Handbook of Experimental Pharmacology

Volume 168

Cannabinoids

Contributors

M.E. Abood, S. Bátkai, T. Bisogno, G.A. Cabral, M.J. Christie, A.A. Coutts, S.N. Davies, R. de Miguel, L. De Petrocellis, V. Di Marzo, M. Egertová, M.R. Elphick, J. Fernández-Ruiz, S.J. Gatley, S.T. Glaser, M. Gómez, S. Gonzáles, M. Guzmán, C.J. Hillard, W.-S.V. Ho, A.G. Hohmann, A.C. Howlett, M.A. Huestis, A.A. Izzo, M. Karsak, G. Kunos, C. Li, A.H. Lichtman, K.P. Lindsey, M. Maccarrone, K. Mackie, A. Makriyannis, B.R. Martin, S.P. Nikas, P. Pacher, R.G. Pertwee, J.A. Ramos, P.H. Reggio, G. Riedel, P. Robson, E. Schlicker, A. Staab, B. Szabo, G.A. Thakur, O. Valverde, C.W. Vaughan, J.M. Walker, T. Wenger, A. Zimmer

Editor
Roger G. Pertwee

Springer

Professor
Dr. Roger Pertwee
School of Medical Sciences
Institute of Medical Science
University of Aberdeen
Foresterhill
Aberdeen AB25 2ZD
Scotland, UK
email: rgp@abdn.ac.uk

With 84 Figures and 26 Tables

ISSN 0171-2004

ISBN 3-540-22565-X Springer Berlin Heidelberg New York

Library of Congress Control Number: 2004109756

Springer is a part of Springer Science + Business Media
springeronline.com

Editor: Dr. R. Lange
Desk Editor: S. Dathe
Cover design: *design&production* GmbH, Heidelberg, Germany
Typesetting and production: LE-TEX Jelonek, Schmidt & Vöckler GbR, Leipzig, Germany
Printed on acid-free paper 27/3150-YL - 5 4 3 2 1 0

In memory of H.J.C. and Bill Paton

– and for Teresa.

Preface

Less than 20 years ago the field of cannabis and the cannabinoids was still considered a minor, somewhat quaint, area of research. A few groups were active in the field, but it was already being viewed as stagnating. The chemistry of cannabis was well known, Δ^9-tetrahydrocannabinol (Δ^9-THC), identified in 1964, being the only major psychoactive constituent and cannabidiol, which is not psychoactive, possibly contributing to some of the effects. These cannabinoids and several synthetic analogs had been thoroughly investigated for their pharmacological effects. Their mode of action was considered to be non-specific. The reasons for this assumption were both technical and conceptual. On the technical side, it had been shown that THC was active in both enantiomeric forms (though with a different level of potency) and this observation was incompatible with action on biological substrates—a receptor, an enzyme, an ion channel—which react with a single stereoisomer only. The conceptual problem related to THC activity. This had been pointed out by several highly regarded research groups that had shown that many of the effects seen with cannabinoids were related to those of biologically active lipophiles, and that many of the effects of THC, particularly chronic ones, were comparable to those seen with anaesthetics and solvents. The technical problems were eliminated when it was found, by several groups, that cannabinoid action is actually stereospecific and most of the previous work, which had pointed to a different conclusion, was based on insufficiently purified samples. The conceptual hurdle was overcome when Allyn Howlett's group in 1988 brought out the first evidence that a specific cannabinoid receptor exists in the brain. This receptor was cloned shortly thereafter and a second receptor, which is not present in the brain, was identified in the periphery. As, presumably, receptors do not exist in mammalian brains for the sake of a plant constituent, several groups went ahead looking for endogenous cannabinoids. The first such endocannabinoid, named anandamide, was reported in 1992, and a second major one, 2-arachidonoylglycerol (2-AG), was discovered in 1995. Several additional, apparently minor ones are now known. A research flood followed. Antagonists to both receptors have been synthesized, specific enzymes, which regulate endocannabinoid levels, have been found, and the biosynthetic and degradation patterns have been established. The endocannabinoid system has turned out to be of major biochemical importance. It is involved in many of our physiological processes—in the nervous, digestive, reproductive, pulmonary and immune systems. Endocannabinoids enhance appetite, reduce pain, act as neuroprotectants and regulators of cytokine production and are somehow involved in the extinction of memories—to mention just a few of their effects.

At cannabinoid meetings in the past, very few representatives of the pharmaceutical companies were present. Now the picture has changed. At least two synthetic cannabinoids are in advanced phase III clinical trials. SR-141716, a CB_1 antagonist, developed by Sanofi, represents a new type of appetite modulator, and HU-211, developed by Pharmos, is a neuroprotectant in head trauma. If the clinical trials are successful, both drugs may represent pharmaceutical breakthroughs in important therapeutic areas. Numerous companies are following in their footsteps. Other clinical conditions apparently are also being looked into. Sleep disorders, inflammatory conditions, neurodegenerative diseases, liver cirrhosis and even cancer represent possible targets.

What can we expect in the future? Compared to the classical neurotransmitters dopamine, serotonin, norepinephrine, and acetylcholine, we still know very little about anandamide and 2-AG. There are strong indications that additional anandamide/cannabinoid receptors exist, but their identification and cloning is still elusive. As both anandamide and 2-AG are arachidonic acid derivatives, their leukotriene-type and prostaglandin-type metabolites may be of biological importance—but, are they? It has been shown that the cannabinoids are rather unique retrograde messengers at the synapse. But the actual messengers have not been identified. Are they anandamide and 2-AG? There are initial indications that the endocannabinoid system is involved in numerous, additional, unrelated biological conditions such as stress, bone formation, aggression, addictive behaviours. We know very little of any possible endocannabinoid involvement. And the list is long.

People smoke cannabis in order to change their mood. The tricyclic cannabinoids (and possibly the endocannabinoids) certainly alter mood, social behaviour and emotions. But we know next to nothing of the chemistry of emotions. Until quite recently the field of emotions was left to the poets and some psychologists and psychiatrists. From the point of view of a chemist or a pharmacologist, unfortunately, we have very few tools to approach problems of emotions. Could the endocannabinoids represent such tools?

The present book is an outstanding summary of many aspects of cannabinoid research. It represents a stepping-stone to many unsolved problems in biochemistry, pharmacology, physiology and the clinic. Perhaps it will help generate novel ideas, such as how to approach the scientific study of emotions.

Spring, 2005

Professor Raphael Mechoulam

Department of Medicinal Chemistry
and Natural Products, Medical Faculty,
Hebrew University,
Jerusalem 91120, Israel
(e-mail: mechou@cc.huji.ac.il)

List of Contributors

(Addresses stated at the beginning of the respective chapters)

List of Contents

HEP (2005) 168:1–51

Pharmacological Actions of Cannabinoids

R.G. Pertwee

School of Medical Sciences, Institute of Medical Sciences, University of Aberdeen, Foresterhill, Aberdeen AB25 2ZD, UK
rgp@abdn.ac.uk

Abstract Mammalian tissues express at least two types of cannabinoid receptor, CB_1 and CB_2, both G protein coupled. CB_1 receptors are expressed predominantly at nerve terminals where they mediate inhibition of transmitter release. CB_2 receptors are found mainly on immune cells, one of their roles being to modulate cytokine release. Endogenous ligands for these receptors (endocannabinoids) also exist. These are all eicosanoids; prominent examples include arachidonoylethanolamide (anandamide) and 2-arachidonoyl glycerol. These discoveries have led to the development of CB_1- and CB_2-selective agonists and antagonists and of bioassays for characterizing such ligands. Cannabinoid receptor antagonists include the CB_1-selective SR141716A, AM251, AM281 and LY320135, and the CB_2-selective SR144528 and AM630. These all behave as inverse agonists, one indication that CB_1 and CB_2 receptors can exist in a constitutively active state. Neutral cannabinoid receptor antagonists that seem to lack inverse agonist properties have recently also been developed. As well as acting on CB_1 and CB_2 receptors, there is convincing evidence that anandamide can activate transient receptor potential vanilloid type 1 (TRPV1) receptors. Certain cannabinoids also appear to have non-CB_1, non-CB_2, non-TRPV1 targets, for example CB_2-like receptors that can mediate antinociception and "abnormal-cannabidiol" receptors that mediate vasorelaxation and promote microglial cell migration. There is evidence too for TRPV1-like receptors on glutamatergic neurons, for α_2-adrenoceptor-like (imidazoline) receptors at sympathetic nerve terminals, for novel G protein-coupled receptors for *R*-(+)-WIN55212 and anandamide in the brain and spinal cord, for novel receptors for Δ^9-tetrahydrocannabinol and cannabinol on perivascular sensory nerves and for novel anandamide receptors in the gastro-intestinal tract. The presence of allosteric sites for cannabinoids on various ion channels and non-cannabinoid receptors has also been proposed. In addition, more information is beginning to emerge about the pharmacological actions of the non-psychoactive plant cannabinoid, cannabidiol. These recent advances in cannabinoid pharmacology are all discussed in this review.

Keywords Cannabinoid receptors · Cannabinoid receptor agonists and antagonists · Abnormal-cannabidiol · Cannabidiol · Inverse agonism

1
Introduction

"Cannabinoid" was originally the collective name given to a set of oxygen-containing C_{21} aromatic hydrocarbon compounds that occur naturally in the plant *Cannabis sativa* (ElSohly 2002; Mechoulam and Gaoni 1967). However, this term is now generally also used for all naturally occurring or synthetic compounds that can mimic the actions of plant-derived cannabinoids or that have structures that closely resemble those of plant cannabinoids. Consequently, a separate term, "phytocannabinoid", has been coined for the cannabinoids produced by cannabis (Pate 1999). One phytocannabinoid, Δ^9-tetrahydrocannabinol (Δ^9-THC; Fig. 1), has attracted particular attention. This is because it is the main psychoactive constituent of cannabis (reviewed in Pertwee 1988) and because it is one of just two cannabinoids to be licensed for medical use, the other being nabilone (Cesamet; Fig. 2), a synthetic analogue of Δ^9-THC (reviewed in the chapter by Robson, this vol-

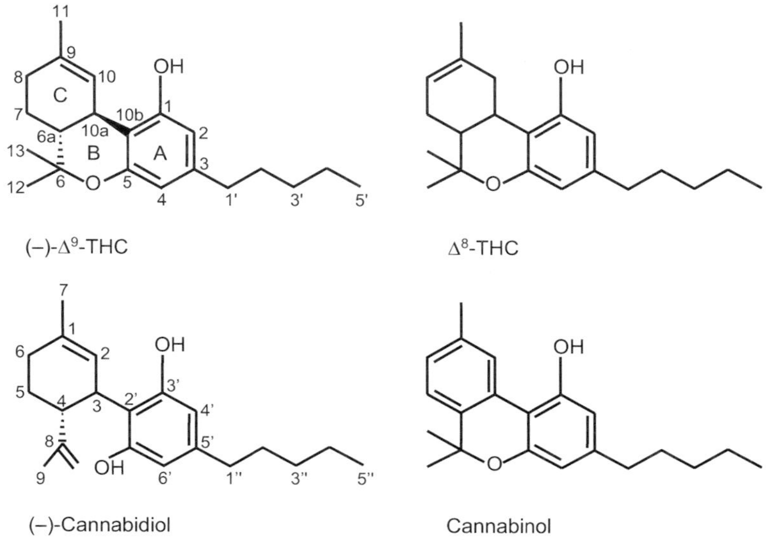

Fig. 1. The structures of four plant cannabinoids, Δ^9-THC, Δ^8-THC, cannabinol and cannabidiol

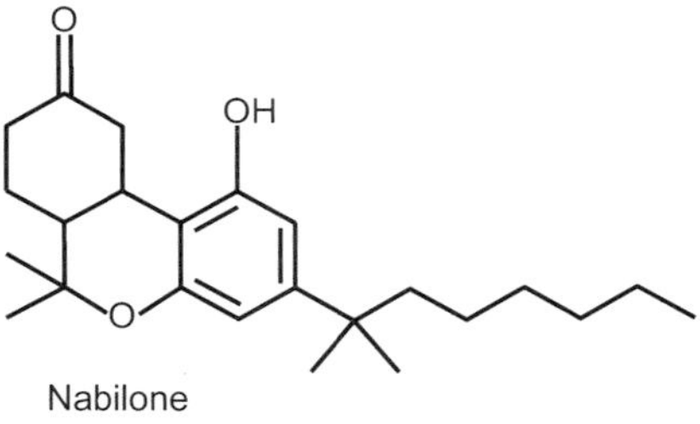

Fig. 2. The structure of nabilone

ume). Because of its high lipid solubility and low water solubility, Δ^9-THC was long thought to owe its pharmacological properties to an ability to perturb the phospholipid constituents of biological membranes (reviewed in Pertwee 1988). However, all this changed in the late 1980s with the discovery in mammalian tissues of specific cannabinoid receptors.

Two types of cannabinoid receptor have so far been identified (reviewed in Howlett et al. 2002). These are CB_1, cloned in Tom Bonner's laboratory in the USA in 1990, and CB_2, cloned by Sean Munro in the UK in 1993. Both these receptors are coupled through $G_{i/o}$ proteins, negatively to adenylate cyclase and positively to mitogen-activated protein kinase. CB_1 receptors are also coupled through $G_{i/o}$ proteins, positively to A-type and inwardly rectifying potassium channels and negatively to N-type and P/Q-type calcium channels and to D-type potassium channels. In addition, there are reports that CB_1 and CB_2 receptors can enhance intracellular free Ca^{2+} concentrations (Fan and Yazulla 2003; Rubovitch et al. 2002; Sugiura et al. 1996, 1997, 2000). It is unclear whether this enhancement is $G_{i/o}$

mediated. In experiments with NG108-15 cells, Sugiura et al. (1996) found CB_1-mediated increases in intracellular free Ca^{2+} levels to be abolished by pretreatment with pertussis toxin, pointing to an involvement of $G_{i/o}$ proteins. However, in experiments with N18TG2 neuroblastoma cells, Rubovich et al. (2002) reported that pertussis toxin failed to prevent CB_1-mediated enhancement of intracellular free Ca^{2+} levels by low concentrations of desacetyl-L-nantradol, a cannabinoid receptor agonist (Sect. 3.1), and instead unmasked a stimulatory effect of higher concentrations of this agonist that in the absence of pertussis toxin did not alter intracellular free Ca^{2+} levels at all. Rubovich et al. (2002) also obtained evidence that the stimulatory effect of desacetyl-L-nantradol on intracellular Ca^{2+} release depended on an ability to delay the inactivation of open L-type voltage-dependent calcium channels and that it was mediated mainly by cyclic AMP-dependent protein kinase (PKA).

Although there is no doubt that $G_{i/o}$ proteins play a major role in cannabinoid receptor signalling, there is also no doubt that transfected and naturally expressed CB_1 receptors can act through G_s proteins to activate adenylate cyclase (Calandra et al. 1999; Glass and Felder 1997; Maneuf and Brotchie 1997). The extent to which CB_1 receptors signal through G_s proteins may be determined by CB_1 receptor location or by cross-talk with colocalized G protein-coupled non-CB_1 receptors (Breivogel and Childers 2000; Calandra et al. 1999; Glass and Felder 1997; Jarrahian et al. 2004). As proposed by Calandra et al. (1999), it is also possible that there are distinct subpopulations CB_1 receptors, one coupled to $G_{i/o}$ proteins and the other to G_s. Additional signalling mechanisms for cannabinoid CB_1 and CB_2 receptors have been proposed and descriptions of these can be found elsewhere (Howlett et al. 2002; see also the chapter by Howlett, this volume).

CB_1 receptors are expressed by central and peripheral neurons and also by some nonneuronal cells (reviewed in Howlett et al. 2002; Pertwee 1997; see also the chapter by Mackie, this volume). Within the central nervous system, the distribution pattern of CB_1 receptors is heterogeneous and can account for several of the characteristic pharmacological properties of CB_1 receptor agonists. For example, the presence of large populations of CB_1 receptors in cerebral cortex, hippocampus, caudate-putamen, substantia nigra pars reticulata, globus pallidus, entopeduncular nucleus and cerebellum, as well as in some areas of the brain and spinal cord that process or modulate nociceptive information, probably accounts for the ability of CB_1 receptor agonists to impair cognition and memory, to alter the control of motor function and to produce antinociception (reviewed in Iversen 2003; Pertwee 2001; see also the chapters by Riedel and Davies, Fernández-Ruiz and González, and Walker and Hohmann, this volume). Some CB_1 receptors are located at central and peripheral nerve terminals. Here they modulate the release of excitatory and inhibitory neurotransmitters when activated (Howlett et al. 2002). Although the effect of CB_1 receptor agonists on release that has been most often observed is one of inhibition, there has been one report that the CB_1/CB_2 receptor agonist, *R*-(+)-WIN55212 (Sect. 3.1), can act through CB_1 receptors to stimulate release of glutamate from primary cultures of rat cerebral cortical neurons (Ferraro et al. 2001). This effect, which disappeared when the concentration of *R*-(+)-WIN55212 was increased from 1 or 10 nM to 100 nM, was most probably triggered by cal-

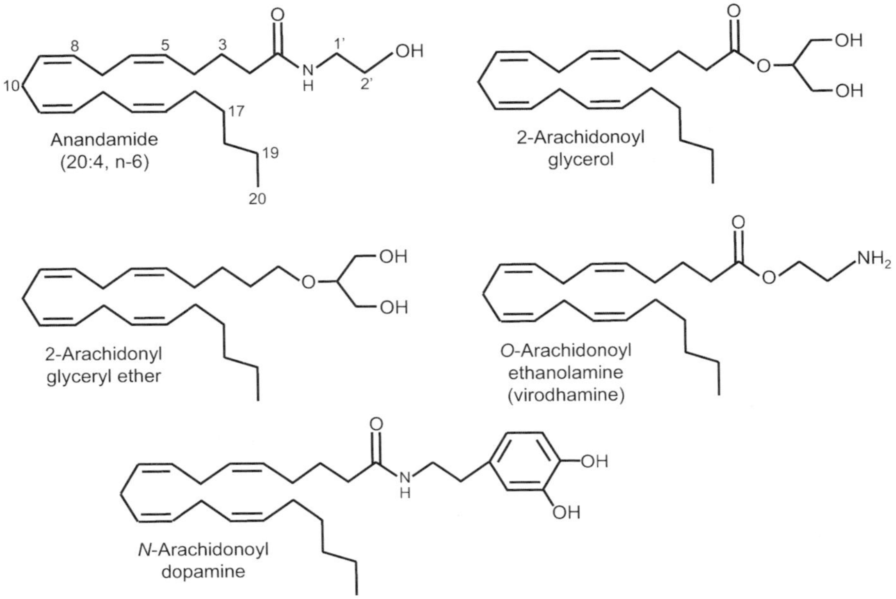

Fig. 3. The structures of five putative endogenous cannabinoids

cium released from inositol 1,4,5-triphosphate-controlled intracellular stores in response to a CB_1 receptor-mediated activation of phospholipase C. CB_2 receptors are expressed mainly by immune cells that include lymphocytes, macrophages, mast cells, natural killer cells, peripheral mononuclear cells and microglia (reviewed in Howlett et al. 2002; Pertwee 1997; see also the chapter by Cabral and Staab, this volume). Less is known about the roles of CB_2 than of CB_1 receptors, although there is good evidence that CB_2 receptors can trigger microglial cell migration (Sect. 4.1.5) and regulate cytokine release. Thus, one property CB_1 and CB_2 receptors share is the ability to modulate ongoing release of chemical messengers.

The discovery of cannabinoid receptors was followed by the demonstration that mammalian tissues can produce endogenous agonists for these receptors, all of which have so far proved to be derivatives of arachidonic acid (reviewed in Di Marzo et al. 1998; Hillard 2000; Mechoulam et al. 1998; see also the chapter by Di Marzo et al., this volume). The most investigated of these "endocannabinoids" have been arachidonoylethanolamide (anandamide) and 2-arachidonoyl glycerol (Fig. 3), both of which are synthesized on demand rather than stored. Other compounds that may be endocannabinoids include 2-arachidonylglyceryl ether (noladin ether), *O*-arachidonoylethanolamine (virodhamine) and *N*-arachidonoyldopamine (Howlett et al. 2002; Porter et al. 2002; Walker et al. 2002). Endocannabinoids together with cannabinoid receptors constitute what is now usually referred to as the "endocannabinoid system". It is likely that endocannabinoids function as both neuromodulators and immunomodulators and indeed, there is already evidence that within the central nervous system they serve as retrograde synaptic messengers (reviewed in the chapter by Vaughan and Christie, this volume). There is also evidence that following their release, anandamide and 2-arachidonoyl glycerol enter cells by a combination of simple diffusion and facilitated, carrier-mediated transport (reviewed in Hillard and Jarrahian 2003) and are then metabolized by intracellular enzymes, anandamide by fatty acid amide hydrolase and 2-arachidonoyl glycerol mainly by monoacylglycerol lipase (monoglyceride lipase) but also by fatty acid amide hydrolase (reviewed in Cravatt and Lichtman 2002; Dinh et al. 2002; Ueda 2002; van der Stelt and Di Marzo 2004; see also the chapter by Di Marzo et al., this volume). Noladin ether also seems to be a substrate for anandamide/2-arachidonoyl glycerol membrane transporter(s) (Fezza et al. 2002). The processes responsible for the production, membrane transport and enzymic inactivation of endocannabinoids are all pharmacological targets through which the activity of the endocannabinoid system can or might be modulated to experimental or therapeutic advantage (reviewed in the chapters by Howlett and by Di Marzo et al., this volume). There is evidence that such modulation may also take place naturally as a result of the co-release of endogenous fatty acid derivatives such as palmitoylethanolamide and oleamide, which can potentiate anandamide, or of 2-linoleyl glycerol and 2-palmitoyl glycerol, which can potentiate 2-arachidonoyl glycerol (Mechoulam et al. 1998). For anandamide, mechanisms through which co-released ligands induce this "entourage effect" include not only inhibition of its metabolism by fatty acid amide hydrolase but also increases in the sensitivity of CB_1 or vanilloid receptors or of other pharmacological targets for anandamide through allosteric or other mechanisms (De Petrocellis et al. 2001b, 2002; Franklin et al. 2003; Mechoulam et al. 1998; Smart et al. 2002).

This chapter describes the in vitro and in vivo bioassays that have been most widely used to characterize ligands for CB_1 and/or CB_2 receptors and reviews the ability of compounds commonly used in cannabinoid research as experimental tools to activate or block these receptors. The likelihood that the most widely used cannabinoid receptor antagonists are inverse agonists rather than neutral antagonists is also discussed, as is evidence for the presence in mammalian tissues of non-CB_1, non-CB_2 pharmacological targets for cannabinoids.

2 Bioassays for Characterizing CB_1 and CB_2 Receptor Ligands

2.1 In Vitro Binding Assays

Several cannabinoid receptor ligands have been radiolabelled with tritium, and these have been used both to determine the CB_1 and CB_2 receptor affinities of unlabelled cannabinoids in displacement assays and to establish the tissue distribution patterns of these receptors (reviewed in Howlett et al. 2002; Pertwee 1999a). As indicated in Tables 1, 2 and 3, some of these compounds bind more readily to CB_1 or to CB_2 receptors, whilst the others bind more or less equally well to both these

Table 1. Typical dissociation constant (K_D) values of radiolabelled ligands at cannabinoid receptor CB_1 and CB_2 binding sites

Radioligand	Source of membranes	Receptor	K_D (nM)	Reference(s)
[^{3}H]SR141716A	Rat brain[a]	rCB_1	0.19–1.20	For references, see Pertwee 1999a
	Guinea-pig forebrain	g-pCB_1	1.24	
[^{123}I]AM251	Rat cerebellum	rCB_1	0.25	
[^{3}H]*R*-(+)-WIN55212	Rat cerebellum	rCB_1	1.89, 4.67, 8.6	
	Guinea-pig forebrain	g-pCB_1	2.34	
	Cultured cells[b]	rCB_1	2.60	
	Cultured cells[b]	hCB_1	16.2, 11.9	
	Cultured cells[b]	hCB_2	3.7, 3.8	
[^{3}H]HU-210 (HU-243)	Rat brain minus brain stem	rCB_1	0.045	
	Cultured cells[b]	hCB_2	0.061	
[^{3}H]CP55940	Cultured cells[b]	hCB_1	0.4 to 3.3	For references, see Pertwee 1997, 1999a
	Cultured cells[b]	rCB_1	4	
	Rat brain[a]	rCB_1	0.07 to 2.3	
	Mouse whole brain	mCB_1	3.4	
	Cultured cells[b]	hCB_2	0.2 to 7.4	
	Cultured cells[b]	mCB_2	0.39	
[^{3}H]CP55940	Rat cerebellum	rCB_1	2.37	Mauler et al. 2002
	Human cerebral cortex	hCB_1	1.29	
	Cultured cells[b]	hCB_1	1.10	
	Cultured cells[b]	hCB_2	4.20	
[^{3}H]BAY 38-7271	Rat cerebellum	rCB_1	1.84	Mauler et al. 2002
	Human cerebral cortex	hCB_1	2.10	
	Cultured cells[b]	hCB_1	2.91	
	Cultured cells[b]	hCB_2	4.24	

g-pCB_1, Guinea-pig CB_1 receptors; hCB_1 and hCB_2, human cannabinoid receptors; mCB_1 and mCB_2, mouse cannabinoid receptors; rCB_1 and rCB_2, rat cannabinoid receptors.

[a]Whole brain or a discrete area.

[b]Cells transfected with CB_1 or CB_2 receptors.

receptor types. It is noteworthy, therefore, that CB_1 or CB_2 selectivity can still be achieved in displacement assays with the non-selective radiolabelled ligands by using membranes obtained from cannabinoid receptor-free cultured cells that have been transfected with CB_1 or CB_2 receptors or membranes obtained from brain (CB_1-rich) or spleen (CB_2-rich). Some care is needed in interpreting binding data obtained with brain or spleen membranes. Thus, whilst there is little evidence that CB_2 receptors are expressed by central neurons, these receptors are expressed by microglial cells (Howlett et al. 2002). Similarly, although it is mainly CB_2 receptors that are present in spleen, this tissue also expresses some CB_1 receptors (reviewed in Howlett et al. 2002; Pertwee 1997). Moreover, there is growing evidence for the presence in brain and other tissues of non-CB_1, non-CB_2 cannabinoid recep-

Table 2. Examples of K_i values of certain cannabinoid CB_1 and/or CB_2 receptor agonists for the in vitro displacement of [^{3}H]CP55940, [^{3}H]HU243 or [^{3}H]BAY-38-7271 from CB_1- and CB_2-specific binding sites (*continued on next page*)

Agonist	CB_1 K_i value (nM)	CB_2 K_i value (nM)	Reference
CB_1-selective agonists in order of decreasing CB_1/CB_2 selectivity			
ACEA	1.4[a,b]	>2,000[a,b]	Hillard et al. 1999
	5.29[a,b]	195[c]	Lin et al. 1998
O-1812	3.4[b]	3,870[b]	Di Marzo et al. 2001
ACPA	2.2[a,b]	715[a,b]	Hillard et al. 1999
2-Arachidonyl glyceryl ether	21.2[b]	>3,000[d]	Hanus et al. 2001
R-(+)-methanandamide	17.9[a,b]	868[c]	Lin et al. 1998
	20[a,b]	815[c]	Khanolkar et al. 1996
	28.3[b]	868[c]	Goutopoulos et al. 2001
Agonists without any marked CB_1 or CB_2 selectivity			
Anandamide	61[a,b]	1,930[c]	Lin et al. 1998
	78.2[a,b]	1,926[c]	Khanolkar et al. 1996
	89[a]	371[a]	Showalter et al. 1996
	543	1,940	Felder et al. 1995
	71.7[a,b]	279[a,b]	Hillard et al. 1999
	252[e,d]	581[e,d]	Mechoulam et al. 1995
BAY 38-7271	1.85[f]	5.96[f]	Mauler et al. 2002
2-Arachidonoyl glycerol	472[e,d]	1,400[e,d]	Mechoulam et al. 1995
	58.3[e,d]	145[e,d]	Ben-Shabat et al. 1998
O-1057	4.4	11.2	Pertwee et al. 2000
HU-210	0.0608	0.524	Felder et al. 1995
	0.1[e,b]	0.17[e]	Rhee et al. 1997
	0.73	0.22	Showalter et al. 1996
CP55940	5	1.8	Ross et al. 1999a
	3.72	2.55	Felder et al. 1995
	1.37[b]	1.37[b]	Rinaldi-Carmona et al. 1994
	0.58	0.69	Showalter et al. 1996
	0.50[a,b]	2.80[a,b]	Hillard et al. 1999
Δ^9-THC	53.3	75.3	Felder et al. 1995
	39.5[e,b]	40[e]	Bayewitch et al. 1996
	40.7	36.4	Showalter et al. 1996
	80.3[e,b]	32.2[e]	Rhee et al. 1997
	35.3[b]	3.9[b]	Rinaldi-Carmona et al. 1994
	5.05	3.13	Iwamura et al. 2001
Nabilone	1.84	2.19	Gareau et al. 1996
Δ^8-THC	47.6[b]	39.3[c]	Busch-Petersen et al. 1996
	44[b]	44	Huffman et al. 1999
Cannabinol	211.2[e,b]	126.4[e]	Rhee et al. 1997
	308	96.3	Showalter et al. 1996
	1,130	301	Felder et al. 1995
CP56667	61.7	23.6	Showalter et al. 1996
R-(+)-WIN55212	9.94[b]	16.2[b]	Rinaldi-Carmona et al. 1994
	4.4[a,b]	1.2[a,b]	Hillard et al. 1999

Table 2. (*continued*)

Agonist	CB_1 K_i value (nM)	CB_2 K_i value (nM)	Reference
	1.89	0.28	Showalter et al. 1996
	62.3	3.3	Felder et al. 1995
	123	4.1	Shire et al. 1996
	9.87	0.29	Iwamura et al. 2001
CB_2-selective agonists in order of increasing CB_2/CB_1 selectivity			
AM1241	280^b	3.4^c	Ibrahim et al. 2003
3-(1′1′-dimethylbutyl)-1-deoxy-Δ^8-THC (JWH-133)	677^b	3.4	Huffman et al. 1999
L-759633	1,043	6.4	Ross et al. 1999a
	15,850	20	Gareau et al. 1996
L-759656	529^b	35	Huffman et al. 1999
	713^b	57	Huffman et al. 2002
	4,888	11.8	Ross et al. 1999a
	>20,000	19.4	Gareau et al. 1996
HU-308	$>10{,}000^{e,b}$	$22.7^{e,d}$	Hanus et al. 1999

See Figs. 1 to 9 for the structures of the compounds listed in this table.
DMH, dimethylheptyl; ND, not determined; THC, tetrahydrocannabinol.
[a]With phenylmethylsulphonyl fluoride (PMSF) in order to inhibit enzymic hydrolysis.
[b]Binding to rat cannabinoid receptors on transfected cells or on brain (CB_1) or spleen tissue (CB_2).
[c]Binding to mouse brain (CB_1) or spleen tissue (CB_2).
[d]Species unspecified. All other data from experiments with human cannabinoid receptors.
[e]Displacement of [^{3}H]HU243 from CB_1- and CB_2-specific binding sites.
[f]Displacement of [^{3}H]BAY-38-7271 from CB_1- and CB_2-specific binding sites.

tors to which at least some CB_1 and/or CB_2 receptor ligands can bind (Sect. 4). Radiolabelled probes for single photon emission computed tomography (SPECT) or positron emission tomography (PET) have also been developed (reviewed in Gifford et al. 2002; see also the chapter by Lindsey et al., this volume).

2.2 In Vitro Functional Bioassays

2.2.1 Assays Using Whole Cells or Cell Membranes

The most commonly employed assays using whole cells or cell membranes are the [^{35}S]guanosine-5′-*O*-(3-thiotriphosphate) ([^{35}S]GTPγS) binding assay and the cyclic AMP assay. The first measures cannabinoid receptor agonist-stimulated binding to G proteins of the hydrolysis-resistant GTP analogue, [^{35}S]GTPγS, whereas the cyclic AMP assay relies on cannabinoid receptor-mediated inhibition (usual effect) or enhancement of basal or drug-induced cyclic AMP production

Table 3. K_i values of cannabinoid receptor antagonists/inverse agonists for the in vitro displacement of [^{3}H]CP55940 from CB_1- and CB_2-specific binding sites

Ligand	CB_1 K_i value (nM)	CB_2 K_i value (nM)	Reference
CB_1-selective antagonists/inverse agonists			
NESS 0327	0.00035[a]	21[a]	Ruiu et al. 2003
SR141716A	11.8	13,200	Felder et al. 1998
	11.8	973	Felder et al. 1995
	12.3	702	Showalter et al. 1996
	5.6	>1,000	Rinaldi-Carmona et al. 1994
	1.98[b]	>1,000[b]	Rinaldi-Carmona et al. 1994
	1.8[a]	514[a]	Ruiu et al. 2003
AM281	12[b]	4,200[a]	Lan et al. 1999a
AM251 (compound 12)	7.49[b]	2,290[a]	Lan et al. 1999b
LY320135	141	14,900	Felder et al. 1998
CB_2-selective antagonists/inverse agonists			
AM 630	5,152	31.2	Ross et al. 1999a
SR144528	437	0.60	Rinaldi-Carmona et al. 1998
	305[b]	0.30[b]	Rinaldi-Carmona et al. 1998
	>10,000	5.6	Ross et al. 1999a
	70[a]	0.28[a]	Ruiu et al. 2003
	50.3	1.99	Iwamura et al. 2001

See Figs. 10 and 11 for the structures of the compounds listed in this table.
[a] Binding to mouse brain (CB_1) or spleen tissue (CB_2).
[b] Binding to rat cannabinoid receptors on transfected cells or on brain (CB_1) or spleen tissue (CB_2).
All other data from experiments with human cannabinoid receptor.

(reviewed in Howlett et al. 2002; Pertwee 1997, 1999a). Both assays can be performed with membranes obtained from brain tissue or from cultured cells that express CB_1 or CB_2 receptors either naturally or after transfection. In addition, the cyclic AMP assay can be performed with whole cells, including primary cultures of central neurons, and the [^{35}S]GTPγS assay can be used in autoradiography experiments with tissue sections (Breivogel et al. 1997; Selley et al. 1996; Sim et al. 1995). The cyclic AMP assay is more sensitive than the [^{35}S]GTPγS assay. Presumably this is because modulation of cyclic AMP production takes place further along the signalling cascade than [^{35}S]GTPγS binding so that there is greater signal amplification. For the [^{35}S]GTPγS assay, it is important to include guanosine diphosphate (GDP) and sodium chloride at appropriate concentrations (Breivogel et al. 1998; Selley et al. 1996; Sim et al. 1995). GDP increases the ratio of agonist-stimulated to basal [^{35}S]GTPγS binding (signal-to-noise ratio) but also decreases the absolute levels of both agonist-stimulated and basal [^{35}S]GTPγS binding. In addition, it magnifies the differences in efficacy exhibited in this assay by full and partial agonists (Savinainen et al. 2001). The signal-to-noise ratio in this bioassay can be further improved by including an adenosine A_1 receptor antagonist (Savinainen et al. 2003). It has also proved possible to assay cannabinoid receptor agonists by exploiting their ability to increase intracellular free Ca^{2+} levels (CB_1 and CB_2 agonists) (Bisogno et al. 2000; Rubovitch et al. 2002; Sugiura et al. 1996, 1997, 2000; Suhara et al. 2001) or to inhibit lipopolysaccharide-induced release of tumour necrosis factor-α (CB_2 agonists) (Wrobleski et al. 2003). Some information about the pharmacological properties of cannabinoid receptor ligands has also been obtained using bioassays performed with cultured neurons that exploit the negative coupling of the CB_1 receptor to N- and P/Q-type calcium channels (reviewed in Pertwee 1997, 1999a).

2.2.2 Isolated Nerve–Smooth Muscle Preparations

Preparations in which cannabinoid receptor agonists can act through neuronal CB_1 receptors to produce a concentration-related inhibition both of electrically-evoked contractile transmitter release (Schlicker et al. 2003; Trendelenburg et al. 2000) and of the contractions caused by this release (reviewed in Howlett et al. 2002; Pertwee 1997; Pertwee et al. 1996a; Schlicker and Kathmann 2001) are called isolated nerve–smooth muscle preparations. The ones most commonly used are the mouse vas deferens and the myenteric plexus-longitudinal muscle preparation of guinea-pig small intestine. However, CB_1 receptor agonists also show activity in other isolated nerve-smooth muscle preparations, for example the rat vas deferens and the mouse urinary bladder. The usual measured response in these bioassays is inhibition of electrically evoked contractions, a response that can also be elicited in these tissues by agonists for several types of non-cannabinoid receptor. Consequently, to establish whether or not the production of such inhibition by a test compound is CB_1 receptor-mediated, it is necessary to measure the susceptibility of this compound to antagonism by a selective CB_1 antagonist. For the mouse vas deferens, an alternative strategy for meeting this objective has been to exploit the ability of a cannabinoid receptor agonist (Δ^9-THC) to induce cannabinoid tolerance without affecting the sensitivity of the twitch response to inhibition by non-cannabinoids (Pertwee 1997).

2.3 In Vivo Bioassays

Probably the most commonly used in vivo bioassay is the mouse tetrad assay, in which the ability of a test compound to produce four effects in the same animal is determined. These effects, hypokinesia, hypothermia, catalepsy in the Pertwee ring test and antinociception in the tail-flick or hot plate test, are usually produced by a CB_1 receptor agonist over a relatively narrow dose range (reviewed in Howlett et al. 2002; Martin et al. 1995). One or other of these effects can be produced by some centrally active non-CB_1 receptor agonists or antagonists. However, when performed together, the tetrad tests provide at least some degree of selectivity since, in contrast to established CB_1 receptor agonists, many other classes of centrally active agent lack activity in at least one of the tests (Wiley and Martin 2003). This feature of the tetrad assay was particularly important when it was first devised, as selective CB_1 receptor antagonists had still to be developed. Now that such antagonists are available (Sect. 3.2), there is less need for a bioassay with CB_1 receptor selectivity. Some non-CB_1 receptor ligands do show activity in all four tetrad tests. These include stearoylethanolamide (Maccarrone et al. 2002), the anandamide analogue, O-2093 (Di Marzo et al. 2002), metabolites of anandamide (reviewed in Pertwee and Ross 2002) and certain anti-psychotic agents (Wiley and Martin 2003). Moreover, although the endocannabinoid anandamide shows cannabimimetic activity in the mouse tetrad assay, it is only antagonized by SR141716A when protected from enzymic hydrolysis (reviewed in Pertwee and Ross 2002). However, other CB_1 receptor agonists do show susceptibility to antagonism by SR141716A in this bioassay (reviewed in Howlett et al. 2002).

Other in vivo bioassays for CB_1 receptor agonists include the dog static ataxia test, the monkey behavioural test, the rat catalepsy test and the drug discrimination test, which is usually carried out with monkeys, rats or pigeons (reviewed in Howlett et al. 2002; Martin et al. 1995). The potencies shown by some cannabinoids in drug discrimination experiments performed with rats have been found to correlate well with their psychoactive potencies in humans (Balster and Prescott 1992). In vivo bioassays that provide measures of other CB_1 receptor-mediated effects in animals, for example changes in memory, have also been developed (reviewed in Howlett et al. 2002; see also the chapter by Riedel and Davies, this volume). However, these have not been used widely for characterizing novel cannabinoid receptor ligands. Methods for evaluating cannabinoids in humans have also been developed (Howlett et al. 2002).

2.4
Cannabinoid Receptor Knockout Mice

One important advance has been the development of transgenic $CB_1^{-/-}$, $CB_2^{-/-}$ and $CB_1^{-/-}/CB_2^{-/-}$ mice that lack CB_1, CB_2 or both CB_1 and CB_2 receptors (reviewed in Howlett et al. 2002; see also the chapters by Abood and by Valverde et al., this volume). The availability of such animals provides a useful additional method for establishing whether or not responses to test compounds are CB_1 and/or CB_2 receptor mediated and, indeed, an important means of detecting the presence of new types of cannabinoid receptor (Sect. 4.1). Cannabinoid receptor knockout mice are also being used to help determine the physiological roles of CB_1 and CB_2 receptors.

3 CB_1 and CB_2 Cannabinoid Receptor Ligands

3.1 Cannabinoid Receptor Agonists

In terms of chemical structure, established cannabinoid receptor agonists fall essentially into four main groups: classical, nonclassical, aminoalkylindole and eicosanoid (reviewed in Howlett et al. 2002; Pertwee 1999a).

- The classical group consists of dibenzopyran derivatives that are either cannabis-derived compounds (phytocannabinoids) or their synthetic analogues. Notable examples are the phytocannabinoids Δ^9-THC, Δ^8-THC and cannabinol (Fig. 1), and the synthetic cannabinoids, 11-hydroxy-Δ^8-THC-dimethylheptyl (HU-210), JWH-133, L-759633, L-759656, L-nantradol and desacetyl-L-nantradol (Figs. 4 and 5).

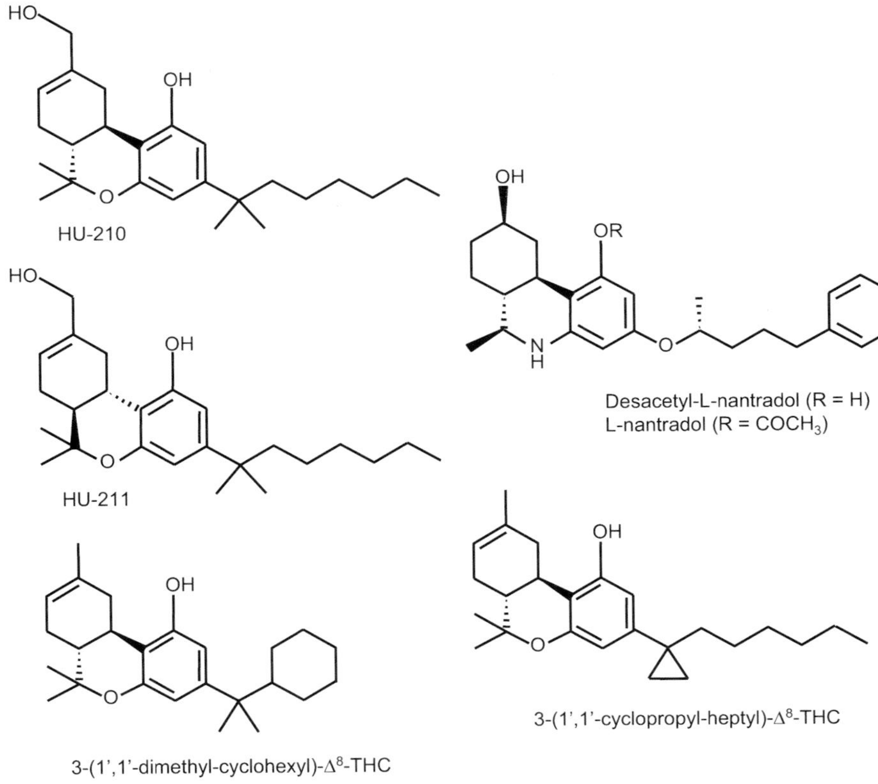

Fig. 4. The structures of five synthetic classical cannabinoids

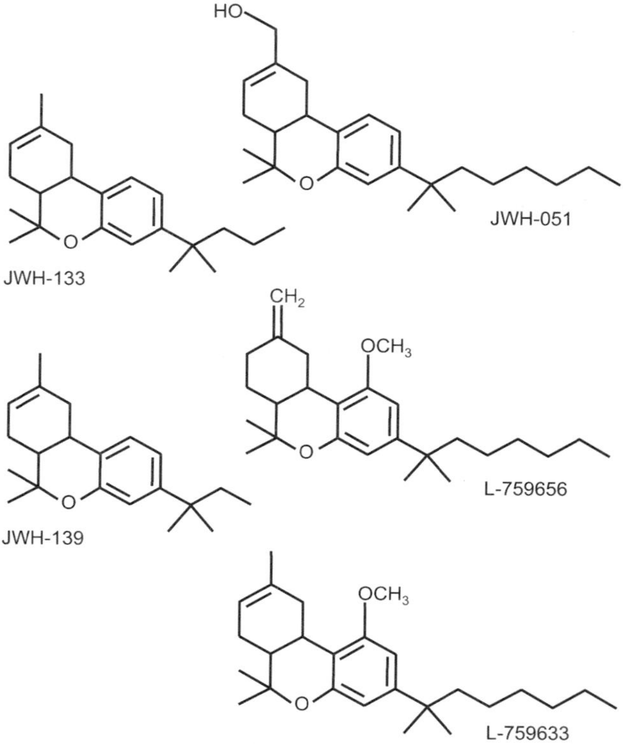

Fig. 5. The structures of four nonclassical cannabinoids

- Nonclassical cannabinoids consist of bicyclic and tricyclic analogues of Δ^9-THC that lack a pyran ring; examples include CP55940, CP47497, CP55244 and HU-308 (Fig. 6). They are, therefore, closely related to the classical cannabinoids.
- In contrast, the aminoalkylindole group of cannabinoid receptor agonists (Fig. 7) have structures that are completely different from those of other cannabinoids. Indeed, results from experiments performed with wild-type and mutant CB_1 receptors (Chin et al. 1998; Petitet et al. 1996; Song and Bonner 1996; Tao and Abood 1998) suggest that *R*-(+)-WIN55212 (WIN55212-2), the most widely investigated of the aminoalkylindoles, binds differently to the CB_1 receptor than classical, nonclassical or eicosanoid cannabinoids, albeit it in a manner that still allows mutual competition between *R*-(+)-WIN55212 and non-aminoalkylindole cannabinoids for binding sites on the wild-type receptor.
- Members of the eicosanoid group of cannabinoid receptor agonists have markedly different structures both from the aminoalkylindoles and from classical and nonclassical cannabinoids. Important members of this group are the endocannabinoids, arachidonoylethanolamide (anandamide), *O*-arachidonoylethanolamine (virodhamine), 2-arachidonoyl glycerol and 2-arachidonyl glyceryl

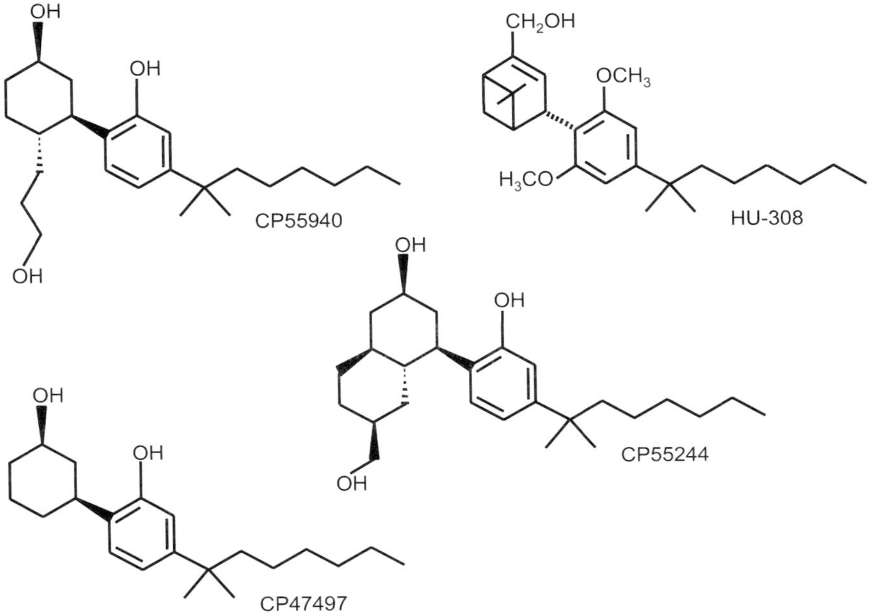

Fig. 6. The structures of four nonclassical cannabinoids. The (+)-enantiomer of CP55940 is CP56667

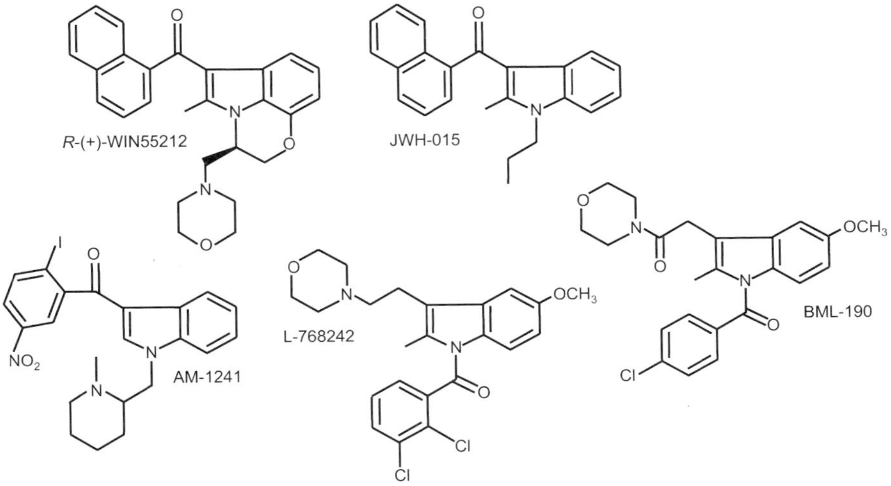

Fig. 7. The structures of *R*-(+)-WIN55212, JWH-015, AM1241, L-768242 and BML-190

ether (noladin ether) (Fig. 3) and several synthetic analogues of anandamide, including *R*-(+)-methanandamide, arachidonyl-2′-chloroethylamide (ACEA), arachidonylcyclopropylamide (ACPA), O-689 and O-1812 (Fig. 8) (Howlett et al. 2002; Pertwee 1999a; Porter et al. 2002).

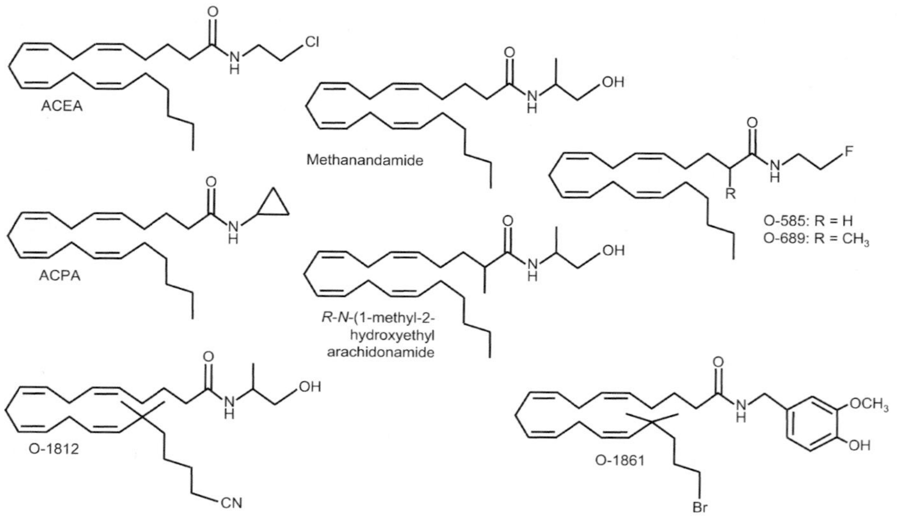

Fig. 8. The structures of eight structural analogues of anandamide

Many cannabinoid receptor agonists exhibit marked stereoselectivity in pharmacological assays, reflecting the presence of chiral centres in these compounds (reviewed in Howlett et al. 2002). Classical and nonclassical cannabinoids with the same absolute stereochemistry as (−)-Δ^9-THC at 6a and 10a, *trans* (6a*R*, 10a*R*), are more active than their *cis* (6a*S*, 10a*S*) enantiomers, whilst *R*-(+)-WIN55212 is more active than *S*-(−)-WIN55212. Although anandamide does not contain any chiral centres, some of its synthetic analogues do. One of these is methanandamide, the *R*-(+)-isomer of which exhibits nine times higher affinity for CB_1 receptors than the *S*-(−)-isomer (Abadji et al. 1994).

Several cannabinoid receptor agonists bind more or less equally well to CB_1 and CB_2 receptors (Table 2), although they do exhibit different relative intrinsic activities at these receptors. Among these are HU-210, CP55940, *R*-(+)-WIN55212, (−)-Δ^9-THC, anandamide and 2-arachidonoyl glycerol (reviewed in Howlett et al. 2002; Pertwee 1999a).

- HU-210 has particularly high affinity for both CB_1 and CB_2 receptors. It also exhibits high relative intrinsic activities at these receptors. Indeed, it is remarkably potent as a cannabinoid receptor agonist and exhibits an exceptionally long duration of action in vivo. The marked affinity and efficacy that HU-210 shows at cannabinoid receptors is due largely to the replacement of the pentyl side chain of Δ^8-THC with a dimethylheptyl group.
- CP55940 and *R*-(+)-WIN55212 have CB_1 and CB_2 relative intrinsic activities of the same order as those of HU-210 and, although they have lower CB_1 and CB_2 affinities than HU-210, are still reasonably potent as they bind to these receptors at concentrations in the low nanomolar range.

- (−)-Δ^9-THC has lower CB_1 and CB_2 affinities and relative intrinsic activities than HU-210, CP55940 or *R*-(+)-WIN55212. Whilst it behaves as a partial agonist at both these receptor types, it exhibits less efficacy at CB_2 than at CB_1 receptors to the extent that in one bioassay system it has been found to behave as a CB_2 receptor antagonist (Bayewitch et al. 1996). (−)-Δ^9-THC can also produce CB_1 receptor antagonism. Thus, it has been found to oppose CB_1 receptor activation by the higher efficacy agonist, 2-arachidonoyl glycerol, in hippocampal cultures that may have contained neurons with rather low CB_1 receptor density (Kelley and Thayer 2004). This it did with an IC_{50} of 42 nM, which is close to its reported CB_1 K_i values (Table 2).
- Anandamide resembles (−)-Δ^9-THC in its affinity for CB_1 receptors, in behaving as a CB_1 and CB_2 receptor partial agonist (Gonsiorek et al. 2000; Hillard 2000; Mackie et al. 1993; Savinainen et al. 2001; Sugiura et al. 1996, 2000) and in having lower CB_2 than CB_1 intrinsic activity (reviewed in Howlett et al. 2002; Pertwee 1999a). It has also been found that, like (−)-Δ^9-THC, anandamide can behave as a CB_2 receptor antagonist in at least one bioassay system (Gonsiorek et al. 2000). In contrast to *R*-(+)-WIN55212, which has slightly higher CB_2 than CB_1 affinity, anandamide binds marginally more readily to CB_1 than to CB_2 receptors.
- 2-Arachidonoyl glycerol is known to activate both CB_1 and CB_2 receptors. It binds about equally well to both receptor types (Table 2) and has been reported to exhibit greater CB_1 intrinsic activity but less CB_1 potency than CP55940 and greater CB_1 intrinsic activity and potency than anandamide (Gonsiorek et al. 2000; Savinainen et al. 2001, 2003; Sugiura et al. 1996). This endocannabinoid also has greater CB_2 potency than anandamide or 1-arachidonoyl glycerol (Gonsiorek et al. 2000; Sugiura et al. 2000).

One recently developed synthetic cannabinoid receptor agonist that interacts almost as well with CB_2 as with CB_1 receptors (Tables 1 and 2) is BAY 38-7271 (De Vry and Jentzsch 2002; Mauler et al. 2002, 2003). This compound has a structure that is not classical, non-classical, aminoalkylindole or eicosanoid (Fig. 9).

Phytocannabinoids other than Δ^9-THC that are known to activate cannabinoid receptors are (−)-Δ^8-THC and cannabinol (reviewed in Pertwee 1999a). Of these, (−)-Δ^8-THC resembles (−)-Δ^9-THC both in its CB_1 and CB_2 receptor affinities (Table 2) and in its relative intrinsic activity at the CB_1 receptor (Gérard et al. 1991; Howlett and Fleming 1984; Matsuda et al. 1990). Cannabinol also behaves as a partial agonist at CB_1 receptors but has even less relative intrinsic activity than (−)-Δ^9-THC (Howlett 1987; Matsuda et al. 1990; Petitet et al. 1997, 1998). Whilst there is one report that cannabinol activates CB_2 receptors in the cyclic AMP assay more effectively than Δ^9-THC (Rhee et al. 1997), there is another that in the GTPγS binding assay, it behaves as a CB_2 receptor inverse agonist (MacLennan et al. 1998).

As to the endocannabinoid virodhamine, Porter et al. (2002) have shown that this activates both CB_1 and CB_2 receptors. Their experiments with transfected cells yielded CB_1 and CB_2 EC_{50} values in the GTPγS binding assay of 1.9 and 1.4 μM, respectively, for this endocannabinoid, indicating it to be less potent than anandamide, 2-arachidonoyl glycerol or *R*-(+)-WIN55212. The CB_2 intrinsic

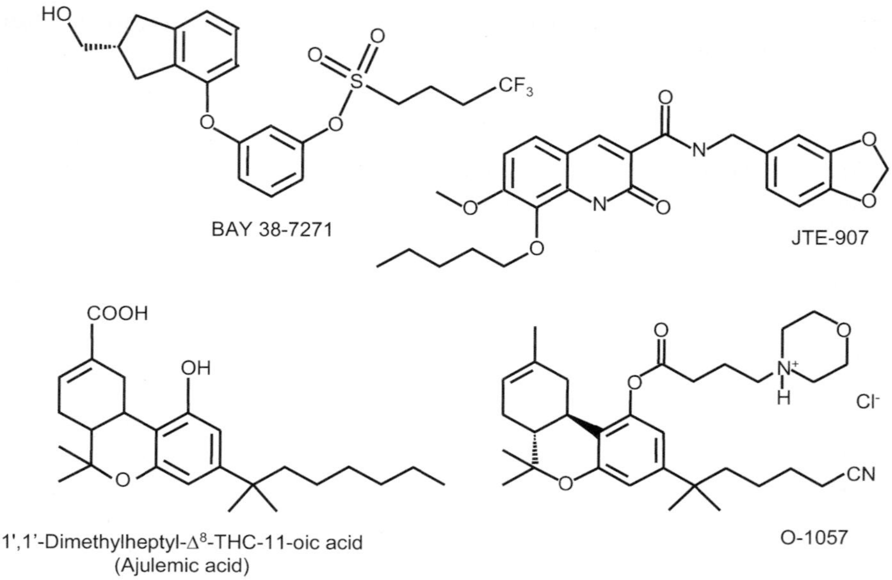

Fig. 9. The structures of BAY 38-7271, JTE-907, ajulemic acid and O-1057

activity of virodhamine matched that of anandamide which, however, behaved as a full agonist in this investigation, suggesting that the CB_2 expression level of the cell line used may have been rather high. In contrast, the CB_1 intrinsic activity of virodhamine was less than that of anandamide, and indeed it was found that virodhamine could attenuate anandamide-induced activation of CB_1 receptors. No binding data are yet available for virodhamine.

Turning now to potent cannabinoid receptor agonists that interact more readily with CB_1 or CB_2 receptors, a number of these have been developed. The starting point for all current CB_1-selective agonists has been anandamide. Thus, results from binding experiments have shown that it is possible to enhance the marginal CB_1 selectivity exhibited by anandamide by replacing a hydrogen atom on the 1′ or 2 carbon with a methyl group to form *R*-(+)-methanandamide or O-689 (Fig. 8) (Abadji et al. 1994; Showalter et al. 1996). As well as increasing CB_1 selectivity, insertion of a methyl group on the 1′ or 2 carbon of anandamide increases resistance to the hydrolytic action of fatty acid amide hydrolase (FAAH) (Abadji et al. 1994; Adams et al. 1995). Anandamide analogues that exhibit particularly marked CB_1-selectivity in binding assays are ACEA, ACPA and a cyano analogue of methanandamide (O-1812) (Table 2; Fig. 8). All three behave as potent CB_1 receptor agonists (Di Marzo et al. 2001; Hillard et al. 1999). O-1812 appears to lack significant susceptibility to hydrolysis by FAAH, presumably because it resembles *R*-(+)-methanandamide in having a methyl group attached to its 1′-carbon. ACEA and ACPA, which do not have the 1′-carbon methyl substituent of *R*-(+)-methanandamide, show no sign of reduced susceptibility to enzymic hy-

Table 4. K_i values of certain other ligands for the in vitro displacement of [^{3}H]CP55940 or [^{3}H]HU243[a] from CB_1- and CB_2-specific binding sites

Ligand	CB_1 K_i value (nM)	CB_2 K_i value (nM)	Reference
CB_1-selective ligands in order of decreasing CB_1/CB_2 selectivity			
R-N-(1-methyl-2-hydroxyethyl)-2-*R*-methyl-arachidonamide	7.42[b,c]	1,952[d]	Goutopoulos et al. 2001
O-585	8.6[b]	324[b]	Showalter et al. 1996
O-689	5.7[b]	132[b]	Showalter et al. 1996
Ligands without any marked CB_1 or CB_2 selectivity			
Ajulemic acid (CT-3)	32.3[a,c]	170.5[a]	Rhee et al. 1997
11-OH-cannabinol-DMH	0.1[a,c]	0.2[a]	Rhee et al. 1997
3-(1′,1′-dimethyl-cyclohexyl)-Δ^8-THC	0.57	0.65	Krishnamurthy et al. 2003
11-OH-cannabinol	38[a,c]	26.6[a]	Rhee et al. 1997
Δ^9-THC-DMH	0.241[a,c]	0.199[a]	Rhee et al. 1997
Cannabinol-DMH	2[a,c]	1.5[a]	Rhee et al. 1997
Cannabidiol	4,350	2,860	Showalter et al. 1996
	>10[a,c]	>10[a,e]	Bisogno et al. 2001
11-OH-Δ^8-THC	25.8[a,c]	7.4[a]	Rhee et al. 1997
1-Deoxy-Δ^8-THC-DMH	23[c]	2.9	Huffman et al. 1996
3-(1′,1′-cyclopropyl-heptyl)-Δ^8-THC	0.44[c]	0.86[d]	Papahatjis et al. 2002
O-1184	5.25	7.41	Ross et al. 1999b
cis (6a*S*, 10a*S*)-3-(1′,1′-DMH)-11-hydroxy-Δ^8-THC (HU-211)	1,990	>10,000	Showalter et al. 1996
Abnormal-cannabidiol	>10,000	>10,000	Showalter et al. 1996
CB_2-selective ligands in order of increasing CB_1/CB_2 selectivity			
JWH-015	383	13.8	Showalter et al. 1996
1-Deoxy-11-hydroxy-Δ^8-THC-DMH (JWH-051)	1.2[c]	0.032	Huffman et al. 1996
JTE-907	2,370	35.9	Iwamura et al. 2001
L-768242	1,917	12	Gallant et al. 1996
3-(1′1′-dimethylpropyl)-1-deoxy-Δ^8-THC (JWH-139)	2,290[c]	14	Huffman et al. 1999
3-(1′1′-dimethylhexyl)-1-methoxy-Δ^8-THC	3,134[c]	18	Huffman et al. 2002
1-Deoxy-Δ^8-THC	>10,000[c]	32	Huffman et al. 1999

See Figs. 1, 4, 5, 7, 8, 9, 11 and 12 for the structures of some of the compounds listed in this table.
DMH, dimethylheptyl; ND, not determined; THC, tetrahydrocannabinol.
[b]With phenylmethylsulphonyl fluoride (PMSF) in order to inhibit enzymic hydrolysis.
[c]Binding to rat cannabinoid receptors on transfected cells or on brain (CB_1) or spleen tissue (CB_2).
[d]Binding to mouse brain (CB_1) or spleen tissue (CB_2).
[e]Species unspecified. All other data from experiments with human cannabinoid receptors.

drolysis. Although insertion of this group into ACEA does markedly reduce the susceptibility of this molecule to FAAH-mediated hydrolysis, it also decreases the affinity of ACEA for CB_1 receptors by about 14-fold (Jarrahian et al. 2000). *R-N*-(1-

methyl-2-hydroxyethyl)-2-*R*-methyl-arachidonamide, which also exhibits marked CB_1-selectivity in binding assays (Table 4), has less metabolic stability than *R*-(+)-methanandamide (Goutopoulos et al. 2001). Another CB_1-selective agonist of note is the endocannabinoid 2-arachidonyl glyceryl ether (Hanus et al. 2001), the CB_1 intrinsic activity of which has been reported to match that of CP55940 and to be less than that of 2-arachidonoyl glycerol. 2-Arachidonyl glyceryl ether exhibits less potency at CB_1 receptors than either CP55940 or 2-arachidonoyl glycerol (Savinainen et al. 2001, 2003; Suhara et al. 2000, 2001).

The best CB_2-selective agonists to have been developed to date are all non-eicosanoid cannabinoids (Howlett et al. 2002; Ibrahim et al. 2003; Pertwee 1999a). They include the classical cannabinoids, L-759633, L-759656 and JWH-133, the non-classical cannabinoid HU-308, and the aminoalkylindole AM1241 (Figs. 5, 6 and 7). All these ligands bind more readily to CB_2 than to CB_1 receptors (Table 2) and have also been shown to behave as potent CB_2-selective agonists in functional bioassays (Hanus et al. 1999; Ibrahim et al. 2003; Pertwee 2000; Ross et al. 1999a).

One other cannabinoid receptor agonist of note is 3-(5′-cyano-1′,1′-dimethylpentyl)-1-(4-*N*-morpholinobutyryloxy)-Δ^8-THC hydrochloride (O-1057). Thus, unlike all established cannabinoid receptor agonists, this is readily soluble in water and yet, compared to CP55940, its potency in the cyclic AMP assay is just 2.9 times less at CB_1 receptors and 6.5 times less at CB_2 receptors (Pertwee et al. 2000). The finding that it is possible to solubilize a cannabinoid and yet retain pharmacological activity has important implications for cannabinoid delivery not only in the laboratory but also in the clinic. As to structure–activity relationships for cannabinoid receptor agonists, the salient features of these have been well described elsewhere (Howlett et al. 2002; Pertwee 1999a). Recent findings of special interest are that the CB_1 and CB_2 affinities of Δ^8-THC can be greatly enhanced both by replacing its C3 pentyl side chain with a 1′,1′-dimethyl-1′-cyclohexyl moiety (Fig. 4; Table 4) (Krishnamurthy et al. 2003) and by changing this side chain from pentyl to heptyl and introducing a cyclopropyl group at the 1′ position (Fig. 4; Table 4) (Papahatjis et al. 2002).

3.2 Cannabinoid CB_1 and CB_2 Receptor Antagonists

3.2.1 Selective CB_1 Receptor Antagonists

The first selective CB_1 receptor antagonist, the diarylpyrazole SR141716A (Fig. 10), was developed by Sanofi Recherche (Rinaldi-Carmona et al. 1994). This readily prevents or reverses effects induced by cannabinoids at CB_1 receptors, both in vitro and in vivo (reviewed in Howlett et al. 2002; Pertwee 1997). It binds with significantly higher affinity to CB_1 than CB_2 receptors (Table 3), lacks significant affinity for a wide range of non-cannabinoid receptors and does not exhibit detectable agonist activity at CB_1 and CB_2 receptors (Hirst et al. 1996; Rinaldi-Carmona et al. 1994, 1996a,b; Shire et al. 1996). Other established CB_1-selective antagonists are

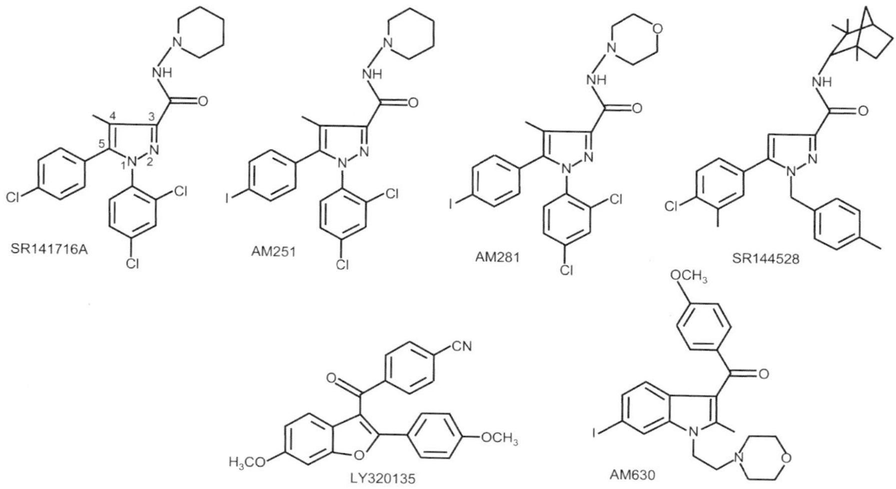

Fig. 10. The structures of several CB_1- or CB_2-selective antagonists/inverse agonists

LY320135, AM251 and AM281 (Fig. 10). LY320135, developed by Eli Lilly, also binds with lower affinity to CB_1 than CB_2 receptors (Table 3). However, its CB_1 affinity is less than that of SR141716A. Moreover, at concentrations in the low micromolar range, LY320135 also binds to muscarinic and 5-hydroxytryptamine $(5\text{-HT})_2$ receptors (K_i<10 μM) and, at higher concentrations, to histamine H_1 receptors (K_I=12.9 μM), α_1- and α_2-adrenoceptors and dopamine D_1 and D_2 receptors (Felder et al. 1998). AM251 and AM281 are both structural analogues of SR141716A. They have been found to displace [^{3}H]SR141716A from binding sites on mouse cerebellar membranes with respectively three and eight times less potency than SR141716A (Gatley et al. 1998), and both compounds have also been shown to bind more readily to CB_1 than CB_2 receptors (Table 3). There are numerous reports that, like SR141716A, AM251 and AM281 can attenuate in vivo or in vitro responses to established cannabinoid receptor agonists (e.g. Cosenza et al. 2000; Gifford et al. 1997; Hájos and Freund 2002a; Lan et al. 1999a; Simoneau et al. 2001).

Although SR141716A is CB_1-selective, it is not CB_1-specific. Thus, results from binding experiments indicate that whilst it may be reasonable to assume that concentrations of this ligand in the low or mid nanomolar range will interact mainly with the CB_1 receptors when it is applied to tissues that contain both CB_1 and CB_2 receptors, this is not so for higher concentrations of SR141716A (Table 3). Results obtained in vitro from functional bioassays also suggest that CB_1 receptors are not the only pharmacological targets with which this compound can interact at micromolar concentrations. For example, it has been found that SR141716A can stimulate extracellular-signal-regulated protein kinase (ERK) at 1 μM (Berdyshev et al. 2001) and antagonize anandamide-induced vasodilation in the mesenteric arteries of $CB_1^{-/-}$ mice at 1 and 5 μM (Járai et al. 1999). In addition there are reports that at concentrations above 1 μM, SR141716A can both block and activate transient receptor potential vanilloid type 1 (TRPV1) receptors (previously known as VR1 receptors), suggesting that it may be a TRPV1 receptor partial agonist (De Petrocellis et al. 2001a; Zygmunt et al. 1999), block adenosine A_1 receptors (as can AM251) (Savinainen et al. 2003), oppose vasorelaxation induced by acetylcholine in ring preparations of rabbit preconstricted isolated superior mesenteric arteries (Chaytor et al. 1999) and by bradykinin in human preconstricted myometrial small arteries (Kenny et al. 2002), and block potassium and L-type calcium channels in rat isolated mesenteric arteries (White and Hiley 1998) and gap junctions between COS-7 cells (Chaytor et al. 1999).

Unexpectedly, in spite of the close similarity between the structures of AM251, AM281 and SR141716A, differences in their pharmacological profiles have been detected in vitro in experiments with cardiovascular tissue (reviewed in Pertwee 2004a). It has also been found that the ability of *R*-(+)-WIN55212 to reduce glutamatergic transmission is opposed by 1 μM SR141716A in $CB_1^{-/-}$ mouse hippocampal slices but not by 2 μM AM251 in rat hippocampal slices (Hájos and Freund 2002a; Hájos et al. 2001).

3.2.2 Selective CB_2 Receptor Antagonists

The most important selective CB_2 receptor antagonists are the diarylpyrazole SR144528 and the aminoalkylindole 6-iodopravadoline (AM630) (Fig. 10). Both bind with markedly higher affinity to CB_2 than CB_1 receptors (Table 3) and prevent or reverse in vitro effects mediated by CB_2 receptors (Portier et al. 1999; Rinaldi-Carmona et al. 1998; Ross et al. 1999a). Evidence also exists that on the one hand, SR144528 lacks significant affinity for a wide range of established non-cannabinoid receptors (Rinaldi-Carmona et al. 1998), and on the other hand it is an antagonist for a putative CB_2-like receptor that is activated by palmitoylethanolamide, a ligand that does not have significant CB_2 receptor affinity (Sect. 4.1.3). Interestingly, it has proved possible to develop diarylpyrazoles with even greater CB_2 selectivity and affinity than SR144528 (Mussinu et al. 2003). This has been achieved by making these molecules less flexible.

Turning now to AM630, particularly with regard to its behaviour at the CB_1 receptor, there are several reports that when administered at concentrations in the micromolar range, it exhibits the mixed agonist-antagonist properties typical of a weak partial agonist for this receptor (reviewed in Pertwee 1999a). However, there are also reports that AM630 can behave as a CB_1 receptor inverse agonist (Landsman et al. 1998; Vásquez et al. 2003).

3.3 Inverse Agonism at Cannabinoid Receptors

There is good evidence that when administered by itself in vivo or in vitro, SR141716A is capable of producing inverse cannabimimetic effects, i.e. effects that are opposite in direction to those produced by the activation of CB_1 receptors (reviewed in Pertwee 2003). There are also reports that such inverse effects can be induced by the other cannabinoid receptor antagonists described in Sect. 3.2: AM251 (Vásquez et al. 2003), AM281 (Cosenza et al. 2000; Gifford et al. 1997; Izzo et al. 2000; Vásquez et al. 2003), LY320135 (Felder et al. 1998) and AM630 (Sect. 3.2.2) at CB_1 receptors and SR144528 (Portier et al. 1999; Rinaldi-Carmona et al. 1998; Ross et al. 1999b), AM630 (New and Wong 2003; Ross et al. 1999a) and AM251 (New and Wong 2003) at CB_2 receptors. These effects include SR141716A- and AM281-induced hyperkinesia in rats and/or mice (Compton et al. 1996; Cosenza et al. 2000; Costa and Colleoni 1999) and the attenuation in vitro of CB_1 or CB_2 receptor signalling. Two other compounds, the CB_2-selective ligands JTE-907 and BML-190 (Figs. 7 and 9), also behave as CB_2 receptor inverse agonists (Iwamura et al. 2001; New and Wong 2003). However, whether JTE-907 or BML-190 produces antagonism at CB_2 receptors has not been reported.

Whereas some inverse cannabimimetic effects of SR141716A may be produced as a result of antagonism of responses to endogenously released endocannabinoids, there is evidence that others are not, prompting the hypothesis that this compound is an inverse agonist that can elicit responses at CB_1 receptors that are opposite in direction from those elicited by conventional agonists. This turn has been taken to indicate that CB_1 receptors can exist in two or more interchangeable conformations (reviewed in Pertwee 2003, 2005). More specifically, it has been proposed that these are (1) a constitutively active “on” state in which the receptors are functionally coupled to their effector mechanisms even in the absence of exogenously added or endogenously produced cannabinoid receptor agonists and (2) one or more “off” states in which the receptors are uncoupled from their effector mechanisms. According to this hypothesis, agonists increase the proportion of receptors in the “on” state, inverse agonists increase the proportion of receptors in the “off” state(s) and neutral antagonists leave the number of receptors in each state unchanged.

There is evidence that SR141716A exhibits greater potency in opposing effects induced by CB_1 agonists than in producing inverse effects at CB_1 receptors by itself (e.g. Sim-Selley et al. 2001). This raises the possibilities, first, that SR141716A may be a neutral CB_1 receptor antagonist at low concentrations that exhibits additional CB_1 inverse agonist activity only at higher concentrations, and secondly, that SR141716A may have two sites of action on the CB_1 receptor, one at which it displaces agonists to produce antagonism and another at which it somehow induces inverse agonism, perhaps through an allosteric mechanism (Sim-Selley et al. 2001).

Although it is likely that at least some of the inverse effects produced by SR144528 or AM630 at CB_2 receptors are also due to inverse agonism, no attempts have been made to establish this conclusively. It is noteworthy, therefore, that the finding that a maximal concentration of SR144528 enhances forskolin-stimulated cyclic AMP production by human (h)CB_2-transfected CHO cells considerably more than a maximal concentration of AM630 (Ross et al. 1999a,b) can be better explained in terms of inverse agonism at the CB_2 receptor than in terms of antagonism of endogenously released endocannabinoids. This is because the simplest explanation for this difference between the maximal inverse effects of these two ligands is that

SR144528 has greater inverse intrinsic activity than AM630. If this interpretation of the data is valid, it is of course an indication that just as the intrinsic activities of CB_1 and CB_2 receptor agonists can vary from compound to compound, so too the (inverse) intrinsic activities of cannabinoid receptor inverse agonists will not be the same for all such ligands.

Whilst there is little doubt that the presence of CB_1 receptors is a prerequisite for the production by SR141716A of many of its inverse cannabimimetic effects, it is noteworthy that this compound has been found to produce an effect on GTPγS binding to whole brain membranes obtained from $CB_1^{-/-}$ mice (enhancement) opposite to that produced by *R*-(+)-WIN55212 or anandamide (inhibition) (Breivogel et al. 2001). This finding supports the hypothesis that at least some apparent inverse effects of SR141716A may be induced at sites that are not located on CB_1 receptors (Sim-Selley et al. 2001). Indeed, it is already known that SR141716A not only binds to CB_2 receptors at concentrations in the high nanomolar range and above (Table 3) but also behaves as a CB_2 receptor inverse agonist at such concentrations, as measured by inhibition of [^{35}S]GTPγS binding to hCB_2 receptors on CHO cell membranes (MacLennan et al. 1998).

3.4 Neutral Antagonism at Cannabinoid Receptors

An important recent pharmacological objective has been the development of cannabinoid receptor ligands for CB_1 and CB_2 receptors that completely lack both inverse agonist and agonist properties (neutral antagonists). One cannabinoid receptor ligand that comes close to being a neutral antagonist is 6′-azidohex-2′-yne-Δ^8-THC (O-1184; Fig. 11 and Table 4), as this behaves as a high-affinity, low-efficacy agonist at CB_1 receptors and as a high-affinity, low-efficacy inverse agonist at CB_2 receptors, and as it produces potent antagonism of *R*-(+)-WIN55212 and CP55940 in the myenteric plexus–longitudinal muscle preparation of guinea-pig small intestine (Ross et al. 1998, 1999b). More recently, an analogue of SR141716A, NESS 0327, has been developed that behaves as a neutral CB_1 receptor antagonist and is markedly more potent and CB_1-selective than SR141716A (Table 3) (Ruiu et al. 2003). This was achieved by reducing the molecule's flexibility through the introduction of a seven-membered ring (Fig. 11). Evidence has also emerged that insertion of a 6″-azidohex-2″-yne side chain into cannabidiol (Fig. 1) converts this molecule into a neutral cannabinoid receptor antagonist (Thomas et al. 2004). This compound, O-2654 (Fig. 11), has markedly higher affinity than cannabidiol for CB_1 receptors and antagonizes *R*-(+)-WIN55212-induced inhibition of electrically evoked contractions of the mouse isolated vas deferens in a competitive, surmountable manner with a K_B (85.7 nM) that is close to its K_i for displacing [^{3}H]CP55940 from CB_1 receptors (114 nM). The conclusion that O-2654 may be a neutral antagonist is based on the observation that at concentrations of up to 10 μM, it exhibits no detectable CB_1 agonist or inverse agonist properties in the mouse isolated vas deferens. Thus, unlike SR141716A (Pertwee et al. 1996b), O-2654 does not increase the amplitude of electrically evoked contractions of this

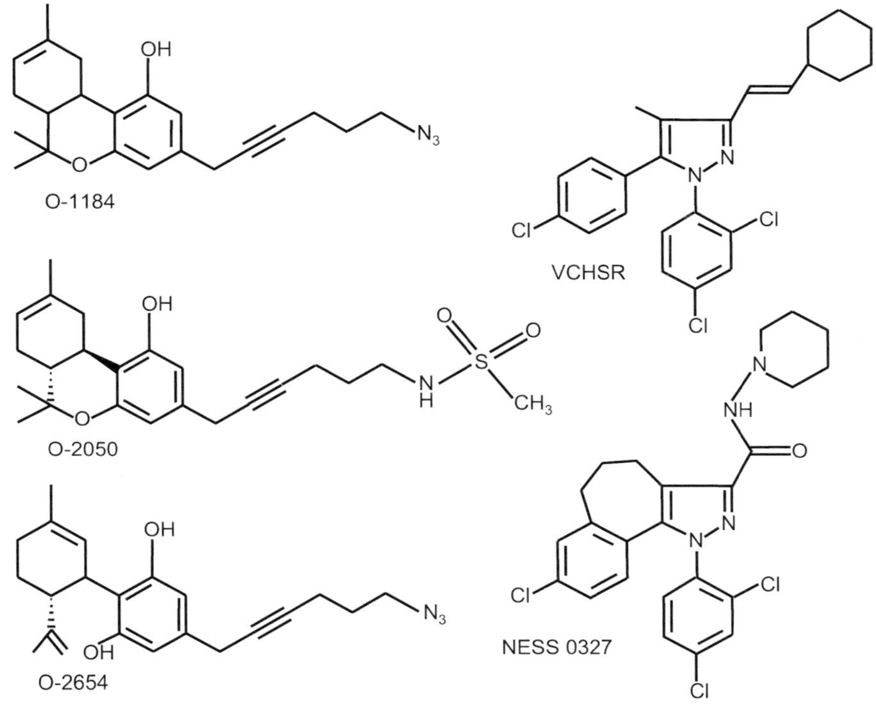

Fig. 11. The structure of O-1184 and of some putative neutral cannabinoid receptor antagonists

preparation. Nor does it share the ability of the CB_1 partial agonist, O-1184, to inhibit these contractions (Ross et al. 1999b). O-2050, a sulphonamide analogue of Δ^8-THC with an acetylenic side chain also behaves as a neutral CB_1 receptor antagonist in the mouse vas deferens (Martin et al. 2002). Another compound that seems to be a neutral CB_1 antagonist is VCHSR (Fig. 11). This is an analogue of SR141716A that lacks hydrogen bonding capability in its C3 substituent region and has a CB_1 K_i value in the low nanomolar range. VCHSR (1 μM) has been found to share the ability of SR141716A to attenuate *R*-(+)-WIN55212-induced inhibition of Ca^{2+} current in rat superior cervical ganglion neurons expressing the human CB_1 receptor but to differ from SR141716A in not affecting Ca^{2+} current in these neurons when administered by itself at 1 or 10 μM (Hurst et al. 2002; Pan et al. 1998). In terms of the two-state model of inverse agonism (see Pertwee 2003, 2005 and Sect. 3.3), this finding suggests that preferential binding by SR141716A to the "off" state of the CB_1 receptor is determined by hydrogen bond formation between the C3 substituent of this molecule and the receptor. Further experiments are required to establish whether putative neutral antagonists, such as NESS 0327, O-2654 and O-2050, resemble SR141716A (Sect. 3.3) in exhibiting inverse agonist properties at concentrations above those at which they behave as neutral antagonists.

4
Other Pharmacological Targets for Cannabinoids in Mammalian Tissues

As discussed in greater detail elsewhere (Hájos and Freund 2002b; Howlett et al. 2002; Pertwee 1999b, 2004a; Pertwee and Ross 2002; Wiley and Martin 2002), evidence is emerging that in addition to CB_1 and CB_2 receptors, there are other pharmacological targets in mammalian tissues with which at least some established CB_1 and/or CB_2 receptor agonists can interact to elicit pharmacological responses.

4.1
Receptors

4.1.1
Vanilloid Receptors

It is now generally accepted that the endogenous CB_1/CB_2 receptor agonist, anandamide, and certain of its analogues are agonists for the TRPV1 receptor (reviewed in Howlett et al. 2002; Pertwee 2004a; Pertwee and Ross 2002; Ross 2003). This receptor is a non-selective cation channel that is present on sensory neurons in tissues such as skin, heart, blood vessels and lung, and an important consequence of its activation is the release of sensory neuropeptides that then produce effects such as pain, tachycardia, vasodilation and bronchoconstriction. It is noteworthy, however, that anandamide has less TRPV1 intrinsic activity than the well-known TRPV1 receptor agonist capsaicin (Ross 2003; Ross et al. 2001). *R*-(+)-methanandamide is even less potent or effective than anandamide at activating TRPV1 receptors (Ross et al. 2001; Zygmunt et al. 1999), whereas lipoxygenase metabolites of anandamide show greater potency at these receptors than their parent compound, at least in guinea-pig bronchus (Craib et al. 2001; Pertwee and Ross 2002). The TRPV1 receptor is not activated by 2-arachidonoyl glycerol or by non-eicosanoid CB_1/CB_2 receptor agonists (Zygmunt et al. 1999), although it is activated by micromolar concentrations of the phytocannabinoid cannabidiol (Bisogno et al. 2001). One compound that behaves as a potent agonist at both TRPV1 and CB_1 receptors is the synthetic anandamide analogue O-1861 (Fig. 8) (Di Marzo et al. 2001). TRPV1 and CB_1 receptors have opposite effects on calcium channel conductance, and there are several reports that in cells such as cultured dorsal root ganglion neurons that co-express these receptors, responses elicited by TRPV1 receptor activation can be opposed by the simultaneous activation of CB_1 receptors (Ahluwalia et al. 2003; Ellington et al. 2002; Millns et al. 2001; Richardson et al. 1998; Ross 2003). Unexpectedly, however, there is also a report that in human embryonic kidney cells co-transfected with CB_1 and TRPV1 receptors, activation of the CB_1 receptors increases the sensitivity of the TRPV1 receptors to subsequent (but not simultaneous) activation (Hermann et al. 2003). Under physiological conditions, TRPV1 receptors on primary sensory neurons are less sensitive to anandamide than CB_1 receptors (Németh et al. 2003; Tognetto et al. 2001). There is also evidence that anandamide production increases during inflammation, raising the possibility that in healthy tissue, one role of anandamide may be to act through CB_1 receptors to oppose any increase in the excitability of sensory neurons, whilst in pathological states such as inflammation, anandamide concentrations and TRPV1 receptor sensitivity increase to the extent that anandamide-induced activation of TRPV1 receptors becomes sufficient to cause an increase in the excitability of sensory neurons (Ahluwalia et al. 2003). Although there is little doubt that anandamide is an endogenous agonist for CB_1 and CB_2 receptors, the question of whether it also serves as an endogenous TRPV1 agonist under normal or pathological conditions has still to be resolved. Also currently uncertain is the extent to which CB_1 and TRPV1 receptors are co-expressed on the same neurons (reviewed in Ross 2003).

4.1.2
CB_1 Receptor Subtypes

Shire et al. (1995) have isolated a spliced variant of CB_1 cDNA (CB_{1A}) from a human lung cDNA library. CB_{1A} mRNA is present in human brain tissue, its distribution pattern matching that of CB_1 mRNA. It has also been detected in peripheral tissues. The spliced variant resembles the CB_1 receptor in its affinity for Δ^9-THC, CP55940 and *R*-(+)-WIN55212, and it also has at least two signal transduction mechanisms in common with the CB_1 receptor (Rinaldi-Carmona et al. 1996a). However, the central and peripheral concentrations of CB_{1A} mRNA are far below those of CB_1 mRNA (Shire et al. 1995). Onaivi et al. (1996) have discovered three distinct CB_1 mRNAs in brain tissue from C57BL/6 mice, although only one CB_1 receptor cDNA. C57BL/6 mice were less sensitive to the hypothermic and antinociceptive effects of Δ^9-THC than two other mouse strains in which only one CB_1 mRNA was detectable.

Results from pharmacological experiments with rats and mice performed by Sandra Welch's group also suggest that there may be more than one subtype of CB_1 receptor (reviewed in Howlett et al. 2002; Pertwee 2001). In mouse experiments, for example, it was found that intraperitoneal SR141716A was more effective in opposing the antinociceptive effects of some CB_1 receptor agonists than of other such agonists when these were administered intrathecally and that intrathecal morphine interacted synergistically with intrathecal THC but not with intrathecal CP55940. Apparent differences between mouse cannabinoid receptors in brain and spinal cord were also detected.

4.1.3
CB_2-Like Receptors

It is possible that palmitoylethanolamide may produce antinociception in rat and mouse models of inflammatory or neuropathic pain by acting on a CB_2-like receptor (Calignano et al. 1998, 2001; Conti et al. 2002; Farquhar-Smith et al. 2002; Farquhar-Smith and Rice 2001; Helyes et al. 2003). The existence of such a receptor is supported by the finding that even though palmitoylethanolamide lacks significant CB_2 receptor affinity or efficacy (Griffin et al. 2000; Lambert et al. 1999;

Sheskin et al. 1997; Showalter et al. 1996), the antinociceptive effects of this fatty acid amide are opposed by SR144528. Evidence for CB_2-like receptors has also been obtained from experiments with the mouse vas deferens (Griffin et al. 1997). Other possibilities, i.e. that palmitoylethanolamide acts through CB_1 or TRPV1 receptors, can be ruled out. Thus, it produces antinociceptive effects that are not opposed by SR141716A (Calignano et al. 1998, 2001; Farquhar-Smith et al. 2002; Farquhar-Smith and Rice 2001) and it has been found not to attenuate nociceptive behaviour induced in mice by intraplantar injection of capsaicin (Calignano et al. 2001). Also, palmitoylethanolamide does not bind to or activate CB_1 receptors at concentrations below 1 or 10 μM (Devane et al. 1992; Felder et al. 1993; Griffin et al. 2000; Lambert et al. 1999; Showalter et al. 1996). Anandamide shares the ability of palmitoylethanolamide to induce antinociception in mice and rats. However, unlike palmitoylethanolamide, it has been found to be susceptible to SR141716A -induced antagonism and resistant to SR144528-induced antagonism in several pain models (Calignano et al. 1998, 2001; Farquhar-Smith and Rice 2001). Also, in contrast to palmitoylethanolamide, anandamide attenuates nociceptive behaviour induced in mice by intraplantar injection of capsaicin (Calignano et al. 2001). Another observation—that palmitoylethanolamide and anandamide interact synergistically rather than additively in the mouse formalin paw and abdominal stretch tests—also supports the hypothesis that they have different antinociceptive mechanisms (Calignano et al. 1998, 2001).

4.1.4
Neuronal Non-CB_1, Non-CB_2, Non-TRPV1 Receptors

Central G Protein-Coupled Receptors for Anandamide and R-(+)-WIN55212

Evidence for the presence of a G protein-coupled non-CB_1, non-CB_2 receptor for anandamide and *R*-(+)-WIN55212 has come from experiments in which it was found that [^{35}S]GTPγS binding to whole-brain membranes from $CB_1^{-/-}$ C57BL/6 mice or to cerebellar homogenates from $CB_1^{-/-}$ CD1 mice could be enhanced by these two cannabinoids (Breivogel et al. 2001; Di Marzo et al. 2000; Monory et al. 2002). Near maximal concentrations of anandamide and *R*-(+)-WIN55212 were not fully additive in their effects on [^{35}S]GTPγS binding to $CB_1^{-/-}$ C57BL/6 brain membranes, supporting the hypothesis that these two agents were acting through a common mechanism (Breivogel et al. 2001). This putative receptor for anandamide and *R*-(+)-WIN55212 appears not to be a TRPV1 receptor (Sect. 4.1.1) or to resemble the proposed abnormal-cannabidiol receptor (Sect. 4.1.5) as neither of these pharmacological targets is *R*-(+)-WIN55212-sensitive and as the TRPV1 receptor is not G protein coupled. However, the possibility does remain that it may be a novel metabotropic "vanilloid-like" receptor (see below). The proposed new receptor also differs from established cannabinoid receptors in several ways.

- It is not sensitive to activation by the established CB_1/CB_2 receptor agonists, Δ^9-THC, CP55940 or HU-210 (Breivogel et al. 2001; Di Marzo et al. 2000; Monory et al. 2002).
- It is not coupled to adenylate cyclase, at least in the cerebellum of $CB_1^{-/-}$ CD1 mice (Monory et al. 2002).
- It differs from the CB_1 receptor in its central distribution pattern (Breivogel et al. 2001; Monory et al. 2002).
- SR141716A and SR144528 do not appear to be competitive antagonists for this putative receptor (Breivogel et al. 2001; Monory et al. 2002).
- There are no specific binding sites for [^{3}H]CP55940 on $CB_1^{-/-}$ C57BL/6 mouse brain membranes (Breivogel et al. 2001).

It has also been found that [^{3}H]*R*-(+)-WIN55212 undergoes selective binding to $CB_1^{-/-}$ C57BL/6 membranes obtained from brain areas in which *R*-(+)-WIN55212 enhances [^{35}S]GTPγS binding (cerebral cortex, hippocampus and brain stem) (Breivogel et al. 2001). Furthermore, $CB_1^{-/-}$ C57BL/6 brain areas that are unresponsive to *R*-(+)-WIN55212-induced enhancement of [^{35}S]GTPγS binding seem to lack [^{3}H]*R*-(+)-WIN55212 binding sites (Breivogel et al. 2001). It is noteworthy, however, that some WIN55212-sensitive brain areas of $CB_1^{-/-}$ C57BL/6 mice (midbrain and diencephalon) and of $CB_1^{-/-}$ CD1 mice (cerebellum) also seem to lack [^{3}H]*R*-(+)-WIN55212 binding sites (Breivogel et al. 2001; Ledent et al. 1999; Monory et al. 2002). Although $CB_1^{-/-}$ C57BL/6 mouse brain does contain specific binding sites for both [^{3}H]SR141716A and [^{3}H]*R*-(+)-WIN55212, these two binding site populations have different distribution patterns (Breivogel et al. 2001). This is further evidence that SR141716A lacks affinity for the proposed *R*-(+)-WIN55212/anandamide receptor.

A pharmacological property that the proposed *R*-(+)-WIN55212/anandamide receptor may share with the CB_1 receptor is the ability to mediate antinociception, catalepsy and hypokinesia. Thus, whilst Δ^9-THC produced these effects only in the wild-type mice, anandamide was essentially as potent and effective in producing these effects in $CB_1^{-/-}$ as in $CB_1^{+/+}$ C57BL/6 mice (Di Marzo et al. 2000). Indeed, this putative new receptor may well prove to be a novel target for anti-spasticity and analgesic drugs (Brooks et al. 2002). The presence of specific binding sites for [^{3}H]SR141716A on $CB_1^{-/-}$ C57BL/6 mouse brain membranes may explain the ability of SR141716A both to inhibit [^{35}S]GTPγS binding to such membranes (Breivogel et al. 2001) and to reduce milk intake and survival of newborn $CB_1^{-/-}$ C57BL/6 mice (Fride et al. 2003).

Central TRPV1-Like Receptors

Evidence has emerged for the presence of G protein-coupled, non-CB_1 receptors on glutamatergic axonal terminals in the hippocampus with which at least some cannabinoid receptor agonists can interact to inhibit glutamate release. More specifically, results from electrophysiological experiments with hippocampal slices obtained from rats or $CB_1^{+/+}$ CD1 mice have shown that *R*-(+)-WIN55212 reduces both excitatory postsynaptic currents (EPSCs) evoked in CA1 pyramidal cells or dentate granule cells and paired pulse facilitation of EPSCs, even though it has not proved possible to detect CB_1 receptor immunostaining on axonal terminals that form glutamatergic synapses in rat hippocampus (Hájos and Freund 2002a; Hájos et al. 2000, 2001). Similar results have been obtained in experiments with $CB_1^{-/-}$ CD1 mouse hippocampal slices (Hájos et al. 2001). *R*-(+)-WIN55212 also inhibits potassium-evoked glutamate release from hippocampal synaptosomes obtained from rats or from $CB_1^{+/+}$ or $CB_1^{-/-}$ mice in an SR141716A- and AM251-independent manner (Köfalvi et al. 2003). Evidence for an involvement of G proteins in the apparent inhibitory effect of *R*-(+)-WIN55212 on glutamate release in mouse hippocampal slices comes from the finding that this effect is pertussis toxin-sensitive (Misner and Sullivan 1999).

The ability of *R*-(+)-WIN55212 to reduce evoked EPSCs in rat hippocampal slices is shared by CP55940 and capsaicin, and all three of these agonists are antagonized by the TRPV1 receptor antagonist capsazepine (Hájos and Freund 2002a). Because the peripheral TRPV1 receptor is neither activated by *R*-(+)-WIN55212 or CP55940 nor coupled to G proteins, it may be that *R*-(+)-WIN55212, CP55940 and capsaicin modulate central glutamate release by acting through a novel metabotropic "vanilloid-like" receptor. Consequently, it would be of interest to establish first whether capsaicin enhances GTPγS binding to brain membranes, and secondly whether *R*-(+)-WIN55212-induced enhancement of GTPγS binding to $CB_1^{-/-}$ mouse brain membranes (see above) can be antagonized by capsazepine.

Evidence for the presence of vanilloid-like receptors in the hippocampus has also been obtained by Al-Hayani et al. (2001). They found paired-pulse depression in the CA1 region of rat hippocampal slices to be increased both by anandamide and by two other TRPV1 receptor agonists, capsaicin and resiniferatoxin, in a manner that was sensitive to antagonism by capsazepine but not by the CB_1 receptor antagonist AM281. Given the results obtained by Hájos et al. (see above), it is possible that these agonists were acting through central vanilloid-like receptors to cause a decrease in excitatory glutamatergic transmission. Alternatively, they may have been acting through these putative receptors to cause an increase in inhibitory γ-aminobutyric acid (GABA)ergic transmission. If anandamide was acting through vanilloid-like receptors, then it apparently activates them more readily than CB_1 receptors, which contrasts with reports that this endocannabinoid interacts less potently with established TRPV1 receptors than with CB_1 receptors (Sect. 4.1.1). In contrast to anandamide, both *R*-(+)-WIN55212 and 2-arachidonoyl glycerol were found to decrease paired-pulse depression in an SR141716A or AM281-sensitive manner (Al-Hayani et al. 2001; Paton et al. 1998). This would suggest that unlike anandamide, these two agonists interact preferentially with CB_1 receptors in this experimental model. There is evidence that anandamide and/or capsaicin can modulate glutamatergic transmission in brain areas other than the hippocampus in a manner that is CB_1-independent and susceptible to antagonism by capsazepine and/or iodoresiniferatoxin. These brain areas include rat locus coeruleus, substantia nigra and medullary dorsal horn (Jennings et al. 2003; Marinelli et al. 2002,

2003). In these experiments, however, glutamatergic transmission was facilitated by anandamide and/or capsaicin.

There is currently no support for the hypothesis that *R*-(+)-WIN55212 inhibits glutamate release in the hippocampus by acting on the non-CB_1, non-CB_2 molecular target that is thought to mediate its enhancement of GTPγS binding to central neuronal membranes (see above). Thus, although *R*-(+)-WIN55212 does suppress evoked EPSCs and paired pulse facilitation in $CB_1^{-/-}$ CD1 mouse hippocampal slices (Hájos et al. 2001), it does not enhance GTPγS binding to $CB_1^{-/-}$ CD1 mouse hippocampal membranes (Monory et al. 2002). Also, whilst CP55940 suppresses evoked EPSCs in rat hippocampal slices (Hájos and Freund 2002a) and potassium-evoked glutamate release from rat hippocampal synaptosomes (Köfalvi et al. 2003), it does not share the ability of *R*-(+)-WIN55212 or anandamide to enhance GTPγS binding to $CB_1^{-/-}$ C57BL/6 mouse brain membranes (Breivogel et al. 2001).

Peripheral Nervous System

Results from experiments with phenylephrine-precontracted rat isolated mesenteric and hepatic arteries suggest that Δ^9-THC can relax these vessels by acting on capsaicin-sensitive perivascular sensory neurons to induce release of calcitonin gene-related peptide (Zygmunt et al. 2002). The underlying mechanism is most probably CB_1 and CB_2 receptor-independent, as this relaxant effect of Δ^9-THC was not prevented by 300 nM SR141716A or by 30 nM AM251 and as the CB_1/CB_2 receptor agonists HU-210 and CP55940 lacked detectable relaxant activity, whereas cannabinol, which has relatively low activity as a cannabinoid receptor agonist (Sect. 3.1), was equipotent with Δ^9-THC. The possibility, that Δ^9-THC was acting through ionotropic or metabotropic glutamate receptors was also excluded. Other observations made in this investigation were that Δ^9-THC- and cannabinol-induced activation of CGRP release from rat arterial segments could be prevented by capsaicin pretreatment and that Δ^9-THC- and cannabinol-induced relaxations of precontracted arterial segments could be attenuated by the noncompetitive TRPV1 antagonist ruthenium red. However, these cannabinoids were most probably not acting through TRPV1 receptors in these experiments. Thus, the competitive TRPV1 antagonist capsazepine did not attenuate the vasorelaxant effects of Δ^9-THC and cannabinol, and in contrast to both capsaicin and anandamide, Δ^9-THC also relaxed phenylephrine-precontracted mesenteric arterial segments that had been obtained from TRPV1$^{-/-}$ mice. In more recent experiments, Jordt et al. (2004) have obtained evidence that Δ^9-THC and cannabinol may have induced vasorelaxation by acting through ANKTM1, another member of the transient receptor potential (TRP) family of ion channels that, unlike the TRPV1 receptor, appears to be insensitive to anandamide and is implicated in the detection of noxious cold. ANKTMI was found to be insensitive to HU-210, CP55940 and 2-arachidonoyl glycerol.

It has also been proposed that the terminals of sympathetic neurons supplying cardiovascular tissue express a non-I_1, non-I_2 subtype of the putative imidazoline receptor that is both CB_1 receptor-like and α_2-adrenoceptor-like and that mediates inhibition of evoked noradrenaline release when activated (reviewed in

Göthert et al. 1999; Molderings and Göthert 1999; Pertwee 2004a). There is evidence that this putative receptor can be activated both by the cannabinoids—CP55940, *R*-(+)-WIN55212 and anandamide—and by non-CB_1, non-CB_2 ligands such as aganodine and clonidine, and that this activation is sensitive to antagonism by SR141716A (1 μM), LY320135 (0.1 or 1 μM) and rauwolscine (30 μM) (reviewed in Pertwee 2004a). It also appears that this proposed receptor may belong to the G protein-coupled receptor family originally known as endothelial differentiation gene (EDG) receptors and that it can be activated by 1-oleoyl-lysophosphatidic acid (Molderings et al. 2002).

Mang et al. (2001) have obtained evidence that anandamide can act on nerve terminals of the myenteric plexus–longitudinal muscle preparation of the guinea-pig ileum to inhibit electrically evoked release of the contractile transmitter acetylcholine through a mechanism that is independent of both TRPV1 and CB_1 receptors. Thus, the inhibitory effects of anandamide on electrically evoked release of [^{3}H]acetylcholine and on electrically evoked contractions of this isolated tissue preparation were insensitive to antagonism by 1 μM capsazepine. They were also much less sensitive to antagonism by SR141716A than expected for CB_1-mediated effects. Results from other experiments with this tissue preparation suggest that anandamide can increase both basal acetylcholine release from neurons and longitudinal muscle tone by acting on neuronal TRPV1 receptors (Mang et al. 2001). Additional support for the presence of a non-CB_1 receptor for anandamide in the gastro-intestinal tract comes from experiments both with the strips of longitudinal muscle obtained from guinea-pig distal colon (Kojima et al. 2002) and with the rat isolated gastric fundus (Storr et al. 2002). In the colon experiments, evidence was obtained that anandamide, possibly after its conversion to active metabolites, can induce contractions by acting through a TRPV1 and CB_1 receptor-independent mechanism (Kojima et al. 2002). 2-Arachidonoyl glycerol also seems to act through such a mechanism to induce contractions of this tissue preparation (Kojima et al. 2002). In the gastric fundus experiments it was found that at 10 μM, the CB_2-selective antagonist AM630 attenuated anandamide- but not *R*-(+)-WIN55212-induced inhibition of electrically evoked contractions (Storr et al. 2002). It is likely that anandamide was acting on prejunctional neurons in this tissue, as it did not affect contractions produced by 5-HT or carbachol. AM630 has also been found to antagonize Δ^9-THC, CP55940, *R*-(+)-WIN55212, methanandamide and anandamide in the mouse isolated vas deferens in an agonist-dependent and competitive manner. However, in this bioassay system, AM630 was less potent as an antagonist of anandamide than of *R*-(+)-WIN55212 (Pertwee et al. 1995). In view of evidence that the mouse vas deferens expresses neuronal CB_2-like receptors that can mediate inhibition of electrically evoked contractions (Griffin et al. 1997; Sect. 4.1.3), it may be that AM630 was producing its antagonism of cannabinoids in this tissue by competing for these putative CB_2-like receptors.

4.1.5
Receptors for Abnormal-Cannabidiol

Cardiovascular System

There is evidence, mainly from in vitro experiments with rat or mouse phenylephrine- or methoxamine-precontracted buffer-perfused isolated mesenteric arterial beds or isolated mesenteric arterial segments, for the presence in these tissues of non-CB_1, non-CB_2 receptors with which anandamide and methanandamide can interact to induce a relaxant effect (reviewed in Howlett et al. 2002; Pertwee 2004a; Wiley and Martin 2002). There are several reasons for believing that these are not CB_1 or CB_2 receptors. First, relaxation is not induced in rat precontracted mesenteric arterial beds by 2-arachidonoyl glycerol or by established non-eicosanoid cannabinoid receptor agonists such as Δ^9-THC or *R*-(+)-WIN55212 (Wagner et al. 1999) but is induced in rat and mouse precontracted mesenteric arterial beds or rat precontracted mesenteric arterial segments by two cannabidiol analogues, abnormal-cannabidiol and O-1602 (Fig. 12), neither of which exhibits significant affinity for CB_1 receptors (Ho and Hiley 2003; Járai et al. 1999; Offertáler et al. 2003; Showalter et al. 1996). Second, anandamide, methanandamide and abnormal-cannabidiol also relax precontracted buffer-perfused mesenteric arterial beds of $CB_1^{-/-}$ knockout or $CB_1^{-/-}/CB_2^{-/-}$ double-knockout C57BL6J mice (Járai et al. 1999). Third, the CB_1-selective antagonist AM281 (1 μM) and the CB_2-selective antagonist AM630 (10 μM) do not attenuate abnormal-cannabidiol-induced relaxations of rat precontracted mesenteric arterial segments (Ho and Hiley 2003). Although SR141716A has been found to oppose the vasorelaxant effects of abnormal-cannabidiol, methanandamide and anandamide in rat or mouse precontracted mesenteric arterial beds or segments, this is generally with a potency lower than expected from its affinity for CB_1 receptors (Ho and Hiley 2003; Járai et al. 1999). Negative results obtained with capsaicin and capsazepine also make it unlikely that the putative "abnormal-cannabidiol" receptor is a TRPV1 receptor (Ho and Hiley 2003; Járai et al. 1999; Offertáler et al. 2003).

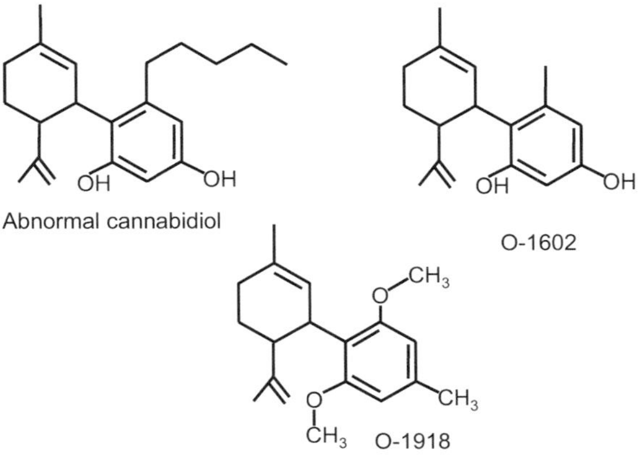

Fig. 12. The structures of abnormal cannabidiol, O-1602 and O-1918

One cannabidiol analogue has been found to behave as a selective abnormal-cannabidiol receptor antagonist. This is O-1918 (Fig. 12), which lacks detectable affinity for CB_1 and CB_2 receptors and, at concentrations of 1 to 30 μM, opposes abnormal-cannabidiol and anandamide-induced relaxations of rat arterial segments and does not reduce vasomotor tone when administered alone (Offertáler et al. 2003). It has also been found to attenuate abnormal-cannabidiol-induced hypotension in anaesthetized mice at doses not affecting hypotension induced by the CB_1/CB_2 receptor agonist HU -210 (Offertáler et al. 2003). Cannabidiol also behaves as a selective abnormal-cannabidiol receptor antagonist in both the rat mesenteric arterial bed and the anaesthetized mouse (Járai et al. 1999). However, in contrast to O-1918, it has been found to share the ability of abnormal-cannabidiol to relax rat precontracted mesenteric arterial segments (Offertáler et al. 2003).

It is likely that there are two sub-types of abnormal-cannabidiol-sensitive receptor in mesenteric arteries capable of mediating a relaxant effect, one expressed by endothelial cells and the second by non-endothelial cells (reviewed in Pertwee 2004a). Activation of the endothelial receptor appears to open large conductance calcium-activated potassium (BK_{Ca}) channels, whereas the non-endothelial receptor seems to signal mainly through inhibition of L-type calcium channels (Begg et al. 2003; Ho and Hiley 2003; Járai et al. 1999; Offertáler et al. 2003). There is also now evidence that abnormal-cannabidiol receptors can mediate stimulation of the migration of vascular endothelial cells through a mechanism that is $G_{i/o}$ protein-coupled and susceptible to antagonism by O-1918 (Mo et al. 2004).

Microglial Cells

Experiments with the mouse microglial cell line BV-2 (Walter et al. 2003) have provided evidence that microglial cells express receptors that have certain properties in common with the putative vascular abnormal-cannabidiol receptor discussed above. These include susceptibility to activation by abnormal-cannabidiol and anandamide and to blockade by O-1918 and lack of sensitivity to activation by Δ^9-THC, at least at concentrations below 3 μM. When activated, these proposed abnormal-cannabidiol-sensitive receptors appear to trigger chemokinetic and chemotaxic migration of microglial cells. Such migration can also be induced by 2-arachidonoyl glycerol (EC_{50}=25 nM). This endocannabinoid seems to act through both microglial CB_2 receptors and microglial abnormal-cannabidiol-sensitive receptors, since it is antagonized by cannabidiol at 300 nM and by SR144528 at 30 nM but not by 30 nM SR141716A (Walter et al. 2003). Indeed, it has been proposed that microglial CB_2 receptors and abnormal-cannabidiol receptors interact in a synergistic manner when triggering the migration of microglial cells (Walter et al. 2003). This could explain why the CB_1-selective agonist ACPA (Sect. 3.1), induces microglial cell migration at concentrations well below those at which it has been reported to bind to CB_2 receptors, as this compound appears to induce migration by acting on both abnormal-cannabidiol-sensitive receptors and CB_2 receptors (Franklin and Stella 2003). By itself, cannabidiol behaves as a weak partial agonist, producing a slight enhancement of basal migration (EC_{50}=250 nM) (Walter et al. 2003). Microglial cells are thought to migrate towards neuroinflammatory lesion sites and to release proinflammatory cytokines and cytotoxic agents at these sites. Consequently, since Walter et al. (2003) also obtained evidence that the production of 2-arachidonoyl glycerol by microglial cells can be increased by a pathological stimulus, it may be that a CB_2 receptor antagonist and/or an antagonist of the putative abnormal-cannabidiol receptor could come to play a part in the clinical management of neuroinflammation. More recently, evidence has emerged that BV-2 microglial cells express non-CB_1, non-CB_2, non-CB_2-like, non-TRPV1, non-abnormal-cannabidiol G_i/G_o-coupled-receptors upon which the endogenous fatty acid amide palmitoylethanolamide can act at concentrations in the low nanomolar range to potentiate anandamide- but not 2-arachidonoyl glycerol-induced migration of these cells (Franklin et al. 2003). There is also evidence for the presence in rat migroglial cells of non-CB_1, non-CB_2, pertussis toxin-insensitive receptors with which *R*-(+)- but not *S*-(−)-WIN55212 can interact to inhibit lipopolysaccharide-induced release of the proinflammatory cytokine tumour necrosis factor-α (Facchinetti et al. 2003).

Mouse Vas Deferens

A finding that abnormal-cannabidiol and cannabidiol can attenuate phenylephrine-induced contractions of the mouse isolated vas deferens points to the presence of abnormal-cannabidiol-sensitive receptors in the smooth muscle cells of this tissue (Pertwee et al. 2002; Thomas et al. 2004). Cannabidiol also decreases methoxamine and noradrenaline-induced contractions of the mouse vas deferens and antagonizes phenylephrine and noradrenaline in an insurmountable manner (Pertwee et al. 2002). It may be, therefore, that cannabidiol, and possibly also abnormal-cannabidiol, are negative allosteric modulators of the α_1-adrenoceptor.

4.2 Allosteric Sites

There is evidence for the presence of allosteric sites for anandamide and/or certain other cannabinoids on several non-cannabinoid receptors (reviewed in Pertwee 2004a). These are 5-HT_2 receptors (Cheer et al. 1999), 5-HT_3 receptors (Barann et al. 2002; Fan 1995; Godlewski et al. 2003; Oz et al. 2002), α_1-adrenoceptors (Sect. 4.1.5), M_1 and M_4 muscarinic receptors (Christopoulos and Wilson 2001) and α-amino-3-hydroxy-5-methyl-4-isoxazolepropionic acid (AMPA) GLU_{A1} and GLU_{A3} glutamate receptors (Akinshola et al. 1999a,b). The functional consequences of occupation of the proposed allosteric sites on 5-HT_2 receptors (by HU-210) and on M_1 and M_4 receptors (by anandamide, methanandamide and SR141716A) have yet to be determined. However, cannabinoids have been found to inhibit currents triggered by the activation of GLU_{A1} and GLU_{A3} receptors (anandamide) or 5-HT_3 receptors (Δ^9-THC, *R*-(+)-WIN55212, anandamide, JWH-015 (Fig. 7), CP55940 and the CB_1 receptor antagonist, LY320135). Cannabinoids have also been found to attenuate the von Bezold-Jarisch reflex induced in urethane-anaesthetized rats by 5-HT_3 receptor activation (CP55940 and *R*-(+)-WIN55212) and to oppose α_1-

adrenoceptor-mediated contractions of the mouse vas deferens (cannabidiol). In addition, there are reports that 2-arachidonoyl glycerol and 5-HT each binds more readily to washed human platelets in the presence of the other compound (Maccarrone et al. 2003) and that 5-HT enhances binding of *R*-(+)-WIN55212 to CB_1 receptors (Devlin and Christopoulos 2002). Importantly, cannabinoids inhibited 5-HT_3 receptor currents in transfected human embryonic kidney cells with a rank order of potency, Δ^9-THC>*R*-(+)-WIN55212>anandamide>JWH-015>LY320135>CP55940 (Barann et al. 2002), that does not correlate with their CB_1 or CB_2 receptor affinities or intrinsic activities (Sect. 3). The IC_{50} values of these ligands were 38, 104, 130, 147, 523 and 648 nM, respectively (Barann et al. 2002). In contrast, the IC_{50} values of anandamide for inhibition of kainate-activated currents in GLU_{A1}- and GLU_{A3}-transfected *Xenopus laevis* oocytes exceeded 100 μM (Akinshola et al. 1999b). In addition, some cannabinoids, including anandamide, methanandamide, *R*-(+)-WIN55212, Δ^9-THC and cannabidiol, may serve as negative modulators of delayed rectifier potassium channels (reviewed in Pertwee 2004a). There is also evidence that nanomolar concentrations of anandamide can block low-voltage-activated (T-type) calcium channels through a mechanism that is independent of CB_1 and CB_2 receptors and of G proteins (Chemin et al. 2001). Evidence has also recently emerged for the presence of an allosteric site on the cannabinoid CB_1 receptor (R. Pertwee, R. Ross and M. Price, unpublished).

4.3 Some CB_1- and CB_2-Independent Actions of Cannabidiol, HU-211 and Other Phenol-Containing Cannabinoids

4.3.1 Neuroprotective Actions

Cannabinoids that contain a phenol group possess anti-oxidant (electron donor) activity that is sufficient to protect neurons against oxidative stress associated, for example, with glutamate-induced excitoxicity. Thus, as discussed in greater detail elsewhere (El-Remessy et al. 2003; Fowler 2003; Hampson et al. 1998, 2000; Marsicano et al. 2002; Mechoulam et al. 2002; Pertwee 2004b; Platt and Drysdale 2004; van der Stelt et al. 2002), this anti-oxidant activity is apparently independent of CB_1 or CB_2 receptors as it is exhibited both by the CB_1/CB_2 agonists Δ^9-THC, HU-210 and CP55940, and by the non-psychoactive phytocannabinoid cannabidiol (Fig. 1) and the *cis* (6a*S*, 10a*S*) enantiomer of 11-hydroxy-Δ^8-THC-dimethylheptyl, HU-211 (Fig. 4), neither of which has significant affinity for CB_1 or CB_2 receptors (Table 4). Moreover, neurons of $CB_1^{-/-}$ mice are no less well protected from oxidative stress by phenolic cannabinoids than neurons of $CB_1^{+/+}$ mice (Marsicano et al. 2002). The neuroprotective properties of HU-211 are also thought to stem from its ability to behave as a non-competitive antagonist at *N*-methyl-D-aspartate (NMDA) receptors and to inhibit tumour necrosis factor-α production (Mechoulam et al. 2002; Darlington 2003), and it is possible that cannabidiol may also protect from glutamate-induced excitotoxicity by opposing metabotropic glutamate receptor-mediated release of calcium from intracellular stores (Drysdale et al. 2004). Non-phenolic cannabinoids have been reported to lack anti-oxidant activity (Marsicano et al. 2002). Even so, some non-phenolic (and phenolic) cannabinoids can protect against glutamate-induced excitotoxicity by acting through receptors to inhibit neuronal glutamate release (possibly putative TRPV1-like receptors; Sect. 4.1.4) and calcium entry into neurons through N- and P/Q-type channels (CB_1 receptors) (Fowler 2003; Mechoulam et al. 2002; van der Stelt et al. 2002).

4.3.2
Other Actions of Cannabidiol

Results from in vitro experiments suggest that cannabidiol has a number of CB_1/CB_2 receptor-independent actions through which it may affect neurotransmission (reviewed in Pertwee 1988, 2004b). For example, there is evidence that at concentrations in the nanomolar or low micromolar range, this cannabinoid enhances spontaneous or evoked release of certain transmitters, antagonizes *R*-(+)-WIN55212- and CP55940-induced inhibition of electrically evoked contractile transmitter release in the mouse isolated vas deferens through a CB_1-independent mechanism and inhibits the uptake of calcium, 5-HT, noradrenaline and dopamine by rat or mouse synaptosomes. Higher concentrations of cannabidiol inhibit anandamide uptake by rat basophilic leukaemia cells, the metabolism of this endocannabinoid by fatty acid amide hydrolase and the synaptosomal uptake of GABA. There is also evidence that cannabidiol is a TRPV1 receptor agonist, a ligand for the putative abnormal-cannabidiol receptor (Sect. 4.1.5) and a negative allosteric modulator of α_1-adrenoceptors (Sect. 4.1.5) and delayed rectifier potassium channels (Sect. 4.2). In addition, cannabidiol inhibits/induces certain cytochrome P450 (CYP450) enzymes, has anti-tumour activity and possesses anti-inflammatory properties that may be due at least in part to inhibition of lipoxygenase activity and cytokine release (Pertwee 2004b).

The CB_1 and CB_2 affinities of cannabidiol can be greatly enhanced both by changing its stereochemistry from (–)-(3*R*, 4*R*) to (+)-(3*S*, 4*S*) and by making certain structural modifications (reviewed in Howlett et al. 2002; Pertwee 2004b). Cannabidiol analogues with particularly high affinities for CB_1 and CB_2 receptors are (+)-(3*S*, 4*S*)-4′-dimethylheptyl-cannabidiol and (+)-(3*S*, 4*S*)-7-hydroxy-4′-dimethylheptyl-cannabidiol (Bisogno et al. 2001). Several (–)-(3*R*, 4*R*)-analogues of cannabidiol with high CB_1 and CB_2 affinities have also been developed, for example O-1660, O-1871 and O-1422 (Wiley et al. 2002). Whether these (+)-(3*S*, 4*S*)- and (–)-(3*R*, 4*R*)-analogues of cannabidiol are agonists or antagonists remains to be established. However, one (–)-(3*R*, 4*R*)-cannabidiol analogue that is already known to be a potent CB_2-selective agonist is HU-308 (Sect. 3.1), whilst another cannabidiol analogue, O-2654, behaves as a reasonably potent CB_1 receptor antagonist (Sect. 3.4).

Finally, there is evidence that cannabidiol can induce apoptosis in cultures of at least some types of human cancer cell: HL-60 myeloblastic leukaemia cells and glioma cells. More specifically, it has been reported to produce signs of apoptosis at 3.2 μM in γ-irradiated HL-60 cells, at 12.7 μM in non-irradiated HL-60 cells and at 25 μM but not 10 μM in U87 and U373 glioma cells (Gallily et al. 2003; Massi et al. 2004). At these or higher concentrations, cannabidiol did not induce detectable apoptosis in γ-irradiated or non-irradiated monocytes obtained from normal individuals (Gallily et al. 2003).

5 CB_1 Receptor Oligomerization

There is some evidence that the CB_1 receptor can exist as a homodimer and also that it may form heterodimers or oligomers with one or more other classes of co-expressed G protein-coupled receptor (e.g. dopamine D_2 and opioid receptors) (Wager-Miller et al. 2002). Resulting cross-talk between CB_1 and non CB_1 receptors may involve the sequestration of G proteins either from other receptor types by CB_1 receptors (reviewed in Pertwee 2003) or conversely, from CB_1 receptors by other receptor types. For example, results obtained from experiments with primary cultures of rat striatal neurons (Glass and Felder 1997) and with human embryonic kidney cells co-transfected with CB_1 and dopamine D_2 receptors (Jarrahian et al. 2004) suggest that D_2 receptors can sequester $G\alpha_{i/o}$ so as to cause co-expressed CB_1 receptors to switch coupling from $G\alpha_{i/o}$ to $G\alpha_s$. Interestingly, Jarrahian et al. (2004) also found that in the human embryonic kidney cells expressing both CB_1 and D_2 receptors, persistent activation of the D_2 receptors promoted the re-establishment of CB_1 receptor coupling with $G\alpha_{i/o}$. Results from other in vitro experiments have provided evidence that in the presence of ongoing $G\alpha_s$-mediated adenylate cyclase stimulation by adenosine A_2 receptor activation, D_2 and CB_1 receptor agonists can interact synergistically through their respective receptors to produce further adenylate cyclase stimulation via $\beta\gamma$-subunits released from $G\alpha_{i/o}$ (Yao et al. 2003).

6 Future Directions

Clearly there is now incontrovertible evidence for the existence of a mammalian endocannabinoid system that consists of at least two types of cannabinoid receptor, CB_1 and CB_2, and of endogenous agonists (endocannabinoids) for these receptors. Agonists that activate both these receptor types with similar potency or that show marked selectivity for one or other receptor type have been discovered, as have potent CB_1- and CB_2-selective cannabinoid receptor antagonists. Quantitative and sensitive in vitro and in vivo bioassays for these ligands are also available, and these have played a crucial role in determining the CB_1 and CB_2 receptor affinities and intrinsic activities of a number of cannabinoids. There is good evidence that the endocannabinoid system can become tonically active and that this is due in some instances to endocannabinoid release and in other instances to the ability of cannabinoid receptors to exist in a constitutively activity state, not only when overexpressed in cultured cells but also when expressed naturally. The existence of such constitutive activity is reflected in the pharmacological properties of established cannabinoid receptor antagonists, all of which appear to be inverse agonists rather than neutral antagonists. Ligands that behave as neutral cannabinoid receptor antagonists are beginning to be described in the literature. These now need to be characterized more fully, as such antagonists would serve as important additional pharmacological tools and might also possess advantages over inverse agonists in the clinic. Evidence for the presence of non-CB_1, non-CB_2 pharmacological targets for at least some cannabinoid receptor agonists is emerging, prompting a need to establish the extent to which these proposed additional targets contribute to the pharmacology of these agonists. For some of these targets, ligands that do not also interact with CB_1 or CB_2 receptors have already been identified, and it will now be important to characterize the actions of these ligands more fully and to investigate the possibility of developing potent and selective non-CB_1, non-CB_2 agonists for all the proposed new targets. This in turn will greatly facilitate a fuller understanding of these targets as well as the discovery of any additional targets. The extent to which cross-talk can occur between identical (e.g. CB_1-CB_1) or different pharmacological targets for cannabinoids (e.g. between CB_2 and abnormal cannabidiol receptors), or between cannabinoid and non-cannabinoid targets (e.g. between CB_1 and dopamine D_2 receptors), and the nature of the mechanisms that underlie such cross-talk also merit further investigation.

References

Abadji V, Lin S, Taha G, Griffin G, Stevenson LA, Pertwee RG, Makriyannis A (1994) (R)-methanandamide: a chiral novel anandamide possessing higher potency and metabolic stability. J Med Chem 37:1889–1893

Adams IB, Ryan W, Singer M, Thomas BF, Compton DR, Razdan RK, Martin BR (1995) Evaluation of cannabinoid receptor binding and in vivo activities for anandamide analogs. J Pharmacol Exp Ther 273:1172–1181

Ahluwalia J, Yaqoob M, Urban L, Bevan S, Nagy I (2003) Activation of capsaicin-sensitive primary sensory neurones induces anandamide production and release. J Neurochem 84:585–591

Akinshola BE, Chakrabarti A, Onaivi ES (1999a) In-vitro and in-vivo action of cannabinoids. Neurochem Res 24:1233–1240

Akinshola BE, Taylor RE, Ogunseitan AB, Onaivi ES (1999b) Anandamide inhibition of recombinant AMPA receptor subunits in Xenopus oocytes is increased by forskolin and 8-bromo-cyclic AMP. Naunyn Schmiedebergs Arch Pharmacol 360:242–248

Al-Hayani A, Wease KN, Ross RA, Pertwee RG, Davies SN (2001) The endogenous cannabinoid anandamide activates vanilloid receptors in the rat hippocampal slice. Neuropharmacology 41:1000–1005

Balster RL, Prescott WR (1992) Δ9-Tetrahydrocannabinol discrimination in rats as a model for cannabis intoxication. Neurosci Biobehav Rev 16:55–62

Barann M, Molderings G, Brüss M, Bönisch H, Urban BW, Göthert M (2002) Direct inhibition by cannabinoids of human 5-HT3A receptors: probable involvement of an allosteric modulatory site. Br J Pharmacol 137:589–596

Bayewitch M, Rhee M-H, Avidor-Reiss T, Breuer A, Mechoulam R, Vogel Z (1996) (–)-Δ9-tetrahydrocannabinol antagonizes the peripheral cannabinoid receptor-mediated inhibition of adenylyl cyclase. J Biol Chem 271:9902–9905

Begg M, Mo F-M, Offertáler L, Bátkai S, Pacher P, Razdan RK, Lovinger DM, Kunos G (2003) G protein-coupled endothelial receptor for atypical cannabinoid ligands modulates a Ca2+-dependent K+ current. J Biol Chem 278:46188–46194

Ben-Shabat S, Fride E, Sheskin T, Tamiri T, Rhee M-H, Vogel Z, Bisogno T, De Petrocellis L, Di Marzo V, Mechoulam R (1998) An entourage effect: inactive endogenous fatty acid glycerol esters enhance 2-arachidonoyl-glycerol cannabinoid activity. Eur J Pharmacol 353:23–31

Berdyshev EV, Schmid PC, Krebsbach RJ, Hillard CJ, Huang CS, Chen N, Dong Z, Schmid HHO (2001) Cannabinoid-receptor-independent cell signalling by N-acylethanolamines. Biochem J 360:67–75

Bisogno T, Melck D, Bobrov MY, Gretskaya NM, Bezuglov VV, De Petrocellis L, Di Marzo V (2000) N-acyl-dopamines: novel synthetic CB1 cannabinoid-receptor ligands and inhibitors of anandamide inactivation with cannabimimetic activity in vitro and in vivo. Biochem J 351:817–824

Bisogno T, Hanus L, De Petrocellis L, Tchilibon S, Ponde DE, Brandi I, Moriello AS, Davis JB, Mechoulam R, Di Marzo V (2001) Molecular targets for cannabidiol and its synthetic analogues: effect on vanilloid VR1 receptors and on the cellular uptake and enzymatic hydrolysis of anandamide. Br J Pharmacol 134:845–852

Breivogel CS, Childers SR (2000) Cannabinoid agonist signal transduction in rat brain: comparison of cannabinoid agonists in receptor binding, G-protein activation, and adenylyl cyclase inhibition. J Pharmacol Exp Ther 295:328–336

Breivogel CS, Sim LJ, Childers SR (1997) Regional differences in cannabinoid receptor/G-protein coupling in rat brain. J Pharmacol Exp Ther 282:1632–1642

Breivogel CS, Selley DE, Childers SR (1998) Cannabinoid receptor agonist efficacy for stimulating [35S]GTPγS binding to rat cerebellar membranes correlates with agonist-induced decreases in GDP affinity. J Biol Chem 273:16865–16873

Breivogel CS, Griffin G, Di Marzo V, Martin BR (2001) Evidence for a new G protein-coupled cannabinoid receptor in mouse brain. Mol Pharmacol 60:155–163

Brooks JW, Pryce G, Bisogno T, Jaggar SI, Hankey DJR, Brown P, Bridges D, Ledent C, Bifulco M, Rice ASC, Di Marzo V, Baker D (2002) Arvanil-induced inhibition of spasticity and persistent pain: evidence for therapeutic sites of action different from the vanilloid VR1 receptor and cannabinoid CB1/CB2 receptors. Eur J Pharmacol 439:83–92

Busch-Petersen J, Hill WA, Fan PS, Khanolkar A, Xie X-Q, Tius MA, Makriyannis A (1996) Unsaturated side chain β-11-hydroxyhexahydrocannabinol analogs. J Med Chem 39:3790–3796

Calandra B, Portier M, Kernéis A, Delpech M, Carillon C, Le Fur G, Ferrara P, Shire D (1999) Dual intracellular signaling pathways mediated by the human cannabinoid CB1 receptor. Eur J Pharmacol 374:445–455

Calignano A, La Rana G, Giuffrida A, Piomelli D (1998) Control of pain initiation by endogenous cannabinoids. Nature 394:277–281

Calignano A, La Rana G, Piomelli D (2001) Antinociceptive activity of the endogenous fatty acid amide, palmitylethanolamide. Eur J Pharmacol 419:191–198

Chaytor AT, Martin PEM, Evans WH, Randall MD, Griffith TM (1999) The endothelial component of cannabinoid-induced relaxation in rabbit mesenteric artery depends on gap junctional communication. J Physiol 520:539–550

Cheer JF, Cadogan A-K, Marsden CA, Fone KCF, Kendall DA (1999) Modification of 5-HT2 receptor mediated behaviour in the rat by oleamide and the role of cannabinoid receptors. Neuropharmacology 38:533–541

Chemin J, Monteil A, Perez-Reyes E, Nargeot J, Lory P (2001) Direct inhibition of T-type calcium channels by the endogenous cannabinoid anandamide. EMBO J 20:7033–7040

Chin C, Lucas-Lenard J, Abadji V, Kendall DA (1998) Ligand binding and modulation of cyclic AMP levels depend on the chemical nature of residue 192 of the human cannabinoid receptor 1. J Neurochem 70:366–373

Christopoulos A, Wilson K (2001) Interaction of anandamide with the M1 and M4 muscarinic acetylcholine receptors. Brain Res 915:70–78

Compton DR, Aceto MD, Lowe J, Martin BR (1996) In vivo characterization of a specific cannabinoid receptor antagonist (SR141716A): inhibition of Δ9-tetrahydrocannabinol-induced responses and apparent agonist activity. J Pharmacol Exp Ther 277:586–594
Conti S, Costa B, Colleoni M, Parolaro D, Giagnoni G (2002) Antiinflammatory action of endocannabinoid palmitoylethanolamide and the synthetic cannabinoid nabilone in a model of acute inflammation in the rat. Br J Pharmacol 135:181–187
Cosenza M, Gifford AN, Gatley SJ, Pyatt B, Liu Q, Makriyannis A, Volkow ND (2000) Locomotor activity and occupancy of brain cannabinoid CB1 receptors by the antagonist/inverse agonist AM281. Synapse 38:477–482
Costa B, Colleoni M (1999) SR141716A induces in rats a behavioral pattern opposite to that of CB1 receptor agonists. Acta Pharmacol Sin 20:1103–1108
Craib SJ, Ellington HC, Pertwee RG, Ross RA (2001) A possible role of lipoxygenase in the activation of vanilloid receptors by anandamide in the guinea-pig bronchus. Br J Pharmacol 134:30–37
Cravatt BF, Lichtman AH (2002) The enzymatic inactivation of the fatty acid amide class of signaling lipids. Chem Phys Lipids 121:135–148
Darlington CL (2003) Dexanabinol: a novel cannabinoid with neuroprotective properties. IDrugs 6:976–979
De Petrocellis L, Bisogno T, Maccarrone M, Davis JB, Finazzi-Agrò A, Di Marzo V (2001a) The activity of anandamide at vanilloid VR1 receptors requires facilitated transport across the cell membrane and is limited by intracellular metabolism. J Biol Chem 276:12856–12863
De Petrocellis L, Davis JB, Di Marzo V (2001b) Palmitoylethanolamide enhances anandamide stimulation of human vanilloid VR1 receptors. FEBS Lett 506:253–256
De Petrocellis L, Bisogno T, Ligresti A, Bifulco M, Melck D, Di Marzo V (2002) Effect on cancer cell proliferation of palmitoylethanolamide, a fatty acid amide interacting with both the cannabinoid and vanilloid signalling systems. Fundam Clin Pharmacol 16:297–302
De Vry J, Jentzsch KR (2002) Discriminative stimulus effects of BAY 38-7271, a novel cannabinoid receptor agonist. Eur J Pharmacol 457:147–152
Devane WA, Hanus L, Breuer A, Pertwee RG, Stevenson LA, Griffin G, Gibson D, Mandelbaum A, Etinger A, Mechoulam R (1992) Isolation and structure of a brain constituent that binds to the cannabinoid receptor. Science 258:1946–1949
Devlin MG, Christopoulos A (2002) Modulation of cannabinoid agonist binding by 5-HT in the rat cerebellum. J Neurochem 80:1095–1102
Di Marzo V, Melck D, Bisogno T, De Petrocellis L (1998) Endocannabinoids: endogenous cannabinoid receptor ligands with neuromodulatory action. Trends Neurosci 21:521–528
Di Marzo V, Breivogel CS, Tao Q, Bridgen DT, Razdan RK, Zimmer AM, Zimmer A, Martin BR (2000) Levels, metabolism, and pharmacological activity of anandamide in CB1 cannabinoid receptor knockout mice: evidence for non-CB1, non-CB2 receptor-mediated actions of anandamide in mouse brain. J Neurochem 75:2434–2444
Di Marzo V, Bisogno T, De Petrocellis L, Brandi I, Jefferson RG, Winckler RL, Davis JB, Dasse O, Mahadevan A, Razdan RK, Martin BR (2001) Highly selective CB1 cannabinoid receptor ligands and novel CB1/VR1 vanilloid receptor "hybrid" ligands. Biochem Biophys Res Commun 281:444–451
Di Marzo V, Griffin G, De Petrocellis L, Brandi I, Bisogno T, Williams W, Grier MC, Kulasegram S, Mahadevan A, Razdan RK, Martin BR (2002) A structure/activity relationship study on arvanil, an endocannabinoid and vanilloid hybrid. J Pharmacol Exp Ther 300:984–991
Dinh TP, Freund TF, Piomelli D (2002) A role for monoglyceride lipase in 2-arachidonoyl-glycerol inactivation. Chem Phys Lipids 121:149–158
Drysdale AJ, Pertwee RG, Platt B (2004) Modulation of calcium homeostasis by cannabidiol in primary hippocampal culture. Proc Br Pharmacol Soc at http://www.pa2online.org/Vol1Issue4abst052P.html

El-Remessy AB, Khalil IE, Matragoon S, Abou-Mohamed G, Tsai N-J, Roon P, Caldwell RB, Caldwell RW, Green K, Liou GI (2003) Neuroprotective effect of (−)Δ9-tetrahydrocannabinol and cannabidiol in N-methyl-D-aspartate-induced retinal neurotoxicity: involvement of peroxynitrite. Am J Pathol 163:1997–2008

Ellington HC, Cotter MA, Cameron NE, Ross RA (2002) The effect of cannabinoids on capsaicin-evoked calcitonin gene-related peptide (CGRP) release from the isolated paw skin of diabetic and non-diabetic rats. Neuropharmacology 42:966–975

ElSohly MA (2002) Chemical constituents of cannabis. In: Grotenhermen F, Russo E (eds) Cannabis and Cannabinoids Pharmacology, Toxicology and Therapeutic Potential. Haworth Press, New York, pp 27–36

Facchinetti F, Del Giudice E, Furegato S, Passarotto M, Leon A (2003) Cannabinoids ablate release of TNFα in rat microglial cells stimulated with lipopolysaccharide. Glia 41:161–168

Fan P (1995) Cannabinoid agonists inhibit the activation of 5-HT3 receptors in rat nodose ganglion neurons. J Neurophysiol 73:907–910

Fan S-F, Yazulla S (2003) Biphasic modulation of voltage-dependent currents of retinal cones by cannabinoid CBI receptor agonist WIN 55212-2. Visual Neurosci 20:177–188

Farquhar-Smith WP, Rice ASC (2001) Administration of endocannabinoids prevents a referred hyperalgesia associated with inflammation of the urinary bladder. Anesthesiology 94:507–513

Farquhar-Smith WP, Jaggar SI, Rice ASC (2002) Attenuation of nerve growth factor-induced visceral hyperalgesia via cannabinoid CB1 and CB2-like receptors. Pain 97:11–21

Felder CC, Briley EM, Axelrod J, Simpson JT, Mackie K, Devane WA (1993) Anandamide, an endogenous cannabimimetic eicosanoid, binds to the cloned human cannabinoid receptor and stimulates receptor-mediated signal transduction. Proc Natl Acad Sci USA 90:7656–7660

Felder CC, Joyce KE, Briley EM, Mansouri J, Mackie K, Blond O, Lai Y, Ma AL, Mitchell RL (1995) Comparison of the pharmacology and signal transduction of the human cannabinoid CB1 and CB2 receptors. Mol Pharmacol 48:443–450

Felder CC, Joyce KE, Briley EM, Glass M, Mackie KP, Fahey KJ, Cullinan GJ, Hunden DC, Johnson DW, Chaney MO, Koppel GA, Brownstein M (1998) LY320135, a novel cannabinoid CB1 receptor antagonist, unmasks coupling of the CB1 receptor to stimulation of cAMP accumulation. J Pharmacol Exp Ther 284:291–297

Ferraro L, Tomasini MC, Gessa GL, Bebe BW, Tanganelli S, Antonelli T (2001) The cannabinoid receptor agonist WIN 55,212-2 regulates glutamate transmission in rat cerebral cortex: an in vivo and in vitro study. Cerebral Cortex 11:728–733

Fezza F, Bisogno T, Minassi A, Appendino G, Mechoulam R, Di Marzo V (2002) Noladin ether, a putative novel endocannabinoid: inactivation mechanisms and a sensitive method for its quantification in rat tissues. FEBS Lett 513:294–298

Fowler CJ (2003) Plant-derived, synthetic and endogenous cannabinoids as neuroprotective agents: non-psychoactive cannabinoids, 'entourage' compounds and inhibitors of N-acyl ethanolamine breakdown as therapeutic strategies to avoid psychotropic effects. Brain Res Rev 41:26–43

Franklin A, Stella N (2003) Arachidonylcyclopropylamide increases microglial cell migration through cannabinoid CB2 and abnormal-cannabidiol-sensitive receptors. Eur J Pharmacol 474:195–198

Franklin A, Parmentier-Batteur S, Walter L, Greenberg DA, Stella N (2003) Palmitoylethanolamide increases after focal cerebral ischemia and potentiates microglial cell motility. J Neurosci 23:7767–7775

Fride E, Foox A, Rosenberg E, Faigenboim M, Cohen V, Barda L, Blau H, Mechoulam R (2003) Milk intake and survival in newborn cannabinoid CB1 receptor knockout mice: evidence for a "CB3" receptor. Eur J Pharmacol 461:27–34

Gallant M, Dufresne C, Gareau Y, Guay D, Leblanc Y, Prasit P, Rochette C, Sawyer N, Slipetz DM, Tremblay N, Metters KM, Labelle M (1996) New class of potent ligands for the human peripheral cannabinoid receptor. Bioorg Med Chem Lett 6:2263–2268

Gallily R, Even-Chen T, Katzavian G, Lehmann D, Dagan A, Mechoulam R (2003) γ-Irradiation enhances apoptosis induced by cannabidiol, a non-psychotropic cannabinoid, in cultured HL-60 myeloblastic leukemia cells. Leuk Lymphoma 44:1767–1773

Gareau Y, Dufresne C, Gallant M, Rochette C, Sawyer N, Slipetz DM, Tremblay N, Weech PK, Metters KM, Labelle M (1996) Structure activity relationships of tetrahydrocannabinol analogues on human cannabinoid receptors. Bioorg Med Chem Lett 6:189–194

Gatley SJ, Lan R, Volkow ND, Pappas N, King P, Wong CT, Gifford AN, Pyatt B, Dewey SL, Makriyannis A (1998) Imaging the brain marijuana receptor: development of a radioligand that binds to cannabinoid CB1 receptors in vivo. J Neurochem 70:417–423

Gérard CM, Mollereau C, Vassart G, Parmentier M (1991) Molecular cloning of a human cannabinoid receptor which is also expressed in testis. Biochem J 279:129–134

Gifford AN, Tang Y, Gatley SJ, Volkow ND, Lan R, Makriyannis A (1997) Effect of the cannabinoid receptor SPECT agent, AM 281, on hippocampal acetylcholine release from rat brain slices. Neurosci Lett 238:84–86

Gifford AN, Makriyannis A, Volkow ND, Gatley SJ (2002) In vivo imaging of the brain cannabinoid receptor. Chem Phys Lipids 121:65–72

Glass M, Felder CC (1997) Concurrent stimulation of cannabinoid CB1 and dopamine D2 receptors augments cAMP accumulation in striatal neurons: evidence for a Gs linkage to the CB1 receptor. J Neurosci 17:5327–5333

Godlewski G, Göthert M, Malinowska B (2003) Cannabinoid receptor-independent inhibition by cannabinoid agonists of the peripheral 5-HT3 receptor-mediated von Bezold-Jarisch reflex. Br J Pharmacol 138:767–774

Gonsiorek W, Lunn C, Fan X, Narula S, Lundell D, Hipkin RW (2000) Endocannabinoid 2-arachidonyl glycerol is a full agonist through human type 2 cannabinoid receptor: antagonism by anandamide. Mol Pharmacol 57:1045–1050

Göthert M, Brüss M, Bönisch H, Molderings GJ (1999) Presynaptic imidazoline receptors: new developments in characterization and classification. Ann N Y Acad Sci 881:171–184

Goutopoulos A, Fan P, Khanolkar AD, Xie X-Q, Lin S, Makriyannis A (2001) Stereochemical selectivity of methanandamides for the CB1 and CB2 cannabinoid receptors and their metabolic stability. Bioorg Med Chem 9:1673–1684

Griffin G, Fernando SR, Ross RA, McKay NG, Ashford MLJ, Shire D, Huffman JW, Yu S, Lainton JAH, Pertwee RG (1997) Evidence for the presence of CB2-like cannabinoid receptors on peripheral nerve terminals. Eur J Pharmacol 339:53–61

Griffin G, Tao Q, Abood ME (2000) Cloning and pharmacological characterization of the rat CB2 cannabinoid receptor. J Pharmacol Exp Ther 292:886–894

Hájos N, Freund TF (2002a) Pharmacological separation of cannabinoid sensitive receptors on hippocampal excitatory and inhibitory fibers. Neuropharmacology 43:503–510

Hájos N, Freund TF (2002b) Distinct cannabinoid sensitive receptors regulate hippocampal excitation and inhibition. Chem Phys Lipids 121:73–82

Hájos N, Katona I, Naiem SS, Mackie K, Ledent C, Mody I, Freund TF (2000) Cannabinoids inhibit hippocampal GABAergic transmission and network oscillations. Eur J Neurosci 12:3239–3249

Hájos N, Ledent C, Freund TF (2001) Novel cannabinoid-sensitive receptor mediates inhibition of glutamatergic synaptic transmission in the hippocampus. Neuroscience 106:1–4

Hampson AJ, Grimaldi M, Axelrod J, Wink D (1998) Cannabidiol and (–)Δ9-tetrahydrocannabinol are neuroprotective antioxidants. Proc Natl Acad Sci USA 95:8268–8273

Hampson AJ, Grimaldi M, Lolic M, Wink D, Rosenthal R, Axelrod J (2000) Neuroprotective antioxidants from marijuana. Ann N Y Acad Sci 899:274–282

Hanus L, Breuer A, Tchilibon S, Shiloah S, Goldenberg D, Horowitz M, Pertwee RG, Ross RA, Mechoulam R, Fride E (1999) HU-308: a specific agonist for CB2, a peripheral cannabinoid receptor. Proc Natl Acad Sci USA 96:14228–14233

Hanus L, Abu-Lafi S, Fride E, Breuer A, Vogel Z, Shalev DE, Kustanovich I, Mechoulam R (2001) 2-Arachidonyl glyceryl ether, an endogenous agonist of the cannabinoid CB1 receptor. Proc Natl Acad Sci USA 98:3662–3665

Helyes Z, Németh J, Thán M, Bölcskei K, Pintér E, Szolcsányi J (2003) Inhibitory effect of anandamide on resiniferatoxin-induced sensory neuropeptide release in vivo and neuropathic hyperalgesia in the rat. Life Sci 73:2345–2353

Hermann H, De Petrocellis L, Bisogno T, Schiano Moriello A, Lutz B, Di Marzo V (2003) Dual effect of cannabinoid CB1 receptor stimulation on a vanilloid VR1 receptor-mediated response. Cell Mol Life Sci 60:607–616

Hillard CJ (2000) Biochemistry and pharmacology of the endocannabinoids arachidonylethanolamide and 2-arachidonylglycerol. Prostaglandins Other Lipid Mediat 61:3–18

Hillard CJ, Jarrahian A (2003) Cellular accumulation of anandamide: consensus and controversy. Br J Pharmacol 140:802–808

Hillard CJ, Manna S, Greenberg MJ, Dicamelli R, Ross RA, Stevenson LA, Murphy V, Pertwee RG, Campbell WB (1999) Synthesis and characterization of potent and selective agonists of the neuronal cannabinoid receptor (CB1). J Pharmacol Exp Ther 289:1427–1433

Hirst RA, Almond SL, Lambert DG (1996) Characterisation of the rat cerebella CB1 receptor using SR141716A, a central cannabinoid receptor antagonist. Neurosci Lett 220:101–104

Ho W-SV, Hiley CR (2003) Vasodilator actions of abnormal-cannabidiol in rat isolated small mesenteric artery. Br J Pharmacol 138:1320–1332

Howlett AC (1987) Cannabinoid inhibition of adenylate cyclase: relative activity of constituents and metabolites of marihuana. Neuropharmacology 26:507–512

Howlett AC, Fleming RM (1984) Cannabinoid inhibition of adenylate cyclase. Pharmacology of the response in neuroblastoma cell membranes. Mol Pharmacol 26:532–538

Howlett AC, Barth F, Bonner TI, Cabral G, Casellas P, Devane WA, Felder CC, Herkenham M, Mackie K, Martin BR, Mechoulam R, Pertwee RG (2002) International Union of Pharmacology. XXVII. Classification of cannabinoid receptors. Pharmacol Rev 54:161–202

Huffman JW, Yu S, Showalter V, Abood ME, Wiley JL, Compton DR, Martin BR, Bramblett RD, Reggio PH (1996) Synthesis and pharmacology of a very potent cannabinoid lacking a phenolic hydroxyl with high affinity for the CB2 receptor. J Med Chem 39:3875–3877

Huffman JW, Liddle J, Yu S, Aung MM, Abood ME, Wiley JL, Martin BR (1999) 3-(1',1'-dimethylbutyl)-1-deoxy-Δ8-THC and related compounds: synthesis of selective ligands for the CB2 receptor. Bioorg Med Chem 7:2905–2914

Huffman JW, Bushell SM, Miller JRA, Wiley JL, Martin BR (2002) 1-methoxy-, 1-deoxy-11-hydroxy- and 11-hydroxy-1-methoxy-Δ8-tetrahydrocannabinols: new selective ligands for the CB2 receptor. Bioorg Med Chem 10:4119–4129

Hurst DP, Lynch DL, Barnett-Norris J, Hyatt SM, Seltzman HH, Zhong M, Song Z-H, Nie J, Lewis D, Reggio PH (2002) N-(Piperidin-1-yl)-5-(4-chlorophenyl)-1-(2,4-dichlorophenyl)-4-methyl-1H-pyrazole-3-carboxamide (SR141716A) interaction with LYS 3.28(192) is crucial for its inverse agonism at the cannabinoid CB1 receptor. Mol Pharmacol 62:1274–1287

Ibrahim MM, Deng H, Zvonok A, Cockayne DA, Kwan J, Mata HP, Vanderah TW, Lai J, Porreca F, Makriyannis A, Malan TP (2003) Activation of CB2 cannabinoid receptors by AM1241 inhibits experimental neuropathic pain: pain inhibition by receptors not present in the CNS. Proc Natl Acad Sci USA 100:10529–10533

Iversen L (2003) Cannabis and the brain. Brain 126:1252–1270

Iwamura H, Suzuki H, Ueda Y, Kaya T, Inaba T (2001) In vitro and in vivo pharmacological characterization of JTE-907, a novel selective ligand for cannabinoid CB2 receptor. J Pharmacol Exp Ther 296:420–425

Izzo AA, Mascolo N, Tonini M, Capasso F (2000) Modulation of peristalsis by cannabinoid CB1 ligands in the isolated guinea-pig ileum. Br J Pharmacol 129:984–990

Járai Z, Wagner JA, Varga K, Lake KD, Compton DR, Martin BR, Zimmer AM, Bonner TI, Buckley NE, Mezey E, Razdan RK, Zimmer A, Kunos G (1999) Cannabinoid-induced mesenteric vasodilation through an endothelial site distinct from CB1 or CB2 receptors. Proc Natl Acad Sci USA 96:14136–14141

Jarrahian A, Manna S, Edgemond WS, Campbell WB, Hillard CJ (2000) Structure-activity relationships among N-arachidonylethanolamine (anandamide) head group analogues for the anandamide transporter. J Neurochem 74:2597–2606

Jarrahian A, Watts VJ, Barker EL (2004) D2 dopamine receptors modulate Gα-subunit coupling of the CB1 cannabinoid receptor. J Pharmacol Exp Ther 308:880–886
Jennings EA, Vaughan CW, Roberts LA, Christie MJ (2003) The actions of anandamide on rat superficial medullary dorsal horn neurons in vitro. J Physiol 548:121–129
Jordt S-E, Bautista DM, Chuang H, McKemy DD, Zygmunt PM, Högestätt ED, Meng ID, Julius D (2004) Mustard oils and cannabinoids excite sensory nerve fibres through the TRP channel ANKTM1. Nature 427:260–265
Kelley BG, Thayer SA (2004) Δ9-tetrahydrocannabinol antagonizes endocannabinoid modulation of synaptic transmission between hippocampal neurons in culture. Neuropharmacology 46:709–715
Kenny LC, Baker PN, Kendall DA, Randall MD, Dunn WR (2002) The role of gap junctions in mediating endothelium-dependent responses to bradykinin in myometrial small arteries isolated from pregnant women. Br J Pharmacol 136:1085–1088
Khanolkar AD, Abadji V, Lin S, Hill WAG, Taha G, Abouzid K, Meng Z, Fan P, Makriyannis A (1996) Head group analogs of arachidonylethanolamide, the endogenous cannabinoid ligand. J Med Chem 39:4515–4519
Köfalvi A, Vizi ES, Ledent C, Sperlágh B (2003) Cannabinoids inhibit the release of [3H]glutmate from rodent hippocampal synaptosomes via a novel CB1 receptor-independent action. Eur J Neurosci 18:1973–1978
Kojima S, Sugiura T, Waku K, Kamikawa Y (2002) Contractile response to a cannabimimetic eicosanoid, 2-arachidonoylglycerol, of longitudinal smooth muscle from the guinea-pig distal colon in vitro. Eur J Pharmacol 444:203–207
Krishnamurthy M, Ferreira AM, Moore BM (2003) Synthesis and testing of novel phenyl substituted side-chain analogues of classical cannabinoids. Bioorg Med Chem Lett 13:3487–3490
Lambert DM, DiPaolo FG, Sonveaux P, Kanyonyo M, Govaerts SJ, Hermans E, Bueb J-L, Delzenne NM, Tschirhart EJ (1999) Analogues and homologues of N-palmitoylethanolamide, a putative endogenous CB2 cannabinoid, as potential ligands for the cannabinoid receptors. Biochim Biophys Acta 1440:266–274
Lan R, Gatley J, Lu Q, Fan P, Fernando SR, Volkow ND, Pertwee R, Makriyannis A (1999a) Design and synthesis of the CB1 selective cannabinoid antagonist AM281: a potential human SPECT ligand. AAPS PharmSci 1:U13–U24
Lan R, Liu Q, Fan P, Lin S, Fernando SR, McCallion D, Pertwee R, Makriyannis A (1999b) Structure-activity relationships of pyrazole derivatives as cannabinoid receptor antagonists. J Med Chem 42:769–776
Landsman RS, Makriyannis A, Deng H, Consroe P, Roeske WR, Yamamura HI (1998) AM630 is an inverse agonist at the human cannabinoid CB1 receptor. Life Sci 62:PL109–113
Ledent C, Valverde O, Cossu G, Petitet F, Aubert J-F, Beslot F, Böhme GA, Imperato A, Pedrazzini T, Roques BP, Vassart G, Fratta W, Parmentier M (1999) Unresponsiveness to cannabinoids and reduced addictive effects of opiates in CB1 receptor knockout mice. Science 283:401–404
Lin S, Khanolkar AD, Fan P, Goutopoulos A, Qin C, Papahadjis D, Makriyannis A (1998) Novel analogues of arachidonylethanolamide (anandamide): affinities for the CB1 and CB2 cannabinoid receptors and metabolic stability. J Med Chem 41:5353–5361
Maccarrone M, Cartoni A, Parolaro D, Margonelli A, Massi P, Bari M, Battista N, Finazzi-Agrò A (2002) Cannabimimetic activity, binding, and degradation of stearoylethanolamide within the mouse central nervous system. Mol Cell Neurosci 21:126–140
Maccarrone M, Bari M, Del Principe D, Finazzi-Agrò A (2003) Activation of human platelets by 2-arachidonoylglycerol is enhanced by serotonin. Thromb Haemost 89:340–347
Mackie K, Devane WA, Hille B (1993) Anandamide, an endogenous cannabinoid, inhibits calcium currents as a partial agonist in N18 neuroblastoma cells. Mol Pharmacol 44:498–503
MacLennan SJ, Reynen PH, Kwan J, Bonhaus DW (1998) Evidence for inverse agonism of SR141716A at human recombinant cannabinoid CB1 and CB2 receptors. Br J Pharmacol 124:619–622

Maneuf YP, Brotchie JM (1997) Paradoxical action of the cannabinoid WIN 55,212-2 in stimulated and basal cyclic AMP accumulation in rat globus pallidus slices. Br J Pharmacol 120:1397–1398

Mang CF, Erbelding D, Kilbinger H (2001) Differential effects of anandamide on acetylcholine release in the guinea-pig ileum mediated via vanilloid and non-CB1 cannabinoid receptors. Br J Pharmacol 134:161–167

Marinelli S, Vaughan CW, Christie MJ, Connor M (2002) Capsaicin activation of glutamatergic synaptic transmission in the rat locus coeruleus in vitro. J Physiol 543:531–540

Marinelli S, Di Marzo V, Berretta N, Matias I, Maccarrone M, Bernardi G, Mercuri NB (2003) Presynaptic facilitation of glutamatergic synapses to dopaminergic neurons of the rat substantia nigra by endogenous stimulation of vanilloid receptors. J Neurosci 23:3136–3144

Marsicano G, Moosmann B, Hermann H, Lutz B, Behl C (2002) Neuroprotective properties of cannabinoids against oxidative stress: role of the cannabinoid receptor CB1. J Neurochem 80:448–456

Martin B, Stevenson LA, Pertwee RG, Breivogel CS, Williams W, Mahadevan A, Razdan RK (2002) Agonists and silent antagonists in a series of cannabinoid sulfonamides. Symposium on the Cannabinoids. Burlington, Vermont, International Cannabinoid Research Society, p 2

Martin BR, Thomas BF, Razdan RK (1995) Structural requirements for cannabinoid receptor probes. In: Pertwee RG (ed) Cannabinoid receptors. Academic Press, London, pp 35–85

Massi P, Vaccani A, Ceruti S, Colombo A, Abbracchio MP, Parolaro D (2004) Antitumor effects of cannabidiol, a nonpsychoactive cannabinoid, on human glioma cell lines. J Pharmacol Exp Ther 308:838–845

Matsuda LA, Lolait SJ, Brownstein MJ, Young AC, Bonner TI (1990) Structure of a cannabinoid receptor and functional expression of the cloned cDNA. Nature 346:561–564

Mauler F, Mittendorf J, Horváth E, De Vry J (2002) Characterization of the diarylether sulfonylester (−)-(R)-3-(2-hydroxymethylindanyl-4-oxy)phenyl-4,4,4-trifluoro-1-sulfonate (BAY 38–7271) as a potent cannabinoid receptor agonist with neuroprotective properties. J Pharmacol Exp Ther 302:359–368

Mauler F, Hinz V, Augstein K-H, Fassbender M, Horváth E (2003) Neuroprotective and brain edema-reducing efficacy of the novel cannabinoid receptor agonist BAY 38–7271. Brain Res 989:99–111

Mechoulam R, Gaoni Y (1967) Recent advances in the chemistry of hashish. Fortschr Chem Org Naturst 25:175–213

Mechoulam R, Ben-Shabat S, Hanus L, Ligumsky M, Kaminski NE, Schatz AR, Gopher A, Almog S, Martin BR, Compton DR, Pertwee RG, Griffin G, Bayewitch M, Barg J, Vogel Z (1995) Identification of an endogenous 2-monoglyceride, present in canine gut, that binds to cannabinoid receptors. Biochem Pharmacol 50:83–90

Mechoulam R, Fride E, Di Marzo V (1998) Endocannabinoids. Eur J Pharmacol 359:1–18

Mechoulam R, Panikashvili D, Shohami E (2002) Cannabinoids and brain injury: therapeutic implications. Trends Mol Med 8:58–61

Millns PJ, Chapman V, Kendall DA (2001) Cannabinoid inhibition of the capsaicin-induced calcium response in rat dorsal root ganglion neurones. Br J Pharmacol 132:969–971

Misner DL, Sullivan JM (1999) Mechanism of cannabinoid effects on long-term potentiation and depression in hippocampal CA1 neurons. J Neurosci 19:6795–6805

Mo FM, Offertáler L, Kunos G (2004) Atypical cannabinoid stimulates endothelial cell migration via a Gi/Go-coupled receptor distinct from CB1, CB2 or EDG-1. Eur J Pharmacol 489:21–27

Molderings GJ, Göthert M (1999) Imidazoline binding sites and receptors in cardiovascular tissue. Gen Pharmacol 32:17–22

Molderings GJ, Bönisch H, Hammermann R, Göthert M, Brüss M (2002) Noradrenaline release-inhibiting receptors on PC12 cells devoid of $\alpha 2$- and CB1 receptors: similarities to presynaptic imidazoline and edg receptors. Neurochem Int 40:157–167

Monory K, Tzavara ET, Lexime J, Ledent C, Parmentier M, Borsodi A, Hanoune J (2002) Novel, not adenylyl cyclase-coupled cannabinoid binding site in cerebellum of mice. Biochem Biophys Res Commun 292:231–235
Mussinu J-M, Ruiu S, Mulè AC, Pau A, Carai MAM, Loriga G, Murineddu G, Pinna GA (2003) Tricyclic pyrazoles. Part 1: synthesis and biological evaluation of novel 1,4-dihydroindeno[1,2-c]pyrazol-based ligands for CB1 and CB2 cannabinoid receptors. Bioorg Med Chem 11:251–263
Németh J, Helyes Z, Thán M, Jakab B, Pintér E, Szolcsányi J (2003) Concentration-dependent dual effect of anandamide on sensory neuropeptide release from isolated rat tracheae. Neurosci Lett 336:89–92
New DC, Wong YH (2003) BML-190 and AM251 act as inverse agonists at the human cannabinoid CB2 receptor: signalling via cAMP and inositol phosphates. FEBS Lett 536:157–160
Offertáler L, Mo F-M, Bátkai S, Liu J, Begg M, Razdan RK, Martin BR, Bukoski RD, Kunos G (2003) Selective ligands and cellular effectors of a G protein-coupled endothelial cannabinoid receptor. Mol Pharmacol 63:699–705
Onaivi ES, Chakrabarti A, Gwebu ET, Chaudhuri G (1996) Neurobehavioral effects of Δ9-THC and cannabinoid (CB1) receptor gene expression in mice. Behav Brain Res 72:115–125
Oz M, Zhang L, Morales M (2002) Endogenous cannabinoid, anandamide, acts as a noncompetitive inhibitor on 5-HT3 receptor-mediated responses in Xenopus oocytes. Synapse 46:150–156
Pan X, Ikeda SR, Lewis DL (1998) SR 141716A acts as an inverse agonist to increase neuronal voltage-dependent Ca2+ currents by reversal of tonic CB1 cannabinoid receptor activity. Mol Pharmacol 54:1064–1072
Papahatjis DP, Nikas SP, Andreou T, Makriyannis A (2002) Novel 1',1'-chain substituted Δ8-tetrahydrocannabinols. Bioorg Med Chem Lett 12:3583–3586
Pate DW (1999) Anandamide structure-activity relationships and mechanisms of action on intraocular pressure in the normotensive rabbit model. Doctoral dissertation. Kuopio University Publications, Kuopio
Paton GS, Pertwee RG, Davies SN (1998) Correlation between cannabinoid mediated effects on paired pulse depression and induction of long term potentiation in the rat hippocampal slice. Neuropharmacology 37:1123–1130
Pertwee R, Griffin G, Fernando S, Li X, Hill A, Makriyannis A (1995) AM630, a competitive cannabinoid receptor antagonist. Life Sci 56:1949–1955
Pertwee RG (1988) The central neuropharmacology of psychotropic cannabinoids. Pharmacol Ther 36:189–261
Pertwee RG (1997) Pharmacology of cannabinoid CB1 and CB2 receptors. Pharmacol Ther 74:129–180
Pertwee RG (1999a) Pharmacology of cannabinoid receptor ligands. Curr Med Chem 6:635–664
Pertwee RG (1999b) Evidence for the presence of CB1 cannabinoid receptors on peripheral neurones and for the existence of neuronal non-CB1 cannabinoid receptors. Life Sci 65:597–605
Pertwee RG (2000) Cannabinoid receptor ligands: clinical and neuropharmacological considerations, relevant to future drug discovery and development. Expert Opin Investig Drugs 9:1553–1571
Pertwee RG (2001) Cannabinoid receptors and pain. Prog Neurobiol 63:569–611
Pertwee RG (2003) Inverse agonism at cannabinoid receptors. In: IJzerman AP (ed) Inverse agonism. Elsevier, Amsterdam, pp 75–86
Pertwee RG (2004a) Novel pharmacological targets for cannabinoids. Curr Neuropharmacol 2:9–29
Pertwee RG (2004b) The pharmacology and therapeutic potential of cannabidiol. In: Di Marzo V (ed) Cannabinoids Kluwer Academic/Plenum Publishers, New York, pp 32–83

Pertwee RG (2005) Inverse agonism and neutral antagonism at cannabinoid CB1 receptors. Life Sci 76:1307–1324

Pertwee RG, Ross RA (2002) Cannabinoid receptors and their ligands. Prostaglandins Leukot Essent Fatty Acids 66:101–121

Pertwee RG, Coutts AA, Griffin G, Fernando SR, McCallion D, Stevenson L (1996a) Presence of cannabinoid CB1 receptors on prejunctional neurones of certain isolated tissue preparations: a brief review. Med Sci Monit 2:840–848

Pertwee RG, Fernando SR, Griffin G, Ryan W, Razdan RK, Compton DR, Martin BR (1996b) Agonist-antagonist characterization of 6'-cyanohex-2'-yne-Δ8-tetrahydrocannabinol in two isolated tissue preparations. Eur J Pharmacol 315:195–201

Pertwee RG, Gibson TM, Stevenson LA, Ross RA, Banner WK, Saha B, Razdan RK, Martin BR (2000) O-1057, a potent water-soluble cannabinoid receptor agonist with antinociceptive properties. Br J Pharmacol 129:1577–1584

Pertwee RG, Ross RA, Craib SJ, Thomas A (2002) (−)-Cannabidiol antagonizes cannabinoid receptor agonists and noradrenaline in the mouse vas deferens. Eur J Pharmacol 456:99–106

Petitet F, Marin L, Doble A (1996) Biochemical and pharmacological characterization of cannabinoid binding sites using [3H]SR141716A. Neuroreport 7:789–792

Petitet F, Jeantaud B, Capet M, Doble A (1997) Interaction of brain cannabinoid receptors with guanine nucleotide binding protein. A radioligand binding study. Biochem Pharmacol 54:1267–1270

Petitet F, Jeantaud B, Reibaud M, Imperato A, Dubroeucq MC (1998) Complex pharmacology of natural cannabinoids: evidence for partial agonist activity of Δ9-tetrahydrocannabinol and antagonist activity of cannabidiol on rat brain cannabinoid receptors. Life Sci 63:PL1–PL6

Platt B, Drysdale AJ (2004) Search and rescue: identification of cannabinoid actions relevant for neuronal survival and protection. Curr Neuropharmacol 2:103–114

Porter AC, Sauer J-M, Knierman MD, Becker GW, Berna MJ, Bao J, Nomikos GG, Carter P, Bymaster FP, Leese AB, Felder CC (2002) Characterization of a novel endocannabinoid, virodhamine, with antagonist activity at the CB1 receptor. J Pharmacol Exp Ther 301:1020–1024

Portier M, Rinaldi-Carmona M, Pecceu F, Combes T, Poinot-Chazel C, Calandra B, Barth F, Le Fur G, Casellas P (1999) SR 144528, an antagonist for the peripheral cannabinoid receptor that behaves as an inverse agonist. J Pharmacol Exp Ther 288:582–589

Rhee M-H, Vogel Z, Barg J, Bayewitch M, Levy R, Hanus L, Breuer A, Mechoulam R (1997) Cannabinol derivatives: binding to cannabinoid receptors and inhibition of adenylyl cyclase. J Med Chem 40:3228–3233

Richardson JD, Kilo S, Hargreaves KM (1998) Cannabinoids reduce hyperalgesia and inflammation via interaction with peripheral CB1 receptors. Pain 75:111–119

Rinaldi-Carmona M, Barth F, Héaulme M, Shire D, Calandra B, Congy C, Martinez S, Maruani J, Néliat G, Caput D, Ferrara P, Soubrié P, Brelière JC, Le Fur G (1994) SR141716A, a potent and selective antagonist of the brain cannabinoid receptor. FEBS Lett 350:240–244

Rinaldi-Carmona M, Calandra B, Shire D, Bouaboula M, Oustric D, Barth F, Casellas P, Ferrara P, Le Fur G (1996a) Characterization of two cloned human CB1 cannabinoid receptor isoforms. J Pharmacol Exp Ther 278:871–878

Rinaldi-Carmona M, Pialot F, Congy C, Redon E, Barth F, Bachy A, Brelière J-C, Soubrié P, Le Fur G (1996b) Characterization and distribution of binding sites for [3H]-SR141716A, a selective brain (CB1) cannabinoid receptor antagonist, in rodent brain. Life Sci 58:1239–1247

Rinaldi-Carmona M, Barth F, Millan J, Derocq J-M, Casellas P, Congy C, Oustric D, Sarran M, Bouaboula M, Calandra B, Portier M, Shire D, Brelière J-C, Le Fur G (1998) SR 144528, the first potent and selective antagonist of the CB2 cannabinoid receptor. J Pharmacol Exp Ther 284:644–650

Ross RA (2003) Anandamide and vanilloid TRPV1 receptors. Br J Pharmacol 140:790–801

Ross RA, Brockie HC, Fernando SR, Saha B, Razdan RK, Pertwee RG (1998) Comparison of cannabinoid binding sites in guinea-pig forebrain and small intestine. Br J Pharmacol 125:1345–1351

Ross RA, Brockie HC, Stevenson LA, Murphy VL, Templeton F, Makriyannis A, Pertwee RG (1999a) Agonist-inverse agonist characterization at CB1 and CB2 cannabinoid receptors of L759633, L759656 and AM630. Br J Pharmacol 126:665–672

Ross RA, Gibson TM, Stevenson LA, Saha B, Crocker P, Razdan RK, Pertwee RG (1999b) Structural determinants of the partial agonist-inverse agonist properties of 6'-azidohex-2'-yne-Δ8-tetrahydrocannabinol at cannabinoid receptors. Br J Pharmacol 128:735–743

Ross RA, Gibson TM, Brockie HC, Leslie M, Pashmi G, Craib SJ, Di Marzo V, Pertwee RG (2001) Structure-activity relationship for the endogenous cannabinoid, anandamide, and certain of its analogues at vanilloid receptors in transfected cells and vas deferens. Br J Pharmacol 132:631–640

Rubovitch V, Gafni M, Sarne Y (2002) The cannabinoid agonist DALN positively modulates L-type voltage-dependent calcium-channels in N18TG2 neuroblastoma cells. Mol Brain Res 101:93–102

Ruiu S, Pinna GA, Marchese G, Mussinu J-M, Saba P, Tambaro S, Casti P, Vargiu R, Pani L (2003) Synthesis and characterization of NESS 0327: a novel putative antagonist of the CB1 cannabinoid receptor. J Pharmacol Exp Ther 306:363–370

Savinainen JR, Järvinen T, Laine K, Laitinen JT (2001) Despite substantial degradation, 2-arachidonoylglycerol is a potent full efficacy agonist mediating CB1 receptor-dependent G-protein activation in rat cerebellar membranes. Br J Pharmacol 134:664–672

Savinainen JR, Saario SM, Niemi R, Järvinen T, Laitinen JT (2003) An optimized approach to study endocannabinoid signaling: evidence against constitutive activity of rat brain adenosine A1 and cannabinoid CB1 receptors. Br J Pharmacol 140:1451–1459

Schlicker E, Kathmann M (2001) Modulation of transmitter release via presynaptic cannabinoid receptors. Trends Pharmacol Sci 22:565–572

Schlicker E, Redmer A, Werner A, Kathmann M (2003) Lack of CB1 receptors increases noradrenaline release in vas deferens without affecting atrial noradrenaline release or cortical acetylcholine. Br J Pharmacol 140:323–328

Selley DE, Stark S, Sim LJ, Childers SR (1996) Cannabinoid receptor stimulation of guanosine-5'-O-(3-[35S]thio)triphosphate binding in rat brain membranes. Life Sci 59:659–668

Sheskin T, Hanus L, Slager J, Vogel Z, Mechoulam R (1997) Structural requirements for binding of anandamide-type compounds to the brain cannabinoid receptor. J Med Chem 40:659–667

Shire D, Carillon C, Kaghad M, Calandra B, Rinaldi-Carmona M, Le Fur G, Caput D, Ferrara P (1995) An amino-terminal variant of the central cannabinoid receptor resulting from alternative splicing. J Biol Chem 270:3726–3731

Shire D, Calandra B, Rinaldi-Carmona M, Oustric D, Pessègue B, Bonnin-Cabanne O, Le Fur G, Caput D, Ferrara P (1996) Molecular cloning, expression and function of the murine CB2 peripheral cannabinoid receptor. Biochim Biophys Acta 1307:132–136

Showalter VM, Compton DR, Martin BR, Abood ME (1996) Evaluation of binding in a transfected cell line expressing a peripheral cannabinoid receptor (CB2): identification of cannabinoid receptor subtype selective ligands. J Pharmacol Exp Ther 278:989–999

Sim LJ, Selley DE, Childers SR (1995) In vitro autoradiography of receptor-activated G proteins in rat brain by agonist-stimulated guanylyl 5'-[γ-[35S]thio]triphosphate binding. Proc Natl Acad Sci USA 92:7242–7246

Sim-Selley LJ, Brunk LK, Selley DE (2001) Inhibitory effects of SR141716A on G-protein activation in rat brain. Eur J Pharmacol 414:135–143

Simoneau II, Hamza MS, Mata HP, Siegel EM, Vanderah TW, Porreca F, Makriyannis A, Malan TP (2001) The cannabinoid agonist WIN55,212-2 suppresses opioid-induced emesis in ferrets. Anesthesiology 94:882–887

Smart D, Jonsson K-O, Vandevoorde S, Lambert DM, Fowler CJ (2002) 'Entourage' effects of N-acyl ethanolamines at human vanilloid receptors. Comparison of effects upon anandamide-induced vanilloid receptor activation and upon anandamide metabolism. Br J Pharmacol 136:452–458

Song Z-H, Bonner TI (1996) A lysine residue of the cannabinoid receptor is critical for receptor recognition by several agonists but not WIN55212-2. Mol Pharmacol 49:891–896

Storr M, Gaffal E, Saur D, Schusdziarra V, Allescher HD (2002) Effect of cannabinoids on neural transmission in rat gastric fundus. Can J Physiol Pharmacol 80:67–76

Sugiura T, Kodaka T, Kondo S, Tonegawa T, Nakane S, Kishimoto S, Yamashita A, Waku K (1996) 2-Arachidonoylglycerol, a putative endogenous cannabinoid receptor ligand, induces rapid, transient elevation of intracellular free Ca2+ in neuroblastoma x glioma hybrid NG108–15 cells. Biochem Biophys Res Commun 229:58–64

Sugiura T, Kodaka T, Kondo S, Nakane S, Kondo H, Waku K, Ishima Y, Watanabe K, Yamamoto I (1997) Is the cannabinoid CB1 receptor a 2-arachidonoylglycerol receptor? Structural requirements for triggering a Ca2+ transient in NG108–15 cells. J Biochem (Tokyo) 122:890–895

Sugiura T, Kondo S, Kishimoto S, Miyashita T, Nakane S, Kodaka T, Suhara Y, Takayama H, Waku K (2000) Evidence that 2-arachidonoylglycerol but not N-palmitoylethanolamine or anandamide is the physiological ligand for the cannabinoid CB2 receptor: comparison of the agonistic activities of various cannabinoid receptor ligands in HL-60 cells. J Biol Chem 275:605–612

Suhara Y, Takayama H, Nakane S, Miyashita T, Waku K, Sugiura T (2000) Synthesis and biological activities of 2-arachidonoylglycerol, an endogenous cannabinoid receptor ligand, and its metabolically stable ether-linked analogues. Chem Pharm Bull (Tokyo) 48:903–907

Suhara Y, Nakane S, Arai S, Takayama H, Waku K, Ishima Y, Sugiura T (2001) Synthesis and biological activities of novel structural analogues of 2-arachidonoylglycerol, an endogenous cannabinoid receptor ligand. Bioorg Med Chem Lett 11:1985–1988

Tao Q, Abood ME (1998) Mutation of a highly conserved aspartate residue in the second transmembrane domain of the cannabinoid receptors, CB1 and CB2, disrupts G-protein coupling. J Pharmacol Exp Ther 285:651–658

Thomas A, Ross RA, Saha B, Mahadevan A, Razdan RK, Pertwee R (2004) 6"-Azidohex-2"-yne-cannabidiol: a potential neutral, competitive cannabinoid CB1 receptor antagonist. Eur J Pharmacol 487:213–221

Tognetto M, Amadesi S, Harrison S, Creminon C, Trevisani M, Carreras M, Matera M, Geppetti P, Bianchi A (2001) Anandamide excites central terminals of dorsal root ganglion neurons via vanilloid receptor-1 activation. J Neurosci 21:1104–1109

Trendelenburg AU, Cox SL, Schelb V, Klebroff W, Khairallah L, Starke K (2000) Modulation of 3H-noradrenaline release by presynaptic opioid, cannabinoid and bradykinin receptors and β-adrenoceptors in mouse tissues. Br J Pharmacol 130:321–330

Ueda N (2002) Endocannabinoid hydrolases. Prostaglandins Other Lipid Mediat 68–69:521–534

van der Stelt M, Di Marzo V (2004) Metabolic fate of endocannabinoids. Curr Neuropharmacol 2:37–48

van der Stelt M, Veldhuis WB, Maccarrone M, Bär PR, Nicolay K, Veldink GA, Di Marzo V, Vliegenthart JFG (2002) Acute neuronal injury, excitotoxicity, and the endocannabinoid system. Mol Neurobiol 26:317–346

Vásquez C, Navarro-Polanco RA, Huerta M, Trujillo X, Andrade F, Trujillo-Hernández B, Hernández L (2003) Effects of cannabinoids on endogenous K+ and Ca2+ currents in HEK293 cells. Can J Physiol Pharmacol 81:436–442

Wager-Miller J, Westenbroek R, Mackie K (2002) Dimerization of G protein-coupled receptors: CB1 cannabinoid receptors as an example. Chem Phys Lipids 121:83–89

Wagner JA, Varga K, Járai Z, Kunos G (1999) Mesenteric vasodilation mediated by endothelial anandamide receptors. Hypertension 33:429–434

Walker JM, Krey JF, Chu CJ, Huang SM (2002) Endocannabinoids and related fatty acid derivatives in pain modulation. Chem Phys Lipids 121:159–172

Walter L, Franklin A, Witting A, Wade C, Xie Y, Kunos G, Mackie K, Stella N (2003) Nonpsychotropic cannabinoid receptors regulate microglial cell migration. J Neurosci 23:1398–1405

White R, Hiley CR (1998) The actions of the cannabinoid receptor antagonist, SR 141716A, in the rat isolated mesenteric artery. Br J Pharmacol 125:689-696

Wiley JL, Martin BR (2002) Cannabinoid pharmacology: implications for additional cannabinoid receptor subtypes. Chem Phys Lipids 121:57–63

Wiley JL, Martin BR (2003) Cannabinoid pharmacological properties common to other centrally acting drugs. Eur J Pharmacol 471:185–193

Wiley JL, Beletskaya ID, Ng EW, Dai Z, Crocker PJ, Mahadevan A, Razdan RK, Martin BR (2002) Resorcinol derivatives: a novel template for the development of cannabinoid CB1/CB2 and CB2-selective agonists. J Pharmacol Exp Ther 301:679–689

Wrobleski ST, Chen P, Hynes J, Lin S, Norris DJ, Pandit CR, Spergel S, Wu H, Tokarski JS, Chen X, Gillooly KM, Kiener PA, McIntyre KW, Patil-Koota V, Shuster DJ, Turk LA, Yang G, Leftheris K (2003) Rational design and synthesis of an orally active indolopyridone as a novel conformationally constrained cannabinoid ligand possessing antiinflammatory properties. J Med Chem 46:2110–2116

Yao L, Fan P, Jiang Z, Mailliard WS, Gordon AS, Diamond I (2003) Addicting drugs utilize a synergistic molecular mechanism in common requiring adenosine and Gi-$\beta\gamma$ dimers. Proc Natl Acad Sci USA 100:14379–14384

Zygmunt PM, Petersson J, Andersson DA, Chuang H, Søgård M, Di Marzo V, Julius D, Högestätt ED (1999) Vanilloid receptors on sensory nerves mediate the vasodilator action of anandamide. Nature 400:452–457

Zygmunt PM, Andersson DA, Högestätt ED (2002) Δ9-tetrahydrocannabinol and cannabinol activate capsaicin-sensitive sensory nerves via a CB1 and CB2 cannabinoid receptor-independent mechanism. J Neurosci 22:4720–4727

HEP (2005) 168:53–79

Cannabinoid Receptor Signaling

A.C. Howlett

Neuroscience/Drug Abuse Research Program, 208 JLC-BBRI, North Carolina Central University, 700 George Street, Durham NC, 27707, USA
ahowlett@nccu.edu

Abstract The cannabinoid receptor family currently includes two types: CB_1, characterized in neuronal cells and brain, and CB_2, characterized in immune cells and tissues. CB_1 and CB_2 receptors are members of the superfamily of seven-transmembrane-spanning (7-TM) receptors, having a protein structure defined by an array of seven membrane-spanning helices with intervening intracellular loops and a C-terminal domain that can associate with G proteins. Cannabinoid receptors are associated with G proteins of the Gi/o family (Gi1,2 and 3, and Go1 and 2). Signal transduction via Gi inhibits adenylyl cyclase in most tissues and cells, although signaling via Gs stimulates adenylyl cyclase in some experimental models. Evidence exists for cannabinoid receptor-mediated Ca^{2+} fluxes and stimulation of phospholipases A and C. Stimulation of CB_1 and CB_2 cannabinoid receptors leads to phosphorylation and activation of p42/p44 mitogen-activated protein kinase (MAPK), p38 MAPK and Jun N-terminal kinase (JNK) as signaling pathways to regulate nuclear transcription factors. The CB_1 receptor regulates K^+ and Ca^{2+} ion channels, probably via Go. Ion channel regulation serves as an important component of neurotransmission modulation by endogenous cannabinoid compounds released in response to neuronal depolarization. Cannabinoid receptor signaling via G proteins results from interactions with the second, third and fourth intracellular loops of the receptor. Desensitization of signal transduction pathways that couple through the G proteins probably entails phosphorylation of critical amino acid residues on these intracellular surfaces.

Keywords Adenylyl cyclase · Aminoalkylindole · Anandamide · Ca^{2+} · Cannabinoid · Cyclic AMP · Depolarization suppression of inhibition or excitation · Desensitization · Endocannabinoid · G proteins · Ion channels · Mitogen activated protein kinases · Neurotransmission · Nitric oxide · Serine/threonine kinases · Seven-transmembrane spanning receptors · Synaptic plasticity · Tyrosine kinases

1 Introduction

Cannabinoid receptors are members of the rhodopsin-like family of seven-transmembrane-spanning (7-TM) receptors that are formed by the interaction of the seven transmembrane helices, and generally couple to G proteins at their intracellular surface as one mechanism for their signal transduction. The cannabinoid receptor family currently includes two types: CB_1, found in neuronal cells and brain, and CB_2, found in immune cells and tissues (see Howlett et al. 2002 for a comprehensive review of cannabinoid receptor pharmacology). Until the discovery of cannabinoid receptors, the mechanism of action of cannabinoid drugs was generally attributed to their lipid solubility properties, with the membrane/buffer partition coefficients for Δ^9-tetrahydrocannabinol (Δ^9-THC) reported to be in the range of 500 to 12,500 (Seeman et al. 1972; Roth and Williams 1979). Δ^9-THC in the 3 μM to 10 μM range could increase fluidity of synaptic plasma membranes (Hillard et al. 1985). The ability of both psychoactive and inactive cannabinoid drugs to influence ATPase and monoamine oxidase activities, hormone and neurotransmitter binding, and synaptosomal uptake of neurotransmitters in in vitro assays was attributed to their ability to intercalate into cellular membranes (for discussion see Martin 1986; Pertwee 1988). The discovery that sub-micromolar concentrations of psychoactive cannabinoid drugs could attenuate cyclic AMP accumulation in cultured neuronal cells and inhibit adenylyl cyclase activity in membranes (Howlett and Fleming 1984; Howlett 1984, 1985) led to the notion that cannabinoid compounds must be working through signal transduction mechanisms comparable to those defined for hormones and neurotransmitters. The involvement of G proteins in the response to active cannabinoid drugs was demonstrated as the characteristic requirement of sub-millimolar Mg^{2+} concentrations and micromolar guanosine triphosphate (GTP) concentrations for Gi-mediated inhibition of adenylyl cyclase (Howlett 1985). The elimination of the response to cannabinoid drugs by pretreatment of the neuronal cells or membranes with pertussis toxin confirmed that a member of the pertussis toxin-sensitive Gi/o family mediated the response (Howlett et al. 1986). The observation that the order of potency for this signal transduction pathway paralleled that for in vivo biological responses of antinociception, immobility, and hypothermia (Howlett et al. 1988; Little et al. 1988; Melvin and Johnson 1987; Howlett 1987) led to the understanding that a cellular receptor was responsible for the effects rather than membrane fluidity changes (Howlett et al. 1989; Thomas et al. 1990).

The development of a high-affinity, stereoselective radioligand, [^{3}H]CP55940, led to the pharmacological characterization of a binding site in brain membranes that could be shown to correlate with the pharmacology of in vivo biological responses (Devane et al. 1988; Howlett et al. 1988). [^{3}H]CP55940 was subsequently used to describe structure–activity relationships for the brain cannabinoid receptor (Howlett et al. 1990; Melvin et al. 1993, 1995) and to define brain regional localization of the receptor (Herkenham et al. 1990, 1991). It was soon determined that high-affinity [^{3}H]CP55940 binding could be attributed to two receptor types: the CB_1 receptor cloned from rat and human brain cDNA libraries (Matsuda et al. 1990; Gerard et al. 1990), and the CB_2 receptor cloned from HL60 promyelocytic cells (Munro et al. 1993).

Cannabinoid pharmacology progressed with the discovery of a number of potent ligands; however, until recently little pharmacological specificity for CB_1 and CB_2 receptors was identified. Increased potency and efficacy for both receptors was found for HU210, a dimethylheptyl analog of Δ^9-THC (Howlett et al. 1990; Felder et al. 1995). A number of non-classical AC-bicyclic (e.g., CP55940) and ACD-tricyclic cannabinoid (e.g., CP55244) compounds also exhibited high potency but limited receptor specificity (Johnson et al. 1981). This class of compounds resembles the classical cannabinoid ABC-tricyclic ring structures with the exception that the pyran "B" ring is eliminated in these structures. WIN55212-2, an aminoalkylindole compound, was discovered as a highly potent, full agonist for both cannabinoid receptor types (Compton et al. 1992; Pacheco et al. 1991). The endogenous agonists for cannabinoid receptors are arachidonic acid metabolites, including arachidonylethanolamide (anandamide), 2-arachidonoylglycerol (2-AG), and 2-arachidonylglyceryl ether (noladin ether) (see Di Marzo et al. 1999; Freund et al. 2003; Giuffrida et al. 2001; Howlett and Mukhopadhyay 2000; Martin et al. 1999; Schmid 2000; Sugiura and Waku 2000; Reggio and Traore 2000 for review). The first specific antagonist for the CB_1 cannabinoid receptor was SR141716 (rimonabant), an aryl pyrazole compound discovered at Sanofi Recherche (Rinaldi-Carmona et al. 1994; Barth and Rinaldi-Carmona 1999). A specific CB_2 receptor antagonist, SR144528, has structural similarities to the CB_1 receptor antagonist (Rinaldi-Carmona et al. 1998). These compounds have been the prevalent ligands utilized in studies of signal transduction pathways for cannabinoid receptors.

2 The Cyclic AMP and Protein Kinase A Signal Transduction Pathway

2.1 Cannabinoid Receptor-Mediated Inhibition of Cyclic AMP Production

Cannabinoid receptor-regulated signal transduction through the cyclic AMP system has been reviewed (Howlett 1995; Pertwee 1997, 1999). For the CB_1 receptor, inhibition of cyclic AMP production is the characteristic response to cannabinoid agonists in brain tissue (Bidaut-Russell and Howlett 1991; Childers et al. 1994). Pharmacological studies have been performed using N18TG2 neuroblastoma cells expressing endogenous CB_1 receptors (Howlett et al. 1988; Pinto et al. 1994) and cell lines expressing recombinant CB_1 receptors (Matsuda et al. 1990; Felder et al. 1993, 1995; Vogel et al. 1993). CB_1 receptor-mediated inhibition of adenylyl cyclase is pertussis toxin-sensitive, indicating the requirement for Gi/o proteins (Howlett et al. 1986; Pacheco et al. 1993; Vogel et al. 1993).

Regulation of cellular activities by cyclic AMP-dependent protein kinase (PKA) is a critical pathway in neuronal responses via the potassium channel A-current (Childers and Deadwyler 1996). In rat hippocampal cells, PKA phosphorylation of the potassium channel produced a negative shift in the voltage-dependence (Deadwyler et al. 1995). CB_1 receptor stimulation resulted in a decrease in intracellular cyclic AMP, net dephosphorylation of the channels, activation of the A-type potassium currents, and hyperpolarization of the membrane (Deadwyler et al. 1995; Hampson et al. 1995). The significance of cannabinoid-mediated hyperpolarization of the axon terminals is that it can cause a depression in the response to depolarizing stimuli and failure in neurotransmitter release at the synapse (Childers and Deadwyler 1996).

Synaptic plasticity and neuronal remodeling can be modified by cannabinoid receptors via the cyclic AMP/PKA pathway. CB_1 receptor agonists induced neurite retraction in a neuroblastoma cell model (Zhou and Song 2001) and inhibition of nerve growth factor (NGF)-induced neurite extension in neural progenitor cells or PC12 pheochromocytoma cells transfected with the CB_1 receptor (Rueda et al. 2002). CB_1 receptors could attenuate the NGF-mediated signaling through p42/p44 MAPK (see below). A cannabinoid receptor-mediated decrease in cyclic AMP and PKA activity was demonstrated to be the mechanism from evidence that this response could be reversed by forskolin or hormone-stimulated cyclic AMP production (Rueda et al. 2002). Cannabinoid receptor-stimulation led to Tyr-phosphorylation of focal adhesion kinase (pp125 FAK) in hippocampal slices, and this response was blocked by SR141716 and pertussis toxin, demonstrating its mediation by CB_1 receptors and Gi/o (Derkinderen et al. 1996; Derkinderen et al. 2001b). Evidence demonstrating that Gi-mediated inhibition of adenylyl cyclase is integral to this pathway comes from studies in which Tyr-phosphorylation of both FAK in brain slices (Derkinderen et al. 1996) and FAK-related non-kinase (FRNK) (Zhou and Song 2002) were reversed by 8-Br-cyclic AMP, and mimicked by PKA inhibitors.

Inhibition of forskolin-stimulated cyclic AMP production has been pharmacologically characterized in human lymphocytes and mouse spleen cells expressing endogenous CB_2 receptors, and in CHO cells expressing recombinant CB_2 receptors (Felder et al. 1995; Gonsiorek et al. 2000; Slipetz et al. 1995). This response was blocked by pertussis toxin, indicating the involvement of Gi/o proteins (Felder et al. 1995). The ramifications of the cellular response to a CB_2 receptor-mediated decrease in cyclic AMP have not been fully characterized in immune cells.

2.2 Cannabinoid Receptor-Mediated Stimulation of Cyclic AMP Production

In contrast to the above studies, stimulation of cyclic AMP production has also been observed in response to cannabinoid drugs. Cannabinoid receptor agonists produced an increase in basal cyclic AMP production in globus pallidus slice preparations (Maneuf and Brotchie 1997). Evidence that the same (CB_1) receptor type mediates both the inhibitory and stimulatory components stems from findings that the order of potency for various agonists was the same, and SR141716 was a competitive inhibitor for both components (Bonhaus et al. 1998). Several mechanisms have been reported that might explain this response. One mechanism might be the cellular production of an endogenous stimulator of adenylyl cyclase. The cannabinoid-mediated production of prostaglandins has been reported (Burstein et al. 1986, 1994), and prostaglandin synthesis has been implicated in cannabinoid-mediated cyclic AMP production (Hillard and Bloom 1983).

A second mechanism for cannabinoid receptor-mediated stimulation of cyclic AMP production could depend upon which isoform of adenylyl cyclase is expressed in target cells and the way that the particular isoform responds to Gi/o-mediated regulation. Inhibition of adenylyl cyclase by recombinant CB_1 or CB_2 receptors was observed in cells that co-express either the isoform 5/6 family or the 1/3/8 family (Rhee et al. 1998) as a result of inhibition by Gi (α subunit). On the other hand, stimulation of adenylyl cyclase was observed in cells coexpressing cannabinoid receptors and the adenylyl cyclase isoform 2/4/7 family, as a result of augmentation of a Gs response by the $G\beta\gamma$ dimers released from Gi due to cannabinoid receptor stimulation (Rhee et al. 1998).

A third mechanism could be the direct interaction between CB_1 receptors and Gs. Evidence for this mechanism has come from findings that pertussis toxin treatment of neurons and CHO cells expressing recombinant CB_1 receptors resulted in cannabinoid agonist stimulation of cyclic AMP accumulation (Glass and Felder 1997; Felder et al. 1998; Bonhaus et al. 1998). In cultured striatal cells, stimulation by combinations of dopamine and cannabinoid agonists resulted in an increase in cyclic AMP production (Glass and Felder 1997). To further investigate this phenomenon, Jarrahian and colleagues (2004) transfected recombinant D_2 dopamine and CB_1 receptors into HEK293 cells, and found that the expression of D_2 dopamine receptors was sufficient to convert the inhibition of forskolin-stimulated cyclic AMP production by CP55940 to a stimulation of cyclic AMP production. Pertussis toxin attenuated the inhibition but not the stimulation of cyclic AMP production, consistent with Gi mediation of the inhibition component and Gs mediation of the stimulation component. The finding that overexpression of Gαi1 could overcome the stimulatory component led these researchers to suggest that the D_2 dopamine receptors could sequester Gi proteins, resembling the response to pertussis toxin treatment, and thereby preclude their coupling to the CB_1 receptors (Jarrahian et al. 2004). The CB_1 receptor-mediated stimulation of cyclic AMP production required greater concentrations of CP55940 than did inhibition (Jarrahian et al. 2004). The efficacies of cannabinoid receptor agonists for regulation of Gs were not as great as for regulation of Gi (Bonhaus et al. 1998). HU210, CP55940, and WIN55212-2 were full agonists to inhibit forskolin-stimulated cyclic AMP accumulation by Gi, and Δ^9-THC and anandamide were partial agonists. Following pertussis toxin treatment, WIN55212-2 was a full agonist to stimulate cyclic AMP accumulation, but HU210, CP55940, Δ^9-THC, and anandamide behaved as partial agonists for this response.

3 Cannabinoid Receptor-Mediated Ca^{2+} Fluxes and Phospholipases C and A

Cannabinoid and endocannabinoid compounds increased intracellular free Ca^{2+} as determined by fura-2 fluorescence in undifferentiated N18TG2 neuroblastoma and NG108-15 neuroblastoma-glioma hybrid cells (Sugiura et al. 1996, 1997a, 1999). The CB_1 receptor and Gi/o proteins were implicated because this response was blocked by SR141716 and pertussis toxin (Sugiura et al. 1996, 1999). From studies directly measuring isotopic Ca^{2+} influx into N18TG2 neuroblastoma cells, the evidence suggests that desacetyllevonantradol stimulated Ca^{2+} uptake via CB_1 receptor coupling to Gs, cyclic AMP production, and PKA activation (Bash et al. 2003). Further evidence suggested that a second component of Ca^{2+} influx was due to CB_1 receptor coupling to Gi/o, leading to receptor Tyr kinase transactivation, PKC phosphorylation, and regulation of MAPK (Rubovitch et al. 2004). Evidence for a CB_1 receptor-mediated Tyr phosphorylation of N18TG2 cell proteins that can be immunoprecipitated with the CB_1 receptor has been reported (Peterson et al. 2004).

Some controversy exists regarding the ability of cannabinoid receptors to signal through the inositol 1,4,5-triphosphate (IP_3)-Ca^{2+} mobilization pathway. In studies using Ca^{2+} reporter fura-2 fluorometry, Ca^{2+} mobilization in N18TG2 neuroblastoma cells was blocked by a phospholipase C (PLC) inhibitor, indicating that PLCβ could be the effector (Sugiura et al. 1996, 1997a). CB_1 receptor activation in cultured cerebellar granule cells resulted in an augmented Ca^{2+} signal in response to depolarization by glutamate receptors or high K^+ (Netzeband et al. 1999). In these cells, Ca^{2+} was mobilized from a caffeine-sensitive and IP_3 receptor-sensitive pool. This Ca^{2+} signal was attenuated by SR141716, pertussis toxin, and a PLC inhibitor (Netzeband et al. 1999), indicative of a CB_1 receptor-mediated PLC mechanism for Ca^{2+} mobilization from endoplasmic reticulum stores.

The primary evidence against a PLC-mediated pathway is that agonist-stimulated CB_1 receptors that were heterologously expressed in competent CHO cells failed to couple to IP_3 or phosphatidic acid release (Felder et al. 1992, 1995). Furthermore, cannabinoid compounds inhibited (rather than augmented) neurotransmitter-stimulated inositol phospholipid production in hippocampal preparations (Nah et al. 1993). Anandamide and WIN55212-2 both failed to activate PLC in competent CHO cells expressing recombinant CB_2 receptors (Felder et al. 1992, 1993, 1995).

Some evidence exists for phospholipase A_2 (PLA_2) activity that could be regulated by cannabinoid receptors. Cannabinoid-induced arachidonic acid release has been observed in several cell culture systems, and this is believed to be mediated both by phospholipase activity and G proteins (Burstein 1991; Burstein et al. 1994; Shivachar et al. 1996).

4 Cannabinoid Receptor-Mediated Regulation of Ion Channels

Studies of the effects of cannabinoid drugs on neurophysiological responses in the years prior to the elucidation of the existence of a cannabinoid receptor were targeted at investigating a mechanism for the anticonvulsant properties of cannabidiol and mixed excitatory properties of Δ^9-THC (for a description and other original references see Karler and Turkanis 1981; Turkanis and Karler 1981). The laboratory of Karler and Turkanis used an in vivo model of cat spinal motor neurons to observe changes in amplitude of excitatory post-synaptic potentials evoked by these cannabinoid compounds (Turkanis and Karler 1983, 1986). These researchers also used cultured neuroblastoma cells to identify Δ^9-THC and 11-OH-Δ^9-THC-induced depression of inward Na^+ currents, suggesting a possible mechanism for CNS depression by these compounds (Turkanis et al. 1991).

4.1 Voltage-Gated Ca^{2+}-Channels

The first reports of the CB_1 receptor and Gi/o protein regulation of Ca^{2+} currents described a cannabinoid agonist-mediated inhibition of N-type voltage-gated Ca^{2+} channels in differentiated N18 neuroblastoma and NG108-15 neuroblastoma-glioma hybrid cells (Caulfield and Brown 1992; Mackie and Hille 1992; Mackie et al. 1993; Priller et al. 1995; Pan et al. 1996). WIN55212-2 and CP55940 elicited a maximal response, anandamide produced agonist/antagonist actions, and SR141716 antagonized this response (Mackie et al. 1993). In studies using fura-2 fluorescence to measure intracellular Ca^{2+} levels, 2-AG and anandamide inhibited the depolarization-evoked intracellular Ca^{2+} increase in differentiated NG108-15 cells (Sugiura et al. 1997b). Further investigations on the mechanism of inhibition of N-type currents have been carried out using neuronal expression systems (Priller et al. 1995; Pan et al. 1996, 1998; Vasquez and Lewis 1999; Guo and Ikeda 2004).

Q-type Ca^{2+} currents were inhibited by WIN55212-2 and anandamide in AtT-20 pituitary cells expressing recombinant CB_1, but not CB_2 receptors (Mackie et al. 1995). Pertussis toxin-sensitivity indicated that Gi/o proteins mediated the response. P/Q-type Ca^{2+} fluxes, detected by fura-2 fluorescence in rat cortical and cerebellar preparations, were inhibited by anandamide (Hampson et al. 1998). This response was blocked by SR141716 and pertussis toxin, indicating mediation by CB_1 receptors and Gi/o proteins.

L-type Ca^{2+} currents were inhibited by anandamide and WIN55212-2 in cat brain arterial smooth muscle cells that endogenously express the CB_1 receptor (Gebremedhin et al. 1999). This response was blocked by SR141716 and pertussis toxin, indicating a critical role for CB_1 receptors and Gi/o. Regulation of L-type Ca^{2+} channels in these smooth muscle cells could be pharmacologically correlated with vascular relaxation in cat cerebral arterial rings (Gebremedhin et al. 1999).

4.2 G Protein-Coupled Inwardly-Rectifying K^+ Channels

In AtT-20 pituitary tumor cells exogenously expressing CB_1 receptors, cannabinoid receptor agonists anandamide and WIN55212-2 activated the inwardly rectifying K^+ currents (K_{ir}). This was a pertussis toxin-sensitive response, indicating the mediation by Gi/o proteins (Mackie et al. 1995; Henry and Chavkin 1995; McAllister et al. 1999). A reduction in cyclic AMP and PKA activity was not required, providing evidence that a direct interaction exists between G protein subunits and the ion channel proteins. Cannabinoid receptor-mediated regulation of these channels was also demonstrated in *Xenopus laevis* oocytes (McAllister et al. 1999) and rat sympathetic neurons (Guo and Ikeda 2004) coexpressing the CB_1 receptor and G Protein-Coupled Inwardly-Rectifying K^+ (GIRK1) and GIRK4 channels.

4.3 Depolarization-Induced Suppression of Inhibition and Excitation

The above-described neurophysiological mechanisms of CB_1 receptor signaling permit a critical function for endocannabinoids as retrograde regulators of neuronal excitability via a mechanism referred to as depolarization-induced suppression of inhibition (DSI) or excitation (DSE) (Wilson and Nicoll 2001, 2002; Wilson et al. 2001). DSI, or DSE, is the feedback mechanism by which a depolarized postsynaptic cell can release a neuromodulator that diffuses to neighboring neurons in the synaptic network to block release of an inhibitory, or excitatory, neurotransmitter (Alger 2002). Wilson, Nicoll, and colleagues showed that DSI could be blocked by the CB_1 antagonist SR141716 and was absent in CB_1 receptor (−/−) knock-out mice, implicating release of endocannabinoids and participation of presynaptic CB_1 receptors (Wilson and Nicoll 2001; Wilson et al. 2001). According to the proposed schema, endocannabinoid production and diffusion from the postsynaptic cell would stimulate CB_1 receptors on presynaptic terminals of a subclass of interneurons in the hippocampus, leading to decreased release of γ-aminobutyric acid (GABA) (Wilson and Nicoll 2001, 2002; Wilson et al. 2001). In addition to hippocampal circuits (Hoffman et al. 2003; Ohno-Shosaku et al. 2002b;

Misner and Sullivan 1999), other brain areas in which neurotransmission appears to be modulated by endocannabinoid release include basal forebrain (Harkany et al. 2003; Steffens et al. 2003), striatum (Gerdeman et al. 2002), and cerebellum (Breivogel et al. 2004; Kreitzer et al. 2002; Maejima et al. 2001b).

In the hippocampus, depolarization-induced opening of pyramidal cell N-type voltage-gated Ca^{2+} channels (Wilson and Nicoll 2002) would lead to release of endocannabinoid neuromodulators (Piomelli 2003). This response did not occur with a high probability in hippocampal cells firing under normal conditions, leading some researchers to suggest that high frequency discharges would be more likely to evoke elevated intracellular Ca^{2+} levels via activated voltage-gated Ca^{2+} channels (Hampson et al. 2003; Zhuang et al. 2003; Alger et al. 1996; Beau and Alger 1998; Morishita et al. 1998). Other synaptic events that might occur concurrently to promote endocannabinoid release in DSI or DSE include convergence of multiple signals that increase intracellular Ca^{2+} (Kim et al. 2002; Brenowitz and Regehr 2003), signal transduction directed by metabotropic glutamate receptors (Galante and Diana 2004; Maejima et al. 2001a; Morishita et al. 1998; Ohno-Shosaku et al. 2002a; Varma et al. 2001), and regulation of post-synaptic transport mechanisms for these retrograde modulators (Ronesi et al. 2004).

The mechanism by which cannabinoid receptors modulate neurotransmitter release is not understood. Some evidence suggests that this could involve K^+ channels (Daniel et al. 2004; Kreitzer et al. 2002). Alternatively, regulation of N or P/Q voltage-gated Ca^{2+} channels might be the mechanism for endocannabinoid agonist action (Shen and Thayer 1998; Guo and Ikeda 2004). Synergism between endocannabinoid-stimulated cellular responses and signal transduction pathways initiated by other synaptic events might be important in the regulation of neurotransmitter release (Netzeband et al. 1999).

5
Cannabinoid Receptor-Mediated Signal Transduction to the Nucleus

5.1
p42/p44 Mitogen-Activated Protein Kinases (Extracellular Signal-Regulated Kinase 1 and 2)

Although in vivo administration of Δ^9-THC can activate brain p42/p44 mitogen-activated protein kinases (MAPK), also known as extracellular signal-regulated kinase 1 and 2 (ERK1 and ERK2), it is likely that this response could reflect multi-synaptic cellular events involving multiple neuromodulators, including dopamine (Valjent et al. 2004). Thus, signal transduction studies have been performed using cultured cell model systems. p42/p44 MAPK activation by an SR141716-sensitive and pertussis toxin-sensitive pathway was first identified in several cell types, including WI-38 fibroblasts, U373MG astrocytic cells, C6 glioma cells and primary astrocytes, and various host cells expressing recombinant CB_1 receptors (Bouaboula et al. 1995b; Guzman and Sanchez 1999; Sanchez et al. 1998; Wartmann et al. 1995).

One mechanism for p42/p44 MAPK activation by CB_1 receptors coupled to Gi/o could utilize the $G\beta\gamma$ dimer to provide a scaffold for proteins in the MAPK activation complex. According to this schema, recruitment of phosphatidylinositol-3-kinase (PI3K) and phosphorylation of membrane inositol phospholipids would recruit protein kinase B (PKB, also known as Akt). This would result in the sequential phosphorylation and activation of the three-kinase module consisting of Raf-1, MAP-ERK kinase (MEK) and p42/p44 MAPK. Evidence for this mechanism comes from studies in which CB_1 receptor-mediated signaling via p42/p44 MAPK was blocked by the PI3K inhibitors wortmannin and LY294002 (Bouaboula et al. 1995b; Galve-Roperh et al. 2002; Wartmann et al. 1995). Δ^9-THC, HU210, and CP55940 produced an SR141716-sensitive activation of the PKB isoform I_B in the human astrocytoma cell line U373MG and in CHO cells expressing recombinant CB_1 receptors (Galve-Roperh et al. 2002; Gomez et al. 2000). Δ^9-THC promoted PI3K and tyrosine phosphorylation of Raf-1 and its translocation to the membrane in rat cortical astrocytes (Sanchez et al. 1998).

An alternative mechanism for regulation of p42/p44 MAPK could be the release of the inhibitory regulation of c-Raf that results from the phosphorylation of Raf by PKA. CB_1 receptor/Gi-mediated inhibition of cyclic AMP production and reduction of PKA activity would promote a net dephosphorylation of c-Raf, thereby permitting the Raf kinase to serve as an activator of MEK in the p42/p44 MAPK activation module. Evidence for this pathway has been described for WIN55212-2-stimulated N1E-115 neuroblastoma cells (Davis et al. 2003) and hippocampal slice preparations (Derkinderen et al. 2003).

Activation of p42/p44 MAPK can be linked to expression of immediate early genes, as has been demonstrated for krox-24 expression induced by CB_1 receptors in U373MG human astrocytoma cells (Bouaboula et al. 1995a). Administration of Δ^9-THC to mice led to the p42/p44 MAPK-dependent expression of c-fos and zif268 in the hippocampus (Derkinderen et al. 2003). These transcription factors modulate the gene expression pattern for proteins involved in cellular functions associated with synaptic plasticity, cell survival, and differentiation.

CB_2 receptors promoted the phosphorylation of 42/p44 MAPK in cultured human promyelocytic-HL60 cells, and in CHO cells expressing recombinant CB_2 receptors (Bouaboula et al. 1996). The mediation by pertussis toxin-sensitive G proteins was demonstrated for HL60 cells (Kobayashi et al. 2001). A PI3K pathway was not the mechanism for regulation by CB_2 receptors in HL60 cells inasmuch as cannabinoid agonists failed to activate PKB/Akt (Gomez del Pulgar et al. 2000). Stimulation of CB_2 receptors by 2-AG in rat RTMGL1 microglial cells led to p42 MAPK activation and cell proliferation (Carrier et al. 2004). It should be pointed out that regulation of p42/p44 MAPK signaling is often by a complex network involving multiple stimuli. For example, sustained p42/p44 MAPK phosphorylation in mouse splenocytes resulted from stimulation of PKC by phorbol esters in addition to calmodulin kinase by Ca^{2+} ionophores (Faubert Kaplan and Kaminski 2003). Under these conditions, cannabinoid compounds were able to block the response (Faubert Kaplan and Kaminski 2003).

Krox-24 expression was induced by CB_2 receptors in HL60 promyelocytes (Bouaboula et al. 1996). A gene expression profile for CB_2 receptor-activated HL60 cells showed an induction of genes involved in cytokine synthesis, regulation of transcription, and cell differentiation (Derocq et al. 2000).

5.2
p38 MAPK and Jun N-Terminal Kinases

p38 MAPK was activated by cannabinoid receptor agonists in CHO cells expressing recombinant CB_1 receptors (Rueda et al. 2000) and in human vein endothelial cells possessing endogenous CB_1 receptors (Liu et al. 2000). Anandamide, 2-AG, and Δ^9-THC activated p38 MAPK via the CB_1 receptor in mouse hippocampal slices (Derkinderen et al. 2001a).

Jun N-terminal kinases (JNK1 and JNK2) were activated in response to cannabinoid receptor agonists in CHO cells expressing recombinant CB_1 receptors, and this was mediated through a pathway that included Gi/o, PI3K and Ras (Rueda et al. 2000). In the CHO cells (a fibroblast cell line), the transactivation of platelet-derived growth factor receptor was implicated in the JNK activation mechanism (Rueda et al. 2000).

Cellular kinase activation and sequelae in the absence of evidence of cannabinoid receptor participation should be interpreted with caution. Mechanisms other than cannabinoid receptor-mediated signal transduction could be possible. For example, anandamide stimulated p38 MAPK and JNK activation in PC12 pheochromocytoma cells (Sarker et al. 2003) and human umbilical vein endothelial cells (Yamaji et al. 2003), and these activated kinases were associated with triggering processes leading to apoptotic cell death. Further investigation indicated that activation of these kinases, leading to apoptosis in a number of cultured cell models (PC12, C6 glioma, Neuro-2A, CHO, HEK, Jurkat, and HL60), is a non-CB_1, non-CB_2 receptor-mediated process that involves anandamide and membrane lipids (Sarker and Maruyama 2003). In a second example, cannabinol and Δ^9-THC at high micromolar concentrations activated p42/p44 MAPK, leading to inhibition of gap junction function in a liver epithelial cell line by an undefined non-CB_1, non-CB_2 receptor-mediated process (Upham et al. 2003).

6
Cannabinoid Receptor-Mediated Nitric Oxide Production

Cannabinoid receptor agonists stimulate the production and release of nitric oxide (NO) by a CB_1 receptor-mediated mechanism utilizing one of the NO synthase (NOS) isoforms in neuronal tissues and model cells (see Fimiani et al. 1999a for review). The signal transduction pathway between CB_1 receptors and neuronal NOS (nNOS) regulation is believed to be important for mediating the effects of Δ^9-THC on hypothermia and locomotor activity (but not antinociception), as determined by the absence of these responses in nNOS (–/–) knock-out mice (Azad et al. 2001). NO production was stimulated by anandamide via SR141716-sensitive CB_1 receptors in rat median eminence slices, but it was not clear from these studies whether NOS in neurons was responsible (Prevot et al. 1998). The presence of Ca^{2+}-dependent constitutive NOS in N18 neuroblastoma homogenates was inferred from a cyclic guanosine monophosphate (cGMP) reporter assay (Simmons and Murphy 1992), and demonstrated by Western blot identification (Mukhopadhyay et al. 2002b; Norford et al. 2002). NO production was stimulated by anandamide and CP55940 in leech or mussel ganglia by an SR141716-sensitive mechanism, implicating the involvement of a CB_1-like receptor (Stefano et al. 1997a,b). Antagonism by the NOS inhibitor L-*N*-arg-methyl ester is evidence that this CB_1-like receptor initiates a signal transduction pathway leading to regulation of one of the isoforms of NOS (Prevot et al. 1998).

It is possible that the CB_1-mediated NO signal transduction pathway may play a role in inhibition of neurotransmitter release by cannabimimetic agonists. Both anandamide and the NO generating agent *S*-nitroso-*N*-acetyl-penicillamine could inhibit the release of preloaded radiolabeled dopamine from invertebrate ganglia, leading Stefano and coworkers to postulate a role for NO in mediating anandamide's effects on neurotransmitter release (Stefano et al. 1997a). Glutamate release from neurons in the rat medulla was blocked by NO donors SIN-1 and spermine NONOate (Huang et al. 2004). This response was blocked by a peroxynitrite decomposition catalyst but not by an NO-stimulated guanylyl cyclase inhibitor, indicating that generation of peroxynitrite was the mechanism (Huang et al. 2004). Further studies indicated that adenosine released in response to the peroxynitrite might mediate the inhibition of glutamatergic neurotransmission (Huang et al. 2004).

In non-neuronal cells, anandamide and HU210 stimulated NO production in human saphenous vein segments (Stefano et al. 1998), cultured human arterial endothelial cells (Fimiani et al. 1999b; Mombouli et al. 1999), cultured human umbilical vein endothelial cells (Maccarrone et al. 2000), and human monocytes (Stefano et al. 1996) in an SR141716-sensitive manner, implicating CB_1 receptors. NO production in cultured human arterial endothelial cells followed a rapid intracellular Ca^{2+} mobilization (Fimiani et al. 1999b; Mombouli et al. 1999). The generation of NO in saphenous vein endothelial cells required extracellular Ca^{2+} (Stefano et al. 1998). Although the isoform(s) of NOS was not identified in these cell lines, these characteristics of NO production are consistent with the stimulation of a Ca^{2+}-regulated constitutive NOS, perhaps endothelial NOS (eNOS).

NO and peroxynitrite in human endothelial cells, human embryonic kidney (HEK) cells, and C6 glioma cells promoted activation of the anandamide and 2-AG transporter(s) (Maccarrone et al. 2000; Bisogno et al. 2001; De Petrocellis et al. 2001). This phenomenon may have ramifications for cellular mechanisms that require anandamide as a regulator. For example, indomethacin is thought to augment anandamide's stimulation of CB_1 receptors in a model of inflammatory hyperalgesia by reducing spinal NO and relieving the activation of the anandamide transporter (Guhring et al. 2002). The net result would be increased extracellular concentrations of anandamide with decreased concentrations of NO, producing an antinociceptive response that was not reversed by prostaglandin E_2 (Guhring et al. 2002). Another example is the potential for NO to activate the anandamide transporter leading to increased intracellular accumulation of anandamide where it can serve as a regulator of transient receptor potential vanilloid type 1 (TRPV1, formerly VR1) (De Petrocellis et al. 2001).

Inhibition of iNOS induction is an important function of cannabinoid receptor agonists in inflammatory reactions, and may be a critical contributor to their anti-inflammatory and neuroprotective effects. Lipopolysaccharide plus interferon-γ induced iNOS expression in saphenous vein endothelium, and this was inhibited by anandamide (Stefano et al. 1998). A similar phenomenon was reported for CP55940 in rat microglial cells (Cabral et al. 2001) and mouse astrocytes (Molina-Holgado et al. 1997; Molina-Holgado et al. 2002), and for WIN55212-2 in rat C6 astrocytoma cells (Esposito et al. 2002). The mechanism could involve feedback by NO, inasmuch as it could be mimicked by NO donors (Esposito et al. 2002; Stefano et al. 1998). The mechanism also appears to involve stimulation of the CB_1 receptor and a reduction in cellular cyclic AMP, presumably via production of NO (Esposito et al. 2002; Molina-Holgado et al. 2002; Stefano et al. 1998). Δ^9-THC inhibited iNOS induction in RAW264.7 macrophage cells by a mechanism that involves CB_2 receptors and a reduction in cyclic AMP (Jeon et al. 1996). A final common pathway for the CB_1- and CB_2-mediated responses is the release of the cytokine interleukin-1 receptor antagonist (IL-1ra), which suppresses iNOS expression (Molina-Holgado et al. 2003).

7 Mechanisms by Which the CB_1 Receptor Signals Through G Proteins

Studies from our own laboratory have investigated domains of the CB_1 receptor that are important for activating selective Gi/o proteins, using strategies that include use of peptides that mimic intracellular domains and co-immunoprecipitation of G proteins to determine selectivity of protein–protein associations. When the CB_1 receptor was immunoprecipitated from detergent-solubilized rat brain membranes, Gαo and various Gαi subtypes were found to be associated with the CB_1 receptor (Houston and Howlett 1998; Mukhopadhyay et al. 2000; Mukhopadhyay and Howlett 2001). Similar immunoprecipitation of CB_1 receptors solubilized from N18TG2 neuroblastoma cell membranes revealed an association with Gαi1, Gαi2, and Gαi3. Pertussis toxin treatment disrupted the CB_1 receptor-Gα association, demonstrating that these complexes represent a functional equilibrium with the receptor–G protein complex as the preferred state (Howlett et al. 1999; Mukhopadhyay et al. 2000).

The domains of the CB_1 receptor that interact with G proteins were studied using peptides representing the juxtamembrane C-terminal region or a series of peptide analogs (Howlett et al. 1998; Mukhopadhyay et al. 1999). Palmitoylation of a *cys* residue anchors the C-terminal domain to the plasma membrane distal to the putative helical intracellular domain (Mukhopadhyay et al. 2002a). Thus, this region is also referred to as the fourth intracellular loop (IC4). The peptide mimicking the juxtamembrane C-terminal domain promoted G protein activation in rat brain membranes and the inhibition of Gs-stimulated or forskolin-activated adenylyl cyclase in N18TG2 membranes (Howlett et al. 1999; Mukhopadhyay et al. 1999; Howlett et al. 1998). In solubilized brain or N18TG2 membrane preparations, the juxtamembrane C-terminal peptide competed for the protein–protein association of the CB_1 receptor with Gαo or Gαi3 (Mukhopadhyay et al. 2000; Mukhopadhyay and Howlett 2001). Because this peptide failed to disrupt the CB_1 receptor interaction with Gαi1 or Gαi2, it is believed that the C-terminal IC4 domain interacts primarily with Gαo or Gαi3 proteins. The IC4 peptide was able to form a helical structure only in a negatively charged environment (Mukhopadhyay et al. 1999), suggesting that changes in the seventh transmembrane helix (TM7) that would alter the positions of critical amino acids could promote activation of Gαo or Gαi3. CB_1 receptor mutants that are truncated two residues distal to the palmitoylated *cys* showed perturbed regulation of Ca^{2+} currents (Nie and Lewis 2001). However, mutants truncated such that the entire IC4 region was deleted were devoid of Ca^{2+} channel regulation (Nie and Lewis 2001), as would be expected if this region were critical for interaction with Go as the transducer of this response.

Three peptides comprising the third intracellular loop (IC3) of the CB_1 receptor were able to disrupt the CB_1 receptor association with Gαi1 or Gαi2 in solubilized preparations of rat brain or N18TG2 membranes (Mukhopadhyay et al. 2000; Mukhopadhyay and Howlett 2001). The C-terminal side of IC3 was considered to be most important for the activation of G proteins, presumed to be Gαi1 or Gαi2 (Howlett et al. 1998). In support of this, a nine-amino acid peptide, mimicking the C-terminal side of IC3 at the membrane-cytosol interface, promoted GTPase activity of a pure preparation of Gαi1 (Ulfers et al. 2002b). The structure of a larger peptide comprising the entire IC3 loop was shown by nuclear magnetic resonance (NMR) analysis to be helical at the N-terminal side distal to TM5 (Ulfers et al. 2002a). The peptide appeared to be amorphous at the middle third except for a turn occurring at an intracellular Gly residue, and exhibited helical structure beginning within the C-terminal third approximately two turns proximal to TM6 (Ulfers et al. 2002a). NMR analysis of a peptide representing this C-terminal region indicated that this peptide was also helical in the presence of Gαi1 (Ulfers et al. 2002b). A Leu-Ala-Lys-Thr sequence at the membrane interface may be critical to Gi interaction because reversal of this Leu-Ala sequence to Ala-Leu in a mutated CB_1 receptor resulted in a loss of coupling to Gi, thereby attenuating inhibition of cyclic AMP production (Abadji et al. 1999). This mutation also promoted coupling to Gs when Gi proteins were inactivated by pertussis toxin (Ulfers et al. 2002b).

Computational modeling studies have made some predictions regarding how the movement of transmembrane helices might be associated with activation of the CB_1 receptor. Shim and colleagues (Shim et al. 2003) have developed a CB_1 cannabinoid receptor homology model based upon the ground-state structure of rhodopsin. A docking site for non-classical cannabinoid ligands was deduced, and included interactions with multiple amino acid residues, including a hydrophobic binding pocket that would accommodate the aromatic A ring and the alkyl side chain of non-classical cannabinoid ligands (Shim et al. 2003). Assuming that the conformation of the ligand that is necessary to conform to the ground-state receptor was not the lowest energy conformation, Shim and Howlett (Shim and Howlett 2004) predicted potential ligand conformations that would release the constrained energy. As the ligand achieved lower free energy states, steric clash with amino acid residues in TM3 and TM6 would be predicted, which may release inter-helical bonds and trigger a conformational change in the CB_1 receptor. Reggio's laboratory has envisioned that helical translocation may occur in a manner similar to what has been predicted for rhodopsin and the β-adrenergic receptor. Starting with a model based on the ground state of rhodopsin, these researchers have modified helical structure to predict a receptor conformation that could represent one of the agonist-activated states of the receptor–G protein cycle (Singh et al. 2002). These modeling studies envision changes in the TM3 and TM6 that might be directed at regulation of movement of the IC3. Future studies will be necessary to test these hypotheses, and to extend them to other intracellular domains that could be important for G protein coupling.

8
Cellular Changes in Signal Transduction upon Chronic Exposure to Agonists

Chronic exposure to Δ^9-THC and other cannabinoid receptor agonists generally leads to biological adaptive mechanisms that may be related to the phenomenon of tolerance. Cellular modifications in response to chronic agonist stimulation have included cannabinoid receptor down-regulation, as well as desensitization of signal transduction pathways. These effects have been recently reviewed in detail (Sim-Selley 2003).

CB_1 cannabinoid receptor numbers in the brain have been reported to decrease after prolonged treatment of animals with agonist drugs (Fan et al. 1996; Oviedo et al. 1993; Rodriguez de Fonseca et al. 1994; Romero et al. 1997). In other studies that used different drugs, concentrations and times of exposure, this decline in CB_1 receptor levels was not observed (Romero et al. 1995; Abood et al. 1993). Differences in the rates and magnitudes of receptor down-regulation across brain regions have been demonstrated (Breivogel et al. 1999). Chronic Δ^9-THC treatment abrogated G protein activation by cannabinoid receptors ([^{35}S]GTPγS binding) in a number of rat brain regions that are expected to be important for cannabinoid effects (Sim et al. 1996). The time course of the decrease in cannabinoid-stimulated [^{35}S]GTPγS binding to G proteins differed between brain regions (Breivogel et al. 1999). More distal responses may not be obviously correlated with the changes in receptor number and coupling to G proteins. Chronic treatment of animals with CP55940 did not produce a measurable change in adenylyl cyclase in cerebellar membranes even though cannabinoid receptor numbers were reduced (Fan et al. 1996). Chronic exposure of rodents to Δ^9-THC increased the MAPK pathway that signals to phosphorylated cyclic AMP response element binding protein (phosphoCREB) and FosB transcription factors in the nucleus (Rubino et al. 2004). These researchers reported evidence that sustained stimulation of the MAPK pathway could be coupled to the development of tolerance to the antinociception and hypomobility responses (Rubino et al. 2004).

Studies of cellular adaptation to cannabinoid drugs have identified cellular changes that could predict the mechanism of synaptic plasticity. Homologous desensitization of adenylyl cyclase inhibition was observed within minutes of exposure to Δ^9-THC and levonantradol in cultured neuroblastoma cells (Dill and Howlett 1988; Shapira et al. 1998). One-way cross-desensitization has been reported, in that chronic exposure to morphine in cultured N18TG2 or NG108-15 cells caused a reduction in the response to acute stimulation of the cannabinoid receptor (Shapira et al. 1998; Eisinger et al. 2002).

8.1 Phosphorylation of the Cannabinoid Receptors as a Mechanism for Desensitization

Phosphorylation of Ser residues on the IC3 and C-terminal of the CB_1 receptor is important for regulation of coupling to G proteins and subsequent signaling. A critical Ser_{317} on the IC3 could be phosphorylated by activation of PKC in a recombinant model system (Garcia et al. 1998). This modification might serve as a heterologous desensitization mechanism by which activation of PKC could lead to the failure of CB_1 receptors to regulate GIRK channels and inhibit P/Q-type Ca^{2+} channels (Garcia et al. 1998). Studies of site-mutations of Ser_{426} and Ser_{430} indicated that these residues were required for desensitization, suggesting the importance of this domain for G protein receptor kinase-3 phosphorylation, and perhaps, association with β-arrestin 2 (Jin et al. 1999). CB_1 receptor mutants that are truncated two residues distal to the palmitoylated *cys* of the C-terminal failed to desensitize the GIRK channel activation response to agonist stimulation of the receptor, demonstrating the importance of the C-terminal tail for desensitization (Jin et al. 1999).

The role of protein kinases in maintaining the tolerant state in rodents was examined by the Welch laboratory (Lee et al. 2003). In those studies, animals were chronically exposed to Δ^9-THC, and then tested for their antinociceptive response to a dose of Δ^9-THC. The tolerance to Δ^9-THC was reversed by prior administration of a PKA inhibitor and a Src family tyrosine kinase inhibitor. These studies suggested that PKA and a tyrosine kinase could be important in maintaining the tolerant state. It is intriguing to speculate on what these findings might imply regarding the signal transduction pathways that might include cannabinoid receptors. However, the complexity of the intact brain and spinal cord in the nociceptive response makes it difficult to assign any particular substrate for PKA, or Src tyrosine kinases, as the target(s) for these phosphorylation-dependent changes.

9 Summary and Predictions

The CB_1 and CB_2 cannabinoid receptors in nervous, immune, and other tissues of the body participate in G protein-mediated signal transduction pathways. Particularly well characterized are those that regulate the second messengers cyclic AMP, Ca^{2+}, and perhaps IP_3. CB_1 receptors are modulators of ion channels, which makes them key players in the control of neurotransmission. These receptors also participate in signal transduction via scaffolding mechanisms, including regulation of MAPK signaling to the nucleus via transcription factors. These receptors promote intercellular signaling via NO, a diffusible ligand that can impact properties of neighboring cells. Chronic administration of cannabinoid receptor agonists can orchestrate pleiotropic changes in cellular signal transduction that contribute to synaptic plasticity in the processes of learning and memory, cognition, nociception, and other responses to CB_1 receptor stimulation.

Future studies should elucidate additional signal transduction pathways in which the cannabinoid receptors can participate. G proteins other than Gi, Go, Gs, and Gq may be important in initiating signal transduction pathways that have not yet been considered for these receptors. Transactivation of alternative signal transduction pathways, with or without the participation of G proteins, may be discovered to be important for cannabinoid receptor-mediated responses. Non-G protein-mediated signal transduction mechanisms may represent alternative cellular signaling pathways. As we continue to learn more about other cellular proteins with which the cannabinoid receptors can potentially interact, we will have a better appreciation of both physiological and pathological processes mediated by endocannabinoid compounds.

References

Abadji V, Lucas-Lenard JM, Chin C, Kendall DA (1999) Involvement of the carboxyl terminus of the third intracellular loop of the cannabinoid CB1 receptor in constitutive activation of Gs. J Neurochem 72:2032–2038

Abood ME, Sauss C, Fan F, Tilton CL, Martin BR (1993) Development of behavioral tolerance to delta 9-THC without alteration of cannabinoid receptor binding or mRNA levels in whole brain. Pharmacol Biochem Behav 46:575–579

Alger BE (2002) Retrograde signaling in the regulation of synaptic transmission: focus on endocannabinoids. Prog Neurobiol 68:247–286

Alger BE, Pitler TA, Wagner JJ, Martin LA, Morishita W, Kirov SA, Lenz RA (1996) Retrograde signalling in depolarization-induced suppression of inhibition in rat hippocampal CA1 cells. J Physiol 496:197–209

Azad SC, Marsicano G, Eberlein I, Putzke J, Zieglgansberger W, Spanagel R, Lutz B (2001) Differential role of the nitric oxide pathway on delta(9)-THC-induced central nervous system effects in the mouse. Eur J Neurosci 13:561–568

Barth F, Rinaldi-Carmona M (1999) The development of cannabinoid antagonists. Curr Med Chem 6:745–755

Bash R, Rubovitch V, Gafni M, Sarne Y (2003) The stimulatory effect of cannabinoids on calcium uptake is mediated by Gs GTP-binding proteins and cAMP formation. Neurosignals 12:39–44

Beau FE, Alger BE (1998) Transient suppression of GABA-A-receptor-mediated IPSPs after epileptiform burst discharges in CB1 pyramidal cells. J Neurophysiol 79:659–669

Bidaut-Russell M, Howlett AC (1991) Cannabinoid receptor-regulated cyclic AMP accumulation in the rat striatum. J Neurochem 57:1769–1773

Bisogno T, Maccarrone M, De Petrocellis L, Jarrahian A, Finazzi-Agro A, Hillard C, Di Marzo, V (2001) The uptake by cells of 2-arachidonoylglycerol, an endogenous agonist of cannabinoid receptors. Eur J Biochem 268:1982–1989

Bonhaus DW, Chang LK, Kwan J, Martin GR (1998) Dual activation and inhibition of adenylyl cyclase by cannabinoid receptor agonists: evidence for agonist-specific trafficking of intracellular responses. J Pharmacol Exp Ther 287:884–888

Bouaboula M, Bourrie B, Rinaldi-Carmona M, Shire D, Le Fur G, Casellas P (1995a) Stimulation of cannabinoid receptor CB1 induces krox-24 expression in human astrocytoma cells. J Biol Chem 270:13973–13980

Bouaboula M, Poinot-Chazel C, Bourrie B, Canat X, Calandra B, Rinaldi-Carmona M, Le Fur G, Casellas P (1995b) Activation of mitogen-activated protein kinases by stimulation of the central cannabinoid receptor CB1. Biochem J 312:637–641

Bouaboula M, Poinot-Chazel C, Marchand J, Canat X, Bourrie B, Rinaldi-Carmona M, Calandra B, Le Fur G, Casellas P (1996) Signaling pathway associated with stimulation of CB2 peripheral cannabinoid receptor. Involvement of both mitogen-activated protein kinase and induction of Krox-24 expression. Eur J Biochem 237:704–711

Breivogel CS, Childers SR, Deadwyler SA, Hampson RE, Vogt LJ, Sim-Selley LJ (1999) Chronic delta9-tetrahydrocannabinol treatment produces a time-dependent loss of cannabinoid receptors and cannabinoid receptor-activated G proteins in rat brain. J Neurochem 73:2447–2459

Breivogel CS, Walker JM, Huang SM, Roy MB, Childers SR (2004) Cannabinoid signaling in rat cerebellar granule cells: G-protein activation, inhibition of glutamate release and endogenous cannabinoids. Neuropharmacology 47:81–91

Brenowitz SD, Regehr WG (2003) Calcium dependence of retrograde inhibition by endocannabinoids at synapses onto Purkinje cells. J Neurosci 23:6373–6384

Burstein S (1991) Cannabinoid induced changes in eicosanoid synthesis by mouse peritoneal cells. Adv Exp Med Biol 288:107–112

Burstein S, Hunter SA, Latham V, Mechoulam R, Melchior DL, Renzulli L, Tefft RE Jr (1986) Prostaglandins and cannabis XV. Comparison of enantiomeric cannabinoids in stimulating prostaglandin synthesis in fibroblasts. Life Sci 39:1813–1823

Burstein S, Budrow J, Debatis M, Hunter SA, Subramanian A (1994) Phospholipase participation in cannabinoid-induced release of free arachidonic acid. Biochem Pharmacol 48:1253–1264

Cabral GA, Harmon KN, Carlisle SJ (2001) Cannabinoid-mediated inhibition of inducible nitric oxide production by rat microglial cells: evidence for CB1 receptor participation. Adv Exp Med Biol 493:207–214

Carrier EJ, Kearn CS, Barkmeier AJ, Breese NM, Yang W, Nithipatikom K, Pfister SL, Campbell WB, Hillard CJ (2004) Cultured rat microglial cells synthesize the endocannabinoid 2-arachidonylglycerol, which increases proliferation via a CB2 receptor-dependent mechanism. Mol Pharmacol 65:999–1007

Caulfield MP, Brown DA (1992) Cannabinoid receptor agonists inhibit Ca current in NG108-15 neuroblastoma cells via a pertussis toxin-sensitive mechanism. Br J Pharmacol 106:231–232

Childers SR, Deadwyler SA (1996) Role of cyclic AMP in the actions of cannabinoid receptors. Biochem Pharmacol 52:819–827

Childers SR, Sexton T, Roy MB (1994) Effects of anandamide on cannabinoid receptors in rat brain membranes. Biochem Pharmacol 47:711–715

Compton DR, Gold LH, Ward SJ, Balster RL, Martin BR (1992) Aminoalkylindole analogs: cannabimimetic activity of a class of compounds structurally distinct from delta 9-tetrahydrocannabinol. J Pharmacol Exp Ther 263:1118–1126

Daniel H, Rancillac A, Crepel F (2004) Mechanisms underlying cannabinoid inhibition of presynaptic Ca2+ influx at parallel fibre synapses of the rat cerebellum. J Physiol 557:159–174

Davis MI, Ronesi J, Lovinger DM (2003) A predominant role for inhibition of the adenylate cyclase/protein kinase A pathway in ERK activation by cannabinoid receptor 1 in N1E-115 neuroblastoma cells. J Biol Chem 278:48973–48980

De Petrocellis L, Bisogno T, Maccarrone M, Davis JB, Finazzi-Agro A, Di Marzo V (2001) The activity of anandamide at vanilloid VR1 receptors requires facilitated transport across the cell membrane and is limited by intracellular metabolism. J Biol Chem 276:12856–12863

Deadwyler SA, Hampson RE, Mu J, Whyte A, Childers S (1995) Cannabinoids modulate voltage sensitive potassium A-current in hippocampal neurons via a cAMP-dependent process. J Pharmacol Exp Ther 273:734–743
Derkinderen P, Toutant M, Burgaya F, Le Bert M, Siciliano JC, de F, V, Gelman M, Girault JA (1996) Regulation of a neuronal form of focal adhesion kinase by anandamide. Science 273:1719–1722
Derkinderen P, Ledent C, Parmentier M, Girault JA (2001a) Cannabinoids activate p38 mitogen-activated protein kinases through CB1 receptors in hippocampus. J Neurochem 77:957–960
Derkinderen P, Toutant M, Kadare G, Ledent C, Parmentier M, Girault JA (2001b) Dual role of Fyn in the regulation of FAK+6,7 by cannabinoids in hippocampus. J Biol Chem 276:38289–38296
Derkinderen P, Valjent E, Toutant M, Corvol JC, Enslen H, Ledent C, Trzaskos J, Caboche J, Girault JA (2003) Regulation of extracellular signal-regulated kinase by cannabinoids in hippocampus. J Neurosci 23:2371–2382
Derocq JM, Jbilo O, Bouaboula M, Segui M, Clere C, Casellas P (2000) Genomic and functional changes induced by the activation of the peripheral cannabinoid receptor CB2 in the promyelocytic cells HL-60. Possible involvement of the CB2 receptor in cell differentiation. J Biol Chem 275:15621–15628
Devane WA, Dysarz FA, III, Johnson MR, Melvin LS, Howlett AC (1988) Determination and characterization of a cannabinoid receptor in rat brain. Mol Pharmacol 34:605–613
Di Marzo V, Bisogno T, De Petrocellis L, Melck D, Martin BR (1999) Cannabimimetic fatty acid derivatives: the anandamide family and other endocannabinoids. Curr Med Chem 6:721–744
Dill JA, Howlett AC (1988) Regulation of adenylate cyclase by chronic exposure to cannabimimetic drugs. J Pharmacol Exp Ther 244:1157–1163
Eisinger DA, Ammer H, Schulz R (2002) Chronic morphine treatment inhibits opioid receptor desensitization and internalization. J Neurosci 22:10192–10200
Esposito G, Ligresti A, Izzo AA, Bisogno T, Ruvo M, Di Rosa M, Di Marzo V, Iuvone T (2002) The endocannabinoid system protects rat glioma cells against HIV-1 Tat protein-induced cytotoxicity. Mechanism and regulation. J Biol Chem 277:50348–50354
Fan F, Tao Q, Abood M, Martin BR (1996) Cannabinoid receptor down-regulation without alteration of the inhibitory effect of CP 55,940 on adenylyl cyclase in the cerebellum of CP 55,940-tolerant mice. Brain Res 706:13–20
Faubert Kaplan BL, Kaminski NE (2003) Cannabinoids inhibit the activation of ERK MAPK in PMA/Io-stimulated mouse splenocytes. Int Immunopharmacol 3:1503–1510
Felder CC, Veluz JS, Williams HL, Briley EM, Matsuda LA (1992) Cannabinoid agonists stimulate both receptor- and non-receptor-mediated signal transduction pathways in cells transfected with and expressing cannabinoid receptor clones. Mol Pharmacol 42:838–845
Felder CC, Briley EM, Axelrod J, Simpson JT, Mackie K, Devane WA (1993) Anandamide, an endogenous cannabimimetic eicosanoid, binds to the cloned human cannabinoid receptor and stimulates receptor-mediated signal transduction. Proc Natl Acad Sci U S A 90:7656–7660
Felder CC, Joyce KE, Briley EM, Mansouri J, Mackie K, Blond O, Lai Y, Ma AL, Mitchell RL (1995) Comparison of the pharmacology and signal transduction of the human cannabinoid CB1 and CB2 receptors. Mol Pharmacol 48:443–450
Felder CC, Joyce KE, Briley EM, Glass M, Mackie KP, Fahey KJ, Cullinan GJ, Hunden DC, Johnson DW, Chaney MO, Koppel GA, Brownstein M (1998) LY320135, a novel cannabinoid CB1 receptor antagonist, unmasks coupling of the CB1 receptor to stimulation of cAMP accumulation. J Pharmacol Exp Ther 284:291–297
Fimiani C, Liberty T, Aquirre AJ, Amin I, Ali N, Stefano GB (1999a) Opiate, cannabinoid, and eicosanoid signaling converges on common intracellular pathways nitric oxide coupling. Prostaglandins Other Lipid Mediat 57:23–34

Fimiani C, Mattocks D, Cavani F, Salzet M, Deutsch DG, Pryor S, Bilfinger TV, Stefano GB (1999b) Morphine and anandamide stimulate intracellular calcium transients in human arterial endothelial cells: coupling to nitric oxide release. Cell Signal 11:189–193
Freund TF, Katona I, Piomelli D (2003) Role of endogenous cannabinoids in synaptic signaling. Physiol Rev 83:1017–1066
Galante M, Diana MA (2004) Group I metabotropic glutamate receptors inhibit GABA release at interneuron-Purkinje cell synapses through endocannabinoid production. J Neurosci 24:4865–4874
Galve-Roperh I, Rueda D, Gomez dP, Velasco G, Guzman M (2002) Mechanism of extracellular signal-regulated kinase activation by the CB(1) cannabinoid receptor. Mol Pharmacol 62:1385–1392
Garcia DE, Brown S, Hille B, Mackie K (1998) Protein kinase C disrupts cannabinoid actions by phosphorylation of the CB1 cannabinoid receptor. J Neurosci 18:2834–2841
Gebremedhin D, Lange AR, Campbell WB, Hillard CJ, Harder DR (1999) Cannabinoid CB1 receptor of cat cerebral arterial muscle functions to inhibit L-type Ca2+ channel current. Am J Physiol 276:H2085–H2093
Gerard C, Mollereau C, Vassart G, Parmentier M (1990) Nucleotide sequence of a human cannabinoid receptor cDNA. Nucleic Acids Res 18:7142
Gerdeman GL, Ronesi J, Lovinger DM (2002) Postsynaptic endocannabinoid release is critical to long-term depression in the striatum. Nat Neurosci 5:446–451
Giuffrida A, Beltramo M, Piomelli D (2001) Mechanisms of endocannabinoid inactivation: biochemistry and pharmacology. J Pharmacol Exp Ther 298:7–14
Glass M, Felder CC (1997) Concurrent stimulation of cannabinoid CB1 and dopamine D2 receptors augments cAMP accumulation in striatal neurons: evidence for a Gs linkage to the CB1 receptor. J Neurosci 17:5327–5333
Gomez del Pulgar T, Velasco G, Guzman M (2000) The CB1 cannabinoid receptor is coupled to the activation of protein kinase B/Akt. Biochem J 347:369–373
Gonsiorek W, Lunn C, Fan X, Narula S, Lundell D, Hipkin RW (2000) Endocannabinoid 2-arachidonyl glycerol is a full agonist through human type 2 cannabinoid receptor: antagonism by anandamide. Mol Pharmacol 57:1045–1050
Guhring H, Hamza M, Sergejeva M, Ates M, Kotalla CE, Ledent C, Brune K (2002) A role for endocannabinoids in indomethacin-induced spinal antinociception. Eur J Pharmacol 454:153–163
Guo J, Ikeda SR (2004) Endocannabinoids modulate N-type calcium channels and G-protein-coupled inwardly rectifying potassium channels via CB1 cannabinoid receptors heterologously expressed in mammalian neurons. Mol Pharmacol 65:665–674
Guzman M, Sanchez C (1999) Effects of cannabinoids on energy metabolism. Life Sci 65:657–664
Hampson AJ, Bornheim LM, Scanziani M, Yost CS, Gray AT, Hansen BM, Leonoudakis DJ, Bickler PE (1998) Dual effects of anandamide on NMDA receptor-mediated responses and neurotransmission. J Neurochem 70:671–676
Hampson RE, Evans GJ, Mu J, Zhuang SY, King VC, Childers SR, Deadwyler SA (1995) Role of cyclic AMP dependent protein kinase in cannabinoid receptor modulation of potassium "A-current" in cultured rat hippocampal neurons. Life Sci 56:2081–2088
Hampson RE, Zhuang SY, Weiner JL, Deadwyler SA (2003) Functional significance of cannabinoid-mediated, depolarization-induced suppression of inhibition (DSI) in the hippocampus. J Neurophysiol 90:55–64
Harkany T, Hartig W, Berghuis P, Dobszay MB, Zilberter Y, Edwards RH, Mackie K, Ernfors P (2003) Complementary distribution of type 1 cannabinoid receptors and vesicular glutamate transporter 3 in basal forebrain suggests input-specific retrograde signalling by cholinergic neurons. Eur J Neurosci 18:1979–1992
Henry DJ, Chavkin C (1995) Activation of inwardly rectifying potassium channels (GIRK1) by co-expressed rat brain cannabinoid receptors in Xenopus oocytes. Neurosci Lett 186:91–94

Herkenham M, Lynn AB, Little MD, Johnson MR, Melvin LS, De Costa BR, Rice KC (1990) Cannabinoid receptor localization in brain. Proc Natl Acad Sci U S A 87:1932–1936
Herkenham M, Lynn AB, Johnson MR, Melvin LS, De Costa BR, Rice KC (1991) Characterization and localization of cannabinoid receptors in rat brain: a quantitative in vitro autoradiographic study. J Neurosci 11:563–583
Hillard CJ, Bloom AS (1983) Possible role of prostaglandins in the effects of the cannabinoids on adenylate cyclase activity. Eur J Pharmacol 91:21–27
Hillard CJ, Harris RA, Bloom AS (1985) Effects of the cannabinoids on physical properties of brain membranes and phospholipid vesicles: fluorescence studies. J Pharmacol Exp Ther 232:579–588
Hoffman AF, Riegel AC, Lupica CR (2003) Functional localization of cannabinoid receptors and endogenous cannabinoid production in distinct neuron populations of the hippocampus. Eur J Neurosci 18:524–534
Houston DB, Howlett AC (1998) Differential receptor-G-protein coupling evoked by dissimilar cannabinoid receptor agonists. Cell Signal 10:667–674
Howlett AC (1984) Inhibition of neuroblastoma adenylate cyclase by cannabinoid and nantradol compounds. Life Sci 35:1803–1810
Howlett AC (1985) Cannabinoid inhibition of adenylate cyclase. Biochemistry of the response in neuroblastoma cell membranes. Mol Pharmacol 27:429–436
Howlett AC (1987) Cannabinoid inhibition of adenylate cyclase: relative activity of constituents and metabolites of marihuana. Neuropharmacology 26:507–512
Howlett AC (1995) Pharmacology of cannabinoid receptors. Annu Rev Pharmacol Toxicol 35:607–634
Howlett AC, Fleming RM (1984) Cannabinoid inhibition of adenylate cyclase. Pharmacology of the response in neuroblastoma cell membranes. Mol Pharmacol 26:532–538
Howlett AC, Mukhopadhyay S (2000) Cellular signal transduction by anandamide and 2-arachidonoylglycerol. Chem Phys Lipids 108:53–70
Howlett AC, Qualy JM, Khachatrian LL (1986) Involvement of Gi in the inhibition of adenylate cyclase by cannabimimetic drugs. Mol Pharmacol 29:307–313
Howlett AC, Johnson MR, Melvin LS, Milne GM (1988) Nonclassical cannabinoid analgetics inhibit adenylate cyclase: development of a cannabinoid receptor model. Mol Pharmacol 33:297–302
Howlett AC, Scott DK, Wilken GH (1989) Regulation of adenylate cyclase by cannabinoid drugs. Insights based on thermodynamic studies. Biochem Pharmacol 38:3297–3304
Howlett AC, Champion TM, Wilken GH, Mechoulam R (1990) Stereochemical effects of 11-OH-delta 8-tetrahydrocannabinol-dimethylheptyl to inhibit adenylate cyclase and bind to the cannabinoid receptor. Neuropharmacology 29:161–165
Howlett AC, Song C, Berglund BA, Wilken GH, Pigg JJ (1998) Characterization of CB1 cannabinoid receptors using receptor peptide fragments and site-directed antibodies. Mol Pharmacol 53:504–510
Howlett AC, Mukhopadhyay S, Shim JY, Welsh WJ (1999) Signal transduction of eicosanoid CB1 receptor ligands. Life Sci 65:617–625
Howlett AC, Barth F, Bonner TI, Cabral G, Casellas P, Devane WA, Felder CC, Herkenham M, Mackie K, Martin BR, Mechoulam R, Pertwee RG (2002) International Union of Pharmacology. XXVII. Classification of cannabinoid receptors. Pharmacol Rev 54:161–202
Huang CC, Chan SH, Hsu KS (2004) 3-Morpholinylsydnonimine inhibits glutamatergic transmission in rat rostral ventrolateral medulla via peroxynitrite formation and adenosine release. Mol Pharmacol 66:492–501
Jarrahian A, Watts VJ, Barker EL (2004) D2 dopamine receptors modulate Galpha-subunit coupling of the CB1 cannabinoid receptor. J Pharmacol Exp Ther 308:880–886
Jeon YJ, Yang KH, Pulaski JT, Kaminski NE (1996) Attenuation of inducible nitric oxide synthase gene expression by delta 9-tetrahydrocannabinol is mediated through the inhibition of nuclear factor-kappa B/Rel activation. Mol Pharmacol 50:334–341

Jin W, Brown S, Roche JP, Hsieh C, Celver JP, Kovoor A, Chavkin C, Mackie K (1999) Distinct domains of the CB1 cannabinoid receptor mediate desensitization and internalization. J Neurosci 19:3773–3780

Johnson MR, Melvin LS, Althuis TH, Bindra JS, Harbert CA, Milne GM, Weissman A (1981) Selective and potent analgetics derived from cannabinoids. J Clin Pharmacol 21:271S–282S

Karler R, Turkanis SA (1981) The cannabinoids as potential antiepileptics. J Clin Pharmacol 21:437S–448S

Kim J, Isokawa M, Ledent C, Alger BE (2002) Activation of muscarinic acetylcholine receptors enhances the release of endogenous cannabinoids in the hippocampus. J Neurosci 22:10182–10191

Kobayashi Y, Arai S, Waku K, Sugiura T (2001) Activation by 2-arachidonoylglycerol, an endogenous cannabinoid receptor ligand, of p42/44 mitogen-activated protein kinase in HL-60 cells. J Biochem (Tokyo) 129:665–669

Kreitzer AC, Carter AG, Regehr WG (2002) Inhibition of interneuron firing extends the spread of endocannabinoid signaling in the cerebellum. Neuron 34:787–796

Lee MC, Smith FL, Stevens DL, Welch SP (2003) The role of several kinases in mice tolerant to delta 9-tetrahydrocannabinol. J Pharmacol Exp Ther 305:593–599

Little PJ, Compton DR, Johnson MR, Melvin LS, Martin BR (1988) Pharmacology and stereoselectivity of structurally novel cannabinoids in mice. J Pharmacol Exp Ther 247:1046–1051

Liu J, Gao B, Mirshahi F, Sanyal AJ, Khanolkar AD, Makriyannis A, Kunos G (2000) Functional CB1 cannabinoid receptors in human vascular endothelial cells. Biochem J 346:835–840

Maccarrone M, Bari M, Lorenzon T, Bisogno T, Di Marzo V, Finazzi-Agro A (2000) Anandamide uptake by human endothelial cells and its regulation by nitric oxide. J Biol Chem 275:13484–13492

Mackie K, Hille B (1992) Cannabinoids inhibit N-type calcium channels in neuroblastoma-glioma cells. Proc Natl Acad Sci U S A 89:3825–3829

Mackie K, Devane WA, Hille B (1993) Anandamide, an endogenous cannabinoid, inhibits calcium currents as a partial agonist in N18 neuroblastoma cells. Mol Pharmacol 44:498–503

Mackie K, Lai Y, Westenbroek R, Mitchell R (1995) Cannabinoids activate an inwardly rectifying potassium conductance and inhibit Q-type calcium currents in AtT20 cells transfected with rat brain cannabinoid receptor. J Neurosci 15:6552–6561

Maejima T, Hashimoto K, Yoshida T, Aiba A, Kano M (2001a) Presynaptic inhibition caused by retrograde signal from metabotropic glutamate to cannabinoid receptors. Neuron 31:463–475

Maejima T, Ohno-Shosaku T, Kano M (2001b) Endogenous cannabinoid as a retrograde messenger from depolarized postsynaptic neurons to presynaptic terminals. Neurosci Res 40:205–210

Maneuf YP, Brotchie JM (1997) Paradoxical action of the cannabinoid WIN 55,212-2 in stimulated and basal cyclic AMP accumulation in rat globus pallidus slices. Br J Pharmacol 120:1397–1398

Martin BR (1986) Cellular effects of cannabinoids. Pharmacol Rev 38:45–74

Martin BR, Mechoulam R, Razdan RK (1999) Discovery and characterization of endogenous cannabinoids. Life Sci 65:573–595

Matsuda LA, Lolait SJ, Brownstein MJ, Young AC, Bonner TI (1990) Structure of a cannabinoid receptor and functional expression of the cloned cDNA. Nature 346:561–564

McAllister SD, Griffin G, Satin LS, Abood ME (1999) Cannabinoid receptors can activate and inhibit G protein-coupled inwardly rectifying potassium channels in a xenopus oocyte expression system. J Pharmacol Exp Ther 291:618–626

Melvin LS, Johnson MR (1987) Structure-activity relationships of tricyclic and nonclassical bicyclic cannabinoids. NIDA Res Monogr 79:31–47

Melvin LS, Milne GM, Johnson MR, Subramaniam B, Wilken GH, Howlett AC (1993) Structure-activity relationships for cannabinoid receptor-binding and analgesic activity: studies of bicyclic cannabinoid analogs. Mol Pharmacol 44:1008–1015
Melvin LS, Milne GM, Johnson MR, Wilken GH, Howlett AC (1995) Structure-activity relationships defining the ACD-tricyclic cannabinoids: cannabinoid receptor binding and analgesic activity. Drug Des Discov 13:155–166
Misner DL, Sullivan JM (1999) Mechanism of cannabinoid effects on long-term potentiation and depression in hippocampal CA1 neurons. J Neurosci 19:6795–6805
Molina-Holgado F, Lledo A, Guaza C (1997) Anandamide suppresses nitric oxide and TNF-alpha responses to Theiler's virus or endotoxin in astrocytes. Neuroreport 8:1929–1933
Molina-Holgado F, Molina-Holgado E, Guaza C, Rothwell NJ (2002) Role of CB1 and CB2 receptors in the inhibitory effects of cannabinoids on lipopolysaccharide-induced nitric oxide release in astrocyte cultures. J Neurosci Res 67:829–836
Molina-Holgado F, Pinteaux E, Moore JD, Molina-Holgado E, Guaza C, Gibson RM, Rothwell NJ (2003) Endogenous interleukin-1 receptor antagonist mediates anti-inflammatory and neuroprotective actions of cannabinoids in neurons and glia. J Neurosci 23:6470–6474
Mombouli JV, Schaeffer G, Holzmann S, Kostner GM, Graier WF (1999) Anandamide-induced mobilization of cytosolic Ca2+ in endothelial cells. Br J Pharmacol 126:1593–1600
Morishita W, Kirov SA, Alger BE (1998) Evidence for metabotropic glutamate receptor activation in the induction of depolarization-induced suppression of inhibition in hippocampal CA1. J Neurosci 18:4870–4882
Mukhopadhyay S, Howlett AC (2001) CB1 receptor-G protein association. Subtype selectivity is determined by distinct intracellular domains. Eur J Biochem 268:499–505
Mukhopadhyay S, Cowsik SM, Lynn AM, Welsh WJ, Howlett AC (1999) Regulation of Gi by the CB1 cannabinoid receptor C-terminal juxtamembrane region: structural requirements determined by peptide analysis. Biochemistry 38:3447–3455
Mukhopadhyay S, McIntosh HH, Houston DB, Howlett AC (2000) The CB_1 cannabinoid receptor juxtamembrane C-terminal peptide confers activation to specific G proteins in brain. Mol Pharmacol 57:162–170
Mukhopadhyay S, Sandiford S, Howlett AC (2002a) Palmitoylation regulates CB1 cannabinoid receptor-G protein interaction. Proc XIV World Congress Pharmacol
Mukhopadhyay S, Shim JY, Assi AA, Norford D, Howlett AC (2002b) CB1 cannabinoid receptor-G protein association: a possible mechanism for differential signaling. Chem Phys Lipids 121:91–109
Munro S, Thomas KL, Abu-Shaar M (1993) Molecular characterization of a peripheral receptor for cannabinoids. Nature 365:61–65
Nah SY, Saya D, Vogel Z (1993) Cannabinoids inhibit agonist-stimulated formation of inositol phosphates in rat hippocampal cultures. Eur J Pharmacol 246:19–24
Netzeband JG, Conroy SM, Parsons KL, Gruol DL (1999) Cannabinoids enhance NMDA-elicited Ca2+ signals in cerebellar granule neurons in culture. J Neurosci 19:8765–8777
Nie J, Lewis DL (2001) The proximal and distal C-terminal tail domains of the CB1 cannabinoid receptor mediate G protein coupling. Neuroscience 107:161–167
Norford DC, Newton D, Jones J, Carney S, Howlett AC (2002) Detection of cannabinoid-induced nitric oxide production in neuronal and glial cells. J Cell Biol
Ohno-Shosaku T, Shosaku J, Tsubokawa H, Kano M (2002a) Cooperative endocannabinoid production by neuronal depolarization and group I metabotropic glutamate receptor activation. Eur J Neurosci 15:953–961
Ohno-Shosaku T, Tsubokawa H, Mizushima I, Yoneda N, Zimmer A, Kano M (2002b) Presynaptic cannabinoid sensitivity is a major determinant of depolarization-induced retrograde suppression at hippocampal synapses. J Neurosci 22:3864–3872
Oviedo A, Glowa J, Herkenham M (1993) Chronic cannabinoid administration alters cannabinoid receptor binding in rat brain: a quantitative autoradiographic study. Brain Res 616:293–302

Pacheco M, Childers SR, Arnold R, Casiano F, Ward SJ (1991) Aminoalkylindoles: actions on specific G-protein-linked receptors. J Pharmacol Exp Ther 257:170–183
Pacheco MA, Ward SJ, Childers SR (1993) Identification of cannabinoid receptors in cultures of rat cerebellar granule cells. Brain Res 603:102–110
Pan X, Ikeda SR, Lewis DL (1996) Rat brain cannabinoid receptor modulates N-type Ca2+ channels in a neuronal expression system. Mol Pharmacol 49:707–714
Pan X, Ikeda SR, Lewis DL (1998) SR 141716A acts as an inverse agonist to increase neuronal voltage-dependent Ca2+ currents by reversal of tonic CB1 cannabinoid receptor activity. Mol Pharmacol 54:1064–1072
Pertwee RG (1988) The central neuropharmacology of psychotropic cannabinoids. Pharmacol Ther 36:189–261
Pertwee RG (1997) Pharmacology of cannabinoid CB1 and CB2 receptors. Pharmacol Ther 74:129–180
Pertwee RG (1999) Pharmacology of cannabinoid receptor ligands. Curr Med Chem 6:635–664
Peterson LJ, McIntosh HH, Howlett AC (2004) Tyrosine phosphorylation of the CB1 receptor. Int Cannab Res Soc 14:127
Pinto JC, Potie F, Rice KC, Boring D, Johnson MR, Evans DM, Wilken GH, Cantrell CH, Howlett AC (1994) Cannabinoid receptor binding and agonist activity of amides and esters of arachidonic acid. Mol Pharmacol 46:516–522
Piomelli D (2003) The molecular logic of endocannabinoid signalling. Nat Rev Neurosci 4:873–884
Prevot V, Rialas CM, Croix D, Salzet M, Dupouy JP, Poulain P, Beauvillain JC, Stefano GB (1998) Morphine and anandamide coupling to nitric oxide stimulates GnRH and CRF release from rat median eminence: neurovascular regulation. Brain Res 790:236–244
Priller J, Briley EM, Mansouri J, Devane WA, Mackie K, Felder CC (1995) Mead ethanolamide, a novel eicosanoid, is an agonist for the central (CB1) and peripheral (CB2) cannabinoid receptors. Mol Pharmacol 48:288–292
Reggio PH, Traore H (2000) Conformational requirements for endocannabinoid interaction with the cannabinoid receptors, the anandamide transporter and fatty acid amidohydrolase. Chem Phys Lipids 108:15–35
Rhee MH, Bayewitch M, Avidor-Reiss T, Levy R, Vogel Z (1998) Cannabinoid receptor activation differentially regulates the various adenylyl cyclase isozymes. J Neurochem 71:1525–1534
Rinaldi-Carmona M, Barth F, Heaulme M, Shire D, Calandra B, Congy C, Martinez S, Maruani J, Neliat G, Caput D (1994) SR141716A, a potent and selective antagonist of the brain cannabinoid receptor. FEBS Lett 350:240–244
Rinaldi-Carmona M, Barth F, Millan J, Derocq JM, Casellas P, Congy C, Oustric D, Sarran M, Bouaboula M, Calandra B, Portier M, Shire D, Breliere JC, Le Fur GL (1998) SR 144528, the first potent and selective antagonist of the CB2 cannabinoid receptor. J Pharmacol Exp Ther 284:644–650
Rodriguez de Fonseca F, Gorriti MA, Fernandez-Ruiz JJ, Palomo T, Ramos JA (1994) Downregulation of rat brain cannabinoid binding sites after chronic delta 9-tetrahydrocannabinol treatment. Pharmacol Biochem Behav 47:33–40
Romero J, Garcia L, Fernandez-Ruiz JJ, Cebeira M, Ramos JA (1995) Changes in rat brain cannabinoid binding sites after acute or chronic exposure to their endogenous agonist, anandamide, or to delta 9-tetrahydrocannabinol. Pharmacol Biochem Behav 51:731–737
Romero J, Garcia-Palomero E, Castro JG, Garcia-Gil L, Ramos JA, Fernandez-Ruiz JJ (1997) Effects of chronic exposure to delta9-tetrahydrocannabinol on cannabinoid receptor binding and mRNA levels in several rat brain regions. Brain Res Mol Brain Res 46:100–108
Ronesi J, Gerdeman GL, Lovinger DM (2004) Disruption of endocannabinoid release and striatal long-term depression by postsynaptic blockade of endocannabinoid membrane transport. J Neurosci 24:1673–1679

Roth SH, Williams PJ (1979) The non-specific membrane binding properties of delta9-tetrahydrocannabinol and the effects of various solubilizers. J Pharm Pharmacol 31:224–230

Rubino T, Forlani G, Vigano D, Zippel R, Parolaro D (2004) Modulation of extracellular signal-regulated kinases cascade by chronic delta 9-tetrahydrocannabinol treatment. Mol Cell Neurosci 25:355–362

Rubovitch V, Gafni M, Sarne Y (2004) The involvement of VEGF receptors and MAPK in the cannabinoid potentiation of Ca2+ flux into N18TG2 neuroblastoma cells. Brain Res Mol Brain Res 120:138–144

Rueda D, Galve-Roperh I, Haro A, Guzman M (2000) The CB(1) cannabinoid receptor is coupled to the activation of c-Jun N-terminal kinase. Mol Pharmacol 58:814–820

Rueda D, Navarro B, Martinez-Serrano A, Guzman M, Galve-Roperh I (2002) The endocannabinoid anandamide inhibits neuronal progenitor cell differentiation through attenuation of the Rap1/B-Raf/ERK pathway. J Biol Chem 277:46645–46650

Sanchez C, Galve-Roperh I, Rueda D, Guzman M (1998) Involvement of sphingomyelin hydrolysis and the mitogen-activated protein kinase cascade in the Delta9-tetrahydrocannabinol-induced stimulation of glucose metabolism in primary astrocytes. Mol Pharmacol 54:834–843

Sarker KP, Maruyama I (2003) Anandamide induces cell death independently of cannabinoid receptors or vanilloid receptor 1: possible involvement of lipid rafts. Cell Mol Life Sci 60:1200–1208

Sarker KP, Biswas KK, Yamakuchi M, Lee KY, Hahiguchi T, Kracht M, Kitajima I, Maruyama I (2003) ASK1-p38 MAPK/JNK signaling cascade mediates anandamide-induced PC12 cell death. J Neurochem 85:50–61

Schmid HH (2000) Pathways and mechanisms of N-acylethanolamine biosynthesis: can anandamide be generated selectively? Chem Phys Lipids 108:71–87

Seeman P, Chau-Wong M, Moyyen S (1972) The membrane binding of morphine, diphenylhydantoin, and tetrahydrocannabinol. Can J Physiol Pharmacol 50:1193–1200

Shapira M, Gafni M, Sarne Y (1998) Independence of, and interactions between, cannabinoid and opioid signal transduction pathways in N18TG2 cells. Brain Res 806:26–35

Shen M, Thayer SA (1998) The cannabinoid agonist Win55,212-2 inhibits calcium channels by receptor-mediated and direct pathways in cultured rat hippocampal neurons. Brain Res 783:77–84

Shim, JY, Howlett AC (2004) Steric trigger as a mechanism for CB1 cannabinoid receptor activation. J Chem Inf Comp Sci 44:1466–1476

Shim JY, Welsh WJ, Howlett AC (2003) Homology model of the CB1 cannabinoid receptor: sites critical for non-classical cannabinoid agonist interaction. Biopolymers 71:169–189

Shivachar AC, Martin BR, Ellis EF (1996) Anandamide- and delta9-tetrahydrocannabinol-evoked arachidonic acid mobilization and blockade by SR141716A [N-(Piperidin-1-yl)-5-(4-chlorophenyl)-1-(2,4-dichlorophenyl)-4-methyl-1H-pyrazole-3-carboximide hydrochloride]. Biochem Pharmacol 51:669–676

Sim LJ, Hampson RE, Deadwyler SA, Childers SR (1996) Effects of chronic treatment with delta9-tetrahydrocannabinol on cannabinoid-stimulated [35S]GTPgammaS autoradiography in rat brain. J Neurosci 16:8057–8066

Sim-Selley LJ (2003) Regulation of cannabinoid CB1 receptors in the central nervous system by chronic cannabinoids. Crit Rev Neurobiol 15:91–119

Simmons ML, Murphy S (1992) Induction of nitric oxide synthase in glial cells. J Neurochem 59:897–905

Singh R, Hurst DP, Barnett-Norris J, Lynch DL, Reggio PH, Guarnieri F (2002) Activation of the cannabinoid CB1 receptor may involve a W6.48/F3.36 rotamer toggle switch. J Pept Res 60:357–370

Slipetz DM, O'Neill GP, Favreau L, Dufresne C, Gallant M, Gareau Y, Guay D, Labelle M, Metters KM (1995) Activation of the human peripheral cannabinoid receptor results in inhibition of adenylyl cyclase. Mol Pharmacol 48:352–361

Stefano GB, Liu Y, Goligorsky MS (1996) Cannabinoid receptors are coupled to nitric oxide release in invertebrate immunocytes, microglia, and human monocytes. J Biol Chem 271:19238–19242

Stefano GB, Salzet B, Rialas CM, Pope M, Kustka A, Neenan K, Pryor S, Salzet M (1997a) Morphine- and anandamide-stimulated nitric oxide production inhibits presynaptic dopamine release. Brain Res 763:63–68

Stefano GB, Salzet B, Salzet M (1997b) Identification and characterization of the leech CNS cannabinoid receptor: coupling to nitric oxide release. Brain Res 753:219–224

Stefano GB, Salzet M, Magazine HI, Bilfinger TV (1998) Antagonism of LPS and IFN-gamma induction of iNOS in human saphenous vein endothelium by morphine and anandamide by nitric oxide inhibition of adenylate cyclase. J Cardiovasc Pharmacol 31:813–820

Steffens M, Szabo B, Klar M, Rominger A, Zentner J, Feuerstein TJ (2003) Modulation of electrically evoked acetylcholine release through cannabinoid CB1 receptors: evidence for an endocannabinoid tone in the human neocortex. Neuroscience 120:455–465

Sugiura T, Waku K (2000) 2-Arachidonoylglycerol and the cannabinoid receptors. Chem Phys Lipids 108:89–106

Sugiura T, Kodaka T, Kondo S, Tonegawa T, Nakane S, Kishimoto S, Yamashita A, Waku K (1996) 2-Arachidonoylglycerol, a putative endogenous cannabinoid receptor ligand, induces rapid, transient elevation of intracellular free Ca2+ in neuroblastoma x glioma hybrid NG108–15 cells. Biochem Biophys Res Commun 229:58–64

Sugiura T, Kodaka T, Kondo S, Nakane S, Kondo H, Waku K, Ishima Y, Watanabe K, Yamamoto I (1997a) Is the cannabinoid CB1 receptor a 2-arachidonoylglycerol receptor? Structural requirements for triggering a Ca2+ transient in NG108-15 cells. J Biochem (Tokyo) 122:890–895

Sugiura T, Kodaka T, Kondo S, Tonegawa T, Nakane S, Kishimoto S, Yamashita A, Waku K (1997b) Inhibition by 2-arachidonoylglycerol, a novel type of possible neuromodulator, of the depolarization-induced increase in intracellular free calcium in neuroblastoma x glioma hybrid NG108–15 cells. Biochem Biophys Res Commun 233:207–210

Sugiura T, Kodaka T, Nakane S, Miyashita T, Kondo S, Suhara Y, Takayama H, Waku K, Seki C, Baba N, Ishima Y (1999) Evidence that the cannabinoid CB1 receptor is a 2-arachidonoylglycerol receptor. Structure-activity relationship of 2-arachidonoylglycerol, ether-linked analogues, and related compounds. J Biol Chem 274:2794–2801

Thomas BF, Compton DR, Martin BR (1990) Characterization of the lipophilicity of natural and synthetic analogs of delta 9-tetrahydrocannabinol and its relationship to pharmacological potency. J Pharmacol Exp Ther 255:624–630

Turkanis SA, Karler R (1981) Electrophysiologic properties of the cannabinoids. J Clin Pharmacol 21:449S–463S

Turkanis SA, Karler R (1983) Effects of delta 9-tetrahydrocannabinol on cat spinal motoneurons. Brain Res 288:283–287

Turkanis SA, Karler R (1986) Cannabidiol-caused depression of spinal motoneuron responses in cats. Pharmacol Biochem Behav 25:89–94

Turkanis SA, Karler R, Partlow LM (1991) Differential effects of delta-9-tetrahydrocannabinol and its 11-hydroxy metabolite on sodium current in neuroblastoma cells. Brain Res 560:245–250

Ulfers AL, McMurry JL, Kendall DA, Mierke DF (2002a) Structure of the third intracellular loop of the human cannabinoid 1 receptor. Biochemistry 41:11344–11350

Ulfers AL, McMurry JL, Miller A, Wang L, Kendall DA, Mierke DF (2002b) Cannabinoid receptor-G protein interactions: G(alphai1)-bound structures of IC3 and a mutant with altered G protein specificity. Protein Sci 11:2526–2531

Upham BL, Rummel AM, Carbone JM, Trosko JE, Ouyang Y, Crawford RB, Kaminski NE (2003) Cannabinoids inhibit gap junctional intercellular communication and activate ERK in a rat liver epithelial cell line. Int J Cancer 104:12–18

Valjent E, Pages C, Herve D, Girault JA, Caboche J (2004) Addictive and non-addictive drugs induce distinct and specific patterns of ERK activation in mouse brain. Eur J Neurosci 19:1826–1836

Varma N, Carlson GC, Ledent C, Alger BE (2001) Metabotropic glutamate receptors drive the endocannabinoid system in hippocampus. J Neurosci 21:RC188
Vasquez C, Lewis DL (1999) The CB1 cannabinoid receptor can sequester G-proteins, making them unavailable to couple to other receptors. J Neurosci 19:9271–9280
Vogel Z, Barg J, Levy R, Saya D, Heldman E, Mechoulam R (1993) Anandamide, a brain endogenous compound, interacts specifically with cannabinoid receptors and inhibits adenylate cyclase. J Neurochem 61:352–355
Wartmann M, Campbell D, Subramanian A, Burstein SH, Davis RJ (1995) The MAP kinase signal transduction pathway is activated by the endogenous cannabinoid anandamide. FEBS Lett 359:133–136
Wilson RI, Nicoll RA (2001) Endogenous cannabinoids mediate retrograde signalling at hippocampal synapses. Nature 410:588–592
Wilson RI, Nicoll RA (2002) Endocannabinoid signaling in the brain. Science 296:678–682
Wilson RI, Kunos G, Nicoll RA (2001) Presynaptic specificity of endocannabinoid signaling in the hippocampus. Neuron 31:453–462
Yamaji K, Sarker KP, Kawahara K, Iino S, Yamakuchi M, Abeyama K, Hashiguchi T, Maruyama I (2003) Anandamide induces apoptosis in human endothelial cells: its regulation system and clinical implications. Thromb Haemost 89:875–884
Zhou D, Song ZH (2001) CB1 cannabinoid receptor-mediated neurite remodeling in mouse neuroblastoma N1E-115 cells. J Neurosci Res 65:346–353
Zhou D, Song ZH (2002) CB1 cannabinoid receptor-mediated tyrosine phosphorylation of focal adhesion kinase-related non-kinase. FEBS Lett 525:164–168
Zhuang SY, Chen Y, Weiner JL, Hampson RE, Deadwyler SA (2003) Lack of functional presynaptic, but putative postsynaptic actions of cannabinoids in hippocampal neurons. Soc Neurosci Abstracts 29:462.1

HEP (2005) 168:81–115

Molecular Biology of Cannabinoid Receptors

M.E. Abood

Forbes Norris MDA/ALS Research, California Pacific Medical Center, 2351 Clay St 416, San Francisco CA, 94115, USA
mabood@cooper.cpmc.org

Abstract To date, two cannabinoid receptors have been isolated by molecular cloning. The CB_1 and CB_2 cannabinoid receptors are members of the G protein-coupled receptor family. There is also evidence for additional cannabinoid receptor subtypes. The CB_1 and CB_2 receptors recognize endogenous and exogenous cannabinoid compounds, which fall into five structurally diverse classes. Mutagenesis and molecular modeling studies have identified several key amino acid residues involved in the selective recognition of these ligands. Numerous residues involved in receptor activation have been elucidated. Regions of the CB_1 receptor mediating desensitization and internalization have also been discovered. The known genetic structures of the CB_1 and CB_2 receptors indicate polymorphisms and multiple exons that may be involved in tissue and species-specific regulation of these genes. The cannabinoid receptors are regulated during chronic agonist exposure, and gene expression is altered in disease states. There is a complex molecular architecture of the cannabinoid receptors that allows a single receptor to recognize multiple classes of compounds and produce an array of distinct downstream effects.

Keywords Cannabinoid receptor · Mutagenesis · Polymorphism · Gene regulation, binding

1
Introduction

Our knowledge of the mechanism of action of cannabinoids has increased greatly in the past several years due to numerous major discoveries. The development of novel synthetic analogs of (–)-Δ^9-tetrahydrocannabinol (Δ^9-THC), the primary psychoactive constituent in marijuana, played a major role in the characterization and cloning of a neuronal cannabinoid receptor, a member of the G protein-coupled receptor family (GPCR) (Matsuda et al. 1990). The identity of the cDNA clone as the cannabinoid receptor (CB_1) was confirmed by transfection into Chinese hamster ovary (CHO) cells and the demonstration of cannabinoid-mediated inhibition of adenylyl cyclase (Gerard et al. 1991; Matsuda et al. 1990). This receptor can also modulate G protein-coupled Ca^{2+} and K^+ channels (Mackie and Hille 1992; McAllister et al. 1999). Five structurally distinct classes of cannabinoid compounds have now been identified: the classical cannabinoids [Δ^9-THC, Δ^8-THC-dimethylheptyl (HU210)]; non-classical cannabinoids (CP 55,940); indoles (WIN 55,212-2), eicosanoids (anandamide, 2-arachidonoylglycerol) and antagonist/inverse agonists (SR141716A, SR145528) (Devane et al. 1992; Eissenstat et al. 1995; Howlett 1995; Mechoulam et al. 1995; Rinaldi-Carmona et al. 1994; Rinaldi-Carmona et al. 1998a; Xie et al. 1996).

The CB_1 receptor gene has been inactivated in mice (by in-frame deletion of most of the coding region) using homologous recombination in two laboratories (Ledent et al. 1999; Zimmer et al. 1999). Significantly, not only did the CB_1 receptor knockout mice lose responsiveness to most cannabinoids, the reinforcing properties of morphine and the severity of the withdrawal syndrome were strongly reduced (Ledent et al. 1999). The CB_1 receptor appears to play a central role in drug addiction.

The existence of a second type of cannabinoid receptor in the spleen was established (Kaminski et al. 1992). The CB_2 receptor was isolated by a polymerase chain reaction (PCR)-based strategy designed to isolate GPCRs in differentiated myeloid cells (Munro et al. 1993). The CB_2 receptor, which has only been found in the spleen and cells of the immune system, has 44% amino acid identity with CB_1, and a distinct yet similar binding profile, and thus represents a receptor subtype. The CB_2 receptor gene has been inactivated by homologous recombination in mice (Buckley et al. 2000); the most notable effect was impairment of immunomodulation by helper T cells.

Another major breakthrough in cannabinoid research was the discovery of endogenous ligands for the cannabinoid receptors; this uncovered a novel neurotransmitter/neuromodulatory system. The first ligand, arachidonoyl ethanolamide (anandamide, AEA) was isolated from porcine brain; it competed for binding to the CB_1 receptor and inhibited electrically stimulated contractions of the mouse vas deferens in the same manner as Δ^9-THC (Devane et al. 1992). The pharmacological properties of anandamide are consistent with its initial identification as an endogenous ligand for the cannabinoid receptor(s). In vivo, anandamide produces many of the same pharmacological effects as the classical cannabinoid ligands, including hypomotility, antinociception, catalepsy, and hypothermia (Fride and Mechoulam 1993). The biosynthetic pathways of anandamide synthesis, release, and removal are under investigation by several laboratories (Deutsch and Chin 1993; Di Marzo et al. 1994; Hilliard and Campbell 1997; Piomelli et al. 1999; Walker et al. 1999). Additional fatty acid ethanolamides with cannabimimetic properties have been isolated, suggesting the existence of a family of endogenous cannabinoids (Hanus et al. 1993). 2-Arachidonoylglycerol (2AG) in several systems acts as a full agonist, whereas anandamide is a partial agonist, suggesting that the CB_1 receptor may in fact be a 2AG receptor (Stella et al. 1997; Sugiura et al. 1997).

Additionally, virodhamine, arachidonic acid and ethanolamine joined by an ester linkage, has been isolated (Porter et al, 2001). Noladin ether, 2-arachidonyl glyceryl ether, is a potent endogenous agonist at the CB_1 receptor (Hanus et al. 2001). *N*-Arachidonoyl-dopamine (NADA), is primarily a vanilloid receptor agonist, but has some activity at CB_1 receptors as well (Huang et al. 2002). Palmitoylethanolamide (PEA) has been suggested as a possible endogenous ligand at the CB_2 receptor (Facci et al. 1995). However, subsequent studies showed no affinity for palmitoylethanolamide at the CB_2 receptor (Griffin et al. 2000; Lambert et al. 1999; Showalter et al. 1996). Instead, PEA seems to increase the potency of AEA, in part by inhibiting fatty acid amide hydrolase (FAAH), the enzyme responsible for breakdown of AEA (Di Marzo et al. 2001).

In addition to actions at cannabinoid receptors, AEA, 2AG, virodhamine, noladin ether, and NADA also act at the vanilloid receptor (transient receptor potential vanilloid type 1 TRPV1; previously know as VR1), a ligand-gated ion channel that is a member of the transient receptor potential (TRP) ion channel family (recently reviewed by Di Marzo et al. 2002). In addition, Δ^9-THC and cannabinol at high (20 μM) concentrations have recently been identified as agonists at another TRP, the ANKTM1 channel (Jordt et al. 2004). These findings raise the possibility that the TRP channels may be ionotropic cannabinoid receptors.

The existence of a family of endogenous ligands suggests the presence of additional cannabinoid receptor subtypes. In addition, some of the diverse effects may result from different receptor conformations. Experimental evidence from several laboratories suggests that cannabinoid receptor ligands can induce different conformations of the CB_1 receptor, which in turn can activate select G proteins (Glass and Northup 1999; Griffin et al. 1998; Kearn et al. 1999; Mukhopadhyay et al. 2000; Selley et al. 1996). This selectivity appears to be driven by distinct molecular interactions that occur between the different classes of cannabinoid compounds and the receptor proteins. These data indicate that receptor "subtypes" may also be observed as a result of activation of distinct second messenger pathways that produce different physiological responses.

This chapter will focus on the molecular biology of the G protein-coupled cannabinoid receptors.

2 General Structure and Distribution

Two cannabinoid receptors have been identified to date; the CB_1 receptor is localized predominantly in the central nervous system (CNS), whereas the CB_2 receptor is located primarily in the immune system. The CB_1 receptor cDNA was isolated from a rat brain library by a homology screen for GPCRs and its identity confirmed by transfecting the clone into CHO cells and demonstrating cannabinoid-mediated inhibition of adenylyl cyclase (Matsuda et al. 1990). Initial identification of the ligand for this "orphan receptor" involved the screening of many candidate ligands, including opioids, neurotensin, angiotensin, substance P, and neuropeptide Y, among others, until cannabinoids were found to act via this molecule. In cells transfected with the clone, CP 55,940, Δ^9-THC and other psychoactive cannabinoids, but not cannabidiol (which lacks CNS activity) were found to inhibit adenylyl cyclase, whereas in untransfected cells no such response was found. Furthermore, the rank order of potency for inhibition of adenylyl cyclase in transfected cells correlated well with cell lines previously shown to possess cannabinoid-inhibited adenylyl cyclase activity. Distribution of the expression of CB_1 mRNA also paralleled that of cannabinoid receptor binding in rat brain. Analysis of the primary amino acid sequence of the CB_1 receptor predicts seven transmembrane (TM) domain regions, typical of GPCRs. Bramblett et al. (1995) have constructed a model of the cannabinoid receptor. A representation of the CB_1 receptor based on their model is shown in Fig. 1.

The CB_2 receptor was also isolated by its homology to other GPCRs, using a PCR-based approach in myeloid cells (Munro et al. 1993). The human CB_2 receptor cDNA was isolated from the human promyelocytic cell line, HL60. The clone has 44% amino acid sequence identity overall with the CB_1 clone, and percentage similarity rises to 68% in the TM domains. The amino acid residues conserved between CB_1 and CB_2 are shaded in Fig. 1. The localization of the CB_2 receptor appears to be mainly in the periphery: in the spleen and in low levels in adrenal, heart, lung, prostate, uterus, pancreas, and testis and in cells of immune origin, including microglia in the CNS (Munro et al. 1993; Galiegue et al. 1995; Walter et al. 2003). An alignment of human CB_1 and CB_2 is shown in Fig. 2. Using the numbering scheme of Ballesteros and Weinstein (Ballesteros and Weinstein 1995), each amino acid is given a number that begins with the helix number followed by

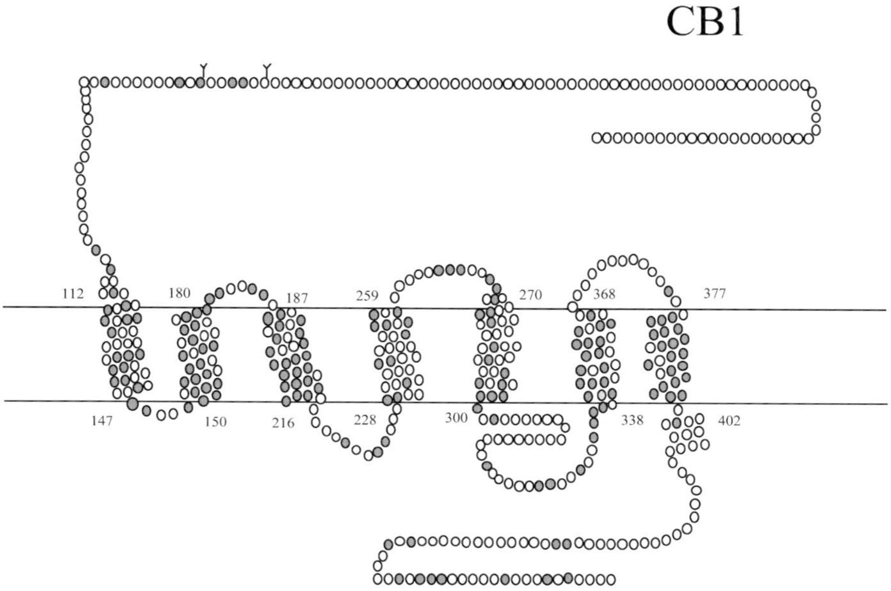

Fig. 1. A helix net representation of the human CB_1 receptor. The amino acids shared with the CB_2 receptor are *shaded*

a two-digit decimal. The most highly conserved residue in each helix is assigned a value of 0.50 and the other residues numbered relative to the conserved residue.

Transfected cell lines expressing the CB_2 receptor have an affinity for CP 55,940 that is similar to those expressing the CB_1 receptor (Felder et al. 1995; Munro et al. 1993; Showalter et al. 1996). Furthermore, the affinities for Δ^9-THC, 11-OH-Δ^9-THC, anandamide and cannabidiol at the CB_2 receptor are comparable to the brain (Showalter et al. 1996) receptor. In contrast, cannabinol (which is known to be ten times less potent than Δ^9-THC at the CB_1 receptor) was found to be equipotent to Δ^9-THC at the CB_2 receptor (Showalter et al. 1996). Based on these binding profiles, it was concluded that the peripheral receptor clone may be a cannabinoid receptor subtype. Indeed, a more extensive characterization of this receptor demonstrates a separation of pharmacological selectivities (Felder et al. 1995; Showalter et al. 1996; Slipetz et al. 1995). The compounds that have been identified as CB_1 and CB_2 selective serve as lead compounds in the design of even more selective ligands. The affinity of SR141716A (the CB_1 receptor antagonist) is at least 50-fold higher at the CB_1 receptor than at the CB_2 receptor (Felder et al. 1995; Rinaldi-Carmona et al. 1994; Showalter et al. 1996) and has provided a starting point for the design of more selective antagonists and agonists.

```
CB1   1 MKSILDGLAD TTFRTITTDL LYVGSNDIQY EDIKGDMASK LGYFPQKFPL   50
CB2     .......... .......... .......... .......... ..........

CB1  51 TSFRGSPFQE KMTAGDNPQL VPADQVNITE FYNKSLSSFK ENEENIQCGE  100
CB2   1 .......... .......... ..MEECWVTE IANGSKDGLD SN........   20

                    111111111 1111111111 1111111111 11111    2
                    223333333 3334444444 4455555555 55566    3
                    890123456 7890123456 7890123456 78901    7
CB1 101 NFMDIECFMV LNPSQQLAIA VLSLTLGTFT VLENLLVLCV ILHSRSLRCR  150
CB2  21 ...PMKDYMI LSGPQKTAVA VLCTLLGLLS ALENVAVLYL ILSSHQLRRK   67

        2222222222 2222222222 2222222222     333333 3333333333
        3344444444 5555555555 5566666666     222222 2223333333
        8901234567 8901234567 8901234567     123456 7890123456
CB1 151 PSYHFIGSLA VADLLGSVIF VYSFIDFHVF HRKDSRNVFL FKLGGVTASF  200
CB2  68 PSYLFIGSLA GADFLASVVF ACSFVNFHVF HGVDSKAVFL LKIGSVTMTF  117

        3333333333 3333333333         44 4444444444 4444444444
        3334444444 4445555555         33 4444444444 5555555555
        7890123456 7890123456         89 0123456789 0123456789
CB1 201 TASVGSLFLT AIDRYISIHR PLAYKRIVTR PKAVVAFCLM WTIAIVIAVL  250
CB2 118 TASVGSLLLT AIDRYLCLRY PPSYKALLTR GRGLVTLGIM WVLSALVSYL  167

        4444                5 5555555555 5555555555 5555555555
        6666                3 3333344444 4444455555 5555566666
        0123                4 5678901234 5678901234 5678901234
CB1 251 PLLGWNCEKL QSVCSDIFPH IDETYLMFWI GVTSVLLLFI VYAYMYILWK  300
CB2 168 PLMGWTCC.. PRPCSELFPL IPNDYLLSWL LFIAFLFSGI IYTYGHVLWK  215

                                          666666666 6666666666
                                          222222333 3333333444
                                          456789012 3456789012
CB1 301 AHSHAVRMIQ RGTQKSIIIH TSEDGKVQVT RPDQARMDIR LAKTLVLILV  350
CB2 216 AHQHVASL.. .......... .SGHQDRQVP GMARMRLDVR LAKTLGLVLA  252

        6666666666 6666666666      77777 7777777777 7777777777
        4444444555 5555555666      33333 3334444444 4445555555
        3456789012 3456789012      23456 7890123456 7890123456
CB1 351 VLIICWGPLL AIMVYDVFGK MNKLIKTVFA FCSMLCLLNS TVNPIIYALR  400
CB2 253 VLLICWFPVL ALMAHSLATT LSDQVKKAFA FCSMLCLINS MVNPVIYALR  302

        7777777777 777
        5555666666 666
        7890123456 789
CB1 401 SKDLRHAFRS MFPSCEGTAQ PLDNSMGDSD CLHKHANNAA SV.HRAAESC  449
CB2 303 SGEIR..... ..SSAHHCLA HWKKCVRGLG SEAKEEAPRS SVTETEADGK  345

CB1 450 IKSTVKIAKV TMSVSTDTSA EAL* 472
CB2 346 ITPWPDSRDL DLSDC*.... .... 360
```

Fig. 2. An alignment of the human CB_1 and CB_2 receptors. The transmembrane domains are *underlined*. The standard single letter amino acid code is used. The numbering system of Ballesteros and Weinstein (1995) is shown *above* each transmembrane domain

3 Gene Structure and Species Diversity

Shortly after the cloning of the rat cannabinoid receptor, isolation of a human CB_1 receptor cDNA was reported (Gerard et al. 1991). The rat and human receptors are highly conserved, 93% identity at the nucleic acid level and 97% at the amino acid level. There is an excellent correlation between binding affinities at the cloned CB_1 receptor as compared to binding in brain homogenates using [^{3}H]CP 55,940 as the radioligand (Felder et al. 1992).

There is evidence for splice variants of the cannabinoid receptors. A PCR amplification product was isolated that lacked 167 base pairs of the coding region of the human CB_1 receptor (Shire et al. 1995). This alternative splice form (CB_{1A}) is unusual in that it is generated from the mRNA encoding CB_1, and not from a separate exon (Shire et al. 1995). When expressed, the CB_{1A} clone would translate to a receptor truncated by 61 amino acid residues with 28 amino acid residues different at the NH_2-terminal. This might lead to a receptor with altered ligand-binding properties. CB_{1A} expression has been detected in many tissues by RT-PCR (Table 1). It will be important to confirm that the CB_{1A} receptor protein is expressed, since splice variants often arise from incomplete splicing during library construction and RT-PCR techniques. The construction of antibodies selective to CB_1 or CB_{1A} peptides would be useful to detect these proteins. The CB_{1A} splice variant is not present in rat or mouse, because the splice consensus sequence is absent in these genes (the invariant GT of the splice donor site becomes a GA in both the rat and mouse) (Bonner 1996).

The mouse CB_1 gene and cDNA sequences have been reported (Abood et al. 1997; Chakrabarti et al. 1995; Ho and Zhao 1996). Sequence analysis of the mouse CB_1 clones also indicates a high degree of conservation among species. The mouse and

Table 1. Amino acid residues important in cannabinoid receptor ligand recognition

CB_1 receptor	**CP 55,940 binding**	**WIN 55,212-2 binding**
SR141716A binding	F3.25(189)	G3.31(195)
K3.28(192)	K3.28(192)	F3.36(201)
F3.36(201)	C174	W5.43(280)
W5.43(280)	C179	V5.46(282)
W6.48(357)		W6.48(357)
Anandamide binding	**CB_2 receptor**	**WIN 55,212-2 binding**
F3.25(190)	SR144528 binding	S3.31(112)
K3.28(192)	S4.53(161)	F5.46(197)
	S4.57(165)	
	C175	
All ligand binding lost (conformational changes)		
Y5.39 (Y275 in CB_1, Y190 in CB_2)	C174 in CB_1	C179 in CB_2
D3.49(130) in CB_2	W4.50(158) in CB_2	W4.64(172) in CB_2
L5.50(201) in CB_2	Y7.53(299) in CB_2	

rat clones have 95% nucleic acid identity (100% amino acid identity). The mouse and human clones have 90% nucleic acid identity (97% amino acid identity). Rat CB_1 probes can be used to detect mouse cannabinoid receptor mRNA (Abood et al. 1993), again indicating conservation among species. However, the human and rat sequences diverge about 60 bp upstream of the translation initiation codon. Furthermore, we have isolated a rat CB_1 clone that is identical to the published sequence in the coding region, but diverges about 60 bp upstream of the translation codon (unpublished data). Examination of the 5′ untranslated sequence of the mouse CB_1 genomic clone indicates a splice junction site approximately 60 bp upstream from the translation start site. This splice junction site is also present in the human CB_1 gene (Shire et al. 1995). These data suggest the existence of splice variants of the CB_1 receptor as well as possible divergence of regulatory sequences between these genes. A third exon is present in the rat and human genes in their 5′ untranslated regions (Bonner 1996). The reported transcription start sites are consistent with the presence of two promoters for the CB_1 genes (Bonner 1996).

The CB_1 receptor has been studied in a molecular phylogenetic analysis of 64 mammalian species (Murphy et al. 2001). The sequence diversity in 62 species examined varied from 0.41% to 27%. In addition to mammals, the CB_1 receptor has been isolated from birds (Soderstrom et al. 2000b), fish (Yamaguchi et al. 1996), amphibia (Cottone et al. 2003; Soderstrom et al. 2000a), and an invertebrate, *Ciona intestitinalis* (Elphick et al. 2003). This deuterostomian invertebrate cannabinoid receptor contains 28% amino acid identity with CB_1, and 24% with CB_2 (Elphick et al. 2003). Since a CB receptor ortholog has not been found in *Drosophila melanogaster* or *Caenorhabditis elegans*, it has been suggested that the ancestor of vertebrate CB_1 and CB_2 receptors originated in a deuterostomian invertebrate (Elphick et al. 2003).

The CB_2 receptor has also been isolated from mouse (Shire et al. 1996b; Valk et al. 1997), rat (Griffin et al. 2000; Brown et al. 2002), and the puffer fish *Fugu rubripes* (Elphick 2002). The CB_2 receptor shows less homology between species than does CB_1; for instance, the human and mouse CB_2 receptors share 82% amino acid identity (Shire et al. 1996b), and the mouse and rat 93% amino acid identity. The human, rat, and mouse sequences diverge at the C-terminus; the mouse sequence is 13 amino acids shorter, whereas the rat clone is 50 amino acids longer than the human CB_2 (Brown et al. 2002).

There is also an intron in the C-terminus of the CB_2 receptor. This intron is also species-specific; it is only present in the rat CB_2 receptor (Brown et al. 2002). This may give rise to rat-specific pharmacology of the CB_2 receptor. We found differences in ligand recognition with a number of compounds at the rat CB_2 receptor compared to the human CB_2 receptor in transfected cells (Griffin et al. 2000). It is important to note, however, that the clone described in these studies was a genomic clone of rat CB_2 and did not contain the edited C-terminus discovered by Brown et al. (2002).

To date, the complete genetic structure including 5′ and 3′ untranslated regions and transcription start sites of the CB_1 and CB_2 genes have not been mapped. From what we know so far, the diversity in the regulatory regions of the CB_1 and CB_2 genes may provide flexibility in gene regulation.

4
Ligand Recognition at the CB_1 Receptor

4.1
The Aminoalklylindole/SR141716A Binding Region

Mutation studies as well as studies with novel ligands have suggested a separation of the binding site for aminoalkylindoles (typified by WIN 55,212-2) from that of the other three classes of cannabinoid agonist ligands (Table 2) (Chin et al. 1998; Song and Bonner 1996; Tao et al. 1999). A K3.28(192)A mutation of CB_1 results in no loss of affinity or efficacy for WIN 55,212-2, but greater than 1,000-fold loss in affinity and efficacy for HU-210, CP 55,940, and anandamide (Chin et al. 1998; Song and Bonner 1996), and a 17-fold loss for SR141716A (Hurst et al. 2002). The CB_2 selectivity of WIN 55,212-2 (Felder et al. 1995; Showalter et al. 1996) may be due to the presence of an additional TM helix (TMH)5 aromatic residue, F5.46 in the CB_2 receptor (Song et al. 1999). Receptor chimera studies of the CB_1 and CB_2 receptors have demonstrated that the region delimited by the fourth and fifth TM domains of the CB_1 receptor is crucial for the binding of the CB_1 receptor antagonist SR141716A, but not CP 55,940, and that this same region in the CB_2 receptor is crucial for the binding of WIN 55,212-2 and the CB_2 receptor antagonist SR144528 (Shire et al. 1996a, 1999). These results reinforce the hypothesis that the aminoalkylindole-binding region at the CB_1 receptor is in the TMH 3-4-5 region and is not identical to that for other CB agonists. Furthermore, these results suggest that SR141716A binding shares the aminoalkylindole binding region but also interacts with K(3.28)192.

In addition, the carbonyl oxygen as well as the morpholino ring of the aminoalkylindoles can be replaced without affecting affinity; therefore hydrogen bonding may not be the primary interaction of these compounds at the CB_1 receptor (Huffman 1999; Huffman et al. 1994; Kumar et al. 1995; Reggio 1999). Huffman et al. (1994) also reported that the replacement of the naphthyl ring of WIN 55,212-2 with an alkyl or alkenyl group resulted in complete loss of CB_1 receptor affinity (K_i>10,000 nM in both cases). The fact that the carbonyl oxygen or the morpholino ring of the aminoalkylindoles can be removed without significant effect, along with evidence that the presence of the carbonyl and morpholino group (in the absence of an aryl substituent) is insufficient to produce CB_1 affinity, suggests that aromatic stacking, rather than hydrogen bonding, may be the primary interaction for aminoalkylindoles at the CB_1 receptor.

Aromatic–aromatic stacking interactions are significant contributors to protein structure stabilization (Burley and Petsko 1985). Modeling studies indicate that in the active state (R*) model of CB_1, there is a patch of aromatic amino acids in the TMH 3-4-5 region with which WIN 55,212-2 can interact (McAllister et al. 2003). There is an upper (extracellular side) stack formed by F3.25(189 in human CB_1, 190 in mouse CB_1), W4.64(255/256), Y5.39(275/276), and W5.43(279/280). When WIN 55,212-2 is computationally docked to interact with this patch, it also can interact with a lower (towards intracellular side) aromatic residue, F3.36(200/201). In this docking position, WIN 55,212-2 creates a continuous aromatic stack over several turns of TMHs 3, 4, and 5 that is likely to be energetically favored. Similarly, studies in the Reggio lab suggested that in the inactive (R) state of CB_1 the amide oxygen of SR141716A interacts with a salt bridge formed by K3.28 and D6.58(366), while the dichlorophenyl ring of SR141716A interacts with F3.36 and W6.48 and the monochlorophenyl ring interacts with F3.36 and W5.43 (Hurst et al. 2002).

In a recent study, McAllister et al. tested the hypothesis that a CB_1 TMH3-4-5-6 aromatic microdomain that includes F3.25, F3.36, W4.64, Y5.39, W5.43, and W6.48, constitutes the binding domain of SR141716A and WIN 55,212-2 (McAllister et al. 2003). Stably transfected cell lines were created for single-point mutations of each aromatic microdomain residue to alanine. The binding of SR141716A and WIN 55,212-2 were found to be affected by the F3.36A, W5.43A, and W6.48A mutations, suggesting that these residues are part of the binding site for these two ligands. In particular, the W5.43A mutation resulted in profound loss of affinity for SR141716A. Mutation of W4.64 to A resulted in loss of ligand binding and signal transduction; however, this was shown to be a result of improper cellular localization; the mutant receptor was not expressed on the cell surface.

Anandamide was used as a control in this study, as aromatic stacking interactions are not key to its binding. However, according to the molecular model, F3.25A is a direct interaction site for anandamide. F3.25A had no effect on WIN 55,212-2 or SR141716A binding, but resulted in a sixfold loss in affinity for anandamide (McAllister et al. 2003).

4.2
The Classical/Non-Classical/Endogenous CB Binding Region

As stated above, the mutation studies of CB_1 demonstrated greater than 1,000-fold loss in affinity and efficacy for HU-210, CP 55,940, and anandamide at K3.28(192)A (Chin et al. 1998; Song and Bonner 1996). This indicated that K3.28(192) is a primary interaction site for the phenolic hydroxyl of HU-210 and other classical cannabinoids, as well as the non-classical cannabinoids (e.g., CP 55,940) in the CB_1 receptor (Huffman et al. 1996). Modeling studies suggested that the alkyl side chain of CP 55,940 resides in a hydrophobic pocket (Tao et al. 1999). In CB_1, the primary interaction is between the phenolic hydroxyl of CP 55,940 and K3.28(192). These considerations suggest that the TMH 3-6-7 region is the binding site for classical and non-classical cannabinoids, and presumably the endogenous cannabinoids.

It should be noted that the two binding regions identified (i.e., TMH 3-4-5 for aminoalkylindoles and TMH 3-6-7 for other agonist classes) overlap spatially such that the binding of a ligand in one region would preclude binding in the other region. This would be detected as competitive inhibition in a binding assay.

Residues in the N-terminus as well as in and near extracellular loop 1 have been shown to be important for binding of CP 55,940 (Murphy and Kendall 2003). Loss of affinity for CP 55,940 was seen when dipeptide insertions were made at residues 113, 181, and 188. Six substitution mutants (to alanine) were constructed around these residues; they showed weaker affinity than the wild-type (WT) receptor, but less of a loss than observed with the corresponding insertion mutant. This pattern suggests that the loop structure itself is important for recognition of CP 55,940.

Interestingly, F189(3.25)A in human CB_1 results in a dramatic reduction of CP 55,940 affinity (Murphy and Kendall 2003), but in mouse CB_1, CP 55,940 binding is not affected, and instead anandamide's affinity is lowered (McAllister et al. 2003). This suggests the minor sequence variation in mouse vs human CB_1 can result in structural differences in ligand recognition.

5 Ligand Recognition at the CB_2 Receptor

5.1 Identification of Amino Acids Which Discriminate CB_1 and CB_2 Receptor Subtypes

The CB_1 and CB_2 receptors (Fig. 2) share only 44% overall amino acid identity, which rises to 68% in the TM domains (Munro et al. 1993). However, most cannabinoid receptor agonists do not discriminate between the receptor subtypes (Felder et al. 1995; Pertwee 1997). There are several ligands which are CB_1- or CB_2-selective (5- to 60-fold), and a few ligands with a greater separation of activity at each receptor (100- to 1,000-fold) (Griffin et al. 1999, 2000; Hanus et al. 1999; Huffman et al. 1996, 1999; Ibrahim et al. 2003; Showalter et al. 1996; Tao et al. 1999). For example, 1-deoxy-Δ^8-THC showed no affinity for the CB_1 receptor but has good affinity (K_i=32 nM) for the CB_2 receptor (Huffman et al. 1999). However, there is a need for more selective agonists to produce specific receptor-mediated effects for in vivo studies.

Structure–activity relationships of Δ^9-THC analogs have revealed three critical points of attachment to a receptor: (1) a free phenolic hydroxyl group; (2) an appropriate substituent at the C9 position and (3) a lipophilic side chain (Howlett et al. 1988). However, compounds with a dimethylheptyl side chain retain affinity for both CB_1 and CB_2 receptors even when they lack a phenolic hydroxyl (Gareau et al. 1996; Huffman et al. 1996). Moreover, these ligands are CB_2-selective (Huffman et al. 1996, 1999).

An alternative approach to traditional structure–activity relationships with synthetic ligands is to map the ligand binding sites of the receptors using in vitro mutagenesis of receptor cDNAs. For example, the lysine residue in the third TM domain of the cannabinoid receptors, which is conserved between the CB_1 and CB_2 receptors, appears to mediate different functional roles in the receptor subtypes. K3.28(192) in the CB_1 receptor is critically important for ligand recognition for several agonists (CP 55,940, HU-210, Δ^9-THC, and anandamide) but not for WIN 55,212-2 (Chin et al. 1998; Song and Bonner 1996). Mutation of the analogous residue in the CB_2 receptor (K109) to alanine or arginine resulted in fully functional CB_2 receptors with all ligands tested (Tao et al. 1999). In this same study a molecular model was generated in order to explain these findings. The model suggested an alternative binding mode could be achieved in the K109A CB_2 mutant in contrast to K192A CB_1. Assuming that ligand binding occurs within the pore formed by the TMH bundle, and the hydrophobic cluster of amino acids on helices 6 and 7 form the hydrophobic pocket with which the dimethylheptyl side chain of CP 55,940 interacts, receptor docking studies indicated that CP 55,940 is oriented differently in the binding pocket in CB_1 vs CB_2. A unique feature identified in the CP 55,940/CB_2 binding site was a hydrogen bonding cluster formed by a serine, threonine, and an asparagine. In the CP 55,940/CB_1 docking studies this cluster is not present. This suggested that when CB_2 K109 was mutated to A, the hydrogen bonding cluster could compensate for receptor binding to CP 55,940, whereas when CB_1 K192 was mutated to A this compensation did not occur. To test this hypothesis the CB_2 hydrogen-bonding cluster was disrupted by generating the double-mutant K109AS112G. When the serine in the hydrogen-bonding cluster was replaced with a glycine, the receptor was not able to recognize several cannabinoid agonists excluding WIN 55,212-2. This was reminiscent of the findings of CB_1 K192A, except only 10% vs full inhibition of cyclic 3′,5′-adenosine monophosphate (cAMP) accumulation could be produce even in the presence of 10 μM WIN 55,212-2. Receptor expression was determined by immunofluorescence. The WT CB_2 protein was expressed in approximately 90% of the cells. Only 30% of the cells expressed the double-mutant K109AS112G, and the pattern of staining exhibited entrapment of the receptor within the perinuclear region. Interestingly, even the expression of the K109A mutant receptor, which exhibited WT receptor characteristics, was expressed less than the WT receptor (50% vs 90% of the cells expressed the protein, respectively). The reduced expression of the double-mutant K109AS112G could explain why only 10% inhibition of cAMP accumulation was observed in the presence of WIN 55,212-2. Regardless, the serine in combination with the lysine in the CB_2 receptor appears to play a crucial role in determining proper function of the receptor.

The K3.28 mutation studies demonstrated that a separate but overlapping receptor binding site must occur with WIN 55,212-2 compared to other cannabinoid ligands in the CB_1 receptors. Another important feature of WIN 55,212-2 is that it has a higher affinity for the CB_2 receptors, albeit only five- to tenfold higher (Showalter et al. 1996). Two groups sought to discover critical residues in the cannabinoid receptors that impart this agonist selectivity. The first used a molecular modeling approach; it indicated that aromatic stacking interactions are important for aminoalkylindole binding (Song et al. 1999). There is a phenylalanine at position 5.46(F197) in CB_2 vs a valine (V282) in CB_1, which could provide greater aromatic stacking and may impart the selectivity of WIN 55,212-2 for CB_2. Therefore, valine and phenylalanine were switched between the receptors. The CB1V282F mutant bound WIN 55,212-2 in a similar fashion to WT CB_2, whereas the CB_2 F197V mutant adopted CB_1 receptor binding affinity for WIN 55,212-2. This data strongly favored the hypothesis that a phenylalanine at position 5.46 is crucial for WIN 55,212-2 selectivity.

At the same time, the role of TM3 in WIN 55,212-2 selectivity was reported (Chin et al. 1999). In this investigation, a CB_1/CB_2 chimera was constructed, $CB_{1/2}$(TM3), in which the TM3 of CB_1 was replaced with the corresponding region of CB_2. The $CB_{1/2}$(TM3) mutant bound WIN 55,212-2, and the other related aminoalkylindole analogs (JWH015 and JWH018) with WT CB_2 affinities. These results suggested that the TM3 of the cannabinoid receptor imparts selectivity of aminoalkylindoles to CB_2. When individual amino acid changes were evaluated, S112(3.31) in CB_2, which corresponded to G195 in CB_1, was the amino acid responsible for CB_2 selectivity of aminoalkylindoles. Tao et al. (1998) also reported that mutation of S112 in the K109AS112G mutation resulted in dramatic effects on ligand binding.

Key differences in the ligand recognition sites of the CB_1 and CB_2 receptors were identified using a combination of receptor chimeras and site-directed mutagenesis (Shire et al. 1996a). This study focused on the SR141716A (CB_1-selective) and CP 55,940 (non-selective) binding sites. Replacing the CB_1 receptor with up to the seventh TM region of the CB_2 receptor, including the third extracellular loop, resulted in a receptor that still exhibited CB_1 receptor properties. Further extending the CB_2 structure into the sixth TM region of the CB_1 altered receptor expression; the mutant was sequestered in the intracellular compartment of the cell and could not be analyzed. Further extending the CB_2 structure into the fifth and then fourth TM region of the CB_1 receptor systematically resulted in a $CB_{1/2}$ chimera that acted like a CB_1 receptor. The fifth TM $CB_{1/2}$ chimera acted as a $CB_{1/2}$ hybrid and the reciprocal mutation fifth TM $CB_{2/1}$ chimera had almost identical properties. The fourth TM $CB_{1/2}$ chimera was similar to the WT CB_2 receptor.

A sandwich chimera was next constructed where the CB_1 receptor TM4-e2-TM5 region was replaced with the CB_2 receptor regions (Shire et al. 1996a). This chimera resembled the WT CB_2 receptor, strengthening the findings that these regions are important for CB_1 receptor selectivity of SR 141716A. A sandwich chimera was then created in which just the CB_1 receptor e2 region was replaced with the CB_2 receptor e2 region; SR141716A binding was almost identical to the WT CB_2, but in this case CP 55,940 binding was lost. A smaller sandwich chimera was also created in which just the CB_1 receptor e2 region between conserved cysteines was replaced with the corresponding CB_2 receptor regions; this mutation resulted in a sequestration of the receptor.

Generation of functional CB_2/CB_1 chimeras proved to be more difficult when trying to study the TM4-e2-TM5 regions. When the CB_2 receptor TM4-e2-TM5 region was replaced with the CB_1 or a sandwich chimera was created in which just the CB_2 receptor e2 region was replaced with CB_1 e2, the receptors were expressed but could not bind CP 55,940 or SR141716A (Shire et al. 1996a).

One notable difference between cannabinoid receptors and many other GPCRs is the lack of conserved cysteines in the second extracellular (EC) domain. However, the third EC domain of both cannabinoid receptors does contain two or more cysteines. These cysteines are thought to form sulfhydryl bonds with cysteines in neighboring TM domains and to stabilize the receptor. When C257 and C264 in the third EC domain of the CB_1 receptor were replaced with serine residues, the mutant receptors were sequestered (Shire et al. 1996a). These residues were then replaced with alanine. In this case the receptors were expressed normally but failed to bind CP 55,940. When cysteine residues (C174 and C179) in the third EC domain of the CB_2 receptor were replaced with serine residues, the mutant receptor, although expressed normally on the cell surface, could not bind CP 55,940. Disruption of a disulfide bridge with the two cysteines in the amino-terminal region of the CB_1 receptor was not the explanation, because the double mutant C98,107S resulted in a receptor with WT properties. Overall, these results suggest the e2 domain and corresponding cysteines are important for CP 55,940 ligand recognition, but not for SR141716A.

5.2
The SR14428 Binding Site

The SR144528 binding site (Table 1) on CB_2 has been analyzed by a combination of site-directed mutagenesis and molecular modeling (Gouldson et al. 2000). Mutation of C175 (in the third EC loop) to serine resulted in a receptor with normal affinity for [^{3}H]CP 55,940, but loss of recognition of SR144528. Consequently, SR144528 did not act as an antagonist at this mutant. An eightfold loss of affinity for WIN 55,212-2 was observed with the C175S mutant. Mutation of S4.53(161) and S4.57(165) to alanines also resulted in the loss of SR144528 binding and functional activity. These serines are alanines in the CB_1 receptor, which supports a direct ligand–residue interaction at CB_2. Several other mutations were analyzed that did not affect SR144528 binding. In the corresponding molecular model of CB_2, SR144528 interacts with residues in TM 3,4, and 5 through a combination of hydrogen bonds and hydrophobic interactions (Gouldson et al. 2000). In particular, W4.64(172) and W5.43(194) form an aromatic stack similar to that proposed for WIN 55,212-2 in the CB_2 receptor (Song et al. 1999) and WIN 55,212-2 and SR141716A in the CB_1 receptor (McAllister et al. 2003).

6
Receptor Conformation

In addition to specific ligand–receptor interactions, several residues have been shown to be keys to maintaining proper receptor conformation for ligand recognition. For example, at the top of the TMH 3-4-5 aromatic cluster in both the CB_1 [Y5.39(275)] and CB_2 [Y5.39(190)] receptors is a tyrosine residue. Creating a tyrosine-to-phenylalanine mutation in both CB_1 and CB_2 resulted in subtle alterations in receptor affinity and signal transduction. In contrast, a tyrosine-to-isoleucine mutation in CB_1 and CB_2 led to receptors that lost ligand-binding capability (McAllister et al. 2002). Evaluation of receptor expression revealed no significant differences between the Y5.39I mutant and the WT receptor. Mutation of Y5.39(275) to A resulted in a receptor which failed to be expressed at the cell surface (Shire et al. 1999). Monte Carlo/stochastic dynamics studies suggested the hypothesis that aromaticity at position 5.39(275) in CB_1 and 5.39(190) in CB_2 is essential to maintain cannabinoid ligand WT affinity; while the CB_1 Y5.39(275)F mutant was very similar to WT, the Y5.39(275)I mutant showed pronounced topology changes in the TMH 3-4-5 region (McAllister et al. 2002).

Two conserved tryptophan residues, W4.50(158) and W4.64(172), are required for proper ligand recognition and signal transduction (Rhee et al. 2000a). W4.50 is conserved among most GPCRs, whereas W4.64 is conserved between CB_1 and CB_2

receptors. Substitutions to aromatic residues phenylalanine or tyrosine as well as to leucine and alanine were evaluated. For both tryptophan residues, the W-to-F mutant retained WT binding and signaling properties and the L and A mutations resulted in loss of ligand binding and signal transduction. In this study, expression of protein was assessed by Western analyses; however, cellular localization was not examined (Rhee et al. 2000a). W4.64 has been suggested to be an interaction site for the aminoalkylindoles and pyrazole antagonists, and in CB_1, the W4.64A mutation resulted in a receptor that did not localize to the cell surface (McAllister et al. 2002).

Absence of a conserved proline is crucial for proper function of the CB_2 receptor (Song and Feng 2002). In most GPCRs, there is a proline residue in the middle of TM5, but in the cannabinoid receptors this residue is a leucine. Substitution of L5.50(201) to proline caused a complete loss of ligand binding and function, probably due to an overall conformational change in the mutant receptor (Song and Feng 2002).

The highly conserved tyrosine in the $NP(X)_nY$ motif in TM7 also plays an important role in the CB_2 receptor's proper conformation for ligand recognition and signal transduction (Feng and Song 2001). The Y7.53(299)A mutation produced a receptor that was correctly targeted to the cell membrane, yet led to a complete loss of ligand binding and functional coupling to adenylyl cyclase. Since the location of Y299 is very close to the cytoplasmic face, it is not postulated to be directly involved in ligand binding; instead these results are probably due to conformational changes in the receptor protein (Feng and Song 2001).

7 CB_1 Receptor Activation

7.1 Constitutive Activity

Overexpression of many GPCRs leads to some degree of constitutive (agonist-independent) activity (Lefkowitz et al. 1993). Experimental evidence for constitutively active CB_1 receptors was first noted when SR141716A, initially described as a CB_1 antagonist, was found to have inverse agonist properties (Bouaboula et al. 1997). In transfected CHO cells expressing CB_1, cannabinoid agonists activated mitogen-activated protein kinase (MAPK) activity (Bouaboula et al. 1997). However, basal MAPK activity was higher in CB_1-transfected cells as compared to untransfected cells, suggesting the presence of autoactivated CB_1 receptors. SR141716A not only antagonized the agonist effect on MAPK, but also reduced basal MAPK activity in CB_1-transfected but not untransfected cells. Similarly, basal cAMP levels were reduced, and SR141716A raised basal cAMP levels in transfected cells. The EC_{50} for SR141716A was similar to its IC_{50}, suggesting that these effects are a result of direct binding to unoccupied (precoupled) CB_1 receptors and not due to the presence of endogenous ligands in the cultures. A significantly higher EC_{50} would be predicted if endogenous agonists were competing with SR141716A. Subsequent studies extended these findings to CB_1 receptor-activated guanosine-5′-*O*-(3-thiotriphosphate) (GTPγS) binding (Landsman et al. 1997) and inhibition of calcium conductance (Pan et al. 1998). Additionally, CB_1 receptors can sequester G proteins, making them unavailable to couple to other receptors (Vasquez and Lewis 1999). SR141716A is also an inverse agonist when CB_1 receptors are co-expressed with G protein-coupled potassium channels in *Xenopus* oocytes (McAllister et al. 1999).

Previously, inverse agonist effects had not been observed in cell lines possessing native CB_1 receptors (Bouaboula et al. 1995), or in primary neuronal cultures (Jung et al. 1997). However, a study in primary cultures of rat cerebellar granule neurons presented evidence for inverse agonism by SR141716A on nitric oxide synthase activity (Hillard et al. 1999). Evidence for inverse agonism was also reported in the guinea pig small intestine (Coutts et al. 2000).

Constitutively active GPCRs can arise from mutations (either naturally occurring or engineered), presumably as a result of transforming the receptor to a constitutively active state. Mutations that result in constitutive activity may provide clues to the key amino acids involved in receptor activation. Generally, constitutively active receptors are also constitutively phosphorylated and desensitized, providing support for a model where a single active state conformation is the target for phosphorylation, internalization and desensitization (Leurs et al. 1998). However, a recent study on the angiotensin II receptor and a series of studies on the CB_1 receptor suggest that GPCRs may possess several transition states, each associated with conformationally distinguishable states of receptor activation and regulation (Houston and Howlett 1998; Hsieh et al. 1999; Jin et al. 1999; Roche et al. 1999; Thomas et al. 2000).

A F3.36/W6.48 interaction is proposed to be key to the maintenance of the CB_1 inactive state (Singh et al. 2002). Previous modeling studies have suggested that a F3.36/W6.48 interaction requires a F3.36 *trans* χ1/W6.48 g+ χ1 rotameric state. SR141716A stabilizes this F3.36/W6.48 aromatic stacking interaction, while WIN55,212-2 favors a F3.36 g+ χ1/W6.48 *trans* χ1 state (Singh et al. 2002). Cannabinoid receptor activation of GIRK1/4 channels in *Xenopus* oocytes was used to assess functional characteristics of the mutant proteins (McAllister et al., 2004). Of five mutant receptors tested, only the F3.36(201)A demonstrated a limited activation profile in the presence of multiple agonists. Ligand-independent receptor activation of GIRK1/4 channels showed that the F3.36A mutant had statistically higher

Table 2. Amino acids important in signal transduction

CB_1 receptor	**CB_2 receptor**
D2.50(163/164)	D2.50(80)
F3.36(201)	R3.50(131)
L6.34(341) and A6.35(342)	Y2.51(132)
C-terminus (401–417)	Y5.58(207)
	A6.34(244)
	C313
	C320

levels of constitutive activity compared to WT CB_1. This result supports the hypothesis of a $\chi1$ rotamer "toggle" switch (W6.48 $\chi1$ g+, F3.36 $\chi1$ *trans*) → (W6.48 $\chi1$ *trans*, F3.36 $\chi1$ g+) for activation of CB_1.

7.2 Residues Involved in Activation of CB_1

Studies to date have indicated that not only are sets of different amino acids involved in the binding of several cannabinoid ligands, but that these ligands promote interactions with different G proteins (Bonhaus et al. 1998; Glass and Northup 1999; Griffin et al. 1998; Kearn et al. 1999; Mukhopadhyay et al. 2000; Selley et al. 1996; Tao et al. 1999). The different sites of ligand–receptor interaction may promote different receptor conformations, which in turn result in selective interaction with different G proteins. Evidence that different receptor conformations can promote distinct G protein interactions is provided by a study in which a mutation produced a constitutively active CB_1 receptor that coupled to G_s in preference to G_i (Abadji et al. 1999). The predominant coupling of the WT CB_1 receptor is to G_i; coupling to G_s can usually only be demonstrated in the presence of pertussis toxin, which uncouples receptors from $G_{i/o}$ proteins (Glass and Felder 1997). A swap of two adjacent residues in the carboxyl terminus of the third intracellular loop/bottom of helix 6, L6.34(341)A/A6.35(342)L, resulted in a receptor that produced minimal inhibition of adenylyl cyclase in the presence of agonist, but instead showed increased basal levels of cAMP in the absence of agonist (Abadji et al. 1999).

Using synthetic peptides derived from the CB_1 receptor, Howlett's laboratory has demonstrated that the amino terminal side of the intracellular (i3) loop can interact with G_i, leading to the inhibition of adenylyl cyclase and that the juxtamembrane portion of the C-terminus is critical for G protein activation (Howlett et al. 1998). As in many other GPCRs, the CB_1 receptor C terminal region may assume a helical structure. In fact, this helical segment is quite clear in the Rho crystal structure (Palczewski et al. 2000). Synthetic peptides derived from this region can autonomously inhibit adenylyl cyclase by regulation of G_i and G_o proteins (Mukhopadhyay et al. 1999, 2000). Residues R400, K402, and C415 have been implicated as potential sites for G protein activation (Mukhopadhyay et al. 1999). Interestingly, the analogous region of CB_2 does not activate G_i (Mukhopadhyay et al. 1999, 2000).

Residues in the C-terminus have also been shown to be important in G protein coupling and sequestration (Nie and Lewis 2001a,b). Truncation of the CB_1 receptor at residue 417 attenuates G protein coupling, and truncation at residue 400 abolishes the inhibition of calcium channels produced by CB_1 receptors expressed in superior cervical ganglia neurons (Nie and Lewis 2001a). Truncation at residue 417 also enhances constitutive activity and G protein sequestration of receptors (Nie and Lewis 2001b). These mutations did not affect trafficking of the receptor to the cell surface.

In contrast, mutation of D2.50(164) to N abolished G protein sequestration and constitutive activity without disrupting agonist activity of CB_1 receptors expressed in neurons (Nie and Lewis 2001b). The consequences of mutation of D2.50, a highly conserved residue present in most GPCRs, appear to depend on the system in which the mutant receptor is expressed. Mutation of human CB_1 D2.50(163) to glutamine or glutamate disrupted G protein coupling but allowed the receptors to retain high affinity for cannabinoid compounds when the mutant receptors were expressed in human embryonic kidney (HEK) 293 cells (Tao and Abood 1998). A subsequent study by Roche et al. (1999) found that rat CB_1 D164N expressed in AtT20 cells retained coupling to adenylyl cyclase and inhibition of calcium currents, but did not couple to GIRK channels internalized following cannabinoid exposure. Interestingly, this same disparity had previously been observed with the α-adrenergic receptor, in that transfection of D2.50N mutant receptors into fibroblasts lacked adenylyl cyclase coupling, but those expressed in AtT20 pituitary cells coupled to adenylyl cyclase (Surprenant et al. 1992). Thus, the cellular background into which the mutant receptors are introduced is also an important determinant of functional coupling. It is possible that this is due to differential localization of the transfected receptors or differential G protein expression.

8 CB_2 Receptor Activation and Constitutive Activity

8.1 Constitutive Activity

The CB_2 receptor has also been shown to be constitutively active (Bouaboula et al. 1999a). Furthermore, CB_2 receptors expressed in CHO cells also sequester G_i proteins; the CB_2 inverse agonist SR144528 inhibits basal G protein activity as well as switching off MAPK activation from receptor tyrosine kinases and the GPCR lysophosphatidic acid (LPA) receptor (Bouaboula et al. 1999a). CB_2 receptors are constitutively phosphorylated and internalized (Bouaboula et al. 1999b). Autophosphorylation as well as agonist-induced phosphorylation occurs on S352 and involves a GPCR kinase (GRK) (Bouaboula et al. 1999b).

8.2 CB_2 Receptor Activation

As with the CB_1 receptor, mutation of the highly conserved aspartate residue in the second TM domain of the CB_2 receptor, D2.50(80) to glutamine or glutamate, disrupted G protein coupling without affecting high-affinity agonist binding (Tao and Abood 1998).

The DRY motif has been shown to be important for activation of a number of GPCRs. This motif has been examined in two separate studies of the CB_2 receptor, with different results (Feng and Song 2003; Rhee et al. 2000b). Both investigations found that mutation of D3.49(130) to A resulted in loss of ligand binding and subsequent signal transduction (Feng and Song 2003; Rhee et al. 2000b). This was proposed to be due to a conformational change in the CB_2 receptor, rather than a direct effect on ligand binding, since this residue is at the cytoplasmic end of TM3. Mutation of Y2.51(132) to A resulted in a loss of signal transduction without affecting ligand recognition (Rhee et al. 2000b). However, Rhee et al. (2000a) demonstrated that mutation of R3.50(131) to A resulted in a slight reduction of signal transduction, whereas Feng and Song (2003) found no evidence for G protein coupling in the mutant receptor, including an abolition of constitutive activity in the mutant cell line. In one case, transient transfection into COS cells was employed (Rhee et al. 2000b), in the other, stable transfection into HEK 293 cells was used (Feng and Song 2003), again suggesting the cellular background plays an important role in the function of these GPCRs. Coupling to different G proteins is one explanation for the disparate results. In fact, a recent study found that 2AG induced a pertussis toxin-sensitive response, whereas CP 55,940 functional responses were unaffected by treatment with pertussis toxin; mutation of R3.50(131) to A resulted in reduction of the 2AG but not the CP 55,940-mediated responses (Alberich Jorda et al. 2004).

Mutation of A6.34(244) to glutamate resulted in a loss of ligand binding, signal transduction and constitutive activity (Feng and Song 2003). The location of this amino acid, at the bottom of helix 6, suggests that it may be important in receptor conformation. Highlighting the differences between CB_1 and CB_2 receptors, this amino acid in the CB_1 receptor was partly responsible for enhancing G protein coupling to G_s (Abadji et al. 1999).

The presence of a tyrosine residue conserved between CB_1 and CB_2,Y5.58(207), is critical for signal transduction in the CB_2 receptor (Song and Feng 2002). The Y5.58A mutant receptor retained ligand binding, albeit with an eightfold reduced affinity for [^{3}H]WIN 55,212-2, and fivefold reduction in HU-210 and anandamide binding. This residue resides at the cytoplasmic end of helix 5, an area which has been demonstrated to be involved in G protein coupling; therefore this conserved tyrosine may play a role in propagation of agonist-induced conformational changes for signal transduction (Song and Feng 2002).

Cysteine residues in the C-terminal domains have been shown to be important in functional coupling in several GPCRs. Mutation of C313 or C320 to alanine in the CB_2 receptor resulted in a mutant that retained WT ligand recognition properties but loss of functional coupling to adenylyl cyclase (Feng and Song 2001). In several other GPCRs, C-terminal cysteine mutations also led to lack of desensitization; this was not the case with the CB_2 receptor (Feng and Song 2001). These data demonstrate the importance of residues in the C-terminal domain to functional coupling in the CB_2 receptor.

9
CB_1 Receptor Polymorphisms in Addiction and Disease

The CB_1 receptor has been shown to regulate cocaine and heroin reinforcement as well as opioid dependence (De Vries et al. 2001; Ledent et al. 1999). When the CB_1 receptor was knocked out by homologous recombination, not only did the mutant mice lose responsiveness to cannabinoids, the reinforcing properties of morphine and the severity of the withdrawal syndrome were strongly reduced (Ledent et al. 1999). Several laboratories have demonstrated that CB_1 receptors regulate mesolimbic dopaminergic transmission in brain areas known to be involved in the reinforcing effects of morphine, and it has now been shown that the CB_1 receptor is critical for this μ-opioid receptor effect (Chen et al. 1990; Mascia et al. 1999; Tanda et al. 1997). In addition to increasing mesolimbic dopamine, Δ^9-THC facilitates brain stimulation reward, an animal model for abuse liability (Gardner and Lowinson 1991). Moreover, genetic variations in the response have been clearly demonstrated in three strains of rats (Lepore et al. 1996). Lewis rats showed the most pronounced Δ^9-THC-induced enhancement of brain reward functions. Sprague-Dawley rats showed an enhancement that was approximately half that seen in Lewis rats and, at the dose tested, brain reward functions in Fischer 344 rats were unaffected. A subsequent study also found a strain-specific facilitatory effect on dopamine efflux in nucleus accumbens (Chen et al. 1991). These data demonstrate that genetic variations to cannabinoid effects exist and suggest that genetic variation influences drug abuse vulnerability. Indeed, differential sensitivity to Δ^9-THC in the elevated plus-maze test of anxiety was also shown in three mouse strains (Onaivi et al. 1995). Two different doses of Δ^9-THC induced aversion to the open arms of the maze in ICR mice, but not in DBA/2 and C57BL/6 mice. Basal locomotor activity was significantly different in the three strains of mice, and may be related to differences in CB_1 receptor function (Basavarajappa and Hungund 2001).

The CB_1 receptor has been cloned and sequenced from two strains of mice, C57BL/6 (Chakrabarti et al. 1995) and 129SJ (Abood et al. 1997) as well as from NG108-15 cells (Ho and Zhao 1996). Additional mouse genomic sequence information has been deposited at NCBI. However, the additional full-length sequences are also from the 129SJ strain. Sequence analysis of the C57BL/6 CB_1 receptor cDNA (accession No. U17985), indicates three amino acid differences compared to that obtained from the 129SJ strain (genomic clones, accession No. U22948 and Abood et al. 1997) and NG108-15 (cDNA clone, accession No. U40709). One of them, T210R, is in the third TM domain, an area found to be critical for ligand recognition in the CB_1 receptor (Chin et al. 1998, 1999; Song and Bonner 1996; Tao et al. 1999). CB_1 receptor polymorphisms may underlie differential sensitivity to Δ^9-THC. In addition, a recent report showed distinct differences in CB_1 receptor binding properties in the brains of C57Bl/6 and DBA/2 mice (Hungund and Basavarajappa 2000). It is possible that naturally occurring mutations confer functional differences in CB_1 responses.

Human CB_1 receptor polymorphisms have been identified. One study found a positive association between a microsatellite polymorphism in the CB_1 gene and intravenous drug abuse (Comings et al. 1997). The initial polymorphism found was a restriction fragment length polymorphism (RFLP) in the intron preceding the coding exon of the receptor (Caenazzo et al. 1991). The CB_1 receptor gene is intronless in its coding region, but possesses an intron 5′ to the coding exon with three putative upstream exons (Abood et al. 1997; Bonner 1996). The first polymorphism in the coding exon was recently reported by Gadzicki et al. (1999).

They identified a silent mutation in T453 (G to A)—a conserved amino acid present in the C-terminal region of the CB_1 and CB_2 receptors—that was a common polymorphism in the German population. While this mutation is silent, analysis of several human sequences present in the database reveals that CB1K5 (accession No. AF107262), a full-length sequence, contains five nucleotide changes, three of which result in amino acid differences. Coincidentally, two amino acid differences are in the third TM domain, F200L and I216V. The third variant is in the fourth TM domain, V246A. A recent report by the group that submitted the sequence to the database revealed that this was a somatic mutation in an epilepsy patient; i.e., DNA obtained from his or her blood was unaltered, but DNA from the hippocampus showed the mutation (Kathmann et al. 2000). The presence of a somatic mutation rather than a polymorphism is generally indicative of the disease process in cancers [e.g. mutant p53 or APC expression in tumors but not normal tissues (Baker et al. 1989; Lamlum et al. 2000)]. CB_1 receptor polymorphisms may affect responsiveness to cannabinoids.

10 The Role of Receptor Regulation in the Development of Cannabinoid Tolerance

Cannabinoid tolerance develops in the absence of pharmacokinetic changes (Martin et al. 1976); therefore, biochemical and/or cellular changes are responsible for this adaptation. The production of tolerance can be associated with a drug's abuse potential (O'Brien 1996); therefore receptor mechanisms contributing to cannabinoid tolerance are of significant interest. One hypothesis for tolerance development is that receptors lose function during chronic agonist treatment, leading to diminished biological responses. Potential cellular mechanisms that might play important roles in tolerance include receptor desensitization, internalization, and downregulation.

Current theories for GPCR regulation predict that activated receptors are phosphorylated by GRKs and/or second messenger-activated kinases (Garcia et al. 1998; Leurs et al. 1998). β-Arrestins bind to phosphorylated receptors and sterically hinder further association of the receptor with G protein, terminating signaling. For some GPCRs, arrestins can serve as adapters to target the receptors for clathrin-mediated internalization and to promote coupling to tyrosine kinase signaling pathways (Luttrell et al. 1999). Also, in the continued presence of agonist, receptors are targeted to lysosomes for degradation (Zastrow and Kobilka 1992). It is this last event that is detected as decreased surface receptor binding.

Early studies of cannabinoid receptor downregulation at the mRNA level in conjunction with ligand binding did not detect changes in either receptor number or mRNA levels in whole brains from mice tolerant to Δ^9-THC (Abood et al. 1993). However, in mice tolerant to CP 55,940, cannabinoid receptor downregulation in cerebella is concomitant with increased levels of receptor mRNA, without alteration of the inhibitory effect of cannabinoid agonists on cAMP accumulation (Fan et al. 1996). Extensive downregulation in cerebellar membranes without any effect on receptor-G protein coupling was subsequently confirmed (Breivogel et al. 1999). Brain region specificity of receptor downregulation has also been demonstrated by several laboratories (Breivogel et al. 1999; Oviedo et al. 1993; Rodriguez-de-Fonseca et al. 1994; Romero et al. 1997). A comprehensive study examining the time course of changes in cannabinoid-stimulated [^{35}S]GTPγS binding and cannabinoid receptor binding in both rat brain sections and membranes, following daily Δ^9-THC treatments for 3, 7, 14, and 21 days, found time-dependent decreases in both [^{35}S]GTPγS and [^{3}H]WIN 55212-2 and [^{3}H]SR141716A binding in cerebellum, hippocampus, caudate-putamen, and globus pallidus, with regional differences in the rate and magnitude of downregulation and desensitization (Breivogel et al. 1999). In a parallel study, the time course and regional specificity of expression of the CB_1 receptor was examined (Zhuang et al. 1998). They found that CB_1 mRNA levels were increased above vehicle control animals at 7 days of treatment (Fan et al. 1996). However, another laboratory found some regions which showed no changes in receptor binding, [^{35}S]GTPγS activation, or mRNA levels following chronic cannabinoid administration (Romero et al. 1998a,b).

Several recent studies in transfected cell systems have implicated regions of the CB_1 receptor involved in receptor regulation following chronic agonist exposure. Rapid internalization of CB_1 receptors was observed after agonist exposure (Hsieh et al. 1999; Rinaldi-Carmona et al. 1998b). In contrast, chronic treatment of cells with the inverse agonist SR141716A caused upregulation of cell surface receptors (Rinaldi-Carmona et al. 1998b). As in other GPCRs, the C-terminal domain is critical for receptor internalization; truncation of the terminal 14 amino acids eliminates receptor internalization (Hsieh et al. 1999). Truncation of the C-terminus at residue 418 abolished desensitization, as did deletion of residues 418–439 (Jin et al. 1999).

On the other hand, phosphorylation of S426 and S430 (tail region) or S317 (third intracellular loop) resulted in CB_1 receptor desensitization; however, these sites had no influence on internalization (Garcia et al. 1998; Jin et al. 1999). While receptor internalization was not affected when G protein signaling was disrupted by treatment with pertussis toxin, a mutation of the highly conserved aspartate residue in the second TM domain in which G protein coupling is altered did block CB_1 receptor internalization (Roche et al. 1999).

Both in vivo and in vitro, different cannabinoid compounds can produce various degrees of tolerance and desensitization, suggesting their actions at cannabinoid receptors may not be identical (Dill and Howlett 1988; Fan et al. 1994). In a comparison of three cannabinoid agonists, the most potent compound (CP 55,940)

Table 3. Amino acids important for desensitization and internalization

Desensitization	Internalization
S317 in CB_1	D2.50(164) in CB_1
S426 in CB_1	C-terminus 458–472 in CB_1
S430 in CB_1	
S352 in CB_2	
C-terminus 418–439	

produced the most tolerance in vivo (Fan et al. 1994). In most in vitro studies, a single cannabinoid agonist has been used; so the cellular basis for this differential tolerance has yet to be determined.

The CB_2 receptor is also desensitized and internalized following agonist treatment in vitro (Bouaboula et al. 1999b). These studies, conducted in CB_2-transfected CHO cells, demonstrated that phosphorylation at S352 appears to play a key role in the loss of responsiveness of the CB_2 receptor. Furthermore, SR144528 could regenerate the desensitized CB_2 receptors by activating a phosphatase that dephosphorylated the receptor. Hence the pharmacological properties and phosphorylation state of the CB_2 receptor can be regulated by both agonists and antagonists.

11 Physiological Receptor Regulation and Disease

Early studies investigated cannabinoid receptor mRNA levels using in situ hybridization (Mailleux and Vanderhaeghen 1993; Mailleux and Vanderhaeghen 1993; Mailleux and Vanderhaeghen 1994). Following adrenalectomy, CB_1 mRNA levels in the striatum increased 50% as compared to control rats (Mailleux and Vanderhaeghen 1993). This increase could be counteracted by dexamethasone treatment, suggesting glucocorticoid downregulation of cannabinoid receptor gene expression in the striatum. A negative dopaminergic influence on CB_1 gene expression has been suggested by studies in which a unilateral 6-hydroxydopamine lesion was associated with 45% increase in mRNA levels in the ipsilateral side; furthermore, treatment with dopamine receptor antagonists mimicked the effect (Mailleux and Vanderhaeghen 1993). Previous experiments had documented the disappearance of CP 55,940 binding following an ibotenic acid lesion of the striatum, but not following a 6-hydroxydopamine lesion, indicating that cannabinoid receptors are not co-localized with dopamine-containing neurons but are probably on axonal terminals of striatal intrinsic neurons (Herkenham et al. 1991). Glutamatergic regulation of cannabinoid receptor mRNA levels in the striatum has also been reported (Mailleux and Vanderhaeghen 1994). Unilateral cerebral decortication resulted in 30% decrease in mRNA levels, and treatment with the *N*-methyl-D-aspartate (NMDA) receptor antagonist MK-801 resulted in an approximate 52% decrease, as compared to control. These data suggest an NMDA receptor-mediated upregulation of cannabinoid receptor mRNA levels. The mechanisms by which these changes occur are not known.

CB_1 receptors are drastically reduced in substantia nigra and lateral globus pallidus in Huntington's disease (Glass et al. 1993; Richfield and Herkenham 1994). The CB_1 receptor agonist nabilone significantly reduced L-dopa-induced dyskinesia in an animal model of Parkinson's disease as well as in Parkinson's patients (Sieradzan et al. 2001; Fox et al. 2002). CB_1 receptor knockout mice displayed increased neuropeptide expression in striatal output pathways and were severely hypoactive in an exploratory test, although their motor coordination was unaltered, suggesting these receptors may be important for initiation of movement (Steiner et al. 1998).

The first report of alteration of CB_2 receptor expression was in the original cloning paper; CB_2 was isolated as a result of its differential expression following treatment with dimethylformamide to produce granulocyte differentiation in the human promyelocytic leukemia line HL60 (Munro et al. 1993). CB_2 transcripts are also elevated when HL60 cells are induced to differentiate into macrophages by tetradecanoylphorbol acetate treatment (Munro et al. 1993). The chromosomal location of CB_2 is in a common virus integration site, and it is overexpressed in retrovirally transformed mouse myeloid leukemias (Valk et al. 1997). Furthermore, CB_2 is aberrantly expressed in several human myeloid cell lines and primary acute myeloid leukemia samples, whereas normal bone marrow precursor cells do not express CB_2 (Alberich Jorda et al. 2004).

Evidence for CB_2 receptor expression has not been found in normal human CNS; however, CB_2 has been found in Alzheimer's brains (Benito et al. 2003). CB_2 immunoreactivity was selectively expressed in microglia associated with neuritic plaques, suggesting that modulation of their activity may have therapeutic implications (Benito et al. 2003).

12
Evidence for Additional Cannabinoid Receptor Subtypes

Not all of the effects of anandamide are mediated through the currently defined cannabinoid receptors. Anandamide inhibits gap-junction conductance and intercellular signaling in striatal astrocytes via a CB-receptor independent mechanism, since the cannabimimetic agents CP 55,940 and WIN 55,212-2 did not mimic the effect of anandamide, nor did the CB_1 receptor antagonist SR141716A reverse anandamide's actions (Venance et al. 1995). Additional fatty acid ethanolamides have been isolated, as well as a 2-arachidonoyl glycerol with cannabimimetic properties, suggesting the existence of a family of endogenous cannabinoids that may interact with additional cannabinoid receptor subtypes (Mechoulam et al. 1995; Mechoulam et al. 1994).

CB_1 receptor knockout mice have now been constructed in four laboratories (Ibrahim et al. 2003; Ledent et al. 1999; Marsicano et al. 2002; Zimmer et al. 1999). In one strain, although CB_1 receptor knockout mice lost responsiveness to most cannabinoids, Δ^9-THC still produced antinociception in the tail-flick test of analgesia (Zimmer et al. 1999). Further characterization of this non-CB_1 Δ^9-THC response suggests the presence of a novel cannabinoid receptor/ion channel in the pain pathway (Zygmunt et al. 2002).

Anandamide produces the full range of behavioral effects (antinociception, catalepsy, and impaired locomotor activity) in CB_1 receptor knockout mice (Di Marzo et al. 2000). Furthermore, anandamide-stimulated GTPγS activity can be elicited in brain membranes from these mice (Breivogel et al. 2001). These effects were not sensitive to inhibition by SR141716A. Interestingly, of all cannabinoid ligands tested, only WIN 55,212-2 elicited GTPγS activity in CB_1 knockout mice. This same phenomenon has also been demonstrated in a second strain of CB_1 receptor knockout mice (Monory et al. 2002).

A cannabinoid receptor subtype has been found in the hippocampus that is responsive to WIN 55,212-2 and CP 55,940 and blocked by capsazepine (Hajos et al. 2001). These receptors are found on excitatory (pyramidal) axon terminals and have been shown to suppress glutamate release in CB_1 receptor knockout animals.

An "abnormal cannabidiol receptor" has also been characterized. Cannabinoids, including anandamide, elicit cardiovascular effects via peripherally located CB_1 receptors (Ishac et al. 1996; Jarai et al. 1999; Wagner et al. 1999). Abnormal cannabidiol (abn-cbd), a neurobehaviorally inactive cannabinoid that does not bind to CB_1 receptors, caused hypotension and mesenteric vasodilation in WT mice and in mice lacking CB_1 receptors or both CB_1 and CB_2 receptors (Jarai et al. 1999). In contrast to the studies described above, these cardiovascular and endothelial effects were SR141716A-sensitive. A stable analog of AEA (methanandamide) also produced SR141716A-sensitive hypotension in CB_1/CB_2 knockout mice. These effects were not due to activation of vanilloid receptors, which also interact with AEA (Zygmunt et al. 1999). A selective antagonist, O-1918, has recently been developed; it inhibits the vasorelaxant effects of abn-cbd and anandamide (Offertaler et al. 2003).

Signal transduction pathways for the abn-cbd receptor have been studied in human umbilical endothelial cells (HUVEC) (Offertaler et al. 2003). Abn-cbd induces phosphorylation of extracellular signal-regulated kinase (ERK) and protein kinase B/Akt via a PI3 kinase-dependent pertussis toxin-sensitive pathway; these effects were blocked by O-1918 (Offertaler et al. 2003). The abn-cbd receptor subtype also appears to be present in microglia (Walter et al. 2003). Anandamide and 2AG triggered migration in BV-2 cells, a microglial cell line; their effects were blocked with O-1918. 2AG also induced phosphorylation of ERK1/2 in BV-2 cells (Walter et al. 2003). These data suggest a common signaling pathway for the abn-cbd receptor in endothelial cells and microglia.

Palmitoylethanolamide has been suggested as a possible endogenous ligand at the CB_2 receptor (Facci et al. 1995). However, it has a low affinity for the cloned human CB_2 receptor (Showalter et al. 1996). This difference suggested that there may be species differences with the CB_2 receptor, as have been found with other GPCRs, but the cloned rat and mouse CB_2 receptors also showed low affinity for palmitoylethanolamide (Griffin et al. 2000). Palmitoylethanolamide has recently been shown produce to a G protein-mediated response in microglial cells that was not affected by CB_1, CB_2, or abn-cbd antagonists, suggesting it acts via its own GPCR (Franklin et al. 2003).

In summary, there is compelling evidence for the existence of additional cannabinoid receptor subtypes. Proof of their existence awaits molecular cloning and expression studies.

13
Conclusion

It is apparent from the growing number of mutagenesis investigations, synthesis of CB_1- and CB_2-selective compounds, and discovery of multiple endogenous agonists, that there is a complex molecular architecture of the cannabinoid receptors. This arrangement allows for a single receptor to recognize multiple classes of compounds and produce an array of distinct downstream effects. Natural polymorphisms and alternative splice variants may also contribute to the pharmacological diversity of the cannabinoid receptors. As our knowledge of the distinct differences grows, we may be able to target select receptor conformations and their corresponding pharmacological responses. Importantly, the basic biology of the endocannabinoid system will continue to be revealed by ongoing investigations.

References

Abadji V, Lucas-Lenard J, Chin C, Kendall D (1999) Involvement of the carboxyl terminus of the third intracellular loop of the cannabinoid CB1 receptor in constitutive activation of Gs. J Neurochem 72:2032–2038

Abood ME, Sauss C, Fan F, Tilton CL, Martin BR (1993) Development of behavioral tolerance to Δ9-THC without alteration of cannabinoid receptor binding or mRNA levels in whole brain. Pharmacol Biochem Behav 46:575–579

Abood ME, Ditto KA, Noel MA, Showalter VM, Tao Q (1997) Isolation and expression of mouse CB1 cannabinoid receptor gene: comparison of binding properties with those of native CB1 receptors in mouse brain and N18TG2 neuroblastoma cells. Biochem Pharmacol 53:207–214

Alberich Jorda M, Rayman N, Tas M, Verbakel SE, Battista N, Van Lom K, Lowenberg B, Maccarrone M, Delwel R (2004) The peripheral cannabinoid receptor Cb2, frequently expressed on AML blasts, either induces a neutrophilic differentiation block or confers abnormal migration properties in a ligand-dependent manner. Blood

Baker SJ, Fearon ER, Nigro JM, Hamilton SR, Preisinger AC, Jessup JM, vanTuinen P, Ledbetter DH, Barker DF, Nakamura Y, et al (1989) Chromosome 17 deletions and p53 gene mutations in colorectal carcinomas. Science 244:217–221

Ballesteros JA, Weinstein H (1995) Integrated methods for the construction of three dimensional models and computational probing of structure function relations in G protein-coupled receptors. In: Conn PM, Sealfon SM (eds) Methods in Neuroscience. Academic Press, San Diego, pp 366–428

Basavarajappa BS, Hungund BL (2001) Cannabinoid receptor agonist-stimulated [35S]guanosine triphosphate γS binding in the brain of C57BL/6 and DBA/2 mice. J Neurosci Res 64:429–446

Benito C, Nunez E, Tolon RM, Carrier EJ, Rabano A, Hillard CJ, Romero J (2003) Cannabinoid CB2 receptors and fatty acid amide hydrolase are selectively overexpressed in neuritic plaque-associated glia in Alzheimer's disease brains. J Neurosci 23:11136–11141

Bonhaus D, Chang L, Kwan J, Martin G (1998) Dual activation and inhibition of adenylyl cyclase by cannabinoid receptor agonists: evidence for agonist-specific trafficking of intracellular responses. J Pharmacol Exp Ther 287:884–888

Bonner T (1996) Molecular biology of cannabinoid receptors. J Neuroimmunol 69:15–23

Bouaboula M, Bourrie B, Rinaldi-Carmona M, Shire D, Fur GL, Casellas P (1995) Stimulation of Cannabinoid Receptor CB1 induces krox-24 expression in human astrocytoma cells. J Biol Chem 270:13973–13980

Bouaboula M, Perrachon S, Milligan L, Canat X, Rinaldi-Carmona M, Portier M, Barth F, Calandra B, Pecceu F, Lupker J, Maffrand J-P, LeFur G, Casellas P (1997) A selective inverse agonist for central cannabinoid receptor inhibits mitogen-activated protein kinase activation stimulated by insulin or insulin-like growth factor 1. J Biol Chem 272:22330–22339

Bouaboula M, Desnoyer N, Carayon P, Combes T, Casellas P (1999a) Gi protein modulation induced by a selective inverse agonist for the peripheral cannabinoid receptor CB2: implication for intracellular signalization cross-regulation. Mol Pharmacol 55:473–480

Bouaboula M, Dussossoy D, Casellas P (1999b) Regulation of peripheral cannabinoid receptor CB2 phosphorylation by the inverse agonist SR 144528. Implications for receptor biological responses. J Biol Chem 274:20397–20405

Breivogel C, Childers S, Deadwyler S, Hampson R, Vogt L, Sim-Selley L (1999) Chronic delta9-tetrahydrocannabinol treatment produces a time-dependent loss of cannabinoid receptors and cannabinoid receptor-activated G proteins in rat brain. J Neurochem 73:2447–2459

Breivogel CS, Griffin G, Di Marzo V, Martin BR (2001) Evidence for a new G protein-coupled cannabinoid receptor in mouse brain. Mol Pharmacol 60:155–163

Brown SM, Wager-Miller J, Mackie K (2002) Cloning and molecular characterization of the rat CB2 cannabinoid receptor. Biochim Biophys Acta 1576:255–264

Buckley NE, McCoy KL, Mezey E, Bonner T, Zimmer A, Felder CC, Glass M (2000) Immunomodulation by cannabinoids is absent in mice deficient for the cannabinoid CB(2) receptor. Eur J Pharmacol 396:141–149

Burley S, Petsko G (1985) Aromatic-aromatic interaction: a mechanism of protein structure stabilization. Science 229:23–28

Caenazzo L, Hoehe M, Hsieh W, Berrettini W, Bonner T, Gershon E (1991) HindIII identifies a two allele DNA polymorphism of the human cannabinoid receptor gene (CNR). Nucleic Acids Res 19:4798

Chakrabarti A, Onaivi ES, Chaudhuri G (1995) Cloning and sequencing of a cDNA encoding the mouse brain-type cannabinoid receptor protein. DNA Seq 5:385–388

Chen J, Paredes W, Li J, Smith D, Lowinson J, Gardner E (1990) Ð9-Tetrahydrocannabinol produces naloxone-blockable enhancement of presynaptic basal dopamine efflux in nucleus accumbens of conscious, freely-moving rats as measured by intracerebral microdialysis. Psychopharmacology (Berl) 102:156–162

Chen J, Paredes W, Lowinson J, Gardner E (1991) Strain-specific facilitation of dopamine efflux by delta 9-tetrahydrocannabinol in the nucleus accumbens of rat: an in vivo microdialysis study. Neurosci Lett 129:136–180

Chin C, Abadji V, Lucas-Lenard J, Kendall D (1998) Ligand binding and modulation of cyclic AMP levels depends on the chemical nature of residue 192 of the human cannabinoid receptor 1. J Neurochem 70:366–373

Chin C, Murphy J, Huffman J, Kendall D (1999) The third transmembrane helix of the cannabinoid receptor plays a role in the selectivity of aminoalkylindoles for CB2, peripheral cannabinoid receptor. J Pharmacol Exp Ther 291:837–844

Comings D, Muhleman D, Gade R, Johnson P, Verde R, Saucier G, MacMurray J (1997) Cannabinoid receptor gene (CNR1): association with i.v. drug use. Mol Psychiatry 2:161–168

Cottone E, Salio C, Conrath M, Franzoni MF (2003) Xenopus laevis CB1 cannabinoid receptor: molecular cloning and mRNA distribution in the central nervous system. J Comp Neurol 464:487–496

Coutts A, Brewster N, Ingram T, Razdan R, Pertwee R (2000) Comparison of novel cannabinoid partial agonists and SR141716A in the guinea-pig small intestine. Br J Pharmacol 129:645–652

De Vries TJ, Shaham Y, Homberg JR, Crombag H, Schuurman K, Dieben J, Vanderschuren LJ, Schoffelmeer AN (2001) A cannabinoid mechanism in relapse to cocaine seeking. Nat Med 7:1151–1154

Deutsch DG, Chin SA (1993) Enzymatic synthesis and degradation of anandamide, a cannabinoid receptor agonist. Biochem Pharmacol 46:791–796

Devane WA, Hanus L, Breuer A, Pertwee RG, Stevenson LA, Griffin G, Gibson D, Mandelbaum A, Etinger A, Mechoulam R (1992) Isolation and structure of a brain constituent that binds to the cannabinoid receptor. Science 258:1946–1949

Di Marzo V, Fontana A, Cadas H, Schinelli S, Cimino G, Schwartz J-C, Piomelli D (1994) Formation and inactivation of endogenous cannabinoid anandamide in central neurons. Nature 372:686–691

Di Marzo V, Breivogel CS, Tao Q, Bridgen DT, Razdan RK, Zimmer AM, Zimmer A, Martin BR (2000) Levels, metabolism, and pharmacological activity of anandamide in CB(1) cannabinoid receptor knockout mice: evidence for non-CB(1), non-CB(2) receptor-mediated actions of anandamide in mouse brain. J Neurochem 75:2434–2444

Di Marzo V, Melck D, Orlando P, Bisogno T, Zagoory O, Bifulco M, Vogel Z, De Petrocellis L (2001) Palmitoylethanolamide inhibits the expression of fatty acid amide hydrolase and enhances the anti-proliferative effect of anandamide in human breast cancer cells. Biochem J 358:249–255

Di Marzo V, De Petrocellis L, Fezza F, Ligresti A, Bisogno T (2002) Anandamide receptors. Prostaglandins Leukot Essent Fatty Acids 66:377–391

Dill JA, Howlett AC (1988) Regulation of adenylate cyclase by chronic exposure to cannabimimetic drugs. J Pharmacol Exp Ther 244:1157–1163

Eissenstat MA, Bell MR, D'Ambra TE, Alexander EJ, Daum SJ, Ackerman JH, Gruett MD, Kumar V, Estep KG, Olefirowicz EM, et al (1995) Aminoalkylindoles: structure-activity relationships of novel cannabinoid mimetics. J Med Chem 38:3094–3105

Elphick MR (2002) Evolution of cannabinoid receptors in vertebrates: identification of a CB(2) gene in the puffer fish Fugu rubripes. Biol Bull 202:104–107

Elphick MR, Satou Y, Satoh N (2003) The invertebrate ancestry of endocannabinoid signalling: an orthologue of vertebrate cannabinoid receptors in the urochordate Ciona intestinalis. Gene 302:95–101

Facci L, Toso RD, Romanello S, Buriani A, Skaper SD, Leon A (1995) Mast cells express a peripheral cannabinoid receptor with differential sensitivity to anandamide and palmitoylethanolamide. Proc Natl Acad Sci USA 92:3376–3380

Fan F, Compton DR, Ward S, Melvin L, Martin BR (1994) Development of cross-tolerance between Δ9-THC, CP 55,940 and WIN 55,212. J Pharmacol Exp Ther 271:1383–1390

Fan F, Tao Q, Abood M, Martin BR (1996) Cannabinoid receptor down-regulation without alteration of the inhibitory effect of CP 55,940 on adenylyl cyclase in the cerebellum of CP 55,940-tolerant mice. Brain Res 706:13–20

Felder CC, Joyce KE, Briley EM, Mansouri J, Mackie K, Blond O, Lai Y, Ma AL, Mitchell RL (1995) Comparison of the pharmacology and signal transduction of the human cannabinoid CB1 and CB2 receptors. Mol Pharmacol 48:443–450

Feng W, Song ZH (2001) Functional roles of the tyrosine within the NP(X)(n)Y motif and the cysteines in the C-terminal juxtamembrane region of the CB2 cannabinoid receptor. FEBS Lett 501:166–170

Feng W, Song ZH (2003) Effects of D3.49A, R3.50A, and A6.34E mutations on ligand binding and activation of the cannabinoid-2 (CB2) receptor. Biochem Pharmacol 65:1077–1085

Fox SH, Henry B, Hill M, Crossman A, Brotchie J (2002) Stimulation of cannabinoid receptors reduces levodopa-induced dyskinesia in the MPTP-lesioned nonhuman primate model of Parkinson's disease. Mov Disord 17:1180–1187

Franklin A, Parmentier-Batteur S, Walter L, Greenberg DA, Stella N (2003) Palmitoylethanolamide increases after focal cerebral ischemia and potentiates microglial cell motility. J Neurosci 23:7767–7775

Fride E, Mechoulam R (1993) Pharmacological activity of the cannabinoid receptor agonist, anandamide, a brain constituent. Eur J Pharmacol 231:313–314

Gadzicki D, Muller-Vahl K, Stuhrmann M (1999) A frequent polymorphism in the coding exon of the human cannabinoid receptor (CNR1) gene. Mol Cell Probes 13:321–323

Galiegue S, Mary S, Marchand J, Dussossoy D, Carriere D, Carayon P, Bouaboula M, Shire D, LeFur G, Casellas P (1995) Expression of central and peripheral cannabinoid receptors in human immune tissues and leukocyte subpopulations. Eur J Biochem 232:54–61

Garcia DE, Brown S, Hille B, Mackie K (1998) Protein kinase C disrupts cannabinoid actions by phosphorylation of the CB1 cannabinoid receptor. J Neurosci 18:2834–2841

Gardner E, Lowinson J (1991) Marijuana's Interaction with Brain Reward Systems: Update 1991. Pharmacol Biochem Behav 40:571–580
Gareau Y, Dufresne C, Gallant M, Rochette C, Sawyer N, Slipetz DM, Tremblay N, Weech PK, Metters KM, Labelle M (1996) Structure activity relationships of tetrahydrocannabinol analogues on human cannabinoid receptors. Bioorg Med Chem Lett 6:189–194
Gerard CM, Mollereau C, Vassart G, Parmentier M (1991) Molecular cloning of a human cannabinoid receptor which is also expressed in testis. Biochem J 279:129–134
Glass M, Felder CC (1997) Concurrent stimulation of cannabinoid CB1 and dopamine D2 receptors augments cAMP accumulation in striatal neurons: evidence for a Gs linkage to the CB1 receptor. J Neurosci 17:5327–5333
Glass M, Northup J (1999) Agonist selective regulation of G proteins by cannabinoid CB(1) and CB(2) receptors. Mol Pharmacol 56:1362–1369
Glass M, Faull R, Dragunow M (1993) Loss of Cannabinoid Receptors in the Substantia Nigra in Huntington's Disease. Neuroscience 56:523–527
Gouldson P, Calandra B, Legoux P, Kerneis A, Rinaldi-Carmona M, Barth F, Le Fur G, Ferrara P, Shire D (2000) Mutational analysis and molecular modelling of the antagonist SR 144528 binding site on the human cannabinoid CB(2) receptor. Eur J Pharmacol 401:17–25
Griffin G, Wray E, Tao Q, McAllister S, Rorrer W, Aung M, Martin B, Abood M (1999) Evaluation of the cannabinoid CB2 receptor-selective antagonist, SR144528: further evidence for cannabinoid CB2 receptor absence in the rat central nervous system. Eur J Pharmacol 377:117–125
Griffin G, Tao Q, Abood M (2000) Cloning and pharmacological characterization of the Rat CB2 cannabinoid receptor. J Pharmacol Exp Ther 292
Griffin GR, Atkinson PJ, Showalter VM, Martin BR, Abood ME (1998) Evaluation of cannabinoid receptor agonists and antagonists using the guanosine-5'-O-(3-[35S]thio)-triphosphate binding assay in rat cerebellar membranes. J Pharmacol Exp Ther 285:553–560
Hajos N, Ledent C, Freund TF (2001) Novel cannabinoid-sensitive receptor mediates inhibition of glutamatergic synaptic transmission in the hippocampus. Neuroscience 106:1–4
Hanus L, Gopher A, Almog S, Mechoulam R (1993) Two new unsaturated fatty acid ethanolamides in brain that bind to the cannabinoid receptor. J Med Chem 36:3032–3034
Hanus L, Breuer A, Tchilibon S, Shiloah S, Goldenberg D, Horowitz M, Pertwee RG, Ross RA, Mechoulam R, Fride E (1999) HU-308: a specific agonist for CB(2), a peripheral cannabinoid receptor. Proc Natl Acad Sci USA 96:14228–14233
Hanus L, Abu-Lafi S, Fride E, Breuer A, Vogel Z, Shalev DE, Kustanovich I, Mechoulam R (2001) 2-arachidonyl glyceryl ether, an endogenous agonist of the cannabinoid CB1 receptor. Proc Natl Acad Sci USA 98:3662–3665
Hillard C, Muthian S, Kearn C (1999) Effects of CB(1) cannabinoid receptor activation on cerebellar granule cell nitric oxide synthase activity. FEBS Lett 459:277–281
Hilliard CJ, Campbell WB (1997) Biochemistry and pharmacology of arachidonylethanolamide, a putative endogenous cannabinoid. J Lipid Res 38:2383–2398
Ho BY, Zhao J (1996) Determination of the cannabinoid receptors in mouse x rat hybridoma NG108–15 cells and rat GH4C1 cells. Neurosci Lett 212:123–126
Houston DB, Howlett AC (1998) Differential receptor-G-protein coupling evoked by dissimilar cannabinoid receptor agonists. Cell Signal 10:667–674
Howlett AC (1995) Pharmacology of cannabinoid receptors. Annu Rev Pharmacol Toxicol 35:607–634
Howlett A, Song C, Berglund B, Wilken G, Pigg J (1998) Characterization of CB1 cannabinoid receptors using receptor peptide fragments and site-directed antibodies. Mol Pharmacol 53:504–510
Howlett AC, Johnson MR, Melvin LS, Milne GM (1988) Nonclassical cannabinoid analgetics inhibit adenylate cyclase: development of a cannabinoid receptor model. Mol Pharmacol 33:297–302

Hsieh C, Brown S, Derleth C, Mackie K (1999) Internalization and recycling of the CB1 cannabinoid receptor. J Neurochem 73:493–501

Huang SM, Bisogno T, Trevisani M, Al-Hayani A, De Petrocellis L, Fezza F, Tognetto M, Petros TJ, Krey JF, Chu CJ, Miller JD, Davies SN, Geppetti P, Walker JM, Di Marzo V (2002) An endogenous capsaicin-like substance with high potency at recombinant and native vanilloid VR1 receptors. Proc Natl Acad Sci USA 99:8400–8405

Huffman JW (1999) Cannabimimetic indoles, pyrroles and indenes. Curr Med Chem 6:705–720

Huffman JW, Dai D, Martin BR, Compton DR (1994) Design, synthesis and pharmacology of cannabimimetic indoles. Bioorg Med Chem Lett 4

Huffman JW, Yu S, Showalter V, Abood ME, Wiley JL, Compton DR, Martin BR, Bramblett RD, Reggio PH (1996) Synthesis and pharmacology of a very potent cannabinoid lacking a phenolic hydroxyl with high affinity for the CB2 receptor. J Med Chem 39:3875–3877

Huffman JW, Liddle J, Yu S, Aung MM, Abood ME, Wiley JL, Martin BR (1999) 3-(1',1'-Dimethylbutyl)-1-deoxy-delta8-THC and related compounds: synthesis of selective ligands for the CB2 receptor. Bioorg Med Chem 7:2905–2914

Hungund BL, Basavarajappa BS (2000) Distinct differences in the cannabinoid receptor binding in the brain of C57BL/6 and DBA/2 mice, selected for their differences in voluntary ethanol consumption. J Neurosci Res 60:122–128

Hurst DP, Lynch DL, Barnett-Norris J, Hyatt SM, Seltzman HH, Zhong M, Song ZH, Nie J, Lewis D, Reggio PH (2002) N-(Piperidin-1-yl)-5-(4-chlorophenyl)-1-(2,4-dichlorophenyl)-4-methyl-1H-p yrazole-3-carboxamide (SR141716A) interaction with LYS 3.28(192) is crucial for its inverse agonism at the cannabinoid CB1 receptor. Mol Pharmacol 62:1274–1287

Ibrahim MM, Deng H, Zvonok A, Cockayne DA, Kwan J, Mata HP, Vanderah TW, Lai J, Porreca F, Makriyannis A, Malan TP Jr (2003) Activation of CB2 cannabinoid receptors by AM1241 inhibits experimental neuropathic pain: pain inhibition by receptors not present in the CNS. Proc Natl Acad Sci USA 100:10529–10533

Ishac EJN, Jiang L, Lake KD, Varga K, Abood ME, Kunos G (1996) Inhibition of exocytotic noradrenaline release by presynaptic cannabinoid CB1 receptors on peripheral sympathetic nerves. Br J Pharmacol 118:2023–2028

Jarai Z, Wagner J, Varga K, Lake K, Compton D, Martin B, Zimmer A, Bonner T, Buckley N, Mezey E, Razdan R, Zimmer A, Kunos G (1999) Cannabinoid-induced mesenteric vasodilation through an endothelial site distinct from CB1 or CB2 receptors. Proc Natl Acad Sci USA 96:14136–14141

Jin W, Brown S, Roche J, Hsieh C, Celver J, Kovoor A, Chavkin C, Mackie K (1999) Distinct domains of the CB1 cannabinoid receptor mediate desensitization and internalization. J Neurosci 19:3773–3780

Jordt SE, Bautista DM, Chuang HH, McKemy DD, Zygmunt PM, Hogestatt ED, Meng ID, Julius D (2004) Mustard oils and cannabinoids excite sensory nerve fibres through the TRP channel ANKTM1. Nature 427:260–265

Jung M, Calassi R, Rinaldi-Carmona M, Chardenot P, LeFur G, Soubrie P, Oury-Donat F (1997) Characterization of CB1 receptors on rat neuronal cell cultures: binding and functional studies using the selective receptor antagonist SR 141716A. J Neurochem 68:402–409

Kaminski NE, Abood ME, Kessler FK, Martin BR, Schatz AR (1992) Identification of a functionally relevant cannabinoid receptor on mouse spleen cells that is involved in cannabinoid-mediated immune modulation. Mol Pharmacol 42:736–742

Kathmann M, Haug K, Heils A, Nothen M, Schlicker E (2000) Exchange of three amino acids in the cannabinoid CB1 receptor (CNR1) of an epilepsy patient 2000 Symposium on the Cannabinoids. International Cannabinoid Research Society, Burlington, Vermont

Kearn C, Greenberg M, DiCamelli R, Kurzawa K, Hillard C (1999) Relationships between ligand affinities for the cerebellar cannabinoid receptor CB1 and the induction of GDP/GTP exchange. J Neurochem 72:2379–2387

Kumar V, Alexander MD, Bell MR, Eissenstat MA, Casiano FM, Chippari SM, Haycock DA, Lutinger DA, Kuster JE, Miller MS, Stevenson JI, Ward SJ (1995) Morpholinoalkylindenes as antinociceptive agents: novel cannabinoid receptor agonists. Bioorg Med Chem Lett 5:381–386

Lambert D, DiPaolo F, Sonveaux P, Kanyonyo M, Govaerts S, Hermans E, Bueb J, Delzenne N, Tschirhart E (1999) Analogues and homologues of N-palmitoylethanolamide, a putative endogenous CB(2) cannabinoid, as potential ligands for the cannabinoid receptors. Biochim Biophys Acta 1440:266–274

Lamlum H, Papadopoulou A, Ilyas M, Rowan A, Gillet C, Hanby A, Talbot I, Bodmer W, Tomlinson I (2000) APC mutations are sufficient for the growth of early colorectal adenomas. Proc Natl Acad Sci USA 97:2225–2228

Landsman RS, Burkey TH, Consroe P, Roeske WR, Yamamura HI (1997) SR141716A is an inverse agonist at the human cannabinoid CB1 receptor. Eur J Pharmacol 334:R1–R2

Ledent C, Valverde O, Cossu G, Petitet F, Aubert J, Beslot F, Bohme G, Imperato A, Pedrazzini T, Roques B, Vassart G, Fratta W, Parmentier M (1999) Unresponsiveness to cannabinoids and reduced addictive effects of opiates in CB1 receptor knockout mice. Science 283:401–404

Lefkowitz RJ, Cotecchia S, Samama P, Costa T (1993) Constitutive activity of receptors coupled to guanine nucleotide regulatory proteins. Trends Pharmacol Sci 14:303–307

Lepore M, Liu X, Savage V, Matalon D, Gardner E (1996) Genetic differences in delta 9-tetrahydrocannabinol-induced facilitation of brain stimulation reward as measured by a rate-frequency curve-shift electrical brain stimulation paradigm in three different rat strains. Life Sci 58:PL365–PL372

Leurs R, Smit M, Alewijnse A, Timmerman H (1998) Agonist-independent regulation of constitutively active G-protein-coupled receptors. Trends Biochem Sci 23:418–422

Luttrell L, Ferguson S, Daaka Y, Miller W, Maudsley S, Rocca GD, Lin F, Kawakatsu H, Owada K, Luttrell D, Caron M, Lefkowitz R (1999) Beta-arrestin-dependent formation of beta2 adrenergic receptor-Src protein kinase complexes. Science 283:655–661

Mackie K, Hille B (1992) Cannabinoids inhibit N-type calcium channels in neuroblastoma-glioma cells. Proc Natl Acad Sci USA 89:3825–3829

Marsicano G, Wotjak CT, Azad SC, Bisogno T, Rammes G, Cascio MG, Hermann H, Tang J, Hofmann C, Zieglgansberger W, Di Marzo V, Lutz B (2002) The endogenous cannabinoid system controls extinction of aversive memories. Nature 418:530–534

Martin BR, Dewey WL, Harris LS, Beckner JS (1976) 3H-Δ9-tetrahydrocannabinol tissue and subcellular distribution in the central nervous system and tissue distribution in peripheral organs of tolerant and nontolerant dogs. J Pharmacol Exp Ther 196:128–144

Mascia M, Obinu M, Ledent C, Parmentier M, Bohme G, Imperato A, Fratta W (1999) Lack of morphine-induced dopamine release in the nucleus accumbens of cannabinoid CB(1) receptor knockout mice. Eur J Pharmacol 383:R1–R2

Matsuda LA, Lolait SJ, Brownstein MJ, Young AC, Bonner TI (1990) Structure of a cannabinoid receptor and functional expression of the cloned cDNA. Nature 346:561–564

McAllister S, Griffin G, Satin L, Abood M (1999) Cannabinoid receptors can activate and inhibit G protein-coupled inwardly rectifying potassium channels in a Xenopus oocyte expression system. J Pharmacol Exp Ther 291:618–626

McAllister SD, Tao Q, Barnett-Norris J, Buehner K, Hurst DP, Guarnieri F, Reggio PH, Nowell Harmon KW, Cabral GA, Abood ME (2002) A critical role for a tyrosine residue in the cannabinoid receptors for ligand recognition. Biochem Pharmacol 63:2121–2136

McAllister SD, Rizvi G, Anavi-Goffer S, Hurst DP, Barnett-Norris J, Lynch DL, Reggio PH, Abood ME (2003) An aromatic microdomain at the cannabinoid CB(1) receptor constitutes an agonist/inverse agonist binding region. J Med Chem 46:5139–5152

McAllister SD, Hurst DP, Barnett-Norris J, Lynch D, Reggio PH, Abood ME (2004) Structural mimicry in class A G protein-coupled receptor rotamer toggle switches: the importance of the F3.36(201)/W6.48(357) interaction in cannabinoid CB1 receptor activation. J Biol Chem 279:48024–48037

Mechoulam R, Hanus L, Ben-Shabat S, Fride E, Weidenfeld J (1994) The anandamides, a family of endogenous cannabinoid ligands—chemical and biological studies. Neuropsychopharmacology 10:145S
Mechoulam R, Ben-Shabat S, Hanus L, Ligumsky M, Kaminski NE, Schatz AR, Gopher A, Almog S, Martin BR, Compton DR, Pertwee RG, Griffin G, Bayewitch M, Barg J, Vogel Z (1995) Identification of an endogenous 2-monoglyceride, present in canine gut, that binds to cannabinoid receptors. Biochem Pharmacol 50:83–90
Monory K, Tzavara ET, Lexime J, Ledent C, Parmentier M, Borsodi A, Hanoune J (2002) Novel, not adenylyl cyclase-coupled cannabinoid binding site in cerebellum of mice. Biochem Biophys Res Commun 292:231–235
Mukhopadhyay S, Cowsik S, Lynn A, Welsh W, Howlett A (1999) Regulation of Gi by the CB1 cannabinoid receptor C-terminal juxtamembrane region: structural requirements determined by peptide analysis. Biochemistry 38:3447–3455
Mukhopadhyay S, McIntosh H, Houston D, Howlett A (2000) The CB(1) cannabinoid receptor juxtamembrane C-terminal peptide confers activation to specific G proteins in brain. Mol Pharmacol 57:162–170
Munro S, Thomas KL, Abu-Shaar M (1993) Molecular characterization of a peripheral receptor for cannabinoids. Nature 365:61–65
Murphy JW, Kendall DA (2003) Integrity of extracellular loop 1 of the human cannabinoid receptor 1 is critical for high-affinity binding of the ligand CP 55,940 but not SR 141716A. Biochem Pharmacol 65:1623–1631
Murphy WJ, Eizirik E, Johnson WE, Zhang YP, Ryder OA, O'Brien SJ (2001) Molecular phylogenetics and the origins of placental mammals. Nature 409:614–618
Nie J, Lewis DL (2001a) The proximal and distal C-terminal tail domains of the CB1 cannabinoid receptor mediate G protein coupling. Neuroscience 107:161–167
Nie J, Lewis DL (2001b) Structural domains of the CB1 cannabinoid receptor that contribute to constitutive activity and G-protein sequestration. J Neurosci 21:8758–8764
O'Brien C (1996) Drug addiction and drug abuse. In: Hardman J, Limbird L (eds) Goodman and Gilman's the pharmacological basis of therapeutics. Mc-Graw Hill, New York, pp 557–577
Offertaler L, Mo FM, Batkai S, Liu J, Begg M, Razdan RK, Martin BR, Bukoski RD, Kunos G (2003) Selective ligands and cellular effectors of a G protein-coupled endothelial cannabinoid receptor. Mol Pharmacol 63:699–705
Onaivi E, Chakrabarti A, Gwebu E, Chaudhuri G (1995) Neurobehavioral effects of delta 9-THC and cannabinoid (CB1) receptor gene expression in mice. Behav Brain Res 72:115–125
Oviedo A, Glowa J, Herkenham M (1993) Chronic cannabinoid administration alters cannabinoid receptor binding in rat brain: a quantitative autoradiographic study. Brain Res 616:293–302
Palczewski K, Kumasaka T, Hori T, Behnke CA, Motoshima H, Fox BA, Le Trong I, Teller DC, Okada T, Stenkamp RE, Yamamoto M, Miyano M (2000) Crystal structure of rhodopsin: A G protein-coupled receptor. Science 289:739–745
Pan X, Ikeda S, Lewis D (1998) SR 141716A acts as an inverse agonist to increase neuronal voltage-dependent Ca2+ currents by reversal of tonic CB1 cannabinoid receptor activity. Mol Pharmacol 54:1064–1072
Pertwee RG (1997) Pharmacology of cannabinoid CB1 and CB2 receptors. Pharmacol Ther 74:129–180
Piomelli D, Beltramo M, Glasnapp S, Lin S, Goutopoulos A, Xie X, Makriyannis A (1999) Structural determinants for recognition and translocation by the anandamide transporter. Proc Natl Acad Sci USA 96:5802–5807
Porter AC, Sauer JM, Knierman MD, Becker GW, Berna MJ, Bao J, Nomikos GG, Carter P, Bymaster FP, Leese AB, Felder CC (2002) Characterization of a novel endocannabinoid, virodhamine, with antagonist activity at the CB1 receptor. J Pharmacol Exp Ther 301:1020–1024

Reggio P (1999) Ligand-ligand and ligand-receptor approaches to modeling the cannabinoid CB1 and CB2 receptors: achievements and challenges. Curr Med Chem 8:665–683
Rhee M-H, Nevo I, Bayewitch ML, Zagoory O, Vogel Z (2000a) Functional role of tryptophan residues in the fourth transmembrane domain of the CB2 cannabinoid receptor. J Neurochem 75:2485–2491
Rhee MH, Nevo I, Levy R, Vogel Z (2000b) Role of the highly conserved Asp-Arg-Tyr motif in signal transduction of the CB2 cannabinoid receptor. FEBS Lett 466:300–304
Richfield E, Herkenham M (1994) Selective vulnerability in Huntington's disease: preferential loss of cannabinoid receptors in lateral globus pallidus. Ann Neurol 36:577–584
Rinaldi-Carmona M, Barth F, Heaulme M, Shire D, Calandra B, Congy C, Martinez S, Maruani J, Neliat G, Caput D, Ferrar P, Soubrie P, Breliere JC, Fur GL (1994) SR141716A, a potent and selective antagonist of the brain cannabinoid receptor. FEBS Lett 350:240–244
Rinaldi-Carmona M, Barth F, Millan J, Derocq JM, Casellas P, Congy C, Oustric D, Sarran M, Bouaboula M, Calandra B, Portier M, Shire D, Breliere JC, Fur GL (1998a) SR 144528, the first potent and selective antagonist of the CB2 cannabinoid receptor. J Pharmacol Exp Ther 284:644–650
Rinaldi-Carmona M, Duigou AL, Oustric D, Barth F, Bouaboula M, Carayon P, Casellas P, Fur GL (1998b) Modulation of CB1 cannabinoid receptor functions after a long-term exposure to agonist or inverse agonist in the Chinese hamster ovary cell expression system. J Pharmacol Exp Ther 287:1038–1047
Roche J, Bounds S, Brown S, Mackie K (1999) A mutation in the second transmembrane region of the CB1 receptor selectively disrupts G protein signaling and prevents receptor internalization. Mol Pharmacol 56:611–618
Rodriguez de Fonseca F, Gorriti M, Fernandez RJ, Palomo T, Ramos JA (1994) Down regulation of rat brain cannabinoid binding sties after chronic Δ9-tetrahydrocannabinol treatment. Pharmacol Biochem Behav 47:33–40
Romero J, Garcia-Palomero E, Castro J, Garcia-Gil L, Ramos J, Fernandez-Ruiz J (1997) Effects of chronic exposure to delta9-tetrahydrocannabinol on cannabinoid receptor binding and mRNA levels in several rat brain regions. Brain Res Mol Brain Res 46:100–108
Romero J, Berrendero F, Garcia-Gil L, Ramos J, Fernandez-Ruiz J (1998a) Cannabinoid receptor and WIN-55,212-2-stimulated [35S]GTP gamma S binding and cannabinoid receptor mRNA levels in the basal ganglia and the cerebellum of adult male rats chronically exposed to delta 9-tetrahydrocannabinol. J Mol Neurosci 11:109–119
Romero J, Berrendero F, Manzanares J, Perez A, Corchero J, Fuentes J, Fernandez-Ruiz J, Ramos J (1998b) Time-course of the cannabinoid receptor down-regulation in the adult rat brain caused by repeated exposure to delta9-tetrahydrocannabinol. Synapse 30:298–308
Selley DE, Stark S, Sim LJ, Childers SR (1996) Cannabinoid receptor stimulation of guanosine-5'-O-(3-[35S]thio)triphosphate binding in rat brain membranes. Life Sci 59:659–668
Shire D, Calandra B, Delpech M, Dumont X, Kaghad M, Fur GL, Caput D, Ferrara P (1996a) Structural features of the central cannabinoid CB1 receptor involved in the binding of the specific CB1 antagonist SR 141716A. J Biol Chem 271:6941–6946
Shire D, Calandra B, Rinaldi-Carmona M, Oustric D, Pessegue B, Bonnin-Cabanne O, Fur GL, Caput D, Ferrara P (1996b) Molecular cloning, expression and function of the murine CB2 peripheral cannabinoid receptor. Biochim Biophys Acta 1307:132–136
Shire D, Calandra B, Bouaboula M, Barth F, Rinaldi-Carmona M, Casellas P, Ferrara P (1999) Cannabinoid receptor interactions with the antagonists SR 141716A and SR 144528. Life Sci 65:627–635
Showalter VM, Compton DR, Martin BR, Abood ME (1996) Evaluation of binding in a transfected cell line expressing a peripheral cannabinoid receptor (CB2): Identification of cannabinoid receptor subtype selective ligands. J Pharmacol Exp Ther 278:989–999
Sieradzan KA, Fox SH, Hill M, Dick JP, Crossman AR, Brotchie JM (2001) Cannabinoids reduce levodopa-induced dyskinesia in Parkinson's disease: a pilot study. Neurology 57:2108–2111

Singh R, Hurst DP, Barnett-Norris J, Lynch DL, Reggio PH, Guarnieri F (2002) Activation of the cannabinoid CB1 receptor may involve a W6 48/F3 36 rotamer toggle switch. J Pept Res 60:357–370

Soderstrom K, Leid M, Moore FL, Murray TF (2000a) Behavioral, pharmacological, and molecular characterization of an amphibian cannabinoid receptor. J Neurochem 75:413–423

Soderstrom K, Leid M, Moore FL, Murray TF (2000b) Behavioral, pharmacological, and molecular characterization of an amphibian cannabinoid receptor. J Neurochem 75:413–423

Song Z, Slowey C-A, Hurst D, Reggio P (1999) The difference between the CB1 and CB2 cannabinoid receptors at position 5.46 is crucial for the selectivity of WIN55212-2 for CB2. Mol Pharmacol 56:834–840

Song Z-H, Bonner TI (1996) A lysine residue of the cannabinoid receptor is critical for receptor recognition by several agonists but not WIN55212-2. Mol Pharmacol 49:891–896

Song ZH, Feng W (2002) Absence of a conserved proline and presence of a conserved tyrosine in the CB2 cannabinoid receptor are crucial for its function. FEBS Lett 531:290–294

Steiner H, Bonner T, Zimmer A, Kita S, Zimmer A (1998) CB1 cannabinoid receptor knockout mice display increased neuropeptide expression in striatal output pathways and are hypoactive in an exploratory test. Soc Neurosci 24:411

Stella N, Schweitzer P, Piomelli D (1997) A second endogenous cannabinoid that modulates long-term potentiation. Nature 388:773–778

Sugiura T, Kodaka T, Kondo S, Nakane S, Kondo H, Waku K, Ishima Y, Watanabe K, Yamamoto I (1997) Is the Cannabinoid CB1 receptor a 2-arachidonylglycerol receptor? Structural requirements for Triggering a Ca2+ transient in NG108–15 cells. J Biochem (Tokyo) 122:890–895

Surprenant A, Horstman DA, Akbarali H, Limbird LE (1992) A point mutation of the alpha 2-adrenoceptor that blocks coupling to potassium but not calcium currents. Science 257:977–980

Tanda G, Pontieri F, Chiara GD (1997) Cannabinoid and heroin activation of mesolimbic dopamine transmission by a common mu1 opioid receptor mechanism. Science 276:2048–2050

Tao Q, Abood ME (1998) Mutation of a highly conserved aspartate residue in the second transmembrane domain of the cannabinoid receptors, CB1 and CB2, disrupts G-protein coupling. J Pharmacol Exp Ther 285:651–658

Tao Q, McAllister S, Andreassi J, Nowell K, Cabral G, Hurst D, Bachtel K, Ekman M, Reggio P, Abood M (1999) Role of a conserved lysine residue in the peripheral cannabinoid receptor (CB2): evidence for subtype specificity. Mol Pharmacol 55:605–613

Thomas W, Qian H, Chang C, Karnik S (2000) Agonist-induced phosphorylation of the angiotensin II (AT(1A)) receptor requires generation of a conformation that is distinct from the inositol phosphate-signaling state. J Biol Chem 275:2893–2900

Valk PJM, Hol S, Vankan Y, Ihle JN, Askew D, Jenkins NA, Gilbert DJ, Copeland NG, deBoth NJ, Lowenberg B, Delwel R (1997) The genes encoding the peripheral cannabinoid receptor and α-L-fucosidase are located near a newly identified common virus integration site. J Virol 71:6796–6804

Vasquez C, Lewis D (1999) The CB1 cannabinoid receptor can sequester G-proteins, making them unavailable to couple to other receptors. J Neurosci 19:9271–9280

Venance L, Piomelli D, Glowinski J, Giaume C (1995) Inhibition by anandamide of gap junctions and intercellular calcium signalling in striatal astrocytes. Nature 376:590–594

Wagner J, Varga K, Jarai Z, Kunos G (1999) Mesenteric vasodilation mediated by endothelial anandamide receptors. Hypertension 33:429–434

Walker J, Huang S, Strangman N, Tsou K, Sanudo-Pena M (1999) Pain modulation by release of the endogenous cannabinoid anandamide. Proc Natl Acad Sci USA 96:12198–12203

Walter L, Franklin A, Witting A, Wade C, Xie Y, Kunos G, Mackie K, Stella N (2003) Nonpsychotropic cannabinoid receptors regulate microglial cell migration. J Neurosci 23:1398–1405

Xie XQ, Melvin LS, Makryiannis A (1996) The conformational properties of the highly selective cannabinoid receptor ligand CP-55,940. J Biol Chem 271:10640–10647

Yamaguchi F, Macrae AD, Brenner S (1996) Molecular cloning of two cannabinoid type-1 receptor genes from the puffer fish Fugu rubripes. Genomics 35:603–605

Zastrow MV, Kobilka BK (1992) Ligand-regulated internalization and recycling of human β2-adrenergic receptors between the plasma membrane and endosomes containing transferrin receptors. J Biol Chem 267:3530–3538

Zhuang S, Kittler J, Grigorenko E, Kirby M, Sim L, Hampson R, Childers S, Deadwyler S (1998) Effects of long-term exposure to delta9-THC on expression of cannabinoid receptor (CB1) mRNA in different rat brain regions. Brain Res Mol Brain Res 62:141–149

Zimmer A, Zimmer A, Hohmann A, Herkenham M, Bonner T (1999) Increased mortality, hypoactivity, and hypoalgesia in cannabinoid CB1 receptor knockout mice. Proc Natl Acad Sci USA 96:5780–5785

Zygmunt P, Petersson J, Andersson D, Chuang H, Sorgard M, Di Marzo V, Julius D, Hogestatt E (1999) Vanilloid receptors on sensory nerves mediate the vasodilator action of anandamide. Nature 400:452–457

Zygmunt PM, Andersson DA, Hogestatt ED (2002) Delta 9-tetrahydrocannabinol and cannabinol activate capsaicin-sensitive sensory nerves via a CB1 and CB2 cannabinoid receptor-independent mechanism. J Neurosci 22:4720–4727

HEP (2005) 168:117–145

Analysis of the Endocannabinoid System by Using CB_1 Cannabinoid Receptor Knockout Mice

O. Valverde[1] · M. Karsak[2] · A. Zimmer[2] (✉)

[1]Laboratori de Neurofarmacologia, Departament de Ciències Experimentals i de la Salut, Universitat Pompeu Fabra, Carrer Dr. Aiguader, 80, 08003 Barcelona, Spain

[2]Laboratory of Molecular Neurobiology, Clinic of Psychiatry, Neurocenter, University of Bonn, Sigmund-Freud-Strasse 25, 53105 Germany

neuro@uni-bonn.de

Abstract The endocannabinoid system has been involved in the control of several neurophysiological and behavioural responses. To date, three lines of CB_1 knockout mice have been established independently in different laboratories. This chapter reviews the main results obtained with these lines of CB_1 knockout mice in several physiological responses that have been previously related to the activity of the endocannabinoid system. Studies using CB_1 knockout mice have demonstrated that this receptor participates in the control of several behavioural responses including locomotion, anxiety- and depressive-like states, cognitive functions such as memory and learning processes, cardiovascular responses and feeding. Furthermore, the CB_1 cannabinoid receptor is involved in the control of pain by acting at peripheral, spinal and supraspinal levels. The involvement of the CB_1 cannabinoid receptor in the behavioural and biochemical processes underlying drug addiction has also been investigated. These CB_1 knockouts have provided new findings to clarify the interactions between cannabinoids and the other drugs of abuse such as opioids, psychostimulants, nicotine and ethanol. Recent studies have demonstrated that endocannabinoids can function as retrograde messengers, modulating the release of different neurotransmitters, including opioids, γ-aminobutyric acid (GABA), and cholecystokinin (CCK), which could explain some of the responses observed after the stimulation of the CB_1 cannabinoid receptor. This review provides an update of the apparently controversial data reported in the literature using the three different lines of CB_1 knockout mice, which seem to be mainly due to the use of different experimental procedures rather than any constitutive alteration in these lines of knockouts.

Keywords CB_1 knockout mice · Locomotion · Emotional-like behaviour · Cognitive functions · Cardiovascular responses · Nociception · Feeding behaviour · Drug addiction · Opioids · Psychostimulants · Nicotine · Ethanol · Retrograde neurotransmitter

In this chapter we will focus on the physiological functions of CB_1 cannabinoid receptors that have been reported in knockout mice, rather than review the general physiology of the CB_1 cannabinoid receptors.

1 Generation of CB_1 Knockout Mice

The murine CB_1 receptor is encoded by the *Cnr1* gene on chromosome 4. Like many other G protein-coupled receptors (GPCRs), the entire CB_1 receptor is encoded by a single large exon. To date three lines of CB_1 knockout mice have been established independently in three different laboratories. In the line generated by Ledent and her co-workers (1999), the first 233 codons were replaced by a phosphoglycerate kinase (PGK)-neo cassette. One of our laboratories (A.Z.) generated a knockout strain by replacing the region between amino acid 32 and 448 with PGK-neo (Zimmer et al. 1999). Both mutations constitutively invalidate the gene. The Ledent line has been crossed to an outbred CD1 genetic background, and thus individual mutant animals from this strain can be expected to have a heterogeneous genetic background. The initial results from the Zimmer line were also obtained with animals from a CD1 genetic background, but it has since been crossed for more than 10 generations to C57BL/6J mice, thus generating a congenic strain in which all animals are genetically homogeneous. Marsicano and colleagues (2002) generated a third line of mice that carries a CB_1 gene flanked by lox sites ("floxed"). These lox sites are recognized by the Cre enzyme, a DNA recombinase derived from P1 bacteriophages. When such mice are bred to a transgenic strain that express Cre, floxed genes will be deleted in all tissues in which the Cre enzyme is active. This strategy is now frequently used for the tissue-specific inactivation of genes (Sauer 1998).

Mice develop apparently normally in the absence of the CB_1 receptor. They are fertile, care for their offspring, and do not show any behavioural abnormalities that would be obvious to the casual observer. However, CB_1-deficient animals have a much higher mortality rate than wild-type animals (Zimmer et al. 1999). Approximately 30% of the mutant animals die of natural causes during the first 6 months, in contrast to less than 5% of the heterozygous and wild-type control animals. The mortality rate in knockout mice is equally high in animals of different age, and death occurs suddenly without prior evidence of illness. Careful examination of dead animals has not yet revealed a cause of death. However, we have frequently observed epileptic seizures in mutant animals and believe that these may have contributed to the increased mortality rate.

2
Neurochemical and Biochemical Adaptive Changes Produced by the Lack of the CB_1 Cannabinoid Receptors

Genetic mutations or deletions can lead to molecular or cellular changes that have been interpreted as an attempt of the organism to compensate for the missing or malfunctioning gene product (Nelson and Young 1998; Pich and Epping-Jordan 1998). CB_1 receptor knockouts have been extensively studied to determine whether such compensatory changes occur in the absence of CB_1 receptors.

Binding of the CB_1-specific agonist CP55,940 was completely abolished in CB_1 knockout mice (Zimmer et al. 1999), and neither CP55,940 nor HU-210 [nor Δ^9-tetrahydrocannabinol (THC)] stimulated [^{35}S]GTP binding in brain tissues from these animals (Breivogel et al. 2001). These results indicated that the CB_1 receptor is the only target for these ligands. A 50% reduction of CB_1 sites was also observed in heterozygous mice when WIN55,212-2 was used. However, the maximal stimulation of [^{35}S]GTP binding was only reduced by 20%–25% in most brain regions, suggesting that there is a small receptor reserve in wild-type animals that was depleted in heterozygous mice (Breivogel et al. 2001). A notable exception was the striatum, where the decrease in stimulation was proportional to the receptor density. Interestingly, some stimulation of [^{35}S]GTP binding by WIN55,212-2 was still observed in homozygous mutant animals, strongly indicating that there is also a non-CB_1 target for this compound. Di Marzo and colleagues analysed anandamide levels in wild-type and CB_1-deficient animals (Di Marzo et al. 2000). They found that, in the absence of CB_1 receptors, anandamide levels were decreased in the hippocampus and to a lesser extent in the striatum. Because fatty acid amide hydrolase (FAAH) activity was unchanged in these animals, the authors argue that the CB_1 receptor may control anandamide biosynthesis. In contrast, Maccarone and co-workers reported that anandamide hydrolysis, mediated by FAAH, was age-dependently increased in CB_1-deficient, but not in wild-type, mice (Maccarone et al. 2001). Old CB_1 knockouts also showed a significantly elevated enzyme activity (V_{max}), in the cerebral cortex. Although the reason for these disparate results are unclear, the different genetic backgrounds of the animals or, more likely, differences in holding conditions may have contributed.

3
CB_1 Cannabinoid Receptors Participate in the Control of Locomotion

Among the most striking behavioural effects of cannabinoids in rodents is a profound dose-dependent induction of catalepsy and reduction of locomotor activity (Rodriguez de Fonseca et al. 1998; Chaperon and Thiebot 1999). In contrast, even high doses of THC (up to 100 mg/kg) have no locomotor effects in CB_1-deficient animals, demonstrating that they are mediated by CB_1 receptors (Zimmer et al. 1999). An endocannabinoid tone in the regulation of locomotor activity has been suggested, because the CB_1 receptor antagonist SR141716A stimulates locomotor activity (Compton et al. 1996) and potentiates the locomotor stimulant effects of amphetamine and apomorphine (Masserano et al. 1999). This idea is supported by the observation of Ledent and co-workers (1999) that locomotor activity is slightly increased in mice without cannabinoid receptors. However, Steiner and colleagues (1999) found a decrease in open-field activity in the Zimmer CB_1 knockout strain. There are two explanations for these differences. First, because cannabinoids have biphasic effects (Chaperon and Thiebot 1999), it is conceivable that abolishing the endocannabinoid tone may lead to different outcomes, depending on the level of the endogenous tone. Secondly, because CB_1 knockout mice apparently have higher levels of anxiety (see below), the results may have been influenced by the experimental conditions. Indeed, Steiner et al. used a relatively large open field apparatus and regular laboratory illumination, whilst Ledent et al. conducted their open field test under low light conditions using a smaller device. The latter conditions are less anxiogenic in mice, thus resulting in a higher locomotor activity.

The locomotor effects of THC are thought to be mediated in part by CB_1 receptors in the basal ganglia (Rodriguez de Fonseca et al. 1998). In the striatum, CB_1 receptors display a distinct medial-to-lateral and dorsal-to-rostral distribution, with the highest receptor densities in the lateral part of the middle striatum (Steiner et al. 1999). The striatum has two distinct output pathways, one to the substantia nigra and one to the globus pallidus (Gerfen 1992, 1993). The primary neurotransmitter of both pathways is γ-aminobutyric acid (GABA), but they have different neuropeptide co-transmitters. Striato-pallidal neurons contain enkephalins, whilst striato-nigral neurons express substance P and dynorphin (Steiner and Gerfen 1998). Steiner and colleagues have shown that dynorphin and substance P mRNA levels were significantly elevated in the medio-lateral striatum of CB_1 knockout mice, which also contained the highest CB_1 receptor densities (Steiner et al. 1999). Enkephalin expression was also elevated in CB_1 knockout mice, but unrelated to CB_1 receptor densities. These results are consistent with a local CB_1 inhibition of striato-nigral neurons, whilst effects on striato-pallidal neurons probably involve network-level alterations.

4 CB_1 Cannabinoid Receptors and Emotional Behaviour

Different evidence suggests that the endocannabinoid system plays an important role in the regulation of emotional-like behaviour. Thus, the CB_1 cannabinoid receptor is widely distributed in limbic and cortical areas involved in the control of emotion. The administration of cannabinoid ligands produces emotional-like responses in different behavioural paradigms. Furthermore, cannabinoids also exert a modulatory role on the activity of the hypothalamic-pituitary adrenal axis (HPA), and these compounds modulate the release of several neurotransmitters involved in emotional behaviour, including CCK and GABA.

Studies using CB_1 knockout mice have supported and clarified the previous data reported by using different pharmacological approaches. Thus, it has been shown that CB_1 knockout animals (on a CD1 genetic background) displayed anxiogenic-like responses in different behavioural models, including the open-field, light-dark box and elevated plus maze (Haller et al. 2002; Maccarrone et al. 2002; Martin et al. 2002; Uriguen et al. 2004). Similar anxiogenic-like responses were exhibited in CB_1 knockout mice with an inbred genetic background (C57BL/6). Thus, an anxiogenic-like response in the elevated plus-maze and impairment in the extinction in auditory fear-conditioning test were revealed in these mice (Marsicano et al. 2002), supporting previous results obtained in the CB_1 knockout mice with a CD1 background. In agreement, the administration of SR141716A mimicked the phenotype of CB_1-deficient mice, supporting the role of the endocannabinoids in the control of emotional-like responses (Marsicano et al. 2002). Furthermore, the anxiogenic-like responses in the CB_1 knockout mice were accompanied by alterations in the HPA axis under basal conditions, as well as a hypersensitivity to stress and an impaired action of anxiolytic drugs (bromazepam and buspirone) in the light-dark box (Uriguen et al. 2004). Indeed, basal corticosterone concentrations in the plasma were lower in mutant CB_1 than in wild-type mice, whereas CB_1 knockout mice showed a greater increase in plasma corticosterone concentrations than wild-type littermates after the exposure to restraint stress, supporting the results obtained in the behavioural models (Uriguen et al. 2004). In addition to the anxiogenic-like profile observed in mice lacking CB_1 cannabinoid receptors, these animals also exhibited an increase in aggressive behaviour when exposed to the resident-intruder paradigm, and an enhanced sensitivity to develop a state of anhedonia (depressive-like state) during the exposure to the chronic unpredictable mild stress paradigm (Martin et al. 2002).

A strong impairment of short-term and long-term extinction in auditory fear-conditioning test has been also reported in CB_1 knockout mice (Marsicano et al. 2002). Thus, tone presentation during extinction trials resulted in elevated levels of endocannabinoids in the basolateral amygdala complex, a region known to control extinction of aversive memories, which indicates that endocannabinoids facilitate extinction of aversive memories through their selective blockade of local inhibitory networks in the amygdala (Marsicano et al. 2002). These authors proposed that the decrease of activity of local inhibitory networks within the basolateral amygdala induced by CB_1 activation leads to a disinhibition of principal neurons and finally to extinction of the freezing response, this being a physiological function impaired in CB_1 knockout mice (Marsicano et al. 2002).

Studies using CB_1 knockout mice also suggest the existence of a novel cannabinoid receptor involved in the control of mood. A recent study has investigated the effects induced by SR141716A on CB_1 knockout mice and wild-type littermates in the elevated plus-maze, showing that surprisingly, the cannabinoid antagonist reduced anxiety in both wild-type and CB_1 knockout mice (Haller et al. 2002). This result shows a discrepancy between genetic and pharmacological blockade of the CB_1 receptor, supporting the hypothesis that a third cannabinoid receptor participates in the responses induced by SR141716A (Haller et al. 2002). Biochemical studies have supported this idea and provided evidence for putative "CB_3" or "CB_x" receptor binding sites in the brain that are sensitive to WIN55,212-2, anandamide and SR141716A (Di Marzo et al. 2000; Breivogel et al. 2001).

In conclusion, pharmacological studies show that cannabinoid agonists induce a broad spectrum of actions in different experimental models of anxiety. Data from knockout mice deficient in the CB_1 cannabinoid receptors demonstrate the existence of an endogenous cannabinoid tonus modulating mood through the stimulation of these CB_1 receptors and also support the possible existence of a third cannabinoid receptor, which seems to play an opposite role to the CB_1 receptor in emotional control. CB_1 cannabinoid receptors modulate the HPA axis activity and the release of several neurotransmitters such as CCK, GABA, serotonin and nicotine, providing a neurochemical substrate for this physiological role. The modulation of several neurotransmitter systems by CB_1 receptors would explain the different effects that cannabinoids can have on anxiety.

5
CB_1 Cannabinoid Receptors Participate in the Control of Cognitive Functions

Cannabinoid ligands produce clear effects on learning and memory that have been widely reported (Dewey 1986; Ameri 1999; Diana and Marty 2004). However, the precise role of the endocannabinoid system on these processes has not yet been completely clarified. In humans, THC administration induces the disruption of short-term recall, as well as disorienting effects (Miller and Branconnier 1983; Chait and Perry 1992). In animals, cannabinoid administration impairs memory and learning processes. In particular, there are reports that cannabinoids impair task acquisition and working memory in different animal species (Molina-Holgado et al. 1995; Lichtman and Martin 1996; Winsauer et al. 1999). The alterations are especially important for spatial memory (Molina-Holgado et al. 1995; Lichtman and Martin 1996) and short-term memory (Molina-Holgado et al. 1995). In rodents, endogenous cannabinoids have been reported to prevent the induction of long-term potentiation in the hippocampus (Stella et al. 1997), and to impair memory in different behavioural tasks, an effect attenuated by SR141716A administration (Mallet and Beninger 1998). On the other hand, the CB_1 antagonist SR141716A can induce an enhancement of memory in some experimental conditions (Hampson and Deadwyler 2000).

In agreement with these pharmacological data, mice lacking CB_1 cannabinoid receptors showed an improved performance in the active avoidance paradigm (Martin et al. 2002), and in the two-trial object recognition test (Reibaud et al. 1999; Bohme et al. 2000). A facilitation of long-term potentiation in the hippocampus was also reported in the same line of CB_1 knockout mice (Böhme et al. 2000). On the other hand, CB_1 knockout mice have been reported to exhibit similar acquisition rates in the Morris water maze as wild-type littermates, whilst CB_1 knockout animals demonstrated deficits in a reversal task in which the hidden platform was located in a different place, also suggesting that the endocannabinoid system has a role in facilitating extinction and/or forgetting processes (Varvel and Lichtman 2002). Indeed, CB_1 cannabinoid receptor-deficient mice exhibited strong impairments in short- and long-term extinction in the auditory fear-conditioning test, indicating that these animals have a prolonged aversive memory (Marsicano et al. 2002).

A recent study has shown that CB_1 knockout mice exhibited an increased acetylcholine release in the hippocampus (Kathmann et al. 2001). Inhibition of acetylcholine activity has been associated with cannabinoid-induced impairment of memory (Braida and Sala 2000). The hippocampus and the neocortex play a crucial role in the control of learning and memory. In both brain structures, CB_1 cannabinoid receptors are expressed in a well-defined subpopulation of GABAergic interneurons (Katona et al. 1999; Marsicano and Lutz 1999; Tsou et al. 1999). Moreover, CB_1 cannabinoid receptor-positive interneurons are distinctive in forming inhibitory synapses with particularly fast kinetics. These GABAergic interneurons seem to control plasticity at excitatory synapses, and thus the blockade of inhibition induced by cannabinoids generally promotes long-term potentiation at excitatory synapses (Wilson and Nicoll 2002; Diana and Marty 2004). This facilitation in the plasticity phenomenon seems to be mediated, at least in part, by extracellular-regulated kinases (ERK). THC has been reported to activate ERK and to induce expression of immediate early genes products in both hippocampal slices and in vivo in this brain structure (Derkinderen et al. 2003). In view of this facilitatory effect induced by cannabinoids in the hippocampal neurons, one may wonder if the endocannabinoid system facilitates learning. However, pharmacological and genetic studies have clearly demonstrated a cannabinoid-induced impairment of memory processes. A possible explanation for this apparent discrepancy has been proposed by Wilson and Nicoll (2002), who suggest that endocannabinoids modulate at a physiological level the activity of interneurons forming fast synapses in the hippocampus to orchestrate fast synchronous oscillations in the gamma range (Banks et al. 2000). The administration of marijuana derivatives might permit promiscuous plasticity, suppressing many hippocampal inhibitory synapses, and cause deficits in cognition and recall (Wilson and Nicoll 2002). Further studies are necessary in order to clarify the complex role of the endocannabinoid system on learning and memory processes and the nature of the changes promoted in the brain by the exogenous administration of cannabinoids.

6
CB_1 Cannabinoid Receptors Participate in the Control of Cardiovascular Responses

It is well known that the acute consumption of THC causes tachycardia in humans without any significant effect on blood pressure, whilst the chronic ingestion of cannabinoids leads to hypotension and bradycardia (Benowitz and Jones 1975). Pharmacological studies using selective CB_1 receptor antagonists (Varga et al. 1995; Lake et al. 1997) have suggested that some of these cardiovascular responses are mediated by CB_1 receptors.

Considering the pharmacological effects of cannabinoids, it was somewhat surprising to see that basal blood pressure and heart rate were normal in CB_1-deficient mice, thus suggesting that endogenous cannabinoids do not exert a tonic control on these cardiovascular parameters. However, when the CB_1 agonists anandamide or WIN55,212-2 were administered to CB_1 knockout animals, they failed to produce the sustained decrease in heart rate and blood pressure that was observed in control littermates (Ledent et al. 1999). A similar result was observed when CB_1-deficient and control mice were treated with 2-arachidonylglyceryl ether, a metabolically stable analogue of 2-arachidonoylglycerol (2-AG). In contrast, 2-AG, which is rapidly metabolized, still produced hypotension and tachycardia in the absence of CB_1 receptors, indicating that a metabolic product of 2-AG elicits cardiovascular effects that are not mediated by CB_1 receptors (Jarai et al. 2000).

Interestingly, "abnormal cannabidiol", a neurobiologically inactive cannabinoid, causes hypotension and mesenteric vasodilation in mice lacking CB_1 and CB_2 receptors that can be blocked by SR141716A (Jarai et al. 1999). These findings suggest the existence of a yet unidentified endothelial cannabinoid receptor. A further line of evidence was obtained when endotoxin lipopolysaccharide (LPS)-induced hypotension was studied in cannabinoid receptor-deficient animals. Intravenous injection of 100 μg/kg LPS caused a similar hypotension in phenobarbital anaesthetised wild-type animals and in mice deficient in CB_1 or both CB_1 and CB_2 receptors (Batkai et al. 2001). This hypotensive effect was also blocked by pretreatment with SR141716A (Batkai et al. 2004), again indicating that this compound exerts some of its effects through non-CB_1 receptors.

7
Participation of the CB_1 Cannabinoid Receptors in the Control of Pain

Cannabinoids produce antinociception through multiple mechanisms at peripheral, spinal and supraspinal levels through CB_1 and CB_2 cannabinoid receptors in several animal species, including mice, rats, rabbits, cats, dogs, monkeys and humans (Pertwee 2001). These responses were revealed in multiple acute nociceptive models using thermal (Buxbaum 1972; Hutcheson et al. 1998; Martin and Lichtman 1998), mechanical (Smith et al. 1998), chemical (Bicher and Mechoulam 1968; Welch et al. 1995) and electrical stimuli (Bicher and Mechoulam 1968; Weissman et al. 1982). Cannabinoid agonists also induce antinociception in inflammatory models of pain, including hyperalgesia induced by carrageenan (Mazzari et al. 1996), capsaicin (Li et al. 1999), formalin (Calignano et al. 1998; Jaggar et al. 1998) or Freund's adjuvant (Martin et al. 1999). Cannabinoid agonists are also effective in visceral models of pain, such as inflammation of the bladder wall induced by turpentine administration (Jaggar et al. 1998), 2,4-dinitrobenzene sulphonic acid (DNBS)-induced colitis (Massa et al. 2004) and also in neuropathic pain models, such as the painful mononeuropathy induced by loose ligature of the sciatic nerve (Herzberg et al. 1997; Mao et al. 2000). Electrophysiological studies also provide evidence that cannabinoids attenuate nociceptive transmission in vivo (Pertwee 2001; Hohmann 2002). Thus, cannabinoids suppress noxious stimulus-evoked neuronal activity in nociceptive neurons in the spinal cord and thalamus (Hohmann et al. 1995; Martin et al. 1996; Tsou et al. 1996).

Several central structures involved in cannabinoid antinociception have been identified. Hence, the local microinjection of cannabinoid agonists in areas such as the periaqueductal grey matter (Martin and Lichtman 1998; Martin et al. 1999), the rostral ventromedial medulla (Martin et al. 1996), the submedius and latero-posterior nuclei of the thalamus (Mailleux and Vanderhaeghen 1992), the superior colliculus and the amygdaloid complex (Martin et al. 1996; Martin et al. 1999) was able to produce antinociceptive responses. All these neuroanatomical structures related to cannabinoid-induced antinociception are involved in pain transmission and constitute the descending system involved in the control of pain (Basbaum and Fields 1984; Fields et al. 1991). At the spinal level, CB_1 cannabinoid receptors are abundant in the dorsal horn responsible for pain transmission. Most primary afferent neurons that express CB_1 receptor mRNA are those with larger diameter fibres involved in the transmission of non-nociceptive-sensitive inputs (Hohmann and Herkenham 1998). However, CB_1 cannabinoid receptors also modulate the transmission of C fibre-evoked responses (Kelly and Chapman 2001), inhibiting the release of neurotransmitters responsible for pain transmission (Wilson and Nicoll 2002). CB_1 cannabinoid receptor mRNA was also highly expressed in dorsal root ganglion cells (Hohmann 2002; Bridges et al. 2003). At this level, CB_1 cannabinoid receptor stimulation seems to produce a presynaptic inhibition of Ca^{2+} channels, attenuating the release of neurotransmitters (Millns et al. 2001).

On peripheral terminals, the activation of CB_1 and CB_2 cannabinoid receptors was shown to inhibit nociceptive transmission, and both receptors seem to be implicated in mediating the existing endogenous cannabinoid tone (Calignano et al. 1998; Strangman et al. 1998; Hanus et al. 1999; Ko and Woods 1999). Thus, behavioural studies support a role for peripheral cannabinoid CB_2 receptors in animal models of persistent pain and the existence of a synergism between CB_1- and CB_2-mediated responses at this level (Malan et al. 2002). However, other studies do not support such a role of peripheral cannabinoid receptors (Di Marzo et al. 2000). CB_2 receptor activation can also inhibit oedema and plasma extravasations produced by inflammation at a peripheral level (Malan et al. 2002). Cannabinoid CB_2 receptors are likely located on non-neuronal cells in inflamed tissues, where they inhibit the release of inflammatory mediators that excite nociceptors (Mazzari et al. 1996).

Recent studies using knockout mice deficient in cannabinoid receptors have provided new and important information on the involvement of the cannabinoid system in nociception. Different results were reported on spontaneous nociceptive perception of CB_1 knockout mice, depending on the genetic construction of the knockout mice. In CB_1 knockout mice with an outbred CD1 genetic background, no changes in the nociceptive threshold were found after the application of thermal (tail-immersion and hot-plate tests), mechanical (tail-pressure) or chemical (writhing test) stimuli (Ledent et al. 1999; Valverde et al. 2000b). However, CB_1 knockout mice on an inbred C57BL/6J genetic background displayed hypoalgesia in the hot-plate and in the formalin test, whereas no difference in the tail-flick test was found (Zimmer et al. 1999). The hypoalgesic phenotype observed in this latter strain was surprising because CB_1 agonists produce similar behavioural effects in wild-type mice. Moreover, intrathecally administered SR141716A or antisense knockdown of spinal CB_1 receptors produced hyperalgesia in the hot-plate test (Richardson et al. 1998). The discrepancies between the two studies performed with knockout mice could be due to the different genetic background of the lines, but also to the different behavioural responses evaluated in the nociceptive test. Thus, Zimmer et al. (1999) measured the first discomfort response exhibited in the hot-plate test (paw lifting, paw shaking, paw licking or jumping), whereas Valverde et al. (2000b) have quantified jumping latency.

A recent study has demonstrated that the endogenous cannabinoid system mediates a protective role during visceral inflammation through the activation of the CB_1 cannabinoid receptors. Thus, CB_1 knockout mice exposed to an experimental colitis, induced by intrarectal DNBS, exhibited a higher sensibility to chemical-induced visceral inflammation. Pharmacological blockade of CB_1 receptors with the selective antagonist SR141716A led to a worsening of colitis similar to that observed in CB_1-deficient mice. Moreover, the cannabinoid agonist HU-210 reduced the severity of experimental colitis, and FAAH-deficient mice showed significant protection against DNBS treatment (Massa et al. 2004).

In mice lacking CB_1 cannabinoid receptors, the antinociceptive properties of THC were abolished in the hot-plate test, and were strongly reduced in the tail-immersion test. In this latter test, a slight antinociceptive response was still observed in mutant mice only at the highest dose of THC used (Ledent et al. 1999; Zimmer et al. 1999). In contrast, morphine-induced antinociception was preserved in these knockout mice in the tail immersion and the hot-plate tests. Furthermore, the antinociceptive effects induced by the selective δ-opioid agonists [D-penicillamine^{2,5}]enkephalin (DPDPE) and deltorphin II and by the selective κ-opioid agonist U-50,488H were unchanged (Valverde et al. 2000b). Therefore, CB_1 receptors do not seem to be involved in the antinociceptive responses induced by exogenous opioids. However, CB_1 receptors participate in the antinociceptive responses produced by non-steroidal anti-inflammatory drugs. Thus, the antinociceptive responses induced by the non-selective cyclooxygenase inhibitor indomethacin in the formalin test were abolished in CB_1 knockout mice (Guhring et al. 2002).

Several studies have shown tolerance to several behavioural responses induced by cannabinoids, including antinociception (Buxbaum 1972; Hutcheson et al. 1998;

Martin and Lichtman 1998; Pertwee 2001). The development of cannabinoid tolerance seems to be mainly due to pharmacodynamic events. Thus, a significant decrease in both CB_1 cannabinoid receptor binding sites and mRNA levels has been observed in different brain areas after a chronic treatment with cannabinoid agonists. Changes in G protein expression and functional activity were also observed in rats chronically treated with cannabinoids (Rodriguez de Fonseca et al. 1994; Rubino et al. 1994, 1998, 2000; Fan et al. 1996; Sim et al. 1996; Romero et al. 1998). Studies using knockout mice deficient in the different components of the endogenous opioid system provide new data concerning the possible mechanisms involved in the development of cannabinoid tolerance. Thus, knockout mice lacking the pre-proenkephalin gene showed a decrease in the development of tolerance to THC antinociceptive effects (Valverde et al. 2000a). A similar decrease in the development of cannabinoid tolerance was also observed in double mutant mice, lacking δ- and κ-opioid receptors (Castañe et al. 2003).

There is increasing evidence to support a role for peripheral CB_2 receptors in the analgesic effects of cannabinoids. Thus, chronic pain induced by peripheral nerve injury, but not that produced by peripheral inflammation, was associated with the enhancement of CB_2 cannabinoid receptor expression, specifically located in the lumbar spinal cord (Malan et al. 2002). Thus, a selective induction of spinal CB_2 expression presumably occurs on activated microglia in regions undergoing neuronal damage.

Taken together, these results show that the endocannabinoid system plays an important role in the physiological modulation of nociceptive transmission and in the development of inflammatory and neuropathic pain. Furthermore, the endocannabinoid system seems to participate in the antinociception induced by anti-inflammatory drugs, and displays an important synergic effect with opioid agonists. These data strongly support the therapeutic potential of cannabinoid receptor agonists for the treatment of chronic pain.

8
CB_1 Cannabinoid Receptors and Addiction

Behavioural and neurochemical studies have now clarified the controversy about the abuse liability of cannabinoids by demonstrating that such drugs fulfil most of the common features attributed to compounds with reinforcing properties. Cannabinoid rewarding properties have been identified using intracranial self-stimulation, conditioned place preference and intravenous self-administration paradigms. Furthermore, a cannabinoid withdrawal syndrome has also been characterized in different animal species (Lichtman and Martin 2002; Maldonado and Rodriguez de Fonseca 2002).

The administration of cannabinoid agonists can produce both rewarding and aversive/dysphoric effects in the place conditioning paradigm, depending on the dose and the experimental conditions. Thus, THC produced place preference in rats when administered at low doses and when animals were exposed to a 24-h washout period between the two THC conditioning sessions (Lepore et al. 1995). THC also produces a clear place preference in mice when a long period of conditioning is used and the possible dysphoric consequences of the first drug exposure are avoided (Valjent and Maldonado 2000). Concerning intracranial self-stimulation, acute administration of THC has been reported to decrease the intracranial self-stimulation threshold in rats, suggesting the activation of central hedonic systems (Gardner et al. 1988; Lepore et al. 1996). In contrast, CP55,940 administration did not modify electrical brain stimulation, supporting the hypothesis that cannabinoids have a relatively modest influence on reward circuits (Arnold et al. 2001).

Different studies have reported that THC is unable to induce self-administration behaviour in any of the animal species studied (Corcoran and Amit 1974; Harris et al. 1974; Carney et al. 1977; Mansbach et al. 1996). However, one study has revealed THC intravenous operant self-administration behaviour in squirrel monkeys that have a previous history of cocaine self-administration (Tanda et al. 2000). Recently, Justinova et al. (2003) reported self-administration of THC by drug-naïve monkeys, demonstrating that THC can act as an effective reinforcer of drug-taking behaviour in monkeys with no history of exposure to other drugs (Justinova et al. 2003). The pharmacokinetic properties of THC seem to be crucial for the behavioural responses observed in the self-administration paradigm. Thus, the synthetic cannabinoid agonists WIN55,212-2 and CP55,940, which have a shorter half-life than THC, are intravenously self-administered by mice (Martellotta et al. 1998) and rats (Braida et al. 2001). A selective involvement of the CB_1 cannabinoid receptors is implicated in the reinforcing properties of all these cannabinoid compounds because the CB_1 receptor antagonist SR141716A completely blocked the self-administration induced by WIN55,212-2 (Martellotta et al. 1998), CP55,940 (Braida et al. 2001) and THC (Tanda et al. 2000). Furthermore, CB_1 knockout mice failed to self-administer WIN55,212-2 in contrast to wild-type animals (Fattore et al. 1999; Ledent et al. 1999).

Administration of the selective CB_1 cannabinoid receptor antagonist SR141716A to animals (mouse, rat and dog) chronically treated with THC has been shown to precipitate different somatic manifestations of cannabinoid withdrawal. In rodents, this cannabinoid withdrawal syndrome is characterized by the presence of a large number of somatic signs and the absence of vegetative manifestations (Lichtman and Martin 2002; Maldonado and Rodriguez de Fonseca 2002). However, the doses of THC required to induce physical dependence in rodents are extremely high, currently from 10 to 100 mg/kg of THC (i.p.), daily for 5 to 10 days (Tsou et al. 1995; Aceto et al. 1996; Cook et al. 1998; Hutcheson et al. 1998). CB_1 cannabinoid receptors are responsible for the somatic manifestations of cannabinoid withdrawal. Indeed, CB_1-deficient mice chronically treated with THC did not exhibit any manifestation of cannabinoid withdrawal (Ledent et al. 1999; Lichtman et al. 2001).

In conclusion, these data clearly demonstrate that the functional activity of the CB_1 cannabinoid receptor is necessary for the manifestation of the rewarding properties of cannabinoids and for the development of cannabinoid physical dependence and withdrawal.

9 Interaction Between Cannabinoid Receptors and Other Addictive Drugs

Different evidence supports the possible existence of functional interactions between cannabinoids and other drugs of abuse including opioids, psychostimulants, ethanol and nicotine. Findings in support of a link between cannabinoids and other drugs of abuse include: (1) the existence of common physiological and pharmacological properties (opioids, ethanol, nicotine); (2) the stimulation of dopamine release after their administration (psychostimulants, opioids, ethanol, nicotine); (3) the existence of interactions at a signal-transduction level (opioids, psychostimulants, ethanol and nicotine); and (4) the observation that many of these drugs are consumed together.

9.1 Interaction Between Cannabinoids and Opioids

The interaction between cannabinoids and opioids has been widely evaluated because of the diverse physiological effects shared by both types of compounds, including antinociception, hypothermia, and control of locomotion, rewarding properties and the ability to induce drug abuse. Interestingly, the interaction between these two systems seems to be bi-directional. Thus, morphine-induced intravenous self-administration (Ledent et al. 1999; Cossu et al. 2001) and conditioned place preference (Martin et al. 2002) was abolished in knockout mice lacking the CB_1 cannabinoid receptors. These studies underlie the relevance of CB_1 cannabinoid receptors for the manifestation of the reinforcing properties of morphine. The ability of cannabinoid agents to reinstate or prevent heroin-seeking behaviour after a period of extinction has been also evaluated. The cannabinoid agonists WIN55,212-2 and CP55,940, but not THC, restored heroin-seeking behaviour in rats, whereas the CB_1 cannabinoid antagonist SR141716A completely prevented the reinstatement of drug-seeking behaviour induced by a priming injection of heroin (Fattore et al. 2003), supporting the cooperation between opioid and cannabinoid systems in the modulation of addictive behaviour.

Different pharmacological and molecular approaches have been used to investigate the interaction between cannabinoids and opioids in physical dependence. For example, administration of the CB_1 cannabinoid antagonist SR141716A can precipitate behavioural and biochemical manifestations of withdrawal in morphine-dependent rats (Navarro et al. 2001). In contrast to these data, SR141716A did not precipitate any behavioural sign of withdrawal in morphine-dependent mice (Lichtman et al. 2001). These discrepancies could be due to the different animal species and/or differences in the experimental procedure. However, studies performed in CB_1 knockout mice clearly demonstrated the important role played by the CB_1 cannabinoid receptors in the physical manifestations of the morphine withdrawal syndrome. Thus, a robust decrease in the severity of naloxone-precipitated morphine withdrawal syndrome was reported in CB_1 knockout mice (Ledent et al. 1999). In agreement, the co-administration of SR141716A and morphine over

5 days produced an important attenuation in the incidence of the morphine withdrawal manifestations (Mas-Nieto et al. 2001). Early studies have also demonstrated that acute administration of cannabinoid agonists strongly attenuated the severity of morphine abstinence (Hine et al. 1975; Bhargava 1976a,b; Bhargava and Way 1976; Vela et al. 1995). Furthermore, a chronic pre-treatment with THC before starting chronic morphine administration reduced the somatic manifestations of naloxone-precipitated morphine withdrawal, without modifying the motivational responses of this opioid compound (Valverde et al. 2000b).

Reciprocally, the endogenous opioid system has been reported to be involved in the motivational responses and withdrawal manifestations induced by cannabinoids. Thus, the rewarding effects induced by THC were abolished in μ-opioid receptor knockout mice (Ghozland et al. 2002). Furthermore, the dysphoric effects induced by a high dose of THC (5 mg/kg) were slightly attenuated in μ-knockout mice and completely blocked in mice lacking κ-opioid receptors (Ghozland et al. 2002). The conditioned place aversion induced by a high dose of THC (5 mg/kg) was also abolished in prodynorphin knockout mice, also supporting the involvement of κ-opioid receptors in the motivational responses induced by cannabinoids (Zimmer et al. 2001). In addition, the rewarding responses induced by THC in the conditioned place paradigm were also abolished in double knockout mice lacking both μ- and δ-opioid receptors (Castañe et al. 2003). There is also evidence to suggest that the endogenous opioid system participates in the reinforcing properties of cannabinoids. Thus, the opioid antagonist naloxone partially blocked self-administration of the cannabinoid agonist CP55,940 (Braida et al. 2001). THC self-administration behaviour was also attenuated by a different opioid antagonist naltrexone (Justinova et al. 2004). Furthermore, naloxone precipitated some behavioural signs of abstinence in rats chronically treated with a cannabinoid agonist (Kaymakcalan et al. 1977; Navarro et al. 2001).

The role of the endogenous opioid peptides in cannabinoid dependence has also been investigated by using knockout mice. The expression of cannabinoid withdrawal was attenuated in THC-dependent knockout mice lacking the preproenkephalin gene (Valverde et al. 2000a). However, THC abstinence was not modified in μ-, δ- or κ-opioid receptor knockout mice (Ghozland et al. 2002). In contrast, another study reported a decrease in the severity of cannabinoid withdrawal syndrome in μ-opioid receptor knockout mice (Lichtman et al. 2001). The different genetic construction of knockout mice and the changes in the experimental conditions can explain these discrepancies. Finally, a significant decrease in the severity of cannabinoid withdrawal syndrome was observed in double μ-, δ-opioid receptor knockout mice (Castañe et al. 2003), suggesting that a cooperative action of μ- and δ-opioid receptors is required for the entire expression of THC dependence.

All these results indicate that the bi-directional interactions between the endogenous cannabinoid and opioid systems are crucial for the motivational properties and the development of physical dependence induced by these two kinds of drugs, and could provide new strategies for a more rational approach to the treatment of drug abuse.

9.2
Interaction Between Cannabinoids and Psychostimulants

The endogenous cannabinoid system has been reported to be involved in the addictive effects induced by other drugs of abuse, such as cocaine and other psychostimulants. Dopaminergic activity in the mesocorticolimbic system is considered a common feature mediating the primary reinforcing effects of most drugs of abuse (Di Chiara 1998). Psychostimulants facilitate this dopaminergic neurotransmission by different mechanisms, including the enhancement of extracellular dopamine concentrations, mainly through inhibition of the dopamine transporter. On the other hand, CB_1 cannabinoid receptors are important modulators of dopaminergic activity in the mesocorticolimbic system, suggesting that the endogenous cannabinoid system may contribute to the reinforcing properties of different drugs of abuse, including psychostimulants. However, the possible mechanisms involved in such an interaction remain controversial, because only a few studies have been performed on this topic and have frequently provided contradictory results.

Several studies suggest that CB_1 cannabinoid receptors do not participate in the acute rewarding properties of psychostimulants. Thus, cocaine-induced conditioned place preference and sensitization to the hyperlocomotor effects produced by chronic administration of the drug were preserved in CB_1 knockout mice (Martin et al. 2000). In addition, acute self-administration of cocaine, performed during a single session, was also maintained in mice lacking CB_1 receptors (Cossu et al. 2001). However, administration of the cannabinoid agonist WIN55,212-2 has been found to decrease the reinforcing actions of cocaine in a brain stimulation paradigm in mice (Vlachou et al. 2003), whereas the blockade of CB_1 receptors by SR141716A treatment decreased the reinforcing value of intracranial self-stimulation in rats (Deroche-Gamonet et al. 2001). These results suggest that the endogenous cannabinoid system could modulate cocaine reward. Other studies have also supported the existence of an interaction between cocaine and cannabinoids in reinforcing responses. Thus, pretreatment with WIN55,212-2 of rats self-administering cocaine reduces cocaine intake in a dose-dependent manner. The CB_1 antagonist SR141716A completely reversed these effects of WIN55,212-2, indicating that the reinforcing effects of CB_1-mediated and cocaine-induced reward mechanisms are additive (Fattore et al. 1999).

Furthermore, the endocannabinoid system plays an important role in the neuronal processes underlying cocaine-seeking behaviour. Thus, the cannabinoid agonist HU-210 induces relapse to cocaine seeking after prolonged withdrawal periods, and the antagonist SR141716A attenuates this response when it is induced by re-exposure to cocaine-associated cues or to cocaine itself (De Vries et al. 2001). It therefore seems necessary to perform further studies by using CB_1 knockout mice to evaluate the contribution of these receptors in processes related to the acquisition, maintenance and extinction of cocaine self-administration, and thus further clarify the nature of the interaction between cocaine and the endocannabinoid system.

Recent studies have also evaluated the interaction between cannabinoids and other psychostimulants such as amphetamine and MDMA (methylenedioxymethamphetamine; ecstasy) (Braida and Sala 2002; Parker et al. 2004). These studies showed that infusion of the cannabinoid agonist CP55,940 decreased intracerebroventricular MDMA self-administration in rats (Braida and Sala 2002). It remains to be determined, however, if cannabinoids modulate the addictive properties of psychostimulant drugs.

9.3 Interaction Between Cannabinoids and Nicotine

The consumption of cannabis is highly associated with tobacco, which contains nicotine, an important psychoactive compound (Nemeth-Coslett et al. 1986; McCambridge and Strang 2004). The administration of THC and nicotine in rodents produces multiple common pharmacological responses including analgesia, hypothermia, impairment of locomotor activity and addiction (Hildebrand et al. 1997; Ameri 1999; Maldonado and Rodriguez de Fonseca 2002). Nicotine responses are mediated by the activation of nicotinic acetylcholine receptors, which have a pentameric structure consisting of different receptor subunits (Grutter and Changeux 2001; Le Novere et al. 2002).

Several studies have suggested a possible functional interaction between cannabinoid and nicotinic systems. The specific behavioural and biochemical consequences of such an interaction are poorly documented in animal models in spite of the high frequency of association of these two substances in humans. Nicotine facilitated THC-induced acute pharmacological and biochemical responses in mice, including hypothermia, antinociception, hypolocomotion and anxiolytic-like responses. Furthermore, the co-administration of sub-threshold doses of THC and nicotine produced conditioned place preference (Valjent et al. 2002). Mice co-treated with nicotine and THC displayed attenuation in THC tolerance and an enhancement in the somatic expression of cannabinoid antagonist-precipitated THC withdrawal (Valjent et al. 2002). These findings showed that low doses of cannabinoids associated with nicotine could have a higher capability to induce behavioural responses related to addictive processes than THC administration alone, and could enhance the somatic consequences of chronic consumption of these drugs.

Some behavioural responses induced by nicotine were modified in mice lacking CB_1 cannabinoid receptors. Thus, whereas the severity of nicotine withdrawal syndrome was not affected in CB_1 knockout mice, the rewarding properties of nicotine, evaluated in the conditioned place preference assay, was abolished in these animals (Castañe et al. 2003). In contrast, the absence of CB_1 cannabinoid receptors did not modify acute self-administration induced by nicotine (Cossu et al. 2001). The effective doses in these two behavioural models (acute intravenous self-administration and conditioned place preference) are different, which makes it difficult to directly compare the results of these studies. However, the interaction between THC and nicotine previously reported by using pharmacological and biochemical approaches (Valjent et al. 2002) are in agreement with the impairment of nicotine rewarding effects in CB_1 knockout mice (Castañe et al. 2002). In addition, the administration of SR141716A decreased nicotine self-administration in rats, and nicotine-induced dopamine release in the nucleus accumbens and the bed nucleus of the stria terminalis, supporting the role of the endocannabinoid system in nicotine rewarding effects (Cohen et al. 2002). SR141716A increased dopamine, noradrenaline and serotonin levels in the cortex and the nucleus accumbens (Tzavara et al. 2003), which could contribute to its ability to reverse nicotine-induced responses. SR141716A could have anti-smoking activity in humans, accordingly to promising findings obtained in a placebo-controlled phase III clinical trial using this compound (Fernandez and Allison 2004).

Studies into the addictive properties of cannabinoids using knockout mice lacking different protein subunits of nicotinic receptors could greatly extend our knowledge of the neurobiological mechanisms involved in the interaction between cannabinoids and nicotine.

9.4 Interaction Between Cannabinoids and Ethanol

There is now considerable evidence to suggest a possible involvement of the cannabinoid CB_1 receptor in the addiction-related effects of ethanol (Mechoulam and Parker 2003). Both, cannabinoids and ethanol produce some similar physiological and behavioural responses including euphoria, motor incoordination and hypothermia. CB_1 ligands are able to modulate ethanol preference and self-administration (Arnone et al. 1997; Freedland et al. 2001; Mechoulam and Parker 2003). Furthermore, chronic ethanol treatment increases the synthesis of endocannabinoids and down-regulates brain CB_1 receptors and their function (Basavarajappa and Hungund 2002), supporting the hypothesis of an interaction between these two drugs. Pharmacological studies reported that blocking the CB_1 receptor with SR141716A reduced ethanol consumption (Arnone et al. 1997; Freedland et al. 2001).

A recent study on a CD1 genetic background showed that ethanol consumption and preference were decreased in CB_1 knockout mice, whereas ethanol sensitivity and withdrawal severity were increased in these mice (Naassila et al. 2004). These observations are similar to those reported in a previous study showing decreased ethanol consumption and increased sensitivity to the acute effects of ethanol in CB_1 knockout mice on a C57BL/6J genetic background (Hungund et al. 2003). Furthermore, ethanol did not cause release of dopamine in the nucleus accumbens in CB_1 knockout mice, in contrast to the effects observed in wild-type littermates. In agreement, SR141716A completely abolished the enhancement of dopamine responses induced by acute ethanol in the nucleus accumbens of wild-type mice (Hungund et al. 2003). Similarly, a reduction in the effects of ethanol on extracellular levels of dopamine in the nucleus accumbens after SR141716A administration has been previously reported, suggesting that cannabinoids modulate the reinforcing properties of ethanol by decreasing the release of dopamine in limbic areas

(Cohen et al. 2002). Another study also supports the hypothesis that endocannabinoids acting on CB_1 receptors contribute to ethanol rewarding effects, albeit in an apparent age-dependent manner (Wang et al. 2003). Thus, a high ethanol preference was found in young (6–10 weeks) C57BL/6J mice that was reduced in CB_1 knockout mice. The administration of the antagonist SR141716A to young wild-type mice reduced ethanol preference to the level exhibited by CB_1 knockout mice. Ethanol preference declined in old wild-type mice (26–48 weeks), and this reached a level similar to that observed in CB_1 knockout mice (similar for young and old animals). Ethanol preference in old CB_1 knockout and wild-type littermates was unaffected by SR141716A (Wang et al. 2003). The age-dependent differences for ethanol preference reported in this study could probably explain some of the discrepancies between results that have been obtained from different studies with CB_1 knockout mice. Thus, Racz et al. (2003) reported that CB_1 knockout mice (on a C57BL/6J genetic background) showed initially an even higher preference for ethanol than wild-type littermates. After 1 week, the ethanol consumption was virtually identical in knockout and wild-type mice. Withdrawal symptoms after the cessation of chronic ethanol administration were completely absent in CB_1 knockout mice (Racz et al. 2003). Activation of the CB_1 receptor promotes alcohol craving and suggests a role of this receptor in excessive ethanol drinking behaviour and the development of alcoholism (Schmidt et al. 2002). Interestingly, this recent clinical study associated a CB_1 cannabinoid receptor gene polymorphism with the severity of withdrawal symptoms in humans (Schmidt et al. 2002).

Recently, a new CB_1 receptor antagonist, namely SR147778, has been developed. This compound is able to reduce both ethanol and sucrose consumption in mice and rats (Rinaldi-Carmona et al. 2004), supporting the involvement of the CB_1 cannabinoid receptor in ethanol consumption. Taken together, these results suggest an involvement of endocannabinoids in the rewarding effects, physical dependence and craving induced by ethanol. Further studies must to be performed in order to clarify the apparent discrepancies observed in the different studies performed with CB_1 knockout mice.

10
CB_1 Receptors in the Control of Feeding Behaviour

The appetite-stimulating effects of marijuana have been known for centuries and constitute one of the established medicinal uses of cannabis preparations. Today THC (dronabinol/Marinol) is clinically used for the treatment of cachexia-anorexia in human immunodeficiency virus (HIV) and palliative care patients. There have also been very promising advances in the development of a cannabinoid receptor antagonist (SR141716A, now named Rimonabant or Acomplia) for the treatment of obesity.

Pharmacological studies in animals are consistent with a role of the endogenous cannabinoid system in the regulation of feeding behaviours and food palatability (Williams and Kirkham 2002a,b; Higgs et al. 2003). Administration of THC to rats produced a significant hyperphagia that was reversed by SR141716A (Williams et al. 1998; Williams and Kirkham 2002b). Since 2-AG is present in the milk of humans and animals, Fride and her collegues asked whether this endocannabinoid might promote appetite and suckling behaviour in newborn animals. Indeed, the administration of SR141716A to newborn mice, within the first 24 h after birth, had a devastating effect on milk ingestion and often led to the death of the treated animals. CB_1 receptor-deficient mice also failed to drink in the first 24 h after birth, but started to display milk bands from day 2. It seems that this delayed onset of milk intake affects the survival rate of CB_1 knockout pups, which was significantly lower than that of wild-type littermates in Fride's studies (Fride et al. 2001, 2003). Our (A.Z.) previous analysis of the distribution of genotypes among offspring of heterozygous matings indicated a small deviation from the expected Mendelian frequency at the time of weaning ($CB_1{}^{+/+}$, 29%; $CB_1{}^{+/-}$, 47,7%; $CB_1{}^{-/-}$, 23.3%; $n = 1{,}439$), thus also suggesting a somewhat reduced viability of homozygous and even heterozygous pups (Zimmer et al. 1999). These results suggest that endocannabinoids in the milk promote suckling behaviour during the early postnatal period.

The body weight of adult CB_1 receptor knockout mice was, however, similar to that of control animals, indicating that the endocannabinoid system is not critical for maintaining regular food intake under normal laboratory conditions (Zimmer et al. 1999). In contrast, when animals were food deprived for 18 h, wild-type mice consumed significantly more food at the end of the fasting period than CB_1-deficient animals (Di Marzo et al. 2001). Wild-type mice that were treated with 3 mg/kg SR141716A 10 min before the start of the testing period also showed a lower food intake, similar to that of CB_1 knockouts. The orexigenic effects of cannabinoids are thought to be mediated by hypothalamic CB_1 receptors, although the CB_1 receptor density in the hypothalamus is lower than in many other brain regions (Marsicano and Lutz 1999; Harrold and Williams 2003). The endocannabinoid system in the hypothalamus seems to be part of a leptin-sensitive regulatory pathway, as leptin decreases hypothalamic endocannabinoid synthesis, whilst defective leptin signalling in obese (ob/ob) or diabetic (db/db) mice is accompanied by elevated endocannabinoid levels (Di Marzo et al. 2001). Fasting also increased 2-AG levels in the hypothalamus and in the limbic forebrain, whilst hypothalamic 2-AG levels declined as animals ate (Kirkham et al. 2002). Together these results are consistent with a role of leptin-regulated endocannabinoids in the control of motivational aspects of feeding behaviour.

11 Endocannabinoid as Retrograde Neurotransmitter

Several recent studies have begun to elucidate the cellular and molecular mechanisms underlying the numerous and profound effects of cannabinoids on the brain. Indeed there is now compelling evidence that endocannabinoids act as activity-dependent retrograde inhibitors of synaptic transmission.

In the hippocampus, CB_1 receptors are localized presynaptically in GABA axon terminals, most of which originate from CCK-positive basket cells (Katona et al.

1999). Endocannabinoids are probably synthesized by Ca^{2+}-dependent postsynaptically localized enzymes (Bisogno et al. 2003). Activation of the presynaptic CB_1 receptors exerts diverse effects on synaptic functions, including the activation of inwardly rectifying K^+ channels, the inhibition of voltage-gated Ca^{2+} channels and the suppression of neurotransmitter release (Di Marzo et al. 1998; Freund et al. 2003). Because of the distribution and function of its various components, the endocannabinoid system seemed ideally suited to mediate a form of activity-dependent modulation of synaptic activity in the hippocampus that has been termed depolarization-induced suppression of inhibition (DSI). DSI describes a phenomenon in which a brief depolarization of a pyramidal neuron transiently suppresses the release of GABA from presynaptic terminals (Pitler and Alger 1992, 1994). A similar phenomenon affecting excitatory glutamatergic synapses has been described in the cerebellum and hippocampus, and is termed depolarization-induced suppression of excitation (DSE). Because DSI and DSE are initiated postsynaptically through an elevation of cytoplasmic Ca^{2+} and expressed presynaptically as an inhibition of neurotransmitter release, a retrograde signal that travels backwards across synapses had been postulated (Wilson and Nicoll 2002). Several studies have now conclusively demonstrated that the retrograde messengers responsible for this signalling are endocannabinoids. In the hippocampus, the CB_1-selective agonist WIN55,212-2 blocked GABA release and suppressed baseline inhibitory post-synaptic current (IPSC) amplitudes (Hajos et al. 2000; Wilson and Nicoll 2001). The CB_1 antagonists SR141716A and AM251 blocked DSI (Wilson and Nicoll 2001). Excitatory hippocampal synapses displayed an analogous reduction: WIN55,212-2 blocked excitatory post-synaptic currents (EPSC) and SR141716A blocked DSE. Importantly, DSI and DSE were completely absent in CB_1 knockout mice from the Zimmer laboratory in the hippocampus and in the cerebellum (Yoshida et al. 2002). However, Hajos and colleagues have pointed out that anatomical studies could not confirm the existence of CB_1 receptors on hippocampal glutamatergic terminals and have reported that CB_1-deficient mice generated by Ledent and co-workers still show a reduction of postsynaptic excitatory currents in hippocampal slices by WIN55,212-2 (Hajos et al. 2001). These authors speculate that the effect of cannabinoids on excitatory hippocampal neurons is mediated by a non-CB_1 receptor. Clearly, further studies are necessary to determine the reason for these contradictory findings.

12 Outlook

Knockout mice have revealed many novel and interesting aspects of the physiological functions of CB_1 receptors in locomotor activity, emotional behaviours, regulation of blood pressure, cognition, pain, reproduction and addiction. In addition, these animals have become invaluable tools for studying the interactions between cannabinoids and other drugs of abuse, i.e. opioids, nicotine, ethanol and cocaine. The multitude of phenotypes that have been observed in these animals reflects the diversity of functions of the endogenous cannabinoid system. Undoubtedly, these results will further the potential medical uses of cannabinoid receptor agonist and antagonists.

Although the phenotype of the different knockout mice is very similar among the individual strains and laboratories involved, small differences do exist. It remains to be determined if these phenotypic differences are due to variations in the genetic background, different holding conditions, or both. Understanding the impact of these epigenetic factors may help us to appreciate the significance of the endocannabinoid system in environmentally and genetically more complex systems.

Whilst most of the research of the endocannabinoid system in the last decade has focussed on the CB_1 and CB_2 receptors, we have also made substantial advances in the identification of endocannabinoid degrading and synthesizing enzymes and the effects of endocannabinoids that are not mediated by these receptors. Future animal models will therefore increasingly address the relevance of non-CB_1 and non-CB_2 endocannabinoid binding sites and the regulation of endocannabinoid levels.

References

Aceto MD, Scates SM, Lowe JA, Martin BR (1996) Dependence on delta 9-tetrahydrocannabinol: studies on precipitated and abrupt withdrawal. J Pharmacol Exp Ther 278:1290–1295

Ameri A (1999) The effects of cannabinoids on the brain. Prog Neurobiol 58:315–348

Arnold JC, Hunt GE, McGregor IS (2001) Effects of the cannabinoid receptor agonist CP 55,940 and the cannabinoid receptor antagonist SR 141716 on intracranial self-stimulation in Lewis rats. Life Sci 70:97–108

Arnone M, Maruani J, Chaperon F, Thiebot MH, Poncelet M, Soubrie P, Le Fur G (1997) Selective inhibition of sucrose and ethanol intake by SR 141716, an antagonist of central cannabinoid (CB_1) receptors. Psychopharmacology (Berl) 132:104–106

Banks MI, White JA, Pearce RA (2000) Interactions between distinct GABA(A) circuits in hippocampus. Neuron 25:449–457

Basavarajappa BS, Hungund BL (2002) Neuromodulatory role of the endocannabinoid signaling system in alcoholism: an overview. Prostaglandins Leukot Essent Fatty Acids 66:287–299

Basbaum AI, Fields HL (1984) Endogenous pain control systems: brainstem spinal pathways and endorphin circuitry. Annu Rev Neurosci 7:309–338

Batkai S, Jarai Z, Wagner JA, Goparaju SK, Varga K, Liu J, Wang L, Mirshahi F, Khanolkar AD, Makriyannis A, Urbaschek R, Garcia N Jr, Sanyal AJ, Kunos G (2001) Endocannabinoids acting at vascular CB_1 receptors mediate the vasodilated state in advanced liver cirrhosis. Nat Med 7:827–832

Batkai S, Pacher P, Jarai Z, Wagner JA, Kunos G (2004) Cannabinoid antagonist SR-141716 inhibits endotoxic hypotension by a cardiac mechanism not involving CB_1 or CB_2 receptors. Am J Physiol Heart Circ Physiol 287:H595–H600

Benowitz NL, Jones RT (1975) Cardiovascular effects of prolonged delta-9-tetrahydrocannabinol ingestion. Clin Pharmacol Ther 18:287–297

Bhargava HN (1976a) Effect of some cannabinoids on naloxone-precipitated abstinence in morphine-dependent mice. Psychopharmacology (Berl) 49:267–270

Bhargava HN (1976b) Inhibition of naloxone-induced withdrawal in morphine dependent mice by 1-trans-delta9-tetrahydrocannabinol. Eur J Pharmacol 36:259–262

Bhargava HN, Way EL (1976) Morphine tolerance and physical dependence: influence of cholinergic agonists and antagonists. Eur J Pharmacol 36:79–88
Bicher HI, Mechoulam R (1968) Pharmacological effects of two active constituents of marihuana. Arch Int Pharmacodyn Ther 172:24–31
Bisogno T, Howell F, Williams G, Minassi A, Cascio MG, Ligresti A, Matias I, Schiano-Moriello A, Paul P, Williams EJ, Gangadharan U, Hobbs C, Di Marzo V, Doherty P (2003) Cloning of the first sn1-DAG lipases points to the spatial and temporal regulation of endocannabinoid signaling in the brain. J Cell Biol 163:463–468
Bohme GA, Laville M, Ledent C, Parmentier M, Imperato A (2000) Enhanced long-term potentiation in mice lacking cannabinoid CB_1 receptors. Neuroscience 95:5–7
Braida D, Sala M (2000) Cannabinoid-induced working memory impairment is reversed by a second generation cholinesterase inhibitor in rats. Neuroreport 11:2025–2029
Braida D, Sala M (2002) Role of the endocannabinoid system in MDMA intracerebral self-administration in rats. Br J Pharmacol 136:1089–1092
Braida D, Pozzi M, Parolaro D, Sala M (2001) Intracerebral self-administration of the cannabinoid receptor agonist CP 55,940 in the rat: interaction with the opioid system. Eur J Pharmacol 413:227–234
Breivogel CS, Griffin G, Di Marzo V, Martin BR (2001) Evidence for a new G protein-coupled cannabinoid receptor in mouse brain. Mol Pharmacol 60:155–163
Bridges D, Rice AS, Egertova M, Elphick MR, Winter J, Michael GJ (2003) Localisation of cannabinoid receptor 1 in rat dorsal root ganglion using in situ hybridisation and immunohistochemistry. Neuroscience 119:803–812
Buxbaum DM (1972) Analgesic activity of 9-tetrahydrocannabinol in the rat and mouse. Psychopharmacologia 25:275–280
Calignano A, La Rana G, Giuffrida A, Piomelli D (1998) Control of pain initiation by endogenous cannabinoids. Nature 394:277–281
Carney JM, Uwaydah IM, Balster RL (1977) Evaluation of a suspension system for intravenous self-administration studies of water-insoluble compounds in the rhesus monkey. Pharmacol Biochem Behav 7:357–364
Castañe A, Valjent E, Ledent C, Parmentier M, Maldonado R, Valverde O (2002) Lack of CB_1 cannabinoid receptors modifies nicotine behavioural responses, but not nicotine abstinence. Neuropharmacology 43:857–867
Castañe A, Robledo P, Matifas A, Kieffer BL, Maldonado R (2003) Cannabinoid withdrawal syndrome is reduced in double mu and delta opioid receptor knockout mice. Eur J Neurosci 17:155–159
Chait LD, Perry JL (1992) Factors influencing self-administration of, and subjective response to, placebo marijuana. Behav Pharmacol 3:545–552
Chaperon F, Thiebot MH (1999) Behavioral effects of cannabinoid agents in animals. Crit Rev Neurobiol 13:243–281
Cohen C, Perrault G, Voltz C, Steinberg R, Soubrie P (2002) SR141716, a central cannabinoid (CB(1)) receptor antagonist, blocks the motivational and dopamine-releasing effects of nicotine in rats. Behav Pharmacol 13:451–463
Compton DR, Aceto MD, Lowe J, Martin BR (1996) In vivo characterization of a specific cannabinoid receptor antagonist (SR141716A): inhibition of delta 9-tetrahydrocannabinol-induced responses and apparent agonist activity. J Pharmacol Exp Ther 277:586–594
Cook SA, Lowe JA, Martin BR (1998) CB_1 receptor antagonist precipitates withdrawal in mice exposed to Delta9-tetrahydrocannabinol. J Pharmacol Exp Ther 285:1150–1156
Corcoran ME, Amit Z (1974) Reluctance of rats to drink hashish suspensions: free-choice and forced consumption, and the effects of hypothalamic stimulation. Psychopharmacologia 35:129–147
Cossu G, Ledent C, Fattore L, Imperato A, Bohme GA, Parmentier M, Fratta W (2001) Cannabinoid CB_1 receptor knockout mice fail to self-administer morphine but not other drugs of abuse. Behav Brain Res 118:61–65

De Vries TJ, Shaham Y, Homberg JR, Crombag H, Schuurman K, Dieben J, Vanderschuren LJ, Schoffelmeer AN (2001) A cannabinoid mechanism in relapse to cocaine seeking. Nat Med 7:1151–1154

Derkinderen P, Valjent E, Toutant M, Corvol JC, Enslen H, Ledent C, Trzaskos J, Caboche J, Girault JA (2003) Regulation of extracellular signal-regulated kinase by cannabinoids in hippocampus. J Neurosci 23:2371–2382

Deroche-Gamonet V, Le Moal M, Piazza PV, Soubrie P (2001) SR141716, a CB_1 receptor antagonist, decreases the sensitivity to the reinforcing effects of electrical brain stimulation in rats. Psychopharmacology (Berl) 157:254–259

Dewey WL (1986) Cannabinoid pharmacology. Pharmacol Rev 38:151–178

Di Chiara G (1998) A motivational learning hypothesis of the role of mesolimbic dopamine in compulsive drug use. J Psychopharmacol 12:54–67

Di Marzo V, Melck D, Bisogno T, De Petrocellis L (1998) Endocannabinoids: endogenous cannabinoid receptor ligands with neuromodulatory action. Trends Neurosci 21:521–528

Di Marzo V, Breivogel CS, Tao Q, Bridgen DT, Razdan RK, Zimmer AM, Zimmer A, Martin BR (2000) Levels, metabolism, and pharmacological activity of anandamide in CB(1) cannabinoid receptor knockout mice: evidence for non-CB(1), non-CB(2) receptor-mediated actions of anandamide in mouse brain. J Neurochem 75:2434–2444

Di Marzo V, Goparaju SK, Wang L, Liu J, Batkai S, Jarai Z, Fezza F, Miura GI, Palmiter RD, Sugiura T, Kunos G (2001) Leptin-regulated endocannabinoids are involved in maintaining food intake. Nature 410:822–825

Diana MA, Marty A (2004) Endocannabinoid-mediated short-term synaptic plasticity: depolarization-induced suppression of inhibition (DSI) and depolarization-induced suppression of excitation (DSE). Br J Pharmacol 142:9–19

Fan F, Tao Q, Abood M, Martin BR (1996) Cannabinoid receptor down-regulation without alteration of the inhibitory effect of CP 55,940 on adenylyl cyclase in the cerebellum of CP 55,940-tolerant mice. Brain Res 706:13–20

Fattore L, Martellotta MC, Cossu G, Mascia MS, Fratta W (1999) CB_1 cannabinoid receptor agonist WIN 55,212-2 decreases intravenous cocaine self-administration in rats. Behav Brain Res 104:141–146

Fattore L, Spano MS, Cossu G, Deiana S, Fratta W (2003) Cannabinoid mechanism in reinstatement of heroin-seeking after a long period of abstinence in rats. Eur J Neurosci 17:1723–1726

Fernandez JR, Allison DB (2004) Rimonabant Sanofi-Synthelabo. Curr Opin Investig Drugs 5:430–435

Fields HL, Heinricher MM, Mason P (1991) Neurotransmitters in nociceptive modulatory circuits. Annu Rev Neurosci 14:219–245

Freedland CS, Sharpe AL, Samson HH, Porrino LJ (2001) Effects of SR141716A on ethanol and sucrose self-administration. Alcohol Clin Exp Res 25:277–282

Freund TF, Katona I, Piomelli D (2003) Role of endogenous cannabinoids in synaptic signaling. Physiol Rev 83:1017–1066

Fride E, Ginzburg Y, Breuer A, Bisogno T, Di Marzo V, Mechoulam R (2001) Critical role of the endogenous cannabinoid system in mouse pup suckling and growth. Eur J Pharmacol 419:207–214

Fride E, Foox A, Rosenberg E, Faigenboim M, Cohen V, Barda L, Blau H, Mechoulam R (2003) Milk intake and survival in newborn cannabinoid CB_1 receptor knockout mice: evidence for a "CB3" receptor. Eur J Pharmacol 461:27–34

Gardner EL, Paredes W, Smith D, Donner A, Milling C, Cohen D, Morrison D (1988) Facilitation of brain stimulation reward by delta 9-tetrahydrocannabinol. Psychopharmacology (Berl) 96:142–144

Gerfen CR (1992) The neostriatal mosaic: multiple levels of compartmental organization. Trends Neurosci 15:133–139

Gerfen CR (1993) Functional organization of the striatum: relevance to actions of psychostimulant drugs of abuse. NIDA Res Monogr 125:82–91

Ghozland S, Matthes HW, Simonin F, Filliol D, Kieffer BL, Maldonado R (2002) Motivational effects of cannabinoids are mediated by mu-opioid and kappa-opioid receptors. J Neurosci 22:1146–1154
Grutter T, Changeux JP (2001) Nicotinic receptors in wonderland. Trends Biochem Sci 26:459–463
Guhring H, Hamza M, Sergejeva M, Ates M, Kotalla CE, Ledent C, Brune K (2002) A role for endocannabinoids in indomethacin-induced spinal antinociception. Eur J Pharmacol 454:153–163
Hajos N, Katona I, Naiem SS, MacKie K, Ledent C, Mody I, Freund TF (2000) Cannabinoids inhibit hippocampal GABAergic transmission and network oscillations. Eur J Neurosci 12:3239–3249
Hajos N, Ledent C, Freund TF (2001) Novel cannabinoid-sensitive receptor mediates inhibition of glutamatergic synaptic transmission in the hippocampus. Neuroscience 106:1–4
Haller J, Bakos N, Szirmay M, Ledent C, Freund TF (2002) The effects of genetic and pharmacological blockade of the CB_1 cannabinoid receptor on anxiety. Eur J Neurosci 16:1395–1398
Hampson RE, Deadwyler SA (2000) Cannabinoids reveal the necessity of hippocampal neural encoding for short-term memory in rats. J Neurosci 20:8932–8942
Hanus L, Breuer A, Tchilibon S, Shiloah S, Goldenberg D, Horowitz M, Pertwee RG, Ross RA, Mechoulam R, Fride E (1999) HU-308: a specific agonist for CB(2), a peripheral cannabinoid receptor. Proc Natl Acad Sci U S A 96:14228–14233
Harris RT, Waters W, McLendon D (1974) Evaluation of reinforcing capability of delta-9-tetrahydrocannabinol in rhesus monkeys. Psychopharmacologia 37:23–29
Harrold JA, Williams G (2003) The cannabinoid system: a role in both the homeostatic and hedonic control of eating? Br J Nutr 90:729–734
Herzberg U, Eliav E, Bennett GJ, Kopin IJ (1997) The analgesic effects of R(+)-WIN 55,212-2 mesylate, a high affinity cannabinoid agonist, in a rat model of neuropathic pain. Neurosci Lett 221:157–160
Higgs S, Williams CM, Kirkham TC (2003) Cannabinoid influences on palatability: microstructural analysis of sucrose drinking after delta(9)-tetrahydrocannabinol, anandamide, 2-arachidonoyl glycerol and SR141716. Psychopharmacology (Berl) 165:370–377
Hildebrand BE, Nomikos GG, Bondjers C, Nisell M, Svensson TH (1997) Behavioral manifestations of the nicotine abstinence syndrome in the rat: peripheral versus central mechanisms. Psychopharmacology (Berl) 129:348–356
Hine B, Torrelio M, Gershon S (1975) Interactions between cannabidiol and delta9-THC during abstinence in morphine-dependent rats. Life Sci 17:851–857
Hohmann AG (2002) Spinal and peripheral mechanisms of cannabinoid antinociception: behavioral, neurophysiological and neuroanatomical perspectives. Chem Phys Lipids 121:173–190
Hohmann AG, Herkenham M (1998) Regulation of cannabinoid and mu opioid receptors in rat lumbar spinal cord following neonatal capsaicin treatment. Neurosci Lett 252:13–16
Hohmann AG, Martin WJ, Tsou K, Walker JM (1995) Inhibition of noxious stimulus-evoked activity of spinal cord dorsal horn neurons by the cannabinoid WIN 55,212-2. Life Sci 56:2111–2118
Hungund BL, Szakall I, Adam A, Basavarajappa BS, Vadasz C (2003) Cannabinoid CB_1 receptor knockout mice exhibit markedly reduced voluntary alcohol consumption and lack alcohol-induced dopamine release in the nucleus accumbens. J Neurochem 84:698–704
Hutcheson DM, Tzavara ET, Smadja C, Valjent E, Roques BP, Hanoune J, Maldonado R (1998) Behavioural and biochemical evidence for signs of abstinence in mice chronically treated with delta-9-tetrahydrocannabinol. Br J Pharmacol 125:1567–1577
Jaggar SI, Hasnie FS, Sellaturay S, Rice AS (1998) The anti-hyperalgesic actions of the cannabinoid anandamide and the putative CB_2 receptor agonist palmitoylethanolamide in visceral and somatic inflammatory pain. Pain 76:189–199

Jarai Z, Wagner JA, Varga K, Lake KD, Compton DR, Martin BR, Zimmer AM, Bonner TI, Buckley NE, Mezey E, Razdan RK, Zimmer A, Kunos G (1999) Cannabinoid-induced mesenteric vasodilation through an endothelial site distinct from CB_1 or CB_2 receptors. Proc Natl Acad Sci U S A 96:14136–14141

Jarai Z, Wagner JA, Goparaju SK, Wang L, Razdan RK, Sugiura T, Zimmer AM, Bonner TI, Zimmer A, Kunos G (2000) Cardiovascular effects of 2-arachidonoyl glycerol in anesthetized mice. Hypertension 35:679–684

Justinova Z, Tanda G, Redhi GH, Goldberg SR (2003) Self-administration of delta9-tetrahydrocannabinol (THC) by drug naive squirrel monkeys. Psychopharmacology (Berl) 169:135–140

Justinova Z, Tanda G, Munzar P, Goldberg SR (2004) The opioid antagonist naltrexone reduces the reinforcing effects of Delta 9 tetrahydrocannabinol (THC) in squirrel monkeys. Psychopharmacology (Berl) 173:186–194

Kathmann M, Weber B, Zimmer A, Schlicker E (2001) Enhanced acetylcholine release in the hippocampus of cannabinoid CB(1) receptor-deficient mice. Br J Pharmacol 132:1169–1173

Katona I, Sperlagh B, Sik A, Kafalvi A, Vizi ES, Mackie K, Freund TF (1999) Presynaptically located CB_1 cannabinoid receptors regulate GABA release from axon terminals of specific hippocampal interneurons. J Neurosci 19:4544–4558

Kaymakcalan S, Ayhan IH, Tulunay FC (1977) Naloxone-induced or postwithdrawal abstinence signs in delta9-tetrahydrocannabinol-tolerant rats. Psychopharmacology (Berl) 55:243–249

Kelly S, Chapman V (2001) Selective cannabinoid CB_1 receptor activation inhibits spinal nociceptive transmission in vivo. J Neurophysiol 86:3061–3064

Kirkham TC, Williams CM, Fezza F, Di Marzo V (2002) Endocannabinoid levels in rat limbic forebrain and hypothalamus in relation to fasting, feeding and satiation: stimulation of eating by 2-arachidonoyl glycerol. Br J Pharmacol 136:550–557

Ko MC, Woods JH (1999) Local administration of delta9-tetrahydrocannabinol attenuates capsaicin-induced thermal nociception in rhesus monkeys: a peripheral cannabinoid action. Psychopharmacology (Berl) 143:322–326

Lake KD, Martin BR, Kunos G, Varga K (1997) Cardiovascular effects of anandamide in anesthetized and conscious normotensive and hypertensive rats. Hypertension 29:1204–1210

Le Novere N, Grutter T, Changeux JP (2002) Models of the extracellular domain of the nicotinic receptors and of agonist- and Ca2+-binding sites. Proc Natl Acad Sci U S A 99:3210–3215

Ledent C, Valverde O, Cossu G, Petitet F, Aubert JF, Beslot F, Bohme GA, Imperato A, Pedrazzini T, Roques BP, Vassart G, Fratta W, Parmentier M (1999) Unresponsiveness to cannabinoids and reduced addictive effects of opiates in CB_1 receptor knockout mice. Science 283:401–404

Lepore M, Vorel SR, Lowinson J, Gardner EL (1995) Conditioned place preference induced by delta 9-tetrahydrocannabinol: comparison with cocaine, morphine, and food reward. Life Sci 56:2073–2080

Lepore M, Liu X, Savage V, Matalon D, Gardner EL (1996) Genetic differences in delta 9-tetrahydrocannabinol-induced facilitation of brain stimulation reward as measured by a rate-frequency curve-shift electrical brain stimulation paradigm in three different rat strains. Life Sci 58:PL365–372

Li J, Daughters RS, Bullis C, Bengiamin R, Stucky MW, Brennan J, Simone DA (1999) The cannabinoid receptor agonist WIN 55,212-2 mesylate blocks the development of hyperalgesia produced by capsaicin in rats. Pain 81:25–33

Lichtman AH, Martin BR (1996) Delta 9-tetrahydrocannabinol impairs spatial memory through a cannabinoid receptor mechanism. Psychopharmacology (Berl) 126:125–131

Lichtman AH, Martin BR (2002) Marijuana withdrawal syndrome in the animal model. J Clin Pharmacol 42:20S–27S

Lichtman AH, Sheikh SM, Loh HH, Martin BR (2001) Opioid and cannabinoid modulation of precipitated withdrawal in delta(9)-tetrahydrocannabinol and morphine-dependent mice. J Pharmacol Exp Ther 298:1007–1014

Maccarrone M, Attina M, Bari M, Cartoni A, Ledent C, Finazzi-Agro A (2001) Anandamide degradation and N-acylethanolamines level in wild-type and CB_1 cannabinoid receptor knockout mice of different ages. J Neurochem 78:339–348

Maccarrone M, Valverde O, Barbaccia ML, Castañe A, Maldonado R, Ledent C, Parmentier M, Finazzi-Agro A (2002) Age-related changes of anandamide metabolism in CB_1 cannabinoid receptor knockout mice: correlation with behaviour. Eur J Neurosci 15:1178–1186

Mailleux P, Vanderhaeghen JJ (1992) Localization of cannabinoid receptor in the human developing and adult basal ganglia. Higher levels in the striatonigral neurons. Neurosci Lett 148:173–176

Malan TP Jr, Ibrahim MM, Vanderah TW, Makriyannis A, Porreca F (2002) Inhibition of pain responses by activation of CB(2) cannabinoid receptors. Chem Phys Lipids 121:191–200

Maldonado R, Rodriguez de Fonseca F (2002) Cannabinoid addiction: behavioral models and neural correlates. J Neurosci 22:3326–3331

Mallet PE, Beninger RJ (1998) The cannabinoid CB_1 receptor antagonist SR141716A attenuates the memory impairment produced by delta9-tetrahydrocannabinol or anandamide. Psychopharmacology (Berl) 140:11–19

Mansbach RS, Rovetti CC, Winston EN, Lowe JA 3rd (1996) Effects of the cannabinoid CB_1 receptor antagonist SR141716A on the behavior of pigeons and rats. Psychopharmacology (Berl) 124:315–322

Mao J, Price DD, Lu J, Keniston L, Mayer DJ (2000) Two distinctive antinociceptive systems in rats with pathological pain. Neurosci Lett 280:13–16

Marsicano G, Lutz B (1999) Expression of the cannabinoid receptor CB_1 in distinct neuronal subpopulations in the adult mouse forebrain. Eur J Neurosci 11:4213–4225

Marsicano G, Wotjak CT, Azad SC, Bisogno T, Rammes G, Cascio MG, Hermann H, Tang J, Hofmann C, Zieglgansberger W, Di Marzo V, Lutz B (2002) The endogenous cannabinoid system controls extinction of aversive memories. Nature 418:530–534

Martellotta MC, Cossu G, Fattore L, Gessa GL, Fratta W (1998) Self-administration of the cannabinoid receptor agonist WIN 55,212-2 in drug-naive mice. Neuroscience 85:327–330

Martin BR, Lichtman AH (1998) Cannabinoid transmission and pain perception. Neurobiol Dis 5:447–461

Martin M, Ledent C, Parmentier M, Maldonado R, Valverde O (2000) Cocaine, but not morphine, induces conditioned place preference and sensitization to locomotor responses in CB_1 knockout mice. Eur J Neurosci 12:4038–4046

Martin M, Ledent C, Parmentier M, Maldonado R, Valverde O (2002) Involvement of CB_1 cannabinoid receptors in emotional behaviour. Psychopharmacology (Berl) 159:379–387

Martin WJ, Hohmann AG, Walker JM (1996) Suppression of noxious stimulus-evoked activity in the ventral posterolateral nucleus of the thalamus by a cannabinoid agonist: correlation between electrophysiological and antinociceptive effects. J Neurosci 16:6601–6611

Martin WJ, Loo CM, Basbaum AI (1999) Spinal cannabinoids are anti-allodynic in rats with persistent inflammation. Pain 82:199–205

Mas-Nieto M, Pommier B, Tzavara ET, Caneparo A, Da Nascimento S, Le Fur G, Roques BP, Noble F (2001) Reduction of opioid dependence by the CB(1) antagonist SR141716A in mice: evaluation of the interest in pharmacotherapy of opioid addiction. Br J Pharmacol 132:1809–1816

Massa F, Marsicano G, Hermann H, Cannich A, Monory K, Cravatt BF, Ferri GL, Sibaev A, Storr M, Lutz B (2004) The endogenous cannabinoid system protects against colonic inflammation. J Clin Invest 113:1202–1209

Masserano JM, Karoum F, Wyatt RJ (1999) SR 141716A, a CB_1 cannabinoid receptor antagonist, potentiates the locomotor stimulant effects of amphetamine and apomorphine. Behav Pharmacol 10:429–432

Mazzari S, Canella R, Petrelli L, Marcolongo G, Leon A (1996) N-(2-hydroxyethyl)hexadecanamide is orally active in reducing edema formation and inflammatory hyperalgesia by down-modulating mast cell activation. Eur J Pharmacol 300:227–236

McCambridge J, Strang J (2004) The efficacy of single-session motivational interviewing in reducing drug consumption and perceptions of drug-related risk and harm among young people: results from a multi-site cluster randomized trial. Addiction 99:39–52

Mechoulam R, Parker L (2003) Cannabis and alcohol—a close friendship. Trends Pharmacol Sci 24:266–268

Miller LL, Branconnier RJ (1983) Cannabis: effects on memory and the cholinergic limbic system. Psychol Bull 93:441–456

Millns PJ, Chapman V, Kendall DA (2001) Cannabinoid inhibition of the capsaicin-induced calcium response in rat dorsal root ganglion neurones. Br J Pharmacol 132:969–971

Molina-Holgado F, Gonzalez MI, Leret ML (1995) Effect of delta 9-tetrahydrocannabinol on short-term memory in the rat. Physiol Behav 57:177–179

Naassila M, Pierrefiche O, Ledent C, Daoust M (2004) Decreased alcohol self-administration and increased alcohol sensitivity and withdrawal in CB_1 receptor knockout mice. Neuropharmacology 46:243–253

Navarro M, Carrera MR, Fratta W, Valverde O, Cossu G, Fattore L, Chowen JA, Gomez R, del Arco I, Villanua MA, Maldonado R, Koob GF, Rodriguez de Fonseca F (2001) Functional interaction between opioid and cannabinoid receptors in drug self-administration. J Neurosci 21:5344–5350

Nelson RJ, Young KA (1998) Behavior in mice with targeted disruption of single genes. Neurosci Biobehav Rev 22:453–462

Nemeth-Coslett R, Henningfield JE, O'Keeffe MK, Griffiths RR (1986) Effects of marijuana smoking on subjective ratings and tobacco smoking. Pharmacol Biochem Behav 25:659–665

Ohno-Shosaku T, Tsubokawa H, Mizushima I, Yoneda N, Zimmer A, Kano M (2002) Presynaptic cannabinoid sensitivity is a major determinant of depolarization-induced retrograde suppression at hippocampal synapses. J Neurosci 22:3864–3872

Parker LA, Burton P, Sorge RE, Yakiwchuk C, Mechoulam R (2004) Effect of low doses of Delta(9)-tetrahydrocannabinol and cannabidiol on the extinction of cocaine-induced and amphetamine-induced conditioned place preference learning in rats. Psychopharmacology (Berl) May 11 [Epub ahead of print]

Pertwee RG (2001) Cannabinoid receptors and pain. Prog Neurobiol 63:569–611

Pich EM, Epping-Jordan MP (1998) Transgenic mice in drug dependence research. Ann Med 30:390–396

Pitler TA, Alger BE (1992) Postsynaptic spike firing reduces synaptic GABAA responses in hippocampal pyramidal cells. J Neurosci 12:4122–4132

Pitler TA, Alger BE (1994) Depolarization-induced suppression of GABAergic inhibition in rat hippocampal pyramidal cells: G protein involvement in a presynaptic mechanism. Neuron 13:1447–1455

Racz I, Bilkei-Gorzo A, Toth ZE, Michel K, Palkovits M, Zimmer A (2003) A critical role for the cannabinoid CB_1 receptors in alcohol dependence and stress-stimulated ethanol drinking. J Neurosci 23:2453–2458

Reibaud M, Obinu MC, Ledent C, Parmentier M, Böhme GA, Imperato A (1999) Enhancement of memory in cannabinoid CB_1 receptor knock-out mice. Eur J Pharmacol 379:R1–R2

Richardson JD, Aanonsen L, Hargreaves KM (1998) Antihyperalgesic effects of spinal cannabinoids. Eur J Pharmacol 345:145–153

Rinaldi-Carmona M, Barth F, Congy C, Martinez S, Oustric D, Perio A, Poncelet M, Maruani J, Arnone M, Finance O, Soubrie P, Le Fur G (2004) SR147778 [5-(4-bromophenyl)-1-(2,4-dichlorophenyl)-4-ethyl-N-(1-piperidinyl)-1H-pyr azole-3-carboxamide], a new potent and selective antagonist of the CB_1 cannabinoid receptor: biochemical and pharmacological characterization. J Pharmacol Exp Ther 310:905–914

Rodriguez de Fonseca F, Gorriti MA, Fernandez-Ruiz JJ, Palomo T, Ramos JA (1994) Down-regulation of rat brain cannabinoid binding sites after chronic delta 9-tetrahydrocannabinol treatment. Pharmacol Biochem Behav 47:33–40
Rodriguez de Fonseca F, Del Arco I, Martin-Calderon JL, Gorriti MA, Navarro M (1998) Role of the endogenous cannabinoid system in the regulation of motor activity. Neurobiol Dis 5:483–501
Romero J, Berrendero F, Garcia-Gil L, Ramos JA, Fernandez-Ruiz JJ (1998) Cannabinoid receptor and WIN-55,212-2-stimulated [35S]GTP gamma S binding and cannabinoid receptor mRNA levels in the basal ganglia and the cerebellum of adult male rats chronically exposed to delta 9-tetrahydrocannabinol. J Mol Neurosci 11:109–119
Rubino T, Massi P, Patrini G, Venier I, Giagnoni G, Parolaro D (1994) Chronic CP-55,940 alters cannabinoid receptor mRNA in the rat brain: an in situ hybridization study. Neuroreport 5:2493–2496
Rubino T, Patrini G, Massi P, Fuzio D, Vigano D, Giagnoni G, Parolaro D (1998) Cannabinoid-precipitated withdrawal: a time-course study of the behavioral aspect and its correlation with cannabinoid receptors and G protein expression. J Pharmacol Exp Ther 285:813–819
Rubino T, Vigano D, Costa B, Colleoni M, Parolaro D (2000) Loss of cannabinoid-stimulated guanosine 5′-O-(3-[(35)S]thiotriphosphate) binding without receptor down-regulation in brain regions of anandamide-tolerant rats. J Neurochem 75:2478–2484
Sauer B (1998) Inducible gene targeting in mice using the Cre/lox system. Methods 14:381–392
Schmidt LG, Samochowiec J, Finckh U, Fiszer-Piosik E, Horodnicki J, Wendel B, Rommelspacher H, Hoehe MR (2002) Association of a CB_1 cannabinoid receptor gene (CNR1) polymorphism with severe alcohol dependence. Drug Alcohol Depend 65:221–224
Sim LJ, Hampson RE, Deadwyler SA, Childers SR (1996) Effects of chronic treatment with delta9-tetrahydrocannabinol on cannabinoid-stimulated [35S]GTPgammaS autoradiography in rat brain. J Neurosci 16:8057–8066
Smith FL, Fujimori K, Lowe J, Welch SP (1998) Characterization of delta9-tetrahydrocannabinol and anandamide antinociception in nonarthritic and arthritic rats. Pharmacol Biochem Behav 60:183–191
Steiner H, Gerfen CR (1998) Role of dynorphin and enkephalin in the regulation of striatal output pathways and behavior. Exp Brain Res 123:60–76
Steiner H, Bonner TI, Zimmer AM, Kitai ST, Zimmer A (1999) Altered gene expression in striatal projection neurons in CB_1 cannabinoid receptor knockout mice [see comments]. Proc Natl Acad Sci U S A 96:5786–5790
Stella N, Schweitzer P, Piomelli D (1997) A second endogenous cannabinoid that modulates long-term potentiation. Nature 388:773–778
Strangman NM, Patrick SL, Hohmann AG, Tsou K, Walker JM (1998) Evidence for a role of endogenous cannabinoids in the modulation of acute and tonic pain sensitivity. Brain Res 813:323–328
Tanda G, Munzar P, Goldberg SR (2000) Self-administration behavior is maintained by the psychoactive ingredient of marijuana in squirrel monkeys. Nat Neurosci 3:1073–1074
Tsou K, Patrick SL, Walker JM (1995) Physical withdrawal in rats tolerant to delta 9-tetrahydrocannabinol precipitated by a cannabinoid receptor antagonist. Eur J Pharmacol 280:R13–15
Tsou K, Lowitz KA, Hohmann AG, Martin WJ, Hathaway CB, Bereiter DA, Walker JM (1996) Suppression of noxious stimulus-evoked expression of Fos protein-like immunoreactivity in rat spinal cord by a selective cannabinoid agonist. Neuroscience 70:791–798
Tsou K, Mackie K, Sanudo-Pena MC, Walker JM (1999) Cannabinoid CB_1 receptors are localized primarily on cholecystokinin-containing GABAergic interneurons in the rat hippocampal formation. Neuroscience 93:969–975

Tzavara ET, Davis RJ, Perry KW, Li X, Salhoff C, Bymaster FP, Witkin JM, Nomikos GG (2003) The CB_1 receptor antagonist SR141716A selectively increases monoaminergic neurotransmission in the medial prefrontal cortex: implications for therapeutic actions. Br J Pharmacol 138:544–553

Uriguen L, Perez-Rial S, Ledent C, Palomo T, Manzanares J (2004) Impaired action of anxiolytic drugs in mice deficient in cannabinoid CB_1 receptors. Neuropharmacology 46:966–973

Valjent E, Maldonado R (2000) A behavioural model to reveal place preference to delta 9-tetrahydrocannabinol in mice. Psychopharmacology (Berl) 147:436–438

Valjent E, Mitchell JM, Besson MJ, Caboche J, Maldonado R (2002) Behavioural and biochemical evidence for interactions between Delta 9-tetrahydrocannabinol and nicotine. Br J Pharmacol 135:564–578

Valverde O, Maldonado R, Valjent E, Zimmer AM, Zimmer A (2000a) Cannabinoid withdrawal syndrome is reduced in pre-proenkephalin knock-out mice. J Neurosci 20:9284–9289

Valverde O, Ledent C, Beslot F, Parmentier M, Roques BP (2000b) Reduction of stress-induced analgesia but not of exogenous opioid effects in mice lacking CB_1 receptors. Eur J Neurosci 12:533–539

Varga K, Lake K, Martin BR, Kunos G (1995) Novel antagonist implicates the CB_1 cannabinoid receptor in the hypotensive action of anandamide. Eur J Pharmacol 278:279–283

Varma N, Carlson GC, Ledent C, Alger BE (2001) Metabotropic glutamate receptors drive the endocannabinoid system in hippocampus. J Neurosci 21:RC188

Varvel SA, Lichtman AH (2002) Evaluation of CB_1 receptor knockout mice in the Morris water maze. J Pharmacol Exp Ther 301:915–924

Vela G, Ruiz-Gayo M, Fuentes JA (1995) Anandamide decreases naloxone-precipitated withdrawal signs in mice chronically treated with morphine. Neuropharmacology 34:665–668

Vlachou S, Nomikos GG, Panagis G (2003) WIN 55,212-2 decreases the reinforcing actions of cocaine through CB_1 cannabinoid receptor stimulation. Behav Brain Res 141:215–222

Wang L, Liu J, Harvey-White J, Zimmer A, Kunos G (2003) Endocannabinoid signaling via cannabinoid receptor 1 is involved in ethanol preference and its age-dependent decline in mice. Proc Natl Acad Sci U S A 100:1393–1398

Weissman A, Milne GM, Melvin LS Jr (1982) Cannabimimetic activity from CP-47,497, a derivative of 3-phenylcyclohexanol. J Pharmacol Exp Ther 223:516–523

Welch SP, Dunlow LD, Patrick GS, Razdan RK (1995) Characterization of anandamide- and fluoroanandamide-induced antinociception and cross-tolerance to delta 9-THC after intrathecal administration to mice: blockade of delta 9-THC-induced antinociception. J Pharmacol Exp Ther 273:1235–1244

Williams CM, Kirkham TC (2002a) Observational analysis of feeding induced by Delta9-THC and anandamide. Physiol Behav 76:241–250

Williams CM, Kirkham TC (2002b) Reversal of delta 9-THC hyperphagia by SR141716 and naloxone but not dexfenfluramine. Pharmacol Biochem Behav 71:333–340

Williams CM, Rogers PJ, Kirkham TC (1998) Hyperphagia in pre-fed rats following oral delta9-THC. Physiol Behav 65:343–346

Wilson RI, Nicoll RA (2001) Endogenous cannabinoids mediate retrograde signalling at hippocampal synapses. Nature 410:588–592

Wilson RI, Nicoll RA (2002) Endocannabinoid signaling in the brain. Science 296:678–682

Winsauer PJ, Lambert P, Moerschbaecher JM (1999) Cannabinoid ligands and their effects on learning and performance in rhesus monkeys. Behav Pharmacol 10:497–511

Yoshida T, Hashimoto K, Zimmer A, Maejima T, Araishi K, Kano M (2002) The cannabinoid CB_1 receptor mediates retrograde signals for depolarization-induced suppression of inhibition in cerebellar Purkinje cells. J Neurosci 22:1690–1697

Zimmer A, Zimmer AM, Hohmann AG, Herkenham M, Bonner TI (1999) Increased mortality, hypoactivity, and hypoalgesia in cannabinoid CB_1 receptor knockout mice [see comments]. Proc Natl Acad Sci U S A 96:5780–5785

HEP (2005) 168:147–185

The Biosynthesis, Fate and Pharmacological Properties of Endocannabinoids

V. Di Marzo[1] (✉) · L. De Petrocellis[2] · T. Bisogno[1]

[1]Endocannabinoid Research Group, Istituto di Chimica Biomolecolare, Via Campi Flegrei 34, Comprensorio Olivetti, Fabbricato 70, 80078 Pozzuoli (Napoli), Italy
vdimarzo@icmib.na.cnr.it

[2]Endocannabinoid Research Group, Istituto di Cibernetica, Consiglio Nazionale delle Ricerche, Via Campi Flegrei 34, Comprensorio Olivetti, Fabbricato 70, 80078 Italy

Abstract The finding of endogenous ligands for cannabinoid receptors, the endocannabinoids, opened a new era in cannabinoid research. It meant that the biological role of cannabinoid signalling could be finally studied by investigating not only the pharmacological actions subsequent to stimulation of cannabinoid receptors by their agonists, but also how the activity of these receptors was regulated under physiological and pathological conditions by varying levels of the endocannabinoids. This in turn meant that the enzymes catalysing endocannabinoid biosynthesis and inactivation had to be identified and characterized, and that selective inhibitors of these enzymes had to be developed to be used as (1) probes to confirm endocannabinoid involvement in health and disease, and (2) templates for the design of new therapeutic drugs. This chapter summarizes the progress achieved in this direction during the 12 years following the discovery of the first endocannabinoid.

Keywords Anandamide · 2-Arachidonoylglycerol · Cannabinoid · Enzyme · Inhibitors

1 Introduction

When the longstanding issue of the mechanism of action of (−)-Δ^9-tetrahydrocannabinol (THC) was solved with the finding of the cannabinoid receptors, studies aimed at finding endogenous ligands for these receptors could be started. These studies culminated in 1992 with the report of the discovery of the first of such ligands, *N*-arachidonoyl-ethanolamine (AEA), which was named anandamide from the Sanskrit word *ananda*, meaning "internal bliss" (Devane et al. 1992). In the following years, the finding of anandamide, which apart from binding to cannabinoid CB_1 (and later also CB_2) receptors could also functionally activate them, led to the revelation that there is a whole endogenous signalling system now known as the *endogenous cannabinoid system*. This comprises, apart from the cannabinoid receptors (Pertwee 1997), other endogenous ligands [named endocannabinoids by our group in 1995 (Di Marzo and Fontana 1995)] and the proteins for their synthesis and inactivation, as well as, possibly, other molecular targets for the endocannabinoids (see Pertwee 2004 for review). First came the finding that a well-known intermediate in phosphoglyceride metabolism, 2-arachidonoyl-glycerol (2-AG), was also able to activate both CB_1 and CB_2 receptors (Mechoulam et al. 1995; Sugiura et al. 1995). The end of the 1990s brought: (1) the finding of the biochemical pathways and the identification of the first enzymes for the formation and inactivation of AEA and 2-AG (Di Marzo et al. 1994; Cravatt et al. 1996; Bisogno et al. 1997b), a breakthrough that was very much facilitated by important similar studies carried out in the 1970s on other lipids belonging to the same families as the two endocannabinoids (Schmid et al. 1990 and Horrocks 1989 for reviews); and (2) the recognition that AEA was a rather promiscuous ligand for several membrane receptors and channels, particularly for vanilloid VR1 receptors (now classified as TRPV1 receptors) (Zygmunt et al. 1999), and as-yet-uncharacterized binding sites in the vascular endothelium (Jarai et al. 1999). Therefore, at the turn of the century it was clear that the endocannabinoid system was going to include new receptors, new ligands and new enzymes. This feeling was confirmed, among other things, by the characterization of: (1) more putative endocannabinoids, all derived from arachidonic acid, i.e. 2-arachidonyl-glyceryl ester (noladin, 2-AGE), *O*-arachidonoyl-ethanolamine (virodhamine, OAE) and *N*-arachidonoyl-dopamine (NADA) (Bisogno et al. 2000;

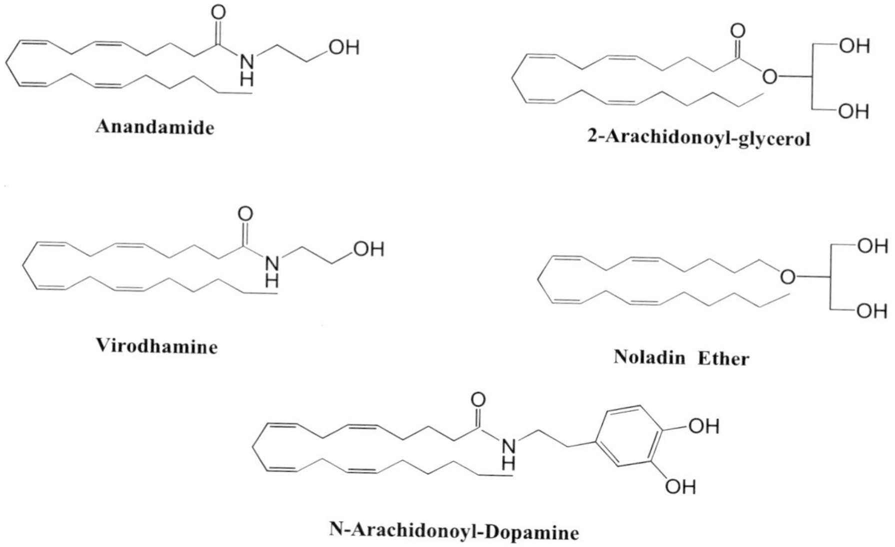

Fig. 1. Established and newly proposed endocannabinoids. Chemical structures of the five endogenous cannabinoid ligands identified so far

Hanus et al. 2001; Huang et al. 2002; Porter et al. 2002); (2) more possible targets for AEA and some synthetic cannabimimetic compounds (Breivogel et al. 2001); and (3) the biosynthetic enzymes for 2-AG and AEA (Bisogno et al. 2003; Okamoto et al. 2004). Clearly, the history of the endocannabinoid system is far from set, but nevertheless the following sections shall attempt at providing the reader with a picture as updated and as complete as possible of the multi-faceted biochemical and pharmacological aspects of the endocannabinoids (Fig. 1).

2 Biosynthesis and Release of Endocannabinoids

The biosynthetic and metabolic pathways of the two best-studied endocannabinoids, AEA and 2-AG, have several features in common. Both compounds are produced from the enzymatic hydrolysis of precursors derived from the remodelling of membrane phospholipids; both appear to be released and then taken up by cells via diffusion through the plasma membrane, possibly facilitated by a membrane carrier protein; and both are inactivated mostly via intracellular enzymatic hydrolysis. Yet, although overlaps are theoretically possible between the biosynthetic pathways of the two endocannabinoids, fundamentally different enzymes are involved in the formation of AEA and 2-AG. This explains why, as is becoming increasingly clear, the two compounds can be produced independently from each other and why their levels can undergo differential and even opposing changes with different physiological and pathological stimuli. For this reason, the biochemical pathways underlying the production of the two major endocannabinoids will be discussed here separately. In general, however, the three following commonalities can be observed:

- Both AEA and that portion of 2-AG acting as endocannabinoid (2-AG is in fact also an important intermediate in phosphoglyceride metabolism), are not stored in secretory vesicles but are, instead, synthesized and released "on demand", often following Ca^{2+} influx, which causes activation of Ca^{2+}-dependent biosynthetic enzymes (Di Marzo et al. 1998b).
- Pharmacological and electrophysiological data have shown that activation of metabotropic (glutamate or muscarinic) receptors, either cooperatively with or independently from Ca^{2+}-influx, can also induce the formation of non-chemically identified endocannabinoids acting as retrograde synaptic signals (Kim et al. 2002; Brenowitz and Regehr 2003; Ohno-Shosaku et al. 2003).
- The formation of both compounds is accompanied by the biosynthesis of cannabinoid-inactive or weakly active congeners, which have been suggested to exert an enhancement of AEA and 2-AG actions via various mechanisms collectively referred to as "entourage" effects (Ben-Shabat et al. 1998; Mechoulam et al. 1998b for review).

2.1 Biosynthesis of AEA and Other *N*-Acylethanolamines

AEA belongs to the family of the *N*-acylethanolamines (NAEs), which have been investigated since the 1960s. Work performed by H. Schmid and co-workers long before the discovery of AEA had shown that these compounds are biosynthesized via a phospholipid-dependent pathway (Fig. 2), i.e. the enzymatic hydrolysis of the corresponding *N*-acyl-phosphatidylethanolamines (NAPEs) (Schmid et al. 1990, 1996, 2002a; Hansen et al. 1998, for reviews). The enzyme catalysing this reaction is a phospholipase D selective for NAPEs (NAPE-PLD), which, in turn, are produced from the transfer to the *N*-position of phosphatidylethanolamine of an acyl group from the *sn*-1 position of phospholipids (PE), catalysed by a Ca^{2+}-dependent *trans*-acylase. Already in these early studies it appeared clear that NAPE-PLD was quite different from other PLD enzymes, and that this enzyme as well as the *trans*-acylase exhibited no selectivity for a particular fatty acid moiety. After the discover of AEA, this route was shown to underlie also the biosynthesis of this endocannabinoid in central neurons after depolarization (Di Marzo et al. 1994). Subsequent studies confirmed the occurrence of *N*-arachidonoyl-phosphatidylethanolamine (NArPE), the NAPE precursor of AEA, in murine brain, testes and leukocytes (Sugiura et al. 1996a,b; Di Marzo et al. 1996a,b; Cadas et al. 1997), and showed that NAPE-PLD lacks the transphosphatidylation activity typical of other PLD enzymes (Petersen and Hansen 1999), is dependent on Ca^{2+} for optimal activity (Ueda et al. 2001a) and is stimulated by polyamines (Liu et al. 2002). In fact, all the previous information gained on roughly purified fractions of NAPE-PLD have been recently confirmed by its cloning, which in addition showed that the enzyme belongs to the zinc

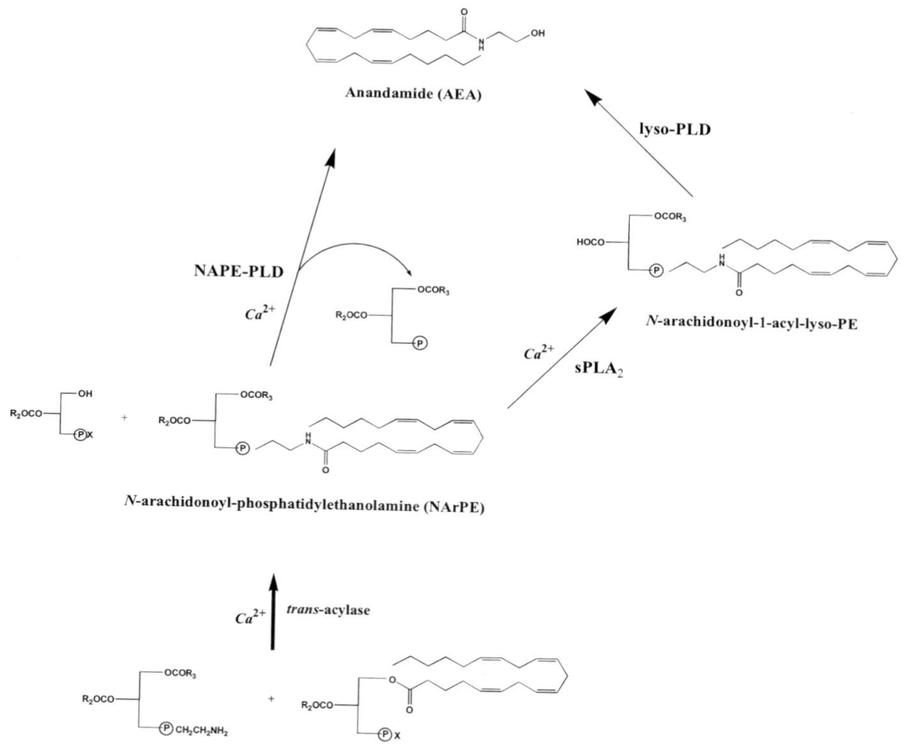

Fig. 2. Major biosynthetic pathways and enzymes for the endocannabinoid anandamide (AEA) and other *N*-acylethanolamines. The *circled P* indicates a phosphate group. *N-ArPE*, *N*-arachidonoyl-phosphatidylethanolamine; *PE*, phosphatidylethanolamine; *PLD*, phospholipase D; *sPLA₂*, secretory phospholipase A_2 of group IB

metalloproteinase family of hydrolases of the β-lactamase fold (Okamoto et al. 2004). Several independent lines of evidence strongly suggest that this pathway is the one mostly responsible for AEA biosynthesis in intact cells, and in particular:

- The finding of a similar distribution of NArPE and AEA in nine different brain areas (Bisogno et al. 1999a), and of increasing levels of both NArPE and AEA in rat brain at different stages of development (Berrendero et al. 1999), confirms a precursor/product relationship for the two compounds.
- The Ca^{2+} sensitivity of both the *trans*-acylase and NAPE-PLD is in agreement with the fact that AEA biosynthesis is triggered by neuronal depolarization and other Ca^{2+} mobilizing stimuli.
- The fact that this biosynthetic pathway is common to other NAEs, and that the percentage fatty acyl chain composition of these compounds in tissues is ultimately dependent on that of the *sn*-1 position of phospholipids (Fig. 2), explains why AEA is the minor of its congeners.

However, a recent study (Sun et al. 2004) also highlights another possible way for NAPEs to be transformed into NAEs, at least in cell-free homogenates, i.e. via the sequential action of a group IB secretory phospholipase A_2 (PLA_2), with the formation of *N*-acyl-1-acyl-lyso-PE, followed by the action of a lyso-PLD enzyme distinct from the known NAPE-PLD (Fig. 2).

2.2
Biosynthesis of 2-Arachidonoylglycerol

Although probably over-estimated due to artefactual production, for example following rat decapitation (Sugiura et al. 2001), the levels of 2-AG in unstimulated tissues and cells, but not in the blood or cerebrospinal fluid (CSF), are usually much higher than those of AEA, and sufficient in principle to permanently activate both cannabinoid receptor subtypes (Sugiura et al. 1995; Stella et al. 1997). This simple observation, and the fact that this compound is at the crossroads of several metabolic pathways and is an important precursor and/or degradation product of phospho-, di- and triglycerides, as well as of arachidonic acid, indicates that the 2-AG found in tissues is not uniquely used to stimulate cannabinoid receptors, although the one measured in extracellular fluids, such as serum and CSF, probably is. While an enhancement of intracellular Ca^{2+} is necessary and sufficient for AEA biosynthesis, 2-AG formation is triggered also, but not only, by Ca^{2+}-mobilizing stimuli (and, hence, also, but not only, following neuronal depolarization). In fact, the most important biosynthetic precursors of 2-AG are the *sn*-1-acyl-2-arachidonoylglycerols (DAGs) (Fig. 3), which, like other diacylglycerols, are produced from phospholipid metabolism and remodelling and, ultimately, by the stimulation of G protein-coupled receptors (GPCRs). This observation raises the possibility that the biosynthesis of 2-AG may be regulated independently from that of AEA, and requires different conditions. Several stimuli have been shown to lead to the formation of 2-AG in intact neuronal and non-neuronal cells, including lipopolysaccharides (in macrophages), ethanol or glutamate (in neurons), carbachol or thrombin (in endothelial cells), endothelin (in astrocytes), platelet-activating factor (in macrophages), etc. (Bisogno et al. 1997b; Stella et al. 1997; Sugiura et al. 1998; Mechoulam et al. 1998a; Bisogno et al. 1999b; Di Marzo et al. 1999a; Basavarajappa et al. 2000; Berdyshev et al. 2001; Stella and Piomelli 2001; Liu et al. 2003; Walter and Stella 2003; and Sugiura et al. 2002, for review), but only seldom have the pathways for 2-AG biosynthesis been investigated. In most cases, the DAGs necessary for 2-AG biosynthesis are obtained from the hydrolysis of 2-arachidonate-containing phosphoinositides (PIs), catalysed by the PI-selective phospholipase C or other phospholipases of this type (Di Marzo et al. 1996b; Stella et al. 1997; Kondo et al. 1998; Berdyshev et al. 2001; Stella and Piomelli 2001; Liu et al. 2003), whereas in the case of ionomycin-stimulated neuroblastoma cells and cultured rat microglial cells, DAGs appear to be formed from the hydrolysis of 2-arachidonate-containing phosphatidic acid (PA), catalysed by a PA phosphohydrolase (Bisogno et al. 1999b; Carrier et al. 2004). Regarding the conversion of DAGs into 2-AG, this requires a *sn*-1-selective DAG lipase (Bisogno et al. 1997b;

Stella et al. 1997). Two *sn*-1 DAG lipase isozymes (DAGLα and DAGLβ) have been cloned and enzymatically characterized (Bisogno et al. 2003). They are located in the plasma membrane, are stimulated by Ca^{2+}, appear to possess a catalytic triad typical of serine hydrolases, and, like NAPE-PLD, do not appear to be particularly selective for 2-arachidonate-containing DAGs. Nevertheless, several lines of evidence (Bisogno et al. 2003) suggest that they are responsible for the formation of the endocannabinoid 2-AG in intact cells:

- Over-expression of DAGLα and DAGLβ in COS cells results in significantly higher levels of 2-AG produced following stimulation with ionomycin, but not in higher 2-AG basal levels.
- The expression of DAGLα and DAGLβ in several cell lines correlates with their ability to produce 2-AG following stimulation with ionomycin.
- Inhibition of DAGLα and DAGLβ activity with tetrahydrolipstatin in COS cells and cell lines stimulated with ionomycin results in the impaired production of 2-AG.
- The distribution of the mRNAs encoding for DAGLα correlates with the relative abundance of 2-AG in rodent tissues and organs (Kondo et al. 1998).
- Finally, the two enzymes exhibit a pattern of subcellular expression in nervous tissues that fits with the proposed role of 2-AG either as a mediator of neurite growth, during brain development (Williams et al. 2003) or as a retrograde signal mediating depolarization-induced suppression of neurotransmission and heterosynaptic plasticity in the adult brain (Chevaleyre and Castillo 2003; Wilson and Nicoll 2002, for review). In fact, the enzymes are located on axons during development and post-synaptically in adult neurons (Bisogno et al. 2003).

However, DAG-independent biosynthetic pathways for 2-AG have also been proposed (Sugiura et al. 2002, for review), although their relevance to the regulation of the endocannabinoid signal has not yet been investigated. Noteworthy is the enzymatic hydrolysis of a particular type of lysophosphatidic acid, 2-arachidonoyl-*sn*-glycero-3-phosphate (Nakane et al. 2002).

2.3 Biosynthesis of Other Putative Endocannabinoids

Very little is known about the biosynthesis of the three most recently proposed endocannabinoids, 2-AGE, virodhamine and NADA. Regarding 2-AGE (noladin ether), this compound was previously identified in pig brain (Hanus et al. 2001) and in some rat tissues and brain areas (Fezza et al. 2002) by using mass-spectrometric (MS) methods coupled to chromatographic separations. However, a recent study cast some doubt on the actual existence of 2-AGE in mammalian brain tissue (Oka et al. 2003). At the time of this study it was already known that (1) the only acyl ethers to have been detected in animals before the discovery of 2-AGE were 2-acyl ethers (e.g. alkenyl ethers such as platelet activating factor and plasmalogens); (2) there was no evidence for the existence of any enzyme catalysing the formation of 2-alkenyl glyceryl ethers from the corresponding fatty acyl alcohols; and (3) although similar enzymes had been previously identified, these had a stringent specificity for the *sn*-1 position of glyceryl ethers with short-medium chain, saturated fatty acids (Nagan and Zoeller 2001). Oka et al. (2003), using MS techniques, could not confirm the presence of 2-AGE in the brain of several mammalian species including pig and rat. These contradictory data might be explained by the use of different extracting procedures, or with the possibility of a "false-positive" MS signal, i.e. an endogenous compound structurally related but not identical to 2-AGE (i.e. with the same molecular weight and similar mass spectrometric fragmentation pattern), which cannot be picked up by all MS techniques. This compound, however, cannot be the 2-AGE isomer 1-arachidonyl glyceryl ether, which can be distinguished from 2-AGE simply on the basis of its chromatographic properties. Clearly, if the existence of 2-AGE were to be confirmed by future studies carried out in other laboratories using exactly the same procedures used by Hanus et al. (2001) and Fezza et al. (2002), some as-yet-unknown biosynthetic pathway, different from that leading to plasmalogens, may exist for this compound. Neuroblastoma N18TG2 intact cells are not capable of converting arachidonate-containing phospholipids into 2-AGE when stimulated with ionomycin, i.e. under conditions where high levels of 2-AG are produced (Fezza et al. 2002). This might suggest a Ca^{2+}-independent or a non-phospholipid-mediated pathway for the formation of this putative endocannabinoid in neurons.

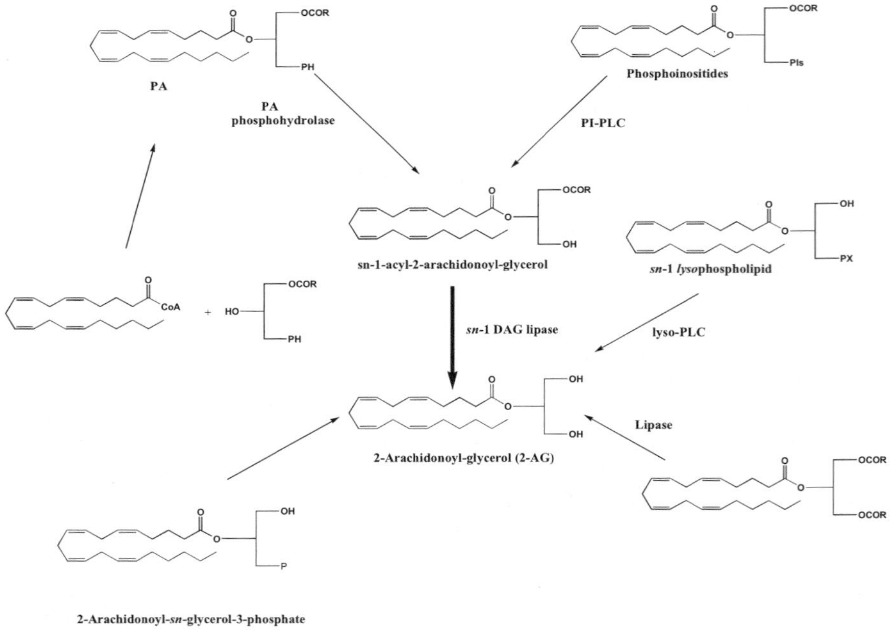

Fig. 3. Major biosynthetic pathways and enzymes for the endocannabinoid 2-arachidonoyl-glycerol (2-AG). *DAG*, di-acyl-glycerol lipase; *PA*, phosphatidic acid; *PI*, phosphoinositide; *PLC*, phospholipase C. P represents a phosphate group

Virodhamine, which seems to accompany AEA in all tissues analysed (Porter et al. 2002), might be biosynthetically related to AEA, since the non-enzymatic transformation of NAEs into the corresponding *O*-acyl esters, and vice versa, in the presence of bases or acids, has been reported (Markey et al. 2000). Given the seemingly opposing activity of the two compounds at their receptors (*virodha* in Sanskrit means "opposing"), with virodhamine being an antagonist at cannabinoid CB_1 receptors and a partial agonist at CB_2 receptors, and anandamide a possible antagonist at cannabinoid CB_2 receptors and a partial agonist at CB_1 receptors, this possibility might give rise to an interesting interplay between the two compounds under those pathological conditions (i.e. inflammation) that cause a local decrease of pH.

Original evidence for the formation of NADA from arachidonic acid and dopamine or tyrosine (Huang et al. 2002) suggested a biosynthetic pathway common to that of the recently discovered arachidonoyl amino acids (Huang et al. 2001), i.e. from the direct condensation between arachidonic acid and dopamine, or, alternatively, from the condensation between arachidonic acid and tyrosine followed by the transformation of *N*-arachidonoyl-tyrosine into NADA by the enzymes catalysing dopamine biosynthesis from tyrosine. Preliminary data have shown, however, that NADA cannot be produced from either *N*-arachidonoyl-tyrosine or *N*-arachidonoyl-L-DOPA either in vitro, in brain homogenates, or in vivo, and that the lipid formed from tyrosine and arachidonic acid is not NADA (M.J. Walker and V. Di Marzo, unpublished observations). Clearly, further studies are needed to understand the biosynthetic mechanism for this putative endocannabinoid.

2.4 Inhibitors of Endocannabinoid Biosynthesis

Partly owing to the fact that the NAPE-PLD for AEA and the two DAGLs for 2-AG have been cloned only very recently, no selective inhibitors of endocannabinoid biosynthesis have been developed to date. However, several non-specific inhibitors have been shown to prevent the formation of either AEA or 2-AG. For the former compound, Cadas et al. (1997) showed that several non-selective hydrolase inhibitors, and particularly the PLA_2 inhibitor (*E*)-6-(bromomethylene)-tetrahydro-3-(1-naphthalenyl)-2H-pyran-2-one (BTNP), could block the activity of crude preparations of the Ca^{2+}-dependent *trans*-acylase. Regarding 2-AG, Bisogno et al. (1999b) found that the PLA_2 inhibitor, oleoyl-oxyethyl-phosphoryl-choline, and the blocker of acylCoA-dependent synthase, thimerosal, could oppose ionomycin-induced formation of 2-AG in intact neurons, possibly by inhibiting the formation of PA precursors. Furthermore, the DAG lipase inhibitor RHC80267 was also found to block 2-AG release from DAGs (Stella et al. 1997; Bisogno et al. 1997b, 1999b). More importantly, the lipase inhibitor tetrahydrolipstatin (orlistat) was recently shown to inhibit the two DAGLs, DAGLα and DAGLβ, at concentrations (IC_{50}=60–250 nM) lower than those previously found to be required to inhibit other lipases (Bisogno et al. 2003). Clearly the chemical structure of this compound (Fig. 3), which is marketed by Roche as an anti-obesity drug, might serve as a template for the development of more selective DAGL inhibitors.

2.5 Endocannabinoid Release

After their biosynthesis, AEA and 2-AG are immediately released into the extracellular medium. This occurs via an unknown mechanism, which, however, several pieces of evidence suggest is one that is dependent on the same putative membrane transporter proposed to facilitate the opposite process, i.e. endocannabinoid cellular uptake (see below). In particular:

- Cells loaded with radiolabelled AEA release this compound through a temperature-dependent and pharmacologically inhibitable mechanism (Hillard et al. 1997).
- AEA biosynthesized de novo inside the cell is released into the extracellular medium via a process that can be inhibited by selective inhibitors of AEA cellular uptake, with subsequent increase of intracellular AEA levels (Ligresti et al. 2004).
- Endocannabinoids have been proposed to act as retrograde messengers for both short- and long-term forms of synaptic plasticity, such as depolarization-induced suppression of excitatory or inhibitory neurotransmission (DSE or DSI) and long-term depression (LTD; see Wilson and Nicoll 2002, for review). It is thought that endocannabinoids are released from the post-synaptic cell following its depolarization, and then act retrogradely on CB_1 receptors on pre-synaptic neurons to inhibit neurotransmitter release. In one model of this phenomenon, inhibitors of the putative endocannabinoid membrane transporter (EMT), injected into the post-synaptic neuron, have been found to inhibit LTD (Ronesi et al. 2004).

Once released, extracellular endocannabinoids act mostly, and with varying selectivity, on cannabinoid receptors, possibly including subtypes other than CB_1 and CB_2. However, endocannabinoids such as AEA and/or NADA may also act, prior to their release, on intracellular sites on ion channels, such as those on vanilloid TRPV1 (transient receptor potential vanilloid type 1) receptors and T-type Ca^{2+} channels (see below). In this case, release represents a possible way to inactivate, rather than facilitate, the action of endocannabinoids.

3 Endocannabinoid Metabolic Fate

3.1 Cellular Uptake

When incubated with intact cells in vitro, all endocannabinoids are rapidly ($t_{1/2} \leq 5$ min) cleared away from the extracellular medium (Di Marzo et al. 1994; Ben-Shabat et al. 1998; Beltramo and Piomelli 2000; Bisogno et al. 2001; Fezza et al. 2002; Huang et al. 2002). It has been suggested that this process depends on the presence of one or more membrane transporters, the putative EMT (see above). This hypothesis is supported by evidence that the transport process is saturable and exhibits sensitivity to temperature, selectivity for unsaturated (particularly polyunsaturated) long-chain fatty acid amides and sensitivity to synthetic inhibitors (Di Marzo et al. 1994; Beltramo et al. 1997; Hillard et al. 1997; Bisogno et al. 1997a). Since this process only transports AEA down transmembrane concentration gradients, it can also: (1) mediate AEA release, and (2) act in the absence of other sources of energy and, therefore, function independently of Na^+- and ATP (Hillard and Jarrahian 2000). However, the putative EMT has not been isolated, and its molecular biology remains uncharacterized. This lack of information, together with the following observations, suggested to some authors that AEA membrane transport might simply occur through passive diffusion driven by intracellular enzymatic hydrolysis:

- Endocannabinoids are lipophilic compounds, and such compounds often do not need a membrane transporter to cross the plasma membrane (although there are several exceptions to this rule).
- The presence in the cell of an active AEA-hydrolysing enzyme, fatty acid amide hydrolase (FAAH) (see below), strongly enhances AEA cellular uptake (Deutsch et al. 2001).
- Inhibitors of AEA intracellular metabolism often (but not always) also inhibit AEA transport into the cell (Deutsch et al. 2001; Glaser et al. 2003).
- Under certain experimental conditions, AEA accumulation into the cell is not saturable (Glaser et al. 2003), whereas, in the absence of a cell monolayer, the plastic ware used in studies of AEA cellular uptake can mimic the AEA sequestration process in terms of temperature sensitivity (Fowler et al. 2004).

However, several observations still strongly, albeit indirectly, support the existence of an EMT, or at least of some specific intracellular process distinct from FAAH for bringing about the cellular uptake of endocannabinoids (for a more detailed review see Hillard and Jarrahian 2003):

- Several cell types can be found that can rapidly take up AEA from the extracellular medium even though they do not express FAAH; furthermore, synaptosomes from transgenic mice lacking FAAH can still take up AEA efficiently and in a saturable manner (Ligresti et al. 2004);
- Several compounds have been developed that are capable of inhibiting AEA cellular uptake without inhibiting AEA enzymatic hydrolysis via FAAH (Di Marzo et al. 2001b, 2002c; De Petrocellis et al. 2000; Lopez-Rodriguez et al. 2001; Ortar et al. 2003); indeed, the chemical prerequisites necessary for fatty acid amide derivatives to inhibit AEA uptake are so stringent that there can be no doubt that this process is mediated by a specific protein (Piomelli et al. 1999; Ligresti et al. 2004).
- FAAH inhibitors enhance, and anandamide uptake inhibitors inhibit, anandamide accumulation into some cells (Kathuria et al. 2003).

- Substances that inhibit the EMT *enhance* responses to exogenous AEA that are elicited at extracellular sites (i.e. at CB_1 receptors) and *inhibit* those that are elicited at intra-cellular targets (i.e. TRPV1 receptors, see De Petrocellis et al. 2001)—if these compounds were simply acting by inhibiting FAAH, they should enhance AEA effects in both cases.
- A selective EMT inhibitor can modify the distribution of de novo biosynthesized AEA between the intracellular and extracellular milieu, without altering its total amounts (Ligresti et al. 2004).
- 2-AGE and NADA, two endocannabinoids that are resistant and refractory to enzymatic hydrolysis, respectively, are still taken up by cells in a temperature-dependent way; their uptake is inhibited competitively by AEA (Huang et al. 2002; Fezza et al. 2002), although none of the specific EMT inhibitors mentioned above has ever been tested on the cellular uptake of these compounds.
- Lipopolysaccharide inhibits FAAH expression without affecting AEA cellular uptake (Maccarrone et al. 2001); conversely, nitric oxide, peroxynitrite and superoxide anions stimulate AEA cellular re-uptake (Maccarrone et al. 2000a), while acute or chronic ethanol inhibits this process (Basavarajappa et al. 2003), without affecting FAAH activity.

These data suggest that, although intracellular hydrolysis does greatly influence the rate of AEA facilitated diffusion, the uptake process is likely to be mediated by a mechanism subject to regulation and distinct from the one catalysing AEA hydrolysis.

3.2 Enzymatic Hydrolysis

3.2.1 Anandamide Hydrolysis

The hydrolysis of AEA is catalysed by FAAH, an enzyme originally purified and cloned from rat liver microsomes (Cravatt et al. 1996), that also catalyses the hydrolysis of other long-chain NAEs and, in vitro, of 2-AG. Since the hydrolysis products do not activate cannabinoid receptors, this reaction represents a true inactivation mechanism. FAAH is probably the same enzyme identified in the 1970s and 1980s as a NAE-hydrolysing enzyme (see Natarajan et al. 1984, for an example). It also catalyses the hydrolysis of arachidonoyl methyl ester, and hence it is possible that virodhamine is also a substrate, although this possibility has not been tested yet. Finally, FAAH also catalyses the hydrolysis of long-chain primary fatty acid amides, such as the putative sleep-inducing factor oleamide (Maurelli et al. 1995; Cravatt et al. 1996). The structural and kinetic properties of FAAH have been widely reviewed in the literature (Bisogno et al. 2002; Cravatt and Lichtman 2003, for recent reviews) and will be described in more detail in other chapters of this volume. In brief, the enzyme has an alkaline optimal pH and is found in intracellular membranes; what was originally thought to be the hydrophobic domain responsible for this localization is instead important for the formation of active oligomers, whereas its localization on intracellular membranes might be regulated by an SH3 (Src homology region 3) consensus proline-rich sequence also necessary for enzymatic activity. Furthermore, judging from the recently elucidated X-ray structure of FAAH crystals in complex with its substrate (Bracey et al. 2002), one more domain may exist conferring the enzyme with the ability to associate with the plasma membrane. The catalytic amino acid of FAAH has been identified as Ser241, and two other residues of the amidase consensus sequence, Ser217 and Cys249, contribute to its enzymatic activity through a catalytic mechanism different from that of other amidases and *Ser* hydrolases (Patricelli and Cravatt 2000). The promoter region on the FAAH gene has been identified (Puffenbarger et al. 2001; Waleh et al. 2002), and is up-regulated by progesterone and leptin (Maccarrone et al. 2003a,b), and down-regulated by estrogens and glucocorticoids (Waleh et al. 2002).

Finally, transgenic FAAH-deficient mice have been developed. They are more responsive to exogenously administered AEA (Cravatt et al. 2001), and their brains contain 15-fold higher levels of AEA than wild-type mice. The phenotype of these mice is characterized also by higher susceptibility to kainate-induced seizures (Clement et al. 2003) and by lower sensitivity to some painful stimuli (Cravatt et al. 2001), which suggests that inhibition of FAAH might lead to the development of novel analgesics.

Another amidase has been characterized whose molecular size, substrate selectivity, optimal pH and tissue distribution are very different from those of FAAH (Ueda et al. 2001b; Ueda 2002, for a review). This enzyme appears to be located in lysosomes and might play a major role in the inactivation not so much of AEA as of its anti-inflammatory and analgesic congener, *N*-palmitoylethanolamine, which lacks activity at both CB_1 and CB_2 receptors (see Lambert et al. 2002, for review).

3.2.2
2-Arachidonoylglycerol Hydrolysis

Although FAAH can catalyse 2-AG hydrolysis both in cell-free homogenates and in some intact cells (Di Marzo et al. 1998a; Ligresti et al. 2003), 2-AG levels are not increased in FAAH knockout mice (Lichtman et al. 2002). This finding, together with previous reports on the existence of additional hydrolases for 2-AG degradation in porcine brain, in rat circulating platelets and macrophages, and in mouse J774 macrophages (Di Marzo et al. 1999a,b; Goparaju et al. 1999), suggests that FAAH may not be uniquely responsible for 2-AG inactivation under physiological conditions in vivo. The additional 2-AG hydrolases are known as monoacylglycerol lipases (MAGLs), are usually found in both membrane and cytosolic fractions, and also recognize other unsaturated monoacylglycerols, such as, for example, mono-oleoyl-glycerol, which is in fact a competitive inhibitor of 2-AG hydrolysis (Ben-Shabat et al. 1998; Di Marzo et al. 1998a). In rat circulating macrophages and platelets, 2-AG hydrolase activity was found to be lower following lipopolysaccharide treatment (Di Marzo et al. 1999a). A MAGL with enzymatic properties and subcellular distribution very similar to these roughly characterized enzymes was cloned in the 1990s from mouse (Karlsson et al. 1997), and more recently from man and rat (Karlsson et al. 2001; Ho et al. 2002; Dinh et al. 2002). Evidence for its participation in 2-AG degradation was provided for the rat enzyme that, as previously found for FAAH (Egertova et al. 1998), is expressed in brain regions with high cannabinoid CB_1 receptor density, such as the hippocampus, but, unlike FAAH, occurs in pre-synaptic neurons and is likely to be expressed in the same neurons as CB_1 receptors (Dinh et al. 2002). This finding supports the role of rat MAGL in the degradation of that pool of 2-AG that acts as an endocannabinoid retrograde synaptic signal.

3.3 Other Metabolic Reactions

3.3.1 Re-esterification

The hydrolysis products of both AEA and 2-AG, i.e. arachidonic acid and ethanolamine or glycerol, are immediately recycled into membrane phospholipids to possibly re-enter the biosynthetic pathways of the two endocannabinoids at a later stage. However, 2-AG can be directly esterified into (phospho)glycerides prior to its hydrolysis, via phosphorylation and/or acylation of its two free hydroxyl groups (for a review see Sugiura et al. 2002). This pathway was suggested to occur in mouse N18TG2 neuroblastoma and rat basophilic leukaemia (RBL-2H3) cells and in mouse J774 macrophages (Di Marzo et al. 1998, 1999a,b). Most importantly, direct esterification into membrane phosphoglyceride fractions, and, to a minor extent, into diacylglycerol and triacylglycerol fractions, occurs for 2-AGE (Fezza et al. 2002), which would otherwise be difficult to metabolize, as its ether bond is refractory to enzymatic hydrolysis.

3.3.2 Oxidation and Methylation

The possible enzymatic oxidation of the arachidonoyl moiety of endocannabinoids was hypothesized shortly after the discovery and definition of endocannabinoids in the early 1990s (Fontana and Di Marzo 1995). Support for this hypothesis was soon obtained in the form of evidence for AEA metabolism by cell-free homogenates expressing various lipoxygenases and cytochrome P450 oxidases (Bornheim et al. 1993; Ueda et al. 1995) and, later, also for AEA metabolism by cyclooxygenase-2, but not cyclooxygenase-1 (Yu et al. 1997). In more recent years it was also found that oxidation products of both AEA and 2-AG could be formed easily in intact cells, and that 2-AG is as good a substrate for cyclooxygenase-2 as arachidonic acid (for a review see Kozak and Marnett 2002). The activity of the oxidation products at cannabinoid receptors depended very much on the type of metabolite formed, with some lipoxygenase products being still capable of binding to both CB_1 and CB_2, and cyclooxygenase-2 products being inactive (Edgemond et al. 1998; Berglund et al. 1999; Maccarrone et al. 2000b; van der Stelt et al. 2002). Indeed, recent pharmacological data point to the existence of distinct, non-cannabinoid receptor, specific molecular targets for *both* prostaglandin-ethanolamides (prostamides), in particular prostamide $F_{2\alpha}$ (Matias et al. 2004), and prostaglandin E_2 glycerol ester (Nirodi et al. 2004). Prostamides, however, are rather stable to further metabolism, except for prostamide E_2, which undergoes slow dehydration/isomerization to prostamide B_2 (Kozak et al. 2001), whereas prostaglandin E_2 glyceryl ester is instead rapidly hydrolysed in rat, but not human, plasma (Kozak et al. 2001). None of these compounds is a substrate for the endocannabinoid transporter or FAAH (Matias et al. 2004; V. Di Marzo and L. Marnett, unpublished data). Regarding lipoxygenase products of AEA and 2-AG, it has been suggested that undefined lipoxygenase products of AEA act via vanilloid TRPV1 receptors (see below) (Craib et al. 2001), although there is no direct evidence for the interaction of hydroxy-anandamides or leukotriene-ethanolamides with these receptors. In contrast, 12- and 5-lipoxygenase products of arachidonic acid are known to interact with TRPV1 receptors (Hwang et al. 2000). The 15-(*S*)-hydroxy-derivative of 2-AG was recently shown to be formed in intact cells and to activate the peroxisome proliferation activator receptor-α (Kozak et al. 2002). Very little data, if any, exist on the further metabolism of AEA and 2-AG lipoxygenase products. Based on evidence available to date, it is possible that oxidation of AEA and 2-AG, while leading to the partial or complete inactivation of their endocannabinoid signal, might produce in some cases compounds active on other molecular targets, and hence represent an unusual example of "agonist functional plasticity".

Apart from its arachidonoyl moiety, the catecholamine moiety of NADA is also likely to be subject to both enzyme-catalysed and non-enzymatic oxidation. However, to date, only the methylation of the 3-hydroxy-group of NADA by catechol-*O*-methyl transferase has been observed (Huang et al. 2002). The reaction product is significantly less active at TRPV1 receptors (Huang et al. 2002), whereas its activity at CB_1 receptors has not been investigated.

3.4
Inhibitors of Endocannabinoid Inactivation

Several selective FAAH inhibitors have been developed (for reviews see Bisogno et al. 2002; Deutsch et al. 2002), some of which have IC_{50} values in the low nanomolar or subnanomolar range of concentrations (Boger et al. 2000; Kathuria et al. 2003) (Fig. 4). The first FAAH inhibitors to be developed, such as the irreversible inhibitor methyl-arachidonoyl-fluoro-phosphonate (MAFP) (Deutsch et al. 1997b; De Petrocellis et al. 1997), and the trifluoromethyl ketones, which are competitive inhibitors, (Koutek et al. 1994), came from the large pool of previously identified PLA_2 inhibitors, and were also found to interfere with CB_1 receptor activity. Others, such as the still widely used palmitylsulphonyl fluoride (AM374) (Deutsch et al. 1997a), appeared to be more selective towards CB_1 receptors but have never

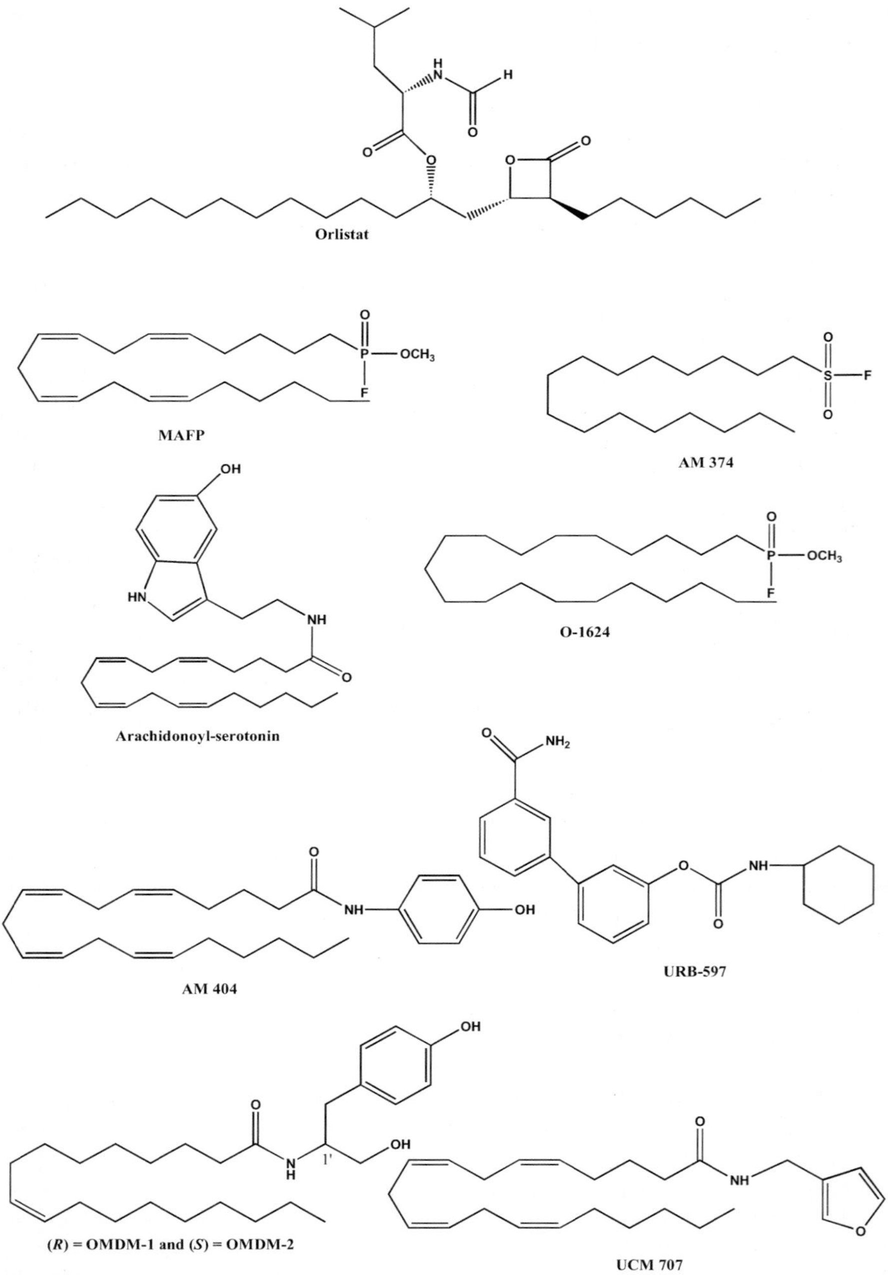

Fig. 4. Chemical structures of inhibitors of endocannabinoid biosynthesis or inactivation

been tested against PLA_2 enzymes. Among the FAAH inhibitors developed so far, particularly noteworthy are:

- *N*-arachidonoyl-serotonin (AA-5-HT, Bisogno et al. 1998), which is not particularly potent (IC_{50} values in the low μM range), but was tested against CB_1 and CB_2 receptors and PLA_2 enzymes and found to be inactive, and is suitable for use in vivo (V. Di Marzo, unpublished observations); so far, it has not been possible to enhance its inhibitory potency by chemical modification (Fowler et al. 2003).
- Several ultra-potent compounds developed by Boger and co-workers (Boger et al. 2000, 2001), whose use in vivo, however, has not been reported as yet.
- A series of MAFP analogues, one of which, O-1624, is quite potent and selective vs CB_1 receptors and was found to enhance anandamide levels after intrathecal administration to mice (Martin et al. 2000).
- A series of alkylcarbamic acid aryl esters, which were found to have very interesting structure–activity relationships against FAAH (Tarzia et al. 2003). One of these compounds, URB-597, is very potent and very selective for FAAH, although it was not tested against PLA_2. It is suitable for in vivo use, as its administration to rats causes a strong elevation of brain AEA levels with corresponding analgesic activity and anxiolytic actions (Kathuria et al. 2003).

With regard to inhibitors of the putative EMT, the development of a very potent and selective inhibitor has been hindered so far by the lack of any molecular data on this elusive protein. The prototypical EMT inhibitor, AM404 (Beltramo et al. 1997, Fig. 4), exhibits IC_{50} values in the 1- to 10-μM range of concentrations and has been widely used in vivo in laboratory animals. However, it has now been established that this compound can also inhibit FAAH and stimulate TRPV1 vanilloid receptors (Jarrahian et al. 2000; Zygmunt et al. 2000; De Petrocellis et al. 2000; Ross et al. 2001) and that both these properties, together with inhibition of EMT, can explain why AM404 can enhance AEA levels in vivo, since TRPV1 stimulation leads to enhanced AEA biosynthesis (Di Marzo et al. 2001d; Ahluwalia et al. 2003a). Therefore, great care is needed when using this compound in vivo. Recently, several compounds have been developed that are more potent as EMT inhibitors than as FAAH inhibitors or TRPV1 agonists:

- VDM11 and VDM13 (De Petrocellis et al. 2000) have been used as pharmacological tools in vitro, for example to demonstrate the action of AEA on TRPV1 at an intracellular site (De Petrocellis et al. 2001; Andersson et al. 2002). VDM11 has also been used to demonstrate anti-proliferative endocannabinoid tone in colorectal carcinoma cells in vitro (Ligresti et al. 2003), and to investigate the role of endocannabinoids in retrograde signalling during long-term depression (Ronesi et al. 2004). Finally, VDM11 has been used successfully in many in vivo studies, for example in the gastrointestinal system following i.p. administration (Pinto et al. 2002; Mascolo et al. 2002; Izzo et al. 2003). Interestingly, VDM11 was recently shown to also block endocannabinoid release (Ligresti et al. 2004; Ronesi et al. 2004). The major drawback of VDM11 and VDM13 is that, like AM404, they are not very stable metabolically and can be hydrolysed to arachidonic acid by brain homogenates (Ortar et al. 2003).

- UCM-707 was developed from several other "head" analogues of AEA and found to be very potent on the EMT on some cells (Lopez-Rodriguez et al. 2003a,b), but not others (Ruiz-Llorente et al. 2004; Fowler et al. 2004). Apart from being more potent as an AEA uptake inhibitor than as a TRPV1 agonist or FAAH inhibitor, this compound is very suitable for in vivo use (de Lago et al. 2002), and has been successfully employed to help demonstrate that AEA plays a role in neuroprotection against kainate-induced seizures in mice (Marsicano et al. 2003).
- OMDM-1 and OMDM-2 are the first selective inhibitors of AEA cellular uptake to be developed from a fatty acid other than arachidonic acid, i.e. oleic acid (Ortar et al. 2003). For this reason, and also because it is more stable to hydrolysis in rat brain homogenates, OMDM-2 appears to exert a more long-lasting inhibition of spasticity in mice with experimental allergic encephalomyelitis (de Lago et al. 2004b), and to improve several motor and immunological parameters of the disorder (C. Guaza and V. Di Marzo, unpublished observations).

Although both basic and applied research with AEA transport inhibitors has already produced several interesting results of relevance to the endocannabinoid system, the isolation and cloning of the putative EMT remains an important objective since, if such a protein really exists, the identification of its molecular features should lead to the development of even more potent inhibitors.

4
Pharmacology of Endocannabinoids

4.1
Endocannabinoid Molecular Targets: Beyond CB_1 and CB_2 Receptors

By definition (Di Marzo and Fontana 1995), endocannabinoids act primarily at cannabinoid CB_1 and/or CB_2 receptors, and they do so with varying affinity and efficacy. AEA, NADA and 2-AGE are more selective for CB_1, with the following rank of affinity: AEA$\geq$2-AGE>NADA. 2-AG has almost the same affinity for both receptors, and its K_i varies considerably according to the experimental conditions. Several assays have been used to examine the functional activity of endocannabinoids and to compare it with that of synthetic cannabinoids and THC (Pertwee 1997), and those used most often are:

- The GTP-γ-S binding assay, which provides an indirect measure of the ability of ligands to induce coupling of receptors to G-proteins.
- The cyclic adenosine monophosphate (cAMP) assay, in which the ability to inhibit forskolin-induced cAMP production is measured.

- Assays that measure the ability of ligands to inhibit voltage-activated Ca^{2+} channels or to stimulate the activity of G protein-coupled inwardly rectifying K^+ channels (GIRK).
- An assay that measures the ability of ligands to inhibit electrically evoked contractions of the mouse vas deferens.

Just as the affinity constant of AEA, 2-AGE and NADA depends on the type of membrane preparation and radioligand used to carry out the binding assay (for an example see Appendino et al. 2003), so too their efficacy depends very much on the type of functional assay used. For example, both noladin and NADA are more potent than AEA at inducing Ca^{2+} transients in neuroblastoma cells via CB_1 receptors (Sugiura et al. 1999; Bisogno et al. 2000). Noladin and 2-AG are equipotent, and much more potent than AEA at inhibiting voltage-activated Ca^{2+} channels in rat sympathetic neurons previously injected with cDNA encoding human CB_1 (Guo and Ikeda 2004). Indeed, AEA behaves as a partial agonist at CB_1 in most assays of functional activity, and is almost functionally inactive on CB_2 (see McAllister and Glass 2002 for review). Virodhamine acts as an antagonist for CB_1 and a partial agonist for CB_2, thus behaving in an opposite way to AEA (Porter et al. 2002). 2-AG appears to be equipotent and a full agonist at both receptor subtypes (McAllister and Glass 2002), although its affinity constants at both targets are lower than those of AEA (Mechoulam et al. 1995).

To add further complexity to this scenario, there is now increasing evidence, based on pharmacological and biochemical data, for the existence of non-CB_1, non-CB_2 GPCRs that respond to physiologically relevant concentrations of endocannabinoids and their congeners, and of AEA in particular (for comprehensive reviews see Di Marzo et al. 2002b and Pertwee 2004). These putative receptors can be grouped into three categories:

- "WIN-55,212-2/AEA/vanillyl-fatty acid amide" receptors: the first example of such sites of action was detected in murine astrocytes (Sagan et al. 1999). Through this, or a very similar receptor, AEA inhibits adenylyl cyclase and, possibly, gap-junction-mediated Ca^{2+} signalling in astrocytes (Venance et al. 1995). Indeed, a GPCR with a distribution different from CB_1 receptors and sensitive to both AEA and WIN-55,212-2, but not to other cannabinoid receptor agonists, was described in several brain areas of CB_1 knockout mice (Di Marzo et al. 2000a; Breivogel et al. 2001; Monory et al. 2002). Still to be clarified is whether this proposed receptor is similar to the putative site of action that mediates the inhibitory effect of WIN-55,212-2 on glutamate release in the mouse hippocampus (Hajos et al. 2001) and is sensitive to capsaicin (Hajos and Freund 2002). This in turn, might be the same receptor as the one that has been postulated to mediate some of the actions of fatty acid–vanillamine amides, such as arvanil and its analogues (Di Marzo et al. 2001b,d; 2002c; Brooks et al. 2002).
- "AEA/abnormal-cannabidiol receptors": another possible GPCR for AEA and for the non-psychotropic cannabinoid, abnormal-cannabidiol (abn-cbd), has been

detected in vascular endothelial cells. This putative receptor mediates the local vasodilator (but not the systemic hypotensive) effects of AEA, and is blocked by both cannabidiol and a synthetic analogue, O-1918 (Jarai et al. 1999; Offertaler et al. 2003). It is coupled to guanylyl cyclase and p42/44 mitogen-activated protein kinase and protein kinase B/Akt. Interestingly, this novel endothelial receptor seems to be activated also by NADA (O'Sullivan et al. 2004). A receptor sensitive to abn-cbd has been proposed to mediate microglial cell migration (Walter et al. 2003), but this site of action, unlike the one in endothelial cells, was also activated by 2-AG.

- "Saturated NAE receptors": one other GPCR, for *N*-palmitoylethanolamine, has been proposed to explain some of the analgesic and anti-inflammatory actions of this AEA congener (Calignano et al. 1998). A receptor for *N*-palmitoylethanolamine has been proposed to occur also in microglial cells (Franklin et al. 2003) and shown to be different from that proposed to mediate the central effects of another saturated AEA congener, *N*-stearoylethanolamine (Maccarrone et al. 2002b).

In addition, several channels that transport Ca^{2+} and K^+ across the cell membrane are targeted directly by sub-micromolar concentrations of AEA (Di Marzo et al. 2002b). These are:

- TASK-1 K^+ channels (Maingret et al. 2001), which are inhibited by AEA.
- T-type Ca^{2+} channels (Chemin et al. 2001), which are also blocked by AEA, apparently acting at an intracellular site.
- Vanilloid TRPV1 receptors, the sites of action of capsaicin, the pungent component of "hot" red peppers (Szallasi and Blumberg 1999), which in contrast are activated by AEA and NADA (Zygmunt et al. 1999; Smart et al. 2000; Huang et al. 2002). In this case the effect clearly requires the activation of an intracellular domain of the protein (De Petrocellis et al. 2001; Jordt and Julius 2002), a mechanism that explains the significantly higher potency with which AEA and NADA induce TRPV1-mediated currents when injected directly into the neuron (Premkumar et al. 2004; Evans et al. 2004).

In heterologous expression systems, the potency of AEA for inducing typical TRPV1-mediated effects (e.g. cation currents, Ca^{2+}-influx and cell depolarization) is at least fivefold lower than its average potency at CB_1 receptors. However, recent data (recently reviewed by Di Marzo et al. 2001a; 2002a; Ross 2003; van der Stelt and Di Marzo 2004) indicate that the potency of AEA and NADA at TRPV1 receptors increases by up to 10- to 15-fold in some pathological states. In fact, the number of TRPV1-mediated pharmacological effects, in vitro and in vivo, being reported in the literature for AEA is increasing by the day. A recent study showed that elevated levels of endocannabinoids acting at TRPV1 cause ileitis in toxin A-treated rats (McVey et al. 2003). Evidence for a role for AEA and TRPV1 in store-operated Ca^{2+}-entry into sensory neurons has also been found (M. van der Stelt and V. Di Marzo, manuscript in preparation). Furthermore, as AEA often exerts opposing actions, depending on whether it acts on CB_1 or TRPV1 receptors, blockade of CB_1 receptors may reveal that TRPV1-mediated effects of AEA can be exerted at concentrations lower than originally thought (Ahluwalia et al. 2003b). Finally, there are in vitro preparations, such as the rat mesenteric artery, where the efficacy and potency of AEA and NADA at TRPV1 are comparable to those of capsaicin (Zygmunt et al. 1999; O'Sullivan et al. 2004). Thus, many authors now agree that TRPV1 and CB_1 receptors may be considered as ionotropic and metabotropic receptors for the same class of endogenous fatty acid amides, including so far AEA and NADA (Di Marzo et al. 2002a,b). A further recent development in this area of research has been the demonstration that THC and a second plant cannabinoid, cannabinol, but not AEA, activate the ANKTM1 receptor, which is another type of transient receptor potential (TRP) channel and appears to be the primary molecular target for mustard oils in some sensory efferents (Jordt et al. 2004). In contrast, AEA and 2-AG, after their hydrolysis to arachidonic acid and conversion to epoxygenase derivatives, activate TRPV4 channels (Watanabe et al. 2003). These TRP channel-mediated actions seem to be important, for example, in the control of small artery dilation (see below), and indicate a partial overlap between the ligand recognition pre-requisites of cannabinoid receptors and some TRP channels.

4.2 Endocannabinoid Pharmacological Actions: Some Major Differences from THC

The pharmacology of endocannabinoids overlaps with that of THC to a great extent. However, important qualitative and quantitative differences have been observed between the pharmacological actions in vivo of THC and, for example, AEA. Together with the high metabolic instability of endocannabinoids, the observation that some of these compounds can activate receptors different from CB_1 and CB_2 can certainly explain some of these differences. This is particularly true for the four behavioural actions that, when assessed together in mice, have been used to characterize a compound as *cannabimimetic* in vivo, i.e. the ability to: (1) inhibit an acute pain response in the "tail flick" or "hot plate" tests; (2) induce immobility on a "ring"; (3) inhibit spontaneous locomotion in an open field; and (4) reduce body temperature. Although activity in this "mouse tetrad" is exhibited by all CB_1 receptor agonists, particularly if they possess a cannabinoid-like chemical structure, it is now accepted that a compound may still exhibit activity in all four tests and yet not act via these receptors (see Wiley and Martin 2003 for a recent critical discussion of this concept). For example, AEA-vanilloid "hybrid" compounds that potently stimulate TRPV1, but not CB_1, receptors are also very potent and efficacious in the tetrad (Di Marzo et al. 2002c), and each of the activities assessed in this way can also be elicited by capsaicin in either mice or rats. Indeed, AEA, unlike THC, is still active in at least three of the four tetrad tests when these are carried out in transgenic mice lacking a functional CB_1 receptor (Di Marzo et al. 2000a), or when these receptors are blocked with SR141716A (rimonabant) (Adams et al. 1998). However, the activity of AEA in these tests has never been assessed using TRPV1-knockout mice. Therefore, the possibility that the effects of this endocannabinoid on the tetrad in CB_1-knockout mice are mediated by these receptors has not been addressed experimentally. Interestingly, a recent study showed that AEA, if administered i.p. to Wistar rats, can cause hypolocomotion via TRPV1 receptors (de Lago et al. 2004a). Indeed, given the ability of AEA to interact with several other receptors (see previous section), and the possible lack of specificity of the tetrad of tests, the fact that this compound can exert central actions even in the absence of CB_1 receptors cannot be regarded any longer as surprising, although the search for the possible alternative target(s) responsible for these actions in vivo is far from being concluded.

The local vasodilator actions, and the effects (or lack of effects) on the release from sensory neurons of nociceptive neuropeptides, represent two other examples of pharmacological differences between THC and endocannabinoids (Randall et al. 2002). THC appears to be either inactive or weakly active in isolated artery preparations, depending on the absence or presence on capsaicin-sensitive perivascular neurons of novel THC receptors (Wagner et al. 1999; Zygmunt et al. 1999; Zygmunt et al. 2002), recently identified as ANKTM1 channels (Jordt et al. 2004). AEA does not activate ANKTM1 but nevertheless produces vasodilation through several complex, concurrent mechanisms (see Ralevic et al. 2003 and Hiley and Ford 2004, for recent reviews) that, for example, involve the participation of endothelial abn-cbd-sensitive receptors, TRPV1 receptors on perivascular neurons and K^+ and Ca^{2+} channels, etc., as well as the possible formation of arachidonate metabolites. The potent vasodilator effect of NADA is also complex (O'Sullivan et al. 2004). Thus, it is mediated by TRPV1 channels, abn-cbd-sensitive receptors and CB_1 receptors, with the relative contribution made by each of these varying according to whether experiments are performed with the superior mesenteric artery or with small mesenteric vessels. Finally, while the vasodilator actions of 2-AG in such preparations have been found to depend solely on its hydrolysis to arachidonic acid and subsequent conversion to cyclooxygenase products (Járai et al. 2000), recent data suggest that 2-AGE (noladin) acts via a novel non CB_1/CB_2 $G_{i/o}$-linked receptor (Ralevic et al. 2004).

Apart from resulting in qualitatively and quantitatively different vasodilator effects, the difference in the abilities of AEA, NADA and THC to stimulate the release of nociceptive/vasodilator neuropeptides (i.e. substance P and calcitonin gene-related peptide) via TRPV1 receptors explains why THC, which does not activate TRPV1, is never pro-nociceptive, whereas AEA and, particularly, NADA can produce hyperalgesic effects (Ahluwalia et al. 2003a; Price et al. 2004). Interestingly, NADA can be anti-nociceptive when administered systemically in vivo, possibly due to its agonist activity at CB_1 receptors (Bisogno et al. 2000), and induces nocifensive reactions when administered locally (Huang et al. 2002; Price et al. 2004).

Finally, neuroprotection is another area in which endocannabinoids and THC produce qualitatively and quantitatively different effects both in vitro and in vivo (see van der Stelt et al. 2003; Walter and Stella 2004, for recent reviews). Apart from its actions on CB_1 receptors, THC, but not anandamide, was also found to behave as an anti-oxidant in vivo (Hampson et al. 1998; Marsicano et al. 2002). Conversely, AEA exerts neuroprotective effects against excitotoxicity that are not uniquely mediated by CB_1 receptors (van der Stelt et al. 2001; Veldhuis et al. 2003).

5 New Drugs from the Endocannabinoid System

5.1 Regulation of Endocannabinoid Levels Under Pathological Conditions

Although we now know that the effects of endocannabinoids and exogenously administered THC can be both qualitatively and quantitatively different, the fact that the symptoms of many ailments have been reported to be alleviated by THC and *Cannabis* provided the rationale to test whether pathological alterations of endocannabinoid signalling can be causative of pathological states, or of their signs. There is now increasing evidence that endocannabinoid levels undergo significant changes in several animal models of both acute and chronic disorders. In particular, they appear to be transiently elevated in specific brain areas during several pathological conditions of the CNS, i.e. following insults or stressful stimuli, such as:

- Glutamate excitotoxicity, in the hippocampus (Marsicano et al. 2003)
- Food deprivation, in the hypothalamus and limbic forebrain (Kirkham et al. 2002)
- Exposure to an aversive memory, in the basolateral amygdala (Marsicano et al. 2002)
- Administration of a painful stimulus, in the periaqueductal grey (Walker et al. 1999)

In these cases, endocannabinoid signalling is enhanced to minimize the impact of the insult or of the stressful stimulus, respectively by:

- Protecting neurons from damage, via feed-back inhibition of glutamatergic neuron activity (Marsicano et al. 2003)
- Reinforcing appetite, via inhibition of anorexic signals (Kirkham et al. 2002; Di Marzo et al. 2001c; Cota et al. 2003)
- Suppressing aversive memories, via inhibition of γ-aminobutyric acid (GABA)-ergic signalling (Marsicano et al. 2002)
- Producing central analgesia, by suppressing the activity of nociceptive circuits (Walker et al. 1999)

Findings that CB_1 receptors appear to contribute significantly to protection from stroke in animals (Parmentier-Batteur et al. 2002), and that 2-AG is protective in a model of head trauma (Panikashvili et al. 2001), support the hypothesis that endocannabinoids have a neuroprotective role. In fact, endocannabinoid signalling is also elevated in animal models of neurodegenerative diseases, such as:

- In reserpine- or 6-hydroxy-dopamine-treated rats (two models of Parkinson's disease) at the level of the basal ganglia (Di Marzo et al. 2000b; Maccarrone et al. 2003c)

- In β-amyloid-treated rats (a model of Alzheimer's disease), in the hippocampus (authors' own unpublished results)
- In mice with chronic relapsing experimental allergic encephalomyelitis (CREAE, a model of multiple sclerosis), in the spinal cord (Baker et al. 2001)

The possible function of this up-regulated signalling, as suggested by pharmacological studies, is presumably to counteract neuronal hyperactivity and local inflammation, and hence damage, or, in the case of multiple sclerosis, to inhibit tremor and spasticity (Baker et al. 2000). However, the progressive nature of some disorders appears to result in a permanent, as opposed to transient, hyperactivation of the endocannabinoid system. This phenomenon appears to even contribute to the development of symptoms typical of Parkinson's disease and Alzheimer's disease, i.e. inhibition of motor activity and loss of memory, respectively, which in fact can be antagonized by CB_1 blockers (Di Marzo et al. 2000b; Mazzola et al. 2003). Furthermore, these effects may result, in some cases, in a compensatory down-regulation of CB_1 receptor expression (Silverdale et al. 2001; Berrendero et al. 2001). In contrast, in animal models of Huntington's chorea there is a loss of CB_1-expressing fibres from the basal ganglia even at the early stages of the disorder, and this results in reduced levels of both endocannabinoids and CB_1 receptors. This decrease in endocannabinoid signalling may contribute to the hyperkinesia typical of the first phase of the disease (Lastres-Becker et al. 2001; Denovan-Wright and Robertson 2000).

The endocannabinoid system is implicated in the physiological control of food intake and energy balance, not only after food deprivation but also in animal models of genetic obesity in which it appears to become overactive at the level of both the hypothalamus and adipocytes (Di Marzo et al. 2001c; Bensaid et al. 2003). This possibly explains why, following treatment of mice and rats with rimonabant, a transient inhibition of food intake and a more persistent reduction of fat mass are observed (Ravinet-Trillou et al. 2003), and why CB_1 "knockout" mice show a reduced susceptibility to obesity in response to a fat diet (Ravinet-Trillou et al. 2004).

Endocannabinoids also participate in pathological conditions of the cardiovascular, immune, gastrointestinal and reproductive systems. Enhanced macrophage and/or platelet endocannabinoid levels are found in rats during hemorrhagic and septic shock, or following liver cirrhosis and experimental myocardial infarction, and cause the hypotensive state typical of these conditions (Wagner et al. 1997; Varga et al. 1998; Batkai et al. 2001; Wagner et al. 2001). Anandamide levels and/or cannabinoid CB_1 receptor expression levels are also enhanced in three mouse models of intestinal disorders, i.e.:

- Small intestine inflammation (Izzo et al. 2001)
- Cholera toxin-induced intestinal hyper-secretion and diarrhoea (Izzo et al. 2003)
- Peritonitis-induced paralytic ileus (Mascolo et al. 2002)

While enhanced signalling at CB_1 receptors contributes to the production of reduced intestinal motility typical of paralytic ileus, in small intestine inflammation and cholera toxin-induced hyper-secretion and diarrhoea it affords tonic protection against the symptoms of the disorders.

Again, by acting preferentially at cannabinoid CB_1 receptors, anandamide plays a dual function in mouse embryo implantation, by stimulating it at low concentrations and inhibiting it at higher ones (Wang et al. 2003). Indeed, impaired anandamide hydrolysis in the blood of pregnant women leads to high levels of this compound correlating with premature abortion or failure of implanted in vitro-fertilized oocytes (Maccarrone et al. 2000c, 2002a).

Finally, enhanced endocannabinoid signalling is found in some human malignancies as compared to the corresponding healthy tissues (Ligresti et al. 2003; Schmid et al. 2002b), as well as in human cancer cells that are exhibiting a high degree of invasiveness (Sanchez et al. 2001; Portella et al. 2003). This observation, together with the finding that stimulation of either CB_1 or CB_2 receptors causes blockage of the proliferation of cancer cells or induction of their apoptosis in vitro (Ligresti et al. 2003; Galve-Roperh et al. 2001), and inhibition of cancer growth, angiogenesis and metastasis in vivo (Galve-Roperh et al. 2001; Bifulco et al. 2001; Portella et al. 2003; Casanova et al. 2003), suggests that endocannabinoids may afford some protection against tumoural growth and spread.

In summary, altered endocannabinoid signalling accompanies several central and peripheral disorders, the effect of this being to counteract symptoms and, maybe, even disease progression. In some cases, a hyperactive or a defective endocannabinoid system contributes to the production of symptoms. In view of the parallelism found between many experimental models and the corresponding clinical disorders (see Di Marzo et al. 2004 for a review), these findings suggest that substances that either prolong the half-life of endocannabinoids or prevent their formation or action may have therapeutic uses.

5.2
Potential Therapeutic Use of Inhibitors of Endocannabinoid Metabolic Fate

As pointed out in this chapter, endocannabinoids appear to be produced "on demand" to play, in many cases, a protective role "when and where needed". This observation provides one more rationale for the design of novel substances that, by retarding the inactivation of endocannabinoids when they are being produced with a protective function, might be exploited therapeutically. Promising results in preclinical studies have already been published with inhibitors of endocannabinoid metabolism in animal models of:

- Acute pain (Martin et al. 2000; Kathuria et al. 2003), particularly with FAAH inhibitors
- Epilepsy (Marsicano et al. 2003), with the uptake inhibitor UCM-707

- Multiple sclerosis (Baker et al. 2001; de Lago et al. 2004b; C. Guaza and V. Di Marzo, unpublished observations), with both uptake and FAAH inhibitors
- Parkinson's disease (Maccarrone et al. 2003c), with both uptake and FAAH inhibitors
- Anxiety (Kathuria et al. 2003), with URB-597, a potent FAAH inhibitor
- Cholera toxin-induced intestinal hypersecretion and diarrhoea (Izzo et al. 2003), with the uptake inhibitor VDM11

Unlike the direct stimulation of cannabinoid receptors with systemic agonists, this approach is likely to influence endocannabinoid levels mostly in those tissues where there is an ongoing production of these compounds, and hence it is less likely to result in undesired side-effects.

6 Concluding Remarks

As can be surmised from the data reviewed in this chapter, considerable progress has been made in little more than 10 years towards the understanding of those mechanisms underlying the regulation of the "cannabinergic" signal, particularly if one takes into consideration the fact that the cloning of the first cannabinoid receptor was only reported in 1990, and the first endocannabinoid identified only 2 years later. Apart from being conserved in all vertebrate phyla, the endocannabinoid system is also present, possibly with some major differences in the structure of receptors and in their function, in most invertebrates (McPartland 2004), thus corroborating the concept of its participation in vital functions. Although great breakthroughs in endocannabinoid biochemistry and pharmacology have been achieved in little more than a decade, several other milestones need to be met. In particular, it will be necessary:

- To understand the regulation at the molecular level of the enzymes catalysing anandamide and 2-AG biosynthesis and inactivation
- To assess the role as endocannabinoids of virodhamine, NADA and 2-AGE, find their biosynthetic pathways and clarify their regulation
- To establish transgenic mice lacking functional genes for endocannabinoid biosynthesis, and to study their phenotype
- To isolate and clone the putative EMT
- To develop selective and potent inhibitors of endocannabinoid biosynthesis and of 2-AG degradation that can be used in vivo
- To clone the novel receptors proposed for AEA and to establish their actual participation in endocannabinoid pharmacological actions in vivo

- To carry on identifying those disorders that can be caused, at least in part, by a malfunctioning endocannabinoid system, or whose onset, progress and/or symptoms are counteracted tonically by the endocannabinoids

Once these further tasks have been achieved, and the regulation of the endocannabinoid system under both physiological and pathological conditions is fully understood, it will be possible to assess whether endocannabinoid-based medicines with clear advantages over other established therapeutic drugs could be developed in the future.

References

Adams IB, Compton DR, Martin BR (1998) Assessment of anandamide interaction with the cannabinoid brain receptor: SR 141716A antagonism studies in mice and autoradiographic analysis of receptor binding in rat brain. J Pharmacol Exp Ther 284:1209–1217

Ahluwalia J, Urban L, Bevan S, Nagy I (2003a) Anandamide regulates neuropeptide release from capsaicin-sensitive primary sensory neurons by activating both the cannabinoid 1 receptor and the vanilloid receptor 1 in vitro. Eur J Neurosci 17:2611–2618

Ahluwalia J, Yaqoob M, Urban L, Bevan S, Nagy I (2003b) Activation of capsaicin-sensitive primary sensory neurones induces anandamide production and release. J Neurochem 84:585–591

Andersson DA, Adner M, Hogestatt ED, Zygmunt PM (2002) Mechanisms underlying tissue selectivity of anandamide and other vanilloid receptor agonists. Mol Pharmacol 62:705–713

Appendino G, Ligresti A, Minassi A, Daddario N, Bisogno T, Di Marzo V (2003) Homologues and isomers of noladin ether, a putative novel endocannabinoid: interaction with rat cannabinoid CB(1) receptors. Bioorg Med Chem Lett 13:43–46

Baker D, Pryce G, Croxford JL, Brown P, Pertwee RG, Huffman JW, Layward L (2000) Cannabinoids control spasticity and tremor in a multiple sclerosis model. Nature 404:84–87

Baker D, Pryce G, Croxford JL, Brown P, Pertwee RG, Makriyannis A, Khanolkar A, Layward L, Fezza F, Bisogno T, Di Marzo V (2001) Endocannabinoids control spasticity in a multiple sclerosis model. FASEB J 15:300–302

Basavarajappa BS, Saito M, Cooper TB, Hungund BL (2000) Stimulation of cannabinoid receptor agonist 2-arachidonylglycerol by chronic ethanol and its modulation by specific neuromodulators in cerebellar granule neurons. Biochim Biophys Acta 1535:78–86

Basavarajappa BS, Saito M, Cooper TB, Hungund BL (2003) Chronic ethanol inhibits the anandamide transport and increases extracellular anandamide levels in cerebellar granule neurons. Eur J Pharmacol 466:73–83

Batkai S, Jarai Z, Wagner JA, Goparaju SK, Varga K, Liu J, Wang L, Mirshahi F, Khanolkar AD, Makriyannis A, Urbaschek R, Garcia N Jr, Sanyal AJ, Kunos G (2001) Endocannabinoids acting at vascular CB1 receptors mediate the vasodilated state in advanced liver cirrhosis. Nat Med 7:827–832

Beltramo M, Piomelli D (2000) Carrier-mediated transport and enzymatic hydrolysis of the endogenous cannabinoid 2-arachidonylglycerol. Neuroreport 11:1231–1235

Beltramo M, Stella N, Calignano A, Lin SY, Makriyannis A, Piomelli D (1997) Functional role of high-affinity anandamide transport, as revealed by selective inhibition. Science 277:1094–1097

Ben-Shabat S, Fride E, Sheskin T, Tamiri T, Rhee MH, Vogel Z, Bisogno T, De Petrocellis L, Di Marzo V, Mechoulam R (1998) An entourage effect: inactive endogenous fatty acid glycerol esters enhance 2-arachidonoyl-glycerol cannabinoid activity. Eur J Pharmacol 353:23–31

Bensaid M, Gary-Bobo M, Esclangon A, Maffrand JP, Le Fur G, Oury-Donat F, Soubrie P (2003) The cannabinoid CB1 receptor antagonist SR141716 increases Acrp30 mRNA expression in adipose tissue of obese fa/fa rats and in cultured adipocyte cells. Mol Pharmacol 63:908–914

Berdyshev EV, Schmid PC, Krebsbach RJ, Schmid HH (2001) Activation of PAF receptors results in enhanced synthesis of 2-arachidonoylglycerol (2-AG) in immune cells. FASEB J 15:2171–2178

Berglund BA, Boring DL, Howlett AC (1999) Investigation of structural analogs of prostaglandin amides for binding to and activation of CB1 and CB2 cannabinoid receptors in rat brain and human tonsils. Adv Exp Med Biol 469:527–533

Berrendero F, Sepe N, Ramos JA, Di Marzo V, Fernandez-Ruiz JJ (1999) Analysis of cannabinoid receptor binding and mRNA expression and endogenous cannabinoid contents in the developing rat brain during late gestation and early postnatal period. Synapse 33:181–191

Berrendero F, Sanchez A, Cabranes A, Puerta C, Ramos JA, Garcia-Merino A, Fernandez-Ruiz J (2001) Changes in cannabinoid CB(1) receptors in striatal and cortical regions of rats with experimental allergic encephalomyelitis, an animal model of multiple sclerosis. Synapse 41:195–202

Bifulco M, Laezza C, Portella G, Vitale M, Orlando P, De Petrocellis L, Di Marzo V (2001) Control by the endogenous cannabinoid system of ras oncogene-dependent tumor growth. FASEB J 15:2745–2747

Bisogno T, Maurelli S, Melck D, De Petrocellis L, Di Marzo V (1997a) Biosynthesis, uptake, and degradation of anandamide and palmitoylethanolamide in leukocytes. J Biol Chem 272:3315–3323

Bisogno T, Sepe N, Melck D, Maurelli S, De Petrocellis L, Di Marzo V (1997b) Biosynthesis, release and degradation of the novel endogenous cannabimimetic metabolite 2-arachidonoylglycerol in mouse neuroblastoma cells. Biochem J 322:671–677

Bisogno T, Melck D, De Petrocellis L, Bobrov MYu, Gretskaya NM, Bezuglov VV, Sitachitta N, Gerwick WH, Di Marzo V (1998) Arachidonoylserotonin and other novel inhibitors of fatty acid amide hydrolase. Biochem Biophys Res Commun 248:515–522

Bisogno T, Berrendero F, Ambrosino G, Cebeira M, Ramos JA, Fernandez-Ruiz JJ, Di Marzo V (1999a) Brain regional distribution of endocannabinoids: implications for their biosynthesis and biological function. Biochem Biophys Res Commun 256:377–380

Bisogno T, Melck D, De Petrocellis L, Di Marzo V (1999b) Phosphatidic acid as the biosynthetic precursor of the endocannabinoid 2-arachidonoylglycerol in intact mouse neuroblastoma cells stimulated with ionomycin. J Neurochem 72:2113–2119

Bisogno T, Melck D, Bobrov MYu, Gretskaya NM, Bezuglov VV, De Petrocellis L, Di Marzo V (2000) N-acyl-dopamines: novel synthetic CB(1) cannabinoid-receptor ligands and inhibitors of anandamide inactivation with cannabimimetic activity in vitro and in vivo. Biochem J 351:817–824

Bisogno T, Maccarrone M, De Petrocellis L, Jarrahian A, Finazzi-Agrò A, Hillard C, Di Marzo V (2001) The uptake by cells of 2-arachidonoylglycerol, an endogenous agonist of cannabinoid receptors. Eur J Biochem 268:1982–1989

Bisogno T, De Petrocellis L, Di Marzo V (2002) Fatty acid amide hydrolase, an enzyme with many bioactive substrates. Possible therapeutic implications. Curr Pharm Des 8:533–547

Bisogno T, Howell F, Williams G, Minassi A, Cascio MG, Ligresti A, Matias I, Paul P, Gangadharan U, Hobbs C, Di Marzo V, Doherty, P (2003) Cloning of the first sn 1-DAG lipases points to the spatial and temporal regulation of endocannabinoid signalling in the brain. J Cell Biol 163:463–468

Boger DL, Sato H, Lerner AE, Hedrick MP, Fecik RA, Miyauchi H, Wilkie GD, Austin BJ, Patricelli MP, Cravatt BF (2000) Exceptionally potent inhibitors of fatty acid amide hydrolase: the enzyme responsible for degradation of endogenous oleamide and anandamide. Proc Natl Acad Sci USA 97:5044–5049

Boger DL, Miyauchi H, Hedrick MP (2001) Alpha-keto heterocycle inhibitors of fatty acid amide hydrolase: carbonyl group modification and alpha-substitution. Bioorg Med Chem Lett 11:1517–1520

Bornheim LM, Kim KY, Chen B, Correia MA (1993) The effect of cannabidiol on mouse hepatic microsomal cytochrome P450-dependent anandamide metabolism. Biochem Biophys Res Commun 197:740–746

Bracey MH, Hanson MA, Masuda KR, Stevens RC, Cravatt BF (2002) Structural adaptations in a membrane enzyme that terminates endocannabinoid signaling. Science 298:1793–1796

Breivogel CS, Griffin G, Di Marzo V, Martin BR (2001) Evidence for a new G protein-coupled cannabinoid receptor in mouse brain. Mol Pharmacol 60:155–163

Brenowitz SD, Regehr WG (2003) Calcium dependence of retrograde inhibition by endocannabinoids at synapses onto Purkinje cells. J Neurosci 23:6373–6384

Brooks JW, Pryce G, Bisogno T, Jaggar SI, Hankey DJ, Brown P, Bridges D, Ledent C, Bifulco M, Rice AS, Di Marzo V, Baker D (2002) Arvanil-induced inhibition of spasticity and persistent pain: evidence for therapeutic sites of action different from the vanilloid VR1 receptor and cannabinoid CB(1)/CB(2) receptors. Eur J Pharmacol 439:83–92

Cadas H, di Tomaso E, Piomelli D (1997) Occurrence and biosynthesis of endogenous cannabinoid precursor, N-arachidonoyl phosphatidylethanolamine, in rat brain. J Neurosci 17:1226–1242

Calignano A, La Rana G, Giuffrida A, Piomelli D (1998) Control of pain initiation by endogenous cannabinoids. Nature 394:277–281

Carrier EJ, Kearn CS, Barkmeier AJ, Breese NM, Yang W, Nithipatikom K, Pfister SL, Campbell WB, Hillard CJ (2004) Cultured rat microglial cells synthesize the endocannabinoid 2-arachidonylglycerol, which increases proliferation via a CB2 receptor-dependent mechanism. Mol Pharmacol 65:999–1007

Casanova ML, Blazquez C, Martinez-Palacio J, Villanueva C, Fernandez-Acenero MJ, Huffman JW, Jorcano JL, Guzman M (2003) Inhibition of skin tumor growth and angiogenesis in vivo by activation of cannabinoid receptors. J Clin Invest 111:43–50

Chemin J, Monteil A, Perez-Reyes E, Nargeot J, Lory P (2001) Direct inhibition of T-type calcium channels by the endogenous cannabinoid anandamide. EMBO J 20:7033–7040

Chevaleyre V, Castillo PE (2003) Heterosynaptic LTD of hippocampal GABAergic synapses: a novel role of endocannabinoids in regulating excitability. Neuron 38:461–472

Clement AB, Hawkins EG, Lichtman AH, Cravatt BF (2003) Increased seizure susceptibility and proconvulsant activity of anandamide in mice lacking fatty acid amide hydrolase. J Neurosci 23:3916–3923

Cota D, Marsicano G, Tschop M, Grubler Y, Flachskamm C, Schubert M, Auer D, Yassouridis A, Thone-Reineke C, Ortmann S, Tomassoni F, Cervino C, Nisoli E, Linthorst AC, Pasquali R, Lutz B, Stalla GK, Pagotto U (2003) The endogenous cannabinoid system affects energy balance via central orexigenic drive and peripheral lipogenesis. J Clin Invest 112:423–431

Craib SJ, Ellington HC, Pertwee RG, Ross, RA (2001) A possible role of lipoxygenase in the activation of vanilloid receptors by anandamide in the guinea-pig bronchus. Br J Pharmacol 134:30–37

Cravatt BF, Lichtman AH (2003) Fatty acid amide hydrolase: an emerging therapeutic target in the endocannabinoid system. Curr Opin Chem Biol 7:469–475

Cravatt BF, Giang DK, Mayfield SP, Boger DL, Lerner RA, Gilula NB (1996) Molecular characterization of an enzyme that degrades neuromodulatory fatty-acid amides. Nature 384:83–87

Cravatt BF, Demarest K, Patricelli MP, Bracey MH, Giang DK, Martin BR, Lichtman AH (2001) Supersensitivity to anandamide and enhanced endogenous cannabinoid signaling in mice lacking fatty acid amide hydrolase. Proc Natl Acad Sci USA 98:9371–9376

de Lago E, Fernandez-Ruiz J, Ortega-Gutierrez S, Viso A, Lopez-Rodriguez ML, Ramos JA (2002) UCM707, a potent and selective inhibitor of endocannabinoid uptake, potentiates hypokinetic and antinociceptive effects of anandamide. Eur J Pharmacol 449:99–103

de Lago E, de Miguel R, Lastres-Becker I, Ramos JA, Fernández-Ruiz J (2004a) Involvement of vanilloid-like receptors in the effects of anandamide on motor behavior and nigrostriatal dopaminergic activity: in vivo and in vitro evidence. Brain Res 1007:152–159
de Lago E, Ligresti A, Ortar G, Morera E, Cabranes A, Pryce G, Bifulco M, Baker D, Fernandez-Ruiz J, Di Marzo V (2004b) In vivo pharmacological actions of two novel inhibitors of anandamide cellular uptake. Eur J Pharmacol 484:249–257
De Petrocellis L, Melck D, Ueda N, Maurelli S, Kurahashi Y, Yamamoto S, Marino G, Di Marzo V (1997) Novel inhibitors of brain, neuronal, and basophilic anandamide amidohydrolase. Biochem Biophys Res Commun 231:82–88
De Petrocellis L, Bisogno T, Davis JB, Pertwee RG, Di Marzo V (2000) Overlap between the ligand recognition properties of the anandamide transporter and the VR1 vanilloid receptor: inhibitors of anandamide uptake with negligible capsaicin-like activity. FEBS Lett 483:52–56
De Petrocellis L, Bisogno T, Maccarrone M, Davis JB, Finazzi-Agrò A, Di Marzo V (2001) The activity of anandamide at vanilloid VR1 receptors requires facilitated transport across the cell membrane and is limited by intracellular metabolism. J Biol Chem 276:12856–12863
Denovan-Wright EM, Robertson HA (2000) Cannabinoid receptor messenger RNA levels decrease in a subset of neurons of the lateral striatum, cortex and hippocampus of transgenic Huntington's disease mice. Neuroscience 98:705–713
Deutsch DG, Lin S, Hill WA, Morse KL, Salehani D, Arreaza G, Omeir RL, Makriyannis A (1997a) Fatty acid sulfonyl fluorides inhibit anandamide metabolism and bind to the cannabinoid receptor. Biochem Biophys Res Commun 231:217–221
Deutsch DG, Omeir R, Arreaza G, Salehani D, Prestwich GD, Huang Z, Howlett A (1997b) Methyl arachidonyl fluorophosphonate: a potent irreversible inhibitor of anandamide amidase. Biochem Pharmacol 53:255–260
Deutsch DG, Glaser ST, Howell JM, Kunz JS, Puffenbarger RA, Hillard CJ, Abumrad N (2001) The cellular uptake of anandamide is coupled to its breakdown by fatty-acid amide hydrolase. J Biol Chem 276:6967–6973
Deutsch DG, Ueda N, Yamamoto S (2002) The fatty acid amide hydrolase (FAAH). Prostaglandins Leukot Essent Fatty Acids 66:201–210
Devane WA, Hanus L, Breuer A, Pertwee RG, Stevenson LA, Griffin G, Gibson D, Mandelbaum A, Etinger A, Mechoulam R (1992) Isolation and structure of a brain constituent that binds to the cannabinoid receptor. Science 258:1946–1949
Di Marzo V, Fontana A (1995) Anandamide, an endogenous cannabinomimetic eicosanoid: 'killing two birds with one stone'. Prostaglandins Leukot Essent Fatty Acids 53:1–11
Di Marzo V, Fontana A, Cadas H, Schinelli S, Cimino G, Schwartz JC, Piomelli D (1994) Formation and inactivation of endogenous cannabinoid anandamide in central neurons. Nature 372:686–691
Di Marzo V, De Petrocellis L, Sepe N, Buono A (1996a) Biosynthesis of anandamide and related acylethanolamides in mouse J774 macrophages and N18 neuroblastoma cells. Biochem J 316:977–984
Di Marzo V, De Petrocellis L, Sugiura T, Waku K (1996b) Potential biosynthetic connections between the two cannabimimetic eicosanoids, anandamide and 2-arachidonoylglycerol, in mouse neuroblastoma cells. Biochem Biophys Res Commun 227:281–288
Di Marzo V, Bisogno T, Sugiura T, Melck D, De Petrocellis L (1998a) The novel endogenous cannabinoid 2-arachidonoylglycerol is inactivated by neuronal- and basophil-like cells: connections with anandamide. Biochem J 331:15–19
Di Marzo V, Melck D, Bisogno T, De Petrocellis L (1998b) Endocannabinoids: endogenous cannabinoid receptor ligands with neuromodulatory action. Trends Neurosci 21:521–528
Di Marzo V, Bisogno T, De Petrocellis L, Melck D, Orlando P, Wagner JA, Kunos G (1999a) Biosynthesis and inactivation of the endocannabinoid 2-arachidonoylglycerol in circulating and tumoral macrophages. Eur J Biochem 264:258–267

Di Marzo V, De Petrocellis L, Bisogno T, Melck D (1999b) Metabolism of anandamide and 2-arachidonoylglycerol: an historical overview and some recent developments. Lipids 34:319–325

Di Marzo V, Breivogel CS, Tao Q, Bridgen DT, Razdan RK, Zimmer AM, Zimmer A, Martin, BR (2000a) Levels, metabolism, and pharmacological activity of anandamide in CB(1) cannabinoid receptor knockout mice: evidence for non-CB(1), non-CB(2) receptor-mediated actions of anandamide in mouse brain. J Neurochem 75:2434–2444

Di Marzo V, Hill MP, Bisogno T, Crossman AR, Brotchie JM (2000b) Enhanced levels of endogenous cannabinoids in the globus pallidus are associated with a reduction in movement in an animal model of Parkinson's disease. FASEB J 14:1432–1438

Di Marzo V, Bisogno T, De Petrocellis L (2001a) Anandamide: some like it hot. Trends Pharmacol Sci 22:346–349

Di Marzo V, Bisogno T, De Petrocellis L, Brandi I, Jefferson RG, Winckler L, Davis JB, Dasse O, Mahadevan A, Razdan RK, Martin BR (2001b) Highly selective CB(1) cannabinoid receptor ligands and novel CB(1)/VR(1) vanilloid receptor "hybrid" ligands. Biochem Biophys Res Commun 281:444–451

Di Marzo V, Goparaju SK, Wang L, Liu J, Batkai S, Jarai Z, Fezza F, Miura GI, Palmiter RD, Sugiura T, Kunos G (2001c) Leptin-regulated endocannabinoids are involved in maintaining food intake. Nature 410:822–825

Di Marzo V, Lastres-Becker I, Bisogno T, De Petrocellis L, Milone A, Davis JB, Fernandez-Ruiz JJ (2001d) Hypolocomotor effects in rats of capsaicin and two long chain capsaicin homologues. Eur J Pharmacol 420:123–131

Di Marzo V, Blumberg PM, Szallasi A (2002a) Endovanilloid signaling in pain. Curr Opin Neurobiol 12:372–379

Di Marzo V, De Petrocellis L, Fezza F, Ligresti A, Bisogno T (2002b) Anandamide receptors. Prostaglandins Leukot Essent Fatty Acids 66:377–391

Di Marzo V, Griffin G, De Petrocellis L, Brandi I, Bisogno T, Williams W, Grier MC, Kulasegram S, Mahadevan A, Razdan RK, Martin BR (2002c) A structure/activity relationship study on arvanil, an endocannabinoid and vanilloid hybrid. J Pharmacol Exp Ther 300:984–991

Di Marzo V, Bifulco M, De Petrocellis L (2004) The endocannabinoid system and its therapeutic exploitation. Nat Rev Drug Discov 3:771–784

Dinh TP, Carpenter D, Leslie FM, Freund TF, Katona I, Sensi S, Kathuria S, Piomelli D (2002) Brain monoglyceride lipase participating in endocannabinoid inactivation. Proc Natl Acad Sci USA 99:10819–10824

Edgemond WS, Hillard CJ, Falck JR, Kearn CS, Campbell, WB (1998) Human platelets and polymorphonuclear leukocytes synthesize oxygenated derivatives of arachidonylethanolamide (anandamide): their affinities for cannabinoid receptors and pathways of inactivation. Mol Pharmacol 54:180–188

Egertova M, Giang DK, Cravatt BF, Elphick MR (1998) A new perspective on cannabinoid signalling: complementary localization of fatty acid amide hydrolase and the CB1 receptor in rat brain. Proc R Soc Lond B Biol Sci 265:2081–2085

Evans RM, Scott RH, Ross RA (2004) Multiple actions of anandamide on neonatal rat cultured sensory neurones. Br J Pharmacol 141:1223–1233

Fezza F, Bisogno T, Minassi A, Appendino G, Mechoulam R, Di Marzo V (2002) Noladin ether, a putative novel endocannabinoid: inactivation mechanisms and a sensitive method for its quantification in rat tissues. FEBS Lett 513:294–298

Fowler CJ, Tiger G, Lopez-Rodriguez ML, Viso A, Ortega-Gutierrez S, Ramos JA (2003) Inhibition of fatty acid amidohydrolase, the enzyme responsible for the metabolism of the endocannabinoid anandamide, by analogues of arachidonoyl-serotonin. J Enzyme Inhib Med Chem 18:225–231

Fowler CJ, Tiger G, Ligresti A, López-Rodríguez ML, Di Marzo V (2004) Selective inhibition of anandamide cellular uptake versus enzymatic hydrolysis—a difficult issue to handle. Eur J Pharmacol 492:1–11

Franklin A, Parmentier-Batteur S, Walter L, Greenberg DA, Stella N (2003) Palmitoylethanolamide increases after focal cerebral ischemia and potentiates microglial cell motility. J Neurosci 23:7767–7775
Galve-Roperh I, Sanchez C, Cortes ML, del Pulgar TG, Izquierdo M, Guzman M (2000) Anti-tumoral action of cannabinoids: involvement of sustained ceramide accumulation and extracellular signal-regulated kinase activation. Nat Med 6:313–319
Glaser ST, Abumrad NA, Fatade F, Kaczocha M, Studholme KM, Deutsch DG (2003) Evidence against the presence of an anandamide transporter. Proc Natl Acad Sci USA 100:4269–4274
Goparaju SK, Ueda N, Taniguchi K, Yamamoto S (1999) Enzymes of porcine brain hydrolyzing 2-arachidonoylglycerol, an endogenous ligand of cannabinoid receptors. Biochem Pharmacol 57:417–423
Guo J, Ikeda SR (2004) Endocannabinoids modulate N-type calcium channels and G-protein-coupled inwardly rectifying potassium channels via CB1 cannabinoid receptors heterologously expressed in mammalian neurons. Mol Pharmacol 65:665–674
Hajos N, Freund TF (2002) Pharmacological separation of cannabinoid sensitive receptors on hippocampal excitatory and inhibitory fibers. Neuropharmacology 43:503–510
Hajos N, Ledent C, Freund TF (2001) Novel cannabinoid-sensitive receptor mediates inhibition of glutamatergic synaptic transmission in the hippocampus. Neuroscience 106:1–4
Hampson AJ, Grimaldi M, Axelrod J, Wink D (1998) Cannabidiol and (-)delta9-tetrahydrocannabinol are neuroprotective antioxidants. Proc Natl Acad Sci USA 95:8268–8273
Hansen HS, Lauritzen L, Moesgaard B, Strand AM, Hansen HH (1998) Formation of N-acyl-phosphatidylethanolamines and N-acetylethanolamines: proposed role in neurotoxicity. Biochem Pharmacol 55:719–725
Hanus L, Abu-Lafi S, Fride E, Breuer A, Vogel Z, Shalev DE, Kustanovich I, Mechoulam R (2001) 2-Arachidonyl glyceryl ether, an endogenous agonist of the cannabinoid CB1 receptor. Proc Natl Acad Sci USA 98:3662–3665
Hiley CR, Ford WR (2004) Cannabinoid pharmacology in the cardiovascular system: potential protective mechanisms through lipid signalling. Biol Rev Camb Philos Soc 79:187–205
Hillard CJ, Jarrahian A (2000) The movement of N-arachidonoylethanolamine (anandamide) across cellular membranes. Chem Phys Lipids 108:123–134
Hillard CJ, Jarrahian A (2003) Cellular accumulation of anandamide: consensus and controversy. Br J Pharmacol 140:802–808
Hillard CJ, Edgemond WS, Jarrahian A, Campbell WB (1997) Accumulation of N-arachidonoylethanolamine (anandamide) into cerebellar granule cells occurs via facilitated diffusion. J Neurochem 69:631–638
Ho SY, Delgado L, Storch J (2002) Monoacylglycerol metabolism in human intestinal Caco-2 cells: evidence for metabolic compartmentation and hydrolysis. J Biol Chem 277:1816–1823
Horrocks LA (1989) Sources for brain arachidonic acid uptake and turnover in glycerophospholipids. Ann N Y Acad Sci 559:17–24
Huang SM, Bisogno T, Petros TJ, Chang SY, Zavitsanos PA, Zipkin RE, Sivakumar R, Coop A, Maeda DY, De Petrocellis L, Burstein S, Di Marzo V, Walker JM (2001) Identification of a new class of molecules, the arachidonyl amino acids, and characterization of one member that inhibits pain. J Biol Chem 276:42639–42644
Huang SM, Bisogno T, Trevisani M, Al-Hayani A, De Petrocellis L, Fezza F, Tognetto M, Petros TJ, Krey JF, Chu CJ, Miller JD, Davies SN, Geppetti P, Walker JM, Di Marzo V (2002) An endogenous capsaicin-like substance with high potency at recombinant and native vanilloid VR1 receptors. Proc Natl Acad Sci USA 99:8400–8405
Hwang SW, Cho H, Kwak J, Lee SY, Kang CJ, Jung J, Cho S, Min KH, Suh YG, Kim D, Oh U (2000) Direct activation of capsaicin receptors by products of lipoxygenases: endogenous capsaicin-like substances. Proc Natl Acad Sci USA 97:6155–6160

Izzo AA, Fezza F, Capasso R, Bisogno T, Pinto L, Iuvone T, Esposito G, Mascolo N, Di Marzo V, Capasso F (2001) Cannabinoid CB1-receptor mediated regulation of gastrointestinal motility in mice in a model of intestinal inflammation. Br J Pharmacol 134:563–570

Izzo AA, Capasso F, Costagliola A, Bisogno T, Marsicano G, Ligresti A, Matias I, Capasso R, Pinto L, Borrelli F, Cecio A, Lutz B, Mascolo N, Di Marzo V (2003) An endogenous cannabinoid tone attenuates cholera toxin-induced fluid accumulation in mice. Gastroenterology 125:765–774

Jarai Z, Wagner JA, Varga K, Lake KD, Compton DR, Martin BR, Zimmer AM, Bonner TI, Buckley NE, Mezey E, Razdan RK, Zimmer A, Kunos G (1999) Cannabinoid-induced mesenteric vasodilation through an endothelial site distinct from CB1 or CB2 receptors. Proc Natl Acad Sci USA 96:14136–14141

Jarai Z, Wagner JA, Goparaju SK, Wang L, Razdan RK, Sugiura T, Zimmer AM, Bonner TI, Zimmer A, Kunos G (2000) Cardiovascular effects of 2-arachidonoyl glycerol in anesthetized mice. Hypertension 35:679–684

Jarrahian A, Manna S, Edgemond WS, Campbell WB, Hillard CJ (2000) Structure-activity relationships among N-arachidonylethanolamine (anandamide) head group analogues for the anandamide transporter. J Neurochem 74:2597–2606

Jordt SE, Julius D (2002) Molecular basis for species-specific sensitivity to "hot" chili peppers. Cell 108:421–430

Jordt SE, Bautista DM, Chuang HH, McKemy DD, Zygmunt PM, Hogestatt ED, Meng ID, Julius D (2004) Mustard oils and cannabinoids excite sensory nerve fibres through the TRP channel ANKTM1. Nature 427:260–265

Karlsson M, Contreras JA, Hellman U, Tornqvist H, Holm C (1997) cDNA cloning, tissue distribution, and identification of the catalytic triad of monoglyceride lipase. Evolutionary relationship to esterases, lysophospholipases, and haloperoxidases. J Biol Chem 272:27218–27223

Karlsson M, Reue K, Xia YR, Lusis AJ, Langin D, Tornqvist H, Holm C (2001) Exon-intron organization and chromosomal localization of the mouse monoglyceride lipase gene. Gene 272:11–18

Kathuria S, Gaetani S, Fegley D, Valino F, Duranti A, Tontini A, Mor M, Tarzia G, La Rana G, Calignano A, Giustino A, Tattoli M, Palmery M, Cuomo V, Piomelli D (2003) Modulation of anxiety through blockade of anandamide hydrolysis. Nat Med 9:76–81

Kim J, Isokawa M, Ledent C, Alger BE (2002) Activation of muscarinic acetylcholine receptors enhances the release of endogenous cannabinoids in the hippocampus. J Neurosci 22:10182–10191

Kirkham TC, Williams CM, Fezza F, Di Marzo V (2002) Endocannabinoid levels in rat limbic forebrain and hypothalamus in relation to fasting, feeding and satiation: stimulation of eating by 2-arachidonoyl glycerol. Br J Pharmacol 136:550–557

Kondo S, Kondo H, Nakane S, Kodaka T, Tokumura A, Waku K, Sugiura T (1998) 2-Arachidonoylglycerol, an endogenous cannabinoid receptor agonist: identification as one of the major species of monoacylglycerols in various rat tissues, and evidence for its generation through Ca2+-dependent and -independent mechanisms. FEBS Lett 429:152–156

Koutek B, Prestwich GD, Howlett AC, Chin SA, Salehani D, Akhavan N, Deutsch DG (1994) Inhibitors of arachidonoyl ethanolamide hydrolysis. J Biol Chem 269:22937–22940

Kozak KR, Marnett LJ (2002) Oxidative metabolism of endocannabinoids. Prostaglandins Leukot Essent Fatty Acids 66:211–220

Kozak KR, Crews BC, Ray JL, Tai HH, Morrow JD, Marnett LJ (2001) Metabolism of prostaglandin glycerol esters and prostaglandin ethanolamides in vitro and in vivo. J Biol Chem 276:36993–36998

Kozak KR, Gupta RA, Moody JS, Ji C, Boeglin WE, DuBois RN, Brash AR, Marnett LJ (2002) 15-Lipoxygenase metabolism of 2-arachidonylglycerol. Generation of a peroxisome proliferator-activated receptor alpha agonist. J Biol Chem 277:23278–23286

Lambert DM, Vandevoorde S, Jonsson KO, Fowler CJ (2002) The palmitoylethanolamide family: a new class of anti-inflammatory agents? Curr Med Chem 9:663–674

Lastres-Becker I, Fezza F, Cebeira M, Bisogno T, Ramos JA, Milone A, Fernandez-Ruiz J, Di Marzo V (2001) Changes in endocannabinoid transmission in the basal ganglia in a rat model of Huntington's disease. Neuroreport 12:2125–2129

Lichtman AH, Hawkins EG, Griffin G, Cravatt BF (2002) Pharmacological activity of fatty acid amides is regulated, but not mediated, by fatty acid amide hydrolase in vivo. J Pharmacol Exp Ther 302:73–79

Ligresti A, Bisogno T, Matias I, De Petrocellis L, Cascio MG, Cosenza V, D'argenio G, Scaglione G, Bifulco M, Sorrentini I, Di Marzo V (2003) Possible endocannabinoid control of colorectal cancer growth. Gastroenterology 125:677–687

Ligresti A, Morera E, Van Der Stelt MM, Monory K, Lutz B, Ortar G, Di Marzo V (2004) Further evidence for the existence of a specific process for the membrane transport of anandamide. Biochem J 380(Pt 1):265–272

Liu J, Batkai S, Pacher P, Harvey-White J, Wagner JA, Cravatt BF, Gao B, Kunos G (2003) Lipopolysaccharide induces anandamide synthesis in macrophages via CD14/MAPK/ phosphoinositide 3-kinase/NF-kappaB independently of platelet-activating factor. J Biol Chem 278:45034–45039

Liu Q, Tonai T, Ueda N (2002) Activation of N-acylethanolamine-releasing phospholipase D by polyamines. Chem Phys Lipids 115:77–84

Lopez-Rodriguez ML, Viso A, Ortega-Gutierrez S, Lastres-Becker I, Gonzalez S, Fernandez-Ruiz JJ, Ramos, JA (2001) Design, synthesis and biological evaluation of novel arachidonic acid derivatives as highly potent and selective endocannabinoid transporter inhibitors. J Med Chem 44:4505–4508

Lopez-Rodriguez ML, Viso A, Ortega-Gutierrez S, Fowler CJ, Tiger G, de Lago E, Fernandez-Ruiz J, Ramos JA (2003a) Design, synthesis and biological evaluation of new endocannabinoid transporter inhibitors. Eur J Med Chem 38:403–412

Lopez-Rodriguez ML, Viso A, Ortega-Gutierrez S, Fowler CJ, Tiger G, de Lago E, Fernandez-Ruiz J, Ramos JA (2003b) Design, synthesis, and biological evaluation of new inhibitors of the endocannabinoid uptake: comparison with effects on fatty acid amidohydrolase. J Med Chem 46:1512–1522

Maccarrone M, Bari M, Lorenzon T, Bisogno T, Di Marzo V, Finazzi-Agrò A (2000a) Anandamide uptake by human endothelial cells and its regulation by nitric oxide. J Biol Chem 275:13484–13492

Maccarrone M, Salvati S, Bari M, Finazzi-Agrò A (2000b) Anandamide and 2-arachidonoylglycerol inhibit fatty acid amide hydrolase by activating the lipoxygenase pathway of the arachidonate cascade. Biochem Biophys Res Commun 278:576–583

Maccarrone M, Valensise H, Bari M, Lazzarin N, Romanini C, Finazzi-Agro' A (2000c) Relation between decreased anandamide hydrolase concentrations in human lymphocytes and miscarriage. Lancet 355:1326–1329

Maccarrone M, De Petrocellis L, Bari M, Fezza F, Salvati S, Di Marzo V, Finazzi-Agrò A (2001) Lipopolysaccharide downregulates fatty acid amide hydrolase expression and increases anandamide levels in human peripheral lymphocytes. Arch Biochem Biophys 393:321–328

Maccarrone M, Bisogno T, Valensise H, Lazzarin N, Fezza F, Manna C, Di Marzo V, Finazzi-Agro' A (2002a) Low fatty acid amide hydrolase and high anandamide levels are associated with failure to achieve an ongoing pregnancy after IVF and embryo transfer. Mol Hum Reprod 8:188–195

Maccarrone M, Cartoni A, Parolaro D, Margonelli A, Massi P, Bari M, Battista N, Finazzi-Agrò A (2002b) Cannabimimetic activity, binding, and degradation of stearoylethanolamide within the mouse central nervous system. Mol Cell Neurosci 21:126–140

Maccarrone M, Bari M, Di Rienzo M, Finazzi-Agrò A, Rossi A (2003a) Progesterone activates Fatty Acid Amide Hydrolase (FAAH) promoter in Human T lymphocytes through the transcription factor Ikaros: evidence for a synergistic effect of leptin. J Biol Chem 278:32726–32732

Maccarrone M, Di Rienzo M, Finazzi-Agrò A, Rossi A (2003b) Leptin activates the anandamide hydrolase promoter in human T lymphocytes through STAT3. J Biol Chem 278:13318–13324
Maccarrone M, Gubellini P, Bari M, Picconi B, Battista N, Centonze D, Bernardi G, Finazzi-Agro' A, Calabresi P (2003c) Levodopa treatment reverses endocannabinoid system abnormalities in experimental parkinsonism. J Neurochem 85:1018–1025
Maingret F, Patel AJ, Lazdunski M, Honore E (2001) The endocannabinoid anandamide is a direct and selective blocker of the background K(+) channel TASK-1. EMBO J 20:47–54
Markey SP, Dudding T, Wang TC (2000) Base- and acid-catalyzed interconversions of O-acyl- and N-acyl-ethanolamines: a cautionary note for lipid analyses. J Lipid Res 41:657–662
Marsicano G, Wotjak CT, Azad SC, Bisogno T, Rammes G, Cascio MG, Hermann H, Tang J, Hofmann C, Zieglgansberger W, Di Marzo V, Lutz B (2002) The endogenous cannabinoid system controls extinction of aversive memories. Nature 418:530–534
Marsicano G, Goodenough S, Monory K, Hermann H, Eder M, Cannich A, Azad SC, Cascio MG, Gutierrez SO, van der Stelt M, Lopez-Rodriguez ML, Casanova E, Schutz G, Zieglgansberger W, Di Marzo V, Behl C, Lutz B (2003) CB1 cannabinoid receptors and on-demand defense against excitotoxicity. Science 302:84–88
Martin BR, Beletskaya I, Patrick G, Jefferson R, Winckler R, Deutsch DG, Di Marzo V, Dasse O, Mahadevan A, Razdan RK (2000) Cannabinoid properties of methylfluorophosphonate analogs. J Pharmacol Exp Ther 294:1209–1218
Mascolo N, Izzo AA, Ligresti A, Costagliola A, Pinto L, Cascio MG, Maffia P, Cecio A, Capasso F, Di Marzo V (2002) The endocannabinoid system and the molecular basis of paralytic ileus in mice. FASEB J 16:1973–1975
Matias I, Chen J, De Petrocellis L, Bisogno T, Ligresti A, Fezza F, Krauss AH, Shi L, Protzman CE, Li C, Liang Y, Nieves AL, Kedzie KM, Burk RM, Di Marzo V, Woodward DF (2004) Prostaglandin-ethanolamides (prostamides): in vitro pharmacology and metabolism. J Pharmacol Exp Ther 309:745–757
Maurelli S, Bisogno T, De Petrocellis L, Di Luccia A, Marino G, Di Marzo V (1995) Two novel classes of neuroactive fatty acid amides are substrates for mouse neuroblastoma 'anandamide amidohydrolase'. FEBS Lett 377:82–86
Mazzola C, Micale V, Drago F (2003) Amnesia induced by beta-amyloid fragments is counteracted by cannabinoid CB1 receptor blockade. Eur J Pharmacol 477:219–225
McAllister SD, Glass M (2002) CB(1) and CB(2) receptor-mediated signalling: a focus on endocannabinoids. Prostaglandins Leukot Essent Fatty Acids 66:161–171
McPartland JM (2004) Phylogenomic and chemotaxonomic analysis of the endocannabinoid system. Brain Res Brain Res Rev 45:18–29
McVey DC, Schmid PC, Schmid HH, Vigna, SR (2003) Endocannabinoids induce ileitis in rats via the capsaicin receptor (VR1). J Pharmacol Exp Ther 304:713–722
Mechoulam R, Ben-Shabat S, Hanus L, Ligumsky M, Kaminski NE, Schatz AR, Gopher A, Almog S, Martin BR, Compton DR (1995) Identification of an endogenous 2-monoglyceride, present in canine gut, that binds to cannabinoid receptors. Biochem Pharmacol 50:83–90
Mechoulam R, Fride E, Ben-Shabat S, Meiri U, Horowitz M (1998a) Carbachol, an acetylcholine receptor agonist, enhances production in rat aorta of 2-arachidonoyl glycerol, a hypotensive endocannabinoid. Eur J Pharmacol 362:R1–R3
Mechoulam R, Fride E, Di Marzo V (1998b) Endocannabinoids. Eur J Pharmacol 359:1–18
Monory K, Tzavara ET, Lexime J, Ledent C, Parmentier M, Borsodi A, Hanoune J (2002) Novel, not adenylyl cyclase-coupled cannabinoid binding site in cerebellum of mice. Biochem Biophys Res Commun 292:231–235
Nagan N, Zoeller RA (2001) Plasmalogens: biosynthesis and functions. Prog Lipid Res 40:199–229
Nakane S, Oka S, Arai S, Waku K, Ishima Y, Tokumura A, Sugiura T (2002) 2-Arachidonoyl-sn-glycero-3-phosphate, an arachidonic acid-containing lysophosphatidic acid: occurrence and rapid enzymatic conversion to 2-arachidonoyl-sn-glycerol, a cannabinoid receptor ligand, in rat brain. Arch Biochem Biophys 402:51–58

Natarajan V, Schmid PC, Reddy PV, Schmid HH (1984) Catabolism of N-acylethanolamine phospholipids by dog brain preparations. J Neurochem 42:1613–1619
Nirodi CS, Crews BC, Kozak KR, Morrow JD, Marnett LJ (2004) The glyceryl ester of prostaglandin E2 mobilizes calcium and activates signal transduction in RAW264.7 cells. Proc Natl Acad Sci USA 101:1840–1845
O'Sullivan SE, Kendall DA, Randall MD (2004) Characterisation of the vasorelaxant properties of the novel endocannabinoid N-arachidonoyl-dopamine (NADA). Br J Pharmacol 141:803–812
Offertaler L, Mo FM, Batkai S, Liu J, Begg M, Razdan RK, Martin BR, Bukoski RD, Kunos G (2003) Selective ligands and cellular effectors of a G protein-coupled endothelial cannabinoid receptor. Mol Pharmacol 63:699–705
Ohno-Shosaku T, Matsui M, Fukudome Y, Shosaku J, Tsubokawa H, Taketo MM, Manabe T, Kano M (2003) Postsynaptic M1 and M3 receptors are responsible for the muscarinic enhancement of retrograde endocannabinoid signalling in the hippocampus. Eur J Neurosci 18:109–116
Oka S, Tsuchie A, Tokumura A, Muramatsu M, Suhara Y, Takayama H, Waku, Sugiura T (2003) Ether-linked analogue of 2-arachidonoylglycerol (noladin ether) was not detected in the brains of various mammalian species. J Neurochem 85:1374–1381
Okamoto Y, Morishita J, Tsuboi K, Tonai T, Ueda N (2004) Molecular characterization of a phospholipase D generating anandamide and its congeners. J Biol Chem 279:5298–5305
Ortar G, Ligresti A, De Petrocellis L, Morera E, Di Marzo V (2003) Novel selective and metabolically stable inhibitors of anandamide cellular uptake. Biochem Pharmacol 65:1473–1481
Panikashvili D, Simeonidou C, Ben-Shabat S, Hanus L, Breuer A, Mechoulam R, Shohami E (2001) An endogenous cannabinoid (2-AG) is neuroprotective after brain injury. Nature 413:527–531
Parmentier-Batteur S, Jin K, Mao XO, Xie L, Greenberg DA (2002) Increased severity of stroke in CB1 cannabinoid receptor knock-out mice. J Neurosci 22:9771–9775
Patricelli MP, Cravatt BF (2000) Clarifying the catalytic roles of conserved residues in the amidase signature family. J Biol Chem 275:19177–19184
Pertwee RG (1997) Pharmacology of cannabinoid CB1 and CB2 receptors. Pharmacol Ther 74:129–180
Pertwee RG (2004) Novel pharmacological targets for cannabinoids. Curr Neuropharmacol 2:9–29
Petersen G, Hansen HS (1999) N-acylphosphatidylethanolamine-hydrolysing phospholipase D lacks the ability to transphosphatidylate. FEBS Lett 455:41–44
Pinto L, Izzo AA, Cascio MG, Bisogno T, Hospodar-Scott K, Brown DR, Mascolo N, Di Marzo V, Capasso F (2002) Endocannabinoids as physiological regulators of colonic propulsion in mice. Gastroenterology 123:227–234
Piomelli D, Beltramo M, Glasnapp S, Lin SY, Goutopoulos A, Xie XQ, Makriyannis A (1999) Structural determinants for recognition and translocation by the anandamide transporter. Proc Natl Acad Sci USA 96:5802–5807
Portella G, Laezza C, Laccetti P, De Petrocellis L, Di Marzo V, Bifulco M (2003) Inhibitory effects of cannabinoid CB1 receptor stimulation on tumor growth and metastatic spreading: actions on signals involved in angiogenesis and metastasis. FASEB J 17:1771–1773
Porter AC, Sauer JM, Knierman MD, Becker GW, Berna MJ, Bao J, Nomikos GG, Carter P, Bymaster FP, Leese AB, Felder CC (2002) Characterization of a novel endocannabinoid, virodhamine, with antagonist activity at the CB1 receptor. J Pharmacol Exp Ther 301:1020–1024
Premkumar LS, Qi ZH, Van Buren J, Raisinghani M (2004) Enhancement of potency and efficacy of NADA by PKC-mediated phosphorylation of vanilloid receptor. J Neurophysiol 91:1442–1449

Price TJ, Patwardhan A, Akopian AN, Hargreaves KM, Flores CM (2004) Modulation of trigeminal sensory neuron activity by the dual cannabinoid-vanilloid agonists anandamide, N-arachidonoyl-dopamine and arachidonyl-2-chloroethylamide. Br J Pharmacol 141:1118–1130
Puffenbarger RA, Kapulin O, Howell JM, Deutsch DG (2001) Characterization of the 5'-sequence of the mouse fatty acid amide hydrolase. Neurosci Lett 314:21–24
Ralevic V, Duncan M, Millns P, Smart D, Wright J, Kendall D (2004) Noladin ether, a putative endocannabinoid, attenuates sensory neurotransmission in the rat isolated mesenteric arterial bed via a non CB1/CB2 Gi/o linked receptor. Br J Pharmacol 142:509–518
Randall MD, Harris D, Kendall DA, Ralevic V (2002) Cardiovascular effects of cannabinoids. Pharmacol Ther 95:191–202
Ravinet Trillou C, Arnone M, Delgorge C, Gonalons N, Keane P, Maffrand JP, Soubrie P (2003) Anti-obesity effect of SR141716, a CB1 receptor antagonist, in diet-induced obese mice. Am J Physiol Regul Integr Comp Physiol 284:R345–R353
Ravinet Trillou C, Delgorge C, Menet C, Arnone M, Soubrie P (2004) CB1 cannabinoid receptor knockout in mice leads to leanness, resistance to diet-induced obesity and enhanced leptin sensitivity. Int J Obes Relat Metab Disord 28:640–648
Ronesi J, Gerdeman GL, Lovinger DM (2004) Disruption of endocannabinoid release and striatal long-term depression by postsynaptic blockade of endocannabinoid membrane transport. J Neurosci 24:1673–1679
Ross RA (2003) Anandamide and vanilloid TRPV1 receptors. Br J Pharmacol 140:790–801
Ross RA, Gibson TM, Brockie HC, Leslie M, Pashmi G, Craib SJ, Di Marzo V, Pertwee RG (2001) Structure-activity relationship for the endogenous cannabinoid, anandamide, and certain of its analogues at vanilloid receptors in transfected cells and vas deferens. Br J Pharmacol 132:631–640
Ruiz-Llorente L, Ortega-Gutiérrez S, Viso A, Sánchez MG, Sánchez AM, Fernández C, Ramos JA, Hillard C, Lasunción MA, López-Rodríguez ML, Díaz-Laviada I (2004) Characterization of an anandamide degradation system in prostate epithelial PC-3 cells: synthesis of new transporter inhibitors as tools for this study. Br J Pharmacol 141:457–467
Sagan S, Venance L, Torrens Y, Cordier J, Glowinski J, Giaume C (1999) Anandamide and WIN 55212-2 inhibit cyclic AMP formation through G-protein-coupled receptors distinct from CB1 cannabinoid receptors in cultured astrocytes. Eur J Neurosci 11:691–699
Sanchez C, de Ceballos ML, del Pulgar TG, Rueda D, Corbacho C, Velasco G, Galve-Roperh I, Huffman JW, Ramon y Cajal S, Guzman M (2001) Inhibition of glioma growth in vivo by selective activation of the CB(2) cannabinoid receptor. Cancer Res 61:5784–5789
Schmid HH, Schmid PC, Natarajan V (1990) N-acylated glycerophospholipids and their derivatives. Prog Lipid Res 29:1–43
Schmid HH, Schmid PC, Natarajan V (1996) The N-acylation-phosphodiesterase pathway and cell signalling. Chem Phys Lipids 80:133–142
Schmid HH, Schmid PC, Berdyshev EV (2002a) Cell signaling by endocannabinoids and their congeners: questions of selectivity and other challenges. Chem Phys Lipids 121:111–134
Schmid PC, Wold LE, Krebsbach RJ, Berdyshev EV, Schmid HH (2002b) Anandamide and other N-acylethanolamines in human tumors. Lipids 37:907–912
Silverdale MA, McGuire S, McInnes A, Crossman AR, Brotchie JM (2001) Striatal cannabinoid CB1 receptor mRNA expression is decreased in the reserpine-treated rat model of Parkinson's disease. Exp Neurol 169:400–406
Smart D, Gunthorpe MJ, Jerman JC, Nasir S, Gray J, Muir AI, Chambers JK, Randall AD, Davis JB (2000) The endogenous lipid anandamide is a full agonist at the human vanilloid receptor (hVR1). Br J Pharmacol 129:227–230
Stella N, Piomelli D (2001) Receptor-dependent formation of endogenous cannabinoids in cortical neurons. Eur J Pharmacol 425:189–196
Stella N, Schweitzer P, Piomelli D (1997) A second endogenous cannabinoid that modulates long-term potentiation. Nature 388:773–778

Sugiura T, Kondo S, Sukagawa A, Nakane S, Shinoda A, Itoh K, Yamashita A, Waku K (1995) 2-Arachidonoylglycerol: a possible endogenous cannabinoid receptor ligand in brain. Biochem Biophys Res Commun 215:89–97

Sugiura T, Kondo S, Sukagawa A, Tonegawa T, Nakane S, Yamashita A, Waku K (1996a) Enzymatic synthesis of anandamide, an endogenous cannabinoid receptor ligand, through N-acylphosphatidylethanolamine pathway in testis: involvement of Ca(2+)-dependent transacylase and phosphodiesterase activities. Biochem Biophys Res Commun 218:113–117

Sugiura T, Kondo S, Sukagawa A, Tonegawa T, Nakane S, Yamashita A, Ishima Y, Waku K (1996b) Transacylase-mediated and phosphodiesterase-mediated synthesis of N-arachidonoylethanolamine, an endogenous cannabinoid-receptor ligand, in rat brain microsomes. Comparison with synthesis from free arachidonic acid and ethanolamine. Eur J Biochem 240:53–62

Sugiura T, Kodaka T, Nakane S, Kishimoto S, Kondo S, Waku K (1998) Detection of an endogenous cannabimimetic molecule, 2-arachidonoylglycerol, and cannabinoid CB1 receptor mRNA in human vascular cells: is 2-arachidonoylglycerol a possible vasomodulator? Biochem Biophys Res Commun 243:838–843

Sugiura T, Kodaka T, Nakane S, Miyashita T, Kondo S, Suhara Y, Takayama H, Waku K, Seki C, Baba N, Ishima Y (1999) Evidence that the cannabinoid CB1 receptor is a 2-arachidonoylglycerol receptor. Structure-activity relationship of 2-arachidonoylglycerol, ether-linked analogues, and related compounds. J Biol Chem 274:2794–2801

Sugiura T, Yoshinaga N, Waku K (2001) Rapid generation of 2-arachidonoylglycerol, an endogenous cannabinoid receptor ligand, in rat brain after decapitation. Neurosci Lett 297:175–178

Sugiura T, Kobayashi Y, Oka S, Waku, K (2002) Biosynthesis and degradation of anandamide and 2-arachidonoylglycerol and their possible physiological significance. Prostaglandins Leukot Essent Fatty Acids 66:173–192

Sun YX, Tsuboi K, Okamoto Y, Tonai T, Murakami M, Kudo I, Ueda N (2004) Biosynthesis of anandamide and N-palmitoylethanolamine by sequential actions of phospholipase A2 and lysophospholipase D. Biochem J 380:749–756

Szallasi A, Blumberg PM (1999) Vanilloid (Capsaicin) receptors and mechanisms. Pharmacol Rev 51:159–212

Tarzia G, Duranti A, Tontini A, Piersanti G, Mor M, Rivara S, Plazzi PV, Park C, Kathuria S, Piomelli D (2003) Design, synthesis, and structure-activity relationships of alkylcarbamic acid aryl esters, a new class of fatty acid amide hydrolase inhibitors. J Med Chem 46:2352–2360

Ueda N (2002) Endocannabinoid hydrolases. Prostaglandins Other Lipid Mediat 68–69:521–534

Ueda N, Yamamoto K, Yamamoto S, Tokunaga T, Shirakawa E, Shinkai H, Ogawa M, Sato T, Kudo I, Inoue K (1995) Lipoxygenase-catalyzed oxygenation of arachidonylethanolamide, a cannabinoid receptor agonist. Biochim Biophys Acta 1254:127–134

Ueda N, Liu Q, Yamanaka K (2001a) Marked activation of the N-acylphosphatidylethanolamine-hydrolyzing phosphodiesterase by divalent cations. Biochim Biophys Acta 1532: 121–127

Ueda N, Yamanaka K, Yamamoto S (2001b) Purification and characterization of an acid amidase selective for N-palmitoylethanolamine, a putative endogenous anti-inflammatory substance. J Biol Chem 276:35552–35557

van der Stelt M, Di Marzo V (2004) Endovanilloids: endogenous ligands of transient receptor potential vanilloid 1 (TRPV1) channels. Eur J Pharmacol 271:1827–1834

van der Stelt M, Veldhuis WB, van Haaften GW, Fezza F, Bisogno T, Bar PR, Veldink GA, Vliegenthart JF, Di Marzo V, Nicolay K (2001) Exogenous anandamide protects rat brain against acute neuronal injury in vivo. J Neurosci 21:8765–8771

van der Stelt M, van Kuik JA, Bari M, van Zadelhoff G, Leeflang BR, Veldink GA, Finazzi-Agrò A, Vliegenthart JF, Maccarrone M (2002) Oxygenated metabolites of anandamide and 2-arachidonoylglycerol: conformational analysis and interaction with cannabinoid receptors, membrane transporter, and fatty acid amide hydrolase. J Med Chem 45:3709–3720

van der Stelt M, Hansen HH, Veldhuis WB, Bar PR, Nicolay K, Veldink GA, Vliegenthart JF, Hansen HS (2003) Biosynthesis of endocannabinoids and their modes of action in neurodegenerative diseases. Neurotox Res 5:183–200

Varga K, Wagner JA, Bridgen DT, Kunos G (1998) Platelet- and macrophage-derived endogenous cannabinoids are involved in endotoxin-induced hypotension. FASEB J 12:1035–1044

Veldhuis WB, van der Stelt M, Wadman MW, van Zadelhoff G, Maccarrone M, Fezza F, Veldink GA, Vliegenthart JF, Bar PR, Nicolay K, Di Marzo V (2003) Neuroprotection by the endogenous cannabinoid anandamide and arvanil against in vivo excitotoxicity in the rat: role of vanilloid receptors and lipoxygenases. J Neurosci 23:4127–4133

Venance L, Piomelli D, Glowinski J, Giaume C (1995) Inhibition by anandamide of gap junctions and intercellular calcium signalling in striatal astrocytes. Nature 376:590–594

Wagner JA, Varga K, Ellis EF, Rzigalinski BA, Martin BR, Kunos G (1997) Activation of peripheral CB1 cannabinoid receptors in haemorrhagic shock. Nature 390:518–521

Wagner JA, Varga K, Jarai Z, Kunos G (1999) Mesenteric vasodilation mediated by endothelial anandamide receptors. Hypertension 33:429–434

Wagner JA, Hu K, Bauersachs J, Karcher J, Wiesler M, Goparaju SK, Kunos G, Ertl G (2001) Endogenous cannabinoids mediate hypotension after experimental myocardial infarction. J Am Coll Cardiol 38:2048–2054

Waleh NS, Cravatt BF, Apte-Deshpande A, Terao A, Kilduff TS (2002) Transcriptional regulation of the mouse fatty acid amide hydrolase gene. Gene 291:203–210

Walker JM, Huang SM, Strangman NM, Tsou K, Sanudo-Pena MC (1999) Pain modulation by release of the endogenous cannabinoid anandamide. Proc Natl Acad Sci USA 96:12198–12203

Walter L, Stella N (2003) Endothelin-1 increases 2-arachidonoyl glycerol (2-AG) production in astrocytes. Glia 44:85–90

Walter L, Stella N (2004) Cannabinoids and neuroinflammation. Br J Pharmacol 141:775–785

Walter L, Franklin A, Witting A, Wade C, Xie Y, Kunos G, Mackie K, Stella N (2003) Nonpsychotropic cannabinoid receptors regulate microglial cell migration. J Neurosci 23:1398–1405

Wang H, Matsumoto H, Guo Y, Paria BC, Roberts RL, Dey SK (2003) Differential G protein-coupled cannabinoid receptor signaling by anandamide directs blastocyst activation for implantation. Proc Natl Acad Sci USA 100:14914–14919

Watanabe H, Vriens J, Prenen J, Droogmans G, Voets T, Nilius B (2003) Anandamide and arachidonic acid use epoxyeicosatrienoic acids to activate TRPV4 channels. Nature 424:434–438

Wiley JL, Martin BR (2003) Cannabinoid pharmacological properties common to other centrally acting drugs. Eur J Pharmacol 471:185–193

Williams EJ, Walsh FS, Doherty P (2003) The FGF receptor uses the endocannabinoid signaling system to couple to an axonal growth response. J Cell Biol 160:481–486

Wilson RI, Nicoll RA (2002) Endocannabinoid signaling in the brain. Science 296:678–682

Yu M, Ives D, Ramesha CS (1997) Synthesis of prostaglandin E2 ethanolamide from anandamide by cyclooxygenase-2. J Biol Chem 272:21181–21186

Zygmunt PM, Petersson J, Andersson DA, Chuang H, Sorgard M, Di Marzo V, Julius D, Hogestatt ED (1999) Vanilloid receptors on sensory nerves mediate the vasodilator action of anandamide. Nature 400:452–457

Zygmunt PM, Chuang H, Movahed P, Julius D, Hogestatt ED (2000) The anandamide transport inhibitor AM404 activates vanilloid receptors. Eur J Pharmacol 396:39–42

Zygmunt PM, Andersson DA, Hogestatt ED (2002) Delta 9-tetrahydrocannabinol and cannabinol activate capsaicin-sensitive sensory nerves via a CB1 and CB2 cannabinoid receptor-independent mechanism. J Neurosci 22:4720–4727

HEP (2005) 168:187–207

Modulators of Endocannabinoid Enzymic Hydrolysis and Membrane Transport

W.-S.V. Ho · C.J. Hillard (✉)

Department of Pharmacology and Toxicology, Medical College of Wisconsin, 8701 Watertown Plank Road, Milwaukee WI, 53226, USA
chillard@mcw.edu

Abstract Tissue concentrations of the endocannabinoids *N*-arachidonoylethanolamine (AEA) and 2-arachidonoylglycerol (2-AG) are regulated by both synthesis and inactivation. The purpose of this review is to compile available data regarding three inactivation processes: fatty acid amide hydrolase, monoacylglycerol lipase, and cellular membrane transport. In particular, we have focused on mechanisms by which these processes are modulated. We describe the in vitro and in vivo effects of inhibitors of these processes as well as available evidence regarding their modulation by other factors.

Keywords Fatty acid amide hydrolase · Monoacylglycerol lipase · Transporter · Carrier · Anandamide · 2-Arachidonoylglycerol · *N*-Arachidonoylethanolamine

1
Introduction

It is becoming clear that like most neuromodulatory molecules, the effective concentrations of the endocannabinoids (eCBs) *N*-arachidonoylethanolamine (AEA) and 2-arachidonoylglycerol (2-AG) are regulated by both synthesis and catabolism (Di Marzo, this volume). Catabolism of both AEA and 2-AG occurs via hydrolysis to arachidonic acid, and ethanolamine and glycerol, respectively. Hydrolysis of AEA is mediated primarily via fatty acid amide hydrolase (FAAH) (Cravatt et al. 2001). 2-AG is also a substrate for FAAH (Goparaju et al. 1998), but monoacylglycerol lipase (MGL) likely plays a more important role in its hydrolysis in vivo (Cravatt and Lichtman 2002). Both of these catabolic enzymes are localized intracellularly (Tsou et al. 1998; Dinh et al. 2002). This compartmentalization of the catabolic enzymes begs the question of whether a mechanism exists by which the eCBs move from the extracellular environment where they are functional signaling molecules into the intracellular environment where they are degraded. Functional studies support the possibility that a transmembrane carrier protein can transport AEA (Hillard and Jarrahian 2003), and perhaps 2-AG (Beltramo and Piomelli 2000; Bisogno et al. 2001), from one side of the plasma membrane to the other. This putative carrier has been suggested to function as an inactivation mechanism, since it would remove the eCBs from extracellular space, effectively sequestering the ligands away from their CB_1 cannabinoid receptor target. Since the putative carrier has the characteristics of a facilitated diffusion process and can also transport AEA from inside to outside (Hillard et al. 1997), it could also play a role in the release of newly synthesized AEA. Indeed, intracellular administration of uptake inhibitors blocks eCB-dependent activation of the CB_1 receptor in striatal slices (Ronesi et al. 2004).

In light of the widespread role of the eCB/CB_1 receptor signaling system in the regulation of CNS function, it is a near certainty that drugs acting on one or more of the three eCB inactivation processes characterized to date (i.e., FAAH, MGL, and cellular uptake) will be useful therapeutic agents in the future. Of the three processes, FAAH is the best characterized, and inhibitor development is the most mature. MGL has been cloned (Karlsson et al. 1997; Dinh et al. 2002), which will allow for clear identification of its role in 2-AG inactivation as well as facilitate inhibitor development. The cellular uptake process is the least characterized of the three at this point. The molecular identities of the proteins involved are not known, with the exception of data suggesting that FAAH can drive cellular uptake in some cell types (Glaser et al. 2003). In spite of the lack of molecular information, inhibitors of the uptake process have been developed and are discussed in this chapter.

2 Fatty Acid Amide Hydrolase

2.1 Characteristics of FAAH

FAAH is an integral-membrane serine hydrolase found in intracellular compartments (predominantly in microsomal fractions) of various cell types in the central nervous system and the periphery. FAAH is widely expressed in the brain, particularly in the neocortex, hippocampal formation, amygdala, and cerebellum (Herkenham et al. 1990; Giang and Cravatt 1997; Thomas et al. 1997; Yazulla et al. 1999). In the periphery, FAAH activity has been reported in the lung, liver, kidney, blood vessels, blood cells, and gastrointestinal tract, as well as the reproductive tract (Deutsch and Chin 1993; Desarnaud et al. 1995; Bisogno et al. 1997a; Giang and Cravatt 1997; Pratt et al. 1998; Maccarrone et al. 2001b).

The FAAH cDNA has been cloned from several mammalian species, and a functional homolog of the mammalian FAAH has also been reported in the plant *Arabidopsis thaliana* (Cravatt et al. 1996; Giang and Cravatt 1997; Shrestha et al. 2003). The rat and mouse FAAH sequences share 91% identity, while the human FAAH shares over 80% sequence identity with rat and mouse FAAHs. Given that human and rodent FAAHs have been shown to display broadly similar substrate selectivity and inhibitor sensitivity profiles (Giang and Cravatt 1997), FAAH activities detected in animal model systems are likely to be relevant to humans.

FAAH belongs to a class of hydrolytic enzymes called the "amidase signature family," which are defined by a conserved serine- and glycine-rich "amidase signature sequence" of approximately 130 amino acids (Cravatt et al. 1996). Its optimal pH is 8 to 9. Site-directed mutagenesis studies and structural determination of FAAH have indicated that the conserved residues Ser-241, Ser-217, Ser-218, Lys-142, and Arg-243 within the signature sequence of FAAH are essential for its catalytic activity (Patricelli and Cravatt 1999; Patricelli et al. 1999; Patricelli and Cravatt 2000; Bracey et al. 2002; McKinney and Cravatt 2003). Ser-241, Ser-217, and Lys-142 are hypothesized to form a catalytic triad. The carbonyl group of AEA or another substrate is believed to react with the hydroxyl group of Ser-241 (the catalytic nucleophile) of FAAH, forming an oxyanion tetrahedral intermediate (the "transition-state"), followed by protonation, facilitated by Ser-217 and Lys-142, of the substrate-leaving group. It has been hypothesized that an almost simultaneous occurrence of the oxyanion formation and subsequent protonation contributes to the unusual ability of FAAH to hydrolyze amides and esters at equivalent rates (McKinney and Cravatt 2003). Interestingly, FAAH with mutated Arg-243, but not the other four critical residues, has differentially reduced amidase over esterase activity (Patricelli and Cravatt 2000), indicating potential separation of the amidase and esterase activities of FAAH.

2.2
Substrate Specificity of FAAH

FAAH hydrolyzes a broad spectrum of long, unsaturated acyl chain amides and esters. Both AEA and 2-AG are hydrolyzed by FAAH at similar concentrations (K_m 3–12 μM for both AEA and 2-AG; Hillard et al. 1995; Goparaju et al. 1998). There is some evidence that other putative eCBs, arachidonoyldopamine and virodhamine, are substrates of FAAH (Bisogno et al. 2000; Huang et al. 2002; Porter et al. 2002). FAAH also hydrolyzes the sleep-inducing factor oleamide, a fatty acid amide, to its corresponding acid (Cravatt et al. 1996), as well as other biologically active fatty acid ethanolamides, including *N*-oleoylethanolamine (a satiety factor), *N*-palmitoylethanolamine (PEA; an anti-inflammatory and analgesic agent), and the lipoamino acid *N*-arachidonoylglycine (a potential analgesic agent) (Cravatt et al. 1996; Huang et al. 2001; Rodriguez de Fonseca et al. 2001; see also Ueda et al. 2000 for review).

2.3
Mechanisms of FAAH Regulation

Expression of FAAH is up-regulated by progesterone and leptin and down-regulated by estrogen and glucocorticoids (Maccarrone et al. 2001a, 2003b; Waleh et al. 2002). Changes in FAAH protein concentrations are paralleled by changes in mRNA levels, consistent with transcription regulation by these factors. Although steroid hormone-response elements have been described in the promoter region of the FAAH gene in rodents and human, the precise mechanisms by which progesterone, estrogen, and glucocorticoids regulate FAAH transcription remain unclear (Maccarrone et al. 2001a, 200, 2003b; Puffenbarger et al. 2001; Waleh et al. 2002). Regulation is tissue- and species-specific; FAAH expression is decreased in mouse uterus, but increased in rat uterus, in response to sex hormones (Maccarrone et al. 2000b; Xiao et al. 2002). The FAAH promoter region also contains a cyclic adenosine monophosphate (cAMP)-response element-like site, which is a transcriptional target of signal transducer and activator of transcription (STAT)3. It has been shown that activation of leptin receptors, probably via activation of STAT3, increases FAAH gene transcription and translation (Maccarrone et al. 2003a).

FAAH contains a type II polyproline sequence that binds Src homology 3 (SH3)-containing proteins. Given that SH3 domains are found in many signal transduction proteins, including phospholipase Cγ and phosphoinositol-3-kinase, and cytoskeletal proteins, these proteins could potentially regulate the activity and subcellular localization of FAAH (Kuriyan and Cowburn 1997; Arreaza and Deutsch 1999). Indeed, ablation of the SH3-binding domain results in loss of enzymatic activity (Arreaza and Deutsch 1999).

AEA hydrolysis by FAAH in vitro is not affected by calcium (Hillard et al. 1995; Maurelli et al. 1995). Interestingly, however, lipoxygenase products appear to inhibit FAAH activity such that inhibition of 5-lipoxygenase enhances AEA hydrolysis in mast cells (Maccarrone et al. 2000c) and neuroblastoma cells (Maccarrone et al. 2000a). Inhibition of FAAH activity in cultured endothelial cells by estrogen seems to require 15-lipoxygenase (Maccarrone et al. 2002). Maccarrone et al. (2004) have recently reported that a yet-to-be-characterized soluble lipid, which is released from blastocytes, increases FAAH activity without affecting its expression.

2.4 FAAH Inhibitors

The characterization of FAAH activity and its role in eCB signaling has been enabled by the development of effective FAAH inhibitors, with diverse structures and affinities for the enzyme (Table 1). Most of the inhibitors target the catalytic site of FAAH and thereby prevent the interaction of the enzyme and its substrates. The first identified inhibitor of FAAH was phenylmethylsulfonyl fluoride (PMSF) an agent widely used to inhibit serine proteases (Deutsch and Chin 1993). PMSF inhibits FAAH irreversibly via sulfonation of serine residues (Hillard et al. 1995; Ueda et al. 1995; Deutsch et al. 1997b). It is commonly included in CB_1 receptor ligand binding studies to inhibit FAAH-mediated catabolism of AEA. Analogs of PMSF with fatty acyl substitutions, such as palmitylsulfonyl fluoride (AM374) and stearylsulfonyl fluoride (AM381) also covalently modify serine residues in FAAH with nanomolar IC_{50} values (Lang et al. 1996; Deutsch et al. 1997b). These acyl sulfonyl fluorides display reasonable separation between FAAH inhibition and CB_1 receptor binding, especially for those with a longer saturated alkyl chain (K_i for CB_1 receptor, AM374:520 nM; AM381:19 μM; Deutsch et al. 1997b).

Another series was derived from inhibitors of phospholipase A_2 (PLA_2) and exploits the preference of FAAH for substrates with long, unsaturated acyl chains. Arachidonoyltrifluoromethylketone (ATFMK) is a reversible inhibitor of AEA hydrolysis at low micromolar range (Maurelli et al. 1995; Ueda et al. 1995; Beltramo et al. 1997a; Deutsch et al. 1997a), probably by forming a stabilized adduct of the trifluoromethylketone and an active-site serine residue (the so-called "transition-state" enzyme inhibitor). However, ATFMK is also a slow- and tight-binding inhibitor of cytosolic PLA_2 with an IC_{50} of 2–15 μM (Street et al. 1993; Riendeau et al. 1994) and it binds to CB_1 receptors in the same concentration range that inhibits AEA degradation (Koutek et al. 1994; Deutsch et al. 1997b). ATFMK also inhibits MGL (see Sect. 3.6). Methyl arachidonoyl fluorophosphonate (MAFP) is another inhibitor of arachidonoyl-selective PLA_2 (Street et al. 1993; Lio et al. 1996). It also interacts with CB_1 receptors in an irreversible manner (Deutsch et al. 1997a; Fernando and Pertwee 1997). The X-ray structure of FAAH crystallized with MAFP has shed light on FAAH substrate recognition and position in the lipid bilayer (Bracey et al. 2002).

Diazomethylarachidonoylketone (DAK) (De Petrocellis et al. 1997; Edgemond et al. 1998) also inhibits FAAH; its carbonyl carbon is likely to bind to an active site serine, whereas the diazomethyl carbon reacts with the imidazolium residue of a histidine, resulting in a very stable complex. In line with this model, three histidine residues are conserved in rodent and human FAAHs (Giang and Cravatt

Table 1. Inhibitors of FAAH activity

Inhibitor	Tissue	IC_{50} (nM)	Reference	Remarks
PMSF	Rat brain homogenates	290	Deutsch et al. 1997b	Irreversible. Unstable in aqueous solution ($t_{1/2}$ <60 min, pH 7)
	Rat forebrain membranes	13,000	Hillard et al. 1995	
	Solubilized from porcine brain	~15,000	Ueda et al. 1995	
AM374	Rat brain homogenates	13	Deutsch et al. 1997b	Irreversible; binds to CB_1 receptors (IC_{50} 520 nM; Deutsch et al. 1997b)
	Brain microsomes	50	Lang et al. 1996	
ATFMK	N18TG2 cells	700	Deutsch et al. 1997a	Reversible; inhibits $cPLA_2$ (IC_{50} 2–15 μM; Street et al. 1993); binds to CB_1 receptor (IC_{50} 0.65–2.5 μM, Koutek et al. 1994)
	Rat brain homogenate	900	Deutsch et al. 1997b	
	Solubilized porcine brain FAAH	1,000	Ueda et al. 1995	
	Mouse neuroblastoma N18TG2 cells	3,000	Maurelli et al. 1995	
	Rat brain microsomes	4,000	Beltramo et al. 1997a	
MAFP	Solubilized from N18TG2 cells, RBL-1 cells, and porcine brain RBL-1 cells, and porcine brain	1–3	De Petrocellis et al. 1997	Irreversible. Also inhibits $cPLA_2$ (Lio et al. 1996). Binds irreversibly to CB_1 receptor IC_{50} 20 nM; Deutsch et al. 1997a
	Rat brain homogenates	2.5	Deutsch et al. 1997a	
	N18TG2 cells	20	Deutsch et al. 1997a	
DAK	Rat brain membranes	500	Edgemond et al. 1998	Irreversible; binds to CB_1 receptors (IC_{50} 1.3 μM; Edgemond et al. 1998)
	Solubilized from N18TG2 cells, RBL-1 cells, and porcine brain	2,000–6,000	De Petrocellis et al. 1997	Reversible
AA-5-HT	RBL-2H3 cells	560	Bisogno et al. 1998	Reversible; inactive at $cPLA_2$, CB_1, or CB_2 receptors
	N18TG2 cells	1,200	Bisogno et al. 1998	
URB597	Rat cortical neurons	0.5	Kathuria et al. 2003	Irreversible; active in vivo at 0.3 mg/kg
	Rat brain membranes	4.6	Kathuria et al. 2003	

AA-5-HT, arachidonoylserotonin; AM374, palmitylsulfonylfluoride; ATFMK, arachidonyltrifluoromethylketone; DAK, diazomethylarachidonylketone; MAFP, methylarachidonylfluorophosphonate; PMSF, phenylmethylsulfonyl fluoride.

1997). Structural alignment of the residues are crucial for the irreversible action of DAK, as indicated by the observation that DAK inhibition of FAAH in detergent-solubilized preparation, but not that in native membranes, is reversed after anion exchange chromatography of the proteins (De Petrocellis et al. 1997; Edgemond et al. 1998). It is unclear if DAK also inhibits PLA_2, but it binds to neuronal CB_1 receptors at concentrations similar to those producing FAAH inhibition (K_i 1.3 μM; Edgemond et al. 1998).

While the inclusion of arachidonic acyl groups in the inhibitors results in high affinity for FAAH, these inhibitors also bind to other arachidonate-binding proteins, such as PLA_2 and the CB_1 receptor. An exception to this is arachidonoylserotonin (AA-5-HT), which is a tight-binding but reversible inhibitor of FAAH that is devoid of activity at CB_1 receptors and $cPLA_2$ (Bisogno et al. 1998). More recently, Boger et al. (2000) reported a new class of α-keto heterocyclic inhibitors of FAAH by combining several features—an optimal fatty acid chain length (C8-C12), *cis*-double bond at the corresponding arachidonoyl location and an α-keto oxazolopyridine ring with a weakly basic nitrogen. These compounds inhibit FAAH reversibly at a picomolar or low nanomolar range, including $\Delta^{9,10}$-octadecynoyl-α-keto-oxazolopyridine, which exhibits a K_i of 140 pM (Boger et al. 2000). However, its pharmacological profile and specificity for FAAH remain to be determined.

Another series of irreversible inhibitors—alkylcarbamic acid aryl esters, with apparent specificity for FAAH—has also been reported (Kathuria et al. 2003; Tarzia et al. 2003). These inhibitors, which do not bind to CB_1 or CB_2 receptors or inhibit MGL or AEA cellular uptake, act by carbamoylation of the active site serine residue. The most potent of the series is URB597 (Kathuria et al. 2003). Of added significance is that these analogs, although difficult to emulsify, are also active as inhibitors of FAAH in vivo, resulting in an elevation of brain AEA content of approximately threefold at a dose of 0.3 mg/kg without an effect on the content of 2-AG (Kathuria et al. 2003).

Several endogenous fatty acid derivatives can inhibit FAAH-mediated catabolism of AEA by virtue of the fact that they function as alternative substrates. For example, *N*-arachidonoylglycine does not bind to CB_1 or CB_2 receptors but, as a substrate of FAAH, can decrease AEA catabolism (Huang et al. 2001; Burstein et al. 2002; Grazia Cascio et al. 2004). FAAH inhibition and the subsequent increases in concentrations of AEA and/or PEA mediate the analgesic effect of *N*-arachidonoylglycine. Oleamide has also been suggested to induce sleep, at least in part, by competing with AEA for FAAH (Mechoulam et al. 1997; Mendelson and Basile 1999), although recent studies have cast doubt over this mechanism (Fedorova et al. 2001; Lichtman et al. 2002; Leggett et al. 2004).

Understanding the effects of endogenous or pharmacological inhibition of FAAH is important for the elucidation of the biological activity of fatty acid-derived substances and the investigation of the therapeutic potentials of selective FAAH inhibitors (Gaetani et al. 2003). Modulation of FAAH activity could play a role in the mechanism of currently used drugs. For example, propofol (2,6-diisopropyl phenol), an intravenous anesthetic that is frequently used for both induction and maintenance of general anesthesia, inhibits FAAH, elevates brain AEA content, and is dependent upon activation of CB_1 receptors for its effect after i.p. administration (Patel et al. 2003). Similarly, the non-steroidal anti-inflammatory drugs indomethacin, ibuprofen, and suprofen have been suggested to inhibit FAAH as well as cyclooxygenase (Fowler et al. 1997a,b). While these compounds are not very potent inhibitors of FAAH (IC_{50} values of 10^{-5}–10^{-4} M), high doses of ibuprofen (400 mg) used by patients with rheumatoid arthritis result in peak plasma concentrations in the range 110 to 150 μM (Karttunen et al. 1990). In addition, FAAH is most sensitive to inhibition by these compounds at acidic pH (Fowler et al. 1997a; Holt et al. 2001; Fowler et al. 2003), which might occur in certain inflammatory conditions including rheumatoid arthritis.

3 Monoacylglycerol Lipase

3.1 Biochemical and Molecular Characteristics of MGL

MGL was first purified and characterized from rat adipose tissue (Tornqvist and Belfrage 1976). This enzyme has a molecular weight of approximately 33 kDa and a pI of 7.2. The purified protein was shown to hydrolyze 1(3)- and 2-monoacylglycerols at equal rates but to have no hydrolytic activity against triacylglycerols or diacylglycerols. Enzymatic activity is inhibited by micromolar concentrations of diisopropylfluorophosphate (DFP), indicating that the enzyme active site contains a reactive serine, and by mercurials, indicating the presence of essential sulfhydryl groups. Cloning of MGL from mouse adipose tissue confirmed and extended these biochemical studies (Karlsson et al. 1997). MGL is a serine hydrolase with a $GXS_{122}XG$ consensus sequence; the other members of the catalytic triad are Asp-239 and His-269. MGL has a ubiquitous tissue distribution, including brain, heart, and spleen. However, Western blot analyses of different mouse tissues reveal protein size differences (Karlsson et al. 2001). In particular, mouse brain MGL exhibits two immunoreactive bands, one migrating at the same molecular weight as the adipose enzyme and another with a slightly larger size. The same doublet has been observed in rat brain tissue (Dinh et al. 2002). The differences in migration on Western blot of MGL-like immunoreactive proteins could be evidence of MGL splice variants, isoforms, or post-translational processing. Interestingly, neuronal nuclei from rabbit cerebral cortex express a 1-monoacylglycerol lipase that has not been well characterized (Baker and Chang 2000). Human (Karlsson et al. 2001; Ho et al. 2002) and rat brain (Dinh et al. 2002) MGL have also been cloned; these two sequences are highly homologous to the mouse adipose clone.

3.2 Brain MGL

Within the brain, the distribution of MGL mRNA is ubiquitous but the expression levels are variable (Dinh et al. 2002). Regions with high transcript expression include cerebellum, cortex, and hippocampus, while low transcript levels are found in the brain stem and hypothalamus. Protein distribution within the hippocampus is consistent with the presence of MGL in axon terminals; no protein is detected in hippocampal principal cells. This distribution agrees with earlier work in which MGL activity was found to be enriched in synaptosomes (Vyvoda and Rowe 1973) and synaptoneurosomes (Farooqui and Horrocks 1997) and contrasts with the distribution of FAAH, which is found predominately within hippocampal pyramidal cell bodies and is absent from presynaptic profiles (Tsou et al. 1998).

3.3 Subcellular Distribution of MGL

The subcellular distribution of MGL has been studied in several tissues and cell types using the distribution of enzymatic activity as an assay. In many tissues and cells, including brain and adipocytes, MGL activity is nearly equivalent in cytosolic and particulate fractions (Sakurada and Noma 1981; Bisogno et al. 1997b; Di Marzo et al. 1999; Goparaju et al. 1999). However, in pancreatic islet cells (Konrad et al. 1994) and erythrocytes (Somma-Delpero et al. 1995), the majority of MGL activity is associated with plasma membrane-enriched fractions and very little activity is seen in the cytosolic fractions. These data suggest that MGL can be associated with membranes but that it is not an intrinsic membrane protein. This conclusion agrees with the lack of obvious transmembrane domains in the MGL amino acid sequence.

A few studies have examined the question of whether different subcellular pools of MGL are kinetically similar. In rat adipocytes, the particulate and cytosolic enzymes are essentially identical with respect to pH dependence and substrate and inhibitor profiles (Sakurada and Noma 1981). Similar inhibitor profiles are also seen in cytosolic and membrane fractions from porcine brain (Goparaju et al. 1999). However, cytosolic MGL activity is reduced by 50% in adipocytes isolated from 24-h fasted rats without any change in membrane MGL activity (Sakurada and Noma 1981). Similarly, treatment of rat macrophages and platelets with lipopolysaccharide results in inhibition of MGL activity in particulate fractions but has no effect on cytosolic enzymatic activity (Di Marzo et al. 1999). The available data suggest that MGL of different subcellular compartments is likely the same enzyme isoform but that the location of the enzyme results in differential regulation by cellular processes.

A study comparing the subcellular distribution of MGL activity between resting and activated human neutrophils suggests that MGL can translocate from one subcellular compartment to another in response to cellular changes (Balsinde et al. 1991). In this study, MGL activity in resting neutrophils was localized primarily in gelatinase-containing granules, but upon activation by A23187, a dramatic shift in activity to the plasma membrane occurred. Interestingly, the enzyme associated with the plasma membrane exhibited an increase in V_{max} for the hydrolysis of 2-AG, suggesting there is a greater pool of substrate available at the plasma membrane.

3.4 Substrate Specificity of MGL

MGL can hydrolyze both 1(3)- and 2-monoacylglycerols and has very little ability to hydrolyze triacylglycerols or diacylglycerols (Tornqvist and Belfrage 1976; Di Marzo et al. 1999; Goparaju et al. 1999). MGL hydrolyzes fatty acyl esters but not fatty acyl amides or ethers. In particular, neither AEA (Bisogno et al. 1997b; Goparaju et al. 1999; Dinh et al. 2002; Saario et al. 2004) nor noladin ether (Saario et al. 2004) are hydrolyzed by MGL. Interestingly, there is one report that AEA (at a concentration of 100 μM) inhibits MGL activity by 77% in macrophage membranes (Di Marzo et al. 1999). The enzyme prefers but does not require a glycerol head group as MGL purified from erythrocytes hydrolyzes oleoylethanol at a rate about 50% of the oleic ester of glycerol (Somma-Delpero et al. 1995) and the ester virodhamine is hydrolyzed to arachidonic acid at a rate about twofold lower than that of 2-AG (Saario et al. 2004).

The glycerol esters of arachidonic acid, oleic acid, and palmitic acid are all hydrolyzed by MGL (Vyvoda and Rowe 1973; Dinh et al. 2002; Saario et al. 2004). Only a few studies have compared the rates of hydrolysis of various monoacylesters. In macrophages the 1(3)-monoglycerols of arachidonic acid, γ-linolenoyl and linolenoyl acid were hydrolyzed at a higher specific activity than the palmitic acid analog (Di Marzo et al. 1999). In another study, the 1(3)-monoglycerol of arachidonic acid was hydrolyzed at a higher rate than the corresponding oleic acid ester (Goparaju et al. 1999). However, none of these differences is large, and when K_m values are compared across studies the values are very similar, in spite of the fact that different substrates and tissue sources were used (Table 2). These data suggest that any possible isoforms of MGL differ little in their affinity for monoacylglycerols and that there is little selectivity for the acyl substituents of the substrates, at least among long chain fatty acid esters.

Table 2. K_m values for monoacylglycerol hydrolysis by MGL-like enzymatic activities

Tissue	Substrate	K_m (μM)	Reference
Rat adipocyte membrane	1(3)- and 2-oleoylglycerol	210	Sakurada and Noma 1981
Rat adipocyte cytosol	1(3)- and 2-oleoylglycerol	370	Sakurada and Noma 1981
Human erythrocyte membranes	1(3)-Oleoylglycerol	270	Somma-Delpero et al. 1995
Human erythrocyte membranes	2-Oleoylglycerol	490	Somma-Delpero et al. 1995
Rat pancreatic islet homogenates	2-Arachidonoylglycerol	0.14	Konrad et al. 1994
J774 macrophage membranes	2-Arachidonoylglycerol	110	Di Marzo et al. 1999
J774 macrophage membranes	2-Palmitoylglycerol	170	Di Marzo et al. 1999
Purified rat adipose enzyme in detergent	1(3)- or 2-oleoylglycerol	200	Tornqvist and Belfrage 1976
Human neutrophil supernatants	2-Arachidonoylglycerol	34	Balsinde et al. 1991

3.5 Regulation of MGL Activity

MGL activity in vitro is not affected by the addition of calcium to the incubation buffer (Sakurada and Noma 1981; Balsinde et al. 1991; Konrad et al. 1994). However, MGL activity could be inhibited by calcium, as two studies have reported an increase in activity following the addition of ethylene glycol-bis(β-aminoethyl ether)N,N,N′,N′,-tetraacetic acid (EGTA) to the assay buffer (Sakurada and Noma 1981; Witting et al. 2004). Very high concentrations of sodium (i.e., 1 M) are inhibitory (Sakurada and Noma 1981) as is zinc (Tornqvist and Belfrage 1976).

Although calcium does not appear to have a direct effect on MGL activity, activation of *N*-methyl-D-aspartate (NMDA) receptors in spinal cord neurons in

Table 3. Inhibitors of MGL activity

Inhibitor	Tissue	IC_{50} or % inhibition (concentration)	Reference
DFP	Purified from adipocyte	100% (10 μM)	Tornqvist and Belfrage 1976
	Rat adipocyte cytosol	81% (2 mM)	Sakurada and Noma 1981
	Rat adipocyte membrane	89% (2 mM)	Sakurada and Noma 1981
	Purified enzyme in CHAPS	88% (5 mM)	Somma-Delpero et al. 1995
PMSF	Purified enzyme in CHAPS	43% (0.5 mM)	Somma-Delpero et al. 1995
	N18 cell cytosol	30% (0.5 mM)	Bisogno et al. 1997b
	N18 cell microsomes	47% (0.5 mM)	Bisogno et al. 1997b
	Platelet membranes	1% (0.1 mM)	Di Marzo et al. 1999
	Macrophage membranes	21% (0.1 mM)	Di Marzo et al. 1999
	Rat cerebellar membranes	155 μM	Saario et al. 2004
MAFP	Porcine brain cytosol	2 nM	Goparaju et al. 1999
	Platelet membranes	0 (50 nM)	Di Marzo et al. 1999
	Macrophage membranes	21% (50 nM)	Di Marzo et al. 1999
	Rat brain cytosol	800 nM	Dinh et al. 2002
	Rat cerebellar membranes	2 nM	Saario et al. 2004
ATFMK	N18 cell cytosol	11% (0.5 mM)	Bisogno et al. 1997b
	N18 cell microsomes	23% (0.5 mM)	Bisogno et al. 1997b
	Porcine brain cytosol	30 μM	Goparaju et al. 1999
	Platelet membranes	7.5% (10 μM)	Di Marzo et al. 1999
	Macrophage membranes	89% (10 μM)	Di Marzo et al. 1999
	Rat brain cytosol	2.5 μM	Dinh et al. 2002
	Rat cerebellar membranes	66 μM	Saario et al. 2004
HDSF	Rat brain cytosol	6.2 μM	Dinh et al. 2002
	Rat cerebellar membranes	241 nM	Saario et al. 2004
URB597	Rat brain cytosol	0 (30 μM)	Kathuria et al. 2003
	Rat cerebellar membranes	30% (1 mM)	Saario et al. 2004

ATFMK, arachidonyltrifluoromethylketone; DFP, diisopropylfluorophosphate; HDSF, hexadecylsulfonyl fluoride; MAFP, methylarachidonylfluorophosphonate; PMSF, phenylmethylsulfonyl fluoride.

culture (Farooqui et al. 1993) and in synaptoneurosomes prepared from young rat brain (Farooqui and Horrocks 1997) results in a very significant activation of MGL activity. While the mechanism of this activation is not known, the time course of activation following NMDA or glutamate treatment is short (onset at 6 min in cells). Since MGL is not activated by calcium directly, it is possible that the regulation involves phosphorylation or another post-translational modification. Interestingly, a recent study by Di Marzo and colleagues revealed that MGL activity in the striatum but not the hippocampus was reduced in rat brain harvested during the dark phase (i.e., active phase) compared to the light phase of the day (Valenti et al. 2004).

3.6 MGL Inhibitors

Since MGL is a serine hydrolase, its sensitivity to many of the available serine hydrolase inhibitors has been explored (Table 3). The results support the hypothesis that MGL can be inhibited by compounds that interact with its reactive serine. On the other hand, the potencies of the inhibitors are quite variable; in some cases, this likely reflects differences in assay methodology (i.e., substrate concentration, pH, form of the enzyme). However, in a few cases, the same assay conditions revealed very different inhibitory potencies (e.g., compare the platelet and macrophage membrane studies by Di Marzo et al. 1999). In any event, studies of these compounds are not likely to yield selective inhibitors of MGL. All of these compounds are inhibitors of FAAH (see above) and many are also inhibitors of PLA_2, diacylglycerol lipase, and acetylcholine esterase, among other hydrolases. By analogy to the development of the URB series of FAAH inhibitors (Kathuria et al. 2003), it is likely that selective inhibitors of MGL will come from other synthetic avenues.

4 Endocannabinoid Transmembrane Movement

4.1 Introduction

While the molecular identities of the proteins involved are not yet understood, it is clear that neurons and other cell types accumulate AEA intracellularly (Hillard and Jarrahian 2003). There are several characteristics of endocannabinoid transmembrane movement that are well supported by data obtained in multiple laboratories. To summarize, the accumulation of AEA by cells does not require sodium or ATP and is moderately temperature dependent. The accumulation exhibits saturation in the micromolar range and is inhibitable by a variety of structural analogs of AEA, suggesting that AEA accumulation involves its interaction with a saturable cellular component. Some data are consistent with the component being a plasma membrane transporter (see for example Hillard and Jarrahian 2000; Ronesi et al. 2004) while other data indicate that, in some cells, the accumulation is driven by

FAAH-mediated catabolism (Deutsch et al. 2001; Glaser et al. 2003). Regardless of the mechanisms involved, inhibitors of the accumulation process have been developed that will help to shed light on the fundamental processes involved in the accumulation as well as the importance of this process in the biological activity of the eCBs.

4.2 AEA Uptake Inhibitors

Many analogs of AEA have been tested as inhibitors of the AEA uptake process (Table 4). The reader is referred to comprehensive papers that include most of the structure–activity profiles of the first generation of inhibitors (Piomelli et al. 1999; Jarrahian et al. 2000; Di Marzo et al. 2002). Of these analogs, the best studied has been AM404 (*N*-(4-hydroxybenzyl)arachidonoylamine), which inhibits AEA uptake into neurons and other cell types with IC_{50} values in the low micromolar range and potentiates the effects of exogenously administered AEA in vivo (Beltramo et al. 1997b). However, AM404 is not an ideal inhibitor because it also inhibits FAAH (Jarrahian et al. 2000) and activates TPRV1 vanilloid receptors (De Petrocellis et al. 2000) at similar concentrations. VDM 11 (*N*-(4-hydroxy-2-methylphenyl)arachidonoylamine) has also been used as an uptake inhibitor in vitro (De Petrocellis et al. 2000) and in vivo (Gubellini et al. 2002). While VDM 11 has the advantage over AM404 of much lower affinity for the TRPV1 receptors, it inhibits FAAH-mediated hydrolysis of AEA at the same concentration range (Fowler et al. 2004).

Another series of analogs of arachidonic acid with furyl substitutions in the head group have been tested (Lopez-Rodriguez et al. 2001). Of this series, UCM707 (*N*-(3-furylmethyl)arachidonoylamine) has the highest affinity for the transporter and exhibits low binding affinities for the CB_1 and TPRV1 receptors. Interestingly, UCM707 has relatively high affinity for the CB_2 receptor (67 nM). UCM707 is hydrolyzed by FAAH with an IC_{50} of 30 μM (Lopez-Rodriguez et al. 2003; Fowler et al. 2004), which makes it metabolically unstable and, although it does not inhibit FAAH as potently as AM404, this feature of the molecule could be responsible for its ability to inhibit uptake in some cell types (Fowler et al. 2004).

Another series of inhibitors, OMDM-1, -2, -3, and-4, like AM404, are fatty acid amides with aromatic head groups but the acyl chain has been changed to oleoyl (18:1) (Ortar et al. 2003). Two members of the series, OMDM-1 and OMDM-2 (*R*- and *S*-1′-4-hydroxybenzyl derivatives of *N*-oleoylethanolamine, respectively) exhibit affinity for inhibition of AEA uptake similar to AM404 in RBL-2H3 cells and are resistant to FAAH. However, both compounds have a small but measurable effect on TPRV1 receptor activity in the same concentration range. The inhibitory potencies and efficacies of these two compounds as uptake inhibitors appear to be cell-specific, having greater potency and, for OMDM-2, greater efficacy in RBL2H3 cells than C6 glioma cells (Fowler et al. 2004). Their affinities for inhibition of uptake in primary neurons have not been determined. In vivo studies of these compounds have been carried out, and they exhibit anti-spasticity efficacy in a mouse model of

Table 4. Inhibitors of AEA cellular uptake

Inhibitor	Cell	IC_{50} (μM)	Reference(s)	Remarks
AM404	Primary neurons	4–5	Beltramo et al. 1997b	Inhibits FAAH, VR1 agonist
VDM 11	RBL-2H3 and C6 glioma	10–11	De Petrocellis et al. 2000	Inhibits FAAH, little effect on VR1
	PC-3 cells	11	Ruiz-Llorente et al. 2004	
UCM-707	U937 cells	0.8	Lopez-Rodriguez et al. 2001	Moderate inhibition of FAAH, little effect on VR1, binds CB_2 receptors
	RBL-2H3	25	Fowler et al. 2004	
	C6 glioma	41 (max. 50% inhibition)	Fowler et al. 2004	
	PC-3 cells	25% inhibition at 100 μM	Ruiz-Llorente et al. 2004	
UMC-119	PC-3 cells	11	Ruiz-Llorente et al. 2004	
OMDM-1	RBL-2H3	2.4	Ortar et al. 2003; Fowler et al. 2004	K_i: CB_1—12 μM; FAAH>50 μM; VR1>10 μM
	C6 glioma	>20	Fowler et al. 2004	
OMDM-2	RBL-2H3	3	Ortar et al. 2003	*K*i: CB_1—5 μM; FAAH>50 μM; VR1—10 μM
	RBL-2H3	3	Fowler et al. 2004	
	C6 glioma	17 (max. 50% inhibition)	Fowler et al. 2004	
AM1172	Cortical neurons	2	Fegley et al. 2004	No inhibition of FAAH at 10 μM; moderate affinity for CB_1 and CB_2; no effect at VR1

Please refer to text for the chemical names of the inhibitors.

multiple sclerosis alone and can produce potentiation of exogenously administered AEA (de Lago et al. 2004). However, the role of the CB_1 receptor in the effects was not determined.

The "reverse" of AM404, i.e., *N*-arachidonoyl-4-hydroxybenzamide (also called AM1172) has been synthesized and studied (Fegley et al. 2004). This compound is not a substrate for FAAH and does not inhibit the hydrolysis of AEA at concentrations up to 10 μM but is equivalent to AM404 in its ability to inhibit AEA uptake into primary cortical neurons. Interestingly, however, this analog has a moderate affinity for both CB_1 and CB_2 receptors and behaves as a partial agonist in biochemical assays of receptor activation.

5
Summary

It is easy to argue from the current eCB literature that pharmacological manipulations of eCB inactivation will be important for human health. It is also important that selective inhibitors of each of the inactivation processes be designed so that mechanistically interpretable studies of these processes can be accomplished. Although significant progress has been made in the development of these agents, it is clear that more selective inhibitors are needed.

References

Arreaza G, Deutsch DG (1999) Deletion of a proline-rich region and a transmembrane domain in fatty acid amide hydrolase. FEBS Lett 454:57–60
Baker RR, Chang H (2000) A metabolic path for the degradation of lysophosphatidic acid, an inhibitor of lysophosphatidylcholine lysophospholipase, in neuronal nuclei of cerebral cortex. Biochim Biophys Acta 1483:58–68
Balsinde J, Diez E, Mollinedo F (1991) Arachidonic acid release from diacylglycerol in human neutrophils. Translocation of diacylglycerol-deacylating enzyme activities from an intracellular pool to plasma membrane upon cell activation. J Biol Chem 266:15638–15643
Beltramo M, Piomelli D (2000) Carrier-mediated transport and enzymatic hydrolysis of the endogenous cannabinoid 2-arachidonylglycerol. Neuroreport 11:1231–1235
Beltramo M, di Tomaso E, Piomelli D (1997a) Inhibition of anandamide hydrolysis in rat brain tissue by (E)-6-(bromomethylene) tetrahydro-3-(1-naphthalenyl)-2H-pyran-2-one. FEBS Lett 403:263–267
Beltramo M, Stella N, Calignano A, Lin SY, Makriyannis A, Piomelli D (1997b) Functional role of high-affinity anandamide transport, as revealed by selective inhibition. Science 277:1094–1097
Bisogno T, Maurelli S, Melck D, De Petrocellis L, Di Marzo V (1997a) Biosynthesis, uptake, and degradation of anandamide and palmitoylethanolamide in leukocytes. J Biol Chem 272:3315–3323
Bisogno T, Sepe N, Melck D, Maurelli S, De Petrocellis L, Di Marzo V (1997b) Biosynthesis, release and degradation of the novel endogenous cannabimimetic metabolite 2-arachidonoylglycerol in mouse neuroblastoma cells. Biochem J 322:671–677
Bisogno T, Melck D, De Petrocellis L, Bobrov M, Gretskaya NM, Bezuglov VV, Sitachitta N, Gerwick WH, Di Marzo V (1998) Arachidonoylserotonin and other novel inhibitors of fatty acid amide hydrolase. Biochem Biophys Res Commun 248:515–522

Bisogno T, Melck D, Bobrov M, Gretskaya NM, Bezuglov VV, De Petrocellis L, Di Marzo V (2000) N-acyl-dopamines: novel synthetic CB(1) cannabinoid-receptor ligands and inhibitors of anandamide inactivation with cannabimimetic activity in vitro and in vivo. Biochem J 351:817–824

Bisogno T, Maccarrone M, De Petrocellis L, Jarrahian A, Finazzi-Agro A, Hillard C, Di Marzo V (2001) The uptake by cells of 2-arachidonoylglycerol, an endogenous agonist of cannabinoid receptors. Eur J Biochem 268:1982–1989

Boger DL, Sato H, Lerner AE, Hedrick MP, Fecik RA, Miyauchi H, Wilkie GD, Austin BJ, Patricelli MP, Cravatt BF (2000) Exceptionally potent inhibitors of fatty acid amide hydrolase: the enzyme responsible for degradation of endogenous oleamide and anandamide. Proc Natl Acad Sci U S A 97:5044–5049

Bracey MH, Hanson MA, Masuda KR, Stevens RC, Cravatt BF (2002) Structural adaptations in a membrane enzyme that terminates endocannabinoid signaling. Science 298:1793–1796

Burstein SH, Huang SM, Petros TJ, Rossetti RG, Walker JM, Zurier RB (2002) Regulation of anandamide tissue levels by N-arachidonylglycine. Biochem Pharmacol 64:1147–1150

Cravatt BF, Lichtman AH (2002) The enzymatic inactivation of the fatty acid amide class of signaling lipids. Chem Phys Lipids 121:135–148

Cravatt BF, Giang DK, Mayfield SP, Boger DL, Lerner RA, Gilula NB (1996) Molecular characterization of an enzyme that degrades neuromodulatory fatty-acid amides. Nature 384:83–87

Cravatt BF, Demarest K, Patricelli MP, Bracey MH, Giang DK, Martin BR, Lichtman AH (2001) Supersensitivity to anandamide and enhanced endogenous cannabinoid signaling in mice lacking fatty acid amide hydrolase. Proc Natl Acad Sci U S A 98:9371–9376

de Lago E, Ligresti A, Ortar G, Morera E, Cabranes A, Pryce G, Bifulco M, Baker D, Fernandez-Ruiz J, Di Marzo V (2004) In vivo pharmacological actions of two novel inhibitors of anandamide cellular uptake. Eur J Pharmacol 484:249–257

De Petrocellis L, Melck D, Ueda N, Maurelli S, Kurahashi Y, Yamamoto S, Marino G, Di Marzo V (1997) Novel inhibitors of brain, neuronal, and basophilic anandamide amidohydrolase. Biochem Biophys Res Commun 231:82–88

De Petrocellis L, Bisogno T, Davis JB, Pertwee RG, Di Marzo V (2000) Overlap between the ligand recognition properties of the anandamide transporter and the VR1 vanilloid receptor: inhibitors of anandamide uptake with negligible capsaicin-like activity. FEBS Lett 483:52–56

Desarnaud F, Cadas H, Piomelli D (1995) Anandamide amidohydrolase activity in rat brain microsomes. Identification and partial characterization. J Biol Chem 270:6030–6035

Deutsch DG, Chin SA (1993) Enzymatic synthesis and degradation of anandamide, a cannabinoid receptor agonist. Biochem Pharmacol 46:791–796

Deutsch DG, Omeir R, Arreaza G, Salehani D, Prestwich GD, Huang Z, Howlett A (1997a) Methyl arachidonyl fluorophosphonate: a potent irreversible inhibitor of anandamide amidase. Biochem Pharmacol 53:255–260

Deutsch DG, Lin S, Hill WA, Morse KL, Salehani D, Arreaza G, Omeir RL, Makriyannis A (1997b) Fatty acid sulfonyl fluorides inhibit anandamide metabolism and bind to the cannabinoid receptor. Biochem Biophys Res Commun 231:217–221

Deutsch DG, Glaser ST, Howell JM, Kunz JS, Puffenbarger RA, Hillard CJ, Abumrad N (2001) The cellular uptake of anandamide is coupled to its breakdown by fatty-acid amide hydrolase. J Biol Chem 276:6967–6973

Di Marzo V, Bisogno T, De Petrocellis L, Melck D, Orlando P, Wagner JA, Kunos G (1999) Biosynthesis and inactivation of the endocannabinoid 2-arachidonoylglycerol in circulating and tumoral macrophages. Eur J Biochem 264:258–267

Di Marzo V, Griffin G, De Petrocellis L, Brandi I, Bisogno T, Williams W, Grier MC, Kulasegram S, Mahadevan A, Razdan RK, Martin BR (2002) A structure/activity relationship study on arvanil, an endocannabinoid and vanilloid hybrid. J Pharmacol Exp Ther 300:984–991

Dinh TP, Carpenter D, Leslie FM, Freund TF, Katona I, Sensi SL, Kathuria S, Piomelli D (2002) Brain monoglyceride lipase participating in endocannabinoid inactivation. Proc Natl Acad Sci U S A 99:10819–10824

Edgemond WS, Greenberg MJ, McGinley PJ, Muthian S, Campbell WB, Hillard CJ (1998) Synthesis and characterization of diazomethylarachidonyl ketone: an irreversible inhibitor of N-arachidonylethanolamine amidohydrolase. J Pharmacol Exp Ther 286:184–190

Farooqui AA, Horrocks LA (1997) Nitric oxide synthase inhibitors do not attenuate diacylglycerol or monoacylglycerol lipase activities in synaptoneurosomes. Neurochem Res 22:1265–1269

Farooqui AA, Anderson DK, Horrocks LA (1993) Effect of glutamate and its analogs on diacylglycerol and monoacylglycerol lipase activities of neuron-enriched cultures. Brain Res 604:180–184

Fedorova I, Hashimoto A, Fecik RA, Hedrick MP, Hanus LO, Boger DL, Rice KC, Basile AS (2001) Behavioral evidence for the interaction of oleamide with multiple neurotransmitter systems. J Pharmacol Exp Ther 299:332–342

Fegley D, Kathuria S, Mercier R, Li C, Goutopoulos A, Makriyannis A, Piomelli D (2004) Anandamide transport is independent of fatty-acid amide hydrolase activity and is blocked by the hydrolysis-resistant inhibitor AM1172. Proc Natl Acad Sci U S A 101:8756–8761

Fernando SR, Pertwee RG (1997) Evidence that methyl arachidonyl fluorophosphonate is an irreversible cannabinoid receptor antagonist. Br J Pharmacol 121:1716–1720

Fowler CJ, Tiger G, Stenstrom A (1997a) Ibuprofen inhibits rat brain deamidation of anandamide at pharmacologically relevant concentrations. Mode of inhibition and structure-activity relationship. J Pharmacol Exp Ther 283:729–734

Fowler CJ, Stenstrom A, Tiger G (1997b) Ibuprofen inhibits the metabolism of the endogenous cannabimimetic agent anandamide. Pharmacol Toxicol 80:103–107

Fowler CJ, Holt S, Tiger G (2003) Acidic nonsteroidal anti-inflammatory drugs inhibit rat brain fatty acid amide hydrolase in a pH-dependent manner. J Enzyme Inhib Med Chem 18:55–58

Fowler CJ, Tiger G, Ligresti A, Lopez-Rodriguez ML, Di Marzo V (2004) Selective inhibition of anandamide cellular uptake versus enzymatic hydrolysis-a difficult issue to handle. Eur J Pharmacol 492:1–11

Gaetani S, Cuomo V, Piomelli D (2003) Anandamide hydrolysis: a new target for anti-anxiety drugs? Trends Mol Med 9:474–478

Giang DK, Cravatt BF (1997) Molecular characterization of human and mouse fatty acid amide hydrolases. Proc Natl Acad Sci U S A 94:2238–2242

Glaser ST, Abumrad NA, Fatade F, Kaczocha M, Studholme KM, Deutsch DG (2003) Evidence against the presence of an anandamide transporter. Proc Natl Acad Sci U S A 100:4269–4274

Goparaju SK, Ueda N, Yamaguchi H, Yamamoto S (1998) Anandamide amidohydrolase reacting with 2-arachidonoylglycerol, another cannabinoid receptor ligand. FEBS Lett 422:69–73

Goparaju SK, Ueda N, Taniguchi K, Yamamoto S (1999) Enzymes of porcine brain hydrolyzing 2-arachidonoylglycerol, an endogenous ligand of cannabinoid receptors. Biochem Pharmacol 57:417–423

Grazia Cascio M, Minassi A, Ligresti A, Appendino G, Burstein S, Di Marzo V (2004) A structure-activity relationship study on N-arachidonoyl-amino acids as possible endogenous inhibitors of fatty acid amide hydrolase. Biochem Biophys Res Commun 314:192–196

Gubellini P, Picconi B, Bari M, Battista N, Calabresi P, Centonze D, Bernardi G, Finazzi-Agro A, Maccarrone M (2002) Experimental parkinsonism alters endocannabinoid degradation: implications for striatal glutamatergic transmission. J Neurosci 22:6900–6907

Herkenham M, Lynn AB, Little MD, Johnson MR, Melvin LS, de Costa BR, Rice KC (1990) Cannabinoid receptor localization in brain. Proc Natl Acad Sci U S A 87:1932–1936

Hillard CJ, Jarrahian A (2000) The movement of N-arachidonoylethanolamine (anandamide) across cellular membranes. Chem Phys Lipids 108:123–134

Hillard CJ, Jarrahian A (2003) Cellular accumulation of anandamide: consensus and controversy. Br J Pharmacol 140:802–808

Hillard CJ, Wilkison DM, Edgemond WS, Campbell WB (1995) Characterization of the kinetics and distribution of N-arachidonylethanolamine (anandamide) hydrolysis by rat brain. Biochim Biophys Acta 1257:249–256

Hillard CJ, Edgemond WS, Jarrahian A, Campbell WB (1997) Accumulation of N-arachidonoylethanolamine (anandamide) into cerebellar granule cells occurs via facilitated diffusion. J Neurochem 69:631–638

Ho SY, Delgado L, Storch J (2002) Monoacylglycerol metabolism in human intestinal Caco-2 cells: evidence for metabolic compartmentation and hydrolysis. J Biol Chem 277:1816–1823

Holt S, Nilsson J, Omeir R, Tiger G, Fowler CJ (2001) Effects of pH on the inhibition of fatty acid amidohydrolase by ibuprofen. Br J Pharmacol 133:513–520

Huang SM, Bisogno T, Petros TJ, Chang SY, Zavitsanos PA, Zipkin RE, Sivakumar R, Coop A, Maeda DY, De Petrocellis L, Burstein S, Di Marzo V, Walker JM (2001) Identification of a new class of molecules, the arachidonyl amino acids, and characterization of one member that inhibits pain. J Biol Chem 276:42639–42644

Huang SM, Bisogno T, Trevisani M, Al-Hayani A, De Petrocellis L, Fezza F, Tognetto M, Petros TJ, Krey JF, Chu CJ, Miller JD, Davies SN, Geppetti P, Walker JM, Di Marzo V (2002) An endogenous capsaicin-like substance with high potency at recombinant and native vanilloid VR1 receptors. Proc Natl Acad Sci USA 99:8400–8405

Jarrahian A, Manna S, Edgemond WS, Campbell WB, Hillard CJ (2000) Structure-activity relationships among N-arachidonylethanolamine (Anandamide) head group analogues for the anandamide transporter. J Neurochem 74:2597–2606

Karlsson M, Contreras JA, Hellman U, Tornqvist H, Holm C (1997) cDNA cloning, tissue distribution, and identification of the catalytic triad of monoglyceride lipase. Evolutionary relationship to esterases, lysophospholipases, and haloperoxidases. J Biol Chem 272:27218–27223

Karlsson M, Reue K, Xia YR, Lusis AJ, Langin D, Tornqvist H, Holm C (2001) Exon-intron organization and chromosomal localization of the mouse monoglyceride lipase gene. Gene 272:11–18

Karttunen P, Saano V, Paronen P, Peura P, Vidgren M (1990) Pharmacokinetics of ibuprofen in man: a single-dose comparison of two over-the-counter, 200 mg preparations. Int J Clin Pharmacol Ther Toxicol 28:251–255

Kathuria S, Gaetani S, Fegley D, Valino F, Duranti A, Tontini A, Mor M, Tarzia G, La Rana G, Calignano A, Giustino A, Tattoli M, Palmery M, Cuomo V, Piomelli D (2003) Modulation of anxiety through blockade of anandamide hydrolysis. Nat Med 9:76–81

Konrad RJ, Major CD, Wolf BA (1994) Diacylglycerol hydrolysis to arachidonic acid is necessary for insulin secretion from isolated pancreatic islets: sequential actions of diacylglycerol and monoacylglycerol lipases. Biochemistry 33:13284–13294

Koutek B, Prestwich GD, Howlett AC, Chin SA, Salehani D, Akhavan N, Deutsch DG (1994) Inhibitors of arachidonoyl ethanolamide hydrolysis. J Biol Chem 269:22937–22940

Kuriyan J, Cowburn D (1997) Modular peptide recognition domains in eukaryotic signaling. Annu Rev Biophys Biomol Struct 26:259–288

Lang W, Qin C, Hill WA, Lin S, Khanolkar AD, Makriyannis A (1996) High-performance liquid chromatographic determination of anandamide amidase activity in rat brain microsomes. Anal Biochem 238:40–45

Leggett JD, Aspley S, Beckett SR, D'Antona AM, Kendall DA (2004) Oleamide is a selective endogenous agonist of rat and human CB1 cannabinoid receptors. Br J Pharmacol 141:253–262

Lichtman AH, Hawkins EG, Griffin G, Cravatt BF (2002) Pharmacological activity of fatty acid amides is regulated, but not mediated, by fatty acid amide hydrolase in vivo. J Pharmacol Exp Ther 302:73–79

Lio YC, Reynolds LJ, Balsinde J, Dennis EA (1996) Irreversible inhibition of Ca(2+)-independent phospholipase A2 by methyl arachidonyl fluorophosphonate. Biochim Biophys Acta 1302:55–60

Lopez-Rodriguez ML, Viso A, Ortega-Gutierrez S, Lastres-Becker I, Gonzalez S, Fernandez-Ruiz J, Ramos JA (2001) Design, synthesis and biological evaluation of novel arachidonic acid derivatives as highly potent and selective endocannabinoid transporter inhibitors. J Med Chem 44:4505–4508

Lopez-Rodriguez ML, Viso A, Ortega-Gutierrez S, Fowler CJ, Tiger G, de Lago E, Fernandez-Ruiz J, Ramos JA (2003) Design, synthesis and biological evaluation of new endocannabinoid transporter inhibitors. Eur J Med Chem 38:403–412

Maccarrone M, Salvati S, Bari M, Finazzi A (2000a) Anandamide and 2-arachidonoylglycerol inhibit fatty acid amide hydrolase by activating the lipoxygenase pathway of the arachidonate cascade. Biochem Biophys Res Commun 278:576–583

Maccarrone M, De Felici M, Bari M, Klinger F, Siracusa G, Finazzi-Agro A (2000b) Down-regulation of anandamide hydrolase in mouse uterus by sex hormones. Eur J Biochem 267:2991–2997

Maccarrone M, Fiorucci L, Erba F, Bari M, Finazzi-Agro A, Ascoli F (2000c) Human mast cells take up and hydrolyze anandamide under the control of 5-lipoxygenase and do not express cannabinoid receptors. FEBS Lett 468:176–180

Maccarrone M, Valensise H, Bari M, Lazzarin N, Romanini C, Finazzi-Agro A (2001a) Progesterone up-regulates anandamide hydrolase in human lymphocytes: role of cytokines and implications for fertility. J Immunol 166:7183–7189

Maccarrone M, De Petrocellis L, Bari M, Fezza F, Salvati S, Di Marzo V, Finazzi-Agro A (2001b) Lipopolysaccharide downregulates fatty acid amide hydrolase expression and increases anandamide levels in human peripheral lymphocytes. Arch Biochem Biophys 393:321–328

Maccarrone M, Bari M, Battista N, Finazzi-Agro A (2002) Estrogen stimulates arachidonoylethanolamide release from human endothelial cells and platelet activation. Blood 100:4040–4048

Maccarrone M, Di Rienzo M, Finazzi-Agro A, Rossi A (2003a) Leptin activates the anandamide hydrolase promoter in human T lymphocytes through STAT3. J Biol Chem 278:13318–13324

Maccarrone M, Bari M, Di Rienzo M, Finazzi-Agro A, Rossi A (2003b) Progesterone activates fatty acid amide hydrolase (FAAH) promoter in human T lymphocytes through the transcription factor Ikaros. Evidence for a synergistic effect of leptin. J Biol Chem 278:32726–32732

Maccarrone M, DeFelici M, Klinger FG, Battista N, Fezza F, Dainese E, Siracusa G, Finazzi-Agro A (2004) Mouse blastocysts release a lipid which activates anandamide hydrolase in intact uterus. Mol Hum Reprod 10:215–221

Maurelli S, Bisogno T, De Petrocellis L, Di Luccia A, Marino G, Di Marzo V (1995) Two novel classes of neuroactive fatty acid amides are substrates for mouse neuroblastoma 'anandamide amidohydrolase'. FEBS Lett 377:82–86

McKinney MK, Cravatt BF (2003) Evidence for distinct roles in catalysis for residues of the serine-serine-lysine catalytic triad of fatty acid amide hydrolase. J Biol Chem 278:37393–37399

Mechoulam R, Fride E, Hanus L, Sheskin T, Bisogno T, Di Marzo V, Bayewitch M, Vogel Z (1997) Anandamide may mediate sleep induction. Nature 389:25–26

Mendelson WB, Basile AS (1999) The hypnotic actions of oleamide are blocked by a cannabinoid receptor antagonist. Neuroreport 10:3237–3239

Ortar G, Ligresti A, De Petrocellis L, Morera E, Di Marzo V (2003) Novel selective and metabolically stable inhibitors of anandamide cellular uptake. Biochem Pharmacol 65:1473–1481

Patel S, Wohlfeil ER, Rademacher DJ, Carrier EJ, Perry LJ, Kundu A, Falck JR, Nithipatikom K, Campbell WB, Hillard CJ (2003) The general anesthetic propofol increases brain

N-arachidonylethanolamine (anandamide) content and inhibits fatty acid amide hydrolase. Br J Pharmacol 139:1005–1013
Patricelli MP, Cravatt BF (1999) Fatty acid amide hydrolase competitively degrades bioactive amides and esters through a nonconventional catalytic mechanism. Biochemistry 38:14125–14130
Patricelli MP, Cravatt BF (2000) Clarifying the catalytic roles of conserved residues in the amidase signature family. J Biol Chem 275:19177–19184
Patricelli MP, Lovato MA, Cravatt BF (1999) Chemical and mutagenic investigations of fatty acid amide hydrolase: evidence for a family of serine hydrolases with distinct catalytic properties. Biochemistry 38:9804–9812
Piomelli D, Beltramo M, Glasnapp S, Lin SY, Goutopoulos A, Xie XQ, Makriyannis A (1999) Structural determinants for recognition and translocation by the anandamide transporter. Proc Natl Acad Sci U S A 96:5802–5807
Porter AC, Sauer JM, Knierman MD, Becker GW, Berna MJ, Bao J, Nomikos GG, Carter P, Bymaster FP, Leese AB, Felder CC (2002) Characterization of a novel endocannabinoid, virodhamine, with antagonist activity at the CB1 receptor. J Pharmacol Exp Ther 301:1020–1024
Pratt PF, Hillard CJ, Edgemond WS, Campbell WB (1998) N-arachidonylethanolamide relaxation of bovine coronary artery is not mediated by CB1 cannabinoid receptor. Am J Physiol 274:H375–381
Puffenbarger RA, Kapulina O, Howell JM, Deutsch DG (2001) Characterization of the 5'-sequence of the mouse fatty acid amide hydrolase. Neurosci Lett 314:21–24
Riendeau D, Guay J, Weech PK LF, Yergey J, Li C, Desmarais S, Perrier H, Liu S, Nicoll-Griffith D, et al (1994) Arachidonyl trifluoromethyl ketone, a potent inhibitor of 85-kDa phospholipase A2, blocks production of arachidonate and 12-hydroxyeicosatetraenoic acid by calcium ionophore-challenged platelets. J Biol Chem 269:15619–15624
Rodriguez de Fonseca F, Navarro M, Gomez R, Escuredo L, Nava F, Fu J, Murillo-Rodriguez E, Giuffrida A, LoVerme J, Gaetani S, Kathuria S, Gall C, Piomelli D (2001) An anorexic lipid mediator regulated by feeding. Nature 414:209–212
Ronesi J, Gerdeman GL, Lovinger DM (2004) Disruption of endocannabinoid release and striatal long-term depression by postsynaptic blockade of endocannabinoid membrane transport. J Neurosci 24:1673–1679
Ruiz-Llorente L, Ortega-Gutierrez S, Viso A, Sanchez MG, Sanchez AM, Fernandez C, Ramos JA, Hillard C, Lasuncion MA, Lopez-Rodriguez ML, Diaz-Laviada I (2004) Characterization of an anandamide degradation system in prostate epithelial PC-3 cells: synthesis of new transporter inhibitors as tools for this study. Br J Pharmacol 141:457–467
Saario SM, Savinainen JR, Laitinen JT, Jarvinen T, Niemi R (2004) Monoglyceride lipase-like enzymatic activity is responsible for hydrolysis of 2-arachidonoylglycerol in rat cerebellar membranes. Biochem Pharmacol 67:1381–1387
Sakurada T, Noma A (1981) Subcellular localization and some properties of monoacylglycerol lipase in rat adipocytes. J Biochem (Tokyo) 90:1413–1419
Shrestha R, Dixon RA, Chapman KD (2003) Molecular identification of a functional homologue of the mammalian fatty acid amide hydrolase in Arabidopsis thaliana. J Biol Chem 278:34990–34997
Somma-Delpero C, Valette A, Lepetit-Thevenin J, Nobili O, Boyer J, Verine A (1995) Purification and properties of a monoacylglycerol lipase in human erythrocytes. Biochem J 312:519–525
Street I, Lin H, Laliberte F, Ghomashchi F, Wang Z, Perrier H, Tremblay N, Huang Z, Weech P, Gelb M (1993) Slow- and tight-binding inhibitors of the 85-kDa human phospholipase A2. Biochemistry 32:5935–5940
Tarzia G, Duranti A, Tontini A, Piersanti G, Mor M, Rivara S, Plazzi PV, Park C, Kathuria S, Piomelli D (2003) Design, synthesis, and structure-activity relationships of alkylcarbamic acid aryl esters, a new class of fatty acid amide hydrolase inhibitors. J Med Chem 46:2352–2360

Thomas EA, Cravatt BF, Danielson PE, Gilula NB, Sutcliffe JG (1997) Fatty acid amide hydrolase, the degradative enzyme for anandamide and oleamide, has selective distribution in neurons within the rat central nervous system. J Neurosci Res 50:1047–1052

Tornqvist H, Belfrage P (1976) Purification and some properties of a monoacylglycerol-hydrolyzing enzyme of rat adipose tissue. J Biol Chem 251:813–819

Tsou K, Nogueron MI, Muthian S, Sanudo-Pena MC, Hillard CJ, Deutsch DG, Walker JM (1998) Fatty acid amide hydrolase is located preferentially in large neurons in the rat central nervous system as revealed by immunohistochemistry. Neurosci Lett 254:137–140

Ueda N, Kurahashi Y, Yamamoto S, Tokunaga T (1995) Partial purification and characterization of the porcine brain enzyme hydrolyzing and synthesizing anandamide. J Biol Chem 270:23823–23827

Ueda N, Puffenbarger RA, Yamamoto S, Deutsch DG (2000) The fatty acid amide hydrolase (FAAH). Chem Phys Lipids 108:107–121

Valenti M, Vigano D, Casico MG, Rubino T, Steardo L, Parolaro D, Di Marzo V (2004) Differential diurnal variations of anandamide and 2-arachidonoyl-glycerol levels in rat brain. Cell Mol Life Sci 61:945–950

Vyvoda OS, Rowe CE (1973) Glyceride lipases in nerve endings of guinea-pig brain and their stimulation by noradrenaline, 5-hydroxytryptamine and adrenaline. Biochem J 132:233–248

Waleh NS, Cravatt BF, Apte-Deshpande A TA, Kilduff TS (2002) Transcriptional regulation of the mouse fatty acid amide hydrolase gene. Gene 291:203–210

Witting A, Walter L, Wacker J, Moller T, Stella N (2004) P2X7 receptors control 2-arachidonoylglycerol production by microglial cells. Proc Natl Acad Sci U S A 101:3214–3219

Xiao AZ, Zhao YG, Duan EK (2002) Expression and regulation of the fatty acid amide hydrolase gene in the rat uterus during the estrous cycle and peri-implantation period. Mol Hum Reprod 8:651–658

Yazulla S, Studholme KM, McIntosh HH, Deutsch DG (1999) Immunocytochemical localization of cannabinoid CB1 receptor and fatty acid amide hydrolase in rat retina. J Comp Neurol 415:80–90

HEP (2005) 168:209–246

Structural Requirements for Cannabinoid Receptor Probes

G.A. Thakur · S.P. Nikas · C. Li · A. Makriyannis (✉)

Center for Drug Discovery, Department of Pharmaceutical Sciences,
University of Connecticut, Storrs CT, 06269, USA
makriyan@uconnvm.uconn.edu

Abstract The discovery and cloning of CB_1 and CB_2, the two known $G_{i/o}$ protein-coupled cannabinoid receptors, as well as the isolation and characterization of two families of endogenous cannabinergic ligands represented by arachidonoylethanolamide (anandamide) and 2-arachidonoylglycerol (2-AG), have opened new horizons in this newly discovered field of biology. Furthermore, a considerable number of cannabinoid analogs belonging to structurally diverse classes of compounds have been synthesized and tested, thus providing substantial information on the structural requirements for cannabinoid receptor recognition and activation. Experiments with site-directed mutated receptors and computer modeling studies have suggested that these diverse classes of ligands may interact with the receptors through different binding motifs. The information about the exact binding site may be obtained with the help of suitably designed molecular probes. These ligands either interact with the receptors in a reversible fashion (reversible probes) or alternatively attach at or near the receptor active site with the formation of covalent bonds (irreversible probes). This review focuses on structural requirements of cannabinoid receptor ligands and highlights their pharmacological and therapeutic potential.

Keywords Cannabinoid receptors · Cannabinoid receptor probes · Structure–activity relationships · Selectivity

1 Introduction

Marijuana (*Cannabis sativa*) is one of the oldest drugs of abuse with a strong social, legal, and medical controversy over its therapeutic utility. Its major psychoactive component, Δ^9-tetrahydrocannabinol (Δ^9-THC), was characterized and synthesized in 1964 and served as a prototype for the synthesis of numerous analogs as potential pharmacological agents (Gaoni and Mechoulam 1964). The next milestone in cannabinoid research was the discovery that cannabinoids produce most of their biochemical and pharmacological effects by interacting with CB_1 and CB_2, the two known $G_{i/o}$ protein-coupled cannabinoid receptors (Devane et al. 1988; Gerard et al. 1990; Matsuda et al. 1990; Munro et al. 1993). CB_1 is found in the central nervous system (CNS) with high density in the cerebellum, hippocampus, and striatum (Gatley et al. 1998; Herkenham 1991, 1990; Mailleux et al. 1992; Matsuda et al. 1993). It is also found in a variety of other organs including the heart, vascular endothelium, vas deferens, testis (Breivogel and Childers 1998; Gerard et al. 1991), small intestine, sperm (Schuel et al. 1999), and uterus (Paria et al. 1998). Conversely, the CB_2 receptor appears to be associated exclusively with the immune system. It is found in the periphery of the spleen and other cells associated with immunochemical functions, but not in neurons in the brain (Munro et al. 1993), and is believed to have an immunomodulatory role. Recent data suggest the presence of a third cannabinoid-like receptor (Begg et al. 2003).

CB_1 and CB_2 share an overall homology of 44% and 68% in the transmembrane domains. The rat (Matsuda et al. 1990), mouse (Abood et al. 1997; Chakrabarti et al. 1995), and human CB_1 receptors (Gerard et al. 1990) have been cloned and show 97%–99% sequence identity across species, while the mouse CB_2 (Shire et al. 1996a,b) exhibits 82% sequence identity with the human clone (Munro et al. 1993). CB_1 and CB_2 share common signal transduction pathways, such as inhibition of adenylyl cyclase and stimulation of mitogen-activated protein kinase. However, unlike CB_1, CB_2 has not been shown to affect ion channels (Pertwee 1997).

The subsequent discovery of the endocannabinoids, arachidonoylethanolamine (anandamide) (Devane et al. 1992b; Hanus et al. 1993) and 2-arachidonoyl glycerol (2-AG) (Di Marzo 1998; Mechoulam et al. 1995; Stella et al. 1997) has led to a better understanding of the physiological and biochemical roles of the endocannabinoid system. 2-Arachidonyl glyceryl ether, also known as noladin ether (Hanus et al. 2001), has been proposed as a representative of a third endocannabinoid class. However, noladin ether's pathway of formation has not been characterized and its occurrence in the normal brain has been questioned (Oka et al. 2003).

Extensive studies on the endocannabinoid system have revealed a number of cannabinergic proteins involved in the inactivation and biosynthesis of endocannabinoids. These include fatty acid amide hydrolase (FAAH) (Di Marzo et al. 1994; Gaetani et al. 2003; Piomelli et al. 1999), monoglyceride lipase (MAG) (Dinh et al. 2002), and the anandamide transporter (ANT) (Beltramo et al. 1997; Di Marzo et al. 1994; Fegley et al. 2004; Hillard et al. 1997). The above three proteins and the two cannabinoid receptors have received considerable attention and show great promise as potential targets for the development of novel medications for various conditions, including pain, immunosuppression, peripheral vascular disease, appetite enhancement or suppression, and motor disorders.

Although both CB_1 and CB_2 have been cloned and their primary sequences are known, their three-dimensional structures and the amino acid residues at the active sites which are involved in ligand recognition, binding, and activation have not been characterized. In the absence of any X-ray crystallographic and nuclear magnetic resonance (NMR) data, information about the structural requirements for ligand–receptor interactions is obtained with the help of suitably designed molecular probes (Khanolkar et al. 2000). These ligands either interact with the receptor in a reversible fashion or, alternatively, attach at or near the receptor active site with the formation of a covalent bond. Information related to ligand binding and receptor activation can also be obtained with the help of receptor mutants (McPartland and Glass 2003; Rhee et al. 2000) and computer modeling (Reggio 1999).

During the last decade, numerous ligands with high affinities and selectivity profiles for cannabinoid receptors (CB_1 and CB_2) evolved from rigorously pursued structure–activity relationship (SAR) studies (for recent reviews see Goutopoulos and Makriyannis 2002; Palmer et al. 2002). These ligands can be classified into six major classes: (1) classical cannabinoids, (2) non-classical cannabinoids (NCCs), (3) hybrid cannabinoids, (4) aminoalkylindoles, (5) diarylpyrazoles, and (6) endocannabinoid-like ligands.

This review focuses on key cannabinoid receptor probes representing the different classes of cannabinergic ligands, their SAR, and therapeutic potentials. The stereoselectivity aspects of interactions between these probes and cannabinoid receptors will also be briefly discussed. Throughout this review we have used the K_i values of individual ligands as measures of their relative abilities to recognize their binding sites. However, it is well known that the K_i values are subject to considerable variability depending on the radioligand used in the binding assays as well as on the experimental details under which the assays were carried out (e.g., albumin concentration, etc.). Direct comparisons hold best within groups of compounds that have been tested under identical experimental conditions. The reader is thus advised to consider the K_i values only as approximate relative measures of a ligand's affinity when interpreting the SAR data and not necessarily a measure of functional potency.

2
Classification of Cannabinoid Receptor Ligands

2.1
Classical Cannabinoids

Classical cannabinoids (CCs) are ABC-tricyclic terpenoid compounds bearing a benzopyran moiety (Figs. 1–3, 5, and 6). This class includes the natural product (–)-Δ^9-THC (**1**, Fig. 1), the more stable and almost equipotent isomer (–)-Δ^8-THC (**2**, Fig. 1), and other pharmacologically active constituents of the plant *Cannabis sativa*. Many CC analogs have been synthesized and evaluated pharmacologically and biochemically (for reviews see Goutopoulos and Makriyannis 2002; Khanolkar et al. 2000; Makriyannis and Goutopoulos 2004; Makriyannis and Rapaka 1990; Mechoulam et al. 1999; Palmer et al. 2002; Razdan 1986). SAR studies recognize four pharmacophores within the cannabinoid prototype: a phenolic hydroxyl (PH), a lipophilic alkyl side chain (SC), a northern aliphatic hydroxyl (NAH), and a southern aliphatic hydroxyl (SAH). The first two are encompassed in the plant-derived cannabinoids, while all four pharmacophores are represented in some of the synthetic NCCs developed by Pfizer (e.g., **25**, Fig. 7). The CC structural features that are important for cannabinoid activity are discussed below.

2.1.1
SAR of Classical Cannabinoids

The Phenolic Hydroxyl This group can be substituted by an amino group, but not by a thiol group (Matsumoto et al. 1977a) while its replacement by a fluorine atom diminishes CB_1 affinity (e.g., **3**, Fig. 2) (Martin et al. 2002). It has also been shown that CCs in which the phenolic hydroxyl is either replaced by a methoxy group (e.g., **4**, Fig. 2) or totally absent (**5** and **6**, Fig. 2) retain some receptor-binding affinity, especially for CB_2 (Gareau et al. 1996; Huffman et al. 2002, 1999, 1996). However, this is not the case for the cannabinol series in which the C-ring is fully aromatized (Khanolkar et al. 2000; Mahadevan et al. 2000).

The Benzopyran Ring This ring is not essential for activity and its expansion to B-ring homocannabinoid derivatives has been considered since the early days of

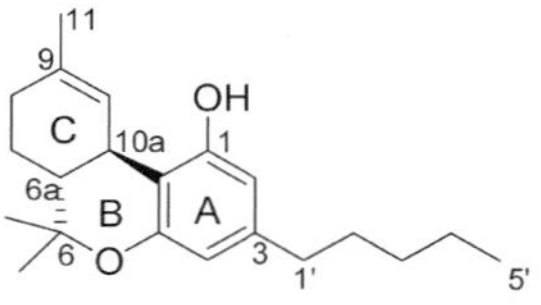

1
(-)-Δ^9-THC (Marinol)
Ki = 39.5 nM (CB1)
= 40.0 nM (CB2)

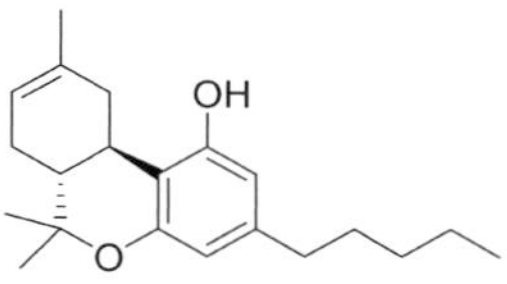

2
(-)-Δ^8-THC
Ki = 47.6 nM (CB1)
= 39.3 nM (CB2)

Fig. 1. The structures of (–)-Δ^9-and (–)-Δ^8-tetrahydrocannabinol (THC)

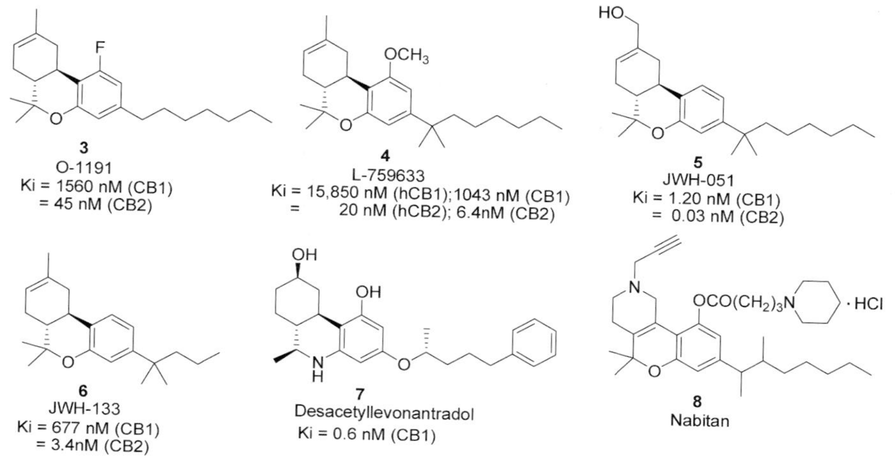

Fig. 2. Phenolic hydroxyl, B- and C-ring modified cannabinoid analogs

cannabinoid structure–activity correlations (Matsumoto et al. 1977b). The pyran oxygen can be substituted by nitrogen as exemplified by compound **7** developed at Pfizer (Fig. 2) (Melvin et al. 1995) or can be eliminated in open phenol or resorcinol analogs. The latter gave rise to the NCC class described in Sect. 2.2.

Neither the double bond nor the 9-methyl at the C-ring is necessary for activity, and this ring may be modified into a heterocyclic system (e.g., **8**, Fig. 2) (Lee et al. 1977, 1983; Osgood et al. 1978; Pars et al. 1976).

C-3 Side Chain This alkyl chain has been recognized as the most critical CC pharmacophoric group. Variation of the *n*-pentyl group of natural cannabinoids can lead to wide variations in potency and selectivity. Optimal activity is obtained with a seven or eight carbon length substituted with 1′,1′-or 1′,2′-dimethyl groups (e.g., **9**, Fig. 3) as was first demonstrated by Adams (Adams et al. 1949; Huffman et al. 2003b; Liddle and Huffman 2001). More recent studies have focused on novel side chains bearing 1′,1′-cyclic moieties (Papahatjis et al. 1998, 2001, 2002, 2003). Some of the synthesized analogs exhibited remarkably high affinities for both CB_1 and CB_2 cannabinoid receptors (e.g., **10**, **11**, **12**, Fig. 3) while in vitro pharmacological testing found the dithiolane analog **10** to be a potent CB_1-selective agonist (Papahatjis et al. 2003). The results of these studies suggest the presence of a subsite within the CB_1 and CB_2 binding domain at the level of the benzylic side carbon in the THC series. In an effort to define the stereochemical limits of this putative subsite, we generated receptor-essential volume maps and receptor-excluded volume maps using molecular modeling approaches (Fig. 4) (Papahatjis et al. 2003).

The observation that the bulky adamantyl Δ^8-THC (**13**, Fig. 3) (Khanolkar et al. 2000; Palmer et al. 2002) exhibits considerable affinity and selectivity for CB_1 points to a greater tolerance for steric bulk in that receptor subsite. Oxygen atoms (ethers) and unsaturations (Busch-Petersen et al. 1996; Papahatjis et al. 1998)

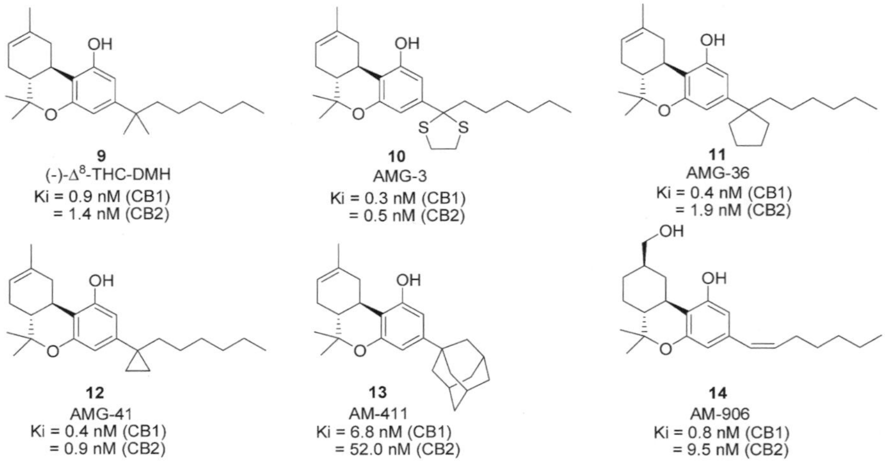

Fig. 3. Representative C-1′ side chain-modified analogs

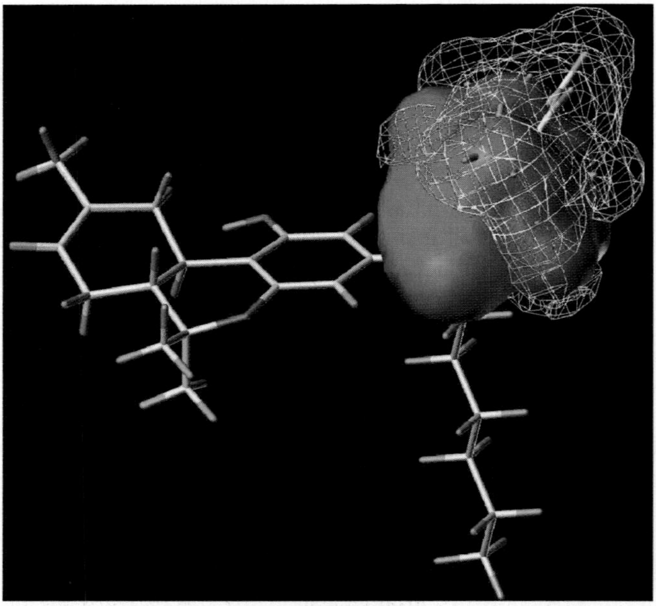

Fig. 4. Molecular modeling of (–)-Δ^8-THC ligands with different substitution in the C-1′ side chain position using molecular mechanics/molecular dynamics. CB_1/CB_2 receptor-excluded volume map (*red contours*) and essential volume map (*white grid*) for the C-1′ subsite in Δ^8-THC series. The *red area* represents the free space within the receptor region that accommodates high-affinity C-1′-substituted ligands, whereas, C-1′ substituents falling within the *white grid* experience unfavorable or less favorable interactions at the binding site

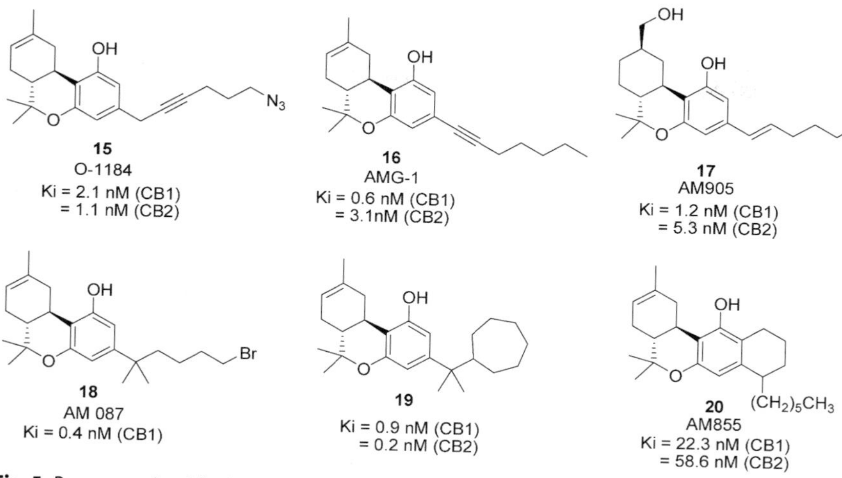

Fig. 5. Representative side chain-modified analogs

within the chain or terminal carboxamido, cyano, azido, and halogen groups are also well tolerated (Charalambous et al. 1991; Crocker et al. 1999; Khanolkar et al. 2000; Martin et al. 1993, 2002; Nikas et al. 2004; Tius et al. 1997, 1993) (e.g., **14**, Fig. 3; **15, 16, 17**, Fig. 5). The side chain seems to be the place of choice for halogen substitution and a considerable enhancement in affinity for CB_1 is observed by halogen substitution at the end carbon of the side chain with the bulkier halogens producing the largest effects (e.g., **18**, Fig. 5). Additionally, naphthyl, phenyl, and cycloalkyl groups have served as side chain substituents (Krishnamurthy et al. 2003; Nadipuram et al. 2003; Papahatjis et al. 1996). Thus, substitution of the 1′,1′-dimethylalkyl side chain with a 1′,1′-dimethylcycloalkyl or 1′,1′-dimethylphenyl group can lead to analogs possessing high affinities for both CB_1 and CB_2 (e.g., **19**, Fig. 5). In another variation, novel tetracyclic analogs of Δ^8-THC in which the alkyl side chain is conformationally more defined by adding a fourth ring in the ABC-tricyclic cannabinoid skeleton fused to the aromatic A-ring have also been reported (e.g., **20**, Fig. 5) (Khanolkar et al. 1999).

Northern Aliphatic Hydroxyl Group It has been shown that introduction of a hydroxyl group at the C-9 or C-11 positions (northern aliphatic hydroxyl; NAH) leads to significant enhancement in affinity and potency for CB_1 and CB_2. Thus, (−)-11-hydroxydimethylheptyl-Δ^8-THC (**21**, Fig. 6), a ligand that has received considerable attention because of its high affinity for both receptors, is more potent than the parent analog with no 11-hydroxy substitution (Mechoulam et al. 1988, 1987). This is also the case for the cannabinol series in which the C-ring is fully aromatized (Rhee et al. 1997) and in the hexahydrocannabinols (HHC, e.g., **22** and **23**, Fig. 6) in which the C-ring is fully saturated. It has also been shown that the relative configuration of C-9 substituents in CCs can have significant effects in the compound's potency (Kriwacki and Makriyannis 1989; Reggio et al. 1989) where

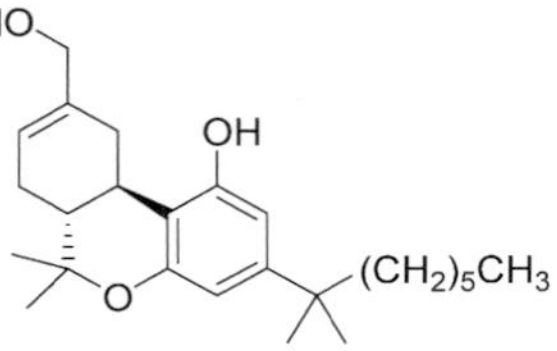

21
HU-210
Ki = 0.7 nM (CB1)
= 0.2 nM (CB2)

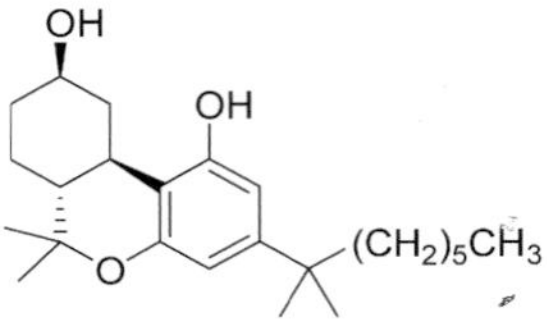

22
Ki = 3.1 nM (CB1)
= 31.4 nM (CB2)

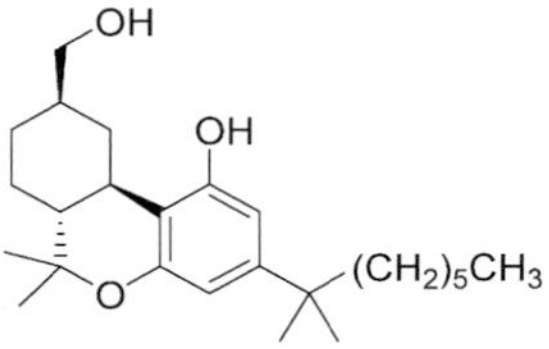

23

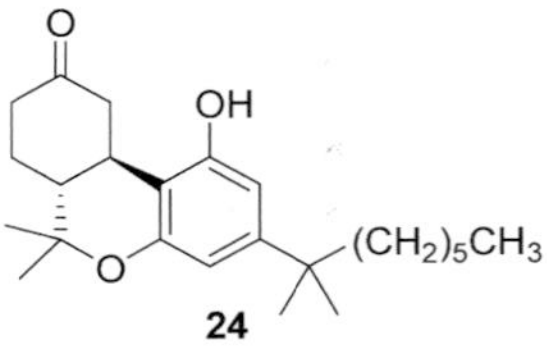

24
Nabilone (Cesamet)
Ki = 2.2 nM (CB1)
= 1.8 nM (CB2)

Fig. 6. Cannabinoid analogs possessing a northern aliphatic hydroxyl (NAH) group

an unfavorable orientation of a C-9 hydroxyl or hydroxymethyl substituent can seriously interfere with this ligand's ability to interact with cannabinoid receptors. Based on the relative configuration at the C-9 position, the HHC encompasses two types of isomers (9α and 9β). Although both isomers are biologically active, the β-epimers in which the C-9 hydroxyl or hydroxymethyl group is equatorial (e.g., **22** and **23**, Fig. 6) have been shown to be more potent than the α-axial isomers (Devane et al. 1992a; Wilson et al. 1976; Yan et al. 1994). The preference for the 9β relative configuration has been used for the design and synthesis of high-affinity photoactivatable probes for the cannabinoid receptors (e.g., AM1708, **70,** Fig. 19) (Khanolkar et al. 2000). Presence of a C-9 carbonyl group encompassed in nabilone (**24**, Fig. 6) is also known to significantly enhance cannabinergic activity (Archer et al. 1986). Although the nature of the substituent at the northern end of the classical cannabinoid structure has an effect on the ligands' potencies, these effects have not yet been fully investigated. Thus, 9-nor-Δ^9-THC, a molecule that lacks a C-9 substituent, exhibits significant cannabinoid activity (Martin et al. 1975).

2.2
Non-classical Cannabinoids

A second class of cannabinergic ligands possessing close similarity with CCs was developed at Pfizer in an effort to simplify the CC structure, while maintaining or improving biological activity (Johnson and Melvin 1986; Little et al. 1988). This group of compounds, generally designated as non-classical cannabinoids (NCCs), includes AC-bicyclic (e.g., **25** and **26**, Fig. 7) and ACD-tricyclic (e.g., **27**, Fig. 7) ligands lacking the pyran B-ring of CCs. Of these the best known is CP-

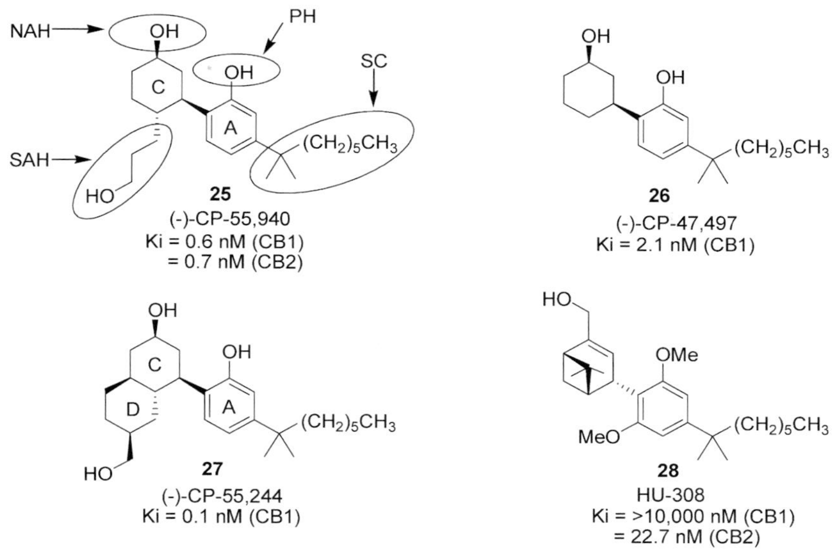

Fig. 7. Non-classical cannabinoid receptor ligands

55,940 (**25**) a crystalline ligand exhibiting high affinity for both CB_1 and CB_2 as well as a high degree of stereoselectivity. [^{3}H]CP-55,940, the tritiated analog, was the key compound that led to the discovery of CB_1 (Devane et al. 1988). This class of compounds shares some of the key pharmacophores of the CCs, namely the phenolic OH, the side chain, and the northern aliphatic hydroxyl groups. Additionally, it encompasses an hydroxypropyl chain on the cyclohexyl ring contiguous and trans to the aromatic phenolic group as with CP-55,940. This important new pharmacophore was designated as the southern aliphatic hydroxyl group (SAH) (Makriyannis and Rapaka 1990) and has been subjected to extensive investigation by the Makriyannis and Tius groups (Chu et al. 2003; Drake et al. 1998; Harrington et al. 2000; Tius et al. 1997, 1994).

The recently introduced ligand HU-308 (**28**, Fig. 7), which has the opposite absolute configuration from all other CC and NCC analogs, is another example of bicyclic cannabinoid receptor ligands (Hanus et al. 1999) and exhibits a high degree of CB_2 selectivity.

2.3 CC/NCC Hybrid Cannabinoids

The southern aliphatic hydroxyl (SAH) pharmacophore is absent in the naturally occurring cannabinoids. To study more precisely the stereochemical requirements of this new pharmacophore, Makriyannis and co-workers designed a group of hybrid ligands that incorporated all of the structural features of both classical and non-classical cannabinoids (Drake et al. 1998; Tius et al. 1995, 1994).

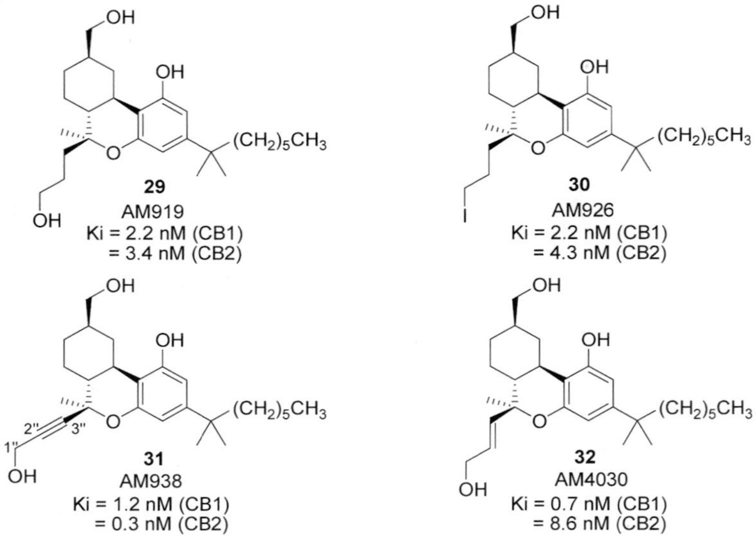

Fig. 8. Hybrid classical/non-classical (CC/NCC) cannabinoids

This new class of analogs (CC/NCC hybrids) had the added advantage of serving as conformationally more defined three-dimensional probes for the CB_1 and CB_2 active sites than their non-classical counterparts. Receptor binding data showed that at C-6 the equatorial β-hydroxypropyl analog had higher affinity than its α-axial epimer (e.g., **29** and **30**, Fig. 8) (Drake et al. 1998; Tius et al. 1994). Further refinement of the CC/NCC hybrid cannabinoids was obtained by imposing restricted rotation around this SAH pharmacophore. This was accomplished through the introduction of double and triple bonds at the C2$''$ position of the 6β-hydroxypropyl chain (e.g., **31** and **32**, Fig. 8).

The affinity data for CB_1/CB_2 receptors shown in Fig. 8 for analogs **31** and **32** refer to the racemic compounds. Enantiomers of **32** were recently separated using chiral AD [amylose tris(3,5-dimethylphenylcarbamate] columns (Thakur et al. 2002) (see Sect. 4). This very promising class of compounds encompassing four asymmetric centers is among the most structurally complex and potent cannabinergic agents synthesized to date.

2.4 Aminoalkylindoles

The fourth chemical class of cannabinergic ligands, the aminoalkylindoles (AAIs) were initially developed at Sterling Winthrop as potential non-ulcerogenic analogs of non-steroidal anti-inflammatory drugs (NSAIDs) (Bell et al. 1991) and bear no structural relationship to the cannabinoids. These analogs also exhibited antinociceptive properties that eventually were attributed to their interactions with the cannabinoid receptors (D'Ambra et al. 1992; Eissenstat et al. 1995). The most widely studied compound of this series is WIN-55,212-2 (**33**, Fig. 9), a potent CB_1 and CB_2 agonist with a slight preference for CB_2. Cannabinergic activity resides principally with only one optical antipode and is more potent than Δ^9-THC in several pharmacological and behavioral assays (Compton et al. 1992; Martin et al. 1991). WIN-55,212-2 has played an important role in the identification and characterization of cannabinoid receptors and their associated functions and is now in standard use as a CB_1/CB_2 radioligand. The four pharmacophores identified for the aminoalkylindoles are: (1) C-3 substituents, (2) the N-1 aminoalkyl side chain, (3) C-2 substituents, and (4) indole ring substituents and modifications. The SAR requirements of this class of compounds are summarized as follows:

2.4.1 SAR of Aminoalkylindoles

C-3 Substituents Pravadoline (**34**, Fig. 9), which carries a *p*-methoxybenzoyl group at C-3, was used as a benchmark ligand to explore structural requirements at this site (Eissenstat et al. 1995). Its *o*-methoxy isomer exhibits higher potency. However, *ortho*-substitution with other groups such as $-CH_3$, –OH, –Cl, –CN, or –F diminishes activity. The presence of an ethyl group at the para position improves potency, but further increase in chain length results in diminished potency. The 1-naphthoyl substitution at C-3 is more potent (IC_{50} = 19 nM) than the 2-napthoyl analog (IC_{50} = 128 nM). Replacement of the naphthyl ring with an alkyl (e.g., CH_3) or alkenyl [$(CH_3)_2C{=}CH$] groups results in complete loss of CB_1 receptor affinity (K_i>10,000 nM) (Huffmann et al. 1994).

NMR and X-ray crystallography studies of **34** and its C-2H congener have revealed that AAIs can exist in two distinct conformations based on the orientation of the C-3 aroyl system (Bell et al. 1991; Reggio et al. 1998). In the *s-trans* conformation, which predominates when the C-2 substitution is hydrogen, the aryl group is proximal to C-2, while the carbonyl oxygen atom is located near C-4. In the *s-cis* conformation, which predominates when the C-2 substituent is a methyl group, the conformational preference shows the aryl ring to be located near C-4, and the carbonyl oxygen near C-2.

Naphthylidene-substituted aminoalkylindenes (e.g., **35**, Fig. 9), a conformationally more rigid version of initial AAIs, were originally designed to circumvent the CNS side effects of pravadoline (Kumar et al. 1995). These analogs were tested as a mixture of *E*- and *Z*-isomers and exhibited higher CB_1 affinity compared to pravadoline. Later, it was shown that the CB_1 and CB_2 affinities and pharmacological potencies were higher for the *E*-geometric isomer (**35**, s-trans, Fig. 9) compared to the *Z*-isomer (Reggio et al. 1998). Removal of the carbonyl oxygen of the C-3 aroyl group in AAIs having unsubstituted C-2 results in moderate reduction in affinity for CB_1 compared to their carbonyl precursors (Huffman et al. 2003a). However, the loss of affinity is larger in the 2-methyl substituted analogs (e.g., **36**, Fig. 9). Both observations support the hypothesis that the *s-trans* conformation of AAI analogs such as **33** is the preferred conformation for interaction at both CB_1

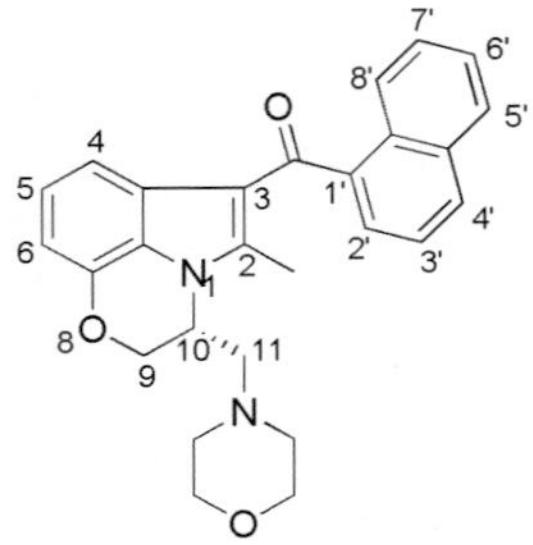

33
R-(+)-WIN 55,212-2
Ki = 1.9 nM (CB1)
= 0.3 nM (CB2)

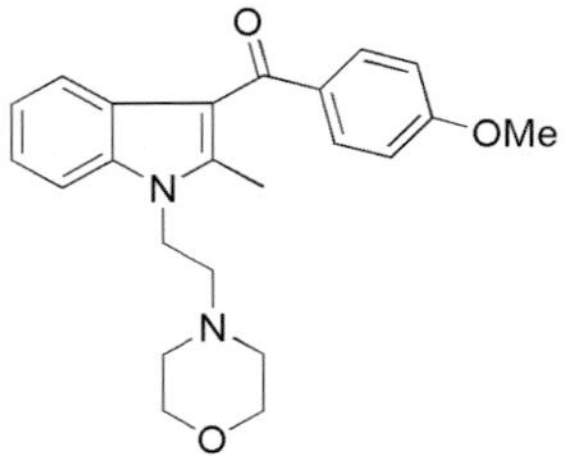

34
Pravadoline
Ki = 2510 nM (CB1)
= 320 nM (CB2)

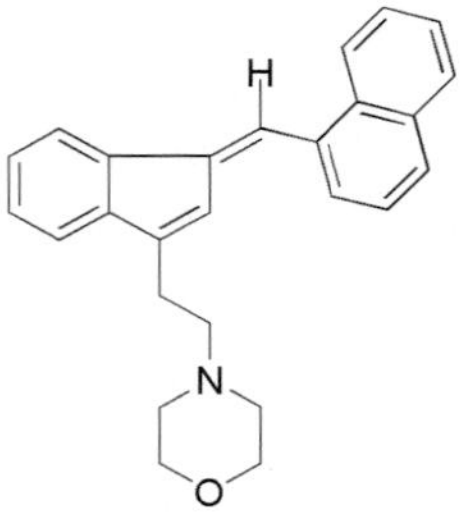

35
Ki = 2.7 nM (CB1)
= 2.7 nM (CB2)

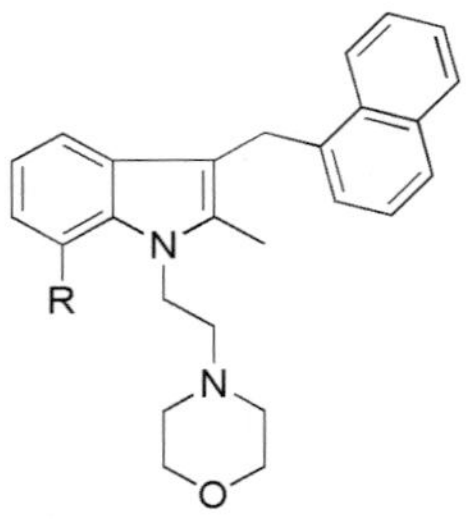

36
R = H; Ki = 113 nM (CB1)
R = CH_3; Ki = 41nM (CB1)
R = OMe; Ki = 20nM (CB1)

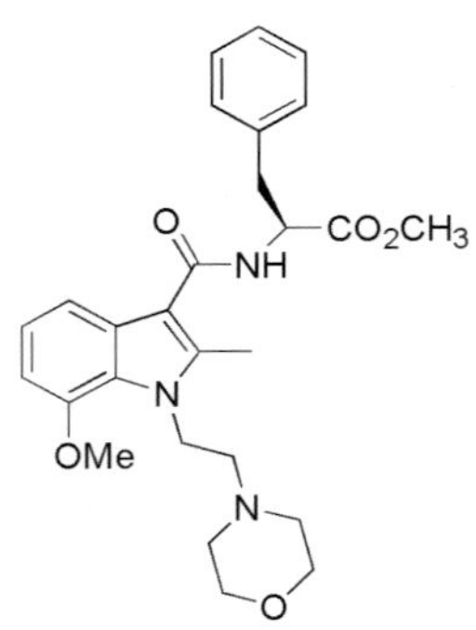

37
Ki = 4000 nM (CB1)
= 8 nM (CB2)

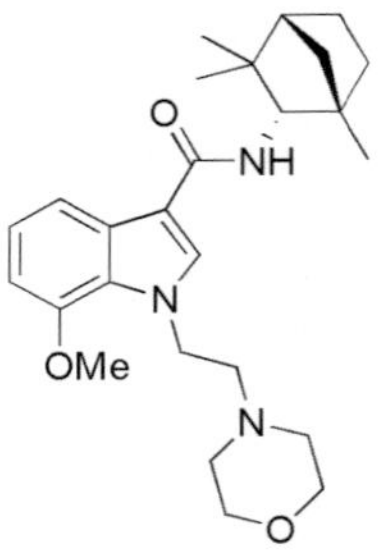

38
Ki = 245 nM (CB1)
= 11 nM (CB2)

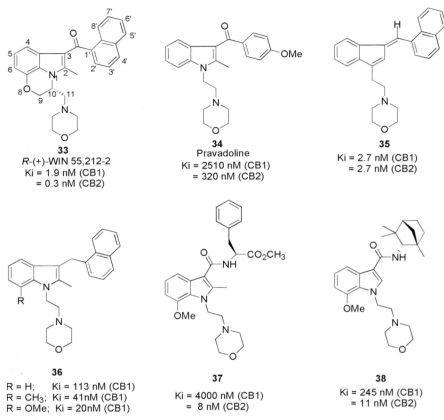

Fig. 9. C-3 modified cannabinergic aminoalkylindoles

and CB_2 receptors and that aromatic stacking of the ligands with aromatic residues in helices 3, 4, and 5 of both receptors may be an important interaction for AAIs at these receptors (Burley and Petsko 1985; Huffman et al. 2003a; Reggio et al. 1998).

The spatial and electronic requirements of the C-3 substituent were further explored by introducing a C-3 amide group (Bristol Myers Squibb). The AAI C-3 amide ligand **37** (Fig. 9) with a methoxy group at C-7, exhibited high CB_2 affinity ($K_i = 8$ nM) and selectivity ($CB_1/CB_2 = 500$) (Hynes et al. 2002). Replacement of the amino acid moiety in **37** with the *S*-fenchylamine component resulted in slightly reduced affinity for the CB_2 receptor ($K_i = 30$ nM). However, in the *S*-fenchyl amide series, when the 2-methyl group in indole was replaced by hydrogen, the resulting ligand (**38**, Fig. 9) showed improved CB_2 affinity ($K_i = 11$ nM).

The 4-alkyloxy indole analogs were derived by translocating the C-3 substituent of AAIs to C-4 via an ether linkage. Some of these exhibited in vivo cannabimimetic activity, but most of them lacked cannabinoid receptor affinity (Dutta et al. 1997).

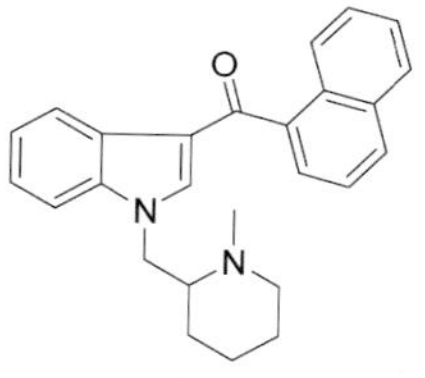

39

IC_{50} (nM) = 1.2 nM (CB1)

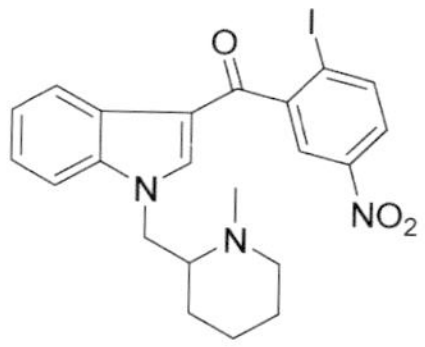

40
AM1241
Ki = 280 nM (CB1)
= 3.4 nM (CB2)

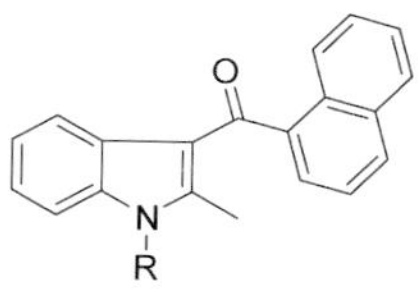

41
JWH-015; R = *n*-propyl;
Ki = 383nM (CB1) and 14 nM (CB2)

JWH-007; R = *n*-pentyl;
Ki = 9.5nM (CB1) and 2.9 nM (CB2)

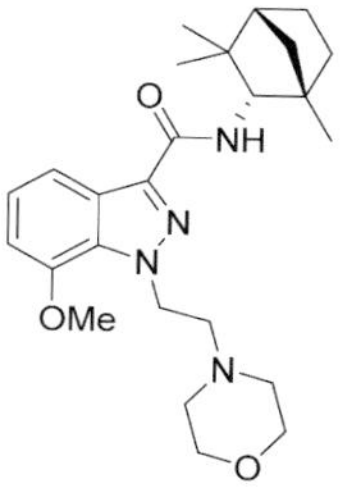

42

Ki = 24 nM (CB1)
= 2.0 nM (CB2)

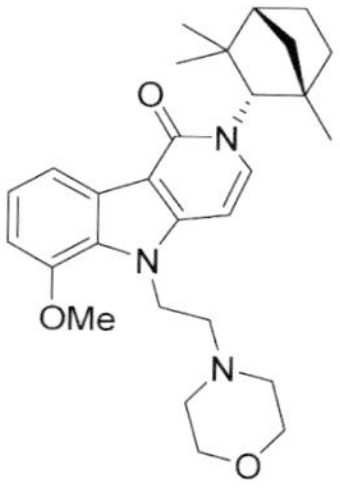

43

Ki = 16.4 nM (CB1)
= 1.0 nM (CB2)

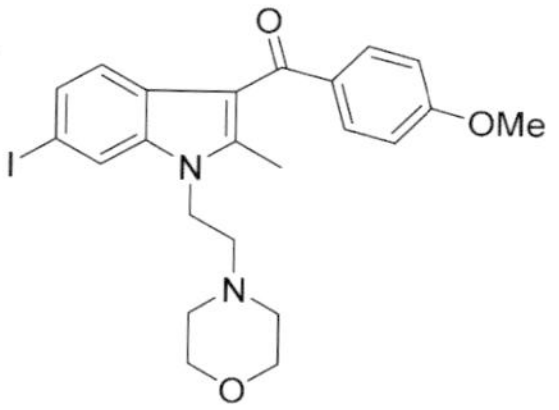

44
AM630
Ki = 5152 nM (CB1)
= 31.2 nM (CB2)

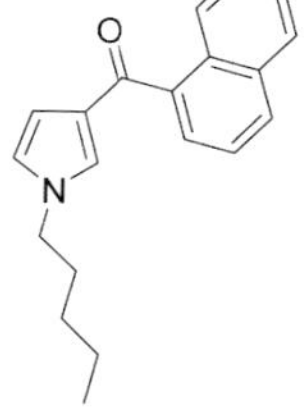

45
JWH-030
Ki = 87 nM (CB1)

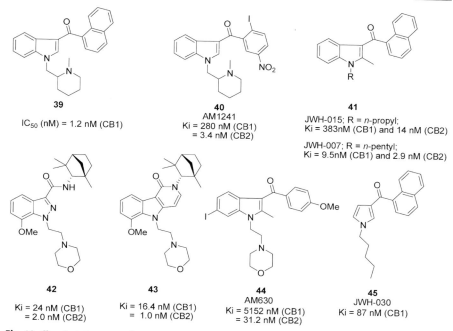

Fig. 10. Chemical structures of some aminoalkylindole-derived analogs

N-1 Aminoalkyl Chain A number of indole analogs bearing different aminoalkyl substituents at N-1 were synthesized (*N*-attached analogs, e.g., **34**, Fig. 9) and tested (Eissenstat et al. 1995). This study found the aminoethyl substitution as an optimal requirement with morpholino, thiomorpholino, and piperidino analogs showing the highest activities. The respective acyclic amine and piperazine analogs were inactive.

The Sterling Winthrop and Makriyannis laboratories further explored structural requirements at the N-1 position by synthesizing novel analogs in which the aminoalkyl chain of the indole ring is attached to a heterocyclic amine through a C–C bond. These analogs are generally more potent compared to the C–N analogs and exhibit more favorable physicochemical properties. Potency was optimum for *N*-methylpiperidinyl-2-methyl substitution at the N-1 position (**39**, Fig. 10), while activity resided predominately in the *R*-enantiomer (D'Ambra et al. 1996).

AM1241 (**40**, Fig. 10), a highly CB_2-selective and potent agonist (Ibrahim et al. 2003; Malan et al. 2001) was recently developed by Makriyannis. Design of this molecule incorporated the *N*-methylpiperidinyl-2-methyl substituent at the N-1 position and a novel 2-iodo-5-nitrobenzoyl group at C-3. AM1241 exhibits remarkably high peripheral analgesia in vivo and does not produce catalepsy, hypothermia, inhibition of spontaneous locomotor activity, or impairment of performance on the rotarod apparatus. The potential use of this CB_2 receptor agonist for the treatment of neuropathic pain is being explored.

Replacement of the aminoalkyl substituent by an alkyl chain results in *N*-alkyl indoles (non-AAIs) (e.g., **41**, Fig. 10). The SAR of cannabimimetic 2-methylindoles indicates that compounds with *N*-alkyl substituents from *n*-propyl to *n*-hexyl have good affinities for both CB_1 and CB_2 receptors with a preference for CB_2. The in vivo potencies of these compounds were reported to be consistent with their receptor affinities (Huffmann et al. 1994; Wiley et al. 1998).

C-2 Substituents Analysis of the effect of C-2 substitution on cannabinoid receptor affinity in AAIs reveals a strong preference for a small substituent at C-2. Thus, hydrogen or methyl groups are well tolerated with the C-2H analogs exhibiting slightly higher affinities for the CB_2 than C-2 methyl analogs (Eissenstat et al. 1995; Hynes et al. 2002; Wrobleski et al. 2003).

Recently, researchers at Bristol Myers Squibb reported their discovery of indazole carboxamides (e.g., **42**, Fig. 10), a new class of cannabimimetics, in which the C-2 carbon of 3-amido AAIs (e.g., **38**, Fig. 9) is replaced by nitrogen. The indazole analog **42** exhibits high affinity for the CB_2 receptor ($K_i = 2.0$ nM) compared to the corresponding AAI analogs **38** (Wrobleski et al. 2003). Indolopyridones (e.g., **43**, Fig. 10), which are conformationally restricted C-3 amido AAIs, exhibit increased affinities for the CB_2 receptor ($K_i = 1.0$ nM) and possess anti-inflammatory properties when administered orally in an in vivo murine inflammation model (Wrobleski et al. 2003).

Indole Ring Substituents and Modifications Introduction of a methyl group at C-4 or various substituents such as $-CH_3$, $-OCH_3$, –F, –Br, or –OH groups at C-5 of pravadoline diminishes affinity. Conversely, C-6 substitution with $-CH_3$, $-OCH_3$, or –Br (WIN-54,461, bromopravadoline) groups improves receptor affinity, but the ligands exhibit diminished agonist properties (Eissenstat et al. 1995). Incorporation of an iodo group at C-6 led to AM630 (**44**, Fig. 10), a ligand that exhibits improved affinity as well as selectivity for CB_2 (Hosohata et al. 1997a,b; Pertwee et al. 1995). This compound was shown to be a potent and selective antagonist/inverse agonist for CB_2 and is a useful pharmacological tool developed before its principal target site was identified (Ross et al. 1999). Substitution at C-7 gives modest improvement in binding affinity. Potent AAI analogs were generated by conformationally restricting the N-1 side chain through the formation of a six-membered ring between the N-1 and C-7 substituents (D'Ambra et al. 1992). In *N*-alkyl indoles, replacement of the indole phenyl ring with a cyclohexyl ring led to an analog with reduced affinities for both CB_1 and CB_2 (Tarzia et al. 2003). Removal of the phenyl ring in AAIs or non-AAIs led to a pyrrole class of cannabimimetics (e.g., **45**, Fig. 10). The SAR of pyrrole cannabinoids has been explored first by Sterling Winthrop and later by Huffman (Wiley et al. 1998) and Tarzia et al. (2003). Most of the pyrrole-derived analogs are less potent than the corresponding indole derivatives. However, the 4-bromopyrrole analog (Tarzia et al. 2003) exhibits high affinity for both CB_1 and CB_2 ($EC_{50} = 13.3$ nM for rCB_1 and 6.8 nM for hCB_2) comparable to WIN-55,212-2.

2.5 Diarylpyrazoles

The most widely studied compound of the diarylpyrazole class is SR141716A (Rimonabant) (**46**, Fig. 11) developed by Rinaldi-Carmona and co-workers at Sanofi (Rinaldi-Carmona et al. 1994) and is currently undergoing clinical trials as an antiobesity medication. This highly potent and selective CB_1 receptor ligand has served as a unique pharmacological and biochemical tool for further characterization of the CB_1 cannabinoid receptor (Lan et al. 1999; Nakamura-Palacios et al. 1999). In vitro, SR141716A antagonizes the inhibitory effects of cannabinoid agonists on both mouse vas deferens (MVD) contractions and adenylyl cyclase activity in rat brain membranes. SR141716A also antagonizes the pharmacological and behavioral effects produced by CB_1 agonists after intraperitoneal (i.p.) or oral administration (Rinaldi-Carmona et al. 1994).

Other diarylpyrazole ligands that have contributed to our understanding of CB_1 pharmacology are AM251 and AM281 (Lan et al. 1999), both of which are CB_1 antagonist/inverse agonists (**47** and **48** respectively, Fig. 11) capable of displacing [^{3}H]SR141716A and [^{3}H]CP-55,940 in CB_1 receptor membrane preparations. Both AM251 and AM281 share the ability of SR141716A to attenuate the responses to

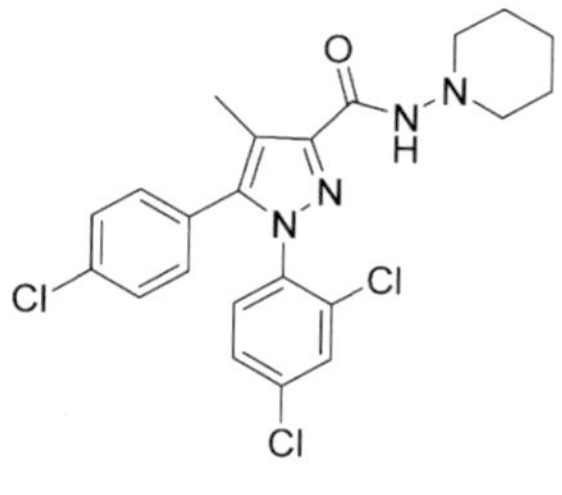

46
SR-141716A
Ki = 11.5nM (CB1)
= 1640 nM (CB2)

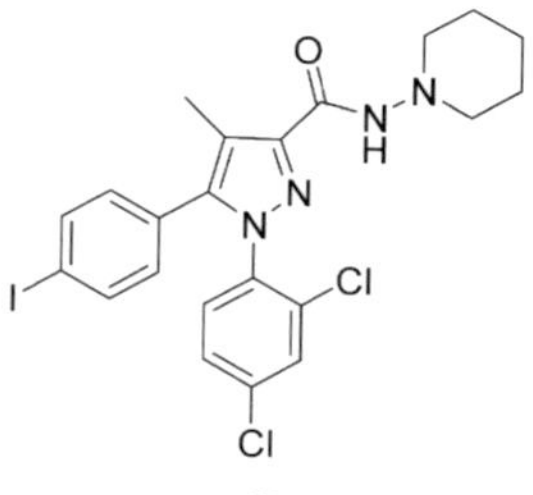

47
AM251
Ki = 7.5 nM (CB1)
= 2290 nM (CB2)

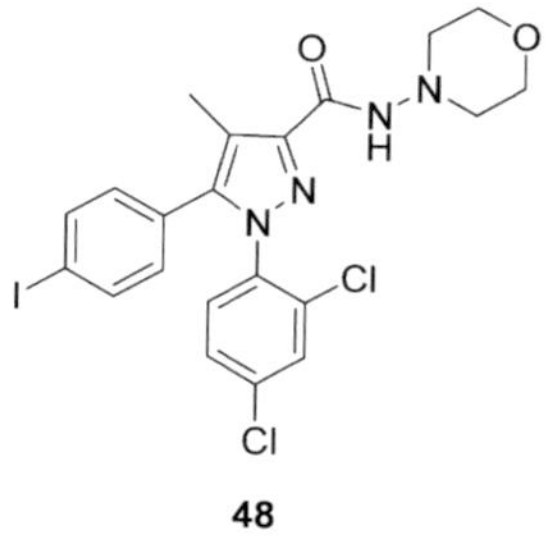

48
AM281
Ki = 12 nM (CB1)
= 4200 nM (CB2)

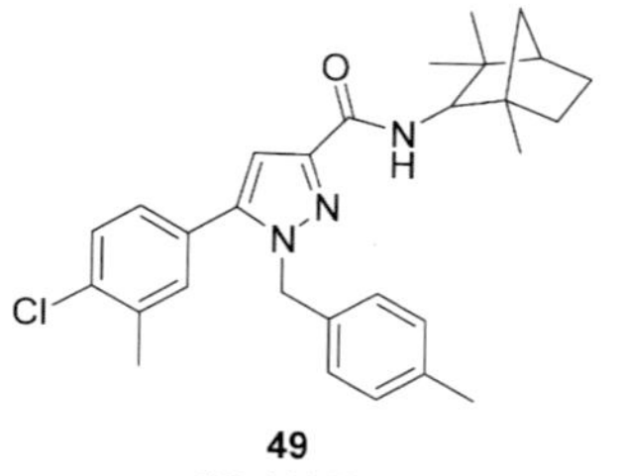

49
SR-144528
Ki = 437nM (CB1)
= 0.6 nM (CB2)

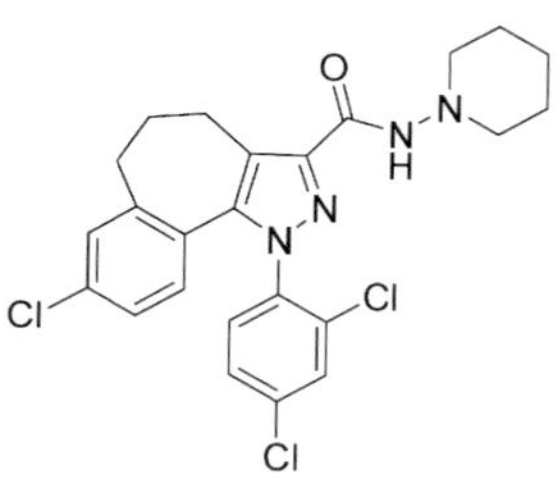

50
Ki = 6.9nM (CB1)

Fig. 11. Representative diarylpyrazole ligands

established cannabinoid receptor agonists like WIN-55,212-2 or CP-55,940. However, recent evidence indicates that AM251 may have a more "CB_1-selective" role than SR141716A (Hajos and Freund 2002). In addition to AM630, the most notable CB_2 receptor antagonist/inverse agonist is SR144528, a diarylpyrazole (**49**, Fig. 11) developed by Sanofi, exhibiting 700-fold selectivity for the CB_2 receptor over CB_1 (Rinaldi-Carmona et al. 1998). Structural requirements for SR141716A-like compounds are summarized below (for earlier reviews see Howlett et al. 2002; Palmer et al. 2002).

2.5.1 SAR of Pyrazole Cannabinoid Receptor Antagonists

N-1 Substituents 2,4-Dichlorophenyl is the optimal substituent for both high CB_1 affinity and subtype selectivity (Barth and Rinaldi-Carmona 1999; Lan et al. 1999). Its replacement with 1-(5-isothiocyanato)-pentyl group decreased CB_1 affinity only by a factor of four (Howlett et al. 2000). The inclusion of 4-butylphenyl, 4-pentylphenyl or a phenyl group at N-1 significantly reduces affinity while *n*-pentyl, *n*-hexyl, *n*-heptyl substitution retains affinity (Shim et al. 2002). Optimal selectivity for CB_2 is contributed by a 4-methylbenzyl group as represented in SR144528 (**49**, Fig. 11) (Rinaldi-Carmona et al. 1998). In the 2,4-dichlorophenyl moiety, elimination of *p*-chloro substitution or replacement of *o*-chloro with *o*-fluoro or *o*-methoxy groups led to low-affinity analogs (Katoch-Rouse et al. 2003). Replacement of the 2,4-dichlorophenyl by unsubstituted cycloalkyl groups decreased both CB_1 and CB_2 affinities, while the 3-methyl and 4-methylcyclohexyl analogs exhibited moderate improvement in CB_2 affinity without any enhancement in selectivity compared to SR141716A (Krishnamurthy et al. 2004).

C-3 Substituents Alkylation of the amide group as well as its replacement by a ketone, alcohol, or ether (Wiley et al. 2001) greatly decreases CB_1 affinity. Replacement of the piperidinyl group with the respective five- or seven-membered heterocyclic rings or by a cyclohexyl group does not alter CB_1 binding affinity, while replacement with a morpholine group or linear alkyl chains leads to reduction in CB_1 affinity (Lan et al. 1999). Alkyl hydrazines, amines, and hydroxyalkylamines of varying lengths were substituted for the aminopiperidinyl moiety to probe the structural and steric requirements of this pharmacophore (Francisco et al. 2002). For alkylamides, hydroxyalkyl amides, and alkyl hydrazides, affinity for CB_1 was found to increase with increasing chain length from ethyl to butyl or pentyl. Further increase in the carbon chain length reduced affinity for both receptors. Alkylamide analogs exhibited enhanced CB_1 selectivity when compared to SR141716A, whereas hydroxyalkyl amide and alkylhydrazide analogs had both decreased affinities and selectivities (Francisco et al. 2002).

C-4 Substituents Compounds with methyl, ethyl, bromo, or iodo substituents in the 4-position of the pyrazole ring are approximately equipotent, whereas replacement of methyl with hydrogen results in a 12-fold decrease in CB_1 affinity (Wiley et al. 2001).

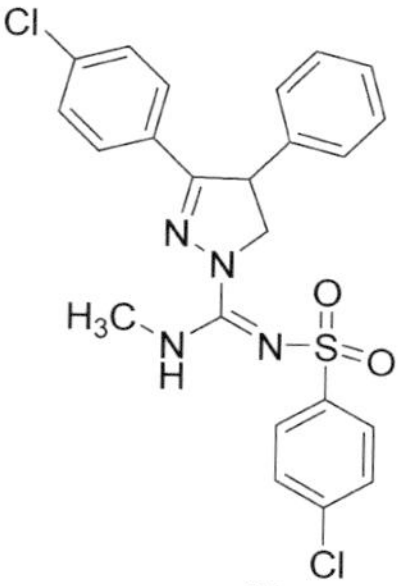

51
Ki = 25.2 nM (CB1)
= >1,000 nM (CB2)

Fig. 12. 3,4-Disubstituted pyrazolines

C-5 Substituents The 4-chloro group of the phenyl ring can be replaced by bromo or alkyl groups but not by nitro or amino groups (Lan et al. 1999; Thomas et al. 1998; Wiley et al. 2001). Replacement of 4-chloro with a 4-iodo substituent (AM251) leads to optimal CB_1 affinity and CB_1/CB_2 selectivity. AM251 has proved to be an excellent CB_1 probe and is widely used as a standard. Conversely, replacement of the aromatic ring with alkyl groups abolishes CB_1 affinity (Lan et al. 1999).

Recently, two research groups independently reported a number of rigid analogs of SR141716A. Solvay (Stoit et al. 2002) first reported some tricyclic CB_1-selective ligands in which the 4- and 5-substituents are conformationally restricted through the formation of a relatively rigid tricyclic system. In these compounds the 4-methyl group is connected with the *ortho* position of the aromatic 5-aryl substituent to form benzocycloheptapyrazole analogs represented by **50** (Fig. 11) that exhibited higher CB_1 affinity than the parent SR141716A (Stoit et al. 2002). However, the compound had poor oral bioavailability. Later Pinna and co-workers (Mussinu et al. 2003) reported similar tricyclic pyrazole analogs in which the above additional 7-membered ring was replaced by a five-membered ring. Interestingly, most ligands in this class had high affinity and selectivity for CB_2 compared to **50** and SR141716A.

Very recently, Solvay Pharmaceuticals (Lange et al. 2004) reported a novel class of 3,4-disubstituted pyrazoline analogs exhibiting high CB_1 selectivity (e.g., **51**, Fig. 12). Another novel class of CB_1 antagonists that has received only limited attention includes the 3-alkyl-5-arylhydantoins (Ooms et al. 2002).

While the search for high affinity/efficacy ligands is ongoing, the development of well-designed radiolabeled ligands has enhanced our understanding of the physiological role of the endocannabinoid system. $[^{123}I]$AM281, an ^{123}I-labeled 1,5-biarylpyrazole, has served as a useful imaging agent in single photon emission computed tomography (SPECT) studies (Gatley et al. 1997, 1998; Gifford et al. 1997).

2.6 Endocannabinoids

In 1992 an arachidonic acid ethanolamide derivative (**52**, AEA, Fig. 13) isolated from porcine brain and characterized as an endogenous ligand for the cannabinoid receptors was named anandamide (Devane et al. 1992b). AEA is a highly lipophilic compound encompassing four non-conjugated cis double bonds and is sensitive to both oxidation and hydrolysis. It was shown to bind to the CB_1 receptor with moderate affinity ($K_i = 61$ nM), has low affinity for the CB_2 receptor ($K_i = 1{,}930$ nM), and behaves as a partial agonist in the biochemical and pharmacological tests used to characterize cannabinoid activity. Its role as a neurotransmitter or neuromodulator is supported by its pharmacological profile as well as by the biochemical mechanisms involved in its biosynthesis and bioinactivation. Two other polyunsaturated fatty acid ethanolamides, homo-γ-linolenoylethanolamide and 7,10,13,16-docosatetraenoylethanolamide, also were isolated subsequently from porcine brain and shown to bind with high affinity to CB_1 (Hanus et al. 1993). Following that, 2-AG (**53**, Fig. 13), a monoglyceride representing a new class of endocannabinoid ligands and capable of binding to both CB_1 and CB_2 receptors was isolated from intestinal and brain tissues and shown to be another endogenous cannabinoid (Mechoulam et al. 1995; Stella et al. 1997) present in brain in concentrations approximately 170-fold higher than anandamide (Di Marzo et al. 1998; Mechoulam et al. 1996; Mechoulam et al. 1995; Stella et al. 1997). Another endogenous agonist for both CB_1 and CB_2 receptors is mead ethanolamide (Priller et al. 1995).

An ether-type endocannabinoid, 2-arachidonyl glyceryl ether (noladin ether, **54**, Fig. 13) was reported to be isolated from porcine brain (Hanus et al. 2001). Noladin ether was found to bind selectively to the CB_1 receptor ($K_i = 21.2$ nM) and cause sedation, hypothermia, intestinal immobility, and mild antinociception in

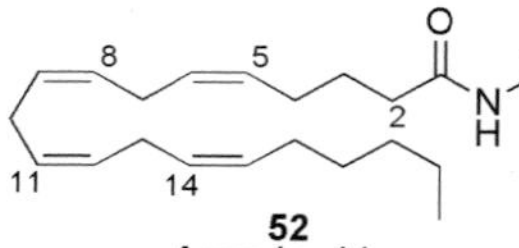

52
Anandamide
Ki = 61 nM (CB1)
= 1930 nM (CB2)

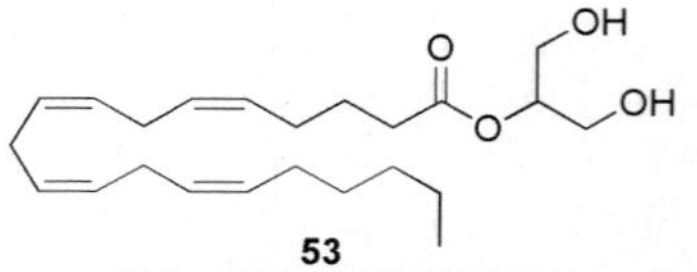

53
2-Arachidonoyl Glycerol
Ki = 472 nM (CB1)
= 1400 nM (CB2)

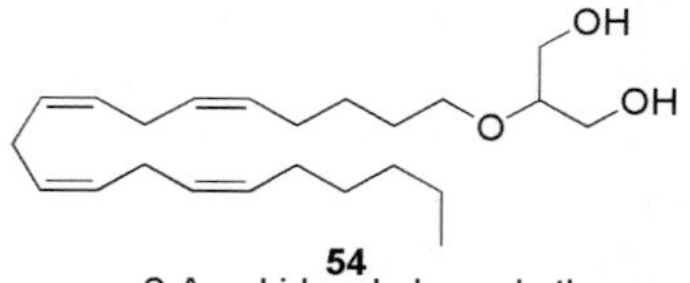

54
2-Arachidonyl glyceryl ether
(Noladin ether)
Ki = 21.2 nM (CB1)
= >3000 nM (CB2)

Fig. 13. Endogenous cannabinoid receptor agonists

mice, effects typically produced by cannabinoid agonists. Synthetic noladin ether was used by Sugiura and co-workers to examine its effects on Ca^{2+} levels in cells (Sugiura et al. 1999; Suhara et al. 2000) and found to exhibit appreciable agonistic activity, although significantly lower than that of 2-AG.

2.6.1
SAR of Endocannabinoids

The chemical structure of anandamide can be divided into two major molecular fragments: (1) a polar ethanolamido head group and (2) a hydrophobic arachidonoyl chain (see Fig. 14). The polar head group is comprised of a secondary amide functionality with an *N*-hydroxyalkyl substituent, while the hydrophobic fragment is a non-conjugated cis tetraolefinic chain and an *n*-pentyl tail reminiscent of the lipophilic side chain found in the classical cannabinoids.

A number of anandamide analogs have been synthesized and tested for their biological activities. These efforts have resulted in the development of several potent metabolically stable analogs some of which are important pharmacological tools useful in elucidating the physiological role of anandamide. Below we summarize the SAR (for previous reviews see Khanolkar and Makriyannis 1999; Palmer et al. 2000; Razdan and Mahadevan 2002; Reggio 2002; Thomas et al. 1996) of anandamide analogs for the currently known high-affinity cannabinergic sites with which anandamide and its analogs are known to interact.

All known arachidonoylethanolamides are primarily CB_1-selective ligands and bind poorly to the peripheral CB_2 receptor. Therefore, the following discussion will focus on the endocannabinoid ligand SAR for the CB_1 receptor.

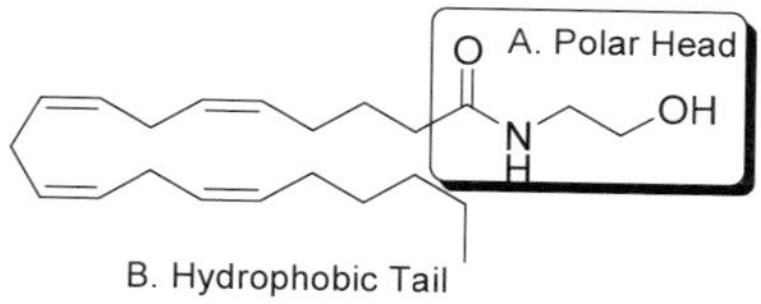

Fig. 14. Structural features of anandamide

Modification of *N*-Hydroxyethyl Group One carbon homologation to the *N*-hydroxypropyl analog increases CB_1 receptor affinity. However, further extension, with or without branching, leads to a decrease in binding affinity (Pinto et al. 1994; Sheskin et al. 1997). Thus, a three-carbon chain separating the amido NH group from the terminal OH appears to be an optimal requirement for a favorable ligand–receptor interaction. However, the hydroxyl group is not a necessary requirement for receptor affinity/potency. *N*-alkyl analogs such as *N*-ethyl, *N*-propyl, and *N*-butyl all show good receptor affinities. *N*-(*n*-Propyl)arachidonamide has a threefold higher CB_1 affinity than anandamide, while the *n*-butyl homolog has about equal affinity (Pinto et al. 1994). Substitution of the ethanolamine head group with an *N*-cyclopropyl group leads to a high-affinity CB_1-selective compound (55,

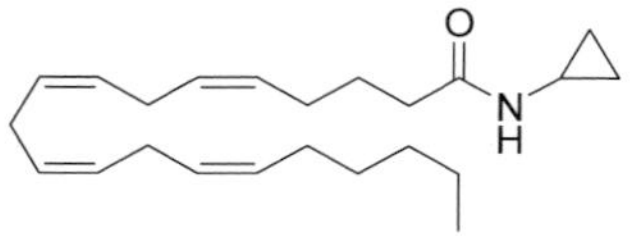

55
ACPA
Ki = 2.2 nM (CB1)
= 715 nM (CB2)

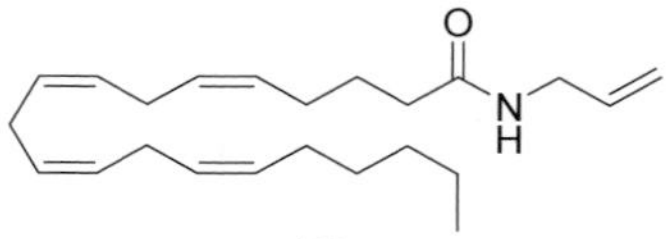

56
AM883
Ki = 9.9 nM (CB1)
= 226 nM (CB2)

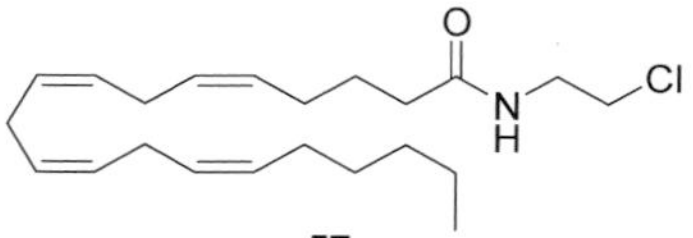

57
AM881, Chloroanandamide
Ki = 5.3 nM (CB1)
= 95 nM (CB2)

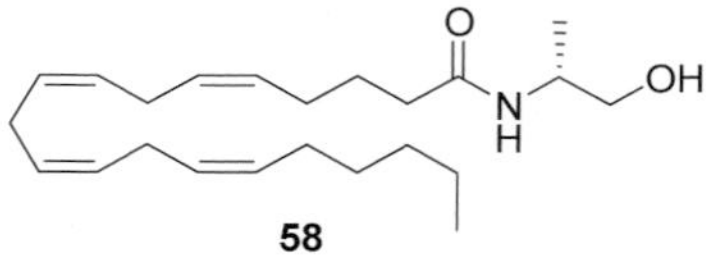

58
AM356, *R*-(+)-methanandamide
Ki =17.9 nM (CB1)
= 868 nM (CB2)

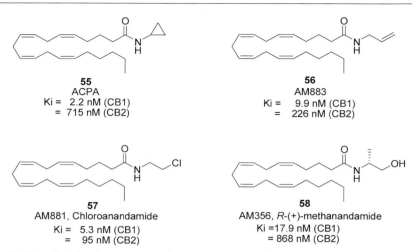

Fig. 15. High-affinity head group analogs of anandamide

Fig. 15). *N*-Allyl (**56**, Fig. 15) and *N*-propargyl analogs also show high CB_1 affinities (Lin et al. 1998). Substitution of the hydroxyl group with a halogen such as F and Cl (**57**, Fig. 15) also increases affinity for CB_1 (Adams et al. 1995a,b; Lin et al. 1998). The above data suggest that anandamide analogs can interact with the CB_1 receptor without the participation of the ethanolamide hydroxyl group.

One of the shortcomings of anandamide as an effective pharmacological tool is its facile in vivo and in vitro enzymatic degradation. It was, thus, important to develop analogs that are resistant to the hydrolytic actions of anandamide amidohydrolase. To address this shortcoming, four chiral anandamide analogs possessing a methyl group at the C-1′ or the C-2′ positions were synthesized (Abadji et al. 1994; Goutopoulos et al. 2001; Lin et al. 1998). The rationale behind the design was to slow down the enzymatic hydrolysis by increasing steric hindrance around the amido group. Of these, the 1′-*R*-methyl isomer [AM356, *R*-(+)-methanandamide **58**, Fig. 15] showed four times higher CB_1 affinity than anandamide while exhibiting excellent metabolic stability. This analog is now being used as an important pharmacological tool in cannabinoid research. Interestingly, an inverse correlation in stereoselectivity between CB_1 receptor affinity and the ability of the ligand to serve as a substrate for FAAH (fatty acid amide hydrolase) was observed. Thus, in the case of 1′-methyl headgroup analogs, the *R*-enantiomer that has higher CB_1 affinity also exhibited lower susceptibility to enzymatic hydrolysis. Introduction of larger alkyl groups, e.g., ethyl or isopropyl, has a detrimental effect on CB_1 affinity (Khanolkar et al. 1996; Khanolkar and Makriyannis 1999).

Substitution of the 2-hydroxyethyl group with a phenolic group results in decreased affinity for CB_1 (Khanolkar et al. 1996). However, *N*-(*o*-hydroxy)phenylarachidonamide (AM403) was found to be an excellent substrate for FAAH (Lang et al. 1999) while a second phenolic analog, *N*-(*p*-hydroxy)phenylarachidonamide (AM404), was found to be an inhibitor for the anandamide transporter (ANT)

(Beltramo et al. 1997). Arachidonamide and arachidonic acid esters (methyl, ethyl, propyl) do not show significant affinity for CB_1 (Sheskin et al. 1997), while cyclization of the head group into an oxazoline ring diminishes affinity (Lin et al. 1998).

Modification of the Amide Group Replacement of the amido group by a thioamido group results in reduced affinity for CB_1. Thus, both thioanandamide and *R*-thiomethanandamide bind weakly to the receptor and show no significant biological activity (Lin et al. 1998). The SAR also indicates that the amide group must be secondary. Primary amides, e.g., arachidonamide, as well as tertiary amides, e.g., *N*-methylanandamide, do not bind to the CB_1 receptor (Lin et al. 1998; Pinto et al. 1994; Sheskin et al. 1997). Reversing the position of the carbonyl and the NH groups slightly decreases receptor affinity. These anandamides, designated as retroanandamides (e.g., **59**, Fig. 16), which were first developed by Makriyannis, exhibit exceptional stability with regard to hydrolysis by FAAH (Lin et al. 1998).

Replacement of the amido group by a carbamate group decreases affinity for CB_1. However, when the amido group is replaced by substituted ureas (**60**, Fig. 16) binding affinity as well as stability towards amidase hydrolysis is increased compared to anandamide (Ng et al. 1999).

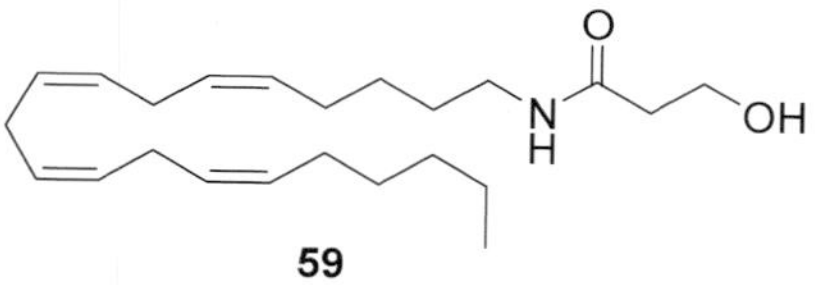

59
AM1174 Retroanandamide
Ki = 114 nM(CB1)
= 3540 nM (CB2)

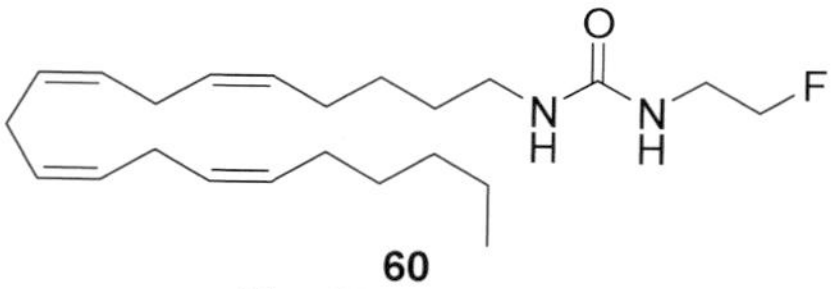

60
Ki = 55nM (CB1)

Fig. 16. Amide group modified analogs of anandamide

Importance of *cis*-Olefinic Bonds for Cannabimimetic Activity Drastic structural modifications of the arachidonyl component, such as complete saturation or replacement of the double bonds with triple bonds, result in complete loss of receptor affinity (Sheskin et al. 1997). Furthermore, ethanolamides of partially unsaturated fatty acids such as linoleic (two double bonds) and oleic (one double bond) acids exhibit considerably diminished affinity for CB_1 and cannabimimetic activity (Sheskin et al. 1997; Lin et al. 1998). From these results it can be argued that the presence of four *cis* olefinic bonds is optimal for activity. Prostaglandins and related analogs, which can be considered as conformationally rigid arachidonic acid analogs, do not bind to the CB_1 receptor (Pinto et al. 1994). Their inability to interact with the receptor may be due to the conformational restriction imposed by the five-member carbocyclic ring, which leads to preferred conformations that are incongruent with those of arachidonoylethanolamide and its analogs. It could also be due to the positions and stereochemistries of their hydroxyl and/or keto groups, which may destabilize their interactions with the receptor. Introduction of a methyl group or *gem*-dimethyl group at the C-2 position results in metabolically stable analogs with concomitant increase in CB_1 affinity as in the case of C-1′ methylation (Adams et al. 1995b; Goutopoulos et al. 2001)

***n*-Pentyl Group Tail Modifications** Although there is no apparent structural similarity between the classical cannabinoids and anandamide, there is considerable evidence suggesting that these two classes of cannabimimetic agents bind similarly to the CB_1 active site (Barnett-Norris et al. 2002; A. Makriyannis and C. Li, unpublished results). There is ample chemical and computational evidence indicating that arachidonic acid, the parent fatty acid of anandamide, favors a bent or looped conformation in which the carbonyl group is proximal to the C14–C15 olefinic bond. The chemical evidence for such a conformation includes the highly regiospecific intramolecular epoxidation of arachidonoyl peracid (Corey et al. 1984) and the facile macrolactonization of C20 hydroxyl methyl arachidonate (Corey et al. 1983). These experimental results are corroborated by molecular dynamics calculations (Rich 1993) that indicate that indeed a bent conformation is thermodynamically favorable. In the case of arachidonoylethanolamides, molecular modeling studies (Barnett-Norris et al. 1998, 2002; Rich 1993) have shown that anandamide and other fatty acid ethanolamides and esters also prefer a hairpin conformation. Additional data (Thomas et al. 1996; Tong et al. 1998) indicate that such a bent conformation is capable of mimicking the three-dimensional structure of tetrahydro- and hexahydrocannabinols.

However, it is unclear whether the hairpin conformation is also the conformation at the CB_1 receptor active site. Recent biophysical work on the conformational properties of anandamide in the membrane provide evidence for a more extended conformation for the C20 chain (A. Makriyannis and X. Tian, unpublished results) and suggest alternative CB_1 pharmacophoric conformations.

As discussed earlier, the SAR for the side chain of classical cannabinoids has been studied extensively, and it is known that a 1′,1′-dimethylheptyl (DMH) substituent generally leads to optimal potency. There is also evidence that classical cannabinoids and anandamides interact with similar residues at the CB_1 binding sites. This it was postulated that a similar substitution in anandamide should result

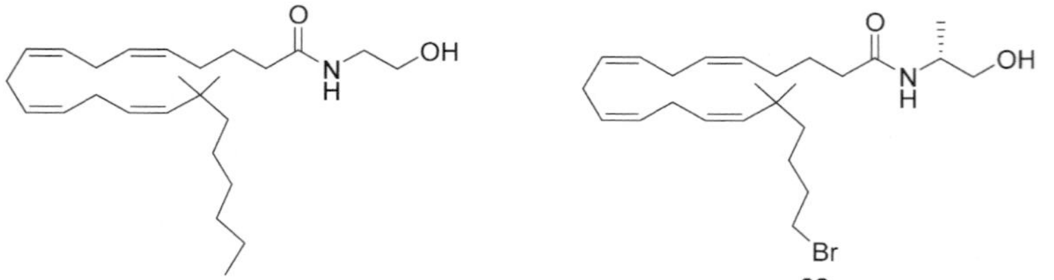

61
Ki = 1.0nM (CB1)with PMSF

62
O-1860
Ki = 2.2nM (CB1)
= >10,000nM (CB2)

Fig. 17. Tail modified analogs of anandamide

in an increase in receptor affinity and potency. To test the hypothesis, dimethylheptyl and other alkyl chain analogs of anandamide were synthesized and tested for their biological activities. As predicted, the dimethylheptyl analogs showed marked increases in receptor affinity and in vivo potency (**61**, Fig. 17) (Ryan et al. 1997; Seltzman et al. 1997; A. Makriyannis and J.K. Kawakami, unpublished results). Also, congruent with classical cannabinoid SAR, introduction of either bromo (**62**, Fig. 17) (Di Marzo et al. 2001) or cyano groups at the C-20 increases CB_1 affinity, whereas a hydroxyl group diminishes CB_1 affinity.

2.7 Other Cannabinergic Classes

A notable CB_1 receptor-selective antagonist that also exhibits inverse CB_1 receptor agonist properties in some assay systems is LY320135 (**63**, Fig. 18). This ligand was developed by Eli Lilly (Felder et al. 1998) and shares the ability of SR141716A to bind preferentially to CB_1. However, it has lower affinity for CB_1 than SR141716A and also binds to muscarinic and 5-HT_2 receptors at low micromolar concentrations (Felder et al. 1998). LY320135 also shares the ability of SR141716A to exhibit inverse agonist activity at some signal transduction pathways of the CB_1 receptor.

Aventis reported (Mignani et al. 2000) a new class of CB_1 receptor antagonists, which are represented by the diarylmethyleneazetidine analog **64** (Fig. 18). Very recently some novel 1,2,4-triazole derivatives were shown to behave as silent cannabinoid antagonists (Jagerovic et al. 2004). Although, these compounds bind

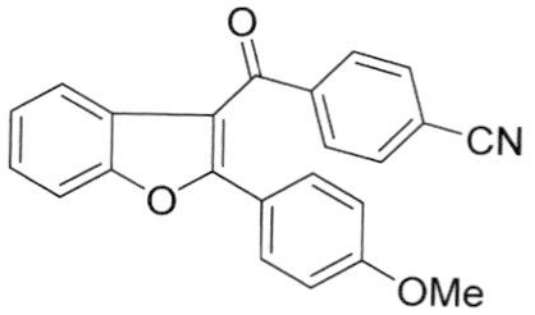

63
LY-320135
Ki = 141nM (CB1)
= 14900nM (CB2)

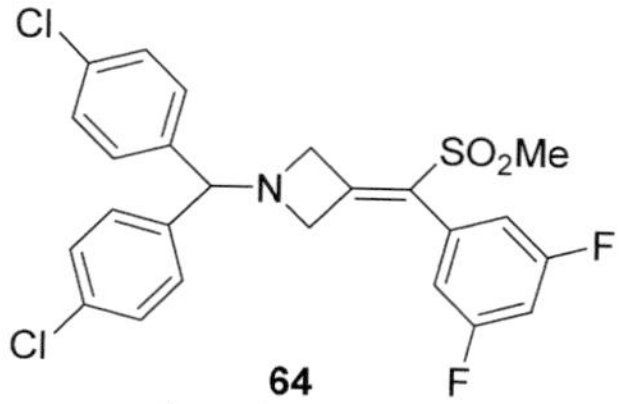

64
Aventis Pharma

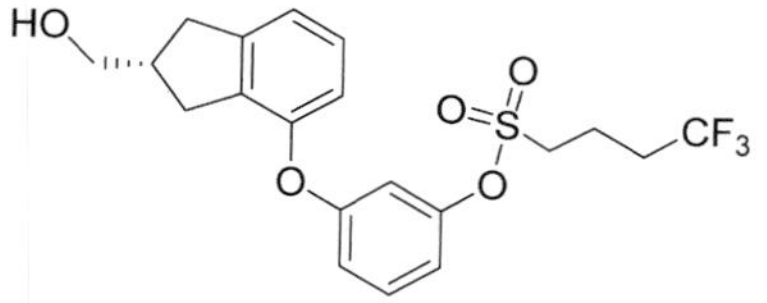

65
BAY 38-7271
Ki = 0.5 nM (CB1)
= 6.0 nM (CB2)

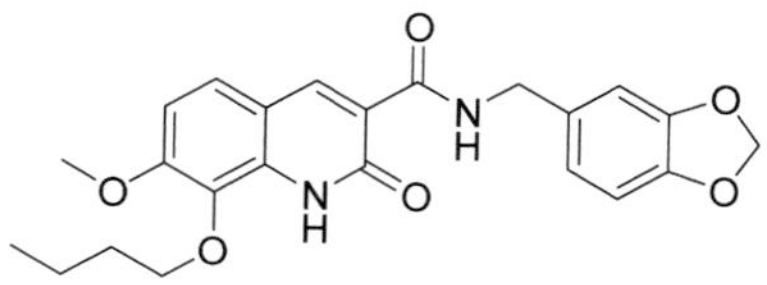

66
JTE-907
Ki = 2760 nM (CB1)
= 1.5 nM (CB2)

Fig. 18. Structurally novel cannabinergic ligands

to the CB_1 receptor with much reduced affinity compared to SR141716A, they exhibit similar antagonist efficacy in functional studies.

Recently, a novel class of diarylether sulfonyl ester cannabinoid agonists possessing neuroprotective properties was reported by Bayer AG (Wuppertal, Germany) (Mauler et al. 2002). The representative agonist, (–)-*R*-3-(2-hydroxy-methyl-indanyl-4-oxy)phenyl-4,4,4-trifluoro-1-sulfonate (**65**, BAY38-7271, Fig. 18), is a high-affinity CB_1 ligand ($K_i = 0.46–1.85$ nM; rat brain, human cortex, and recombinant human CB_1 receptor) (Mauler et al. 2003).

Researchers at Japan Tobacco (Osaka, Japan) reported the CB_2 selective inverse agonist JTE-907, whose structure is characterized by the presence of a carboxamide group in the 3-position of a quinolone nucleus (**66**, Fig. 18) (Iwamura et al. 2001) with anti-inflammatory in vivo activity. Naphthyridine derivatives sharing some structural features of JTE-907 were recently reported as cannabinoid receptor ligands with a preference for the CB_2 receptor (Ferrarini et al. 2004).

3 Covalent Binding Probes

Makriyannis and co-workers have developed several novel cannabinoid receptor affinity ligands (for recent reviews see Khanolkar et al. 2000; Palmer et al. 2002) that encompass reactive groups at judiciously chosen positions within the classical cannabinoid structure and can be used as probes for obtaining information on the receptor binding domain. Two types of reactive groups were incorporated: (1) electrophilic isothiocyanate group (NCS) that target nucleophilic amino acid residues such as lysine, histidine, and cysteine at or near the active site and (2) a photoactivatable aliphatic azido groups (N_3) capable of labeling the amino acid residues at the active site via a highly reactive nitrene intermediate. Both types of probes were shown to successfully label the cannabinoid receptors (Picone et al. 2002). The first photoaffinity label for the cannabinoid receptor, (–)-5′-azido-Δ^8-THC (**67**, Fig. 19) was reported in 1992 and was shown to covalently attach to CB_1 (Charalambous et al. 1992).

Second generation covalent probes carrying isothiocyanato or azido groups with improved affinities for both CB_1 and CB_2 were also reported and shown to label these receptors. The best known of these are (–)-11-hydroxy-7′-isothiocyanato-1′,1′-dimethylheptyl-Δ^8-THC (**68**, Fig. 19) and (–)-11-hydroxy-7′-azido-1′,1′-dimethylheptyl-Δ^8-THC (**69**, Fig. 19) (Yan et al. 1994).

A significant improvement in the design of these new probes was the introduction of a ^{125}I-substituent in the ligand without compromising its high receptor affinity (e.g., AM1708, **70**, Fig. 19) (Khanolkar et al. 2000; A.D. Khanolkar, G.A. Thakur, and A. Makriyannis, unpublished). These radio-iodinated probes have served as valuable tools for receptor purification and characterization of the CB_1 and CB_2 receptors (A. Makriyannis and W. Xu unpublished). Currently, a variety of mono- and bifunctional covalent ligands with hybrid cannabinoid structures (**71**, Fig. 19) (Chu et al. 2003), as well as endocannabinoid-like compounds (C. Li and A. Makriyannis, unpublished) are being used to elucidate the binding motifs

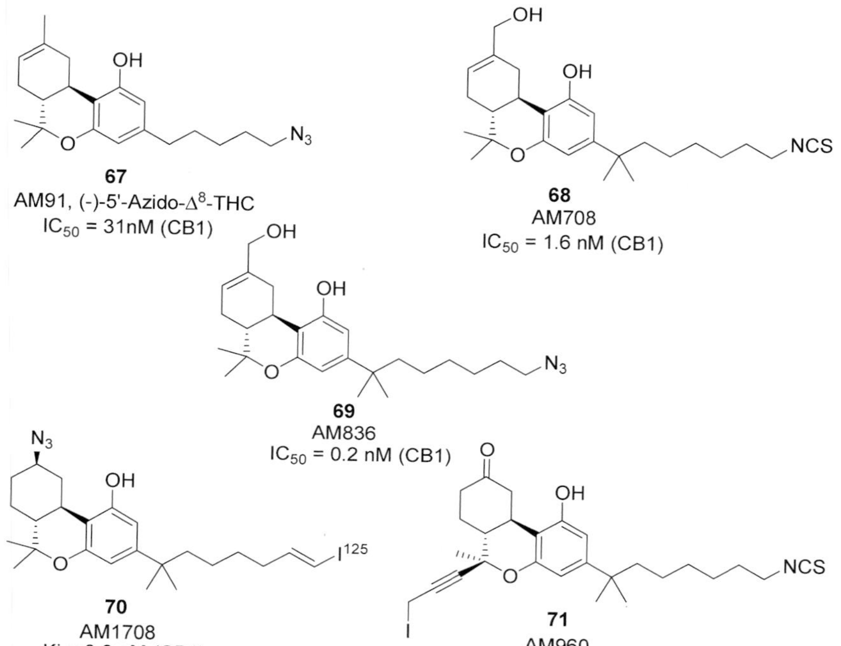

67 AM91, (-)-5'-Azido-Δ8-THC IC_{50} = 31nM (CB1)

68 AM708 IC_{50} = 1.6 nM (CB1)

69 AM836 IC_{50} = 0.2 nM (CB1)

70 AM1708 Ki = 0.8 nM (CB1) = 0.9 nM (CB2)

71 AM960 IC_{50} = 25 nM (CB1)

Fig. 19. Covalent probes for cannabinoid receptors

of the various classes of cannabinergics for the CB_1 and CB_2 receptors. This ligand-based approach in structural biology can serve as a useful avenue for studying the active sites of membrane-bound structural proteins that are not easily amenable to a crystallization approach.

4 Enantioselective Cannabinergic Ligands

Ligand enantioselectivity is often an important criterion in the characterization of drug receptors and in the development of biochemical and pharmacological assays. Thus, a highly enantioselective enantiomer can be a radioligand in a binding assay in which its much-less-potent enantiomer can be used to determine non-specific binding. Similarly, the less active enantiomer can serve as a control in in vitro or in vivo drug evaluations.

The cannabinergic ligand library includes a number of key enantiomeric pairs that have found substantial use in laboratories engaged in cannabinoid research. A careful examination of the literature reveals striking discrepancies in reported bioenantioselectivities. These are generally attributable to inadequate chiral resolution leading to a chirally impure enantiomer. Variation in enantioselectivity can

Table 1. Stereoselectivity ratios of cannabinergic ligands[a]

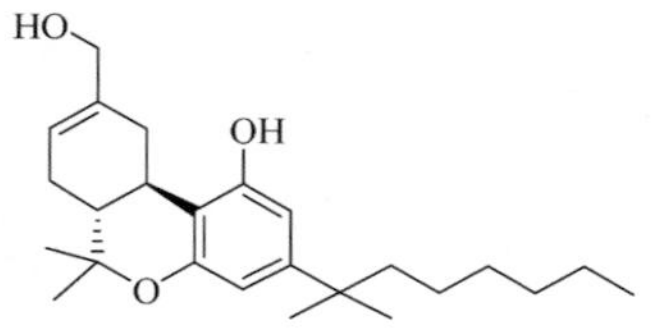

HU-210

	K_i (nM)	
	CB_1	CB_2
(6a*R*,10a*R*) (−)	0.7	0.2
(6a*S*,10a*S*) (+)	Does not bind significantly	

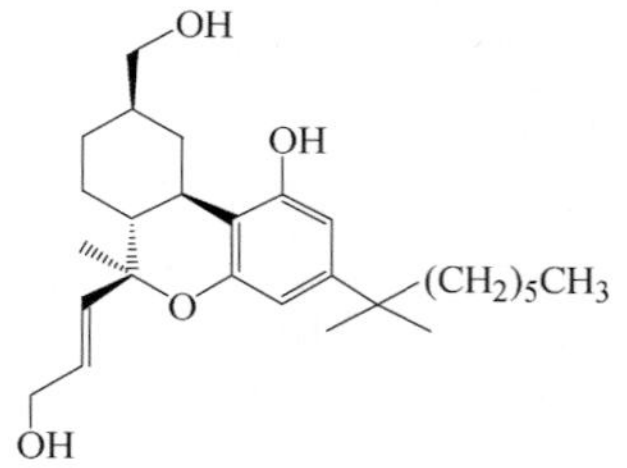

AM4030

	K_i (nM)	
	CB_1	CB_2
(6S,6aR,9R,10aR) (−)	0.6	1.1
(6R,6aS,9S,10aS) (+)	94.8	124.8

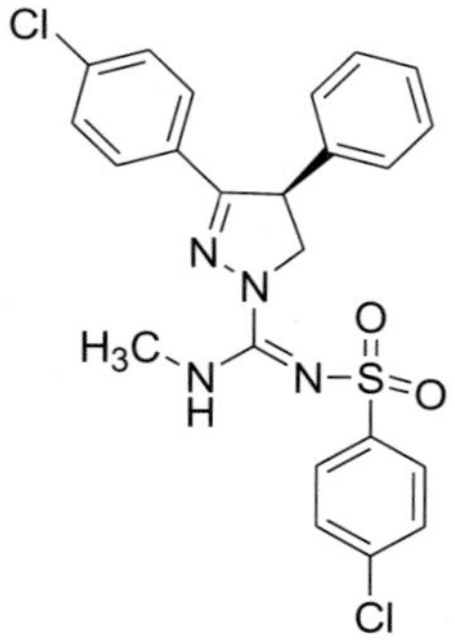

SLV-319

	K_i (nM)	
	CB_1	CB_2
4S(−)	7.8	7,943
4R(+)	894	>1000

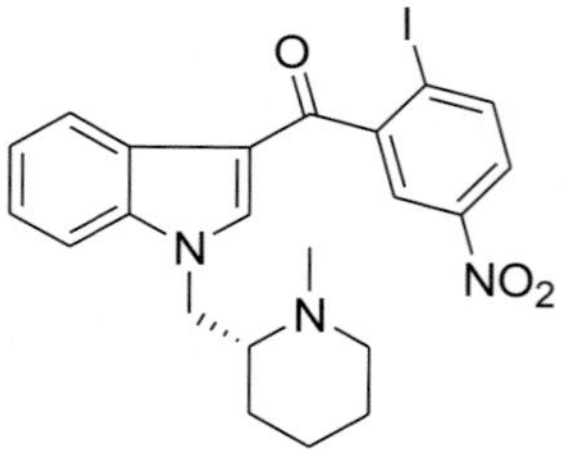

AM1241

	K_i (nM)	
	CB_1	CB_2
R (+)	139.7	1.4
S (−)	2049	160.5

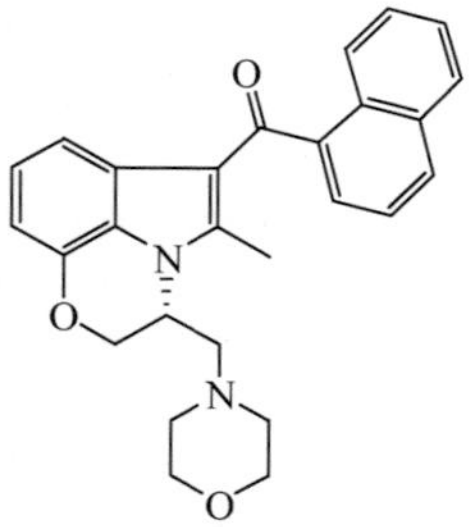

WIN-55,212-2

	K_i (nM)	
	CB_1	CB_2
R (+)	1.9	0.3
S (−)	6300	>1000

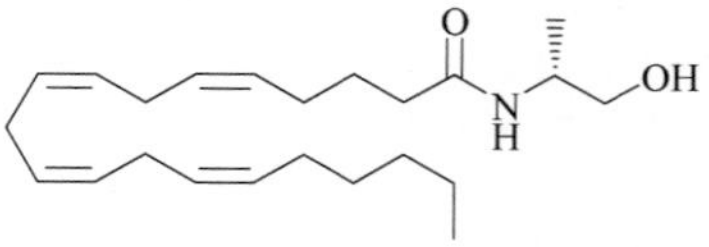

AM356

	K_i (nM)	
	CB_1	CB_2
R (+)	17.9	868
S (−)	309	8220

[a] The structures shown in this table represent the most active enantiomer.

be seen depending on the target protein or for the corresponding protein among different species, the CB_2 receptor being a case in point where the homology between the commonly used mouse spleen CB_2 preparation and that of expressed human receptor is only 82%. Discrepancies between in vitro and in vivo enantioselectivities may also be due to metabolic or bioavailability factors where the two enantiomers of a chiral ligand can be metabolized by the same enzyme but at different rates or exhibit different rates of uptake. Below we list some key chiral cannabinergic ligands currently used in cannabinoid research (Table 1).

(−)-Δ^9-THC, the active constituent of marijuana, which has a 6a*R*, 10a*R* stereochemistry, was found to be 5 to 100 times more potent than its synthetic (+)-enantiomer in producing static ataxia in dogs, depressing schedule-controlled responding in monkeys, and in producing hypothermia and inhibiting spontaneous activity in mice (Dewey et al. 1984; Martin et al. 1981). Similarly, Hollister and co-workers (Hollister et al. 1987) showed enantioselectivity of THC enantiomers in human studies using indices of the subjective experience, or "high," while May's group found enantioselectivity in a series of structurally modified Δ^9-THC analogs in tests of motor depression and analgesia (Wilson and May 1975; Wilson et al. 1976, 1979).

Pfizer's levonantradol (CP-50,556-1) is 30 times as potent as (−)-Δ^9-THC in several in vivo tests, whereas its (+)-enantiomer, dextronantradol (CP-53,870-1) is inactive (Little et al. 1988). (−)-CP-55,244 (NCCs with ACD ring) and (−)-CP-55,940 analogs are 30 to 2,000 times more potent than their respective (+)-enantiomers (Little et al. 1988).

(−)-Cannabidiol (CBD) is a non-psychotropic component of cannabis with possible therapeutic use as an anti-inflammatory drug. Recent studies on both enantiomers of CBD showed enantioselectivity in their interaction with cannabinoid and vanniloid (VR1) receptors as well as on the cellular uptake and enzymatic hydrolysis of anandamide (Bisogno et al. 2001).

HU210 [(−)-*R*,*R*-11-hydroxy-1′,1′-dimethylhepthyl-Δ^8-THC] is one of the most potent cannabinoids known. It acts through CB_1 and CB_2 receptors and is a potent inhibitor of forskolin-stimulated cyclic adenosine monophosphate (cAMP) production. Both the affinity and potency of HU210 are much higher than those of its synthetic (+)-*S*, *S*-enantiomer HU211 (also called dexanabinol). HU-211 is devoid of cannabinoid activity but has other interesting in vivo properties, including its action as an NMDA (*N*-methyl-D-aspartate) antagonist, antioxidant, and inhibitor of the synthesis of tumor-necrosis factor (TNF). It has found utility as a potential neuroprotective agent, and after favorable results in animal models (Shohami and Mechoulam 2000), it is now undergoing phase III clinical trials in Europe and Israel for traumatic brain injury (Knoller et al. 2002; Agranat et al. 2002).

The classical/non-classical cannabinoid hybrid AM4030 was resolved using chiral AD columns (Thakur et al. 2002).The (−)-isomer AM4030a has the (6*S*, 6a*R*, 9*R*, 10a*R*) stereochemistry and binds to CB_1 with subnanomolar affinity. The affinity of AM4030a was 158 times higher than that of its (+)-isomer AM4030b.

In the class of 3,4-diarylpyrazolines, SLV-319, the (−)-enantiomer, was found to bind to CB_1 with high affinity and selectivity ($CB_1 = 7.8$ nM, $CB_2 = 7{,}943$ nM) and ~100-fold higher potency than its (+)-isomer (Lange et al. 2004).

WIN-55,212-2, the (+)-enantiomer binds with high affinity to CB_1 (1.9 nM) and CB_2 (0.3 nM) whereas its (–)-isomer, WIN-55,212-3 does not bind significantly to CB_1 and CB_2 (both >1000 nM) (Pertwee 1997; Xie et al. 1995). The aminoalkylindole AM1241 exhibits high CB_2 selectivity (Ibrahim et al. 2003; Malan et al. 2001). Enantiomeric resolution of this ligand using chiral AD column gave the eutomer *R*-(+)-AM1241, which shows higher CB_2 affinity and selectivity ($CB_1 = 139.7$ nM; $CB_2 = 1.4$ nM) than *S*-(–)-AM1241 ($CB_1 = 2049$ nM; $CB_2 = 160.5$ nM). Recently, the asymmetric synthesis of *R*-(+)-AM1241 was carried out (A. Zvonok and A. Makriyannis, unpublished results).

AM356, *R*-(+) methanandamide, (Abadji et al. 1994; Lin et al. 1998) showed 4 times higher affinity ($CB_1 = 17.9$ nM) for CB_1 receptor than that of anandamide and 17 times higher than that of *S*-(–) methanandamide ($CB_1 = 309$ nM). Conversely, the *S*-enantiomer is a considerably better substrate of FAAH.

5 Present and Future

Currently, the field of cannabinoid research is at a very exciting phase. Understanding of the structural–activity relationships (SARs) of cannabinergic ligands has led to the development of highly selective and potent agonists, antagonists, and inverse agonists that in turn have assisted in the biochemical and pharmacological characterization of the cannabinoid receptors. These potent and selective compounds are now playing a major role in unraveling the physiological functions of the endocannabinoid system and the signaling mechanisms associated with it. Furthermore, some of these ligands are being evaluated for their potential therapeutic usefulness. In parallel with the above work, the binding motifs of the different classes of cannabinergic ligands are being elucidated with the help of receptor mutants and suitably designed high-affinity covalent binding probes.

Recent results describing the effects of some cannabinergic ligands in CB_1/CB_2 knockout mice suggest the presence of more cannabinoid-like receptors. One such receptor has been characterized pharmacologically in the vascular endothelium. The prospect of such novel cannabinoid or cannabinoid-like receptors offers excellent opportunities for future SAR work and the development of suitable probes for these new systems. Similarly, the recognition that the endocannabinoid system is closely linked biochemically to a number of key lipid modulators offers additional opportunities for the development of novel lipidomimetic ligand probes and potential therapeutic agents.

Acknowledgements. Supported by grants from National Institutes on Drug Abuse (DA9158, DA03801, and DA07215).

References

Abadji V, Lin S, Taha G, Griffin G, Stevenson LA, Pertwee RG, Makriyannis A (1994) (R)-Methanandamide: a chiral novel anandamide possessing higher potency and metabolic stability. J Med Chem 37:1889–1893

Abood ME, Ditto KE, Noel MA, Showalter VM, Tao Q (1997) Isolation and expression of a mouse CB1 cannabinoid receptor gene. Comparison of binding properties with those of native CB1 receptors in mouse brain and N18TG2 neuroblastoma cells. Biochem Pharmacol 53:207–214

Adams IB, Ryan W, Singer M, Razdan RK, Compton DR, Martin BR (1995a) Pharmacological and behavioral evaluation of alkylated anandamide analogs. Life Sci 56:2041–2048

Adams IB, Ryan W, Singer M, Thomas BF, Compton DR, Razdan RK, Martin BR (1995b) Evaluation of cannabinoid receptor binding and in vivo activities for anandamide analogs. J Pharmacol Exp Ther 273:1172–1181

Adams R, Harfenist M, Loewe S (1949) New analogs of tetrahydrocannabinol. XIX. J Am Chem Soc 71:1624–1628

Agranat I, Caner H, Caldwell J (2002) Putting chirality to work: the strategy of chiral switches. Nat Rev Drug Discov 1:753–768

Archer RA, Stark P, Lemberger L (1986) Nabilone. In: Mechoulam R (ed) Cannabinoids as therapeutic agents. CRC Press, Boca Raton, pp 85–103

Barnett-Norris J, Guarnieri F, Hurst DP, Reggio PH (1998) Exploration of biologically relevant conformations of anandamide, 2-arachidonylglycerol, and their analogues using conformational memories. J Med Chem 41:4861–4872

Barnett-Norris J, Hurst DP, Lynch DL, Guarnieri F, Makriyannis A, Reggio PH (2002) Conformational memories and the endocannabinoid binding site at the cannabinoid CB1 receptor. J Med Chem 45:3649–3659

Barth F, Rinaldi-Carmona M (1999) The development of cannabinoid antagonists. Curr Med Chem 6:745–755

Begg M, Mo FM, Offertaler L, Batkai S, Pacher P, Razdan RK, Lovinger DM, Kunos G (2003) G protein-coupled endothelial receptor for atypical cannabinoid ligands modulates a Ca2+-dependent K+ current. J Biol Chem 278:46188–46194

Bell MR, D'Ambra TE, Kumar V, Eissenstat MA, Herrmann JLJ, Wetzel JR, Rosi D, Philion RE, Daum SJ, Hlasta DJ, Kullnig RK, Ackerman JH, Haubrich BR, Luttinger DA, Baizman ER, Miller MS, Ward SJ (1991) Antinociceptive (aminoalkyl)indoles. J Med Chem 34:1099–1110

Beltramo M, Stella N, Calignano A, Lin SY, Makriyannis A, Piomelli D (1997) Functional role of high-affinity anandamide transport, as revealed by selective inhibition. Science 277:1094–1097

Bisogno T, Hanus L, De Petrocellis L, Tchilibon S, Ponde DE, Brandi I, Moriello AS, Davis JB, Mechoulam R, Di Marzo V (2001) Molecular targets for cannabidiol and its synthetic analogues: effect on vanilloid VR1 receptors and on the cellular uptake and enzymatic hydrolysis of anandamide. Br J Pharmacol 134:845–852

Breivogel CS, Childers SR (1998) The functional neuroanatomy of brain cannabinoid receptors. Neurobiol Dis 5:417–431

Burley SK, Petsko GA (1985) Aromatic-aromatic interaction: a mechanism of protein structure stabilization. Science 229:23–28

Busch-Petersen J, Hill WA, Fan P, Khanolkar A, Xie XQ, Tius MA, Makriyannis A (1996) Unsaturated side chain β-11-hydroxyhexahydrocannabinol analogs. J Med Chem 39:3790–3796

Chakrabarti A, Onaivi ES, Chaudhuri G (1995) Cloning and sequencing of a cDNA encoding the mouse brain-type cannabinoid receptor protein. DNA Seq 5:385–388

Charalambous A, Lin S, Marciniak G, Banijamali A, Friend FL, Compton DR, Martin BR, Makriyannis A (1991) Pharmacological evaluation of halogenated Δ8-THC analogs. Pharmacol Biochem Behav 40:509–512

Charalambous A, Yan G, Houston DB, Howlett AC, Compton DR, Martin BR, Makriyannis A (1992) 5′-Azido-Δ8-THC: a novel photoaffinity label for the cannabinoid receptor. J Med Chem 35:3076–3079
Chu C, Ramamurthy A, Makriyannis A, Tius MA (2003) Synthesis of covalent probes for the radiolabeling of the cannabinoid receptor. J Org Chem 68:55–61
Compton DR, Gold LH, Ward SJ, Balster RL, Martin BR (1992) Aminoalkylindole analogs: cannabimimetic activity of a class of compounds structurally distinct from Δ9-tetrahydrocannabinol. J Pharmacol Exp Ther 263:1118–1126
Corey EJ, Iguchi S, Albright J, De B (1983) Studies on the conformational mobility of arachidonic acid. Facile macrolactonization of 20-hydroxyarachidonic acid. Tetrahedron Lett 24:37–40
Corey EJ, Cashman JR, Kantner SS, Wright SW (1984) Rationally designed, potent competitive inhibitors of leukotriene biosynthesis. J Am Chem Soc 106:1503–1504
Crocker PJ, Saha B, Ryan WJ, Wiley JL, Martin BR, Ross RA, Pertwee RG, Razdan RK (1999) Development of agonists, partial agonists and antagonists in the Δ8-tetrahydrocannabinol series. Tetrahedron 55:13907–13926
D'Ambra TE, Estep KG, Bell MR, Eissenstat MA, Josef KA, Ward SJ, Haycock DA, Baizman ER, Casiano FM, Beglin NC, Chippari SM, Grego JD, Kullnig RK, Daley GT (1992) Conformationally restrained analogues of pravadoline: nanomolar potent, enantioselective, (aminoalkyl) indole agonists of the cannabinoid receptor. J Med Chem 35:124–135
D'Ambra TE, Eissenstat MA, Abt J, Ackerman JH, Bacon ER, Bell MR, Carabateas PM, Josef KA, Kumar V, Weaver JDI, Arnold R, Casiano FM, Chippari SM, Haycock DA, Kuster JE, Luttinger DA, Stevenson LA, Ward SJ, Hill WA, Khanolkar AD, Makriyannis A (1996) C-attached aminoalkylindoles: potent cannabinoid mimetics. Bioorg Med Chem Lett 6:17–22
Devane WA, Dysarz FA, Johnson RM, Melvin LS, Howlett AC (1988) Determination and characterization of a cannabinoid receptor in rat brain. Mol Pharmacol 34:605–613
Devane WA, Breuer A, Sheskin T, Jarbe TU, Eisen MS, Mechoulam R (1992a) A novel probe for the cannabinoid receptor. J Med Chem 35:2065–2069
Devane WA, Hanus L, Breuer A, Pertwee RG, Stevenson LA, Griffin G, Gibson D, Mandelbaum A, Etinger A, Mechoulam R (1992b) Isolation and structure of a brain constituent that binds to the cannabinoid receptor. Science 258:1946–1949
Dewey WL, Martin BR, May EL (1984) Cannabinoid stereoisomers: pharmacological effects. In: Smith DF (ed) CRC Handbook. Stereoisomers: drugs psychopharmacology. CRC Press, Boca Raton, pp 317–326
Di Marzo V (1998) 2-Arachidonoyl-glycerol as an "endocannabinoid": limelight for a formerly neglected metabolite. Biochemistry (Mosc) 63:13–21
Di Marzo V, Fontana A, Cadas H, Schinelli S, Cimino G, Schwartz JC, Piomelli D (1994) Formation and inactivation of endogenous cannabinoid anandamide in central neurons. Nature 372:686–691
Di Marzo V, Bisogno T, Sugiura T, Melck D, De Petrocellis L (1998) The novel endogenous cannabinoid 2-arachidonoylglycerol is inactivated by neuronal- and basophil-like cells: connections with anandamide. Biochem J 331:15–19
Di Marzo V, Bisogno T, De Petrocellis L, Brandi I, Jefferson RG, Winckler RL, Davis JB, Dasse O, Mahadevan A, Razdan RK, Martin BR (2001) Highly selective CB1 cannabinoid receptor ligands and novel CB1/VR1 vanilloid receptor "hybrid" ligands. Biochem Biophys Res Commun 281:444–451
Dinh TP, Carpenter D, Leslie FM, Freund TF, Katona I, Sensi SL, Kathuria S, Piomelli D (2002) Brain monoglyceride lipase participating in endocannabinoid inactivation. Proc Natl Acad Sci USA 99:10819–10824
Drake DJ, Jensen RS, Busch-Petersen J, Kawakami JK, Fernandez-Garcia MC, Fan P, Makriyannis A, Tius MA (1998) Classical/nonclassical hybrid cannabinoids: southern aliphatic chain-functionalized C-6β methyl, ethyl and propyl analogues. J Med Chem 41:3596–3608

Dutta AK, Ryan W, Thomas BF, Singer M, Compton DR, Martin BR, Razdan RK (1997) Synthesis, pharmacology, and molecular modeling of novel 4-alkyloxy indole derivatives related to cannabimimetic aminoalkyl indoles (AAIs). Bioorg Med Chem 5:1591–1600

Eissenstat MA, Bell MR, D'Ambra TE, Alexander EJ, Daum SJ, Ackerman JH, Gruett MD, Kumar V, Estep KG, Olefirowicz EM, Wetzel JR, Alexander EJ, Weaver JDI, Haycock DA, Luttinger DA, Casiano FM, Chippari SM, Kuster JE, Stevenson LA, Ward SJ (1995) Aminoalkylindoles: structure-activity relationships of novel cannabinoid mimetics. J Med Chem 38:3094–3105

Fegley D, Kathuria S, Mercier R, Li C, Goutopoulos A, Makriyannis A, Piomelli D (2004) Anandamide transport is independent of fatty-acid amide hydrolase activity and is blocked by the hydrolysis-resistant inhibitor AM1172. Proc Natl Acad Sci USA 101:8756–8761

Felder CC, Joyce KE, Briley EM, Glass M, Mackie KP, Fahey KJ, Cullinan GJ, Hunden DC, Johnson DW, Chaney MO, Koppel GA, Brownstein M (1998) LY320135, a novel cannabinoid CB1 receptor antagonist, unmasks coupling of the CB1 receptor to stimulation of cAMP accumulation. J Pharmacol Exp Ther 284:291–297

Ferrarini PL, Calderone V, Cavallini T, Manera C, Saccomanni G, Pani L, Ruiu S, Gessa GL (2004) Synthesis and biological evaluation of 1,8-naphthyridin-4(1H)-on-3-carboxamide derivatives as new ligands of cannabinoid receptors. Bioorg Med Chem 12:1921–1933

Francisco MEY, Seltzman HH, Gilliam AF, Mitchell RA, Rider SL, Pertwee RG, Stevenson LA, Thomas BF (2002) Synthesis and structure-activity relationships of amide and hydrazide analogues of the cannabinoid CB1 receptor antagonist N-(piperidinyl)-5-(4-chlorophenyl)-1-(2,4-dichlorophenyl)-4-methyl-1H-pyrazole-3-carboxamide (SR141716). J Med Chem 45:2708–2719

Gaetani S, Cuomo V, Piomelli D (2003) Anandamide hydrolysis: a new target for anti-anxiety drugs? Trends Mol Med 9:474–478

Gaoni Y, Mechoulam R (1964) Hashish. III. Isolation, structure, and partial synthesis of an active constituent of hashish. J Am Chem Soc 86:1646–1647

Gareau Y, Dufresne C, Gallant M, Rochette C, Sawyer N, Slipetz DM, Tremblay N, Weech PK, Metters KM, Labelle M (1996) Structure activity relationships of tetrahydrocannabinol analogs on human cannabinoid receptors. Bioorg Med Chem Lett 6:189–194

Gatley SJ, Lan R, Pyatt B, Gifford AN, Volkow ND, Makriyannis A (1997) Binding of the non-classical cannabinoid CP-55,940, and the diarylpyrazole AM251 to rodent brain cannabinoid receptors. Life Sci 61:L191–197

Gatley SJ, Lan R, Volkow ND, Pappas N, King P, Wong CT, Gifford AN, Pyatt B, Dewey SL, Makriyannis A (1998) Imaging the brain marijuana receptor: development of a radioligand that binds to cannabinoid CB1 receptors in vivo. J Neurochem 70:417–423

Gerard C, Mollereau C, Vassart G, Parmentier M (1990) Nucleotide sequence of a human cannabinoid receptor cDNA. Nucleic Acids Res 18:7142

Gerard CM, Mollereau C, Vassart G, Parmentier M (1991) Molecular cloning of a human brain cannabinoid receptor which is also expressed in testis. Biochem J 279:129–134

Gifford AN, Tang Y, Gatley SJ, Volkow ND, Lan R, Makriyannis A (1997) Effect of the cannabinoid receptor SPECT agent, AM 281, on hippocampal acetylcholine release from rat brain slices. Neurosci Lett 238:84–86

Goutopoulos A, Fan P, Khanolkar AD, Xie XQ, Lin S, Makriyannis A (2001) Stereochemical selectivity of methanandamides for the CB1 and CB2 cannabinoid receptors and their metabolic stability. Bioorg Med Chem 9:1673–1684

Goutopoulos A, Makriyannis A (2002) From cannabis to cannabinergics new therapeutic opportunities. Pharmacol Ther 95:103–117

Hajos N, Freund TF (2002) Pharmacological separation of cannabinoid sensitive receptors on hippocampal excitatory and inhibitory fibers. Neuropharmacology 43:503–510

Hanus L, Gopher A, Almog S, Mechoulam R (1993) Two new unsaturated fatty acid ethanolamides in brain that bind to the cannabinoid receptor. J Med Chem 36:3032–3034

Hanus L, Breuer A, Tchilibon S, Shiloah S, Goldenberg D, Horowitz M, Pertwee RG, Ross RA, Mechoulam R, Fride E (1999) HU-308: a specific agonist for CB(2), a peripheral cannabinoid receptor. Proc Natl Acad Sci USA 96:14228–14233

Hanus L, Abu-Lafi S, Fride E, Breuer A, Vogel Z, Shalev DE, Kustanovich I, Mechoulam R (2001) 2-Arachidonyl glyceryl ether, an endogenous agonist of the cannabinoid CB1 receptor. Proc Natl Acad Sci U S A 98:3662–3665

Harrington PE, Stergiades IA, Erickson J, Makriyannis A, Tius MA (2000) Synthesis of functionalized cannabinoids. J Org Chem 65:6576–6582

Herkenham M (1991) Characterization and localization of cannabinoid receptors in brain: an in vitro technique using slide-mounted tissue sections. NIDA Res Monogr 112:129–145

Herkenham M, Lynn AB, Little MD, Johnson MR, Melvin LS, de Costa BR, Rice KC (1990) Cannabinoid receptor localization in brain. Proc Natl Acad Sci USA 87:1932–1936

Hillard CJ, Edgemond WS, Jarrahian A, Campbell WB (1997) Accumulation of N-arachidonoylethanolamine (anandamide) into cerebellar granule cells occurs via facilitated diffusion. J Neurochem 69:631–638

Hollister LE, Gillespie HK, Mechoulam R, Srebnik M (1987) Human pharmacology of 1S and 1R enantiomers of Δ-3-tetrahydrocannabinol. Psychopharmacology (Berl) 92:505–507

Hosohata K, Quock RM, Hosohata Y, Burkey TH, Makriyannis A, Consroe P, Roeske WR, Yamamura HI (1997a) AM630 is a competitive cannabinoid receptor antagonist in the guinea pig brain. Life Sci 61:PL115–PL118

Hosohata Y, Quock RM, Hosohata K, Makriyannis A, Consroe P, Roeske WR, Yamamura HI (1997b) AM630 antagonism of cannabinoid-stimulated [35S]GTPgS binding in the mouse brain. Eur J Pharmacol 321:R1–R3

Howlett AC, Wilken GH, Pigg JJ, Houston DB, Lan R, Liu Q, Makriyannis A (2000) Azido- and isothiocyanato-substituted aryl pyrazoles bind covalently to the CB1 cannabinoid receptor and impair signal transduction. J Neurochem 74:2174–2181

Howlett AC, Barth F, Bonner TI, Cabral G, Casellas P, Devane WA, Felder CC, Herkenham M, Mackie K, Martin BR, Mechoulam R, Pertwee RG (2002) International Union of Pharmacology. XXVII. Classification of cannabinoid receptors. Pharmacol Rev 54:161–202

Huffmann JW, Dai D, Martin BR, Compton DR (1994) Design, synthesis and pharmacology of cannabimimetic indoles. Bioorg Med Chem Lett 4:563–566

Huffman JW, Yu S, Showalter V, Abood ME, Wiley JL, Compton DR, Martin BR, Bramblett RD, Reggio PH (1996) Synthesis and pharmacology of a very potent cannabinoid lacking a phenolic hydroxyl with high affinity for the CB2 receptor. J Med Chem 39:3875–3877

Huffman JW, Liddle J, Yu S, Aung MM, Abood ME, Wiley JL, Martin BR (1999) 3-(1′,1′-Dimethylbutyl)-1-deoxy-Δ8-THC and related compounds: synthesis of selective ligands for the CB2 receptor. Bioorg Med Chem 7:2905–2914

Huffman JW, Bushell SM, Miller JRA, Wiley JL, Martin BR (2002) 1-Methoxy-, 1-deoxy-11-hydroxy- and 11-hydroxy-1-methoxy-Δ8-tetrahydrocannabinols: new selective ligands for the CB2 receptor. Bioorg Med Chem 10:4119–4129

Huffman JW, Mabon R, Wu M-J, Lu J, Hart R, Hurst DP, Reggio PH, Wiley JL, Martin BR (2003a) 3-Indolyl-1-naphthylmethanes: new cannabimimetic indoles provide evidence for aromatic stacking interactions with the CB1 cannabinoid receptor. Bioorg Med Chem 11:539–549

Huffman JW, Miller JRA, Liddle J, Yu S, Thomas BF, Wiley JL, Martin BR (2003b) Structure-activity relationships for 1′,1′-dimethylalkyl-Δ8-tetrahydrocannabinols. Bioorg Med Chem 11:1397–1410

Hynes JJ, Leftheris K, Wu H, Pandit CR, Chen P, Norris DJ, Chen BC, Zhao R, Kiener PA, Chen X, Turk LA, Patil-koota V, Gillooly KM, Shuster DJ, McIntyre KW (2002) C-3 Amido-indole cannabinoid receptor modulators. Bioorg Med Chem Lett 12:2399–2402

Ibrahim MM, Deng H, Zvonok A, Cockayne DA, Kwan J, Mata HP, Vanderah TW, Lai J, Porreca F, Makriyannis A, Malan TPJ (2003) Activation of CB2 cannabinoid receptors by AM1241 inhibits experimental neuropathic pain: pain inhibition by receptors not present in the CNS. Proc Natl Acad Sci U S A 100:10529–10533

Iwamura H, Suzuki H, Ueda Y, Kaya T, Inaba T (2001) In vitro and in vivo pharmacological characterization of JTE-907, a novel selective ligand for cannabinoid CB2 receptor. J Pharmacol Exp Ther 296:420–425

Jagerovic N, Hernandez-Folgado L, Alkorta I, Goya P, Navarro M, Serrano A, Rodriguez de Fonseca F, Dannert MT, Alsasua A, Suardiaz M, Pascual D, Martin MI (2004) Discovery of 5-(4-chlorophenyl)-1-(2,4-dichlorophenyl)-3-hexyl-1H-1,2,4-triazole, a novel in vivo cannabinoid antagonist containing a 1,2,4-triazole motif. J Med Chem 47:2939–2942

Johnson MR, Melvin LS (1986) The discovery of non-classical cannabinoid analgesics. In: Mechoulam R (ed) Cannabinoids as therapeutic agents. CRC Press, Boca Raton, pp 121–145

Katoch-Rouse R, Pavlova OA, Caulder T, Hoffman AF, Mukhin AG, Horti AG (2003) Synthesis, structure-activity relationship, and evaluation of SR141716 analogues: development of central cannabinoid receptor ligands with lower lipophilicity. J Med Chem 46:642–645

Khanolkar AD, Abadji V, Lin S, Hill WA, Taha G, Abouzid K, Meng Z, Fan P, Makriyannis A (1996) Head group analogs of arachidonylethanolamide, the endogenous cannabinoid ligand. J Med Chem 39:4515–4519

Khanolkar AD, Makriyannis A (1999) Structure-activity relationships of anandamide, an endogenous cannabinoid ligand. Life Sci 65:607–616

Khanolkar AD, Lu D, Fan P, Tian X, Makriyannis A (1999) Novel conformationally restricted tetracyclic analogs of Δ8-tetrahydrocannabinol. Bioorg Med Chem Lett 9:2119–2124

Khanolkar AD, Palmer SL, Makriyannis A (2000) Molecular probes for the cannabinoid receptors. Chem Phys Lipids 108:37–52

Knoller N, Levi L, Shoshan I, Reichenthal E, Razon N, Rappaport ZH, Biegon A (2002) Dexanabinol (HU-211) in the treatment of severe closed head injury: a randomized, placebo-controlled, phase II clinical trial. Crit Care Med 30:548–554

Krishnamurthy M, Ferreira AM, Moore BM (2003) Synthesis and testing of novel phenyl substituted side-chain analogues of classical cannabinoids. Bioorg Med Chem Lett 13:3487–3490

Krishnamurthy M, Li W, Moore BM (2004) Synthesis, biological evaluation, and structural studies on N1 and C5 substituted cycloalkyl analogues of the pyrazole class of CB1 and CB2 ligands. Bioorg Med Chem 12:393–404

Kriwacki RW, Makriyannis A (1989) The conformational analysis of Δ9- and Δ9,11-tetrahydrocannabinols in solution using high resolution nuclear magnetic resonance spectroscopy. Mol Pharmacol 35:495–503

Kumar V, Alexander MD, Bell MR, Eissenstat MA, Casiano FM, Chippari SM, Haycock DA, Luttinger DA, Kuster JE, Miller MS, Stevenson LA, Ward SJ (1995) Morpholinoalkylindenes as antinociceptive agents: novel cannabinoid receptor agonists. Bioorg Med Chem Lett 5:381–386

Lan R, Liu Q, Fan P, Lin S, Fernando SR, McCallion D, Pertwee RG, Makriyannis A (1999) Structure-activity relationship of pyrazole derivatives as cannabinoid receptor antagonists. J Med Chem 42:776–779

Lang W, Qin C, Lin S, Khanolkar AD, Goutopoulos A, Fan P, Abouzid K, Meng Z, Biegel D, Makriyannis A (1999) Substrate specificity and stereoselectivity of rat brain microsomal anandamide amidohydrolase. J Med Chem 42:896–902

Lange JHM, Coolen HKAC, Van Stuivenberg HH, Dijksman JAR, Herremans AHJ, Ronken E, Keizer HG, Tipker K, McCreary AC, Veerman W, Wals HC, Stork B, Verveer PC, den Hartog AP, de Jong NMJ, Adolfs TJP, Hoogendoorn J, Kruse CG (2004) Synthesis, biological properties, and molecular modeling investigations of novel 3,4-diarylpyrazolines as potent and selective CB1 cannabinoid receptor antagonists. J Med Chem 47:627–643

Lee CM, Michaels RJ, Zaugg HE, Dren AT, Plotnikoff NP, Young PR (1977) Cannabinoids. Synthesis and central nervous system activity of 8-substituted 10-hydroxy-5,5-dimethyl-5H-[1]benzopyrano[4,3-c]pyridine and derivatives. J Med Chem 20:1508–1511

Lee CM, Zaugg HE, Michaels RJ, Dren AT, Plotnikoff NP, Young PR (1983) New azacannabinoids highly active in the central nervous system. J Med Chem 26:278–280
Liddle J, Huffman JW (2001) Enantioselective synthesis of 11-hydroxy-(1′S,2′R)-dimethylheptyl-Δ8-THC, a very potent CB1 agonist. Tetrahedron 57:7607–7612
Lin S, Khanolkar AD, Fan P, Goutopoulos A, Qin C, Papahadjis D, Makriyannis A (1998) Novel analogues of arachidonylethanolamide (anandamide): affinities for the CB1 and CB2 cannabinoid receptors and metabolic stability. J Med Chem 41:5353–5361
Little PJ, Compton DR, Johnson MR, Melvin LS, Martin BR (1988) Pharmacology and stereoselectivity of structurally novel cannabinoids in mice. J Pharmacol Exp Ther 247:1046–1051
Mahadevan A, Siegel C, Martin BR, Abood ME, Beletskaya I, Razdan RK (2000) Novel cannabinol probes for CB1 and CB2 cannabinoid receptors. J Med Chem 43:3778–3785
Mailleux P, Parmentier M, Vanderhaeghen JJ (1992) Distribution of cannabinoid receptor messenger RNA in the human brain: an in situ hybridization histochemistry with oligonucleotides. Neurosci Lett 143:200–204
Makriyannis A, Rapaka RS (1990) The molecular basis of cannabinoid activity. Life Sci 47:2173–2184
Makriyannis A, Goutopoulos A (2004) Cannabinergics: old and new therapeutic possibilities. In: Makriyannis A, Biegel D (eds) Drug discovery strategies and methods. Marcel Dekker, New York, pp 89–128
Malan TPJ, Ibrahim MM, Deng H, Liu Q, Mata HP, Vanderah T, Porreca F, Makriyannis A (2001) CB2 cannabinoid receptor-mediated peripheral antinociception. Pain 93:239–245
Martin BR, Dewey WL, Harris LS, Beckner J (1975) Marihuana-like activity of new synthetic tetrahydrocannabinols. Pharmacol Biochem Behav 3:849–853
Martin BR, Balster RL, Razdan RK, Harris LS, Dewey WL (1981) Behavioral comparisons of the stereoisomers of tetrahydrocannabinols. Life Sci 29:565–574
Martin BR, Compton DR, Thomas BF, Prescott WR, Little PJ, Razdan RK, Johnson MR, Melvin LS, Mechoulam R, Ward SJ (1991) Behavioral, biochemical, and molecular modeling evaluations of cannabinoid analogs. Pharmacol Biochem Behav 40:471–478
Martin BR, Compton DR, Semus SF, Lin S, Marciniak G, Grzybowska J, Charalambous A, Makriyannis A (1993) Pharmacological evaluation of iodo and nitro analogs of Δ8-THC and Δ9-THC. Pharmacol Biochem Behav 46:295–301
Martin BR, Jefferson RG, Winckler R, Wiley JL, Thomas BF, Crocker PJ, Williams W, Razdan RK (2002) Assessment of structural commonality between tetrahydrocannabinol and anandamide. Eur J Pharmacol 435:35–42
Matsuda LA, Lolait SJ, Brownstein MJ, Young AC, Bonner TI (1990) Structure of a cannabinoid receptor and functional expression of the cloned cDNA. Nature 346:561–564
Matsuda LA, Bonner TI, Lolait SJ (1993) Localization of cannabinoid receptor mRNA in rat brain. J Comp Neurol 327:535–550
Matsumoto K, Stark P, Meister RG (1977a) Cannabinoids. 1. 1-Amino- and 1-mercapto-7,8,9,10-tetrahydro-6H-dibenzo [b,d]pyrans. J Med Chem 20:17–24
Matsumoto K, Stark P, Meister RG (1977b) Synthesis and central nervous system activities of some B-ring homocannabinoid derivatives and related lactones. J Med Chem 20:25–30
Mauler F, Mittendorf J, Horvath E, De Vry J (2002) Characterization of the diarylether sulfonylester (−)-(R)-3-(2-hydroxymethylindanyl-4-oxy)phenyl-4,4,4-trifluoro-1-sulfonate (BAY 38-7271) as a potent cannabinoid receptor agonist with neuroprotective properties. J Pharmacol Exp Ther 302:359–368
Mauler F, Horvath E, de Vry J, Jaeger R, Schwarz T, Sandmann S, Weinz C, Heinig R, Boettcher M (2003) BAY 38-7271: a novel highly selective and highly potent cannabinoid receptor agonist for the treatment of traumatic brain injury. CNS Drug Rev 9:343–358
McPartland JM, Glass M (2003) Functional mapping of cannabinoid receptor homologs in mammals, other vertebrates, and invertebrates. Gene 312:297–303
Mechoulam R, Lander N, Srebnik M, Breuer A, Segal M, Feigenbaum JJ, Jarbe TU, Consroe P (1987) Stereochemical requirements for cannabimimetic activity. NIDA Res Monogr 79:15–30

Mechoulam R, Feigenbaum JJ, Lander N, Segal M, Jarbe TU, Hiltunen AJ, Consroe P (1988) Enantiomeric cannabinoids: stereospecificity of psychotropic activity. Experientia 44:762–764

Mechoulam R, Ben-Shabat S, Hanus L, Ligumsky M, Kaminski NE, Schatz AR, Gopher A, Almog S, Martain BR, Comton DR (1995) Identification of an endogenous 2-monoglyceride, present in canine gut, that binds to cannabinoid receptors. Biochem Pharmacol 50:83–90

Mechoulam R, Ben Shabat S, Hanus L, Fride E, Vogel Z, Bayewitch M, Sulcova AE (1996) Endogenous cannabinoid ligands-chemical and biological studies. J Lipid Mediat Cell Signal 14:45–49

Mechoulam R, Devane WA, Glaser R (1999) Cannabinoid geometry and biological activity. In: Nahas GG, Sutin KM, Agurell S (eds) Marijuana and medicine. Humana Press, Totowa, pp 65–90

Melvin LS, Milne GM, Johnson MR, Wilken GH, Howlett AC (1995) Structure-activity relationships defining the ACD-tricyclic cannabinoids: cannabinoid receptor binding and analgesic activity. Drug Des Discov 13:155–166

Mignani S, Hittinger A, Achard D, Bouchard H, Bouquerel J, Capet M, Grisoni S, Malleron Jl (2000) Preparation of 1-bis(aryl)methyl-3-(alkylsulfonyl)arylmethyleneazetidines as cannabinoid CB1 receptor antagonists. USA Patent No. WO 0015609, pp 1–239

Munro S, Thomas KL, Abu-Shaar M (1993) Molecular characterization of a peripheral receptor for cannabinoids. Nature 365:61–65

Mussinu JM, Ruiu S, Mule AC, Pau A, Carai MAM, Loriga G, Murineddu G, Pinna GA (2003) Tricyclic pyrazoles. Part 1: Synthesis and biological evaluation of novel 1,4-dihydroindeno[1,2-c]pyrazol-based ligands for CB1and CB2 cannabinoid receptors. Bioorg Med Chem 11:251–263

Nadipuram AK, Krishnamurthy M, Ferreira AM, Li W, Moore BM (2003) Synthesis and testing of novel classical cannabinoids: exploring the side chain ligand binding pocket of the CB1 and CB2 receptors. Bioorg Med Chem Lett 11:3121–3132

Nakamura-Palacios EM, Moerschbaecher JM, Barker LA (1999) The pharmacology of SR 141716A: a review. CNS Drug Rev 5:43–58

Ng EW, Aung MM, Abood ME, Martin BR, Razdan RK (1999) Unique analogues of anandamide: arachidonyl ethers and carbamates and norarachidonyl carbamates and ureas. J Med Chem 42:1975–1981

Nikas SP, Grzybovska J, Papahatjis DP, Charalambous A, Banijamali AR, Chari R, Fan P, Kourouli T, Lin S, Nitowski AJ, Marciniak G, Guo Y, Li X, Wang C-LJ, Makriyannis A (2004) The role of halogen substitution in classical cannabinoids: a CB1 pharmacophore model. AAPS Journal 6(4): Article 30 (http://www.aapsj.org)

Oka S, Tsuchie A, Tokumura A, Muramatsu M, Suhara Y, Takayama H, Waku K, Sugiura T (2003) Ether-linked analogue of 2-arachidonoylglycerol (noladin ether) was not detected in the brains of various mammalian species. J Neurochem 85:1374–1381

Ooms F, Wouters J, Oscari O, Happaerts T, Bouchard G, Carrupt PA, Testa B, Lambert DM (2002) Exploration of the pharmacophore of 3-alkyl-5-arylimidazolidinediones as new CB1 cannabinoid receptor ligands and potential antagonists: synthesis, lipophilicity, affinity, and molecular modeling. J Med Chem 45:1748–1756

Osgood PF, Howes JF, Razdan RK, Pars HG (1978) Drugs derived from cannabinoids. 7. Tachycardia and analgesia structure-activity relationships in Δ9-tetrahydrocannabinol and some synthetic analogues. J Med Chem 21:809–811

Palmer SL, Khanolkar AD, Makriyannis A (2000) Natural and synthetic endocannabinoids and their structure-activity relationships. Curr Pharm Des 6:1381–1397

Palmer SL, Thakur GA, Makriyannis A (2002) Cannabinergic ligands. Chem Phys Lipids 121:3–19

Papahatjis D, Kourouli T, Makriyannis A (1996) Pharmacophoric requirements for cannabinoid side-chains. Naphthoyl and naphthylmethyl substituted Δ8-tetrahydro-cannabinol analogs. J Heterocycl Chem 33:559–562

Papahatjis DP, Kourouli T, Abadji V, Goutopoulos A, Makriyannis A (1998) Pharmacophoric requirements for cannabinoid side chains: multiple bond and C1′-substituted Δ8-tetrahydrocannabinols. J Med Chem 41:1195–1200

Papahatjis DP, Nikas S, Tsotinis A, Vlachou M, Makriyannis A (2001) A new ring-forming methodology for the synthesis of conformationally constrained bioactive molecules. Chem Lett 3:192–193

Papahatjis DP, Nikas SP, Andreou T, Makriyannis A (2002) Novel 1′,1′-chain substituted Δ8-tetrahydrocannabinols. Bioorg Med Chem Lett 12:3583–3586

Papahatjis DP, Nikas SP, Kourouli T, Chari R, Xu W, Pertwee RG, Makriyannis A (2003) Pharmacophoric requirements for the cannabinoid side chain. Probing the cannabinoid receptor subsite at C1′. J Med Chem 46:3221–3229

Paria BC, Ma W, Andrenyak DM, Schmid PC, Schmid HHO, Moody DE, Deng H, Makriyannis A, Dey SK (1998) Effects of cannabinoids on preimplantation mouse embryo development and implantation are mediated by brain-type cannabinoid receptors. Biol Reprod 58:1490–1495

Pars HG, Granchelli FE, Razdan RK, Keller JK, Teiger DG, Rosenberg FJ, Harris LS (1976) Drugs derived from cannabinoids. 1. Nitrogen analogs, benzopyranopyridines and benzopyranopyrroles. J Med Chem 19:445–454

Pertwee R, Griffin G, Fernando S, Li X, Hill A, Makriyannis A (1995) AM630, a competitive cannabinoid antagonist. Life Sci 56:1949–1955

Pertwee RG (1997) Pharmacology of cannabinoid CB1 and CB2 receptors. Pharmacol Ther 74:129–180

Picone RP, Fournier DJ, Makriyannis A (2002) Ligand based structural studies of the CB1 cannabinoid receptor. J Pept Res 60:348–356

Pinto JC, Potie F, Rice KC, Boring D, Johnson MR, Evans DM, Wilken GH, Cantrell CH, Howlett AC (1994) Cannabinoid receptor binding and agonist activity of amides and esters of arachidonic acid. Mol Pharmacol 46:516–522

Piomelli D, Beltramo M, Glasnapp S, Lin SY, Goutopoulos A, Xie X-Q, Makriyannis A (1999) Structural determinants for recognition and translocation by the anandamide transporter. Proc Natl Acad Sci U S A 96:5802–5807

Priller J, Briley EM, Mansouri J, Devane WA, Mackie K, Felder CC (1995) Mead ethanolamide, a novel eicosanoid, is an agonist for the central (CB1) and peripheral (CB2) cannabinoid receptors. Mol Pharmacol 48:288–292

Razdan RK (1986) Structure-activity relationships in cannabinoids. Pharmacol Rev 38:75–149

Razdan RK, Mahadevan A (2002) Recent advances in the synthesis of endocannabinoid related ligands. Chem Phys Lipids 121:21–33

Reggio PH (1999) Ligand-ligand and ligand-receptor approaches to modeling the cannabinoid CB1 and CB2 receptors: achievements and challenges. Curr Med Chem 6:665–683

Reggio PH (2002) Endocannabinoid structure-activity relationships for interaction at the cannabinoid receptors. Prostaglandins Leukot Essent Fatty Acids 66:143–160

Reggio PH, Basu-Dutt S, Barnett-Norris J, Castro MT, Hurst DP, Seltzman HH, Roche MJ, Gilliam AF, Thomas BF, Stevenson LA, Pertwee RG, Abood ME (1998) The bioactive conformation of aminoalkylindoles at the cannabinoid CB1 and CB2 receptors: insights gained from (E)- and (Z)-naphthylidene indenes. J Med Chem 41:5177–5187

Reggio PH, Greer KV, Cox SM (1989) The importance of the orientation of the C9 substituent to cannabinoid activity. J Med Chem 32:1630–1635

Rhee MH, Vogel Z, Barg J, Bayewitch M, Levy R, Hanus L, Breuer A, Mechoulam R (1997) Cannabinol derivatives: binding to cannabinoid receptors and inhibition of adenylylcyclase. J Med Chem 40:3228–3233

Rhee MH, Nevo I, Bayewitch ML, Zagoory O, Vogel Z (2000) Functional role of tryptophan residues in the fourth transmembrane domain of the CB(2) cannabinoid receptor. J Neurochem 75:2485–2491

Rich MR (1993) Conformational analysis of arachidonic and related fatty acids using molecular dynamics simulations. Biochim Biophys Acta 1178:87–96

Rinaldi-Carmona M, Barth F, Heaulme M, Shire D, Calandra B, Congy C, Martinez S, Maruani J, Neliat G, Caput D (1994) SR141716A, a potent and selective antagonist of the brain cannabinoid receptor. FEBS Lett 350:240–244

Rinaldi-Carmona M, Barth F, Millan J, Derocq J-M, Casellas P, Congy C, Oustric D, Sarran M, Bouaboula M, Calandra B, Portier M, Shire D, Breliere J-C, LeFur G (1998) SR144528, the first potent and selective antagonist for the CB2 cannabinoid receptor. J Pharmacol Exp Ther 284:644–650

Ross RA, Brockie HC, Stevenson LA, Murphy VL, Templeton F, Makriyannis A, Pertwee RG (1999) Agonist-inverse agonist characterization at CB1 and CB2 cannabinoid receptors of L759633, L759656, and AM630. Br J Pharmacol 126:665–672

Ryan WJ, Banner WK, Wiley JL, Martin BR, Razdan RK (1997) Potent anandamide analogs: the effect of changing the length and branching of the end pentyl chain. J Med Chem 40:3617–3625

Schuel H, Chang MC, Burkman LJ, Picone RP, Makriyannis A, Zimmerman AM, Zimmerman S (1999) Cannabinoid receptors in sperm. In: Nahas GG, Sutin KM, Agurell S (eds) Marihuana and medicine. Humana Press, Totowa, pp 335–345

Seltzman HH, Fleming DN, Thomas BF, Gilliam AF, McCallion DS, Pertwee RG, Compton DR, Martin BR (1997) Synthesis and pharmacological comparison of dimethylheptyl and pentyl analogs of anandamide. J Med Chem 40:3626–3634

Sheskin T, Hanus L, Slager J, Vogel Z, Mechoulam R (1997) Structural requirements for binding of anandamide-type compounds to the brain cannabinoid receptor. J Med Chem 40:659–667

Shim JY, Welsh WJ, Cartier E, Edwards JL, Howlett AC (2002) Molecular interaction of the antagonist N-(Piperidin-1-yl)-5-(4-chlorophenyl)-1- (2,4-dichlorophenyl)-4-methyl-1H-pyrazole-3-carboxamide with the CB1 cannabinoid receptor. J Med Chem 45:1447–1459

Shire D, Calandra B, Delpech M, Dumont X, Kaghad M, LeFur G, Caput D, Ferrar P (1996a) Structural features of the central cannabinoid CB1 receptor involved in the binding of the specific CB1 antagonist SR141716A. J Biol Chem 271:6941–6946

Shire D, Calandra B, Rinaldi Carmona M, Oustric D, Pessegue B, Bonnin Cabanne O, Le Fur G, Caput D, Ferrara P (1996b) Molecular cloning, expression and function of the murine CB2 peripheral cannabinoid receptor. Biochim Biophys Acta 1307:132–136

Shohami E, Mechoulam R (2000) Dexanabinol (HU-211): a nonpsychotropic cannabinoid with neuroprotective properties. Drug Dev Res 50:211–215

Stella N, Schweitzer P, Piomelli D (1997) A second endogenous cannabinoid that modulates long-term potentiation. Nature 388:773–778

Stoit AR, Lange JHM, den Hartog AP, Ronken E, Tipker K, van Stuivenberg HH, Dijksman JAR, Wals HC, Kruse CG (2002) Design, synthesis and biological activity of rigid cannabinoid CB1 receptor antagonists. Chem Pharm Bull (Tokyo) 50:1109–1113

Sugiura T, Kodaka T, Nakane S, Miyashita T, Kondo S, Suhara Y, Takayama H, Waku K, Seki C, Baba N, Ishima Y (1999) Evidence that the cannabinoid CB1 receptor is a 2-arachidonoylglycerol receptor. Structure-activity relationship of 2-arachidonoylglycerol, ether-linked analogues, and related compounds. J Biol Chem 274:2794–2801

Suhara Y, Takayama H, Nakane S, Miyashita T, Waku K, Sugiura T (2000) Synthesis and biological activities of 2-arachidonoylglycerol, an endogenous cannabinoid receptor ligand, and its metabolically stable ether-linked analogues. Chem Pharm Bull (Tokyo) 48:903–907

Tarzia G, Duranti A, Tontini A, Spadoni G, Mor M, Rivara S, Vincenzo Plazzi P, Kathuria S, Piomelli D (2003) Synthesis and structure-activity relationships of a series of pyrrole cannabinoid receptor agonists. Bioorg Med Chem 11:3965–3973

Thakur GA, Palmer SL, Harrington PE, Stergiades IA, Tius MA, Makriyannis A (2002) Enantiomeric resolution of a novel chiral cannabinoid receptor ligand. J Biochem Biophys Methods 54:415–422

Thomas BF, Adams IB, Mascarella SW, Martin BR, Razdan RK (1996) Structure-activity analysis of anandamide analogs: relationship to a cannabinoid pharmacophore. J Med Chem 39:471–479

Thomas BF, Gilliam AF, Burch DF, Roche MJ, Seltzman HH (1998) Comparative receptor binding analyses of cannabinoid agonists and antagonists. J Pharmacol Exp Ther 285:285–292
Tius MA, Kannangara GSK, Kerr MA, Grace KJS (1993) Halogenated cannabinoid synthesis. Tetrahedron 49:3291–3304
Tius MA, Makriyannis A, Long Zou X, Abadji V (1994) Conformationally restricted hybrids of CP-55,940 and HHC: stereoselective synthesis and activity. Tetrahedron 50:2671–2680
Tius MA, Hill WA, Zou XL, Busch-Petersen J, Kawakami JK, Fernandez Garcia MC, Drake DJ, Abadji V, Makriyannis A (1995) Classical/non-classical cannabinoid hybrids; stereochemical requirements for the southern hydroxyalkyl chain. Life Sci 56:2007–2012
Tius MA, Busch-Petersen J, Marris AR (1997) Synthesis of a bifunctional cannabinoid ligand. J Chem Soc Chem Commun 19:1867–1868
Tong W, Collantes ER, Welsh WJ, Berglund BA, Howlett AC (1998) Derivation of a pharmacophore model for anandamide using constrained conformational searching and comparative molecular field analysis. J Med Chem 41:4207–4215
Wiley JL, Compton DR, Dai D, Lainton JA, Phillips M, Huffman JW, Martin BR (1998) Structure-activity relationships of indole- and pyrrole-derived cannabinoids. J Pharmacol Exp Ther 285:995–1004
Wiley JL, Jefferson RG, Grier MC, Mahadevan A, Razdan RK, Martin BR (2001) Novel pyrazole cannabinoids: insights into CB1 receptor recognition and activation. J Pharmacol Exp Ther 296:1013–1022
Wilson RS, May EL (1975) Analgesic properties of the tetrahydrocannabinols, their metabolites, and analogs. J Med Chem 18:700–703
Wilson RS, May EL, Martin BR, Dewey WL (1976) 9-Nor-9-hydroxyhexahydrocannabinols. Synthesis, some behavioral and analgesic properties, and comparison with the tetrahydrocannabinols. J Med Chem 19:1165–1167
Wilson RS, May EL, Dewey WL (1979) Some 9-hydroxycannabinoid-like compounds. Synthesis and evaluation of analgesic and behavioral properties. J Med Chem 22:886–888
Wrobleski ST, Chen P, Hynes JJ, Lin S, Norris DJ, Pandit CR, Spergel S, Wu H, Tokarski JS, Chen X, Gillooly KM, Kiener PA, McIntyre KW, Patil-koota V, Shuster DJ, Turk LA, Yang G, Leftheris K (2003) Rational design and synthesis of an orally active indolopyridone as a novel conformationally constrained cannabinoid ligand possessing antiinflammatory properties. J Med Chem 46:2110–2116
Xie XQ, Eissenstat M, Makriyannis A (1995) Common cannabimimetic pharmacophoric requirements between aminoalkyl indoles and classical cannabinoids. Life Sci 56:1963–1970
Yan G, Yin D, Khanolkar AD, Compton DR, Martin BR, Makriyannis A (1994) Synthesis and pharmacological properties of 11-hydroxy-3-(1′,1′-dimethylheptyl)hexahydrocannabinol: a high-affinity cannabinoid agonist. J Med Chem 37:2619–2622

HEP (2005) 168:247–281

Cannabinoid Receptors and Their Ligands: Ligand–Ligand and Ligand–Receptor Modeling Approaches

P.H. Reggio

Department of Chemistry and Biochemistry, University of North Carolina Greensboro, P.O. Box 26170, Greensboro NC, 27402, USA
phreggio@uncg.edu

Abstract The cannabinoid CB_1 and CB_2 receptors belong to the class A, rhodopsin-like family of GPCRs. Antagonists for each receptor sub-type, as well as four structural classes of agonists that bind to both receptors, have been identified.

An extensive amount of structure–activity relationship information (SAR) has been developed for agonists and antagonists that bind at CB_1, while the SAR of CB_2 ligands is only now emerging in the literature. This chapter focuses both on recent CB_1 and CB_2 SAR and on the pharmacophores for ligand recognition at the CB_1 receptor that have been developed using ligand–ligand or ligand–receptor approaches. In a ligand–ligand approach, the structure of the binding site of the ligand is not directly considered. This approach is an attempt to infer information about the macromolecular binding site, and/or modes of binding interactions from a correlation between experimentally determined biological activities and the structural and electronic features of a series of small molecules. In a ligand–receptor approach, cannabinoid (CB) receptor models are probed for ligand binding sites and binding sites can be screened using energetic criteria, as well as ligand SAR and the CB mutation literature. This chapter discusses the factors that control the quality of the results emanating from each of these approaches and identifies areas of agreement and of disagreement in the existing CB literature. Challenges for future SAR and pharmacophore development are also identified.

Keywords Cannabinoid SAR · Modeling · Receptor modeling

1 Introduction

Both the CB_1 and the CB_2 receptors belong to the class A rhodopsin-like family of G protein-coupled receptors (GPCRs). The cloning and expression of a complementary DNA from a rat cerebral cortex cDNA library that encoded the first cannabinoid receptor subtype (CB_1) was reported by Matsuda and co-workers (1990). Subsequently, the primary amino acid sequences of an amino terminus variant CB_1 receptor (Shire et al. 1995), as well as the CB_1 sequence in human brain and in mouse were reported (Abood et al. 1997; Gerard et al. 1991). A helix net representation of the human CB_1 receptor sequence is presented in Fig. 1. In addition to being found in the central nervous system (CNS), mRNA for CB_1 has also been identified in testis (Gerard et al. 1991). The CB_1 receptor has been shown to have a high level of ligand-independent activation (i.e., constitutive activity) in transfected cell lines, as well as in cells that naturally express the CB_1 receptor (Bouaboula et al. 1997; Pan et al. 1998; Mato et al. 2002; Meschler et al. 2000). Kearn and co-workers (1999) have estimated that in a population of wild-type (WT) CB_1 receptors, 70% exist in the inactive state (R) and 30% exist in the activated state (R*).

The second cannabinoid receptor sub-type, CB_2, was derived from a human promyelocytic leukemia cell HL60 cDNA library (Munro et al. 1993). The human CB_2 receptor exhibits 68% identity to the human CB_1 receptor within the transmembrane regions, 44% identity throughout the whole protein. The CB_2 receptor in both rat (Griffin et al. 2000) and mouse (Shire et al. 1996) has been cloned as well. A helix net representation of the human CB_2 receptor sequence is presented in Fig. 2. Unlike the CB_1 receptor, which is highly conserved across human, rat, and

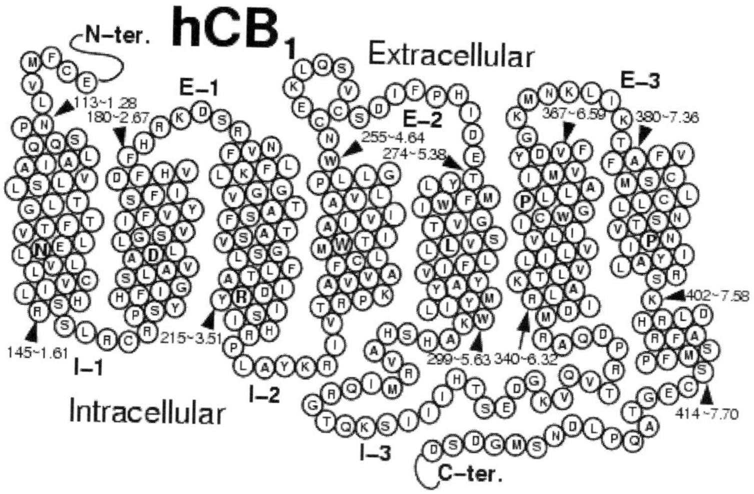

Fig. 1. A helix net representation of the human CB_1 receptor sequence. (Gerard et al. 1991)

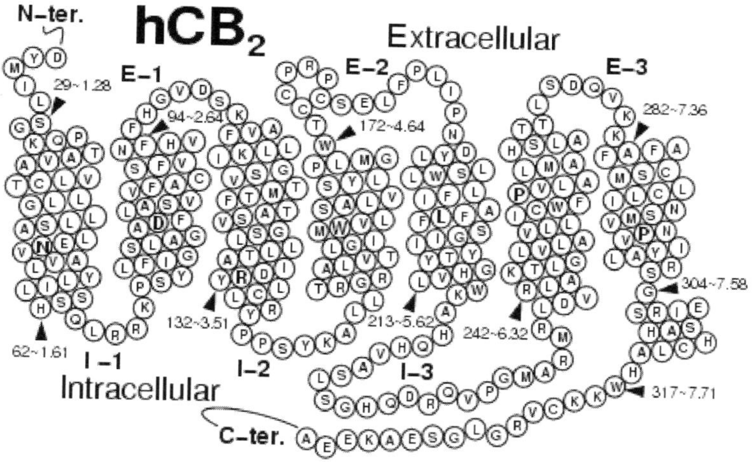

Fig. 2. A helix net representation of the human CB_2 receptor sequence. (Munro et al. 1993)

mouse, the CB_2 receptor is much more divergent. Sequence analysis of the coding region of the rat CB_2 genomic clone indicates 93% amino acid identity between rat and mouse and 81% amino acid identity between rat and human. CB_2 receptor-transfected CHO cells exhibit high constitutive activity (Bouaboula et al. 1999). Evidence for other cannabinoid receptors is mounting in the literature (Breivogel et al. 2001; Di Marzo et al. 2000; Fride et al. 2003; Jarai et al. 1999; Wagner et al. 1999). However, no new CB receptor subtypes have yet been cloned.

1.1 Cannabinoid Receptor Agonists

The CB_1 receptor transduces signals in response to CNS-active constituents of *Cannabis sativa*, such as the classical cannabinoid (CB) (–)-*trans*-Δ^9-tetrahydrocannabinol [(–)-Δ^9-THC (**1**)] and to three other structural classes of ligands, the non-classical CBs typified by (1R,3R,4R)-3-[2-hydroxy-4-(1,1-dimethylheptyl)phenyl]-4-(3-hydroxypropyl) cyclohexan-1-ol [CP-55,940 (**2**)] (Devane et al. 1988; Melvin et al. 1995), the aminoalkylindoles (AAIs) typified by *R*-[2,3-dihydro-5-methyl-3-[(4-morpholinyl)methyl]pyrrolo[1,2,3-de]-1,4-benzoxazin-6-yl](1-naphthalenyl)methanone [WIN55,212-2 (**3**)] (Compton et al. 1992; D'Ambra et al. 1992; Ward et al. 1991), and the endogenous CBs. The non-classical CBs clearly share many structural features with the classical CBs, e.g., a phenolic hydroxyl at C-1 (C2′), and alkyl side chain at C-3 (C-4′), as well as the ability to adopt the same orientation of the carbocyclic ring as that in classical CBs (Reggio et al. 1993). The AAIs, on the other hand, bear no obvious structural similarities with the classical/non-classical CBs.

The first endogenous CB was isolated from porcine brain by Mechoulam and co-workers (Devane et al. 1992). The endogenous CB ligands are unsaturated

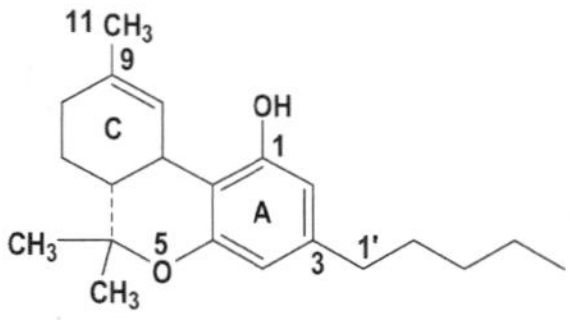

delta-9--THC
1

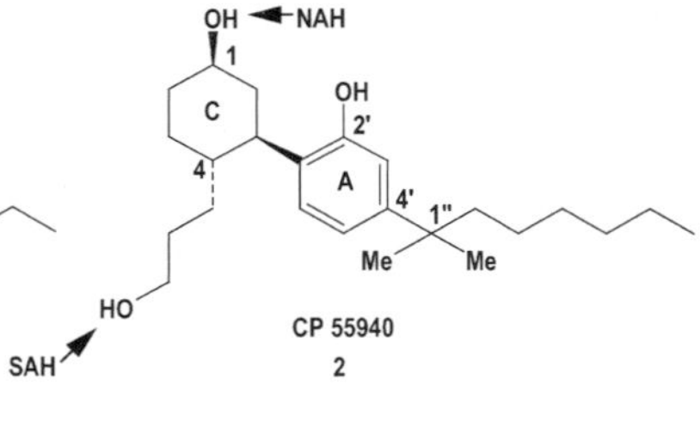

CP 55940
2

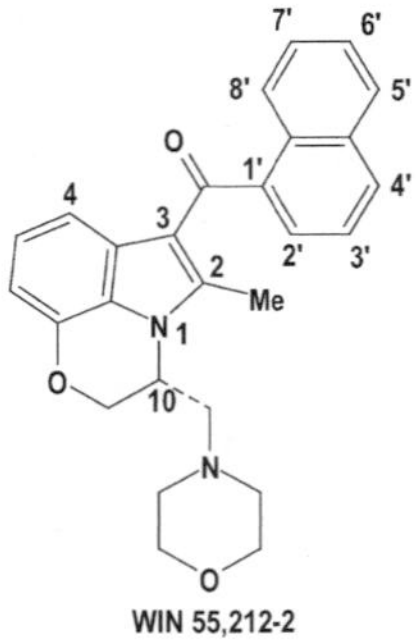

WIN 55,212-2
3

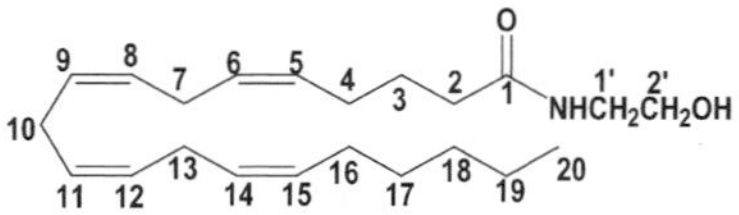

Anandamide
AEA
4

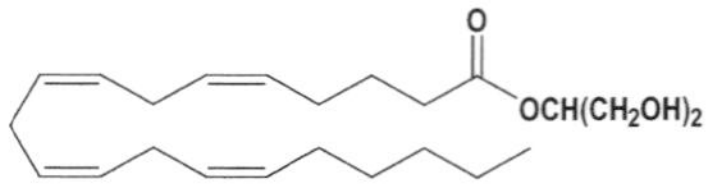

2-AG
5

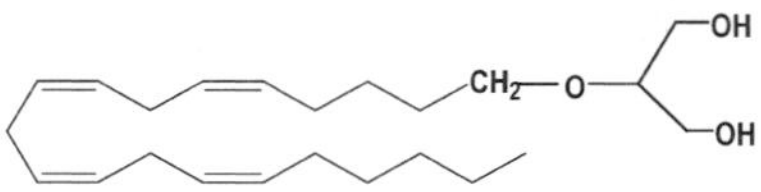

Noladin Ether
6

fatty-acid ethanolamides. The first identified ligand of this class was arachidonoylethanolamide (AEA, also called anandamide, **4**) (Devane et al. 1992). AEA has been shown to be synthesized from lipid in neurons and to be degraded by fatty acid amide hydrolase (FAAH), an integral membrane protein (Bracey et al. 2002). 2-Arachidonoylglycerol (2-AG; **5**) was isolated from intestinal tissue and shown to be a second endogenous CB ligand (CB_1 $K_i = 472 \pm 55$ nM; CB_2 $K_i = 1400 \pm 172$ nM) (Mechoulam et al. 1995). 2-AG has been found present in the brain at concentrations 170 times greater than anandamide (Stella et al. 1997). In addition, a fatty acid glycerol ether, 2-arachidonyl glyceryl ether, called noladin ether (**6**) has been identified as another endogenous CB ligand (Hanus et al. 2001).

1.2 Cannabinoid CB_1 Receptor Antagonists/Inverse Agonists

The first CB_1 antagonist, *N*-(piperidin-1-yl)-5-(4-chlorophenyl)-1-(2,4-dichlorophenyl)-4-methyl-1H-pyrazole-3-carboxamide [SR141716A (**7**)] was developed by Rinaldi-Carmona and co-workers at Sanofi Recherche (Rinaldi-Carmona et al. 1994). SR141716A displays nanomolar CB_1 affinity ($K_i = 1.98 \pm 13$ nM), but very low affinity for CB_2. In vitro, SR141716A antagonizes the inhibitory effects of CB agonists on both mouse vas deferens contractions and adenylyl cyclase activity in rat brain membranes. SR141716A also antagonizes the pharmacological and behavioral effects produced by CB_1 agonists after intraperitoneal (IP) or oral administration (Rinaldi-Carmona et al. 1994). Several other CB_1 antagonists have been reported: LY-320135 (Felder et al. 1998), O-1184 (Ross et al. 1998), CP-27,2871 (Meschler et al. 2000), a class of benzocycloheptapyrazoles (Stoit et al. 2002) and, most recently, a novel series of 3,4-diarylpyrazolines (Lange et al. 2004).

SR141716A (**7**) has been shown to act as a competitive antagonist and inverse agonist in host cells transfected with exogenous CB_1 receptor, as well as in biological preparations endogenously expressing CB_1. Bouaboula and co-workers

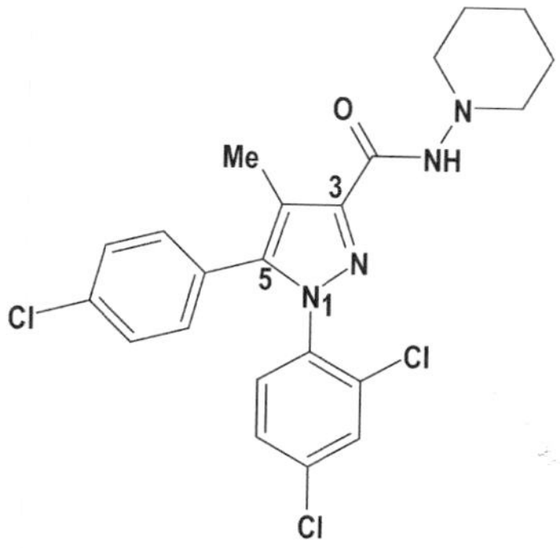

SR 141716A
(CB1)

7

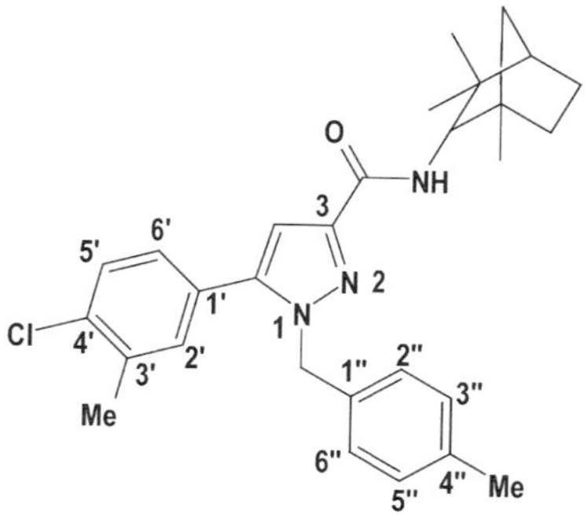

SR 144528
(CB2)

8

(1997) reported that CHO cells transfected with human CB_1 receptor exhibit high constitutive activity at the level of both mitogen-activated protein (MAP) kinase and adenylyl cyclase. Guanine nucleotides enhanced the binding of SR141716A, a property of inverse agonists. Lewis and co-workers (Pan et al. 1998) demonstrated constitutive activity of CB_1 receptors in inhibiting Ca^{2+} currents that was not due to endogenous agonist. These investigators reported that SR141716A antagonized the Ca^{2+} current inhibition induced by the CB agonist, WIN55,212-2, in neurons heterologously expressing either rat or human CB_1 receptors. Further, when applied alone, SR141716A increased the Ca^{2+} current, with an EC_{50} of 32 nM, via a pertussis toxin-sensitive pathway, indicating that SR141716A can act as an inverse agonist by reversal of tonic CB_1 receptor activity. Howlett and co-workers (Meschler et al. 2000) demonstrated that constitutive activity is demonstrable in neuronal cells that endogenously express CB_1 (N18TG2 cells) and that SR141716A acts as a competitive antagonist and reduces basal activity in the manner of an inverse agonist in these cells.

In some experiments, SR141716A has been found to be more potent in blocking the actions of CB_1 agonists than in eliciting inverse responses by itself. For example, in their study that focused upon rat brain membrane and brain sections, Sim-Selley et al. (2001) suggested that SR141716A may bind to two sites on the CB receptor, a high-affinity site at which it exerts its competitive antagonism and a lower affinity site at which it exerts its inverse agonism.

1.3 CB_2 Antagonists

The first CB_2 antagonist, SR144528 (**8**), was reported by Rinaldi-Carmona and co-workers at Sanofi Recherche (Rinaldi-Carmona et al. 1998). SR144528 displays sub-nanomolar affinity for both the rat spleen and cloned human CB_2 receptors ($K_i = 0.60 \pm 0.13$ nM). SR144528 displays a 700-fold lower affinity for both the rat brain and cloned human CB_1 receptors. CB_2 receptor-transfected CHO cells exhibit high constitutive activity, and this activity can be blocked by SR144528, working as an inverse agonist (Bouaboula et al. 1999). More recently, JTE-907 has also been identified as an inverse agonist at CB_2 (Iwamura et al. 2001) and AM630 has been reported to be a CB_2-selective antagonist (Ross et al. 1999a).

2 CB Pharmacophore Development: Ligand–Ligand and Ligand–Receptor Approaches

Pharmacophore development can be approached from a ligand–ligand perspective or from a ligand–receptor perspective. In a ligand–ligand approach, the structure of the binding site of the ligand is not directly considered. This approach is an attempt to infer information about the macromolecular binding site, and/or modes of binding interactions from a correlation between experimentally determined biological activities and the structural and electronic features of a series of small molecules (Nakanishi et al. 1995).

There are many computer modeling/QSAR (quantitative structure–activity relationship) techniques that can be used to deduce information about a receptor binding site based upon ligand SAR. Among these, conformational analysis, molecular electrostatic potential mapping, receptor steric and receptor essential volume mapping, and the comparative molecular field analysis (CoMFA) QSAR method have been used in the literature to gain indirect information about the CB receptors. Central to many of these techniques is a structural superimposition using hypothesized key pharmacophoric features as molecular alignment guides. The quality of results emanating from this approach is highly dependent on these chosen alignments with template molecules. Much of the driving force for structural superpositions in the CB literature has been the fact that the four structural classes of CB agonist ligands and the CB antagonist ligands for a particular receptor sub-type displace one another in radioligand binding experiments. This fact has led to the assumption that key molecular features must superimpose because the molecules must interact with the same key amino acids of the receptor and share the same binding site to be able to displace one another. However, ligands do not necessarily have to occupy exactly the same space nor interact with the same key amino acids in order to displace one another. The presence of steric overlap between binding sites is sufficient to account for ligand displacement data. This means that alignments that incorporate structurally diverse classes of CB ligands may not lead to the best results.

In a ligand–receptor approach, CB receptor models are probed for binding sites for ligand classes and binding sites can be screened using energetic criteria, as well as ligand SAR and the CB mutation literature. The quality of the research emanating from this approach depends heavily on the quality of the receptor model, including the state that this model represents. The reliability of ligand binding sites identified based on energetic criteria is completely dependent on the model itself. If this model is far from the true receptor structure, then the identification of low energy binding sites will have little relevance for the CB field. Since the publication of the 2.8 Å X-ray crystal structure of bovine rhodopsin (Rho) in 2000 (Palczewski et al. 2000), most models of GPCRs, including the CB receptors (Barnett-Norris et al. 2002b; Hurst et al. 2002; McAllister et al. 2003; Salo et al. 2004; Shim et al. 2003; Xie et al. 2003), have been based upon this crystal structure. It is important to note that this structure represents the dark (inactive) state structure of Rho in which the inverse agonist, 11-*cis*-retinal is covalently bound. For ligand–receptor studies that employ a homology model of Rho, the relevant conformational state of the receptor should be taken into consideration because the inactive and active states of a GPCR are fundamentally different in conformation (Ghanouni et al. 2001b; Hulme et al. 1999; Jensen et al. 2001). Therefore, the state of the receptor for which an inverse agonist has high affinity (the inactive state) is not the state for which agonists have high affinity (the activated state).

In the next section, the use of both ligand–ligand and ligand–receptor approaches in the CB field will be discussed. This discussion is organized around individual structural classes of CB ligands.

3 Classical/Non-classical CB Pharmacophores

Prior to the discovery of the cannabinoid CB_1 receptor, CB SARs were developed by those who hypothesized that at least some of the effects produced by CBs may be receptor mediated. The early SAR that emerged has been reviewed comprehensively by Razdan (1986) and by Makriyannis and Rapaka 1990). These reviews consider both classical and non-classical CB compounds. Because there is structural/conformational similarity between the classical and non-classical CBs (Lagu et al. 1995; Reggio et al. 1993; Xie et al. 1994, 1996, 1998), unified pharmacophores developed for these two classes have agreed with one another and have led to a consensus pharmacophore that involves the existence of the following at the CB_1 receptor:

1. A hydrophobic binding pocket of limited depth into which the C-3 (C-4′) alkyl chain fits such that the chain is nearly perpendicular to the aromatic ring (Howlett et al. 1988; Melvin and Johnson 1987; Xie et al. 1998). Analogs with side chains of less than five carbons have no affinity for CB_1. Highest affinity is associated with the 1′,1′-dimethylheptyl side chain.
2. Hydrogen bonding sites for the phenolic hydroxyl of ring A (see **1**), the C-9/C-11 hydroxyls of the carbocyclic ring (ring C; see **1**) (Howlett et al. 1988; Melvin and Johnson 1987) (Huffman et al. 1996; Song and Bonner 1996) and the southern aliphatic hydroxyl (SAH) group (see **2**) (Drake et al. 1998; Tius et al. 1994).
3. An occluded region behind C-9 (classical), C-1 (non-classical) of the carbocyclic ring (ring C; see **1**) occupied by residues of the receptor itself (Reggio et al. 1993).
4. A large hydrophobic pocket that accommodates SAH hydrophobic analogs and a smaller hydrophilic pocket that accommodates the SAH group (see **2**) (Drake et al. 1998; Tius et al. 1994).

3.1 Ligand–Ligand Studies: CoMFA Pharmacophores for Classical/Non-classical CBs

Thomas and co-workers presented the first CoMFA QSAR model of the CB receptor that employed Martin multiple paradigm activity data as the biological activity and considered both classical and non-classical CBs (Thomas et al. 1991). Compounds were superimposed at the aromatic ring and alkyl side chain. The n-propyl alcohol chain (SAH) of CP-55,940 (**2**) was aligned with respect to its restricted analog CP-55,243. Results indicated steric repulsion behind the C-ring (see **2**) is associated with decreased predicted binding affinity and pharmacological potency. The steric bulk of the C-4′ side chain of **2** that is extended up to seven carbons was found to contribute to predictions of increased binding affinity and potency. The electrostatic fields of the CB analogs that correlated with increased predicted potency were predominantly seen around the C11 position of Δ^9-THC (**1**). Results indicated that the protons of hydroxyl groups at positions corresponding to the C11 position of **1** may interact with an electronegative acceptor atom.

3.1.1 Side Chain SAR

One of the molecular regions of the classical/non-classical CBs that has been the focus of recent interest is the C-3 alkyl side chain. Several groups have looked at the effect of the introduction of unsaturation or functionality in the alkyl side chain of classical CBs. 1′,1′-Cyclopropyl side chain substituents were found to enhance the affinities of (−)-Δ^8-tetrahydrocannabinol (Δ^8-THC) and respective cannabidiol analogs for the CB_1 and CB_2 cannabinoid receptors (Papahatjis et al. 2002). For novel analogs of Δ^8-THC (**9**) in which the conformation of the side chain was restricted by incorporating the first one or two carbons into a six-membered ring fused with the aromatic phenolic ring, results indicated that the "southbound" chain conformer retained the highest affinity for both receptors (Khanolkar et al. 1999). Papahatjis and co-workers (1998) published a study involving side chain-constrained analogs of Δ^8-THC, including a 3-(1-heptynyl) analog synthesized in a β-11-HHC (**10**) series and a potent 1′-dithiolane derivative. No analog had the side chain in a fully restricted conformation. However, the authors concluded from their binding data, and in particular the increased potency of the 1′-dithiolane and the 1′-methylene analogs, that a hydrophobic subsite of the CB pharmacophore exists in both CB_1 and CB_2 at the level of the benzylic side chain carbon. To study the

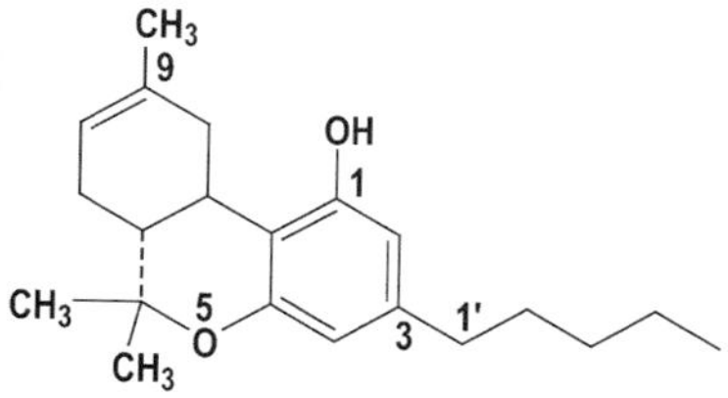

delta-8-THC

9

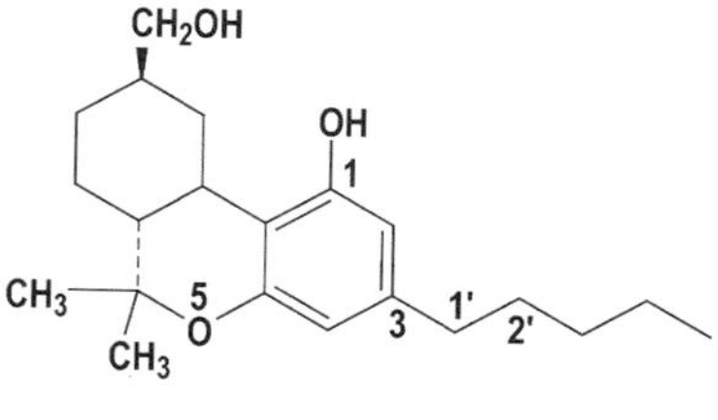

10

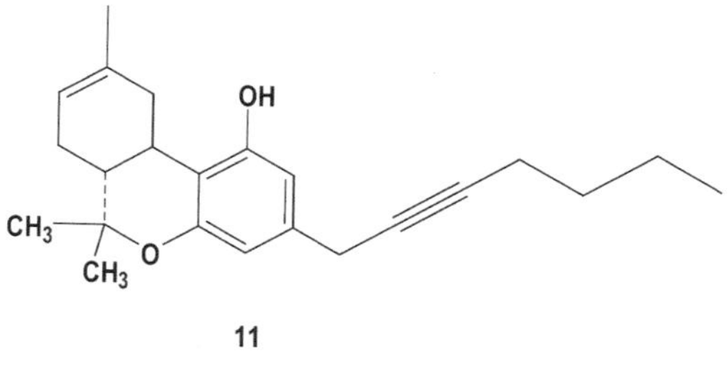

11

stereochemical requirements of the side chain, Busch-Petersen (1996) synthesized a series of β-11-hydroxyhexahydrocannabinol (**10**) CBs in which rotation around the C1′-C2′ bond is blocked by the introduction of a double (*cis* or *trans*) or triple bond. All the analogs tested showed nanomolar affinity for the receptors, the *cis*-hept-1-ene side chain having the highest affinity for CB_1 ($K_i = 0.89$ nM) and showing the widest separation between CB_1 and CB_2 affinities (Busch-Petersen et al. 1996).

Razdan and co-workers have also pursued the effects of unsaturation or added functionality in the C-3 side chain of classical CBs. These investigators have found that manipulations of the side chain can produce high-affinity ligands with either antagonist, partial agonist, or full agonist effect. In particular, antagonists such as **11** were developed through strategic placement of a triple bond (Griffin et al. 1999; Martin et al. 1999; Ross et al. 1998, 1999b; Ryan et al. 1995; Singer et al. 1998). It is possible that the reason that this ligand functions as an antagonist is that such substitution reduces the flexibility of the side chain and leads to loss of efficacy.

Nadipuram and co-authors synthesized a series of C3 cyclic side-chain analogs of Δ^8-THC in which ring substituents were attached at the 1′ position. Substitution of a dithiolane ring at the 1′ position and a carbocyclic ring at the 1′ position led to compounds that retained very good affinity for CB_1 and CB_2, suggesting that the binding pocket for the classical CB side chain may be ellipsoidal rather than elongated (Nadipuram et al. 2003). Substitution of a phenyl ring at the C-1′ position along with a dithiolane ring at C-1′ or a 1′,1′-dimethyl group led to compounds with high affinities for both CB_1 and CB_2, while affinity was reduced when the phenyl ring was attached to a C-1′ CH2 or carbonyl group. The dimethyl and ketone analogs displayed selectivity for the CB_2 receptor (Krishnamurthy et al. 2003)

Pharmacophore for Classical CB Side Chain

Thomas and co-workers used the extensive side chain SAR generated by Razdan to develop a novel QSAR for the side chain region of Δ^8-THC (**9**) (Keimowitz et al. 2000). A series of 36 side chain-substituted Δ^8-THCs with a wide range of pharmacological potency and CB_1 receptor affinity was investigated using computational molecular modeling and QSAR analyses. The conformational mobility of each compound's side chain was characterized using a quenched molecular dynamics approach. The QSAR techniques included a modified active analog approach (MAA), multiple linear regression analyses (MLR), and CoMFA studies. Results obtained support the hypothesis that for optimum affinity and potency, the side chain must have conformational freedom that allows its terminus to fold back and come into proximity with the phenolic ring (Keimowitz et al. 2000). This result fits very well with those of Razdan and co-workers mentioned above who produced classical CB antagonists, such as **11**, by restricting the conformational freedom of the side chain (Griffin et al. 1999; Martin et al. 1999; Ross et al. 1998, 1999b; Ryan et al. 1995; Singer et al. 1998).

3.2
Ligand–Receptor Studies for Classical/Non-classical CB Binding to CB_1

Shim and co-workers recently published a ligand–receptor study in which a molecular docking approach that combined Monte Carlo and molecular dynamics simulations was used to identify putative binding conformations of non-classical CB agonists, including AC-bicyclic CP-47,497 and CP-55,940 (**2**), and ACD-tricyclic CP-55,244 (Shim et al. 2003). These investigators used an inactive state model of CB_1 for these docking studies based upon the X-ray crystal structure of rhodopsin (Palczewski et al. 2000). Ligand placement was based upon the assumption of a critical hydrogen bond between the A-ring OH and the side chain N of Lys192 in transmembrane helix (TMH) 3. Two alternative binding conformations were considered, with a conformation in which the C-3 side chain pointed inside the receptor chosen as the binding site conformation for which the ligand could achieve more interactions. Key hydrogen bonds were identified between both K3.28(192) and E(258) and the A-ring OH (see **2**), and between Q(261) and the C-ring C-12 hydroxypropyl.

3.3
CB_2 Selective Classical/Non-classical CBs Break CB_1 SAR rules

In recent years, it has become clear that one way to develop CB_2-selective compounds is to violate long accepted pharmacophore requirements for binding to CB_1 (see consensus pharmacophore list above). In particular, CB_2-selective compounds have emerged through changes in the phenolic hydroxyl region and through shortening of the alkyl side chain.

3.3.1
Phenolic Hydroxyl

Huffman was first to show that removal of the phenolic hydroxyl of 11-hydroxy-Δ^8-tetrahydrocannabinol-1′, 1′-dimethylheptyl (HU-210, **12**) results in a CB_2-selective compound (**13**) with high affinity for both the CB_1 and CB_2 receptors (CB_1 $K_i = 1.2 \pm 0.1$ nM, CB_2 $K_i = 0.032 \pm 0.019$ nM) (Huffman et al. 1996). In addition, Gareau reported that the conversion of the C-1 phenolic hydroxyl of a classical CB to a methoxy group (i.e., etherification) also produced a CB_2-selective ligand (Gareau et al. 1996). In both of these studies, analogs possessed a longer side chain than natural CBs, a 1′, 1′-dimethylheptyl (DMH) side chain at C-3. Huffman and co-workers also showed that removal of the 11-hydroxy group of deoxy-HU-210 to produce deoxy-Δ^8-THC-DMH still resulted in a ligand with good CB_1 affinity (CB_1 $K_i = 23 \pm 7$ nM) and CB_2 selectivity (CB_2 $K_i = 2.9 \pm 1.6$ nM) (Huffman et al. 1996). O,2-propano-9β-OH-11-nor-HHC (**14**), a rigidified C-1 ether, has also been reported to have good CB_1 affinity ($K_i = 26 \pm 2$ nM) and a 4.5-fold CB_2 selectivity ($K_i = 5.8 \pm 2.9$ nM) (Reggio et al. 1997).

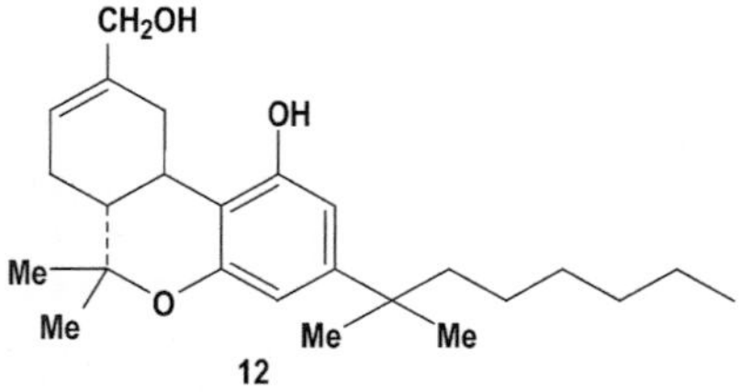

12

HU 210

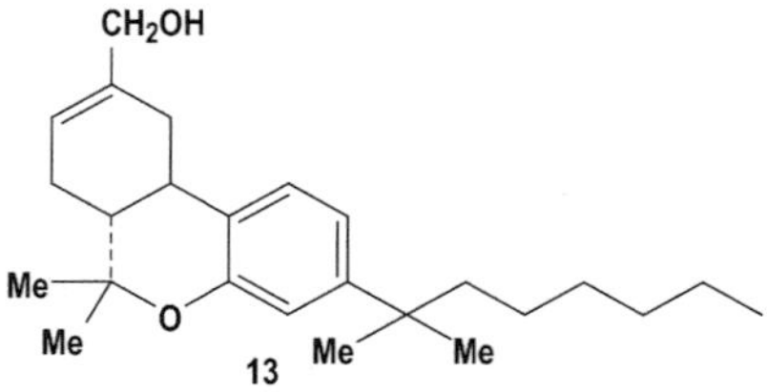

13

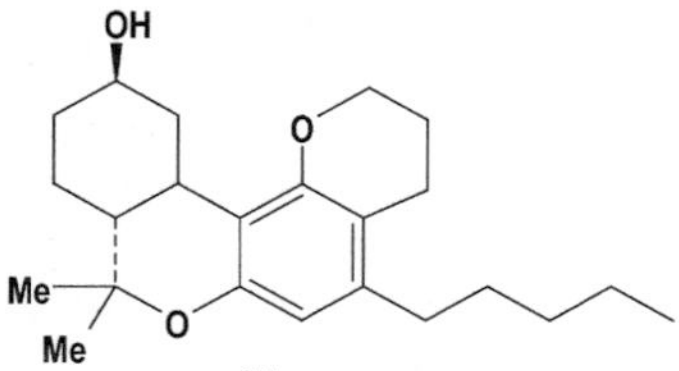

14

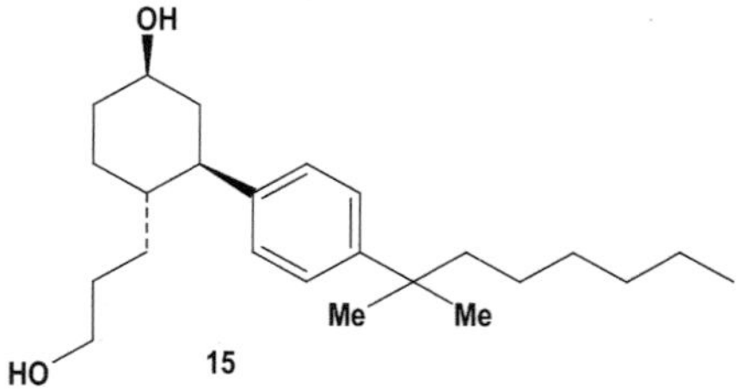

15

In the non-classical CBs, Melvin has also shown that the phenolic hydroxyl at C-2′ (see drawing of **2**) is not necessary as 2′-deoxy-CP-55,940 (**15**) possessed a $K_i = 40.2 \pm 13.5$ nM at the CB_1 receptor (Melvin et al. 1993). In this case, however, although CB_1 affinity is retained, the deoxy analog does have an attenuated affinity relative to that of CP-55,940 (**2**) whose CB_1 $K_i = 0.137 \pm 0.038$ nM. The affinity difference between 1-deoxy-11-hydroxy-Δ^8-THC-DMH (**13**) and 2′-deoxy-CP-55,940 (**15**) may be due to the greater entropic expense incurred by **15** upon binding, as it is a more flexible molecule.

Razdan and co-workers (Wiley et al. 2002) recently reported a series of resorcinol derivatives that exhibited varying affinities for CB_1 and CB_2. When the free phenols at C1 and C3 in this series (with DMH side chains) were etherified, CB_2-selective compounds resulted [e.g., O-1966A (**16**), CB_1 $K_i = 5{,}055 \pm 984$ nM; CB_2 $K_i = 23 \pm 2.1$ nM]. These results are consistent with results reported by Mechoulam and co-workers (Hanus et al. 1999) for HU-308 (**17**), which also has etherification at the same positions as O-1966A and is highly CB_2-selective (CB_1 K_i>10μM; CB_2 $K_i = 22.7 \pm 3.9$ nM).

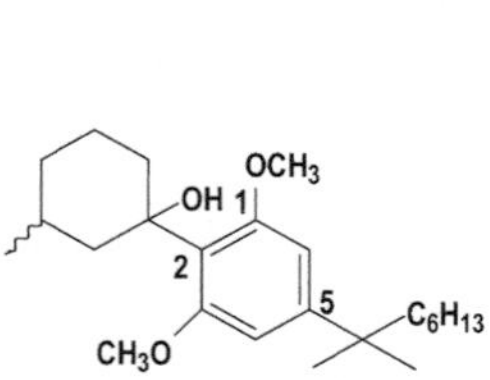

O-1966A

16

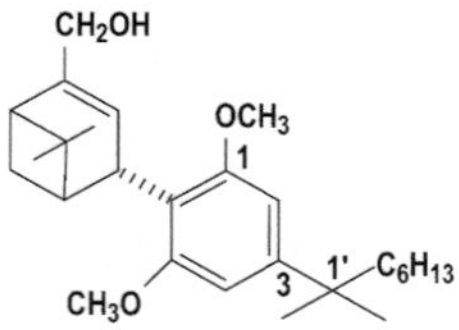

HU-308

17

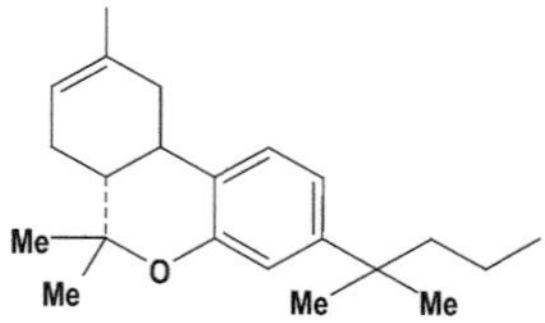

18

3.3.2 Side Chain

In traditional CB SAR, a 1′, 1′-dimethylheptyl C-3 side chain has been a "magic bullet," improving the CB_1 affinity and efficacy of nearly every molecule to which it has been attached (Razdan 1986). In a series of papers, Huffman and co-workers have shown that the CB_2 receptor clearly can accommodate shorter alkyl side chains than can CB_1. This group reported that 1-deoxy-3(1′-1′-dimethylbutyl)-Δ^8-THC (**18**) had high CB_2 affinity [$K_i = 3.4 \pm 1.0$ nM], but 200-fold lower affinity for CB_1 ($K_i = 677 \pm 132$ nM) (Huffman et al. 1998, 1999). 1-deoxy-3(n-butyl)-Δ^8-THC (CB_1 $K_i = 2791 \pm 820$ nM; CB_2 $K_i = 53.8 \pm 8.0$ nM) and 1-deoxy-Δ^8-THC (CB_1 $K_i > 10{,}000$ nM; CB_2 $K_i = 31.6 \pm 8.7$ nM) were also CB_2-selective. 1-deoxy-Δ^8-THCs with 1′, 1′-dimethyl-pentyl and hexyl side chains at C-3 bound weakly to CB_1 ($K_i = 338 \pm 76$ nM, $K_i = 295 \pm 52$ nM), but maintained high CB_2 affinity (CB_2 $K_i = 10 \pm 2$ nM to CB_2 $K_i = 19 \pm 4$ nM) with the octyl and nonyl analogs showing lower affinity (Huffman et al. 1998, 1999).

More recently, this group prepared three series of new CBs: 1-methoxy-3-(1′,1′-dimethylalkyl)-Δ^8-tetrahydrocannabinol, 1-deoxy-11-hydroxy-3-(1′,1′-dimethylalkyl)-Δ^8-tetrahydrocannabinol and 11-hydroxy-1-methoxy-3-(1′,1′-dimethylalkyl)-Δ^8-tetrahydrocannabinol, which contain alkyl chains from dimethylethyl to dimethylheptyl appended to C-3 of the CB. All of these compounds had greater affinity for the CB_2 receptor than for the CB_1 receptor; however, only 1-methoxy-3-(1′,1′-dimethylhexyl)-Δ^8-THC has effectively no affinity for the CB_1 receptor ($K_i = 3134 \pm 110$ nM) and high affinity for CB_2 ($K_i = 18 \pm 2$ nM) (Huffman et al. 2002).

4 Endogenous CB Pharmacophores

The arachidonic acid (AA) moiety in anandamide and congeners confers the molecule with what could be called "dynamic plasticity." The arachidonic acid acyl chain contains four homoallylic double bonds (i.e., *cis* double bonds separated by methylene carbons). Rabinovich and Ripatti (1991) reported that polyunsaturated acyl chains in which double bonds are separated by one methylene group are characterized by the highest equilibrium flexibility compared with other unsaturated acyl chains. Rich (1993) reports that a broad domain of low-energy conformational freedom exists for these C–C bonds. Results of the Biased Sampling phase from Conformational Memories calculations of arachidonic acid are consistent with Rich's and with Rabinovich and Ripatti's results (Barnett-Norris et al. 1998), as they revealed a relatively broad distribution of populated torsional space about the classic skew angles of 119°(s) and −119°(s′) for the C8-C9-C10-C11 torsion angle in anandamide, for example (see **4** for numbering system).

4.1 Endocannabinoid SAR

4.1.1 Acyl Chain SAR

While the fatty acid literature indicates that unsaturated fatty acids that possess multiple homoallylic double bonds, such as AA, exhibit a high degree of flexibility, this literature also indicates that saturated fatty acids tend to be significantly less flexible and adopt primarily extended conformations. Fatty acids with decreasing amounts of unsaturation tend to show a decreasing tendency to form folded structures, but still tend to curve in acyl chain regions in which unsaturation is present (Reggio and Traore 2000). A correlation has been drawn between this acyl chain conformation trend and the SAR of the anandamide (AEA) acyl chain (Barnett-Norris et al. 2002b). Endocannabinoid SAR indicates that the CB_1 receptor recognizes ethanolamides whose fatty acid acyl chains have 20 or 22 carbons, with at least three homoallylic double bonds and saturation in at least the last five carbons of the acyl chain (Sheskin et al. 1997). Reggio and co-workers have suggested that this acyl chain unsaturation SAR requirement is an outgrowth of the shape of the AEA binding pocket at CB_1 which may require tightly folded conformations, conformations not possible for AEA analogs with less than three homoallylic double bonds (Barnett-Norris et al. 2002b).

An analogy has been drawn in the literature between the C16–C20 portion of AEA (see **4**) and the C-3 pentyl side chain of the classical CB, Δ^9-THC (see **1**). Consistent with this hypothesis, replacement of the pentyl tail of AEA with a dimethylheptyl chain results in enhanced affinity (although not to the same degree as seen in the classical CBs) (Ryan et al. 1997; Seltzman et al. 1997).

Although initially it seemed that the development of rigid anandamide analogs that mimic the AA conformations discussed above would help to identify the receptor-appropriate conformation of AEA, all attempts at rigidifying AEA have

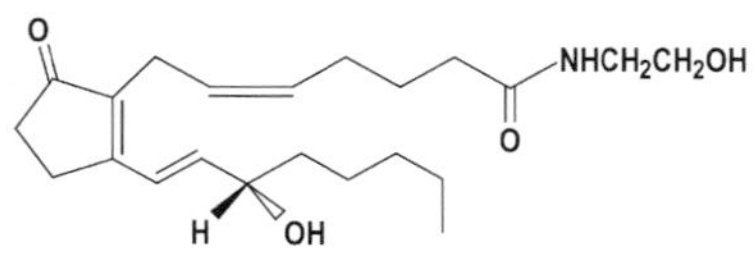

19

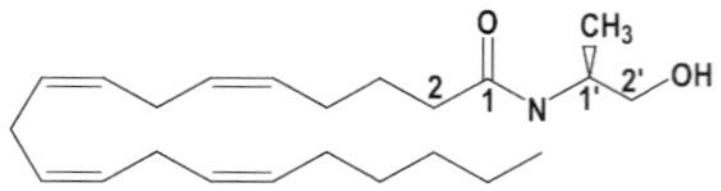

(R)-(+)-methanandamide

20

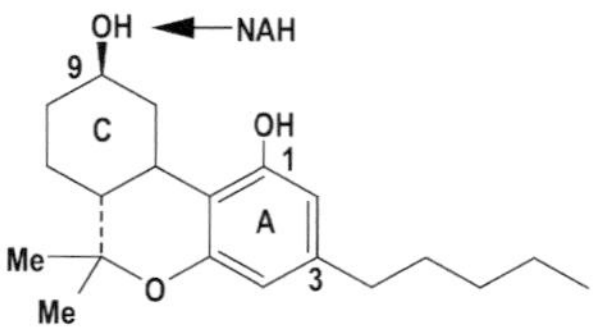

9-nor-9β–OH-HHC

21

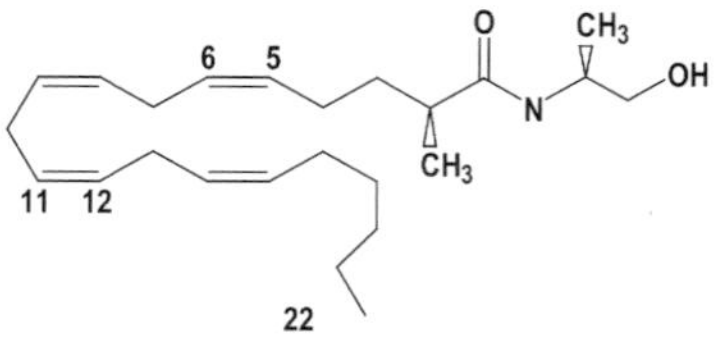

22

been met with little success (Berglund et al. 1999, 2000; Pinto et al. 1994) Pinto and co-workers investigated a series of arachidonyl amides and esters in addition to a series of "rigid hairpin" conformations typified by *N*-(2-hydroxyethyl)-prostaglandin amides to determine the structural requirements for binding to the CB_1 receptor. 2D drawings of anandamide and PGB_2-EA (**19**) make the shapes of these two compounds look similar. However, all of the rigid prostaglandin analogs synthesized by Pinto et al. (1994) failed to alter [^{3}H]CP-55,940 binding to CB_1 in concentrations as great as 100 μM. Barnett-Norris and co-workers 1998) reported conformational memories (CM) results for PGB_2-EA (**19**) which showed an attenuated ability for the prostaglandin ethanolamide to adopt extended conformations or to form U-shaped conformations like AEA and 2-AG. Instead, the CM results showed that the conjugation of the acyl chain with the ring double bond introduces "stiffness" into this part of the molecule, resulting in a predominantly folded L-shaped conformation.

4.1.2 Head Group SAR

In order for high-affinity binding to the CB_1 receptor to occur and for agonist binding to activate G proteins, the carbonyl group of the AEA amide head group must be present (Berglund et al. 1998). Arachidonamide and simple alkyl esters of arachidonic acid did not show significant CB_1 affinity (Pinto et al. 1994). Cyclization of the head group into an oxazoline ring diminished affinity (Lin et al. 1998). Arachidonylethers, carbamates, and norarachidonlycarbamates had poor CB_1 affinity (Ng et al. 1999). However, norarachidonyl ureas showed generally good binding affinities for the CB_1 receptor ($K_i = 55–746$ nM). Some of the weaker affinity analogs in this series produced potent pharmacological activity. These analogs showed hydrolytic stability toward amidase enzymes as well (Ng et al. 1999).

Methylation at the C-1′ position in the AEA (see **4** for numbering system) head group resulted in an 1′-*R*-methyl isomer (*R*-methanandamide, **20**) which had fourfold higher CB_1 affinity than AEA, while the 1′-*S*-methyl isomer had two-fold lower CB_1 affinity than AEA. *R*-Methanandamide (**20**) also was found to be resistant to enzymatic breakdown (Abadji et al. 1994). Methylation at the 2′ position also produced some stereoselectivity, as the *S*(+) isomer was found to have twofold to fivefold higher CB_1 affinity than the *R*(−)-isomer (Abadji et al. 1994; Berglund et al. 1998). Introduction of larger alkyl groups had a detrimental effect on CB_1 affinity (Adams et al. 1995a). A series of C1′-C2 dimethyl anandamide analogs revealed stereochemical requirements of the CB_1 binding pocket, as only the *R*,*R* isomer, (*R*)-N-(1-methyl-2-hydroxyethyl)-2-(*R*)-methyl-arachidonamide, had significant affinity for CB_1 ($K_i = 7.42 \pm 0.86$ nM) (Goutopoulos et al. 2001).

Enlargement of the ethanolamine head group by insertion of methylene groups revealed that the N-propanol analog had slightly higher CB_1 affinity than AEA, while higher homologs had reduced CB_1 affinity (Pinto et al. 1994; Sheskin et al. 1997). Alkyl branching of the alcoholic head group led to lower affinity analogs (Sheskin et al. 1997). N-(Propyl) arachidonylamide possessed higher CB_1 affinity

($K_i = 7.3$ nM) than anandamide itself ($K_i = 22$ nM) (Pinto et al. 1994; Sheskin et al. 1997). Substitution of an N-cyclopropyl group for the ethanolamine head group of AEA led to a very high CB_1-affinity compound (Hillard et al. 1999). These results suggest that there may exist a hydrophobic sub-site for the AEA head group such that the hydroxyl of AEA may not be necessary for receptor interaction (Lin et al. 1998). Replacement of the hydroxyl group of AEA with a halogen such as F or CI increased CB_1 affinity as well (Adams et al. 1995b; Hillard et al. 1999; Lin et al. 1998). Substitution of the 2-hydroxyethyl group of AEA with a phenolic group, however, greatly decreased affinity for CB_1 (Edgemond et al. 1995, 1998; Khanolkar et al. 1996; Lang et al. 1999).

Taken together, all of these results suggest that the hydroxyl in the anandamide head group is not essential for receptor interaction, but that the CB receptor can accommodate both hydrophobic and hydrophilic head groups, possibly in two different subsites. The size(s) of the cavity(ies) in which the head group binds, however, is (are) small as only relatively small variations on the head group permit the retention of high-affinity binding.

4.2 Ligand–Ligand Studies of Endocannabinoids

CoMFA models for endocannabinoids have been developed using rigid classical CBs as templates. However, the inherent flexibility of the arachidonic acyl chain of anandamide has been an obstacle to the identification of an unambiguous overlay needed for COMFA studies.

Thomas et al. were first to report a CoMFA QSAR pharmacophore model for anandamide (**4**) and its analogs (Thomas et al. 1996). These authors used molecular dynamics studies to explore conformations of **4** that present pharmacophoric similarities with Δ^9-THC (**1**). A J-shaped or looped conformation of **4** was identified that had good molecular volume overlap with **1** when (1) the carboxyamide of **4** was overlaid with the pyran oxygen (O-5) in **1**; (2) the head group hydroxyl of **4** was overlaid with the C-1 phenolic hydroxyl group of **1**, (3) the five terminal carbons of the **4** fatty acid acyl chain were overlaid with the C-3 pentyl side chain of **1**; and (4) the polyolefin loop of **4** was overlaid with the tricyclic ring system of **1**. These authors supported their use of a J-shaped conformation for **4** by citing synthetic results for the internal epoxidation undergone by peroxyarachidonic acid, which point to the J shape as necessary for such a reaction (Corey et al. 1979).

Tong and co-workers reported a pharmacophore model for anandamide (**4**) using constrained conformational searching and CoMFA (Tong et al. 1998) and a different alignment between key elements of the pharmacophore. 9-nor-9β-OH-HHC (**21**) was used as the template to which **4** and its analogs were fitted. The training set for the CoMFA model contained 29 classical and non-classical CBs. The conformation identified for **4** was a helical conformation in which (1) the oxygen of the carboxyamide overlaid the C-1 phenolic hydroxyl group of **21**; (2) the head group hydroxyl overlaid the C-9 hydroxyl of **21**; (3) the alkyl tail of **4** overlaid the C-3 alkyl side chain of **21**, and (4) the polyolefin loop overlaid the tricyclic ring structure of **21**. These authors supported their use of a helix-shaped **4** by citing a recent X-ray crystallographic structure that shows that arachidonic acid adopts a helical conformation when it is a substrate for cyclooxygenase (Stegeman et al. 1998).

4.2.1
Tests of CoMFA Models

Howlett and co-workers (Berglund et al. 2000) used the Tong pharmacophore (Tong et al. 1998) to design a series of monocyclic and bicyclic alkyl amides. The bend in the U-shaped conformation of AEA was approximated with incorporation of a phenyl or naphthyl ring, and the importance of a flat ring was tested by incorporation of a cyclohexyl ring. Aspects of the Tong pharmacophore that were reported to be important (i.e., the alkyl tail and carbonyl of the amide) were included or excluded in the series. Highest affinity was associated with phenyl analogs, and among these analogs, meta substitution on the phenyl ring yielded the highest affinity compounds, presumably because it places the amide group and the alkyl side chain at the best distance. Using the pharmacophoric elements proposed to be important in the Tong model, the investigators calculated the distances between the pharmacophoric elements [carbocyclic ring (ring C)-hydroxyl, phenolic hydroxyl and C-3 alkyl side chain] of a series of high-affinity, moderate-affinity and low-affinity analogs relative to these distances in the high-affinity non-classical CB, CP-55,244. However, the authors found it difficult to establish a clear relationship between relative binding affinities and their corresponding pharmacophoric distances due to the high flexibility of the compounds.

Van der Steldt and co-workers (van der Stelt et al. 2002) evaluated a series of anandamide and 2-AG lipoxygenase products for their CB_1 and CB_2 affinities, as well as their ability to inhibit AEA hydrolysis at the FAAH enzyme and to inhibit AEA transport. Several of these have previously been reported by Hillard and co-workers (Edgemond et al. 1998). Conformational analysis was performed using nuclear magnetic resonance (NMR) solution studies, as well as molecular dynamics calculations. Conformational analysis results for the hydroxylated AEAs were probed for consistency of placement of the key pharmacophoric elements identified by Tong and co-workers (Tong et al. 1998) vs a CP-55,940 template. However, the overlapping regions of CP-55,940 and the hydroxylated-AEA series did not reveal great differences between analogs with high or low CB_1 affinities. Taken together, the Howlett (Berglund et al. 2000) and van der Stelt et al. (2002) studies illustrate the limited utility of a CoMFA pharmacophore model based on structural superpositions of AEA analogs with a classical CB template (Tong et al. 1998).

4.3
Ligand–Receptor Modeling Studies of Endocannabinoid Binding

Endocannabinoid SAR indicates that the CB_1 receptor recognizes ethanolamides whose fatty acid acyl chains have 20 or 22 carbons, with at least three homoallylic double bonds and saturation in at least the last five carbons of the acyl chain (Reggio and Traore 2000). Endocannabinoid SAR also indicates that the CB_1 receptor does not tolerate large endocannabinoid head groups; however, it does recognize both polar and non-polar moieties in the head group region (Reggio and Traore 2000). Reggio and co-workers (Barnett-Norris et al. 1998, 2002b) have taken a different approach to the development of an endocannabinoid pharmacophore by focusing on sets of AEA analogs with variation in one region of the molecule at a time and by using the conformational memories (CM) method. CM is a Monte Carlo/simulated annealing based approach (Guarnieri and Weinstein 1996) that generates 100 low free energy structures of each compound at 310 K. In adopting this approach, all possible endocannabinoid conformations can be considered, rather than considering a smaller region of conformational space as is necessitated by working hypotheses of required overlap of key regions with a rigid template (see the CoMFA studies discussed above) (Thomas et al. 1996; Tong et al. 1998).

In order to probe the molecular basis for these acyl chain requirements, Barnett-Norris and co-workers used the CM method to study the conformations available to an n-6 series of ethanolamide fatty acid acyl chain congeners, 22:4,n-6 ($K_i = 34.4 \pm 3.2$ nM), 20:4,n-6 ($K_i = 39.2 \pm 5.7$ nM), 20:3,n-6 ($K_i = 53.4 \pm 5.5$ nM); and 20:2,n-6 ($K_i > 1500$ nM) (Sheskin et al. 1997). CM studies indicated that each analog could form both U/J-shaped (Cls 1) and extended (Cls 2) families of conformers. However, for the low-affinity 20:2,n-6 ethanolamide, the higher populated family was the extended conformer family, while for the other analogs in the series, the U/J-shaped family had the higher population. In addition, the 20:2,n-6 ethanolamide U-shaped family was not as tightly curved as were those of the other analogs studied. In order to quantitate this variation in curvature, the radius of curvature (in the C-3 to C-17 region) of each member of each U/J-shaped family was measured. The average radii of curvature (with their 95% confidence intervals) were found to be 5.8 Å (5.3–6.2) for 20:2,n-6; 4.4 Å (4.1–4.7) for 20:3,n-6; 4.0 Å (3.7–4.2) for 20:4,n-6; and 4.0 Å (3.6–4.5) for 22:4,n-6. These results suggest that higher CB_1 affinity is associated with endocannabinoids that can form tightly curved structures.

In order to identify a head group orientation that results in high CB_1 affinity, Barnett-Norris and co-workers studied a series of dimethyl anandamide analogs (*R*)-*N*-(1-methyl-2-hydroxyethyl)-2-(*R*)-methyl-arachidonamide ($K_i = 7.42 \pm 0.86$ nM; **22**), (*R*)-*N*-(1-methyl-2-hydroxyethyl)-2-(*S*)-methyl-arachidonamide ($K_i = 185 \pm 12$ nM), (*S*)-*N*-(1-methyl-2-hydroxyethyl)-2-(*S*)-methyl-arachidonamide ($K_i = 389 \pm 72$ nM), and (*S*)-*N*-(1-methyl-2-hydroxyethyl)-2-(*R*)-methyl-arachidonamide ($K_i = 233 \pm 69$ nM) (Goutopoulos et al. 2001) using CM and computer receptor docking studies in an active state (R*) model of CB_1 (Barnett-Norris et al. 2002b). These studies suggested that the high CB_1 affinity of the *R,R* stereoisomer (**22**) is due to the ability of the head group to form an intramolecular hydrogen bond between the carboxamide oxygen and the head group hydroxyl that orients the C2 and C1′ methyl groups to have hydrophobic interactions with valine 3.32(196), while the carboxamide oxygen forms a hydrogen bond with lysine 3.28(192) at CB_1. In this position in the CB_1 binding pocket, F2.57(170) and F3.25(189) have C-Hπ interactions with the C5–C6 and C11–C12 acyl chain double bonds, respectively. This binding site is supported by: NMR solution studies of AEA that have shown the persistence of this same AEA headgroup intramolecular hydrogen bond (Bonechi et al. 2001); CB_1 K3.28A mutation studies that show that K3.28 is critical for the binding of AEA at CB_1 (Song and Bonner 1996); and recent CB_1 F3.25A mutation studies that suggest that F3.25 is an interaction site for AEA at CB_1 (McAllister et al. 2003). Taken together, these studies suggest that anandamide and its congeners must adopt tightly curved U/J-shaped conformations at CB_1, and suggest that the TMH 2–3–6–7 region is the endocannabinoid-binding region at CB_1.

Finally, it is important to mention that the binding site model proposed by Reggio and co-workers (Barnett-Norris et al. 2002b) does not address one last aspect of endocannabinoid acyl chain SAR, the requirement for an acyl chain of 20–22 carbons. These investigators have hypothesized that this length requirement originates not from the requirements of the final binding site itself, but from requirements for endocannabinoid entry into the binding pocket from lipid. Recently, this group showed that alkyl tail interaction with V6.43(351)/I6.46(354) (which form a groove on CB_1 TMH 6 into which an alkyl tail can fit) results in the induction of an active state conformation for TMH 6 (Barnett-Norris et al. 2002a). Simulations of TMH 6/endocannabinoid interaction in a lipid environment are currently underway in this laboratory to test the hypothesis that only endocannabinoids with 20 to 22 carbon acyl chains (with at least 3 homoallylic double bonds and at least 5 saturated carbons at their ends) extend to the proper depth in the lipid membrane to access the V6.43/I6.46 groove.

5
Aminoalkylindole Pharmacophores

Of all CB agonists, the aminoalkylindoles (AAIs) are the most structurally dissimilar to the classical CBs. This class of compounds also has been found to differ significantly in the set of amino acids important for its binding as revealed by mutation studies (Chin et al. 1998; McAllister et al. 2003; Song and Bonner 1996). It is no wonder, then, that attempts to construct pharmacophores that include WIN55,212-2 in structural superpositions with classical CB agonists have led to the greatest ambiguity.

Adding another layer of complexity to the use of structural superpositions is the fact that the AAIs, as typified by WIN55,212-2 are not rigid compounds, but can adopt several low-energy conformations. AM1 conformational analysis revealed two general classes of accessible conformers at biological temperature, s-*cis* and s-*trans* conformations (see drawings **23** and **24**) for a 2D representation of these conformers (Reggio et al. 1998). This leads to the question: What is the bioactive conformation of the AAIs at CB receptors? It is clear in **23** and **24** that the s-*cis* vs the s-*trans* conformations of WIN55,212-2 place their naphthyl rings in very different regions of space and that these conformers will differ in surface area accessible for intermolecular interactions. As will become more evident in the discussion which follows, the existence of both s-*cis* and s-*trans* conformers of WIN55,212-2 also permits more than one superposition upon a classical CB template.

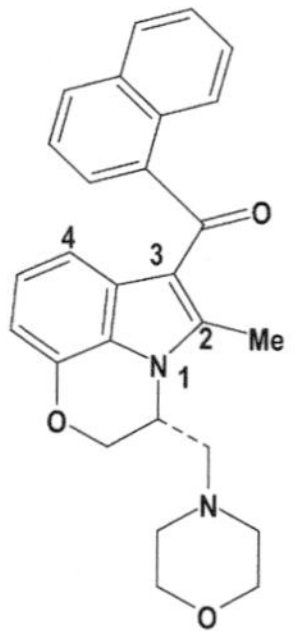

s-cis

23

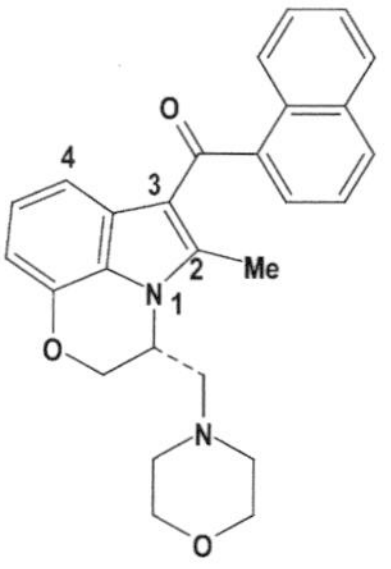

s-trans

24

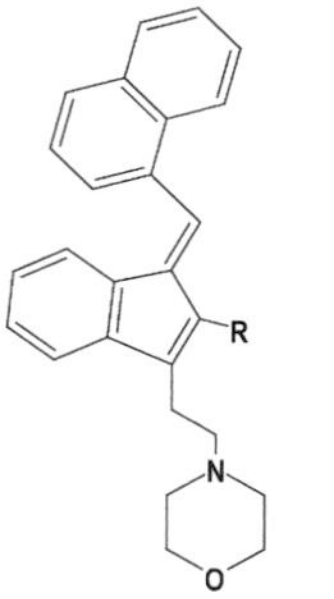

Z-naphthylidene indene

25

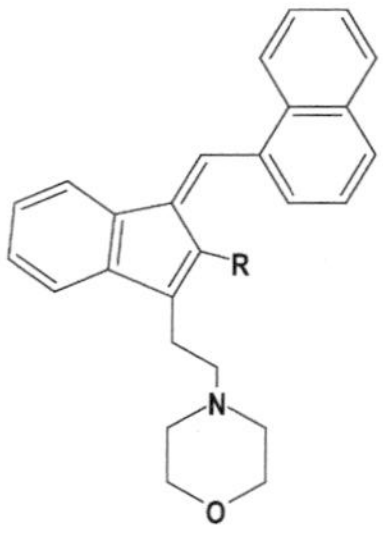

E-naphthylidene indene

26

One way to resolve the ambiguity concerning the bioactive conformation of the AAIs is by the development of rigid AAI analogs. The Z,E-naphthylidene indene AAI analogs (**25**, **26**) synthesized as a mixture by Kumar are rigidified compounds that lack the carbonyl oxygen of the AAIs, but still exhibit high CB_1 affinity (Kumar et al. 1995). Reggio and co-workers extended the work of Kumar by synthesizing each naphthylidene indene geometric isomer. AM1 conformational analysis revealed that the indene E-isomer (**26**) mimics the s-*trans* conformation of WIN55,212-2, while the Z isomer (**25**) mimics the s-*cis* conformation (Reggio et al. 1998). CB_1/CB_2 binding assays revealed that the E-naphthylidene indene (**26**) has significantly higher CB_1 and CB_2 affinity (R = H CB_1 $K_i = 2.72 \pm 0.22$ nM, CB_2 $K_i = 2.72 \pm 0.32$ nM; R = Me CB_1 $K_i = 2.89 \pm 0.41$ nM, CB_2 $K_i = 2.05 \pm 0.22$ nM) than the corresponding Z isomer (**25**) (R = H CB_1 $K_i = 148 \pm 29$ nM, CB_2 $K_i = 132.0 \pm 45.6$ nM; R = Me CB_1 $K_i = 1945 \pm 94$ nM, CB_2 $K_i = 658 \pm 206$ nM) (Reggio et al. 1998). These results point to the s-*trans* AAI conformer (corresponds to the E-indene) as the AAI bioactive conformation at the CB_1 and CB_2 receptors. Detailed below are pharmacophores that have been developed for the AAIs.

5.1
Ligand–Ligand Studies of the Aminoalkylindoles and Related Compounds

Eissenstat and co-workers were first to develop a pharmacophore for AAI binding at CB_1. These investigators presented two pharmacophoric models developed using mouse vas deferens (MVD) data (Eissenstat et al. 1995). The first model was an independent pharmacophore with three key structural features (see compound **3** for numbering system): (1) the nitrogen atom in the amino alkyl side chain; (2) the C-3-aroyl ring, represented by a dummy atom placed at its centroid; and (3) a heterocyclic nucleus represented by a dummy atom placed at the end of a 3-Å normal passing through its centroid. No AAI agonists were identified that did not conform to these pharmacophoric requirements; but not every molecule that fit the pharmacophore was active in the MVD assay. A second approach taken by

Eissenstat et al. involved the potential commonality between AAIs and classical CBs (Eissenstat et al. 1995). The amine of the AAIs was considered to mimic the C-1 phenolic hydroxyl of classical CBs—both engaging in a hydrogen bond with the receptor. Furthermore, these investigators proposed that the amine of the AAIs is similar to the amide functionality of anandamide. They also equated the AAI indole ring with the dibenzopyran ring of classical CBs and the naphthyl ring of the AAIs with the C-3 alkyl side chain of classical CBs.

Shim et al. (1998) reported two CoMFA models for AAI interaction at the CB_1 receptor based on pK_i values measured using radioligand binding assays for [^{3}H]CP-55,940 and [^{3}H]WIN55,212-2. Both models exhibited a strong correlation between the calculated steric-electrostatic fields and the observed biological activity for the respective training set compounds. CoMFA models with AAIs protonated at the morpholino nitrogen were also developed. Comparison of the statistical parameters resulting from these CoMFA models, however, failed to provide unequivocal evidence as to whether the AAIs are protonated or neutral as receptor-bound species. When experimental pK_i values for the training set compounds to displace [^{3}H]WIN55,212-2 were plotted against pK_i values predicted for the same compounds to displace [^{3}H]CP-55,940, the correlation was moderately strong ($R^2 = 0.73$). These authors found that the variation in binding affinity among AAIs was dominated by steric interactions at the receptor site. For CoMFA model 2, the presence of a lipophilic aroyl group promoted increased binding inside a presumed large hydrophobic pocket within the receptor cavity. A sterically forbidden region surrounded C2′ and C3′ of the naphthyl moiety. A region of enhanced binding was found surrounding the C4′ of the naphthyl moiety in 4′-substituted AAIs. Although these CoMFA models were developed using only AAIs, these investigators propose that the CB_1 receptor binding sites of the classical CBs and AAIs may partly overlap and that the two distinct classes of compounds share some common structural features to allow association with the CB_1 receptor (Shim et al. 1998).

Xie and co-workers used a combination of NMR solution studies and molecular modeling to study WIN55,212-2 (**3**; see numbering system). Their results suggest that the minimum energy conformations of the WIN55,212-2 have distinct pharmacophoric features: (1) the naphthyl ring is oriented off the plane of the benzoxazine ring by approximately 59 degrees with the carbonyl C = O group pointing toward the C-2 methyl group, and (2) at the C10-position, the axial morpholinomethyl conformation is preferred over the equatorial in order to relieve a steric interaction with the C-2 methyl group. The preferred conformer as defined by the three key pharmacophores, naphthyl, morpholino, and 3-keto groups, shows that the morpholinyl ring of the molecule WIN55,212-2 deviates from the plane of the benzoxazine ring by about 32 degrees and orients in the left molecular quadrant. This model supports the hypothesis that a certain deviation of the morpholino group from the plane of the indole ring in WIN55,212-2 is essential for cannabimimetic activity. These authors have postulated that such an alignment by the respective pharmacophores allows them to interact optimally with the receptor (Xie et al. 1999).

Dutta and co-workers (1997) reported results for 4-alkyloxy indole AAI derivatives. These investigators aligned the naphthoyl group of WIN55,212-2 (**3**) with the

C-3 side chain of Δ^9-THC (**1**) and the morpholino group of **3** with the carbocyclic ring system of **1**. A similar alignment was suggested by Xie et al. (1995) and by Eissenstat and co-workers (1995; see above). The C-2 methyl group of **3** was aligned with the phenolic hydroxyl of **1**. Because of the conformational flexibility of **3**, these investigators studied the conformational energy of **3** as a function of volume difference with **1**. No specific AAI conformation that could be superimposed with **1** was found to be preferable. Because no specific AAI conformation could be identified as best to overlay with **1**, these investigators proposed that a unique AAI pharmacophore may need to be developed. In addition, Dutta et al. proposed that the keto group of the AAIs may be important for interaction with the CB_1 receptor (Dutta et al. 1997).

In their initial studies of the AAIs, Huffman and co-workers aligned the carbonyl oxygen of WIN55,212-2 with the phenolic hydroxyl of Δ^9-THC (**1**); the naphthyl moiety, with the cyclohexyl and pyran ring of **1**; and the indole nitrogen and the substituent extending from it with C-3 and its alkyl tail in **1** (Huffman et al. 1994). Huffman demonstrated that elimination of the AAI naphthyl substituent led to inactive compounds that failed to bind to the CB_1 receptor (Huffman et al. 1994). Huffman also showed that the morpholino ring can be replaced by an alkyl side chain and retain good CB_1 affinity (Huffman et al. 1994).

On the basis of this initial classical CB/AAI alignment, Huffman and co-workers have synthesized a series of indole- and pyrrole-derived CBs in which the morpholinoethyl group was replaced with another cyclic structure or with a carbon chain that more directly corresponded to the side chain of Δ^9-THC (**1**). Receptor affinity and potency of these novel CBs were related to the length of the carbon chain. Short side chains resulted in inactive compounds, whereas chains with four to six carbons produced optimal in vitro and in vivo activity. Pyrrole-derived CBs were consistently less potent than were the corresponding indole derivatives. These results suggest that, whereas the site of the morpholinoethyl group in these CBs seems crucial for attachment to CB_1 receptors, the exact structural constraints on this part of the molecule are not as strict as previously thought (Wiley et al. 1998).

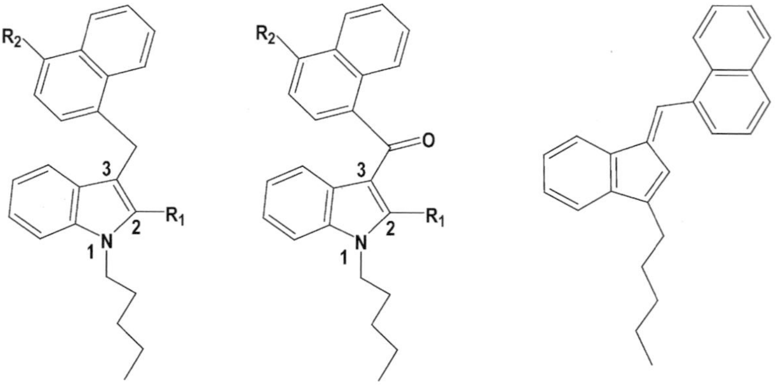

27 **28** **29**

Most recently, Huffman synthesized a series of 1-pentyl-1*H*-indol-3-yl-(1-naphthyl)methanes and 2-methyl-1-pentyl-1*H*-indol-3-yl-(1-naphthyl)methanes to investigate the hypothesis that cannabimimetic 3-(1-naphthoyl)indoles interact with the CB_1 receptor by hydrogen bonding to the carbonyl group. Indoles (**27**) for which $R_1 = H$ and $R_2 = H$, CH_3, or OCH_3 were to found have significant (CB_1 $K_i = 17–23$ nM) receptor affinity, somewhat less than that of the corresponding naphthoylindoles (**28**; $R_1 = H$, $R_2 = H$, CH_3, or OCH_3). A cannabimimetic E-indene hydrocarbon (**29**), which lacks any hydrogen bonding capability, was synthesized and found to have a CB_1 $K_i = 26 \pm 4$ nM. These results suggest that hydrogen bonding of the AAI carbonyl to CB_1 is not crucial for binding.

5.2 Ligand–Receptor Studies of Aminoalkylindole Binding

Reggio and co-workers constructed a pharmacophore for AAI binding at CB_1/CB_2 based on their earlier experimental work that suggested that the s-*trans* conformation of WIN55,212-2 was its bioactive conformation (Reggio et al. 1998). Their work suggested that aromatic stacking was the primary interaction for AAIs at CB_1 and that the aromatic residue-rich TMH 3–4–5–6 region in CB_1 and CB_2 constitutes the binding pocket for AAIs at the CB receptors (Song et al. 1999). These investigators used their CB_1 and CB_2 receptor models to identify F5.46 in CB_2 (a residue that is aromatic only in CB_2) as the residue responsible for the higher affinity of WIN55,212-2 for CB_2. This prediction was confirmed by mutation studies (Song et al. 1999).

Support for the TMH 3–4–5–6 region as the AAI binding region has come from Shire and colleagues' mutation/chimera studies of the CB receptors that suggest that the TMH 4-E-2 loop–TMH 5 region of the CB receptors contains residues important to the binding of the AAI, WIN55,212-2 (Shire et al. 1999). Subsequent modeling studies in the Reggio lab of the CB_1 R* (active state) identified direct stacking interactions between WIN55,212-2 and F3.36, W5.43 and W6.48, with W4.64 and Y5.39 forming part of the extended ligand–CB_1 aromatic cluster. Results of CB_1 F3.36A, W5.43A, and W6.48A mutation studies were consistent with this binding site model (McAllister et al. 2003). A recent modeling study reported by Salo and co-workers identified aromatic stacking interactions of WIN55,212-2 with F3.36, Y5.39, and W5.43 (Salo et al. 2004).

Molecular modeling and receptor docking studies of naphthoylindole (**28**; $R_1 = H$, $R_2 = OCH_3$) and its 2-methyl congener (**28**; $R_1 = CH_3$, $R_2 = OCH_3$) vs indolyl-1-naphthylmethanes (**27**; $R_1 = H$, $R_2 = OCH_3$) and (**27**; $R_1 = CH_3$, $R_2 = OCH_3$), combined with the receptor affinities of these cannabimimetic indoles, strongly suggested that these CB receptor ligands bind primarily by aromatic stacking interactions in the TMH 3–4–5–6 region of the CB_1 receptor (Huffman et al. 2003).

In summary, there is a great divergence between pharmacophores established for AAI binding at CB_1. This divergence can be attributed to the use of different conformations of WIN55,212-2 in superpositions with classical or non-classical

CB templates and the identification of different AAI functional groups as key for interaction at CB_1. Importance in some pharmacophores has been placed on the morpholino nitrogen or the carbonyl oxygen. The fact that the morpholino ring can be replaced by an alkyl chain with no loss in CB_1 affinity (Huffman et al. 1994) and that the carbonyl group can be replaced with a non-hydrogen bonding isostere (i.e., the indenes) with little loss in CB_1 affinity (Huffman et al. 2003; Reggio et al. 1998) argues against the morpholino nitrogen or carbonyl oxygen serving as key interaction sites at CB_1. Instead, it appears that the aromatic rings are key to the receptor interactions of the AAIs at CB_1 (McAllister et al. 2003). Compound **29** ($K_i = 26 \pm 4$ nM) (Huffman et al. 2003) is a good example of this.

6 SR141716A Pharmacophores

SR141716A (**7**) has been shown to act as a competitive antagonist and inverse agonist in host cells transfected with exogenous CB_1 receptor, as well as in biological preparations endogenously expressing CB_1 (Bouaboula et al. 1997; Meschler et al. 2000; Pan et al. 1998). In some experiments, SR141716A has been found to be more potent in blocking the actions of CB_1 agonists than in eliciting inverse responses by itself (Sim-Selley et al. 2001).

6.1 Ligand–Ligand Studies of SR141716A

Thomas and co-workers (Thomas et al. 1998) developed an SAR for SR141716A (**6**) using a set of seven halogenated SR141716A analogs. They concluded that this SR141716A SAR was consistent with a pharmacophoric alignment in which the mono-chloro ring of **7** is overlaid with the C-3 alkyl side chain of Δ^9-THC (**1**); the pyrazole nitrogen of **7** is overlaid with the C-1 phenolic hydroxyl of **1**; and, the carbonyl oxygen of **7** is overlaid with the pyran oxygen (O-5) of **1**. In this superposition, the dichloro ring of SR141716A represents a region unique to SR141716A. This region was hypothesized to be the antagonist-conferring moiety of SR141716A.

More recently, Thomas and co-workers reported an SAR study of the amino-piperidine region (C-3 substituent) of SR141716 (Francisco et al. 2002). Structural modifications made in this study include the substitution of alkyl hydrazines, amines, and hydroxyalkylamines of varying lengths for the aminopiperidinyl moiety. In general, it was found that increasing the length and bulk of the C-3 substituent was associated with increased receptor affinity and efficacy (as measured in a guanosine 5′-triphosphate-γ-[^{35}S] assay). However, in most instances, receptor affinity and efficacy increases were no longer observed after a certain chain length was reached. A quantitative SAR study was carried out to characterize the pharmacophoric requirements of the aminopiperidine region. This model indicated that ligands that exceed 3 Å in length would have reduced potency and affinity with respect to SR141716A and that substituents with a positive charge density in the aminopiperidine region would be predicted to possess increased pharmacological activity (Francisco et al. 2002).

Makriyannis and co-workers designed and synthesized a series of pyrazole derivatives to aid in the characterization of the CB receptor binding sites and also to serve as potentially useful pharmacological probes. Structural requirements for potent and selective brain cannabinoid CB_1 receptor antagonistic activity included (1) a para-substituted phenyl ring at the 5-position, (2) a carboxamido group at the 3-position, and (3) a 2,4-dichlorophenyl substituent at the 1-position of the pyrazole ring (Lan et al. 1999).

Razdan and co-workers synthesized and evaluated a series of SR141716A analogs that retained the central pyrazole structure of SR141716A with replacement of the 1-, 3-, 4-, and/or 5-substituents by alkyl side chains or other substituents known to impart potent agonist activity in traditional tricyclic CB compounds (Wiley et al. 2001). Although none of the analogs alone produced the profile of cannabimimetic effects seen with full agonists, several of the 3-substituent analogs with higher binding affinities showed partial agonism for one or more measures. Cannabimimetic activity was most noted when the 3-substituent of SR141716A was replaced with an alkyl amide or ketone group. None of the 3-substituted analogs produced antagonist effects when tested in combination with 3 mg/kg Δ^9-THC (**1**). In contrast, antagonism of Δ^9-THC's effects without accompanying agonist or partial agonist effects was observed with substitutions at positions 1, 4, and 5. These results suggest that the structural properties of 1- and 5-substituents are primarily responsible for the antagonist activity of SR141716A.

Shim and co-workers used the semi-empirical AM1 method to perform a conformational analysis of SR141716A in its unprotonated and protonated forms. Results from conformational analyses, superimposition models, and 3D-QSAR models suggested that the N1 aromatic ring moiety of SR141716A dominates the steric binding interaction with the receptor in much the same way as does the C3 alkyl side chain of CB agonists and the C-3 aroyl ring of the AAI agonists. Several of the conformers considered in this study were found to possess the proper spatial orientation and distinct electrostatic character to bind to the CB_1 receptor. The authors proposed that the unique region in space occupied by the C-5 aromatic ring of SR141716A might contribute to conferring antagonist activity and that the pyrazole C-3 substituent of SR141716A might contribute to conferring either neutral antagonist or inverse agonist activity, depending upon the interaction with the receptor (Shim et al. 2002).

Lambert and co-workers synthesized a set of 3-alkyl 5-arylimidazolidinediones (hydantoins) with affinity for the human cannabinoid CB_1 receptor (Kanyonyo et al. 1999; Ooms et al. 2002). At least three of these compounds were found to act as neutral antagonists. Using a set of selected compounds, experimental lipophilicity was measured by reversed-phase high-pressure liquid chromatography (RP-HPLC) and calculated by a fragmental method (CLOGP) and a conformation-dependent method (CLIP) based on the molecular lipophilicity potential. The CB agonist 9β-OH-HHC (**21**) was used as a template to which both polar and non-polar hydantoins were fit. For the polar hydantoins, optimal alignment with **21** was achieved by matching the oxygen atom of the morpholino ring or hydroxy moiety with the northern aliphatic hydroxyl (NAH) of **21**, the oxygen atom of the carboxyl amide with the phenolic hydroxyl oxygen of **21**, and the pro-R phenyl ring with the side chain of **21**. For the non-polar hydantoins, two pairs of atoms were used for alignment with **21**: the hydantoin carbonyl oxygens and the oxygen atoms in the carbocyclic and aromatic rings of **21**. No discussion of the basis for the antagonist properties of these ligands was offered (Ooms et al. 2002).

6.2 Ligand–Receptor Models for SR141716A Binding

Reggio and co-workers recently used a combination of synthesis, mutation, electrophysiology, and modeling to identify a binding site for SR141716A at CB_1 (Hurst et al. 2002). A mutant thermodynamic cycle was used to show that K3.28 is a direct interaction site for SR141716A in the inactive state of CB_1. Modeling studies of the CB_1 inactive R state indicated that aromatic stacking interactions were also crucial for SR141716A binding. Direct stacking interactions were identified with F3.36, Y5.39, and W5.43, while W4.64 and W6.48 were part of the larger ligand/aromatic cluster. CB_1 F3.36A, W5.43A, and W6.48A mutation study results were found to be consistent with this binding site model (McAllister et al. 2003). Furthermore, these modeling studies suggested that at the SR141716A binding site, the interaction between the dichlorophenyl ring and F3.36, which in turn interacts with W6.48, helps to maintain the receptor in its inactive state.

A recent modeling study of CB_1 reported by Salo and co-workers identified aromatic stacking interactions for SR141716A in the same aromatic cluster region of CB_1, with direct aromatic stacking interactions identified between SR141716A and Y5.39 and W5.43 (Salo et al. 2004).

6.3 SAR of Other Recently Synthesized CB_1 Antagonists

Mussinu and co-workers (2003) recently reported a new series of rigid 1-aryl-1,4-dihydroindeno[1,2-c]pyrazole-3-carboxamides that are conformationally restricted analogs of SR141716A. These investigators found that conformational restriction resulted in markedly improved CB_2 affinity and selectivity. These compounds were not screened for agonism/antagonism.

Stoit and colleagues reported that benzocycloheptapyrazoles constitute a class of very potent CB_1 receptor antagonists in vitro (Stoit et al. 2002), while Mignani and co-workers have reported that diarylmethyleneazetidine compounds also act as CB_1 antagonists (Mignani et al. 2000). Ruiu and colleagues recently reported that the antagonist NESS0327 (*N*-piperidinyl-[8-chloro-1-(2,4-dichlorophenyl)-1,4,5,6 tetrahydrobenzo[6,7]cyclo-hepta [1,2-*c*]pyrazole-3-carboxamide] is more then 60,000-fold selective for CB_1 (Ruiu et al. 2003). Lange and co-workers reported a series of novel 3,4-diarylpyrazolines which elicited potent in vitro CB_1

antagonistic activities and in general exhibited high CB_1 vs CB_2 receptor subtype selectivities (Lange et al. 2004). The binding affinities of these new compounds were rationalized using the binding site model proposed by Hurst and co-workers (2002)

7 Conclusions

This chapter clearly shows that there has been great growth in our knowledge of the CB receptors and their ligands in the past decade. Pharmacophores have been developed for the CB_1 antagonist SR141716A and for every structural class of CB_1 agonist. Attempts at creating a unified pharmacophore for all CB_1 ligands have not met with success. This is likely because CB_1 ligands do not share a single binding site. CB_2-selective ligand development has been one of the major advances in the CB field in the past decade, and information about GPCR structure combined with mutation studies has permitted refinement of CB_1 and CB_2 receptor models.

Many future challenges for SAR and pharmacophore development still exist. No pharmacophores have yet been developed for agonist or antagonist recognition at CB_2. Another major frontier in modeling studies of CB receptors will be the elucidation of how the CB receptors are activated by agonists and how they are inactivated by inverse agonists. Methodologically, this will be a challenge because the timescale over which activation takes place [milliseconds for receptors with diffusible ligands (Ghanouni et al. 2001b)] is orders of magnitude longer than the timescales currently accessible computationally (nanoseconds). In addition, there is growing evidence that there may be more than one activated state for a GPCR, with the activated conformational state dependent on the agonist that induced it (Ghanouni et al. 2001a). This is very likely the situation in the CB field as well (Glass and Northup 1999) and could present us with the opportunity ultimately to develop highly selective CB ligands that couple through a specific G protein subtype.

Acknowledgements.
This work was supported by the National Institute on Drug Abuse (Grants DA03934 and DA000489).

References

Abadji V, Lin S, Taha G, Griffin G, Stevenson LA, Pertwee RG, Makriyannis A (1994) (R)-Methanandamide: a chiral novel anandamide possessing higher potency and metabolic stability. J Med Chem 37:1889–1893

Abood ME, Ditto KE, Noel MA, Showalter VM, Tao Q (1997) Isolation and expression of a mouse CB1 cannabinoid receptor gene. Comparison of binding properties with those of native CB1 receptors in mouse brain and N18TG2 neuroblastoma cells. Biochem Pharmacol 53:207–214

Adams IB, Ryan W, Singer M, Razdan RK, Compton DR, Martin BR (1995a) Pharmacological and behavioral evaluation of alkylated anandamide analogs. Life Sci 56:2041–2048

Adams IB, Ryan W, Singer M, Thomas BF, Compton DR, Razdan RK, Martin BR (1995b) Evaluation of cannabinoid receptor binding and in vivo activities for anandamide analogs. J Pharmacol Exp Ther 273:1172–1181

Barnett-Norris J, Guarnieri F, Hurst DP, Reggio PH (1998) Exploration of biologically relevant conformations of anandamide, 2-arachidonylglycerol, and their analogues using conformational memories. J Med Chem 41:4861–4872

Barnett-Norris J, Hurst DP, Buehner K, Ballesteros JA, Guarnieri F, Reggio PH (2002a) Agonist alkyl tail interaction with cannabinoid CB1 receptor V6.43/I646 groove induces a helix 6 active conformation. Int J Quantum Chem 88:76–86

Barnett-Norris J, Hurst DP, Lynch DL, Guarnieri F, Makriyannis A, Reggio PH (2002b) Conformational memories and the endocannabinoid binding site at the cannabinoid CB1 receptor. J Med Chem 45:3649–3659

Berglund BA, Boring DL, Wilken GH, Makriyannis A, Howlett AC, Lin S (1998) Structural requirements for arachidonylethanolamide interaction with CB1 and CB2 cannabinoid receptors: pharmacology of the carbonyl and ethanolamide groups. Prostaglandins Leukot Essent Fatty Acids 59:111–118

Berglund BA, Boring DL, Howlett AC (1999) Investigation of structural analogs of prostaglandin amides for binding to and activation of CB1 and CB2 cannabinoid receptors in rat brain and human tonsils. Adv Exp Med Biol 469:527–533

Berglund BA, Fleming PR, Rice KC, Shim JY, Welsh WJ, Howlett AC (2000) Development of a novel class of monocyclic and bicyclic alkyl amides that exhibit CB1 and CB2 cannabinoid receptor affinity and receptor activation. Drug Des Discov 16:281–294

Bonechi C, Brizzi A, Brizzi V, Francoli M, Donati A, Rossi C (2001) Conformational analysis of N-arachidonylethanolamide (anandamide) using nuclear magnetic R\resonanceand theoretical calculations. Magn Reson Chem 39:432–437

Bouaboula M, Perrachon S, Milligan L, Canat X, Rinaldi-Carmona M, Portier M, Barth F, Calandra B, Pecceu F, Lupker J, Maffrand JP, Le Fur G, Casellas P (1997) A selective inverse agonist for central cannabinoid receptor inhibits mitogen-activated protein kinase activation stimulated by insulin or insulin-like growth factor 1. Evidence for a new model of receptor/ligand interactions. J Biol Chem 272:22330–22339

Bouaboula M, Desnoyer N, Carayon P, Combes T, Casellas P (1999) Gi protein modulation induced by a selective inverse agonist for the peripheral cannabinoid receptor CB2: implication for intracellular signalization cross-regulation. Mol Pharmacol 55:473–480

Bracey MH, Hanson MA, Masuda KR, Stevens RC, Cravatt BF (2002) Structural adaptations in a membrane enzyme that terminates endocannabinoid signaling. Science 298:1793–1796

Breivogel CS, Griffin G, Di Marzo V, Martin BR (2001) Evidence for a new G protein-coupled cannabinoid receptor in mouse brain. Mol Pharmacol 60:155–163

Busch-Petersen J, Hill WA, Fan P, Khanolkar A, Xie XQ, Tius MA, Makriyannis A (1996) Unsaturated side chain beta-11-hydroxyhexahydrocannabinol analogs. J Med Chem 39:3790–3796

Chin C, Lucas-Lenard J, Abadji V, Kendall DA (1998) Ligand binding and modulation of cyclic AMP levels depend on the chemical nature of residue 192 of the human cannabinoid receptor 1. J Neurochem 70:366–373

Compton DR, Gold LH, Ward SJ, Balster RL, Martin BR (1992) Aminoalkylindole analogs: cannabimimetic activity of a class of compounds structurally distinct from delta-9-tetrahydrocannabinol. J Pharmacol Exp Ther 263:1118–1126

Corey EJ, Niwa H, Falck JR (1979) Selective epoxidation of ecosa-cis-5, 8, 11, 14-tetraenoic (arachidonic) acid and eicosa-cis-8, 11,14-trienoic acid. J Am Chem Soc 101:1586–1587

D'Ambra TE, Estep KG, Bell MR, Eissenstat MA, Josef KA, Ward SJ, Haycock DA, Baizman ER, Casiano FM, Beglin NC, et al (1992) Conformationally restrained analogues of pravadoline: nanomolar potent, enantioselective, (aminoalkyl)indole agonists of the cannabinoid receptor. J Med Chem 35:124–135

Devane WA, Dysarz FA, 3rd, Johnson MR, Melvin LS, Howlett AC (1988) Determination and characterization of a cannabinoid receptor in rat brain. Mol Pharmacol 34:605–613

Devane WA, Hanus L, Breuer A, Pertwee RG, Stevenson LA, Griffin G, Gibson D, Mandelbaum A, Etinger A, Mechoulam R (1992) Isolation and structure of a brain constituent that binds to the cannabinoid receptor. Science 258:1946–1949

Di Marzo V, Breivogel CS, Tao Q, Bridgen DT, Razdan RK, Zimmer AM, Zimmer A, Martin BR (2000) Levels, metabolism, and pharmacological activity of anandamide in CB(1) cannabinoid receptor knockout mice: evidence for non-CB(1), non-CB(2) receptor-mediated actions of anandamide in mouse brain. J Neurochem 75:2434–2444

Drake DJ, Jensen RS, Busch-Petersen J, Kawakami JK, Concepcion Fernandez-Garcia M, Fan P, Makriyannis A, Tius MA (1998) Classical/nonclassical hybrid cannabinoids: southern aliphatic chain-functionalized C-6beta methyl, ethyl, and propyl analogues. J Med Chem 41:3596–3608

Dutta AK, Ryan W, Thomas BF, Singer M, Compton DR, Martin BR, Razdan RK (1997) Synthesis, pharmacology, and molecular modeling of novel 4-alkyloxy indole derivatives related to cannabimimetic aminoalkyl indoles (AAIs). Bioorg Med Chem 5:1591–1600

Edgemond WS, Campbell WB, Hillard CJ (1995) The binding of novel phenolic derivatives of anandamide to brain cannabinoid receptors. Prostaglandins Leukot Essent Fatty Acids 52:83–86

Edgemond WS, Hillard CJ, Falck JR, Kearn CS, Campbell WB (1998) Human platelets and polymorphonuclear leukocytes synthesize oxygenated derivatives of arachidonylethanolamide (anandamide): their affinities for cannabinoid receptors and pathways of inactivation. Mol Pharmacol 54:180–188

Eissenstat MA, Bell MR, D'Ambra TE, Alexander EJ, Daum SJ, Ackerman JH, Gruett MD, Kumar V, Estep KG, Olefirowicz EM, et al (1995) Aminoalkylindoles: structure-activity relationships of novel cannabinoid mimetics. J Med Chem 38:3094–3105

Felder CC, Joyce KE, Briley EM, Glass M, Mackie KP, Fahey KJ, Cullinan GJ, Hunden DC, Johnson DW, Chaney MO, Koppel GA, Brownstein M (1998) LY320135, a novel cannabinoid CB1 receptor antagonist, unmasks coupling of the CB1 receptor to stimulation of cAMP accumulation. J Pharmacol Exp Ther 284:291–297

Francisco ME, Seltzman HH, Gilliam AF, Mitchell RA, Rider SL, Pertwee RG, Stevenson LA, Thomas BF (2002) Synthesis and structure-activity relationships of amide and hydrazide analogues of the cannabinoid CB(1) receptor antagonist N-(piperidinyl)-5-(4-chlorophenyl)-1-(2,4-dichlorophenyl)-4-methyl-1H-pyrazole-3-carboxami de (SR141716). J Med Chem 45:2708–2719

Fride E, Foox A, Rosenberg E, Faigenboim M, Cohen V, Barda L, Blau H, Mechoulam R (2003) Milk intake and survival in newborn cannabinoid CB1 receptor knockout mice: evidence for a "CB3" receptor. Eur J Pharmacol 461:27–34

Gareau Y, Dufresne C, Gallant M, Rochette C, Sawyer N, Slipetz DM, Tremblay N, Weech PK, Metters KM, Labelle M (1996) Structure activity relationships of tetrahydrocannabinol analogues on human cannabinoid receptors. Bioorg Med Chem Lett 6:189–194

Gerard CM, Mollereau C, Vassart G, Parmentier M (1991) Molecular cloning of a human cannabinoid receptor which is also expressed in testis. Biochem J 279:129–134

Ghanouni P, Gryczynski Z, Steenhuis JJ, Lee TW, Farrens DL, Lakowicz JR, Kobilka BK (2001a) Functionally different agonists induce distinct conformations in the G protein coupling domain of the beta 2 adrenergic receptor. J Biol Chem 276:24433–24436

Ghanouni P, Steenhuis JJ, Farrens DL, Kobilka BK (2001b) Agonist-induced conformational changes in the G-protein-coupling domain of the beta 2 adrenergic receptor. Proc Natl Acad Sci U S A 98:5997–6002

Glass M, Northup JK (1999) Agonist selective regulation of G proteins by cannabinoid CB(1) and CB(2) receptors. Mol Pharmacol 56:1362–1369

Goutopoulos A, Fan P, Khanolkar AD, Xie XQ, Lin S, Makriyannis A (2001) Stereochemical selectivity of methanandamides for the CB1 and CB2 cannabinoid receptors and their metabolic stability. Bioorg Med Chem 9:1673–1684

Griffin G, Wray EJ, Rorrer WK, Crocker PJ, Ryan WJ, Saha B, Razdan RK, Martin BR, Abood ME (1999) An investigation into the structural determinants of cannabinoid receptor ligand efficacy. Br J Pharmacol 126:1575–1584

Griffin G, Tao Q, Abood ME (2000) Cloning and pharmacological characterization of the rat CB(2) cannabinoid receptor. J Pharmacol Exp Ther 292:886–894

Guarnieri F, Weinstein H (1996) Conformational memories and the exploration of biologically relevant peptide conformations: an illustration for the gonadotropin-releasing hormone. J Am Chem Soc 118:5580–5589

Hanus L, Breuer A, Tchilibon S, Shiloah S, Goldenberg D, Horowitz M, Pertwee RG, Ross RA, Mechoulam R, Fride E (1999) HU-308: a specific agonist for CB(2), a peripheral cannabinoid receptor. Proc Natl Acad Sci U S A 96:14228–14233

Hanus L, Abu-Lafi S, Fride E, Breuer A, Vogel Z, Shalev DE, Kustanovich I, Mechoulam R (2001) 2-Arachidonyl glyceryl ether, an endogenous agonist of the cannabinoid CB1 receptor. Proc Natl Acad Sci U S A 98:3662–3665

Hillard CJ, Manna S, Greenberg MJ, DiCamelli R, Ross RA, Stevenson LA, Murphy V, Pertwee RG, Campbell WB (1999) Synthesis and characterization of potent and selective agonists of the neuronal cannabinoid receptor (CB1). J Pharmacol Exp Ther 289:1427–1433

Howlett AC, Johnson MR, Melvin LS, Milne GM (1988) Nonclassical cannabinoid analgetics inhibit adenylate cyclase: development of a cannabinoid receptor model. Mol Pharmacol 33:297–302

Huffman JW, Dai D, Martin BR, Compton DR (1994) Design, synthesis and pharmacology of cannabimimetic indoles. Bioorg Med Chem Lett 4:563–566

Huffman JW, Yu S, Showalter V, Abood ME, Wiley JL, Compton DR, Martin BR, Bramblett RD, Reggio PH (1996) Synthesis and pharmacology of a very potent cannabinoid lacking a phenolic hydroxyl with high affinity for the CB2 receptor. J Med Chem 39:3875–3877

Huffman JW, Yu S, Liddle J, Wiley JL, Abood M, Martin BR, Aung MM (1998) 1998 Symposium on the Cannabinoids; International Cannabinoid Research Society; Burlington, VT. 10 [ISBN: 09658053-2-096580538]

Huffman JW, Liddle J, Yu S, Aung MM, Abood ME, Wiley JL, Martin BR (1999) 3-(1′,1′-Dimethylbutyl)-1-deoxy-delta-8-THC and related compounds: synthesis of selective ligands for the CB2 receptor. Bioorg Med Chem 7:2905–2914

Huffman JW, Bushell SM, Miller JR, Wiley JL, Martin BR (2002) 1-Methoxy-, 1-deoxy-11-hydroxy- and 11-hydroxy-1-methoxy-delta-8-tetrahydrocannabinols: new selective ligands for the CB(2) receptor. Bioorg Med Chem 10:4119–4129

Huffman JW, Mabon R, Wu MJ, Lu J, Hart R, Hurst DP, Reggio PH, Wiley JL, Martin BR (2003) 3-Indolyl-1-naphthylmethanes: new cannabimimetic indoles provide evidence for aromatic stacking interactions with the CB(1) cannabinoid receptor. Bioorg Med Chem 11:539–549

Hulme EC, Lu ZL, Ward SD, Allman K, Curtis CA (1999) The conformational switch in 7-transmembrane receptors: the muscarinic receptor paradigm. Eur J Pharmacol 375:247–260

Hurst DP, Lynch DL, Barnett-Norris J, Hyatt SM, Seltzman HH, Zhong M, Song ZH, Nie J, Lewis D, Reggio PH (2002) N-(piperidin-1-yl)-5-(4-chlorophenyl)-1-(2,4-dichlorophenyl)-4-methyl-1H-p yrazole-3-carboxamide (SR141716A) interaction with LYS 3.28(192) is crucial for its inverse agonism at the cannabinoid CB1 receptor. Mol Pharmacol 62:1274–1287

Iwamura H, Suzuki H, Ueda Y, Kaya T, Inaba T (2001) In vitro and in vivo pharmacological characterization of JTE-907, a novel selective ligand for cannabinoid CB2 receptor. J Pharmacol Exp Ther 296:420–425

Jarai Z, Wagner JA, Varga K, Lake KD, Compton DR, Martin BR, Zimmer AM, Bonner TI, Buckley NE, Mezey E, Razdan RK, Zimmer A, Kunos G (1999) Cannabinoid-induced mesenteric vasodilation through an endothelial site distinct from CB1 or CB2 receptors. Proc Natl Acad Sci U S A 96:14136–14141

Jensen AD, Guarnieri F, Rasmussen SG, Asmar F, Ballesteros JA, Gether U (2001) Agonist-induced conformational changes at the cytoplasmic side of transmembrane segment 6

in the beta 2 adrenergic receptor mapped by site-selective fluorescent labeling. J Biol Chem 276:9279–9290
Kanyonyo M, Govaerts SJ, Hermans E, Poupaert JH, Lambert DM (1999) 3-Alkyl-(5,5′-diphenyl)imidazolidineiones as new cannabinoid receptor ligands. Bioorg Med Chem Lett 9:2233–2236
Kearn CS, Greenberg MJ, DiCamelli R, Kurzawa K, Hillard CJ (1999) Relationships between ligand affinities for the cerebellar cannabinoid receptor CB1 and the induction of GDP/GTP exchange. J Neurochem 72:2379–2387
Keimowitz AR, Martin BR, Razdan RK, Crocker PJ, Mascarella SW, Thomas BF (2000) QSAR analysis of delta-8-THC analogues: relationship of side-chain conformation to cannabinoid receptor affinity and pharmacological potency. J Med Chem 43:59–70
Khanolkar AD, Abadji V, Lin S, Hill WA, Taha G, Abouzid K, Meng Z, Fan P, Makriyannis A (1996) Head group analogs of arachidonylethanolamide, the endogenous cannabinoid ligand. J Med Chem 39:4515–4519
Khanolkar AD, Lu D, Fan P, Tian X, Makriyannis A (1999) Novel conformationally restricted tetracyclic analogs of delta-8-tetrahydrocannabinol. Bioorg Med Chem Lett 9:2119–2124
Krishnamurthy M, Ferreira AM, Moore BM, 2nd (2003) Synthesis and testing of novel phenyl substituted side-chain analogues of classical cannabinoids. Bioorg Med Chem Lett 13:3487–3490
Kumar V, Alexander MD, Bell MR, Eissenstat MA, Casiano FM, Chippari SM, Haycock DA, Luttinger DA, Kuster JE, Miller MS, Stevenson JI, Ward SJ (1995) Morpholinoalkylindenes as antinociceptive agents: novel cannabinoid receptor agonists. Bioorg Med Chem Lett 5:381–386
Lagu SG, Varona A, Chambers JD, Reggio PH (1995) Construction of a steric map of the binding pocket for cannabinoids at the cannabinoid receptor. Drug Des Discov 12:179–192
Lan R, Liu Q, Fan P, Lin S, Fernando SR, McCallion D, Pertwee R, Makriyannis A (1999) Structure-activity relationships of pyrazole derivatives as cannabinoid receptor antagonists. J Med Chem 42:769–776
Lang W, Qin C, Lin S, Khanolkar AD, Goutopoulos A, Fan P, Abouzid K, Meng Z, Biegel D, Makriyannis A (1999) Substrate specificity and stereoselectivity of rat brain microsomal anandamide amidohydrolase. J Med Chem 42:896–902
Lange JH, Coolen HK, van Stuivenberg HH, Dijksman JA, Herremans AH, Ronken E, Keizer HG, Tipker K, McCreary AC, Veerman W, Wals HC, Stork B, Verveer PC, den Hartog AP, de Jong NM, Adolfs TJ, Hoogendoorn J, Kruse CG (2004) Synthesis, biological properties, and molecular modeling investigations of novel 3,4-diarylpyrazolines as potent and selective CB(1) cannabinoid receptor antagonists. J Med Chem 47:627–643
Lin S, Khanolkar AD, Fan P, Goutopoulos A, Qin C, Papahadjis D, Makriyannis A (1998) Novel analogues of arachidonylethanolamide (anandamide): affinities for the CB1 and CB2 cannabinoid receptors and metabolic stability. J Med Chem 41:5353–5361
Makriyannis A, Rapaka RS (1990) The molecular basis of cannabinoid activity. Life Sci 47:2173–2184
Martin BR, Jefferson R, Winckler R, Wiley JL, Huffman JW, Crocker PJ, Saha B, Razdan RK (1999) Manipulation of the tetrahydrocannabinol side chain delineates agonists, partial agonists, and antagonists. J Pharmacol Exp Ther 290:1065–1079
Mato S, Pazos A, Valdizan EM (2002) Cannabinoid receptor antagonism and inverse agonism in response to SR141716A on cAMP production in human and rat brain. Eur J Pharmacol 443:43–46
Matsuda LA, Lolait SJ, Brownstein MJ, Young AC, Bonner TI (1990) Structure of a cannabinoid receptor and functional expression of the cloned cDNA. Nature 346:561–564
McAllister SD, Rizvi G, Anavi-Goffer S, Hurst DP, Barnett-Norris J, Lynch DL, Reggio PH, Abood ME (2003) An aromatic microdomain at the cannabinoid CB(1) receptor constitutes an agonist/inverse agonist binding region. J Med Chem 46:5139–5152
Mechoulam R, Ben-Shabat S, Hanus L, Ligumsky M, Kaminski NE, Schatz AR, Gopher A, Almog S, Martin BR, Compton DR, et al (1995) Identification of an endogenous

2-monoglyceride, present in canine gut, that binds to cannabinoid receptors. Biochem Pharmacol 50:83–90
Melvin LS, Johnson MR (1987) Structure-activity relationships of tricyclic and nonclassical bicyclic cannabinoids. NIDA Res Monogr 79:31–47
Melvin LS, Milne GM, Johnson MR, Subramaniam B, Wilken GH, Howlett AC (1993) Structure-activity relationships for cannabinoid receptor-binding and analgesic activity: studies of bicyclic cannabinoid analogs. Mol Pharmacol 44:1008–1015
Melvin LS, Milne GM, Johnson MR, Wilken GH, Howlett AC (1995) Structure-activity relationships defining the ACD-tricyclic cannabinoids: cannabinoid receptor binding and analgesic activity. Drug Des Discov 13:155–166
Meschler JP, Kraichely DM, Wilken GH, Howlett AC (2000) Inverse agonist properties of N-(piperidin-1-yl)-5-(4-chlorophenyl)-1-(2, 4-dichlorophenyl)-4-methyl-1H-pyrazole-3-carboxamide HCl (SR141716A) and 1-(2-chlorophenyl)-4-cyano-5-(4-methoxyphenyl)-1H-pyrazole-3-carboxylic acid phenylamide (CP-272871) for the CB(1) cannabinoid receptor. Biochem Pharmacol 60:1315–1323
Mignani S, Hittinger O, Archard D, Bouchard H, Bouquerel J, Capet M, Grisoni S, Malleron JL (2000) Preparation of 1-bis(aryl)methyl-3-(alkylsulfonyl)arylmethyleneazetidines as cannabinoid CB1 receptor antagonists. Chemical Abstracts. Aventis Pharma SA, France, p 236982s
Munro S, Thomas KL, Abu-Shaar M (1993) Molecular characterization of a peripheral receptor for cannabinoids. Nature 365:61–65
Mussinu JM, Ruiu S, Mule AC, Pau A, Carai MA, Loriga G, Murineddu G, Pinna GA (2003) Tricyclic pyrazoles. Part 1: synthesis and biological evaluation of novel 1,4-dihydroindeno[1,2-c]pyrazol-based ligands for CB(1)and CB(2)cannabinoid receptors. Bioorg Med Chem 11:251–263
Nadipuram AK, Krishnamurthy M, Ferreira AM, Li W, Moore BM (2003) Synthesis and testing of novel classical cannabinoids: exploring the side chain ligand binding pocket of the CB1 and CB2 receptors. Bioorg Med Chem 11:3121–3132
Nakanishi K, Zhang H, Lerro KA, Takekuma S, Yamamoto T, Lien TH, Sastry L, Back D-J, Moquin-Pattey C, Boehm MF, Derguini F, Gawinowicz MA (1995) Photoaffinity labeling of rhodopsin and bacteriorhodopsin. Biophys Chem 56:13–22
Ng EW, Aung MM, Abood ME, Martin BR, Razdan RK (1999) Unique analogues of anandamide: arachidonyl ethers and carbamates and norarachidonyl carbamates and ureas. J Med Chem 42:1975–1981
Ooms F, Wouters J, Oscari O, Happaerts T, Bouchard G, Carrupt PA, Testa B, Lambert DM (2002) Exploration of the pharmacophore of 3-alkyl-5-arylimidazolidinediones as new CB(1) cannabinoid receptor ligands and potential antagonists: synthesis, lipophilicity, affinity, and molecular modeling. J Med Chem 45:1748–1756
Palczewski K, Kumasaka T, Hori T, Behnke CA, Motoshima H, Fox BA, Le Trong I, Teller DC, Okada T, Stenkamp RE, Yamamoto M, Miyano M (2000) Crystal structure of rhodopsin: A G protein-coupled receptor. Science 289:739–745
Pan X, Ikeda SR, Lewis DL (1998) SR 141716A acts as an inverse agonist to increase neuronal voltage-dependent Ca2+ currents by reversal of tonic CB1 cannabinoid receptor activity. Mol Pharmacol 54:1064–1072
Papahatjis DP, Kourouli T, Abadji V, Goutopoulos A, Makriyannis A (1998) Pharmacophoric requirements for cannabinoid side chains: multiple bond and C1′-substituted delta-8-tetrahydrocannabinols. J Med Chem 41:1195–1200
Papahatjis DP, Nikas SP, Andreou T, Makriyannis A (2002) Novel 1′,1′-chain substituted delta-8-tetrahydrocannabinols. Bioorg Med Chem Lett 12:3583–3586
Pinto JC, Potie F, Rice KC, Boring D, Johnson MR, Evans DM, Wilken GH, Cantrell CH, Howlett AC (1994) Cannabinoid receptor binding and agonist activity of amides and esters of arachidonic acid. Mol Pharmacol 46:516–522
Rabinovich AL, Ripatti PO (1991) On the conformational, physical properties and function of polyunsaturated AcylChains. Biochim Biophys Acta 1085:53–56

Razdan RK (1986) Structure-activity relationships in cannabinoids. Pharmacol Rev 38:75–149

Reggio PH, Traore H (2000) Conformational requirements for endocannabinoid interaction with the cannabinoid receptors, the anandamide transporter and fatty acid amidohydrolase. Chem Phys Lipids 108:15–35

Reggio PH, Panu AM, Miles S (1993) Characterization of a region of steric interference at the cannabinoid receptor using the active analog approach. J Med Chem 36:1761–1771

Reggio PH, Wang T, Brown AE, Fleming DN, Seltzman HH, Griffin G, Pertwee RG, Compton DR, Abood ME, Martin BR (1997) Importance of the C-1 substituent in classical cannabinoids to CB2 receptor selectivity: synthesis and characterization of a series of O,2-propano-delta-8-tetrahydrocannabinol analogs. J Med Chem 40:3312–3318

Reggio PH, Basu-Dutt S, Barnett-Norris J, Castro MT, Hurst DP, Seltzman HH, Roche MJ, Gilliam AF, Thomas BF, Stevenson LA, Pertwee RG, Abood ME (1998) The bioactive conformation of aminoalkylindoles at the cannabinoid CB1 and CB2 receptors: insights gained from (E)- and (Z)-naphthylidene indenes. J Med Chem 41:5177–5187

Rich MR (1993) Conformational analysis of arachidonic and related fatty acids using molecular dynamics simulations. Biochim Biophys Acta 1178:87–96

Rinaldi-Carmona M, Barth F, Heaulme M, Shire D, Calandra B, Congy C, Martinez S, Maruani J, Neliat G, Caput D, et al (1994) SR141716A, a potent and selective antagonist of the brain cannabinoid receptor. FEBS Lett 350:240–244

Rinaldi-Carmona M, Barth F, Millan J, Derocq JM, Casellas P, Congy C, Oustric D, Sarran M, Bouaboula M, Calandra B, Portier M, Shire D, Breliere JC, Le Fur GL (1998) SR 144528, the first potent and selective antagonist of the CB2 cannabinoid receptor. J Pharmacol Exp Ther 284:644–650

Ross RA, Brockie HC, Fernando SR, Saha B, Razdan RK, Pertwee RG (1998) Comparison of cannabinoid binding sites in guinea-pig forebrain and small intestine. Br J Pharmacol 125:1345–1351

Ross RA, Brockie HC, Stevenson LA, Murphy VL, Templeton F, Makriyannis A, Pertwee RG (1999a) Agonist-inverse agonist characterization at CB1 and CB2 cannabinoid receptors of L759633, L759656 and AM630. Br J Pharmacol 126:665–672

Ross RA, Gibson TM, Stevenson LA, Saha B, Crocker P, Razdan RK, Pertwee RG (1999b) Structural determinants of the partial agonist-inverse agonist properties of 6′-azidohex-2′-yne-delta-8-tetrahydrocannabinol at cannabinoid receptors. Br J Pharmacol 128:735–743

Ruiu S, Pinna GA, Marchese G, Mussinu JM, Saba P, Tambaro S, Casti P, Vargiu R, Pani L (2003) Synthesis and characterization of NESS 0327: a novel putative antagonist of the CB1 cannabinoid receptor. J Pharmacol Exp Ther 306:363–370

Ryan W, Singer M, Razdan RK, Compton DR, Martin BR (1995) A novel class of potent tetrahydrocannabinols (THCS): 2′-yne-delta-8- and delta-9-THCs. Life Sci 56:2013–2020

Ryan WJ, Banner WK, Wiley JL, Martin BR, Razdan RK (1997) Potent anandamide analogs: the effect of changing the length and branching of the end pentyl chain. J Med Chem 40:3617–3625

Salo OM, Lahtela-Kakkonen M, Gynther J, Jarvinen T, Poso A (2004) Development of a 3D model for the human cannabinoid CB1 receptor. J Med Chem 47:3048–3057

Seltzman HH, Fleming DN, Thomas BF, Gilliam AF, McCallion DS, Pertwee RG, Compton DR, Martin BR (1997) Synthesis and pharmacological comparison of dimethylheptyl and pentyl analogs of anandamide. J Med Chem 40:3626–3634

Sheskin T, Hanus L, Slager J, Vogel Z, Mechoulam R (1997) Structural requirements for binding of anandamide-type compounds to the brain cannabinoid receptor. J Med Chem 40:659–667

Shim JY, Collantes ER, Welsh WJ, Subramaniam B, Howlett AC, Eissenstat MA, Ward SJ (1998) Three-dimensional quantitative structure-activity relationship study of the cannabimimetic (aminoalkyl)indoles using comparative molecular field analysis. J Med Chem 41:4521–4532

Shim JY, Welsh WJ, Cartier E, Edwards JL, Howlett AC (2002) Molecular interaction of the antagonist N-(piperidin-1-yl)-5-(4-chlorophenyl)-1-(2,4-dichlorophenyl)-4-methyl-1H-pyrazole-3-carboxamide with the CB1 cannabinoid receptor. J Med Chem 45:1447–1459
Shim JY, Welsh WJ, Howlett AC (2003) Homology model of the CB1 cannabinoid receptor: sites critical for nonclassical cannabinoid agonist interaction. Biopolymers 71:169–189
Shire D, Carillon C, Kaghad M, Calandra B, Rinaldi-Carmona M, Le Fur G, Caput D, Ferrara P (1995) An amino-terminal variant of the central cannabinoid receptor resulting from alternative splicing. J Biol Chem 270:3726–3731
Shire D, Calandra B, Rinaldi-Carmona M, Oustric D, Pessegue B, Bonnin-Cabanne O, Le Fur G, Caput D, Ferrara P (1996) Molecular cloning, expression and function of the murine CB2 peripheral cannabinoid receptor. Biochim Biophys Acta 1307:132–136
Shire D, Calandra B, Bouaboula M, Barth F, Rinaldi-Carmona M, Casellas P, Ferrara P (1999) Cannabinoid receptor interactions with the antagonists SR 141716A and SR 144528. Life Sci 65:627–635
Sim-Selley LJ, Brunk LK, Selley DE (2001) Inhibitory effects of SR141716A on G-protein activation in rat brain. Eur J Pharmacol 414:135–143
Singer M, Ryan WJ, Saha B, Martin BR, Razdan RK (1998) Potent cyano and carboxamido side-chain analogues of 1′, 1′-dimethyl-delta-8-tetrahydrocannabinol. J Med Chem 41:4400–4407
Song ZH, Bonner TI (1996) A lysine residue of the cannabinoid receptor is critical for receptor recognition by several agonists but not WIN55212-2. Mol Pharmacol 49:891–896
Song ZH, Slowey CA, Hurst DP, Reggio PH (1999) The difference between the CB(1) and CB(2) cannabinoid receptors at position 5.46 is crucial for the selectivity of WIN55212-2 for CB(2). Mol Pharmacol 56:834–840
Stegeman R, Pawlitz J, Stevens A, Gierse J, Stallings W, Kurumbail R (1998) Mechanism of cyclooxygenase reactions: structure of arachidonic acid bound to cyclooxygenase-2. American Crystallographic Association. Abstract
Stella N, Schweitzer P, Piomelli D (1997) A second endogenous cannabinoid that modulates long-term potentiation. Nature 388:773–778
Stoit AR, Lange JHM, den Hartog AP, Ronken E, Tipker K, van Stuivenberg HH, Dijksman JAR, Wals HC, Kruse CG (2002) Design, synthesis and biological activity of rigid cannabinoid CB1 receptor antagonists. Chem Pharm Bull (Tokyo) 50:1109–1113
Thomas BF, Compton DR, Martin BR, Semus SF (1991) Modeling the cannabinoid receptor: a three-dimensional quantitative structure-activity analysis. Mol Pharmacol 40:656–665
Thomas BF, Adams IB, Mascarella SW, Martin BR, Razdan RK (1996) Structure-activity analysis of anandamide analogs: relationship to a cannabinoid pharmacophore. J Med Chem 39:471–479
Thomas BF, Gilliam AF, Burch DF, Roche MJ, Seltzman HH (1998) Comparative receptor binding analyses of cannabinoid agonists and antagonists. J Pharmacol Exp Ther 285:285–292
Tius MA, Makriyannis A, Zoua XL, Abadji V (1994) Conformationally restricted hybrids of CP-55,940 and HHC: stereoselective synthesis and activity. Tetrahedron 50:2671–2680
Tong W, Collantes ER, Welsh WJ, Berglund BA, Howlett AC (1998) Derivation of a pharmacophore model for anandamide using constrained conformational searching and comparative molecular field analysis. J Med Chem 41:4207–4215
van der Stelt M, van Kuik JA, Bari M, van Zadelhoff G, Leeflang BR, Veldink GA, Finazzi-Agro A, Vliegenthart JF, Maccarrone M (2002) Oxygenated metabolites of anandamide and 2-arachidonoylglycerol: conformational analysis and interaction with cannabinoid receptors, membrane transporter, and fatty acid amide hydrolase. J Med Chem 45:3709–3720
Wagner JA, Varga K, Jarai Z, Kunos G (1999) Mesenteric vasodilation mediated by endothelial anandamide receptors. Hypertension 33:429–434

Ward SJ, Baizman E, Bell M, Childers S, D'Ambra T, Eissenstat M, Estep K, Haycock D, Howlett A, Luttinger D, et al (1991) Aminoalkylindoles (AAIs): a new route to the cannabinoid receptor? NIDA Res Monogr 105:425–426

Wiley JL, Compton DR, Dai D, Lainton JA, Phillips M, Huffman JW, Martin BR (1998) Structure-activity relationships of indole- and pyrrole-derived cannabinoids. J Pharmacol Exp Ther 285:995–1004

Wiley JL, Jefferson RG, Grier MC, Mahadevan A, Razdan RK, Martin BR (2001) Novel pyrazole cannabinoids: insights into CB(1) receptor recognition and activation. J Pharmacol Exp Ther 296:1013–1022

Wiley JL, Beletskaya ID, Ng EW, Dai Z, Crocker PJ, Mahadevan A, Razdan RK, Martin BR (2002) Resorcinol derivatives: a novel template for the development of cannabinoid CB(1)/CB(2) and CB(2)-selective agonists. J Pharmacol Exp Ther 301:679–689

Xie XQ, Yang DP, Melvin LS, Makriyannis A (1994) Conformational analysis of the prototype nonclassical cannabinoid CP-47,497, using 2D NMR and computer molecular modeling. J Med Chem 37:1418–1426

Xie XQ, Eissenstat M, Makriyannis A (1995) Common cannabimimetic pharmacophoric requirements between aminoalkyl indoles and classical cannabinoids. Life Sci 56:1963–1970

Xie XQ, Melvin LS, Makriyannis A (1996) The conformational properties of the highly selective cannabinoid receptor ligand CP-55,940. J Biol Chem 271:10640–10647

Xie XQ, Pavlopoulos S, DiMeglio CM, Makriyannis A (1998) Conformational studies on a diastereoisomeric pair of tricyclic nonclassical cannabinoids by NMR spectroscopy and computer molecular modeling. J Med Chem 41:167–174

Xie XQ, Han XW, Chen JZ, Eissenstat M, Makriyannis A (1999) High-resolution NMR and computer modeling studies of the cannabimimetic aminoalkylindole prototype WIN-55212-2. J Med Chem 42:4021–4027

Xie XQ, Chen JZ, Billings EM (2003) 3D structural model of the G-protein-coupled cannabinoid CB2 receptor. Proteins 53:307–319

HEP (2005) 168:283–297

The Phylogenetic Distribution and Evolutionary Origins of Endocannabinoid Signalling

M.R. Elphick (✉) · M. Egertová

School of Biological Sciences, Queen Mary, University of London, London E1 4NS, UK
M.R.Elphick@qmul.ac.uk

Abstract The endocannabinoid signalling system in mammals comprises several molecular components, including cannabinoid receptors (e.g. CB_1, CB_2), putative endogenous ligands for these receptors [e.g. anandamide, 2-arachidonoylglycerol (2-AG)] and enzymes involved in the biosynthesis and inactivation of anandamide (e.g. NAPE-PLD, FAAH) and 2-AG (e.g. DAG lipase, MGL). In this review we examine the occurrence of these molecules in non-mammalian organisms (in particular, animals and plants) by surveying published data and by basic local alignment search tool (BLAST) analysis of the GenBank database and of genomic sequence data from several vertebrate and invertebrate species. We conclude that the ability of cells to synthesise molecules that are categorised as "endocannabinoids" in mammals is an evolutionarily ancient phenomenon that may date back to the unicellular common ancestor of animals and plants. However, exploitation of these molecules for intercellular signalling may have occurred independently in different lineages during the evolution of the eukaryotes. The CB_1- and CB_2-type receptors that mediate effects of endocannabinoids in mammals occur throughout the vertebrates, and an orthologue of vertebrate cannabinoid receptors was recently identified in the deuterostomian invertebrate *Ciona intestinalis* (CiCBR). However, orthologues of the vertebrate cannabinoid receptors are not found in protostomian invertebrates (e.g. *Drosophila, Caenorhabditis elegans*). Therefore, it is likely that a CB_1/CB_2-type cannabinoid receptor originated in a deuterostomian invertebrate. This phylogenetic information provides a basis for exploitation of selected non-mammalian organisms as model systems for research on endocannabinoid signalling.

Keywords Cannabinoid · Anandamide · 2-Arachidonoylglycerol · Deuterostome · Protostome

1 Introduction

Cannabinoid receptors are activated by Δ^9-tetrahydrocannabinol, the main psychoactive constituent of the drug cannabis (Howlett et al. 2002). Two G protein-coupled cannabinoid receptors have been identified in humans and other mammals and are known as CB_1 and CB_2 (Matsuda et al. 1990; Munro et al. 1993). CB_1 is expressed by neurons and mediates effects of cannabis on the central nervous system (CNS) whereas CB_2 is associated with cells in the immune system. Following the discovery of CB_1 and CB_2, putative endogenous ligands for these receptors were isolated from mammalian tissues and identified as derivatives of arachidonic acid. The first "endocannabinoid" to be characterised was arachidonoylethanolamide ("anandamide"; Devane et al. 1992) followed by 2-arachidonoylglycerol (2-AG; Mechoulam et al. 1995; Sugiura et al. 1995). With these discoveries the concept of an endocannabinoid signalling system in mammals has emerged. Moreover, the physiological roles of the endocannabinoid signalling system in mammals are beginning to be elucidated. Recently it was established that endocannabinoids and the CB_1 receptor mediate retrograde signalling at synapses in the brain (Wilson and Nicoll 2002; Kreitzer and Regehr 2002), confirming a hypothesis first put forward by Egertová et al. (1998) and elaborated on by Elphick and Egertová (2001).

The purpose of this article is not to review research on endocannabinoid signalling in mammals, as this topic is covered in detail in other chapters of this volume and in other recent reviews (Freund et al. 2003; Piomelli 2003). The aim here is to examine the phylogenetic distribution and evolutionary origins of the molecular components that are recognised as constituents of the endocannabinoid signalling system in mammals. This is not the first article to discuss the evolution of endocannabinoid signalling; several reviews on comparative aspects of cannabinoid biology have been published in recent years, including: Salzet et al. (2000), Elphick and Egertová (2001), Salzet and Stefano (2002) and McPartland and Pruitt (2002). What then justifies writing another? First, important discoveries have been made since the last review appeared. Second, there are conflicting views on interpretation of some published data. For this review we will largely restrict our analysis to eukaryotes and in particular animals and plants, although in doing so we do not presume that some elements of the endocannabinoid signalling system in mammals might not have their origins in more ancient prokaryotic organisms.

To investigate the phylogenetic distribution of proteins that could mediate the biosynthesis, inactivation and physiological effects of endocannabinoids in non-mammalian organisms, in addition to surveying published papers, we have employed the basic local alignment search tool (BLAST; Altschul et al. 1990) to analyse the GenBank database and databases specifically associated with genome sequencing projects, using mammalian endocannabinoid-related proteins as search sequences. The primary focus for these searches were several non-mammalian animal species where complete or near complete genome sequence data are available. These include the vertebrate species *Fugu rubripes* (puffer fish; Aparicio et al. 2002), *Danio rerio* (zebrafish), *Xenopus laevis* (African clawed toad) and *Gallus gallus* (chicken), and the invertebrate species *Caenorhabditis elegans* (nematode worm; The *C. elegans* sequencing consortium 1998), *Drosophila elegans* (fruit fly; Adams et al. 2000), and *Ciona intestinalis* (sea-squirt; Dehal et al. 2002).

Interpretation of the significance of results obtained from BLAST analysis of genome sequence data from different species requires knowledge of animal phylogeny, and therefore a brief introduction is necessary here. Comparative analysis of extant animals based on both morphological and molecular data indicates that the animal kingdom comprises two main clades: (1) the deuterostomes, which include vertebrates, cephalochordates, urochordates (e.g. *Ciona*), hemichordates and echinoderms and (2) the protostomes, which are further sub-divided into two assemblages: (a) the ecdysozoa, which include nematodes (e.g. *C. elegans*) and arthropods (e.g. *Drosophila*) and (b) the lophotrochozoa, which include molluscs and annelids. Basal to the deuterostomes and protostomes are the cnidarians (e.g. *Hydra*), which are the most primitive animals with nervous systems (Adoutte et al. 2000).

2
The Phylogeny of Endocannabinoids

2.1
The Phylogenetic Distribution of Anandamide and Enzymes Involved in Anandamide Biosynthesis

Anandamide (arachidonoylethanolamide) is just one of a family of lipids known as *N*-acylethanolamines (NAEs), which are generated from membrane phospholipids via a common enzymatic pathway (see below). The occurrence of anandamide in an organism is dependent on: (1) the presence of the fatty acid arachidonic acid as a component of membrane phospholipids and (2) the presence of enzymes that can catalyse formation of NAEs from membrane phospholipids. Therefore, the phylogenetic distribution of anandamide is likely to reflect a combination of both the phylogenetic distribution of arachidonic acid as a fatty acid component of membrane lipids and the phylogenetic distribution of the enzymes that can catalyse formation of NAEs.

The presence of arachidonic acid in an organism is determined by diet and/or the presence of enzymes that catalyse formation of arachidonic acid from other fatty acids. In mammals, arachidonic acid is synthesised from linoleic acid through the sequential activity of $\Delta 6$ fatty acid desaturase, $\Delta 6$ fatty acid elongase and $\Delta 5$ fatty acid desaturase (Nakamura and Nara 2003). Interestingly, zebrafish have a single gene encoding an enzyme with both $\Delta 5$ and $\Delta 6$ fatty acid desaturase activities, whereas the nematode *C. elegans*, like mammals, has two genes encoding a $\Delta 5$ fatty acid desaturase and a $\Delta 6$ fatty acid desaturase (Hastings et al. 2001; Napier and Michaelson 2001). These findings indicate that vertebrate and invertebrate species can generate arachidonic acid, although this fatty acid is not necessarily ubiquitous in animals. For example, arachidonic acid is not found as a component of phospholipids in *Drosophila* heads (Yoshioka et al. 1985), which may reflect lack of expression of genes encoding fatty acid desaturases and elongases or loss of genes encoding these enzymes. However, in animal species that lack genes encoding fatty acid desaturases and/or elongases, arachidonic acid may be a dietary constituent. Thus, determination of an organism's potential for generating anandamide from arachidonic acid, as a component of membrane phospholipids, may require assessment of both molecular genetic and dietary information. Consequently, there are unlikely to be discrete phylogenetic patterns in the distribution of arachidonic acid, and hence the potential to generate anandamide.

Anandamide and other NAEs are synthesised in mammalian tissues through the sequential action of two enzymes: (1) a *N*-acyltransferase that generates *N*-acylphosphatidylethanolamine (NAPE) from phosphatidylcholine and phosphatidylethanolamine and (2) a NAPE-phospholipase D (NAPE-PLD) that generates anandamide and other NAEs by cleavage of NAPE (Schmid et al. 1990; Di Marzo et al. 1994; Piomelli 2003). The presence of enzymes that catalyse these reactions has also been reported in invertebrate animals and in plant species (Bisogno et al. 1997; Chapman 2000), indicating that the enzymatic machinery for formation of NAEs may be evolutionarily ancient. Unfortunately, phylogenetic analysis of the distribution of these enzymes has been hindered by lack of sequence data for genes that encode these enzymes. An important breakthrough was reported recently, however, with the cloning and sequencing of cDNAs encoding NAPE-PLD in human, rat and mouse (Okamoto et al. 2004). Thus, it is now possible to investigate the occurrence of related proteins in non-mammalian species. Analysis of genome sequence data for the puffer fish *Fugu rubripes* reveals the presence of a gene encoding a protein that shares a high level of sequence identity (60%) with mammalian NAPE-PLDs. This protein is likely to be a fish orthologue of mammalian NAPE-PLDs, and therefore NAPE-PLDs probably occur throughout the vertebrates. However, genes encoding proteins resembling NAPE-PLD do not appear to be present in two of the invertebrate species for which there are complete genome sequence data available, the insect *Drosophila melanogaster* and the sea-squirt *Ciona intestinalis*. Proteins sharing approximately 40% sequence identity with mammalian NAPE-PLDs are present in the nematode worm *C. elegans* and in numerous bacterial species. However, experimental studies are required to determine if these proteins actually function as NAPE-PLDs.

2.2
The Phylogenetic Distribution of Fatty Acid Amide Hydrolase

The existence of an enzyme in mammalian tissues that catalyses hydrolysis of anandamide to arachidonic acid and ethanolamide was established soon after the identification of anandamide as an endogenous cannabinoid (Deutsch and Chin 1993; Di Marzo et al. 1994; Ueda et al. 1995). Molecular characterisation of this enzyme was accomplished by Cravatt et al. (1996) with the cloning and sequencing of a rat cDNA encoding a protein that is now known as fatty acid amide hydrolase (FAAH). Genes encoding orthologues of rat FAAH have been identified in human and mouse (Giang and Cravatt 1997), but relatively little is known about the occurrence of FAAH in non-mammalian animals. There are, however, several reports of FAAH activity in homogenates of tissues from a variety of invertebrate species. For example, FAAH-like activity has been detected in whole-animal homogenates of the cnidarian *Hydra viridis* (De Petrocellis et al. 1999), in the nervous system of the leech *Hirudo medicinalis* (Matias et al. 2001) and in the ovaries of the sea urchin *Paracentrotus lividus* (Bisogno et al. 1997). Moreover, FAAH-like activity has also been detected in plant tissues (Shrestha et al. 2002), indicating that FAAH may be an evolutionarily ancient enzyme.

An important recent discovery has been the identification of a FAAH gene in the plant species *Arabidopsis* (Shrestha et al. 2003). *Arabidopsis* FAAH is a 607 amino acid protein that shares only 18% overall sequence identity with rat FAAH, although this rises to 37%–60% in the catalytic domain, depending on the length of sequence compared. Analysis of the enzymatic properties of heterologously expressed *Arabidopsis* FAAH reveals that, like mammalian FAAHs, it catalyses hydrolysis of anandamide and other NAEs. Therefore, it appears that FAAH is an evolutionarily ancient enzyme whose ancestry dates back at least as far as the unicellular eukaryotic common ancestor of plants and animals. Moreover, the discovery of a plant gene encoding a protein that functions as a FAAH enzyme, but which shares relatively little sequence similarity with mammalian FAAHs, suggests that related genes in non-mammalian animal species may also encode enzymes that have FAAH activity. For example, genes encoding FAAH-like proteins that share much higher levels of sequence similarity with mammalian FAAHs than with *Arabidopsis* FAAH are present in the genomes of the bird *Gallus gallus* (chicken), the puffer fish *Fugu rubripes*, the urochordate *Ciona intestinalis* and the nematode *C. elegans*. Further studies are now required to characterise the properties of the enzymes encoded by these putative non-mammalian FAAH genes.

2.3
The Phylogenetic Distribution of 2-AG and Enzymes Involved in 2-AG Biosynthesis

2-AG was originally identified as a potential endogenous cannabinoid in mammals by Mechoulam et al. (1995) and Sugiura et al. (1995) and subsequent studies indicate that 2-AG is also present in several non-mammalian species, including the insect *Drosophila melanogaster* (McPartland et al. 2001) and the annelid *Hirudo medicinalis* (Matias et al. 2001). These findings suggest that 2-AG may have a broad phylogenetic distribution. However, as with anandamide, the ability of organisms to generate 2-AG will depend on the presence of arachidonic acid as a component of phospholipids.

Two enzymatic pathways have been proposed as potential mechanisms for 2-AG biosynthesis in mammalian cells (Piomelli 2003). First, a pathway in which phosphatidylinositol is cleaved by phospholipase C (PLC) to generate 1,2-diacylglycerol (DAG), which is then converted to 2-AG through the action of DAG lipase. Second, a pathway in which phosphatidylinositol is cleaved by phospholipase A_1 to generate 2-arachidonoyl-lysophospholipid, which is then converted to 2-AG through the action of lyso-PLC. For the purposes of this review we will focus on the first pathway because: (1) PLC is a ubiquitous effector for G protein-coupled receptors throughout the animal kingdom and therefore a potentially important and evolutionarily ancient mediator of 2-AG formation and (2) genes encoding mammalian DAG lipases have recently been identified, opening up new opportunities for analysis of the molecular and cellular biology of 2-AG formation in cells.

Analysis of human genome sequence data revealed the presence of two genes that encode *sn*1-DAG lipases and which are now known as DAGLα and DAGLβ (Bisogno et al. 2003). Importantly, heterologous expression of DAGLα or DAGLβ conferred increased formation of 2-AG from *sn*-1-stearoyl-2-arachidonoyl-glycerol as a substrate, demonstrating that these enzymes can catalyse synthesis of 2-AG in cells. Therefore, expression of DAGLα or DAGLβ in cells and tissues may serve as molecular markers for cells that generate 2-AG in vivo. Consistent with this notion, DAGLα is expressed in the dendrites of cerebellar Purkinje cells, neurons which are sources of endocannabinoids that act as retrograde signalling molecules by activating presynaptic CB_1 receptors located on the axons of cerebellar granule cells (Kreitzer and Regehr 2002).

Genes encoding orthologues of DAGLα and DAGLβ are present in other mammals (e.g. mouse) and, more importantly for purposes of this review, in non-mammalian vertebrates that include the bird *Gallus gallus* (chicken) and the zebrafish *Danio rerio* (Bisogno et al. 2003). Moreover, a DAG lipase-like gene (CG33174) is also present in an invertebrate species, the insect *Drosophila melanogaster* (Adams et al. 2000).

2.4 The Phylogenetic Distribution of Monoglyceride Lipase

The inactivation of 2-AG in mammals is thought to be mediated by the enzyme monoglyceride lipase (MGL). However, molecular characterisation of MGL was not driven by an interest in 2-AG but by research directed at identification of the enzymes involved in the sequential hydrolysis of stored triglycerides. A mouse cDNA encoding this enzyme was cloned and sequenced by Karlsson et al. (1997) and found to encode a 302 amino acid protein that is expressed in a wide range of tissues, including brain. Subsequently, Dinh et al. (2002) demonstrated that rat MGL catalyses hydrolysis of 2-AG when expressed in cells. Interestingly, 2-AG is also a substrate for the enzyme FAAH in vitro (Goparaju et al. 1998), but in mice lacking FAAH, the 2-AG content of the brain is not significantly different from that in wild-type mice (Lichtman et al. 2002). Therefore, it is thought that MGL is the primary physiological mediator of 2-AG inactivation in the mammalian brain (Dihn et al. 2002).

Analysis of the occurrence of MGL-like proteins in non-mammalian organisms by BLAST analysis reveals closely related proteins in the zebrafish *Danio rerio* and the chicken *Gallus gallus* . It is likely, therefore, that MGL occurs throughout the vertebrates. However, genes encoding proteins resembling MGL do not appear to be present in any of the invertebrate species for which complete genome sequence data are available (i.e. *Drosophila, C. elegans, Ciona*). Genes encoding related proteins are, however, present in the genomes of plant, bacterial and viral species. This is an unusual pattern of phylogenetic distribution that raises questions about the evolutionary origin of vertebrate MGL proteins. Relevant to this issue, it is interesting to note that a cowpox virus gene encodes a protein that shares 40% sequence identity with mammalian MGL proteins (Karlsson et al. 1997). Therefore, perhaps an ancestral MGL gene was introduced into the vertebrate genome by horizontal gene transfer mediated by a virus.

3
The Phylogeny of Cannabinoid Receptors and Other Endocannabinoid Receptors

What our survey of the phylogenetic distribution of endocannabinoids and associated enzymes indicates is that the ability of cells to produce and inactivate the molecules that we classify as endocannabinoids in mammals is an evolutionarily ancient phenomenon. Moreover, some components of the endocannabinoid system may date back as far as the common ancestor of all eukaryotic organisms. However, the ability of cells to produce these molecules does not necessarily imply that they function as signalling molecules in all eukaryotes. In assessing the evolution of endocannabinoid signalling, we should not assume that because endocannabinoids activate CB_1/CB_2-type G protein-coupled receptors in mammals that receptors of this type necessarily mediate effects of these molecules in other eukaryotes. Some organisms may have independently evolved their “own” endocannabinoid receptors unrelated to the mammalian cannabinoid receptors. Other organisms may be able to produce the chemicals that we, with our mammalian bias, refer to as “endocannabinoids” but lack receptors for these molecules.

3.1
Receptors Related to Mammalian CB_1 and CB_2 Cannabinoid Receptors

Genes encoding orthologues of the mammalian CB_1 and CB_2 receptors have been identified in the puffer fish *Fugu rubripes* (Yamaguchi et al. 1996; Elphick 2002). This indicates that the existence of CB_1 and CB_2 receptors in vertebrates can be traced back at least as far as the common ancestor of teleost fish like *Fugu* and the amphibians, reptiles, birds and mammals. Accordingly, CB_1-type genes have also been identified in birds and amphibians (Soderstrom and Johnson 2000, 2001; Soderstrom et al. 2000). Thus far, the *Fugu* CB_2 gene is the only one reported for a non-mammalian vertebrate (Elphick 2002). However, BLAST analysis of genome sequence data for the bird *Gallus gallus* (chicken) reveals the presence of both CB_1-and CB_2-type genes in this species.

An interesting feature of the puffer fish *Fugu rubripes* is that it has one CB_2 gene (Elphick 2002) but two CB_1-type genes (CB_{1A} and CB_{1B}; Yamaguchi et al. 1996). The occurrence of duplicated genes, with respect to other vertebrates, is a feature of teleost fish that is thought to be a legacy of a whole-genome duplication event that occurred in an ancestral species (Taylor et al. 2001). However, duplicates of some genes will have been lost with the passage of evolutionary time, which probably explains the existence of only one CB_2 gene in *Fugu*.

Although both CB_1 and CB_2 genes have been found in the "higher" vertebrates, it remains to be established if CB_1 and CB_2 genes are also present in cartilaginous fish (e.g. sharks, rays) and in primitive agnathan vertebrates (e.g. hagfish, lamprey). However, progress has been made recently in investigating the occurrence of cannabinoid receptors in invertebrate chordates. The extant invertebrates that are most closely related to the vertebrates are the cephalochordates (e.g. Amphioxus), based on both morphological and molecular evidence (Adoutte et al. 2000). Unfortunately, relatively little is known about the physiology and biochemistry of these animals. However, because of the important phylogenetic position of these animals with respect to vertebrates, there are plans to sequence the genome of a cephalochordate species.

An invertebrate chordate species that has had its genome sequenced recently is the urochordate (sea-squirt) *Ciona intestinalis* (Dehal et al. 2002). As adults, these animals exhibit little similarity with other chordates (vertebrates and cephalochordates), but as larvae *Ciona* have several morphological characters that distinguish them as chordates. Moreover, urochordates are the most primitive of the extant chordates, and thus these animals are of particular interest for evolutionary studies. Analysis of the *Ciona* genome sequence has revealed the presence of a putative cannabinoid receptor gene (*CiCBR*) encoding a 423 amino acid protein that shares 28% and 24% sequence identity with the human CB_1 and CB_2 receptor, respectively (Elphick et al. 2003). These are relatively low levels of sequence similarity, but analysis of the relationship of CiCBR with cannabinoid receptors and other G protein-coupled receptors, by construction of a phylogenetic tree based on sequence alignments, demonstrated that CiCBR is an orthologue of the vertebrate cannabinoid receptors CB_1 and CB_2 (Elphick et al. 2003). Thus, CiCBR is the first putative cannabinoid receptor to be identified in an invertebrate species. Moreover, phylogenetic analysis indicates that the common ancestor of CiCBR and vertebrate CB_1 and CB_2 receptors predates a duplication event that gave rise to CB_1 and CB_2 in vertebrates. In this respect, cannabinoid receptor genes conform to a pattern seen in other gene families, where for each invertebrate gene there are often two or more related genes in vertebrates. This feature is thought to reflect a whole-genome duplication event that occurred in the invertebrate ancestor of the vertebrates (Furlong and Holland 2002).

The discovery of CiCBR indicates that the evolutionary history of cannabinoid receptors that are related to the vertebrate CB_1 and CB_2 receptors extends back at least as far as the common ancestor of vertebrates and the invertebrate chordates (urochordates, cephalochordates). What remains to be established is whether other, more distantly invertebrate animals also have orthologues of the vertebrate cannabinoid receptors. The invertebrate animals that are most closely related to the chordates are the hemichordates and echinoderms (Adoutte et al. 2000). Hemichordates are a relatively obscure group of animals (e.g. acorn worms) that have not been studied in great detail. There has, however, been a surge of interest in these animals recently with the advent of molecular techniques for research on developmental and evolutionary biology (e.g. Lowe et al. 2003). Moreover, there are plans to sequence the genome of a hemichordate species, the acorn worm *Saccoglossus kowalevskii*. Therefore, as with Amphioxus, there may be opportunities to investigate the occurrence of a cannabinoid receptor in hemichordates in the near future.

The echinoderms (e.g. sea urchins and starfish) are an invertebrate group that has been studied extensively, in particular for research on early stages of development. Moreover, at the time of writing, a genome sequencing project for the sea urchin species *Strongylocentrotus purpuratus* is ongoing (Cameron et al. 2000) and due to be completed during 2004. This is of special interest for research on cannabinoid receptors because this species has been the subject of a detailed study on the effects of cannabinoids. Herbert Schuel and colleagues demonstrated that cannabinoids block the acrosome reaction in sea urchin sperm, indicating that endocannabinoids may have a physiological role in preventing polyspermy (Schuel et al. 1991, 1994). Moreover, Chang et al. (1993) demonstrated that cannabinoid binding sites are present on sea urchin sperm. The molecular properties of these cannabinoid binding sites and their relationship to vertebrate cannabinoid receptors are currently unknown. However, analysis of sea urchin genome sequence data, when it is available, may provide new opportunities for further research on this issue.

Having considered the deuterostomian invertebrates, we will now turn our attention to the protostomian clade of the animal kingdom. First we will consider the ecdysozoa, which include two well-studied species for which complete genome sequence data are available—the insect *Drosophila melanogaster* and the nematode *C. elegans*. Analysis of the genome sequences of both of these species has revealed, however, that orthologues of cannabinoid receptors are not present (Elphick and Egertová 2001). Moreover, these species also do not have orthologues of the G protein-coupled receptors in vertebrates that are most closely related to CB_1 and CB_2—lysophospholipid receptors and melanocortin receptors (Elphick and Egertová 2001). These data indicate, therefore, that the group of G protein-coupled receptors that include cannabinoid receptors may have originated in the deuterostomian branch of the animal kingdom, after the deuterostomian-protostomian split. Consistent with these conclusions based on genome sequence data, biochemical analysis of insect species has not revealed the presence of cannabinoid binding sites (Egertová 1999; Elphick and Egertová 2001; McPartland et al. 2001).

Turning now to the lophotrochozoan phyla, here there have been a few studies that have reported detection of cannabinoid binding sites. Stefano et al. (1996) reported the presence of binding sites for anandamide on haemocytes from the bivalve mollusc *Mytilus edulis* , whilst Stefano et al. (1997) reported anandamide binding sites in the nervous system of the leech *Hirudo medicinalis* (Phylum Annelida). Interestingly, the latter study was accompanied by a partial leech cDNA sequence that shared sequence similarity with vertebrate CB_1 receptors. However, subsequent detailed analysis of this sequence revealed that it is chimeric with one region that shares 98% amino acid identity with the bovine adrenocorticotropic hormone receptor and two regions that share 68% and 65% amino acid identity with mammalian CB_1 receptors (Elphick 1998). It is unlikely, therefore, that this sequence represents part of a *bone fide* leech cannabinoid receptor cDNA. How, then, can the discovery of this unusual sequence be explained. One possibility is that leech cDNA was contaminated with bovine DNA derived from blood that leeches had fed on. Clearly, further work is required, but thus far there have been no follow-up studies to confirm the existence of a full-length cannabinoid receptor cDNA in the leech or in any other protostomian invertebrate.

The detection of cannabinoid binding sites in *Mytilus* and *Hirudo*, but not in insects, has been explained by some authors as a consequence of loss of cannabinoid receptor genes in the ecdysozoan lineage but not in the lophotrochozoan lineage (McPartland and Pruitt 2002). However, as highlighted above, both *Drosophila* and *C. elegans* also lack orthologues of the vertebrate G protein-coupled receptors that are most closely related to cannabinoid receptors (lysophospholipid and melanocortin receptors). Therefore, a more parsimonious explanation is that this group of receptors originated in the deuterostomian branch of the animal kingdom after the protostomian–deuterostomian split.

If orthologues of cannabinoid receptors are not present in protostomian invertebrates, as proposed above and in previous reports (Elphick and Egertová 2001; Elphick et al. 2003), how then can the existence of cannabinoid binding sites in *Mytilus* and *Hirudo* be explained? Detection of these binding sites may reflect interaction of cannabinoids with membrane proteins in these species that are unrelated to the vertebrate CB_1/CB_2-type cannabinoid receptors but which have evolved independently. However, demonstrating that these binding sites equate to functional receptors that mediate physiological effects of endocannabinoids in these organisms will require detailed molecular characterisation of the putative receptors, and thus far this has yet to be accomplished. The same applies to cannabinoid binding sites detected in the primitive cnidarian species *Hydra viridis* (De Petrocellis et al. 1999). If these putative receptors can be characterised then they may provide fascinating examples of convergent evolution in signalling mechanisms.

3.2
Other Endocannabinoid Receptors and Cannabinoid Receptors

Although the CB_1 and CB_2 cannabinoid receptors are by far the most well characterised receptors for endocannabinoids in vertebrates, it is important to recognise that there are also other receptor types that may mediate physiological effects of anandamide and 2-AG. For example, there is evidence of a third G protein-coupled receptor in mammals that is activated by endocannabinoids (Breivogel et al. 2001). Without molecular characterisation of this putative receptor it is impossible to investigate its phylogenetic distribution. However, the possibility remains that this receptor may have more widespread phylogenetic distribution than CB_1/CB_2-related receptors and thereby account for cannabinoid binding sites that have been reported in some invertebrate species.

Another receptor that has been implicated as a mediator of physiological effects of the endocannabinoid anandamide in mammals is the vanilloid receptor VR1, more recently referred to as transient receptor potential vanilloid type 1 (TRPV1) (Zygmunt et al. 1999). However, VR1 is not activated by "classical" Δ^9-tetrahydrocannabinol-like cannabinoid agonists. Therefore, VR1 is an endocannabinoid receptor but not a cannabinoid receptor. Unlike CB_1 and CB_2, VR1 is not a G protein-coupled receptor but belongs to the TRP family of ligand-gated cation channels (Montell et al. 2002). Genes encoding proteins that are closely related to the mammalian VR1 receptor have been identified in *Drosophila* (nan) and in *C. elegans* (OSM-9) (Montell 2003). However, to the best of our knowledge, it is not known if these invertebrate VR1-like channels are activated by anandamide. Therefore, it remains to be determined if the ability of anandamide to activate TRP-type channels is an evolutionarily ancient phenomenon.

Another interesting member of the TRP channel family that has been characterised recently is ANKTM1, which is activated by Δ^9-tetrahydrocannabinol as well as being implicated in the detection of noxious cold (Jordt et al. 2004). However, the physiological relevance of the effect of Δ^9-tetrahydrocannabinol on ANKTM1 is unclear because the endocannabinoids anandamide and 2-AG do not activate this TRP channel (Jordt et al. 2004). Nevertheless, it is possible that other as-yet-unidentified endocannabinoids act as endogenous ligands for ANKTM1.

In conclusion, there is now an emerging concept of TRP-type ion channels that are receptors for cannabinoids and/or endocannabinoids, and an interesting area for future research will be to investigate the occurrence of invertebrate TRP-type channels that are also activated by cannabinoid-related molecules.

4 The Evolutionary Origins of Endocannabinoid Signalling

What can we conclude from our survey of the phylogenetic distribution of (1) endocannabinoids, (2) enzymes involved in endocannabinoid biosynthesis and inactivation and (3) cannabinoid/endocannabinoid receptors? It is clear that many of the components of the enzymatic machinery that are used for biosynthesis and inactivation of endocannabinoids in mammals are evolutionarily ancient. For example, there is evidence that enzymes involved in biosynthesis and inactivation of anandamide occur in animals and plants. Most notable in this respect has been the recent discovery and enzymatic characterisation of a FAAH-like enzyme in the plant *Arabidopsis* (Shrestha et al. 2003). Therefore, it appears that the ability of organisms to synthesise endocannabinoids such as anandamide and 2-AG may date back at least as far as the unicellular eukaryotic common ancestor of plants and animals. However, exploitation of these molecules for intercellular signalling may have occurred independently in different lineages during the evolution of the eukaryotes. For example, there is evidence that plants may also have receptors for anandamide and/or related NAEs, because Tripathy et al. (2003) have detected binding sites for NAEs in cell membranes from the plant species *Nicotiana tabacum* (tabacco) and *Arabidopsis*. If molecular characterisation of a putative NAE receptor in plants can be accomplished, this may provide a fascinating example of how plants have independently exploited NAEs as signalling molecules.

So far, the best-characterised example of endocannabinoid signalling in eukaryotes is CB_1/CB_2-mediated processes in vertebrates. Moreover, our phylogenetic analysis of the occurrence of CB_1/CB_2 receptors in invertebrates indicates that the ancestor of these receptors originated in a deuterostomian invertebrate, and in accordance with this view receptors of this type have so far not been found in protostomian invertebrates. The CiCBR gene that was recently identified in the invertebrate chordate *Ciona intestinalis* (Elphick et al. 2003) is an example of a receptor in a deuterostomian invertebrate that may resemble the putative ancestor of the vertebrate CB_1 and CB_2 receptors. Therefore, analysis of CiCBR function in *Ciona* is now of particular interest.

Looking ahead, we hope that this review may stimulate scientists with an interest in endocannabinoid signalling to exploit not only the familiar mammalian model species (rats, mice) but also the rich diversity of non-mammalian animals where the existence of endocannabinoid receptors has been established.

References

Adams MD, Celniker SE, Holt RA, et al (2000) The genome sequence of Drosophila melanogaster. Science 287:2185–2195

Adoutte A, Balavoine G, Lartillot N, Lespinet O, Prud'homme B, De Rosa R (2000) The new animal phylogeny: reliability and implications. Proc Natl Acad Sci USA 97:4453–4456

Altschul SF, Gish W, Miller W, Myers EW, Lipman DJ (1990) Basic local alignment search tool. J Mol Biol 215:403–410

Aparicio S, Chapman J, Stupka E, et al (2002) Whole-genome shotgun assembly and analysis of the genome of Fugu rubripes. Science 297:1301–1310

Bisogno T, Ventriglia M, Milone A, Mosca M, Cimino G, Di Marzo V (1997) Occurrence and metabolism of anandamide and related acyl-ethanolamides in ovaries of the sea urchin Paracentrotus lividus. Biochim Biophys Acta 1345:338–348

Bisogno T, Howell F, Williams G, Minassi A, Cascio MG, Ligresti A, Matias I, Schiano-Moriello A, Paul P, Williams EJ, Gangadharan U, Hobbs C, Di Marzo V, Doherty P (2003) Cloning of the first sn1-DAG lipases points to the spatial and temporal regulation of endocannabinoid signaling in the brain. J Cell Biol 163:463–468

Breivogel CS, Griffin G, Di Marzo V, Martin BR (2001) Evidence for a new G protein-coupled cannabinoid receptor in mouse brain. Mol Pharmacol 60:155–163

Cameron RA, Mahairas G, Rast JP, Martinez P, Biondi TR, Swartzell S, Wallace JC, Poustka AJ, Livingston BT, Wray GA, Ettensohn CA, Lehrach H, Britten RJ, Davidson EH, Hood L (2000) A sea urchin genome project: sequence scan, virtual map, and additional resources. Proc Natl Acad Sci USA 97:9514–9518

Chang MC, Berkery D, Schuel R, Laychock SG, Zimmerman AM, Zimmerman S, Schuel H (1993) Evidence for a cannabinoid receptor in sea urchin sperm and its role in blockade of the acrosome reaction. Mol Reprod Dev 36:507–516

Chapman KD (2000) Emerging physiological roles for N-acylphosphatidylethanolamine metabolism in plants: signal transduction and membrane protection. Chem Phys Lipids 108:221–229

Cravatt BF, Giang DK, Mayfield SP, Boger DL, Lerner RA, Guilula NB (1996) Molecular characterization of an enzyme that degrades neuromodulatory fatty-acid amides. Nature 384:83–87

De Petrocellis L, Melck D, Bisogno T, Milone A, Di Marzo V (1999) Finding of the endocannabinoid signalling system in Hydra, a very primitive organism: possible role in the feeding response. Neuroscience 92:377–387

Dehal P, Satou Y, Campbell RK, et al (2002) The draft genome of Ciona intestinalis: insights into chordate and vertebrate origins. Science 298:2157–2167

Deutsch DG, Chin SA (1993) Enzymatic synthesis and degradation of anandamide, a cannabinoid receptor agonist. Biochem Pharmacol 46:791–796

Devane WA, Dysarz FAI, Johnson MR, Melvin LS, Howlett AC (1988) Determination and characterization of a cannabinoid receptor in rat brain. Mol Pharmacol 34:605–613

Devane WA, Hanus L, Breuer A, Pertwee RG, Stevenson LA, Griffin G, Gibson D, Mandelbaum A, Etinger A, Mechoulam R (1992) Isolation and structure of a brain constituent that binds to the cannabinoid receptor. Science 258:1946–1948

Di Marzo V, Fontana A, Cadas H, Schinelli S, Cimino G, Schwartz J-C, Piomelli D (1994) Formation and inactivation of endogenous cannabinoid anandamide in central neurons. Nature 372:686–691

Dinh TP, Carpenter D, Leslie FM, Freund TF, Katona I, Sensi SL, Kathuria S, Piomelli D (2002) Brain monoglyceride lipase participating in endocannabinoid inactivation. Proc Natl Acad Sci USA 99:10819–10824

Egertová M (1999) Neuroanatomy and phylogeny of cannabinoid signalling. Ph.D. Thesis, University of London, UK

Egertová M, Giang DK, Cravatt BF, Elphick MR (1998) A new perspective on cannabinoid signalling: complementary localization of fatty acid amide hydrolase and the CB1 receptor in rat brain. Proc R Soc Lond B Biol Sci 265:2081–2085

Elphick MR (1998) An invertebrate G-protein coupled receptor is a chimeric cannabinoid/melanocortin receptor. Brain Res 780:170–173

Elphick MR (2002) Evolution of cannabinoid receptors in vertebrates: identification of a CB2 gene in the puffer fish Fugu rubripes. Biol Bull 202:104–107

Elphick MR, Egertová M (2001) The neurobiology and evolution of cannabinoid signalling. Philos Trans R Soc Lond B Biol Sci 356:381–408

Elphick MR, Satou Y, Satoh N (2003) The invertebrate ancestry of endocannabinoid signalling: an orthologue of vertebrate cannabinoid receptors in the urochordate Ciona intestinalis. Gene 302:95–101

Freund TF, Katona I, Piomelli D (2003) Role of endogenous cannabinoids in synaptic signaling. Physiol Rev 83:1017–1066

Furlong RF, Holland PWH (2002) Were vertebrates octoploid? Philos Trans R Soc Lond B Biol Sci 357:531–544

Giang DK, Cravatt BF (1997) Molecular characterization of human and mouse fatty acid amide hydrolases. Proc Natl Acad Sci USA 94:2238–2242

Goparaju SK, Ueda N, Yamaguchi H, Yamamoto S (1998) Anandamide amidohydrolase reacting with 2-arachidonoylglycerol, another cannabinoid receptor ligand. FEBS Lett 422:69–73

Hastings N, Agaba M, Tocher DR, Leaver MJ, Dick JR, Sargent JR, Teale AJ (2001) A vertebrate fatty acid desaturase with delta 5 and delta 6 activities. Proc Natl Acad Sci USA 98:14304–14309

Howlett AC, Barth F, Bonner TI, Cabral G, Casellas P, Devane WA, Felder CC, Herkenham M, Mackie K, Martin BR, Mechoulam R, Pertwee RG (2002) International Union of

Pharmacology. XXVII. Classification of cannabinoid receptors. Pharmacol Rev 54:161–202

Jordt SE, Bautista DM, Chuang HH, McKemy DD, Zygmunt PM, Hogestatt ED, Meng ID, Julius D (2004) Mustard oils and cannabinoids excite sensory nerve fibres through the TRP channel ANKTM1. Nature 427:260–265

Karlsson M, Contreras JA, Hellman U, Tornqvist H, Holm C (1997) cDNA cloning, tissue distribution, and identification of the catalytic triad of monoglyceride lipase. Evolutionary relationship to esterases, lysophospholipases, and haloperoxidases. J Biol Chem 27:27218–27223

Kreitzer AC, Regehr WG (2002) Retrograde signaling by endocannabinoids. Curr Opin Neurobiol 12:324–330

Lichtman AH, Hawkins EG, Griffin G, Cravatt BF (2002) Pharmacological activity of fatty acid amides is regulated, but not mediated, by fatty acid amide hydrolase in vivo. J Pharmacol Exp Ther 302:73–79

Lowe CJ, Wu M, Salic A, Evans L, Lander E, Stange-Thomann N, Gruber CE, Gerhart J, Kirschner M (2003) Anteroposterior patterning in hemichordates and the origins of the chordate nervous system. Cell 113:853–865

Matias I, Bisogno T, Melck D, Vandenbulcke F, Verger-Bocquet M, De Petrocellis L, Sergheraert C, Breton C, Di Marzo V, Salzet M (2001) Evidence for an endocannabinoid system in the central nervous system of the leech Hirudo medicinalis. Brain Res Mol Brain Res 87:145–159

Matsuda LA, Lolait SJ, Brownstein MJ, Young AC, Bonner TI (1990) Structure of a cannabinoid receptor and functional expression of the cloned cDNA. Nature 346:561–564

McPartland JM, Pruitt P (2002) Sourcing the code: searching for the evolutionary origins of cannabinoid receptors, vanilloid receptors, and anandamide. J Cannabis Ther 2:73–103

McPartland J, Di Marzo V, De Petrocellis L, Mercer A, Glass M (2001) Cannabinoid receptors are absent in insects. J Comp Neurol 436:423–429

Mechoulam R, Ben-Shabat S, Hanus L, Ligumsky M, Kaminski NE, Schatz AR, Gopher A, Almog S, Martin BR, Compton DR, Pertwee RG, Griffin G, Bayewitch M, Barg J, Vogel ZVI (1995) Identification of an endogenous 2-monoglyceride, present in canine gut that binds to cannabinoid receptors. Biochem Pharmacol 50:83–90

Montell C (2003) The venerable inveterate invertebrate TRP channels. Cell Calcium 33:409–417

Montell C, Birnbaumer L, Flockerzi V (2002) The TRP channels, a remarkably functional family. Cell 108:595–598

Munro S, Thomas KL, Abu-Shaar M (1993) Molecular characterization of a peripheral receptor for cannabinoids. Nature 365:61–65

Nakamura MT, Nara TY (2003) Essential fatty acid synthesis and its regulation in mammals. Prostaglandins Leukot Essent Fatty Acids 68:145–150

Napier J, Michaelson LV (2001) Genomic and functional characterisation of polyunsaturated fatty acid biosynthesis in Caenorhabditis elegans. Lipids 36:761–766

Okamoto Y, Morishita J, Tsuboi K, Tonai T, Ueda N (2004) Molecular characterization of a phospholipase D generating anandamide and its congeners. J Biol Chem 279:5298–5305

Piomelli D (2003) The molecular logic of endocannabinoid signalling. Nat Rev Neurosci 4:873–884

Salzet M, Stefano G (2002) The endocannabinoid system in invertebrates. Prostaglandins Leukot Essent Fatty Acids 66:353–361

Salzet M, Breton C, Bisogno T, Di Marzo V (2000) Comparative biology of the endocannabinoid system. Possible role in the immune response. Eur J Biochem 15:4917–4927

Schmid HH, Schmid PC, Natarajan V (1990) N-Acylated glycerophospholipids and their derivatives. Prog Lipid Res 29:1–43

Schuel H, Berkery D, Schuel R, Chang MC, Zimmerman AM, Zimmerman S (1991) Reduction of the fertilizing capacity of sea urchin sperm by cannabinoids derived from marihuana. I. Inhibition of the acrosome reaction induced by egg jelly. Mol Reprod Dev 29:51–59

Schuel H, Goldstein E, Mechoulam R, Zimmerman AM, Zimmerman S (1994) Anandamide (arachidonylethanolamide), a brain cannabinoid receptor agonist, reduces sperm fertilizing capacity in sea urchins by inhibiting the acrosome reaction. Proc Natl Acad Sci USA 91:7678–7682

Shrestha R, Noordermeer MA, van der Stelt M, Veldink GA, Chapman KD (2002) N-Acylethanolamines are metabolised by lipoxygenase and amidohydrolase in competing pathways during cottonseed imbibition. Plant Physiol 130:391–401

Shrestha R, Dixon RA, Chapman KD (2003) Molecular identification of a functional homologue of the mammalian fatty acid amide hydrolase in Arabidopsis thaliana. J Biol Chem 278:34990–34997

Soderstrom K, Johnson F (2000) CB1 cannabinoid receptor expression in brain regions associated with zebra finch song control. Brain Res 857:151–157

Soderstrom K, Johnson F (2001) Zebra finch CB1 cannabinoid receptor: pharmacology and in vivo and in vitro effects of activation. J Pharmacol Exp Ther 297:189–197

Soderstrom K, Leid M, Moore FL, Murray TF (2000) Behavioural, pharmacological and molecular characterization of an amphibian cannabinoid receptor. J Neurochem 75:413–423

Stefano GB, Liu Y, Goligorsky MS (1996) Cannabinoid receptors are coupled to nitric oxide release in invertebrate immunocytes, microglia and human monocytes. J Biol Chem 271:19238–19242

Stefano GB, Salzet B, Salzet M (1997) Identification and characterization of the leech CNS cannabinoid receptor: coupling to nitric oxide release. Brain Res 753:219–224

Sugiura T, Kondo S, Sukagawa A, Nakane S, Shinoda A, Itoh K, Yamashita A, Waku K (1995) 2-Arachidonoylglycerol: a possible endogenous cannabinoid receptor ligand in brain. Biochem Biophys Res Commun 215:89–97

Taylor JS, Van de Peer Y, Braasch I, Meyer A (2001) Comparative genomics provides evidence for an ancient genome duplication event in fish. Philos Trans R Soc Lond B Biol Sci 356:1661–1679

The C. elegans Sequencing Consortium (1998) Genome sequence of the nematode C. elegans: a platform for investigating biology. Science 282:2012–2018

Tripathy S, Kleppinger-Sparace K, Dixon RA, Chapman KD (2003) N-acylethanolamine signaling in tobacco is mediated by a membrane-associated, high-affinity binding protein. Plant Physiol 131:1781–1791

Ueda N, Kurahashi Y, Yamamoto S, Tokunaga T (1995) Partial purification and characterization of the porcine brain enzyme hydrolyzing and synthesizing anandamide. J Biol Chem 270:23823–23827

Wilson RI, Nicoll RA (2002) Endocannabinoid signaling in the brain. Science 296:678–682

Yamaguchi F, Macrae AD, Brenner S (1996) Molecular cloning of two cannabinoid type 1-like receptor genes from the puffer fish Fugu rubripes. Genomics 35:603–605

Yoshioka T, Inoue H, Kasama T, Seyama Y, Nakashima S, Nozawa Y, Hotta Y (1985) Evidence that arachidonic acid is deficient in phosphatidylinositol of Drosophila heads. J Biochem (Tokyo) 98:657–662

Zygmunt PM, Petersson J, Andersson DA, Chuang HH, Sorgard M, DiMarzo V, Julius D, Hogestatt ED (1999) Vanilloid receptors on sensory nerves mediate the vasodilator action of anandamide. Nature 400:452–457

HEP (2005) 168:299–325

Distribution of Cannabinoid Receptors in the Central and Peripheral Nervous System

K. Mackie

University of Washington, Box 356540, Seattle WA, 98195-6540, USA
kmackie@u.washington.edu

Abstract CB_1 cannabinoid receptors appear to mediate most, if not all of the psychoactive effects of delta-9-tetrahydrocannabinol and related compounds. This G protein-coupled receptor has a characteristic distribution in the nervous system: It is particularly enriched in cortex, hippocampus, amygdala, basal ganglia outflow tracts, and cerebellum—a distribution that corresponds to the most prominent behavioral effects of cannabis. In addition, this distribution helps to predict neurological and psychological maladies for which manipulation of the endocannabinoid system might be beneficial. CB_1 receptors are primarily expressed on neurons, where most of the receptors are found on axons and synaptic terminals, emphasizing the important role of this receptor in modulating neurotransmission at specific synapses. While our knowledge of CB_1 localization in the nervous system has advanced tremendously over the past 15 years, there is still more to learn. Particularly pressing is the need for (1) detailed anatomical studies of brain regions important in the therapeutic actions of drugs that modify the endocannabinoid system and (2) the determination of the localization of the enzymes that synthesize, degrade, and transport the endocannabinoids.

Keywords Immunocytochemistry · In situ hybridization · Autoradiography · Cholecystokinin · Synapse

1 Introduction

1.1 Background

The CB_1 cannabinoid receptor is the major mediator of the psychoactive effects of cannabis and its derivatives. In addition, this G protein-coupled receptor transduces many of the effects of the endogenous cannabinoids. Understanding the distribution of CB_1 receptors has proved helpful to both predict and understand the effects of cannabinoids. For example, the high CB_1 receptor levels found in cortex, basal ganglia, and cerebellum coincide with the prominent effects cannabinoids have on functions subserved by these brain regions. By comparison, the low levels present in the medullary nuclei responsible for regulating respiration are consistent with the modest effects cannabinoids have on respiratory drive. Furthermore, the strong presynaptic localization of the receptor found in ultrastructural studies underscores its major role as a modulator of neurotransmitter release.

The distribution of cannabinoid receptors has been extensively mapped by quantitative autoradiography, in situ hybridization, and immunocytochemistry. Each of these techniques has its strengths and weaknesses. Properly calibrated, quantitative autoradiography provides the best measure of absolute receptor density. Nonetheless, its spatial resolution is limited and specificity depends on the ligand used. In situ hybridization identifies the cells synthesizing CB_1 mRNA. However, mRNA levels and protein levels may not necessarily correlate. Immunocytochemistry provides outstanding spatial resolution; however, fixation artifacts and unanticipated antibody crossreactivity must be assiduously avoided. For the most part, the results obtained from these three approaches have provided complementary and logically consistent results. In addition to these anatomical approaches, it is possible to obtain a measure of CB_1 receptor function by guanosine triphosphate (GTP)γS binding, giving spatial resolution similar to quantitative autoradiography. Finally, the results of experiments using regionally or neuron specific CB_1 knockout mice can give additional insight into receptor localization.

1.2
Autoradiography

Miles Herkenham performed the first CB_1 receptor distribution studies using autoradiography with the tritiated CB_1 agonist, CP55,940. Examples of his results from human brain are shown in Fig. 1. A striking feature of the autoradiographic studies was the extraordinarily high levels of CB_1 receptors found in substantia nigra, globus pallidus, hippocampus, cerebellum, and cortex. The levels of CB_1 receptors found in these brain regions in the rat approached 6 pmole/mg (Herkenham et al. 1991). To give a sense of the magnitude of CB_1 receptor expression, CB_1 receptors are tenfold denser than D_2 receptors in the basal ganglia and have a density similar to cortical ionotropic glutamate receptors. The specificity of these results was verified by Virginia Seybold and her colleagues, who performed a systematic autoradiographic study of rat brain using tritiated WIN55,212-2, a structurally distinct CB_1 agonist (Jansen et al. 1992). These thorough studies in rodents have been complemented by autoradiographic studies in human brain (Glass et al. 1997; Mato et al. 2003). The results of the human and rodent studies are qualitatively similar once the evolutionary changes associated with the development of the human brain are considered.

1.3
In Situ Hybridization

Cloning the CB_1 receptor (Matsuda et al. 1990) made it possible to identify CB_1 synthesizing cells by in situ hybridization (Mailleux and Vanderhaeghen 1992; Matsuda et al. 1993). Correlating the results of the autoradiographic and in situ hybridization studies reveals several common themes of the CB_1 system. The first was that in some brain regions, particularly forebrain (for example, cortex, amygdala, and hippocampus), CB_1 receptors are expressed at very high levels in a very restricted set of neurons. These neurons then project widely, resulting in a dense network of CB_1-positive axons. The second was that CB_1 receptors were primarily found on axons and terminals. For example, high levels of CB_1 are present in the striatonigral pathway and substantia nigra, yet nigral neurons express no CB_1 mRNA. These findings strongly suggest CB_1 receptors are synthesized in the striatal projection neurons (medium spiny neurons—which contain moderate levels of CB_1 mRNA) and are trafficked to their axons. The axonal and terminal localization of CB_1 receptors, coupled with the observation that CB_1 receptors inhibit presynaptic calcium channels, implied that a major function of CB_1 receptors would be to

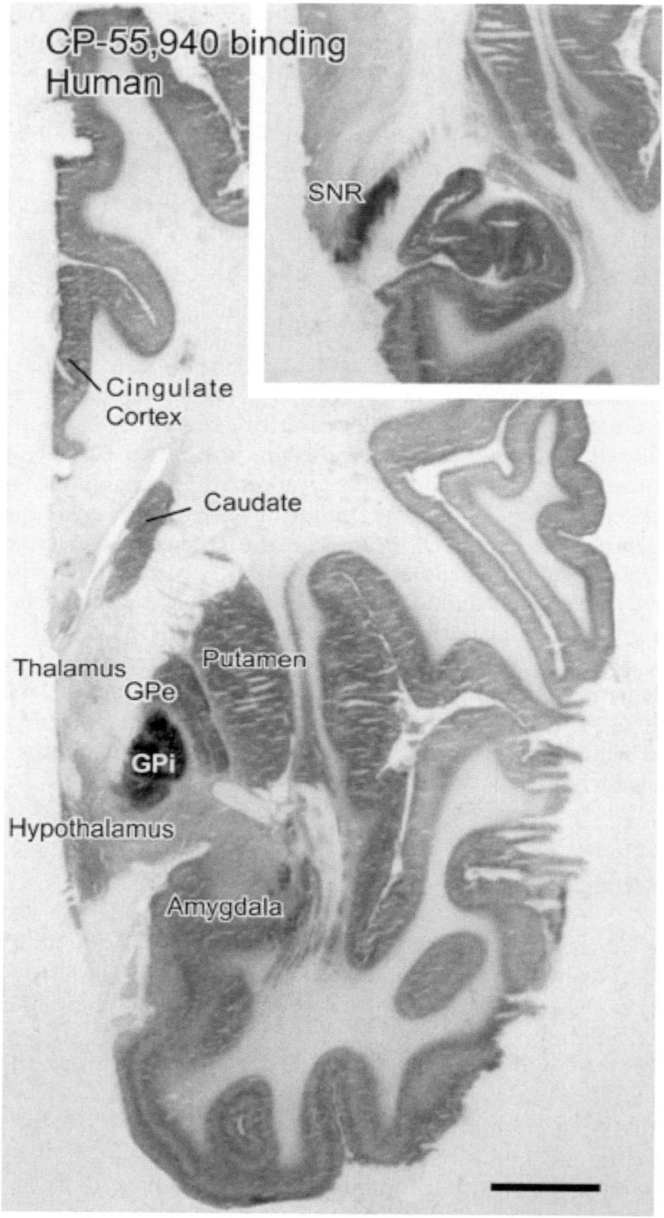

Fig. 1. CB_1 expression in human brain. CB_1 receptors were detected by quantitative autoradiography using tritiated CP55,940. Strikingly high levels are found in the substantia nigra pars reticulata (*SNR*) and the internal segment of the globus pallidus (*GPi*). Moderate levels are present in the caudate, putamen, the external segment of the globus pallidus (*GPe*), amygdala, and cortex. Lesser levels are present in hypothalamus, and very low expression is apparent in most areas of the thalamus. The laminar nature of CB_1 expression is apparent in the most rostral parts of the cortex. *Scale bar* = 1 cm. (Original figure provided by Miles Herkenham)

inhibit neurotransmitter release. The third theme was that in a few brain regions (for example, anterior olfactory nucleus, caudate nucleus and cerebellum) CB_1 receptors are uniformly expressed at moderate levels on a single class of neurons.

1.4 Immunocytochemistry

Elucidation of the primary sequence of the CB_1 receptor allowed for production of numerous CB_1 receptor antibodies. There have been two thorough immunocytochemical mapping studies in rodent brain (Tsou et al. 1998a; Egertová and Elphick 2000) and one in spinal cord (Farquhar-Smith et al. 2000). These generally support the results from the autoradiographic studies, with some differences in relative intensity of staining. These variations may be due to differences in antibody access to specific epitopes, variable post-translational modification of an epitope (e.g., phosphorylation), or fixation conditions. There is little evidence for alternative splicing in the coding region of rodent CB_1 receptors (Matsuda 1997; Lutz 2002), despite the report of alternatively splicing of the human CB_1 receptor (Shire et al. 1995; Matsuda 1997; Lutz 2002); so alternative splicing is less likely to explain the reported differences.

The immunocytochemical studies have led to additional insights into cannabinoid action. The first is that rigorous electron microscopic studies in the hippocampus demonstrated that in this region CB_1 is undetectable on somatic cell membranes and dendrites, yet is very highly expressed in axon terminals and preterminal segments (Hajos et al. 2000; Katona et al. 2000, 2001). An example of this is shown in Fig. 2, with the labeling of four consecutive ultrathin sections of a cortical axodendritic synapse. The second is that double-label immunostaining experiments demonstrated that in forebrain there is a striking correlation between cholecystokinin (CCK) and CB_1 receptor expression (Katona et al. 1999, 2001; Tsou et al. 1999). These findings have been confirmed and extended with double-label in situ hybridization studies (Marsicano and Lutz 1999). Thus, the cells expressing the highest levels of CB_1 receptors in forebrain are γ-aminobutyric acid (GABA)ergic, CCK-positive interneurons. Although inhibition of GABA release is measured in the in vitro electrophysiological studies, activation of CB_1 receptors in vivo will attenuate both inhibitory transmission (generally fast, mediated by GABA A receptors) (Wilson et al. 2001) as well as the slow, excitatory actions mediated by CCK receptors (Beinfeld and Connolly 2001). Thus, the localization of CB_1 receptors on CCK-containing neurons suggests that CB_1 receptors are well positioned to modulate complex network behaviors (Freund 2003).

Once antibodies to the anandamide-degrading enzyme, namely fatty acid amide hydrolase (FAAH), became available, it was apparent that in many regions FAAH and CB_1 expression is reciprocal in nature (Egertová et al. 1998, 2003; Tsou et al. 1998b). For example, FAAH, but not CB_1 is highly expressed in the somata and proximal dendrites of hippocampal pyramidal cells and cerebellar Purkinje neurons. These neurons are, in turn, densely innervated by CB_1-positive fibers. Thus, it has been proposed that anandamide, despite its possible presynaptic site

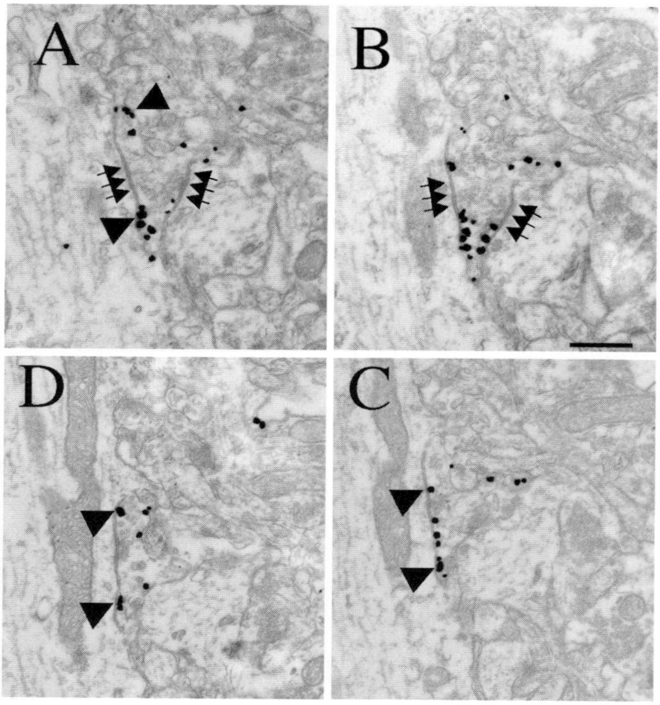

Fig. 2A–C. CB_1 expression in serial sections of a γ-aminobutyric acidergic (GABAergic) terminal synapsing onto an apical dendrite in cortex. CB_1 receptors (*arrowheads*) were detected with an antibody directed against the C terminus of rat CB_1 using pre-embedding immunogold with silver enhancement. The boutons are forming symmetric synapses (*arrows*), characteristic of GABAergic axon terminals, onto the apical dendrite of a cortical pyramidal cell. *Scale bar* = 0.5 μm. (Original photomicrograph provided by Tamas Freund and Agnes Bodor)

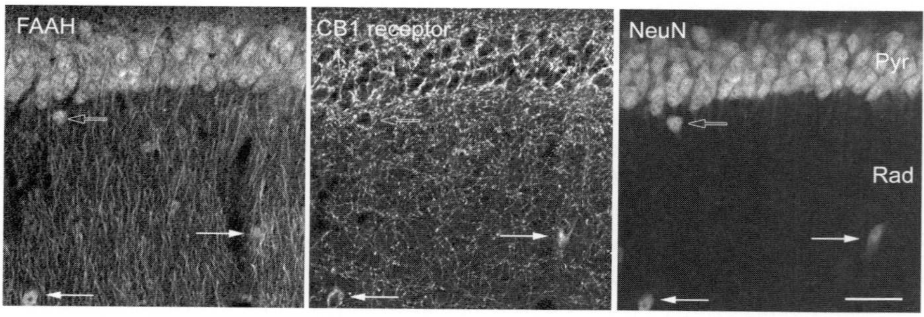

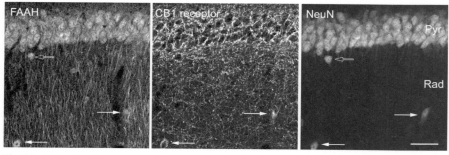

Fig. 3A–C. Reciprocal expression of CB_1 and FAAH in mouse hippocampus. FAAH was detected using an antibody raised against the last 200 residues of FAAH, CB_1 receptors were detected by an antibody directed against its C terminus, and neuronal nuclear antigen (NeuN) was detected using a mouse monoclonal antibody from Chemicon. FAAH is expressed uniformly by pyramidal neurons (*Pyr*) including the apical dendrites. FAAH is also expressed in interneurons (*open and filled arrows*). CB_1 receptors are present in axons investing the pyramidal cell layer and also some interneuron cell bodies (*filled arrows*), but not in others (*open arrow*). Staining of neurons by NeuN identifies neuronal nuclei in the field. *Scale bar* = 18 μm. (Figure provided by Tibor Harkany)

of action, is synthesized and degraded in the postsynaptic neuron. An example of this reciprocal localization in the CA1 region of mouse hippocampus is shown in Fig. 3. The situation for monoacylglycerol (MAG) lipase, the major 2-arachidonoyl glycerol-degrading enzyme, is still being clarified. However, a recent paper suggests that MAG lipase, in contrast to FAAH is predominately localized presynaptically (Gulyas et al. 2004). As the majority of CB_1 receptors a presynaptic, location of MAG lipase near these receptors would mean the endogenous cannabinoid 2-AG would be metabolized at its likely site of action, rather than having to diffuse back across the synapse. Thorough studies on the anatomical distribution of the endocannabinoid-synthesizing enzymes, diacylglycerol lipase (Bisogno et al. 2003) and the *N*-acyl phosphatidylethanolamine-preferring phospholipase D (Okamoto et al. 2004), remain to be done.

1.5 Functional Studies

Functional studies have provided another dimension in cannabinoid receptor localization. The most pertinent studies for this chapter are GTPγS studies and results inferred from studies with CB_1 knockout mice. The chapter by Lindsey et al. (this volume) will consider advances in positron emission tomography (PET), single-photon emission computed tomography (SPECT) and 2-deoxy-glucose imaging of CB_1 receptors and their activation. GTPγS studies provide a measure of regional CB_1 receptor activation of G proteins with a spatial resolution similar to other autoradiographic studies. Informative results from these studies include the observation that CB_1 receptors are relatively inefficient activators of G protein (for example, sevenfold less efficient than μ- or δ-opioid receptors) and that activation of G proteins by CB_1 receptors desensitizes strongly with chronic tetrahydrocannabinol (THC) treatment (Sim et al. 1996a,b). As mentioned below, the region-specific CB_1 knockout mice experiments support the contention that some CB_1 receptors may be expressed on hippocampal pyramidal neurons.

2 CB_1 Expression in Specific CNS Regions

2.1 Olfactory Areas

The highest levels of CB_1 receptors in olfactory bulb are in the inner granule cell layer, followed by the inner plexiform layer. The external plexiform layer, the mitral cell (glomerular) layer, and the accessory olfactory bulb have few CB_1 receptors (Herkenham et al. 1991; Tsou et al. 1998a; Egertová and Elphick 2000). The anterior olfactory nucleus and anterior commissure, which connects the olfactory bulbs, both contain high levels of CB_1 receptor. In contrast to neighboring regions, CB_1 receptors are expressed uniformly by most neurons in the anterior olfactory

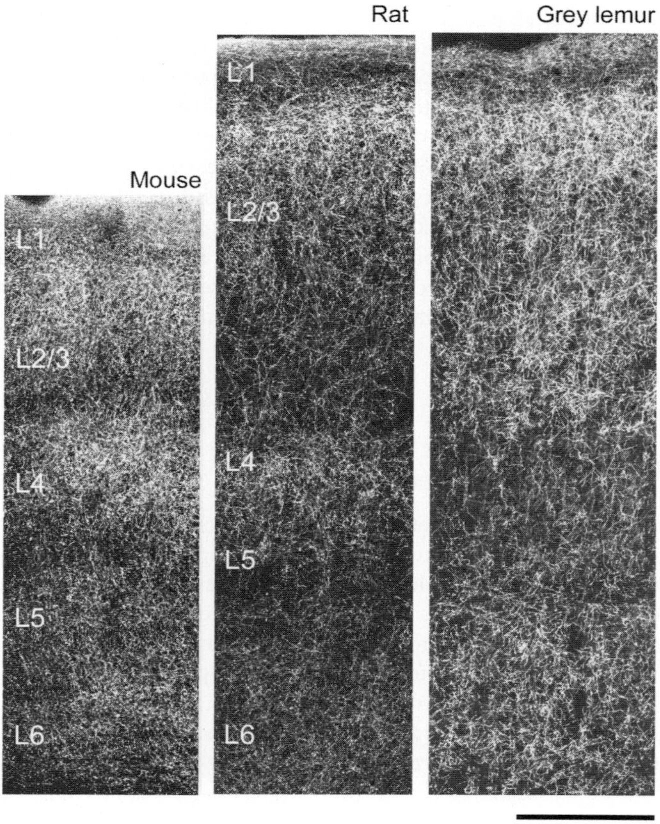

Fig. 4. Laminar CB_1 expression in cortex of three mammals. CB_1 receptors were detected with an antibody directed against the C terminus of rat CB_1. Particularly high levels of CB_1 are found in lamina layers II, upper III (*L2/3*), IV (*L4*), and VI (*L6*). *Scale bar* = 250 μm. (Figure provided by Tibor Harkany)

nucleus (Herkenham et al. 1991; Mailleux and Vanderhaeghen 1992; Matsuda et al. 1993; Tsou et al. 1998a; Marsicano and Lutz 1999; Egertová and Elphick 2000). CB_1 receptors are also found in the supporting cells of the olfactory epithelium as well as axon bundles of the lamina propria (M. Caillol, personal communication).

2.2 Neocortex

CB_1 receptors are densely expressed in all regions of the cortex (Herkenham et al. 1991; Mailleux and Vanderhaeghen 1992; Matsuda et al. 1993; Glass et al. 1997; Tsou et al. 1998a; Egertová and Elphick 2000). The variation in CB_1 expression across cortical regions has been examined most extensively in human brain using receptor autoradiography. Here there is variation between regions, with higher

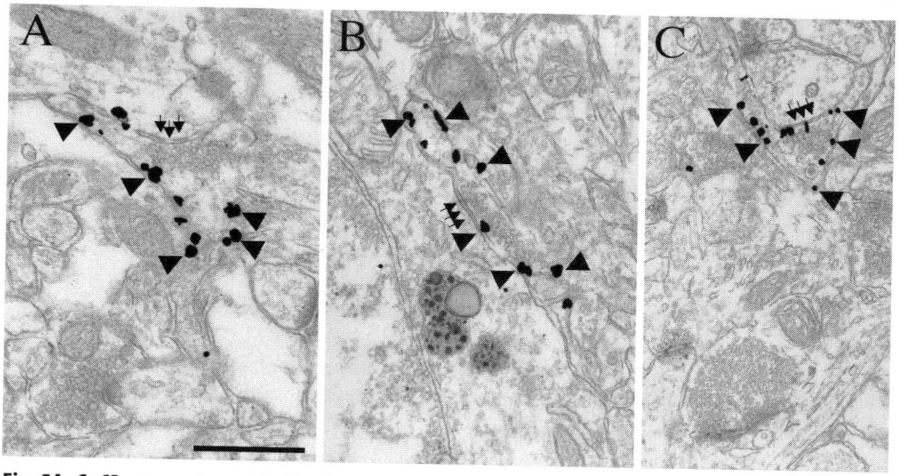

Fig. 5A–C. CB_1 expression on GABAergic terminals in rat somatosensory cortex. CB_1 receptors (*arrowheads*) were detected with an antibody directed against the C terminus of rat CB_1 using pre-embedding immunogold with silver enhancement. The boutons are forming symmetric synapses (*arrows*), characteristic of cortical GABAergic axon terminals. CB_1-positive terminals form synapses with pyramidal cell bodies (**A**), main apical dendrites (**B**), and fine-caliber dendrite branches (**C**). *Scale bar* = 0.5 μm. (Original photomicrograph provided by Tamas Freund and Agnes Bodor)

levels found in cingulate gyrus, frontal cortex, and secondary somatosensory and motor cortex. Lesser levels are found in primary somatosensory and motor cortex (Glass et al. 1997). The laminar nature of CB_1 expression within the neocortex is striking. The relative levels of expression between regions vary (Glass et al. 1997). However, as an example, in rat somatosensory cortex, CB_1 levels are relatively higher in layers II, upper III, IV, and VI. In contrast, CB_1 receptor expression is relatively less in deeper layer III and layer V (Freund et al. 2003). Layer I appears almost devoid of CB_1 receptors. Examples of CB_1 immunoreactivity in mouse, rat, and mouse lemur cortex are shown in Fig. 4. While the general laminar pattern between species is preserved, the amount of CB_1 expression appears to increase, particularly in layers III and V in the primate. Ultrastructural studies reveal that in cortex, CB_1-positive terminals synapse onto pyramidal cell bodies, apical dendrites, and smaller caliber branches (Fig. 5).

In neocortex, almost all neurons expressing CB_1 at high or moderate levels are likely to be inhibitory due to the tight correlation between GAD65 and CB_1 mRNA expression (Marsicano and Lutz 1999). However, there appear to be CB_1-mediated actions on glutamatergic transmission in cortex (Sjostrom et al. 2003). The localization and nature of these cannabinoid receptors remain to be identified. As in most other forebrain areas, the majority of strongly CB_1-positive axons in the cortex appear to arise from CCK-expressing interneurons (Marsicano and Lutz 1999). However, among cortical neurons, those expressing lower levels of CB_1 receptors represent a more heterogeneous population, with 20% of the CB_1-positive cells not expressing detectable levels of CCK mRNA (Marsicano and Lutz

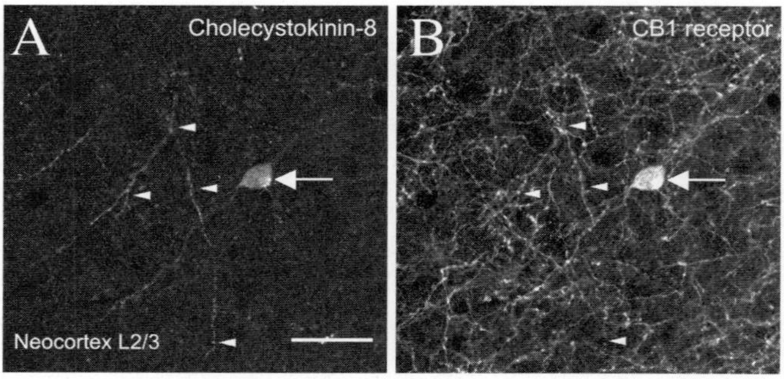

Fig. 6A, B. Co-localization of CCK with CB_1 in neocortex. **A** Expression of CCK in a cortical interneuron (*arrow*) and CCK-positive processes (*arrowheads*). **B** CB_1 is widely expressed in cortical axons. CCK-positive processes are often CB_1 positive as well (*arrowheads*). *Scale bar* = 25 μm. (Figure provided by Tibor Harkany)

1999). An example of this for layer II/III cortex is shown in Fig. 6. Here, a strongly CB_1-expressing neuron co-localizes with CCK immunoreactivity, and most CCK-containing fibers also are immunopositive for CB_1. However, there are also many CB_1-positive fibers that do not appear to contain CCK.

2.3 Hippocampal Formation

2.3.1 Hippocampus

The hippocampus expresses high levels of cannabinoid receptors. Because of the cognitive effects of cannabinoids, this brain region has received much attention as a site of action of endogenous and exogenous cannabinoids. The first autoradiographic studies found very high levels of CB_1 receptors in all subfields of the hippocampus as well as the dentate gyrus (Herkenham et al. 1991; Jansen et al. 1992). In situ hybridization studies revealed that most of this CB_1 receptor expression arose from a restricted subset of interneurons (Matsuda et al. 1990, 1993; Mailleux and Vanderhaeghen 1992). Immunocytochemical studies identified a dense plexus of CB_1-containing axon terminals surrounding the pyramidal cell layer (perisomatic labeling), consistent with CB_1 receptor expression on basket cell axons (Tsou et al. 1998a, 1999; Katona et al. 1999; Egertová and Elphick 2000). This is illustrated in Figs. 3 and 7.

Basket cells can be conveniently separated into two groups distinguished by CCK or parvalbumin expression (Freund and Buzsaki 1996; Freund 2003). Double-label immunocytochemistry has shown that high levels of CB_1 receptor expression are restricted to the CCK-expressing interneurons (Katona et al. 1999; Tsou et al. 1999). Given that the CCK-expressing interneurons may be involved in the more subjective (emotional and motivational) aspects of information processing, it is likely that endocannabinoids are involved in the normal function of these circuits, and exogenous cannabinoids may serve to disrupt them in some fashion. This

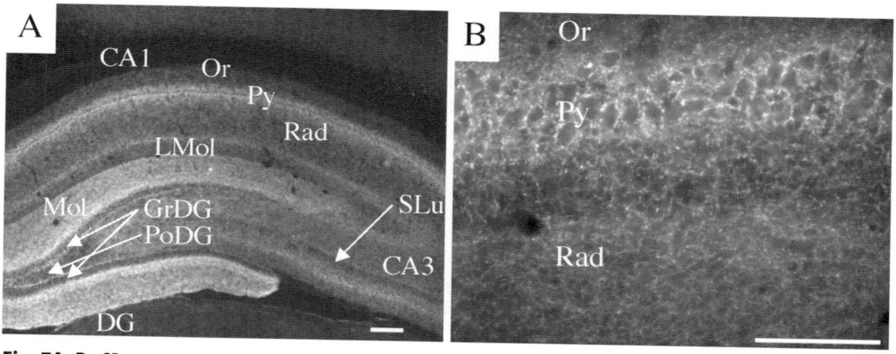

Fig. 7A, B. CB_1 expression in rat hippocampal formation. **A** CB_1 cannabinoid receptors were detected with an antibody raised against the C terminus of rat CB_1. Receptor levels are particularly high in the pyramidal cell layer (*Py*), the molecular layer (*Mol*) of the dentate gyrus (*DG*), and at the base of the granule cell layer (*GrDG*) of the dentate gyrus. Lesser levels are found in the stratum oriens (*Or*), stratum radiatum (*Rad*), stratum lucidum (*SLu*), and the polymorphic layer of the dentate gyrus (*PoDG*). *CA1*, field CA1 of the hippocampus; *CA3*, field CA3 of the hippocampus. **B** CB_1-positive fibers surround the somata of pyramidal cells (*Py*) in CA1. Numerous varicosities, corresponding to terminals, are apparent. CB_1 receptors are also seen on axon fibers, although at lower levels, in stratum oriens (*Or*) and stratum radiatum (*Rad*). For both images, *scale bar* = 100 μm. (Original photomicroph provided by Marja Van Sickle and Keith Sharkey)

pattern of selective interneuron and axonal CB_1 receptor expression is preserved at all stages of postnatal development in the rat (Morozov and Freund 2003).

Tight functional separation of GABAergic input onto CA1 pyramidal cells has also been demonstrated in an elegant electrophysiological study where only large, fast GABAergic inhibitory postsynaptic currents (IPSCs) mediated by inhibitory terminals expressing N-type [(Cav1.2); but not P-type (Cav1.1)] calcium channels were subject to depolarization-induced suppression of inhibition (Wilson et al. 2001). These electrophysiological results are satisfyingly consistent with the anatomical localization of the CB_1 receptor on perisomatic GABAergic terminals.

The expression of CB_1 receptors on principal cells of the hippocampus is a source of some controversy (as reviewed by Freund et al. 2003). On one hand, careful electron microscopic immunocytochemical studies with specific and sensitive CB_1 receptor antibodies have consistently failed to find CB_1 receptor expression in pyramidal cells (Katona et al. 1999, 2000; Hajos et al. 2000; Chen et al. 2003). On the other hand, in situ hybridization studies consistently show low levels of CB_1 mRNA in the stratum pyramidale (Mailleux and Vanderhaeghen 1992; Matsuda et al. 1993; Marsicano and Lutz 1999). Complicating interpretation of these studies are the observations that several drugs acting at CB_1 receptors (for example, WIN55,212-2 and SR141716) also inhibit glutamate release from pyramidal neurons in a CB_1 receptor-independent fashion [that is, they inhibit release in CB_1 knockout mice (Hajos et al. 2001; Hajos and Freund 2002)]. The electrophysiological and in situ data could conceivably be reconciled by crossreactivity of the in situ probes with a receptor closely related to the CB_1 receptor. However, this does not seem to be the case, as targeted deletion of CB_1 receptors from hippocampal pyramidal neurons (sparing CB_1 receptors in the interneurons) eliminates cannabinoid-mediated protection in a kainate neurotoxicity model (Marsicano et al. 2003). Although this issue is not yet resolved, a parsimonious explanation of experimental results thus far is that hippocampal pyramidal neurons may express CB_1 receptors, albeit at far lower levels than the CCK-containing basket cells.

2.3.2
Dentate Gyrus

As in the hippocampus, CB_1 receptors in dentate gyrus are primarily found in CCK-containing basket cells—parvalbumin-positive basket cells and the granule cells do not express CB_1 (Mailleux and Vanderhaeghen 1992; Matsuda et al. 1993; Katona et al. 1999; Marsicano and Lutz 1999; Tsou et al. 1999). This results in high levels of CB_1 receptors in the inner third of the molecular layer and at the base of the granule cell layer in the dentate gyrus (Fig. 7). While it has not been studied anatomically, functional studies suggest the glutamatergic terminals of the perforant path may express CB_1 receptors (Kirby et al. 1995).

2.4
Amygdala

CB_1 receptor distribution in the amygdala is markedly heterogeneous (Katona et al. 2001; McDonald and Mascagni 2001). High levels are found in the basolateral complex (comprising the lateral, basal, and accessory basal nucleus), nucleus of the lateral olfactory tract, the periamygdaloid cortex, and amygdalohippocampal areas. In contrast, CB_1 receptors are sparsely expressed in the medial, central, and intercalated nuclei (Fig. 1). As in other regions of the forebrain, CB_1 receptors are primarily expressed on large, GABAergic, CCK-containing axon terminals (Katona et al. 2001; McDonald and Mascagni 2001). Activation of these CB_1 receptors by cannabinoids decreases GABA release from these terminals, which may disinhibit the basolateral glutamatergic pyramidal cells (Katona et al. 2001). As in other forebrain regions, there is also a relatively high concordance between CB_1 and serotonin-3 (5-HT3) receptor expression in amygdala (Morales et al. 2004). Compelling evidence suggests that endocannabinoids play a role in modulating fear conditioning at the level of the amygdala (Marsicano et al. 2002), and amygdaloid CB_1 receptors may play a role in the panic states occasionally seen following consumption of prodigious quantities of cannabis.

2.5
Subcortical CB_1 Receptors

2.5.1
Basal Forebrain

Moderate levels of CB_1 receptors are present in the basal forebrain. Autoradiographic studies found CB_1 in the medial and lateral septum and the intermediate nucleus of the lateral septum (Herkenham et al. 1991). CB_1 mRNA is present at moderate levels in many cells of the medial septum and the nucleus of the diagonal band (Mailleux and Vanderhaeghen 1992; Matsuda et al. 1993). A recent immunocytochemical study in mouse revealed that the tenia tecta, ventral pallidum, and substantia innominata all contained a dense network of CB_1-positive fibers. In contrast, a fine meshwork of CB_1 receptor-containing fibers was present in the medial septum, diagonal bands, and nucleus basalis (Harkany et al. 2003). No CB_1 immunoreactivity was detected in basal forebrain cholinergic cells; instead these cells contained high levels of FAAH (Harkany et al. 2003). These results are in contrast to a report in monkey, which found CB_1 expression in cholinergic forebrain neurons (Lu et al. 1999). This discrepancy may be due to a difference between species or methodologies.

2.5.2
Basal Ganglia

The subcortical structures with the highest level of CB_1 receptor expression are the basal ganglia. In fact, the highest levels of CB_1 receptors in the brain detected in autoradiography studies were found in the substantia nigra (Herkenham et al. 1991). In situ hybridization studies demonstrated that many striatal medium spiny neurons express CB_1 receptors (Matsuda et al. 1993; Julian et al. 2003). In contrast, adult pallidal and nigral neurons contain little or no CB_1 mRNA (Matsuda et al. 1993; Julian et al. 2003). Rather, CB_1 receptors in the globus pallidus and substantia nigra are localized to the axons traversing or terminating in these structures (Tsou et al. 1998a; Egertová and Elphick 2000). Thus, the high levels of pallidal and nigral CB_1 receptor binding and protein observed in autoradiographic and immunocytochemical studies mostly arise from GABAergic neurons projecting from the caudate putamen. Figure 8 illustrates the intense immunostaining of CB_1 receptors that begins at the border between the caudate putamen and globus pallidus. It is possible that dopaminergic neurons may transiently express CB_1 receptors during development, as CB_1 co-localizes with tyrosine hydroxylase in cultured mesencephalic neurons (Hernandez et al. 2000).

Both autoradiographic and immunocytochemical studies show a gradient of CB_1 expression in the rodent caudate putamen with the highest levels found dorsolaterally (Tsou et al. 1998a; Egertová and Elphick 2000). Both the matrix and patch structures of the caudate putamen contain CB_1 receptors, where they partially overlap with μ-opioid receptors (Rodriguez et al. 2001). CB_1 receptors are present on both the striatonigral (prodynorphin or preprotachykinin A positive) and striatopallidal (proenkephalin positive) projection pathways (Hohmann and Herkenham 2000). Thus, CB_1 receptors are positioned to modulate both the direct and indirect striatal output pathways.

In addition to medium spiny neurons, anatomical and functional studies identified CB_1 receptors on the terminals of the corticostriatal pathway (Gerdeman and Lovinger 2001; Huang et al. 2001; Rodriguez et al. 2001) and GABAergic aspiny interneurons (Hohmann and Herkenham 2000). In contrast, CB_1 receptors

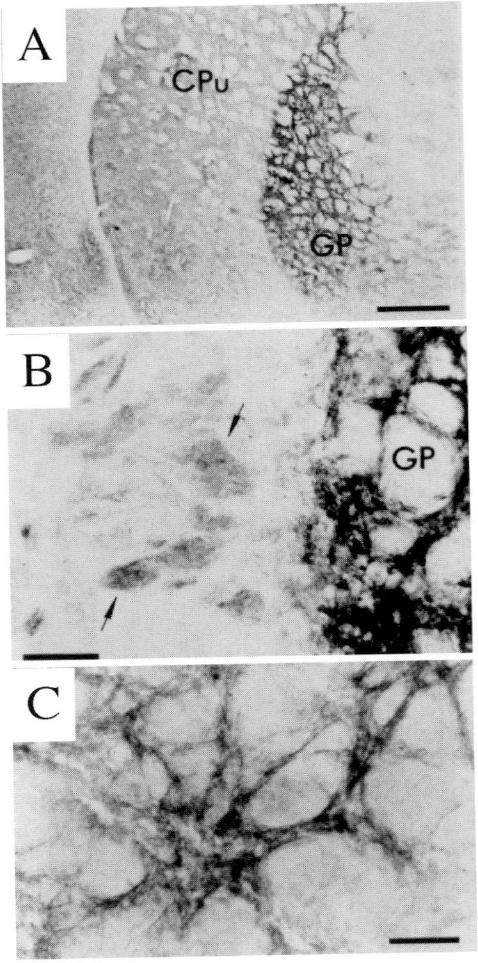

Fig. 8A–C. CB_1 expression in basal ganglia detected by an antibody raised against the amino terminus of rat CB_1. **A** Low-power view showing moderate levels of CB_1 in caudate putamen (*CPu*) and very high levels in the globus pallidus (*GP*). The sharp demarcation between the two structures is evident. **B** Boundary of CPu and GP. Two moderately stained fiber bundles are indicated by the *arrows*. **C** High-power view of globus pallidus with fine, strongly immunoreactive, non-varicose processes corresponding to medium spiny neuron axons. *Scale bars* = 500 μm (**A**); 50 μm (**B** and **C**). (Modified from a photomicrograph provided by Kang Tsou)

do not appear to be expressed in the large aspiny cholinergic interneurons or somatostatin-containing interneurons (Hohmann and Herkenham 2000). CB_1 receptors are also present on the neurons in the subthalamic nucleus (Matsuda et al. 1993). Taken together, the presence of CB_1 receptors on diverse neuronal populations in the basal ganglia can account for the complex effects of cannabinoids on motor behaviors (Sanudo-Pena et al. 1999b; Romero et al. 2002).

2.5.3 Nucleus Accumbens

CB_1 receptors are also expressed at low to moderate levels in the nucleus accumbens. Here CB_1 receptors are found in a pattern reminiscent of the striatum. CB_1 receptors are expressed on terminals of the glutamatergic prefrontal cortex accumbens pathways (Robbe et al. 2001). They are also present on the accumbens medium spiny neurons. They appear to be absent from the dopaminergic terminals projecting to the accumbens from the ventral tegmentum. Consequently, cannabinoid stimulation of dopamine release in nucleus accumbens (Tanda et al. 1997) appears to be an indirect effect, perhaps mediated by inhibition of GABA release (Szabo et al. 1999, 2002).

2.5.4 Thalamus

Expression of CB_1 receptors in the thalamus is low (Herkenham et al. 1991; Jansen et al. 1992; Matsuda et al. 1993; Glass et al. 1997; Tsou et al. 1998a; Egertová and Elphick 2000). Regions of the thalamus with some CB_1 expression include the (lateral) habenular nucleus, the anterior dorsal thalamic nucleus, and the reticular thalamic nucleus (Herkenham et al. 1991; Mailleux and Vanderhaeghen 1992; Tsou et al. 1998a).

2.5.5 Hypothalamus

Given the marked effects of CB_1 receptor agonists on body temperature and antagonists on consumptive behavior, it is not surprising that CB_1 receptors are present in the hypothalamus. Low to moderate levels of CB_1 immunoreactivity are found in the paraventricular nucleus, ventral medial hypothalamic nucleus, infundibular stem, and lateral hypothalamic area (Tsou et al. 1998a). There are in situ data suggesting CB_1 receptors in the hypothalamus are primarily present on glutamatergic neurons (Marsicano and Lutz 1999). Intriguingly, although the levels of CB_1 receptors in hypothalamus are fairly low, functional studies with GTPγS suggests these CB_1 receptors are more strongly coupled to G proteins than are most CB_1 receptors (Breivogel et al. 1997). A careful and detailed anatomical study of CB_1 expression in hypothalamus is needed because of the likely involvement of this region in the anti-appetitive actions of CB_1 antagonists.

2.6 Midbrain

2.6.1 Substantia Nigra

A striking feature of CB_1 receptor expression is the high number of CB_1 receptors found in the substantia nigra (Fig. 9A and 9B). As mentioned above, these receptors

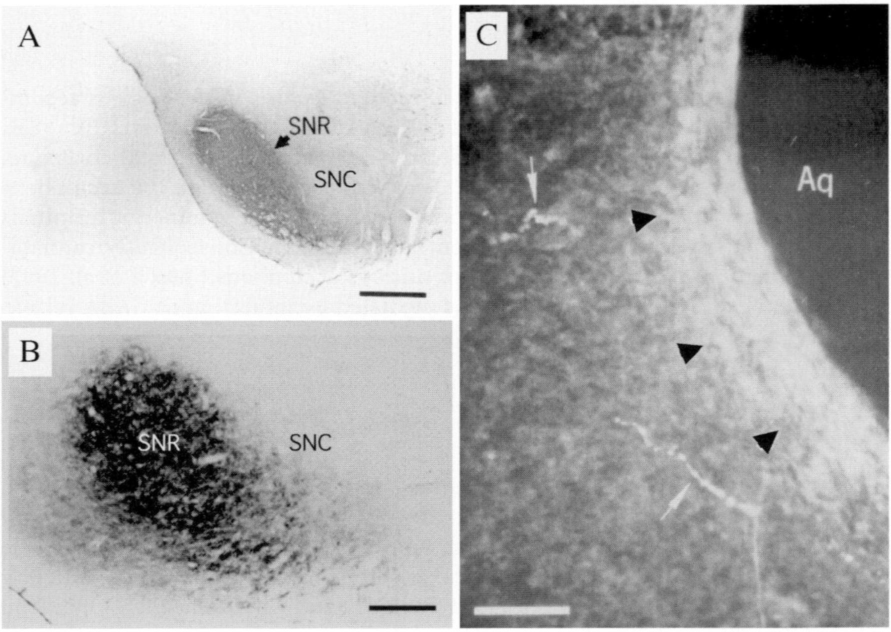

Fig. 9A–C. CB_1 receptor expression in midbrain structures detected by an antibody against the amino terminus of rat CB_1. **A** CB_1 immunostaining is very strong in substantia nigra pars reticulata (*SNR*) but virtually absent in substantia nigra pars compacta (*SNC*). **B** Higher magnification view of SNR. When the plane of the section is perpendicular to the striatonigral pathway, immunoreactivity is apparent as puncta, from the high levels of axonal CB_1 expression. **C** In caudal periaqueductal gray, CB_1-positive fibers (*arrows*) and intensely labeled neuropil (*arrowheads*) are apparent. *Aq*, lumen of the aqueduct. *Scale bars* = 500 μm (**A**), 50 μm (**B**), and 20 μm (**C**). (Modified from a photomicrograph provided by Kang Tsou)

appear to be restricted to the GABAergic axons of the putamen medium spiny neurons—the nigral dopaminergic neurons appear to be devoid of CB_1 receptors (Matsuda et al. 1993; Julian et al. 2003). Anatomical and functional evidence also suggests that the excitatory glutamatergic projection from the subthalamic nucleus to the substantia nigra contains CB_1 receptors (Mailleux and Vanderhaeghen 1992; Sanudo-Pena and Walker 1997; Sanudo-Pena et al. 1999b).

2.6.2
Ventral Tegmentum

CB_1 expression and function in the ventral tegmental area (VTA) is of interest because of the euphoric and reinforcing properties of cannabinoids—evident in carefully conducted studies. There is no evidence for CB_1 receptor expression on the tegmental dopamine neurons (Herkenham et al. 1991). Emerging functional evidence (detailed immunocytochemical studies remain to be done) suggests that CB_1 receptors are present on intrinsic GABAergic terminals, GABAergic terminals present on accumbens neurons projecting to VTA, and glutamatergic terminals (Szabo et al. 2002; Riegal et al. 2003; Melis et al. 2004). These findings suggest that CB_1 receptor activation may play a role in the reinforcing effects of cannabinoids and, more provocatively, that disorders in endocannabinoid-mediated synaptic plasticity may be important in a broader range of addictive disorders.

2.6.3 Periaqueductal Gray

Moderate levels of CB_1 receptor are also found in several other regions of the midbrain. One of these is the periaqueductal gray (PAG) (Fig. 9C). Here CB_1 receptors are found on the terminals of GABAergic neurons. In contrast to opiate receptors on GABAergic aqueductal neurons, CB_1 receptors are preferentially localized in the dorsal portion of the PAG (Tsou et al. 1998a). Autoradiographic studies indicate that CB_1 receptors are also found at moderate levels in the reticular formation and raphe nucleus (Glass et al. 1997).

2.7 Brainstem

Expression of CB_1 receptors in brainstem is relatively low. In contrast to the opioid receptors, few cannabinoid receptors are found in the medullary respiratory control centers (Herkenham et al. 1991; Glass et al. 1997). This likely underlies

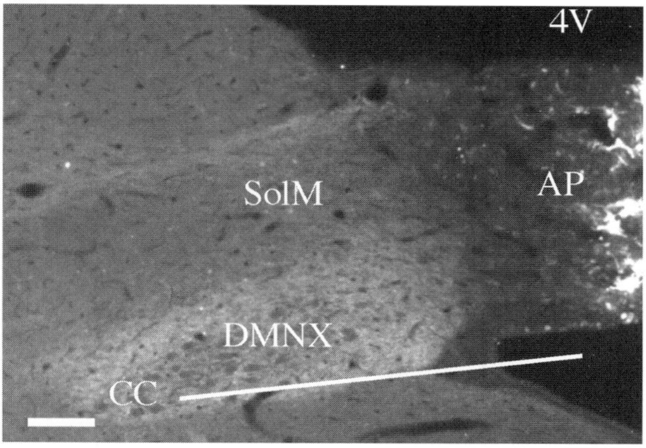

Fig. 10. CB_1 expression in emetic centers. CB_1 is prominently expressed in the ferret area postrema (*AP*), dorsal vagal complex (*DMNX*), and associated regions involved in emesis as detected with a C-terminal CB_1 receptor antibody. Particularly strong immunostaining is present in a restricted group of cells in the area postrema as well as diffusely through the dorsal motor nucleus of the vagus (notice the lack of staining of cell bodies in DMNX), and the medial nucleus of the solitary tract (*SolM*). *4V*, fourth ventricle; *CC*, central canal. *Scale bar* = 100 μm. (Original photomicrograph provided by Marja Van Sickle and Keith Sharkey)

the low lethality of high doses of cannabinoids. One exception to the low levels of cannabinoid receptor in the brainstem is the medullary nuclei associated with emesis (Van Sickle et al. 2001). Here, as illustrated in Fig. 10, relatively high levels of CB_1 receptor are found in the dorsal motor nucleus of the vagus and the medial subnucleus of the nucleus of the solitary tract. Moderate levels are present in the subnucleus gelatinosus of the solitary tract (Fig. 10). Occasional, very strongly stained cells are evident in the area postrema (Fig. 10). In most cases, CB_1 receptors appear to be localized to terminal structures. Interestingly, FAAH immunoreactivity was restricted to the cell bodies invested by the CB_1-positive fibers (Van Sickle et al. 2001), continuing the theme of complementary expression of CB_1 receptors and FAAH. Compelling evidence suggests that a major portion of the antiemetic actions of cannabinoids is a consequence of CB_1 receptor activation in these nuclei (Van Sickle et al. 2001, 2003).

2.8 Cerebellum

CB_1 receptor expression in the cerebellum follows a striking and very predictable pattern. Autoradiographic and immunocytochemical studies show very strong labeling of the molecular layer (Fig. 11A), while in situ hybridization studies show robust expression in the granule cell layer (Matsuda et al. 1990; Herkenham et al. 1991; Glass et al. 1997; Tsou et al. 1998a; Egertová and Elphick 2000). Combining these results with functional studies suggests CB_1 receptors are expressed in climbing fibers and parallel fibers, as well as the basket cells, particularly at the basket cell–Purkinje cell synapse (Fig. 11A, B). In contrast, there is little evidence that Purkinje neurons express CB_1 receptors (Matsuda et al. 1990). Thus, both ma-

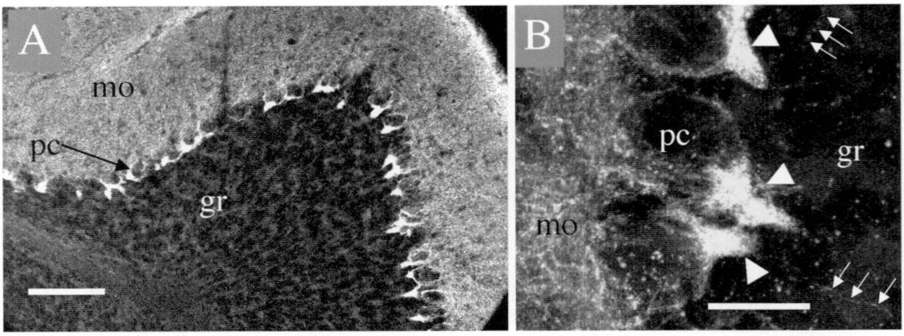

Fig. 11. CB_1 is highly expressed in the molecular layer and on the basket cell–Purkinje neuron synapse of the mouse cerebellum. **A** Using an antibody directed against the C terminus of the CB_1 receptor, strikingly high levels of CB_1 receptors are apparent at basket cell synapses onto the Purkinje neurons (*pc*) as well as diffusely high levels in the molecular layer (*mo*), corresponding to the parallel fiber–Purkinje neuron synapse. **B** Higher magnification view showing intense labeling of basket cell synapses (*arrowheads*), labeled fibers in the granule cell layer (*gr*) (*arrows*), and diffuse labeling in the molecular layer. *Scale bars* = 150 μm (**A**), and 15 μm (**B**). (Modified from a photomicrograph provided by Jane Lauckner)

jor glutamatergic inputs and at least some of the GABAergic input onto Purkinje neurons are subject to modulation by cannabinoids. These anatomical observations are supported by several elegant electrophysiological studies demonstrating a role for endogenous cannabinoid inhibition of glutamatergic and GABAergic neurotransmission onto Purkinje neurons (Kreitzer and Regehr 2001; Maejima et al. 2001; Diana et al. 2002; Kreitzer et al. 2002; Brenowitz and Regehr 2003). While most of the actions of exogenous and endogenous cannabinoids can be interpreted as effects on presynaptic CB_1 receptors, there is also solid evidence for somatic expression of CB_1 receptors. This comes from experiments by the Regehr lab showing that the release of endocannabinoids from Purkinje neurons can slow the firing rate of basket cells, consistent with an activation of somatic potassium channels (Kreitzer et al. 2002).

2.9 Spinal Cord

Because of the efficacy of intrathecal cannabinoids in various pain models, it is not surprising that moderate levels of CB_1 receptor are found in the regions of the spinal cord associated with analgesia. In particular, the superficial layers of the dorsal horn, the dorsolateral funiculus, and lamina X all have moderate levels of CB_1 receptor (Farquhar-Smith et al. 2000). Cannabinoids inhibit glutamate release from afferents in lamina I of the dorsal horn in a CB_1 receptor-dependent fashion (Jennings et al. 2001; Morisset and Urban 2001). Providing anatomical support for these functional studies, CB_1 receptors are found in the dorsal horn in a characteristic twin band corresponding to lamina I and the inner portion of lamina II (Farquhar-Smith et al. 2000).

The source of CB_1 receptors in the dorsal horn remains controversial. One immunocytochemical study found little decrease in CB_1 receptor immunoreactivity following dorsal rhizotomy or hemisection of the spinal cord, suggesting CB_1 receptors are primarily expressed on interneurons (Farquhar-Smith et al. 2000). In contrast, another study using autoradiography to quantify CB_1 expression found a 50% decrease in CB_1 expression following dorsal rhizotomy, suggesting that approximately 50% of CB_1 receptors are found on primary afferents while the balance are on interneurons and descending pathways (Hohmann et al. 1999). Additional evidence supporting functionally significant levels of CB_1 expression on primary afferents includes the findings that CB_1 receptor activation inhibits glutamate release in lamina I (Jennings et al. 2001; Morisset and Urban 2001), only low levels of CB_1 mRNA are present in spinal cord (Mailleux and Vanderhaeghen 1992), and CB_1 receptor mRNA and protein are both expressed in dorsal root ganglia cells (Hohmann et al. 1999; Hohmann and Herkenham 1999b; Bridges et al. 2003). Despite the presence of CB_1 receptors on some C fibers, many more are present on large, myelinated fibers (Abeta and Adelta) (Hohmann and Herkenham 1998, 1999b; Bridges et al. 2003; Price et al. 2003). In balance, it is likely that the analgesic effects of CB_1 receptor activation in the spinal cord are due to interplay between cannabinoid actions on primary afferents, interneurons, and descending pathways.

Emerging evidence suggests that CB_2 agonists are analgesic in a number of neuropathic and inflammatory pain models (Ibrahim et al. 2003; Nackley et al. 2003; Hohmann et al. 2004). There is little evidence for CB_2 expression in normal spinal cord (for example, Buckley et al. 1998). However, CB_2 expression is induced in the spinal cord, likely in microglial cells, following nerve injury and the development of a neuropathic state (Zhang et al. 2003). Precise localization of these receptors using immunocytochemistry remains to be performed. Intriguingly, CB_2 receptor expression was not increased in an inflammatory pain model, despite the efficacy of CB_2 agonists as analgesics in this model. This suggests that CB_2 receptors are selectively upregulated only after specific forms of nerve injury. It also implies that peripherical CB_2 receptors mediate some of the effects of CB_2 agonists, at least some inflammatory pain states.

While expression of CB_1 in the dorsal horn is well established, its expression in spinal cord areas associated with movement is less certain. However, some immunocytochemical evidence suggests CB_1 receptors are found in the ventral horn (Tsou et al. 1998a; Sanudo-Pena et al. 1999a). Interestingly, FAAH is also found in the cell bodies of ventral horn neurons (Tsou et al. 1998b). The localization of CB_1 receptors and FAAH in neuronal circuits associated with movement may underlie the antispastic effects of cannabinoids.

3 Peripheral Nervous System

3.1 Peripheral Nerves

There is strong evidence for CB_1 receptor expression in the periphery. For example, ligation of the sciatic nerve leads to accumulation of CB_1 receptors proximal to the ligation (Hohmann and Herkenham 1999a) and peripherally administered, but systemically inactive, doses of CB_1 agonists can be analgesic (Calignano et al. 1998). To date, no studies have been published examining CB_1 receptors in the periphery beyond major nerves (e.g., sciatic). The development of sufficiently sensitive techniques to study CB_1 and CB_2 expression in the periphery is needed to thoroughly understand the peripheral actions of these compounds. Cannabinoids also regulate autonomic nervous system function. Examples include cannabinoid inhibition of neurotransmitter release in ileum (Roth 1978; Pertwee et al. 1992; Croci et al. 1998) and vas deferens (Nicolau et al. 1978; Pertwee et al. 1992).

3.2 Enteric Nervous System

CB_1 receptors are richly distributed throughout the enteric nervous system; their function has been the focus of reviews (Pertwee 2001; Pinto et al. 2002). Cannabis and its psychoactive extracts inhibit intestinal motility (Shook and Burks 1989; Izzo et al. 1999). Detailed anatomical studies have found high levels of CB_1 receptor in specific populations of nerves innervating the gut (Kulkarni-Narla and Brown 2000; Coutts et al. 2002; MacNaughton et al. 2004). Studies of guinea pig ileum suggest that CB_1 receptors are localized, in part to the cholinergic myenteric motor neurons (Coutts et al. 2002). Activation of these presynaptic CB_1 receptors inhibits acetylcholine release, decreasing longitudinal muscle contractions. Intestinal motility mediated by non-adrenergic, non-cholinergic (NANC) neurotransmission is also decreased by CB_1 agonists (Izzo et al. 1998); likewise, CB_1 receptors are also found on some NANC neurons (MacNaughton et al. 2004). Activation of CB_1 receptors also decreases fluid secretion in the stomach and intestine. Consistent with this, CB_1 receptors are present in both cholinergic and non-cholinergic sensorimotor submucosal neurons (Tyler et al. 2000; Adami et al. 2002; MacNaughton et al. 2004). CB_1 receptors are also present on some vagal afferents, where their expression is decreased by food intake and CCK (Burdyga et al. 2004).

3.3 Pelvic Viscera

Several studies suggest CB_1 receptor activation has effects on bladder, vas deferens, and uterine function, in both normal and pathophysiological states (Nicolau et al. 1978; Pertwee et al. 1992; Pertwee and Fernando 1996; Dmitrieva and Berkley 2002; Farquhar-Smith et al. 2002). While CB_1 receptors are expressed on tyrosine hydroxylase (noradrenaline)-positive pelvic neurons (Pan et al. 1998), detailed studies on CB_1 receptor distribution to these organs remains to be performed.

4 Summary

The pattern of CB_1 expression in the brain generally correlates with its function both at the macroscopic and microscopic levels. High levels of cannabinoid receptors are found in brain regions implicated in the behavioral effects of cannabinoids, particularly cortex, hippocampus, amygdala, basal ganglia, cerebellum, and the emetic centers of the brainstem. Conversely, low levels are found in other regions, such as the thalamus, pons, and the remainder of the brainstem. Correspondingly, these areas have generally not been implicated in playing a major role in the actions of cannabis or cannabinoids. Undoubtedly, the future will bring further refinement in the localization of CB_1 receptors as well as the badly needed details on where endocannabinoid synthesizing and degrading enzymes are found. Together, this information will aid in our understanding of the role of CB_1 receptors in the function of the CNS, both in normal physiology as well as in pathological states.

Acknowledgements. I would like to dedicate this review to the memory of Professor Kang Tsou, who first introduced me to the intricacies, beauty, and logic of CB_1 immunocytochemistry in the CNS. I am indebted to my collaborators and colleagues who have provided images to use as figures in this review: Tibor Harkany, Tamas Freund and Agnes Bodor, Kang Tsou, Marja Van Sickle and Keith Sharkey, Miles Herkenham, and Jane Lauckner. In addition, I am grateful to these individuals and many others for stimulating discussions on the insight that CB_1 receptor localization gives us to understand the role of endogenous and exogenous cannabinoids in CNS function. The writing of this chapter has been supported in part by grants from the NIH, DA00286 and DA11322.

References

Adami M, Frati P, Bertini S, Kulkarni-Narla A, Brown DR, de Caro G, Coruzzi G, Soldani G (2002) Gastric antisecretory role and immunohistochemical localization of cannabinoid receptors in the rat stomach. Br J Pharmacol 135:1598–1606

Beinfeld MC, Connolly K (2001) Activation of CB1 cannabinoid receptors in rat hippocampal slices inhibits potassium-evoked cholecystokinin release, a possible mechanism contributing to the spatial memory defects produced by cannabinoids. Neurosci Lett 301:69–71

Bisogno T, Howell F, Williams G, Minassi A, Cascio MG, Ligresti A, Matias I, Schiano-Moriello A, Paul P, Williams EJ, Gangadharan U, Hobbs C, Di Marzo V, Doherty P (2003) Cloning of the first sn1-DAG lipases points to the spatial and temporal regulation of endocannabinoid signaling in the brain. J Cell Biol 163:463–468

Breivogel CS, Sim LJ, Childers SR (1997) Regional differences in cannabinoid receptor/G-protein coupling in rat brain. J Pharmacol Exp Ther 282:1632–1642

Brenowitz SD, Regehr WG (2003) Calcium dependence of retrograde inhibition by endocannabinoids at synapses onto Purkinje cells. J Neurosci 23:6373–6384

Bridges D, Rice AS, Egertová M, Elphick MR, Winter J, Michael GJ (2003) Localisation of cannabinoid receptor 1 in rat dorsal root ganglion using in situ hybridisation and immunohistochemistry. Neuroscience 119:803–812

Buckley NE, Hansson S, Harta G, Mezey E (1998) Expression of the CB1 and CB2 receptor messenger RNAs during embryonic development in the rat. Neuroscience 82:1131–1149

Burdyga G, Lal S, Varro A, Dimaline R, Thompson DG, Dockray GJ (2004) Expression of cannabinoid CB1 receptors by vagal afferent neurons is inhibited by cholecystokinin. J Neurosci 24:2708–2715

Calignano A, La Rana G, Giuffrida A, Piomelli D (1998) Control of pain initiation by endogenous cannabinoids. Nature 394:277–281

Chen K, Ratzliff A, Hilgenberg L, Gulyas A, Freund TF, Smith M, Dinh TP, Piomelli D, Mackie K, Soltesz I (2003) Long-term plasticity of endocannabinoid signaling induced by developmental febrile seizures. Neuron 39:599–611

Coutts AA, Irving AJ, Mackie K, Pertwee RG, Anavi-Goffer S (2002) Localisation of cannabinoid CB(1) receptor immunoreactivity in the guinea pig and rat myenteric plexus. J Comp Neurol 448:410–422

Croci T, Manara L, Aureggi G, Guagnini F, Rinaldi-Carmona M, Maffrand JP, Le Fur G, Mukenge S, Ferla G (1998) In vitro functional evidence of neuronal cannabinoid CB1 receptors in human ileum. Br J Pharmacol 125:1393–1395

Diana MA, Levenes C, Mackie K, Marty A (2002) Short-term retrograde inhibition of GABAergic synaptic currents in rat Purkinje cells is mediated by endogenous cannabinoids. J Neurosci 22:200–208

Dinh TP, Carpenter D, Leslie FM, Freund TF, Katona I, Sensi SL, Kathuria S, Piomelli D (2002) Brain monoglyceride lipase participating in endocannabinoid inactivation. Proc Natl Acad Sci U S A 99:10819–10824

Dmitrieva N, Berkley KJ (2002) Contrasting effects of WIN 55212-2 on motility of the rat bladder and uterus. J Neurosci 22:7147–7153

Egertová M, Elphick MR (2000) Localisation of cannabinoid receptors in the rat brain using antibodies to the intracellular C-terminal tail of CB. J Comp Neurol 422:159–171

Egertová M, Giang DK, Cravatt BF, Elphick MR (1998) A new perspective on cannabinoid signalling: complementary localization of fatty acid amide hydrolase and the CB1 receptor in rat brain. Proc R Soc Lond B Biol Sci 265:2081–2085

Egertová M, Cravatt BF, Elphick MR (2003) Comparative analysis of fatty acid amide hydrolase and cb(1) cannabinoid receptor expression in the mouse brain: evidence of a widespread role for fatty acid amide hydrolase in regulation of endocannabinoid signaling. Neuroscience 119:481–496

Farquhar-Smith WP, Egertová M, Bradbury EJ, McMahon SB, Rice AS, Elphick MR (2000) Cannabinoid CB(1) receptor expression in rat spinal cord. Mol Cell Neurosci 15:510–521

Farquhar-Smith WP, Jaggar SI, Rice AS (2002) Attenuation of nerve growth factor-induced visceral hyperalgesia via cannabinoid CB(1) and CB(2)-like receptors. Pain 97:11–21

Freund TF (2003) Interneuron diversity series: rhythm and mood in perisomatic inhibition. Trends Neurosci 26:489–495

Freund TF, Buzsaki G (1996) Interneurons of the hippocampus. Hippocampus 6:347–470

Freund TF, Katona I, Piomelli D (2003) Role of endogenous cannabinoids in synaptic signaling. Physiol Rev 83:1017–1066

Gerdeman G, Lovinger DM (2001) CB1 cannabinoid receptor inhibits synaptic release of glutamate in rat dorsolateral striatum. J Neurophysiol 85:468–471

Glass M, Dragunow M, Faull RL (1997) Cannabinoid receptors in the human brain: a detailed anatomical and quantitative autoradiographic study in the fetal, neonatal and adult human brain. Neuroscience 77:299–318

Gulyas AI, Cravatt BF, Bracey MH, Dinh TP, Piomelli D, Boscia F, Freud TF (2004) Segregation of two endocannabinoid-hydrolyzing enzymes into pre- and postsynaptic compartements in the rat hippocamus, cerebellum, and amygdala. Eur J Neurosci 20:441–458

Hajos N, Freund TF (2002) Pharmacological separation of cannabinoid sensitive receptors on hippocampal excitatory and inhibitory fibers. Neuropharmacology 43:503–510

Hajos N, Katona I, Naiem SS, MacKie K, Ledent C, Mody I, Freund TF (2000) Cannabinoids inhibit hippocampal GABAergic transmission and network oscillations. Eur J Neurosci 12:3239–3249

Hajos N, Ledent C, Freund TF (2001) Novel cannabinoid-sensitive receptor mediates inhibition of glutamatergic synaptic transmission in the hippocampus. Neuroscience 106:1–4

Harkany T, Hartig W, Berghuis P, Dobszay MB, Zilberter Y, Edwards RH, Mackie K, Ernfors P (2003) Complementary distribution of type 1 cannabinoid receptors and vesicular glutamate transporter 3 in basal forebrain suggests input-specific retrograde signalling by cholinergic neurons. Eur J Neurosci 18:1979–1992

Herkenham M, Lynn AB, Johnson MR, Melvin LS, de Costa BR, Rice KC (1991) Characterization and localization of cannabinoid receptors in rat brain: a quantitative in vitro autoradiographic study. J Neurosci 11:563–583

Hernandez M, Berrendero F, Suarez I, Garcia-Gil L, Cebeira M, Mackie K, Ramos JA, Fernandez-Ruiz J (2000) Cannabinoid CB(1) receptors colocalize with tyrosine hydroxylase in cultured fetal mesencephalic neurons and their activation increases the levels of this enzyme. Brain Res 857:56–65

Hohmann AG, Herkenham M (1998) Regulation of cannabinoid and mu opioid receptors in rat lumbar spinal cord following neonatal capsaicin treatment. Neurosci Lett 252:13–16

Hohmann AG, Herkenham M (1999a) Cannabinoid receptors undergo axonal flow in sensory nerves. Neuroscience 92:1171–1175

Hohmann AG, Herkenham M (1999b) Localization of central cannabinoid CB1 receptor messenger RNA in neuronal subpopulations of rat dorsal root ganglia: a double-label in situ hybridization study. Neuroscience 90:923–931

Hohmann AG, Herkenham M (2000) Localization of cannabinoid CB(1) receptor mRNA in neuronal subpopulations of rat striatum: a double-label in situ hybridization study. Synapse 37:71–80

Hohmann AG, Briley EM, Herkenham M (1999) Pre- and postsynaptic distribution of cannabinoid and mu opioid receptors in rat spinal cord. Brain Res 822:17–25
Hohmann AG, Farthing JN, Zvonok AM, Makriyannis A (2004) Selective activation of cannabinoid CB2 receptors suppresses hyperalgesia evoked by intradermal capsaicin. J Pharmacol Exp Ther 308:446–453
Huang CC, Lo SW, Hsu KS (2001) Presynaptic mechanisms underlying cannabinoid inhibition of excitatory synaptic transmission in rat striatal neurons. J Physiol 532:731–748
Ibrahim MM, Deng H, Zvonok A, Cockayne DA, Kwan J, Mata HP, Vanderah TW, Lai J, Porreca F, Makriyannis A, Malan TP Jr (2003) Activation of CB2 cannabinoid receptors by AM1241 inhibits experimental neuropathic pain: pain inhibition by receptors not present in the CNS. Proc Natl Acad Sci U S A 100:10529–10533
Izzo AA, Mascolo N, Borrelli F, Capasso F (1998) Excitatory transmission to the circular muscle of the guinea-pig ileum: evidence for the involvement of cannabinoid CB1 receptors. Br J Pharmacol 124:1363–1368
Izzo AA, Mascolo N, Pinto L, Capasso R, Capasso F (1999) The role of cannabinoid receptors in intestinal motility, defaecation and diarrhoea in rats. Eur J Pharmacol 384:37–42
Jansen EM, Haycock DA, Ward SJ, Seybold VS (1992) Distribution of cannabinoid receptors in rat brain determined with aminoalkylindoles. Brain Res 575:93–102
Jennings EA, Vaughan CW, Christie MJ (2001) Cannabinoid actions on rat superficial medullary dorsal horn neurons in vitro. J Physiol 534:805–812
Julian MD, Martin AB, Cuellar B, Rodriguez De Fonseca F, Navarro M, Moratalla R, Garcia-Segura LM (2003) Neuroanatomical relationship between type 1 cannabinoid receptors and dopaminergic systems in the rat basal ganglia. Neuroscience 119:309–318
Katona I, Sperlagh B, Sik A, Kafalvi A, Vizi ES, Mackie K, Freund TF (1999) Presynaptically located CB1 cannabinoid receptors regulate GABA release from axon terminals of specific hippocampal interneurons. J Neurosci 19:4544–4558
Katona I, Sperlagh B, Magloczky Z, Santha E, Kofalvi A, Czirjak S, Mackie K, Vizi ES, Freund TF (2000) GABAergic interneurons are the targets of cannabinoid actions in the human hippocampus. Neuroscience 100:797–804
Katona I, Rancz EA, Acsady L, Ledent C, Mackie K, Hajos N, Freund TF (2001) Distribution of CB1 cannabinoid receptors in the amygdala and their role in the control of GABAergic transmission. J Neurosci 21:9506–9518
Kirby MT, Hampson RE, Deadwyler SA (1995) Cannabinoids selectively decrease paired-pulse facilitation of perforant path synaptic potentials in the dentate gyrus in vitro. Brain Res 688:114–120
Kreitzer AC, Regehr WG (2001) Retrograde inhibition of presynaptic calcium influx by endogenous cannabinoids at excitatory synapses onto Purkinje cells. Neuron 29:717–727
Kreitzer AC, Carter AG, Regehr WG (2002) Inhibition of interneuron firing extends the spread of endocannabinoid signaling in the cerebellum. Neuron 34:787–796
Kulkarni-Narla A, Brown DR (2000) Localization of CB1-cannabinoid receptor immunoreactivity in the porcine enteric nervous system. Cell Tissue Res 302:73–80
Lu XR, Ong WY, Mackie K (1999) A light and electron microscopic study of the CB1 cannabinoid receptor in monkey basal forebrain. J Neurocytol 28:1045–1051
Lutz B (2002) Molecular biology of cannabinoid receptors. Prostaglandins Leukot Essent Fatty Acids 66:123–142
MacNaughton WK, Van Sickle MD, Keenan CM, Cushing K, Mackie K, Sharkey KA (2004) Distribution and function of the cannabinoid-1 receptor in the modulation of ion transport in the guinea pig ileum: relationship to capsaicin-sensitive nerves. Am J Physiol Gastrointest Liver Physiol 286:G863–G871
Maejima T, Ohno-Shosaku T, Kano M (2001) Endogenous cannabinoid as a retrograde messenger from depolarized postsynaptic neurons to presynaptic terminals. Neurosci Res 40:205–210

Mailleux P, Vanderhaeghen JJ (1992) Distribution of neuronal cannabinoid receptor in the adult rat brain: a comparative receptor binding radioautography and in situ hybridization histochemistry. Neuroscience 48:655–668

Marsicano G, Lutz B (1999) Expression of the cannabinoid receptor CB1 in distinct neuronal subpopulations in the adult mouse forebrain. Eur J Neurosci 11:4213–4225

Marsicano G, Wotjak CT, Azad SC, Bisogno T, Rammes G, Cascio MG, Hermann H, Tang J, Hofmann C, Zieglgansberger W, Di Marzo V, Lutz B (2002) The endogenous cannabinoid system controls extinction of aversive memories. Nature 418:530–534

Marsicano G, Goodenough S, Monory K, Hermann H, Eder M, Cannich A, Azad SC, Cascio MG, Gutierrez SO, van der Stelt M, Lopez-Rodriguez ML, Casanova E, Schutz G, Zieglgansberger W, Di Marzo V, Behl C, Lutz B (2003) CB1 cannabinoid receptors and on-demand defense against excitotoxicity. Science 302:84–88

Mato S, Del Olmo E, Pazos A (2003) Ontogenetic development of cannabinoid receptor expression and signal transduction functionality in the human brain. Eur J Neurosci 17:1747–1754

Matsuda LA (1997) Molecular aspects of cannabinoid receptors. Crit Rev Neurobiol 11:143–166

Matsuda LA, Lolait SJ, Brownstein MJ, Young AC, Bonner TI (1990) Structure of a cannabinoid receptor and functional expression of the cloned cDNA. Nature 346:561–564

Matsuda LA, Bonner TI, Lolait SJ (1993) Localization of cannabinoid receptor mRNA in rat brain. J Comp Neurol 327:535–550

McDonald AJ, Mascagni F (2001) Localization of the CB1 type cannabinoid receptor in the rat basolateral amygdala: high concentrations in a subpopulation of cholecystokinin-containing interneurons. Neuroscience 107:641–652

Melis M, Pistis M, Perra S, Muntoni AL, Pillolla G, Gessa GL (2004) Endocannabinoids mediate presynaptic inhibition of glutamatergic transmission in rat ventral tegmental area dopamine neurons through activation of CB1 receptors. J Neurosci 24:53–62

Morales M, Wang SD, Diaz-Ruiz O, Jho DH (2004) Cannabinoid CB1 receptor and serotonin 3 receptor subunit A (5-HT3A) are co-expressed in GABA neurons in the rat telencephalon. J Comp Neurol 468:205–216

Morisset V, Urban L (2001) Cannabinoid-induced presynaptic inhibition of glutamatergic EPSCs in substantia gelatinosa neurons of the rat spinal cord. J Neurophysiol 86:40–48

Morozov YM, Freund TF (2003) Post-natal development of type 1 cannabinoid receptor immunoreactivity in the rat hippocampus. Eur J Neurosci 18:1213–1222

Nackley AG, Makriyannis A, Hohmann AG (2003) Selective activation of cannabinoid CB(2) receptors suppresses spinal fos protein expression and pain behavior in a rat model of inflammation. Neuroscience 119:747–757

Nicolau M, Lapa AJ, Valle JR (1978) The inhibitory effect induced by delta9-tetrahydrocannabinol on the contractions of the isolated rat vas deferens. Arch Int Pharmacodyn Ther 236:131–136

Okamoto Y, Morishita J, Tsuboi K, Tonai T, Ueda N (2004) Molecular characterization of a phospholipase D generating anandamide and its congeners. J Biol Chem 279:5298–5305

Pan X, Ikeda SR, Lewis DL (1998) SR 141716A acts as an inverse agonist to increase neuronal voltage-dependent Ca2+ currents by reversal of tonic CB1 cannabinoid receptor activity. Mol Pharmacol 54:1064–1072

Pertwee RG (2001) Cannabinoids and the gastrointestinal tract. Gut 48:859–867

Pertwee RG, Fernando SR (1996) Evidence for the presence of cannabinoid CB1 receptors in mouse urinary bladder. Br J Pharmacol 118:2053–2058

Pertwee RG, Stevenson LA, Elrick DB, Mechoulam R, Corbett AD (1992) Inhibitory effects of certain enantiomeric cannabinoids in the mouse vas deferens and the myenteric plexus preparation of guinea-pig small intestine. Br J Pharmacol 105:980–984

Pinto L, Capasso R, Di Carlo G, Izzo AA (2002) Endocannabinoids and the gut. Prostaglandins Leukot Essent Fatty Acids 66:333–341

Price TJ, Helesic G, Parghi D, Hargreaves KM, Flores CM (2003) The neuronal distribution of cannabinoid receptor type 1 in the trigeminal ganglion of the rat. Neuroscience 120:155–162

Riegal AC, Williams JT, Lupica CR (2003) Cananbionid CB1 receptors inhibit GABA-B-mediated synaptic currents in midbrain dopaminergic neurons. Soc Neurosci Abstr 33:462466

Robbe D, Alonso G, Duchamp F, Bockaert J, Manzoni OJ (2001) Localization and mechanisms of action of cannabinoid receptors at the glutamatergic synapses of the mouse nucleus accumbens. J Neurosci 21:109–116

Rodriguez JJ, Mackie K, Pickel VM (2001) Ultrastructural localization of the CB1 cannabinoid receptor in mu-opioid receptor patches of the rat Caudate putamen nucleus. J Neurosci 21:823–833

Romero J, Lastres-Becker I, de Miguel R, Berrendero F, Ramos JA, Fernandez-Ruiz J (2002) The endogenous cannabinoid system and the basal ganglia. biochemical, pharmacological, and therapeutic aspects. Pharmacol Ther 95:137–152

Roth SH (1978) Stereospecific presynaptic inhibitory effect of delta9-tetrahydrocannabinol on cholinergic transmission in the myenteric plexus of the guinea pig. Can J Physiol Pharmacol 56:968–975

Sanudo-Pena MC, Walker JM (1997) Role of the subthalamic nucleus in cannabinoid actions in the substantia nigra of the rat. J Neurophysiol 77:1635–1638

Sanudo-Pena MC, Strangman NM, Mackie K, Walker JM, Tsou K (1999a) CB1 receptor localization in rat spinal cord and roots, dorsal root ganglion, and peripheral nerve. Zhongguo Yao Li Xue Bao 20:1115–1120

Sanudo-Pena MC, Tsou K, Walker JM (1999b) Motor actions of cannabinoids in the basal ganglia output nuclei. Life Sci 65:703–713

Shire D, Carillon C, Kaghad M, Calandra B, Rinaldi-Carmona M, Le Fur G, Caput D, Ferrara P (1995) An amino-terminal variant of the central cannabinoid receptor resulting from alternative splicing. J Biol Chem 270:3726–3731

Shook JE, Burks TF (1989) Psychoactive cannabinoids reduce gastrointestinal propulsion and motility in rodents. J Pharmacol Exp Ther 249:444–449

Sim LJ, Hampson RE, Deadwyler SA, Childers SR (1996a) Effects of chronic treatment with delta9-tetrahydrocannabinol on cannabinoid-stimulated [35S]GTPgammaS autoradiography in rat brain. J Neurosci 16:8057–8066

Sim LJ, Selley DE, Xiao R, Childers SR (1996b) Differences in G-protein activation by mu- and delta-opioid, and cannabinoid, receptors in rat striatum. Eur J Pharmacol 307:97–105

Sjostrom PJ, Turrigiano GG, Nelson SB (2003) Neocortical LTD via coincident activation of presynaptic NMDA and cannabinoid receptors. Neuron 39:641–654

Szabo B, Muller T, Koch H (1999) Effects of cannabinoids on dopamine release in the corpus striatum and the nucleus accumbens in vitro. J Neurochem 73:1084–1089

Szabo B, Siemes S, Wallmichrath I (2002) Inhibition of GABAergic neurotransmission in the ventral tegmental area by cannabinoids. Eur J Neurosci 15:2057–2061

Tanda G, Pontieri FE, Di Chiara G (1997) Cannabinoid and heroin activation of mesolimbic dopamine transmission by a common mu1 opioid receptor mechanism. Science 276:2048–2050

Tsou K, Brown S, Sanudo-Pena MC, Mackie K, Walker JM (1998a) Immunohistochemical distribution of cannabinoid CB1 receptors in the rat central nervous system. Neuroscience 83:393–411

Tsou K, Nogueron MI, Muthian S, Sanudo-Pena MC, Hillard CJ, Deutsch DG, Walker JM (1998b) Fatty acid amide hydrolase is located preferentially in large neurons in the rat central nervous system as revealed by immunohistochemistry. Neurosci Lett 254:137–140

Tsou K, Mackie K, Sanudo-Pena MC, Walker JM (1999) Cannabinoid CB1 receptors are localized primarily on cholecystokinin-containing GABAergic interneurons in the rat hippocampal formation. Neuroscience 93:969–975

Tyler K, Hillard CJ, Greenwood-Van Meerveld B (2000) Inhibition of small intestinal secretion by cannabinoids is CB1 receptor-mediated in rats. Eur J Pharmacol 409:207–211
Van Sickle MD, Oland LD, Ho W, Hillard CJ, Mackie K, Davison JS, Sharkey KA (2001) Cannabinoids inhibit emesis through CB1 receptors in the brainstem of the ferret. Gastroenterology 121:767–774
Van Sickle MD, Oland LD, Mackie K, Davison JS, Sharkey KA (2003) Delta9-tetrahydrocannabinol selectively acts on CB1 receptors in specific regions of dorsal vagal complex to inhibit emesis in ferrets. Am J Physiol Gastrointest Liver Physiol 285:G566–G576
Wilson RI, Kunos G, Nicoll RA (2001) Presynaptic specificity of endocannabinoid signaling in the hippocampus. Neuron 31:453–462
Zhang J, Hoffert C, Vu HK, Groblewski T, Ahmad S, O'Donnell D (2003) Induction of CB2 receptor expression in the rat spinal cord of neuropathic but not inflammatory chronic pain models. Eur J Neurosci 17:2750–2754

HEP (2005) 168:327–365

Effects of Cannabinoids on Neurotransmission

B. Szabo[1] (✉) · E. Schlicker[2]

[1]Institut für Experimentelle und Klinische Pharmakologie und Toxikologie, Albert-Ludwigs-Universität, Albertstrasse 25, 79104 Freiburg, Germany
szabo@pharmakol.uni-freiburg.de

[2]Institut für Pharmakologie und Toxikologie, Rheinische Friedrich-Wilhelms-Universität, Reuterstrasse 2b, 53113 Bonn, Germany

Abstract The CB_1 cannabinoid receptor is widely distributed in the central and peripheral nervous system. Within the neuron, the CB_1 receptor is often localised in axon terminals, and its activation leads to inhibition of transmitter release. The consequence is inhibition of neurotransmission via a presynaptic mechanism. Inhibition of glutamatergic, GABAergic, glycinergic, cholinergic, noradrenergic and serotonergic neurotransmission has been observed in many regions of the central nervous system. In the peripheral nervous system, CB_1 receptor-mediated inhibition of adrenergic, cholinergic and sensory neuroeffector transmission has been frequently observed. It is characteristic for the ubiquitous operation of CB_1 receptor-mediated presynaptic inhibition that antagonistic components of functional systems (for example, the excitatory and inhibitory inputs of the same neuron) are simultaneously inhibited by cannabinoids. Inhibition of voltage-dependent calcium channels, activation of potassium channels and direct interference with the synaptic vesicle release mechanism are all implicated in the cannabinoid-evoked inhibition of transmitter release. Many presynaptic CB_1 receptors are subject to an endogenous tone, i.e. they are constitutively active and/or are continuously activated by endocannabinoids. Compared with the abundant data on presynaptic inhibition by cannabinoids, there are only a few examples for cannabinoid action on the somadendritic parts of neurons in situ.

Keywords Acetylcholine · Axon terminal · CB_1 cannabinoid receptor · GABA · Glutamate · Neurotransmission · Noradrenaline · Presynaptic inhibition · Transmitter release

1 Introduction

As described in the chapter by Mackie (this volume), the CB_1 cannabinoid receptor is widely distributed in the central and peripheral nervous system. One of the primary consequences of activation of CB_1 receptors is the inhibition or activation of ion channels. For example, voltage-dependent calcium channels are typically inhibited by cannabinoids, whereas several kinds of potassium channels are activated. Theoretically, due to their influence on ion channels, cannabinoids can change the function of neurons in several ways. By acting in the dendrites, they can interfere with the conduction of synaptic currents to the soma of the neuron. By acting in the soma, they can interfere with the generation of action potentials. By acting on ion channels in axon terminals, they can inhibit transmitter release from the terminals; the consequence is inhibition of neurotransmission with a presynaptic mechanism. Inhibition of neurotransmission appears to be, at present, the best-characterised electrophysiological effect of cannabinoids, and this review focuses on this effect. Before analysing the presynaptic effect, we describe cannabinoid effects on ion channels and the anatomical evidence for the presence of cannabinoid receptors in axon terminals. Presynaptic inhibition by endogenous cannabinoids released by postsynaptic neurons—retrograde signaling—is described in the chapter by Vaughan and Christie (this volume).

2 Effects of Cannabinoids on Ion Channels

The somadendritic region of most neurons is accessible for electrophysiological studies. In contrast, direct electrophysiological recording from axon terminals of mammals is either impossible or extremely difficult. Accordingly, we know relatively well how cannabinoids change the function of ion channels in the somadendritic region. Our knowledge on electrophysiological changes in axon terminals is limited; we can only assume that ion channels are influenced similarly as in the somadendritic region. In this section, effects on the somadendritic region are dealt with.

2.1
Effects of Cannabinoids on Voltage-Gated Ion Channels

2.1.1
Calcium Channels

In the majority of studies, cannabinoids depressed voltage-dependent calcium channels. According to the first observations, activation of CB_1 receptors inhibits N-type voltage-dependent calcium channels in neuronal cell lines (Caulfield and Brown 1992; Mackie and Hille 1992; Mackie et al. 1993). No inhibition occurred in pertussis toxin-treated cells, indicating the involvement of G proteins containing $G\alpha_{i/o}$ subunits. Later, this observation was extended to isolated rat hippocampal neurons and cerebellar granule cells (Twitchell et al. 1997; Nogueron et al. 2001). In isolated rat sympathetic ganglion neurons that previously had been injected with CB_1 receptor cRNA, cannabinoids also inhibited N-type calcium channels (Pan et al. 1996). Q-type calcium channels were also inhibited in CB_1 receptor-transfected AtT20 cells (Mackie et al. 1995). The endogenous cannabinoid (endocannabinoid) anandamide inhibits T-type calcium channels; this effect is, however, not mediated by CB_1 receptors (Chemin et al. 2001).

There are at least two examples for stimulation of calcium channels by cannabinoids: L-type calcium currents in a neuronal cell line (Rubovitch et al. 2002) and in retinal rods of the tiger salamander (Straiker and Sullivan 2003) were enhanced by cannabinoids.

2.1.2
Potassium Channels

Activated CB_1 receptors can also change the function of several types of potassium channels. In oocytes and AtT20 cells artificially expressing the CB_1 receptor, stimulation of inwardly rectifying potassium channels was repeatedly observed (Henry and Chavkin 1995; Mackie et al. 1995; Garcia et al. 1998; McAllister et al. 1999). Potassium A currents in cultured hippocampal neurons are stimulated by cannabinoids (Deadwyler et al. 1995; Mu et al. 2000). The effects of cannabinoids on potassium M currents in hippocampal brain slices have also been studied; M currents were inhibited, which means an enhancement of neuronal excitability (Schweitzer 2000). The potassium K current is inhibited by cannabinoids in cultured hippocampal neurons (Hampson et al. 2000). As in the case of calcium channels, anandamide can elicit a CB_1 receptor-independent effect on potassium channels, i.e. it inhibits the acid-sensitive background potassium channel TASK-1 (Maingret et al. 2001).

2.1.3 Sodium Channels

In an early study, Turkanis et al. (1991) showed that Δ^9-tetrahydrocannabinol inhibits voltage-dependent sodium channels; the involved primary receptor was not identified in this study. More recently, it was observed that anandamide and the synthetic CB_1/CB_2 receptor agonist WIN55212-2 inhibited voltage-dependent sodium channels in synaptosomes prepared from mouse brain (Nicholson et al. 2003). Since the effects were not attenuated by the CB_1 receptor antagonist AM251, the involvement of CB_1 receptors can be excluded.

2.2 Effects of Cannabinoids on Ligand-Gated Ion Channels

The function of several types of ligand-gated ion channels is changed by cannabinoids—as a rule, these effects are not mediated by CB_1 receptors. In isolated rat nodose ganglion neurons, cannabinoids inhibited serotonin-3 ($5\text{-}HT_3$) receptor-mediated currents (Fan 1995). This observation was verified and extended in a recent study. In HEK293 cells expressing the human $5\text{-}HT_{3A}$ receptor, several cannabinoids inhibited the 5-HT-evoked current (Barann et al. 2002). CB_1 receptors could not be involved in this effect, since HEK293 cells do not express CB_1 receptors.

The function of AMPA-type glutamate receptors (Akinshola et al. 1999) and nicotinic acetylcholine receptors (Oz et al. 2003), expressed in oocytes, was inhibited by anandamide. These effects are, again, CB_1 and CB_2 receptor-independent.

2.3 What Is the Functional Consequence of the Inhibition of Somadendritic Ion Channels?

The majority of the experiments in which the effect of cannabinoids on somadendritic ion channels was studied were carried out on cell lines, on cells artificially expressing the CB_1 receptor or on isolated neurons. It is not known whether the effects also occur under natural conditions. For example, cannabinoid receptor agonists did not influence voltage-dependent calcium channels in caudate-putamen medium spiny neurons (Szabo et al. 1998), although these neurons are known to synthesise CB_1 receptors. It is conceivable that in neurons under physiological conditions, the density of somadendritic CB_1 receptors is too low for modulation of certain ion channels. Alternatively, the coupling mechanism between receptor and ion channel may not be functional.

Another important question also remains unanswered. We basically do not know how modulation of somadendritic ion channels by cannabinoids affects the excitability or integrative capacity of neurons. There are only a few experiments in which neurons were studied in situ (in brain slices), and cannabinoid effects were restricted to the somadendritic region of the neurons (by blockade of the synaptic input of the neurons), and cannabinoids elicited an effect. One such experiment was carried out by Kreitzer et al. (2002): cannabinoids lowered the firing rate of cerebellar interneurons and this was attributed to the activation of barium-sensitive potassium channels. In the experiments of Himmi et al. (1998), cannabinoids changed the firing rate of nucleus tractus solitarii neurons in brain slices; since the synaptic input was not blocked, it is not known whether the change in firing rate was due to an effect on the neurons themselves, or to an effect on their synaptic input.

3 Anatomical Evidence for the Presence of CB_1 Cannabinoid Receptors in Axon Terminals

The wide distribution of the CB_1 receptor in the nervous system is described in the chapter by Mackie (this volume). The prerequisite for presynaptic inhibition of neurotransmission is that the receptor is localised in axon terminals. The following paragraph lists known examples for localisation of CB_1 receptors in axon terminals.

In the cerebellum, CB_1 receptors in terminals of basket cells can be seen at the light microscopic level (Tsou et al. 1998; Diana et al. 2002). Electron microscopical studies have indicated that a great portion of CB_1 receptors in the caudate-putamen (Rodriguez et al. 2001), hippocampus (Katona et al. 1999, 2000; Hájos et al. 2000) and amygdala (Katona et al. 2001) is in axon terminals. Comparison of the site of CB_1 receptor synthesis (which was determined by in situ hybridisation) with the distribution of receptor protein (which was determined with receptor autoradiography and immunohistochemistry) indicates localisation of CB_1 receptors in terminals of parallel fibres in the cerebellum and in terminals of striatonigral neurons in the substantia nigra pars reticulata (compare, for example, Mailleux and Vanderhaeghen 1992; Matsuda et al. 1993; Tsou et al. 1998). The changes in the CB_1 receptor distribution pattern during neurodegeneration accompanying Huntington's disease and experimentally elicited neurodegeneration also suggest that CB_1 receptors in the substantia nigra pars reticulata are localised in striatonigral axon terminals (Herkenham et al. 1991; Glass et al. 2000).

In a few instances, it was shown that CB_1 receptors are not uniformly distributed in a neuron, but are preferentially localised in the axon terminal. For example, CB_1 receptors were well visible in cerebellar basket cell terminals, but not in the somata of these neurons (Diana et al. 2002). Preferential localisation of CB_1 receptors in axon terminals was also observed in hippocampal neurons (Twitchell et al. 1997; Irving et al. 2000).

4 Effects of Cannabinoids on Neurotransmission in the Central Nervous System

Two methods were used to study the effect of cannabinoids on presynaptic axon terminals. The more frequently used electrophysiological approach measures neurotransmission. In brain slices or neuronal cultures, electrical currents in postsynaptic neurons are recorded with patch-clamp or microelectrode techniques. Presynaptic axon terminals are electrically stimulated and the postsynaptic current resulting from stimulation of ligand-gated ion channels of postsynaptic neurons by the released transmitter is determined. The change in the postsynaptic current amplitude is a measure of the change in synaptic transmission.

In the other method, the release of endogenous or radiolabelled neurotransmitters from presynaptic axon terminals is determined chemically. Although this latter method shows directly what happens at the level of axon terminals, it does not measure "neurotransmission".

In electrophysiological experiments, cannabinoids inhibited neurotransmission. The inhibition was always due to inhibition of transmitter release from axon terminals and never to interference of cannabinoids with the postsynaptic effects of the neurotransmitters. The experiments in which transmitter release was determined neurochemically also indicated that cannabinoids inhibit transmitter release from axon terminals. In most instances the presynaptic cannabinoid receptors can be classified as CB_1 receptors (but some exceptions are given in Tables 1 and 2). Effects of cannabinoids on the release of individual transmitters are discussed below. Effects of cannabinoids on neurotransmission have also been reviewed by Schlicker and Kathmann (2001).

4.1 Fast Excitatory Neurotransmission

Activation of CB_1 receptors inhibits the release of the excitatory neurotransmitter glutamate in many brain regions and in the spinal cord (Table 1).

Inhibition was seen in nuclei belonging to the extrapyramidal motor control system: caudate-putamen, globus pallidus and substantia nigra pars reticulata (Fig. 1 shows an example of presynaptic inhibition of glutamatergic neurotransmission in the substantia nigra pars reticulata; see Fig. 6 for an overview of cannabinoid effects on neurotransmission in the extrapyramidal motor control system). Inhibition of neurotransmission was also observed in the ventral tegmental area, hippocampus and the nucleus accumbens—these regions are parts of the limbic system. Inhibition of the excitatory synaptic transmission in the hippocampus could contribute to the anticonvulsive effect of cannabinoids. Purkinje cells in the cerebellar cortex receive excitatory inputs from parallel fibres and climbing fibres; both kinds of excitatory inputs are inhibited by activated CB_1 receptors (see Fig. 7 for an overview of cannabinoid effects on neurotransmission in the cerebellar cortex). Moreover, cannabinoids depress the glutamatergic neurotransmission

Table 1. Inhibition of glutamatergic neurotransmission (*continued on next page*)

Neurotransmitter	Species	Region	Method	Mechanism of presynaptic inhibition	Reference(s)
Glutamate	Rat	Cortex (brain slice)	Electrophysiology (patch clamp)	Vesicle release machinery inhibited	Auclair et al. 2000
Glutamate[a]	Rat	Cortex (cell culture)	Endogenous glutamate chemically determined	IP_3 is involved	Ferraro et al. 2001
Glutamate	Rat	Caudate-putamen (brain slice)	Electrophysiology (patch clamp)	Vesicle release machinery inhibited	Gerdeman and Lovinger 2001
Glutamate	Rat	Caudate-putamen (brain slice)	Electrophysiology (patch clamp, extracell. recording)	The inhibition is pertussis toxin sensitive Vesicle release machinery not inhibited Ca^{2+} channels involved	Huang et al. 2001, 2002
Glutamate	Rat	Caudate-putamen (brain slice)	Electrophysiology (extracell. recording), [^{3}H]glutamate release	Glutamate uptake is inhibited, glutamate causes presynaptic inhibition via mGluR	Brown et al. 2003b
Glutamate	Mouse	Nucleus accumbens (brain slice)	Electrophysiology (patch clamp, extracell. recording)	Vesicle release machinery inhibited Ca^{2+} channels not involved K^+ channels involved cAMP and PKA not involved	Robbe et al. 2001
Glutamate	Mouse	Globus pallidus (brain slice)	Electrophysiology (patch clamp)	No details given	Wallmichrath and Szabo 2003
Glutamate	Rat	Ventral tegmental area	Electrophysiology (patch clamp)	Vesicle release machinery inhibited	Melis et al. 2004
Glutamate	Rat	Substantia nigra pars reticulata (brain slice)	Electrophysiology (patch clamp)	No details given	Szabo et al. 2000
Glutamate	Rat	Hippocampus (brain slice)	Electrophysiology (extracell. recording)	No details given	Ameri et al. 1999; Ameri and Simmet 2000
Glutamate	Rat	Hippocampus (brain slice)	Electrophysiology (extracell. recording)	No details given	Al-Hayani and Davies 2000

Table 1. *(continued)*

Neurotransmitter	Species	Region	Method	Mechanism of presynaptic inhibition	Reference(s)
Glutamate	Rat	Hippocampus (cell culture)	Electrophysiology (patch clamp)	Vesicle release machinery inhibited Ca^{2+} channels involved	Sullivan 1999
Glutamate	Rat	Hippocampus (cell culture)	Electrophysiology (patch clamp), microfluorometry (Ca^{2+} spikes determined)	The inhibition is pertussis toxin sensitive	Shen et al. 1996, Shen and Thayer 1999, Kouznetsova et al. 2002
Glutamate? GABA?	Rat	Hippocampus (cell culture)	Release of the vesicle marker FM1–43	No details given	Kim and Thayer 2000
Glutamate	Rat	Hippocampus (synaptosomes)	Endogenous glutamate chemically determined (4-AP evoked release)	The inhibition is pertussis toxin sensitive Ca^{2+} channels involved Vesicle release machinery not inhibited PKA not involved	Wang 2003
Glutamate	Mouse, rat	Hippocampus (brain slice)	Electrophysiology (patch clamp)	No CB_1 receptors are involved	Hájos et al. 2001; Hájos and Freund 2002
Glutamate	Rat, mouse	Hippocampus (synaptosomes)	[^{3}H]Glutamate release	No CB_1 receptors are involved	Köfalvi et al. 2003
Glutamate	Mouse	Hippocampus (brain slice)	Electrophysiology (patch clamp)	The inhibition is pertussis toxin sensitive Vesicle release machinery inhibited	Misner and Sullivan 1999
Glutamate	Mouse	Amygdala	Electrophysiology (patch clamp, extracell. recording)	The inhibition is pertussis toxin sensitive Vesicle release machinery inhibited K^+ channels involved Ca^{2+} channels not involved	Azad et al. 2003
Glutamate	Rat	Cerebellum (brain slice)	Electrophysiology (patch clamp)	Vesicle release machinery inhibited K^+ channels involved Ca^{2+} channels only indirectly involved	Levenes et al. 1998, Daniel and Crepel 2001
Glutamate	Rat	Cerebellum (brain slice)	Electrophysiology (patch clamp)	Vesicle release machinery not inhibited	Takahashi and Linden 2000

Table 1. *(continued)*

Neurotransmitter	Species	Region	Method	Mechanism of presynaptic inhibition	Reference(s)
Glutamate	Rat	Cerebellum (brain slice)	Electrophysiology (patch clamp)	No details given	Kreitzer and Regehr 2001
Glutamate	Mouse	Cerebellum (brain slice)	Electrophysiology (patch clamp)	No details given	Maejima et al. 2001
Glutamate	Mouse	Synaptosomes from whole brain	Endogenous glutamate chemically determined (veratridine-evoked release)	No CB_1 receptors are involved Na^+ channels inhibited	Nicholson et al. 2003
Glutamate	Rat	Spinal cord (cord slice)	Electrophysiology (patch clamp)	Vesicle release machinery inhibited	Morisset and Urban 2001

4-AP, 4-aminopyridine; cAMP, cyclic adenosine monophosphate; extracell., extracellular; GABA, γ-aminobutyric acid; PKA, protein kinase A.
[a]In this exceptional study, cannabinoids did not decrease glutamate release, but increased it.

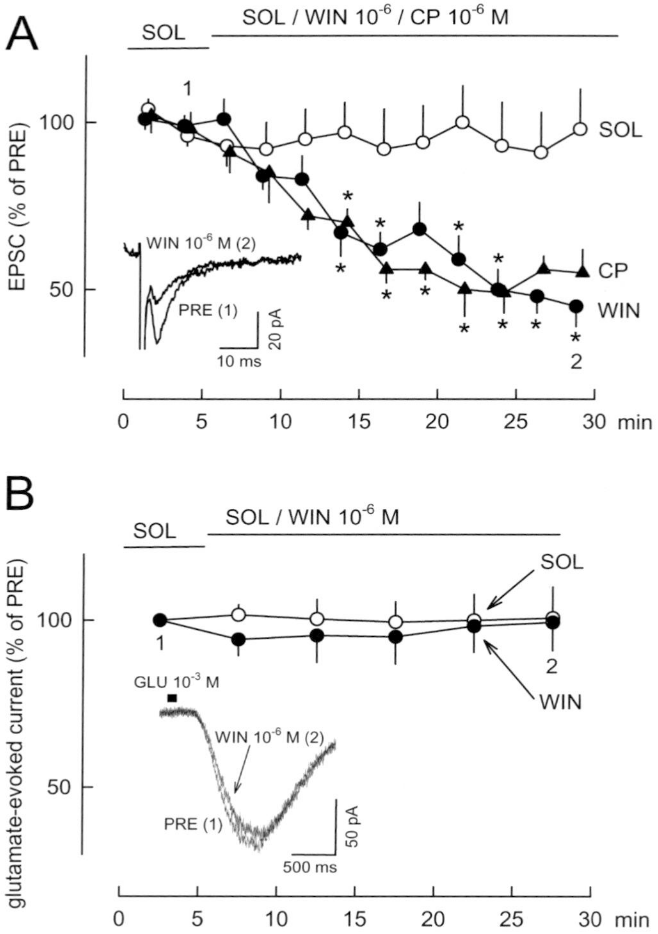

Fig. 1A, B. Cannabinoids inhibit glutamatergic synaptic transmission in the substantia nigra pars reticulata (SNR) of the rat via a presynaptic mechanism. The major glutamatergic afferent input of SNR neurons originates in the subthalamic nucleus. **A** SNR neurons were patch-clamped and their glutamatergic afferent axons electrically stimulated. The stimulation elicited excitatory postsynaptic currents (*EPSCs*) in SNR neurons. EPSCs remained stable in solvent (*SOL*)-superfused slices. The synthetic cannabinoid agonists WIN55212-2 (*WIN*) and CP55940 (*CP*) inhibited the EPSCs. **B** SNR neurons were patched-clamped and glutamate (*GLU*) was pressure-ejected from a pipette in their vicinity. Glutamate-evoked currents remained stable in SOL-superfused slices. Superfusion of WIN also did not change the glutamate-evoked currents. This observation indicates that cannabinoids do not interfere with the postsynaptic effect of glutamate; thus, the inhibition of neurotransmission seen in panel **A** is due to presynaptic inhibition of glutamate release from axon terminals. In both panels, a typical original recording obtained in a WIN experiment (*inset*) and the statistical analysis are shown. *PRE*, initial reference period. See Szabo et al. (2000) for details of the experiments. *, Significant difference from SOL ($p<0.05$)

between primary sensory fibres and neurons in the dorsal horn of the spinal cord: this effect could be the basis of the spinal analgesia produced by cannabinoids.

According to recent observations, some effects of cannabinoids on glutamatergic transmission in the hippocampus are not mediated by CB_1 receptors (and also not by CB_2 receptors). Synthetic cannabinoids depressed excitatory neurotransmission also in brain slices from CB_1 receptor-knockout mice and in the presence of some CB_1 antagonists (Hájos et al. 2001; Hájos and Freund 2002). Similarly, cannabinoid-evoked glutamate release from hippocampal synaptosomes was resistant to CB_1 antagonists and persisted in CB_1 receptor-knockout mice (Köfalvi et al. 2003; but in a similar preparation, effects were sensitive to a CB_1 antagonist; Wang 2003). Based on such observations, the existence of a new cannabinoid receptor was postulated. It must be noted that the involvement of known non-cannabinoid receptors or ion channels—for which cannabinoids might possess a hitherto unrecognised affinity—was not excluded in these studies.

Prolonged exposure of G protein-coupled receptors to their agonists leads to desensitisation due to diminished coupling of the receptors with G proteins and receptor internalisation. This phenomenon was observed also in the case of CB_1 receptor-mediated inhibition of neurotransmission. Cannabinoid-evoked inhibition of glutamatergic and γ-aminobutyric acid (GABA)ergic neurotransmission in the nucleus accumbens was diminished by treatment of animals for 1 week with natural and synthetic cannabinoids (Hoffman et al. 2003a). Cannabinoid-evoked inhibition of excitatory neurotransmission between cultured hippocampal neurons was also strongly desensitised by a 24-h treatment of the neurons with a cannabinoid (Kouznetsova et al. 2002).

4.2 Fast Inhibitory Neurotransmission

CB_1 receptor-mediated inhibition of GABAergic neurotransmission has been observed in many brain regions, belonging to different functional systems (Table 2).

Thus, cannabinoids depress cerebral cortical GABAergic neurotransmission. Neurotransmission is also depressed in nuclei belonging to the extrapyramidal motor control system: caudate-putamen, globus pallidus and substantia nigra pars reticulata (Fig. 6 also shows cannabinoid effects on inhibitory neurotransmission in the extrapyramidal motor control system). GABAergic synaptic transmission in the cerebellum, a major brain region involved in motor control, is inhibited as well (Fig. 7 also shows cannabinoid effects on inhibitory neurotransmission in the cerebellar cortex). Figure 2 shows inhibition of GABAergic neurotransmission in the cerebellar cortex, and Fig. 3 shows that the inhibition is due to the inhibition of GABA release from presynaptic axon terminals. In several nuclei belonging to the limbic system (e.g. hippocampus and amygdala), activation of CB_1 receptors leads to depression of inhibitory neurotransmission. Inhibition of GABA release in the ventral tegmental area—where the mesolimbic reward pathway originates—could explain the euphoria produced by cannabinoids. The rostral ventromedial medulla oblongata and the periaqueductal grey in the midbrain are involved in nocicep-

Table 2. Inhibition of GABAergic neurotransmission (*continued on next page*)

Neurotransmitter	Species	Region	Method	Mechanism of presynaptic inhibition	Reference(s)
GABA	Mouse	Cortex (brain slice)	Electrophysiology (patch clamp)	Vesicle release machinery not inhibited	Trettel and Levine 2002, 2003
GABA	Rat	Caudate-putamen (brain slice)	Electrophysiology (patch clamp)	No details given	Szabo et al. 1998
GABA	Mouse	Nucleus accumbens (brain slice)	Electrophysiology (patch clamp)	Vesicle release machinery inhibited	Manzoni and Bockaert 2001
GABA	Rat	Nucleus accumbens (brain slice)	Electrophysiology (patch clamp)	Vesicle release machinery weakly inhibited	Hoffman and Lupica 2001; Hoffman et al. 2003a
GABA	Mouse	Globus pallidus (brain slice)	Electrophysiology (patch clamp)	Vesicle release machinery not inhibited	Engler and Szabo 2003, 2004
GABA	Rat	Substantia nigra pars reticulata (brain slice)	Electrophysiology (patch clamp)	No details given	Chan et al. 1998, Chan and Yung 1998
GABA	Rat	Substantia nigra pars reticulata (brain slice)	Electrophysiology (patch clamp)	No details given	Wallmichrath and Szabo 2002a
GABA	Mouse	Substantia nigra pars reticulata (brain slice)	Electrophysiology (patch clamp)	Vesicle release machinery not inhibited	Wallmichrath and Szabo 2002b
GABA	Rat	Ventral tegmental area (brain slice)	Electrophysiology (patch clamp)	Vesicle release machinery not inhibited	Szabo et al. 2002
GABA ($GABA_B$-mediated transmission)	Rat	Ventral tegmental area (brain slice)	Electrophysiology (patch clamp)	No details given	Riegel et al. 2003
GABA	Rat	Hippocampus (brain slice)	Electrophysiology (patch clamp)	Vesicle release machinery not inhibited	Hájos et al. 2000
GABA	Rat	Hippocampus (brain slice)	Electrophysiology (patch clamp)	Vesicle release machinery not inhibited K^+ channels not involved	Hoffman and Lupica 2000

Table 2. *(continued)*

Neurotransmitter	Species	Region	Method	Mechanism of presynaptic inhibition	Reference(s)
GABA	Rat	Hippocampus (brain slice)	Electrophysiology (patch clamp)	No details given	Hoffman et al. 2003b
GABA	Rat	Hippocampus (brain slice)	Electrophysiology (patch clamp)	Vesicle release machinery inhibited	Wilson and Nicoll 2001
GABA	Rat	Hippocampus (cell culture)	Electrophysiology (patch clamp)	Vesicle release machinery inhibited	Irving et al. 2000
GABA	Rat	Hippocampus (cell culture)	Electrophysiology (patch clamp)	No details given	Ohno-Shosaku et al. 2001
GABA	Rat	Hippocampus (brain slice)	[^{3}H]GABA release	No details given	Katona et al. 1999
GABA	Mouse	Hippocampus (brain slice)	Electrophysiology (patch clamp)	No details given	Hájos et al. 2001
GABA	Man	Hippocampus (brain slice)	[^{3}H]GABA release	No details given	Katona et al. 2000
GABA	Man	Hippocampus (brain slice)	Electrophysiology (patch clamp)	Vesicle release machinery not inhibited	Nakatsuka et al. 2003
GABA	Rat, mouse	Amygdala (brain slice)	Electrophysiology (patch clamp)	Vesicle release machinery not inhibited	Katona et al. 2001
GABA	Mouse	Amygdala (brain slice)	Electrophysiology (patch clamp)	Vesicle release machinery inhibited	Azad et al. 2003
GABA	Rat	Periaqueductal grey (brain slice)	Electrophysiology (patch clamp)	Vesicle release machinery inhibited	Vaughan et al. 2000
GABA	Rat	Rostral ventro-medial medulla oblongata (brain slice)	Electrophysiology (patch clamp)	Vesicle release machinery inhibited	Vaughan et al. 1999
GABA	Rat	Cerebellum (brain slice)	Electrophysiology (patch clamp)	Intracellular Ca^{2+} concentration decreased	Kreitzer and Regehr 2001

Table 2. (*continued*)

Neurotransmitter	Species	Region	Method	Mechanism of presynaptic inhibition	Reference(s)
GABA	Rat	Cerebellum (brain slice)	Electrophysiology (patch clamp)	Vesicle release machinery inhibited Intracellular Ca^{2+} concentration decreased	Diana et al. 2002
GABA	Rat	Cerebellum (brain slice)	Electrophysiology (patch clamp)	Vesicle release machinery inhibited	Szabo et al. 2004
GABA	Mouse	Cerebellum (brain slice)	Electrophysiology (patch clamp)	No details given	Yoshida et al. 2002
GABA	Mouse	Synaptosomes from whole brain	Endogenous GABA chemically determined, Veratridine-induced release	No CB_1 receptors are involved Na^+ channels inhibited	Nicholson et al. 2003
GABA, glycine	Rat	Superficial medullary dorsal horn	Electrophysiology (patch clamp)	Vesicle release machinery inhibited	Jennings et al. 2001

GABA, γ-aminobutyric acid.

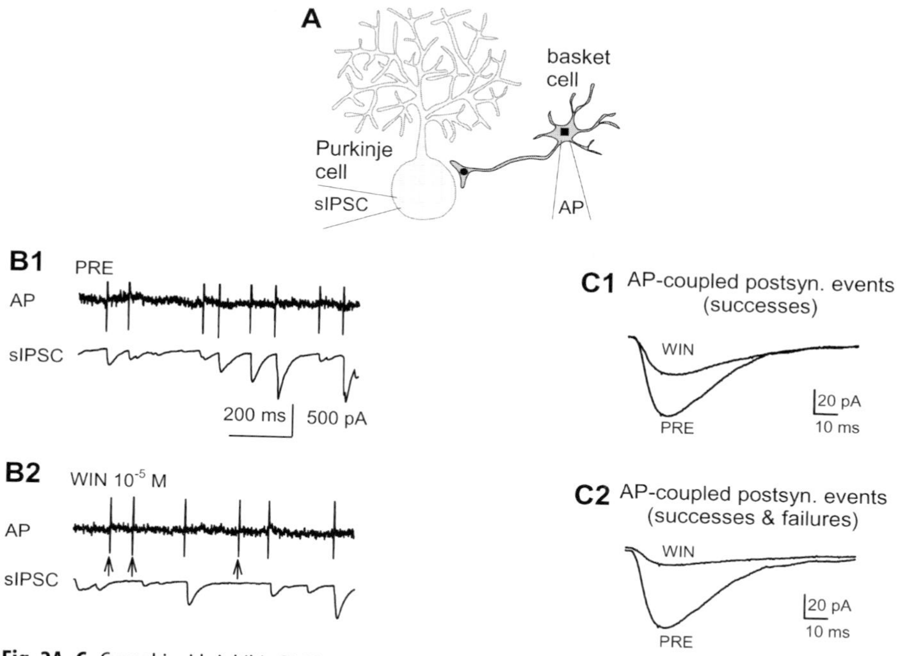

Fig. 2A–C. Cannabinoids inhibit GABAergic synaptic transmission between basket and Purkinje cells in the cerebellar cortex of the rat. **A** The basket cell synthesises CB_1 receptor mRNA (■) and the CB_1 receptor protein (•) is localised in the axon terminal. Action potentials (*APs*) of a basket cell and spontaneous inhibitory postsynaptic currents (*sIPSCs*) in a synaptically coupled Purkinje cell were recorded simultaneously. **B1, B2** APs and sIPSCs were recorded during the initial reference period (*PRE*) and during superfusion with WIN55212-2 (*WIN*). During PRE (**B1**), every presynaptic AP was accompanied by a postsynaptic IPSC: synaptic transmission was always successful. During WIN superfusion (**B2**), synaptic failures appear (marked by *arrows*). Enhancement of synaptic failure is typical for drugs that decrease probability of transmitter release from the presynaptic axon terminal. **C1** AP-coupled postsynaptic currents were averaged only if transmission was successful. The decrease in amplitude indicates inhibition of neurotransmission by WIN. **C2** All AP-coupled postsynaptic currents were averaged (successes and failures). The WIN-evoked inhibition is greater (than in **C1**), because WIN also increased the number of failures. The figure represents five experiments with a similar outcome. See Szabo et al. (2004) for details of the experiments

tive information processing; in both regions, GABAergic synaptic transmission is inhibited by cannabinoids.

In the above-mentioned experiments, cannabinoids inhibited fast GABAergic transmission by inhibiting GABA release from axon terminals. It is expected that if GABA release is inhibited, then $GABA_B$ receptor-mediated slow inhibitory transmission will be inhibited as well. This was indeed observed in the ventral tegmental area (Riegel et al. 2003).

In addition to GABA, glycine is also involved in fast inhibitory neurotransmission. Activation of CB_1 receptors inhibits both GABAergic and glycinergic synaptic transmission in the medulla oblongata (Jennings et al. 2001; see Table 2).

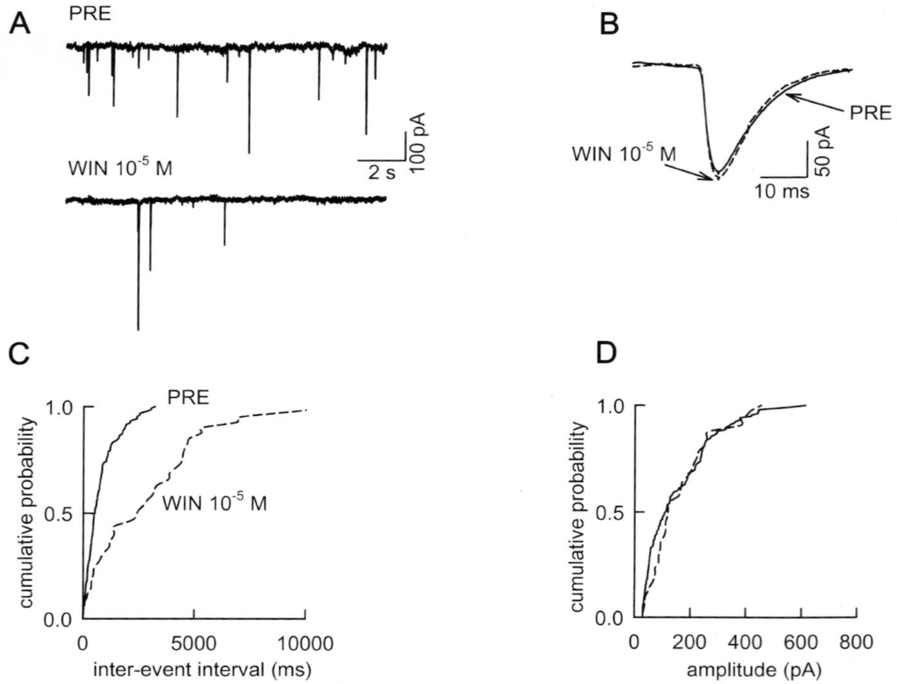

Fig. 3A–D. Cannabinoids inhibit GABAergic synaptic transmission between basket and Purkinje cells in the cerebellar cortex of the rat via a presynaptic mechanism. Miniature inhibitory postsynaptic currents (mIPSCs) were recorded in Purkinje cells in the presence of tetrodotoxin (3×10^{-7} M) during an initial reference period (*PRE*) and during superfusion with WIN55212-2 (*WIN*). **A** Original tracings from an experiment with WIN: WIN obviously lowers the frequency of mIPSCs. **B** Averaged mIPSCs from the experiment shown in **A**: WIN does not change the amplitude of mIPSCs. **C, D** Cumulative probability distribution plots of inter-event intervals and amplitudes of mIPSCs (same experiment as in **A**): the inhibitory effect of WIN on the frequency of mIPSCs and its lack of effect on the amplitude of mIPSCs is evident. Lack of effect of WIN on the amplitude of mIPSCs indicates that the cannabinoid does not interfere with the effect of GABA on the postsynaptic neuron—this is an indication that WIN inhibited neurotransmission between basket and Purkinje cells (see Fig. 2) via a presynaptic mechanism. Lowering the frequency of mIPSCs by WIN suggests that WIN directly interferes with the vesicle release machinery. The figure represents six experiments with a similar outcome. See Szabo et al. (2004) for details of the experiments

4.3 Neurotransmission via Monoamines and Acetylcholine

A synopsis of the inhibitory effects of cannabinoids on the release of the monoamines noradrenaline, dopamine and serotonin and of acetylcholine in the brain and the retina is given in Table 3. Noradrenaline release is inhibited via CB_1 receptors in the hippocampus of guinea-pig and man but not in the hippocampus of rat and mouse (Table 3, Fig. 4; Van Vliet et al. 2000). Although CB_1 receptors inhibit the release of dopamine from amacrine cells of the retina, contradictory results were obtained with respect to the modulation of dopamine release from

Table 3. Inhibition of the release of monoamines and acetylcholine in the brain and the retina

Neurotransmitter	Species	Region	Method	References
Noradrenaline	Guinea-pig	Cortex, hippocampus, hypothalamus, cerebellum (brain slice)	[^{3}H]Noradrenaline release	Schlicker et al. 1997
	Man	Hippocampus (brain slice)		
Dopamine	Rat	Caudate-putamen	[^{3}H]Dopamine release	Cadogan et al. 1997
Dopamine	Rat	Caudate-putamen	NMDA-stimulated [^{3}H]dopamine release	Kathmann et al. 1999
Dopamine	Guinea-pig	Retina	[^{3}H]Noradrenaline release	Schlicker et al. 1996
Serotonin	Mouse	Cortex (brain slice)	[^{3}H]Serotonin release	Nakazi et al. 2000
Acetylcholine	Rat	Hippocampus (brain slice)	[^{3}H]Acetylcholine release	Gifford and Ashby 1996
		Cortex, hippocampus (synaptosomes)		Gifford et al. 2000
Acetylcholine	Mouse	Cortex, hippocampus (brain slice)	[^{3}H]Acetylcholine release	Kathmann et al. 2001a
Acetylcholine	Mouse Man	Cortex (brain slice) Cortex (brain slice)	[^{3}H]Acetylcholine release	Steffens et al. 2003

NMDA, *N*-methyl-D-aspartate

the terminals of the striatonigral axons in the caudate-putamen. Dopamine release was depressed in some studies (Table 3), but not, however, in a study using voltammetry to measure dopamine release (Szabo et al. 1999). Serotonin release was slightly inhibited in the cortex of mice but not affected at all in the cortex of rats (Table 3; Van Vliet et al. 2000). Moreover, cannabinoids inhibit acetylcholine release in the hippocampus and cortex; inhibition also occurs in human cortex (Table 3). However, not all cholinergic neurons are affected by cannabinoids: e.g. acetylcholine release from the cholinergic interneurons of the caudate-putamen is not changed by cannabinoids (Gifford et al. 1997a; Kathmann et al. 2001a).

The papers listed in Table 3 and discussed in the preceding paragraph represent in vitro studies, and the question arises whether similar results are also obtained in vivo. This was examined in a series of studies on rats subjected to in vivo microdialysis; the ligands under study were administered intraperitoneally or intravenously. Cannabinoids indeed decrease acetylcholine release in the dorsal hippocampus (Mishima et al. 2002). In the studies by Tzavara et al. (2001, 2003a), in which cannabinoid agonists were not studied themselves, the CB_1 receptor inverse agonist SR 141716, which elicits effects opposite in direction to those of cannabi-

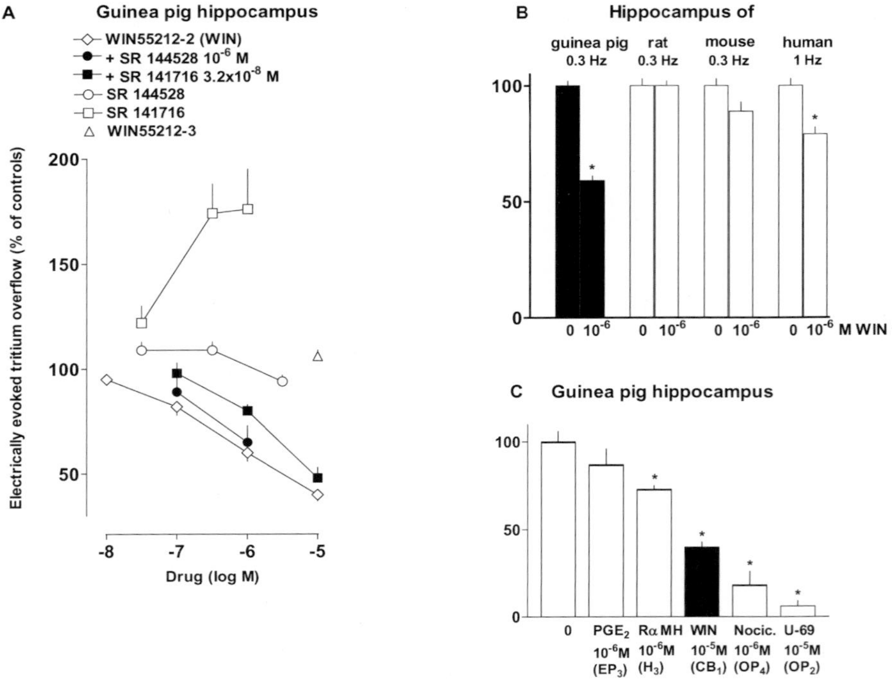

Fig. 4A–C. Cannabinoids inhibit noradrenaline release in the brain. **A** Guinea-pig hippocampal slices were preincubated with [^{3}H]noradrenaline and superfused. The electrically (0.3 Hz) evoked tritium overflow (which represents quasi-physiological noradrenaline release) was inhibited by WIN55212-2 but not affected by its enantiomer WIN55212-3. The concentration–response curve of WIN55212-2 (WIN) was shifted to the right by a low concentration of the CB_1 receptor antagonist SR 141716 (pA_2 8.2) but hardly affected by a high concentration of the CB_2 receptor antagonist SR 144528. Given alone, SR 141716 facilitated, whereas SR 144528 did not affect, noradrenaline release. In another series of experiments, not shown here, slices were superfused with K^+-rich (2.5×10^{-2} M) Ca^{2+}-free medium containing tetrodotoxin 10^{-6} M; under this experimental condition WIN inhibited tritium overflow evoked by re-introduction of Ca^{2+} 1.3×10^{-3} M (in a manner sensitive to SR 141716 3.2×10^{-7}), suggesting that the CB_1 receptors are located presynaptically on the noradrenergic axon terminals. **B** WIN inhibited noradrenaline release also in human hippocampus but failed to do so in rat and mouse hippocampus. Although SR 141716 3.2×10^{-7} M counteracted the effect of WIN in human hippocampus, it did not affect noradrenaline release by itself (not shown). (Since noradrenaline release is relatively low in human hippocampus we used a higher stimulation frequency than in hippocampal slices from the three animal species.) **C** In guinea-pig hippocampus, the inhibitory effect of WIN is higher than that of prostaglandin E_2 (*PGE_2*; acting via prostaglandin EP_3 receptors) and R-α-methylhistamine (*RαMH*; acting via histamine H_3 receptors), but lower than that of nociceptin (*Nocic.*; acting via opioid OP_4 receptors) and U-69,593 (*U-69*; acting via OP_2 receptors). Note that the concentrations of the five agonists cause maximum or near maximum effects at the respective presynaptic inhibitory receptors. *, Significant difference from control ($p<0.001$). See Schlicker et al. (1997) and Timm et al. (1998) for details of the experiments (some of the data shown here are unpublished)

noid agonists under a variety of conditions (for a more detailed discussion, see Sect. 7), increases noradrenaline release in the prefrontal cortex and anterior hypothalamus, dopamine release in the prefrontal cortex and serotonin release in the prefrontal cortex and nucleus accumbens. On the other hand, cannabinoids increase rather than decrease striatal dopamine release (Malone and Taylor 1999) and acetylcholine release in the frontal cortex (Verrico et al. 2003). The situation is even more complicated with respect to the effects of cannabinoids on acetylcholine release in the medial prefrontal cortex and hippocampus. Low doses of cannabinoids increase (Acquas et al. 2000, 2001), whereas high doses decrease (Gessa et al. 1998; Carta et al. 1998), the release of this transmitter.

The fact that cannabinoids when given systemically increase rather than decrease transmitter release in various paradigms in vivo is in all likelihood not related to the fact that there are also facilitatory cannabinoid receptors. Inhibitory CB_1 receptors occur both on facilitatory and inhibitory neurons of complex neuronal networks and cannabinoids may therefore elicit inhibitory or facilitatory effects on transmitter release, depending on the exact site(s) where they act. Two typical networks in which presynaptic inhibitory CB_1 receptors occur on various sites are depicted in Figs. 6 and 7. The recent study by Tzavara et al. (2003b) shows that the differential effects of cannabinoids on hippocampal acetylcholine release (Gessa et al. 1998; Carta et al. 1998; Acquas et al. 2000, 2001) are due to the fact that the cannabinoids, depending on the dose, act on different pathways, involving dopamine D_1 or D_2 receptors.

5 Effects of Cannabinoids on Neurotransmission in the Peripheral Nervous System

Effects of cannabinoids on the sympathetic nervous system have been studied in isolated tissues and in pithed animals (Table 4). Sympathetic neurons were usually activated by electrical stimulation. Activation of CB_1 receptors led to inhibition of noradrenaline and/or ATP release and, consequently, to inhibition of the effector responses in the heart, in mesenteric and renal blood vessels and in the vas deferens. Figure 5A shows that cannabinoids inhibit sympathetic neuroeffector transmission in the heart. Sympathetically mediated vasoconstriction was inhibited in many tissues of pithed rats and rabbits. Sympathetic tone is depressed during long-term Δ^9-tetrahydrocannabinol administration in humans; the presynaptic inhibitory effect of cannabinoids on sympathetic axon endings may be the basis of this effect.

Cannabinoids also inhibit transmitter release from cholinergic autonomic neurons (Table 4). As an example, the bradycardia elicited by vagal nerve stimulation is depressed. Figure 5B shows that cannabinoids inhibit parasympathetic neuroeffector transmission in the heart. Electrically evoked contractions of the ileum and urinary bladder can also be inhibited by activation of CB_1 receptors (Table 4).

Finally, cannabinoids inhibit the release of neuropeptides like calcitonin gene-related peptide (CGRP), substance P and somatostatin from sensory neurons (Table 4). Capsaicin or electrical stimulation was used to evoke neuropeptide release. In some of these studies, the endocannabinoid anandamide was used, which has a dual effect on neuropeptide release from sensory neurons. Anandamide possesses an inhibitory effect mediated via CB_1 receptors at low concentrations and

Table 4. Inhibition of neuroeffector transmission in the peripheral nervous system (*continued on next page*)

Neurotransmitter	Species	Tissue	Method	Reference(s)
Noradrenaline	Mouse	Sympathetic neurons (cell culture)	Electrically evoked [^{3}H]noradrenaline release	Göbel et al. 2000
Noradrenaline	Rat	Vas deferens	Electrically evoked contraction	Christopoulos et al. 2001
Noradrenaline	Rat	Vas deferens, heart atrium	Electrically evoked [^{3}H]noradrenaline release	Ishac et al. 1996
Noradrenaline, ATP	Mouse	Vas deferens	Electrically evoked contraction	Pertwee et al. 1992, 2002
ATP	Mouse	Vas deferens	Electrically evoked contraction	Lay et al. 2000
Noradrenaline	Mouse	Vas deferens	Electrically evoked [^{3}H]noradrenaline release	Trendelenburg et al. 2000; Schlicker et al. 2003
Noradrenaline	Rat	Heart	Electrically evoked cardioaccelerator response	Malinowska et al. 2001
Noradrenaline	Rabbit	Heart	Electrically evoked cardioaccelerator response	Szabo et al. 2001
Acetylcholine			Bradycardia evoked by electrical stimulation of the vagus nerve	
Noradrenaline	Man	Heart atrial appendages	Electrically evoked [^{3}H]noradrenaline release	Molderings et al. 1999
Noradrenaline	Rat	Sympathetically innervated blood vessels of many organs	Electrically evoked increase in blood pressure in pithed rats	Malinowska et al. 1997
Noradrenaline	Rat	Sympathetically innervated tissues of many organs	Electrically evoked increase in blood pressure and plasma noradrenaline in pithed rats	Niederhoffer et al. 2003
Noradrenaline	Rabbit	Sympathetically innervated tissues of many organs	Electrically evoked increase in blood pressure and plasma noradrenaline in pithed rabbits	Niederhoffer and Szabo 1999
Noradrenaline	Rat	Mesenterial vessels	Electrically evoked noradrenaline release	Ralevic and Kendall 2002
Noradrenaline	Rat	Renal arteries	K^+-evoked [^{3}H]noradrenaline release	Deutsch et al. 1997
Noradrenaline	Guinea-pig	Lung (bronchi)	Electrically evoked [^{3}H]noradrenaline release	Vizi et al. 2001
Adrenaline	Rabbit	Adrenal medulla	Electrically evoked increase in plasma adrenaline in pithed rabbits Electrically evoked adrenaline release in isolated adrenal medullary slices	Niederhoffer et al. 2001

Table 4. *(continued)*

Neurotransmitter	Species	Tissue	Method	Reference(s)
Acetylcholine, ATP	Mouse	Urinary bladder	Electrically evoked contraction	Pertwee and Fernando 1996
Acetylcholine, NANC-transmitter	Mouse	Colon	Electrically evoked cholinergic and NANC postsynaptic potentials	Storr et al. 2004
Acetylcholine	Guinea-pig	Ileum	Electrically evoked contraction, acetylcholine release, cholinergic postsynaptic potentials	Pertwee et al. 1992, 1996b; Lopez-Redondo et al. 1997; Mang et al. 2001
Acetylcholine, NANC-transmitter	Guinea-pig	Ileum	Electrically evoked contraction	Izzo et al. 1998
Acetylcholine	Man	Ileum	Electrically evoked contraction	Croci et al. 1998
Adenosine	Guinea-pig	Ileum	Electrically evoked adenosine release	Begg et al. 2002
CGRP	Rat	Primary sensory neurons (cell culture)	Basal and capsaicin-evoked CGRP release	Ahluwalia et al. 2003
CGRP, substance P, somatostatin	Rat	Trachea	Capsaicin-evoked CGRP, substance P and somatostatin release	Nemeth et al. 2003
CGRP	Rat	Hind paw skin	Capsaicin-evoked CGRP release	Richardson et al. 1998; Ellington et al. 2002
CGRP	Rat	Spinal cord slices	Electrically evoked CGRP release	Tognetto et al. 2001

CGRP, calcitonin gene-related peptide; NANC, non-adrenergic-non-cholinergic

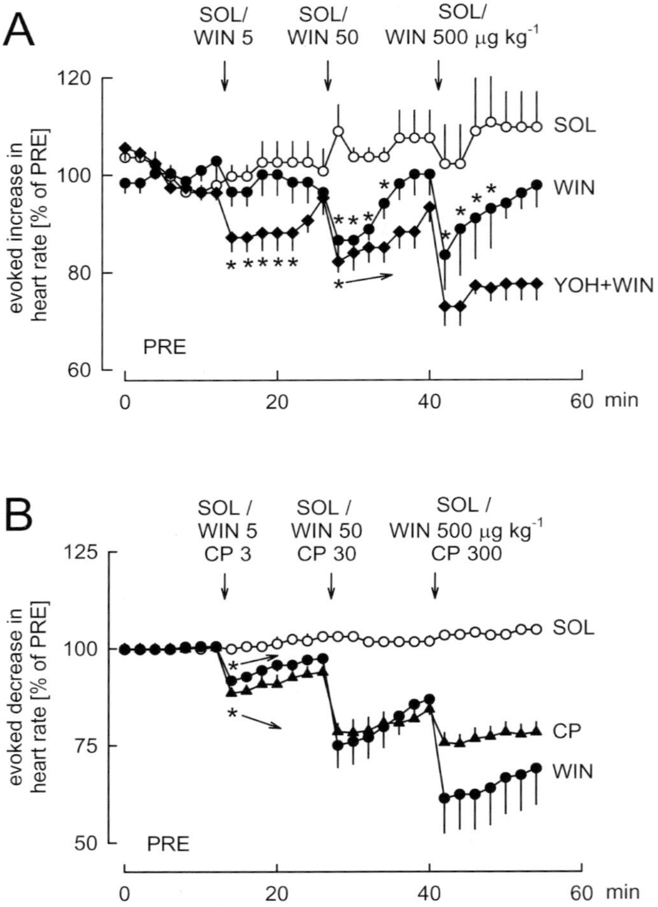

Fig. 5A, B. Cannabinoids inhibit sympathetic and parasympathetic neuroeffector transmission in the heart. **A** Cardiac sympathetic nerves in pithed rabbits were stimulated at a frequency of 1 Hz for 30 s. Solvent (*SOL*) and WIN55212-2 (*WIN*) were administered i.v. as indicated by the *arrows*. One of the WIN groups (*YOH+WIN*) was pretreated with the α_2-adrenoceptor antagonist yohimbine (0.5 mg/kg^{-1}; i.v.) at $t = -14$ min. Cardioaccelerator responses are given as percentages of the initial reference value (*PRE*). WIN inhibited the cardioaccelerator response more strongly in the presence of YOH, probably because YOH prevented concurrent inhibition by endogenous noradrenaline. **B** The right vagus nerve was stimulated at a frequency of 10 Hz for 5 s. SOL, WIN and CP55940 (*CP*) were administered i.v. as indicated by the *arrows*. Cardiodecelerator responses are given as percentages of the initial reference value PRE. *, Significant difference from SOL ($p<0.05$). See Szabo et al. (2001) for details of the experiments

a stimulatory effect mediated via vanilloid receptors (TRPV1, transient receptor potential V1 channel) at high concentrations (Zygmunt et al. 1999; Tognetto et al. 2001; Ahluwalia et al. 2003; Nemeth et al. 2003).

The effect of cannabinoids on peripheral autonomic transmission has been extensively reviewed by Ralevic (2003).

6
Mechanism of the Presynaptic Inhibition

Information regarding the mechanism of presynaptic inhibition is included in Tables 1 and 2. As expected for a $G\alpha_{i/o}$ protein-coupled receptor, the presynaptic inhibition mediated by the CB_1 receptor was sensitive to pertussis toxin in the few cases where this interaction was studied. Moreover, in isolated hippocampal neurons, presynaptic inhibition of excitatory neurotransmission elicited by CB_1 receptor activation could be mediated by several subtypes of $G\alpha_{i/o}$ proteins: $G\alpha_{o1}$, $G\alpha_{i2}$ and $G\alpha_{i3}$ (Straiker et al. 2002). Information on the involvement of second messengers in the presynaptic inhibition by cannabinoids is sparse. For example, the role of $G\beta\gamma$ proteins is not known. Data on the role of the cyclic adenosine monophosphate (cAMP)–protein kinase A messenger system are contradictory (see Tables 1 and 2). After activation of $G\alpha_{i/o}$ protein-coupled receptors, several final mechanisms may lead to inhibition of transmitter release (for review see Thompson et al. 1993; Wu and Saggau 1997; see Fig. 8). Most often, presynaptic inhibition is attributed to inhibition of voltage-dependent calcium channels. In addition, activation of potassium channels and direct interference with the vesicle release machinery can also play a role in presynaptic inhibition. It seems that cannabinoids can use all three mechanisms for producing presynaptic depression (see Tables 1 and 2). Since it is extremely difficult to obtain electrophysiological access to mammalian axon terminals, direct evidence for cannabinoid-evoked modulation of axon terminal ion channels is lacking. Therefore, most of the evidence regarding the mechanism of cannabinoid-evoked presynaptic inhibition is indirect.

6.1
Inhibition of Calcium Channels

As mentioned above, cannabinoids inhibit voltage-dependent calcium channels in somadendritic regions of neurons (see Sect. 2.1.1). It is assumed that such an inhibition also operates in axon terminals and is responsible for presynaptic inhibition. Using microfluorometric methods, it was indeed shown that the action potential-evoked increase in axon terminal calcium concentration is depressed by exogenous and endogenous cannabinoids (Kreitzer and Regehr 2001; Diana et al. 2002; Daniel and Crepel 2001; Brown et al. 2003a; Diana and Marty 2003). Based on the interaction between cannabinoids and calcium channel blockers, Sullivan (1999) concluded that calcium channel inhibition is responsible for the cannabinoid-evoked depression of synaptic transmission.

6.2
Activation of Potassium Channels

As mentioned above, cannabinoids activate several types of potassium channels in the somadendritic region of neurons (see Sect. 2.1.2). Cannabinoid-evoked opening of potassium channels will hyperpolarise axon terminals and shorten action potentials. As a consequence, invasion of axon terminals by action potentials and the activation of calcium channels can be impeded. The duration of calcium influx during the action potential may also decrease. Evidence for the involvement of potassium channels in presynaptic inhibition was obtained by using potassium channel blockers. Thus, potassium channel blockers prevented cannabinoid-evoked presynaptic inhibition (Daniel and Crepel 2001; Robbe et al. 2001; Diana and Marty 2003; Azad et al. 2003) and cannabinoid-evoked inhibition of the action potential-triggered increase in axon terminal calcium concentration (Daniel and Crepel 2001). In contrast, since potassium channel blockers did not affect cannabinoid-evoked presynaptic inhibition, Hoffman and Lupica (2000) excluded a role of potassium channels in presynaptic inhibition.

6.3 Direct Inhibition of the Vesicle Release Machinery

In most nerve terminals, spontaneous and asynchronous quantal transmitter release occurs also in the absence of calcium influx through voltage-dependent calcium channels. Such release events are recorded in electrophysiological experiments either in the presence of tetrodotoxin or calcium channel blockers. The recorded postsynaptic events are called miniature excitatory or inhibitory postsynaptic currents (mEPSCs or mIPSCs). There are many examples for the lowering of the frequency of mEPSCs and mIPSCs by cannabinoids (Tables 1 and 2), including GABAergic synaptic transmission between basket and Purkinje cells in the rat cerebellar cortex (Fig. 3). These observations indicate that cannabinoids are capable of inhibiting neurotransmitter release at a site downward of calcium entry into axon terminals, most probably at the level of the vesicular release machinery. However, it is also clear from Tables 1 and 2 that at many synapses cannabinoids produce presynaptic inhibition without directly interfering with vesicular release.

In conclusion, there are examples for presynaptic inhibition by all three mechanisms: inhibition of voltage-dependent calcium channels, activation of potassium channels and inhibition of the vesicle-release machinery. The inhibitory mechanisms vary in different types of axon terminals. One axon terminal can possess several inhibitory mechanisms (for example, calcium channels and vesicle release can be inhibited simultaneously).

7 Endogenous Tone at Presynaptic Cannabinoid Receptors

There is now increasing evidence that cannabinoid receptors involved in the inhibition of neuroeffector transmission are subject to an endogenous tone (Table 5). A typical example is the presynaptic CB_1 receptors on GABAergic neurons synapsing with the pyramidal neurons in the rat hippocampus (Wilson and Nicoll 2001). Depolarisation of the latter neurons causes an increase in formation of endo-

Table 5. Endogenous tone at CB_1 receptors inhibiting neuroeffector transmission (*continued on next page*)

Neurotransmitter	Species	Tissue	Method	Identification of endogenous tone[1]	Reference(s)
Glutamate	Rat	Cortex	Electrophysiology (patch clamp)	Inverse agonist (SR 141716)	Auclair et al. 2000
Glutamate	Rat	Caudate-putamen	Electrophysiology (patch clamp)	Inverse agonist (AM 281)	Huang et al. 2001
Glutamate	Rat	Hippocampus	Electrophysiology (extracellular recording)	Inverse agonist (SR 141716)	Ameri et al. 1999
GABA	Rat	Hippocampus	Electrophysiology (patch clamp)	Inhibitor of endocannabinoid reuptake (AM 404)	Wilson and Nicoll 2001
Noradrenaline	Guinea-pig	Hippocampus	[^{3}H]Noradrenaline release	Inverse agonists (SR 141716, AM 251, AM 281)	Schlicker et al. 1997, 2002
Noradrenaline	Rat	Vas deferens	Contraction	Inverse agonists (SR 141716, LY 320135), inhibitor of endocannabinoid degradation (phenylmethylsulfonyl fluoride, PMSF)	Christopoulos et al. 2001
Noradrenaline, ATP	Mouse	Vas deferens	Contraction	Inverse agonist (SR 141716)	Pertwee et al. 1996a
Noradrenaline	Mouse	Vas deferens	[^{3}H]Noradrenaline release	Inverse agonist (SR 141716), knockout mouse generated by Zimmer et al. 1999	Schlicker et al. 2003
Dopamine	Guinea-pig	Retina	[^{3}H]Dopamine release	Inverse agonist (SR 141716)	Schlicker et al. 1996
Acetylcholine	Man	Cortex	[^{3}H]Acetylcholine release	Inverse agonist (SR 141716), partial agonist (O-1184), inhibitor of endocannabinoid reuptake (AM 404)	Steffens et al. 2003
Acetylcholine	Rat	Hippocampus	[^{3}H]Acetylcholine release	Inverse agonists (SR 141716, AM 281)	Gifford and Ashby 1996; Gifford et al. 1997b, 2000
Acetylcholine	Mouse	Hippocampus	[^{3}H]Acetylcholine release	Inverse agonist (SR 141716), knockout mouse generated by Zimmer et al. 1999	Kathmann et al. 2001a,b
Acetylcholine	Guinea-pig	Hippocampus	[^{3}H]Acetylcholine release	Inverse agonist (SR 141716)	Schultheiß et al. 2004
Acetylcholine, ATP	Mouse	Urinary bladder	Contraction	Inverse agonist (SR 141716)	Pertwee and Fernando 1996

Table 5. *(continued)*

Neurotransmitter	Species	Tissue	Method	Identification of endogenous tone[1]	Reference(s)
Acetylcholine	Mouse	Colon	Electrophysiology (intracellular recording)	Inverse agonist (SR 141716), knockout mouse generated by Marsicano et al. 2002	Storr et al. 2004
Acetylcholine, substance P	Guinea-pig	Small intestine	Contraction	Inverse agonist (SR 141716)	Pertwee et al. 1996b; Coutts and Pertwee 1997; Izzo et al. 1998
Acetylcholine	Guinea-pig	Ileum	[^{3}H]Acetylcholine release	Inverse agonist (SR 141716)	Mang et al. 2001
Substance P	Mouse	Spinal cord	Substance P release	Inverse agonist (SR 141716)	Lever and Malcangio 2002

[1] Inverse and partial agonists and CB_1 receptor disruption increase, whereas inhibition of endocannabinoid reuptake or degradation decreases, transmitter release.

cannabinoids, which in turn activate the presynaptic inhibitory CB_1 receptors on the GABAergic neurons (Fig. 8; see also the chapter by Vaughan and Christie, this volume). The inhibitory effect is mimicked by a blocker of endocannabinoid reuptake, i.e. AM 404 (in a manner sensitive to the CB_1 receptor inverse agonist SR 141716), suggesting that endocannabinoids are accumulating. This has also been shown in some other paradigms (Table 5) and even in human tissue (Steffens et al. 2003). The same conclusion was reached from experiments in which a blocker of the degradation of the endocannabinoids, i.e. phenylmethylsulfonyl fluoride (PMSF), mimicked the inhibitory effect of the endocannabinoids (Table 5). The third approach was the use of a partial CB_1 receptor agonist, O-1184, which led to an increase in transmitter release, probably by interrupting the inhibition caused by accumulating endocannabinoids (Steffens et al. 2003).

In many studies, SR 141716 or other antagonists/inverse agonists increased transmitter release (Fig. 4; Table 5). Although the reason for their facilitatory effect might be the same as in the case of O-1184, an entirely different explanation has to be considered as well. Thus, presynaptic CB_1 receptors may be constitutively active, i.e. inhibit transmitter release even if they are not activated by endocannabinoids, and in this case inverse agonists would be expected to increase transmitter release as well. Constitutive activity frequently occurs with G protein-coupled receptors expressed in high densities (Seifert and Wenzel-Seifert 2002) and CB_1 receptors are expressed in relatively high densities when compared to other G protein-coupled receptors (Wilson and Nicoll 2002). In at least one of the paradigms shown in Table 5, constitutive activity seems to be the only possible explanation. Thus, SR 141716 increased the Ca^{2+}-induced [^{3}H]acetylcholine release in rat hippocampal synaptosomes (Gifford et al. 2000). In synaptosomes as opposed to isolated tissues (used in most of the other studies shown in Table 5), accumulation of endogenously released ligands cannot occur, since the latter are efficiently removed by the superfusion stream (Starke et al. 1989). For further clarification, neutral CB_1 receptor antagonists (which have become available only recently; Hurst et al. 2002; Ruiu et al. 2003) will be useful, since they are expected to facilitate transmitter release if endocannabinoids are accumulating but should be without effect if CB_1 receptors are constitutively active.

The facilitatory effect of inverse agonists on transmitter release was mimicked in some paradigms by the disruption of CB_1 receptors, i.e. transmitter release was higher in tissues from CB_1 receptor-deficient mice when compared to wild-type animals (Table 5). This experimental approach does not allow one to reach a conclusion as to whether the endogenous tone is related to accumulation of endocannabinoids or constitutively active CB_1 receptors; yet it is remarkable that blockade of, or inverse agonism at, CB_1 receptors during the course of the experiment and complete lack of CB_1 receptors have the same consequence.

The fact that presynaptic CB_1 receptors at many sites are activated by endogenous compounds lends further support to the view that the cannabinoid system plays an important regulatory role. It has also great practical relevance since CB_1 receptor antagonists/inverse agonists may be used for therapeutic purposes (for further discussion, see the chapter by Robson, this volume).

8
What Is the Functional Role of Presynaptic Cannabinoid Receptors?

It is evident from Sects. 4 and 5 that presynaptic CB_1 receptors are ubiquitous in the central and peripheral nervous system. Even within one functional system, several components of the neuronal circuitry are equipped with CB_1 receptors. This will be illustrated in two functional systems: the extrapyramidal motor control system (Fig. 6) and the cerebellum (Fig. 7).

Figure 6 shows the most important glutamatergic, GABAergic and dopaminergic neuronal connections within the extrapyramidal motor control system. Glutamatergic and GABAergic neurotransmission is inhibited at several sites by cannabinoids. In contrast, dopaminergic transmission may not be influenced. A typical motor effect of high doses of cannabinoids is catalepsy (Compton et al. 1996; Sanudo-Pena et al. 1999). Catalepsy is thought to occur if the GABAergic neurons in the output nucleus of the basal ganglia, the substantia nigra pars reticulata, are firing at a high rate (Kolasiewicz et al. 1988). Among the 11 sites where cannabinoids can act presynaptically, an action at 5 sites would indirectly enhance the firing rate of substantia nigra pars reticulata neurons, and thus would lead to

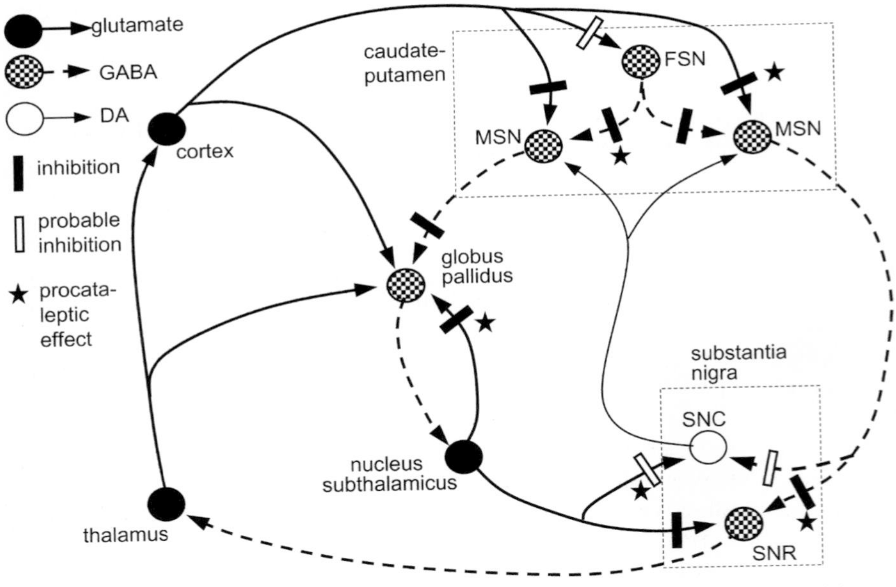

Fig. 6. Effects of cannabinoids on synaptic transmission in nuclei belonging to the extrapyramidal motor control system. *DA*, dopamine; *FSN*, fast spiking neuron; *MSN*, medium spiny neuron; *SNC*, substantia nigra pars compacta; *SNR*, substantia nigra pars reticulata. CB_1 receptor-mediated inhibition of neurotransmission was demonstrated at many synapses of this motor control system. In addition to the proved sites of inhibition, inhibition is very probable at additional sites (based on the localisation of the CB_1 receptor). For the sake of simplicity, the pathway including the entopeduncular nucleus (globus pallidus medialis/internus) is not shown

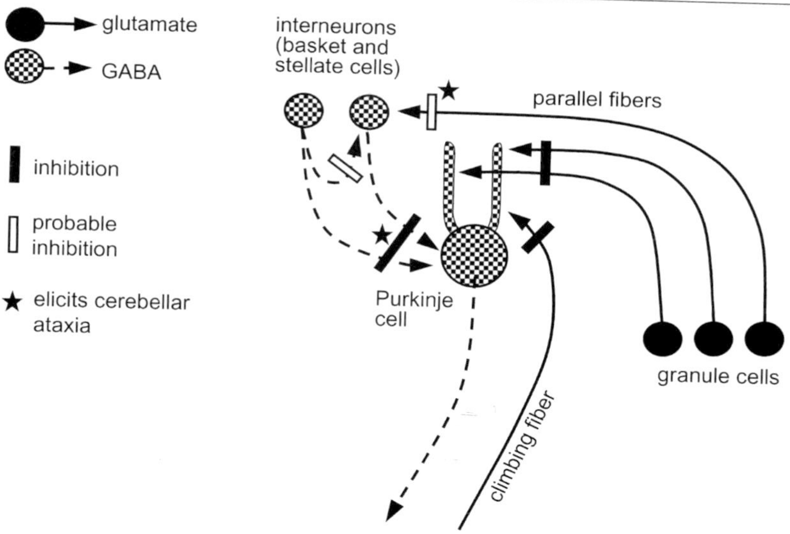

Fig. 7. Effects of cannabinoids on synaptic transmission in the cerebellar cortex. CB_1 receptor-mediated inhibition of neurotransmission was demonstrated at several synapses in the cerebellar cortex. In addition to the proven sites of inhibition, inhibition is very probable at additional sites (based on the localisation of the CB_1 receptor). In addition to synaptic inhibition, activation of CB_1 receptors can also directly decrease the firing rate of interneurons (not shown)

catalepsy. Action at the remaining sites would lead to the opposite effect, i.e. the firing rate of substantia nigra pars reticulata neurons would decrease, which would be an "anticataleptic" effect. In vivo, the balance of all effects obviously favours catalepsy.

Figure 7 shows neuronal connections in the cortex of the cerebellum and the action of cannabinoids on these connections. Activation of CB_1 receptors inhibits glutamatergic as well as GABAergic neurotransmission at altogether five sites. Cannabinoids cause static and gait ataxia, and this is attributed to cerebellar dysfunction (Fränkel 1903; Patel and Hillard 2001). It is thought that the firing rate of Purkinje cells is increased during cerebellar ataxia. Two of the presynaptic cannabinoid effects shown in Fig. 7 would indirectly enhance the firing rate of Purkinje cells; these effects could be the primary events behind cerebellar ataxia. As in the extrapyramidal motor control system, however, inhibitory CB_1 presynaptic receptors are also localised on neurons that play opposite roles in the function of the cortex of the cerebellum.

Further examples for the simultaneous inhibitory effects of cannabinoids on antagonistic components of functional systems can be easily found. For example, cannabinoids inhibit the glutamatergic as well as the GABAergic input of ventral tegmental area dopaminergic neurons (Szabo et al. 2002; Melis et al. 2004) and the sympathetic as well as the parasympathetic input of the heart (Szabo et al. 2001).

What is the physiological role of CB_1 receptors—receptors that are so widely distributed and that simultaneously influence antagonistic components of a given

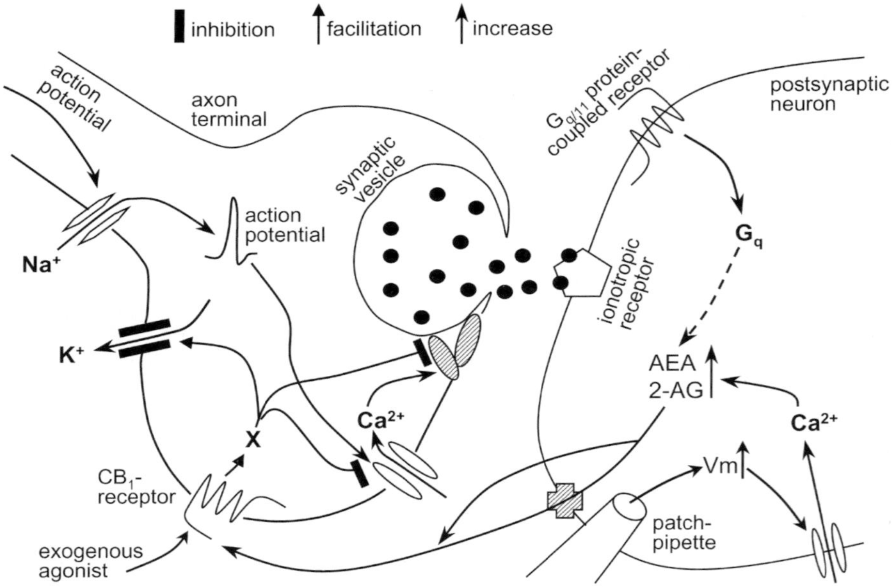

Fig. 8. Effects of cannabinoids on synaptic transmission. Activation of the CB_1 receptor at the presynaptic axon terminal inhibits transmitter release from the synaptic vesicle. Three mechanisms can be involved in presynaptic inhibition (*X* refers to unknown second messengers): inhibition of voltage-dependent calcium channels, activation of potassium channels and direct interference with the vesicle release machinery. The CB_1 receptor can be activated by exogenous agonists, but also by the endocannabinoids anandamide (*AEA*) and 2-arachidonoylglycerol (*2-AG*), which are released from the postsynaptic neuron by passive and/or facilitated diffusion. The synthesis of endocannabinoids is triggered by a depolarisation-induced (V_m, membrane potential) calcium influx or by activation of $G_{q/11}$ protein-coupled receptors

functional system? One functional role of CB_1 receptors is their participation in retrograde signalling, at least with respect to fast excitatory and inhibitory transmission. Endogenous cannabinoids released from postsynaptic neurons can diffuse to presynaptic axon terminals where they produce presynaptic inhibition (Ohno-Shosaku et al. 2001; Wilson and Nicoll 2001). The trigger for synthesis of endocannabinoids is depolarisation of postsynaptic neurons or activation of $G\alpha_{q/11}$ protein-coupled receptors of postsynaptic neurons (see Fig. 8). This phenomenon is called depolarisation-induced suppression of inhibition (DSI; if inhibitory neurotransmission is suppressed by endocannabinoids) or depolarisation-induced suppression of excitation (DSE; if excitatory neurotransmission is suppressed by endocannabinoids). This new research field was reviewed by Wilson and Nicoll (2002) and Freund et al. (2003) and is also reviewed in the chapter by Vaughan and Christie (this volume).

References

Acquas E, Pisanu A, Marrocu P, Di Chiara G (2000) Cannabinoid CB_1 receptor agonists increase rat cortical and hippocampal acetylcholine release in vivo. Eur J Pharmacol 401:179–185

Acquas E, Pisanu A, Marrocu P, Goldberg SR, Di Chiara G (2001) Δ^9-Tetrahydrocannabinol enhances cortical and hippocampal acetylcholine release in vivo: a microdialysis study. Eur J Pharmacol 419:155–161

Ahluwalia J, Urban L, Bevan S, Nagy I (2003) Anandamide regulates neuropeptide release from capsaicin-sensitive primary sensory neurons by activating both the cannabinoid 1 receptor and the vanilloid receptor 1 in vitro. Eur J Neurosci 17:2611–2618

Akinshola BE, Taylor RE, Ogunseitan AB, Onaivi ES (1999) Anandamide inhibition of recombinant AMPA receptor subunits in Xenopus oocytes is increased by forskolin and 8-bromo-cyclic AMP. Naunyn–Schmiedeberg's Arch Pharmacol 360:242–248

Al-Hayani A, Davies SN (2000) Cannabinoid receptor mediated inhibition of excitatory synaptic transmission in the rat hippocampal slice is developmentally regulated. Br J Pharmacol 131:663–665

Ameri A, Simmet T (2000) Effects of 2-arachidonoylglycerol, an endogenous cannabinoid, on neuronal activity in rat hippocampal slices. Naunyn–Schmiedeberg's Arch Pharmacol 361:265–272

Ameri A, Wilhelm A, Simmet T (1999) Effects of the endogenous cannabinoid, anandamide, on neuronal activity in rat hippocampus slices. Br J Pharmacol 126:1831–1839

Auclair N, Otani S, Soubrie P, Crepel F (2000) Cannabinoids modulate synaptic strength and plasticity at glutamatergic synapses of rat prefrontal cortex pyramidal neurons. J Neurophysiol 83:3287–3293

Azad SC, Eder M, Marsicano G, Lutz B, Zieglgänsberger W, Rammes G (2003) Activation of the cannabinoid receptor type 1 decreases glutamatergic and GABAergic synaptic transmission in the lateral amygdala of the mouse. Learn Mem 10:116–128

Barann M, Molderings G, Brüss M, Bönisch H, Urban BW, Göthert M (2002) Direct inhibition by cannabinoids of human $5\text{-}HT_{3A}$ receptors: probable involvement of an allosteric modulatory site. Br J Pharmacol 137:589–596

Begg M, Dale N, Llaudet E, Molleman A, Parsons ME (2002) Modulation of the release of endogenous adenosine by cannabinoids in the myenteric plexus-longitudinal muscle preparation of the guinea-pig ileum. Br J Pharmacol 137:1298–1304

Brown SP, Brenowitz SD, Regehr WG (2003a) Brief presynaptic bursts evoke synapse-specific retrograde inhibition mediated by endogenous cannabinoids. Nat Neurosci 6:1048–1057

Brown TM, Brotchie JM, Fitzjohn SM (2003b) Cannabinoids decrease corticostriatal synaptic transmission via an effect on glutamate uptake. J Neurosci 23:11073–11077

Cadogan A-K, Alexander SPH, Boyd AE, Kendall DA (1997) Influence of cannabinoids on electrically evoked dopamine release and cyclic AMP generation in the rat striatum. J Neurochem 69:1131–1137

Carta G, Nava F, Gessa GL (1998) Inhibition of hippocampal acetylcholine release after acute and repeated Δ^9-tetrahydrocannabinol in rats. Brain Res 809:1–4

Caulfield MP, Brown DA (1992) Cannabinoid receptor agonists inhibit Ca current in NG108-15 neuroblastoma cells via a pertussis toxin-sensitive mechanism. Br J Pharmacol 106:231–232

Chan PKY, Yung W-H (1998) Occlusion of the presynaptic action of cannabinoids in rat substantia nigra pars reticulata by cadmium. Neurosci Lett 249:57–60

Chan PKY, Chan SCY, Yung W-H (1998) Presynaptic inhibition of GABAergic inputs to rat substantia nigra pars reticulata neurones by a cannabinoid agonist. Neuroreport 9:671–675

Chemin J, Monteil A, Perez-Reyes E, Nargeot J, Lory P (2001) Direct inhibition of T-type calcium channels by the endogenous cannabinoid anandamide. EMBO J 20:7033–7040

Christopoulos A, Coles P, Lay L, Lew MJ, Angus JA (2001) Pharmacological analysis of cannabinoid receptor activity in the rat vas deferens. Br J Pharmacol 132:1281–1291

Compton DR, Harris LS, Lichtman AH Martin BR (1996) Marihuana. Pharmacological aspects of drug dependence. In: Schuster CR, Kuhar MJ (eds) Handbook of Experimental Pharmacology. Springer, Heidelberg, pp 83–158

Coutts AA, Pertwee RG (1997) Inhibition by cannabinoid receptor agonists of acetylcholine release from the guinea-pig myenteric plexus. Br J Pharmacol 121:1557–1566

Croci T, Manara L, Aureggi G, Guagnini F, Rinaldi-Carmona M, Maffrand J-P, Le Fur G, Mukenge S, Ferla G (1998) In vitro functional evidence of neuronal cannabinoid CB1 receptors in human ileum. Br J Pharmacol 125:1393–1395

Daniel H, Crepel F (2001) Control of Ca^{2+} influx by cannabinoid and metabotropic glutamate receptors in rat cerebellar cortex requires K^{+} channels. J Physiol (Lond) 537:793–800

Deadwyler SA, Hampson RE, Mu J, Whyte A, Childers S (1995) Cannabinoids modulate voltage sensitive potassium A-current in hippocampal neurons via a cAMP-dependent process. J Pharmacol Exp Ther 273:734–743

Deutsch DG, Goligorsky MS, Schmid PC, Krebsbach RJ, Schmid HHO, Das SK, Dey SK, Arreaza G, Thorup C, Stefano G, Moore L (1997) Production and physiological actions of anandamide in the vasculature of the rat kidney. J Clin Invest 100:1538–1546

Diana MA, Marty A (2003) Characterization of depolarization-induced suppression of inhibition using paired interneuron-Purkinje cell recordings. J Neurosci 23:5906–5918

Diana MA, Levenes C, Mackie K, Marty A (2002) Short-term retrograde inhibition of GABAergic synaptic currents in rat Purkinje cells is mediated by endogenous cannabinoids. J Neurosci 22:200–208

Ellington HC, Cotter MA, Cameron NE, Ross RA (2002) The effect of cannabinoids on capsaicin-evoked calcitonin gene-related peptide (CGRP) release from the isolated paw skin of diabetic and non-diabetic rats. Neuropharmacology 42:966–975

Engler B, Szabo B (2003) Cannabinoids inhibit striatopallidal neurotransmission in mice. Naunyn–Schmiedeberg's Arch Pharmacol 367:R85

Engler B, Szabo B (2004) Characterization of the effects of cannabinoids on synaptic transmission between caudate-putamen and globus pallidus. Naunyn–Schmiedeberg's Arch Pharmacol 369:R80

Fan P (1995) Cannabinoid agonists inhibit the activation of 5-HT_3 receptors in rat nodose ganglion neurons. J Neurophysiol 73:907–910

Ferraro L, Tomasini MC, Gessa GL, Bebe BW, Tanganelli S, Antonelli T (2001) The cannabinoid receptor agonist WIN 55,212-2 regulates glutamate transmission in rat cerebral cortex: an in vivo and in vitro study. Cerebral Cortex 11:728–733

Fränkel S (1903) Chemie und Pharmakologie des Haschisch. Naunyn–Schmiedeberg's Arch Exp Pathol Pharmakol 49:266–284

Freund TF, Katona I, Piomelli D (2003) Role of endogenous cannabinoids in synaptic signaling. Physiol Rev 83:1017–1066

Garcia DE, Brown S, Hille B, Mackie K (1998) Protein kinase C disrupts cannabinoid actions by phosphorylation of the CB1 cannabinoid receptor. J Neurosci 18:2834–2841

Gerdeman G, Lovinger DM (2001) CB1 cannabinoid receptor inhibits synaptic release of glutamate in rat dorsolateral striatum. J Neurophysiol 85:468–471

Gessa GL, Casu MA, Carta G, Mascia MS (1998) Cannabinoids decrease acetylcholine release in the medial-prefrontal cortex and hippocampus, reversal by SR 141716A. Eur J Pharmacol 355:119–124

Gifford AN, Ashby Jr CR (1996) Electrically evoked acetylcholine release from hippocampal slices is inhibited by the cannabinoid receptor agonist, WIN 55212-2, and is potentiated by the cannabinoid antagonist, SR 141716A. J Pharmacol Exp Ther 277:1431–1436

Gifford AN, Samiian L, Gatley JS, Ashby Jr CR (1997a) Examination of the effect of the cannabinoid receptor agonist, CP 55,940, on electrically evoked transmitter release from rat brain slices. Eur J Pharmacol 324:187–192

Gifford AN, Tang Y, Gatley SJ, Volkow ND, Lan R, Makriyannis A (1997b) Effect of the cannabinoid receptor SPECT agent, AM 281, on hippocampal acetylcholine release from rat brain slices. Neurosci Lett 238:84–86

Gifford AN, Bruneus M, Gatley SJ, Volkow ND (2000) Cannabinoid receptor-mediated inhibition of acetylcholine release from hippocampal and cortical synaptosomes. Br J Pharmacol 131:645–650

Glass M, Dragunow M, Faull RLM (2000) The pattern of neurodegeneration in Huntington's disease: a comparative study of cannabinoid, dopamine, adenosine and $GABA_A$ receptor alterations in the human basal ganglia in Huntington's disease. Neuroscience 97:505–519

Göbel I, Trendelenburg AU, Cox SL, Meyer A, Starke K (2000) Electrically evoked release of [^{3}H]noradrenaline from mouse cultured sympathetic neurons: release-modulating heteroreceptors. J Neurochem 75:2087–2094

Hájos N, Freund TF (2002) Pharmacological separation of cannabinoid sensitive receptors on hippocampal excitatory and inhibitory fibers. Neuropharmacology 43:503–510

Hájos N, Katona I, Naiem SS, Mackie K, Ledent C, Mody I, Freund TF (2000) Cannabinoids inhibit hippocampal GABAergic transmission and network oscillations. Eur J Neurosci 12:3239–3249

Hájos N, Ledent C, Freund TF (2001) Novel cannabinoid-sensitive receptor mediates inhibition of glutamatergic synaptic transmission in the hippocampus. Neuroscience 106:1–4

Hampson RE, Mu J, Deadwyler SA (2000) Cannabinoid and kappa opioid receptors reduce potassium K current via activation of G_s proteins in cultured hippocampal neurons. J Neurophysiol 84:2356–2364

Henry DJ, Chavkin C (1995) Activation of inwardly rectifying potassium channels (GIRK1) by co-expressed rat brain cannabinoid receptors in Xenopus oocytes. Neurosci Lett 186:91–94

Herkenham M, Lynn AB, De Costa BR, Richfield EK (1991) Neuronal localization of cannabinoid receptors in the basal ganglia of the rat. Brain Res 547:267–274

Himmi T, Perrin J, El Ouazzani T, Orsini J-C (1998) Neuronal responses to cannabinoid receptor ligands in the solitary tract nucleus. Eur J Pharmacol 359:49–54

Hoffman AF, Lupica CR (2000) Mechanisms of cannabinoid inhibition of GABAA synaptic transmission in the hippocampus. J Neurosci 20:2470–2479

Hoffman AF, Lupica CR (2001) Direct actions of cannabinoids on synaptic transmission in the nucleus accumbens: a comparison with opioids. J Neurophysiol 85:72–83

Hoffman AF, Oz M, Caulder T, Lupica CR (2003a) Functional tolerance and blockade of long-term depression at synapses in the nucleus accumbens after chronic cannabinoid exposure. J Neurosci 23:4815–4820

Hoffman AF, Riegel AC, Lupica CR (2003b) Functional localization of cannabinoid receptors and endogenous cannabinoid production in distinct neuron populations of the hippocampus. Eur J Neurosci 18:524–534

Huang C-C, Lo S-W, Hsu K-S (2001) Presynaptic mechanisms underlying cannabinoid inhibition of excitatory synaptic transmission in rat striatal neurons. J Physiol (Lond) 532:731–748

Huang C-C, Chen Y-L, Lo S-W, Hsu K-S (2002) Activation of cAMP-dependent protein kinase suppresses the presynaptic cannabinoid inhibition of glutamatergic transmission at corticostriatal synapses. Mol Pharmacol 61:578–585

Hurst DP, Lynch DL, Barnett-Norris J, Hyatt SM, Seltzman HH, Zhong M, Song ZH, Nie J, Lewis D, Reggio PH (2002) N-(Piperidin-1-yl)-5-(4-chlorophenyl)-1-(2,4-dichlorophenyl)-4-methyl-1H-pyrazole-3-carboxamide (SR141716A) interaction with LYS 3.28 (192) is crucial for its inverse agonism at the cannabinoid CB1 receptor. Mol Pharmacol 62:1274–1287

Irving AJ, Coutts AA, Harvey J, Rae MG, Mackie K, Bewick GS, Pertwee RG (2000) Functional expression of cell surface cannabinoid CB_1 receptors on presynaptic inhibitory terminals in cultured rat hippocampal neurons. Neuroscience 98:253–262

Ishac EJN, Jiang L, Lake KD, Varga K, Abood ME, Kunos G (1996) Inhibition of exocytotic noradrenaline release by presynaptic cannabinoid CB_1 receptors on peripheral sympathetic nerves. Br J Pharmacol 118:2023–2028

Izzo AA, Mascolo N, Borrelli F, Capasso F (1998) Excitatory transmission to the circular muscle of the guinea-pig ileum: evidence for the involvement of cannabinoid CB_1 receptors. Br J Pharmacol 124:1363–1368
Jennings EA, Vaughan CW, Christie MJ (2001) Cannabinoid actions on rat superficial medullary dorsal horn neurons in vitro. J Physiol 534:805–812
Kathmann M, Bauer U, Schlicker E, Göthert M (1999) Cannabinoid CB_1 receptor-mediated inhibition of NMDA- and kainate-stimulated noradrenaline and dopamine release in the brain. Naunyn–Schmiedeberg's Arch Pharmacol 359:466–470
Kathmann M, Weber B, Schlicker E (2001a) Cannabinoid CB_1 receptor-mediated inhibition of acetylcholine release in the brain of NMRI, CD-1 and C57BL/6 J mice. Naunyn–Schmiedeberg's Arch Pharmacol 363:50–56
Kathmann M, Weber B, Zimmer A, Schlicker E (2001b) Enhanced acetylcholine release in the hippocampus of cannabinoid CB_1 receptor-deficient mice. Br J Pharmacol 132:1169–1173
Katona I, Sperlagh B, Sik A, Köfalvi A, Vizi ES, Mackie K, Freund TF (1999) Presynaptically located CB1 cannabinoid receptors regulate GABA release from axon terminals of specific hippocampal interneurons. J Neurosci 19:4544–4558
Katona I, Sperlagh B, Magloczky Z, Santha E, Köfalvi A, Czirjak S, Mackie K, Vizi ES, Freund TF (2000) GABAergic interneurons are the targets of cannabinoid actions in the human hippocampus. Neuroscience 100:797–804
Katona I, Rancz EA, Acsady L, Ledent C, Mackie K, Hajos N, Freund TF (2001) Distribution of CB1 cannabinoid receptors in the amygdala and their role in the control of GABAergic transmission. J Neurosci 21:9506–9518
Kim DJ, Thayer SA (2000) Activation of CB1 cannabinoid receptors inhibits neurotransmitter release from identified synaptic sites in rat hippocampal cultures. Brain Res 852:398–405
Köfalvi A, Vizi ES, Ledent C, Sperlagh B (2003) Cannabinoids inhibit the release of $[^3H]$glutamate from rodent hippocampal synaptosomes via a novel CB_1 receptor-independent action. Eur J Neurosci 18:1973–1978
Kolasiewicz W, Wolfarth S, Ossowska K (1988) The role of the ventromedial thalamic nucleus in the catalepsy evoked from the substantia nigra pars reticulata in rats. Neurosci Lett 90:219–223
Kouznetsova M, Kelley B, Shen M, Thayer SA (2002) Desensitization of cannabinoid-mediated presynaptic inhibition of neurotransmission between rat hippocampal neurons in culture. Mol Pharmacol 61:477–485
Kreitzer AC, Regehr WG (2001) Retrograde inhibition of presynaptic calcium influx by endogenous cannabinoids at excitatory synapses onto Purkinje cells. Neuron 29:717–727
Kreitzer AC, Carter AG, Regehr WG (2002) Inhibition of interneuron firing extends the spread of endocannabinoid signaling in the cerebellum. Neuron 34:787–796
Lay L, Angus JA, Wright CE (2000) Pharmacological characterisation of cannabinoid CB_1 receptors in the rat and mouse. Eur J Pharmacol 391:151–161
Levenes C, Daniel H, Soubrie P, Crepel F (1998) Cannabinoids decrease excitatory synaptic transmission and impair long-term depression in rat cerebellar Purkinje cells. J Physiol (Lond) 510:867–879
Lever IJ, Malcangio M (2002) CB1 receptor antagonist SR141716A increases capsaicin-evoked release of Substance P from the adult mouse spinal cord. Br J Pharmacol 135:21–24
Lopez-Redondo F, Lees GM, Pertwee RG (1997) Effects of cannabinoid receptor ligands on electrophysiological properties of myenteric neurones of the guinea-pig ileum. Br J Pharmacol 122:330–334
Mackie K, Hille B (1992) Cannabinoids inhibit N-type calcium channels in neuroblastoma-glioma cells. Proc Natl Acad Sci USA 89:3825–3829
Mackie K, Devane WA, Hille B (1993) Anandamide, an endogenous cannabinoid, inhibits calcium currents as a partial agonist in N 18 neuroblastoma cells. Mol Pharmacol 44:498–503

Mackie K, Lai Y, Westenbroek R, Mitchell R (1995) Cannabinoids activate an inwardly rectifying potassium conductance and inhibit Q-type calcium currents in AtT20 cells transfected with rat brain cannabinoid receptor. J Neurosci 15:6552–6561
Maejima T, Hashimoto K, Yoshida T, Aiba A, Kano M (2001) Presynaptic inhibition caused by retrograde signal from metabotropic glutamate to cannabinoid receptors. Neuron 31:463–475
Mailleux P, Vanderhaeghen J-J (1992) Distribution of neuronal cannabinoid receptor in the adult rat brain: a comparative receptor binding radioautography and in situ hybridization histochemistry. Neuroscience 48:655–668
Maingret F, Patel AJ, Lazdunski M, Honore E (2001) The endocannabinoid anandamide is a direct and selective blocker of the background K+ channel TASK-1. EMBO J 20:47–54
Malinowska B, Godlewski G, Bucher B, Schlicker, E (1997) Cannabinoid CB_1 receptor-mediated inhibition of the neurogenic vasopressor response in the pithed rat. Naunyn–Schmiedeberg's Arch Pharmacol 356:197–202
Malinowska B, Piszcz J, Koneczny B, Hryniewicz A, Schlicker E (2001) Modulation of the cardiac autonomic transmission of pithed rats by presynaptic opioid OP_4 and cannabinoid CB_1 receptors. Naunyn–Schmiedeberg's Arch Pharmacol 364:233–241
Malone DT, Taylor DA (1999) Modulation by fluoxetine of striatal dopamine release following Δ^9-tetrahydrocannabinol: a microdialysis study in conscious rats. Br J Pharmacol 128:21–26
Mang CF, Erbelding D, Kilbinger H (2001) Differential effects of anandamide on acetylcholine release in the guinea-pig ileum mediated via vanilloid and non-CB_1 receptors. Br J Pharmacol 134:161–167
Manzoni OJ, Bockaert J (2001) Cannabinoids inhibit GABAergic synaptic transmission in mice nucleus accumbens. Eur J Pharmacol 412:R3–R5
Marsicano G, Wotjak CT, Azad SC, Bisogno T, Rammes G, Cascio MG, Hermann H, Tang J, Hofmann C, Zieglgänsberger W, Di Marzo V, Lutz B (2002) The endogenous cannabinoid system controls extinction of aversive memories. Nature 418:530–534
Matsuda LA, Bonner TI, Lolait SJ (1993) Localization of cannabinoid receptor mRNA in rat brain. J Comp Neurol 327:535–550
McAllister SD, Griffin G, Satin LS, Abood ME (1999) Cannabinoid receptors can activate and inhibit G protein-coupled inwardly rectifying potassium channels in a Xenopus oocyte expression system. J Pharmacol Exp Ther 291:618–626
Melis M, Pistis M, Perra S, Muntoni AL, Pillolla G, Gessa GL (2004) Endocannabinoids mediate presynaptic inhibition of glutamatergic transmission in rat ventral tegmental area dopamine neurons through activation of CB1 receptors. J Neurosci 24:53–62
Mishima K, Egashira N, Matsumoto Y, Iwasaki K, Fujiwara M (2002) Involvement of reduced acetylcholine release in Δ^9-tetrahydrocannabinol-induced impairment of spatial memory in the 8-arm radial maze. Life Sci 72:397–407
Misner DL, Sullivan JM (1999) Mechanism of cannabinoid effects on long-term potentiation and depression in hippocampal CA1 neurons. J Neurosci 19:6795–6805
Molderings GJ, Likungu J, Göthert M (1999) Presynaptic cannabinoid and imidazoline receptors in the human heart and their potential relationship. Naunyn–Schmiedeberg's Arch Pharmacol 360:157–164
Morisset V, Urban L (2001) Cannabinoid-induced presynaptic inhibition of glutamatergic EPSCs in substantia gelatinosa neurons of the rat spinal cord. J Neurophysiol 86:40–48
Mu J, Zhuang S-Y, Hampson RE, Deadwyler SA (2000) Protein kinase-dependent phosphorylation and cannabinoid receptor modulation of potassium A current (I_A) in cultured rat hippocampal neurons. Pflügers Arch 439:541–546
Nakatsuka T, Chen H-X, Roper SN, Gu JG (2003) Cannabinoid receptor-1 activation suppresses inhibitory synaptic activity in human dentate gyrus. Neuropharmacology 45:116–121
Nakazi M, Bauer M, Nickel T, Kathmann M, Schlicker E (2000) Inhibition of serotonin release in the mouse brain via presynaptic cannabinoid CB_1 receptors. Naunyn–Schmiedeberg's Arch Pharmacol 361:19–24

Nemeth J, Helyes Z, Than M, Jakab B, Pinter E, Szolcsanyi J (2003) Concentration-dependent dual effect of anandamide on sensory neuropeptide release from isolated rat tracheae. Neurosci Lett 336:89–92

Nicholson RA, Liao C, Zheng J, David LS, Coyne L, Errington AC, Singh G, Lees G (2003) Sodium channel inhibition by anandamide and synthetic cannabimimetics in brain. Brain Res 978:194–204

Niederhoffer N, Szabo B (1999) Effect of the cannabinoid receptor agonist WIN55212-2 on sympathetic cardiovascular regulation. Br J Pharmacol 126:457–466

Niederhoffer N, Hansen HH, Fernandez-Ruiz JJ, Szabo B (2001) Effects of cannabinoids on adrenaline release from adrenal medullary cells. Br J Pharmacol 134:1319–1327

Niederhoffer N, Schmid K, Szabo B (2003) The peripheral sympathetic nervous system is the major target of cannabinoids in eliciting cardiovascular depression. Naunyn-Schmiedeberg's Arch Pharmacol 367:434–443

Nogueron IM, Porgilsson B, Schneider WE, Stucky CL, Hillard CJ (2001) Cannabinoid receptor agonists inhibit depolarization-induced calcium influx in cerebellar granule neurons. J Neurochem 79:371–381

Ohno-Shosaku T, Maejima T, Kano M (2001) Endogenous cannabinoids mediate retrograde signals from depolarized postsynaptic neurons to presynaptic terminals. Neuron 29:729–738

Oz M, Ravindran A, Diaz-Ruiz O, Zhang L, Morales M (2003) The endogenous cannabinoid anandamide inhibits $\alpha 7$ nicotinic acetylcholine receptor-mediated responses in Xenopus oocytes. J Pharmacol Exp Ther 306:1003–1010

Pan X, Ikeda SR, Lewis DL (1996) Rat brain cannabinoid receptor modulates N-type Ca^{2+} channels in a neuronal expression system. Mol Pharmacol 49:707–714

Patel S, Hillard CJ (2001) Cannabinoid CB1 receptor agonists produce cerebellar dysfunction in mice. J Pharmacol Exp Ther 297:629–637

Pertwee RG, Fernando SR (1996) Evidence for the presence of cannabinoid CB_1 receptors in mouse urinary bladder. Br J Pharmacol 118:2053–2058

Pertwee RG, Stevenson LA, Elrick DB, Mechoulam R, Corbett AD (1992) Inhibitory effects of certain enantiomeric cannabinoids in the mouse vas deferens and the myenteric plexus preparation of guinea-pig small intestine. Br J Pharmacol 105:980–984

Pertwee RG, Fernando SR, Griffin G, Ryan W, Razdan RK, Compton DR, Martin BR (1996a) Agonist-antagonist characterization of 6'-cyanohex-2'-yne-Δ^8-tetrahydrocannabinol in two isolated tissue preparations. Eur J Pharmacol 315:195–201

Pertwee RG, Fernando SR, Nash JE, Coutts AA (1996b) Further evidence for the presence of cannabinoid CB_1 receptors in guinea-pig small intestine. Br J Pharmacol 118:2199–2205

Pertwee RG, Ross RA, Craib S, Thomas A (2002) (–)-Cannabidiol antagonizes cannabinoid receptor agonists and noradrenaline in the mouse vas deferens. Eur J Pharmacol 456:99–106

Ralevic V (2003) Cannabinoid modulation of peripheral autonomic and sensory neurotransmission. Eur J Pharmacol 472:1–21

Ralevic V, Kendall DA (2001) Cannabinoid inhibition of capsaicin-sensitive sensory neurotransmission in the rat mesenteric arterial bed. Eur J Pharmacol 418:117–125

Ralevic V, Kendall DA (2002) Cannabinoids inhibit pre- and postjunctionally sympathetic neurotransmission in rat mesenteric arteries. Eur J Pharmacol 444:171–181

Richardson JD, Kilo S, Hargreaves KM (1998) Cannabinoids reduce hyperalgesia and inflammation via interaction with peripheral CB1 receptors. Pain 75:111–119

Riegel AC, Williams JT, Lupica CR (2003) Cannabinoid receptor activation depresses $GABA_B$-mediated synaptic responses in dopamine neurons. In: 2003 Symposium on the cannabinoids. International Cannabinoid Research Society, Burlington, p 23

Robbe D, Alonso G, Duchamp F, Bockaert J, Manzoni OJ (2001) Localization and mechanisms of action of cannabinoid receptors at the glutamatergic synapses of the mouse nucleus accumbens. J Neurosci 21:109–116

Rodriguez JJ, Mackie K, Pickel VM (2001) Ultrastructural localization of the CB1 cannabinoid receptor in μ-opioid receptor patches of the rat caudate putamen nucleus. J Neurosci 21:823–833

Rubovitch V, Gafni M, Sarne Y (2002) The cannabinoid agonist DALN positively modulates L-type voltage-dependent calcium-channels in N18TG2 neuroblastoma cells. Mol Brain Res 101:93–102

Ruiu S, Pinna GA, Marchese G, Mussinu JM, Saba P, Tambaro S, Casti P, Vargiu R, Pani L (2003) Synthesis and characterization of NESS 0327: a novel putative antagonist of the CB_1 cannabinoid receptor. J Pharmacol Exp Ther 306:363–370

Sanudo-Pena MC, Tsou K, Walker JM (1999) Motor actions of cannabinoids in the basal ganglia output nuclei. Life Sci 65:703–713

Schlicker E, Kathmann M (2001) Modulation of transmitter release via presynaptic cannabinoid receptors. Trends Pharmacol Sci 22:565–572

Schlicker E, Timm J, Göthert M (1996) Cannabinoid receptor-mediated inhibition of dopamine release in the retina. Naunyn–Schmiedeberg's Arch Pharmacol 354:791–795

Schlicker E, Timm J, Zentner J, Göthert M (1997) Cannabinoid CB_1 receptor-mediated inhibition of noradrenaline release in the human and guinea-pig hippocampus. Naunyn–Schmiedeberg's Arch Pharmacol 356:583–589

Schlicker E, Liedtke S, Flau K, Kathmann M (2002) Further evidence that the cannabinoid receptor inhibiting noradrenaline release in the guinea-pig brain belongs to the CB_1 subtype and is subject to an endogenous tone. Pharmacologist 44(Suppl 1):A112

Schlicker E, Redmer A, Werner A, Kathmann M (2003) Lack of CB_1 receptors increases noradrenaline release in vas deferens without affecting atrial noradrenaline release or cortical acetylcholine release. Br J Pharmacol 140:323–328

Schultheiß T, Flau K, Redmer A, Kathmann M, Reggio PH, Seltzman HH, Schlicker E (2004) The facilitatory effect of SR141716 on transmitter release in guinea-pig hippocampus is due to its inverse agonist activity at cannabinoid CB_1 receptors. Naunyn–Schmiedeberg's Arch Pharmacol 369 (Suppl 1):R84

Schweitzer P (2000) Cannabinoids decrease the K^+ M-current in hippocampal CA1 neurons. J Neurosci 20:51–58

Seifert R, Wenzel-Seifert K (2002) Constitutive activity of G-protein-coupled receptors: cause of disease and common property of wild-type receptors. Naunyn–Schmiedeberg's Arch Pharmacol 366:381–416

Shen M, Thayer SA (1999) Δ^9-Tetrahydrocannabinol acts as a partial agonist to modulate glutamatergic synaptic transmission between rat hippocampal neurons in culture. Mol Pharmacol 55:8–13

Shen M, Piser TM, Seybold VS, Thayer SA (1996) Cannabinoid receptor agonists inhibit glutamatergic synaptic transmission in rat hippocampal cultures. J Neurosci 16:4322–4334

Starke K, Göthert M, Kilbinger H (1989) Modulation of neurotransmitter release by presynaptic autoreceptors. Physiol Rev 69:864–989

Steffens M, Szabo B, Klar M, Rominger A, Zentner J, Feuerstein TJ (2003) Modulation of electrically evoked acetylcholine release through cannabinoid CB_1 receptors: evidence for an endocannabinoid tone in the human neocortex. Neuroscience 120:456–465

Storr M, Sibaev A, Marsicano G, Lutz B, Schusdziarra V, Timmermans JP, Allescher HD (2004) Cannabinoid receptor type 1 modulates excitatory and inhibitory neurotransmission in mouse colon. Am J Physiol Gastrointest Liver Physiol 286:G110–G117

Straiker A, Sullivan JM (2003) Cannabinoid receptor activation differentially modulates ion channels in photoreceptors of the tiger salamander. J Neurophysiol 89:2647–2654

Straiker AJ, Borden CR, Sullivan JM (2002) G-protein alpha subunit isoforms couple differentially to receptors that mediate presynaptic inhibition at rat hippocampal synapses. J Neurosci 22:2460–2468

Sullivan JM (1999) Mechanisms of cannabinoid-receptor-mediated inhibition of synaptic transmission in cultured hippocampal pyramidal neurons. J Neurophysiol 82:1286–1294

Szabo B, Dörner L, Pfreundtner C, Nörenberg W, Starke K (1998) Inhibition of GABAergic inhibitory postsynaptic currents by cannabinoids in rat corpus striatum. Neuroscience 85:395–403

Szabo B, Müller T, Koch H (1999) Effects of cannabinoids on dopamine release in the corpus striatum and the nucleus accumbens in vitro. J Neurochem 73:1084–1089

Szabo B, Wallmichrath I, Mathonia P, Pfreundtner C (2000) Cannabinoids inhibit excitatory neurotransmission in the substantia nigra pars reticulata. Neuroscience 97:89–97

Szabo B, Nordheim U, Niederhoffer N (2001) Effects of cannabinoids on sympathetic and parasympathetic neuroeffector transmission in the rabbit heart. J Pharmacol Exp Ther 297:819–826

Szabo B, Siemes S, Wallmichrath I (2002) Inhibition of GABAergic neurotransmission in the ventral tegmental area by cannabinoids. Eur J Neurosci 15:2057–2061

Szabo B, Than M, Wallmichrath I, Thorn D (2004) Analysis of the effects of cannabinoids on synaptic transmission between basket and Purkinje cells in the cerebellar cortex of the rat. J Pharmacol Exp Ther 310:915–925

Takahashi KA, Linden DJ (2000) Cannabinoid receptor modulation of synapses received by cerebellar Purkinje cells. J Neurophysiol 83:1167–1180

Thompson SM, Capogna M, Scanziani M (1993) Presynaptic inhibition in the hippocampus. Trends Neurosci 16:222–227

Timm J, Marr I, Werthwein S, Elz S, Schunack W, Schlicker E (1998) H_2 receptor-mediated facilitation and H3 receptor-mediated inhibition of noradrenaline release in the guinea-pig brain. Naunyn–Schmiedeberg's Arch Pharmacol 357:232–239

Tognetto M, Amadesi S, Harrison S, Creminon C, Trevisani M, Carreras M, Matera M, Geppetti P, Bianchi A (2001) Anandamide excites central terminals of dorsal root ganglion neurons via vanilloid receptor-1 activation. J Neurosci 21:1104–1109

Trendelenburg AU, Cox SL, Schelb V, Klebroff W, Khairallah L, Starke K (2000) Modulation of 3H-noradrenaline release by presynaptic opioid, cannabinoid and bradykinin receptors and ß-adrenoceptors in mouse tissues. Br J Pharmacol 130:321–330

Trettel J, Levine ES (2002) Cannabinoids depress inhibitory synaptic inputs received by layer 2/3 pyramidal neurons of the neocortex. J Neurophysiol 88:534–539

Trettel J, Levine ES (2003) Endocannabinoids mediate rapid retrograde signaling at interneuron→pyramidal neuron synapses of the neocortex. J Neurophysiol 89:2334–2338

Tsou K, Brown S, Sanudo-Pena MC, Mackie K, Walker JM (1998) Immunohistochemical distribution of cannabinoid CB1 receptors in the rat central nervous system. Neuroscience 83:393–411

Turkanis SA, Partlow LM, Karler R (1991) Delta-9-tetrahydrocannabinol depresses inward sodium current in mouse neuroblastoma cells. Neuropharmacology 30:73–77

Twitchell W, Brown S, Mackie K (1997) Cannabinoids inhibit N- and P/Q-type calcium channels in cultured rat hippocampal neurons. J Neurophysiol 78:43–50

Tzavara ET, Perry KW, Rodriguez DE, Bymaster F, Nomikos GG (2001) The cannabinoid CB1 receptor antagonist SR141716A increases norepinephrine outflow in the rat anterior hypothalamus. Eur J Pharmacol 426:R3–R4

Tzavara ET, Davis RJ, Perry KW, Li X, Salhoff C, Witkin JM, Bymaster F, Nomikos GG (2003a) The CB_1 receptor antagonist SR141716A selectively increases monoaminergic neurotransmission in the medial prefrontal cortex: implications for therapeutic actions. Br J Pharmacol 138:544–553

Tzavara ET, Wade M, Nomikos GG (2003b) Biphasic effects of cannabinoids on acetylcholine release in the hippocampus: site and mechanism of action. J Neurosci 23:9374–9384

Van Vliet BJ, Nievelstein HNMW, Long SK, Kruse CG (2000) CB_1 receptor-mediated effects on brain neurotransmitter systems. Eur Neuropsychopharmacol 10 (Suppl 3):S182–S183

Vaughan CW, McGregor IS, Christie McDJ (1999) Cannabinoid receptor activation inhibits GABAergic neurotransmission in rostral ventromedial medulla neurons in vitro. Br J Pharmacol 127:935–940

Vaughan CW, Connor M, Bagley EE, Christie MJ (2000) Actions of cannabinoids on membrane properties and synaptic transmission in rat periaqueductal gray neurons in vitro. Mol Pharmacol 57:288–295

Verrico CD, Jentsch JD, Dazzi L, Roth RH (2003) Systemic, but not local, administration of cannabinoid CB1 receptor agonists modulate prefrontal cortical acetylcholine efflux in the rat. Synapse 48:178–183

Vizi ES, Katona I, Freund TF (2001) Evidence for presynaptic cannabinoid CB_1 receptor-mediated inhibition of noradrenaline release in the guinea pig lung. Eur J Pharmacol 431:237–244

Wallmichrath I, Szabo B (2002a) Analysis of the effect of cannabinoids on GABAergic neurotransmission in the substantia nigra pars reticulata. Naunyn–Schmiedeberg's Arch Pharmacol 365:326–334

Wallmichrath I, Szabo B (2002b) Cannabinoids inhibit striatonigral GABAergic neurotransmission in the mouse. Neuroscience 113:671–682

Wallmichrath I, Szabo B (2003) Effects of cannabinoids on the glutamatergic neurotransmission between nucleus subthalamicus and globus pallidus. Naunyn–Schmiedeberg's Arch Pharmacol 367:R83

Wang S-J (2003) Cannabinoid CB_1 receptor-mediated inhibition of glutamate release from rat hippocampal synaptosomes. Eur J Pharmacol 469:47–55

Wilson RI, Nicoll RA (2001) Endogenous cannabinoids mediate retrograde signalling at hippocampal synapses. Nature 410:588–592

Wilson RI, Nicoll RA (2002) Endocannabinoid signaling in the brain. Science 296:678–682

Wu L-G, Saggau P (1997) Presynaptic inhibition of elicited neurotransmitter release. Trends Neurosci 20:204–212

Yoshida T, Hashimoto K, Zimmer A, Maejima T, Araishi K, Kano M (2002) The cannabinoid CB1 receptor mediates retrograde signals for depolarization-induced suppression of inhibition in cerebellar Purkinje cells. J Neurosci 22:1690–1697

Zimmer A, Zimmer AM, Hohmann AG, Herkenham M, Bonner TI (1999) Increased mortality, hypoactivity, and hypoalgesia in cannabinoid CB1 receptor knockout mice. Proc Natl Acad Sci USA 96:5780–5785

Zygmunt PM, Petersson J, Andersson DA, Chuang H, Sorgard M, Di Marzo V, Julius D, Högestätt ED (1999) Vanilloid receptors on sensory nerves mediate the vasodilator action of anandamide. Nature 400:452–457

HEP (2005) 168:367–383

Retrograde Signalling by Endocannabinoids

C.W. Vaughan · M.J. Christie (✉)

Pain Management Research Institute, Northern Clinical School,
University of Sydney at Royal North Shore Hospital, 2006 NSW,Sydney, Australia
chrisv@med.usyd.edu.au

Abstract The cannabinoid neurotransmitter system comprises cannabinoid G protein-coupled membrane receptors (CB_1 and CB_2), endogenous cannabinoids (endocannabinoids), as well as mechanisms for their synthesis, membrane transport and metabolism. Within the brain the marijuana constituent Δ^9-tetrahydro-cannabinol (THC) produces its pharmacological actions by acting on cannabinoid CB_1 receptors. THC modulates neuronal excitability by inhibiting synaptic transmission via presynaptic CB_1-mediated mechanisms. More recently, it has been established that physiological stimulation of neurons can induce the synthesis of endocannabinoids, which also modulate synaptic transmission via cannabinoid CB_1 and other receptor systems. These endogenously synthesised endocannabinoids appear to act as retrograde signalling agents, reducing synaptic inputs onto the stimulated neuron in a highly selective and restricted manner. In this review we describe the cellular mechanisms underlying retrograde endocannabinoid signalling.

Keywords Endocannabinoid · Synaptic transmission · Retrograde signalling · TRP · mGluR

1 Endocannabinoids

The main psychoactive ingredient of cannabis, Δ^9-tetrahydrocannabinol (THC), is known to produce its actions in the central nervous system by acting on the body's own cannabinoid neurotransmitter system, predominantly via interaction with cannabinoid G protein-coupled CB_1 receptors. Like other neurotransmitter systems, the components of the cannabinoid signalling system also include endogenous cannabinoids (endocannabinoids), as well as mechanisms for their synthesis, membrane transport and metabolism (for recent reviews see Freund et al. 2003; Piomelli 2003; Petrocellis et al. 2004). Some of the endocannabinoids identified to date include anandamide, 2-arachidonoyl glycerol (2-AG), noladin ether and virodhamine. Within the central nervous system, endocannabinoids produce their biological effects by acting at least in part on cannabinoid CB_1 receptors. In this section we briefly describe some of the factors involved in the production and degradation of endocannabinoids and their locus of action, which are relevant to retrograde endocannabinoid signalling.

1.1 Endocannabinoid Synthesis, Release and Degradation

The production and degradation of endocannabinoids proceeds via a number of discrete steps that remain to be fully elucidated (Schmid 2000; Sugiura et al. 2002; Cravatt and Lichtman 2003; Glaser et al. 2003; Hillard and Jarrahian 2003; Piomelli 2003; Petrocellis et al. 2004). Endocannabinoids are thought to be produced on demand from membrane-bound phospholipids as a result of specific stimuli, and to be released in a non-vesicular manner. Briefly, anandamide is thought to be formed in a Ca^{2+}-dependent manner by a specific isoform of the enzyme phospholipase D (PLD) (Okamoto et al. 2004). 2-AG, on the other hand is thought to be formed via the phospholipase C (PLC)/DAG (*sn*-1-acyl-2-arachidonoylglycerol) lipase cascade in a Ca^{2+}-dependent manner (Bisogno et al. 1997; Stella et al. 1997). However, there are other potential synthetic pathways for these endocannabinoids. The biological activity of anandamide and 2-AG is terminated by uptake and subsequent degradation. It has been proposed that these endocannabinoids are removed from the extracellular space by an anandamide membrane transporter (AMT). However, the AMT remains to be identified, and it has been suggested that uptake occurs by passive diffusion maintained through intracellular degradation (Glaser et al. 2003; Hillard and Jarrahian 2003). Once within the cell, the degradation of anandamide and 2-AG appears to be catalysed by at least two distinct enzymes, fatty acid amide hydrolase (FAAH) and monoacylglycerol lipase (MAGL). The biosynthetic and degradation pathways of the other identified endocannabinoids have not been examined.

1.2
Endocannabinoids Act via Presynaptic Cannabinoid CB_1 Receptors to Inhibit Transmitter Release

THC and a number of synthetic non-selective cannabinoid receptor agonists (such as WIN 55,212-2 and HU-210) modulate neuronal excitability by presynaptic CB_1-mediated short-term inhibition of glutamatergic and γ-aminobutyric acid (GABA)ergic synaptic transmission (for more detailed reviews see Schlicker and Kathmann 2001; Alger 2002). Release studies have demonstrated that cannabinoids also modulate other neurotransmitter systems. In accordance with the electrophysiological evidence, anatomical studies have demonstrated that cannabinoid CB_1 receptors are located presynaptically on nerve terminals in numerous brain structures. Exogenous application of the endocannabinoids anandamide and 2-AG also modulates synaptic transmission within a number of regions throughout the central nervous system, including the hippocampus, midbrain periaqueductal grey, spinal and medullary dorsal horn, and the substantia nigra (Shen et al. 1996; Vaughan et al. 2000; Morisset et al. 2001; Jennings et al. 2003; Marinelli et al. 2003). While the inhibitory effects of endocannabinoids are mediated by presynaptic cannabinoid CB_1 receptors, endocannabinoids have complex effects on synaptic transmission that are also mediated by presynaptic vanilloid TRPV1 (transient receptor potential vanilloid type 1) receptors and potentially other mechanisms (Di Marzo et al. 2001, see also Sect. 5).

In addition to their short-term effects, cannabinoids also modulate the induction of long-term synaptic plasticity. Administration of THC and the endocannabinoids anandamide and 2-AG inhibits the induction of long-term potentiation (LTP) in the hippocampus (Nowicky et al. 1987; Terranova et al. 1995; Stella et al. 1997) and long-term depression (LTD) within the cerebellum and nucleus accumbens (Levenes et al. 1998; Hoffman et al. 2003a).

2
Endocannabinoid as Retrograde Transmitters

The importance of endocannabinoid signalling elements remained uncertain for some time because of a lack of direct evidence for physiologically relevant synthesis,

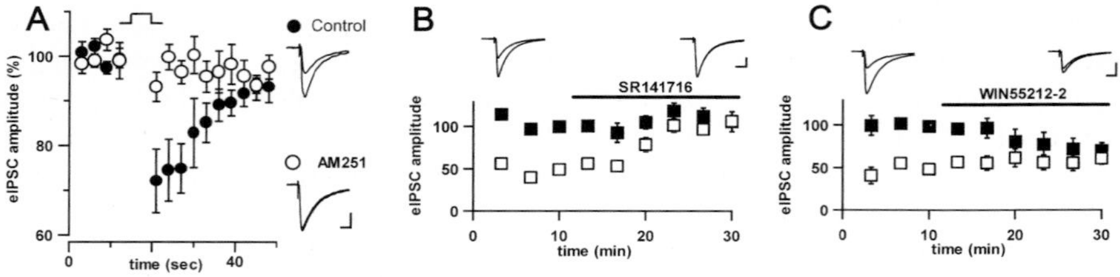

Fig. 1A–C DSI is mediated by endogenous cannabinoids. **A** A 5-s depolarising step produces a transient reduction in the amplitude of evoked inhibitory postsynaptic currents (*eIPSC*, control, *filled circles*) in hippocampal pyramidal neurons that is abolished by the CB_1 antagonist AM251 (2 μM, *open circles*). **B** and **C** Time plots of the eIPSC amplitude just before (*filled squares*) and just after (*open squares*) a 5-s depolarisation over 30 min. **B** DSI is blocked by the CB_1 antagonist SR141716 (2 μM). **C** The CB agonist WIN55212-2 (800 nM) inhibits eIPSCs and occludes DSI. The *insets* show average eIPSCs for the 10 s before and 10 s after the depolarising step (**A**), and in the same at 0 min and 30 min (**B** and **C**). *Scale bars* in inserts are 200 pA, 20 ms. (Modified from Wilson and Nicoll 2001, by permission)

release and activity of endocannabinoids on CB_1 receptors at the level of the synapse. Such evidence has recently arisen from the prior in vitro observation that strong depolarisation of hippocampal pyramidal and cerebellar Purkinje neurons produces a subsequent transient (from <10 s to 120 s), short-term inhibition of GABAergic synaptic inputs impinging upon these cells, observed as a decrease in the amplitude of evoked synaptic currents in the depolarised neuron (Llano et al. 1991; Pitler and Alger 1992) (Fig. 1A). This depolarisation-induced suppression of inhibition (DSI) was likely to be mediated by a retrograde messenger because it had a *postsynaptic* origin, but a *presynaptic* locus of action. However, the specific retrograde transmitter involved in DSI remained elusive (for review see Alger 2002).

2.1 Depolarisation-Induced Transient, Short-Term Retrograde Endocannabinoid Signalling

Wilson and Nicoll (2001) and Ohno-Shosahu et al. (2001) independently established that physiologically relevant stimulation of single hippocampal pyramidal neurons produces an endocannabinoid or endocannabinoids that diffuse onto the terminals of presynaptic GABAergic interneurons, where they act upon cannabinoid CB_1 receptors to produce transient, short-term inhibition of neurotransmitter release. Endocannabinoids also mediate depolarisation-induced retrograde signalling of interneuronal inhibitory GABAergic synapses onto cerebellar Purkinje cells (Kreitzer and Regehr 2001a; Diana et al. 2002; Yoshida et al. 2002), onto neocortical pyramidal cells (Trettel and Levine 2003) and onto substantia nigra neurons (Yoshida et al. 2002). In addition, retrograde endocannabinoid signalling-mediated depolarisation-induced suppression of excitation (DSE) has been demonstrated for excitatory glutamatergic synaptic inputs onto cerebellar Purkinje cells from climbing fibre and parallel fibre inputs (Kreitzer and Regehr 2001b; Maejima et al. 2001) and onto dopaminergic neurons within the ventral tegmental area (Melis et al. 2004). Endocannabinoids thus belong to a small but growing club of retrograde neurotransmitters that act back to modulate presynaptic inputs impinging upon a neuron (Alger 2002).

A role for endocannabinoids in DSI and DSE has been established indirectly by pharmacological antagonism, and mimicry/occlusion with agonists (e.g. Kreitzer and Regehr 2001a,b; Ohno-Shosaku et al. 2001; Varma et al. 2001; Wilson and Nicoll 2001; Diana et al. 2002; Yoshida et al. 2002; Hampson et al. 2003; Yanovsky et al. 2003; Melis et al. 2004). In these studies, the transient inhibition of evoked glutamatergic excitatory and GABAergic inhibitory postsynaptic currents (IPSCs and EPSCs) produced by postsynaptic depolarisation is abolished by the cannabinoid receptor antagonists SR141716, AM251 and AM281 (Fig. 1A and B). The non-selective cannabinoid receptor agonist WIN 55,212-2 inhibits synaptic transmission and occludes DSI and DSE (Fig. 1C). In addition, DSI is absent in mice with a CB_1 receptor deletion (Varma et al. 2001; Wilson et al. 2001; Kim et al. 2002; Yoshida et al. 2002). Thus, DSI and DSE are mediated by a yet-to-be-identified endocannabinoid(s) that acts via cannabinoid CB_1 receptors.

It has long been known that DSI satisfies the criteria for retrograde signalling. First, both DSI and DSE are induced in the postsynaptic cell because they are caused by depolarisation-induced increases in postsynaptic cytoplasmic Ca^{2+} (see Sect. 3.1). Second, DSI and DSE are expressed presynaptically (Kreitzer and Regehr 2001a,b; Ohno-Shosaku et al. 2001; Wilson and Nicoll 2001; Diana et al. 2002; Kreitzer et al. 2002; Yoshida et al. 2002; Diana and Marty 2003; Trettel and Levine 2003). In these studies presynaptic inhibition has been demonstrated using standard electrophysiological techniques, including an increase in the paired-pulse ratio of electrically evoked synaptic currents, an increase in failure rate and variance of evoked synaptic currents using paired recordings and a reduction in the rate, but not the amplitude of tetrodotoxin (TTX)-resistant miniature synaptic currents and of Sr^{2+}-induced evoked asynchronous synaptic currents. Thus, postsynaptic elevations in cytoplasmic Ca^{2+} produce a reduction in the probability of transmitter release from presynaptic terminals impinging upon the depolarised neuron. Retrogradely released endocannabinoids might act directly on the cell bodies of interneurons (Kreitzer et al. 2002; Diana and Marty 2003).

Thus, DSI and DSE satisfy the three criteria of retrograde endocannabinoid signalling: they are (1) mediated by endocannabinoid(s), (2) induced postsynaptically and (3) expressed presynaptically. It might be noted that other transmitters have been implicated in retrograde signalling (e.g. Yanovsky et al. 2003).

2.2 Activation of Postsynaptic Metabotropic Receptors Induces Short-Term Retrograde Endocannabinoid Signalling

Some studies have also shown that retrograde endocannabinoid signalling can also be induced by postsynaptic activation of metabotropic glutamatergic receptors (mGluRs) and muscarinic acetylcholine receptors (mAChRs). Like the cannabinoid receptors, metabotropic glutamate receptors (mGluR) belong to the G protein-coupled membrane receptor superfamily and have been classified into three main groups on the basis of their sequence homology and their biochemical and pharmacological profiles (Conn and Pin 1997). Group II and III mGluRs are located on presynaptic nerve terminals and act as autoreceptors to inhibit glutamate release in numerous brain regions. While group I mGluRs are located mainly on the cell bodies of neurons, activation of group I mGluRs (either with agonists or indirectly by high frequency stimulation of glutamatergic inputs) produces short- and long-term presynaptic inhibition of glutamatergic and GABAergic synaptic transmission within a number of brain regions (for review see Doherty and Dingledine 2003).

The divergence between the anatomical localisation (postsynaptic) and electrophysiological locus of action (presynaptic) of group I mGluRs can be reconciled by retrograde signalling. A role for endocannabinoids in postsynaptic mGluR-induced retrograde signalling has been demonstrated using similar pharmacological and knockout techniques to those described for DSI and DSE. In these studies, the inhibition of evoked synaptic currents produced by the selective group I mGluR agonist 3,5-dihydroxyphenylglycine (DHPG) (and by stimulation-evoked release from glutamatergic inputs) is abolished by cannabinoid CB_1 antagonism and deletion, and is mimicked/occluded by cannabinoid receptor agonists (Maejima et al. 2001; Varma et al. 2001; Ohno-Shosaku et al. 2002; Robbe et al. 2002; but see Chevaleyre and Castillo 2003; Rouach and Nicoll 2003). Similar techniques have demonstrated that endocannabinoids mediate retrograde signalling produced by postsynaptic mAChRs activation (Kim et al. 2002).

Like DSI and DSE, mGluR/stimulation-induced short-term depression is mediated by a retrograde signalling process. The group I mGluR-mediated inhibition is induced at a postsynaptic site because the presynaptic inhibition produced by the group I mGluR agonist DHPG (but not group II agonists) and by high frequency stimulation is abolished by disrupting postsynaptic G protein coupling with guanosine triphosphate (GTP)-γS, or guanosine diphosphate (GDP)-βS (Maejima et al. 2001; Ohno-Shosaku et al. 2002). However, group I mGluR-evoked presynaptic inhibition is not dependent upon postsynaptic elevations in intracellular Ca^{2+} (Maejima et al. 2001; Ohno-Shosaku et al. 2002, and see Sect. 3.2). Also, evidence similar to that described in Sect. 2.2 has demonstrated that mGluR-induced endocannabinoid-mediated inhibition is expressed presynaptically (Maejima et al. 2001; Ohno-Shosaku et al. 2002; Chevaleyre and Castillo 2003).

2.3 Activation of Postsynaptic Metabotropic Receptors Potentiates Depolarisation-Induced Retrograde Endocannabinoid Signalling

Do mGluR-induced and depolarisation-induced retrograde endocannabinoid signalling work independently? While depolarisation-induced DSI does not require activation of mGluRs (e.g. Wilson and Nicoll 2001), there is an interaction between the postsynaptic depolarisation and group I mGluR-induced presynaptic inhibition. Low concentrations of mGluR and mAChR agonists, which alone produce little presynaptic inhibition, potentiate DSI (Varma et al. 2001; Kim et al. 2002; Hoffman et al. 2003b). Given that the two forms of retrograde endocannabinoid signalling are mediated by distinct postsynaptic mechanisms (see Sects. 3.1 and 3.2), there is the potential for synergistic presynaptic inhibition.

2.4 A Role for Retrograde Endocannabinoids in Long-Term Synaptic Plasticity

It is becoming apparent that, in addition to short-term modulation of synaptic transmission, retrograde endocannabinoid signalling is involved in some forms of long-term synaptic plasticity because both short- and long-term modulation are abolished by cannabinoid CB_1 antagonists and are absent in mice with a CB_1 receptor deletion (Fig. 2A). Endocannabinoids mediate high-frequency stimulation-induced LTD of glutamatergic synaptic transmission in the striatum (Gerdeman et al. 2002) and nucleus accumbens (Robbe et al. 2002), stimulation-induced LTD of GABAergic synaptic transmission in the amygdala and hippocampus (Marsicano et al. 2002; Chevaleyre and Castillo 2003), timing-dependent LTD within the neocortex (Sjostrom et al. 2003) and amphetamine-induced LTD of glutamatergic synaptic transmission in the amygdala (Huang et al. 2003). However, endocannabinoids are not involved in all forms of long-term synaptic plasticity (e.g. Carlson et al. 2002; Marsicano et al. 2002; Rouach and Nicoll 2003) and it is therefore likely to be region- and synapse-specific.

Stimulation-induced LTD is induced at a postsynaptic site and is expressed presynaptically. LTD is induced postsynaptically because it is abolished by disrupting postsynaptic G protein coupling (Chevaleyre and Castillo 2003) (Figure 2C). Some forms of stimulation-induced LTD are caused by endogenously released glutamate acting on postsynaptic group I mGluR receptors because they are abolished by group I antagonists (Fig. 2B) and can be mimicked by the application of

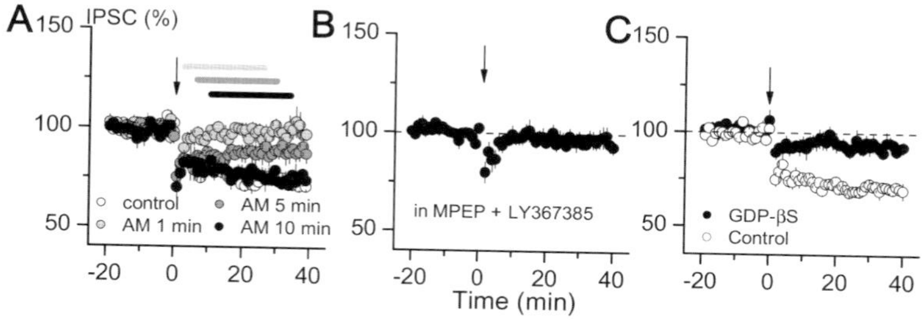

Fig. 2A–C LTD is mediated by postsynaptic mGluR-induced release of endocannabinoids that act on presynaptic CB_1 receptors. **A** High-frequency stimulation (HFS)-induced LTD of evoked IPSCs (*eIPSCs*) in hippocampal pyramidal neurons is abolished by the CB_1 antagonist AM251 (*AM*, 4 μM) when applied prior to, or within 10 min of LTD induction. The HFS-induced LTP is abolished by pre-application of (**B**) group I mGluR antagonists *MPEP* (4 μM) plus LY367385 (100 μM), and by (**C**) disrupting postsynaptic G protein coupling with intracellular *GDP-βS*. (Modified from Chevaleyre and Castillo 2003, by permission)

group I mGluR agonists (Robbe et al. 2002; Chevaleyre and Castillo 2003). Evidence in favour of presynaptic expression of LTD is similar to that described above for short-term depression (Choi and Lovinger 1997; Gerdeman et al. 2002; Marsicano et al. 2002; Robbe et al. 2002; Chevaleyre and Castillo 2003).

Endocannabinoid activation of CB_1 receptors is likely to be involved in the induction, but not in the maintenance of long-term synaptic plasticity because CB_1 antagonists impair long-term depression only when applied either prior to, or within 5 to 10 min of LTD induction (Chevaleyre and Castillo 2003; Ronesi et al. 2004) (Fig. 2A). The difference between short- and long-term synaptic effects of endocannabinoids may be due to the duration (or type) of endocannabinoid release, and sustained (for several minutes) activation of presynaptic CB_1 receptors (note DSI has a shorter time course). The mechanisms by which CB_1 activation induces long-term changes remain to be resolved, but the induction is unlikely to be solely due to CB_1 activation (Ronesi et al. 2004).

3 Production and Release of Endocannabinoids in Retrograde Endocannabinoid Signalling

While the identity of the endocannabinoid(s) involved in retrograde signalling remain to be determined directly, some clues have come from our understanding of the biosynthetic pathways involved in their production and degradation. Below we discuss the mechanisms underlying the postsynaptic production and release of endocannabinoids in relation to retrograde signalling. Depolarisation and mGluR-induced retrograde signalling are mediated by distinct Ca^{2+}-dependent and -independent intracellular cascades that provide some clues to the endocannabinoids involved in these processes. However, many issues remain to be resolved.

3.1 Ca^{2+}-Dependent and Ca^{2+}-Independent Endocannabinoid Production

Depolarisation-induced retrograde endocannabinoid signalling is dependent upon an increase in postsynaptic Ca^{2+}. First, uncaging postsynaptic Ca^{2+} from a photolabile chelator (with flash photolysis) in single neurons produces endocannabinoid-mediated DSI in the absence of depolarisation (Wilson and Nicoll 2001). Second, DSI and DSE are abolished by chelating postsynaptic Ca^{2+} with ethyleneglycoltetraacetic acid (EGTA), or ethylenedioxybis(o-phenylenenitrilo)tetraacetic acid (BAPTA) (Pitler and Alger 1992; Kreitzer and Regehr 2001b; Ohno-Shosaku et al. 2001; Kim et al. 2002; Trettel and Levine 2003).

The postsynaptic intracellular processes involved in mGluR-induced endocannabinoid-mediated retrograde signalling are diverse. Short-term mGluR-induced presynaptic inhibition does not require postsynaptic Ca^{2+} increases because postsynaptic BAPTA does not abolish group I mGluR-induced short-term inhibition of GABAergic and glutamatergic synaptic transmission in the hippocampus and nucleus accumbens, respectively (Maejima et al. 2001; Varma et al. 2001; Kim et al. 2002; Ohno-Shosaku et al. 2002). In contrast, both Ca^{2+}-dependent and Ca^{2+}-independent endocannabinoid-mediated long-term plasticity has been observed. Postsynaptic BAPTA abolishes stimulation-induced LTD of glutamatergic synaptic transmission in the striatum and nucleus accumbens (Gerdeman et al. 2002; Robbe et al. 2002; see also Huang et al. 2003), but not of GABAergic synaptic transmission in the hippocampus (Chevaleyre and Castillo 2003).

3.2 Depolarisation and Stimulation/mGluR-Induced Depression: Distinct Intracellular Cascades and Endocannabinoids?

A major question to be addressed is which endocannabinoids mediate retrograde endocannabinoid signalling? Both anandamide and 2-AG are potential candidates as retrograde signallers because their exogenous application produces presynaptic inhibition, albeit with reduced potency and efficacy compared to synthetic agonists (see Sect. 1.2). The Ca^{2+}-dependent depolarisation-induced short-term retrograde endocannabinoid signalling is not mediated by postsynaptic G protein-coupled processes and the PLC/DAG lipase cascade (Kim et al. 2002; Chevaleyre and Castillo 2003). It might be speculated that anandamide mediates depolarisation-induced short-term depression because anandamide formation/release can be induced by neuronal activation and Ca^{2+} influx. However, a role for 2-AG cannot be entirely ruled out because cellular Ca^{2+} influx also promotes its formation.

Stimulation-induced LTD within the hippocampus and nucleus accumbens is likely to be mediated by activation of group I mGluRs (see Sect. 2.4). Group I mGluRs are coupled via G_q-proteins/PLC to produce DAG and to increase intracellular Ca^{2+} via IP_3 and ryanodine sensitive stores (Conn and Pin 1997). The Ca^{2+}-independent endocannabinoid-mediated LTD within the hippocampus is mediated by the PLC/DAG cascade because it is abolished by PLC and DAG lipase inhibitors (Chevaleyre and Castillo 2003). On the other hand, the Ca^{2+}-dependent endocannabinoid-mediated LTD in the nucleus accumbens is mediated by the PLC mobilisation of intracellular Ca^{2+} because it is abolished thapsigargin and ryanodine (Robbe et al. 2002). It might be speculated that 2-AG mediates stimulation/mGluR-induced depression because 2-AG formation is activated via the PLC cascade (see Sect. 1.1) and high-frequency stimulation-induced hippocampal LTP is associated with an increase in 2-AG, but not anandamide production (Stella et al. 1997). However, a role for anandamide cannot be entirely ruled out because neuronal activation, Ca^{2+} influx and G protein-coupled receptor activation can also trigger anandamide (Piomelli 2003). In addition, striatal LTD is mimicked by loading the postsynaptic cell with either anandamide, or 2-AG (Gerdeman et al. 2002; Ronesi et al. 2004).

In summary, the endocannabinoids involved in the types of retrograde endocannabinoid signalling described to date are not fully understood. This will only be resolved by more fully comparing the intracellular cascades involved in the processes of endocannabinoid production and retrograde signalling. Ultimately, the issue of which endocannabinoid(s) mediates retrograde signalling will only be resolved by direct identification.

4
Spread of Retrograde Endocannabinoid Signalling

Under normal conditions, retrograde endocannabinoid signalling is spatially restricted in a manner likely to be determined by diffusion, uptake and enzymatic degradation.

4.1
Endocannabinoid Signalling Is Spatially Restricted

Inhibition of synaptic inputs onto a cell is dominated by endocannabinoids generated by that cell. There is little heterosynaptic 'spill-over' of endocannabinoids from adjacent cells because retrograde signalling is abolished by disrupting postsynaptic endocannabinoid production (see Sects. 2 and 3). In contrast, other studies have demonstrated that there is some heterosynaptic endocannabinoid 'spill-over'. Using paired patch-clamp recordings, Wilson and Nicoll (2001) and Kreitzer et al. (2002) have exploited the parallel alignment of principal hippocampal and cerebellar neurons to estimate the range of influence of the endocannabinoid(s) released by a single neuron. In these studies, depolarisation of one neuron produces heterosynaptic inhibition of GABAergic synaptic inputs onto an adjacent neuron. Heterosynaptic effects are only observed for inter-electrode distances less than 20–75 μm from the stimulated neuron (the separation between cells is likely to be much less because this distance includes the radius of both cells). The heterosynaptic presynaptic depression is likely to be solely due to diffusion because it is independent of whether the two neurons receive common presynaptic inputs.

4.2
Factors Influencing Endocannabinoid Spread

The spatial restriction of endocannabinoid signalling is likely to be influenced by a number of factors including temperature and the mode of retrograde endocannabinoid induction. Differences in temperature between studies are likely to affect uptake (via the AMT). Diffusion is spatially restricted by uptake/degradation at physiological temperatures (in comparison to room temperature), although this is likely to be influenced by synaptic geometry and diffusion barriers (Kreitzer et al. 2002). Another factor likely to influence the spatial extent of retrograde endocannabinoid signalling is the mode and strength of induction, both of which are likely to affect endocannabinoid production.

4.3 Inhibitors of Uptake and Metabolism

External application of the AMT transport inhibitor AM404 alone causes CB_1-mediated presynaptic inhibition (Wilson and Nicoll 2001; Robbe et al. 2002; Huang et al. 2003) and potentiates amphetamine-induced LTD in the amygdala, but has no effect on stimulation-induced LTD in the striatum (Gerdeman et al. 2002; Huang et al. 2003; Ronesi et al. 2004). Blockade of uptake might alter the spatial influence of endogenously released endocannabinoids. Indeed this seems to be the case for some forms of retrograde endocannabinoid signalling. External AM404 and VDM-11 restore Ca^{2+}-sensitive LTD in EGTA, or BAPTA loaded cells in both the striatum and amygdala. Thus, while endocannabinoid signalling is normally spatially restricted, blockade of uptake allows significant endocannabinoid 'spillover'. It might be noted that while the molecular components involved in endocannabinoid metabolism have been more fully characterised (see Sect. 1.1), their role in retrograde endocannabinoid signalling has not been examined. There is a strong possibility that transport or metabolism inhibitors will increase the range or intensity of retrograde endocannabinoid signalling at cannabinoid sensitive synapses throughout the nervous system. Experimental manipulation of potential enzymes involved in breakdown of endocannabinoids, e.g. FAAH and MAGL inhibitors, may also help to narrow the range of candidate endocannabinoids in future studies.

5 Other Endocannabinoid Targets: TRP Channels

In all of the above studies, endocannabinoid signalling is mediated largely via presynaptic activation of cannabinoid CB_1 receptors. However, endocannabinoids such as anandamide have some affinity for other receptors, such as the TRPV1 'noxious heat/capsaicin' receptor (Di Marzo et al. 2001). Besides their predicted primary afferent localisation, TRPV1 receptors are present within discrete brain regions (Sasamura et al. 1998; Mezey et al. 2000). Thus, in addition to CB_1-mediated presynaptic effects, exogenously applied anandamide acts via TRPV1 receptors to increase spontaneous glutamatergic synaptic transmission not only within the spinal and medullary dorsal horn (Morisset et al. 2001; Jennings et al. 2003) but also within brain regions such as the hippocampus and substantia nigra (Al-Hayani et al. 2001; Marinelli et al. 2003) (Fig. 3A).

Recently, it has been demonstrated that the TRPV1 antagonists capsazepine and iodoresiniferatoxin increase glutamatergic synaptic transmission within the substantia nigra (Marinelli et al. 2003) (Fig. 3B). This suggests that an endogenously released endocannabinoid, possibly anandamide, can produce both inhibitory CB_1 and excitatory TRPV1-mediated presynaptic effects. However, activation of presynaptic ligand-gated ion channels has complex effects on synaptic transmission (Engelman and MacDermott 2004). It also remains to be determined whether this presynaptic TRPV1 action is produced via a retrograde signalling process.

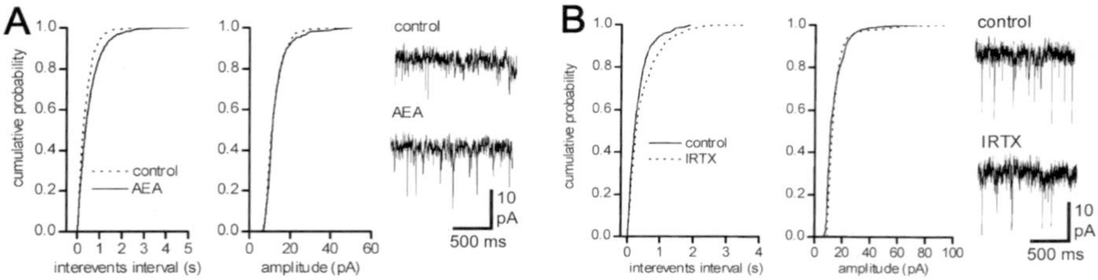

Fig. 3A, B Tonic release of anandamide presynaptically facilitates glutamatergic synaptic transmission in substantia nigra by activating TRPV1 receptors. **A** Anandamide (*AEA*, 30 μM) increases the rate (leftward shift in the cumulative distribution of the inter-event interval), but has no effect on the amplitude distribution of spontaneous EPSCs (sEPSCs). **B** The TRPV1 antagonist iodoresiniferatoxin (*IRTX*, 300 nM) decreases the rate (rightward shift in the cumulative distribution of the inter-event interval), but has no effect on the amplitude distribution of sEPSCs. *Insets* show raw traces of sEPSCs before (control) and during AEA and IRTX. (Modified from Marinelli et al. 2003, by permission)

6 What Is the Functional Significance of Retrograde Endocannabinoid Signalling?

It might be asked whether retrograde endocannabinoid signalling is physiologically relevant. It has been suggested that physiologically relevant action potential firing patterns do not induce DSI in the hippocampus (Hampson et al. 2003). While the depolarisation protocols used in DSI/DSE studies are unphysiological, it has been demonstrated that physiological action potential firing and electrical stimulation in the hippocampus is sufficient to produce DSI, which in turn has a significant disinhibitory effect on LTP (Ohno-Shosaku et al. 2001; Chevaleyre and Castillo 2003). In addition, synaptic activation of cerebellar parallel fibres at physiological rates induces retrograde endocannabinoid signalling of glutamatergic synaptic inputs onto Purkinje cells (Maejima et al. 2001; Brown et al. 2003). However, physiological relevance will only ultimately be determined by establishing retrograde endocannabinoid signalling in vivo using 'natural' stimuli.

7 Summary and Implications

There is compelling evidence to indicate that endocannabinoids are involved in retrograde signalling within a number of brain regions, including those involved in memory, motor control and reward/addiction. It is possible that retrograde endocannabinoid signalling occurs in other 'cannabinoid-sensitive' regions of the central nervous system and might prove to be a general phenomenon. Briefly, the events underlying retrograde endocannabinoid signalling can be described as follows (Fig. 4). Distinct stimuli, including postsynaptic depolarisation and activation of metabotropic receptors (group I mGluRs, mAChRs) appear to activate distinct intracellular postsynaptic cascades which induce de novo production and

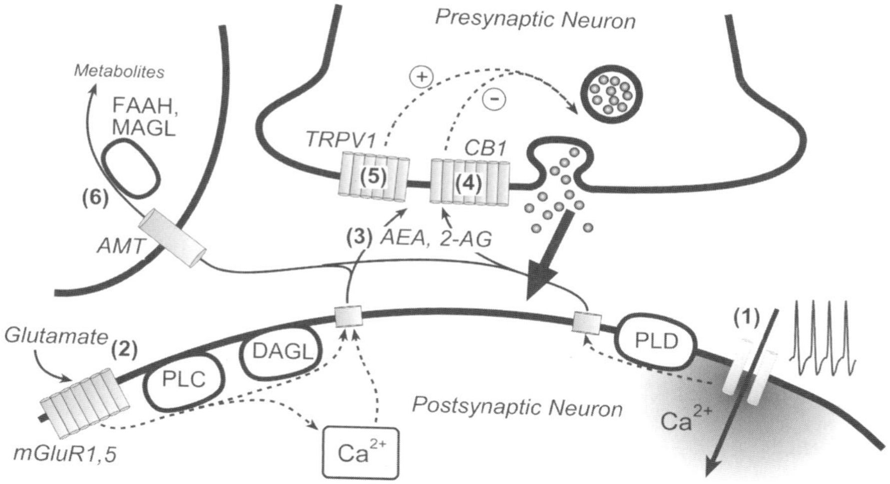

Fig. 4. Model depicting how endogenously released cannabinoids act as retrograde messengers within the brain. *(1)* Neuronal depolarisation induces an influx of Ca^{2+} via voltage dependent Ca^{2+} channels that stimulates phospholipase D (*PLD*). *(2)* Endogenously released glutamate, or synthetic agonist-induced activation of group I mGluRs, stimulates phospholipase C (*PLC*) to activate Ca^{2+} release from internal stores and to activate DAG-lipase in a Ca^{2+}-independent manner. *(3)* These postsynaptic events cause cleavage of membrane lipid precursors to induce de novo synthesis and release of endocannabinoids such as anandamide (*AEA*) and 2-arachidonoyl glycerol (*2-AG*) into the synaptic cleft. *(4)* These endocannabinoids activate cannabinoid CB_1 receptors located on presynaptic terminals of neurons which reduces release of neurotransmitters (such as GABA and glutamate) onto the postsynaptic neuron. *(5)* Endogenously released cannabinoids might also act via TRP ligand gated ion channels (e.g. *TPRV1*) to increase transmitter release. *(6)* Endocannabinoids are taken back up into neuronal and glial cells, possibly by a selective carrier-mediated transporter (*AMT*), and then degraded by enzymes such as fatty acid amide hydrolase (*FAAH*) and MAG-lipase (*MAGL*)

release of endocannabinoid(s). These endocannabinoids diffuse from the postsynaptic cell and act upon presynaptic cannabinoid CB_1 receptors to suppresses specific synaptic inputs impinging upon that cell, producing either short- and/or long-term changes. The action of endocannabinoid(s) is terminated by uptake and degradation. However, a number of issues remain to be resolved, such as the endocannabinoid(s) involved in these processes, the postsynaptic mechanisms of endocannabinoid production/release, other presynaptic endocannabinoid targets (e.g. TRP-like ion channels), and the precise mechanisms involved in uptake and degradation.

Increased understanding of the organisation of endocannabinoid signalling has raised hopes that therapeutic agents without the unwanted side-effects of cannabis could be developed. Natural and synthetic cannabinoids produce a range of pharmacological effects with potential therapeutic applications in the treatment of pain, migraine, muscle spasticity associated with multiple sclerosis, glaucoma, nausea and vomiting, and stimulation of appetite. Unfortunately, the CB_1 receptor is widely distributed throughout the brain and accounts for almost all of the effects of cannabis, including non-therapeutic effects on memory, cognition and motor coordination. The possibility that drugs selective for one cannabinoid receptor type could overcome the unwanted psychotropic actions of cannabis is therefore limited. Another therapeutic possibility relates to the finding that retrograde endocannabinoid signalling is restricted by uptake and degradation. There is a strong possibility that transport or metabolism inhibitors will increase the range/intensity of retrograde endocannabinoid signalling. Transport and degradation inhibitors may provide novel therapeutic agents in a manner analogous to the clinically useful inhibitors associated with other neurotransmitter systems (e.g. Kathuria et al. 2003). In doing so, endocannabinoid transport or metabolism inhibitors might intensify retrograde signalling at specific synapses, in contrast to the disruptive effects of globally acting cannabinoid CB_1 receptor agonists.

References

Al-Hayani A, Wease KN, Ross RA, Pertwee RG, Davies SN (2001) The endogenous cannabinoid anandamide activates vanilloid receptors in the rat hippocampal slice. Neuropharmacology 41:1000–1005

Alger BE (2002) Retrograde signaling in the regulation of synaptic transmission: focus on endocannabinoids. Prog Neurobiol 68:247–286

Bisogno T, Sepe N, Melck D, Maurelli S, De Petrocellis L, Di Marzo V (1997) Biosynthesis, release and degradation of the novel endogenous cannabimimetic metabolite 2-arachidonoylglycerol in mouse neuroblastoma cells. Biochem J 322:671–677

Brown SP, Brenowitz SD, Regehr WG (2003) Brief presynaptic bursts evoke synapse-specific retrograde inhibition mediated by endogenous cannabinoids. Nat Neurosci 6:1048–1057

Carlson G, Wang Y, Alger BE (2002) Endocannabinoids facilitate the induction of LTP in the hippocampus. Nat Neurosci 5:723–724

Chevaleyre V, Castillo PE (2003) Heterosynaptic LTD of hippocampal GABAergic synapses: a novel role of endocannabinoids in regulating excitability. Neuron 38:461–472

Choi S, Lovinger DM (1997) Decreased frequency but not amplitude of quantal synaptic responses associated with expression of corticostriatal long-term depression. J Neurosci 17:8613–8620

Conn PJ, Pin JP (1997) Pharmacology and functions of metabotropic glutamate receptors. Annu Rev Pharmacol Toxicol 37:205–237

Cravatt BF, Lichtman AH (2003) Fatty acid amide hydrolase: an emerging therapeutic target in the endocannabinoid system. Curr Opin Chem Biol 7:469–475

Di Marzo V, Bisogno T, De Petrocellis L (2001) Anandamide: some like it hot. Trends Pharmacol Sci 22:346–349

Diana MA, Marty A (2003) Characterization of depolarization-induced suppression of inhibition using paired interneuron-Purkinje cell recordings. J Neurosci 23:5906–5918

Diana MA, Levenes C, Mackie K, Marty A (2002) Short-term retrograde inhibition of GABAergic synaptic currents in rat Purkinje cells is mediated by endogenous cannabinoids. J Neurosci 22:200–208

Doherty J, Dingledine R (2003) Functional interactions between cannabinoid and metabotropic glutamate receptors in the central nervous system. Curr Opin Pharmacol 3:46–53

Engelman HS, MacDermott AB (2004) Presynaptic ionotropic receptors and control of transmitter release. Nat Rev Neurosci 5:135–145

Freund TF, Katona I, Piomelli D (2003) Role of endogenous cannabinoids in synaptic signaling. Physiol Rev 83:1017–1066

Gerdeman GL, Ronesi J, Lovinger DM (2002) Postsynaptic endocannabinoid release is critical to long-term depression in the striatum. Nat Neurosci 5:446–451

Glaser ST, Abumrad NA, Fatade F, Kaczocha M, Studholme KM, Deutsch DG (2003) Evidence against the presence of an anandamide transporter. Proc Natl Acad Sci USA 100:4269–4274

Hampson RE, Zhuang SY, Weiner JL, Deadwyler SA (2003) Functional significance of cannabinoid-mediated, depolarization-induced suppression of inhibition (DSI) in the hippocampus. J Neurophysiol 90:55–64

Hillard CJ, Jarrahian A (2003) Cellular accumulation of anandamide: consensus and controversy. Br J Pharmacol 140:802–808

Hoffman AF, Oz M, Caulder T, Lupica CR (2003a) Functional tolerance and blockade of long-term depression at synapses in the nucleus accumbens after chronic cannabinoid exposure. J Neurosci 23:4815–4820

Hoffman AF, Riegel AC, Lupica CR (2003b) Functional localization of cannabinoid receptors and endogenous cannabinoid production in distinct neuron populations of the hippocampus. Eur J Neurosci 18:524–534

Huang YC, Wang SJ, Chiou LC, Gean PW (2003) Mediation of amphetamine-induced long-term depression of synaptic transmission by CB1 cannabinoid receptors in the rat amygdala. J Neurosci 23:10311–10320

Jennings EA, Vaughan CW, Roberts LA, Christie MJ (2003) The actions of anandamide on rat superficial medullary dorsal horn neurons in vitro. J Physiol (Lond) 548:121–129

Kathuria S, Gaetani S, Fegley D, Valino F, Duranti A, Tontini A, Mor M, Tarzia G, La Rana G, Calignano A, Giustino A, Tattoli M, Palmery M, Cuomo V, Piomelli D (2003) Modulation of anxiety through blockade of anandamide hydrolysis. Nat Med 9:76–81

Kim J, Isokawa M, Ledent C, Alger BE (2002) Activation of muscarinic acetylcholine receptors enhances the release of endogenous cannabinoids in the hippocampus. J Neurosci 22:10182–10191

Kreitzer AC, Regehr WG (2001a) Cerebellar depolarization-induced suppression of inhibition is mediated by endogenous cannabinoids. J Neurosci 21:RC174

Kreitzer AC, Regehr WG (2001b) Retrograde inhibition of presynaptic calcium influx by endogenous cannabinoids at excitatory synapses onto Purkinje cells. Neuron 29:717–727

Kreitzer AC, Carter AG, Regehr WG (2002) Inhibition of interneuron firing extends the spread of endocannabinoid signaling in the cerebellum. Neuron 34:787–796

Levenes C, Daniel H, Soubrie P, Crepel F (1998) Cannabinoids decrease excitatory synaptic transmission and impair long-term depression in rat cerebellar Purkinje cells. J Physiol (Lond) 510:867–879

Llano I, Leresche N, Marty A (1991) Calcium entry increases the sensitivity of cerebellar Purkinje cells to applied GABA and decreases inhibitory synaptic currents. Neuron 6:565–574

Maejima T, Hashimoto K, Yoshida T, Aiba A, Kano M (2001) Presynaptic inhibition caused by retrograde signal from metabotropic glutamate to cannabinoid receptors. Neuron 31:463–475

Marinelli S, Di Marzo V, Berretta N, Matias I, Maccarrone M, Bernardi G, Mercuri NB (2003) Presynaptic facilitation of glutamatergic synapses to dopaminergic neurons of the rat substantia nigra by endogenous stimulation of vanilloid receptors. J Neurosci 23:3136–3144

Marsicano G, Wotjak CT, Azad SC, Bisogno T, Rammes G, Cascio MG, Hermann H, Tang JR, Hofmann C, Zieglgansberger W, Di Marzo V, Lutz B (2002) The endogenous cannabinoid system controls extinction of aversive memories. Nature 418:530–534

Melis M, Pistis M, Perra S, Muntoni AL, Pillolla G, Gessa GL (2004) Endocannabinoids mediate presynaptic inhibition of glutamatergic transmission in rat ventral tegmental area dopamine neurons through activation of CB1 receptors. J Neurosci 24:53–62

Mezey E, Toth ZE, Cortright DN, Arzubi MK, Krause JE, Elde R, Guo A, Blumberg PM, Szallasi A (2000) Distribution of mRNA for vanilloid receptor subtype 1 (VR1), and VR1-like immunoreactivity, in the central nervous system of the rat and human. Proc Natl Acad Sci USA 97:3655–3660

Morisset V, Ahluwalia J, Nagy I, Urban L (2001) Possible mechanisms of cannabinoid-induced antinociception in the spinal cord. Eur J Pharmacol 429:93–100
Nowicky AV, Teyler TJ, Vardaris RM (1987) The modulation of long-term potentiation by delta-9-tetrahydrocannabinol in the rat hippocampus, in vitro. Brain Res Bull 19:663–672
Ohno-Shosaku T, Maejima T, Kano M (2001) Endogenous cannabinoids mediate retrograde signals from depolarized postsynaptic neurons to presynaptic terminals. Neuron 29:729–738
Ohno-Shosaku T, Shosaku J, Tsubokawa H, Kano M (2002) Cooperative endocannabinoid production by neuronal depolarization and group I metabotropic glutamate receptor activation. Eur J Neurosci 15:953–961
Okamoto Y, Morishita J, Tsuboi K, Tonai T, Ueda N (2004) Molecular characterization of a phospholipase D generating anandamide and its congeners. J Biol Chem 279:5298–5305
Petrocellis LD, Cascio MG, Marzo VD (2004) The endocannabinoid system: a general view and latest additions. Br J Pharmacol 141:765–774
Piomelli D (2003) The molecular logic of endocannabinoid signalling. Nat Rev Neurosci 4:873–884
Pitler TA, Alger BE (1992) Postsynaptic spike firing reduces synaptic GABAA responses in hippocampal pyramidal cells. J Neurosci 12:4122–4132
Robbe D, Kopf M, Remaury A, Bockaert J, Manzoni OJ (2002) Endogenous cannabinoids mediate long-term synaptic depression in the nucleus accumbens. Proc Natl Acad Sci USA 99:8384–8388
Ronesi J, Gerdeman GL, Lovinger DM (2004) Disruption of endocannabinoid release and striatal long-term depression by postsynaptic blockade of endocannabinoid membrane transport. J Neurosci 24:1673–1679
Rouach N, Nicoll RA (2003) Endocannabinoids contribute to short-term but not long-term mGluR-induced depression in the hippocampus. Eur J Neurosci 18:1017–1020
Sasamura T, Sasaki M, Tohda C, Kuraishi Y (1998) Existence of capsaicin-sensitive glutamatergic terminals in rat hypothalamus. Neuroreport 9:2045–2048
Schlicker E, Kathmann M (2001) Modulation of transmitter release via presynaptic cannabinoid receptors. Trends Pharmacol Sci 22:565–572
Schmid HHO (2000) Pathways and mechanisms of N-acylethanolamine biosynthesis: can anandamide be generated selectively? Chem Phys Lipids 108:71–87
Shen M, Piser TM, Seybold VS, Thayer SA (1996) Cannabinoid receptor agonists inhibit glutamatergic synaptic transmission in rat hippocampal cultures. J Neurosci 16:4322–4334
Sjostrom PJ, Turrigiano GG, Nelson SB (2003) Neocortical LTD via coincident activation of presynaptic NMDA and cannabinoid receptors. Neuron 39:641–654
Stella N, Schweitzer P, Piomelli D (1997) A second endogenous cannabinoid that modulates long-term potentiation. Nature 388:773–778
Sugiura T, Kobayashi Y, Oka S, Waku K (2002) Biosynthesis and degradation of anandamide and 2-arachidonoylglycerol and their possible physiological significance. Prostaglandins Leukot Essent Fatty Acids 66:173–192
Terranova JP, Michaud JC, Le Fur G, Soubrie P (1995) Inhibition of long-term potentiation in rat hippocampal slices by anandamide and WIN55212-2: reversal by SR141716 A, a selective antagonist of CB1 cannabinoid receptors. Naunyn Schmiedebergs Arch Pharmacol 352:576–579
Trettel J, Levine ES (2003) Endocannabinoids mediate rapid retrograde signaling at interneuron→pyramidal neuron synapses of the neocortex. J Neurophysiol 89:2334–2338
Varma N, Carlson GC, Ledent C, Alger BE (2001) Metabotropic glutamate receptors drive the endocannabinoid system in hippocampus. J Neurosci 21:RC188
Vaughan CW, Connor M, Bagley EE, Christie MJ (2000) Actions of cannabinoids on membrane properties and synaptic transmission in rat periaqueductal gray neurons in vitro. Mol Pharmacol 57:288–295

Wilson RI, Nicoll RA (2001) Endogenous cannabinoids mediate retrograde signalling at hippocampal synapses. Nature 410:588–592

Wilson RI, Kunos G, Nicoll RA (2001) Presynaptic specificity of endocannabinoid signaling in the hippocampus. Neuron 31:453–462

Yanovsky Y, Mades S, Misgeld U (2003) Retrograde signaling changes the venue of postsynaptic inhibition in rat substantia nigra. Neuroscience 122:317–328

Yoshida T, Hashimoto K, Zimmer A, Maejima T, Araishi K, Kano M (2002) The cannabinoid CB1 receptor mediates retrograde signals for depolarization-induced suppression of inhibition in cerebellar Purkinje cells. J Neurosci 22:1690–1697

HEP (2005) 168:385–423

Effects on the Immune System

G.A. Cabral (✉) · A. Staab

Department of Microbiology and Immunology, Virginia Commonwealth University, School of Medicine, 1101 E. Marshall St., Richmond VA, 23298-0678, USA
gacabral@hsc.vcu.edu

Abstract Marijuana and other exogenous cannabinoids alter immune function and decrease host resistance to microbial infections in experimental animal models and in vitro. Two modes of action by which Δ^9-tetrahydrocannabinol (THC) and other cannabinoids affect immune responses have been proposed. First, cannabinoids may signal through the cannabinoid receptors CB_1 and CB_2. Second, at sites of direct exposure to high concentrations of cannabinoids, such as the lung, membrane perturbation may be involved. In addition, endogenous cannabinoids or endocannabinoids have been identified and have been proposed as native modulators of immune functions through cannabinoid receptors. Exogenously introduced cannabinoids may disturb this homoeostatic immune balance. A mode by which cannabinoids may affect immune responses and host resistance may be by perturbing the balance of T helper $(Th)_1$ pro-inflammatory versus Th_2 anti-inflammatory cytokines. While marijuana and various cannabinoids have been documented to alter immune functions in vitro and in experimental animals, no controlled longitudinal epidemiological studies have yet definitively correlated immunosuppressive effects with increased incidence of infections or immune disorders in humans. However, cannabinoids by virtue of their immunomodulatory properties have the potential to serve as therapeutic agents for ablation of untoward immune responses.

Keywords B lymphocytes · Cannabinoid receptors · Cannabinoids · Cytokines · Endocannabinoids · Human immunodeficiency virus · Immunity · Infections · Macrophages · Mast cells · Microglia · Natural killer cells · THC · Therapeutics · T lymphocytes

1 Introduction

Marijuana, or *Cannabis sativa*, has been valued for its medicinal as well as its psychotropic properties dating back to ancient times. However, reports from as early as the 1960s have indicated that marijuana and select components also could compromise human health, including the ability to resist infections. Included among these components is a class of compounds collectively known as cannabinoids. At least 60 have been identified. These include cannabidiol (CBD), cannabinol (CBN), cannabigerol (CBG), and Δ^9-tetrahydrocannabinol (THC), the major psychoactive ingredient in marijuana that has been implicated as the major immunomodulatory substance.

Early studies on the effects of marijuana on the immune system attributed these to the ability of THC to perturb cellular membranes since it was highly lipophilic. However, it was soon recognized that THC also exhibited specificity of action at the physiological and pharmacological levels as well as in distribution in organs and cells. Studies performed on various rodents demonstrated that THC produced a characteristic tetrad of behavioral effects that consists of catalepsy, antinociception, hypothermia, and hypomobility (see Wiley and Martin 2003). These centrally mediated effects could be elicited following intravenous, intrathecal, and intraperitoneal administration. Furthermore, use of radiolabeled THC in hybridization studies revealed a distribution of binding in rodent brain slices that was consistent with that attributed to areas in the brain that correlated with specified behavioral activities.

A series of experiments that drove the field of cannabinoid pharmacology forward as it relates to the recognition of the existence of a functionally relevant cannabinoid receptor—and which impacted studies on effects of cannabinoids on immunity—was conducted by Howlett and coworkers. Using a pharmacological approach, it was demonstrated that THC, Δ^8-tetrahydrocannabinol, levonantradol, and desacetyllevonantradol inhibited adenylate cyclase activity in plasma membranes of neuroblastoma cells (Howlett and Fleming 1984). This inhibition was not blocked by atropine, yohimbine, or naloxone, suggesting that muscarinic, α-2-adrenergic and opiate receptors were not required for the response. Furthermore, the inhibition of adenylate cyclase appeared to be specific for cannabinoids that were psychoactive, since CBD and CBN produced minimal effects. In addition, the inhibition of adenylate cyclase activity was stereoselective, since dextronantradol as compared to its stereoisomer levonantradol did not produce the response. Finally, inhibition was observed as concentration-dependent over a nanomolar range for both THC and its synthetic analog, desacetyllevonantradol. In a subsequent set of experiments, it was shown that the inhibitory effect of these psychoactive compounds was related to the ability of adenylate cyclase to be regulated by divalent cations and guanine nucleotides (Howlett 1985). Howlett et al. (1986) also demonstrated that pertussis toxin treatment abolished the cannabimimetic inhibition of adenylate cyclase activity, but only for intact neuroblastoma cells, neuroblastoma/glioma hybrid cells, or their derivative membranes. These results were consistent with the existence of a "cannabinoid" receptor, since receptor-mediated inhibition of adenylate cyclase requires the presence of a guanine nucleotide-binding protein complex, G_i, which can be functionally inactivated as a result of an adenosine diphosphate (ADP)-ribosylation modification catalyzed by pertussis toxin. Devane et al. (1988) used a tritium-labeled biologically active synthetic bicyclic cannabinoid (CP 55,940) to identify and characterize specific ligand binding in rat brain. Collectively, the data fulfilled the criteria for the existence of a high-affinity and stereoselective, pharmacologically distinct cannabinoid receptor in brain tissue.

Matsuda and colleagues (Matsuda et al. 1990) reported on the cloning and expression of a complementary DNA (cDNA) from rat brain that encoded a G protein-coupled receptor that exhibited all of the properties described by Howlett and colleagues. Its messenger RNA (mRNA) was found in cell lines and regions of the brain that contained cannabinoid receptors based on radioligand binding analysis. Thus, by the early 1990s, a framework of rigorous cellular, molecular, pharmacological, and physiological methodology had been established that allowed for the systematic characterization and definition of a neuronal or central cannabinoid receptor, currently referred to as the CB_1 receptor. It was within this framework of definable criteria for assessing effects of cannabinoids that a second cannabinoid receptor was discovered that was associated in distribution and functional relevance with the immune system

In 1992, Kaminski and colleagues (Kaminski et al. 1992) demonstrated through an equilibrium binding assay that membranes from mouse spleen had specific binding sites for cannabinoids. In addition, using specific primers for the cannabinoid receptor identified in brain, they amplified from splenic RNA using RNA transcriptase-polymerase chain reaction (RT-PCR), an 854-kb product that hybridized with brain cannabinoid receptor cDNA. These studies demonstrated that a cannabinoid receptor identical to, or that shared homology with that in brain, was present in cells of the immune system. In 1993, Munro and colleagues (Munro et al. 1993) reported on the molecular characterization of a peripheral cannabinoid receptor from a human promyelocytic leukemia cell line HL-60 that was distinctive from the CB_1 receptor. The cloned receptor was not expressed in the brain but was localized in macrophages in the marginal zone of the spleen. Subsequent studies have confirmed and extended these observations and have indicated that gene products for this receptor, designated CB_2, are localized primarily in cells of the immune system, including B lymphocytes, macrophages and monocytes, natural killer (NK) cells, and T lymphocytes (Galiègue et al. 1995; Bouaboula et al. 1993). The CB_2 receptor also is expressed in vitro by microglia (Carlisle et al. 2002; Sinha et al. 1998), a population of resident macrophages of the brain and eye.

Since the discovery of the CB_1 and CB_2 receptors, major advances in cannabinoid pharmacology and physiology have been made and novel insights into the functional role of cannabinoids and their cognate receptors in the immune system have been obtained. The discovery of receptors for exogenous cannabinoids implied the existence of an endogenous system of receptors and cognate ligands, or endocannabinoids. The role of these putative endogenous cannabinoids and their receptors in immune cell function has yet to be fully explored, although it has been postulated that they may be involved in homoeostatic regulation of immune responses. Studies to assess their functional relevance may yield novel insights into mechanisms and modalities of action by which exogenous cannabinoids such as THC may perturb the native functionality of these endogenous systems. In addition, cannabinoid receptor type-specific agonists and antagonists have been developed, as have synthetic compounds of varying affinity for the CB_1 and CB_2 receptors, that are applicable to assessment of structure–activity relationships in immune functional responses. The availability of these pharmacological tools has afforded novel opportunities for evaluating cannabinoid analogs as therapeutic agents for management of select untoward or aberrant immune responses. Furthermore, availability of cannabinoid receptor type-specific agonists and antagonists, and of chemically engineered cannabinoid compounds, may allow for the identification, characterization, and isolation of additional cannabinoid receptors that may be operative in activities linked to immune function.

2 Cannabinoids and the Immune System

2.1 Early Studies

Early in vitro and in vivo studies indicated that marijuana as well as cannabinoids derivative of this plant, particularly THC, had immunosuppressive properties. These compounds were found to exert a wide range of effects on a variety of immune functions from a diverse array of immune cell types. Such effects were observed for immune cells derivative of various rodents and for cell lines exhibiting phenotypic and functional attributes that mimicked those of native immune cells, including those of human origin. Because cannabinoids were recognized as highly lipophilic, their effects on humans and animals originally were considered non-specific and were attributed to perturbing effects on plasma membranes and other lipid-containing structures of the cell (Wing et al. 1985; Friedman et al. 1991). Light and electron microscopy studies performed by Raz and Goldman (1976) demonstrated extensive vacuolation in murine peritoneal macrophages exposed to either THC or CBD. It was indicated that these cytoplasmic alterations had been observed also in alveolar macrophages of hashish smokers. Similar observations, related to disruption and perturbation of cell membranes following exposure to relatively high levels of THC, have been made by Meyers and Heath (1979) and Cabral et al. (1987a). Cabral and Fischer-Stenger (1994) postulated that THC at high concentrations (i.e., 10^{-5} M), as a consequence of membrane perturbation, has a generalized effect on immune cell functions. THC was shown in vitro to have a differential inhibitory effect on murine P388D_1 macrophage-like cell inducible protein expression in response to priming and activating signals such as interferon (IFN)-γ and bacterial lipopolysaccharide (LPS). It was postulated that interaction of the highly lipophilic cannabinoid with cell membranes resulted in alterations in membrane fluidity and selective permeability and the attendant increase in intracellular sodium resulted in shutdown of protein synthesis. A similar process has been proposed for virus infection of cells in which insertion of virus-specified glycoproteins into the cell surface membrane alters selective permeability and affects a shutdown of cell-specified macromolecular synthesis (Carrasco and Smith 1976; Garry et al. 1979, 1982). Nahas et al. (1977) reported that CBD and CBN exerted greater inhibitory effects on phytohemagglutinin (PHA)-induced human lymphocyte transformation as measured by ^{3}H-thymidine incorporation when compared to THC. These effects were exerted at drug concentrations ranging from 10^{-4} M to 10^{-6} M, which are relatively high.

With the discovery of cannabinoid receptors in the 1990s, novel insights were obtained regarding modalities by which cannabinoids affect immune functions. Indeed, a general picture emerged that cannabinoids exhibited a duality of action. Effects on immune functions exerted by cannabinoids at concentrations below the micromolar level may be through the activation of cannabinoid receptors. However, cannabinoids also can exert non-receptor-mediated effects (Felder et al. 1992; Derocq et al. 1998). Such non-receptor-mediated effects may be exerted at high concentrations (i.e., micromolar levels). Such levels are achievable in the lung as a result of exposure to marijuana smoke or through therapeutic application of purified cannabinoids.

2.2
Effects on the Immune System Using In Vivo Models

Experimental animal models, using guinea pigs and mice, have been used for nearly a century to document effects of various toxic and infectious agents on host resistance. These in vivo models have offered unique advantages for assessment of effects of drugs on infection and immunity due to their well-defined immune systems. Furthermore, use of animal models has allowed for the definition of factors of host resistance that are targeted by drugs under stringently controlled conditions. As a result, acquisition of statistically significant data with minimal confounding variables has been possible, a condition that is difficult to attain for human populations as a result of potential environmental toxic exposures and multiple drug use. The collective data that have been obtained through studies involving experimental animals indicate that cannabinoids alter humoral and cellular immunity and can compromise resistance to a variety of infectious agents including bacteria, protozoa, and viruses. Rosenkrantz et al. (1975) indicated that THC suppressed the primary immune response in rats. Klykken et al. (1977) extended these studies and demonstrated that 8,9-epoxyhexahydrocannabinol (EHHC) inhibited both humoral and cell-mediated immunity in BDG1 mice administered *Corynebacterium parvum* (*C. parvum*). Zimmerman et al. (1977) compared the effects of THC, CBD, and CBN on the immune response of immature mice. THC elicited a dose-dependent depression of immune responsiveness. However, the impairment of humoral immunity was specific to THC in that CBD or CBN did not have an effect. These results served as an indicator that the effects of cannabinoids on immune function were selective in that a psychotropic cannabinoid exerted an immunomodulatory effect, while a non-psychotropic cannabinoid at comparable doses did not. Baczynsky et al. (1983a) examined the effects of THC, CBD, and CBN on the primary humoral immune response, the secondary humoral immune response, and the memory aspect of humoral immunity following sheep red blood cell immunization. Mice treated with THC during the primary immunization period exhibited a suppression of the primary humoral immune response while those treated during the secondary immunization period showed no measurable suppression of the secondary humoral immune response. However, CBD and CBN had no effect.

Studies using animal models have suggested also that cannabinoid immunological tolerance differs from pharmacological tolerance. Luthra et al. (1980) administered THC orally to Fischer rats of both sexes at doses that produced pharmacological tolerance. In order to assess for immunological tolerance, the primary immune response was evaluated by determining splenic antibody-forming cells, hemagglutinin titers, and/or hemolysin titers. Modalities of action that were operative in the establishment of pharmacological tolerance appeared to differ from those associated with tolerance at the immunological level. The effects of cannabinoids on immune function appear also to differ with age. Pross et al. (1990) examined the suppression of murine lymphoid cell blastogenesis by marijuana components in adult versus juvenile mice of various ages. Differences in susceptibility to THC- and 11-OH-THC-induced suppression were observed for in vitro proliferative responses of murine lymphoid cells to the mitogens concanavalin A (ConA) and PHA. Thymus cells from adults were suppressed more readily than splenocytes, while splenocytes from mice under 2 weeks of age were suppressed much more readily than those from older mice. Snella et al. (1995) evaluated the relationship of aging and THC-mediated immunomodulation of murine lymphoid cells. Cells from 2- and 18-month-old mice, in contrast to those from adult mice, were resistant to THC-mediated proliferation when stimulated by their CD3 receptor. The THC-induced enhancement appeared to be related in part to levels of interleukin (IL)-2 since addition of this cytokine modified the THC-induced up-regulation of cell proliferation. In contrast, cells from 18-month-old mice remained resistant to modulation by THC. It was concluded that the difference in immune responsiveness to THC related to the age of mice correlated, at least in part, to IL-2 levels in the 2-week-old and young adult mice. Ramarathinam et al. (1997) demonstrated that lymphoid cells from young and old mice exhibited different immunological potential in terms of ability to produce cytokines following stimulation with either ConA or anti-CD3 antibody. Levels of the anti-inflammatory cytokines IL-4 and IL-10 were up-regulated in spleen cell cultures from the older animals. Furthermore, in vivo administration of THC resulted in an up-regulation of the proliferative response of lymphoid cells from young adult mice. Such enhancement was not evident for cells from older animals.

2.3 Effects on the Immune System Using In Vitro Models

2.3.1 Effects on Mixed Cell Populations

Early experiments involved the use of mixed cell populations, since these more closely replicated the in vivo condition of a complex mixture of distinctive cell types cross-talking through soluble mediators as well as interacting with each other through cell-contact-dependent modalities. Furthermore, the use of mixed cell populations lent itself to the application of depletion and reconstitution studies for the definition of specific cell subpopulations affected by cannabinoids. Lefkowitz et al. (1978) examined the effect of THC on the in vitro sensitization of mouse splenic lymphocytes with sheep erythrocytes (SRBC) using a plaque-forming cell (PFC) assay. Splenic lymphocytes from mice injected with THC showed a depressed immunological response when compared with those from control animals. A similar alteration in the immunological response was obtained when THC was added directly to the culture medium as demonstrated by a reduction in the number of plaque-forming centers. Baczynsky et al. (1983b) reported that cannabinoids acted differentially in their suppressant effect on immunocytes in vitro. They examined the effects of THC, CBD, and CBN on the primary-like immune response in cultures of mouse spleen cells. THC and CBD, but not CBN, depressed the primary-like immune response of stimulated mouse splenocytes. It was noted that THC exerted its maximal suppression of immune responses when administered antecedent to antigenic stimulation. Pross et al. (1992) found that, when the mitogens ConA or PHA were used to stimulate THC-treated splenocytes, a down-regulation of lymphocyte proliferation occurred. This down-regulation was accompanied by lower T cell numbers in general and Ly2-positive cells specifically. In contrast, when splenocytes were stimulated directly with anti-CD3 antibody, low concentrations of THC enhanced lymphocyte proliferation that was accompanied by greater numbers of T cells in general and Ly2 cells specifically. In a subsequent study, Pross et al. (1992) indicated that THC suppressed IL-2 and IL-2 receptor expression. The T cell mitogen, anti-CD3, produced the opposite effect when combined with THC in that it increased proliferation and the IL-2 response. This modulation pattern for IL-2 by THC after stimulation of murine spleen cells with ConA, PHA, or anti-CD3 antibody was reproduced and extended by Nakano et al. (1992). It was shown that the THC-related modulation of IL-2 activity corresponded not only to changes in blastogenic activity, but also to variations in numbers of Tac-positive cells. Collectively, these studies indicated that THC could exert differential effects on spleen cell populations dependent upon the stimulators used.

2.3.2 Effects on Mononuclear Cells, Macrophages, and Macrophage-Like Cells

THC and other cannabinoids have been shown to suppress macrophage functions such as phagocytosis, bactericidal activity, and spreading (Friedman et al. 1991; Klein and Friedman 1990). Sacerdote et al. (2000) reported that in vivo and in vitro treatment with CP 55,940 decreased the in vitro migration of macrophages in the rat and that this effect involved both CB_1 and CB_2 receptors. THC also has been reported to alter the gene expression, processing, and secretion of an array of macrophage pro-inflammatory and anti-inflammatory factors. Zheng et al. (1992) indicated that THC caused a significant decrease in tumor necrosis factor (TNF)-α production by BALB/c mouse peritoneal macrophages in response to LPS and IFN-γ. A similar set of responses was obtained for human peripheral blood lymphocyte (PBL) adherent cells treated with THC. Fisher-Stenger et al. (1993) also examined the effects of THC on TNF-α production by murine RAW264.7 macrophage-like cells and reported that it affected TNF-α production by altering its conversion from a 26-kDa presecretory form to the 17-kDa secretory product. THC also has been reported to alter the expression of other cytokines. Klein and Friedman (1990) indicated that IL-1 activity increased in supernatants of mouse macrophage cultures treated with LPS and THC. Studies to assess for its intracellular fate in relation to drug treatment indicated that the THC-induced higher levels in supernatants were due to an increase, and prolongation, of release of the promature IL-1α and mature IL-1β forms. The processing and release of IL-1 in macrophages appeared to be due partly to increased activity of the IL-1 converting enzyme (i.e., caspase), since THC had been shown to induce an augmentation in caspase activity and other markers of apoptosis (Zhu et al. 1998).

Burnette-Curley and Cabral (1995) reported that THC was able to inhibit macrophage-like cell contact-dependent cytolysis of tumor cells and that this inhibition was effected by selective targeting of TNF-dependent pathways versus L-arginine-dependent reactive nitrogen intermediates. In addition, the effect of the enantiomeric pairs (−)CP 55,940/(+)CP 56,667 or HU-210/HU-211 on macrophage cell contact-dependent killing was assessed. Inhibition of macrophage tumoricidal activity against TNF-sensitive murine L929 cells was effected by both isomers of THC analogs. Coffey et al. (1996) confirmed and extended these studies using mouse peritoneal macrophages. They indicated that an early step in NO production, such as NO synthase (NOS) gene transcription or NOS synthesis, rather than NOS activity was affected by THC. A structure–activity order of effectiveness in inhibition was noted for various THC analogues with potency being highest for Δ^8-THC and decreasing in order for THC > CBD ≥ 11-OH-THC > CBN. Furthermore, THC attenuated the cyclic adenosine monophosphate (cAMP) response in the macrophage cultures. The investigators concluded that inhibition of NO was mediated by a process that depended partly on a stereoselective cannabinoid receptor/cAMP pathway and partly on a nonselective molecular process. Jeon et al. (1996), using the murine RAW264.7 macrophage cell line, demonstrated that THC inhibited NOS transcription factors such as nuclear factor (NF)-κB/RelA suggesting a mode by which this cannabinoid affected NO production. Waksman et al. (1999) reported that CP 55,940 mediated inhibition of inducible NOS (iNOS) produced by neonatal rat brain cortical microglia in a mode that was linked functionally to the CB_1 receptor. On the other hand, Stephano et al. (2000) indicated that the synthetic cannabinoid WIN 55,212-2 had an opposite effect on constitutive NO and increased its release from human monocytes and vascular tissues through the CB_1 receptor. They reported also that the endocannabinoid 2-arachidonoylglycerol (2-AG) stimulated NO release from human monocytes and vascular tissues and immunocytes of the invertebrate *Mytilus edulis* (Stephano et al. 2000). This effect was mediated through the CB_1 receptor in human cells and through an apparent "cannabinoid" receptor in the invertebrate immunocytes. Furthermore, in both the monocytes and the immunocytes, NO release elicited in response to 2-AG was blocked by a CB_1, but not by a CB_2, antagonist. In contrast, Gross et al. (2000) reported that inhibition of NO production by WIN 55,212-2, but not palmitoylethanolamide, was attenuated significantly by the CB_2 antagonist SR144528. Their results suggested that inhibition of RAW264.7 macrophage-like cell LPS-induced iNOS expression by WIN 55,212-2 was mediated by the CB_2 receptor.

Cannabinoids have been shown to alter macrophage functions in addition to those related to cytokine and NO production. It has been reported that THC affects macrophage processing of antigens that is necessary for the activation of $CD4^+$ T lymphocytes (McCoy et al. 1995). The T cell response to hen egg lysozyme was dramatically reduced after pretreatment of a macrophage hybridoma with THC. In contrast, THC exposure did not alter the capacity of the macrophage hybridoma to process chicken ovalbumin and augmented their presenting cell function for a pigeon cytochrome *c* response. The level of T cell activation with peptides of lysozyme and cytochrome *c*, which do not require processing, was inhibited only at the highest concentrations of THC, suggesting that THC mainly affected antigen processing. Peritoneal macrophages exposed to THC during an antigen pulse and fixed with paraformaldehyde showed similar effects on the subsequent T cell responses to lysozyme and cytochrome *c* in the absence of THC, arguing against a possible influence of THC on the T cells. The investigators concluded that THC differentially modulated the capacity of macrophages to process antigens that is necessary for the activation of $CD4^+$ T cells. Follow-up studies on effects of THC on processing of intact lysozyme by macrophages provided evidence for a CB_2 receptor participation (McCoy et al. 1999). These observations were confirmed by Buckley et al. (2000) using CB_2 receptor knockout mice.

2.3.3
Effects on B Lymphocytes

B lymphocytes have been reported to express relatively high levels of the CB_2 receptor (Carayon et al. 1998; Galiègue et al. 1995; Lynn and Herkenham 1994). Thus, it is not surprising that cannabinoid agonists should exert major effects on their functional activities. Klein et al. (1985) noted that addition of THC to mouse splenocyte cultures suppressed not only T lymphocyte proliferation in response to the mitogens ConA and PHA, but also that of B lymphocytes induced by LPS, a B cell mitogen. The hydroxylated metabolite of THC, 11-hydroxy-THC, was observed to be much less potent in this inhibition.

Additional reports have confirmed that cannabinoids suppress the antibody response of humans and animals (Friedman et al. 1991; Klein et al. 1998a). Kaminski et al. (1994) reported that suppression of the humoral immune response by cannabinoids was mediated, at least in part, through the inhibition of adenylate cyclase by a pertussis toxin-sensitive G protein-coupled mechanism. THC and CP 55,940 inhibited murine splenocyte proliferative responses to phorbol myristate acetate (PMA) plus ionomycin. The SRBC IgM antibody-forming cell (AFC) response was abrogated by low concentrations of dibutyryl-cAMP. Inhibition of the SRBC AFC response by both THC and CP 55,940 also was abrogated when splenocytes were preincubated with pertussis toxin that also was found to directly abrogate cannabinoid inhibition of adenylate cyclase. Collectively, the results suggested that inhibition of the SRBC AFC response by cannabinoids was mediated, at least in part, by inhibition of adenylate cyclase through a pertussis toxin-sensitive G_i protein-coupled cannabinoid receptor.

On the other hand, Derocq et al. (1995) reported that cannabinoids at low nanomolar concentrations had an enhancing effect on human tonsillar B cell growth. The cannabinoids CP 55,940 and WIN 55,212-2, as well as THC, exerted a dose-dependent increase of B cell proliferation for which the EC_{50} was at low nanomolar concentrations. The cannabinoid-induced enhancing activity was inhibited by pertussis toxin, suggesting a functional linkage to a G protein-coupled receptor. The CB_1 specific antagonist SR141716A had no antagonistic effect on this augmentation. These results, together with the demonstration that human B cells displayed large amounts of CB_2 mRNA, led the investigators to propose that the growth-enhancing activity observed on B cells at very low concentrations of cannabinoids was mediated through the CB_2 receptor. Carayon et al. (1998) reported that the CB_2 receptor was down-regulated at the mRNA and protein levels during B cell differentiation. Lowest levels of expression were observed in germinal center proliferating centroblasts of tonsillar tissue. The potent cannabinoid agonist CP 55,940 enhanced CD40-mediated proliferation of both virgin and germinal center B cell subsets. This enhanced proliferation could be blocked by the CB_2 antagonist SR144528 but not by the CB_1 antagonist SR141716A. These observations, taken together with the observation that SR144528 antagonized the stimulating effects of CP 55,940 on human tonsillar B cell activation evoked by cross-linking of surface immunoglobulins (Rinaldi-Carmona et al. 1998) suggested a functional involvement of CB_2 receptors during B cell differentiation.

2.3.4
Effects on T Lymphocytes

Studies as early as 1977 indicated that THC alters human T lymphocyte functions (Nahas et al. 1977). THC has since been reported to suppress a variety of functions of T cells from a variety of sources, including cytolytic activity and proliferation responses to T cell mitogens. Klein et al. (1991) examined the effect of cannabinoids on the activity of murine cytotoxic T lymphocytes (CTLs). The cytolytic activity of CTLs generated by co-cultivation with either allospecific or trinitrophenol (TNP)-modified-self stimulators was suppressed by THC and 11-hydroxy-THC. Allospecific CTLs generated in vivo also were inhibited by in vitro exposure to these cannabinoids. Drug treatment of mature CTLs appeared to have a minimal effect on binding of these cells to their targets. In addition, THC inhibited the proliferation of lymphocytes responding to an allogeneic stimulus as well as the maturation of these lymphocytes to mature CTLs. Similarly, THC was shown to inhibit CTL activity developing in vivo. It was proposed that CTL functionality was inhibited by cannabinoids by at least two modes. First, the cytolytic activity of mature CTLs was suppressed at a step beyond the binding to the target cell. Second, cannabinoids appeared to suppress the normal development of mature effector cells from the less mature precursor state. Fischer-Stenger et al. (1992) examined the effect of THC on CTL response to herpes simplex virus (HSV)-1 infection. It was indicated that THC decreased CTL activity against virus-infected cells by inhibiting CTL cytoplasmic polarization toward the virus-infected target cell. Granule reorientation toward the effector cell-target cell interface following cell–cell conjugation occurred at a lower frequency in co-cultures containing CTLs from drug-treated mice. Yerbra et al. (1992) examined effects of THC on one of the earliest events in T cell activation, the mobilization of cytosolic free calcium $[Ca^{2+}]$. It was reported that a portion of the proliferation defect in THC-treated lymphocytes could be related to a drug induced inhibition of $[Ca^{2+}]$ mobilization that normally occurs following mitogen treatment.

Schatz et al. (1993) proposed that THC selectively inhibited T cell-dependent humoral immune responses through direct inhibition of accessory T cell function. Oral administration of THC to mice produced a selective and dose-related inhibition of primary humoral immune responses to the T cell-dependent antigen, SRBC, as measured by the AFC response. No inhibitory effect on humoral responses to the T cell-independent antigen, dinitrophenyl (DNP)-Ficoll was obtained. A similar profile of immune inhibition was observed following direct in vitro addition of THC to naïve spleen cell cultures sensitized with defined antigens. In addition, THC produced a marked and dose-related inhibition of anti-CD3 monoclonal antibody-induced T cell proliferation. More recently, Condie et al. (1996) studied the effects of cannabinoids on adenylate cyclase-mediated signal transduction and IL-2 expression in the murine thymoma-derived T cell line EL4.IL-2. Treatment of cells with CBN or THC disrupted the adenylate cyclase signaling cascade by inhibiting forskolin-stimulated cAMP accumulation. This inhibition led to a decrease in protein kinase A activity and binding of transcription factors to a cAMP-response element (CRE) consensus sequence. Likewise, an inhibition of PMA/ionomycin-induced IL-2 protein secretion, which correlated to decreased IL-2 gene transcription, was induced by both CBN and THC. Cannabinoid treatment also decreased PMA/ionomycin-induced NF binding to the activator protein (AP)-1 proximal site of the IL-2 promoter. These findings suggested that inhibition of signal transduction via the adenylate cyclase/cAMP pathway induces T cell dysfunction by diminution in IL-2 gene transcription.

2.3.5 Effects on Natural Killer Cells

Kawakami et al. (1988) indicated that IL-2-induced killing activity and proliferation on NKB61A2 cells, a cell line derived from mouse that contains many morphological and functional similarities to primary NK cells (Warner and Dennert 1982), was suppressed by THC and 11-hydroxy-THC. Similarly, THC suppressed proliferation of murine spleen cells stimulated with recombinant human IL-2 and the appearance of the lymphokine-activated killer (LAK) cell phenomenon in IL-2-treated spleen cells (Kawakami et al. 1988). In addition, spleen cells previously stimulated in culture with IL-2, and then incubated with THC prior to target cell addition, displayed suppressed cytolytic activity against both yeast artificial chromosome (YAC)-1 and murine thymoma (EL)-4 tumor targets. These results suggested that THC could suppress IL-2-linked functions, including clonal expansion of lymphocytes, expansion of killer cell populations, and stimulation of killer cell cytotoxic activity. Additional studies have indicated that THC and other cannabinoids not only suppress the killing activity of mouse NK cells but also those from humans (Klein et al. 1998a, 1998b). The mechanism of this suppression was attributed as due partly to a drug-induced decrease in the number of high- and intermediate-affinity IL-2 binding sites, suggesting suppression in the expression of the IL-2 receptor (IL-2R) proteins (Zhu et al. 1993). Subsequent studies demonstrated that THC increased cellular levels of IL-2Rα and β proteins but decreased levels of the γ-protein and the function of the IL-2R (Zhu et al. 1995). It was concluded that drug treatment disturbed the relative expression of the various IL-2R chains, resulting in overall receptor dysfunction and poor responsiveness to IL-2. Daaka et al. (1997) extended these studies using NKB61A2 NK-like cells, established a link to cannabinoid receptors for these effects, and implicated involvement of the transcription factor NF-κB. These investigators concluded that, in the NK-like cell line used in the studies, a signaling pathway existed that was composed of CB_1, NF-κB, and the IL-2Rα gene. Other immune cell models have been used to demonstrate linkage of cannabinoid receptors to NF-κB-mediated gene activity (Herring et al. 1998; Jeon et al. 1996). Massi et al. (2000) reported that in vivo administration of THC to mice significantly inhibited NK cytolytic activity without affecting ConA-induced splenocyte proliferation. The parallel measurement of IFN-γ revealed that THC significantly reduced production of this cytokine, and the CB_1 and CB_2 antagonists completely reversed this reduction. These results suggested that both cannabinoid receptors were involved in the network mediating NK cytolytic activity.

2.4
Effects on Cytokines

A mode by which cannabinoids may exert their multiplicity of effects may be through the modulation of the expression of chemokines and cytokines which cross-signal among immune cells and play a critical role in pro-inflammatory versus anti-inflammatory activities. Blanchard et al. (1986) and Cabral et al. (1986a) reported that induction of IFN-α/β was suppressed by chronic treatment of mice with THC. Watzl et al. (1991) indicated that cytokine activity also was modulated in human peripheral blood mononuclear cell cultures by THC. However, the non-psychoactive CBD also modulated cytokine production and/or secretion, suggesting that a non-cannabinoid receptor-mediated mode of action could also be involved. The investigators indicated that a possible explanation for the capacity of cannabinoids to act through cannabinoid receptors so as to exert a broad spectrum of immune function effects was that exposure to these compounds resulted in the expression of a differential profile of cytokines.

Srivastava et al. (1998) examined the effect of THC and CBD on cytokine production in vitro by human leukemic T, B, eosinophilic, and $CD8^+$ natural killer lines. THC was found to decrease the constitutive production of IL-8, macrophage inflammatory protein (MIP)-1 α, MIP-1 β, and RANTES (regulated upon activation normal T cell expressed and secreted) protein. Phorbol ester-stimulated production of TNF-α, granulocyte-macrophage colony-stimulating factor (GM-CSF), and IFN-γ produced by NK cells also was affected. These results indicated that THC and CBD could alter production of a multiplicity of cytokines across a diverse array of immune cell lineages. Smith et al. (2000) evaluated the effects of cannabinoid receptor agonists and antagonists on the production of inflammatory cytokines and the anti-inflammatory cytokine IL-10 in endotoxemic mice. WIN 55,212-2 and HU-210 decreased serum levels of TNF-α and IL-12 and increased those for IL-10 when administered to mice before LPS exposure. The cannabinoids also protected *C. parvum* mice (but not unprimed mice) against the lethal effects of LPS. The protection afforded to *C. parvum* could not be attributed to the higher levels of IL-10 present in the mice after agonist treatment. The WIN 55,212-2- and HU-210-mediated changes in the responsiveness of mice to LPS were antagonized by SR141716A, the CB_1 antagonist, but not by SR144528, the CB_2 antagonist. It was concluded that both cannabinoid agonists modulated LPS responses through the CB_1 receptor. It was noted, also, that SR141716A itself modulated cytokine responses in a manner identical with that of WIN 55,212-2 and HU-210. The agonist-like effects of SR141716A were more striking in unprimed mice, suggesting that the antagonist could also function as a partial agonist at the CB_1 receptor. Zhu et al. (2000) reported that THC inhibited anti-tumor immunity by a CB_2 receptor-mediated, cytokine-dependent pathway. In their studies they used two different weakly immunogenic murine lung cancer models. THC decreased tumor immunogenicity, as indicated by the limited capacity for tumor-immunized, THC-treated mice to withstand tumor rechallenge. However, in contrast to the findings in immunocompetent mice, THC did not affect tumor growth in tumor-bearing severe combined immunodeficiency (SCID) mice. Levels of the immune inhibitory T helper $(Th)_2$ cytokines IL-10 and transforming growth factor (TGF) were augmented while those of the immune stimulatory Th_1 cytokine IFN-γ were down-regulated at both the tumor site and in spleens of THC-treated mice. Administration of either anti-IL-10 or anti-TGF-β neutralizing antibodies prevented the THC-induced enhancement in tumor growth. In vivo administration of the CB_2 antagonist SR144528 blocked the effects of THC. These findings suggested that THC promoted tumor growth by inhibiting anti-tumor immunity by a CB_2 receptor-mediated, cytokine-dependent pathway. Furthermore, this cytokine-dependent pathway correlated with a shift in a Th_1 pro-inflammatory to a Th_2 anti-inflammatory cytokine profile.

Berdyshev et al. (1997) examined the effects of anandamide, palmitoylethanolamide and THC on the production of TNF-α, IL-4, IL-6, IL-8, IL-10, IFN-γ, p55, and p75 TNF-α soluble receptors expressed by stimulated human peripheral blood mononuclear cells as well as [^{3}H]-arachidonic acid release by non-stimulated and *N*-formyl-Met-Leu-Phe (fMLP)-stimulated human monocytes. Anandamide diminished IL-6 and IL-8 production at low nanomolar concentrations and inhibited the production of TNF-α, IFN-γ, IL-4, and p75 TNF-α soluble receptors at higher concentrations (i.e., micromolar levels). Palmitoylethanolamide inhibited IL-4, IL-6, and IL-8 synthesis and the production of p75 TNF-α soluble receptors at concentrations similar to those of anandamide but did not affect TNF-α and IFN-γ production. Neither anandamide nor palmitoylethanolamide influenced IL-10 synthesis. THC, on the other hand, exerted a biphasic effect on pro-inflammatory cytokine production. TNF-α, IL-6, and IL-8 synthesis was inhibited maximally by 3 nM THC but stimulated by 3 μM THC. A similar effect was observed for IL-8 and IFN-γ. The level of IL-4, IL-10, and p75 TNF-α soluble receptors was diminished by 3 μM THC. [^{3}H]-Arachidonate release was stimulated only by high THC and anandamide concentrations. Based on these observations, the investigators suggested that the inhibitory properties of anandamide, palmitoylethanolamide, and THC are determined by the activation of peripheral-type cannabinoid receptors (i.e., CB_2) and that various endogenous fatty acid ethanolamides also participate in the regulation of the immune response.

Molena-Holgado et al. (1998) assessed the effects of cannabinoids in the context of a Theiler's murine encephalomyelitis virus (TMEV) infectivity model. In this model, murine cortical astrocyte cultures produce robust levels of IL-6 following infection in vitro. Treatment of cultures with anandamide resulted in increased production of IL-6. Cannabinoid receptors were implicated in these events because the enhancing effect was attenuated by the CB_1 antagonist SR141716A. Since it has been suggested that IL-6 may have a palliative effect in CNS diseases such as multiple sclerosis (MS), the investigators speculated that the increased levels of this cytokine could be related to a protective effect of cannabinoids in such diseases.

3
Cannabinoids and Infections

3.1
In Vitro Infections

A variety of mammalian cellular systems have been used as experimental models for documenting the in vitro effects of cannabinoids on immune responsiveness to viruses, bacteria, and amoebae. Blevins and Dumic (1980) indicated that THC had a protective effect against HSV infection in vitro. It was found that both HSV-1 and HSV-2 failed to replicate and produce extensive cytopathic effect (c.p.e.) in human cell monolayer cultures exposed before infection, at infection, or post infection to various concentrations of THC. In contrast, other studies indicate that THC compromises resistance to virus infection. It has been reported that THC inhibits macrophage extrinsic anti-viral activity (Cabral and Vásquez 1991; Cabral and Vásquez 1992) whereby macrophages normally suppress virus replication in cells to which they attach (Morahan et al. 1980; Stohlman et al. 1982). Noe et al. (1998) reported that a variety of cannabinoid receptor agonists enhanced syncytia formation in human T cell leukemia virus-I (HTLV-I)-transformed human T (MT-2) cells infected with cell free human immunodeficiency virus (HIV-1MN). It was found that CP 55,940, THC, WIN 55,212-2, and WIN 55,212-3 significantly increased syncytia formation, a phenomenon that has been reported to serve as an indicator of HIV infection and cytopathicity.

In addition to exerting modulatory effects on virus replication, cannabinoids have been reported to affect macrophage interaction with bacteria. Arata and colleagues (Arata et al. 1991, 1992) reported that THC could overcome the restriction of the growth of *Legionella pneumophila* , a facultative intracellular pathogen that replicates readily in human and guinea pig macrophages and in peritoneal exudate macrophages from A/J mice. Pretreatment of macrophages with THC did not affect ingestion or replication of *Legionella*. However, treatment with THC following infection with *Legionella* resulted in increased numbers of intracellular bacteria. Stimulation of macrophages with LPS resulted in a reduction in *Legionella* growth within macrophages. In contrast, treatment of these LPS-activated macrophages with THC resulted in greater growth of *Legionella*, indicating that the drug abolished the LPS-induced enhanced resistance. Gross et al. (2000) suggested a role for the CB_1 receptor in THC-mediated ablation of infection of macrophages by the intracellular pathogen *Brucella suis* , a gram-negative bacterium. Multiplication of *Brucella* within macrophages was inhibited by the CB_1 antagonist SR141716A but not by the CB_2 antagonist SR144528. THC has been shown also to alter the capacity of macrophages in vitro to kill *Naegleria fowleri* (Burnette-Curley et al. 1993), free-living amoebae that can cause a fatal disease in humans known as primary amoebic meningoencephalitis (PAME) (Marciano-Cabral 1988).

3.2 In Vivo Infections

Guinea pigs and mice have been used extensively as experimental in vivo models for documenting the effects of cannabinoids on host resistance. One of the earliest studies that indicated cannabinoids exacerbated host resistance to microbes was reported by Bradley et al. (1977), who demonstrated enhanced susceptibility of mice to combinations of THC and live or killed gram-negative bacteria. Morahan et al. (1979) demonstrated that mice exposed to THC were compromised in their ability to resist infection to viral and bacterial agents. BALB/c mice administered THC intraperitoneally exhibited decreased resistance to infection with either *Listeria monocytogenes* or HSV-2. Mishkin and Cabral (1985) and Cabral et al. (1986a,b) confirmed and extended these studies. THC was shown to increase in a dose-related fashion the susceptibility to HSV-2 genital infection in guinea pigs and mice. Animals treated with THC exhibited greater severity of herpes genitalis, higher mortalities, and higher mean titers of virus shed from the vagina. Suppression of antibody production to HSV-2 and a delay in the onset of the delayed hypersensitivity response to HSV-2 were observed. Cabral et al. (1987b) also noted that THC caused a reduction of the splenocyte proliferative response to HSV-2. It was suggested that THC inhibited immune responsiveness of $(B_6C_3)F_1$ mice to homotypic challenge with HSV-2.

Specter et al. (1991) reported that THC augmented murine retroviral-induced immunosuppression and infection. It was noted that THC in vitro administration to spleen cells from mice infected with Friend leukemia virus (FLV) resulted in a decrease, beyond that seen with virus or THC alone, in lymphocyte blastogenesis and NK cell activity. Moreover, when both FLV and THC were administered to mice concurrently infected with HSV, mortality attributed to FLV infection occurred significantly more rapidly than in the absence of HSV or THC. Paradise and Friedman (1993), using a hamster model, indicated that THC enhanced infection with *T. pallidum* , the causative agent of syphilis in humans. A greater degree of enhancement was exhibited also in rabbits in that treponemes proliferated more readily during treatment with THC than in control animals. In addition, Marciano-Cabral et al. (2001) reported that THC exacerbated brain infection in mice by *Acanthamoeba* , free-living amoebae that act as opportunistic pathogens.

There is accumulating data that alterations in cytokine expression play a critical role in enhanced mortality and morbidity in experimental animals. Klein et al. (1993) reported that THC induced cytokine-mediated mortality of mice infected with *L. pneumophila* . Mice administered two injections of THC, one before and one after a sublethal dose of *Legionella* experienced acute collapse and death. The THC-induced mortality resembled cytokine-mediated shock, and acute-phase serum from these animals contained significantly elevated levels of TNF and IL-6. The investigators concluded that THC increased the blood levels of acute-phase cytokines in the infected animals and that these elevated levels, at least in part, accounted for mortalities induced by THC. Newton et al. (1994) demonstrated that drug treatment of mice suppressed Th_1 anti-*Legionella* immunity as demonstrated by reduced production of IFN-γ and antibodies to IgG_{2a}. BALB/c mice, infected with a primary sublethal dose of *L. pneumophila* developed resistance to a larger challenge infection. Intravenous injection of THC 1 day prior to primary infection resulted in increased mortalities after a challenge infection. Furthermore, the level of anti-*L. pneumophila* IgG_1 antibodies in THC-treated animals, which are stimulated by Th_2 cells, was elevated. In contrast, levels of Th_1-regulated, IgG_{2a} were depressed. The investigators suggested that THC decreases the development of anti-*L. pneumophila* immunity by causing a change in the balance of Th_1 and Th_2 functional activities. In a follow-up study, Klein et al. (1997) examined the effects of CBD and CBN, as well as CP 55,940, on sublethal infection of inbred BALB/c mice that are relatively resistant to infection with *L. pneumophila*. CBD and CBN did not affect mortality of mice sublethally infected with *Legionella* to as great an extent as THC. However, CP 55,940 yielded levels of mortalities that were comparable to those induced by THC consistent with augmented lethality. Furthermore, mice receiving THC before and after infection exhibited higher levels of bacteria in their lung compared to sublethally infected mice not receiving cannabinoid. In addition, lung levels of mRNA for IL-6 were increased markedly following treatment of infected animals with THC. More recently, Klein et al. (2000) reported that THC treatment of BALB/c mice results in suppression of not only IFN-γ but also IL-12 and IL-12 receptor $\beta2$ in response to *L. pneumophila* infection. Studies using receptor antagonists suggested that both the CB_1 and CB_2 receptors were linked functionally to the suppression of Th_1 immunity to *Legionella*, resulting in a decrease in IFN-γ and IL-12. In addition, Cabral and Marciano-Cabral (2004) noted that cannabinoid-mediated exacerbation of brain infection with *Acanthamoeba* involved alterations in levels of cytokines. It was shown that mice administered THC and infected with *Acanthamoeba* exhibited dose-related higher mortalities than infected vehicle controls. The greater severity of disease for THC-treated mice was accompanied by decreased accumulation of macrophage-like cells at focal sites of infection in the brain. Furthermore, THC resulted in decreased levels of mRNA for the pro-inflammatory cytokines IL1-α, IL1-β, and TNF-α for neonatal rat microglia co-cultured with *Acanthamoeba*, implicating these resident macrophages of the brain as targets of cannabinoid-mediated decreased resistance.

The studies that have been performed on experimental animals have utilized THC doses in the range of 0.2 to 100 mg/kg. These doses have been administered by different routes to guinea pigs and mice or other rodents. The concentrations of THC measured in the circulation of these animals are achievable in human marijuana smokers following appropriate extrapolation for mass/surface ratio (Rachelefsky and Opelz 1977).

4 Effects of Marijuana and Cannabinoids on Human Health

4.1 Effects Related to Infections Other Than with the Human Immunodeficiency Virus

To date, there is no direct evidence that marijuana smoking or therapeutic administration of cannabinoids leads to an increased incidence of infectious disease in humans. Cohen (1976) reported that based on a University of California at Los Angeles (UCLA) 94-day cannabis study, exposure to marijuana did not alter immune responses. Hollister (1986) reported that no clinical consequences were noted from the effects of marijuana on the immune system. Sidney et al. (1997) examined the relationship of marijuana use to mortality for a population of 65,171 Kaiser Permanente Medical Care Program enrollees, aged 15 through 49 years, who completed questionnaires about smoking habits, including marijuana use, between 1979 and 1985. Mortality follow-up was conducted through 1991. Marijuana use in a prepaid health care-based study cohort was found to have little effect on non-AIDS mortality in men and on total mortality in women. Also, in a study to evaluate the relationship between marijuana use and sexually transmitted diseases in pregnant women, antenatal marijuana use was found to be unrelated to sexually transmitted infections during pregnancy (Miller et al. 2000).

In contrast, there are reports that cannabinoids and marijuana exert deleterious effects on immune function and host resistance. Juel-Jensen (1972) indicated anecdotally that individuals infected with HSV who were marijuana smokers had an increased recurrence of genital viral lesions. Also, Harkess et al. (1989) reported on six unrelated outbreaks of hepatitis A among users of marijuana and intravenously administered methamphetamine. Although the exact mode of transmission could not be determined, it was indicated that practices associated with illicit drug use could have facilitated transmission of hepatitis A. Gross et al. (1991) reported that marijuana use altered responsiveness of human papillomavirus to systemic recombinant IFN-α2a treatment and suggested that THC could be a cofactor influencing the outcome of infection. Liau et al. (2002) investigated the association between biologically confirmed marijuana use and laboratory-confirmed sexually transmitted diseases and condom use among African-American female adolescents. Among the 522 study subjects, 5.4% tested positive for marijuana. It was concluded that the adolescents were more likely to test positive for *Neisseria gonorrhoeae* and *Chlamydia trachomatis*, to have never used condoms in the previous 30 days and to have not used condoms consistently in the previous 6 months, and that sexually transmitted disease and sexual risk behavior may co-occur with marijuana use. Crosby et al. (2002) identified psychosocial predictors of *Trichomonas vaginalis* infection among low-income African American adolescent females living in a high-risk urban area of the United States. The strongest multivariate predictor of *T. vaginalis* infection was biologically confirmed marijuana use. Kagen et al. (1983) studied the possible role of marijuana in inducing sensitization to *Aspergillus* organisms in 28 marijuana smokers. It was reported that the use of marijuana was associated with risks of both fungal exposure and infection, as well as the possible induction of a variety of immunologic lung disorders. Kusher et al. (1994) demonstrated that THC suppressed the functional activities of human large granular lymphocytes (LGL). Exposure of LGL to THC at physiologically relevant concentrations resulted in down-regulation of TNF-α production and diminished LGL cytolytic activity against K562 tumor cells. The investigators suggested that, since the natural killer /polymorphonuclear neutrophil axis represents an important early defense against the opportunistic fungus *Candida albicans* , repression of this system could contribute to susceptibility to opportunistic infections.

In addition to effects on infectious agents, habitual marijuana use may elicit histopathological alterations and anti-inflammatory processes in the lung and respiratory tract. Guarisco et al. (1988) reported on the development of isolated uvulitis secondary to heavy marijuana use in three individuals. Barsky et al. (1998) examined bronchoscopy specimens from groups of smokers and nonsmokers and assessed them for incidence of molecular markers that antedate the development of lung cancer. It was found that smokers of marijuana, cocaine, or tobacco exhibited more molecular and histopathological alterations than did nonsmokers. The investigators concluded that marijuana and cocaine smoking, comparable to tobacco smoking, placed subjects at increased risk of developing lung cancer. Baldwin et al. (1997) indicated that marijuana and cocaine impaired lung alveolar macrophage function and cytokine production. Alveolar macrophages were deficient in their ability to phagocytose *Staphylococcus aureus* and severely limited in their ability to kill both bacteria and tumor cells. Alveolar macrophages of marijuana smokers were not able to use NO as an antibacterial effector molecule and produced less than normal amounts of TNF-α, GM-CSF, and IL-6 when stimulated in culture with LPS. Tashkin et al. (2002) extended these studies and indicated that regular use of marijuana was associated with ultrastructural abnormalities in human alveolar macrophages along with impairment of their cytokine production, antimicrobial activity, and tumoricidal function.

4.2 Effects Related to Infection with Human Immunodeficiency Virus and AIDS

There have been a limited number of studies which have addressed the issue of effects of marijuana or cannabinoids on HIV infection and the acquired immunodeficiency syndrome (AIDS). No conclusive data have been obtained as to potential risks and/or hazards associated with HIV infection and the use of marijuana or administration of cannabinoids in a therapeutic mode. Kaslow et al. (1989), in a report from the Multicenter AIDS Cohort Study, indicated that there was no evidence for a role of alcohol or other psychoactive drugs such as marijuana in accelerating immunodeficiency in HIV-1-positive individuals. Coates et al. (1990) analyzed cofactors of disease progression in a cohort of 249 male sexual contacts of men with AIDS or an AIDS-related condition. No significant association with risk of progression to AIDS was noted for use of various recreational drugs, history of specific infections, age at enrollment, or smoking and drinking status at enrollment.

Difranco et al. (1996), through the San Francisco Men's Health Study (SFMHS), evaluated in a 6-year follow-up study the association of specific recreational drugs and alcohol with laboratory predictors of AIDS. No association with progression to AIDS was observed for marijuana use. Wallace et al. (1998) examined risk factors and outcomes associated with identification of *Aspergillus* in respiratory specimens from individuals with HIV disease as part of a study to evaluate pulmonary complications of HIV infection. Cigarette and marijuana use was found not to be associated with *Aspergillus* respiratory infection. Persaud et al. (1999) conducted a cross-sectional survey among 124 street- and brothel-based female commercial sex workers in Georgetown, Guyana, to determine the seroprevalence of HIV infection and describe the sexual practices and drug use patterns. No statistically significant association was found between HIV infection and marijuana use. Miller and Goodridge (2000) evaluated in a retrospective study the relationship between marijuana use and sexually transmitted diseases in pregnant women. The prevalence of gonorrhea, Chlamydia, syphilis, HIV, hepatitis B surface antigen, human papilloma virus, and HSV was determined. No significant differences were found in the prevalence of any single—or more than one—sexually transmitted disease between pregnant women who used marijuana and drug-free pregnant women. Bredt et al. (2002) examined the short-term effects of cannabinoids on immune phenotype and function in HIV-1-infected patients. A randomized, prospective, controlled trial comparing the use of marijuana cigarettes (3.95% THC), dronabinol (2.5 mg), and oral placebo in HIV-infected adults taking protease inhibitor-containing highly active antiretroviral therapy (HAART) was undertaken. Few statistically significant effects on immune system phenotypes or functions were found in this patient population. Struwe et al. (1993) studied the effect of dronabinol (THC) on nutritional status in HIV infection and found that, in a selected group of HIV-infected patients with weight loss, short-term treatment with dronabinol resulted in improvement in nutritional status and symptom distress. Recently, Abrams et al. (2003) conducted a randomized, placebo-controlled 21-day clinical trial to examine the short-term effects of smoked marijuana on the viral load in 67 HIV-infected patients. The study was conducted in an inpatient setting at the General Clinical Research Center at the San Francisco General Hospital, San Francisco, California. Participants were randomly assigned to a 3.95%-THC marijuana cigarette, a 2.5 mg dronabinol (THC) capsule, or a placebo capsule three times daily before meals. It was concluded that smoked and oral cannabinoids did not appear to present a risk in individuals with HIV infection with respect to HIV RNA levels, $CD4^+$ and $CD8^+$ cell counts, or protease inhibitor levels.

On the other hand, there are reports that marijuana use is associated with compromised health status among HIV-infected individuals. Lozada et al. (1983) assessed oral manifestations of tumor and opportunistic infections in 53 AIDS-affected men with Kaposi's sarcoma (KS). Twenty-seven had biopsy-proved oral KS, the palate being the most common site. Past or present infections with cytomegalovirus, hepatitis, venereal diseases, and gastrointestinal microorganisms occurred in more than 70%. Oral candidiasis was confirmed in 57%. Heavy marijuana smoking was identified as the most common habit among these individuals. Newell et al. (1985) reported that marijuana use was a risk factor among men referred for possible AIDS in their analysis of responses to a lifestyle questionnaire among 13 patients with Kaposi's sarcoma and 18 with an opportunistic infection as compared with those of 29 symptom-free referred individuals. In addition, immuno-epidemiological studies using univariant and multivariant analyses have indicated an association between marijuana use and progression of HIV seropositivity to development of symptomatic AIDS (Tindall et al. 1988). Whitfield et al. (1997) examined the impact of ethanol and Marinol/marijuana on HIV+/AIDS patients undergoing azidothymidine (AZT), azidothymidine/dideoxycytidine (AZT/DDC), or dideoxyinosine (DDI) therapy. Marinol/marijuana was reported to be associated with declining health status in both the AZT and AZT/DDC groups. However, in HIV+/AIDS patients with the lowest $CD4^+$ counts undergoing DDI monotherapy, utilization of Marinol/marijuana did not seem to have a deleterious impact. Caiaffa et al. (1994) indicated that smoking illicit drugs such as marijuana was one of several factors that increased risk of bacterial pneumonia in HIV-seropositive drug users.

5 Distribution of Cannabinoid Receptors in the Immune System

5.1 Native Distribution

Two cannabinoid receptor types have been identified in various cells and tissues of the immune system (Table 1). The first of these, the CB_1 receptor, was cloned originally from a cDNA rat library by Matsuda et al. (1990). Munro et al. (1993) reported on the cloning of the second receptor for cannabinoids, designated the CB_2 receptor, which was not expressed in the brain but rather in the immune system. Recent studies suggest the existence of a third cannabinoid receptor, tentatively designated as non-CB_1, non-CB_2 (Breivogel et al. 2001; Fride et al. 2003; Wiley and Martin 2002). Cannabinoid receptor CB_1 mRNA is found primarily in brain and neural tissue but can be found also at lower levels in peripheral tissues including the adrenal gland, bone marrow, heart, lung, prostate, testis, thymus, tonsils, and spleen (Bouaboula et al. 1993; Galiègue et al. 1995; Kaminski et al. 1992; Noe et al. 2000). Messenger RNA for the CB_1 receptor has been identified also in microglia from the brain (Sinha et al. 1998; Waksman et al. 1999). Transcripts (i.e., mRNAs) for the CB_2 receptor are abundant in spleen and tonsils and are found at levels equivalent to those for CB_1 mRNA in the CNS (Galiègue et al. 1995; Munro et al. 1993). However, in other tissues of the immune system mRNA levels for the CB_2 receptor, while exceeding those for the CB_1 receptor (Bouaboula et al. 1993; Galiègue et al. 1995), are relatively low. The distribution pattern of CB_2 mRNA displays major variation in human blood cell populations with a rank order of B lymphocytes>NK cells>>monocytes>polymorphonuclear neutrophils>T8 lymphocytes>T4 lymphocytes (Bouaboula et al. 1993; Galiègue et al. 1995). Lee et al. (2001b) reported a comparable pattern of distribution for murine immune cells.

Table 1. Distribution of cannabinoid receptors in the immune system

Cell type/tissue	Species	Receptor	Reference
B lymphocytes	Human	CB_2	Galiègue et al. 1995 Carayon et al. 1998
T4 lymphocytes	Human	CB_2	Galiègue et al. 1995
T8 lymphocytes	Human	CB_2	Galiègue et al. 1995
Leukocytes	Human	CB_2	Bouaboula et al. 1993
Macrophages	Human, mouse	CB_2	Galiègue et al. 1995 Lee et al. 2001a,b
Microglia	Rat	CB_1, CB_2	Sinha et al. 1998 Waksman et al. 1999 Carlisle et al. 2002
Mononuclear cells	Human, rat	CB_2	Galiègue et al. 1995 Facci et al 1995
Mast cells	Rat	CB_2	Facci et al. 1995
Natural killer (NK) cells	Human	CB_2	Galiègue et al. 1995
Peyer's patches	Rat	CB[a]	Lynn and Herkenham 1994
Spleen	Human Mouse, rat	CB_1, CB_2	Kaminski et al. 1992 Munro et al. 1993 Galiègue et al. 1995 Facci et al. 1995 Lynn and Herkenham 1994
Thymus	Human	CB_2	Galiègue et al. 1995
Tonsils	Human	CB_2	Galiègue et al. 1995
Lymph nodes	Rat	CB[a]	Lynn and Herkenham 1994

Cannabinoid receptor type not specified.

In addition to the identification of cannabinoid receptor mRNA in immune cells (Fig. 1), cognate protein has been demonstrated in rat lymph nodes, Peyer's patches, and spleen (Fig. 2). Lynn et al. (1994) used a radiolabeled high-affinity cannabinoid receptor ligand (i.e., [^{3}H]CP 55,940) for in vitro binding and autoradiography and found that specific cannabinoid receptor binding was restricted to components of the immune system at peripheral sites. Cannabinoid receptor protein in the immune system was found to be confined to B lymphocyte-enriched areas such as the marginal zone of the spleen, cortex of the lymph nodes, and nodular corona of Peyer's patches. Galiègue et al. (1995), using anti-human CB_2 IgG, localized CB_2 receptors within B lymphocyte-enriched areas of the mantle of secondary lymphoid follicles in sections of human tonsil. Protein for the CB_2 receptor has been identified also by immunohistochemistry in B lymphocyte-enriched areas of the mantle of secondary lymphoid follicles in human tonsil sections (Bouaboula et al. 1993). Carayon et al. (1998) employed immunopurified polyclonal antibody to investigate the expression of CB_2 receptors in leukocytes and showed that peripheral blood and tonsillar B cells were the leukocyte subsets expressing the

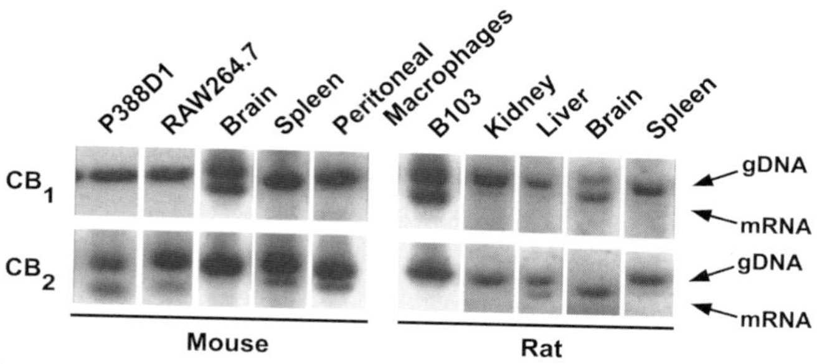

Fig. 1. Identification of cannabinoid receptor mRNA in mouse and rat tissues and cells. For detection of CB_1 or CB_2 mRNA, total nucleic acid was subjected to reverse mutagenic reverse transcription-polymerase chain reaction (MRT-PCR) as described by Carlisle et al. (2002). The bands designated *gDNA* represent amplified genomic DNA used as an internal quantitation standard. The bands designated *mRNA* represent product amplified from receptor message. The *P388D1* and *RAW264.7* designate murine macrophage-like cells, while *B103* designates rat neuroblastoma cells

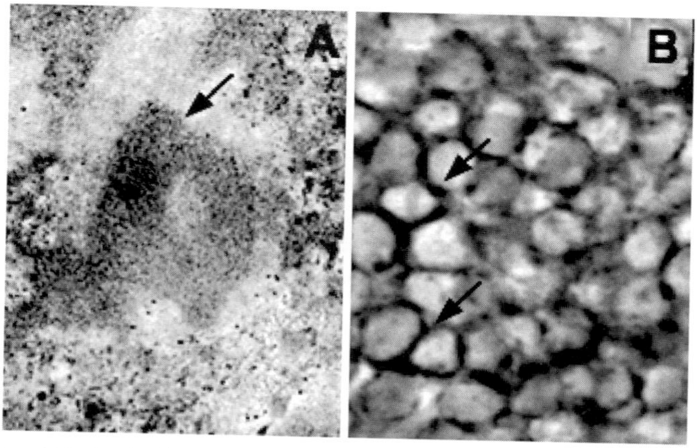

Fig. 2A, B Immunohistochemical localization of CB_2 receptors in germinal centers in mouse spleen. An affinity-purified antibody to the amine terminal domain of the murine CB_2 was used in concert with immunoperoxidase staining. **A** Immunoreactive product is localized in germinal centers enriched for B lymphocytes (*arrow*). Magnification: X50. **B** Immunoreactive product for the CB_2 receptor in B lymphocytes is concentrated at the outer periphery of the cytoplasmic compartment (*arrows*). ×500

highest amount of CB_2 protein. Dual color confocal microscopy performed on human tonsillar tissue demonstrated a marked expression of CB_2 receptors in mantle zones of secondary follicles, whereas germinal centers were weakly stained, suggesting a modulation of this receptor during B lymphocyte differentiation stages from virgin B lymphocytes to memory B cells. In addition, protein for the CB_1 and CB_2 receptors has been identified in neonatal rat microglia maintained in vitro (Carlisle et al. 2002; Sinha et al. 1998; Waksman et al. 1999).

Levels of cannabinoid receptors on cells of the immune system may vary during cell differentiation, activation, or response to external stimuli. Noe et al. (2000) reported that anti-CD40, anti-CD3, and IL-2 stimulation induced contrasting changes in CB_1 mRNA expression in mouse splenocytes. It was suggested that signaling pathways activated by T cell mitogens led to decreased CB_1 gene activation while pathways activated by B cell mitogens and IL-2 led to increased levels of CB_1. Lee et al. (2001a,b) examined the expression pattern of CB_2 mRNA in mouse peritoneal macrophages. CB_2 was expressed in thioglycolate-elicited macrophages, but not in resident macrophages. LPS stimulation down-regulated CB_2 mRNA expression in splenocyte cultures, while stimulation through CD40 using anti-CD40 antibody up-regulated the response. In addition, co-stimulation with IL-4 attenuated the anti-CD40 response. Collectively, the results suggested that the signaling pathways activated by LPS and anti-CD40 exerted differential effects on CB_2 mRNA expression. Carlisle et al. (2002) indicated that the CB_2 receptor was differentially expressed by rodent peritoneal macrophages and macrophage-like cells in relation to their cell activation state. The CB_2 receptor was undetectable in resident macrophages, present at high levels in thioglycolate-elicited inflammatory and IFN-γ-primed macrophages, and detected at significantly diminished levels in LPS fully activated macrophages. A comparable pattern of differential expression of the CB_2 receptor was noted for murine macrophage-like cells and neonatal rat brain cortex microglia.

5.2
Distribution in Cell Lines

Messenger RNA for CB_1 has been found in immune cell lines, including human THP-1 monocytic cells, human Raji B cells, murine NKB61A2 NK-like cells, and murine CTLL2 IL-2-dependent T cells (Daaka et al. 1995). Sinha et al. (1998) employed RT-PCR to detect message and affinity-purified polyclonal antiserum to demonstrate that MHC class II^+, macrophage-like, glial cells contained message and cognate protein for CB_1. A glioma cell line and a B lymphoblastoid cell line also were positive for this protein, which was estimated at 58 kDa. Daaka et al. (1996) indicated that the Jurkat, human T cell line was weakly positive for the CB_1 receptor but mitogen activation increased message levels. Valk et al. (1997) reported the presence of CB_2 mRNA in 45 of 51 cell lines of distinct hematopoietic lineages, including myeloid, macrophage, mast, B lymphoid, T lymphoid, and erythroid cells. A rank order for levels of CB_2 transcripts similar to that for primary human cell types has been recorded for human cell lines belonging to the myeloid, monocytic, and lymphoid lineages (Bouaboula et al. 1993).

6
Mode of Action in the Immune System

6.1
Exogenous Cannabinoids

Smith et al. (1978) were among the first to note structure–activity relationships of natural and synthetic cannabinoids in suppression of humoral and cell-mediated immunity. They indicated that THC, Δ^8-THC, 1-methyl Δ^8-THC and abnormal Δ^8-THC caused immunosuppression, based on reduction of the humoral immune response to SRBC as measured by spleen plaque-forming cells. Furthermore, this suppression was not related to CNS activity, since 1-methyl Δ^8-THC and abnormal Δ^8-THC had minimal CNS activity. Titishov et al. (1989) injected mice with (+) and (−)enantiomers of the dimethyl heptyl derivative of THC, HU-210 and HU-211, and reported that they exerted stereospecific effects on the immune system. It was concluded, however, that immune suppression was effected both by receptor and non-receptor-mediated modes.

Schatz et al. (1992) reported that inhibition of adenylate cyclase by THC constituted a potential mechanism for cannabinoid-mediated immunosuppression. Diaz et al. (1993) treated human peripheral blood mononuclear cell cultures with THC and a variety of cAMP stimulators. Lymphocyte cAMP levels were stimulated using three hormone receptor stimulators, isoproterenol, histamine, or *N*-ethylcarboxamide adenosine (NECA), each of which utilizes a different receptor to enhance cAMP production. THC suppressed cAMP levels independently of the hormone and receptor utilized. It was suggested that THC exerted its effects on second messenger systems at the lymphocyte membrane level, and that a pertussis toxin-sensitive G_i protein was involved. Kaminski et al. (1994) also reported that suppression of the humoral immune response by cannabinoids was mediated partially through inhibition of adenylate cyclase by a pertussis toxin-sensitive G protein-coupled mechanism. More direct evidence for a role of a cannabinoid receptor as linked to cannabinoid mediation of immune responses was provided by Kaminski et al. (1992) when they identified a functionally relevant cannabinoid receptor on mouse spleen cells. It was concluded that a cannabinoid receptor similar, if not identical, to the CB_1 receptor was linked to immune modulation by cannabimimetic agents. With the cloning of the CB_2 receptor from the human promyelocytic cell line HL-60 by Munro et al (1993), a more complete picture concerning functional relevance for cannabinoid receptors in the immune system was obtained. Mckallip et al. (2002) suggested recently that THC-induced apoptosis in the thymus and spleen of mice, mediated through the CB_2 receptor, serves as a mechanism for immunosuppression in vitro and in vivo.

Several investigators have addressed the effects of non-psychotropic cannabinoids on immune function. Herring et al. (1998) demonstrated that CBN, a ligand that exhibits higher binding affinity for the CB_2 receptor in comparison to the CB_1 receptor, modulated immune responses and cAMP-mediated signal transduction in mouse lymphoid cells. The decrease in intracellular cAMP levels resulted in a reduction of protein kinase A activity, leading to an inhibition of transcription factor binding to the cAMP response element and κB motifs. Jan et al. (2002) reported that CBN enhanced IL-2 expression by T cells that was associated with an increase in IL-2 distal nuclear factor of activated T cell activity (NF-AT). It was suggested that this increase was mediated through a CB_1/CB_2-independent mechanism. Enhancement of IL-2 also was demonstrated with CP 55,940, THC, and CBD, suggesting that the phenomenon was not unique to CBN. Luo et al. (1992) examined the effects of THC, Δ^8-THC, and cocaine on the in vitro mitogen-induced transformation of lymphocytes of human and mouse origin. The two cannabinoids exerted a biphasic effect on mitogen-induced transformation. Both stimulated lymphocyte transformation at low concentrations but inhibited mitogenesis at high concentrations. In contrast, cocaine neither affected mitogen-induced lymphocyte transformation nor altered the effect of THC when added together with this cannabinoid. Human lymphocytes and mouse splenocytes appeared to respond in similar patterns.

6.2 Endogenous Cannabinoids (Endocannabinoids)

The recognition that exogenous cannabinoids could alter immune functional activities through cannabinoid receptors implicated the existence of an endogenous functionally relevant ligand-receptor system. Devane et al. (1992) isolated from porcine brain an arachidonic derivative, anandamide, in a screen for endogenous ligands for the cannabinoid receptor with properties suggestive that it acted as a natural ligand for the cannabinoid receptor in the brain. The discovery and characterization of anandamide served as a catalyst for studies to assess its role in signaling through the CB_1 receptor in the brain as well as in the immune system. Valk et al. (1997) reported that anandamide acted through the CB_2 receptor as a synergistic growth factor for hematopoietic cells. Derocq et al. (1998), in a similar study using IL-3-dependent and IL-6-dependent murine cell lines, postulated that anandamide exerted a growth-promotion effect. However, it was indicated that this growth-promoting effect was cannabinoid receptor-independent.

De Petrocellis et al. (1998) reported that anandamide potently and selectively inhibited the proliferation of human breast cancer cells in vitro. Anandamide dose-dependently suppressed the proliferation of MCF-7 and EFM-19 human breast carcinoma cells but did not affect the proliferation of several nonmammary tumoral cell lines. The anti-proliferative effect of anandamide was apparently not due to toxicity or to apoptosis of cells but was accompanied by a reduction of cells in the S phase of the cell cycle. The stable analog of anandamide *R*-methanandamide and the synthetic cannabinoid HU-210 also inhibited EFM-19 cell proliferation. The drug effects were blocked with the CB_1 antagonist SR141716A, suggesting that anandamide blocked human breast cancer cell proliferation through a CB_1-like receptor-mediated inhibition of endogenous prolactin action at the level of the prolactin receptor. Facci et al. (1995) reported that mast cells, multifunctional bone marrow-derived cells found in mucosal and connective tissues and in the nervous system that play an important role in tissue inflammation and neuroimmune interactions, expressed a peripheral cannabinoid receptor with differential sensitivity to anandamide and palmitoylethanolamide. It was found that they expressed the CB_2 receptor and that this receptor exerted negative regulatory effects on mast cell activation. Although palmitoylethanolamide and anandamide bound to the CB_2 receptor, only the former down-modulated mast cell activation in vitro. It was proposed that palmitoylethanolamide, unlike anandamide, behaved as an endogenous agonist for the CB_2 receptor on mast cells. The existence of an autacoid local inflammation antagonism (ALIA) process was proposed. However, more recent experiments have shown that palmitoylethanolamide has very low affinity for CB_2 receptors and little CB_2 receptor efficacy (Griffin et al. 2000; Lambert et al. 1999; Sheskin et al. 1997; Showalter et al. 1996).

In 1995, Mechoulam et al. (1995) identified an endogenous 2-monoglyceride from canine gut that bound to cannabinoid receptors which was designated 2-AG. Studies using 2-AG have indicated that it is more active than anandamide in the immune system. Lee et al. (1995) reported that 2-AG suppressed lymphoproliferation of splenocytes to LPS and anti-CD3 at concentrations greater than 10 μM. Proliferation due to alloantigen stimulation also was suppressed, but no suppression of PMA/ionomycin-induced proliferation was observed. In addition, the in vitro PFC response to SRBC was increased by 2-AG. Sugiura et al. (2000) examined the effect of 2-AG on intracellular free Ca^{2+} concentrations in the human macrophage-like cells HL-60 and found that it induced a rapid transient increase in levels of intracellular free Ca^{2+}. The Ca^{2+} transient induced by 2-AG was blocked by pretreatment of the HL-60 cells with SR144528, the CB_2 antagonist, but not with SR141716A, the CB_1 antagonist, indicating the involvement of the CB_2 receptor in this cellular response. Anandamide and palmitoylethanolamide, other putative endogenous ligands, were found to be a weak partial agonist and an inactive ligand, respectively. Based on these results, Sugiura et al. (2000) proposed that the CB_2 receptor was originally a 2-AG receptor, and that 2-AG constituted its native cognate ligand.

Diaz et al. (1994) examined mechanisms of action of THC in inducing immunosuppression contextual to transductional activities mediated through lipid bioeffector molecule derivatives of arachidonic acid, since THC was known to affect arachidonic acid metabolism in non-lymphoid cells. It was indicated that THC increased the production of the eicosanoid 12-hydroxyeicosateraenoic acid (12-HETE) from peripheral blood mononuclear cells (PBMC). To determine if other eicosanoid metabolites were affected by THC, levels of leukotriene B4 were measured. THC was shown to increase markedly the production of LTB4 from PBMC stimulated with the calcium ionophore A23187. The collective results indicated that THC altered arachidonic acid metabolism in lymphocytes by increasing the production of lipoxygenase products, biological effectors with known immunosuppressive properties (reviewed in Lawrence et al. 2002).

7
Cannabinoids as Immune Therapeutic Agents

Cannabinoids, as immunosuppressive compounds, have been proposed as having therapeutic potential in chronic inflammatory disorders and neurodegenerative disease triggered by inflammatory attack. Lyman et al. (1989) inoculated Lewis rats and strain 13 guinea pigs with myelin-basic protein emulsified in complete Freund's adjuvant to induce experimental autoimmune encephalomyelitis (EAE) to mimic MS and indicated that THC-treated animals had either no clinical signs or exhibited mild signs with delayed onset and greater survival. Examination of CNS tissue revealed a marked reduction of inflammation in the THC-treated animals. Wirguin et al. (1994) examined the effect of Δ^8-THC, a more stable and less psychotropic analog of THC, on EAE using two strains of rats. Δ^8-THC significantly reduced the incidence and severity of neurological deficit in both strains. It was suggested that suppression of EAE by cannabinoids was related to their effect on corticosterone secretion. Pryce et al. (2003), using the EAE model, also reported that cannabinoids could inhibit neurodegeneration. In addition, exogenously introduced CB_1 agonists provided significant neuroprotection from the consequences of inflammatory CNS disease in an animal model of experimental allergic uveitis.

Molina-Holgado et al. (1998) utilized Theiler's murine encephalomyelitis virus (TMEV) to produce persistent brain infection in mice with attendant chronic primary immune-mediated demyelination resembling MS. The effects of anandamide on astrocytes infected with TMEV were examined, since these glial cells in the brain are potent producers of pro-inflammatory cytokines upon virus infection. Astrocytes from susceptible (SJL/J) and resistant (BALB/c) strains of mice infected with TMEV exhibited increased IL-6 release that was enhanced by anandamide. Treatment of TMEV-infected astrocytes with arachidonyl trifluoromethyl ketone, a potent inhibitor of the amidase that degrades anandamide, potentiated this anandamide effect. SR141617A, the CB_1 antagonist, blocked the enhancing effects of anandamide on IL-6 release by TMEV-infected astrocytes, suggesting a cannabinoid receptor-mediated pathway. The investigators indicated that, while the physiological implications of these results were unknown, they could be related to the postulated protective effects of cannabinoids on neurological disorders such as MS. Arevalo-Martin et al. (2003), using the TMEV model, demonstrated that treatment with WIN 55,212-2, the potent highly selective CB_1 agonist ACEA, or the CB_2 receptor high-affinity cannabimimetic JWH-015 during established disease resulted in significant long-term improvement of neurological deficits. Similarly, Croxford et al. (2003) demonstrated that WIN 55,212-2 ameliorated progression of clinical disease symptoms in mice with preexisting Theiler murine encephalomyelitis virus-induced demyelinating disease (TMEV-IDD). Ablation of disease was associated with down-regulation of virus and myelin epitope-specific Th_1 effector functions (i.e., delayed-type hypersensitivity and IFN-γ production) and the inhibition of CNS mRNA expression for the pro-inflammatory cytokines TNF-α, IL-1β, and IL-6. Killestein et al. (2003) assessed the immunomodulatory effects of orally administered cannabinoids in 16 MS patients. A modest increase of TNF-α in LPS-stimulated whole blood was found during cannabis plant-extract treatment, but changes in levels of other cytokines were not observed. In patients with high adverse event scores, it was found that an increase in plasma IL-12p40 occurred. The investigators suggested that cannabinoids had a potential for modifying MS in humans. Li et al. (2001) examined the immunosuppressive effects of THC in streptozotocin (STZ)-induced autoimmune diabetes. THC administered orally to CD-1 mice attenuated, in a transient manner, the STZ-induced elevation in serum glucose and loss of pancreatic insulin. STZ-induced insulitis and increases in IFN-γ, TNF-α, and IL-12 mRNA levels were reduced by coadministration of THC. Studies performed using $(B_6C_3)F_1$ mice showed a moderate hyperglycemia and a significant reduction in pancreatic insulin by STZ in the absence of insulitis. The investigators suggested that THC was capable of attenuating the severity of autoimmune responses in this experimental model of autoimmune diabetes.

In addition, it has been proposed that cannabinoids may protect against septic shock and brain trauma. Bass et al. (1996) suggested that Dexanabinol, the nonpsychotropic cannabinoid HU-211, had potential for use in treatment based on use of an experimental rat model of meningitis in which rats were inoculated with *Streptococcus pneumoniae* . HU-211 was efficacious when used in combination with antimicrobial therapy in reducing brain damage, especially when given concomitantly with antibiotics. Shohami et al. (1997) developed an experimental rat model for closed head injury (CHI), in which edema, blood–brain barrier disruption, and motor and memory dysfunctions were demonstrated. Using this model, spatial and temporal induction of IL-1, IL-6, and TNF-α gene mRNA transcription and TNF-α and IL-6 activity in rat brain after CHI were demonstrated. HU-211 acted as an effective cerebroprotectant in that it suppressed TNF-α production. HU-211, pentoxifylline and TNF-binding protein improved the outcome of CHI. These studies were extended by Gallily et al. (1997) who demonstrated that HU-211 not only suppressed TNF-α production but also rescued mice and rats from endotoxic shock after LPS inoculation.

It has been proposed, also, that cannabinoids have therapeutic potential in the management of select microbial infections. Nok et al. (1994) examined the effect of *Cannabis sativa* on trypanosome-infected rats. It was reported that an aqueous extract of the seeds cured animals infected with *Trypanosome brucei brucei* of blood stream parasites. Berdyshev et al. (1998) investigated the effects of WIN 55,212-2, THC, anandamide, and palmitoylethanolamide on LPS-induced bronchopulmonary inflammation in mice. WIN 55,212-2 and THC induced a concentration-dependent decrease in TNF-α levels in bronchoalveolar lavage fluid. This effect was accompanied by moderately reduced neutrophil recruitment. Palmitoylethanolamide diminished levels of TNF-α in bronchoalveolar lavage fluid but had no effect on neutrophil recruitment. Anandamide did not influence the inflammatory process but TNF-α levels and neutrophil recruitment were decreased. Gross et al. (2000) reported that the CB_1 antagonist SR141716A was a potent inhibitor of macrophage infection by the intracellular gram-negative bacterial pathogen *Brucella suis* . These investigators assessed the influence of the CB_1 or CB_2 antagonists SR141716A and SR144528, respectively, as well as the nonselective CB_1/CB_2 cannabinoid receptor agonists CP55,940 or WIN 55,212-2 on macrophage infection. The intracellular multiplication of *Brucella* was dose-dependently inhibited in cells treated with SR141716A, which exerted a potent microbicidal effect. The involvement of CB_1 receptors in the protective effect was proposed. Furthermore, SR141716A was able to pre-activate macrophages and to trigger an activation signal that inhibited *Brucella* development. Collectively, the results indicated that SR141716A up-regulated the antimicrobial properties of macrophages in vitro and that it might serve as a pharmaceutical compound for counteracting the propagation of intra-macrophagic gram-negative bacteria.

8
Summary and Conclusions

Marijuana and other exogenous cannabinoids, primarily those which possess psychotropic activity, alter immune functionality and decrease host resistance to bacterial, protozoan, and viral infections in experimental animal models and in vitro systems. The main substance in marijuana that exerts these immuno depressive effects is Δ^9-tetrahydrocannabinol (THC). This cannabinoid alters the function of an array of immune cells including lymphocytes, natural killer (NK) cells, and macrophages, thereby affecting their capacity to exert anti-microbial activities. Two modes of action by which THC and other cannabinoids affect immune responsiveness have been proposed. First, these compounds may signal transduce through cannabinoid receptors CB_1 and CB_2. Second, at sites of direct exposure to marijuana or high concentrations of cannabinoids, such as the lung, membrane perturbation may be involved. In addition, endogenous cannabinoids or endocannabinoids have been identified and have been proposed as native modulators of immune functions through cannabinoid receptors. Exogenously introduced cannabinoids may disturb this homoeostatic immune balance. A mode by which cannabinoids may alter immune responsiveness is by driving the expression of cytokines from a Th_1 pro-inflammatory pattern to that of a Th_2 anti-inflammatory pattern. While marijuana and various cannabinoids have been documented to alter immune functions in vitro and in experimental animals, no controlled longitudinal epidemiological studies have yet definitively correlated immunosuppressive effects with increased incidence of infections or immune disorders in different segments of the human population. However, cannabinoids by virtue of their immunomodulatory properties have the potential to serve as therapeutic agents for ablation of untoward immune responses.

References

Abood ME, Martin BR (1996) Molecular neurobiology of the cannabinoid receptor. Int Rev Neurobiol 39:197–221

Abrams DI, Hilton JF, Leiser RJ, Shade SB, Elbeik TA, Aweeka FT, Benowitz NL, Bredt BM, Kosel B, Aberg JA, Deeks SG, Mitchell TF, Mulligan K, Bacchetti P, McCune JM, Schambelan M (2003) Short-term effects of cannabinoids in patients with HIV-1 infection: a randomized, placebo-controlled clinical trial. Ann Intern Med 139:258–266

Arata S, Klein TW, Newton C, Friedman H (1991) Tetrahydrocannabinol treatment suppresses growth restriction of Legionella pneumophila in murine macrophage cultures. Life Sci 49:473–479

Arata S, Newton C, Klein T, Friedman H (1992) Enhanced growth of Legionella pneumophila in tetrahydrocannabinol-treated macrophages. Proc Soc Exp Biol Med 199:65–67

Arevalo-Martin A, Vela JM, Molina-Holgado E, Borrell J, Guaza C (2003) Therapeutic action of cannabinoids in a murine model of multiple sclerosis. J Neurosci 23:2511–2516

Ashfaq MK, Watson ES, elSohly HN (1987) The effect of subacute marijuana smoke inhalation on experimentally induced dermonecrosis by S. aureus infection. Immunopharmacol Immunotoxicol 9:319–331

Baczynsky WO, Zimmerman AM (1983a) Effects of delta-9-tetrahydrocannabinol, cannabinol and cannabidiol on the immune system in mice. I. In vivo investigation of the primary and secondary immune response. Pharmacology 26:1–11

Baczynsky WO, Zimmerman AM (1983b) Effects of delta-9-tetrahydrocannabinol, cannabinol and cannabidiol on the immune system in mice. II. In vitro investigation using cultured mouse splenocytes. Pharmacology 26:12–19

Baldwin GC, Tashkin DP, Buckley DM, Park AN, Dubinett SM, Roth MD (1997) Marijuana and cocaine impair alveolar macrophage function and cytokine production. Am J Respir Crit Care Med 156:1606–1613

Barsky SH, Roth MD, Kleerup EC, Simmons M, Tashkin DP (1998) Histopathologic and molecular alterations in bronchial epithelium in habitual smokers of marijuana, cocaine, and/or tobacco. J Natl Cancer Inst 90:1198–1205

Bass R, Engelhard D, Trembovler V, Shohami E (1996) A novel nonpsychotropic cannabinoid, HU-211, in the treatment of experimental pneumococcal meningitis. J Infect Dis 173:735–738

Berdyshev E, Boichot E, Corbel M, Germain N, Lagente V (1998) Effects of cannabinoid receptor ligands on LPS-induced pulmonary inflammation in mice. Life Sci 63:L125–L129

Berdyshev EV, Boichot E, Germain N, Allain N, Anger JP, Lagente V (1997) Influence of fatty acid ethanolamides and delta-9-tetrahydrocannabinol on cytokine and arachidonate release by mononuclear cells. Eur J Pharmacol 330:231–240

Blanchard DK, Newton C, Klein TW, Stewart WE, Friedman H (1986) In vitro and in vivo suppressive effects of delta-9-tetrahydrocannabinol on interferon production by murine spleen cells. Int J Immunopharmacol 8:819–824

Blevins RD, Dumic MP (1980) The effect of delta-9-tetrahydrocannabinol on herpes simplex virus replication. J Gen Virol 49:427–431

Bouaboula M, Rinaldi M, Carayon P, Carillon C, Delpech B, Shire D, Le Fur G, Casellas P (1993) Cannabinoid-receptor expression in human leukocytes. Eur J Biochem 214:173–180

Bradley SG, Munson AE, Dewey WL, Harris LS (1977) Enhanced susceptibility of mice to combinations of delta-9-tetrahydrocannabinol and live or killed gram-negative bacteria. Infect Immun 17:325–329

Bredt BM, Higuera-Alhino D, Shade SB, Hebert SJ, McCune JM, Abrams DI (2002) Short-term effects of cannabinoids on immune phenotype and function in HIV-1-infected patients. J Clin Pharmacol 42:82S–89S

Breivogel CS, Griffin G, Di Marzo V, Martin BR (2001) Evidence for a new G protein-coupled cannabinoid receptor in mouse brain. Mol Pharmacol 60:155–163

Buckley NE, McCoy KL, Mezey E, Bonner T, Zimmer A, Felder CC, Glass M, Zimmer A (2000) Immunomodulation by cannabinoids is absent in mice deficient for the cannabinoid CB(2) receptor. Eur J Pharmacol 396:141–149

Burnette-Curley D, Cabral GA (1995) Differential inhibition of RAW264.7 macrophage tumoricidal activity by delta-9-tetrahydrocannabinol. Proc Soc Exp Biol Med 210:64–76

Burnette-Curley D, Marciano-Cabral F, Fischer-Stenger K, Cabral GA (1993) Delta-9-tetrahydrocannabinol inhibits cell contact-dependent cytotoxicity of Bacillus Calmétte-Guerin-activated macrophages. Int J Immunopharmacol 15:371–382

Cabral GA, Dove Pettit DA (1998) Drugs and immunity: cannabinoids and their role in decreased resistance to infectious disease. J Neuroimmunol 83:116–123

Cabral GA, Fischer-Stenger K (1994) Inhibition of macrophage inducible protein expression by delta-9-tetrahydrocannabinol. Life Sci 54:1831–1844

Cabral GA, Vasquez R (1991) Effects of marijuana on macrophage function. Adv Exp Med Biol 288:93–105

Cabral GA, Vasquez R (1992) Delta-9-tetrahydrocannabinol suppresses macrophage extrinsic antiherpesvirus activity. Proc Soc Exp Biol Med 199:255–263

Cabral GA, Lockmuller JC, Mishkin EM (1986a) Delta-9-tetrahydrocannabinol decreases alpha/beta interferon response to herpes simplex virus type 2 in the B6C3F1 mouse. Proc Soc Exp Biol Med 181:305–311

Cabral GA, Mishkin EM, Marciano-Cabral F, Coleman P, Harris L, Munson AE (1986b) Effect of delta-9-tetrahydrocannabinol on herpes simplex virus type 2 vaginal infection in the guinea pig. Proc Soc Exp Biol Med 182:181–186

Cabral GA, McNerney PJ, Mishkin EM (1987a) Interaction of delta-9-tetrahydrocannabinol with rat B103 neuroblastoma cells. Arch Toxicol 60:438–449

Cabral GA, McNerney PJ, Mishkin EM (1987b) Delta-9-tetrahydrocannabinol inhibits the splenocyte proliferative response to herpes simplex virus type 2. Immunopharmacol Immunotoxicol 9:361–370

Caiaffa WT, Vlahov D, Graham NM, Astemborski J, Solomon L, Nelson KE, Munoz A (1994) Drug smoking, Pneumocystis carinii pneumonia, and immunosuppression increase risk of bacterial pneumonia in human immunodeficiency virus-seropositive injection drug users. Am J Respir Crit Care Med 150:1493–1498

Carayon P, Marchand J, Dussossoy D, Derocq JM, Jbilo O, Bord A, Bouaboula M, Galiegue S, Mondière P, Penarier G, Le Fur GL, Defrance T, Casellas P (1998) Modulation and functional involvement of CB2 peripheral cannabinoid receptors during B-cell differentiation. Blood 92:3605–3615

Carlisle SJ, Marciano-Cabral F, Staab A, Ludwick C, Cabral GA (2002) Differential expression of the CB2 cannabinoid receptor by rodent macrophages and macrophage-like cells in relation to cell activation. Int Immunopharmacol 2:69–82

Carrasco L, Smith AE (1976) Sodium ions and the shut-off of host cell protein synthesis by picornaviruses. Nature 264:807–809

Coates RA, Farewell VT, Raboud J, Read SE, MacFadden DK, Calzavara LM, Johnson JK, Shepherd FA, Fanning MM (1990) Cofactors of progression to acquired immunodeficiency syndrome in a cohort of male sexual contacts of men with human immunodeficiency virus disease. Am J Epidemiol 132:717–722

Coffey RG, Yamamoto Y, Snella E, Pross S (1996) Tetrahydrocannabinol inhibition of macrophage nitric oxide production. Biochem Pharmacol 52:743–751

Cohen S (1976) The 94-day cannabis study. Ann N Y Acad Sci 282:211–220

Condie R, Herring A, Koh WS, Lee M, Kaminski NE (1996) Cannabinoid inhibition of adenylate cyclase-mediated signal transduction and interleukin 2 (IL-2) expression in the murine T-cell line, EL4.IL-2. J Biol Chem 271:13175–13183

Crosby R, Diclemente RJ, Wingood GM, Harrington K, Davies SL, Hook EW, III, Oh MK (2002) Predictors of infection with Trichomonas vaginalis: a prospective study of low income African-American adolescent females. Sex Transm Infect 78:360–364

Croxford JL, Miller SD (2003) Immunoregulation of a viral model of multiple sclerosis using the synthetic cannabinoid R+WIN55,212. J Clin Invest 111:1231–1240

Daaka Y, Klein TW, Friedman H (1995) Expression of cannabinoid receptor mRNA in murine and human leukocytes. Adv Exp Med Biol 373:91–96

Daaka Y, Friedman H, Klein TW (1996) Cannabinoid receptor proteins are increased in Jurkat, human T-cell line after mitogen activation. J Pharmacol Exp Ther 276:776–783

Daaka Y, Zhu W, Friedman H, Klein TW (1997) Induction of interleukin-2 receptor alpha gene by delta-9-tetrahydrocannabinol is mediated by nuclear factor kappaB and CB1 cannabinoid receptor. DNA Cell Biol 16:301–309

De Petrocellis L, Melck D, Palmisano A, Bisogno T, Laezza C, Bifulco M, Di Marzo V (1998) The endogenous cannabinoid anandamide inhibits human breast cancer cell proliferation. Proc Natl Acad Sci U S A 95:8375–8380

Derocq JM, Segui M, Marchand J, Le Fur G, Casellas P (1995) Cannabinoids enhance human B-cell growth at low nanomolar concentrations. FEBS Lett 369:177–182

Derocq JM, Bouaboula M, Marchand J, Rinaldi-Carmona M, Segui M, Casellas P (1998) The endogenous cannabinoid anandamide is a lipid messenger activating cell growth via a cannabinoid receptor-independent pathway in hematopoietic cell lines. FEBS Lett 425:419–425

Devane WA, Dysarz FA, III, Johnson MR, Melvin LS, Howlett AC (1988) Determination and characterization of a cannabinoid receptor in rat brain. Mol Pharmacol 34:605–613
Devane WA, Hanus L, Breuer A, Pertwee RG, Stevenson LA, Griffin G, Gibson D, Mandelbaum A, Etinger A, Mechoulam R (1992) Isolation and structure of a brain constituent that binds to the cannabinoid receptor. Science 258:1946–1949
Di Franco MJ, Sheppard HW, Hunter DJ, Tosteson TD, Ascher MS (1996) The lack of association of marijuana and other recreational drugs with progression to AIDS in the San Francisco Men's Health Study. Ann Epidemiol 6:283–289
Diaz S, Specter S, Coffey RG (1993) Suppression of lymphocyte adenosine 3':5'-cyclic monophosphate (cAMP) by delta-9-tetrahydrocannabinol. Int J Immunopharmacol 15:523–532
Diaz S, Specter S, Vanderhoek JY, Coffey RG (1994) The effect of delta-9-tetrahydrocannabinol on arachidonic acid metabolism in human peripheral blood mononuclear cells. J Pharmacol Exp Ther 268:1289–1296
Facci L, Dal Toso R, Romanello S, Buriani A, Skaper SD, Leon A (1995) Mast cells express a peripheral cannabinoid receptor with differential sensitivity to anandamide and palmitoylethanolamide. Proc Natl Acad Sci U S A 92:3376–3380
Felder CC, Veluz JS, Williams HL, Briley EM, Matsuda LA (1992) Cannabinoid agonists stimulate both receptor- and non-receptor-mediated signal transduction pathways in cells transfected with and expressing cannabinoid receptor clones. Mol Pharmacol 42:838–845
Fischer-Stenger K, Updegrove AW, Cabral GA (1992) Delta-9-tetrahydrocannabinol decreases cytotoxic T lymphocyte activity to herpes simplex virus type 1-infected cells. Proc Soc Exp Biol Med 200:422–430
Fischer-Stenger K, Dove Pettit DA, Cabral GA (1993) Delta-9-tetrahydrocannabinol inhibition of tumor necrosis factor-alpha: suppression of post-translational events. J Pharmacol Exp Ther 267:1558–1565
Fride E, Foox A, Rosenberg E, Faigenboim M, Cohen V, Barda L, Blau H, Mechoulam R (2003) Milk intake and survival in newborn cannabinoid CB1 receptor knockout mice: evidence for a "CB3" receptor. Eur J Pharmacol 461:27–34
Friedman H, Klein T, Specter S (1991) Immunosuppression by marijuana components. In: Ader R, Felten DL, Cohen N (eds) Psychoneuroimmunology. Academic Press, San Diego, pp 931–953
Friedman H, Newton C, Klein TW (2003) Microbial infections, immunomodulation, and drugs of abuse. Clin Microbiol Rev 16:209–219
Galiègue S, Mary S, Marchand J, Dussossoy D, Carrière D, Carayon P, Bouaboula M, Shire D, Le Fur G, Casellas P (1995) Expression of central and peripheral cannabinoid receptors in human immune tissues and leukocyte subpopulations. Eur J Biochem 232:54–61
Gallily R, Yamin A, Waksmann Y, Ovadia H, Weidenfeld J, Bar-Joseph A, Biegon A, Mechoulam R, Shohami E (1997) Protection against septic shock and suppression of tumor necrosis factor alpha and nitric oxide production by dexanabinol (HU-211), a nonpsychotropic cannabinoid. J Pharmacol Exp Ther 283:918–924
Garry RF, Bishop JM, Parker S, Westbrook K, Lewis G, Waite MR (1979) Na+ and K+ concentrations and the regulation of protein synthesis in Sindbis virus-infected chick cells. Virology 96:108–120
Garry RF, Ulug ET, Bose HR Jr (1982) Membrane-mediated alterations of intracellular Na+ and K+ in lytic-virus-infected and retrovirus-transformed cells. Biosci Rep 2:617–623
Goldsmith MA, Greene WC (1996) Interleukin-2 and interleukin-2 receptor. In: Thomas A (ed) The cytokine handbook. Academic Press, New York, pp 57–80
Griffin G, Tao Q, Abood ME (2000) Cloning and pharmacological characterization of the rat CB2 cannabinoid receptor. J Pharmacol Exp Ther 292:886–894
Gross A, Terraza A, Marchant J, Bouaboula M, Ouahrani-Bettache S, Liautard JP, Casellas P, Dornand J (2000) A beneficial aspect of a CB1 cannabinoid receptor antagonist: SR141716A is a potent inhibitor of macrophage infection by the intracellular pathogen Brucella suis. J Leukoc Biol 67:335–344

Gross G, Roussaki A, Ikenberg H, Drees N (1991) Genital warts do not respond to systemic recombinant interferon alpha-2a treatment during cannabis consumption. Dermatologica 183:203–207

Guarisco JL, Cheney ML, LeJeune FE Jr, Reed HT (1988) Isolated uvulitis secondary to marijuana use. Laryngoscope 98:1309–1312

Harkess J, Gildon B, Istre GR (1989) Outbreaks of hepatitis A among illicit drug users, Oklahoma, 1984–87. Am J Public Health 79:463–466

Herring AC, Koh WS, Kaminski NE (1998) Inhibition of the cyclic AMP signaling cascade and nuclear factor binding to CRE and kappaB elements by cannabinol, a minimally CNS-active cannabinoid. Biochem Pharmacol 55:1013–1023

Hollister LE (1986) Health aspects of cannabis. Pharmacol Rev 38:1–20

Howlett AC (1985) Cannabinoid inhibition of adenylate cyclase. Biochemistry of the response in neuroblastoma cell membranes. Mol Pharmacol 27:429–436

Howlett AC (1995) Pharmacology of cannabinoid receptors. Annu Rev Pharmacol Toxicol 35:607–634

Howlett AC, Fleming RM (1984) Cannabinoid inhibition of adenylate cyclase. Pharmacology of the response in neuroblastoma cell membranes. Mol Pharmacol 26:532–538

Howlett AC, Qualy JM, Khachatrian LL (1986) Involvement of Gi in the inhibition of adenylate cyclase by cannabimimetic drugs. Mol Pharmacol 29:307–313

Howlett AC, Barth F, Bonner TI, Cabral G, Casellas P, Devane WA, Felder CC, Herkenham M, Mackie K, Martin BR, Mechoulam R, Pertwee RG (2002) International Union of Pharmacology. XXVII. Classification of cannabinoid receptors. Pharmacol Rev 54:161–202

Jan TR, Rao GK, Kaminski NE (2002) Cannabinol enhancement of interleukin-2 (IL-2) expression by T cells is associated with an increase in IL-2 distal nuclear factor of activated T cell activity. Mol Pharmacol 61:446–454

Jeon YJ, Yang KH, Pulaski JT, Kaminski NE (1996) Attenuation of inducible nitric oxide synthase gene expression by delta-9-tetrahydrocannabinol is mediated through the inhibition of nuclear factor- kappa B/Rel activation. Mol Pharmacol 50:334–341

Juel-Jensen BE (1972) Cannabis and recurrent herpes simplex. Br Med J 4:296

Kagen SL, Kurup VP, Sohnle PG, Fink JN (1983) Marijuana smoking and fungal sensitization. J Allergy Clin Immunol 71:389–393

Kaminski NE (1998) Regulation of the cAMP cascade, gene expression and immune function by cannabinoid receptors. J Neuroimmunol 83:124–132

Kaminski NE, Abood ME, Kessler FK, Martin BR, Schatz AR (1992) Identification of a functionally relevant cannabinoid receptor on mouse spleen cells that is involved in cannabinoid-mediated immune modulation. Mol Pharmacol 42:736–742

Kaminski NE, Koh WS, Yang KH, Lee M, Kessler FK (1994) Suppression of the humoral immune response by cannabinoids is partially mediated through inhibition of adenylate cyclase by a pertussis toxin-sensitive G-protein coupled mechanism. Biochem Pharmacol 48:1899–1908

Kaslow RA, Blackwelder WC, Ostrow DG, Yerg D, Palenicek J, Coulson AH, Valdiserri RO (1989) No evidence for a role of alcohol or other psychoactive drugs in accelerating immunodeficiency in HIV-1-positive individuals. A report from the Multicenter AIDS Cohort Study. JAMA 261:3424–3429

Kawakami Y, Klein TW, Newton C, Djeu JY, Dennert G, Specter S, Friedman H (1988a) Suppression by cannabinoids of a cloned cell line with natural killer cell activity. Proc Soc Exp Biol Med 187:355–359

Kawakami Y, Klein TW, Newton C, Djeu JY, Specter S, Friedman H (1988b) Suppression by delta-9-tetrahydrocannabinol of interleukin 2-induced lymphocyte proliferation and lymphokine-activated killer cell activity. Int J Immunopharmacol 10:485–488

Killestein J, Hoogervorst EL, Reif M, Blauw B, Smits M, Uitdehaag BM, Nagelkerken L, Polman CH (2003) Immunomodulatory effects of orally administered cannabinoids in multiple sclerosis. J Neuroimmunol 137:140–143

Klein TW, Friedman H (1990) Modulation of murine immune cell function by marijuana components. In: Watson R (ed) Drugs of abuse and immune function. CRC Press, Boca Raton, pp 87–111
Klein TW, Newton CA, Widen R, Friedman H (1985) The effect of delta-9-tetrahydrocannabinol and 11-hydroxy-delta-9-tetrahydrocannabinol on T-lymphocyte and B-lymphocyte mitogen responses. J Immunopharmacol 7:451–466
Klein TW, Kawakami Y, Newton C, Friedman H (1991) Marijuana components suppress induction and cytolytic function of murine cytotoxic T cells in vitro and in vivo. J Toxicol Environ Health 32:465–477
Klein TW, Newton C, Widen R, Friedman H (1993) Delta-9-tetrahydrocannabinol injection induces cytokine-mediated mortality of mice infected with Legionella pneumophila. J Pharmacol Exp Ther 267:635–640
Klein TW, Friedman H, Specter S (1998a) Marijuana, immunity and infection. J Neuroimmunol 83:102–115
Klein TW, Newton C, Friedman H (1998b) Cannabinoid receptors and immunity. Immunol Today 19:373–381
Klein TW, Newton CA, Nakachi N, Friedman H (2000) Delta-9-tetrahydrocannabinol treatment suppresses immunity and early IFN-gamma, IL-12, and IL-12 receptor beta 2 responses to Legionella pneumophila infection. J Immunol 164:6461–6466
Klein TW, Newton C, Snella E, Friedman H (2001) Marijuana, the cannabinoid system, and immunomodulation. In: Ader R, Felten DL, Cohen N (eds) Psychoneuroimmunology. Academic Press, San Diego, pp 415–432
Klykken PC, Smith SH, Levy JA, Razdan R, Munson AE (1977) Immunosuppressive effects of 8,9-epoxyhexahydrocannabinol (EHHC). J Pharmacol Exp Ther 201:573–579
Kusher DI, Dawson LO, Taylor AC, Djeu JY (1994) Effect of the psychoactive metabolite of marijuana, delta-9-tetrahydrocannabinol (THC), on the synthesis of tumor necrosis factor by human large granular lymphocytes. Cell Immunol 154:99–108
Lambert DM, DiPaolo FG, Sonveaux P, Kanyonyo M, Govaerts SJ, Hermans E, Bueb J, Delzenne NM, Tschirhart EJ (1999) Analogues and homologues of N-palmitoylethanolamide, a putative endogenous CB2 cannabinoid, as potential ligands for the cannabinoid receptors. Biochim Biophys Acta 1440:266–274
Lawrence T, Willoughby DA, Gilroy DW (2002) Anti-inflammatory lipid mediators and insights into the resolution of inflammation. Nat Rev Immunol 2:787–795
Lee M, Yang KH, Kaminski NE (1995) Effects of putative cannabinoid receptor ligands, anandamide and 2-arachidonyl-glycerol, on immune function in B6C3F1 mouse splenocytes. J Pharmacol Exp Ther 275:529–536
Lee SF, Newton C, Widen R, Friedman H, Klein TW (2001a) Downregulation of cannabinoid receptor 2 (CB2) messenger RNA expression during in vitro stimulation of murine splenocytes with lipopolysaccharide. Adv Exp Med Biol 493:223–228
Lee SF, Newton C, Widen R, Friedman H, Klein TW (2001b) Differential expression of cannabinoid CB(2) receptor mRNA in mouse immune cell subpopulations and following B cell stimulation. Eur J Pharmacol 423:235–241
Lefkowitz SS, Klager K (1978) Effect of delta-9-tetrahydrocannabinol on in vitro sensitization of mouse splenic lymphocytes. Immunol Commun 7:557–566
Li X, Kaminski NE, Fischer LJ (2001) Examination of the immunosuppressive effect of delta-9-tetrahydrocannabinol in streptozotocin-induced autoimmune diabetes. Int Immunopharmacol 1:699–712
Liau A, Diclemente RJ, Wingood GM, Crosby RA, Williams KM, Harrington K, Davies SL, Hook EW, III, Oh MK (2002) Associations between biologically confirmed marijuana use and laboratory-confirmed sexually transmitted diseases among African American adolescent females. Sex Transm Dis 29:387–390
Lozada F, Silverman S Jr, Migliorati CA, Conant MA, Volberding PA (1983) Oral manifestations of tumor and opportunistic infections in the acquired immunodeficiency syndrome (AIDS): findings in 53 homosexual men with Kaposi's sarcoma. Oral Surg Oral Med Oral Pathol 56:491–494

Luo YD, Patel MK, Wiederhold MD, Ou DW (1992) Effects of cannabinoids and cocaine on the mitogen-induced transformations of lymphocytes of human and mouse origins. Int J Immunopharmacol 14:49–56

Luthra YK, Esber HJ, Lariviere DM, Rosenkrantz H (1980) Assessment of tolerance to immunosuppressive activity of delta-9-tetrahydrocannabinol in rats. J Immunopharmacol 2:245–256

Lyman WD, Sonett JR, Brosnan CF, Elkin R, Bornstein MB (1989) Delta-9-tetrahydrocannabinol: a novel treatment for experimental autoimmune encephalomyelitis. J Neuroimmunol 23:73–81

Lynn AB, Herkenham M (1994) Localization of cannabinoid receptors and nonsaturable high-density cannabinoid binding sites in peripheral tissues of the rat: implications for receptor-mediated immune modulation by cannabinoids. J Pharmacol Exp Ther 268:1612–1623

Marciano-Cabral F (1988) Biology of Naegleria spp. Microbiol Rev 52:114–133

Marciano-Cabral F, Ferguson T, Bradley SG, Cabral G (2001) Delta-9-tetrahydrocannabinol (THC), the major psychoactive component of marijuana, exacerbates brain infection by Acanthamoeba. J Eukaryot Microbiol Suppl:4S–5S

Massi P, Fuzio D, Vigano D, Sacerdote P, Parolaro D (2000) Relative involvement of cannabinoid CB(1) and CB(2) receptors in the Delta(9)-tetrahydrocannabinol-induced inhibition of natural killer activity. Eur J Pharmacol 387:343–347

Matsuda LA, Lolait SJ, Brownstein MJ, Young AC, Bonner TI (1990) Structure of a cannabinoid receptor and functional expression of the cloned cDNA. Nature 346:561–564

McCoy KL, Gainey D, Cabral GA (1995) Delta-9-tetrahydrocannabinol modulates antigen processing by macrophages. J Pharmacol Exp Ther 273:1216–1223

McCoy KL, Matveyeva M, Carlisle SJ, Cabral GA (1999) Cannabinoid inhibition of the processing of intact lysozyme by macrophages: evidence for CB2 receptor participation. J Pharmacol Exp Ther 289:1620–1625

McKallip RJ, Lombard C, Martin BR, Nagarkatti M, Nagarkatti PS (2002) Delta(9)-tetrahydrocannabinol-induced apoptosis in the thymus and spleen as a mechanism of immunosuppression in vitro and in vivo. J Pharmacol Exp Ther 302:451–465

Mechoulam R, Ben Shabat S, Hanus L, Ligumsky M, Kaminski NE, Schatz AR, Gopher A, Almog S, Martin BR, Compton DR, Pertwee RG, Griffin G, Bayewitch M, Barg J, Vogel Z (1995) Identification of an endogenous 2-monoglyceride, present in canine gut, that binds to cannabinoid receptors. Biochem Pharmacol 50:83–90

Meyers WA, III, Heath RG (1979) Cannabis sativa: ultrastructural changes in organelles of neurons in brain septal region of monkeys. J Neurosci Res 4:9–17

Miller JM Jr, Goodridge C (2000) Antenatal marijuana use is unrelated to sexually transmitted infections during pregnancy. Infect Dis Obstet Gynecol 8:155–157

Mishkin EM, Cabral GA (1985) Delta-9-tetrahydrocannabinol decreases host resistance to herpes simplex virus type 2 vaginal infection in the B6C3F1 mouse. J Gen Virol 66:2539–2549

Molina-Holgado F, Molina-Holgado E, Guaza C (1998) The endogenous cannabinoid anandamide potentiates interleukin-6 production by astrocytes infected with Theiler's murine encephalomyelitis virus by a receptor-mediated pathway. FEBS Lett 433: 139–142

Morahan PS, Klykken PC, Smith SH, Harris LS, Munson AE (1979) Effects of cannabinoids on host resistance to Listeria monocytogenes and herpes simplex virus. Infect Immun 23:670–674

Morahan PS, Morse SS, McGeorge MG (1980) Macrophage extrinsic antiviral activity during herpes simplex virus infection. J Gen Virol 46:291–300

Munro S, Thomas KL, Abu-Shaar M (1993) Molecular characterization of a peripheral receptor for cannabinoids. Nature 365:61–65

Nahas GG, Morishima A, Desoize B (1977) Effects of cannabinoids on macromolecular synthesis and replication of cultured lymphocytes. Fed Proc 36:1748–1752

Nakano Y, Pross SH, Friedman H (1992) Modulation of interleukin 2 activity by delta-9-tetrahydrocannabinol after stimulation with concanavalin A, phytohemagglutinin, or anti-CD3 antibody. Proc Soc Exp Biol Med 201:165–168
Newell GR, Mansell PW, Wilson MB, Lynch HK, Spitz MR, Hersh EM (1985) Risk factor analysis among men referred for possible acquired immune deficiency syndrome. Prev Med 14:81–91
Newton C, Klein T, Friedman H (1998) The role of macrophages in THC-induced alteration of the cytokine network. Adv Exp Med Biol 437:207–214
Newton CA, Klein TW, Friedman H (1994) Secondary immunity to Legionella pneumophila and Th1 activity are suppressed by delta-9-tetrahydrocannabinol injection. Infect Immun 62:4015–4020
Noe SN, Nyland SB, Ugen K, Friedman H, Klein TW (1998) Cannabinoid receptor agonists enhance syncytia formation in MT-2 cells infected with cell free HIV-1MN. Adv Exp Med Biol 437:223–229
Noe SN, Newton C, Widen R, Friedman H, Klein TW (2000) Anti-CD40, anti-CD3, and IL-2 stimulation induce contrasting changes in CB1 mRNA expression in mouse splenocytes. J Neuroimmunol 110:161–167
Nok AJ, Ibrahim S, Arowosafe S, Longdet I, Ambrose A, Onyenekwe PC, Whong CZ (1994) The trypanocidal effect of Cannabis sativa constituents in experimental animal trypanosomiasis. Vet Hum Toxicol 36:522–524
Paradise LJ, Friedman H (1993) Syphilis and drugs of abuse. Adv Exp Med Biol 335:81–87
Persaud NE, Klaskala W, Tewari T, Shultz J, Baum M (1999) Drug use and syphilis. Co-factors for HIV transmission among commercial sex workers in Guyana. West Indian Med J 48:52–56
Pross SH, Klein TW, Newton C, Smith J, Widen R, Friedman H (1990) Age-related suppression of murine lymphoid cell blastogenesis by marijuana components. Dev Comp Immunol 14:131–137
Pross SH, Nakano Y, Widen R, McHugh S, Newton CA, Klein TW, Friedman H (1992) Differing effects of delta-9-tetrahydrocannabinol (THC) on murine spleen cell populations dependent upon stimulators. Int J Immunopharmacol 14:1019–1027
Pryce G, Ahmed Z, Hankey DJ, Jackson SJ, Croxford JL, Pocock JM, Ledent C, Petzold A, Thompson AJ, Giovannoni G, Cuzner ML, Baker D (2003) Cannabinoids inhibit neurodegeneration in models of multiple sclerosis. Brain 126:2191–2202
Rachelefsky GS, Opelz G (1977) Normal lymphocyte function in the presence of delta-9-tetrahydrocannabinol. Clin Pharmacol Ther 21:44–46
Ramarathinam L, Pross S, Plescia O, Newton C, Widen R, Friedman H (1997) Differential immunologic modulatory effects of tetrahydrocannabinol as a function of age. Mech Ageing Dev 96:117–126
Raz A, Goldman R (1976) Effect of hashish compounds on mouse peritoneal macrophages. Lab Invest 34:69–76
Rinaldi-Carmona M, Barth F, Millan J, Derocq JM, Casellas P, Congy C, Oustric D, Sarran M, Bouaboula M, Calandra B, Portier M, Shire D, Brelière JC, Le Fur GL (1998) SR 144528, the first potent and selective antagonist of the CB2 cannabinoid receptor. J Pharmacol Exp Ther 284:644–650
Rosenkrantz H, Miller AJ, Esber HJ (1975) Delta-9-tetrahydrocannabinol suppression of the primary immune response in rats. J Toxicol Environ Health 1:119–125
Sacerdote P, Massi P, Panerai AE, Parolaro D (2000) In vivo and in vitro treatment with the synthetic cannabinoid CP 55,940 decreases the in vitro migration of macrophages in the rat: involvement of both CB1 and CB2 receptors. J Neuroimmunol 109:155–163
Schatz AR, Kessler FK, Kaminski NE (1992) Inhibition of adenylate cyclase by delta-9-tetrahydrocannabinol in mouse spleen cells: a potential mechanism for cannabinoid-mediated immunosuppression. Life Sci 51:L25–L30
Schatz AR, Koh WS, Kaminski NE (1993) Delta-9-tetrahydrocannabinol selectively inhibits T-cell dependent humoral immune responses through direct inhibition of accessory T-cell function. Immunopharmacology 26:129–137

Schatz AR, Lee M, Condie RB, Pulaski JT, Kaminski NE (1997) Cannabinoid receptors CB1 and CB2: a characterization of expression and adenylate cyclase modulation within the immune system. Toxicol Appl Pharmacol 142:278–287

Sheskin T, Hanus L, Slager J, Vogel Z, Mechoulam R (1997) Structural requirements for binding of anandamide-type compounds to the brain cannabinoid receptor. J Med Chem 40:659–667

Shohami E, Gallily R, Mechoulam R, Bass R, Ben-Hur T (1997) Cytokine production in the brain following closed head injury: dexanabinol (HU-211) is a novel TNF-alpha inhibitor and an effective neuroprotectant. J Neuroimmunol 72:169–177

Showalter VM, Compton DR, Martin BR, Abood ME (1996) Evaluation of binding in a transfected cell line expressing a peripheral cannabinoid receptor (CB2): identification of cannabinoid receptor subtype selective ligands. J Pharmacol Exp Ther 278:989–999

Sidney S, Beck JE, Tekawa IS, Quesenberry CP, Friedman GD (1997) Marijuana use and mortality. Am J Public Health 87:585–590

Sinha D, Bonner TI, Bhat NR, Matsuda LA (1998) Expression of the CB1 cannabinoid receptor in macrophage-like cells from brain tissue: immunochemical characterization by fusion protein antibodies. J Neuroimmunol 82:13–21

Smith MS, Yamamoto Y, Newton C, Friedman H, Klein T (1997) Psychoactive cannabinoids increase mortality and alter acute phase cytokine responses in mice sublethally infected with Legionella pneumophila. Proc Soc Exp Biol Med 214:69–75

Smith SH, Harris LS, Uwaydah IM, Munson AE (1978) Structure-activity relationships of natural and synthetic cannabinoids in suppression of humoral and cell-mediated immunity. J Pharmacol Exp Ther 207:165–170

Smith SR, Terminelli C, Denhardt G (2000) Effects of cannabinoid receptor agonist and antagonist ligands on production of inflammatory cytokines and anti-inflammatory interleukin-10 in endotoxemic mice. J Pharmacol Exp Ther 293:136–150

Snella E, Pross S, Friedman H (1995) Relationship of aging and cytokines to the immunomodulation by delta-9-tetrahydrocannabinol on murine lymphoid cells. Int J Immunopharmacol 17:1045–1054

Specter S, Lancz G, Westrich G, Friedman H (1991) Delta-9-tetrahydrocannabinol augments murine retroviral induced immunosuppression and infection. Int J Immunopharmacol 13:411–417

Srivastava MD, Srivastava BI, Brouhard B (1998) Delta-9-tetrahydrocannabinol and cannabidiol alter cytokine production by human immune cells. Immunopharmacology 40:179–185

Stefano GB, Bilfinger TV, Rialas CM, Deutsch DG (2000) 2-Arachidonyl-glycerol stimulates nitric oxide release from human immune and vascular tissues and invertebrate immunocytes by cannabinoid receptor 1. Pharmacol Res 42:317–322

Stohlman SA, Woodward JG, Frelinger JA (1982) Macrophage antiviral activity: extrinsic versus intrinsic activity. Infect Immun 36:672–677

Struwe M, Kaempfer SH, Geiger CJ, Pavia AT, Plasse TF, Shepard KV, Ries K, Evans TG (1993) Effect of dronabinol on nutritional status in HIV infection. Ann Pharmacother 27:827–831

Sugiura T, Kondo S, Kishimoto S, Miyashita T, Nakane S, Kodaka T, Suhara Y, Takayama H, Waku K (2000) Evidence that 2-arachidonoylglycerol but not N-palmitoylethanolamine or anandamide is the physiological ligand for the cannabinoid CB2 receptor. Comparison of the agonistic activities of various cannabinoid receptor ligands in HL-60 cells. J Biol Chem 275:605–612

Tashkin DP, Baldwin GC, Sarafian T, Dubinett S, Roth MD (2002) Respiratory and immunologic consequences of marijuana smoking. J Clin Pharmacol 42:71S–81S

Tindall B, Cooper DA, Donovan B, Barnes T, Philpot CR, Gold J, Penny R (1988) The Sydney AIDS Project: development of acquired immunodeficiency syndrome in a group of HIV seropositive homosexual men. Aust N Z J Med 18:8–15

Titishov N, Mechoulam R, Zimmerman AM (1989) Stereospecific effects of (–)- and (+)-7-hydroxy-delta-6-tetrahydrocannabinol-dimethylheptyl on the immune system of mice. Pharmacology 39:337–349

Valk P, Verbakel S, Vankan Y, Hol S, Mancham S, Ploemacher R, Mayen A, Lowenberg B, Delwel R (1997) Anandamide, a natural ligand for the peripheral cannabinoid receptor is a novel synergistic growth factor for hematopoietic cells. Blood 90:1448–1457

Vethanayagam D, Pugsley S, Dunn EJ, Russell D, Kay JM, Allen C (2000) Exogenous lipid pneumonia related to smoking weed oil following cadaveric renal transplantation. Can Respir J 7:338–342

Waksman Y, Olson JM, Carlisle SJ, Cabral GA (1999) The central cannabinoid receptor (CB1) mediates inhibition of nitric oxide production by rat microglial cells. J Pharmacol Exp Ther 288:1357–1366

Wallace JM, Lim R, Browdy BL, Hopewell PC, Glassroth J, Rosen MJ, Reichman LB, Kvale PA (1998) Risk factors and outcomes associated with identification of Aspergillus in respiratory specimens from persons with HIV disease. Pulmonary Complications of HIV Infection Study Group. Chest 114:131–137

Warner JF, Dennert G (1982) Effects of a cloned cell line with NK activity on bone marrow transplants, tumour development and metastasis in vivo. Nature 300:31–34

Watzl B, Scuderi P, Watson RR (1991) Marijuana components stimulate human peripheral blood mononuclear cell secretion of interferon-gamma and suppress interleukin-1 alpha in vitro. Int J Immunopharmacol 13:1091–1097

Whitfield RM, Bechtel LM, Starich GH (1997) The impact of ethanol and Marinol/marijuana usage on HIV+/AIDS patients undergoing azidothymidine, azidothymidine/dideoxycytidine, or dideoxyinosine therapy. Alcohol Clin Exp Res 21:122–127

Wiley JL, Martin BR (2002) Cannabinoid pharmacology: implications for additional cannabinoid receptor subtypes. Chem Phys Lipids 121:57–63

Wiley JL, Martin BR (2003) Cannabinoid pharmacological properties common to other centrally acting drugs. Eur J Pharmacol 471:185–193

Wirguin I, Mechoulam R, Breuer A, Schezen E, Weidenfeld J, Brenner T (1994) Suppression of experimental autoimmune encephalomyelitis by cannabinoids. Immunopharmacology 28:209–214

Yebra M, Klein TW, Friedman H (1992) Delta-9-tetrahydrocannabinol suppresses concanavalin A induced increase in cytoplasmic free calcium in mouse thymocytes. Life Sci 51:151–160

Zheng ZM, Specter S, Friedman H (1992) Inhibition by delta-9-tetrahydrocannabinol of tumor necrosis factor alpha production by mouse and human macrophages. Int J Immunopharmacol 14:1445–1452

Zhu LX, Sharma S, Stolina M, Gardner B, Roth MD, Tashkin DP, Dubinett SM (2000) Delta-9-tetrahydrocannabinol inhibits antitumor immunity by a CB2 receptor-mediated, cytokine-dependent pathway. J Immunol 165:373–380

Zhu W, Igarashi T, Qi ZT, Newton C, Widen RE, Friedman H, Klein TW (1993) Delta-9-tetrahydrocannabinol (THC) decreases the number of high and intermediate affinity IL-2 receptors of the IL-2 dependent cell line NKB61A2. Int J Immunopharmacol 15:401–408

Zhu W, Newton C, Daaka Y, Friedman H, Klein TW (1994) Delta-9-tetrahydrocannabinol enhances the secretion of interleukin 1 from endotoxin-stimulated macrophages. J Pharmacol Exp Ther 270:1334–1339

Zhu W, Igarashi T, Friedman H, Klein TW (1995) Delta-9-tetrahydrocannabinol (THC) causes the variable expression of IL2 receptor subunits. J Pharmacol Exp Ther 274:1001–1007

Zhu W, Friedman H, Klein TW (1998) Delta-9-tetrahydrocannabinol induces apoptosis in macrophages and lymphocytes: involvement of Bcl-2 and caspase-1. J Pharmacol Exp Ther 286:1103–1109

Zimmerman S, Zimmerman AM, Cameron IL, Laurence HL (1977) Delta-1-tetrahydrocannabinol, cannabidiol and cannabinol effects on the immune response of mice. Pharmacology 15:10–23

HEP (2005) 168:425–443

Imaging of the Brain Cannabinoid System

K.P. Lindsey[1] (✉) · S.T. Glaser · S.J. Gatley

Center for Translational Neuroimaging, Brookhaven National Laboratory, 30 Bell Avenue, Upton NY, 11973, USA
klindsey@mclean.harvard.edu

[1] Present address: Harvard Medical School, Mclean Hospital, 115 Mill Street, ADARC/Oaks 328, Belmont, MA 02478, USA

Abstract This review covers two major strategies for imaging of the brain cannabinoid system: autoradiography and in vivo neuroimaging. Cannabinoid receptors can be imaged directly with autoradiography in brain slices using radiolabeled cannabinoid receptor ligands. In addition, the effects of pharmacologic doses of unlabeled cannabinoid drugs can be autoradiographically imaged using indicators of blood flow or indicators of metabolism such as glucose analogs. Although cannabinoid imaging is a relatively new topic of research compared to imaging of other drugs of abuse, autoradiographic strategies have produced high-quality information about the distribution of brain cannabinoid receptors and the effects of cannabinoid drugs on brain metabolism. In vivo neuroimaging, in contrast to autoradiography, utilizes noninvasive techniques such as positron emission tomography (PET), single photon emission computed tomography (SPECT), and magnetic resonance imaging (MRI) to image both the binding and the effects of drugs within living brain. These techniques are well developed; however, in vivo imaging of cannabinoid systems is in a very preliminary state. Early results have been promising yet hard to generalize. Definitive answers to some of the most important questions about cannabinoid drugs and their effects await development of suitable in vivo neuroimaging ligands for cannabinoid systems.

Keywords Cannabinoids · Imaging · Metabolism · Blood flow · MRI (magnetic resonance imaging) · PET (positron emission tomography) · Autoradiography

1 Introduction

Tetrahydrocannabinol (Δ^9-THC; the main psychoactive ingredient of cannabis) acts at G protein-coupled receptors (CB_1 receptors) that are abundant in specific brain areas including the cerebellum, hippocampus and outflow nuclei of the basal ganglia. Investigators have imaged these receptors in cryostat sections of brain and also in the living brains of humans and animals. Additionally, the effects of cannabinoid agonists and antagonists on cerebral metabolism and blood flow have been visualized. Two major strategies are reviewed, ex vivo autoradiographical imaging, and in vivo imaging using positron emission tomography (PET), single photon emission computed tomography (SPECT), or magnetic resonance imaging (MRI).

Autoradiography is a relatively old technique, first used in a biological context by Lacassagne in 1924 (Rogers 1973). Radioactivity incorporated into tissues can be used to generate cumulative, spatially accurate representations of the isotope's distribution by placing tissue samples in close proximity to a recording medium, usually, but not always, a photographic emulsion. Due to their invasive nature, autoradiographic strategies necessitate a preclinical or postmortem focus and between-subjects experimental designs.

Over the last quarter century, in vivo imaging modalities have been developed that allow living brains, including the human brain, to be studied in a non-invasive manner. These modalities include PET which utilizes positron emitting radiotracers that are now available for a growing range of neurotransmitter receptors as well as for blood flow and glucose metabolism. PET, as well as the related modality SPECT, have been used to perform the same general kinds of experiments that are possible using ex vivo autoradiography (see Sect. 2). However, in addition, longitudinal and within-subjects experimental designs are possible.

The non-radionuclide modality functional magnetic resonance imaging (fMRI) is also becoming an important tool for human neuropharmacological studies. These include evaluation of acute effects of drugs on control subjects and drug-dependent individuals. Chronic drug use can be evaluated by means of comparing dependent and non-dependent subjects.

Both autoradiographical and in vivo studies of drugs of abuse such as cannabis may help our understanding of mental diseases, and produce useful leads for the development of drug therapies for these illnesses as well as helping us understand mechanisms of addiction. A better understanding of the cannabinoid receptor system might produce more useful therapeutic drugs with less abuse potential.

2 Overview: Five Major Experimental Strategies

Both autoradiography and PET/SPECT neuroimaging are used to visualize the behavior of drug molecules in complex biological systems. Although these techniques are in some ways quite different, they are used to answer the same types of questions by means of the following five types of studies:

1. Biodistribution. Abused substances such as Δ^9-THC can be directly labeled with radioisotopes. This permits quantitative determination and direct measurement of the distribution of drugs throughout the body. This type of information is invaluable when the substrates of drug action are not known. It also has utility when evaluating new synthetic analogs, yielding information about penetration of the blood–brain barrier, and kinetics. Biodistribution studies are discussed in a recent journal special issue (Gatley and Carroll 2003).

2. Receptor mapping. This technique provides highly detailed information about regional drug distribution within a receptor-rich tissue such as the brain.

3. Competition. The ability of an abused drug or therapeutic drug molecule to compete with or to displace a receptor-mapping radioligand for the same binding sites can be measured using imaging techniques. This information allows calculation of the amount of receptor occupancy provided by the given dose of drug. Measurement of the degree of receptor occupancy achieved by a drug potentially allows evaluation of the relationships between receptor occupancy and physiological, behavioral, and subjective effects of the drug.

4. Metabolism and flow. Imaging strategies may be used to measure regional values of cerebral blood flow (rCBF) (Sakurada et al. 1978) and rates of regional cerebral glucose metabolism (rCGM) (Sokoloff et al. 1977). This strategy allows effects of abused drugs on regional and global brain function to be evaluated, since flow and metabolism are correlated with nerve terminal activity.

5. Effects of drugs on diverse neurotransmitter systems. Radioligands, which bind to different sites from the drug of interest, may be used to examine effects of abused drugs on other neurotransmitter systems. For example, the in vivo binding of the dopamine D_2 receptor PET radioligand [^{11}C]raclopride has been shown to be sensitive to alterations in levels of endogenous dopamine (Dewey et al. 1993). Using this technique, the indirect impact of pharmacological doses of a drug of interest, for instance, Δ^9-THC, on other neurotransmitter systems, such

as the dopaminergic system, can be assessed. This strategy has been utilized to date only accidentally in the context of cannabinoid imaging in one human subject discussed in a case report (Voruganti et al. 2001) and is reviewed in Sect. 4.4.

2.1 In Vitro and Ex Vivo Neuroimaging Using Autoradiography

2.1.1 Technique Overview: Autoradiography

In in vitro receptor autoradiography, frozen human or animal brains are processed to form thin (20 μm) sections fixed onto glass slides. The sections are incubated in receptor radioligand solution to allow labeling of receptor-rich areas and washed to remove unbound radioligand. Subsequent exposure of sections using film, imaging plates, or a beta imager yields maps of the receptor distribution. The images produced by autoradiography have spatial resolution of approximately 50 μm, although specialized applications of this technique can yield images with resolution of 0.05 μm. Similar methodology using labeled polynucleotide probes (in situ hybridization) yields maps of gene transcription. In contrast, ex vivo autoradiography involves preparation of postmortem sections after injection of radiotracer into the living animal. Ex vivo autoradiographs have some dependency on physiological factors such as blood flow, as well as on receptor density. Primary foci of research include imaging of labeled cannabinoid receptors directly, and visualization of metabolic effects of cannabinoid drugs via imaging their effects on neuronal metabolism.

2.1.2 Autoradiographic Tracers and Their Substrates

Autoradiographic Mapping of Cannabinoid Receptors Δ^9-THC binds to cannabinoid receptors with only moderate affinity. Synthetic molecules with higher affinities include the non-classical cannabinoid receptor agonist CP 55,940 (Compton et al. 1992b), the aminoalkylindole agonist WIN 55,212-2 (Compton et al. 1992a), and the antagonist SR141716A (Rinaldi-Carmona et al. 1994). Tritiated versions of each of these three drugs have been used in vitro for autoradiographic imaging studies of cannabinoid receptor density with essentially identical results. Since CB_2-type receptors are nearly absent from normal brain, labeling of brain sections primarily reflects the distribution of CB_1 receptors, even with non-specific radioligands.

Autoradiographic Tracers for Neuronal Activation Studies Ex vivo autoradiography is also commonly used with the glucose analog [^{14}C] 2-deoxyglucose (2-DG) to provide maps of the metabolic demands of neurons (Sokoloff et al. 1977). [^{14}C]2-DG is a substrate for facilitated glucose carriers in the blood–brain barrier and neuronal cell membranes, and also for hexokinase. Since it cannot proceed further down the glycolytic pathway, its local accumulation as 2-deoxyglucose-6-phosphate reflects local rates of glucose consumption. Radiolabeled 2-DG can be used to assess the metabolic impact of chronic or acute treatments with drugs. Blood flow, another correlate of neuronal activation, can be measured autoradiographically using labeled iodoantipyrine ([^{125}I]IAP and [^{14}C]IAP). This lipophilic, blood flow-dependent ligand has a uniform brain–blood equilibrium partition coefficient throughout the brain, and washes out slowly from the brain allowing enough time for preparation of brain sections.

2.2
Noninvasive Neuroimaging Techniques: PET, SPECT, and MRI

2.2.1
Technique Overview: PET

PET scanners are able to measure the regional and temporal concentrations of positron emitting nuclides in small (4×4×4 mm) volumes of the human body (Phelps 1991). They therefore allow the extension of radioligand binding studies to the living human brain, provided that suitable labeled compounds are available (Gatley 1996). For use in PET, carbon, nitrogen, and oxygen all have positron-emitting isotopes: ^{11}C ($t_{1/2} = 20$ min), ^{13}N ($t_{1/2} = 10$ min) and ^{15}O ($t_{1/2} = 2$ min). Of these, ^{11}C is perhaps most useful because it can be used to label organic compounds, including drugs of abuse, without altering their pharmacokinetics or binding profiles. In addition, many compounds labeled with ^{18}F ($t_{1/2} = 110$ min) are useful PET tracers, because fluorine can often replace –H or –OH with retention of activity.

The high sensitivity of PET scanners allows detection of microgram quantities of radiolabeled molecules in vivo—amounts so small that they do not exert measurable pharmacologic effects and only occupy a small fraction of drug binding sites. Kinetic modeling approaches permit quantitative visualization of the distribution of receptor sites or enzymes within brain or other organs using suitable tracers. Using PET, the behavior of radiolabeled drugs within living systems can be evaluated either without perturbing the system or in the presence of other drugs.

2.2.2
Technique Overview: SPECT

SPECT scanners offer poorer resolution, sensitivity, and quantification than PET scanners but are more widely available because of greater clinical use. Iodine-123 ($t_{1/2} = 13$ h) has the best properties for labeling low molecular weight organic compounds for SPECT while retaining biological activity (Gatley 1993), since an iodine atom is isosteric with a methyl group. Many ^{123}I-labeled radioligands are available. The tracer [^{123}I]iodobenzamide ([^{123}I]IBZM) can be used, like raclopride, to assess changes in synaptic dopamine (Laruelle 2000). SPECT can also be used to visualize relative perfusion patterns using clinical radiopharmaceuticals labeled with ^{99m}Tc ($t_{1/2} = 6$ h) (Iyo et al. 1997). Cerebral blood flow can be measured quantitatively from washout kinetics of ^{133}Xe, though this method yields poor resolution images with relatively high radiation exposure to the airways.

2.2.3
In Vivo Imaging Radioligands

PET/SPECT Mapping of Cannabinoid Receptors The properties of Δ^9-THC make it unsuitable as a PET CB_1 radioligand. These include extremely high lipophilicity, only moderate affinity for CB_1 receptors, and a structure difficult to label in the time constraints imposed by ^{11}C. In an early study, THC was modified by labeling with ^{18}F in the hydrocarbon side chain. Unfortunately, [^{18}F]THC showed poor brain uptake with a homogeneous distribution in baboon brain (Charalambous et al. 1991), and there was uptake of radioactivity in the skull, suggesting catabolic loss of labeled fluoride ion. It is likely, therefore, that the PET images represented only non-specific uptake of the tracer with a negligible component due to specific binding to cannabinoid receptors. Later radioligand development efforts have largely focused on pyrazole antagonists, as discussed in Sect. 3.1.

PET/SPECT Ligands for Neuronal Activation Studies PET studies are commonly used with the glucose analog [^{18}F]fluorodeoxyglucose (FDG) to provide maps of the metabolic demands of neurons in a manner analogous to the use of 2-DG in autoradiography (Reivich et al. 1977). Like 2-DG, FDG is a substrate for facilitated glucose carriers in the blood–brain barrier. Its accumulation within neurons reflects local rates of glucose consumption and is correlated with neuronal activation. FDG can be used to assess the metabolic impact of chronic or acute treatments with drugs. No SPECT equivalent of the PET tracer FDG is available for brain metabolic studies (Gatley 2003). Blood flow, another correlate of neuronal activation, can be measured with PET using radiolabeled water (Raichle et al. 1983).

2.2.4
Technique Overview: MRI

MRI is another noninvasive strategy that can be used to visualize the effects of drugs of abuse, such as cannabis. Although MRI can be used to answer pharmacologic questions about drugs of abuse, its technique and applications are somewhat different from those of both autoradiography and other in vivo imaging strategies such as PET and SPECT. Rather than detection of radioactivity, MRI involves detection of spin properties of hydrogen nuclei, "protons," which depend on their physical-chemical environments.

Although MRI is not a particularly sensitive technique, it does produce images with excellent spatial resolution (resolution $\ll 1$ mm). Thus anatomical differences between control subjects and drug abusers can be measured. Furthermore, high-resolution images can be used to more clearly define small brain regions for subsequent analysis using a lower resolution imaging strategy such as PET, or fMRI (see below) in the same individual.

Even though drug concentrations and kinetics can rarely be directly measured within living systems using MRI, because of sensitivity issues, the effect of drugs can be inferred by using MRI to measure the effect of the drug on imaging parameters thought to be correlated with neuronal activity, a technique known as functional MRI (fMRI). The most common fMRI technique is BOLD (blood oxygenation level dependent) scanning, which has excellent time resolution on the order of seconds. The kinetics of changes in neuronal activation caused by drug or by performance of a cognitive or behavioral task can be resolved using MRI with much faster time resolution than using PET, though the spatial resolution of fMRI is similar. Very little cannabinoid research utilizing MRI techniques has yet been reported.

3
Major Topics of Investigations Using Autoradiography

3.1
Measurement of Cannabinoid Receptor Density

CB_1 Receptor Mapping Autoradiographic studies with high-affinity THC analogs both in rat brain tissue (Herkenham et al. 1990) and in postmortem human brain tissue (Thomas et al. 1992; Biegon and Kerman 2001) have demonstrated high concentrations of cannabinoid receptors in the basal ganglia and especially in its outflow nuclei, the globus pallidus, and substantia nigra. High concentrations are also found in hippocampus and cerebellum. The cerebral cortex, especially the cingulate gyrus, also has a fairly high CB_1 receptor density. Some other regions, including most of the brainstem and the thalamus, contain few CB_1 receptors.

Drug Effects on Cannabinoid Receptor Density A number of other studies have assessed the effect of chronic treatments with cannabimimetic drugs on cannabinoid receptor binding. An early study, investigating the mechanism of locomotor tolerance to treatments consisting of 2 weeks of daily i.p. Δ^9-THC or CP 55,940 in rats, reported dose-dependent reductions in binding of radiolabeled CP 55,940. This was attributed to agonist-induced downregulation of CB_1 receptors in striatal brain sections (Oviedo et al. 1993). A later study comparing receptor binding alterations after chronic Δ^9-THC in rats reported that 5 days i.p. administration of Δ^9-THC decreased cannabinoid receptor binding in all brain areas studied, including cerebellum, hippocampus, basal ganglia, limbic nuclei, and cerebral cortex, among others (Romero et al. 1997). CB_1 mRNA levels in these regions were also measured, but did not show reductions in parallel to reductions in receptor binding (Romero et al. 1997). Further research by this group focused on the time-course of receptor down-regulation. Rats were treated with i.p. Δ^9-THC for 1, 3, 7, or

14 days before receptor autoradiography. Downregulation was largely progressive, yet highly variable between brain regions studied. For instance, rapid and robust reductions in [^{3}H]WIN 55,212-2 binding were observed in the dentate gyrus, yet in the basal ganglia reductions in binding were slower in onset and more moderate in degree (Romero et al. 1998); alterations in CB_1 mRNA, where present, occurred after changes in receptor binding. Similar reductions in the binding of [^{3}H]CP 55,940 in rat brain were reported after i.p. Δ^9-THC twice daily for 6 days (Rubino et al. 2000), a regimen that produced maximal reductions. Furthermore, these reductions were accompanied by increases in cyclic adenosine monophosphate (cAMP) and protein kinase A (PKA) activity in the same regions (cerebellum, striatum, globus pallidus, and cortex).

Most autoradiographic studies using postmortem human brain tissue have involved schizophrenic subjects, since a link has been suggested between brain cannabinoid systems and schizophrenia (Arseneault et al. 2004), based, in part, on a high density of CB_1 receptors in brain areas implicated in schizophrenia (Biegon and Kerman 2001). Dean and colleagues studied postmortem tissue from schizophrenic subjects, some of whom had cannabis exposure shortly before death. They associated increases in [^{3}H]CP 55,940 binding in the dorsolateral prefrontal cortex with schizophrenia, but increases in binding in the caudate-putamen with cannabis use (Dean et al. 2001). Another postmortem study of the anterior cingulate cortex revealed a 64% increase in [^{3}H]SR141716A binding in tissue from schizophrenic subjects compared to normal matched controls (Zavitsanou et al. 2004).

Competition for Cannabinoid Receptor Binding Autoradiographic studies in rat brain indicate that anandamide, SR141716A, and CP 55,940 compete for the same cannabinoid receptor, despite the fact that some effects of anandamide are not blocked by the selective CB_1 receptor antagonist SR141716A (Adams et al. 1998). This may partly reflect binding of anandamide to vanilloid receptors (Zygmunt et al. 1999). There are hints that one or more additional central non-vanilloid cannabinoid-type receptors may exist (Calignano et al. 1998; Di Marzo et al. 2000).

3.2
Measurement of Cannabinoid Effects on Neuronal Metabolism

Several studies have examined the effects of Δ^9-THC upon regional cerebral glucose metabolism (rCGM). Biphasic dose-related alterations in glucose utilization in rat have been found. Slight increases in rCGM in limbic and cortical areas but not in diencephalic or brainstem areas were reported after low doses (0.2 mg/kg i.v.) of Δ^9-THC in rats (Margulies and Hammer 1991). Larger doses (2 or 10 mg/kg) decreased rCGM in the cortical and limbic areas. Another paper has shown evidence of increases in rCGM with small doses of Δ^9-THC and decreases with larger doses of Δ^9-THC. Small increases in overall CGM were found using a dose of 1 mg/kg and overall decreases after 5 mg/kg (Brett et al. 2001). Significant reductions in rCGM were seen in rat hippocampal, limbic, sensory, and sensorimotor processing regions, consistent with the effects of Δ^9-THC on memory, sensory perception, and motor control (Brett et al. 2001). Autoradiography has also been used to examine the time-course of Δ^9-THC effects on rCGM. A single dose (2.5 mg/kg i.p.) of Δ^9-THC resulted in an immediate widespread depression of CGM including limbic areas, and motor and sensory systems. Metabolism returned to baseline in 8 of the 17 structures originally affected after 6 h, and in all but 6 structures after 24 h (Whitlow et al. 2002).

Other researchers have not found biphasic effects of Δ^9-THC on rCGM, although they confirm time- and dose-dependent effects of Δ^9-THC on rCGM in rat brain. Acute administrations of low doses of i.p. Δ^9-THC produced dose-dependent reductions in glucose utilization in rat brain. A dose of 1.0 mg/kg i.p. Δ^9-THC was associated with decreases in rCGM occurring primarily in the limbic and sensory systems, with more widespread reductions occurring after a 2.5 mg/kg dose. If rats were pre-treated with the CB_1 antagonist SR141716A prior to Δ^9-THC administration, no significant reductions in rCGM were observed in 34 of the 38 regions examined (Freedland et al. 2002). Freedland and colleagues reconcile the discrepancy between their findings of decreased rCGM after 1 mg/kg i.p. Δ^9-THC, with the findings of Margulies and others cited above (increased rCGM after 1 mg/kg i.v. Δ^9-THC) by pointing out the different routes of Δ^9-THC administration used in the studies. Intraperitoneal administration of drugs leads to slower onset kinetics compared to intravenous administration, a factor that has been found to have a significant impact on the effects of other drugs of abuse (Porrino 1993). Rats with a tolerance to Δ^9-THC exhibited an altered pattern of glucose metabolism relative to drug-naïve animals. While naïve rats exhibited large global decreases in CGM following a single 10 mg/kg i.p. dose of Δ^9-THC, behaviorally tolerant rats had a more localized reduction of CGM, primarily in regions associated with memory, reward, and stress, such as the hippocampus, nucleus accumbens, mediodorsal thalamus, basolateral amygdala and median raphe (Whitlow et al. 2003).

The effects of synthetic cannabinoid drugs have also been studied using the 2-DG strategy. One study found that the effects of the synthetic cannabinoid agonist WIN 55,212-2 on brain glucose metabolism are largely consistent with the effects of Δ^9-THC. Doses between 0.15–0.30 mg/kg WIN 55,212-2 reduced 2-DG accumulation in the hippocampus and ventrolateral thalamic nucleus of rats (Pontieri et al. 1999). Another study found that acute administration of the CB_1 antagonist SR141716A decreased rCGM in limbic areas thought to be involved in motivated behavior. These findings were accompanied by reduced rates of food-reinforced responding in the same animals. Reductions in both responding and rCGM in limbic systems after SR141617A were more pronounced in animals that had been made tolerant to the behavioral effects of Δ^9-THC (Freedland et al. 2003).

3.3
Measurement of Cannabinoid Effects on Blood Flow

Δ^9-THC and its active metabolite 11-OH-THC both reduce rCBF, as measured using autoradiography with [^{14}C]iodoantipyrine (IAP) . Rats injected with Δ^9-

THC at doses ranging from 0.5 to 16 mg/kg 30 min prior to sacrifice had variably decreased rCBF in 16 brain areas, including the CA1 region of the hippocampus, the frontal and medial prefrontal cortex, the nucleus accumbens, and the claustrum. Other regions, such as the medial septum, caudate, cerebellum, and several other cortex regions, remained unaffected. Similar results were obtained when rats were injected with 11-OH-THC. Most of the regions affected by Δ^9-THC in this study express CB_1, with the notable exceptions of the CA3 region of the hippocampus and the cerebellum, which had no change in cerebral blood flow (Bloom et al. 1997).

Anandamide dose-dependently reduces rCBF in the rat, although it has a short duration of action. At 15 min after 10 mg/kg anandamide, flow was reduced in the amygdala, cingulate, frontal prepyriform, sensorimotor, and claustrocortex. A maximal effect, with 16 additional brain regions involved, was observed after 30 mg/kg anandamide. In most areas, reductions were persistent for 60 min following this larger dose. Anandamide at 3 mg/kg had no effect (Stein et al. 1998).

The limited studies examining the effects of both exogenous and putative endogenous cannabinoids upon regional cerebral blood flow complement studies measuring local brain metabolic activity with 2-DG. They both suggest a strong dose- and time-dependent response to a stimulation of the endocannabinoid system. Together, they support the notion of heterogeneous effects of cannabinoids on brain metabolism.

4 Major Topics of In Vivo Investigations Using PET and SPECT

4.1 Measurement of Cannabinoid Receptor Density

Four types of cannabinoid receptor ligands are currently known: the plant cannabinoids, such as Δ^9-THC and their synthetic relatives such as CP55,940; the endocannabinoids, such as anandamide; the pyrazole ligands, such as SR141716A; and the aminoalkylindole ligands, such as WIN55,2 12-2 (Fig. 1). As mentioned previously, radioligands for the mapping of cannabinoid receptors in vivo are still under development. We have recently reviewed efforts at Brookhaven in collaboration with Dr Alexandros Makriyannis' group (Gatley et al. 2004). Our starting point was to replace the chlorophenyl group of SR141716A with an ^{123}I-iodophenyl group. This produced a tracer (AM251) that was evaluated as a SPECT ligand. Although AM251 gave promising results in mice, and in in vitro autoradiography, it failed to enter the brains of baboons in SPECT experiments (Gatley et al. 1998). A further structural modification of AM251 was performed—insertion of an oxygen into the piperidine ring—yielding AM281, which has lower lipophilicity. In ex vivo experiments in rodents, [^{123}I]AM281 yielded brain autoradiographs similar to those obtained using tritiated ligands in in vitro experiments. Using [^{123}I]AM281 in SPECT experiments, we were able to image CB_1 receptors for the first time in the

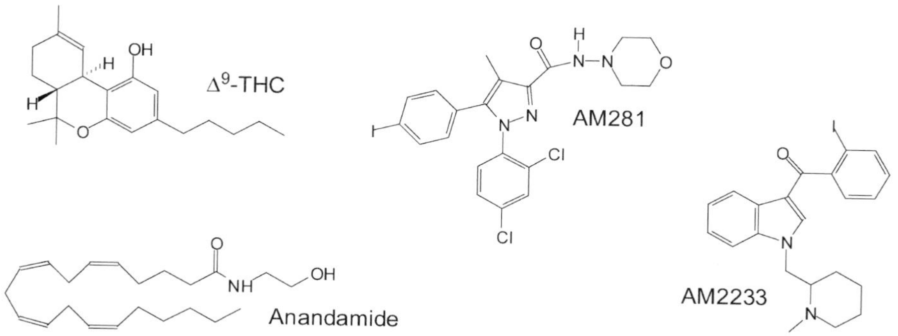

Fig. 1. Representative structures of the different cannabinoid receptor ligand classes: the plant cannabinoid, Δ^9-tetrahydrocannabinol; the endocannabinoid, arachidonoyl ethanolamide (anandamide); the synthetic pyrazole inverse agonist AM281 and the potent aminoalkylindole agonist AM2233. Both AM281 and AM2233 contain an iodine atom that has been labeled with radioiodine for in vitro and in vivo binding experiments

living primate brain (Gatley et al. 1998). The first human brain [^{123}I]AM281 SPECT images of CB_1 receptors have recently been reported by Berding et al. (2004). As anticipated, the extent of specific binding was rather low, and extensive clinical imaging research on CB_1 receptors will probably await development of either a SPECT radioligand with superior properties to [^{123}I]AM281, or, taking advantage of the higher sensitivity of PET cameras, a PET radioligand with at least equivalent brain penetration and receptor affinity to [^{123}I]AM281. Although several candidate PET imaging agents have been synthesized and evaluated biologically by ourselves (Gatley et al. 2004; Gifford et al. 2003) and others (Mathews et al. 2000, 2002; Katoch-Rouse et al. 2003), none has yet been satisfactory.

4.2 Measurement of Cannabinoid Effects on Brain Metabolism

Acute Effects of Δ^9-THC on Brain Glucose Metabolism Two papers have been published utilizing PET to assess the effects of cannabinoids on rCGM in human subjects. (Volkow et al. 1991, 1996). The most consistent observation both in normal controls and habitual marijuana users was an increase in relative metabolic rate in the cerebellum after i.v. Δ^9-THC. This increase was positively correlated both with concentrations of Δ^9-THC in the plasma and with the intensity of self-reported ratings of intoxication. However, the average increase in cerebellar metabolism after Δ^9-THC administration was less in marijuana users than in controls. The FDG/PET studies also demonstrated that marijuana users, but not controls, responded to Δ^9-THC administration with increased metabolic activity in the prefrontal cortex, orbitofrontal cortex, and basal ganglia. Unlike the consistent effects of Δ^9-THC on *relative* metabolic rates, absolute *global* changes were quite variable, as were subjective responses to marijuana or Δ^9-THC. In the studies of Volkow et al. (1991, 1996), Δ^9-THC apparently behaved dissimilarly to acutely administered cocaine, alcohol, morphine, amphetamine, and benzodiazepines (see references in Gatley and Volkow 1998) in that single acute administrations of Δ^9-THC did not reduce overall metabolic rates. These results are largely consistent with results reported using rCBF as an indicator of neuronal activation (discussed below in Sect. 4.3).

Effects of Chronic Δ^9-THC on Brain Glucose Metabolism Marijuana users had lower baseline cerebellar metabolism than controls (Volkow et al. 1996). This finding, coupled with the finding of increased rCGM after acute exposure to cannabinoids suggests that the decreases in basal cerebellar metabolic rates found in habitual marijuana users may reflect a compensatory response to chronic exposure to the drug. Functions known to be associated with the cerebellum, such as motor coordination, proprioception and learning, are adversely affected both during acute marijuana intoxication and in habitual users of the drug (Varma et al. 1988). The PET scanner used in these investigations lacked sufficient resolution to examine metabolic rates in other brain areas, such as hippocampus, substantia nigra, and caudate nucleus, which contain high concentrations of cannabinoid receptors. However, increased rCGM has not been seen in these areas in rodents using autoradiographic imaging with 2-DG. Studies using modern PET cameras will allow more detailed examination of regional changes in human rCGM induced by acute and chronic Δ^9-THC.

4.3 Measurement of Cannabinoid Effects on Blood Flow

Changes in regional neuronal metabolism are coupled to corresponding regional changes in blood flow. Early ^{133}Xe investigations found bilateral hemispheric increases (right>left) in rCBF 30 min after smoking marijuana compared to smoking placebo in 32 normal human males with a history of exposure to marijuana (Mathew et al. 1992). Increases were greatest in the frontal lobes and were proportional to the reported degree of intoxication. More recent blood flow studies have employed PET and ^{15}O water. An advantage of this tracer is the very short (2 min) physical half-life, which unlike FDG allows repeated measurements in a scanning session. Consequently, ^{15}O water studies can detect brief alterations in rCBF, whereas FDG studies measure accumulation of ^{18}F over a period of about one hour. An ^{15}O water PET study by the Mathew group in 32 normal volunteers found that increases in subject-reported intoxication after i.v. Δ^9-THC doses of 0.15 or 0.25 mg/min over 20 min (total doses of 3 or 5 mg) correlated most markedly with rCBF increases in the right frontal regions, specifically frontal cortex, insula, and cingulate gyrus (Mathew et al. 1997). A subsequent paper showed that the earlier reported increases in rCBF were not present in all 46 subjects studied, and that some subjects showed a decrease in rCBF in the cerebellum that was associated with subject-reported disturbances of time sense (Mathew et al. 1998). A third paper by this group (59 normal subjects) confirmed earlier reports of increased rCBF (right>left hemisphere) and also found increased rCBF not only in frontal lobes, but also in anterior cingulate. The increased rCBF in frontal lobes and anterior cingulate was associated with subject-reported sensations of depersonalization (Mathew et al. 1999). Most recently, this group has published the time-course of changes in rCBF and behavior after the same doses of Δ^9-THC in 47 normal subjects. This study again confirms earlier findings and additionally reports that blood flow to the cerebellum is increased after the high Δ^9-THC dose (Mathew et al. 2002).

Self-administration of Δ^9-THC by smoking a marijuana cigarette, where the subject controls the dose and the rate of dosing to achieve a desired effect, might be expected to affect rCBF or rCGM differently from experimenter-controlled i.v. Δ^9-THC. PET studies using ^{15}O water, in fact, have shown increases in similar regions (orbitofrontal lobes, insula, and temporal poles), in addition to increased rCBF in the cerebellum, in normal subjects with previous histories of marijuana use ($n=5$) after smoking marijuana (O'Leary et al. 2000). Rather than measuring blood flow when the subject is in a resting state, this study used a dichotic listening task to measure marijuana effects on task-related changes in rCBF. Large decreases in rCBF were reported during the listening task in temporal lobe areas sensitive to auditory attention effects (O'Leary et al. 2000). A similar PET study from this group using the same technique in 12 experienced subjects assessed the effect of smoking marijuana on task-related rCBF during an auditory attention task (O'Leary et al. 2002). In addition to replication of their earlier findings, this study noted that anterior increases in rCBF occur primarily in paralimbic regions and postulated that these may be related to marijuana's mood-related effects. Also in this study, reduced rCBF was found in several brain regions, including parietal lobe, frontal lobe, and thalamus, which may form part of an attentional network. No rCBF changes were noted in the nucleus accumbens or in any other region thought to be associated with "reward circuitry". Interestingly, brain regions having high densities of cannabinoid receptors, such as basal ganglia and hippocampus, also did not show changes in rCBF (O'Leary et al. 2002). The most recent study from this group assessed the effect of smoked marijuana in 12 occasional and 12 chronic users during counting and finger-tapping tasks accompanied by PET with ^{15}O water. Both counting rate and finger-tapping rate were acutely increased after marijuana smoking and both these effects were correlated with increased blood flow in the cerebellum (O'Leary et al. 2003).

Another group of studies using PET with ^{15}O water have examined the effects of chronic marijuana on rCBF. In contrast to findings after acute marijuana use, chronic users had decreased blood flow to a region of posterior cerebellum after more than 26 h of monitored abstinence from marijuana, compared to control subjects (Block et al. 2000b). Additionally, a later publication using a similar design evaluated effects of chronic marijuana on memory-related blood flow. Decreases in prefrontal cortex, altered lateralization in hippocampus, and increased rCBF in memory-related regions of cerebellum were documented (Block et al. 2002).

The cerebellum is likely to be involved in the psychoactive effects of marijuana. The effects of cannabinoids on rCBF in the cerebellum are consistent with interactions between cannabinoids and the high concentration of CB_1 receptors in this brain area. Both acute marijuana intoxication and habitual use have been shown to affect parameters such as motor coordination, proprioception, and learning, in which the cerebellum plays a key role (Varma et al. 1988). The lack of significant changes in blood flow in the basal ganglia is somewhat surprising given the facts that firstly, high densities of CB_1 receptors are present in this region, and, secondly, that marijuana serves as a reinforcer in at least a large subset of humans. The endocannabinoid system's modulatory/inhibitory actions on presynaptic neurotransmitter release may complicate the interpretation of regional changes in blood flow after exogenous cannabinoid receptor agonists.

4.4 Other Topics

Stimulant drugs and reinforcers have been shown to be associated with elevated synaptic dopamine that can be monitored in PET studies using the D_2 radioligand [^{11}C]raclopride, whose in vivo binding is sensitive to alterations in extracellular dopamine. We are not aware of any published systematic study of cannabis smoking or Δ^9-THC using this experimental paradigm. However, a study of a single individual was inadvertently conducted with the SPECT tracer [^{123}I]IBZM, whose binding is also sensitive to competition with synaptic dopamine. This study was designed to measure alterations in dopaminergic function in schizophrenics. During the scanning protocol, the subject reported feeling anxious and requested a break. During this break he surreptitiously smoked marijuana. This behavior was revealed the next day during a follow-up mental status exam, which showed significant worsening of psychotic symptoms. Comparison of the subject's scans taken before and immediately after marijuana showed a 20% reduction of D_2 binding potential, attributed to increased synaptic dopaminergic activity (Voruganti et al. 2001). This anecdotal report illustrates some of the challenges of clinical drug abuse research.

5 Major Topics of Investigation using MRI

5.1 Cannabinoid Research Utilizing MRI

Anatomical Studies An early paper using the comparatively primitive technique air encephalography to evaluate neuroanatomical changes in chronic cannabis users reported cerebral atrophy (Campbell et al. 1971), and sparked a debate in the field. Other studies using more advanced techniques such as computerized axial tomography have not substantiated these results (Co et al. 1977; Kuehnle et al. 1977). Two more recent papers utilizing MRI to assess anatomical changes in marijuana using subjects have reported conflicting findings. A study combining both PET with ^{15}O water and structural MRI to evaluate alterations in blood flow as well as structural changes in the brains of 57 subjects found that marijuana users who started using marijuana before the age of 17 had smaller brains than either subjects who began using marijuana later, or control subjects (Wilson et al.

2000). Additionally, the subjects who began using marijuana earlier had smaller gray matter/white matter ratios than the other subjects, and were smaller in overall body size. Finally, the early marijuana users had higher global CBF. In contrast, another MRI research group found no significant structural changes in the brains of 18 frequent users of marijuana (Block et al. 2000a). Although the issue of anatomical alterations in brain after marijuana use remains to be definitively resolved, large studies using combinations of more than one imaging modality are likely to provide the best answers about relationships between the presence or the absence of these changes and the functional impact of marijuana use.

Functional Studies As of March 2004, there have been no peer-reviewed papers utilizing BOLD fMRI to examine functional changes in brain after acute or chronic marijuana. However, a recent meeting abstract hints at the wealth of information that remains to be gathered using this powerful technique. This preliminary study reported that marijuana users had decreased brain activation in cerebellar vermis and dorsal parietal cortex compared to normal control subjects during a visual attention task. Additionally, marijuana users were reported to show exposure-dependent decreases in relative BOLD signal in the cerebellar vermis (Chang et al. 2003). A pilot study of working memory in cannabis smokers, tobaco smokers and non-smoking controls has recently appeared (Jacobsen et al. 2004).

6 Summary

Compared with investigations of other drugs of abuse such as cocaine and opioids, imaging research on brain cannabinoid systems is still in its infancy. Although significant progress has been made using autoradiographic techniques, a great deal of work remains to be done with in vivo imaging. The near future will see clinical studies using fMRI, high-resolution FDG/PET, and PET studies with CB_1 receptor radioligands, and with radioligands for other neuroreceptors. The development of small animal imaging technologies, in the form of microPET cameras and small-bore high-field MRI scanners will allow very tightly controlled studies of cannabinoid drugs and their effects in living animal subjects, which may help resolve the conflicting results that have been reported in the human cannabinoid imaging literature.

Acknowledgements. This work was conducted at the Brookhaven National Laboratory under Contract DE-AC02-98CH10886 from the U.S. Department of Energy. KPL and STG thank the National Institute on Drug Abuse for support under award number 1 T32 DA 07316. The authors thank Drs Alexandros Makriyannis, Nora Volkow and Andrew Gifford for advice and encouragement.

References

Adams IB, Compton DR, Martin BR (1998) Assessment of anandamide interaction with the cannabinoid brain receptor: SR 141716A antagonism studies in mice and autoradiographic analysis of receptor binding in rat brain. J Pharmacol Exp Ther 284:1209–1217

Arseneault L, Cannon M, Witton J, Murray RM (2004) Causal association between cannabis and psychosis: examination of the evidence. Br J Psychiatry 184:110–117

Berding G, Muller-Vahl K, Schneider U, Gielow P, Fitschen J, Stuhrmann M, Harke H, Buchert R, Donnerstag F, Hofmann M, Knoop BO, Brooks DJ, Emrich HM, Knapp WH (2004). [(123)I]AM281 single-photon emission computed tomography imaging of central cannabinoid CB(1) receptors before and after Delta(9)-tetrahydrocannabinol therapy and whole-body scanning for assessment of radiation dose in tourette patients. Biol Psychiatry. 55:904–915

Biegon A, Kerman IA (2001) Autoradiographic study of pre- and postnatal distribution of cannabinoid receptors in human brain. Neuroimage 14:1463–1468

Block RI, O'Leary DS, Ehrhardt JC, Augustinack JC, Ghoneim MM, Arndt S, Hall JA (2000a) Effects of frequent marijuana use on brain tissue volume and composition. Neuroreport 11:491–496

Block RI, O'Leary DS, Hichwa RD, Augustinack JC, Ponto LL, Ghoneim MM, Arndt S, Ehrhardt JC, Hurtig RR, Watkins GL, Hall JA, Nathan PE, Andreasen NC (2000b) Cerebellar hypoactivity in frequent marijuana users. Neuroreport 11:749–753

Block RI, O'Leary DS, Hichwa RD, Augustinack JC, Boles Ponto LL, Ghoneim MM, Arndt S, Hurtig RR, Watkins GL, Hall JA, Nathan PE, Andreasen NC (2002) Effects of frequent marijuana use on memory-related regional cerebral blood flow. Pharmacol Biochem Behav 72:237–250

Bloom AS, Tershner S, Fuller SA, Stein EA (1997) Cannabinoid-induced alterations in regional cerebral blood flow in the rat. Pharmacol Biochem Behav 57:625–631

Brett R, MacKenzie F, Pratt J (2001) Delta 9-tetrahydrocannabinol-induced alterations in limbic system glucose use in the rat. Neuroreport 12:3573–3577

Calignano A, La Rana G, Giuffrida A, Piomelli D (1998) Control of pain initiation by endogenous cannabinoids. Nature 394:277–281

Campbell AM, Evans M, Thomson JL, Williams MJ (1971) Cerebral atrophy in young cannabis smokers. Lancet 2:1219–1224

Chang L, Leckova K, Cloak C, Arnold S, Yakupov R, Lozar C, Warren K, Ernst T (2003) Decreased BOLD activation during visual attention tasks in marijuana abusers. International Society of Magnetic Resonance in Medicine. Toronto, ON, Canada

Charalambous A, Marciniak G, Shiue CY, Dewey SL, Schlyer DJ, Wolf AP, Makriyannis A (1991) PET studies in the primate brain and biodistribution in mice using (–)-5'-18F-delta 8-THC. Pharmacol Biochem Behav 40:503–507

Co BT, Goodwin DW, Gado M, Mikhael M, Hill SY (1977) Absence of cerebral atrophy in chronic cannabis users. Evaluation by computerized transaxial tomography. JAMA 237:1229–1230

Compton DR, Gold LH, Ward SJ, Balster RL, Martin BR (1992a) Aminoalkylindole analogs: cannabimimetic activity of a class of compounds structurally distinct from delta 9-tetrahydrocannabinol. J Pharmacol Exp Ther 263:1118–1126

Compton DR, Johnson MR, Melvin LS, Martin BR (1992b) Pharmacological profile of a series of bicyclic cannabinoid analogs: classification as cannabimimetic agents. J Pharmacol Exp Ther 260:201–209

Dean B, Sundram S, Bradbury R, Scarr E, Copolov D (2001) Studies on [3H]CP-55940 binding in the human central nervous system: regional specific changes in density of cannabinoid-1 receptors associated with schizophrenia and cannabis use. Neuroscience 103:9–15

Dewey SL, Smith GS, Logan J, Brodie JD, Fowler JS, Wolf AP (1993b) Striatal binding of the PET ligand 11C-raclopride is altered by drugs that modify synaptic dopamine levels. Synapse 13:350–356

Di Marzo V, Breivogel CS, Tao Q, Bridgen DT, Razdan RK, Zimmer AM, Zimmer A, Martin BR (2000) Levels, metabolism, and pharmacological activity of anandamide in CB(1) cannabinoid receptor knockout mice: evidence for non-CB(1), non-CB(2) receptor-mediated actions of anandamide in mouse brain. J Neurochem 75:2434–2444

Freedland CS, Whitlow CT, Miller MD, Porrino LJ (2002) Dose-dependent effects of delta9-tetrahydrocannabinol on rates of local cerebral glucose utilization in rat. Synapse 45:134–142

Freedland CS, Whitlow CT, Smith HR, Porrino LJ (2003) Functional consequences of the acute administration of the cannabinoid receptor antagonist, SR141716A, in cannabinoid-naive and -tolerant animals: a quantitative 2-[14C]deoxyglucose study. Brain Res 962:169–179

Gatley SJ (1996) Positron radiopharmaceutical agents and their chemistry. In: Henkin RE, Boles MA, Dillehay GL, Halama JR, Karesh SM, Wagner RH, Zimmer AM (eds) Nuclear medicine. Mosby, St Louis, pp 429–444

Gatley SJ (2003) Labeled glucose analogs in the genomic era. J Nucl Med 44:1082–1086

Gatley SJ, Volkow ND (1998) Addiction and imaging of the living human brain. Drug Alcohol Depend 51:97–108

Gatley SJ, Lan R, Volkow ND, Pappas N, King P, Wong CT, Gifford AN, Pyatt B, Dewey SL, Makriyannis A (1998) Imaging the brain marijuana receptor: development of a radioligand that binds to cannabinoid CB1 receptors in vivo. J Neurochem 70:417–423

Gatley SJ, Gifford AN, Ding YS, Lan R, Liu Q, Volkow ND, Makriyannis A (2004) Development of PET and SPECT radioligands for cannabinoid receptors. In: Makriyannis A, Biegel D, Dekker M (eds) Drug discovery strategies and methods. Marcel Dekker, New York, pp 129–146

Gifford AN, Makriyannis A, Volkow ND, Gatley SJ (2002) In vivo imaging of the brain cannabinoid receptor. Chem Phys Lipids 121:65–72

Herkenham M, Lynn AB, Little MD, Johnson MR, Melvin LS, de Costa BR, Rice KC (1990) Cannabinoid receptor localization in brain. Proc Natl Acad Sci U S A 87:1932–1936

Iyo M, Namba H, Yanagisawa M, Hirai S, Yui N, Fukui S (1997) Abnormal cerebral perfusion in chronic methamphetamine abusers: a study using 99MTc-HMPAO and SPECT. Prog Neuropsychopharmacol Biol Psychiatry 21:789–796

Jacobsen LK, Mencl WE, Westerveld M, Pugh KR (2004). Impact of cannabis use on brain function in adolescents. Ann N Y Acad Sci. 1021:384–390

Katoch-Rouse R, Pavlova OA, Caulder T, Hoffman AF, Mukhin AG, Horti AG (2003) Synthesis, structure-activity relationship, and evaluation of SR141716 analogues: development of central cannabinoid receptor ligands with lower lipophilicity. J Med Chem 46:642–645

Kuehnle J, Mendelson JH, Davis KR, New PF (1977) Computed tomographic examination of heavy marijuana smokers. JAMA 237:1231–1232

Laruelle M (2000) Imaging synaptic neurotransmission with in vivo binding competition techniques: a critical review. J Cereb Blood Flow Metab 20:423–451

Margulies JE, Hammer RP Jr (1991) Delta 9-tetrahydrocannabinol alters cerebral metabolism in a biphasic, dose-dependent manner in rat brain. Eur J Pharmacol 202:373–378

Mathew RJ, Wilson WH, Humphreys DF, Lowe JV, Wiethe KE (1992) Regional cerebral blood flow after marijuana smoking. J Cereb Blood Flow Metab 12:750–758

Mathew RJ, Wilson WH, Coleman RE, Turkington TG, DeGrado TR (1997) Marijuana intoxication and brain activation in marijuana smokers. Life Sci 60:2075–2089

Mathew RJ, Wilson WH, Turkington TG, Coleman RE (1998) Cerebellar activity and disturbed time sense after THC. Brain Res 797:183–189

Mathew RJ, Wilson WH, Chiu NY, Turkington TG, Degrado TR, Coleman RE (1999) Regional cerebral blood flow and depersonalization after tetrahydrocannabinol administration. Acta Psychiatr Scand 100:67–75
Mathew RJ, Wilson WH, Turkington TG, Hawk TC, Coleman RE, DeGrado TR, Provenzale J (2002) Time course of tetrahydrocannabinol-induced changes in regional cerebral blood flow measured with positron emission tomography. Psychiatry Res 116:173–185
Mathews WB, Scheffel U, Finley P, Ravert HT, Frank RA, Rinaldi-Carmona M, Barth F, Dannals RF (2000) Biodistribution of [18F] SR144385 and [18F] SR147963: selective radioligands for positron emission tomographic studies of brain cannabinoid receptors. Nucl Med Biol 27:757–762
Mathews WB, Scheffel U, Rauseo PA, Ravert HT, Frank RA, Ellames GJ, Herbert JM, Barth F, Rinaldi-Carmona M, Dannals RF (2002) Carbon-11 labeled radioligands for imaging brain cannabinoid receptors. Nucl Med Biol 29:671–677
O'Leary DS, Block RI, Flaum M, Schultz SK, Boles Ponto LL, Watkins GL, Hurtig RR, Andreasen NC, Hichwa RD (2000) Acute marijuana effects on rCBF and cognition: a PET study. Neuroreport 11:3835–3841
O'Leary DS, Block RI, Koeppel JA, Flaum M, Schultz SK, Andreasen NC, Ponto LB, Watkins GL, Hurtig RR, Hichwa RD (2002) Effects of smoking marijuana on brain perfusion and cognition. Neuropsychopharmacology 26:802–816
O'Leary DS, Block RI, Turner BM, Koeppel J, Magnotta VA, Ponto LB, Watkins GL, Hichwa RD, Andreasen NC (2003) Marijuana alters the human cerebellar clock. Neuroreport 14:1145–1151
Oviedo A, Glowa J, Herkenham M (1993) Chronic cannabinoid administration alters cannabinoid receptor binding in rat brain: a quantitative autoradiographic study. Brain Res 616:293–302
Phelps ME (1991) PET: a biological imaging technique. Neurochem Res 16:929–940
Pontieri FE, Conti G, Zocchi A, Fieschi C, Orzi F (1999) Metabolic mapping of the effects of WIN 55212-2 intravenous administration in the rat. Neuropsychopharmacology 21:773–776
Porrino LJ (1993) Functional consequences of acute cocaine treatment depend on route of administration. Psychopharmacology (Berl) 112:343–351
Raichle ME, Martin WR, Herscovitch P, Mintun MA, Markham J (1983) Brain blood flow measured with intravenous H2(15)O. II. Implementation and validation. J Nucl Med 24:790–798
Reivich M, Kuhl D, Wolf A, Greenberg J, Phelps M, Ido T, Casella V, Fowler J, Gallagher B, Hoffman E, Alavi A, Sokoloff L (1977) Measurement of local cerebral glucose metabolism in man with 18F-2-fluoro-2-deoxy-d-glucose. Acta Neurol Scand Suppl 64:190–191
Rinaldi-Carmona M, Barth F, Heaulme M, Shire D, Calandra B, Congy C, Martinez S, Maruani J, Neliat G, Caput D, et al (1994) SR141716A, a potent and selective antagonist of the brain cannabinoid receptor. FEBS Lett 350:240–244
Rogers A (1973) Techniques of autoradiography, 2nd edn. Elsevier Scientific Publishing Company, Amsterdam
Romero J, Garcia-Palomero E, Castro JG, Garcia-Gil L, Ramos JA, Fernandez-Ruiz JJ (1997) Effects of chronic exposure to delta9-tetrahydrocannabinol on cannabinoid receptor binding and mRNA levels in several rat brain regions. Brain Res Mol Brain Res 46:100–108
Romero J, Berrendero F, Manzanares J, Perez A, Corchero J, Fuentes JA, Fernandez-Ruiz JJ, Ramos JA (1998) Time-course of the cannabinoid receptor down-regulation in the adult rat brain caused by repeated exposure to delta9-tetrahydrocannabinol. Synapse 30:298–308
Rubino T, Vigano D, Massi P, Spinello M, Zagato E, Giagnoni G, Parolaro D (2000) Chronic delta-9-tetrahydrocannabinol treatment increases cAMP levels and cAMP-dependent protein kinase activity in some rat brain regions. Neuropharmacology 39:1331–1336
Sakurada O, Kennedy C, Jehle J, Brown JD, Carbin GL, Sokoloff L (1978) Measurement of local cerebral blood flow with iodo [14C] antipyrine. Am J Physiol 234:H59–66

Smart D, Gunthorpe MJ, Jerman JC, Nasir S, Gray J, Muir AI, Chambers JK, Randall AD, Davis JB (2000) The endogenous lipid anandamide is a full agonist at the human vanilloid receptor (hVR1). Br J Pharmacol 129:227–230

Sokoloff L, Reivich M, Kennedy C, Des Rosiers MH, Patlak CS, Pettigrew KD, Sakurada O, Shinohara M (1977) The [14C]deoxyglucose method for the measurement of local cerebral glucose utilization: theory, procedure, and normal values in the conscious and anesthetized albino rat. J Neurochem 28:897–916

Stein EA, Fuller SA, Edgemond WS, Campbell WB (1998) Selective effects of the endogenous cannabinoid arachidonylethanolamide (anandamide) on regional cerebral blood flow in the rat. Neuropsychopharmacology 19:481–491

Thomas BF, Wei X, Martin BR (1992) Characterization and autoradiographic localization of the cannabinoid binding site in rat brain using [3H]11-OH-delta 9-THC-DMH. J Pharmacol Exp Ther 263:1383–1390

Varma VK, Malhotra AK, Dang R, Das K, Nehra R (1988) Cannabis and cognitive functions: a prospective study. Drug Alcohol Depend 21:147–152

Volkow ND, Gillespie H, Mullani N, Tancredi L, Grant C, Ivanovic M, Hollister L (1991) Cerebellar metabolic activation by delta-9-tetrahydro-cannabinol in human brain: a study with positron emission tomography and 18F-2-fluoro-2-deoxyglucose. Psychiatry Res 40:69–78

Volkow ND, Gillespie H, Mullani N, Tancredi L, Grant C, Valentine A, Hollister L (1996) Brain glucose metabolism in chronic marijuana users at baseline and during marijuana intoxication. Psychiatry Res 67:29–38

Voruganti LN, Slomka P, Zabel P, Mattar A, Awad AG (2001) Cannabis induced dopamine release: an in-vivo SPECT study. Psychiatry Res 107:173–177

Whitlow CT, Freedland CS, Porrino LJ (2002) Metabolic mapping of the time-dependent effects of delta 9-tetrahydrocannabinol administration in the rat. Psychopharmacology (Berl) 161:129–136

Whitlow CT, Freedland CS, Porrino LJ (2003) Functional consequences of the repeated administration of Delta9-tetrahydrocannabinol in the rat. Drug Alcohol Depend 71:169–177

Wilson W, Mathew R, Turkington T, Hawk T, Coleman RE, Provenzale J (2000) Brain morphological changes and early marijuana use: a magnetic resonance and positron emission tomography study. J Addict Dis 19:1–22

Zavitsanou K, Garrick T, Huang XF (2004) Selective antagonist [3H]SR141716A binding to cannabinoid CB1 receptors is increased in the anterior cingulate cortex in schizophrenia. Prog Neuropsychopharmacol Biol Psychiatry 28:355–360

Zygmunt PM, Petersson J, Andersson DA, Chuang H, Sorgard M, Di Marzo V, Julius D, Hogestatt ED (1999) Vanilloid receptors on sensory nerves mediate the vasodilator action of anandamide. Nature 400:452–457

HEP (2005) 168:445–477

Cannabinoid Function in Learning, Memory and Plasticity

G. Riedel (✉) · S.N. Davies

School of Medical Sciences, University of Aberdeen, Foresterhill, Aberdeen AB25 2ZD, UK
g.riedel@abdn.ac.uk

Abstract Marijuana and its psychoactive constituents induce a multitude of effects on brain function. These include deficits in memory formation, but care needs to be exercised since many human studies are flawed by multiple drug abuse, small sample sizes, sample selection and sensitivity of psychological tests for subtle differences. The most robust finding with respect to memory is a deficit in working and short-term memory. This requires intact hippocampus and prefrontal cortex, two brain regions richly expressing CB_1 receptors. Animal studies, which enable a more controlled drug regime and more constant behavioural testing, have confirmed human results and suggest, with respect to hippocampus, that exogenous cannabinoid treatment selectively affects encoding processes. This may be different in other brain areas, for instance the amygdala, where a predominant involvement in memory consolidation and forgetting has been firmly established. While cannabinoid receptor agonists impair memory formation, antagonists reverse these deficits or act as memory enhancers. These results are in good agreement with data obtained from electrophysiological recordings, which reveal reduction in neural plasticity following cannabinoid treatment, and increased plasticity following antagonist exposure. The mixed receptor properties of the pharmacological tool, however, make it difficult to define the exact role of any CB_1 receptor population in memory processes with any certainty. This makes it all the more important that behavioural studies use selective administration of drugs to specific brain areas, rather than global administration to whole animals. The emerging role of the endogenous cannabinoid system in the hippocampus may be to facilitate the induction of long-term potentiation/the encoding of information. Administration of exogenous selective CB_1 agonists may therefore disrupt hippocampus-dependent learning and memory by 'increasing the noise', rather than 'decreasing the signal' at potentiated inputs.

Keywords Endocannabinoids · CB_1 receptors · Perception · Cognition · Memory formation · Hippocampus · In vitro slice · Synaptic plasticity · LTP · LTD · DSI

1 Introduction

The resin made from flowers and leaves of the hemp plant *Cannabis sativa*, commonly known as cannabis or marijuana, comprises approximately 60 terpenophenolic compounds, which are referred to as plant cannabinoids. The primary psychoactive constituent is Δ^9-tetrahydrocannabinol (Δ^9THC) (Gaoni and Mechoulam 1964). Marijuana has been used for hundreds of years all over the world for both recreational and medicinal purposes, but has always been known for both positive effects, including relaxation, calming and stress relief, and negative effects, such as nausea, sickness, vomiting, dizziness and headaches. Like most drugs of abuse, cannabis is known for its ability to induce euphoria, lethargy, confusion, depersonalisation, altered time sense, impaired motor performance, memory defects, paranoia, depression, fear, anxiety and hallucinations. Given that most of these effects are mediated through specific receptors, it makes cannabinoids and their receptors an interesting target for the development of treatment strategies in the clinical setting.

In this chapter we will focus on the cognitive effects that have been described after cannabinoid use in humans and more recently in animals. We rely on many other chapters of this handbook in which details about the pharmacology and physiology, molecular and cell biology, and medicinal chemistry of cannabinoids are reviewed in detail (see chapters by Pertwee, Howlett, Abood, Reggio, Mackie, and Szabo and Schlicker). In the first part we will summarise effects of marijuana smoking on cognition in humans, and this will be followed by a brief introduction to the physiology of these effects. This will mainly concentrate on functional imaging data and electroencephalographic (EEG) recordings in humans. The main focus, however, will review work on animals and cognition (learning and memory) as this provides, to date, the best and most detailed data of the basic behavioural pharmacology of cannabinoids and cannabinoid receptors. Finally, we will try to explain the behavioural results in terms of ex vivo and in vitro physiology and synaptic plasticity.

2 Marijuana and Cognition in Man

Cannabis use alters both motor and cognition-based behaviour in man. Collectively, data strongly indicate acute intoxication to be more effective in disrupting memory than chronic use, probably due to long-term habituation and related changes in brain function. While simple cognitive tasks can be performed normally, the severity of cognitive impairment correlates with task difficulty, and this may be the direct consequence of deficits in attention and goal-directed learning. Importantly, there are few if any gross motor impairments, even after chronic cannabis smoking over many years.

Acute cannabis intoxication leads to multiple effects, including changes in reaction time and perception. Simple reaction times are recorded such that test subjects have to press a button in response to a tone or light. This merely requires motor execution; such tasks are devoid of complex cognitive processing. Several studies have reported that reaction times increase after marijuana use (Borg et al. 1975; Dornbush et al. 1971), but this has not been confirmed by others (Braden et al. 1974; Evans et al. 1976) despite comparable sample sizes and drug doses. Increasing the complexity of the task (pressing different buttons in response to different stimuli) consistently leads to up to 50% longer reaction times in users relative to controls, and there seems to be a strong correlation between task complexity and cannabis-induced impairment (Clark and Nakashima 1968; Chait and Pierri 1992).

Stronger evidence supports the notion that cannabis use alters perception, such as taste, smell, hearing and vision. In users there are clear problems of colour discrimination (Adams et al. 1976) and identification of figures hidden in pictures (Pearl et al. 1973). Perceptual changes also pertain to time sense, which is generally altered in cannabis users. As they estimate time to pass more slowly than control subjects (Tart 1971; Chait and Pierri 1992), this could explain why they are prepared to take greater risks, for example in driving faster and more dangerously (Bech et al. 1973). Such drug effects are of importance when investigating complex behaviours. Hasty reactions and perceptual deficits could easily explain impairments in memory tasks and need to be excluded.

Despite perceptual effects, it is still possible to identify cannabis-induced memory problems. A type of memory highly sensitive to marijuana intoxication is recognition memory. Typically, test subjects are presented with a series of words. After a delay period, a second series is presented containing some words from the original series, but also some new ones. Cannabis users have no problem identifying the words from the original list, but they often recognise some words that are actually new (Dornbush 1974). Such memory intrusions may reflect problems in distinguishing between relevant and irrelevant words, a hypothesis that is supported by observations on free recall. Here, participants write down as many words as they remember from the original list without being primed. This is a more complex paradigm, and users not only remember fewer words than controls (Dornbush et al. 1971), they also have memory intrusions, inserting words that were not presented (Miller and Cornett 1978).

Determinations of cognitive alterations in *chronic* marijuana users are more difficult. Classical studies of Jamaican (Bowman and Pihl 1973) and Costa Rican (Satz et al. 1976) subjects did not reveal any cognitive impairment, despite a battery of psychological tests and the fact that chronic users had been smoking more than nine joints per day for more that 10 years. These results were confirmed in a recent report on 1,300 residents in Baltimore that had been followed in a longitudinal study over 11 years. Mini-Mental State Examination was applied to investigate any changes in mental functioning, yet no significant difference was observed between chronic marijuana consumption and controls (Lyketsos et al. 1999).

Cognitive differences were revealed in a study on 1,600 Egyptian prisoners (Soueif 1976). In this study, 16 different measures were recorded, of which 10 revealed impairment in the user group, while 2 showed better performance. However, the selected groups were not well controlled and many of the critiques listed below apply to this investigation. Similarly, deficits in IQ, memory, time estimation and reaction times were reported in several studies performed in India (Wig and Varma 1977; Menhiratta et al. 1978). Finally, investigations on college students with at least twice weekly marijuana consumption revealed deficits in memory formation, specifically deficits in information transfer into long-term memory (Gianutsos and Litwack 1976; Entin and Glodzung 1973). However, a later study did not confirm these memory impairments (Rochford et al. 1977). More recent studies on cognitive deficits in marijuana users collectively suggest that impairments are (1) predominant for the attentional/executive system related to prefrontal cortex, and (2) increase with the length of cannabis use (Pope and Yurgelun-Todd 1996; Fletcher et al. 1996; Elwan et al. 1997). Such deficits can readily explain impairments in short-term memory, which are frequently reported for cannabis users (Schwartz et al. 1989).

Many of these studies, however, are flawed and do not reveal the true extent to which long-term cannabis use affects human cognition. Especially, early studies from the 1970s and 1980s were conducted on small sample sizes, and it has been calculated that an $n = 25$ per group is necessary to attain reliable results (Cohen 1990). Moreover, subjects that feel less affected by drug use are more likely to sign up for trials, while those experiencing severe problems may feel less eager to participate, even for pay (Strohmetz et al. 1990). As a consequence, the finding of subtle differences may not be a true reflection of the effects of chronic cannabis use. Psychological testing has seen considerable refinement, and the emergence of novel, increasingly sensitive tasks has helped to reveal differences between long-term marijuana smokers and controls. This suggests that tests used in the original studies, which have not found differences between test and control groups, were insensitive and might have been too simple.

Another critique frequently raised with respect to chronic use is the idea that users may have already been different from non-users prior to ever smoking marijuana. This is a valid point, as one might argue that (1) people of lower IQ may be more prone to drug use and (2) any intellectual difference may have preceded any cannabis smoking habits. Randomised control studies in which non-users are signed up for chronic smoking, however, are ethically difficult to justify. Another potential confounder is the use of multiple drugs. Many marijuana smokers are likely to use other and more drugs than controls (Earleywine and Newcomb 1997). Multi-drug effects can only be assessed in the context of each drug alone. Subjects who meet this criterion do not normally form part of studies. Consequently, multi-drug use will make the sample group heterogeneous so that results may not reflect the typical cannabis user.

In contrast, animal research is devoid of many of the above critiques and results are thus not confounded by, for example, polydrug use, low sample sizes, pre-treatment differences, etc. Consequently, the main focus of this chapter rests on such animal models and the effects of acute and chronic cannabis administration on learning, memory, and related brain physiology.

3 Effects of Cannabinoids on the Brain

To date, there is no evidence for gross morphological and structural changes in brain following short-term or long-term marijuana smoking. Although this has been investigated over many years, of particular interest here are studies that have utilised modern imaging techniques such as magnetic resonance imaging (MRI). There were no regional or global changes in brain tissue volume or composition in cannabis users (Block et al. 2000). More subtle changes can be determined through post-mortem analysis using radiolabelled compounds, or measurement of endocannabinoid levels. Such work showed reduced cannabinoid binding in caudate and hippocampus of Alzheimer's brains (Westlake et al. 1994), and in normal ageing (Biegon and Kerman 1995). No such studies have been reported on chronic marijuana smokers yet.

Alterations in brain function following acute and chronic use of cannabis is nevertheless detectable using cerebral blood flow (CBF) measurements such as positron emission tomography (PET) and multi-site EEG. Although very important for the understanding of brain regions associated with behavioural changes, CBF measurements are not ideal for determination of marijuana-induced changes in brain function (Mathew and Wilson 1991; also see chapter by Lindsey et al. in this handbook). This is mainly due to contaminating effects of cannabis on vascular smooth muscles and altered vasomotor tone, but also due to alterations in respiration and general circulation (for details see chapter by Pacher et al. in this volume). Since such circulation-related effects cannot be controlled for properly, they may lead to increased variability and make interpretations of CBF studies in marijuana users difficult. Fortunately, there is no conclusive evidence to suggest these peripheral effects impact significantly on blood circulation in brain. PET, for instance, makes use of a radiotracer (^{11}C, ^{13}N or ^{15}O) followed by reconstruction of tomographic slices depicting isotope concentrations in different brain regions. A shortcoming of PET is, however, its low resolution. Areas of less than 2 mm cannot be resolved properly. As with psychological testing, results of PET tests have been ambiguous; some reported decreased CBF, others increased CBF; some found no difference (see Wilson and Mathew, 2002 for review). Yet, it remains elusive as to why this variability is observed.

Collectively, data from CBF studies confirm the contention that alterations in brain function predominate in areas with high levels of cannabinoid receptor sites (Pertwee 1997). However, global marijuana intoxication will induce multiple effects at the same time, making it difficult to correlate any particular effect and CBF change. Overall, it has been found that chronic cannabis users have a lower resting level of brain blood flow than controls and that marijuana smoking or intravenous administration increases CBF in most cortical areas in a dose- and time-dependent manner (Wilson and Mathew 2002). Increases in CBF peaked at 30 min and returned to near-baseline levels 2 h after smoking. Subcortical areas including basal ganglia, thalamus, hippocampus and amygdala showed reduced CBF relative to placebo and both hemispheres were affected to the same extent. In addition, cerebellar blood flow increased by at least 1 standard error of the mean in about 60% of subjects. It should be obvious from these results that systemic administration of marijuana may not help to resolve the question as to what the function of individual subpopulations of receptors located in specific brain areas might be. CBF will provide important information as to global changes related to drug treatment.

An interesting approach in utilising PET is its combination with cognitive tasks. While subjects perform verbal memory recall tasks, they are monitored in the scanner. Relative to controls, frequent marijuana users presented with reduced memory-related blood flow in prefrontal cortex, but increased CBF in hippocampus and cerebellum (Block et al. 2002). These alterations were paralleled by an increased recency effect, suggesting that users rely on short-term memory and thus fail in multiple trial learning tasks, while control subjects encode and retrieve episodic memory. Consequently, it may be argued that chronic marijuana use leads to a reconfiguration of memory processing. Reductions in prefrontal CBF are consistent with deficits in working memory.

Another functional approach is the use of multiple recording sites on the skull to detect global changes in cortical activity through EEG. Event-related potentials (ERPs) derived from EEGs recorded during complex cognitive tasks have been recorded in a number of studies by Solowij and colleagues (see Solowij 1998 for review). Collectively, data in this area can be summarised as follows. Independent of frequency of marijuana smoking, ERPs in frontal regions progressively decline with the number of years of use. This suggests a physiological mechanism for the reduced ability to focus attention and filter out irrelevant information. Interestingly, the deficit was maintained even after several months of abstinence. The speed of information processing can be measured as positive wave at 300 ms (P300) of the ERP. Similar to reaction times, P300 was impaired with increasing frequency and length of marijuana use. Long-term marijuana use manifests in elevated absolute power and interhemispheric coherence of alpha and theta rhythm of the EEG (Struve et al.1994) and reduces the P50 auditory sensory gating response (Patrick et al. 1999).

4 Cannabinoids Modulate Cognition in Animal Models

Guided by the older work from humans, research into the behavioural effects of cannabinoids concentrated on the disruption of working and short-term memory formation. This is in agreement with data suggesting marijuana-induced increases in CBF in paralimbic regions of the frontal lobes and the cerebellum, but reduced blood flow in the temporal lobe (O'Leary et al. 2002). Hypoactivity in the temporal lobe may constitute the neural basis of cognitive alterations seen in cannabis users and has prompted the search for the underlying mechanisms using behavioural paradigms that specifically activate the medial temporal lobe, or using electrophysiological recording protocols in medial temporal lobe structures. It is also in line with reductions of the cortical P300 amplitude in marijuana addicts. The P300 is an ERP reflecting attentional resource allocation and active working memory (Johnson et al. 1997). Similarly, monkeys treated with Δ^9THC chronically have predominantly slow-wave EEGs (1–2 Hz) in hippocampus, amygdala and septum (Stadnicki et al. 1974) and present with similar deficits as human subjects (Aiger 1988; Branch et al. 1980; Evans and Wenger 1992; Gluck et al. 1973; Nakamura-Palacios et al. 2000; Schulze et al. 1988; Winsauer et al. 1999). Increased sophistication in pharmacological and physiological techniques applicable to rodents has now considerably increased our understanding of cannabinoid mechanisms in different types of memory, suggesting a modulatory role of cannabinoids and cannabinoid receptors in encoding, memory consolidation and even forgetting.

4.1 Spatial Learning

4.1.1 Water Maze

With respect to rodents such as rats and mice, training in the open-field water maze, in which animals search for a submerged and non-visible platform, is probably the most popular learning paradigm tackling spatial and thus hippocampus-dependent memory (Morris 1984). Animals learn to find the submerged platform in opaque water in relation to distal cues; a progressive reduction of the latency to swim to and climb onto the platform is an index for learning. If the platform is kept in the same place, this is a reference memory task while changing the platform location to a new position every day reflects a working memory paradigm.

Despite the paradigm's popularity as a spatial learning task, reports on the effects of cannabinoids are relatively recent. The initial report by Ferrari and colleagues (1999) revealed that HU210 induced a learning deficit in rats trained in a reference memory task. Animals treated with doses of up to 100 mg/kg i.p. were unable to acquire the spatial location of the submerged platform on four consecutive days. By contrast, learning to swim to a visible platform was not different between drug groups and controls, thus excluding sensory perception as a factor to explain the deficit. This has recently been confirmed with Δ^9THC and Δ^8THC in rats and mice (Da Silva and Takahashi 2002; Mishima et al. 2001; Varvel et al. 2001; Diana et al. 2003), but not with nabilone (Diana et al. 2003). Once spatial memory is acquired, consolidation and recall is no longer sensitive to cannabinoid treatment (unless drug doses are extremely high and cause considerable motor side-effects). However, the learning deficit in the water maze may not be due to memory problems. Robinson and co-workers (2003) revealed that Δ^9THC induced place aversion in a novel place preference/aversion task conducted in the water maze. This aversion would in itself account for the observed spatial learning deficits in the drug groups.

When exposed to a working memory paradigm, in which the location of the platform was changed on a daily basis, Δ^9THC-treated mice were impaired in finding the platform despite extensive pre-training over weeks. Consequently, Varvel and co-workers (2001; Lichtman et al. 2002) claimed that spatial working memory in mice is more sensitive to cannabinoid treatment. Despite extensive pre-training of the mice to the working memory task, animals were unable to remember the new platform location when under Δ^9THC. However, mice in the reference memory paradigm were also extensively pre-trained, and a lack of deficit with low doses of Δ^9THC may simply be due to the fact that cannabinoid receptor-dependent mechanisms are not active during memory recall. If animals are naïve as to the exact platform location, acquisition learning is still impaired with Δ^9THC (Da Silva and Takahashi 2002).

Interestingly, Varvel et al. (2001) used a working memory paradigm, in which animals were released from the same location on each day; but this release site was altered between days. This protocol, which was also used more recently for the testing of CB_1-null mutants (Varvel and Lichtman 2002), therefore has a strong egocentric, and thus hippocampus-independent (Jarrard 1993), component; animals may have acquired the task without the use of distal cues and allocentric strategies. It remains uncertain whether cannabinoids selectively interfere with egocentric spatial tasks or whether the deficit is ubiquitous for all forms of spatial acquisition.

Unexpected was the finding that CB_1 knockout mice acquired a spatial reference memory task in the water maze normally. This was unexpected since pharmacological studies had predicted that CB_1 knockout would facilitate learning and memory formation (for radial maze, see Lichtman 2000). Mice were, however, impaired in reversal learning, suggesting a deficit in task flexibility (Varvel and Lichtman 2002). $CB_1^{-/-}$ mice were not different from wild-type littermates in a working memory version of the task, and were also insensitive to Δ^9THC, WIN55,212-2 or methanandamide treatment. By contrast, all these cannabinoids disrupted working memory in wild-type littermates in a manner that was sensitive to rimonabant (SR141716A).

4.1.2
Radial Arm Maze

Radial arm mazes come in different shapes and forms. The most common one consists of eight arms radiating from an octagonal central platform. Animals kept on 80%–85% of their free-feeding body weights learn to retrieve a food reward from the distal end of each arm or, in some cases, a predetermined selection of arms. Use of the eight-arm radial maze has significantly contributed to our understanding of cannabis effects on cognition in rodents, but has some complications. For example, cannabis, Δ^9THC or synthetic analogues can depress locomotor activity (see DeSanty and Dar 2001a,b; Rodriguez de Fonseca et al. 1998; chapter by Fernández-Ruiz and González in this volume for review). This may impact on task performance such that longer latencies to reach and consume the reward may decrease attention processes due to longer test sessions. Such an effect is in line with cannabis-induced alterations in human cognition and thus may not be ideally suited to measure learning/memory. Furthermore, cannabinoids are widely known for their stimulatory effects on appetite (for review, see chapter by Maccarrone and Wenger, this volume). This may lead to a difference in motivation of already hungry animals to perform in this food-rewarded task (Di Marzo et al. 2001; Mechoulam and Friede 2001).

Despite such limitations, numerous reports suggest that cannabinoids impair performance in the eight-arm radial maze, especially when all arms are baited and revisits to the same arms are recorded as working memory errors. Animals were trained to criterion performance with all eight arms baited. Systemic administration of cannabinoids increased the number of working memory errors with low doses not affecting the amount of time required to complete the visits. This was originally reported in chronic experiments with Δ^9THC administered for 3 or 6 months (Stiglick and Kalant 1982), and has been confirmed more recently for acute infusions of Δ^9THC (Hernandez-Tristan et al. 2000; Lichtman and Martin 1996; Lichtman et al. 1995; Mishima et al. 2001; Molina-Holgado et al. 1995; Nakamura et al.1991), synthetic CB_1 receptor agonists WIN55,212-2 and CP55,940, or local infusion of CP55,940 directly into the hippocampus (Lichtman et al. 1995). To increase task difficulty, some researchers have introduced a short delay between visits to arms 1–4 and 5–8 in these experiments. Cannabinoids also disrupt performance in this short-term memory task when delays were 5 s (Molina-Holgado et al. 1995), 30 s (Hernandez-Tristan et al. 2000), or 1 h long (Nakamura et al. 1991). It is, however, unclear whether animals prefer to revisit arms 1–4 or 5–8. In accordance with drug effects on locomotor activity, post-delay performance was prolonged in the Δ^9THC group (Hernandez-Tristan et al. 2000; Nakamura et al. 1991). Mishima and co-workers (2001) employed a mixed reference and working memory protocol. Four predetermined arms were rewarded while four others were not. Entry into an arm that was never baited was counted as a reference memory error, re-entry into a previously baited arm as a working memory error. Δ^9THC (6 mg/kg) significantly impaired working memory but not reference memory. However, inspection of their data suggests that reference memory may have also been affected by drug treatment.

Confirmation that drug effects were mediated via CB_1 receptor activation was obtained through co-administration of the selective receptor antagonist rimonabant, which reversed deficits induced by Δ^9THC (Lichtman and Martin 1997; Mishima et al. 2001). Rimonabant alone had no effect when delays between entries 1–4 and 5–8 were short (Lichtman 2000; Lichtman and Martin 1997; Mishima et al. 2001), but there was a memory enhancement for delays of several hours (Lichtman 2000). This result suggests that blockade of CB_1 receptors may aid the development of short-term memory, and receptor antagonists may become important as cognitive enhancers not only for future animal research but also with respect to treatment of cognitive impairment in humans.

4.1.3
T- and Y-Maze Procedures

Rodents show spontaneous alternation behaviour when tested in simple T- or Y-shaped mazes. They can also be forced to alternate, i.e. when food reward is placed at the end of the goal arm of the T/Y. As pointed out for the eight-arm radial maze, this paradigm uses food or drinking of juice as reward and is contraindicated when using cannabinoids. Systemic administration of Δ^9THC prior to daily testing decreased the alternation score (Nava et al. 2000). In control animals, in vivo brain dialysis confirmed an alternation-induced release of acetylcholine in hippocampus, which was smaller in the Δ^9THC group. In addition, the alternation impairment and the acetylcholine release depression persisted in animals treated with Δ^9THC twice daily with 5 mg/kg Δ^9THC i.p. for up to 1 week (Nava et al. 2001). Both effects were fully reversed by rimonabant, suggesting that no tolerance developed after chronic 5-day Δ^9THC exposure.

Delayed alternation is another possible training protocol for the T/Y-maze, in which animals are rewarded for choosing any goal box in trial one. They are then returned to the start box and released after an inter-trial interval. In trial two, they have to enter the arm not visited in trial one (non-match) and are rewarded. Typically, animals acquire a criterion of 80% correct responses after a short training period. When tested in the presence of Δ^9THC, there was a significant drop in performance coupled with a reduction in monoamine turnover in their prefrontal cortex (Jentsch et al. 1997). Animals treated with a similar dose (5 mg/kg) Δ^9THC, however, were not impaired in brightness discrimination (Jentsch et al. 1996) or visual discrimination of forms procedures (Mishima et al. 2001) administered in the same apparatus.

4.1.4 Delayed Match-to-Position Tasks

Different from the short-term memory tested in the radial arm maze are the delayed match-to-position (DMTP) or delayed match-to-sample (DMTS) tasks that employ standard conditioning chambers. In the most frequently used version, rats learn to press a lever during the sample phase and press the same (match) or opposite (non-match) lever during the choice phase. Task difficulty is modulated by the introduction of a delay between sample and choice phase (0–45 s); performance falls to chance at delays exceeding 45 s (Deadwyler et al. 1996). The laboratories of Hampson and Deadwyler have extensively studied hippocampal involvement in this task (Hampson et al. 1999a) and determined learning-related single unit activity of ensembles of CA3 and CA1 neurones during the different phases of the task (Deadwyler and Hampson 1999; Deadwyler et al. 1996; Hampson and Deadwyler 1996; Hampson et al. 1993, 1999b). In summary, distinct hippocampal pyramidal cells fire during the sample, delay and match phases, respectively. In a series of elegant studies, Hampson, Deadwyler and their colleagues have provided compelling evidence for a modulatory role of cannabinoids in delayed match-to-sample performance. Acutely injected Δ^9THC (0.75–2 mg/kg i.p.) prior to testing resulted in dose- and delay-dependent performance deficits. Animals were able to perform the task at 0–5 s delays, but the severity of impairment increased with the inter-trial interval, suggesting that short-term memory had been compromised (Heyser et al. 1993). At the same time, hippocampal firing during the sample phase was greatly diminished, leading to ensemble miscodes that increase the probability for the occurrence of errors especially at long, but not very short delays. Behavioural tolerance to Δ^9THC developed after 35 days of daily Δ^9THC exposure and was followed by a short withdrawal period of 2 days (Deadwyler et al. 1995). Behavioural sensitisation, however, developed within 4 days of repeated treatment (Miyamoto et al. 1995). These initial results both in terms of behavioural performance and physiological responses have been confirmed for delayed nonmatch-to-position protocols (Hampson and Deadwyler 1998; Mallet and Beninger 1996) and extended to other cannabinoids such as WIN55,212-2 (Hampson and Deadwyler 1999, 2000; Han et al. 2000). A similar performance deficit was reported for exogenous administration of the endocannabinoid anandamide (Mallet and Beninger 1996), and deficits were reversed by the CB_1 receptor antagonist rimonabant (Hampson and Deadwyler 1999, 2000; Mallet and Beninger 1998). Rimonabant alone had no effect (Hampson and Deadwyler 2000; Mallet and Beninger 1998).

In their studies, Hampson and Deadwyler have made a strong case for a role of CB_1 receptors in encoding, since pyramidal cell firing was diminished during encoding in animals exposed to cannabinoids. Firing during delay and matching phases remained unchanged, suggesting that errors occur only when there is an encoding deficit. A non-memory-related explanation for this behavioural deficit could be derived from results obtained by Han and Robinson (2001), who showed that cannabinoids such as WIN55,212-2 or Δ^9THC can shorten time estimation in the rat. This, however, would suggest an increase in response rate during the delay, which is difficult to observe given that the levers are not present during the delay. Also, Hampson and Deadwyler have not reported alterations in the amount or length of delay-related firing in cannabis-treated rats.

4.2 Conditioning of Fear

Auditory fear conditioning is a standard procedure used in animal research (for review, see Crawley 2000). In a typical experiment, the animal is placed in a small chamber and a tone is presented after a short habituation period. The tone co-terminates with a mild footshock delivered through the grid floor, which the animals cannot escape. Consequently, this procedure is also called 'learned helplessness'. It results in a typical freezing reaction consisting of a crouching posture and immobility. One trial often is sufficient for induction of lasting memory, which can be tested hours or days later by measuring the freezing response upon re-exposure to the chamber (contextual fear conditioning) and presentation of the tone (cued fear conditioning). Mice treated with the CB_1 receptor antagonist rimonabant or CB_1 knockout mice readily acquire this fear-conditioning paradigm (Marsicano et al. 2002). While 6 days of extinction training, in which no shock is delivered, reduced the amount of freezing in wild-type littermates, knockout mice or wild-types treated with rimonabant throughout extinction maintain their freezing levels. These data suggest that the endocannabinoid system is highly active during forgetting and the extinction of aversive memories (Marsicano et al. 2002). Re-exposure to the tone during extinction also induced release of the endocannabinoids anandamide and 2-arachidonoyl glycerol in the basolateral amygdala, and this not only confirms the importance of the amygdala in fear conditioning, but also that on-demand release of endocannabinoids controls extinction of the fear response (Marsicano et al. 2002).

The acoustic startle response is based on a naturally occurring startle reaction to loud noise. If this loud tone is presented 30–500 ms after a 20-ms pre-pulse (pure tone), the startle reaction is considerably reduced. This is termed pre-pulse inhibition, reflects a measure of sensory-motor gating and involves a multitude of brain stem areas and transmitter systems (Koch 1999). Rats injected with the synthetic cannabinoids WIN55,212-2 (1.2 mg/kg) or CP55,940 (0.1 mg/kg) show little if any pre-pulse inhibition relative to controls (Mansback et al. 1996; Schneider and Koch 2002). The CP55,940-induced deficit in sensory-motor gating was fully reversed by 10 mg/kg rimonabant, but the antagonist had no effect when given alone (Mansbach et al. 1996). Although these data strongly suggest that cannabinoids can modulate sensory-motor gating, they have not resolved the issue of whether acquisition and execution of a normal startle response, either to a single loud noise or in a pre-pulse paradigm, is under the control of the endocannabinoid system.

4.3
Avoidance Tasks

Similar to the fear conditioning paradigms described above, shock is used as a reinforcer when animals are trained to avoid certain compartments. Two different training procedures are widely used. Passive or inhibitory avoidance refers to the active inhibition of a response in order to avoid a footshock; by contrast, active avoidance refers to behavioural escape from a dangerous area in order to avoid punishment.

Passive or inhibitory avoidance can be performed in two protocols. In step-through passive avoidance tasks, animals are released into a brightly lit chamber and a door is opened to allow entry into a dark compartment. Once entering the dark compartment, the door is closed and a foot shock is delivered. Upon re-exposure, animals will prefer to stay in the bright chamber. Step-down inhibitory avoidance, on the other hand, uses an elevated platform, from which animals will step down onto a grid floor. This triggers shock delivery and escape back onto the platform. Memory is assessed in a test session as an increase in step-down latency relative to the latency observed during acquisition training. As with most shock-motivated tasks, only a few trials are given to induce long-lasting memory traces.

Mice and rats treated with the endocannabinoid anandamide (1.5–6 mg/kg) immediately post-training presented with a significant memory impairment (Castellano et al. 1997, 1999) when tested in a step-through variant. The block of memory formation is due to an effect of anandamide on memory consolidation, since injections 2 h post-training had no effect. Interestingly, this effect was specific to the CD1 and DBA strains, but memory facilitation was observed in C57Bl/6 mice (Castellano et al. 1999). It is likely that modulation of memory strength with cannabinoids affected the monoaminergic transmitter system, since both D_1 and D_2 receptor antagonists reversed deficits induced by anandamide. The memory deficit was also reversed by naltrexone, an opioid receptor antagonist, suggesting cross-talk between cannabinoid and opioid system.

Memory for the shock tested 15 min or 24 h later was also reduced in rats injected intracerebroventricularly with anandamide or arachidonic acid (3.6 nmol in 5 µl) immediately post-training (Rodriguez de Fonseca et al. 1998). While anandamide administration enhanced the amount of slow-wave and rapid eye movement sleep in the period between training and retention testing, arachidonic acid led to a reduction in slow wave sleep. It therefore remains uncertain whether drug-induced changes in sleep pattern have any bearing on the consolidation and expression of an inhibitory avoidance response. A more detailed time course with systemic pre-training, post-training and pre-test injection of Δ^9THC in rats revealed memory deficits independent of injection time (Mishima et al. 2001). Since no deficit in acquisition learning was reported, it is safe to assume that cannabinoids have modulated consolidation and retention processes of emotional memories. Any functional role for the endocannabinoid system, however, still remains elusive. Site-direct infusion of anandamide (100 µmol/0.5 µl) into hippocampal CA1 also induced anterograde amnesia in step-down inhibitory avoidance learning (Barros et al. 2004).

Similarly, an acquisition impairment was reported for rats tested in an active avoidance paradigm (Izquierdo and Nasello 1973). The weak CB_1 receptor ligand cannabidiol (3.5 mg/kg) reduced conditioned responding, but was not effective when administered immediately post-training. This result is in agreement with more recent active avoidance training in CB_1-null mutant mice, which showed increased conditioned responding consistent with memory enhancement (Martin et al. 2002). At odds with these results is the finding that rats chronically treated with Δ^9THC (20 mg/kg) for 3 months and subsequently left untreated for 30 or 118 days before training in a shuttle box outperformed controls (Stiglick et al. 1984). Animals that had been exposed to Δ^9THC attained asymptotic performance levels faster than controls. It remains to be shown whether this effect is mediated by the cannabinoid system. An interesting comparison can be made with hippocampally lesioned animals. Such rats also show enhanced active avoidance and outperform controls (Isaacson et al. 1961), suggesting that systemically administered Δ^9THC may induce a functional lesion of the hippocampus (Hampson and Deadwyler 1998).

4.4
Other Memory Paradigms

Olfactory memory traces can be assessed in a social recognition task in which an adult animal is brought together with a juvenile conspecific. Exploration of the juvenile's anus can be monitored as a dependent variable for periods of up to 5 min. Memory is assessed through re-exposure and a reduction in anogenital sniffing is taken as an index of recognition memory. Results obtained with cannabinoid agonists and antagonists are straightforward.

In the first investigation into the effects of rimonabant, Terranova and coworkers (1996) reported enhancement of social recognition memory at doses of 0.1–3 mg/kg administered subcutaneously within 5 min post-training. Memory was assessed 2 h post-presentation of the juvenile, and the enhancement was present in both aged rats and mice. Reversal of this enhancement was attained by simultaneous administration of scopolamine, suggesting a tight interaction between cholinergic and cannabinoid system. Terranova and colleagues' finding was the first to suggest a memory-related function of the endocannabinoid system, which is particularly important during consolidation and forgetting (Marsicano et al. 2002). In contrast to rimonabant, administration of the CB_1 receptor agonist WIN55,212-2 (0.6–1.2 mg/kg) impaired short-term (30 min) social recognition memory in a dose-dependent manner without affecting anogenital exploration per se (Schneider and Koch 2002).

Another test for short-term memory, but with a different psychological quality, is object recognition (Ennaceur and Delacour 1988). During acquisition, animals are exposed to a novel environment containing object A, and are tested during re-exposure to object A plus first-time exposure to novel object B, after minutes or hours. The time spent exploring each object in the test session is an index of short-term memory for A. Good memory is characterised by preferential exploration of B, bad memory by exploration of A. Brain structures involved in object recognition include entorhinal and perirhinal cortex. Intraperitoneal injection of Δ^9THC (10 mg/kg: Ciccocioppo et al. 2002), WIN55,212-2 (0.6–1.2 mg/kg: Schneider and Koch 2002), CP55,940 (0.025–2.5 mg/kg: Kosiorek et al. 2003) or *R*-(+)-methanandamide (0.25–2.5 mg/kg: Kosiorek et al. 2003) impaired object recognition memory, and this deficit was reversed by 1 mg/kg rimonabant (Ciccocioppo et al. 2002). While this was interpreted as a CB_1 receptor-mediated action, caution should be exercised, since rimonabant alone was not tested, and this allows for the alternative interpretation that the endocannabinoid system might limit the length of the recognition memory. In agreement with this notion, CB_1-null mutant mice show normal exploration during acquisition, but enhanced memory when tested against object B (Maccarone et al. 2002; Reibaud et al. 1999).

A different, more complex learning protocol was used by Brodkin and Moerschbacher (1997). In a specifically modified conditioning box, rats were trained for 14 weeks to respond to a sequence of lights by pressing appropriate keys. Once asymptotic performance criteria were met, drugs like cannabidiol (100 mg/kg) and anandamide (18 mg/kg) were injected, but had no effect on performance. By contrast, Δ^9THC (3.2–18 mg/kg) and *R*-(+)-methanandamide (1–18 mg/kg) impaired performance in a dose-related manner. This impairment was reversed by rimonabant (1 mg/kg), but the antagonist had no effect on its own.

4.5 Summary

Collectively, the behavioural data suggest a modulatory role of cannabinoids in learning and formation of different forms of memory. In view of the numerous side-effects of both natural and synthetic cannabinoids, however, hard proof for this notion is difficult to obtain. Reinforcer modulation may include cannabinoid-induced increases in food consumption (see chapter by Maccarrone and Wenger, this volume); activity-related changes of Δ^9THC or anandamide include the reduction of ambulations in the open field (Järbe and Hiltunen 1987; Järbe et al. 2002; Navarro et al. 1993; see Fernández-Ruiz and González, this volume), induction of catalepsy (Teng and Craft 2004 for a recent example), and suppression of conditioned responding in a lever-press task (Arizzi et al. 2004); anxiogenic properties of cannabinoids would lead to higher levels of emotionality (Onaivi et al. 1990). Consequently, the observed deficit may not be a result of an impairment in acquisition or consolidation, but may be due to unrelated side-effects of drugs, such as reductions in reaction time or even signal detection (Presburger and Robinson 1999).

Despite all these problems, there is now strong evidence for a role of CB_1 receptors in memory formation. For delay-dependent short-term memory tasks, CB_1 receptors may be able to modulate the encoding processes. By contrast, CB_1 receptors may play a role in consolidation and even recall in memory formation of avoidance tasks. These effects are likely to be mediated by different CB_1 receptor populations located in different brain regions, and a better understanding of their function requires more localised administration of selective CB_1 agonists and antagonists.

5
Synaptic Plasticity

The anecdotal reports of effects of smoking cannabis on cognitive processes has naturally prompted investigation into the effects of cannabinoids on synaptic plasticity, and in particular long-term potentiation (LTP). To summarise 17 years of effort, it has been easy to show that cannabinoids have effects on LTP, and most commonly suppress it, but far more difficult to define the mechanisms by which this occurs. Emerging themes are that the commonly reported inhibition of LTP by synthetic cannabinoids may be mediated by non-CB_1 receptors, and that the function of the endogenous cannabinoid system may be to facilitate induction of LTP. The vast majority of studies have focussed on synaptic plasticity in the CA1 region of the rat hippocampal slice and we will start our review there.

5.1
LTP in the Hippocampus

The first reported investigation was that of Nowicky et al. (1987) who showed that pre-incubation of adult rat hippocampal slices with Δ^9THC could either inhibit or potentiate high-frequency stimulation-induced LTP, depending on the concentration used. High-frequency stimulation (200 Hz for 500 ms) induced a potentiation of the CA1 population spike amplitude that had a decay half-time of 280 min. Pre-incubation with 10 pM Δ^9THC increased this to 350 min, whereas pre-incubation with 100 or 1,000 pM Δ^9THC reduced it to 91 and 31 min, respectively. These changes in synaptic plasticity were associated with corresponding changes in the baseline population spikes, so that 10 pM Δ^9THC increased the population spike amplitude, and 100 or 1,000 pM Δ^9THC decreased it. Before delivering the high-frequency train though, stimulus strength was adjusted to counter any effect on baseline transmission. These experiments were performed before the dawn of the age of the cannabinoid receptor and few pharmacological tools were available. It is therefore possible that the effects could be a result of some non-specific mechanism of membrane perturbation.

5.1.1
Effects of Synthetic Cannabinoids on LTP Are Stereoselective

One of the cardinal features of a receptor-mediated effect is stereoselectivity. It was therefore significant that cannabinoids were shown to be stereoselective as inhibitors of LTP. HU-210 and HU-211 were the first to be tested, since their stereochemical purity is particularly high, and only HU-210 is psychoactive. The experiments by Collins et al. (1994) showed that pre-incubation of adult rat hippocampal slices with 100 nM HU-210, but not HU-211, blocked LTP of the CA1 field excitatory postsynaptic potential (fEPSP) induced by high-frequency stimulation (100 Hz for 500 ms). Slices were pre-incubated with the drugs so there was no internal control to determine if HU-210 depressed the baseline fEPSP. Stereoselectivity of the effect has subsequently been confirmed (Paton et al. 1998) for another cannabinoid, where perfusion of 5 μM WIN55,212-2 but not WIN55,212-3 blocked LTP of the CA1 population spike induced by high-frequency stimulation (100 Hz for 500 ms). This inhibition occurred in the absence of any effect on the amplitude of the baseline population spike, but was associated with a decrease in paired pulse depression.

5.1.2
Effects of Cannabinoids Are Blocked by CB_1 Receptor Antagonism

A second cardinal feature of receptor-mediated actions is that they should be blocked by a suitable antagonist. The tools that have been used for the investigation of LTP have been rimonabant, AM251 and AM281. The properties of these, and specifically whether they are antagonists or inverse agonists at CB_1 receptors are discussed elsewhere (see the chapter by Pertwee, this volume). Rimonabant was the first to become available and has been shown to prevent the blockade of LTP in the CA1 region by HU-210 (Collins et al. 1995), and WIN55212-2 (Terranova et al. 1995; Paton et al. 1998; Misner and Sullivan 1999).

An alternative approach to the pharmacological manipulation of receptors is to use knockout animals. CB_1 knockout mice have been produced, and high-frequency stimulation (100 Hz for 250 ms) induced significantly larger LTP of the CA1 field EPSP in hippocampal slices prepared from $CB_1^{-/-}$ mice compared to wild-type controls (Bohme et al. 2000). This was not associated with any detectable change in the amplitude of the half-maximal field EPSP evoked in the two groups.

5.1.3
Prenatal Exposure to Cannabinoids Inhibits LTP

Chronic effects of administration of cannabinoids have been studied in slices prepared from 40-day-old rats born to mothers who received daily subcutaneous injections of WIN55,212-2 throughout gestation (Mereu et al. 2003). LTP in these slices was reduced as compared to control slices prepared from rats born to untreated mothers. The slices also showed impaired basal and K^+-stimulated glutamate release.

5.1.4
Putative Endogenous Cannabinoids Inhibit LTP

The range of putative endogenous cannabinoids have been outlined previously (see Di Marzo et al., this volume), and several studies have investigated the effects of application of these on LTP. In rat hippocampal slices, perfusion of 20 μM sn-2 arachidonylglycerol (2-AG) blocked LTP of CA1 field EPSPs induced by high-frequency stimulation (100 Hz for 1 s), with little or no effect on the baseline field

EPSP (Stella et al. 1997; Schweitzer et al. 1999), and the effect of 2-AG was blocked by 2 μM rimonabant (Stella et al. 1997). Similarly, perfusion of anandamide (3–10 μM) blocked LTP of CA1 population spikes, and this was prevented by rimonabant (10 μM) (Terranova et al. 1995).

5.1.5
Do Cannabinoids Suppress Baseline Excitatory Transmission in the CA1 Region?

Any drug that reduces excitatory drive would be expected to impair high-frequency stimulus-induced LTP by limiting the level of postsynaptic depolarisation achieved during the induction train. This would in turn reduce the relief of the voltage-dependent block of *N*-methyl-D-aspartate (NMDA) receptor-gated channels by Mg^{2+} and hence reduce the postsynaptic Ca^{2+} entry that is required to trigger the processes that lead to synaptic potentiation. The question of whether cannabinoids inhibit baseline excitatory transmission is therefore an important one. The suppression of inhibitory transmission by cannabinoids in the hippocampus has been well documented, but whether cannabinoids also suppress excitatory glutamatergic transmission (as they do in the cerebellum, see the chapter by Szabo and Schlicker, this volume) is less clear cut. Thus, there are reports stating explicitly that WIN55,212-2 either inhibits (Misner and Sullivan 1999; Al-Hayani and Davies 2000; Ameri and Simmet 2000; Hajos et al. 2001), or does not inhibit (Terranova et al. 1995; Paton et al. 1998; Al-Hayani and Davies 2000), excitatory synaptic transmission in the CA1 region. This apparent discrepancy has now been resolved by the demonstration that the most commonly used CB_1 receptor agonist, WIN55,212-2, at tenfold higher concentrations, also activates a TRPV1-like receptor, which is also sensitive to rimonabant (Hajos et al. 2001; Hajos and Freund 2002). Thus, in slices prepared from $CB_1{}^{+/+}$ mice, perfusion of WIN55,212-2 inhibited pharmacologically isolated excitatory postsynaptic currents (EPSCs) with an EC_{50} of 2.01 μM, and pharmacologically isolated inhibitory postsynaptic currents (IPSCs) with an EC_{50} of 0.24 μM (Hajos et al. 2001). In slices prepared from $CB_1{}^{-/-}$ mice, WIN55,212-2 no longer inhibited evoked IPSCs, but still inhibited evoked EPSCs. This inhibition of excitatory transmission was mimicked by the TRPV1 agonist capsaicin (10 μM), and was blocked by the TRPV1 antagonist, capsazepine (10 μM). The fact that the suppression of EPSCs (as well as IPSCs) by WIN55,212-2 is blocked by rimonabant is significant, since this is the criterion by which an effect would previously have been judged to be CB_1 receptor mediated. The tenfold concentration difference in the EC_{50} of WIN55,212-2 in blocking inhibitory, as opposed to excitatory, transmission also explains why the drug has been reported to selectively block paired-pulse depression of population spikes (an effect dependent on feedback inhibitory transmission) but not baseline synaptic transmission (Paton et al. 1998).

Whether this central TRPV1-like receptor has the same properties as the better characterised peripheral TRPV1 receptors (Szallasi and Di Marzo 2001), and whether it represents the same non-CB_1 non-CB_2 receptor characterised by Breivogel et al. (2001; see chapter by Pertwee, this volume) remains to be determined. Equally, it is hard to resolve whether the effects of other cannabinoids are likely to be mediated by the TRPV1-like receptor. Available evidence suggests that the receptor is also activated by CP55,940, but that it is not blocked by AM251 (Hajos and Freund 2002). The answer is therefore that some cannabinoids do indeed inhibit excitatory transmission in the hippocampus, but via a TRPV1-like receptor. This in turn raises the question: Is inhibition of the induction of LTP by cannabinoids secondary to suppression of baseline excitatory transmission?

Several of the studies on LTP reported above have commented in passing on any associated effects of cannabinoids on baseline field EPSP or population spike responses, but answers have been contradictory. Thus, some have found no change (Terranova et al. 1995; Stella et al. 1997; Paton et al. 1988; Schweitzer et al. 1999), whereas others have found an inhibition of baseline excitatory responses (Nowicky et al. 1987; Misner and Sullivan 1999). The question was addressed directly by Misner and Sullivan (1999) who found that the EPSC was reduced by about 50% in slices prepared from neonatal rats. Perfusion of 5 μM WIN55,212-2 blocked the induction of LTP by high-frequency stimulation (100 Hz for 200 ms), but this could be overcome by manipulations designed to overcome the Mg^{2+}-block of NMDA receptor-gated channels, i.e. reducing the concentration of Mg^{2+} in the perfusion medium, or by slightly depolarising the recorded cell during the high-frequency train. They therefore concluded that the effect of WIN55,212-2 was to reduce the excitatory drive and therefore the extent of activation of NMDA receptors.

Note that lack of an effect of cannabinoids on the low-frequency-evoked synaptic responses in some of the experiments described above does not exclude the possibility that the drugs may have an effect on the high-frequency response required to induce LTP. It would therefore be important to establish the effects of cannabinoid receptor ligands on the response to repetitive high-frequency stimulation, as well as on the response to low-frequency stimulation. Though some of the experiments described above suggest that cannabinoids might inhibit induction of LTP by suppressing excitatory drive, none of them distinguishes the possibility that cannabinoids block LTP via an action on the TRPV1-like receptor rather than the CB_1 receptor. WIN55,212-2 is an agonist at both, and rimonabant inhibits both. Resolution of this question must await the development of ligands that will reliably differentiate the two receptors, and/or investigation of the effects of cannabinoids on the induction of LTP in $CB_1^{-/-}$ animals.

5.1.6
Long-Lasting Effects of Perfusion of Cannabinoid Receptor Ligands

Diana et al. (2002) argue that the apparent blockade of LTP by perfusion of WIN55,212-2 is in fact due to the very slow and gradual inhibition of the baseline synaptic response. Thus, when the potentiation of a high-frequency-stimulated pathway is compared to the gradual decay of a control un-tetanised pathway, it appears that administration of WIN55,212-2 does not inhibit LTP. From inspection of individual figures it is hard to see that this could explain the results in all other reports; however, the interpretation of this experiment illustrates an important point. Cannabinoid ligands are very lipophilic, which makes them notoriously difficult to work with. They require use of some vehicle (e.g. DMSO, TWEEN 80, ethanol) to disperse them, they stick to glassware and tubing, they produce relatively slow effects, and they are very difficult to wash off. The last of these points makes it very difficult to distinguish between long-term effects [e.g. LTP, long-term depression (LTD)] caused by transient application of the drug, and a short-term effect that persists because the drug is still present in the tissue. Furthermore, the free concentration of drug available to the receptors is liable to vary widely between different preparations (e.g. cultured neurones vs slices) and different recording conditions. For instance, the observation that WIN55,212-2 inhibits excitatory transmission in slices prepared from neonatal, but not adult, rats (Al-Hayani and Davies 2000), may be due to improved drug access in the neonatal tissue. These considerations make any meaningful comparison of effective concentrations between reports very difficult.

5.1.7 Cannabinoid Involvement in Depolarisation-Induced Suppression of Inhibition

Depolarisation-induced suppression of inhibition (DSI) is a form of short-term plasticity which merits mention here. In the hippocampus, depolarisation of pyramidal neurones induces a short-term suppression of γ-aminobutyric acid (GABA)ergic IPSCs (Pitler and Alger 1992). This DSI is blocked by perfusion of CB_1 receptor antagonists, and it is absent in $CB_1{}^{-/-}$ mice (Ohno-Shosaku et al. 2001; Wilson and Nicoll 2001). These results cemented the role of cannabinoids as retrograde messengers in the nervous system (see chapter by Vaughan and Christie, this volume). Specifically, they are consistent with the notion that increased Ca^{2+} entry during the depolarising pulse triggers synthesis and release of an endocannabinoid from the pyramidal cell, which then activates CB_1 receptors on local inhibitory terminals and suppresses GABAergic inhibition of that cell. An additional trigger to the synthesis and release of endocannabinoids may be activation of group I metabotropic glutamate (mGlu) receptors (Varma et al. 2001; Ohno-Shosaku et al. 2002) and muscarinic acetylcholine receptors (Kim et al. 2002). Note that depolarisation-induced suppression of excitatory synaptic transmission (DSE) has been reliably demonstrated in cerebellar tissue (Kreitzer and Regehr 2001: Maejima et al. 2001), but not in the hippocampus (but see Ohno-Shosaku et al. 2002). This correlates with immunocytochemical studies that reveal CB_1 receptors are located on excitatory glutamate-containing terminals in the cerebellum, but not in the hippocampus.

DSI expressed in the hippocampus is transient, and suppression of the resulting inhibition could not account for long-term plasticity of synapses. However, it may be important in facilitating depolarisation, and therefore induction of LTP, during a high-frequency stimulus.

5.1.8
Release of Endogenous Cannabinoids During DSI Facilitates the Induction of LTP

Carlson et al. (2002) demonstrated that delivery of a subthreshold high-frequency train (i.e. one that would not normally induce LTP) during the phase of DSI could induce LTP. The subthreshold train (50 Hz for 400 ms) usually failed to induce LTP of the fEPSP recorded from a population of cells in the CA1 region, or of the EPSC recorded from a single CA1 pyramidal cell. However, a depolarising pulse (to 0 mV for 1 s) delivered to the single cell 3 or 13 s before the subthreshold train, resulted in LTP of the EPSC, but not of the fEPSP. Furthermore, perfusion of AM251 (2 μM) prevented this facilitation. It is therefore apparent that the release of endocannabinoids (in this instance by the depolarising pulse) causes a local facilitation of the induction of LTP.

It is intriguing to speculate on the consequences of extending this paradigm. High-frequency stimulation (100 Hz for 1 s) also induces release of endocannabinoids, specifically 2-AG (Stella et al. 1997). If synthesis and release of cannabinoids is sufficiently rapid, then the local release of endocannabinoids caused by the inducing train would tend to facilitate the induction of LTP at that particular site and therefore increase the signal-to-noise ratio of any potentiating synapses in that area. The role of endogenous cannabinoids in the CA1 region of the hippocampus may therefore be to cause a local facilitation of LTP. Assuming that LTP is an important neural basis of at least some forms of learning and memory, the physiological role of the endocannabinoid system would be to enhance hippocampal-dependent learning and memory. In contrast, smoking cannabis may impair learning and memory due to the inappropriate global facilitation of LTP at synapses throughout the brain, rather than at discrete local sites, leading to an elevation of the background noise and a reduction in the signal-to-noise ratio of potentiated synapses.

The findings of Carlson et al. (2002) also prompt the question of why administration of CB_1 receptor agonists is almost universally observed to cause suppression, rather than facilitation of high-frequency stimulus induced LTP. This may relate first to the experimental conditions, which have generally used supra-threshold induction protocols, which are liable to saturate LTP and would not allow any potentiation to be observed. Second, the synthetic agonist most commonly used (i.e. WIN55,212-2) may suppress excitatory transmission, and therefore LTP, via its action on the TRPV1-like rather than the CB_1 receptor. While this could also explain why anandamide (another agonist at the TRPV1-like receptor) blocks LTP (Terranova et al. 1995), note that it cannot explain why 2-AG (which is not an antagonist at the TRPV1-like receptor) also does (Stella et al. 1997).

5.2
LTD in the Hippocampus

Misner and Sullivan (1999) described that 5 μM WIN55,212-2 not only blocked the induction of LTP of CA1 field EPSPs (and EPSCs) by high-frequency stimulation

(100 Hz for 200 ms), but also that it blocked LTD induced by lower frequency stimulation (1 Hz for 15 min). This block of LTD was again relieved by reducing the concentration of Mg^{2+} in the perfusion medium, or by slightly depolarising the recorded cell during the low-frequency train. It was therefore interpreted as being secondary to a reduced excitatory drive resulting in any postsynaptic Ca^{2+} changes being insufficient to reach threshold to induce LTD.

There is also intriguing evidence that release of endogenous cannabinoids may contribute to the maintenance of LTD of IPSCs, and hence contribute to the E–S coupling that is observed after the induction of LTP. High-frequency stimulation (100 Hz for 1 s) or theta burst stimulation (10x100 Hz for 50 ms) caused LTD of pharmacologically isolated IPSCs recorded from the CA1 region of rat hippocampal slices (Chevaleyre and Castillo 2003). AM251 blocked the induction of this LTD, but had no effect if it was applied 10 min after the high-frequency train. LTD was blocked by group I mGlu receptor antagonists, and mimicked by group I mGlu receptor agonists. This suggests that activation of group I mGlu receptors during the high- frequency train stimulates transient release of an endocannabinoid that causes a persistent reduction of GABA release. In this scenario, endocannabinoids have a role, not only in regulating the induction of synaptic plasticity, but also in the expression of synaptic plasticity, during the maintenance phase. Specifically, they may contribute to the increased E–S potentiation, i.e. the ability of a given-sized EPSP to evoke an action potential. The mechanism by which transient release of an endocannabinoid can evoke long-lasting depression of the IPSC has not been determined.

5.3
LTP and LTD in Other Brain Regions

The differences in expression of CB_1 receptors in various brain regions suggest that cannabinoid receptor ligands may have different effects on synaptic plasticity in different synaptic pathways.

5.3.1
Cortex

Results from prefrontal cortex are generally consistent with those from the hippocampus, but carry the same caveat that they may reflect activity of cannabinoids at TRPV1-like receptors, rather than CB_1 receptors. High-frequency stimulation (100 Hz for 1 s) induced LTD, LTP, or no change in pharmacologically isolated EPSCs in approximately equal proportions of cells recorded. WIN55,212-2 (1 μM) increased the proportion of cells exhibiting LTD and decreased the proportion exhibiting LTP, whereas rimonabant increased the proportion of cells exhibiting LTP and decreased the proportion exhibiting LTD (Auclair et al. 2000; Barbara et al. 2003). These effects on synaptic plasticity could be explained as a secondary consequence of changes in the baseline EPSC, since WIN55,212-2 decreased, and rimonabant increased, the baseline response. In this region, the effects of cannabinoids on pairing-induced LTP have also been investigated. LTP was induced between pairs of whole-cell recorded layer 5 cells by coincident pairing of strong pre- and postsynaptic depolarisation, and in the presence of AM251 this potentiation was significantly increased (213% vs 162% of control) (Sjöström et al. 2003).

5.3.2
Amygdala

In the amygdala, immunocytochemical labelling shows that CB_1 receptors are, just as in the hippocampus, selectively localised to the terminals of a subset of inhibitory neurones (Katona et al. 2001). In slices of mouse basolateral amygdala, LTP of the population spike induced by high-frequency trains (100 Hz for 1 s) was significantly greater in slices prepared from $CB_1^{-/-}$ mice than that of wild-type controls (147% vs 117% of control population spike amplitude, respectively). Rimonabant did not affect the baseline synaptic response in slices prepared from wild-type mice, suggesting that tonic activity of the endocannabinoid system normally inhibits high-frequency stimulus induction of LTP, but not via an inhibition of the baseline response (Marsciano et al. 2002). Low-frequency stimulation (1 Hz for 900 s) induced LTD of the population spike, but this was comparable in slices prepared from $CB_1^{-/-}$ and wild-type controls (depressed to 75% vs 80% of control population spike amplitude, respectively). The same study also investigated the effect of low-frequency stimulation on pharmacologically isolated IPSCs. Low-frequency stimulation (1 Hz for 100 s) induced LTD of the IPSC, which was blocked in wild-type mice by rimonabant (5 μM), and was absent in $CB_1^{-/-}$ mice. This study therefore indicates that CB_1 receptors are involved in synaptic plasticity in the amygdala, but does not determine whether the effects of synthetic cannabinoids on synaptic plasticity are mediated by CB_1 receptors.

5.3.3
Nucleus Accumbens

In the nucleus accumbens immunocytochemical labelling shows that, in contrast to the hippocampus, CB_1 receptors are located on glutamatergic nerve terminals (Robbe et al. 2001). In a mouse brain slice preparation, low-frequency stimulation (13 Hz for 10 min) evoked LTD of field EPSPs or EPSCs recorded in the nucleus accumbens (Robbe et al. 2002, 2003). This LTD apparently depended on the activity of endocannabinoids, since it was occluded by 0.3 μM WIN55,212-2, blocked by 0.1–1 μM rimonabant or 2 μM AM251, and critically, it was absent in $CB_1^{-/-}$ mice. While the use of $CB_1^{-/-}$ mice shows that CB_1 receptors are essential for this form of synaptic plasticity in the nucleus accumbens, the cannabinoid ligands were not tested in $CB_1^{-/-}$ animals, and so although it is likely, it is not certain that they produced their effects via the CB_1 receptor. In a similar study, low-frequency stimulation (10 Hz for 5 min) induced LTD of the population spike recorded in slices of rat nucleus accumbens to about 70% of control. This was blocked by perfusion of rimonabant (1 μM), and by desensitisation of CB_1 receptors induced by chronic pre-treatment of the rats with WIN55,212-2 (Hoffman et al. 2003).

5.3.4 Striatum

In the striatum, CB_1 receptors are located on excitatory terminals. High-frequency stimulation (100 Hz for 1 s) induced LTD of EPSCs (to 60% of control) in striatal slices prepared from wild-type mice, but not in slices prepared from $CB_1^{-/-}$ mice (Gerdeman et al. 2002). HU-210 (1 μM) inhibited evoked EPSCs in slices prepared from wild-type mice, but had no effect in slices prepared from $CB_1^{-/-}$ mice. Preincubation of slices prepared from wild-type animals in rimonabant (3 μM) for 90 min blocked the induction of LTD by high-frequency stimulation. In the striatum, therefore, there is compelling evidence that HU-210 blocks LTP via an action on CB_1 receptors.

5.3.5 Cerebellum

In slices of rat cerebellum, LTD of parallel fibre inputs to Purkinje cells can be induced by pairing low-frequency stimulation (1 Hz for 5 min) with post-synaptic depolarisation. Both WIN55, 212-2 (1 μM) and CP55,940 (400 nM) reduced the EPSC by about 50%, and impaired the induction of LTD, an effect which was blocked by rimonabant (1 μM) (Levenes et al. 1998). In this pathway, therefore, it appears that cannabinoid effects on synaptic plasticity may be secondary to changes in baseline responses.

5.4 Future Questions

The finding that the most common pharmacological tools used to probe CB_1 receptors, WIN55,212-2 and rimonabant, also have actions on a TRPV1-like receptor requires that much of the existing literature be re-evaluated. It is often not clear whether effects are mediated by CB_1 receptors or the TRPV1-like receptor. The dearth of information on the pharmacological properties of the TRPV1-like receptor means that even results obtained using related drugs, e.g. CP55,940, AM251 and AM281, must also be treated with caution. Notwithstanding this, the current information suggests that the suppression of high-frequency stimulation-induced LTP in the CA1 hippocampal region by WIN55,212-2 is likely to be secondary to suppression of excitatory drive and is mediated by the TRPV1-like receptor. A big step towards resolving the receptors involved would be to establish whether WIN55,212-2 can inhibit high-frequency stimulation-induced LTP in the CA1 region of hippocampal slices prepared from $CB_1^{-/-}$ mice. Note that the situation may be very different in other brain areas, such as nucleus accumbens where CB_1 receptors are also present on excitatory terminals, and so modulation of synaptic plasticity in these regions is more likely to be CB_1 receptor mediated. This makes it all the more important that behavioural studies use selective administration of drugs to specific brain areas, rather than global administration to whole animals.

In contrast to our original preconceptions, it is emerging that the role of the endogenous cannabinoid system in the hippocampus may be to facilitate the induction of LTP. Administration of exogenous-selective CB_1 agonists may therefore disrupt hippocampus-dependent learning and memory by 'increasing the noise', rather than 'decreasing the signal' at potentiated inputs.

An understanding of the role of the endogenous cannabinoid system in the regulation of synaptic plasticity will depend on identification of the physiological stimuli that trigger the synthesis and release of the endocannabinoid(s). There may be a background level of endocannabinoid activity (controlled by, for example, metabotropic, cholinergic and glutamatergic inputs) that sets the threshold for induction of LTP throughout the hippocampus. Alternatively, the very firing patterns that induce LTP may trigger synthesis and release of endocannabinoids and the time scale over which this occurs will determine any impact on the induction of LTP.

The situation is made more complicated by the diverse pharmacology of some of the putative endocannabinoids. 2-AG appears to exert its effects via the CB_1 receptor, but anandamide may activate both CB_1 and TRPV1-like receptors. The question of which endocannabinoid is released is therefore functionally important.

6
Where to Go from Here?

An obvious feature of this chapter is that, despite the increasing knowledge of the physiology and pharmacology of cannabinoid function in the brain, little transfer as to behavioural consequences has been achieved. While it is clear from the in vitro approach that there are numerous stimulation patterns able to trigger release of endocannabinoids, such conditions have yet to be identified in the behavioural setting. This must involve not only the definition of the psychological circumstances that lead to endogenous cannabinoid release, but also the characterisation of the cellular mechanisms that aid this release. It has also become clear from the in vitro approach that cannabinoid receptors are present only in particular neuronal cell types and that the types of neurone that express these receptors vary between brain areas in a manner that is expected to affect the outcome of cannabinoid receptor activation in vivo. In order to understand this variation in terms of behavioural outcome, it will be essential to apply cannabinoid agonists and antagonists in behaving animals in a more restricted and site-directed manner. In addition, local administration and behavioural testing should be combined with physiological assessment of neural changes induced in response to drug-treatment. This is achievable either by using ex vivo slice preparations of the area of interest (for example cerebellum, hippocampus, visual cortex, etc.) or by chronic implantation of electrodes into the area of interest and in vivo recording of either EEG or single units during behavioural testing. Although this latter approach requires a high level of technical sophistication, especially when conducted using mice, it may yet lead to an unprecedented gain of information concerning the function of the endocannabinoid system in the brain.

References

Adams AJ, Brown B, Haegerstrom-Portnoy G, Flom MC (1976) Evidence for acute effects of alcohol and marijuana on color discrimination. Percept Psychophys 20:119–124

Aiger TG (1988) Delta-9-tetrahydrocannabinol impairs visual recognition memory but not discrimination learning in rhesus monkeys. Psychopharmacology (Berl) 95:507–511

Al-Hayani A, Davies SN (2000) Cannabinoid receptor mediated inhibition of excitatory synaptic transmission in the rat hippocampal slice is developmentally regulated. Br J Pharmacol 131:663–665

Ameri A, Simmet T (2000) Effects of 2-arachidonylglycerol, an endogenous cannabinoid, on neuronal activity in rat hippocampal slices. Naunyn Schmiedebergs Arch Pharmacol 361:265–272

Arizzi MN, Cervone KM, Aberman JE, Betz A, Liu Q, Lin S, Markiyannis A, Salamone JD (2004) Behavioral effects of inhibition of cannabinoid metabolism: The amidase inhibitor AM374 enhances the suppression of lever pressing produced by exogenously administered anandamide. Life Sci 74:1001–1011

Auclair N, Oyani S, Soubrie P, Crepel F (2000) Cannabinoids modulate synaptic strength and plasticity at glutamatergic synapses of rat prefrontal cortex pyramidal neurons. J Neurphysiol 83:3287–3293

Barabara J-G, Auclair N, Roisin M-P, Otani S, Valjent E, Caboche J, Soubrie P, Crepel F (2003) Direct and indirect interactions between cannabinoid CB1 receptor and group II metabotropic glutamate receptor signalling in layer V pyramidal neurons from the rat prefrontal cortex. Eur J Neurosci 17:981–990

Barros DM, Carlis V, Maidana M, Silva AS, Baisch ALM, Ramirez MR, Izquierdo I (2004) Interactions between anandamide-induced anterograde amnesia and post-training memory modulatory systems. Brain Res 1016:66–71

Bech P, Rafaelsen L, Rafaelsen OJ (1973) Cannabis and alcohol: Effects on estimation of time and distance. Psychopharmacologia 32:373–381

Biegon A, Kerman I (1995) Quantitative autoradiography of cannabinoid receptors in the human brain post-mortem. In: Biegon A, Volkow ND (eds) Sites of drug action in the human brain. CRC Press, Boca Raton, pp 65–74

Block RI, O'Leary DS, Ehrhardt JC, Augustinack JC, Ghonheim MM, Arndt S, Hall JA (2000) Effects of frequent marijuana use on brain tissue volume and composition. NeuroReport 11:491–496

Block RI, O'Leary DS, Hichwa RD, Augustinack JC, Boles Ponto LL, Ghonheim MM, et al (2002) Effects of frequent marijuana use on memory-related regional cerebral blood flow. Pharmacol Biochem Behav 72:237–250

Bohme GA, Laville M, Ledent C, Parmentier M, Imperato A (2000) Enhanced long-term potentiation in mice lacking cannabinoid CB1 receptors. Neuroscience 95:5–7

Borg J, Gershon S, Alpert M (1975) Dose effects of smoked marijuana on human cognitive and motor functions. Psychopharmacologia 42:211–218

Bowman M, Pihl RO (1973) Cannabis: psychological effects of chronic heavy use: a controlled study of intellectual functioning in chronic users of high potency cannabis. Psychopharmacologia 29:159–190

Braden W, Stillman RC, Wyatt RJ (1974) Effects of marijuana on contingent negative variation and reaction time. Arch Gen Psychiatry 31:537–541

Branch MN, Dearing ME, Lee DM (1980) Acute and chronic effects of Δ9-tetrahydrocannabinol on complex behavior in squirrel monkeys. Psychopharmacology (Berl) 71:247–256

Breivogel CS, Griffin G, Di Marzo V, Martin BR (2001) Evidence for a new G protein-coupled cannabinoid receptor in mouse brain. Mol Pharmacol 60:155–163

Brodkin J, Moerschbacher JM (1997) SR141716A antagonizes the disruptive effects of cannabinoid ligands on learning in rats. J Pharmacol Exp Ther 282:1526–1532

Carlson G, Wang Y, Alger BE (2002) Endocannabinoids facilitate the induction of LTP in the hippocampus. Nat Neurosci 5:723–724

Castellano C, Cabib S, Palmisano A, et al (1997) The effects of anandamide on memory consolidation in mice involve both D1 and D2 dopamine receptors. Behav Pharmacol 8:707–712

Castellano C, Ventura R, Caibb S, et al (1999) Strain-dependent effects of anandamide on memory consolidation in mice are antagonized by naltrexone. Behav Pharmacol 10:453–457

Chait LD, Pierri J (1992) Effects of smoked marijuana on human performance: a critical review. In: Murphy L, Bartke A (eds) Marijuana/cannabinoids: Neurobiology and neurophysiology. CRC Press, Boca Raton, pp 387–423

Chevaleyre V, Castillo PE (2003) Heterosynaptic LTD of hippocampal GABAergic synapses: a novel role of endocannabinoids in regulating excitability. Neuron 38:461–472

Ciccocioppo R, Antonelli L, Biondini M, et al (2002) Memory impairment following combined exposure to Δ9-tetrahydrocannabinol and ethanol in rats. Eur J Pharmacol 449:245–252

Clark LD, Nakashima EN (1968) Experimental studies of marijuana. Am J Psychiatry 125:379–384

Cohen J (1990) Statistical power analysis for the behavioural sciences. Lawrence Earlbaum Publishers, Hillsdale, NJ

Collins DR, Pertwee RG, Davies SN (1995) Prevention by the cannabinoid antagonist, SR141716A, of cannabinoid-mediated blockade of long-term potentiation in the rat hippocampal slice. Br J Pharmacol 115:869–870

Collins DR, Pertwee RG, Davies SN (1994) The actions of synthetic cannabinoids on the induction of long-term potentiation in the rat hippocampal slice. Eur J Pharmacol 259:R7–R8

Crawley JN (2000) What is wrong with my mouse? Wiley-Liss, New York

Da Silva GE, Takahashi RN (2002) SR141716A prevents Δ9-tetrahydrocannabinol-induced spatial learning deficit in a Morris-type water maze in mice. Prog Neuropsychopharmacol Biol Psychiatry 26:321–325

Deadwyler SA, Hampson RE (1999) Anatomic model of hippocampal encoding of spatial information. Hippocampus 9:397–412

Deadwyler SA, Heyser CJ, Hampson RE (1995) Complete adaptation to the memory disruptive effects of delta-9-THC following 35 days of exposure. Neurosci Res Commun 17:9–18

Deadwyler SA, Bunn T, Hampson RE (1996) Hippocampal ensemble activity during spatial delayed-nonmatch-to-sample performance in rats. J Neurosci 16:354–372

DeSanty KP, Dar MS (2001a) Cannabinoid-induced motor incoordination through the cerebellar CB1 receptor in mice. Pharmacol Biochem Behav 69:251–259

DeSanty KP, Dar MS (2001b) Involvement of the cerebellar adenosine A1 receptor in cannabinoid-induced motor incoordination in the acute and tolerant state in mice. Brain Res 905:178–187

Di Marzo V, Goparaju SK, Wang L, et al (2001) Leptin-regulated endocannabinoids are involved in maintaining food intake. Nature 410:822–825

Diana G, Pieri M, Valentini G (2002) Effects of WIN55,212-2 on hippocampal CA1 long-term potentiation in experiments controlled for basal glutamatergic synaptic transmission. Eur J Pharmacol 453:251–254

Diana G, Malloni, M, Pieri M (2003) Effects of the synthetic cannabinoid nabilone on spatial learning and hippocampal transmission. Pharmacol Biochem Behav 75:585–591

Dornbush RL (1974) Marijuana and memory: effects of smoking on storage. Trans N Y Acad Sci 36:94–100
Dornbush RL, Fink M, Freedman AM (1971) Marijuana, memory and perception. Am J Psychiatry 128:194–197
Earleywine M, Newcomb M (1997) Concurrent versus simultaneous polydrug use: prevalence, correlates, discriminant validity, and prospective effects on health outcomes. Exp Clin Psychopharmacol 5:353–364
Elwan O, Hassan AAH, Naseer MA, Elwan F, et al (1997) Brain aging in a sample of normal Egyptians: cognition, education, addiction and smoking. J Neurol Sci 148:79–86
Ennaceur A, Delacour J (1988) A new one-trial test for neurobiological studies of memory in rats. I. Behavioral data. Behav Brain Res 31:47–59
Entin EE, Glodzung PJ (1973) Residual effects of marijuana use on learning and memory. Psychol Rec 23:169–178
Evans EB, Wenger GR (1992) Effects of drugs of abuse on acquisition of behavioral chains in squirrel monkeys. Psychopharmacology (Berl) 107:55–60
Evans MA, Martz R, Rodda BE, Lemberger L, Forney RB (1976) Effects of marihuana-dextroamphetamine combination. Clin Pharmacol Ther 20:350–361
Ferrari F, Ottani A, Vivoli R, et al (1999) Learning impairment produced in rats by the cannabinoid agonist HU210 in a water-maze task. Pharmacol Biochem Behav 64:555–561
Fletcher JM, Page JB, Francis DJ, et al (1996) Cognitive correlates of long-term cannabis use in Costa Rican men. Arch Gen Psychiatry 53:1051–1057
Gaoni Y, Mechoulam R (1964) Isolation, structure and partial synthesis of an active constituent of hashish. J Am Chem Soc 86:1646–1647
Gerdeman GL, Ronesi J, Lovinger DM (2002) Postsynaptic endocannabinoid release is critical to long-term depression in the striatum. Nature Neurosci 5:446–451
Gianutsos R, Litwack AR (1976) Chronic marijuana smokers show reduced coding into long-term storage. Bull Psychon Soc 7:277–279
Gluck JP, Ferraro DP, Marriott RG (1973) Retardation of discrimination reversal by delta-9-tetrahydrocannabinol in monkeys. Pharmacol Biochem Behav 1:605–608
Hajos N, Freund TF (2002) Pharmacological separation of cannabinoid sensitive receptors on hippocampal excitatory and inhibitory fibres. Neuropharmacology 43:503–510
Hajos N, Ledent C, Freund TF (2001) Novel cannabinoid-sensitive receptor mediates inhibition of glutamatergic synaptic transmission in the hippocampus. Neuroscience 106:1–4
Hampson RE, Deadwyler S (2000) Cannabinoids reveal the necessity of hippocampal neural encoding for short-term memory in rats. J Neurosci 20:8932–8942
Hampson RE, Deadwyler SA (1996) Ensemble codes involving hippocampal neurons are at risk during delayed performance tests. Proc Natl Acad Sci USA 93:13487–13493
Hampson RE, Deadwyler SA (1998) Role of cannabinoid receptors in memory storage. Neurobiol Dis 5:474–482
Hampson RE, Deadwyler SA (1999) Cannabinoids, hippocampal function and memory. Life Sci 65:715–723
Hampson RE, Heyser CJ, Deawyler SA (1993) Hippocampal cell firing correlates of delayed-match-to-sample performance in the rat. Behav Neurosci 107:715–739
Hampson RE, Jarrard LE, Deadwyler SA (1999a) Effects of ibotenate hippocampal destruction on delayed matching and nonmatching to sample behavior in rats. J Neurosci 19:1492–1507
Hampson RE, Simeral JD, Deadwyler SA (1999b) Distribution of spatial and nonspatial information in dorsal hippocampus. Nature 402:610–614
Han CJ, Robinson JK (2001) Cannabinoid modulation of time estimation in the rat. Behav Neurosci 115:243–246
Han CJ, Pierre-Louis J, Scheff A, et al (2000) A performance-dependent adjustment of the retention interval in the delayed non-matching-to-position paradigm differentiates effects of amnestic drugs in rats. Eur J Pharmacol 403:87–93

Hernandez-Tristan R, Arevalo C, Canals S, et al (2000) The effects of acute treatment with Delta(9)-THC on exploratory behaviour and memory in the rat. J Physiol Biochem 56:17–24
Heyser CJ, Hampson RE, Deadwyler SA (1993) Effects of delta-9-tetrahydrocannabinol on delayed match to sample performance in rats. Alterations in short-term memory associated with changes in task specific firing of hippocampal cells. J Pharmacol Exp Ther 264:294–307
Hoffman AF, Oz M, Caulder T, Lupicia CR (2003) Functional tolerance and blockade of long-term depression at synapses in the nucleus accumbens after chronic cannabinoid exposure. J Neurosci 23:4815–4820
Isaacson RL, Douglas RJ, Moore RY (1961) The effect of radical hippocampal ablation on acquisition of avoidance response. J Comp Physiol Psychol 54:625–628
Izquierdo I, Nasello AG (1973) Effects of cannabidiol and diphenylhydantoin on the hippocampus and learning. Psychopharmacologia 31:167–175
Järbe TUC, Hiltunen AJ (1987) Cannabimimetic activity of cannabinol in rats and pigeons. Neuropharmacology 26:219–228
Järbe TUC, Andrzejewski ME, DiPatrizio NV (2002) Interaction between the CB1 receptor agonist Δ9-THC and the CB1 receptor antagonist SR-141716 in rats. Open field revisited. Pharmacol Biochem Behav 73:911–919
Jarrard LE (1993) On the role of the hippocampus in learning and memory in the rat. Behav Neural Biol 60:9–26
Jentsch JD, Andrusiak E, Tran A, et al (1996) HA966 prevents increases in prefrontal dopamine turnover and impairments in spatial working memory induced by psychomimetic drugs. FASEB J 20:A716
Jentsch JD, Andrusiak E, Tran A, et al (1997) Δ9-Tetrahydrocannabinol increases prefrontal cortical catecholaminergic utilisation and impairs spatial working memory in the rat: blockade of dopaminergic effects by HA966. Neuropsychopharmacology 16:426–432
Johnson JP, Muhleman D, MacMurray J, et al (1997) Association between the cannabinoid receptor gene (CNR1) and the P300 event-related potential. Mol Psychiatry 2:169–171
Katona I, Rancz EA, Acsady L, Ledent C, Mackie K, Hajos N, Freund TF (2001) Distribution of CB1 cannabinoid receptors in the amygdala and their role in the control of GABAergic transmission. J Neurosci 21:9506–9518
Kim J, Isokawa M, Ledent C, Alger BE (2002) Activation of muscarinic acetylcholine receptors enhances the release of endogenous cannabinoids in the hippocampus. J Neurosci 22:10182–10191
Koch M (1999) The neurobiology of startle. Prog Neurobiol 59:107–128
Kosiorek P, Hryniewicz A, Bialuk I, Zawadzka A, Winnicka A (2003) Cannabinoids alter recognition memory in rats. Pol J Pharmacol 55:903–910
Kreitzer AC, Regehr WG (2001) Retrograde inhibition of presynaptic calcium influx by endogenous cannabinoids at excitatory synapses onto Purkinje cells. Neuron 29:717–727
Levenes C, Daniel H, Soubrie P, Crepel F (1998) Cannabinoids decrease excitatory synaptic transmission and impair long-term depression in rat cerebellar Purkinje cells. J Physiol 510:867–879
Lichtman AH (2000) SR141716A enhances spatial memory assessed in a radial-arm maze task in rats. Eur J Pharmacol 404:175–179
Lichtman AH, Martin BR (1996) Δ9-Tetrahydrocannabinol impairs spatial memory through cannabinoid receptor mechanism. Psychopharmacology (Berl) 126:125–131
Lichtman AH, Martin BR (1997) The selective cannabinoid antagonist SR 141716A blocks cannabinoid-induced antinociception in rats. Pharmacol Biochem Behav 57:7–12
Lichtman AH, Dimen KR, Martin BR (1995) Systemic or intrahippocampal cannabinoid administration impairs spatial memory in rats. Psychopharmacology (Berl) 119:282–290
Lichtman AH, Varvel SA, Martin BR (2002) Endocannabinoids in cognition and dependence. Prostaglandins Leukot Essent Fatty Acids 66:269–285

Lyketsos CG, Garret E, Liang KY, Anthony JC (1999) Cannabis use and cognitive decline in persons under 65 years of age. Am J Epidemiol 149:794–800

Maccarrone M, Valverde O, Barbaccia ML, et al (2002) Age-related changes of anandamide metabolism in CB1 cannabinoid receptor knockout mice: correlation with behaviour. Eur J Neurosci 15:1178–1186

Maejima T, Hashimoto K, Yoshida T, Aiba A, Kano M (2001) Presynaptic inhibition caused by retrograde signal from metabotropic glutamate to cannabinoid receptors. Neuron 31:463–475

Mallet PE, Beninger RJ (1996) The endogenous cannabinoid receptor agonist impairs memory in rats. Behav Pharmacol 7:276–284

Mallet PE, Beninger RJ (1998) The cannabinoid CB1 receptor antagonist SR141716A attenuates the memory impairment produced by Δ9-tetrahydrocannabinol or anandamide. Psychopharmacology (Berl) 140:11–19

Mansbach RS, Rovetti CC, Winston EN, et al (1996) Effects of cannabinoid receptor antagonist SR141716A on the behavior of pigeons and rats. Psychopharmacology (Berl) 124:315–322

Marsciano G, Wotjack CT, Azas SC, Bisogno T, Rammes G, Cascio MG, Hermann H, Tang J, Hofmann C, Zieglgänsberger W, Di Marzo V, Lutz B (2002) The endogenous cannabinoid system controls extinction of aversive memories. Nature 418:530–534

Martin M, Ledent C, Parmentier M, Maldonado R, Valverde O (2002) Involvement of CB1 cannabinoid receptors in emotional behaviour. Psychopharmacology (Berl) 159:379–387

Mathew RJ, Wilson WH (1991) Substance abuse and cerebral blood flow. Am J Psychiatry 148:292

Mechoulam R, Fride E (2001) A hunger for cannabinoids. Nature 410:763–765

Menhiratta SS, Wig NN, Verma SK (1978) Some psychological correlates of long-term heavy cannabis users. Br J Psychiatry 132:482–486

Mereu G, Fà M, Ferraro L, Cagiano R, Antonelli T, Tattoli M, Ghiglieri V, Tanganelli S, Gessa GL, Cuomo V (2003) Prenatal exposure to a cannabinoid agonist produces memory deficits linked to dysfunction in hippocampal long-term potentiation and glutamate release. Proc Natl Acad Sci USA 100:4915–4920

Miller L, Cornett T (1978) Marijuana: dose-response effects on pulse rate, subjective estimates of intoxication, free recall and recognition memory. Pharmacol Biochem Behav 9:573–579

Mishima K, Egashira N, Hirosawa N, et al (2001) Characteristics of learning and memory impairment induced by Δ9-tetrahydrocannabinol in rats. Jpn J Pharmacol 87:297–308

Misner DL, Sullivan JM (1999) Mechanism of cannabinoid effects on long-term potentiation and depression in hippocampal CA1 neurons. J Neurosci 19:6795–6805

Miyamoto A, Yamamoto T, Watanabe S (1995) Effect of repeated administration of Δ9-tetrahydrocannabinol on delayed matching-to-sample performance in rats. Neurosci Lett 201:139–142

Molina-Holgado F, Gonzalez MI, Leret ML (1995) Effect of Δ9-tetrahydocannabinol on short-term memory in the rat. Physiol Behav 57:177–179

Morris RGM (1984) Developments of a water-maze procedure for studying spatial learning in that rat. J Neurosci Methods 11:47–60

Nakamura EM, da Silva EA, Concilio GV, et al (1991) Reversible effects of acute and long-term administration of Δ-9-tetrahydrocannabinol (THC) on memory in the rat. Drug Alcohol Depend 28:167–175

Nakamura-Palacios EM, Winsauer PJ, Moerschbacher JM (2000) Effects of the cannabinoid ligand SR141716A alone or in combination with Δ9-tetrahydrocannabinol or scopolamine on learning in squirrel monkeys. Behav Pharmacol 11:377–389

Nava F, Carta G, Battesi AM, et al (2000) D2 dopamine receptors enable Δ9tetrahydrocannabinol induced memory impairment and reduction of hippocampal extracellular acetylcholine concentration. Br J Pharmacol 130:1201–1210

Nava F, Carta G, Colombo G, et al (2001) Effects of chronic Δ9tetrahydrocannabinol treatment on hippocampal extracellular acetylcholine concentration and alternation performance in the T-maze. Neuropharmacology 41:392–399

Navarro M, Fernandez Ruiz JJ, de Miguel R, et al (1993) Motor disturbances induced by an acute dose of Δ9-tetrahydrocannabinol: possible involvement of nigrostriatal dopaminergic alterations. Pharmacol Biochem Behav 45:291–298

Nowicky AV, Teyler TJ, Vardaris RM (1987) The modulation of long-term potentiation by delta-9-tetrahydrocannabinol in the rat hippocampus, in vitro. Brain Res Bull 19:663–672

O'Leary DS, Block RI, Koeppel JA, et al (2002) Effects of smoking marijuana on brain perfusion and cognition. Neuropsychopharmacology 26:802–816

Ohno-Shosaku T, Maejima T, Kano M (2001) Endogenous cannabinoids mediate retrograde signals from depolarised postsynaptic neurons to presynaptic terminals. Neuron 29:729–738

Ohno-Shosaku T, Tsubokawa H, Mizushima I, Yoneda N, Zimmer A, Kano M (2002) Presynaptic cannabinoid sensitivity is a major determinant of depolarization-induced retrograde suppression at hippocampal synapses. J Neurosci 22:3864–3872

Onaivi ES, Green MR, Martin BR (1990) Pharmacological characterisation of cannabinoids in the elevated plus-maze. J Pharmacol Exp Ther 253:1002–1009

Paton GS, Pertwee RG, Davies SN (1998) Correlation between cannabinoid mediated effects on paired pulse depression and induction of long term potentiation in the rat hippocampal slice. Neuropharmacology 37:1123–1130

Patrick G, Straumanis JJ, Struve FA, Fitz-Gerald MJ, Leavitt J, Manno JE (1999) Reduced P50 auditory gating response in psychiatrically normal chronic marihuana users: a pilot study. Biol Psychiatry 45:1307–1312

Pearl J, Domino E, Rennick P (1973) Short-term effects of marijuana smoking on cognitive behavior in experienced male users. Psychopharmacologia 31:13–24

Pertwee RG (1997) Pharmacology of cannabinoid CB1 and CB2 receptors. Pharmacol Ther 74:129–180

Pitler TA, Alger BE (1992) Postsynaptic spike firing reduces synaptic GABA(A) responses in hippocampal pyramidal cells. J Neurosci 12:4122–4132

Pope HG, Yurgelun-Todd D (1996) The residual cognitive effects of heavy marijuana use in college students. J Am Med Assoc 275:521–527

Porter AC, Felder CC (2001) The endocannabinoid nervous system: Unique opportunities for therapeutic intervention. Pharmacol Ther 90:45–60

Presburger G, Robinson JK (1999) Spatial signal detection in rats is differentially disrupted by Δ-9-tetrahydrocannabinol, scopolamine and MK-801. Behav Brain Res 99:27–34

Reibaud M, Obinu MC, Ledent C, et al (1999) Enhancement of memory in cannabinoid CB1 receptor knock-out mice. Eur J Pharmacol 379:R1–R2

Robbe D, Alonso G, Duchamp F, Bockaert J, Manzoni JM (2001) Localization and mechanisms of action of cannabinoid receptors at the glutamatergic synapses of the mouse nucleus accumbens. J Neurosci 21:109–116

Robbe D, Kopf M, Remaury A, Bockaert J, Manzoni O (2002) Endogenous cannabinoids mediate long-term synaptic depression in the nucleus accumbens. Proc Natl Acad Sci USA 99:8384–8388

Robbe D, Alonso G, Manzoni O (2003) Exogenous and endogenous cannabinoids control synaptic transmission in mice nucleus accumbens. Ann NY Acad Sci 1003:212–225

Robinson L, Hinder L, Pertwee RG, Riedel G (2003) Effects of Δ9-THC and WIN-55,212-2 on place preference in the water maze in rats. Psychopharmacology (Berl) 166:40–50

Rochford J, Grant I, LaVigne G (1977) Medical students and drugs: Further neuropsychological and use pattern considerations. Int J Addict 12:1057–1065

Rodriguez de Fonseca F, Del Arco I, Martin-Calderon JL, et al (1998) Role of the endogenous cannabinoid system in the regulation of motor activity. Neurobiol Dis 5:483–501

Satz P, Fletcher JM, Sutker LS (1976) Neuropsychologic, intellectual and personality correlates in chronic marihuana use in native Costa Ricans. Ann NY Acad Sci 282:266–306
Schneider M, Koch M (2002) The cannabinoid agonist WIN 55,212–2 reduces sensorimotor gating and recognition memory in rats. Behav Pharmacol 13:29–37
Schulze GE, McMillan DE, Bailey JR, et al (1988) Acute effects of delta-9-tetrahydrocannabinol in rhesus monkeys as measured by performance in a battery of complex operant tests. J Pharmacol Exp Ther 245:178–186
Schwartz RH, Gruenewald PJ, Klitzner M, et al (1989) Short-term memory impairments in cannabis-dependent adolescents. Am J Dis Child 143:1214–1219
Schweitzer P, Siggins GR, Madamba SG (1999) Cannabinoid modulation of neuronal activity in adult rat hippocampus. Adv Exp Med Biol 469:547–552
Sjöström PJ, Turrigano GG, Nelson SB (2003) Neocortical LTD via coincident activation of presynaptic NMDA and cannabinoid receptors. Neuron 39:641–654
Solowij N (1998) Cannabis and cognitive functioning. Cambridge University Press, Cambridge
Soueif MI (1976) Some determinant of psychological deficits associated with chronic cannabis consumption. Bull Narc 28:25–42
Stadnicki SW, Schaeppi U, Rosenkrantz H, et al (1974) Crude marihuana extract: EEG and behavioral effects of chronic oral administration in rhesus monkeys. Psychopharmacologia 37:225–233
Stella N, Schweitzer P, Piomelli D (1997) A second endogenous cannabinoid that modulates long-term potentiation. Nature 388:773–778
Stiglick A, Kalant H (1982) Learning impairment in the radial-arm maze following prolonged cannabis treatment in rats. Psychopharmacology (Berl) 77:117–123
Stiglick A, Llewellyn ME, Kalant H (1984) Residual effects of prolonged cannabis treatment on shuttle-box avoidance in the rat. Psychopharmacology (Berl) 84:476–479
Strohmetz DB, Alterman AI, Walter D (1990) Subject selection bias in alcoholics volunteering for a treatment study. Alcohol Clin Exp Res 14:736–738
Struve FA, Straumanis JJ, Patrick G (1994) Persistent topographic quantitative EEG sequelae of chronic marijuana use: A replication study and initial discriminant function analysis. Clin Electroencephalogr 25:63–75
Szallasi A, Di Marzo V (2000) New perspectives on enigmatic vanilloid receptors. Trends Neurosci 23:491–497
Tart CT (1971) On being stoned. Science and Behavior Books,Palo Alto
Teng AH, Craft RM (2004) CB1 receptor mediation of cannabinoid behavioural effects in male and female rats. Psychopharmacology (Berl) 172:25–30
Terranova J-P, Michaud J-C, Le Fur G, Soubrie P (1995) Inhibition of long-term potentiation in rat hippocampal slices by anandamide and WIN55,212-2: reversal by SR141716A, a selective antagonist of CB1 cannabinoid receptors. Naunyn Schmiedebergs Arch Pharmacol 352:576–579
Terranova J-P, Storme J-J, Lafon N, et al (1996) Improvement of memory in rodents by the selective CB1 cannabinoid receptor antagonist, SR141716. Psychopharmacology (Berl) 126:165–172
Varma N, Carlson GC, Ledent C, Alger BE (2001) Metabotropic glutamate receptors drive the endocannabinoid system in hippocampus. J Neurosci 21:RC188
Varvel SA, Lichtman AH (2002) Evaluation of CB1 receptor knockout mice in the Morris water maze. J Pharmacol Exp Ther 301:915–924
Varvel SA, Hamm RJ, Martin BR, et al (2001) Differential effects of Δ9THC on spatial reference and working memory in mice. Psychopharmacology (Berl) 157:142–150
Westlake GA, Howlett AC, Bonner TI, Matsuda LA, Herkenham M (1994) Cannabinoid receptor binding and messenger RNA expression in human brain: an in vitro receptor autoradiography and in situ hybridization histochemistry study of normal and Alzheimer's brains. Neuroscience 63:637–652

Wig NN, Verma SK (1977) Patterns of long-term heavy cannabis use in north India and its effects on cognitive functions: A preliminary report. Drug Alcohol Depend 2:211–219
Wilson RI, Nicoll RA (2001) Endogenous cannabinoids mediate retrograde signalling at hippocampal synapses. Nature 410:588–592
Wilson RI, Nicoll RA (2002) Endocannabinoid signalling in the brain. Science 296:678–682
Wilson WH, Mathew RJ (2002) Effects of marijuana on brain: Function and structure. In: Onaivi ES (ed) Biology of marijuana: from gene to behavior. Taylor and Francis, London, pp 234–281
Winsauer PJ, Lambert P, Moerschbacher JM (1999) Cannabinoid ligands and their effects on learning and performance in rhesus monkeys. Behav Pharmacol 10:497–511

HEP (2005) 168:479–507

Cannabinoid Control of Motor Function at the Basal Ganglia

J. Fernández-Ruiz (✉) · S. González

Departamento de Bioquímica y Biología Molecular III, Facultad de Medicina,
Universidad Complutense, Ciudad Universitaria s/n, 28040 Madrid, Spain
jjfr@med.ucm.es

Abstract Classic and novel data strengthen the idea of a prominent role for the endocannabinoid signaling system in the control of movement. This finding is supported by three-fold evidence: (1) the abundance of the cannabinoid CB_1 receptor subtype, but also of CB_2 and vanilloid VR1 receptors, as well as of endocannabinoids in the basal ganglia and the cerebellum, the areas that control movement; (2) the demonstration of a powerful action, mostly of an inhibitory nature, of plant-derived, synthetic, and endogenous cannabinoids on motor activity, exerted by modulating the activity of various classic neurotransmitters; and (3) the occurrence of marked changes in endocannabinoid transmission in the basal ganglia of humans affected by several motor disorders, an event corroborated in animal models of these neurological diseases. This three-fold evidence has provided support to the idea that cannabinoid-based compounds, which act at key steps of the endocannabinoid transmission [receptors, transporter, fatty acid amide hydrolase (FAAH)], might be of interest because of their potential ability to alleviate motor symptoms and/or provide neuroprotection in a variety of neurological pathologies directly affecting basal ganglia structures, such as Parkinson's disease and Huntington's chorea, or indirectly, such as multiple sclerosis and Alzheimer's disease. The present chapter will review the knowledge on this issue, trying to establish future lines for research into the therapeutic potential of the endocannabinoid system in motor disorders.

Keywords Cannabinoids · Cannabinoid receptors · Movement · Basal ganglia · Motor disorders

1
Function of the Endocannabinoid Signaling System in Motor Regions

The finding that the endocannabinoid system might be involved in the regulation of motor behavior is based on three series of different but complementary studies. A first group of studies, mainly dealing with pharmacological aspects, addressed the motor effects of plant-derived, synthetic, and endogenous cannabinoids in humans and laboratory animals (for reviews see Consroe 1998; Romero et al. 2002). In general, these studies demonstrated that cannabinoid agonists have powerful actions, mostly inhibitory effects, on motor activity (Crawley et al. 1993; Fride and Mechoulam 1993; Wickens and Pertwee 1993; Smith et al. 1994; Romero et al. 1995a and 1995b; for reviews see Sañudo-Peña et al. 1999; Romero et al. 2002). These studies also demonstrated that there exist differences in magnitude and duration for motor effects of the different cannabinoids, but these are mostly attributable to their differences in receptor affinity, potency, and/or metabolic stability. These motor effects are likely the consequence of the capability of cannabinoids to interact with specific neurotransmitters in the basal ganglia structures (for reviews see Sañudo-Peña et al. 1999; Romero et al. 2002; Fernández-Ruiz et al. 2002).

A second set of studies addressed the location and quantification of diverse elements of the endocannabinoid system in motor regions. They demonstrated that endocannabinoids and their receptors, in comparison with other brain structures, are abundant in the basal ganglia and also in the cerebellum, two brain structures directly involved in the control of movement (Herkenham et al. 1991a,b; Mailleux and Vanderhaeghen 1992a; Tsou et al. 1998a,b; Bisogno et al. 1999). Finally, in a third group of studies, possibly the most recent ones, the objective was to examine whether CB_1 receptors or other key proteins of the endocannabinoid system are altered in the basal ganglia of humans affected by several neurological diseases directly or indirectly related to motor function (Glass et al. 1993, 2000; Richfield and Herkenham 1994; Lastres-Becker et al. 2001a; for reviews see Consroe 1998; Fernández-Ruiz et al. 2002). These observations have been corroborated in different animal models of these motor disorders (Zeng et al. 1999; Romero et al.

2000; Page et al. 2000; Lastres-Becker et al. 2001a, 2002a,b). The present chapter will consider all of this previous pharmacological, biochemical, anatomical, and pathological evidence, and also the extent to which existing data support the hypothesis that modulators of the endocannabinoid system have therapeutic potential for the treatment of motor disorders.

1.1 Motor Effects of Cannabinoid-Based Compounds

Among a variety of effects, the consumption of cannabis by humans affects psychomotor activity, reflected by a global impairment of performance (especially in complex and demanding tasks) and resulting in an increased motor activity followed by inertia and incoordination, ataxia, tremulousness, and weakness (for reviews see Dewey 1986; Consroe 1998). Similar results were obtained in experiments with laboratory animals where the administration of plant-derived, synthetic, or endogenous cannabinoids, in particular (−)-Δ^9-tetrahydrocannabinol (Δ^9-THC), the prototypical tricyclic cannabinoid derived from *Cannabis sativa*, produced dose-dependent impairments in a variety of motor tests (open-field, ring test, actimeter, rotarod), thus stressing the relevance of endocannabinoid transmission in the control of motor function by the basal ganglia (for reviews see Di Marzo et al. 1998; Sañudo-Peña et al. 1999; Romero et al. 2002; Fernández-Ruiz et al. 2002).

1.1.1 Effects of Plant-Derived, Synthetic, or Endogenous Cannabinoid Agonists

Among the most notable effects, the administration of Δ^9-THC produced a reduction of spontaneous activity and induction of catalepsy in mice (Pertwee et al. 1988), whereas in rats it reduced ambulation, and spontaneous or induced stereotypic behaviors (Navarro et al. 1993; Romero et al. 1995a), increased inactivity (Rodríguez de Fonseca et al. 1994; Romero et al. 1995a), potentiated reserpine-induced hypokinesia (Moss et al. 1981) while reducing amphetamine-induced hyperlocomotion (Gorriti et al. 1999), increased circling behavior (Jarbe et al. 1998), and disrupted fine motor control (McLaughlin et al. 2000). Many other effects have been also documented (see Table 1 for a summary). Other plant-derived cannabinoids also produced motor inhibition (Hiltunen et al. 1988), although their effects were weak compared with those caused by Δ^9-THC in concordance with their lower affinity for the cannabinoid CB_1 receptor. By contrast, synthetic cannabinoids produced powerful inhibitory effects in a variety of motor tests and animal models (for reviews, see Consroe 1998; Sañudo-Peña et al. 1999; Romero et al. 2002; Table 1 for a summary).

The inhibitory effects reported for plant-derived or synthetic cannabinoids were, in general, mimicked by endocannabinoids, mainly anandamide (see Table 1 for a summary). Thus, Fride and Mechoulam (1993) reported a decrease in rearing behavior and immobility in mice, results that were reproduced by Crawley et al. (1993) and Smith et al. (1994). In addition, Wickens and Pertwee (1993) found that

Table 1. Motor effects of cannabinoid-related compounds in laboratory animals

Compound(s)		Motor effects
Plant-derived cannabinoids	Δ^9-THC	Reduction of spontaneous and stereotypic activity in rats (Navarro et al. 1993; Romero et al. 1995a,b; Jarbe et al. 1998)
		Reduction of spontaneous motor activity and induction of catalepsy (basal and muscimol-induced) in mice (Pertwee et al. 1988; Wickens and Pertwee 1993)
		Increase in reserpine-induced hypokinesia in rats (Moss et al. 1981)
		Increase in inactivity in rats (Rodríguez de Fonseca et al. 1994; Romero et al. 1995a,b: Jarbe et al. 1998)
		Reduction of amphetamine-induced hyperlocomotion in rats (Gorriti et al. 1999)
		Disruption of fine motor control in rats (McLaughlin et al. 2000)
		Increase in motor activity at low doses (Sañudo-Peña et al. 2000)
	CBD and CBN	Motor inhibition, but of lesser magnitude than Δ^9-THC (Hiltunen et al. 1988)
Synthetic cannabinoids	CP 55,940, HU210, WIN 55,212-2	Powerful effects causing motor inhibition in rats (reviewed by Romero et al. 2002)
		Turning behavior at low doses in mice (Souilhac et al. 1995)
Endocannabinoids	Anandamide	Immobility and decreased rearing behavior in rodents (Fride and Mechoulam 1993; Crawley et al. 1993; Smith et al. 1994)
		Reduction of ambulation and stereotypy, and increase of inactivity in rats (as with Δ^9-THC, but its effects were of shorter duration) (Romero et al. 1995a and 1995b)
		Potentiation of muscimol-induced catalepsy (Wickens and Pertwee 1993)
		Turning behavior at low doses in mice (Souilhac et al. 1995)
Endocannabinoid analogs	*R*-methanandamide (AM356)	Decreased ambulation and stereotypy and increased inactivity in rats (effects of longer duration than anandamide, similar to Δ^9-THC) (Romero et al. 1996; Jarbe et al. 1998)
Transporter inhibitors	AM404, VDM11, OMDM2, UCM707	Decreased ambulation and increased inactivity in rats (González et al. 1999; Beltramo et al. 2000)
		Potentiated anandamide-induced motor inhibition (de Lago et al. 2002 and 2004a)
Receptor antagonists	SR141716A (CB_1)	Antagonized motor effects of cannabinoid agonists (Souilhac et al. 1995; Di Marzo et al. 2001)
		Induction of stereotypies and hyperlocomotion (Compton et al. 1996)
	Capsazepine (VR1)	Antagonized motor effects of anandamide (de Lago et al. 2004b)

muscimol-induced catalepsy in rats was potentiated by anandamide and by Δ^9-THC. In our laboratory, we found that anandamide inhibited motor and stereotypic behaviors in a dose-related manner as did Δ^9-THC (Romero et al. 1995a), but, unlike the time-course for the response exhibited by this plant-derived cannabinoid, the time-course of the response to anandamide showed a biphasic pattern that probably reflected its conversion to active metabolite(s) (Romero et al. 1995b). *R*-(+)-methanandamide, a more stable analog of anandamide, produced a dose-dependent motor inhibition in the open-field test with almost the same potency as Δ^9-THC and with a duration of action longer than that of anandamide and almost comparable to that of Δ^9-THC (Romero et al. 1996a; Jarbe et al. 1998). In contrast with the above studies that used a range of doses producing exclusively hypokinetic effects, there are also a few studies showing that lower doses of anandamide, Δ^9-THC, or other cannabinoids increase motor behavior in mice (Souilhac et al. 1995) or rats (Sañudo-Peña et al. 2000).

1.1.2
Effects of Inhibitors of Endocannabinoid Inactivation

The above evidence was obtained with compounds that act directly at the CB_1 receptor, the cannabinoid receptor subtype involved in the psychoactive effects of cannabis derivatives. Similar results were observed with inhibitors of endocannabinoid inactivation, so-called indirect agonists, which act by prolonging the action of endocannabinoids at their receptors. These reuptake inhibitors were AM404 (González et al. 1999; Beltramo et al. 2000), VDM11 (de Lago et al. 2004a), UCM707 (de Lago et al. 2002), and OMDM2 (de Lago et al. 2004a) (see Table 1 for details). The most interesting aspect of the motor effects of these compounds was their ability to potentiate the hypokinetic effects of subeffective doses of anandamide, an effect that was particularly notable in experiments with UCM707 (de Lago et al. 2002). The use of these compounds in the clinic might represent an interesting option for those diseases, such as Huntington's disease (HD) or other hyperkinetic disorders, where a hypofunction of endocannabinoid transmission has been documented (see Fernández-Ruiz et al. 2002 for a review).

1.1.3
Effects of Cannabinoid Receptor Antagonists

The motor effects of cannabinoid agonists are usually prevented by SR141716, a selective CB_1 receptor antagonist (Souilhac et al. 1995; Di Marzo et al. 2001; for a review see Consroe 1998), thus suggesting that they are CB_1 receptor-mediated (see Table 1). However, the administration of SR141716 by itself can cause hyper-locomotion (Compton et al. 1996). All these data are compatible with the idea that the pharmacological blockade of CB_1 receptors might be of value for the treatment of hypokinetic signs of the sort that occur in Parkinson's disease (PD) and related disorders (see Fernández-Ruiz et al. 2002 for a review), an issue that will be discussed in detail below.

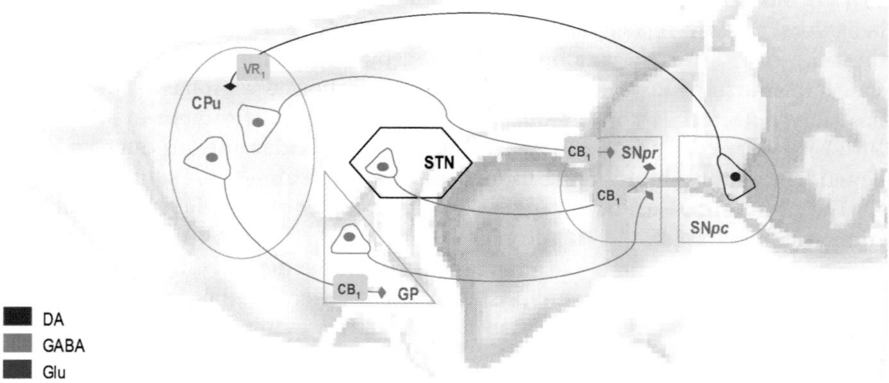

Fig. 1. Distribution of CB_1 and VR1 receptors in the basal ganglia circuitry in rats. *CPU*, caudate-putamen; *GP*, globus pallidus; *STN*, subthalamic nucleus; *SNpr*, substantia nigra pars reticulata; *SNpc*, substantia nigra pars compacta

In addition, recent evidence has demonstrated that vanilloid VR1 receptors might also be involved in the motor effects of certain cannabinoids that include those with an eicosanoid structure but exclude classical cannabinoids (de Lago et al. 2004b). Thus, motor inhibition produced by anandamide has been found to be reversed by capsazepine but not by SR141716 (de Lago et al. 2004b; see Table 1), which agrees with previous observations that: (1) the activation of VR1 receptors with their classic agonist, capsaicin, also produced hypokinesia (Di Marzo et al. 2001), and (2) the antihyperkinetic activity of AM404 in rat models of HD depends on its ability to directly activate VR1 receptors rather than to block the endocannabinoid transporter (Lastres-Becker et al. 2002a, 2003a). These pharmacological data suggest that the VR1 receptor may be another target through which anandamide and its analogs are able to affect motor function and provide therapeutic benefits in motor disorders. The recent detection of VR1 receptors in the basal ganglia (Mezey et al. 2000) supports this hypothesis.

1.2
Control of Different Neurotransmitters by Cannabinoids in Motor Regions

As indicated above, the administration of different cannabinoids impairs movement in rodents and humans. It is expected that this effect depends on the direct or indirect action of cannabinoids on the levels of several neurotransmitters that have been classically involved in the control of basal ganglia function. Three neurotransmitters seem to be influenced by cannabinoids in this circuitry, dopamine, γ-aminobutyric acid (GABA), and glutamate. In the case of the last two neurotransmitters, a direct action is possible since GABAergic and glutamatergic neurons in the basal ganglia contain CB_1 receptors located presynaptically (see Fig. 1). This enables endocannabinoids to directly influence presynaptic events, such as synthesis, release, or reuptake (see Fernández-Ruiz et al. 2002 for a review). In contrast, dopaminergic neurons do not contain CB_1 receptors (Herkenham et al. 1991b). However, these receptors are abundantly expressed in the caudate-putamen, which is innervated by dopamine-releasing neurons (Herkenham et al. 1991a; Mailleux and Vanderhaeghen 1992a; Tsou et al. 1998a), thus allowing an indirect interaction. In addition, recent data showing that VR1 receptors present in the basal ganglia are likely located in nigrostriatal dopaminergic neurons (Mezey et al. 2000) open up the possibility that some cannabinoids may have a direct effect on dopaminergic transmission (de Lago et al. 2004b).

1.2.1 γ-Aminobutyric Acid

The involvement of GABAergic transmission in motor effects of cannabinoids has been documented in several studies (for a review see Fernández-Ruiz et al. 2002). We reported that the blockade of $GABA_B$, but not $GABA_A$, receptors attenuated most of the signs of motor inhibition caused by the administration of cannabinoid agonists in rats (Romero et al. 1996b). This is consistent with results obtained by Miller and Walker (1995, 1996) in a series of electrophysiological studies. These indicated that cannabinoids can modulate GABA release in vivo in the globus pallidus and substantia nigra. However, the effects were very modest. More recently, neurochemical studies demonstrated that the administration of cannabinoids did not affect GABA synthesis or release in the basal ganglia of naïve animals (Maneuf et al. 1996; Romero et al. 1998a; Lastres-Becker et al. 2002a), although cannabinoids were effective in increasing both processes in animals with lesions of striatal GABAergic neurons of the sort that occur in HD (Lastres-Becker et al. 2002a). In addition, the stimulation of CB_1 receptors located on axonal terminals of striatal GABAergic neurons resulted in an inhibition of GABA reuptake in globus pallidus slices (Maneuf et al. 1996) or substantia nigra synaptosomes (Romero et al. 1998a), and hence in the potentiation of GABAergic transmission. These observations are concordant with the finding by Gueudet et al. (1995) that the blockade of CB_1 receptors in striatal projection neurons with SR141716 reduced inhibitory GABAergic tone, thereby allowing the firing of nigrostriatal dopaminergic neurons. The authors concluded that endocannabinoid transmission might increase the action of striatal GABAergic neurons in the substantia nigra, producing a decrease of the stimulation of nigral dopaminergic neurons (Gueudet et al. 1995). However, other studies have reported opposite effects. Thus, Tersigni and Rosenberg (1996) reported an increase by cannabinoids in the activity of nigral neurons without any alteration of GABAergic activity. Other authors have observed inhibition rather than stimulation of GABAergic neurons by cannabinoid agonists via a presynaptic action in the substantia nigra (Chan et al. 1998) or the striatum (Szabo et al. 1998). Therefore, further studies will be required to elucidate the complex interaction of cannabinoids with GABAergic transmission in the basal ganglia.

1.2.2 Glutamate

Cannabinoids may also exert a direct action on glutamate-releasing neurons in the basal ganglia due to the location of CB_1 receptors in subthalamonigral glutamatergic neurons (Mailleux and Vanderhaeghen 1992a; see Fig. 1). This has been demonstrated in a series of electrophysiological studies showing a modification by cannabinoid agonists of the activity of pallidal and nigral neurons, which was exerted by inhibiting glutamate release from subthalamonigral terminals (Sañudo-Peña and Walker 1997; Szabo et al. 2000). The involvement of CB_1 receptors in these effects seems very likely, since they were reversed by SR141716 (Sañudo-Peña et al. 1999; Szabo et al. 2000). In behavioral studies, reductions in motor activity have been observed that probably resulted from a glutamate-lowering effect of cannabinoids (Miller et al. 1998). In addition, a recent electrophysiological study by Gerdeman and Lovinger (2001) demonstrated that cannabinoids were also able to inhibit glutamate release from afferent terminals in the striatum, this effect being also blocked by SR141716. This points to the possibility that CB_1 receptors are also located in cortical afferents projecting to the caudate-putamen which are glutamatergic. In contrast, Herkenham et al. (1991b) found that excitotoxic lesions of the striatum led to an almost complete disappearance of CB_1 receptors. Therefore, it remains to be demonstrated whether the inhibitory effect of cannabinoids on striatal glutamate release is caused by the activation of CB_1 receptors located presynaptically on afferent terminals in the striatum, or whether it is an indirect effect mediated by CB_1 receptors that are located elsewhere.

1.2.3 Dopamine

Dopamine transmission is also affected by cannabinoids in the basal ganglia circuitry, as revealed by the findings that cannabinoids potentiated reserpine-induced hypokinesia (Moss et al. 1981), while reducing amphetamine-induced hyperactivity (Gorriti et al. 1999). Despite the lack of selectivity of reserpine and amphetamine, it appears likely that both acted in the basal ganglia circuitry to modulate dopaminergic transmission. There is also evidence from several neurochemical studies that cannabinoids reduce the activity of nigrostriatal dopaminergic neurons (Romero et al. 1995a,b; Cadogan et al. 1997; see Romero et al. 2002; van der Stelt and Di Marzo 2003 for recent reviews), an effect that is consistent with the ability of cannabinoid receptor agonists to produce hypokinesia. However, in other studies cannabinoids have been found to increase rather than decrease dopaminergic transmission (Sakurai-Yamashita et al. 1989; see also Romero et al. 2002; van der Stelt and Di Marzo 2003 for recent reviews).

We have reported that anandamide and AM404 reduced the activity of tyrosine hydroxylase in the caudate-putamen and the substantia nigra (Romero et al. 1995a,b; González et al. 1999). However, these effects were small and transient possibly because CB_1 receptors are not located on nigrostriatal dopaminergic neurons

(Herkenham et al. 1991b). In concordance with this idea, cannabinoid agonists and antagonists failed to inhibit electrically evoked dopamine release in the striatum (Szabo et al. 1999), although the matter has still to be clarified, since other studies have shown opposite effects of cannabinoids on striatal dopamine release in vitro (increases rather than decreases) (see van der Stelt and Di Marzo 2003 for details). The absence of CB_1 receptors from nigrostriatal dopaminergic neurons would support the hypothesis that the changes in the activity of these neurons produced by classical cannabinoids in vivo were caused indirectly through an effect on GABAergic transmission (Maneuf et al. 1996; Romero et al. 1998a). However, two additional mechanisms are also possible. First, CB_1 receptors might interact with D_1 or D_2 dopaminergic receptors at the level of G protein/adenylyl cyclase signal transduction mechanisms (Giuffrida et al. 1999; Meschler and Howlett 2001), since they colocalize in striatal projection neurons (Herkenham et al. 1991b). Second, certain cannabinoid agonists, such as anandamide and some analogs, but not classical cannabinoids, would be able to directly influence dopaminergic transmission through the activation of vanilloid VR1 receptors, which have been detected in nigrostriatal dopaminergic neurons (Mezey et al. 2000; see Fig. 1). In support of this, we have recently reported that the hypokinetic action and the dopamine-lowering effect of anandamide were both reversed by capsazepine, an antagonist of VR1 receptors, and, more importantly, we have found a direct effect of anandamide on dopamine release in vitro, an effect that was also reversed by capsazepine (de Lago et al. 2004b). Classical cannabinoids, such as Δ^9-THC, that do not bind to vanilloid-like receptors were not able to produce this effect (de Lago et al. 2004b). This is in concordance with the observation that anandamide reduced dopamine release from striatal slices (Cadogan et al. 1997), although these authors also found a dopamine-lowering effect after application of the classical cannabinoid CP 55,940 (Cadogan et al. 1997).

1.3 Presence of Elements of the Endocannabinoid System in Motor Regions

Several studies have addressed the identification and quantification of diverse elements of the endocannabinoid signaling system in the basal ganglia, as a way to establish the importance of the role played by this system in the control of motor function (for a review see Romero et al. 2002). Most of the studies focused on cannabinoid receptors, mainly the CB_1 subtype and more recently the functionally related receptor subtype, VR1, but some studies have dealt with other key proteins of the endocannabinoid system (for review see Romero et al. 2002).

1.3.1 Cannabinoid and Vanilloid Receptors

Autoradiographic studies have demonstrated conclusively that the basal ganglia are among the brain structures containing the highest levels of both binding sites and mRNA expression for the CB_1 receptor (for a review see Romero et al. 2002). In particular, the three nuclei that receive striatal efferent outputs (globus pallidus, entopeduncular nucleus, and substantia nigra pars reticulata), contain high levels of cannabinoid receptor binding sites (Herkenham et al. 1991a), whereas CB_1 receptor-mRNA transcripts are present in the caudate-putamen, which lacks striatal outflow nuclei (Mailleux and Vanderhaeghen 1992a). This observation is compatible with the idea that CB_1 receptors are presynaptically located in striatal projection neurons (see Fig. 1), a notion that has been supported by a series of anatomical studies in which specific neuronal subpopulations in the basal ganglia were lesioned (Herkenham et al. 1991b), and, more recently, by analysis of the cellular distribution of this receptor subtype in the basal ganglia using immunohistochemical techniques (Tsou et al. 1998a). CB_1 receptors are located in both striatonigral (the so-called "direct" striatal efferent pathway) and striatopallidal (the so-called "indirect" striatal efferent pathway) projection neurons, which use GABA as a neurotransmitter. In both pathways, CB_1 receptors are co-expressed with other markers, such as glutamic acid decarboxylase, prodynorphin, substance P, or proenkephalin, as well as D_1 or D_2 dopaminergic receptors (Hohmann and Herkenham 2000). In contrast, intrinsic striatal neurons, that contain somatostatin or acetylcholine, do not contain CB_1 receptors (Hohmann and Herkenham 2000). Another subpopulation of CB_1 receptors in the basal ganglia is located on subthalamopallidal and/or subthalamonigral glutamatergic terminals (see Fig. 1), as revealed by the presence of measurable levels of mRNA for this receptor in the subthalamic nucleus, together with the absence of detectable levels of cannabinoid receptor binding in that structure (Mailleux and Vanderhaeghen 1992a). Finally, CB_1 receptors are also located in GABAergic and glutamatergic neurons in the cerebellum, another brain structure involved in motor function (Herkenham et al. 1991a). These neurons are most likely associated with the effects of cannabinoids on posture and balance (Consroe 1998), although the neurochemical basis for these effects has been poorly explored (see Iversen 2003 for review).

These anatomical data reinforce the notion that CB_1 receptors play an important role in the mediation of motor effects of cannabinoid agonists, an idea that is supported by results obtained when studying motor function in mice deficient in CB_1 receptor gene expression (Ledent et al. 1999; Zimmer et al. 1999). These knockout mice exhibited significant motor disturbances, although the two models developed so far have yielded conflicting results, since a trend to hyperlocomotion was observed in one of these two models (Ledent et al. 1999) and hypoactivity was evident in the other (Zimmer et al. 1999).

CB_2 receptors are not present in motor regions in basal conditions, except in the cerebellum where Nuñez et al. (2004) recently demonstrated immunoreactivity for this receptor subtype in perivascular microglial cells of healthy human brains, but not in rat brain. This is concordant with previous data published by Skaper et al. (1996) working with mouse cerebellar cultures, and suggests that this receptor subtype might play a role in various cerebellar processes in normal conditions, although it most likely takes on a more important role when a neurodegenerative event takes place. Thus, recent studies have demonstrated that CB_2 receptors are significantly induced in different brain structures, including the basal ganglia, in response to different types of insults, including injury or inflammation (Benito et al. 2003; Aroyo et al. 2005). In these conditions, they are possibly located in glial cells (activated astrocytes, reactive microglia) rather than in neurons, playing a role in events related to the protective and/or cytotoxic influences that the different glial cells exert on neuronal survival (see Chen and Swanson 2003 for a review).

Finally, we must also mention the importance of the recent report of vanilloid VR1 receptors in the basal ganglia (Mezey et al. 2000). These receptors are molecular integrators of nociceptive stimuli, abundant on sensory neurons, but they have also been located in the basal ganglia circuitry colocalized with tyrosine hydroxylase, which means that they are located in nigrostriatal dopaminergic neurons (Mezey et al. 2000; see Fig. 1). As mentioned above, recent pharmacological and neurochemical studies have established the involvement of these receptors in the control of motor function (Di Marzo et al. 2001) and in the production of motor effects by certain cannabinoid receptor agonists (de Lago et al. 2004b).

1.3.2 Endocannabinoid Ligands

Endogenous cannabinoid receptor ligands, anandamide and 2-arachidonoylglycerol, are also present in the basal ganglia (Bisogno et al. 1999; Di Marzo et al. 2000a) in concentrations that are in general higher than those measured in the whole brain.

Two key regions involved in the control of movement, the globus pallidus and the substantia nigra, contain not only the highest densities of CB_1 receptors in the brain (Herkenham et al. 1991a) but also the highest levels of endocannabinoids, particularly of anandamide (Di Marzo et al. 2000a). The phenotype of the nerve cells that produce endocannabinoids in the basal ganglia is presently unknown, although the precursor of anandamide, *N*-arachidonoylphosphatidylethanolamine, has been found in the basal ganglia (Di Marzo et al. 2000b), which supports the existence of in situ synthesis for this endocannabinoid. The synthesis of anandamide seems sensitive to dopamine. Thus, Giuffrida et al. (1999) reported that, in the striatum, it is regulated by dopaminergic D_2 receptors, which was interpreted by these authors as an indication that the endocannabinoid system serves as an inhibitory feedback mechanism that counteracts dopamine-induced facilitation of psychomotor activity (Giuffrida et al. 1999).

1.3.3 Endocannabinoid Inactivation

Despite the fact that the endocannabinoid transporter has not yet been isolated or cloned, a situation that has led to some controversy about its existence (Glaser et al. 2003), there are several anandamide analogs that behave in vitro as endocannabinoid transport inhibitors (Giuffrida et al. 2001) and that, in vivo, produce significant effects on motor function (for review see Fernández-Ruiz et al. 2002).

These include compounds such as AM404 (González et al. 1999; Beltramo et al. 2000), VDM11 (de Lago et al. 2004a), and UCM707 (de Lago et al. 2002). Based on this pharmacological evidence, it is to be expected that the transporter for endocannabinoids is abundantly concentrated in the basal ganglia and other motor regions.

Fatty acid amide hydrolase (FAAH), the enzyme involved in the degradation of anandamide, is also present in high levels in all regions of the basal ganglia, in particular in the globus pallidus and the substantia nigra (Desarnaud et al. 1995; Tsou et al. 1998b). As to monoacylglycerol-lipase, the enzyme involved in the degradation of the other important endocannabinoid, 2-arachidonoylglycerol, this has also been detected in the basal ganglia and, to a greater extent, in the cerebellum (see Dinh et al. 2002 for a review). However, these enzymes accept as substrates various *N*-acylethanolamines or mono-acylglycerols, respectively, and so lack the specificity that would allow them to be used as selective markers of endocannabinoid transmission.

2 Potential Therapeutic Applications of Cannabinoids in Motor Disorders

From what has been stated above, it can be hypothesized that compounds affecting endocannabinoid transmission might be useful for reducing motor deterioration in both hyper- and hypokinetic disorders (for reviews see Consroe 1998; Müller-Vahl et al. 1999c; Fernández-Ruiz et al. 2002). To date, much research has been directed at the search for compounds able to alleviate motor symptoms in these disorders (see Fernández-Ruiz et al. 2002; van der Stelt and Di Marzo 2003), but evidence has also been obtained that cannabinoid-related compounds might be neuroprotectant substances (Grundy 2002; Romero et al. 2002). In this chapter, we review the evidence supporting the first of these potential clinical applications, because the potential of cannabinoids to influence cell viability is addressed in another chapter of this book (see contribution by Guzmán).

2.1 General Aspects

Senescence is a physiological process, characterized, in part, by a slow but progressive impairment of motor function, but with no evident signs of a disease state (Schut 1998). This correlates with a decrease in the activity of most of the neurotransmitters acting in the basal ganglia, particularly dopamine and GABA (for a review see Francis et al. 1993). Endocannabinoid transmission is also influenced by normal senescence, since the population of CB_1 receptors was reduced in the basal ganglia of aged rats with no signs of neurological disease (Mailleux and Vanderhaeghen 1992b; Romero et al. 1998b). However, the changes observed in CB_1 receptors in the postmortem basal ganglia of humans affected by several neurodegenerative motor diseases, as well as in animal models of these disorders, are much more dramatic (for reviews see Fernández-Ruiz et al. 2002; Lastres-

Becker et al. 2003b). Among these disorders, PD and HD are the two diseases directly related to the control of movement that have attracted most interest in terms of a potential application of cannabinoids for both alleviation of symptoms and delay/arrest of neurodegeneration (for a review see Fernández-Ruiz et al. 2002). Another interesting motor disorder in which cannabinoids might be effective is Gilles de la Tourette's syndrome (Müller-Vahl 2003 for a review). Finally, together with these classic motor disorders, other diseases not directly related to the control of movement in origin but exhibiting strong motor symptoms, such as Alzheimer's disease (Pazos et al. 2004 for a review) or multiple sclerosis (Baker and Pryce 2003 for a review), have also been examined as potential therapeutic targets for cannabinoid-based compounds.

2.2 Huntington's Disease

HD is an inherited neurodegenerative disorder caused by an unstable expansion of a CAG repeat in exon 1 of the human huntingtin gene. Translation through the CAG span results in a polyglutamine tract near the N-terminus of this protein, which leads to toxicity predominantly of striatal projection neurons (for a recent review see Cattaneo et al. 2001). The symptoms of this disease are primarily characterized by motor disturbances, such as chorea and dystonia, a consequence of the progressive degeneration of the striatum due to the selective death of striatal projection neurons (Berardelli et al. 1999). Secondarily, patients are also affected by cognitive decline (Reddy et al. 1999).

2.2.1 Changes in Endocannabinoid Transmission

Studies in postmortem human tissue have clearly demonstrated that, in HD, there is an almost complete disappearance of CB_1 receptors in the substantia nigra, in the lateral part of the globus pallidus and, to a lesser extent, in the putamen (Glass et al. 1993, 2000; Richfield and Herkenham 1994). This loss of CB_1 receptors is concordant with the characteristic neuronal loss observed in HD that predominantly affects medium-spiny GABAergic neurons, which contain the major population of CB_1 receptors present in basal ganglia structures (Herkenham et al. 1991b; Hohmann and Herkenham 2000). It is also consistent with the finding that other phenotypic markers for these neurons, such as substance P, enkephalin, calcineurin, calbindin, and receptors for neurotransmitters, such as adenosine or dopamine, are also depleted in HD (Hersch and Ferrante 1997). However, recent experiments have revealed that the reduction of CB_1 receptors occurs in advance of other receptor losses and even before the appearance of major HD symptomatology, when the incidence of cell death is still low (Glass et al. 2000). This suggests that losses of CB_1 receptors might be involved in the pathogenesis and/or progression of neurodegeneration in HD.

Studies with animal models of HD have validated the data obtained with postmortem human tissues (see Lastres-Becker et al. 2003b for a review), and also indicate that these models may predominantly reflect partial aspects or specific phases of striatal degeneration. For instance, decreases of CB_1 receptors in the basal ganglia have also been found in various transgenic mouse models that express mutated forms of the human huntingtin gene (Denovan-Wright and Robertson 2000; Lastres-Becker et al. 2002c). In these genetic models, cell dysfunction rather than cell death is the major change that takes place, so the observation of reduced CB_1 receptors in these animals might be equivalent to the reductions of these receptors reported by Glass et al. (2000) in early stages of the human disease. CB_1 receptors were reduced to a greater extent in rat models of HD generated by selective lesions of striato-efferent GABAergic neurons caused by mitochondrial or excitotoxic toxins (Page et al. 2000; Lastres-Becker et al. 2001b, 2002a,b). These toxins, in particular 3-nitropropionic acid, reproduce in animals the same changes that have been proposed to be associated with the human disease, i.e., failure of energy metabolism, glutamate excitotoxicity, and, to a lesser extent, oxidative stress, leading to progressive neuronal death (for reviews see Alexi et al. 1998; Brouillet et al. 1999). However, they are more representative of the pattern of profound neuronal loss that occurs in advanced states of the human disease (Brouillet et al. 1999). In these conditions, the losses of CB_1 receptors might be a mere side effect caused by the progressive and selective destruction of striatal GABAergic projection neurons, neurons on which these receptors are located. In this rat model, the losses of CB_1 receptors were accompanied by a decrease in the content of both anandamide and 2-arachidonoylglycerol in the caudate-putamen (Lastres-Becker et al. 2001b). Therefore, all the data collected from humans and from animal models indicate that endocannabinoid transmission becomes progressively hypofunctional in the basal ganglia in HD. This might contribute to some extent to the hyperkinesia typical of this disorder and so support a therapeutic usefulness of cannabinoid agonists for alleviating motor deterioration, as will be described below.

2.2.2
Therapeutic Usefulness of Cannabinoids

Medicines used for the treatment of HD include mainly antidopaminergic drugs to reduce the hyperkinesia characteristic of the first phases of the disease (Factor and Firedman 1997) and antiglutamatergic agents to reduce excitotoxicity (Kieburtz 1999). However, the outcome of both strategies has been poor in terms of improving quality of life for HD patients, despite the progress in the elucidation of molecular events involved in the pathogenesis of HD (Cattaneo et al. 2001). In this context, cannabinoid agonists might be a reasonable alternative, since they combine both antihyperkinetic and neuroprotective effects (for review see Fernández-Ruiz et al. 2002; Lastres-Becker et al. 2003b). As mentioned above, we will not address here the neuroprotective potential of cannabinoids in HD, because this has been addressed in the chapter by Guzmán (this volume), but we will address the potential antihyperkinetic action of substances that can elevate endocannabinoid activity in

Table 2. Potential therapeutic effects of cannabinoid-related compounds in basal ganglia disorders (*continued on next page*)

Compound	Disease	Therapeutic application
Plant-derived cannabinoids		
Δ^9-THC	Huntington's disease	Reduction of striatal injury in 3NP rat model (Lastres-Becker et al. 2004b)
		Divergent effects on striatal injury in the malonate rat model (Lastres-Becker et al. 2003c; Aroyo et al. 2005)
	Parkinson's disease	Reduction of dopaminergic injury in the 6-hydroxydopamine rat model (Lastres-Becker et al. 2004a)
		Failure to alleviate symptoms in PD patients (reviewed by Consroe 1998)
	Tourette's syndrome	Reduction of tics and obsessive-compulsive behaviors (reviewed by Müller-Vahl 2003)
Cannabidiol	Huntington's disease	Failure to reduce hyperkinetic movements in HD patients (reviewed by Consroe 1998)
		Poor neuroprotective action in the malonate rat model (Aroyo et al. 2005)
	Parkinson's disease	Reduction of dopaminergic injury in the 6-hydroxydopamine rat model (Lastres-Becker et al. 2004a)
Synthetic cannabinoids		
CP 55,940	Huntington's disease	Certain antihyperkinetic activity in 3NP-lesioned rats (Lastres-Becker et al. 2003a)
	Parkinson's disease	Potential reduction of tremor by reducing the overactivity of the subthalamic nucleus in the 6-hydroxydopamine rat model (Sañudo-Peña et al. 1998)
Nabilone	Huntington's disease	Increase of hyperkinesia (choreic movements) in HD patients (Müller-Vahl et al. 1999b)
	Parkinson's disease	Reduction of L-dopa-induced dyskinesia in PD patients (Sieradzan et al. 2001)
	Dystonia	No effects in patients with generalized and segmental primary dystonia (Fox et al. 2002b)
WIN 55,212-2	Parkinson's disease	Reduction of L-dopa-induced dyskinesia in rat models of PD (Segovia et al. 2003; Ferrer et al. 2003)
	Dystonia	Antidystonic effects in mutant dystonic hamsters (Richter and Löscher 1994, 2002)
HU308	Huntington's disease	Reduction of GABAergic injury in the malonate rat model; reversed by SR144528 (Aroyo et al. 2005)
Endogenous cannabinoids		
Anandamide	Huntington's disease	Certain antihyperkinetic activity in 3NP-lesioned rats (possibly VR1-mediated effect) (Lastres-Becker et al. 2002a)

Table 2. *(continued)*

Compound	Disease	Therapeutic application
Inhibitors of endocannabinoid inactivation		
AM404	Huntington's disease	Antihyperkinetic activity and recovery from neurochemical deficits in 3NP-lesioned rats (involvement of VR1 receptors) (Lastres-Becker et al. 2002a, 2003a)
	Parkinson's disease	Unable to reduce L-dopa-induced dyskinesia in the reserpine rat model (Segovia et al. 2003)
VDM11	Huntington's disease	Not effective in 3NP-lesioned rats (Lastres-Becker et al. 2003a)
UCM707	Huntington's disease	Certain antihyperkinetic activity in 3NP-lesioned rats (de Lago et al. 2004c)
AM374	Huntington's disease	Not effective in 3NP-lesioned rats (Lastres-Becker et al. 2003a)
Receptor antagonists		
SR141716 (CB1)	Huntington's disease	Increased striatal damage in the malonate rat model (Lastres-Becker et al. 2003c)
		Unable to reverse antihyperkinetic effects of AM404 in 3NP-lesioned rats (Lastres-Becker et al. 2003a)
		Unable to reduce late akinesia in 3NP-lesioned rats (Lastres-Becker et al. 2002b)
	Parkinson's disease	Effective to reduce L-dopa-induced dyskinesia in MPTP-treated marmosets and the reserpine rat model (Brotchie 1998, 2000; Segovia et al. 2003)
		Unable to reverse bradykinesia and rigidity in MPTP-treated primates (Meschler et al. 2001)
		Able to restore locomotion in the reserpine model of PD (Di Marzo et al. 2000a)
SR144528 (CB2)	Huntington's disease	Able to reverse neuroprotective effect of HU308 in the malonate rat model (Aroyo et al. 2005)
Capsazepine (VR1)	Huntington's disease	Able to reverse antihyperkinetic effects of AM404 in 3NP-lesioned rats (Lastres-Becker et al. 2003a)

3NP, 3-nitropropionic acid; HD, Huntington's disease; MPTP, 1-methyl-4-phenyl-1,2,3,6-tetrahydropyridine; PD, Parkinson's disease.

the basal ganglia. Thus, we have recently demonstrated that the endocannabinoid transporter inhibitor AM404 was able to reduce hyperkinesia and induce recovery from GABAergic and dopaminergic deficits in rats with striatal lesions caused by local application of 3-nitropropionic acid (Lastres-Becker et al. 2002a, 2003a), while direct agonists of CB_1 receptors, such as CP 55,940, only produced very modest effects (Lastres-Becker et al. 2003a). AM404 was also able to normalize motor activity in genetically hyperactive rats without causing overt cannabimimetic effects (Beltramo et al. 2000). However, in view of the fact that a progressive decrease of CB_1 receptors has been recorded in this disease, the efficacy of this compound might a priori be extended only to the early or intermediate hyperkinetic phases, when cell death is still moderate, but not to the late akinetic stages of the disease characterized by high neuronal death (see Lastres-Becker et al. 2003b for a review). These results, however, contrast with some clinical data that indicate that the administration of plant-derived cannabinoids (Consroe 1998), or some of their synthetic analogs (Müller-Vahl et al. 1999b), increased choreic movements in HD patients. It is possible that this is related to the lack of VR1 receptor activity of these cannabinoid agonists, since recent studies carried out in our laboratory (see details in Table 2) in rats with striatal lesions have revealed that only those cannabinoid-based compounds having an additional profile as VR1 receptor agonists were really effective in alleviating hyperkinetic signs (Lastres-Becker et al. 2003a). This was so for AM404, which, in addition to its ability to block the endocannabinoid transporter, also exhibits affinity for the VR1 receptor (Zygmunt et al. 2000). Interestingly, inhibitors of endocannabinoid inactivation that are not active at the VR1 receptor, such as VDM11 or AM374, did not have any antihyperkinetic action in HD rats (Lastres-Becker et al. 2003a), whereas UCM707, the most potent inhibitor to date, only produced modest effects (de Lago et al. 2004c) (see Table 2). Therefore, our data suggest that VR1 receptors alone, or better in combination with CB_1 receptors, might represent novel targets through which the hyperkinetic symptoms of HD could be alleviated. Possibly, the best option might be to develop "hybrid" compounds with the dual capability of activating both VR1 and CB_1 receptors, although the relative contribution made by each of these targets is likely to change during the course of the disease due to a progressive loss of CB_1 receptors without any concomitant loss of VR1 receptors (see Lastres-Becker et al. 2003b for a review).

2.3 Parkinson's Disease

PD is a progressive neurodegenerative disorder in which the capacity of executing voluntary movements is lost gradually. The major clinical symptomatology in PD includes tremor, rigidity, and bradykinesia (slowness of movement). The pathological hallmark of this disease is the degeneration of melanin-containing dopaminergic neurons of the substantia nigra pars compacta that leads to severe dopaminergic denervation of the striatum (for a recent review see Blandini et al. 2000).

2.3.1 Changes in the Endocannabinoid Transmission

Compared with HD, much less data exist on the status of CB_1 receptors in the postmortem basal ganglia of humans affected by PD. Only recently we have found that CB_1 receptor binding and the activation of G proteins by cannabinoid agonists were significantly increased in the basal ganglia as a consequence of the selective degeneration of nigrostriatal dopaminergic neurons (Lastres-Becker et al. 2001a). These increases were not related to the chronic dopaminergic replacement therapy with L-dopa that these patients were undergoing, since they were also seen in 1-methyl-4-phenyl-1,2,3,6-tetrahydropyridine (MPTP)-treated marmosets, a primate PD model, and disappeared after chronic L-dopa administration in these animals (Lastres-Becker et al. 2001a). It has also been found that endocannabinoid levels increase in a rat model of PD and that this increase can be reversed by L-dopa (Gubellini et al. 2002; Macarrone et al. 2003). This suggests the existence of an imbalance between dopamine and endocannabinoids in the basal ganglia in PD (see Fernández-Ruiz et al. 2002 for a review).

As in HD, a change in CB_1 receptor density might also be an early event in the pathogenesis of PD. This is supported by data obtained from individuals affected by incidental Lewy body disease, an early and presymptomatic phase of PD. These individuals, who did not receive any therapy as they presented Lewy bodies and a low degree of nigral pathology without any neurological symptoms, exhibited a trend towards an increase in CB_1 receptors in some basal ganglia structures (Lastres-Becker et al. 2001a). Moreover, preliminary experiments with a genetic model of PD, the parkin-2 knockout mouse (Itier et al. 2003), have yielded data showing an increase in CB_1 receptor binding in the substantia nigra of the knockout mice that occurs in the absence of neuronal death (González S, Lastres-Becker I, Ramos JA, Fernández-Ruiz J, unpublished results).

Overactivity of endocannabinoid transmission (as measured by increases in CB_1 receptor or endocannabinoid levels) has also been observed in the basal ganglia in different rat models of PD (Mailleux and Vanderhaeghen 1993; Romero et al. 2000; Di Marzo et al. 2000a; Gubellini et al. 2002), although the data are not consistent, with some authors reporting no changes (Herkenham et al. 1991b), reductions in CB_1 receptor levels (Silverdale et al 2001), or a dependency on chronic L-dopa co-treatment (Zeng et al. 1999). Despite these conflicts, we consider that most of the data indicate that endocannabinoid transmission becomes overactive in the basal ganglia in PD, a conclusion that is compatible with the hypokinesia that characterizes this disease. This would also support the suggestion that CB_1 receptor antagonists, rather than agonists, might be useful for alleviating motor deterioration in PD, or for reducing the development of dyskinesia caused by prolonged replacement therapy with L-dopa (Brotchie 2000).

2.3.2 Therapeutic Usefulness of Cannabinoids

Dopaminergic replacement therapy represents a useful remedy for rigidity and bradykinesia in PD patients (Carlsson 2002), at least in the early and middle phases of this disease. Later on, the chronic use of L-dopa therapy results in a loss of efficacy and even in the appearance of an irreversible dyskinetic state. Cannabinoid-based compounds might also be useful in PD. In this disorder, CB_1 receptor agonists or antagonists have both been proposed, for their use alone or as coadjuvants, against different signs of the complex motor pathology developed by PD patients (Brotchie 2000; Romero et al. 2000; Di Marzo et al. 2000a; Lastres-Becker et al. 2001a; Fox et al. 2002a; see Table 2). For instance, it has been reported that CB_1 receptor agonists: (1) are able to interact with dopaminergic agonists to improve motor impairments (Anderson et al. 1995; Maneuf et al. 1997; Brotchie 1998; Sañudo-Peña et al. 1998), (2) reduce tremor associated with an overactivity of the subthalamic nucleus (Sañudo-Peña et al. 1998, 1999), and (3) decrease and/or delay the occurrence of dyskinesia associated with long-term dopaminergic replacement (Sierazdan et al. 2001). Cannabinoids, particularly classical cannabinoids with antioxidant properties, have also been reported to provide protection against dopaminergic cell death (Lastres-Becker et al. 2004a; see Table 2).

However, because of the hypokinetic profile of cannabinoid agonists, it is unlikely that these compounds would be useful for alleviating bradykinesia in PD patients. This is confirmed by results obtained with humans or with MPTP-lesioned primates, as these indicated that the administration of plant-derived cannabinoid receptor agonists enhanced motor disability (for reviews see Consroe 1998; Müller-Vahl et al. 1999c). Indeed, it has been proposed that the blockade of CB_1 receptors may be a better strategy for reducing both bradykinesia (see Fernández-Ruiz et al. 2002 for review) and L-dopa-induced dyskinesia (Brotchie 2000, 2003) (see Table 2 for details). In support of this possibility, dysfunction of nigrostriatal dopaminergic neurons is associated with an overactivity of endocannabinoid transmission in the basal ganglia. Such overactivity has been observed after administration of reserpine (Di Marzo et al. 2000a) or dopaminergic antagonists (Mailleux and Vanderhaeghen 1993), or during degeneration of these neurons caused by the local application of 6-hydroxydopamine (Mailleux and Vanderhaeghen 1993; Romero et al. 2000; Gubellini et al. 2002; Fernández-Espejo et al. 2004) or MPTP (Lastres-Becker et al. 2001a). In theory, CB_1 receptor blockade would avoid the excessive inhibition of GABA uptake produced by the increased activation of CB_1 receptors in striatal projection neurons (Maneuf et al. 1996; Romero et al. 1998a), thus allowing a faster removal of this inhibitory neurotransmitter from the synaptic cleft, which would reduce hypokinesia. Despite this evidence, the first pharmacological studies that have examined the capability of rimonabant (SR141716) to reduce hypokinesia in animal models of PD have yielded conflicting resulted (Di Marzo et al. 2000a; Meschler et al. 2001; see Table 2 for more details). It is possible that the blockade of CB_1 receptors might be effective only at very advanced phases of the disease. Indeed, recent evidence obtained by Fernández-Espejo and coworkers (2004) is in favor of this option, which presents an additional advantage since it would make it possible to give an antiparkinsonian compound at a stage of the disease at which classic dopaminergic therapy generally fails. In addition, in view of the recently demonstrated role of VR1 receptors in the regulation of dopamine release from nigral neurons (de Lago et al. 2004b), the potential of VR1 receptor ligands for the treatment of hypokinetic signs of this disease must also be considered.

2.4 Other Motor Disorders

To our knowledge, no data exist on the role(s) of cannabinoid receptors in other basal ganglia disorders in the human, such as tardive dyskinesia, Gilles de la Tourette's syndrome, dystonia, and others. Even so, cannabinoids might be of interest for the treatment of at least some of these diseases (for reviews see Consroe 1998; Fernández-Ruiz et al. 2002; Table 2 for more details). Thus, a relationship between cannabis use and the incidence of tardive dyskinesia has been described in psychiatric patients that were being chronically treated with neuroleptic drugs (Zarestky et al. 1993). A few studies have also addressed this issue for dystonia in humans (Fox et al. 2002b) or animal models (Richter and Löscher 1994, 2002), by demonstrating that cannabinoids have antidystonic effects (for reviews see Consroe 1998; Müller-Vahl et al. 1999c). In addition, plant-derived cannabinoids might have the potential to reduce tics and also to improve behavioral problems in patients with Tourette's syndrome (Hemming and Yellowlees 1993; Consroe 1998; Müller-Vahl et al. 1998, 1999a, 2002; for review see Müller-Vahl 2003). However, there are no data on the status of endocannabinoid signaling in patients or in animal models of this disease, and also no information on the neurochemical pathways mediating the beneficial effects of cannabinoids.

Another relevant disease in which cannabinoids might improve motor deterioration is multiple sclerosis. This is a disease of immune origin, but it progresses with neurological deterioration that affects mainly the motor system. Studies in laboratory animals have convincingly demonstrated that both direct and indirect cannabinoid receptor agonists are useful in this disease, in particular for the management of motor-related symptoms such as spasticity, tremor, dystonia, and others (for reviews see Pertwee 2002; Baker and Pryce 2003). These effects seem to be mediated by CB_1 and, to a lesser extent, CB_2 receptors (Baker et al. 2000). This pharmacological evidence explains previous anecdotal, uncontrolled, or preclinical data that suggested a beneficial effect of marijuana when smoked by multiple sclerosis patients to alleviate some of their symptoms, mainly spasticity and pain (for review see Consroe 1998). In line with these data, a clinical trial, recently completed in the UK, has demonstrated that although cannabis and Δ^9-THC did not have a beneficial effect on spasticity when this was measured objectively, these drugs did increase the patients' perception of improvement of different symptoms of this disease (Zajicek et al. 2003).

In contrast with the numerous pharmacological studies in this area of research, there are no data on possible changes in CB_1 and CB_2 receptors in the postmortem brains of patients with multiple sclerosis, and only a few studies have examined the status of endocannabinoid transmission in animal models of this disease (Baker et al. 2001; Berrendero et al. 2001). Using a rat model of multiple sclerosis, we recently reported a decrease in central CB_1 receptors (Berrendero et al. 2001). This decrease was restricted to basal ganglia structures, which is consistent with the fact that motor deterioration is one of the most prominent neurological signs in these rats and also in the human disease (for review see Baker and Pryce 2003). This decrease was accompanied by a reduction in endocannabinoid levels that also occurred in brain structures other than the basal ganglia (Cabranes et al. 2005). This finding led us to hypothesize that the changes in CB_1 receptors and their ligands in the basal ganglia might be associated with disturbances in several neurotransmitter systems. If this were the case, it follows that the well-known effects of cannabinoid agonists on these systems might underlie their ability to ameliorate the motor symptoms of multiple sclerosis (see Fernández-Ruiz et al. 2002 for review). However, there is no support for this hypothesis. Thus, although we have detected reductions in CB_1 receptors (Berrendero et al. 2001) and endocannabinoid levels (Cabranes et al. 2005) in the basal ganglia of the lesioned rats, we were unable to detect any changes in dopamine, serotonin, GABA, or glutamate. Because of this finding, we recently tested the effects of various inhibitors of endocannabinoid transport that are capable of elevating endocannabinoid levels. We found that although these inhibitors were able to reduce the neurological decline typically exhibited by the lesioned rats, this reduction seemed to depend on the activation of VR1 receptors (Cabranes et al. 2005).

One other disorder worthy of mention is Alzheimer's disease, which, like multiple sclerosis, is not a disorder of the basal ganglia, and yet frequently gives rise to extrapyramidal signs and symptoms that are possibly caused by the degeneration of glutamatergic cortical afferents to the caudate-putamen (for review see Kurlan et al. 2000). Studies in postmortem brain regions of patients affected by this disease have revealed a significant loss of CB_1 receptors in the basal ganglia (Westlake et al. 1994). However, it is important to remark that the authors considered that their results related more to old age than to an effect selectively associated with the pathology characteristic of Alzheimer's disease (Westlake et al. 1994). Also using postmortem tissue from Alzheimer's patients, Benito et al. (2003) reported the induction of CB_2 receptors in activated microglia that surround senile plaques. This would suggest a role of this receptor subtype in the pathogenesis of this disease and a therapeutic potential for compounds that selectively target this receptor (see recent studies by Milton 2002; Iuvone et al. 2004).

3 Concluding Remarks and Future Perspectives

The studies reviewed here are all concordant with the view that control of movement is a key function for endocannabinoid transmission. We have collected the pharmacological and biochemical evidence that supports this hypothesis. We have also shown that endocannabinoid transmission is altered in motor disorders, in parallel with changes in classic neurotransmitters such as GABA, dopamine, or glutamate. This provides the basis for the development of novel pharmacotherapies with compounds selective for the different target proteins that form the endocannabinoid system. However, only a few studies have examined, hitherto, the potential contribution these compounds might make to the management of motor disorders in the clinic. The importance of this novel system demands further investigation and the development of novel promising compounds for the symptomatic and/or neuroprotectant treatment of basal ganglia pathology.

Acknowledgements.
Studies included in this review have been made possible by grants from CAM-PRI (08.5/0029/98 and 08.5/0063/2001) and MCYT (SAF2003-08269).

References

Alexi T, Hughes PE, Faull RLM, Williams LE (1998) 3-Nitropropionic acid's lethal triplet: cooperative pathways of neurodegeneration. Neuroreport 9:57–64

Anderson LA, Anderson JJ, Chase TN, Walters JR (1995) The cannabinoid agonists WIN55, 212-2 and CP55,940 attenuate rotational behaviour induced by a dopamine D1 but not D2 agonist in rats with unilateral lesions of the nigrostriatal pathway. Brain Res 691:106–114

Aroyo I, González S, Nuñez E, Lastres-Becker I, Sagredo O, Mechoulam R, Romero J, Ramos JA, Brouillet E, Fernández-Ruiz J (2005) Involvement of CB2 receptors in the neuroprotective effects of cannabinoids in rats with striatal atrophy induced by local application of malonate, an experimental model of Huntington's disease. J Neurosci (submitted)

Baker D, Pryce G (2003) The therapeutic potential of cannabis in multiple sclerosis. Expert Opin Investig Drugs 12:561–567

Baker D, Pryce G, Croxford JL, Brown P, Pertwee RG, Huffman JW, Layward L (2000) Cannabinoids control spasticity and tremor in a multiple sclerosis model. Nature 404:84–87

Baker D, Pryce G, Croxford JL, Brown P, Pertwee RG, Makriyannis A, Khanolkar A, Layward L, Fezza F, Bisogno T, Di Marzo V (2001) Endocannabinoids control spasticity in a multiple sclerosis model. FASEB J 15:300–302

Beltramo M, Rodríguez de Fonseca F, Navarro M, Calignano A, Gorriti MA, Grammatikopoulos G, Sadile AG, Giuffrida A, Piomelli D (2000) Reversal of dopamine D2 receptor responses by an anandamide transport inhibitor. J Neurosci 20:3401–3407

Benito C, Nunez E, Tolon RM, Carrier EJ, Rabano A, Hillard CJ, Romero J (2003) Cannabinoid CB2 receptors and fatty acid amide hydrolase are selectively overexpressed in neuritic plaque-associated glia in Alzheimer's disease brains. J Neurosci 23:11136–11141

Berardelli A, Noth J, Thompson PD, Bollen EL, Curra A, Deuschl G, van Dijk JG, Topper R, Schwarz M, Roos RA (1999) Pathophysiology of chorea and bradykinesia in Huntington's disease. Mov Disord 14:398–403

Berrendero F, Sánchez A, Cabranes A, Puerta C, Ramos JA, García-Merino A, Fernández-Ruiz J (2001) Changes in cannabinoid CB1 receptors in striatal and cortical regions of rats with experimental allergic encephalomyelitis, an animal model of multiple sclerosis. Synapse 41:195–202

Bisogno T, Berrendero F, Ambrosino G, Cebeira M, Ramos JA, Fernández-Ruiz JJ, Di Marzo V (1999) Brain regional distribution of endocannabinoids: implications for their biosynthesis and biological function. Biochem Biophys Res Commun 256:377–380

Blandini F, Nappi G, Tassorelli C, Martignoni E (2000) Functional changes in the basal ganglia circuitry in Parkinson's disease. Prog Neurobiol 62:63–88

Brotchie JM (1998) Adjuncts to dopamine replacement: a pragmatic approach to reducing the problem of dyskinesia in Parkinson's disease. Mov Disord 13:871–876
Brotchie JM (2000) The neural mechanisms underlying levodopa-induced dyskinesia in Parkinson's disease. Ann Neurol 47:S105–S114
Brotchie JM (2003) CB1 cannabinoid receptor signalling in Parkinson's disease. Curr Opin Pharmacol 3:54–61
Brouillet E, Conde F, Beal MF, Hantraye P (1999) Replicating Huntington's disease phenotype in experimental animals. Prog Neurobiol 59:427–468
Cabranes A, Venderova K, de Lago E, Fezza F, Valenti M, Sánchez A, García-Merino A, Ramos JA, Di Marzo V, Fernández-Ruiz J (2005) Decreased endocannabinoid levels in the brain and beneficial effects of certain endocannabinoid uptake inhibitors in a rat model of multiple sclerosis: involvement of vanilloid TRPV1 receptors. Neurobiol Dis (in press)
Cadogan AK, Alexander SP, Boyd EA, Kendall DA (1997) Influence of cannabinoids on electrically evoked dopamine release and cyclic AMP generation in the rat striatum. J Neurochem 69:1131–1137
Carlsson A (2002) Treatment of Parkinson's with L-DOPA. The early discovery phase, and a comment on current problems. J Neural Transm 109:777–787
Cattaneo E, Rigamonti D, Goffredo D, Zuccato C, Squitieri F, Sipione S (2001) Loss of normal huntingtin function: new developments in Huntington's disease research. Trends Neurosci 24:182–188
Chan PK, Chan SC, Yung WH (1998) Presynaptic inhibition of GABAergic inputs to rat substantia nigra pars reticulata neurones by a cannabinoid agonist. Neuroreport 9:671–675
Chen Y, Swanson RA (2003) Astrocytes and brain injury. J Cereb Blood Flow Metab 23:137–149
Compton DR, Aceto MD, Lowe J, Martin BR (1996) In vivo characterization of a specific cannabinoid antagonis (SR141716A): inhibition of Δ9-tetrahydrocannabinol-induced responses and apparent agonist activity. J Pharmacol Exp Ther 277:586–594
Consroe P (1998) Brain cannabinoid systems as targets for the therapy of neurological disorders. Neurobiol Dis 5:534–551
Crawley JN, Corwin RL, Robinson JK, Felder ChC, Devane WA, Axelrod J (1993) Anandamide, an endogenous ligand of the cannabinoid receptor, induces hypomotility and hypothermia in vivo in rodents. Pharmacol Biochem Behav 46:967–972
de Lago E, Fernandez-Ruiz J, Ortega-Gutierrez S, Viso A, Lopez-Rodriguez ML, Ramos JA (2002) UCM707, a potent and selective inhibitor of endocannabinoid uptake, potentiates hypokinetic and antinociceptive effects of anandamide. Eur J Pharmacol 449:99–103
de Lago E, Ligresti A, Ortar G, Morera E, Cabranes A, Pryce G, Bifulco M, Baker D, Fernandez-Ruiz J, Di Marzo V (2004a) In vivo pharmacological actions of two novel inhibitors of anandamide cellular uptake. Eur J Pharmacol 484:249–257
de Lago E, de Miguel, Lastres-Becker I, Ramos JA, Fernández-Ruiz J (2004b) Involvement of vanilloid-like receptors in the effects of anandamide on motor behavior and nigrostriatal dopaminergic activity: in vivo and in vitro evidence. Brain Res 1007:152–159
de Lago E, Ortega S, López-Rodríguez ML, Ramos JA, Fernández-Ruiz J (2004c) Therapeutic potential of UCM707, an inhibitor of the endocannabinoid transport, in animal models of various neurological diseases. Mov Disord (submitted)
Denovan-Wright EM, Robertson HA (2000) Cannabinoid receptor messenger RNA levels decrease in subset neurons of the lateral striatum, cortex and hippocampus of transgenic Huntington's disease mice. Neuroscience 98:705–713
Desarnaud F, Cadas H, Piomelli D (1995) Anandamide amidohydrolase activity in rat brain microsomes. Identification and partial purification. J Biol Chem 270:6030–6035
Dewey WL (1986) Cannabinoid pharmacology. Pharmacol Rev 38:151–178
Di Marzo V, Melck D, Bisogno T, De Petrocellis L (1998) Endocannabinoids: endogenous cannabinoid receptor ligands with neuromodulatory action. Trends Neurosci 21:521–528

Di Marzo V, Hill MP, Bisogno T, Crossman AR, Brotchie JM (2000a) Enhanced levels of endocannabinoids in the globus pallidus are associated with a reduction in movement in an animal model of Parkinson's disease. FASEB J 14:1432–1438
Di Marzo V, Berrendero F, Bisogno T, Gonzalez S, Cavaliere P, Romero J, Cebeira M, Ramos JA, Fernandez-Ruiz J (2000b) Enhancement of anandamide formation in the limbic forebrain and reduction of endocannabinoid contents in the striatum of Δ9-tetrahydrocannabinol-tolerant rats. J Neurochem 74:1627–1635
Di Marzo V, Lastres-Becker I, Bisogno T, De Petrocellis L, Milone A, Davis JB, Fernandez-Ruiz J (2001) Hypolocomotor effects in rats of capsaicin and two long chain capsaicin homologues. Eur J Pharmacol 420:123–131
Dinh TP, Freund TF, Piomelli D (2002) A role for monoglyceride lipase in 2-arachidonoyl-glycerol inactivation. Chem Phys Lipids 121:149–158
Factor SA, Friedman JH (1997) The emerging role of clozapine in the treatment of movement disorders. Mov Disord 12:483–496
Fernández-Espejo E, Caraballo I, Rodríguez de Fonseca F, El Banoua F, Ferrer B, Flores JA, Galán-Rodríguez B (2004) Homeostatic changes of anandamide synthesis, and functional effects of cannabinoid CB1 antagonists in rats with severe hemiparkinsonism. Neurobiol Dis (in press)
Fernández-Ruiz J, Lastres-Becker I, Cabranes A, González S, Ramos JA (2002) Endocannabinoids and basal ganglia functionality. Prostaglandins Leukot Essent Fatty Acids 66:263–273
Ferrer B, Asbrock N, Kathuria S, Piomelli D, Giuffrida A (2003) Effects of levodopa on endocannabinoid levels in rat basal ganglia: implications for the treatment of levodopa-induced dyskinesias. Eur J Neurosci 18:1607–1614
Fox SH, Henry B, Hill M, Crossman A, Brotchie J (2002a) Stimulation of cannabinoid receptors reduces levodopa-induced dyskinesia in the MPTP-lesioned nonhuman primate model of Parkinson's disease. Mov Disord 17:1180–1187
Fox SH, Kellett M, Moore AP, Crossman AR, Brotchie JM (2002b) Randomised, double-blind, placebo-controlled trial to assess the potential of cannabinoid receptor stimulation in the treatment of dystonia. Mov Disord 17:145–149
Francis PT, Webster MT, Chesell IP, Holmes C, Stratmann GC, Procter AW, Cross AJ, Green AR, Bouen DM (1993) Neurotransmitters and second messengers in aging and Alzheimer's disease. Ann NY Acad Sci 695:19–26
Fride E, Mechoulam R (1993) Pharmacological activity of the cannabinoid receptor agonist, anandamide, a brain constituent. Eur J Pharmacol 231:313–314
Gerdeman G, Lovinger DM (2001) CB1 cannabinoid receptor inhibits synaptic release of glutamate in rat dorsolateral striatum. J Neurophysiol 85:468–471
Giuffrida A, Parsons LH, Kerr TM, Rodríguez de Fonseca F, Navarro M, Piomelli D (1999) Dopamine activation of endogenous cannabinoid signaling in dorsal striatum. Nat Neurosci 2:358–363
Giuffrida A, Beltramo M, Piomelli D (2001) Mechanisms of endocannabinoid inactivation: biochemistry and pharmacology. J Pharmacol Exp Ther 298:7–14
Glaser ST, Abumrad NA, Fatade F, Kaczocha M, Studholme KM, Deutsch DG (2003) Evidence against the presence of an anandamide transporter. Proc Natl Acad Sci USA 100:4269–4274
Glass M, Faull RLM, Dragunow M (1993) Loss of cannabinoid receptors in the substantia nigra in Huntington's disease. Neuroscience 56:523–527
Glass M, Dragunow M, Faull RLM (2000) The pattern of neurodegeneration in Huntington's disease: a comparative study of cannabinoid, dopamine, adenosine and GABA-A receptor alterations in the human basal ganglia in Huntington's disease. Neuroscience 97:505–519
González S, Romero J, de Miguel R, Lastres-Becker I, Villanúa MA, Makriyannis A, Ramos JA, Fernández-Ruiz J (1999) Extrapyramidal and neuroendocrine effects of AM404, an inhibitor of the carrier-mediated transport of anandamide. Life Sci 65:327–336

Gorriti MA, Rodríguez de Fonseca F, Navarro M, Palomo T (1999) Chronic (-) Δ9-tetrahydrocannabinol treatment induces sensitization to the psychomotor effects of amphetamine in rats. Eur J Pharmacol 365:133–142

Grundy RI (2002) The therapeutic potential of the cannabinoids in neuroprotection. Expert Opin Investig Drugs 11:1365–1374

Gubellini P, Picconi B, Bari M, Battista N, Calabresi P, Centonze D, Bernardi G, Finazzi-Agro A, Maccarrone M (2002) Experimental parkinsonism alters endocannabinoid degradation: implications for striatal glutamatergic transmission. J Neurosci 22:6900–6907

Gueudet C, Santucci V, Rinaldi-Carmona M, Soubrie P, Le Fur G (1995) The CB1 cannabinoid receptor antagonist SR141716A affects A9 dopamine neuronal activity in the rats. Neuroreport 6:1421–1425

Hemming M, Yellowlees PM (1993) Effective treatment of Tourette's syndrome with marijuana. J Psychopharmacol 7:389–391

Herkenham M, Lynn AB, Little MD, Melvin LS, Johnson MR, de Costa DR, Rice KC (1991a) Characterization and localization of cannabinoid receptors in rat brain: a quantitative in vitro autoradiographic study. J Neurosci 11:563–583

Herkenham M, Lynn AB, de Costa BR, Richfield EK (1991b) Neuronal localization of cannabinoid receptors in the basal ganglia of the rat. Brain Res 547:267–274

Hersch SM, Ferrante RJ (1997) Neuropathology and pathophysiology of Huntington's disease. In: Watts RL, Koller WC (eds) Movement disorders. Neurologic principles and practice. McGraw-Hill, New York, pp 503–518

Hiltunen AJ, Jarbe TU, Wangdahl K (1988) Cannabinol and cannabidiol in combination: temperature, open-field activity, and vocalization. Pharmacol Biochem Behav 30:675–678

Hohmann AG, Herkenham M (2000) Localization of cannabinoid CB1 receptor mRNA in neuronal subpopulations of rat striatum: a double-label in situ hybridization study. Synapse 37:71–80

Itier JM, Ibanez P, Mena MA, Abbas N, Cohen-Salmon C, Bohme GA, Laville M, Pratt J, Corti O, Pradier L, Ret G, Joubert C, Periquet M, Araujo F, Negroni J, Casarejos MJ, Canals S, Solano R, Serrano A, Gallego E, Sanchez M, Denefle P, Benavides J, Tremp G, Rooney TA, Brice A, Garcia de Yebenes J (2003) Parkin gene inactivation alters behaviour and dopamine neurotransmission in the mouse. Hum Mol Genet 12:2277–2291

Iuvone T, Esposito G, Esposito R, Santamaria R, Di Rosa M, Izzo AA (2004) Neuroprotective effect of cannabidiol, a non-psychoactive component from Cannabis sativa, on beta-amyloid-induced toxicity in PC12 cells. J Neurochem 89:134–141

Iversen L (2003) Cannabis and the brain. Brain 126:1252–1270

Jarbe TU, Sheppard R, Lamb RJ, Makriyannis A, Lin S, Goutopoulos A (1998) Effects of Δ9-tetrahydrocannabinol and (R)-methanandamide on open-field behavior in rats. Behav Pharmacol 9:169–174

Kieburtz K (1999) Antiglutamate therapies in Huntington's disease. J Neural Transm Suppl 55:97–102

Kurlan R, Richard IH, Papka M, Marshall F (2000) Movement disorders in Alzheimer's disease: more rigidity of definitions is needed. Mov Disord 15:24–29

Lastres-Becker I, Cebeira M, de Ceballos M, Zeng B-Y, Jenner P, Ramos JA, Fernández-Ruiz J (2001a) Increased cannabinoid CB1 receptor binding and activation of GTP-binding proteins in the basal ganglia of patients with Parkinson's syndrome and of MPTP-treated marmosets. Eur J Neurosci 14:1827–1832

Lastres-Becker I, Fezza F, Cebeira M, Bisogno T, Ramos JA, Milone A, Fernández-Ruiz JJ, Di Marzo V (2001b) Changes in endocannabinoid transmission in the basal ganglia in a rat model of Huntington's disease. Neuroreport 12:2125–2129

Lastres-Becker I, Hansen HH, Berrendero F, de Miguel R, Pérez-Rosado A, Manzanares J, Ramos JA, Fernández-Ruiz J (2002a) Alleviation of motor hyperactivity and neurochemical deficits by endocannabinoid uptake inhibition in a rat model of Huntington's disease. Synapse 44:23–35

Lastres-Becker I, Gomez M, De Miguel R, Ramos JA, Fernández-Ruiz J (2002b) Loss of cannabinoid CB1 receptors in the basal ganglia in the late akinetic phase of rats with experimental Huntington's disease. Neurotox Res 4:601–608

Lastres-Becker I, Berrendero F, Lucas JJ, Martín-Aparicio E, Yamamoto A, Ramos JA, Fernández-Ruiz J (2002c) Loss of mRNA levels, binding and activation of GTP-binding proteins for cannabinoid CB1 receptors in the basal ganglia of a transgenic model of Huntington's disease. Brain Res 929:236–242

Lastres-Becker I, de Miguel R, De Petrocellis L, Makriyannis A, Di Marzo V, Fernández-Ruiz J (2003a) Compounds acting at the endocannabinoid and/or endovanilloid systems reduce hyperkinesia in a rat model of Huntington's disease. J Neurochem 84:1097–1109

Lastres-Becker I, De Miguel R, Fernández-Ruiz J (2003b) The endocannabinoid system and Huntington's disease. Curr Drug Target CNS Neurol Disord 2:335–347

Lastres-Becker I, Bizat N, Boyer F, Hantraye P, Brouillet E, Fernández-Ruiz J (2003c) Effects of cannabinoids in the rat model of Huntington's disease generated by an intrastriatal injection of malonate. Neuroreport 14:813–816

Lastres-Becker I, Molina-Holgado F, Ramos JA, Mechoulam R, Fernández-Ruiz J (2004a) Cannabinoids provide neuroprotection in experimental models of Parkinson's disease: involvement of their antioxidant properties and/or of glial cell-mediated effects. Neurobiol Dis (in press)

Lastres-Becker I, Bizat N, Boyer F, Hantraye P, Fernández-Ruiz J, Brouillet E (2004b) Potential involvement of cannabinoid receptors in 3-nitropropionic acid toxicity in vivo: implication for Huntington's disease. Neuroreport 15:2375–2379

Ledent C, Valverde O, Cossu G, Petitet F, Aubert JF, Beslot F, Böhme GA, Imperato A, Pedrazzini T, Roques BP, Vassart G, Fratta W, Parmentier M (1999) Unresponsiveness to cannabinoids and reduced addictive effects of opiates in CB1 receptor knockout mice. Science 283:401–404

Maccarrone M, Gubellini P, Bari M, Picconi B, Battista N, Centonze D, Bernardi G, Finazzi-Agro A, Calabresi P (2003) Levodopa treatment reverses endocannabinoid system abnormalities in experimental parkinsonism. J Neurochem 85:1018–1025

Mailleux P, Vanderhaeghen JJ (1992a) Distribution of neuronal cannabinoid receptor in the adult rat brain: a comparative receptor binding radioautography and in situ hybridization histochemistry. Neuroscience 48:655–668

Mailleux P, Vanderhaeghen JJ (1992b) Age-related loss of cannabinoid of cannabinoid receptor binding sites and mRNA in the rat striatum. Neurosci Lett 147:179–181

Mailleux P, Vanderhaeghen JJ (1993) Dopaminergic regulation of cannabinoid receptor mRNA levels in the rat caudate-putamen: an in situ hybridization study. J Neurochem 61:1705–1712

Maneuf YP, Nash JE, Croosman AR, Brotchie JM (1996) Activation of the cannabinoid receptor by Δ9-THC reduces GABA uptake in the globus pallidus. Eur J Pharmacol 308:161–164

Maneuf YP, Croosman AR, Brotchie JM (1997) The cannabinoid receptor agonist WIN 55,212-2 reduces D2, but not D1, dopamine receptor-mediated alleviation of akinesia in the reserpine-treated rat model of Parkinson's disease. Exp Neurol 148:265–270

McLaughlin PJ, Delevan CE, Carnicom S, Robinson JK, Brener J (2000) Fine motor control in rats is disrupted by Δ9-tetrahydrocannabinol. Pharmacol Biochem Behav 66:803–809

Meschler JP, Howlett AC (2001) Signal transduction interactions between CB1 cannabinoid and dopamine receptors in the rat and monkey striatum. Neuropharmacology 40:918–926

Meschler JP, Howlett AC, Madras BK (2001) Cannabinoid receptor agonist and antagonist effects on motor function in normal and 1-methyl-4-phenyl-1,2,5,6-tetrahydropyridine (MPTP)-treated non-human primates. Psychopharmacology (Berl) 156:79–85

Mezey E, Toth ZE, Cortright DN, Arzubi MK, Krause JE, Elde R, Guo A, Blumberg PM, Szallasi A (2000) Distribution of mRNA for vanilloid receptor subtype 1 (VR1), and VR1-like immunoreactivity, in the central nervous system of the rat and human. Proc Natl Acad Sci USA 97:3655–3660

Miller A, Walker JM (1995) Effects of a cannabinoid on spontaneous and evoked neuronal activity in the substantia nigra pars reticulata. Eur J Pharmacol 279:179–185
Miller A, Walker JM (1996) Electrophysiological effects of a cannabinoid on neural activity in the globus pallidus. Eur J Pharmacol 304:29–35
Miller A, Sañudo-Peña MC, Walker JM (1998) Ipsilateral turning behavior induced by unilateral microinjections of a cannabinoid into the rat subthalamic nucleus. Brain Res 793:7–11
Milton NG (2002) Anandamide and noladin ether prevent neurotoxicity of the human amyloid-beta peptide. Neurosci Lett 332:127–130
Moss DE, McMaster SB, Rogers J (1981) Tetrahydrocannabinol potentiates reserpine-induced hypokinesia. Pharmacol Biochem Behav 15:779–783
Müller-Vahl KR (2003) Cannabinoids reduce symptoms of Tourette's syndrome. Expert Opin Pharmacother 4:1717–1725
Müller-Vahl KR, Kolbe H, Schneider U, Emrich HM (1998) Cannabinoids: possible role in the pathophysiology of Gilles de la Tourette-syndrome. Acta Psychiatr Scand 98:502–506
Müller-Vahl KR, Schneider U, Kolbe H, Emrich HM (1999a) Treatment of Tourette-syndrome with Δ9-tetrahydrocannabinol. Am J Psychiatry 156:495
Müller-Vahl KR, Schneider U, Emrich HM (1999b) Nabilone increases choreatic movements in Huntington's disease. Mov Disord 14:1038–1040
Müller-Vahl KR, Kolbe H, Schneider U, Emrich HM (1999c) Cannabis in movement disorders. Forsch Komplementarmed 6:23–27
Müller-Vahl KR, Schneider U, Koblenz A, Jobges M, Kolbe H, Daldrup T, Emrich HM (2002) Treatment of Tourette's syndrome with Δ9-tetrahydrocannabinol (THC): a randomized crossover trial. Pharmacopsychiatry 35:57–61
Navarro M, Fernández-Ruiz JJ, de Miguel R, Hernández ML, Cebeira M, Ramos JA (1993) Motor disturbances induced by an acute dose of Δ9-tetrahydrocannabinol: possible involvement of nigrostriatal dopaminergic alterations. Pharmacol Biochem Behav 45:291–298
Nuñez E, Benito C, Pazos MR, Barbachano A, Fajardo O, González S, Tolón R, Romero J (2004) Cannabinoid CB2 receptors are expressed by perivascular microglial cells in the human brain: an immunohistochemical study. Synapse 53:208–213
Page KJ, Besret L, Jain M, Monaghan EM, Dunnett SB, Everitt BJ (2000) Effects of systemic 3-nitropropionic acid-induced lesions of the dorsal striatum on cannabinoid and mu-opioid receptor binding in the basal ganglia. Exp Brain Res 130:142–150
Pazos MR, Nuñez E, Benito C, Tolón R, Romero J (2004) Role of the endocannabinoid system in Alzheimer's disease: new perspectives. Life Sci 75:1907–1915
Pertwee RG (2002) Cannabinoids and multiple sclerosis. Pharmacol Ther 95:165–174
Pertwee RG, Greentree SG, Swift PA (1988) Drugs which stimulate or facilitate central GABAergic transmission interact synergistically with Δ9-tetrahydrocannabinol to produce marked catalepsy in mice. Neuropharmacology 27:1265–1270
Reddy PH, Williams M, Tagle DA (1999) Recent advances in understanding the pathogenesis of Huntington's disease. Trends Neurosci 22:248–255
Richfield EK, Herkenham M (1994) Selective vulnerability in Huntington's disease: preferential loss of cannabinoid receptors in lateral globus pallidus. Ann Neurol 36:577–584
Richter A, Loscher W (1994) (+)-WIN 55,212-2, a novel cannabinoid receptor agonist, exerts antidystonic effects in mutant dystonic hamsters. Eur J Pharmacol 264:371–377
Richter A, Loscher W (2002) Effects of pharmacological manipulations of cannabinoid receptors on severity of dystonia in a genetic model of paroxysmal dyskinesia. Eur J Pharmacol 454:145–151
Rodriguez de Fonseca F, Gorriti MA, Fernández-Ruiz J, Palomo T, Ramos JA (1994) Down-regulation of rat brain cannabinoid binding sites after chronic Δ9-tetrahydrocannabinol treatment. Pharmacol Biochem Behav 47:33–40
Romero J, García L, Cebeira M, Zadrozny D, Fernández-Ruiz J, Ramos JA (1995a) The endogenous cannabinoid receptor ligand, anandamide, inhibits the motor behaviour: role of nigrostriatal dopaminergic neurons. Life Sci 56:2033–2040

Romero J, de Miguel R, García-Palomero E, Fernández-Ruiz J, Ramos JA (1995b) Time-course of the effects of anandamide, the putative endogenous cannabinoid receptor ligand, on extrapyramidal function. Brain Res 694:223–232
Romero J, García-Palomero E, Lin SY, Ramos JA, Makriyannis A, Fernández-Ruiz J (1996a) Extrapyramidal effects of methanandamide, an analog of anandamide, the endogenous CB1 receptor ligand. Life Sci 58:1249–1257
Romero J, García-Palomero E, Fernández-Ruiz J, Ramos JA (1996b) Involvement of GABA-B receptors in the motor inhibition produced by agonists of brain cannabinoid receptors. Behav Pharmacol 7:299–302
Romero J, de Miguel R, Ramos JA, Fernández-Ruiz J (1998a) The activation of cannabinoid receptors in striatonigral neurons inhibited GABA uptake. Life Sci 62:351–363
Romero J, Berrendero F, García-Gil L, de la Cruz P, Ramos JA, Fernández-Ruiz J (1998b) Loss of cannabinoid receptor binding and messenger RNA levels and cannabinoid agonist-stimulated [35S]-GTPγS binding in the basal ganglia of aged rats. Neuroscience 84:1075–1083
Romero J, Berrendero F, Pérez-Rosado A, Manzanares J, Rojo A, Fernández-Ruiz J, de Yébenes JG, Ramos JA (2000) Unilateral 6-hydroxydopamine lesions of nigrostriatal dopaminergic neurons increased CB1 receptor mRNA levels in the caudate-putamen. Life Sci 66:485–494
Romero J, Lastres-Becker I, de Miguel R, Berrendero F, Ramos JA, Fernández-Ruiz J (2002) The endogenous cannabinoid system and the basal ganglia. biochemical, pharmacological, and therapeutic aspects. Pharmacol Ther 95:137–152
Sakurai-Yamashita Y, Kataoka Y, Fujiwara M, Mine K, Ueki S (1989) Δ9-Tetrahydrocannabinol facilitates striatal dopaminergic transmission. Pharmacol Biochem Behav 33:397–400
Sañudo-Peña MC, Walker JM (1997) Role of the subthalamic nucleus in cannabinoid actions in the substantia nigra of the rat. J Neurophysiol 77:1635–1638
Sañudo-Peña MC, Patrick SL, Khen S, Patrick RL, Tsou K, Walker JM (1998) Cannabinoid effects in basal ganglia in a rat model of Parkinson's disease. Neurosci Lett 248:171–174
Sañudo-Peña MC, Tsou K, Walker JM (1999) Motor actions of cannabinoids in the basal ganglia output nuclei. Life Sci 65:703–713
Sañudo-Peña MC, Romero J, Seale GE, Fernández-Ruiz J, Walker JM (2000) Activational role of cannabinoids on movement. Eur J Pharmacol 391:269–274
Schut LJ (1998) Motor system changes in the aging brain: what is normal and what is not. Geriatrics 53:S16–S19
Segovia G, Mora F, Crossman AR, Brotchie JM (2003) Effects of CB1 cannabinoid receptor modulating compounds on the hyperkinesia induced by high-dose levodopa in the reserpine-treated rat model of Parkinson's disease. Mov Disord 18:138–149
Sieradzan KA, Fox SH, Hill M, Dick JP, Crossman AR, Brotchie JM (2001) Cannabinoids reduce levodopa-induced dyskinesia in Parkinson's disease: a pilot study. Neurology 57:2108–2111
Silverdale MA, McGuire S, McInnes A, Crossman AR, Brotchie JM (2001) Striatal cannabinoid CB1 receptor mRNA expression is decreased in the reserpine-treated rat model of Parkinson's disease. Exp Neurol 169:400–406
Skaper SD, Buriani A, Dal Toso R, Petrelli L, Romanello S, Facci L, Leon A (1996) The ALIAmide palmitoylethanolamide and cannabinoids, but not anandamide, are protective in a delayed postglutamate paradigm of excitotoxic death in cerebellar granule neurons. Proc Natl Acad Sci USA 93:3984–3989
Smith PB, Compton DR, Welch SP, Razdan RK, Mechoulam R, Martin BR (1994) The pharmacological activity of anandamide, a putative endogenous cannabinoid, in mice. J Pharmacol Exp Ther 270:219–227
Souilhac J, Poncelet M, Rinaldi-Carmona M, Le-Fur G, Soubrie P (1995) Intrastriatal injection of cannabinoid receptor agonists induced turning behavior in mice. Pharmacol Biochem Behav 51:3–7

Szabo B, Dorner L, Pfreundtner C, Norenberg W, Starke K (1998) Inhibition of GABAergic inhibitory postsynaptic currents by cannabinoids in rat corpus striatum. Neuroscience 85:395–403
Szabo B, Muller T, Koch H (1999) Effects of cannabinoids on dopamine release in the corpus striatum and the nucleus accumbens in vitro. J Neurochem 73:1084–1089
Szabo B, Wallmichrath I, Mathonia P, Pfreundtner C (2000) Cannabinoids inhibit excitatory neurotransmission in the substantia nigra pars reticulata. Neuroscience 97:89–97
Tersigni TJ, Rosenberg HC (1996) Local pressure application of cannabinoid agonists increases spontaneous activity of rat substantia nigra pars reticulata neurons without affecting response to iontophoretically-applied GABA. Brain Res 733:184–192
Tsou K, Brown S, Sañudo-Peña MC, Mackie K, Walker JM (1998a) Immunohistochemical distribution of cannabinoid CB1 receptors in the rat central nervous system. Neuroscience 83:393–411
Tsou K, Nogueron MI, Muthian S, Sañudo-Peña M, Hillard CJ, Deutsch DG, Walker JM (1998b) Fatty acid amide hydrolase is located preferentially in large neurons in the rat central nervous system as revealed by immunohistochemistry. Neurosci Lett 254:137–140
van der Stelt M, Di Marzo V (2003) The endocannabinoid system in the basal ganglia and in the mesolimbic reward system: implications for neurological and psychiatric disorders. Eur J Pharmacol 480:133–150
Westlake TM, Howlett AC, Bonner TI, Matsuda LA, Herkenham M (1994) Cannabinoid receptor binding and messenger RNA expression in human brain: an in vitro receptor autoradiography and in situ hybridization histochemistry study of normal aged and Alzheimer's brains. Neuroscience 63:637–652
Wickens AP, Pertwee RG (1993) Δ9-Tetrahydrocannabinol and anandamide enhance the ability of muscimol to induce catalepsy in the globus pallidus of rats. Eur J Pharmacol 250:205–208
Zajicek J, Fox P, Sanders H, Wright D, Vickery J, Nunn A, Thompson A (2003) Cannabinoids for treatment of spasticity and other symptoms related to multiple sclerosis (CAMS study): multicentre randomised placebo-controlled trial. Lancet 362:1517–1526
Zaretsky A, Rector NA, Seeman MV, Fornazzari X (1993) Current cannabis use and tardive dyskinesia. Schizophr Res 11:3–8
Zeng BY, Dass B, Owen A, Rose S, Cannizzaro C, Tel BC, Jenner P (1999) Chronic L-DOPA treatment increases striatal cannabinoid CB1 receptor mRNA expression in 6-hydroxydopamine-lesioned rats. Neurosci Lett 276:71–74
Zimmer A, Zimmer AM, Hohmann AG, Herkenham M, Bonner TI (1999) Increased mortality, hypoactivity, and hypoalgesia in cannabinoid CB1 receptor knockout mice. Proc Natl Acad Sci USA 96:5780–5785
Zygmunt PM, Chuang H, Movahed P, Julius D, Hogestatt ED (2000) The anandamide transport inhibitor AM404 activates vanilloid receptors. Eur J Pharmacol 396:39–42

HEP (2005) 168:509–554

Cannabinoid Mechanisms of Pain Suppression

J.M. Walker[1] · A.G. Hohmann[2] (✉)

[1]Department of Psychology, Indiana University Bloomington, IN, 47405-7007, USA
[2]Neuroscience and Behavior Program, Department of Psychology,
The University of Georgia, Athens GA, 30602, USA
ahohmann@uga.edu

Abstract A large body of literature indicates that cannabinoids suppress behavioral responses to acute and persistent noxious stimulation in animals. This review examines neuroanatomical, behavioral, and neurophysiological evidence supporting a role for cannabinoids in suppressing pain at spinal, supraspinal, and peripheral levels. Localization studies employing receptor binding and quantitative autoradiography, immunocytochemistry, and in situ hybridization are reviewed to examine the distribution of cannabinoid receptors at these levels and provide a neuroanatomical framework with which to understand the roles of endogenous cannabinoids in sensory processing. Pharmacological and transgenic approaches that have been used to study cannabinoid antinociceptive mechanisms are described. These studies provide insight into the functional roles of cannabinoid CB_1 (CB1R) and CB_2 (CB2R) receptor subtypes in cannabinoid antinociceptive mechanisms, as revealed in animal models of acute and persistent pain. The role of endocannabinoids and related fatty acid amides that are implicated in endogenous mechanisms for pain suppression are discussed. Human studies evaluating therapeutic potential of cannabinoid pharmacotherapies in experimental and clinical pain syndromes are evaluated. The potential of exploiting cannabinoid antinociceptive mechanisms in novel pharmacotherapies for pain is discussed.

Keywords Endocannabinoid · Spinal cord · Periaqueductal gray · Supraspinal · Peripheral · CB1 · CB2 · THC · Hyperalgesia · Clinical pain

The study of the role of endocannabinoids in pain is founded in research on pain mechanisms, a vital field that is steadily evolving on many fronts. A brief overview of the current thinking on the neural basis of pain is thus provided as background to discussing how cannabinoids and endocannabinoids modulate pain sensation. Extensive reviews of pain mechanisms may be found in a relatively recent volume edited by Wall and Melzack (1999). This review will focus on preclinical studies that evaluate evidence from neuroanatomical, behavioral, electrophysiological, and neurochemical approaches that provide insight into the roles of cannabinoids and endocannabinoids in suppressing pain. Peripheral, spinal, and supraspinal sites of cannabinoid actions are discussed, as well as the endogenous ligands implicated in endocannabinoid mechanisms of pain suppression. This review will also present results from clinical studies that provide insight into the therapeutic potential for cannabinoid pharmacotherapies for pain in man.

1 Brief Overview of Pain Mechanisms

Pain is a complex psychological phenomenon comprising sensory, emotional, and motivational components. The negative emotion and the motivation to escape from the stimulus are essential features of pain—without them the experience would be non-painful tactile stimulation. In the early twentieth century, "labeled-line" theory dominated thinking about pain. In this conception, specific nociceptors in the periphery transmit signals about noxious stimuli to the spinal cord, which relays the information to a pain center in the brain, which in turn gives rise to the sensation of pain. This notion has broken down, first with the realization that while there are specific nociceptors in the periphery, activity from incoming non-nociceptive fibers interacts with that from nociceptors, changing the spinal transmission properties of the nociceptive fibers. Hence, increased activity in larger, non-nociceptive fibers lessens the impact of activity in nociceptors (typically smaller unmyelinated C fibers and finely myelinated Aδ fibers). The second and perhaps even more significant finding was the discovery that the brain contains circuits that modulate the ability of nociceptors to excite ascending pain-transmission pathways. These circuits can either dampen or facilitate pain. The observations by Beecher (1959), who observed soldiers in World War II who felt no pain despite serious injuries, were important in rethinking the labeled line theory, leading to the more sophisticated view that the experience of pain is regulated by the relative activity in peripheral, spinal, and brain networks of pro- and anti-nociceptive circuits. These networks, described in more detail below, provide substrates for actions of cannabinoids on pain.

1.1 Nociceptors

The term nociceptor refers to sensory receptors that respond to noxious stimuli (see Kruger et al. 2003 for review). A variety of cutaneous primary afferent nociceptors have been described, primary among them are the unmyelinated C fibers that are characterized by free nerve endings. The C-polymodal nociceptor responds to mechanical, heat, and chemical stimuli. Primary afferents have a unique morphology. The cell bodies, which are found in the dorsal root ganglion, lack dendrites and synapses and are encased in satellite cells that insulate them. The axon bifurcates, sending a branch to the spinal cord and branch to the periphery. Hence, the sensory apparatus is found on an axon terminal, and indeed action potentials in the peripheral nerve lead to secretion of neurotransmitter at both the peripheral and central terminals. The biochemical machinery of nociceptors includes a variety of molecular transduction elements such as transient receptor potential (TRP) channels, acid sensing channels, and P2X3 receptors, as well as particular neurotransmitters including glutamate, substance P, and calcitonin gene-related peptide (CGRP). On the central terminals are found presynaptic receptors that modulate neurotransmitter release.

1.2 Ascending Pain Pathways

Upon activation of spinal neurons by nociceptors, information about noxious stimuli is carried to the brain by ascending pathways. Multiple pathways have been described (for review see Millan 1999, especially Table 4 therein).

1.2.1 Spinothalamic Tract

The classical ascending pathway (Fig. 1) is the spinothalamic tract, a contralaterally projecting fiber bundle that ascends in the anterolateral aspect of the spinal white matter to the ventral posterolateral thalamus with extensive collateralization to brainstem structures prominent among these being the periaqueductal gray (PAG).

1.2.2 Dorsal Column Visceral Pain Pathway

The dorsal column pathway may be of major importance for visceral pain (Berkley and Hubscher 1995; Willis et al. 1995). This pathway originates from the visceral processing circuitry in the gray matter surrounding central canal of the spinal cord and ascends ipsilaterally in the dorsal columns, the white matter areas adjacent to the midline on the dorsal aspect of the spinal cord. The putative involvement of the dorsal column in visceral pain is noteworthy for two reasons, the first being that it supercedes the classical understanding of the dorsal columns as being the trajectory for discriminative touch sensations, and second that it provides a new understanding of complex neural pathways for visceral compared to somatosensory pain. Rather than operating in isolation, the dorsal column and spinal routes cooperate to produce the many perceptions of touch and pain (Berkley and Hubscher 1995). This ensemble view encourages the development of novel, integrative pharmacotherapies and treatments (Berkley and Hubscher 1995).

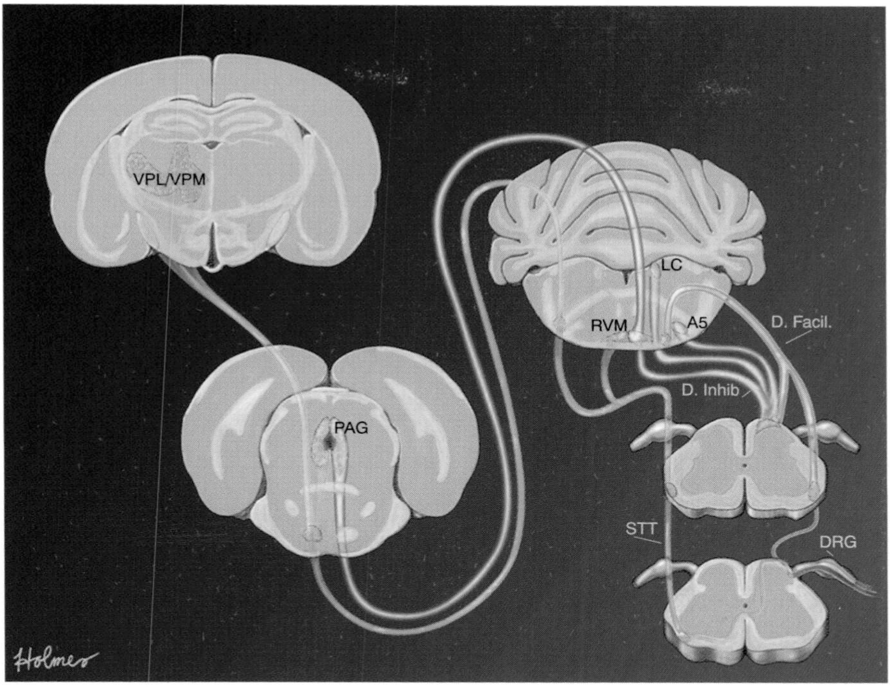

Fig. 1. Schematic of neural pathways that process and modulate the transmission of information about nociceptive signals. In *orange*, the spinothalamic tract is shown, with signals originating in the peripheral nerve, crossing the midline, and ascending the anterolateral white matter of the spinal cord with many collateral outputs to the brainstem shown for the RVM and PAG. This tract terminates in the VPL/VPM thalamus. In *green*, descending pain inhibitory pathways are shown, which connect the PAG to the RVM, and from there makes connections in the spinal cord. Other descending inhibitory pathways originating in the LC and noradrenergic nucleus A5 are also shown. In *red*, pathways that facilitate pain are shown originating in the RVM and descending to the spinal cord. Abbreviations: *A5*, noradrenergic nucleus A5; *D. Facil.*, descending facilitation pathway; *D. Inhib*, descending inhibitory pathways; *DRG*, dorsal root ganglion; *LC*, locus coeruleus; *PAG*, periaqueductal gray; *RVM*, rostral ventromedial medulla; *STT*, spinothalamic tract; *VPL*, ventroposterolateral nucleus; *VPM*, ventral posteromedial nucleus

1.3 Descending Modulation of Pain

1.3.1 Descending Pain Inhibition

With the observation by Kang Tsou (Tsou and Jang 1964) of the potent analgesic effects of morphine applied by microinjection to the periaqueductal gray came the early realization that the brain plays an active role in determining whether pain is felt following noxious stimulation. Later, it was observed by Reynolds (1969) that electrical stimulation of this region in the rat produced sufficient analgesia for a pain-free laparotomy without additional anesthesia. Akil et al. (1976) noted that this analgesic phenomenon could be reversed by naloxone, suggesting that the electrical stimulation releases an endogenous opiate-like substance that led to analgesia. These observations set the stage for extensive studies of how the PAG can entirely block pain sensations (reviewed by Fields et al. 1991). It became clear that this occurs through projections from the PAG to the rostral ventromedial medulla (RVM), and from there to the spinal cord. Specific on- and off-cells in the RVM were found to control the excitability of ascending spinal pathways. On-cells fire just before a nocifensive flexion reflex and off-cells, which are spontaneously active, stop firing just before a nocifensive flexion reflex. This pathway is activated by certain forms of stress and appears to naturally serve to control the organism's response to noxious stimuli, being able to entirely suppress pain under certain conditions.

1.3.2
Descending Pain Facilitation

More recently, it has become clear that the RVM can facilitate as well as dampen pain (reviewed by Porreca et al. 2002). Stimulation of the RVM at relatively low current intensities increases the responses of spinal dorsal horn neurons to noxious stimuli. The role of this facilitation in chronic pain is suggested by studies showing that blockade of the RVM with lidocaine reduces abnormal tactile responses in rats with neuropathic pain (Pertovaara et al. 1996). Other studies of inflammatory and neuropathic pain converge in showing that descending facilitation is an important component of pathological pain.

1.4
Implications for Understanding Cannabinoid Actions in Pain

The above outline of current understanding of the neural processing that underlies pain provides a foundation for understanding the effects of exogenous and endogenous cannabinoids in pain. Cannabinoids act at all of the sites discussed above, i.e., the periphery, spinal cord, and central circuits for pain facilitation and pain modulation. In the following sections, we review the current understanding of the systemic effects of cannabinoids and their sites of action within pain processing circuits from anatomical, physiological, and behavioral perspectives.

2
Antinociception and Suppression of Pain Neurotransmission by Systemically Administered Cannabinoids

Cannabinoid antinociception is observed in preclinical behavioral studies employing different modalities of noxious stimulation including thermal, mechanical, and chemical (see Walker et al. 2001 for review). Perhaps the earliest recorded scientific demonstration of cannabinoid antinociception was provided by one of the fathers of modern pharmacology, Ernest Dixon (1899). He observed that dogs that inhaled cannabis smoke failed to react to pin pricks. Early studies by Bicher and Mechoulam (1968) and Kosersky et al. (1973) provided a foundation for subsequent work that verified the ability of cannabinoids to profoundly suppress behavioral reactions to acute noxious stimuli and inflammatory and nerve injury-induced pain. In these early studies, it was noted that the potency and efficacy of cannabinoids rival that of morphine (Bloom et al. 1977; Buxbaum 1972). However, cannabinoids also produce profound motor effects [e.g., immobility, catalepsy; (Martin et al. 1991)], a potential confound for behavioral studies, which inevitably employ motor responses to noxious stimuli as a measure of pain sensitivity.

In part to address this potential confound, subsequent electrophysiological and neurochemical studies examined the question of whether cannabinoids suppress activity within pain circuits. These studies provided convincing evidence that cannabinoids suppress nociceptive transmission in vivo (see Hohmann 2002 for review). Walker's laboratory first demonstrated that cannabinoids suppress noxious stimulus-evoked neuronal activity in nociceptive neurons in the spinal cord (Fig. 2)

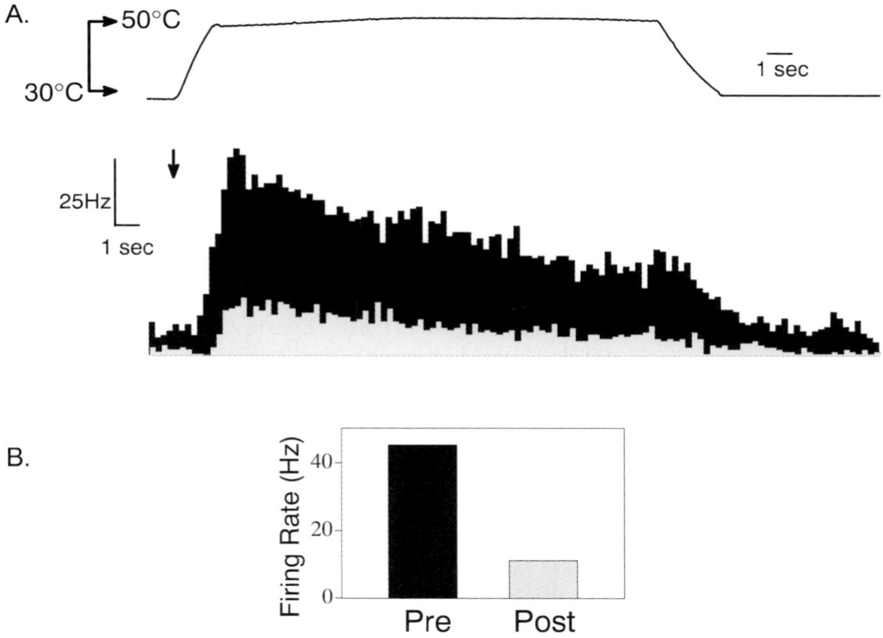

Fig. 2. Example of inhibition of noxious heat-evoked activity in a lumbar dorsal horn neuron by the cannabinoid WIN55,212-2. The responses of the neuron to a 50°C stimulus were examined during 16 stimulus trials. **A** The noxious stimulus, illustrated by the temperature waveform (*top center*), was administered at 2.5-min intervals. The *black peristimulus time histogram* represents baseline firing prior to injection of the synthetic CB1R/CB2R agonist WIN55,212-2 (125 µg/kg i.v.).The *gray peristimulus time histogram* represents the firing rate for the first five post-injection trials. **B** Comparison of the mean firing rate during the stimulus for the five baseline trials to the firing rate during the stimulus for the first five post-injection trials illustrating, approximately, a 75% decrease in responsiveness. (Redrawn from Hohmann et al. 1999b)

and thalamus (Hohmann et al. 1995, 1998, 1999b; Martin et al. 1996; Strangman and Walker 1999). This suppression is observed in nociceptive neurons, generalizes to different modalities of noxious stimulation (mechanical, thermal, chemical), is mediated by cannabinoid receptors (CBRs), and correlates with the antinociceptive effects of cannabinoids (Hohmann et al. 1995, 1998, 1999b,c; Martin et al. 1996). Cannabinoids suppress C fiber-evoked responses in spinal dorsal horn neurons recorded in normal and inflamed rats (Drew et al. 2000; Kelly and Chapman 2001; Strangman and Walker 1999). Spinal Fos protein expression, a neurochemical marker of sustained neuronal activation (Hunt et al. 1987), is also suppressed by cannabinoids in animal models of persistent pain (Farquhar-Smith et al. 2002; Hohmann et al. 1999c; Martin et al. 1999b; Nackley et al. 2003a, 2003b; Tsou et al. 1996). This suppression occurs through cannabinoid CB_1 receptor (CB1R)- and cannabinoid CB_2 receptor (CB2R)-selective mechanisms. These studies provided a foundation for subsequent work, which has identified the sites of action of cannabinoids within pain circuits and the actions of specific endocannabinoids within these circuits.

3 CB1R-Mediated Antinociception: Peripheral, Spinal, and Supraspinal Actions

3.1 Methodological Considerations

The distribution of CBRs in brain was first mapped by Herkenham et al. (1991) using receptor binding and autoradiographic methods. This approach permits quantitative evaluation of the density and distribution of receptors, but lacks cellular resolution. The development of specific antibodies for CBRs has permitted characterization of the cellular distribution of CBRs (Egertová et al. 2003; Egertová et al. 1998; Tsou et al. 1998a). Immunocytochemical approaches, however, are suited to qualitative rather than quantitative evaluation of CBR densities.

CBRs have been studied in rat spinal cord using autoradiographic (Herkenham et al. 1991; Hohmann et al. 1999a; Hohmann and Herkenham 1998) and immunocytochemical (Farquhar-Smith et al. 2000; Morisset et al. 2001; Salio et al. 2002b; Salio et al. 2001; Sanudo-Pena et al. 1999; Tsou et al. 1998a) techniques. It is important to note that localization studies employing antibodies raised against the N-terminal of the CB1R protein may reveal different patterns of immunostaining from antibodies raised against the C-terminal tail and support different conclusions regarding the anatomical localization of CBRs. Antibodies recognizing the intracellular C-terminal domain of CB1R might be expected to behave differently depending on the level of tissue fixation and receptor internalization. It is possible that N-terminal antibodies underestimate localization of CB1R to plasma membrane and primarily reflect synthesis, storage, or transport sites; detection of CB1R at the plasma membrane would require an antibody recognizing the N terminus to penetrate the extracellular space (Salio et al. 2002b). Moreover, N-terminal antibodies are unable to recognize a splice variant of CB_1, CB_{1A} (Shire et al. 1995), because the splice variant bears a truncated N terminus (Salio et al. 2002b). However, it is unclear whether these isoforms are differentially distributed in the spinal dorsal horn. This review will compare the distribution of CB1R mRNA and CB1R immunoreactivity in rat dorsal root ganglion cells. The distribution of CBRs in rat spinal cord revealed by receptor binding and quantitative autoradiography will subsequently be compared with patterns of CB1R immunostaining revealed by immunocytochemistry using antibodies recognizing different epitopes of CB1R.

3.2 Antinociception Mediated by CB1Rs in the Periphery

The distribution of CBRs outside the central nervous system is consistent with behavioral and neurochemical data that implicate a role for peripheral CB1Rs in cannabinoid antinociception. The distribution of CB1Rs in dorsal root ganglia and peripheral nerve is therefore reviewed here. The role of CB2Rs in cannabinoid antinociceptive mechanisms is reviewed in Sect. 4.

Traditionally, the dorsal root ganglion (DRG) has been used as a model of the peripheral nerve because of its more convenient size, location, and the ability to correlate cell size and neurochemical phenotype with peripheral axon caliber. Hohmann and Herkenham (1999a,b) used in situ hybridization to test the hypothesis that dorsal root ganglion cells, the source of primary afferent input to the

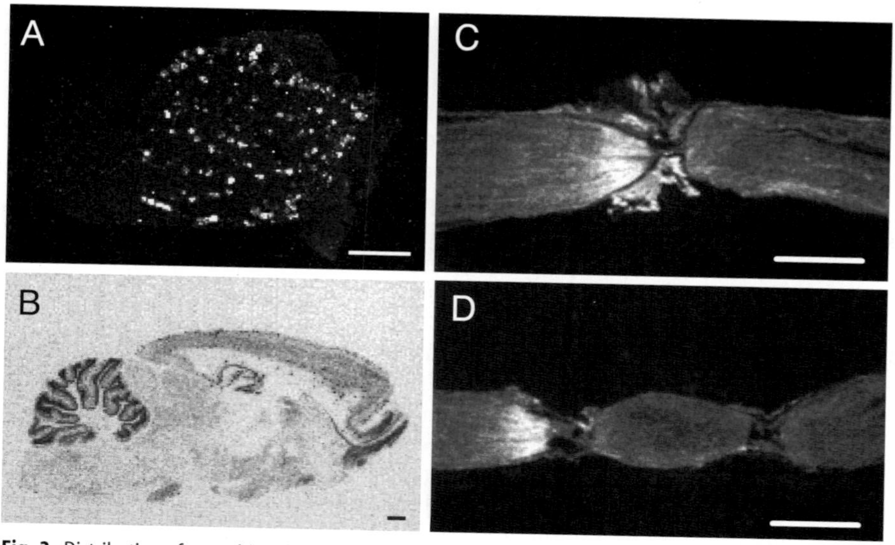

Fig. 3. Distribution of cannabinoid CB_1 receptor (CB1R) mRNA in rat (**A**) dorsal root ganglia and (**B**) brain. Cannabinoid binding sites accumulate proximal to a tight ligation of the sciatic nerve. [^{3}H]CP55,940 binding and high-resolution emulsion autoradiography was used to demonstrate flow of cannabinoid receptors to the periphery. Dark-field photomicrographs show damming of cannabinoid receptors proximal as opposed to distal to (**C**) a single tight ligation and (**D**) the more proximal of two separate ligatures applied to the sciatic nerve. *Scale bars* = 1 mm. (From Hohmann and Herkenham 1999a)

spinal cord, synthesize cannabinoid CB1Rs (Fig. 3A, B). CB1R mRNA was highly expressed in dorsal root ganglion cells of heterogeneous cell size, and predominant in intermediate-sized neurons. These data are consistent with immunocytochemical studies using an N-terminal antibody in native DRG that confirmed the presence of CB1Rs in small, medium, and large cells of rat dorsal root ganglia (Salio et al. 2002a).

Both CB1Rs and CB2Rs have been identified in primary cultures of dorsal root ganglion cells derived from neonatal rats (Ross et al. 2001a). The location and phenotypes of cells expressing CB2Rs in dorsal root ganglion likely represent an important topic of future investigation. It is unclear if CB2Rs are expressed in satellite glial cells, the main glial cells in sensory ganglia, that have recently been shown to be histologically altered in animal models of nociception (Hanani et al. 2002; Li and Zhou 2001). Neuronal expression of CB2R mRNA in native DRG (Hohmann and Herkenham 1999a) and trigeminal ganglia (Price et al. 2003) was similar to background under conditions in which CB1R mRNA was clearly demonstrated. These data suggest that: (1) a high-affinity low-capacity CB2R site may be synthesized in the DRG and contribute peripheral cannabinoid actions, (2) a CB_2-like receptor may mediate the observed effects, and/or (3) a CB2R mechanism exerts its actions indirectly (e.g., by inhibiting the release of inflammatory mediators that excite nociceptors).

3.2.1
Phenotypes of Dorsal Root Ganglion Cells Expressing CB1Rs and CB1R mRNA

To better understand the role of cannabinoids in sensory processing, phenotypes of dorsal root ganglion cells that synthesize CB1Rs have been investigated by several laboratories (Ahluwalia et al. 2000, 2002; Bridges et al. 2003; Hohmann and Herkenham 1999b; Price et al. 2003). Small-diameter cells in the dorsal root ganglia, in general, correspond to nociceptors and thermoreceptors, respond to high-threshold stimuli, and have unmyelinated or thinly myelinated axons. The small-diameter cells fall into two categories—the nerve growth factor-sensitive population of cells that synthesize neuropeptides and express trkA (Averill et al. 1995; Molliver et al. 1995) and those that are sensitive to glial cell-derived neurotrophic factor, contain the enzyme fluoride-resistant acid phosphatase (Nagy and Hunt 1982), bind isolectin B4 (IB4) (Silverman and Kruger 1990), and do not express trks. We evaluated localization of CB1Rs to dorsal root ganglion cells that synthesize preprotachykinin A (a precursor for substance P) and α-CGRP (Hohmann and Herkenham 1999b) using double-label in situ hybridization. In native dorsal root ganglia, only small subpopulations of cells expressing CB1R mRNA colocalized mRNAs for neuropeptide markers of primary afferents preprotachykinin A and α-CGRP (Hohmann and Herkenham 1999b). Neurons expressing mRNA for somatostatin were CB_1-mRNA negative (Hohmann and Herkenham 1999b).

Quantification of double-labeled cells in this study revealed that less than 9% and 13% of cells containing mRNA for precursors of CGRP or substance P mRNA, respectively, expressed CB1R mRNA (Hohmann and Herkenham 1999b). Moreover, the vast majority of CB1R mRNA-expressing cells (75%) in the dorsal root ganglia of naive rats failed to colocalize these neuropeptides. Direct support for localization of CB1Rs to dorsal root ganglion cells bearing myelinated fibers has recently been demonstrated (Bridges et al. 2003; Price et al. 2003). These observations indicate that under normal conditions, CB1Rs are localized mainly to non-nociceptive primary afferent fibers. Inflammation and axotomy induce marked changes in peptide phenotypes of dorsal root ganglia cells (Calza et al. 1998; Donaldson et al. 1994; Galeazza et al. 1995; Hanesch et al. 1995; Ji et al. 1994, 1995; Leslie et al. 1995; Neumann et al. 1996), indicating that different coexpression levels may also exist in chronic pain states.

In native DRG, CB1R is largely associated with myelinated A fibers. Bridges et al. (2003) demonstrated that the majority (69%–80%) of CB1R-immunoreactive cells (labeled using an antibody directed against the C-terminal of CB1R) coexpress neurofilament 200 (Bridges et al. 2003). This marker is largely restricted to primary afferent A fibers. A modest degree of colocalization of CB1R immunoreactivity was observed with IB4 (17%–26%) and CGRP (10%) immunoreactive cells in DRG (Bridges et al. 2003), markers of nociceptors. In addition, 10% of mRNA expressing cells were immunoreactive for transient receptor potential vanilloid family ion channel 2 (TRPV2), the noxious heat-transducing channel found in medium and large lightly myelinated Aδ fibers. Moreover, this study demonstrated that only 11%–20% of CB1R mRNA expressing cells were immunoreactive for TRPV1, a marker of nociceptive C fibers. Similar results are observed in native trigeminal ganglia, where only minor colocalization of CB1R is observed with markers of nociceptors (TRPV1, substance P, CGRP, IB4) and high levels of colocalization (75%) of CB1R with N52, a maker of myelinated non-nociceptive fibers, were observed (Price et al. 2003).

The phenotypes of cells expressing CB1Rs in native DRG differs from that reported in cultured DRG, where colocalization of CB1Rs with markers of nociceptors is more prevalent. CB1Rs have been identified in small-diameter cells expressing capsaicin-sensitive TRPV1 (VR1) receptors in cultured DRG cells (Ahluwalia et al. 2000, 2002). In contrast to observations in native DRG, approximately 80% of the CB1R-like immunopositive cells showed TRPV1-like immunoreactivity, while 98% of the TRPV1-like immunolabeled neurons showed CB_1-like immunostaining (Ahluwalia et al. 2000). A further study demonstrated that CB1R-immunoreactive cells colocalized immunoreactivity for CGRP and IB4 (Ahluwalia et al. 2002). In this study, approximately 20% of CB1R immunostained neurons did not show either CGRP or IB4 immunoreactivity, indicating that they were non-nociceptive. These data support localization of CB1Rs to nociceptive neurons as well as non-nociceptive neurons in dorsal root ganglion cells raised in culture, in contrast with the modest colocalization of CB1Rs with markers of nociceptors observed in native dorsal root (Bridges et al. 2003; Hohmann and Herkenham 1999b) and trigeminal (Price et al. 2003) ganglia. However, in cultured DRG neurons, cannabinoids attenuate depolarization-dependent Ca^{++} influx in intermediate-sized (800–1500 μm^2) dorsal root ganglion cells raised in cultures derived from adult rats, but these effects were largely absent in small ($< 800\ \mu m^2$) neurons (Khasabova et al. 2002).

The differences in colocalization reported here may be attributed to differences between native and cultured dorsal root ganglion cells and/or the use of different antibodies recognizing different epitopes of CB1R. Lower numbers of TRPV1 immunoreactive cells are observed in DRG cultures raised in the absence of neurotrophic factors, but no changes are observed in the number of CB_1-expressing cells under these same conditions (Ahluwalia et al. 2002). Elimination of neurotrophic factors from culture media is also associated with a modest but significant shift in the distribution of the size of CB1R-immunoreactive cells to larger diameters (Ahluwalia et al. 2002).

3.2.2
Axonal Transport of CBRs to the Periphery

We used [^{3}H]CP55,940 binding and high-resolution emulsion autoradiography to test the hypothesis that CBRs synthesized in dorsal root ganglion cells are transported to the periphery. Transport of CBRs to the periphery was occluded by tight ligation of the sciatic nerve (Hohmann and Herkenham 1999a). These data suggest that CBRs synthesized in the DRG are likely to undergo anterograde transport and be inserted on terminals in the peripheral direction (Fig. 3C, D). This observation is also consistent with the observation of CB1R immunoreactivity in rat peripheral nerve and in ventral roots (Sanudo-Pena et al. 1999). More work is necessary to determine if CBRs synthesized in the DRG are differentially transported to peripheral vs central terminals and whether transport of these receptors is modulated by persistent pain states.

3.2.3
Peripheral CB1R-Mediated Antinociception: Acute and Persistent Pain States

Behavioral and neurochemical studies implicate a role for peripheral CB1Rs in cannabinoid antinociception in models of acute, inflammatory, and neuropathic pain states.

Peripheral CB1R-Modulation of Inflammatory Nociception

Richardson and colleagues first demonstrated that activation of peripheral CB1Rs suppresses thermal hyperalgesia and edema in the carrageenan model of inflammation (Richardson et al. 1998c). Hyperalgesia refers to a lowering of the pain threshold or increase in sensitivity to a normally painful stimulus. Anandamide, administered to the site of injury, suppressed the development and maintenance of carrageenan-evoked thermal hyperalgesia (Richardson et al. 1998c). The same dose administered to the noninflamed contralateral paw was inactive, suggesting that antihyperalgesia occurred at low doses that do not produce antinociception. Antihyperalgesia induced by anandamide was blocked by the CB1R-competitive antagonist/inverse agonist SR141716A, demonstrating mediation by CB1R. Intraplantar administration of the mixed CB_1/CB_2 agonist WIN55,212-2 also attenuates the development of carrageenan-evoked mechanical hyperalgesia, allodynia, and spinal Fos protein expression (Nackley et al. 2003b); these latter actions were completely blocked by coadministration of either a CB1R or CB2R antagonist.

Peripheral CB1R in Acute Antinociception and Antinociceptive Synergism

Cannabinoids induce a site-specific topical antinociception to thermal stimulation (Dogrul et al. 2003; Johanek and Simone 2004; Ko and Woods 1999; Yesilyurt et al. 2003). This local antinociceptive effect synergizes with spinal cannabinoid antinociception, as reflected by a 15-fold leftward shift in the dose–response curve (Dogrul et al. 2003), and also synergizes with topical morphine antinociception (Yesilyurt et al. 2003). The latter effects were blocked by a CB1R antagonist (Yesilyurt et al. 2003).

Peripheral CB1R Modulation of Formalin-Evoked Nocifensive Behavior

Intraplantar administration of formalin induces a biphasic pain response that is characterized by an early acute period (phase 1), a brief quiescent period, and a second phase of sustained "tonic" pain behavior (phase 2). The early phase reflects formalin-activation of Aβ, Aδ, and C-primary afferent fibers (McCall et al. 1996; Puig and Sorkin 1996). The late phase also activates Aδ and C fibers not activated during phase 1 (Puig and Sorkin 1996) and involves inflammation and long-term changes in the central nervous system (Coderre and Melzack 1992). Intraplantar administration of exogenous anandamide produces antinociception in the formalin test (Calignano et al. 1998), an effect blocked by systemic administration of the CB1R antagonist SR141716A. Anandamide produced antinociception only during phase 1, which likely reflects the short duration of action of anandamide, as the metabolically stable analog methanandamide suppressed pain behavior during both phase 1 and 2 (Calignano et al. 1998).

Peripheral CB1R Modulation of Capsaicin-Evoked Hyperalgesia

Intradermal administration of capsaicin to rats or humans induces hyperalgesia. Primary hyperalgesia, especially that elicited by noxious thermal stimulation, is mediated partly by sensitization of C-polymodal nociceptors (Baumann et al. 1991; Kenins 1982; LaMotte et al. 1992; Simone et al. 1987; Szolcsanyi et al. 1988; Torebjork et al. 1992). Secondary hyperalgesia is elicited in surrounding uninjured tissue and involves central nervous system sensitization rather than sensitization of peripheral nociceptors (Baumann et al. 1991; LaMotte et al. 1992; LaMotte et al. 1991; Simone et al. 1989) and requires conduction in primary afferent A fibers (Torebjork et al. 1992). Systemic administration of the mixed CB_1/CB_2 agonist WIN55,212-2, but not its receptor-inactive enantiomer, suppresses capsaicin-evoked thermal and mechanical hyperalgesia and nocifensive behavior (Li et al. 1999), demonstrating that the actions of WIN55,212-2 were receptor-mediated. A peripheral CB1R mechanism is implicated in the attenuation of capsaicin-evoked heat hyperalgesia by locally administered cannabinoids in nonhuman primates (Ko and

Woods 1999). Topical administration of the cannabinoid agonist HU210 to human skin also suppresses capsaicin-evoked thermal hyperalgesia and touch-evoked allodynia (Rukwied et al. 2003), although pharmacological specificity has not been assessed. Cannabinoid modulation of capsaicin-evoked hyperalgesia involves peripheral and central mechanisms. A CB1R mechanism is also implicated in the attenuation of hyperalgesia induced by locally administered cannabinoids following intradermal capsaicin (Johanek et al. 2001) or cutaneous heat injury (Johanek and Simone 2004).

The efficacy of peripheral cannabinoid mechanisms in suppressing neuronal activation evoked by corneal application of the small-fiber excitant mustard oil has been documented at the level of the lower brainstem. Corneal nociceptor activity, assessed using mustard oil-evoked Fos protein expression at the trigeminal interpolaris/caudalis (Vi/Vc) transition, was suppressed by direct corneal application of WIN55,212-2, and these effects were blocked by systemic administration of SR141716A (Bereiter et al. 2002), but CB2R mechanisms were not assessed. These suppressions occurred in the absence of changes in Fos at the subnucleus caudalis junction, thereby suggesting a role for CB1R mechanisms, at least in part, in regulating reflexive aspects of nociception and/or contributing to homeostasis of the anterior eye. More work is necessary to determine if CB2R mechanisms are implicated in regulation of corneal nociceptor activity.

Peripheral CB1R Modulation of Capsaicin-Evoked Neuropeptide Release

Anandamide suppressed capsaicin-evoked plasma extravasation in vivo through a peripheral CB1R mechanism (Richardson et al. 1998c) and inhibits capsaicin-evoked CGRP release in rat dorsal horn (Richardson et al. 1998a) and peripheral paw skin in vitro (Richardson et al. 1998c). Although pharmacological specificity was not assessed in the in vitro superfusion studies, these effects occurred at low concentrations [100 nM; (Richardson et al. 1998c)], consistent with mediation by CB1R.

Capsaicin-evoked CGRP release is enhanced in paw skin derived from rats with diabetic neuropathy induced by streptozotocin (Ellington et al. 2002). The mixed CB_1/CB_2 agonist CP55,940 attenuated capsaicin-evoked CGRP release in diabetic and nondiabetic animals, and these effects were blocked by a CB1R but not a CB2R antagonist (Ellington et al. 2002). Interestingly, anandamide inhibited capsaicin-evoked CGRP release in nondiabetic but not in diabetic rat skin, but neither the CB1R nor the CB2R antagonist attenuated these effects. Functional changes following diabetic neuropathy may have prevented these inhibitory effects of anandamide on capsaicin-evoked CGRP release. Anandamide also increased capsaicin-evoked CGRP release at high concentrations, possibly through a TRPV1 mechanism, although susceptibility to blockade by TRPV1 antagonists would be required to establish pharmacological specificity. Anandamide also inhibits in vivo release of CGRP and somatostatin induced by systemically administered resiniferatoxin, a potent TRPV1 ligand; the inhibitory effects of anandamide on plasma neuropeptide levels were blocked by a CB1R antagonist (Helyes et al. 2003).

Peripheral CB1R Modulation of Nerve Injury-Induced Nociception

A role for CB1Rs in suppressing hyperalgesia and allodynia induced by nerve injury has been demonstrated in multiple models of neuropathic pain (Bridges et al. 2001; Fox et al. 2001; Herzberg et al. 1997; Mao et al. 2000). Fox and colleagues demonstrated that intraplantar administration of WIN55,212-2 suppresses mechanical hyperalgesia following partial ligation of the sciatic nerve; these effects were blocked by the CB1R antagonist/inverse agonist SR141716A administered systemically (Fox et al. 2001). These data suggest a peripheral CB1R action in neuropathic pain, although CB2R mechanisms were not assessed. WIN55,212-2 also suppresses thermal hyperalgesia as well as mechanical and cold allodynia following spinal nerve ligation (Bridges et al. 2001). These latter effects were blocked by systemic administration of a CB1R but not a CB2R antagonist (Bridges et al. 2001), suggesting that the antihyperalgesic effects of systemically administered WIN55,212-2 were mediated by CB1R (Bridges et al. 2001; Herzberg et al. 1997).

3.3 Antinociception Mediated by CB1R in Spinal Cord

3.3.1 Distribution of CBRs on Central Terminals of Primary Afferents

Receptors are typically bidirectionally transported from the soma to central and peripheral terminals (Young et al. 1980). To identify afferents likely to contain CBRs, Hohmann and Herkenham assessed their pre- and postsynaptic distributions in the spinal cord using receptor binding and quantitative autoradiography (Hohmann et al. 1999a; Hohmann and Herkenham 1998). Destruction of sensory C fibers with neonatal capsaicin treatment produced only modest (16%) decreases in cannabinoid binding sites in the superficial dorsal horn, as measured by receptor binding and quantitative autoradiography (Hohmann and Herkenham 1998). These data suggest that a majority of spinal CBRs is not localized to central terminals of primary afferent C fibers. Multisegment unilateral cervical dorsal rhizotomy (C3-T1 or T2) produced time-dependent losses in cannabinoid binding densities in the dorsal horn (Hohmann et al. 1999a) of larger magnitude than that induced by neonatal capsaicin treatment. This observation is unsurprising because rhizotomy destroys the central terminals of both small- and large-diameter fibers. Rhizotomy suppressed [^{3}H]CP55,940 binding in the superficial and neck region of the dorsal horn as well as in the nucleus proprius without affecting binding in lamina X or the ventral horn. By contrast, massive losses in μ-opioid binding sites were observed in lamina I and II in adjacent sections following either neonatal capsaicin or rhizotomy (Hohmann et al. 1999a; Hohmann and Herkenham 1998), consistent with previous reports (Besse et al. 1990; Nagy et al. 1980). These data support the conclusion that CB1Rs occur both pre- and postsynaptically in the spinal dorsal horn, with the majority of receptors occurring postsynaptically. This conclusion is consistent with the observation of CB1R-immunoreactive fibers in dorsal roots (Sanudo-Pena et al. 1999) and in axons of Lissauer's tract (Salio et al. 2002b), and immunocytochemical studies showing that CB1R and vanilloid receptor (TRPV1) immunostaining is reduced in parallel in the superficial dorsal horn following neonatal capsaicin treatment (Morisset et al. 2001). Of course, postsynaptic changes that occur subsequent to an extensive rhizotomy (Hohmann et al. 1999a) can also contribute to the pattern of receptor changes observed.

By contrast, Farquhar-Smith and colleagues, using an antibody directed against the C-terminal of CB1R, demonstrated that lumbar dorsal rhizotomy induced a minor, though significant, reduction in CB1R immunoreactivity (Farquhar-Smith et al. 2000). Consistent with these observations, CB1R immunoreactivity in the superficial dorsal horn showed a laminar overlap with markers of thin primary afferents, as identified by immunoreactivity for CGRP, substance P, isolectin B4 (IB4), and TRPV1, but very little colocalization of CB_1 was observed with any of these markers at the single-fiber level (Farquhar-Smith et al. 2000). Similarly, minimal colocalization of CB1Rs was observed with these markers in dorsal root ganglion cells, using the same antibody (Bridges et al. 2003). These data collectively suggest that the majority of CB1Rs are not localized to central terminals of nociceptive primary afferents, but rather are localized on postsynaptic sites, and provide indirect support for the hypothesis that CB1Rs in spinal cord are localized predominantly to fibers of intrinsic spinal neurons.

3.3.2
Distribution of CB1R mRNA and CB1R Immunoreactivity in Spinal Cord

The presence of CB1R mRNA in rat dorsal horn has been reported (Mailleux and Vanderhaeghen 1992). Hohmann (2002) characterized the laminar distribution of CB1R mRNA-expressing cells in rat lumbar spinal cord using a highly sensitive cRNA probe. CB1R mRNA was found in all spinal laminae except lamina IX; motoneurons in this region, which are immunoreactive for fatty acid amide hydrolase (FAAH) (Tsou et al. 1998b), were CB1R mRNA negative. Expression was dense in lamina X and sparsest in III and IV. CB1R mRNA was highly expressed in lamina V and VI and the medial part of IV. These laminae contained many large cells with high levels of expression. In general, primary afferents that project to deeper parts of the dorsal horn (here III–VI) include coarser caliber fibers than those projecting to the superficial laminae, although small diameter fibers are also observed. Small-diameter fibers from viscera also project to lamina V, VII, and X (see Grant 1995 for review). By contrast, the superficial dorsal horn (lamina I and II) had many small cells with low levels of expression compared to cells observed in deeper lamina. Lamina I and II neurons receive inputs from unmyelinated as well as finely myelinated primary afferents (see Grant 1995 for review). Thus, in situ hybridization studies demonstrate that spinal neurons synthesize CB1Rs, although they do not address putative localization of these receptors to spinal interneurons and/or terminals of supraspinally projecting efferents.

Immunocytochemical studies have provided information about the cellular elements expressing CB1Rs in the spinal cord. The C-terminal antibody employed by Farquhar-Smith et al. (2000) exclusively labeled fibers and terminals, whereas the antibody employed by Tsou and colleagues (Sanudo-Pena et al. 1999) additionally labeled cell bodies. Tsou and colleagues, using an antibody raised against the first 77 residues of the N terminus of CB1R, identified beaded immunoreactive fibers throughout the spinal dorsal horn and in lamina X surrounding the central canal (Tsou et al. 1998a). Further work by this group also revealed the presence of lightly stained cells throughout the spinal cord gray matter (Sanudo-Pena et al. 1999). Farquhar-Smith and colleagues, using an antibody directed against the C-terminal 13 amino acids of CB1R, demonstrated immunoreactivity for CB1Rs in fibers and terminals with no consistent immunoreactivity observed in any cell bodies (Farquhar-Smith et al. 2000).

3.3.3
Evidence for CB1Rs on Spinal Interneurons

There is considerable support for localization of CBRs in rat spinal cord postsynaptic to primary afferents at both light and electron microscope levels. Direct evidence for postsynaptic localization of CB_1 in spinal dorsal horn is derived from the observation that intrinsic excitatory interneurons in lamina IIi that expressed protein kinase C isoform γ showed high levels of colocalization with CB_1 (Farquhar-Smith et al. 2000); this pattern may suggest an anatomical basis for the efficacy of cannabinoids in ameliorating inflammatory and neuropathic pain (Bridges et al. 2001; Fox et al. 2001; Herzberg et al. 1997; Malmberg et al. 1997; Mao et al. 2000).

CB1R immunoreactivity has also been localized to dorsal horn interneurons containing γ-aminobutyric acid (GABA) (Salio et al. 2002b). GABA presynaptically inhibits primary afferent input to the spinal cord. The observation of GABAergic dendrites postsynaptic to primary afferents also suggests that primary afferents are anatomically positioned to activate GABAergic inhibitory circuits. GABAergic interneurons can also synapse directly on dorsal horn neurons to reduce excitatory input. The demonstrated colocalization of CB1R with GABA is consistent with functional studies demonstrating a CB1R-mediated presynaptic inhibition of GABAergic and glycinergic transmission in recordings performed in rat medullary dorsal horn in vitro (Jennings et al. 2001). By contrast, postsynaptic effects on medullary substantia gelatinosa neurons were not observed (Jennings et al. 2001). These data suggest that cannabinoids act through a disinhibitory action on lamina II neurons by inhibiting GABAergic transmission.

Immunoreactivity for CB1R and μ-opioid receptors (MOR) is also colocalized on lamina II interneurons at the ultrastructural level (Salio et al. 2001). In this work, CB1R was predominantly localized postsynaptically in dendrites and cell bodies, but immunoreactive axons and axon terminals were also observed (Salio et al. 2001). Both species showed rare labeling of the plasma membrane. Since MOR1 is not colocalized with GABA (Gong et al. 1997; Kemp et al. 1996), these data support the presence of CB1R in distinct populations of intrinsic spinal neurons (Salio et al. 2001). By contrast, colocalization of CB1R with MOR1 in thin primary afferent terminals could not be convincingly demonstrated in this work (Salio et al. 2001).

3.3.4
Evidence for CB1Rs on Afferents Originating Supraspinally

CB1R immunoreactivity is highly expressed at all spinal levels in fibers of the dorsolateral funiculus (DLF) and in the intermediolateral nucleus (Farquhar-Smith et al. 2000). Interruption of descending pathways (and ascending pathways from lamina I) that course in the DLF produced only a 5% change in CB1R immunoreactivity (Farquhar-Smith et al. 2000). These data suggest that CB1R immunoreactivity, in general, is not localized on terminals of neurons originating supraspinally and suggest localization of CB1R to intrinsic spinal neurons and/or ascending projections (Farquhar-Smith et al. 2000). Because visceral primary afferents project to the nucleus of the DLF, CB1Rs are appropriately positioned to influence visceral afferent input as well as viscero-somatic integration (Farquhar-Smith et al. 2000). These observations are consistent with cannabinoid modulation of visceral hyperalgesia (see Hohmann 2002 for review). Ascending projections to the brainstem, hypothalamus, and thalamus have been shown to originate in lamina X (Molander and Grant 1995). The presence of CB1R immunoreactivity in lamina X and in the intermediolateral nucleus may also reflect interaction of CB1R with neurons of the sympathetic nervous system (Farquhar-Smith et al. 2000).

3.3.5
Evidence for CB1Rs on Nonneuronal Cells at the Spinal Level

CB1R has recently been demonstrated in astrocytes in laminae I and II of the spinal dorsal horn using multiple antibodies directed against the C-terminal tail of CB_1 (Salio et al. 2002a). By contrast, astrocytes were not labeled in rat spinal cord when an N-terminal-specific anti-CB1R antibody was employed (Salio et al. 2002b). The functional roles of putative CB1R subtypes in spinal glial cells require further investigation (Salio et al. 2002a).

3.3.6
Antinociceptive and Electrophysiological Effects of Spinally Administered Cannabinoids

Antinociceptive effects of cannabinoids are mediated, in part, at the spinal level. Spinal reflexive responses to noxious stimuli are inhibited by cannabinoids in spinally transected dogs (Gilbert 1981). Support for spinal mechanisms of cannabinoid analgesic action is also derived from the ability of intrathecally administered cannabinoids to produce antinociception (Smith and Martin 1992; Welch et al. 1995; Yaksh 1981). The behavioral data are consistent with the ability of spinally administered cannabinoids to suppress noxious heat-evoked and after-discharge firing (Hohmann et al. 1998) and noxious stimulus-evoked Fos protein expression in the spinal dorsal horn neurons (Hohmann et al. 1999c). Spinal administration of a CB1R-selective agonist also inhibits C fiber and Aδ fiber-evoked responses of wide dynamic range (WDR) neurons through a CB1R mechanism with only minor effects on A-β fiber-evoked responses (Kelly and Chapman 2001). Systemic and intrathecally administered cannabinoids retain a weak but long-lasting antinociceptive effect in spinally transected rats (Lichtman and Martin 1991b; Smith and Martin 1992), providing compelling evidence for spinal mechanisms of cannabinoid antinociception.

Spinal administration of a cannabinoid (HU210) also suppresses C fiber-mediated post-discharge responses, a measure of neuronal hyperexcitability, in carrageenan-inflamed and noninflamed rats (Drew et al. 2000); these effects were blocked by a CB1R antagonist. Spinal administration of anandamide produced CB1R-mediated effects in carrageenan-inflamed rats that were similar to that reported for HU210, but only inconsistent effects were observed in noninflamed rats (Harris et al. 2000). Upregulation of CB1Rs is also observed in the spinal cord following nerve injury, suggesting that regulation of spinal CB1Rs may contribute to the therapeutic efficacy of cannabinoids in neuropathic pain states (Lim et al. 2003). These data implicate involvement of spinal CB1Rs in both acute and persistent pain states.

3.4 Antinociception Mediated by CB1Rs in Supraspinal Pain Circuits

Support for supraspinal sites of cannabinoid antinociceptive action is derived from the antinociceptive effects of cannabinoids following intracerebroventricular administration (Hohmann et al. 1999b; Martin et al. 1993) and the attenuation of cannabinoid antinociception following disruption of communication between brain and spinal cord. Both the antinociceptive (Lichtman and Martin 1991b) and electrophysiological (Hohmann et al. 1999b) effects of systemically administered cannabinoids are attenuated following spinal transection, suggesting the involvement of supraspinal sites of cannabinoid analgesic action. Intrathecal administration of the α_2 antagonist yohimbine but not the serotonin antagonist methysergide also blocks the antinociceptive effect of systemically administered Δ^9-tetrahydrocannabinol (Δ^9-THC) (Lichtman and Martin 1991a). Furthermore, the antinociceptive efficacy of systemically administered cannabinoids is markedly attenuated following neurotoxic destruction of descending noradrenergic projections to the spinal cord (Gutierrez et al. 2003). These data collectively implicate a role for descending noradrenergic systems in cannabinoid antinociceptive mechanisms.

Direct evidence for supraspinal sites of cannabinoid antinociception is derived from studies employing intracranial administration of cannabinoids. Site-specific injections of cannabinoid agonists to various brain regions have permitted the identification of brain loci implicated in cannabinoid antinociception. The active sites included the dorsolateral periaqueductal gray, dorsal raphe nucleus, RVM, amygdala, lateral posterior and submedius regions of the thalamus, superior colliculus, and noradrenergic A5 region (Martin et al. 1995, 1998, 1999a). These studies suggest that endocannabinoid actions at these sites are sufficient to produce antinociception.

3.4.1 Role of the Periaqueductal Gray

Studies of metabolically stable anandamide analogs and the effects of anandamide in FAAH knockout mice lead to the conclusion that anandamide would produce antinociceptive effects upon release in the appropriate brain, spinal, or peripheral sites. Electrical stimulation of the dorsal aspect of the periaqueductal gray (PAG) caused CB1R-mediated analgesia evidenced by a markedly reduced effect following administration of SR141716A (Walker et al. 1999). This work suggested that the dorsal PAG serves as a substrate for cannabinoid antinociception. Exogenously applied cannabinoids have been shown to inhibit GABAergic and glutamatergic neurons in rat PAG neurons through presynaptic mechanisms (Vaughan et al. 2000). These effects occurred in the absence of direct postsynaptic actions on PAG neurons, thus providing a neurophysiological basis for cannabinoid modulation of nociceptive transmission through presynaptic actions.

Metabotropic glutamate and *N*-methyl-D-aspartate (NMDA) receptors are required for cannabinoid antinociception at the level of the PAG. Infusion of the CBR agonist WIN55,212-2 into the PAG produced dose-dependent increases in paw withdrawal latencies to a noxious thermal stimulus (Palazzo et al. 2001). This effect was blocked by pretreatment with SR141716A. Blockade of $mGlu_5$ metabotropic glutamate receptors but not $mGlu_1$ receptors blocked the effects of WIN55,212-2. Both $mGlu_5$ and $mGlu_1$ receptors belong to group I class of metabotropic glutamate receptors that are G protein-coupled and positively coupled to phospholipase C. Pretreatment with antagonists for group II (which includes $mGlu_2$ and $mGlu_3$) and group III (which includes $mGlu_4$, $mGlu_6$, $mGlu_7$, and $mGlu_8$) metabotropic glutamate receptors also suppressed WIN55,212-2-induced analgesia. This latter class of receptors is negatively coupled to adenylate cyclase and preferentially localized to presynaptic active zones associated with autoreceptors. In addition to these metabotropic receptors, a selective antagonist for ionotropic glutamate (NMDA) receptors also blocked the antinociceptive effects of WIN55,212-2.

It has been postulated that the effect of antagonism of group II and III metabotropic receptors on cannabinoid antinociception is attributable to an increased release of GABA in the PAG (Palazzo et al. 2001). Because GABAergic interneurons within the PAG tonically inhibit descending antinociceptive pathways (Moreau and Fields 1986), an inhibition of PAG descending pathways may underlie the observed blockade of cannabinoid antinociception through modulation of GABAergic interneurons. In vitro studies demonstrate that cannabinoids inhibit GABA and glutamate release presynaptically in the PAG in the absence of direct postsynaptic effects on PAG neurons (Vaughan et al. 2000). By contrast, antagonists for $mGlu_5$ and NMDA, which are localized postsynaptically, could reduce the tonic excitatory control of glutamate on descending antinociceptive pathways with cells of origin in the PAG (Palazzo et al. 2001), thereby modulating cannabinoid antinociception through a distinct mechanism.

3.4.2 Role of Rostral Ventral Medulla

Researchers have targeted synthetic cannabinoids at other brainstem nuclei including the RVM (Martin et al. 1998; Monhemius et al. 2001; Vaughan et al. 1999) and the nucleus reticularis gigantocellularis (Monhemius et al. 2001) to better characterize sites of cannabinoid-mediated antinociception. Site-specific administration of cannabinoids (WIN55,212-2 and HU210) in the RVM produced significant antinociception in the tail-flick test (Martin et al. 1998). Mediation by CBRs was established because the antinociceptive effects of HU210 were blocked by the CB1R antagonist SR141716A, and the receptor-inactive enantiomer WIN55,212-3 failed to induce antinociception following microinjection to the same site (Martin et al. 1998).

Electrophysiological studies have provided insight into the mechanisms mediating these antinociceptive effects. Cannabinoids modulate on- and off-cells in the RVM (Meng et al. 1998), demonstrating their ability to control descending pain modulatory signaling in a manner similar to that of morphine. Pharmacological inactivation of the RVM with site-specific administration of the $GABA_A$ receptor agonist muscimol blocked the antinociceptive effects but not the motor deficits of systemically administered WIN55,212-2 (Meng et al. 1998). At the cellular level, it appears that cannabinoids exert their physiological effects in the RVM by presynaptic inhibition of GABAergic neurotransmission (Vaughan et al. 1999).

3.4.3 Role of the Basolateral Amygdala

The amygdala is a nuclear complex located in the limbic forebrain that plays a key role in the coordination of fear and defensive reactions. The amygdala is optimally positioned anatomically to receive and integrate sensory information from multiple modalities and, in turn, to mediate emotional, autonomic, and somatic motor reactions to salient stimuli (especially threatening stimuli) (Davis and Whalen 2001). Within the amygdala, CB1R immunoreactivity has been detected in a subset of GABAergic interneurons in the basolateral complex (Marsicano et al. 2002), a site implicated in the formation and storage of aversive memories (Medina et al. 2002). Endocannabinoids are elevated in the basolateral amygdala in a conditioned fear-aversion paradigm (Marsicano et al. 2002), supporting the hypothesis that endocannabinoids serve naturally to inhibit extinction of aversive memories. Presentation of the conditioned aversive stimulus during extinction trials elicited elevated levels of the endocannabinoids 2-arachidonoylglycerol (2-AG) and anandamide in the basolateral nucleus of the amygdala but not the medial prefrontal cortex (another brain area implicated in memory formation) of mice. Marsicano et al. (2002) reported that endocannabinoids and CB1Rs in the basolateral nucleus of the amygdala are crucial to the long-term depression of GABAergic inhibitory currents, positing that endocannabinoids regulate aversive memory extinction via selective inhibition of local inhibitory networks in the amygdala.

The amygdala also plays a critical role in modulating antinociception. Microinjection of cannabinoids into the basolateral nucleus of the amygdala produces antinociception in the tail-flick test (Martin et al. 1999a). Microinjection of μ-opioid agonists into the basolateral nucleus of the amygdala similarly results in marked antinociceptive responding in the radiant heat tail-flick (Helmstetter et al. 1993, 1995) and formalin tests (Manning and Mayer 1995). Moreover, bilateral lesions of the amygdala rendered nonhuman primates less sensitive to the antinociceptive effects of the potent synthetic cannabinoid WIN55,212-2 (Manning et al. 2001). In rodents, microinjection of the $GABA_A$ agonist muscimol into the central nucleus of the amygdala but not into the basolateral nucleus of the amygdala, reduced the antinociceptive effects of systemic WIN55,212-2 (Manning et al. 2003). Moreover, the endocannabinoid-degrading enzyme FAAH is localized in the basolateral and lateral amygdala (Egertová et al. 2003; Tsou et al. 1998b). These data indicate that a mechanism exists for inactivation of endocannabinoid actions in the basolateral amygdala. Both conditioned (Helmstetter 1992; Helmstetter and Bellgowan 1993) and unconditioned (Bellgowan and Helmstetter 1996) stress-induced analgesia depend on intact functioning of the amygdala. These observations, together with the demonstration of cannabinoid-mediated antinociceptive effects following site-specific administration to the basolateral nucleus of the amygdala (Martin et al. 1999a), suggest that endocannabinoids may serve naturally to suppress environmentally induced pain by actions in the amygdala.

4 Antinociception Mediated by CB2Rs

In clinical trials of THC and other cannabinoid agonists for pain pharmacotherapy, unwanted, negative psychotropic effects limit dosing to levels that are probably below those producing maximal analgesic efficacy. These effects are caused by actions of the compounds at CB1Rs in the brain. However, CB2Rs are either absent or expressed in low levels by neural tissues (Munro et al. 1993; Zimmer et al. 1999). This distribution has led to evaluation and validation of CB2Rs as targets for novel pharmacotherapies for pain.

4.1 Localization of CB2Rs that Contribute to Cannabinoid Antinociception

CB2Rs are expressed by cells that are involved in inflammation and thereby pain. Among them are monocytes, polymorphonuclear neutrophils, mast cells, B cells, T cells, and natural killer cells (see Cabral and Staab, this volume). CB2Rs are also found on microglia (Walter et al. 2003), which play an important role in pathological pain states (Zhang et al. 2003). Recent pharmacological evidence also supports the presence of CB2Rs in human and guinea pig vagus nerve (Patel et al. 2003). CB2R immunoreactivity has been detected in dorsal root ganglion cells (Ross et al. 2001a) in cultures derived from neonatal rats. More work is necessary to identify the phenotypes of cells expressing CB2Rs, especially since levels of CB2R mRNA in neurons of dorsal root (Hohmann and Herkenham 1999a) and trigeminal (Price et al. 2003) ganglia are near background under conditions in which CB_1 mRNA is clearly demonstrated.

4.2
CB2R-Mediated Antinociceptive Effects

CB2R agonists are antinociceptive in models of acute (Malan et al. 2001) and persistent pain (Clayton et al. 2002; Hanus et al. 1999; Hohmann et al. 2004; Ibrahim et al. 2003; Nackley et al. 2003a). Direct evidence of CB2R-mediated antinociceptive effects was reported by Hanus et al. (1999) using HU-308, a highly selective CB2R agonist ($K_i = 22.7$ CB2R vs >10 μM CB1R). They found that HU-308 (50 mg/kg) produced marked decreases in pain behavior in rats receiving hindpaw injections of dilute formalin. This effect occurred without any change in motor function, a centrally mediated effect of CB1R agonists that may predict psychoactivity in humans. HU-308 also reduced the swelling produced by arachidonic acid. The CB2R-selective cannabinoid antagonist SR144528 blocked these effects. Another CB2R agonist, AM1241, has also been shown to induce a CB2R-mediated antinociceptive effect in otherwise untreated rats while failing to elicit centrally mediated side effects such as hypothermia, catalepsy, and hypoactivity (Malan et al. 2001). AM1241 also induces CB2R-mediated suppression of carrageenan and capsaicin-evoked thermal and mechanical hyperalgesia and allodynia (Hohmann et al. 2004; Nackley et al. 2003b; Quartilho et al. 2003) and suppresses carrageenan-evoked Fos protein expression (Nackley et al. 2003a). These effects were blocked by the CB2R-selective antagonist but not by a CB1R-selective antagonist.

Electrophysiological studies also support a role for CB2Rs in suppressing nociception. AM1241 induced CB2R-mediated suppression of C fiber-evoked responses and windup in spinal WDR neurons; this suppression was observed in both the absence and presence of carrageenan inflammation and following local and systemic drug administration (Nackley et al. 2004). The suppressive effects of AM1241 were more pronounced in the presence compared to the absence of inflammation. By contrast, low threshold, purely non-nociceptive spinal neurons did not show sensitization during the development of inflammation and were not altered by AM1241 actions in the periphery (Nackley et al. 2004). Intraplantar administration of anandamide also suppresses mechanically evoked responses in spinal dorsal horn neurons in the carrageenan model of inflammation; these effects were blocked by a CB2R-selective antagonist (Sokal et al. 2003). These data demonstrate that activation of peripheral cannabinoid CB2Rs is sufficient to suppress neuronal activity at central levels of processing in the spinal dorsal horn. Sensory hypersensitivity in animals with nerve injury was also reduced by a CB2R agonist (Ibrahim et al. 2003). In light of the induction of CB2Rs in the spinal dorsal horn by neuropathic pain states, coincident with the appearance of activated microglia, it appears likely that these latter effects are mediated, at least in part, by nonneuronal cells.

The main effect of inflammatory cells in nociception is to sensitize neurons. This occurs in the periphery when the immune response stimulates peripheral cells to secrete mediators that sensitize primary afferent neurons. Substances released by immune cells that sensitize nociceptors include histamine, serotonin, eicosanoids, interleukin 1, tumor necrosis factor-α, and nerve growth factor (Dray 1995; McMahon 1996; Tracey and Walker 1995). Sensitization also occurs in the CNS, and centrally located microglia, which express CB2Rs, may be involved in the sensitization of central nociceptive neurons during inflammation (reviewed by DeLeo et al. 2004).

CB2R agonists reduce the secretion of inflammatory mediators from immune cells. For example, cannabinoids inhibit lipopolysaccharide (LPS)-inducible cytokine mRNA expression in rat microglial cells (Puffenbarger et al. 2000) and cytotoxicity and release of inflammatory mediators from monocytic cells (Klegeris et al. 2003). Activation of CB2Rs localized to mast cells or other immune cells also attenuates the release of inflammatory mediators, including nerve growth factor (Rice et al. 2002) and cytokines (Klegeris et al. 2003) that in turn sensitize nociceptors (Mazzari et al. 1996). In the presence of inflammation, CB2R agonists could thus act locally on immune cells in the periphery and suppress C fiber sensitization. These observations suggest that the effects of CB2R ligands occur via the decreased release of inflammatory mediators from peripheral immune cells in the periphery and microglia in the CNS. However, CB2R modulation of immune responses does not readily account for the effects of AM1241 on windup and C fiber responses in the absence of inflammation and local antinociceptive effects of this compound that are observed in otherwise untreated rats (Malan et al. 2001). Direct effects on CB2Rs localized to primary afferents (Griffin et al. 1997; Patel et al. 2003; Ross et al. 2001a; see also Hohmann and Herkenham 1999a; Price et al. 2003) could provide a parsimonious explanation for the antinociceptive and electrophysiological actions of CB2R agonists observed in the absence of inflammation. Malan's group has recently identified a potential mechanism of action for AM1241; AM1241 is likely to suppress primary afferent activation indirectly by stimulating local release of β-endorphin in peripheral tissue through a CB2R-specific mechanism (Malan et al. 2004).

Besides suggesting a novel pharmacotherapy for pain, these findings suggest that CB2R activation by endocannabinoids would promote anti-inflammatory and antinociceptive effects, some of which may be mediated by non-neuronal cells in the CNS.

5 Pain Modulation by Endocannabinoids

Seven putative endocannabinoids have been identified: (1) anandamide, (2) dihomo-γ-linolenoylethanolamide (HEA), (3) docosatetraenoylethanolamide (DEA), (4) 2-AG, (5) noladin ether, (6) virodhamine, and (7) *N*-arachidonoyldopamine (NADA). The roles of these novel putative endogenous compounds in pain and inflammation have been a recent focus of investigations. The sections above, which described the relationship between pain circuits, exogenous drugs, and CBRs provide a foundation for understanding how these putative endocannabinoids may operate physiologically to modify pain perception. Proving that a particular endocannabinoid plays such a role requires first the demonstration that it can produce antinociception within the proposed site of action, then the demonstration that it is formed and released in the proposed site under conditions where pain sensitivity is altered. In the following, we review the data for each endocannabinoid with these criteria in mind.

5.1 Anandamide

Anandamide was the first putative endocannabinoid to be identified (Devane et al. 1992) and has therefore been the focus of the majority of investigations of endocannabinoid mechanisms of pain suppression.

5.1.1 Effects of Exogenous Anandamide on Pain Sensitivity

In studies of physiological pain (i.e., pain induced by noxious stimuli in animals free of inflammation, nerve injury, or other pathology), anandamide typically produced antinociceptive effects, but these effects were not blocked by cannabinoid antagonists (Adams et al. 1998; Vivian et al. 1998). This effect was likely due to the rapid metabolism of anandamide by FAAH, since FAAH knockout mice exhibit marked CB1R-mediated analgesic responses to anandamide (Cravatt et al. 2001). However, in animals with nerve injury, at doses of 10 and 100 μg i.v., anandamide reversed neuropathic mechanical hyperalgesia, and this effect was antagonized by the CB1R and CB2R antagonists SR141716A and SR144528.

These findings above are in good agreement with electrophysiological and neurochemical studies of the effects of anandamide on sensory neurons. In 64% of neurons examined, anandamide (10 μM) depressed Aδ fiber-evoked excitatory postsynaptic currents (EPSCs) (Luo et al. 2002). By contrast, an inhibitory action of anandamide on C fiber-evoked EPSCs was observed in only 31% of neurons tested. Anandamide also inhibited the release of neuropeptides evoked by a TRPV1 agonist (Helyes et al. 2003). These findings are consistent with studies of the localization of CBRs (see Sect. 3.2.1) and suggest that anandamide acts primarily on larger caliber peripheral afferent fibers and cells (see Sect. 3.2.1).

5.1.2 Effects of Inhibition of the Putative Anandamide Transporter

Another approach to examining the role of endogenous anandamide in pain has been to employ transport inhibitors such as AM404. Blocking transport would be expected to block the reuptake of anandamide and cause increased levels to occur in the vicinity of CBRs, with both processes leading to increased occupation of CBRs. Beltramo et al. (1997) showed that administration of AM404 caused the accumulation of anandamide in cultures of cortical neurons and enhanced the hot-plate analgesia produced by systemically administered anandamide. AM404 alone did not alter pain sensitivity, suggesting that anandamide does not act tonically to maintain pain thresholds for thermal stimuli. The paper did not address whether environmentally produced analgesia was affected by AM404 (but see Hohmann et al. 2001).

5.1.3
Modulation of Pain by Endogenous Anandamide

Anandamide appears to participate in endogenous pain modulation by actions in the PAG. Blocking the CB1R with the antagonist SR141716A produced hyperalgesia in the formalin test (Calignano et al. 1998; Strangman et al. 1998) and prevented the analgesia produced by electrical stimulation of the dorsolateral PAG (Walker et al. 1999). These pro-nociceptive actions of the antagonist are reasonable evidence for an antinociceptive action of one or more endocannabinoids, but conclusions along this line are limited by the possible confound with the proposed inverse-agonist activity of current CBR antagonists (Landsman et al. 1997). In order to address directly the questions regarding the role of endocannabinoids that were made inferentially from the actions of an antagonist, the release of anandamide in the PAG was studied using microdialysis (Walker et al. 1999). This method permits collection of neurotransmitters/modulators from the extracellular space, and is therefore an indicator of the release of these modulators. Microdialysis coupled with liquid chromatography/mass spectrometry established that the analgesia producing electrical stimulation or injections of the chemical irritant formalin into the hindpaws of anesthetized rats induced the release of anandamide in the PAG. Thus, it appears that either pain itself, or electrical stimulation leads to the release of anandamide, which acts on CB1Rs in the PAG to inhibit nociception.

5.2
Dihomo-γ-Linolenoylethanolamide and Docosatetraenylethanolamide

HEA and DEA were reported together by Hanus et al. (1993) as cannabinoids similar in structure to anandamide but with different fatty acyl chains: 20:3 (n-6) and 22:4 (n-6) for HEA and DEA, respectively. As they have been studied together often and produce similar results, they are considered together here. Koga et al. (1997) verified the occurrence of these compounds as endogenous to a variety of mammalian tissues by using liquid chromatography/mass spectrometry. A recent study indicates that, along with anandamide, these two compounds are formed in astrocytes, suggestive of a potential role in inflammatory pain (Walter et al. 2002). These compounds possess binding affinities for CB1Rs that are similar to that of anandamide (Felder et al. 1993; Hanus et al. 1993; Vogel et al. 1994). They also inhibit forskolin-stimulated cyclic AMP (cAMP) and electrically evoked contractions of the mouse vas deferens with potencies similar to that of anandamide (Felder et al. 1993; Pertwee et al. 1994; Vogel et al. 1994). Piomelli et al. (1999) reported that HEA effectively competes against anandamide for the putative anandamide transporter. As with anandamide, DEA exhibits weak activity at the TRPV1 (Ross et al. 2001b). Taken together, the findings indicate that DEA and HEA are naturally occurring compounds in mammals and exhibit a pharmacology that is very similar to that of anandamide. Systemic administration of DEA and HEA causes analgesia to acute thermal stimulation in mice (Fride and Mechoulam 1993), and tolerance develops to this effect (Fride 1995). Whether this effect is CBR-mediated is currently unknown. More work with these poorly studied compounds is warranted.

5.3 2-Arachidonoylglycerol

2-AG was the second endocannabinoid to be identified (Mechoulam et al. 1995; Sugiura et al. 1995). Compared to anandamide, less is known as to what role it may play in pain modulation and whether its effects on nociceptive processing are indeed CB1R-mediated. Intravenous administration of 2-AG caused a suppression of pain behavior in the tail-flick test (Mechoulam et al. 1995). However, the investigators did not test whether the effects could be blocked by CBR antagonists. This leaves open the possibility that active non-CB metabolites may have produced the effect, as was apparently the case with anandamide discussed above. Ben-Shabat et al. (1998) showed that at doses of 2-AG that fail to produce analgesic effects in the hot plate test, the addition of two cannabinoid-inactive endogenous congeners of 2-AG, 2-lineoylglycerol and 2-palmitoylglycerol, caused significant analgesia. These effects were referred to as "entourage effects," a reference to the notion that endogenous mediators of similar structure are often released together and act in concert.

5.4 Noladin Ether

The novel endocannabinoid noladin ether was recently identified by Hanus et al. (2001). Subsequently, its existence in brain was reported by Fezza et al. (2002), but Oka et al. (2003) were unable to detect the compound in the brains of any of several mammalian species by gas chromatography/mass spectrometry. Noladin ether was reported to occur in relatively high amounts in dissected thalamus, but its localization to somatosensory areas of thalamus has not been established. It was reported to occur in much lower amounts in spinal cord (Fezza et al. 2002). Hanus et al. (2001) showed that the compound produces analgesic effects in the hot plate test following systemic administration in mice (20 mg/kg, i.p.). However, as with 2-AG, experiments have not been carried out to determine whether its effects were due to an action at CBRs. More work is needed to verify the formation of this compound in vivo and its potential role in pain modulation.

5.5 Virodhamine

O-Arachidonoylethanolamine was identified in rat brain and named virodhamine (Porter et al. 2002). This compound is similar to anandamide in being formed from arachidonic acid and ethanolamine, but virodhamine contains an ester linkage rather than anandamide's amide linkage. Like anandamide, it appears to act as a partial agonist. However, a microdialysis study suggested that while its tissue concentrations are similar to anandamide, it is released in much higher amounts. The existence of this compound has not been independently verified, and this author has been unable to detect it in rat brain extracts using ultrasensitive LC/MS/MS (liquid chromatography/tandem mass spectrometry) methods developed using the synthetic compound (J.M. Walker, unpublished observations). Additional confirmatory studies of the existence of virodhamine are needed upon which further study of its potential role in pain modulation would be warranted.

5.6 *N*-Arachidonoyldopamine

Another molecule with the arachidonic acid backbone, NADA was recently identified in rat and bovine brain (Huang et al. 2002). It activates CB1Rs and elicits cannabimimetic effects (which include analgesia following systemic administration but not tested with a cannabinoid antagonist) (Bisogno et al. 2000; De Petrocellis et al. 2000; Huang et al. 2002). NADA significantly inhibited innocuous (8, 10 g) mechanically evoked responses of dorsal horn neurons, and these effects were blocked by intraplantar injection of SR141716A (Sagar et al. 2004). In addition, NADA activates TRPV1 receptors and causes hyperalgesia when administered peripherally (Huang et al. 2002). This effect is in contrast to anandamide, which also activates TRPV1 (Smart et al. 2000; Zygmunt et al. 1999), though administration of anandamide typically causes analgesia. The distribution of endogenous NADA in various brain areas differs from that of anandamide, with the highest levels found in the striatum and hippocampus (Huang et al. 2002). It also occurs in the DRG in low levels. Given that NADA is capable of eliciting analgesia upon systemic administration and hyperalgesia upon intradermal injection, it is possible that endogenous NADA activates either TRPV1 or CB1Rs, depending upon location and circumstance.

5.7 Regulation of Endocannabinoids by Fatty Acid Amide Hydrolase

Three putative endogenous cannabinoids, anandamide, 2-AG, and NADA, appear to be susceptible to degradation by FAAH (Cravatt et al. 1996; Deutsch and Chin 1993; Di Marzo et al. 1998; Huang et al. 2002). Immunohistochemical studies show that FAAH is present in the ventral posterior lateral nucleus of the thalamus (Egertová et al. 1998, 2003; Tsou et al. 1998b), the termination zone of the spinothalamic tract. FAAH is also found in Lissauer's tract, which comprises primary afferent fibers entering the spinal cord, and in small neurons in the superficial dorsal horn, which is the termination zone of nociceptive primary afferents. These observations demonstrate that a mechanism capable of inactivating anandamide, 2-AG, and NADA is present in regions of the CNS related to nociceptive processing and thus suggest a role for these ligands in pain modulation. Of course, the presence of FAAH does not necessarily identify that cell as a site of synthesis of endocannabinoids, as FAAH is a catabolic enzyme and also metabolizes fatty acid amides that act through CBR-independent mechanisms.

5.7.1 Pain Sensitivity and Inflammatory Responses in FAAH Knockout Mice

Cravatt and colleagues (2001; Lichtman et al. 2004) developed transgenic mice lacking FAAH and observed in these mutants enhanced analgesic effects of exogenously administered anandamide (Fig. 4). These effects were reversed by the selective CB1R antagonist/inverse agonist SR141716A. Moreover, these animals exhibit tonic CB1R-mediated analgesia, apparently due to the decreased metabolism of FAAH-susceptible endocannabinoids. These findings support the hypothesis that endocannabinoids susceptible to hydrolysis by FAAH serve to naturally suppress pain sensitivity. The development of FAAH and CB1R knockouts and pharmacological approaches employing subtype selective antagonists or antisense knockdown have been used to evaluate a role of endocannabinoids in pain modulation.

In a subsequent study, mice were generated that expressed FAAH in the nervous system but not in peripheral tissues. These mice exhibited normal pain sensi-

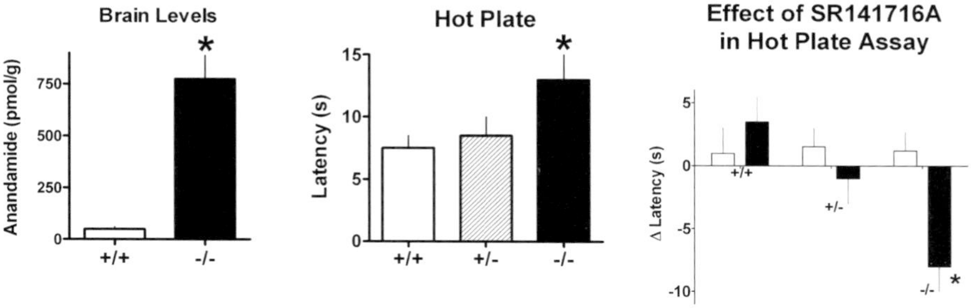

Fig. 4 Marked changes in anandamide levels, hot plate sensitivity, and basal effects of the CB1R antagonist SR141716A in animals lacking the enzyme fatty acid amide hydrolase (FAAH). Wild-type mice (+/+, *left panel*) exhibit relatively low levels of anandamide (50 pmol/g) in brain compared to FAAH knockout mice (–/–) which exhibit 775 pmol/g, indicating that FAAH is the principal mechanism for the metabolism of anandamide. FAAH knockout mice (–/–, *middle panel*) exhibit significantly reduced pain sensitivity under basal conditions compared to wild-type (+/+) and heterozygous (+/-) mice, raising the possibility that the increased levels of anandamide in the knockouts produce a constant state of hypoalgesia. The tonic hypoalgesia observed in the FAAH knockout mice (–/–, *right panel*) is eliminated by the CB1R antagonist SR141716A (*black bars*) compared to vehicle (*white bars*), whereas no significant effect of the antagonist is observed in wild-type (+/+) or heterozygous (–/–) mice. Redrawn from Cravatt et al. (2001)

tivity but a reduced inflammatory response (edema) to carrageenan via a non-cannabinoid mechanism (Cravatt et al. 2004). These findings indicate that the elevated levels of anandamide and other fatty acid conjugates susceptible to FAAH in the nervous system mediate the analgesia observed in FAAH knockouts, while the reduced susceptibility to inflammation is mediated by peripherally elevated lipids acting via a non-CBR mechanism. These data suggest that the central and peripheral FAAH signaling systems regulate discrete phenotypes that may be separately targeted for distinct therapeutic needs.

5.8 Role of Endocannabinoids in the Antinociceptive Actions of Cyclooxygenase Inhibitors

Anandamide is metabolized by cyclooxygenase 2 (COX-2) to form prostaglandin (PG) E2 ethanolamide, PGD_2 ethanolamide, and $PGF_{2\alpha}$ ethanolamide (Kozak et al. 2002; Yu et al. 1997). Ross et al. (2002) demonstrated that PGE_2 ethanolamide binds with nanomolar affinity to prostaglandin EP1, EP2, EP3, and EP4 receptors (K_i (nM) = 5.61 ± 0.1, 6.33 ± 0.01, 6.70 ± 0.13, and 6.29 ± 0.06, respectively; receptor subtypes reviewed by Breyer et al. 2001). Anandamide is not the only derivative of arachidonic acid that is oxygenated by COX-2. The predicted glycerol adduct of PGE_2 is formed upon exposure of 2-AG to recombinant COX-2 (Kozak et al. 2000; Prusakiewicz et al. 2002). The glycerol ester of PGE_2 was recently shown to produce proinflammatory-like effects in macrophage cell line (Nirodi et al. 2004). The above findings indicate that when COX-2 is induced by inflammation, endocannabinoids may be converted from antinociceptive/anti-inflammatory compounds to pro-nociceptive/proinflammatory compounds. This possibility was addressed by Gühring et al. (2002) with the demonstration that the reduction of pain behavior following formalin injection in the hindpaw produced by the COX-2 inhibitor indomethacin was reversed by the CB1R antagonist AM251 but not by PGE_2. This effect was absent in CB1R knockout mice. AM251 also reversed the antihyperalgesic effect of indomethacin subsequent to zymosan-induced inflammation. These findings suggest that COX inhibitors suppress pain, at least in part, by preventing the metabolism of antinociceptive endocannabinoids to pro-nociceptive prostanoids.

5.9 Evidence for Tonic Modulation of Pain via CB1Rs

5.9.1 Pain Sensitivity in CB1R Knockout Mice

Knockouts of the CB1R provided mixed results. Ledent et al. (1999) found that CB1R knockout mice failed to exhibit any of the usual changes produced by exposure to cannabinoids including analgesia. In the absence of any treatment, the basal responses to noxious stimuli in the −/− mice were similar to those of the wild-type mice, in contrast to another study published the same year on a different CB1R knockout (Zimmer et al. 1999), in which a higher pain threshold in the –/– mice compared to wild-type was observed. The surprising finding of analgesia-like effects of the knockouts in this study are at variance with the other study and are difficult to explain, except to hypothesize different patterns of developmental organization of the pain system in the absence of CB1Rs in the groups of mice used by the two laboratories. It is possible that regulatory changes in other receptor systems occur during development subsequent to the knockout of the CB1R gene and contribute to the behavioral phenotype observed in the transgenic mice.

5.9.2 Effects of Endocannabinoids Assessed with CBR antagonists

Studies of the effects of SR141716A, a specific cannabinoid CB1R antagonist (reviewed by Walker et al. 2000) suggest that endocannabinoids participate in endogenous pain modulation and that this action involves the PAG. Blocking the cannabinoid CB1R with SR141716A produced hyperalgesia in the formalin test (Calignano et al. 1998; Strangman et al. 1998) and blocked the analgesia produced by electrical stimulation of the dorsolateral PAG (Walker et al. 1999). These findings are in line with previous studies (Richardson et al. 1997; Richardson et al. 1998b) that demonstrated hyperalgesia following intrathecal administration of this cannabinoid antagonist or CB1R knockdown with an antisense oligonucleotide. Chapman (1999) found that spinal nociceptive neurons exhibit markedly greater C fiber-mediated responses following low doses of SR141716A (0.1–1 ng in 50 ml applied to spinal cord). The authors of these studies posited that the pain-enhancement by the antagonist results from the blockade of endocannabinoids. However, the conclusions from these and other experiments that use SR141716A in this manner are limited by three factors. First, several reports have suggested that SR141716A acts as an inverse agonist, an effect that would mimic that of blocking endocannabinoids (reviewed by Walker et al. 2000). Second, these studies do not identify any particular endocannabinoid that might be involved in the proposed suppression of pain. Third, not all investigators have observed the pain-enhancing effect of SR141716A (Beaulieu et al. 2000), perhaps due to differences in experimental procedures or baseline differences in activation of the endocannabinoid system. For example, ceiling effects in pain behavior could contribute to failures to observe hyperalgesia in the cited work, which used twice the concentration of formalin that was used by Strangman et al. (1998).

6 Effects of Cannabinoids on Pain in Humans

The human trials of cannabis and Δ^9-THC are few in number and typically small in size. These studies differ in important ways. There are marked differences between studies in dose and dose regimens, and the drug preparations differ, with some using smoked marijuana and some using Δ^9-THC by the oral or intravenous routes. Some studies used healthy volunteers whereas others used patients with clinical pain of various origins. Therefore, it is important to note that (1) some negative results may have arisen from ineffective doses; (2) the oral route of administration adds variability due to the unpredictable absorption of Δ^9-THC; (3) smoked marijuana contains additional constituents that likely contribute to any observed effects; (4) clinical pain is very different from experimental pain due to plasticity in the neuronal circuits that mediate pain. In light of the fact that the extant materials do not permit one to reach solid conclusions about the utility of direct-acting full cannabinoid agonists as therapeutic agents in pain, it seems best to examine this literature with an eye toward uncovering whatever therapeutic potential exists.

6.1 Experimental Pain

One approach in studying the effects of cannabinoids in pain perception in humans is through paradigms that involve administering controlled painful stimuli to healthy volunteers. An interesting approach used in two papers (Clark et al. 1981; Zeidenberg et al. 1973) aimed at distinguishing between response bias (often referred to as B, β, or Lx) and sensitivity [often referred to as P(A) or d′] to painful stimuli, using the methods of sensory decision theory. In this approach, response bias refers to the tendency of a particular subject to rate events in a more positive or negative direction. This variable is related to cognitive processes reflecting factors such as a person's temperament. Sensitivity refers to the detectability of a stimulus and the subject's ability to distinguish stimuli that are of similar but slightly different intensities. Sensory decision analysis requires a variety of statistical assumptions, which make interpretation of the results more difficult.

Zeidenberg et al. (1973) administered 5 mg (p.o.) of Δ^9-THC to healthy male volunteers between the ages of 25 and 29, and tested them for thermal pain perception to a radiant heat source before and after administration of the drug. They found that d′ or the ability to distinguish between stimuli of different intensities dropped, and this drop occurred both during the period of subjective effects of the drug and, in 3 of 4 subjects, for the subsequent testing period. Response bias exhibited more intersubject variability. The authors noted that the analgesic effects of the drug remained at a time when effects on memory and psycholinguistic parameters were returning to normal levels, suggesting a longer time course for the drug's effect on pain sensitivity.

A second study that used sensory decision theory reached opposite conclusions (Clark et al. 1981). However, in this study tolerance to cannabinoids is confounded with the pain tests. Healthy volunteers were permitted to smoke increasing quantities of marijuana cigarettes (2%, 20 mg Δ^9-THC content per cigarette, supplied by the U.S. National Institute on Drug Abuse). The total number of cigarettes consumed was very high for both the moderate and high consumption groups (average 19.4 cigarettes per day for high consumption, 13.1 for moderate users), which undoubtedly induced tolerance in the subjects. This confound is so deeply embedded in the experimental design that it is virtually impossible to interpret the data from this experiment.

Raft et al. (1977) used two doses of Δ^9-THC administered intravenously (0.022 and 0.044 mg/kg) in 10 males (ages 18–28) and measured pain induced by two types of noxious stimuli, pressure and electrical. These investigators took the approach of examining the pain threshold (the lowest intensity of stimulation that gives rise to pain) and pain tolerance (the intensity at which pain becomes unbearable). At both doses and for both stimuli the threshold for pain was increased, whereas pain tolerance was not affected. In this and other studies conducted around the same time, the use of threshold and tolerance measures is unfortunate. Clinical pain is normally somewhere in between the two, and it is difficult to assess from the present data what happens in this middle range. Modern approaches would likely use a range of noxious stimuli coupled with ratings of pain intensity, allowing the construction of stimulus–response functions. What is clear from the results of the study by Raft et al. (1977) is that the sensation of pain was entirely absent at some levels of noxious stimulation, but whether this would extend to the clinically relevant levels cannot be assessed from these data. An interesting result from this paper stems from patient reports on pain severity overall. Although the largest decrease in pain threshold occurred with the pressure stimulus at the 0.44 mg/kg dose, most patients rated this condition as the least desirable. It appears that dysphoric effects of Δ^9-THC heightened the overall negativity of the pain. Thus, there is a dissociation between the sensory phenomena and the overall pain experience such that the negative psychotropic effects of Δ^9-THC at the higher dose range overrides the positive effects of the drug on sensory threshold.

Hill et al. (1974) also measured pain thresholds and tolerance. In this single-dose study, healthy male volunteers (ages 21–30, $n = 26$) inhaled marijuana smoke using an apparatus that caused nearly complete combustion of the plant while the subject practiced inhalation in a timed manner. Subjects experienced ascending intensities of electrical stimulation and were asked to report when the stimulation became painful and when it became intolerable. The strength of stimulation was then reversed and the subjects were asked to report when the pain disappeared. The authors found that marijuana smoking lowered the pain threshold as well as pain tolerance. A drawback of this study is the inability to state the dose with any accuracy, a possible basis for the fact that it is at variance with the results of Raft et al. (1977).

A recent study employing topical administration of the cannabinoid agonist HU210 has demonstrated its effectiveness in reducing the magnitude estimation of pain induced in human volunteers following intradermal administration of capsaicin (Rukwied et al. 2003). HU210 also increased the mean heat threshold for pain and reduced tactile allodynia elicited by stimulation with a cotton pad following capsaicin administration. Although pharmacological specificity was not assessed in this work, it is consistent with preclinical studies where mediation by CBRs was confirmed with competitive antagonists (see Sects. 3.2.3 and 4). These data collectively suggest that local administration of a cannabinoid may be employed in humans to suppress pain without psychomimetic side effects.

6.2 Clinical Pain

The studies discussed in this section are the most compelling because the subject population was drawn from patients suffering from significant chronic clinical pain. Chronic pain takes on features that distinguish it from acute pain due to neural plasticity. The changes in sensory processes that take place during periods of prolonged pain serve mainly to amplify the pain. Ongoing painful stimulation leads to peripheral and central sensitization, a process in which the responses to stimulation are enhanced. This leads to allodynia (a painful sensation pursuant to mild tactile stimulation), hyperalgesia (a greater than normal pain sensation to a noxious stimulus), and spontaneous pain. The peripheral mechanisms for different classes of pain (e.g., inflammatory pain versus neuropathic or nerve injury pain) differ. Consequently, different analgesics exhibit different degrees of efficacy in chronic pain of different etiologies. For example, morphine is an excellent analgesic for inflammatory pain, whereas it frequently lacks efficacy in neuropathic pain (Arner and Meyerson 1988). Therefore, studies of clinical pain of different types are necessary precursors to drawing sound conclusions about the possible role of cannabinoids in the pharmacotherapy for pain.

Positive results of cannabinoids have been found in the studies of cancer pain conducted by Noyes and colleagues (Noyes et al. 1975a,b). The larger of the two studies used 36 subjects (26 women and 10 men, mean age 51). These patients reported continuous pain of moderate intensity. In a double-blind random pattern, patients received on successive days placebo, 10 and 20 mg Δ^9-THC, and 60 and 120 mg of codeine. Pain ratings by the patients were used to estimate pain relief and pain reduction scores. The results indicated that 20 mg Δ^9-THC was roughly equivalent to 120 mg codeine. Five of the 36 patients experienced adverse reactions to Δ^9-THC, one following 10 mg Δ^9-THC, four following 20 mg. These side effects undoubtedly limit the amount of analgesia that can be produced by Δ^9-THC. Another report by Noyes (1975) reached similar conclusions with a smaller sample.

Neuropathic pain is a potential target for cannabinoid pharmacotherapies that have been validated at preclinical as well as clinical levels. The cannabinoid Δ^9-THC (dronabinol) has recently been evaluated in multiple sclerosis patients with central neuropathic pain in a double-blind placebo-controlled crossover design (Svendsen et al. 2004). Orally administered dronabinol (10 mg daily for three weeks) lowered median spontaneous pain intensity scores and increased the median pain relief scores relative to placebo treatment. The modest but clear therapeutic effect was associated with improvements on the SF-36 quality-of-life scale with no change in the functional ability of the multiple sclerosis patients. During the first week of treatment, adverse side effects of dronabinol treatment (dizziness, light-headedness) were more frequent with dronabinol than placebo, but the adverse effects decreased over the therapeutic course, possibly due to tolerance (Svendsen et al. 2004). Nonetheless, the clinical relevance of dronabinol for pain management may be limited by unwanted psychoactive side effects (Svendsen et al. 2004). Results of a randomized, placebo-controlled 21-day intervention trial suggest that smoked and oral cannabinoids do not appear to be unsafe [with respect to human immunodeficiency virus (HIV) RNA levels, $CD4^+$ and $CD8^+$ cell counts, or protease inhibitor levels] in individuals with HIV infection (Abrams et al. 2003). Cannabinoids also represent a promising therapeutic target in acquired immunodeficiency syndrome (AIDS) and cancer patients where the antiemetic effects of cannabinoids represent a useful therapeutic adjunct in patient populations for whom the emetic effects of opioids are poorly tolerated.

Recent work has aimed at developing cannabinoids that lack psychotropic side effects, which limit dosing. One example of this may be found in the THC and cannabidiol acid derivatives ajulemic acid (CT-3) and HU-320. These compounds were reported to produce anti-inflammatory effects with a reduced side effect profile (Burstein et al. 1998; Burstein et al. 2004; Sumariwalla et al. 2004), perhaps because they possess either poor (ajulemic acid) or virtually no (HU-320) affinity for either CB1R or CB2R. Consequently, the mechanism by which they produce analgesic effects is not clear. In a recent clinical trial of patients suffering from neuropathic pain, ajulemic acid possessed some efficacy (Karst et al. 2003). While many questions about these and similar compounds are awaiting further research, this appears to be an important line of inquiry.

7 Conclusions

Although cannabinoids have been used for pain relief for centuries, the basis for their analgesic effects were poorly understood until recently. During the last decade a prodigious output of research papers from many laboratories has elucidated many of the major features of cannabinoid analgesia. These studies have not only provided a detailed understanding of the network of neural and inflammatory cells that serve as the targets of cannabinoids, the literature has also begun to address the more difficult question of the physiological role of endocannabinoids in pain regulatory circuits. The low levels of CBRs in brainstem regions that control vital heart rate and respiratory function provided an anatomical basis for the low toxicity of cannabinoids (Herkenham et al. 1991). However, the psychoactivity of direct-acting CB1R agonists proved to be a major barrier to their use as therapeutic tools in the pharmacotherapy of chronic pain. More encouraging results have arisen from a number of studies showing positive effects of CB2R agonists, locally administered cannabinoids, inhibitors of the anandamide-degrading enzyme or the putative anandamide transporter, or the use of atypical cannabinoids such as HU-320. Such novel targets for pain pharmacotherapy represent important future directions for research in this field.

References

Abrams DI, Hilton JF, Leiser RJ, Shade SB, Elbeik TA, Aweeka FT, Benowitz NL, Bredt BM, Kosel B, Aberg JA, Deeks SG, Mitchell TF, Mulligan K, Bacchetti P, McCune JM, Schambelan M (2003) Short-term effects of cannabinoids in patients with HIV-1 infection: a randomized, placebo-controlled clinical trial. Ann Intern Med 139:258–266

Adams IB, Compton DR, Martin BR (1998) Assessment of anandamide interaction with the cannabinoid brain receptor: SR 141716A antagonism studies in mice and autoradiographic analysis of receptor binding in rat brain. J Pharmacol Exp Ther 284:1209–1217

Ahluwalia J, Urban L, Capogna M, Bevan S, Nagy I (2000) Cannabinoid 1 receptors are expressed in nociceptive primary sensory neurons. Neuroscience 100:685–688

Ahluwalia J, Urban L, Bevan S, Capogna M, Nagy I (2002) Cannabinoid 1 receptors are expressed by nerve growth factor- and glial cell-derived neurotrophic factor-responsive primary sensory neurones. Neuroscience 110:747–753

Akil H, Mayer DJ, Liebeskind JC (1976) Antagonism of stimulation-produced analgesia by naloxone, a narcotic antagonist. Science 191:961–962

Arner S, Meyerson BA (1988) Lack of analgesic effect of opioids on neuropathic and idiopathic forms of pain [see comments]. Pain 33:11–23

Averill S, McMahon SB, Clary DO, Reichardt LF, Priestley JV (1995) Immunocytochemical localization of trkA receptors in chemically identified subgroups of adult rat sensory neurons. Eur J Neurosci 7:1484–1494

Baumann TK, Simone DA, Shain CN, LaMotte RH (1991) Neurogenic hyperalgesia: the search for the primary cutaneous afferent fibers that contribute to capsaicin-induced pain and hyperalgesia. J Neurophysiol 66:212–227

Beaulieu P, Bisogno T, Punwar S, Farquhar-Smith WP, Ambrosino G, Di Marzo V, Rice AS (2000) Role of the endogenous cannabinoid system in the formalin test of persistent pain in the rat. Eur J Pharmacol 396:85–92

Beecher HK (1959) The measurement of subjective responses: quantitative effects of drugs. Oxford University Press, New York, pp 164–166

Bellgowan PS, Helmstetter FJ (1996) Neural systems for the expression of hypoalgesia during nonassociative fear. Behav Neurosci 110:727–736

Beltramo M, Stella N, Calignano A, Lin SY, Makriyannis A, Piomelli D (1997) Functional role of high-affinity anandamide transport, as revealed by selective inhibition. Science 277:1094–1097

Ben-Shabat S, Fride E, Sheskin T, Tamiri T, Rhee MH, Vogel Z, Bisogno T, De Petrocellis L, Di Marzo V, Mechoulam R (1998) An entourage effect: inactive endogenous fatty acid glycerol esters enhance 2-arachidonoyl-glycerol cannabinoid activity. Eur J Pharmacol 353:23–31

Bereiter DA, Bereiter DF, Hirata H (2002) Topical cannabinoid agonist, WIN55,212-2, reduces cornea-evoked trigeminal brainstem activity in the rat. Pain 99:547–556

Berkley KJ, Hubscher CH (1995) Are there separate central nervous system pathways for touch and pain? Nat Med 1:766–773

Besse D, Lombard MC, Zajac JM, Roques BP, Besson JM (1990) Pre- and postsynaptic distribution of μ, δ and κ opioid receptors in the superficial layers of the cervical dorsal horn of the rat spinal cord. Brain Res 521:15–22

Bicher HI, Mechoulam R (1968) Pharmacological effects of two active constituents of marihuana. Arch Int Pharmacodyn Ther 172:24–31

Bisogno T, Melck D, Bobrov M, Gretskaya NM, Bezuglov VV, De Petrocellis L, Di Marzo V (2000) N-Acyl-dopamines: novel synthetic CB(1) cannabinoid-receptor ligands and inhibitors of anandamide inactivation with cannabimimetic activity in vitro and in vivo. Biochem J 351:817–824

Bloom AS, Dewey WL, Harris LS, Brosius KK (1977) 9-nor-9beta-hydroxyhexahydrocannabinol, a cannabinoid with potent antinociceptive activity: comparisons with morphine. J Pharmacol Exp Ther 200:263–270

Breyer RM, Bagdassarian CK, Myers SA, Breyer MD (2001) Prostanoid receptors: subtypes and signaling. Annu Rev Pharmacol Toxicol 41:661–690

Bridges D, Ahmad K, Rice AS (2001) The synthetic cannabinoid WIN55,212-2 attenuates hyperalgesia and allodynia in a rat model of neuropathic pain. Br J Pharmacol 133:586–594

Bridges D, Rice AS, Egertová M, Elphick MR, Winter J, Michael GJ (2003) Localisation of cannabinoid receptor 1 in rat dorsal root ganglion using in situ hybridisation and immunohistochemistry. Neuroscience 119:803–812

Burstein SH, Friderichs E, Kogel B, Schneider J, Selve N (1998) Analgesic effects of 1′,1′ dimethylheptyl-delta8-THC-11-oic acid (CT3) in mice. Life Sci 63:161–168

Burstein SH, Karst M, Schneider U, Zurier RB (2004) Ajulemic acid: a novel cannabinoid produces analgesia without a "high". Life Sci 75:1513–1522

Buxbaum DM (1972) Analgesic activity of Δ^9-tetrahydrocannabinol in the rat and mouse. Psychopharmacologia 25:275–280

Calignano A, La Rana G, Giuffrida A, Piomelli D (1998) Control of pain initiation by endogenous cannabinoids. Nature 394:277–281

Calza L, Pozza M, Zanni M, Manzini CU, Manzini E, Hokfelt T (1998) Peptide plasticity in primary sensory neurons and spinal cord during adjuvant-induced arthritis in the rat: an immunocytochemical and in situ hybridization study. Neuroscience 82:575–589

Chapman V (1999) The cannabinoid CB1 receptor antagonist, SR141716A, selectively facilitates nociceptive responses of dorsal horn neurones in the rat. Br J Pharmacol 127:1765–1767

Clark WC, Janal MN, Zeidenberg P, Nahas GG (1981) Effects of moderate and high doses of marihuana on thermal pain: a sensory decision theory analysis. J Clin Pharmacol 21:299S–310S

Clayton N, Marshall FH, Bountra C, O'Shaughnessy CT (2002) CB1 and CB2 cannabinoid receptors are implicated in inflammatory pain. Pain 96:253–260

Coderre TJ, Melzack R (1992) The contribution of excitatory amino acids to central sensitization and persistent nociception after formalin-induced tissue injury. J Neurosci 12:3665–3670

Cravatt BF, Giang DK, Mayfield SP, Boger DL, Lerner RA, Gilula NB (1996) Molecular characterization of an enzyme that degrades neuromodulatory fatty-acid amides. Nature 384:83–87

Cravatt BF, Demarest K, Patricelli MP, Bracey MH, Giang DK, Martin BR, Lichtman AH (2001) Supersensitivity to anandamide and enhanced endogenous cannabinoid signaling in mice lacking fatty acid amide hydrolase. Proc Natl Acad Sci U S A 98:9371–9376

Cravatt BF, Saghatelian A, Hawkins EG, Clement AB, Bracey MH, Lichtman AH (2004) Functional disassociation of the central and peripheral fatty acid amide signaling systems. Proc Natl Acad Sci U S A 101:10821–10826

Davis M, Whalen PJ (2001) The amygdala: vigilance and emotion. Mol Psychiatry 6:13–34

De Petrocellis L, Bisogno T, Davis JB, Pertwee RG, Di Marzo V (2000) Overlap between the ligand recognition properties of the anandamide transporter and the VR1 vanilloid receptor: inhibitors of anandamide uptake with negligible capsaicin-like activity. FEBS Lett 483:52–56

DeLeo JA, Tanga FY, Tawfik VL (2004) Neuroimmune activation and neuroinflammation in chronic pain and opioid tolerance/hyperalgesia. Neuroscientist 10:40–52

Deutsch DG, Chin SA (1993) Enzymatic synthesis and degradation of anandamide, a cannabinoid receptor agonist. Biochem Pharmacol 46:791–796

Devane WA, Hanus L, Breuer A, Pertwee RG, Stevenson LA, Griffin G, Gibson D, Mandelbaum A, Etinger A, Mechoulam R (1992) Isolation and structure of a brain constituent that binds to the cannabinoid receptor. Science 258:1946–1949

Di Marzo V, Bisogno T, Sugiura T, Melck D, De Petrocellis L (1998) The novel endogenous cannabinoid 2-arachidonoylglycerol is inactivated by neuronal- and basophil-like cells: connections with anandamide. Biochem J 331:15–19

Dixon WE (1899) The pharmacology of cannabis. Indica Br Med J 2:1354–1357

Dogrul A, Gul H, Akar A, Yildiz O, Bilgin F, Guzeldemir E (2003) Topical cannabinoid antinociception: synergy with spinal sites. Pain 105:11–16

Donaldson LF, McQueen DS, Seckl JR (1994) Local anaesthesia prevents acute inflammatory changes in neuropeptide messenger RNA expression in rat dorsal root ganglia neurons. Neurosci Lett 175:111–113

Dray A (1995) Inflammatory mediators of pain. Br J Anaesth 75:125–131
Drew LJ, Harris J, Millns PJ, Kendall DA, Chapman V (2000) Activation of spinal cannabinoid 1 receptors inhibits C-fibre driven hyperexcitable neuronal responses and increases [35S]GTPgammaS binding in the dorsal horn of the spinal cord of noninflamed and inflamed rats. Eur J Neurosci 12:2079–2086
Egertová M, Giang DK, Cravatt BF, Elphick MR (1998) A new perspective on cannabinoid signalling: complementary localization of fatty acid amide hydrolase and the CB1 receptor in rat brain. Proc R Soc Lond B Biol Sci 265:2081–2085
Egertová M, Cravatt BF, Elphick MR (2003) Comparative analysis of fatty acid amide hydrolase and CB(1) cannabinoid receptor expression in the mouse brain: evidence of a widespread role for fatty acid amide of endocannabinoid signaling. Neuroscience 119:481–496
Ellington HC, Cotter MA, Cameron NE, Ross RA (2002) The effect of cannabinoids on capsaicin-evoked calcitonin gene-related peptide (CGRP) release from the isolated paw skin of diabetic and non-diabetic rats. Neuropharmacology 42:966–975
Farquhar-Smith WP, Egertova M, Bradbury EJ, McMahon SB, Rice AS, Elphick MR (2000) Cannabinoid CB(1) receptor expression in rat spinal cord. Mol Cell Neurosci 15:510–521
Farquhar-Smith WP, Jaggar SI, Rice AS (2002) Attenuation of nerve growth factor-induced visceral hyperalgesia via cannabinoid CB(1) and CB(2)-like receptors. Pain 97:11–21
Felder CC, Briley EM, Axelrod J, Simpson JT, Mackie K, Devane WA (1993) Anandamide, an endogenous cannabimimetic eicosanoid, binds to the cloned human cannabinoid receptor and stimulates receptor-mediated signal transduction. Proc Natl Acad Sci U S A 90:7656–7660
Fezza F, Bisogno T, Minassi A, Appendino G, Mechoulam R, Di Marzo V (2002) Noladin ether, a putative novel endocannabinoid: inactivation mechanisms and a sensitive method for its quantification in rat tissues. FEBS Lett 513:294–298
Fields HL, Heinricher MM, Mason P (1991) Neurotransmitters in nociceptive modulatory circuits. Annu Rev Neurosci 14:219–245
Fox A, Kesingland A, Gentry C, McNair K, Patel S, Urban L, James I (2001) The role of central and peripheral cannabinoid1 receptors in the antihyperalgesic activity of cannabinoids in a model of neuropathic pain. Pain 92:91–100
Fride E (1995) Anandamides: tolerance and cross-tolerance to delta 9-tetrahydrocannabinol. Brain Res 697:83–90
Fride E, Mechoulam R (1993) Pharmacological activity of the cannabinoid receptor agonist, anandamide, a brain constituent. Eur J Pharmacol 231:313–314
Galeazza MT, Garry MG, Yost HJ, Strait KA, Hargreaves KM, Seybold VS (1995) Plasticity in the synthesis and storage of substance P and calcitonin gene-related peptide in primary afferent neurons during peripheral inflammation. Neuroscience 66:443–458
Gilbert PE (1981) A comparison of THC, nantradol, nabilone, and morphine in the chronic spinal dog. J Clin Pharmacol 21:311S–319S
Gong LW, Ding YQ, Wang D, Zheng HX, Qin BZ, Li JS, Kaneko T, Mizuno N (1997) GABAergic synapses on mu-opioid receptor-expressing neurons in the superficial dorsal horn: an electron microscope study in the cat spinal cord. Neurosci Lett 227:33–36
Grant G (1995) Primary afferent projections to the spinal cord. In: Paxinos G (ed) The Rat Nervous System, 2nd edn. Academic Press, San Diego, pp 61–65
Griffin G, Fernando SR, Ross RA, McKay NG, Ashford ML, Shire D, Huffman JW, Yu S, Lainton JA, Pertwee RG (1997) Evidence for the presence of CB2-like cannabinoid receptors on peripheral nerve terminals. Eur J Pharmacol 339:53–61
Guhring H, Hamza M, Sergejeva M, Ates M, Kotalla CE, Ledent C, Brune K (2002) A role for endocannabinoids in indomethacin-induced spinal antinociception. Eur J Pharmacol 454:153–163
Gutierrez T, Nackley AG, Neely MH, Freeman KG, Edwards GL, Hohmann AG (2003) Effects of neurotoxic destruction of descending noradrenergic pathways on cannabinoid antinocicepetion in models of acute and tonic nociception. Brain Res 987:176–185

Hanani M, Huang TY, Cherkas PS, Ledda M, Pannese E (2002) Glial cell plasticity in sensory ganglia induced by nerve damage. Neuroscience 114:279–283

Hanesch U, Blecher F, Stiller RU, Emson PC, Schaible HG, Heppelmann B (1995) The effect of a unilateral inflammation at the rat's ankle joint on the expression of preprotachykinin-A mRNA and preprosomatostatin mRNA in dorsal root ganglion cells—a study using non-radioactive in situ hybridization. Brain Res 700:279–284

Hanus L, Gopher A, Almog S, Mechoulam R (1993) Two new unsaturated fatty acid ethanolamides in brain that bind to the cannabinoid receptor. J Med Chem 36:3032–3034

Hanus L, Breuer A, Tchilibon S, Shiloah S, Goldenberg D, Horowitz M, Pertwee RG, Ross RA, Mechoulam R, Fride E (1999) HU-308: a specific agonist for CB(2), a peripheral cannabinoid receptor. Proc Natl Acad Sci U S A 96:14228–14233

Hanus L, Abu-Lafi S, Fride E, Breuer A, Vogel Z, Shalev DE, Kustanovich I, Mechoulam R (2001) 2-Arachidonyl glyceryl ether, an endogenous agonist of the cannabinoid CB1 receptor. Proc Natl Acad Sci U S A 98:3662–3665

Harris J, Drew LJ, Chapman V (2000) Spinal anandamide inhibits nociceptive transmission via cannabinoid receptor activation in vivo. Neuroreport 11:2817–2819

Helmstetter FJ (1992) The amygdala is essential for the expression of conditional hypoalgesia. Behav Neurosci 106:518–528

Helmstetter FJ, Bellgowan PS (1993) Lesions of the amygdala block conditional hypoalgesia on the tail flick test. Brain Res 612:253–257

Helmstetter FJ, Bellgowan PS, Tershner SA (1993) Modulation of spinal nociceptive reflexes by the microinjection of morphine into the amygdala. NeuroReport 4:471–474

Helmstetter FJ, Bellgowan PS, Poore LH (1995) Microinfusion of mu, but not delta or kappa opioid agonists into the basolateral amygdala results in inhibition of the tail flick reflex in pentobarbital-anesthetized rats. J Pharmacol Exp Ther 275:381–388

Helyes Z, Nemeth J, Than M, Bolcskei K, Pinter E, Szolcsanyi J (2003) Inhibitory effect of anandamide on resiniferatoxin-induced sensory neuropeptide release in vivo and neuropathic hyperalgesia in the rat. Life Sci 73:2345–2353

Herkenham M, Lynn AB, Johnson MR, Melvin LS, de Costa BR, Rice KC (1991) Characterization and localization of cannabinoid receptors in rat brain: a quantitative in vitro autoradiographic study. J Neurosci 11:563–583

Herzberg U, Eliav E, Bennett GJ, Kopin IJ (1997) The analgesic effects of R(+)-WIN 55,212-2 mesylate, a high affinity cannabinoid agonist, in a rat model of neuropathic pain. Neurosci Lett 221:157–160

Hill SY, Schwin R, Goodwin DW, Powell BJ (1974) Marihuana and pain. J Pharmacol Exp Ther 188:415–418

Hohmann AG (2002) Spinal and peripheral mechanisms of cannabinoid antinociception: behavioral, neurophysiological and neuroanatomical perspectives. Chem Phys Lipids 121:173–190

Hohmann AG, Herkenham M (1998) Regulation of cannabinoid and mu opioid receptor binding sites following neonatal capsaicin treatment. Neurosci Lett 252:13–16

Hohmann AG, Herkenham M (1999a) Cannabinoid receptors undergo axonal flow in sensory nerves. Neuroscience 92:1171–1175

Hohmann AG, Herkenham M (1999b) Localization of central cannabinoid CB1 receptor messenger RNA in neuronal subpopulations of rat dorsal root ganglia: a double-label in situ hybridization study. Neuroscience 90:923–931

Hohmann AG, Martin WJ, Tsou K, Walker JM (1995) Inhibition of noxious stimulus-evoked activity of spinal cord dorsal horn neurons by the cannabinoid WIN 55,212-2. Life Sci 56:2111–2118

Hohmann AG, Tsou K, Walker JM (1998) Cannabinoid modulation of wide dynamic range neurons in the lumbar dorsal horn of the rat by spinally administered WIN55,212-2. Neurosci Lett 257:119–122

Hohmann AG, Briley EM, Herkenham M (1999a) Pre- and postsynaptic distribution of cannabinoid and mu opioid receptors in rat spinal cord. Brain Res 822:17–25

Hohmann AG, Tsou K, Walker JM (1999b) Cannabinoid suppression of noxious heat-evoked activity in wide dynamic range neurons in the lumbar dorsal horn of the rat. J Neurophysiol 81:575–583

Hohmann AG, Tsou K, Walker JM (1999c) Intrathecal cannabinoid administration suppresses noxious-stimulus evoked Fos protein-like immunoreactivity in rat spinal cord: comparison with morphine. Acta Pharmacol Sin 20:1132–1136

Hohmann AG, Neely MH, Suplita RL, Nackley AG, Holmes PV, Crystal JD (2001) Endocannabinoid mechanisms of stress-induced analgesia. Soc Neurosci Abstr 27:716–719

Hohmann AG, Farthing JN, Zvonok AM, Makriyannis A (2004) Selective activation of cannabinoid CB2 receptors suppresses hyperalgesia evoked by intradermal capsaicin. J Pharmacol Exp Ther 308:446–453

Huang SM, Bisogno T, Trevisani M, Al-Hayani A, De Petrocellis L, Fezza F, Tognetto M, Petros TJ, Krey JF, Chu CJ, Miller JD, Davies SN, Geppetti P, Walker JM, Di Marzo V (2002) An endogenous capsaicin-like substance with high potency at recombinant and native vanilloid VR1 receptors. Proc Natl Acad Sci U S A 99:8400–8405

Hunt SP, Pini A, Evan G (1987) Induction of c-fos-like protein in spinal cord neurons following sensory stimulation. Nature 328:632–634

Ibrahim MM, Deng H, Zvonok A, Cockayne DA, Kwan J, Mata HP, Vanderah TW, Lai J, Porreca F, Makriyannis A, Malan TP (2003) Activation of CB2 cannabinoid receptors by AM1241 inhibits experimental neuropathic pain: pain inhibition by receptors not present in the CNS. Proc Natl Acad Sci USA 100:10529–10533

Jennings EA, Vaughan CW, Christie MJ (2001) Cannabinoid actions on rat superficial medullary dorsal horn neurons in vitro. J Physiol 534:805–812

Ji RR, Zhang X, Wiesenfeld-Hallin Z, Hokfelt T (1994) Expression of neuropeptide Y and neuropeptide Y (Y1) receptor mRNA in rat spinal cord and dorsal root ganglia following peripheral tissue inflammation. J Neurosci 14:6423–6434

Ji RR, Zhang X, Zhang Q, Dagerlind A, Nilsson S, Wiesenfeld-Hallin Z, Hokfelt T (1995) Central and peripheral expression of galanin in response to inflammation. Neuroscience 68:563–576

Johanek LM, Simone DA (2004) Activation of peripheral cannabinoid receptors attenuates cutaneous hyperalgesia produced by a heat injury. Pain 109:432–442

Johanek LM, Heitmiller DR, Turner M, Nader N, Hodges J, Simone DA (2001) Cannabinoids attenuate capsaicin-evoked hyperalgesia through spinal and peripheral mechanisms. Pain 93:303–315

Karst M, Salim K, Burstein S, Conrad I, Hoy L, Schneider U (2003) Analgesic effect of the synthetic cannabinoid CT-3 on chronic neuropathic pain: a randomized controlled trial. JAMA 290:1757–1762

Kelly S, Chapman V (2001) Selective cannabinoid CB1 receptor activation inhibits spinal nociceptive transmission in vivo. J Neurophysiol 86:3061–3064

Kemp T, Spike RC, Watt C, Todd AJ (1996) The mu-opioid receptor (MOR1) is mainly restricted to neurons that do not contain GABA or glycine in the superficial dorsal horn of the rat spinal cord. Neuroscience 75:1231–1238

Kenins P (1982) Responses of single nerve fibres to capsaicin applied to the skin. Neurosci Lett 29:83–88

Khasabova IA, Simone DA, Seybold VS (2002) Cannabinoids attenuate depolarization-dependent Ca2+ influx in intermediate-size primary afferent neurons of adult rats. Neuroscience 115:613–625

Klegeris A, Bissonnette CJ, McGeer PL (2003) Reduction of human monocytic cell neurotoxicity and cytokine secretion by ligands of the cannabinoid-type CB2 receptor. Br J Pharmacol 139:775–786

Ko MC, Woods JH (1999) Local administration of delta9-tetrahydrocannabinol attenuates capsaicin-induced thermal nociception in rhesus monkeys: a peripheral cannabinoid action. Psychopharmacology (Berl) 143:322–326

Koga D, Santa T, Fukushima T, Homma H, Imai K (1997) Liquid chromatographic-atmospheric pressure chemical ionization mass spectrometric determination of anandamide and its analogs in rat brain and peripheral tissues. J Chromatogr B Biomed Sci Appl 690:7–13

Kosersky DS, Dewey WL, Harris LS (1973) Antipyretic, analgesic and anti-inflammatory effects of delta 9-tetrahydrocannabinol in the rat. Eur J Pharmacol 24:1–7

Kozak KR, Rowlinson SW, Marnett LJ (2000) Oxygenation of the endocannabinoid, 2-arachidonylglycerol, to glyceryl prostaglandins by cyclooxygenase-2. J Biol Chem 275: 33744–33749

Kozak KR, Crews BC, Morrow JD, Wang LH, Ma YH, Weinander R, Jakobsson PJ, Marnett LJ (2002) Metabolism of the endocannabinoids, 2-arachidonylglycerol and anandamide, into prostaglandin, thromboxane, and prostacyclin glycerol esters and ethanolamides. J Biol Chem 277:44877–44885

Kruger L, Light AR, Schweizer FE (2003) Axonal terminals of sensory neurons and their morphological diversity. J Neurocytol 32:205–216

LaMotte RH, Shain CN, Simone DA, Tsai EF (1991) Neurogenic hyperalgesia: psychophysical studies of underlying mechanisms. J Neurophysiol 66:190–211

LaMotte RH, Lundberg LE, Torebjörk HE (1992) Pain, hyperalgesia and activity in nociceptive C units in humans after intradermal injection of capsaicin. J Physiol 448:749–764

Landsman RS, Burkey TH, Consroe P, Roeske WR, Yamamura HI (1997) SR141716A is an inverse agonist at the human cannabinoid CB1 receptor. Eur J Pharmacol 334:R1–R2

Ledent C, Valverde O, Cossu G, Petitet F, Aubert JF, Beslot F, Bohme GA, Imperato A, Pedrazzini T, Roques BP, Vassart G, Fratta W, Parmentier M (1999) Unresponsiveness to cannabinoids and reduced addictive effects of opiates in CB1 receptor knockout mice. Science 283:401–404

Leslie TA, Emson PC, Dowd PM, Woolf CJ (1995) Nerve growth factor contributes to the up-regulation of growth- associated protein 43 and preprotachykinin A messenger RNAs in primary sensory neurons following peripheral inflammation. Neuroscience 67:753–761

Li J, Daughters RS, Bullis C, Bengiamin R, Stucky MW, Brennan J, Simone DA (1999) The cannabinoid receptor agonist WIN 55,212-2 mesylate blocks the development of hyperalgesia produced by capsaicin in rats. Pain 81:25–33

Li L, Zhou XF (2001) Pericellular Griffonia simplicifolia I isolectin B4-binding ring structures in the dorsal root ganglia following peripheral nerve injury in rats. J Comp Neurol 439:259–274

Lichtman AH, Martin BR (1991a) Cannabinoid-induced antinociception is mediated by a spinal α2-noradrenergic mechanism. Brain Res 559:309–314

Lichtman AH, Martin BR (1991b) Spinal and supraspinal components of cannabinoid-induced antinociception. J Pharmacol Exp Ther 258:517–523

Lichtman AH, Shelton CC, Advani T, Cravatt BF (2004) Mice lacking fatty acid amide hydrolase exhibit a cannabinoid receptor-mediated phenotypic hypoalgesia. Pain 109:319–327

Lim G, Sung B, Ji RR, Mao J (2003) Upregulation of spinal cannabinoid-1-receptors following nerve injury enhances the effects of Win 55,212-2 on neuropathic pain behaviors in rats. Pain 105:275–283

Luo C, Kumamoto E, Furue H, Chen J, Yoshimura M (2002) Anandamide inhibits excitatory transmission to rat substantia gelatinosa neurones in a manner different from that of capsaicin. Neurosci Lett 321:17–20

Mailleux P, Vanderhaeghen JJ (1992) Distribution of neuronal cannabinoid receptor in the adult rat brain: a comparative receptor binding radioautography and in situ hybridization histochemistry. Neuroscience 48:655–668

Malan TP, Ibrahim MM, Makriyannis A, Porreca F (2004) CB2 cannabinoid receptors may produce peripheral analgesia by stimulating local release of endogenous opioids. 2004 Symposium on the Cannabinoids. International Cannabinoid Research Society, p 52

Malan TP Jr, Ibrahim MM, Deng H, Liu Q, Mata HP, Vanderah T, Porreca F, Makriyannis A (2001) CB2 cannabinoid receptor-mediated peripheral antinociception. Pain 93:239–245

Malmberg AB, Chen C, Tonegawa S, Basbaum AI (1997) Preserved acute pain and reduced neuropathic pain in mice lacking PKCgamma. Science 278:279–283

Manning BH, Mayer DJ (1995) The central nucleus of the amygdala contributes to the production of morphine antinociception in the formalin test. Pain 63:141–152

Manning BH, Merin NM, Meng ID, Amaral DG (2001) Reduction in opioid- and cannabinoid-induced antinociception in rhesus monkeys after bilateral lesions of the amygdaloid complex. J Neurosci 21:8238–8246

Manning BH, Martin WJ, Meng ID (2003) The rodent amygdala contributes to the production of cannabinoid-induced antinociception. Neuroscience 120:1157–1170

Mao J, Price DD, Lu J, Keniston L, Mayer DJ (2000) Two distinctive antinociceptive systems in rats with pathological pain. Neurosci Lett 280:13–16

Marsicano G, Wotjak CT, Azad SC, Bisogno T, Rammes G, Cascio MG, Hermann H, Tang J, Hofmann C, Zieglgansberger W, Di Marzo V, Lutz B (2002) The endogenous cannabinoid system controls extinction of aversive memories. Nature 418:530–534

Martin BR, Compton DR, Thomas BF, Prescott WR, Little PJ, Razdan RK, Johnson MR, Melvin LS, Mechoulam R, Ward SJ (1991) Behavioral, biochemical, and molecular modeling evaluations of cannabinoid analogs. Pharmacol Biochem Behav 40:471–478

Martin WJ, Lai NK, Patrick SL, Tsou K, Walker JM (1993) Antinociceptive actions of cannabinoids following intraventricular administration in rats. Brain Res 629:300–304

Martin WJ, Patrick SL, Coffin PO, Tsou K, Walker JM (1995) An examination of the central sites of action of cannabinoid-induced antinociception in the rat. Life Sci 56:2103–2109

Martin WJ, Hohmann AG, Walker JM (1996) Suppression of noxious stimulus-evoked activity in the ventral posterolateral nucleus of the thalamus by a cannabinoid agonist: correlation between electrophysiological and antinociceptive effects. J Neurosci 16:6601–6611

Martin WJ, Tsou K, Walker JM (1998) Cannabinoid receptor-mediated inhibition of the rat tail-flick reflex after microinjection into the rostral ventromedial medulla. Neurosci Lett 242:33–36

Martin WJ, Coffin PO, Attias E, Balinsky M, Tsou K, Walker JM (1999a) Anatomical basis for cannabinoid-induced antinociception as revealed by intracerebral microinjections. Brain Res 822:237–242

Martin WJ, Loo CM, Basbaum AI (1999b) Spinal cannabinoids are anti-allodynic in rats with persistent inflammation. Pain 82:199–205

Mazzari S, Canella R, Petrelli L, Marcolongo G, Leon A (1996) N-(2-hydroxyethyl)hexadecanamide is orally active in reducing edema formation and inflammatory hyperalgesia by down-modulating mast cell activation. Eur J Pharmacol 300:227–236

McCall WD, Tanner KD, Levine JD (1996) Formalin induces biphasic activity in C-fibers in the rat. Neurosci Lett 208:45–48

McMahon SB (1996) NGF as a mediator of inflammatory pain. Philos Trans R Soc Lond B Biol Sci 351:431–440

Mechoulam R, Ben-Shabat S, Hanus L, Ligumsky M, Kaminski NE, Schatz AR, Gopher A, Almog S, Martin BR, Compton DR, et al (1995) Identification of an endogenous 2-monoglyceride, present in canine gut, that binds to cannabinoid receptors. Biochem Pharmacol 50:83–90

Medina JF, Repa CJ, Mauk MD, LeDoux JE (2002) Parallels between cerebellum- and amygdala-dependent conditioning. Nature Rev. Neuroscience 3:122–131

Meng ID, Manning BH, Martin WJ, Fields HL (1998) An analgesia circuit activated by cannabinoids. Nature 395:381–384

Millan MJ (1999) The induction of pain: an integrative review. Prog Neurobiol 57:1–164

Molander C, Grant G (1995) Spinal cord cytoarchitecture. In: Paxinos G (ed) The rat nervous system. Academic Press, San Diego, pp 39–45

Molliver DC, Radeke MJ, Feinstein SC, Snider WD (1995) Presence or absence of TrkA protein distinguishes subsets of small sensory neurons with unique cytochemical characteristics and dorsal horn projections. J Comp Neurol 361:404–416

Monhemius R, Azami J, Green DL, Roberts MH (2001) CB1 receptor mediated analgesia from the nucleus reticularis gigantocellularis pars alpha is activated in an animal model of neuropathic pain. Brain Res 908:67–74

Moreau JL, Fields HL (1986) Evidence for GABA involvement in midbrain control of medullary neurons that modulate nociceptive transmission. Brain Res 397:37–46

Morisset V, Ahluwalia J, Nagy I, Urban L (2001) Possible mechanisms of cannabinoid-induced antinociception in the spinal cord. Eur J Pharmacol 429:93–100

Munro S, Thomas KL, Abu-Shaar M (1993) Molecular characterization of a peripheral receptor for cannabinoids. Nature 365:61–65

Nackley AG, Makriyannis A, Hohmann AG (2003a) Selective activation of cannabinoid CB2 receptors suppresses spinal Fos protein expression and pain behavior in a rat model of inflammation. Neuroscience 119:747–757

Nackley AG, Suplita RL 2nd, Hohmann AG (2003b) A peripheral cannabinoid mechanism suppresses spinal fos protein expression and pain behavior in a rat model of inflammation. Neuroscience 117:659–670

Nackley AG, Zvonok A, Makriyannis A, Hohmann AG (2004) Activation of cannabinoid CB2 receptors suppresses C-fiber responses and windup in spinal wide dynamic range neurons in the absence and presence of inflammation. J Neurophysiol 92:3562–3574

Nagy JI, Hunt SP (1982) Fluoride-resistant acid phosphatase-containing neurones in dorsal root ganglia are separate from those containing substance P or somatostatin. Neuroscience 7:89–97

Nagy JI, Vincent SR, Staines WA, Fibiger HC, Reisine TD, Yamamura HI (1980) Neurotoxic action of capsaicin on spinal substance P neurons. Brain Res 186:435–444

Neumann S, Doubell TP, Leslie T, Woolf CJ (1996) Inflammatory pain hypersensitivity mediated by phenotypic switch in myelinated primary sensory neurons. Nature 384:360–364

Nirodi CS, Crews BC, Kozak KR, Morrow JD, Marnett LJ (2004) The glyceryl ester of prostaglandin E2 mobilizes calcium and activates signal transduction in RAW264.7 cells. Proc Natl Acad Sci U S A 101:1840–1845

Noyes R Jr, Brunk SF, Avery DA, Canter AC (1975a) The analgesic properties of delta-9-tetrahydrocannabinol and codeine. Clin Pharmacol Ther 18:84–89

Noyes R Jr, Brunk SF, Baram DA, Canter A (1975b) Analgesic effect of delta-9-tetrahydrocannabinol. J Clin Pharmacol 15:139–143

Oka S, Tsuchie A, Tokumura A, Muramatsu M, Suhara Y, Takayama H, Waku K, Sugiura T (2003) Ether-linked analogue of 2-arachidonoylglycerol (noladin ether) was not detected in the brains of various mammalian species. J Neurochem 85:1374–1381

Palazzo E, Marabese I, de Novellis V, Oliva P, Rossi F, Berrino L, Maione S (2001) Metabotropic and NMDA glutamate receptors participate in the cannabinoid-induced antinociception. Neuropharmacology 40:319–326

Patel HJ, Birrell MA, Crispino N, Hele DJ, Venkatesan P, Barnes PJ, Yacoub MH, Belvisi MG (2003) Inhibition of guinea-pig and human sensory nerve activity and the cough reflex in guinea-pigs by cannabinoid (CB2) receptor activation. Br J Pharmacol 140:261–268

Pertovaara A, Wei H, Hamalainen MM (1996) Lidocaine in the rostroventromedial medulla and the periaqueductal gray attenuates allodynia in neuropathic rats. Neurosci Lett 218:127–130

Pertwee R, Griffin G, Hanus L, Mechoulam R (1994) Effects of two endogenous fatty acid ethanolamides on mouse vasa deferentia. Eur J Pharmacol 259:115–120

Piomelli D, Beltramo M, Glasnapp S, Lin SY, Goutopoulos A, Xie XQ, Makriyannis A (1999) Structural determinants for recognition and translocation by the anandamide transporter. Proc Natl Acad Sci U S A 96:5802–5807

Porreca F, Ossipov MH, Gebhart GF (2002) Chronic pain and medullary descending facilitation. Trends Neurosci 25:319–325

Porter AC, Sauer JM, Knierman MD, Becker GW, Berna MJ, Bao J, Nomikos GG, Carter P, Bymaster FP, Leese AB, Felder CC (2002) Characterization of a novel endocannabi-

noid, virodhamine, with antagonist activity at the CB1 receptor. J Pharmacol Exp Ther 301:1020–1024
Price TJ, Helesic G, Parghi D, Hargreaves KM, Flores CM (2003) The neuronal distribution of cannabinoid receptor type 1 in the trigeminal ganglion of the rat. Neuroscience 120:155–162
Prusakiewicz JJ, Kingsley PJ, Kozak KR, Marnett LJ (2002) Selective oxygenation of N-arachidonylglycine by cyclooxygenase-2. Biochem Biophys Res Commun 296:612–617
Puffenbarger RA, Boothe AC, Cabral GA (2000) Cannabinoids inhibit LPS-inducible cytokine mRNA expression in rat microglial cells. Glia 29:58–69
Puig S, Sorkin LS (1996) Formalin-evoked activity in identified primary afferent fibers: systemic lidocaine suppresses phase-2 activity. Pain 64:345–355
Quartilho A, Mata HP, Ibrahim MM, Vanderah TW, Porreca F, Makriyannis A, Malan TP Jr (2003) Inhibition of inflammatory hyperalgesia by activation of peripheral CB2 cannabinoid receptors. Anesthesiology 99:955–960
Raft D, Gregg J, Ghia J, Harris L (1977) Effects of intravenous tetrahydrocannabinol on experimental and surgical pain. Psychological correlates of the analgesic response. Clin Pharmacol Ther 21:26–33
Reynolds DV (1969) Surgery in the rat during electrical analgesia induced by focal brain stimulation. Science 164:444–445
Rice AS, Farquhar-Smith WP, Nagy I (2002) Endocannabinoids and pain: spinal and peripheral analgesia in inflammation and neuropathy. Prostaglandins Leukot Essent Fatty Acids 66:243–256
Richardson JD, Aanonsen L, Hargreaves KM (1997) SR 141716A, a cannabinoid receptor antagonist, produces hyperalgesia in untreated mice. Eur J Pharmacol 319:R3–R4
Richardson JD, Aanonsen L, Hargreaves KM (1998a) Antihyperalgesic effects of spinal cannabinoids. Eur J Pharmacol 345:145–153
Richardson JD, Aanonsen L, Hargreaves KM (1998b) Hypoactivity of the spinal cannabinoid system results in NMDA-dependent hyperalgesia. J Neurosci 18:451–457
Richardson JD, Kilo S, Hargreaves KM (1998c) Cannabinoids reduce hyperalgesia and inflammation via interaction with peripheral CB1 receptors. Pain 75:111–119
Ross RA, Coutts AA, McFarlane SM, Anavi-Goffer S, Irving AJ, Pertwee RG, MacEwan DJ, Scott RH (2001a) Actions of cannabinoid receptor ligands on rat cultured sensory neurones: implications for antinociception. Neuropharmacology 40:221–232
Ross RA, Gibson TM, Brockie HC, Leslie M, Pashmi G, Craib SJ, Di Marzo V, Pertwee RG (2001b) Structure-activity relationship for the endogenous cannabinoid, anandamide, and certain of its analogues at vanilloid receptors in transfected cells and vas deferens. Br J Pharmacol 132:631–640
Ross RA, Craib SJ, Stevenson LA, Pertwee RG, Henderson A, Toole J, Ellington HC (2002) Pharmacological characterization of the anandamide cyclooxygenase metabolite: prostaglandin E2 ethanolamide. J Pharmacol Exp Ther 301:900–907
Rukwied R, Watkinson A, McGlone F, Dvorak M (2003) Cannabinoid agonists attenuate capsaicin-induced responses in human skin. Pain 102:283–288
Sagar DR, Smith PA, Millns PJ, Smart D, Kendall DA, Chapman V (2004) TRPV1 and CB(1) receptor-mediated effects of the endovanilloid/endocannabinoid N-arachidonoyl-dopamine on primary afferent fibre and spinal cord neuronal responses in the rat. Eur J Neurosci 20:175–184
Salio C, Fischer J, Franzoni MF, Mackie K, Kaneko T, Conrath M (2001) CB1-cannabinoid and mu-opioid receptor co-localization on postsynaptic target in the rat dorsal horn. Neuroreport 12:3689–3692
Salio C, Doly S, Fischer J, Franzoni M, Conrath M (2002a) Neuronal and astrocytic localization of the cannabinoid receptor-1 in the dorsal horn of the rat spinal cord. Neurosci Lett 329:13
Salio C, Fischer J, Franzoni MF, Conrath M (2002b) Pre- and postsynaptic localizations of the CB1 cannabinoid receptor in the dorsal horn of the rat spinal cord. Neuroscience 110:755–764

Sanudo-Pena MC, Strangman NM, Mackie K, Walker JM, Tsou K (1999) CB1 receptor localization in rat spinal cord and roots, dorsal root ganglion, and peripheral nerve. Acta Pharmacol Sin 20:1115–1120

Shire D, Carillon C, Kaghad M, Calandra B, Rinaldi-Carmona M, Le Fur G, Caput D, Ferrara P (1995) An amino-terminal variant of the central cannabinoid receptor resulting from alternative splicing [published erratum appears in J Biol Chem 1996 Dec 27;271(52):33706]. J Biol Chem 270:3726–3731

Silverman JD, Kruger L (1990) Selective neuronal glycoconjugate expression in sensory and autonomic ganglia: relation of lectin reactivity to peptide and enzyme markers. J Neurocytol 19:789–801

Simone DA, Ngeow JY, Putterman GJ, LaMotte RH (1987) Hyperalgesia to heat after intradermal injection of capsaicin. Brain Res 418:201–203

Simone DA, Baumann TK, LaMotte RH (1989) Dose-dependent pain and mechanical hyperalgesia in humans after intradermal injection of capsaicin. Pain 38:99–107

Smart D, Gunthorpe MJ, Jerman JC, Nasir S, Gray J, Muir AI, Chambers JK, Randall AD, Davis JB (2000) The endogenous lipid anandamide is a full agonist at the human vanilloid receptor (hVR1). Br J Pharmacol 129:227–230

Smith PB, Martin BR (1992) Spinal mechanisms of Δ^9-tetrahydrocannabinol-induced analgesia. Brain Res 578:8–12

Sokal DM, Elmes SJR, Kendall DA, Chapman V (2003) Intraplantar injection of anandamide inhibits mechanically-evoked responses of spinal neurons via activation of CB2 receptors in anesthetized rats. Neuropharmacology 45:404–411

Strangman NM, Walker JM (1999) The cannabinoid WIN 55,212-2 inhibits the activity-dependent facilitation of spinal nociceptive responses. J Neurophysiol 81:472–477

Strangman NM, Patrick SL, Hohmann AG, Tsou K, Walker JM (1998) Evidence for a role of endogenous cannabinoids in the modulation of acute and tonic pain sensitivity. Brain Res 813:323–328

Sugiura T, Kondo S, Sukagawa A, Nakane S, Shinoda A, Itoh K, Yamashita A, Waku K (1995) 2-Arachidonoylglycerol: a possible endogenous cannabinoid receptor ligand in brain. Biochem Biophys Res Commun 215:89–97

Sumariwalla PF, Gallily R, Tchilibon S, Fride E, Mechoulam R, Feldmann M (2004) A novel synthetic, nonpsychoactive cannabinoid acid (HU-320) with antiinflammatory properties in murine collagen-induced arthritis. Arthritis Rheum 50:985–998

Svendsen KB, Jensen TS, Bach FW (2004) Does the cannabinoid dronabinol reduce central pain in multiple sclerosis? Randomised double blind placebo controlled crossover trial. BMJ 329:253

Szolcsanyi J, Anton F, Reeh PW, Handwerker HO (1988) Selective excitation by capsaicin of mechano-heat sensitive nociceptors in rat skin. Brain Res 446:262–268

Torebjork HE, Lundberg LE, LaMotte RH (1992) Central changes in processing of mechanoreceptive input in capsaicin- induced secondary hyperalgesia in humans. J Physiol 448:765–780

Tracey DJ, Walker JS (1995) Pain due to nerve damage: are inflammatory mediators involved? Inflamm Res 44:407–411

Tsou K, Jang CS (1964) Studies on the site of analgesic action of morphine by intracerebral micro-injection. Sci Sin 13:1099–1109

Tsou K, Lowitz KA, Hohmann AG, Martin WJ, Hathaway CB, Bereiter DA, Walker JM (1996) Suppression of noxious stimulus-evoked expression of FOS protein-like immunoreactivity in rat spinal cord by a selective cannabinoid agonist. Neuroscience 70:791–798

Tsou K, Brown S, Sanudo-Peña MC, Mackie K, Walker JM (1998a) Immunohistochemical distribution of cannabinoid CB1 receptors in the rat central nervous system. Neuroscience 83:393–411

Tsou K, Nogueron MI, Muthian S, Sanudo-Pena MC, Hillard CJ, Deutsch DG, Walker JM (1998b) Fatty acid amide hydrolase is located preferentially in large neurons in the rat central nervous system as revealed by immunohistochemistry. Neurosci Lett 254:137–140

Vaughan CW, McGregor IS, Christie MJ (1999) Cannabinoid receptor activation inhibits GABAergic neurotransmission in rostral ventromedial medulla neurons in vitro. Br J Pharmacol 127:935–940

Vaughan CW, Connor M, Bagley EE, Christie MJ (2000) Actions of cannabinoids on membrane properties and synaptic transmission in rat periaqueductal gray neurons in vitro. Mol Pharmacol 57:288–295

Vivian JA, Kishioka S, Butelman ER, Broadbear J, Lee KO, Woods JH (1998) Analgesic, respiratory and heart rate effects of cannabinoid and opioid agonists in rhesus monkeys: antagonist effects of SR 141716A. J Pharmacol Exp Ther 286:697–703

Vogel Z, Bayewitch M, Levy R, Matus-Leibovitch N, Hanus L, Ben-Shabat S, Mechoulam R, Avidor-Reiss T, Barg J (1994) Binding and functional studies with the peripheral and neuronal cannabinoid receptors. Regul Pept 54:313–314

Walker JM, Huang SM, Strangman NM, Tsou K, Sanudo-Pena MC (1999) Pain modulation by release of the endogenous cannabinoid anandamide. Proc Natl Acad Sci U S A 96:12198–12203

Walker JM, Huang SM, Sanudo-Pena (2000) Identification of the role of endogenous cannabinoids in pain modulation: strategies and pitfalls. J Pain 1:20–32

Walker JM, Strangman NM, Huang SM (2001) Cannabinoids and pain. Pain Res Manag 6:74–79

Walter L, Franklin A, Witting A, Moller T, Stella N (2002) Astrocytes in culture produce anandamide and other acylethanolamides. J Biol Chem 277:20869–20876

Walter L, Franklin A, Witting A, Wade C, Xie Y, Kunos G, Mackie K, Stella N (2003) Nonpsychotropic cannabinoid receptors regulate microglial cell migration. J Neurosci 23:1398–1405

Welch SP, Thomas C, Patrick GS (1995) Modulation of cannabinoid-induced antinociception after intracerebroventricular versus intrathecal administration to mice: possible mechanisms for interaction with morphine. J Pharmacol Exp Ther 272:310–321

Willis WD, Westlund KN, Carlton SM (1995) Pain. In: Paxinos G (ed) The Rat Nervous System, 2nd edn. Academic Press, New York, pp 725–750

Yaksh TL (1981) The antinociceptive effects of intrathecally administered levonantradol and desacetyllevonantradol in the rat. J Clin Pharmacol 21:334S–340S

Yesilyurt O, Dogrul A, Gul H, Seyrek M, Kusmez O, Ozkan Y, Yildiz O (2003) Topical cannabinoid enhances topical morphine antinociception. Pain 105:303–308

Young WSd, Wamsley JK, Zarbin MA, Kuhar MJ (1980) Opioid receptors undergo axonal flow. Science 210:76–78

Yu M, Ives D, Ramesha CS (1997) Synthesis of prostaglandin E2 ethanolamide from anandamide by cyclooxygenase-2. J Biol Chem 272:21181–21186

Zeidenberg P, Clark WC, Jaffe J, Anderson SW, Chin S, Malitz S (1973) Effect of oral administration of delta9 tetrahydrocannabinol on memory, speech, and perception of thermal stimulation: results with four normal human volunteer subjects. Preliminary report. Compr Psychiatry 14:549–556

Zhang J, Hoffert C, Vu HK, Groblewski T, Ahmad S, O'Donnell D (2003) Induction of CB2 receptor expression in the rat spinal cord of neuropathic but not inflammatory chronic pain models. Eur J Neurosci 17:2750–2754

Zimmer A, Zimmer AM, Hohmann AG, Herkenham M, Bonner TI (1999) Increased mortality, hypoactivity, and hypoalgesia in cannabinoid CB1 receptor knockout mice. Proc Natl Acad Sci U S A 96:5780–5785

Zygmunt PM, Petersson J, Andersson DA, Chuang H, Sorgard M, Di Marzo V, Julius D, Hogestatt ED (1999) Vanilloid receptors on sensory nerves mediate the vasodilator action of anandamide. Nature 400:452–457

HEP (2005) 168:555–571

Effects of Cannabinoids on Hypothalamic and Reproductive Function

M. Maccarrone[1] (✉) · T. Wenger[2]

[1]Department of Biomedical Sciences, University of Teramo, Piazza A. Moro 45, 64100 Teramo, Italy
Maccarrone@vet.unite.it

[2]Department of Human Morphology and Developmental Biology, Semmelweis University, PO Box 95, 1450 Budapest, Hungary

Abstract Marijuana and cannabinoids have been shown to exert profound effects on hypothalamic regulatory functions and reproduction in both experimental animals and humans. Here we review the role of (endo)cannabinoids in the regulation of appetite and food intake. There is converging evidence that the hypothalamic endocannabinoid system changes after leptin treatment. Cannabinoid administration decreases heat production by altering hypothalamic neurotransmitter production. Experimental and human data have also shown that the endocannabinoid system is involved in the regulation of reproductive function at both central and peripheral levels. We discuss also the role of fatty acid amide hydrolase (FAAH) in gestation, and in particular the regulation of the activity of FAAH by progesterone and leptin. We show that endocannabinoids inhibit the release of leukaemia inhibitory factor (LIF) from peripheral T lymphocytes. Taken together, endocannabinoids not only help to maintain neuroendocrine homeostasis, but also take part in immunological changes occurring during early pregnancy.

Keywords Appetite · Cytokines · Endocannabinoids · Hypothalamus · Lymphocytes · Pituitary · Pregnancy · Reproduction · Sex hormones · Thermoregulation

1 Historical Background

Cannabis was used as a drug as long ago as 2000 B.C. Hemp is mentioned in the Atharva Veda approx. 2000 B.C. (veda: saint book of Hindi religion). The ancient Hindus credited it as giving "vital energy", and Pliny the Elder first mentions its effect on the reproductive system (cited by Butrica 2002): "*...semen eius extinguere genitarum uirorum dicitur...* (Its seed is said to extinguish men's semen)". Aetius mentioned (sixth century A.D.) that it could be used on women as well, although he did not specify the conditions of use. Cannabis has been widely known since the first millennium in the Middle East, and the physician al-Badri (middle of the thirteenth century A.D.) already recommended hashish to stimulate appetite (cited by Peters et al. 1999).

As early as in the tenth century in Middle Eastern medicine, the hemp was used as an antipyretic agent (cited by Lozano 2001). Moreau (1845) also mentioned the hypothermic effect of marijuana.

Cannabis was introduced into the modern Western world as a medicine by O'Shaughnessy in 1830, who recommended it to cure menstrual disorders (in Crawford 2002), probably not because of its effects on hormonal secretion but as an anticonvulsive smooth muscle relaxant.

It was not until 1970 that marijuana was extensively investigated, as a result of the identification of the major psychoactive component of cannabis, Δ^9-tetrahydrocannabinol (THC), by Mechoulam et al. (1965).

2 General Anatomical Features

The hypothalamus is a subdivision of the diencephalon. It is a multifunctional centre for the control of visceromotor and endocrine activity. The hypothalamus integrates and modulates responses to changes in temperature or osmolality or in the level of specific hormones in the general circulation. Anteriorly it is bordered by the lamina terminalis. The third cerebral ventricle is its medial boundary and the lateral border is formed by basal forebrain structures (Fig. 1). The fornix divides the hypothalamus into medial and lateral region. The hypothalamus has varied and complex connections with several other CNS areas (Levine 2000). It receives information from sensory nerves, peripheral hormone secretions, and pathways originating in limbic and cortical structures. The output structures control brainstem autonomic centres like gastrointestinal (appetite, vomiting) regulatory areas. The hypothalamus plays a major role in emotional behaviour, and is sensitive to changes of blood temperature. As such, it plays a role in the regulation of body

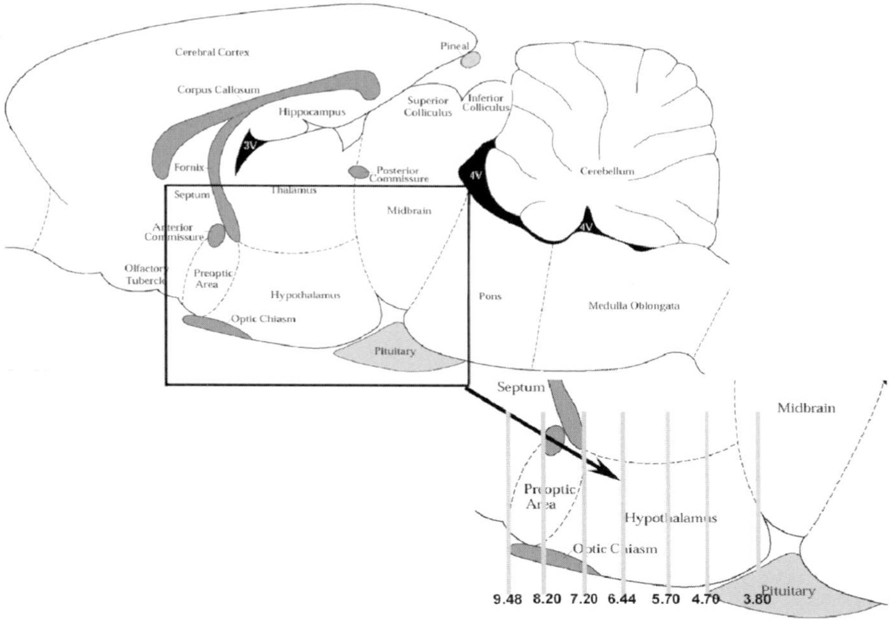

Fig. 1. Schematic drawing of the sagittal section of rat brain 0.4 mm lateral to the midline, according to the atlas of Paxinos-Watson (1997). The *box* shows a part of the forebrain. The *parallel lines* indicate frontal sections seen in Fig. 2. The *numbers* indicate the distance from the interaural line according to the Paxinos-Watson atlas. Only the main structures are labelled (for orientation)

temperature. Indeed, the hypothalamus is a brain area that is generally considered to be particularly important in maintaining homeostasis.

3 Cannabinoids in the Hypothalamus and Pituitary

The presence of endocannabinoids has been shown in the hypothalamus (Herkenham 1995) and in the anterior pituitary (Gonzales 1999). The central cannabinoid receptor (CB_1 receptor) is also present is these structures. The hypothalamus contains fewer cannabinoid binding sites than other areas of the CNS. Nevertheless the effects caused by the activation of CB_1 receptors in the hypothalamus are important, maybe because the receptors are more or less concentrated within specific hypothalamic nuclei-areas (Fig. 2). CB_1 receptors seem to be located on intrinsic hypothalamic neurons rather than on neurons with cell bodies located outside the hypothalamus, since hypothalamic deafferentation is not followed by any reduction in the number of cannabinoid receptor binding sites within this brain area (Romero 1998).

Unlike the hypothalamus, the anterior pituitary, which is regulated by hypothalamic releasing and inhibiting factors, contains a large number of CB_1 recep-

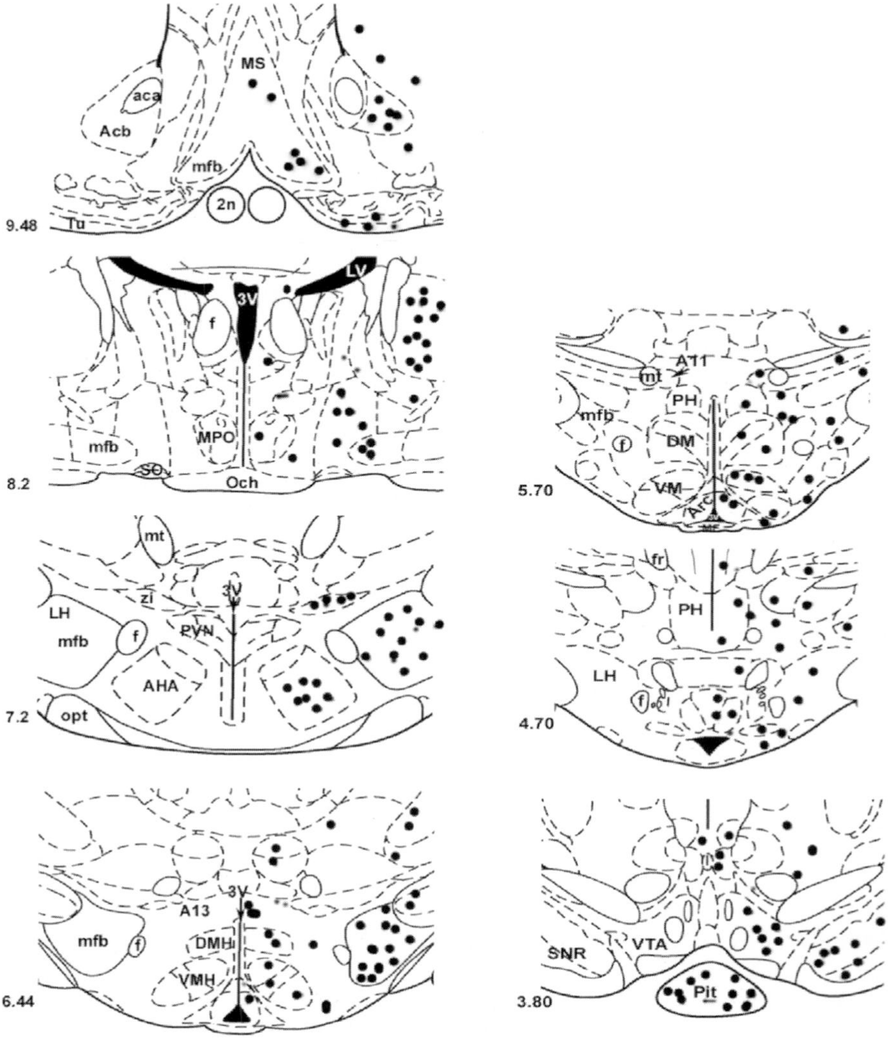

Fig. 2. Schematic drawing of different forebrain and midbrain areas to show the presence of cannabinoid receptor immunoreactivity (*dark spots*). Differences between fibres and cell bodies are not shown. Also not shown are quantitative differences between different areas in CB_1 receptor density. Only structures expressing CB_1 receptor immunoreactivity are labelled. The *numbers* indicate the distance from the interaural line (according to the Paxinos-Watson atlas, see also Fig. 1 for explanations). Abbreviations of structures labelled in this and in consecutive figures (in alphabetical order): *2n*, optic nerve; *3V*, third ventricle; *A11*, A11 dopamine cells; *A13*, A13 dopamine cells; *aca*, anterior commissure; *Acb*, accumbens nucleus; *AHA*, anterior hypothalamic area; *DMH*, dorsomedial hypothalamic nucleus; *f*, fornix; *fr*, fasciculus retroflexus; *LH*, lateral hypothalamic area; *LV*, lateral ventricle; *mfb*, median forebrain bundle; *MPO*, medial preoptic nucleus; *MS*, medial septal nucleus; *mt*, mamillothalamic tract; *Och*, optic chiasma; *opt*, optic tract; *Pit*, pituitary gland; *PVN*, paraventricular nucleus; *SNR*, substantia nigra; *SO*, supraoptic nucleus; *Tu*, olfactory tubercle; *VMH* ventromedial hypothalamic nucleus; *zi*, zona incerta. (Details from Moldrich and Wenger 2000, with the kind permission of Elsevier Science Publishing)

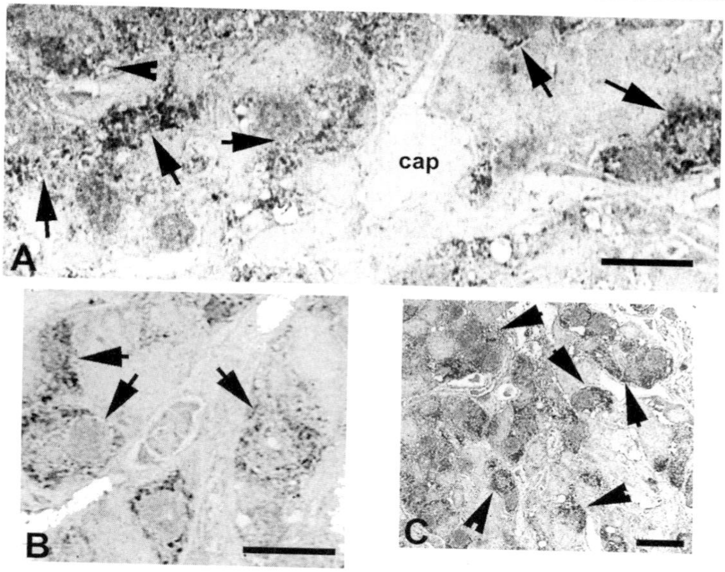

Fig. 3A–C. Immunohistochemical expression of CB_1 receptors in the anterior pituitary of rat. The *arrows* show intensely stained cells. **A** and **B** Gonadotrope cells. **C** Lactotropes. Note that the immunoreactive granules are present mainly at the periphery of the cells. *cap*, sinusoid capillary; *scale bars* = 35 μm in **A** and **B**, 25 μm in **C**

tors, mainly on lactotropes and gonadotropes (Wenger et al. 1999) (Fig. 3). *N*-Arachidonoylethanolamine (AEA) is also present in the pituitary (Gonzales et al. 1999).

No cannabinoid receptors have been found in pituitary corticotrope cells (Wenger at al. 1999).

3.1 Cannabinoids and Appetite and Feeding

Leptin, the 16-kDa product of the *obese* gene, has been implicated in the maintenance of feeding behaviour and energy balance (Campfield et al. 1995). Leptin is regarded as an "appetite-reducing" protein, and as the primary signal through which the hypothalamus regulates food intake and energy balance (Friedman and Halaas 1998). It is known that neurons in the ventromedial hypothalamic nucleus (VMH) and in the lateral hypothalamus (LHY) play a central role in the regulation of feeding and energy homeostasis (Oomura et al. 1969). The leptin receptor (LR) was first demonstrated in the choroid plexus and hypothalamus (Tartaglia 1995). Strong LR immunoreactivity was described in the hypothalamic arcuate nucleus (ARC), VMH and dorsomedial nucleus (DMN), and moderate immunoreactivity in the LHY (Maruta et al. 1999; Funahashi et al. 1999) (Fig. 4). It is interesting that leptin does not only regulate appetite and feeding, but also takes part in hypothalamic neuroendocrine regulation (Takashi et al. 2002).

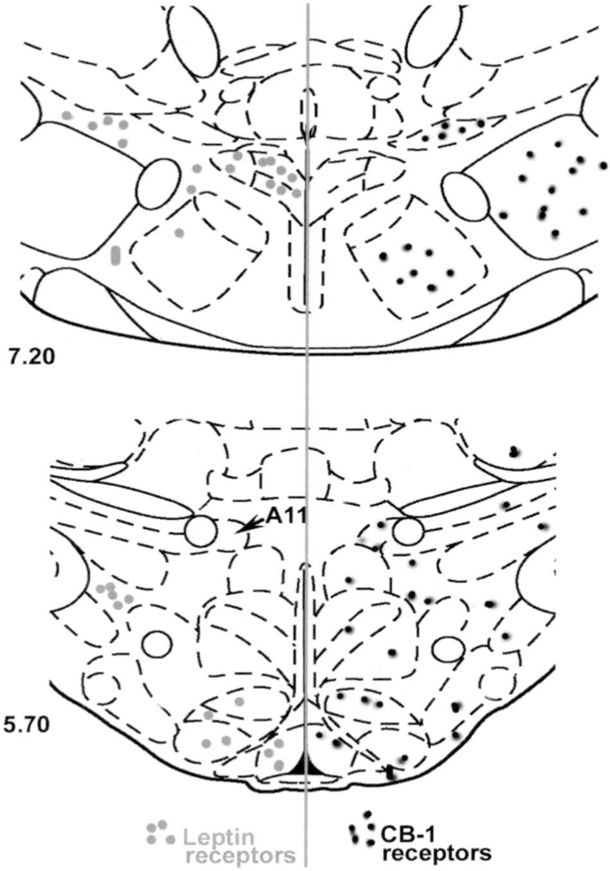

Fig. 4. Schematic drawing of two different forebrain areas to show the presence of CB_1 receptors (*right side*) and leptin receptors (*left side*). The structures are labelled as in Fig. 2. Note that there are sites such as the zona incerta, anterior hypothalamic area, ventromedial nucleus, and dorsomedial nucleus where both receptors are present. (Drawings concerning leptin were made using the data of Maruta et al. 1999)

On the other hand, studies confirmed that THC and AEA might cause overeating (Williams et al. 1998; Williams and Kirkham 1999). Administration of AEA caused hyperphagia and overeating in rats. Attenuation of this effect by the CB_1 receptor antagonist SR 141716A was dose dependent (Williams and Kirkham 1999). Di Marzo et al. (2001) reported that hypothalamic endocannabinoid signalling is constitutively stimulated in obese mice and Zucker rats. They also observed that food intake is lower in CB_1 receptor knockout (KO) mice than in their wild-type littermates. Hypothalamic endocannabinoids appear to be under negative control by leptin. AEA content is reduced in the hypothalamus after leptin treatment (Di Marzo et al. 2001), and AEA-hydrolase (fatty acid amide hydrolase, FAAH) activity is enhanced by leptin in human T cells (Maccarrone et al. 2003a).

It can be concluded that endocannabinoids contribute to the stimulation of appetite by activating the CB_1 receptors present in the hypothalamus (the presence of both leptin and CB_1 receptors in the hypothalamus is shown on Fig. 4) and that the CB_1 receptor antagonist SR 141716A can be considered to be an appetite-suppressing drug.

3.2 Cannabinoids and Thermoregulation

One of the characteristic pharmacological properties of CB_1 receptor agonists is an ability to induce hypothermia (Pertwee 1985). The changes of body temperature caused by cannabinoids are dose dependent. According to Pertwee, higher doses of THC cause hypothermia by lowering the thermoregulatory "set point", while lower doses are hyperthermic. It has been postulated that differential G_s and G_i protein activation by CB_1 receptors could explain these findings (Sulcova et al. 1998).

Cannabinoid-induced hypothermia is mediated by dopaminergic pathways (Pertwee 1992). It was proposed that AEA might not produce all of its effect on thermoregulation by a direct interaction on CB_1 receptors present in hypothalamic thermoregulatory centres. SR 141716A did not block hypothermia caused by AEA (Adams et al. 1998), although this CB_1 receptor antagonist reversed the hypothermia caused by WIN 55,212-2. The endocannabinoids *N*-arachidonoyl-dopamine and 2-arachidonoylglycerol (2-AG)-ether both caused hypothermia (Bisogno et al. 2000; Hanus et al. 2001), supporting the involvement of CB_1 receptors in this process.

On the other hand, *N*-vanillyl-arachidonyl-amide (arvanil), a VR1 receptor agonist, was 100 times more potent than AEA in producing hypothermia (Di Marzo et al. 2000), which indicates that hypothermia caused by cannabimimetic compounds may not (only) be due to the activation of CB_1 receptors.

It is possible that the endocannabinoid system is taking part in thermoregulation, too. However, it is still questionable whether this effect occurs by the activation of cannabinoid receptors and/or vanilloid receptor, or by other mechanisms.

3.3 Cannabinoids and Regulation of the Hypothalamo-Pituitary-Adrenal Cortical Axis

Both exogenous and endogenous CB_1 receptor agonists stimulate adrenocorticotrophic hormone (ACTH) and corticosterone secretion (Dewey 1986; Weidenfield et al. 1994; Wenger et al. 1997; Manzanares et al. 1999).

Chronic administration of THC increased corticotropin-releasing hormone (CRH) and proopiomelanocortin (POMC) gene expression in the rat hypothalamus (Corchero et al. 2001). Circulating gonadal steroids facilitate the latter effect.

AEA activates the CRH-producing neurons in the hypothalamic paraventricular nucleus (Wenger et al. 1997). This effect of AEA may be mediated by a different and as-yet-uncharacterized G protein-coupled cannabinoid receptor (CB_X), the presence of which in the CNS has been proposed (Wenger et al. 1997; Di Marzo et al. 2002; Wenger et al. 2003).

4 Cannabinoids and Reproduction

In the 1980s a great number of papers dealt with the effects of THC on reproduction and on neuroendocrine function. THC increased gonadotropin-releasing hormone (GnRH) (Collu 1976). Kumar et al. (1983) found that hypothalamic GnRH content was increased in ovariectomized rats after a single dose of THC. An accumulation of GnRH in dense-core vesicles was observed in the hypothalamic median eminence after THC treatment, (Doms et al. 1981). Ayalon et al. (1977) and Tyrey (1984) postulated that THC acted primarily through central neuroendocrine mechanisms, since its effects could be reversed by administration of exogenous GnRH. In contrast, Wenger et al. (1987) found no changes in GnRH content in the (anterior) hypothalamus after THC administration, and in in vitro studies it was demonstrated that THC did not alter the release or storage of gonadotropins and did not modify the responsiveness of cultured anterior pituitary cells to GnRH.

Since the early studies by Marks (1973) on the inhibitory effects of THC on pituitary luteinizing hormone (LH) secretion, a number of papers reported similar effects (Smith et al. 1980; Steger et al. 1980; Wenger et al. 1987). THC suppresses the tonic circulating level of LH in male rats (Chakravarty et al. 1982) and episodic LH secretion in female animals (Tyrey 1980).

Studies by Chakravarty et al. (1975) in intact female rats, by Kramer (1974) in male rats, by Dalterio et al. (1981) in male mice, and by Rettori et al. (1988) in male rats in vitro, have shown that administration of THC can lower prolactin (PRL) release.

AEA has similar effects on reproductive hormones as THC. AEA temporarily decreases serum LH level, and this effect lasts up to 2–3 h (Gonzales et al. 2000; Wenger et al. 1999). PRL levels can also be decreased by endocannabinoid treatment (Wenger et al. 1999). Sexual differences in CB_1 receptor density have been detected in the medial basal hypothalamus (MBH) (Rodriguez de Fonseca 1994). The density was higher in diestrus and decreased in oestrus. Gonzales et al. (2000) reported that AEA content in both the hypothalamus and anterior pituitary might be controlled by circulating sex steroids. AEA effects on the control of the regulation of reproduction are mediated by CB_1 receptors located in the hypothalamus and in the anterior pituitary (Fernandez-Ruiz et al. 1997; Romero et al. 1998; Wenger et al. 1997). Recently it has been demonstrated that AEA changes dopaminergic turnover, thus altering inhibitory dopaminergic effects on PRL secretion (Scorticati et al. 2003).

CB_1 receptor inactivation suppresses reproductive hormone secretion (Wenger et al. 2001). Serum LH and testosterone (T) levels significantly decreased in mutant ($CB_1^{-/-}$) mice (Table 1). Results from this investigation also indicated that cannabinoids regulate neuroendocrine function through the activation of CB_1

Table 1. Luteinizing hormone (LH) and testosterone (T) content in central cannabinoid receptor (CB_1 receptor) knockout mice (data from Wenger et al. 2001)

	LH mg/pituitary	LH ng/ml serum	T nmol/testis
$CB_1^{+/+}$	0.71±0.24	5.15±0.8	39.57±4.23
$CB_1^{-/-}$	0.76±0.3	2.6±0.24*	19.89±3.2**

$^{+/+}$, Wild-type mice; $^{-/-}$, CB_1 receptor knockout mice.
n=8–10 in all groups.
*$p<0.01$ vs +/+ (±SEM).
**$pLT0.001$ vs +/+ (±SEM).

Table 2. Anterior pituitary hormone content changes after AEA administration[1]

LH	FSH	PRL	ACTH
↓↓	–	↓↓	↑↑

↓↓, Significant decrease ($p<0.01$ or higher); ↑↑, significant increase ($p<0.01$ or higher); ACTH, adrenocorticotrophic hormone; AEA, anandamide; LH, luteinizing hormone; FSH, follicle-stimulating hormone; PRL, prolactin.
[1] One dose (0.1 mg/kg), i.p. administration.

receptors. This regulation seems to be mainly through inhibition of hormone release at the pituitary level and may or may not also involve the hypothalamus. Table 2 summarizes the effects of endocannabinoids on anterior pituitary hormone content. Interestingly, cannabinoids do not affect the secretion/release of follicle-stimulating hormone (FSH), and it remains to be ascertained whether or not they may modulate the purported FSH-releasing factor (Samson et al. 1980).

A direct regulatory role for endocannabinoids in normal human anterior pituitary gland and pituitary adenomas has also been postulated (Pagotto et al. 2001). Pituitary adenomas had higher AEA and 2-AG concentrations, pointing to a role for endocannabinoids in the development of pituitary adenomas too.

4.1 The Endocannabinoid System and Female Reproductive Function

Adverse effects of cannabinoids, and in particular of THC, on reproductive functions include retarded embryo development, foetal loss and pregnancy failure. They have been known for a long time (Geber and Schramm 1969; Kolodney et al. 1974; Das et al. 1995; Ness et al. 1999), and were recently reviewed (Paria and Dey 2000; Maccarrone et al. 2002).

THC has been reported to account for the majority of the reproductive hazards of marijuana use, and in males it leads to impotence by suppressing spermatogenesis, reducing the weight of reproductive organs, and decreasing the plasma concentration of circulating hormones like testosterone (Kolodney et al. 1974). In females, THC inhibits ovulation by prolonging the oestrous cycle and decreasing the pro-oestrous surge of luteinizing hormone. In addition, exposure to natural cannabis extracts during pregnancy has been linked to embryotoxicity and to the production of specific teratological malformations in rats, hamsters and rabbits (Geber and Schramm 1969).

Also, AEA has been shown to impair pregnancy and embryo development in mice (Paria et al. 1996), suggesting that endocannabinoids might regulate fertility in mammals. Consistently, down-regulation of AEA levels in mouse uterus has been associated with increased uterine receptivity, which instead decreased when AEA was up-regulated (Yang et al. 1996; Schmid et al. 1997). The higher level of AEA in the nonreceptive uterus correlates well with the embryotoxic effect of the nonreceptive uterine environment, and also with the in vitro observation that AEA inhibits embryo development and zona-hatching of blastocysts (Paria et al. 1996; Yang et al. 1996; Schmid et al. 1997). In the mouse, mRNAs of AEA-binding CB_1 and CB_2 receptors are expressed in the preimplantation embryos, and the levels of CB_1 receptors are much higher than those found in brain (Das et al. 1995; Yang et al. 1996; Schmid et al. 1997). A recent study has also shown cross-talk between cannabinoid receptors and progesterone receptors in THC-induced modulation of sexual receptivity (Mani et al. 2001), further demonstrating that dysregulation of cannabinoid signalling disrupts uterine receptivity for embryo implantation (Paria et al. 2001).

4.2 The Endocannabinoid System and Male Reproductive Function

Despite the knowledge that chronic administration of THC to animals lowers testosterone secretion and reduces the production, motility and viability of sperm (Hall and Solowij 1998), it is not yet known whether the endocannabinoid system has any role in the control of male fertility in mammals. The binding of AEA to a cannabinoid receptor present on spermatozoa of sea urchin (*Strongylocentrotus purpuratus*) has been shown to reduce their fertilizing capacity (Chang et al. 1993; Schuel et al. 1994), and evidence that AEA regulates human sperm functions required for fertilization has been recently reviewed (Schuel et al. 2002a). In addition, a recent in vitro study has demonstrated that *N*-palmitoylethanolamine, a homologue of AEA, may affect the time-course of capacitation of human spermatozoa by modulating the properties of their membranes (Ambrosini et al. 2003). On the other hand, rat testis is able to synthesize AEA (Sugiura et al. 1996), and this compound has been detected in human seminal plasma at nanomolar (10 nM) concentrations (Schuel et al. 2002b). More recently, the presence of CB_1 receptors in Leydig cells and their involvement in testosterone secretion have been demonstrated in mice (Wenger et al. 2001). Also, the function of Sertoli cells has been shown to be altered by THC, though the molecular basis for this alteration has not been established (Newton et al. 1993). As Sertoli cells of the mammalian seminiferous epithelium are involved in the regulation of germ cell development by providing nutrients and hormonal signals needed for spermatogenesis (Griswold

1995), we recently sought to investigate whether Sertoli cells were able to bind and degrade AEA, and whether this endocannabinoid might induce apoptosis in these cells. In this context, the effect of FSH was also checked, because it dramatically impacts fetal and early neonatal Sertoli cell proliferation and is critical in determining spermatogenic capacity in the adult mammal (Orth et al. 1998).

We found that Sertoli cells have the biochemical machinery to degrade AEA and express functional CB_2 receptors on their surface (Maccarrone et al. 2003b). In addition, FSH dose-dependently inhibited apoptosis induced by AEA in these cells through a remarkable (four- to fivefold) increase in FAAH activity (Maccarrone et al. 2003b). Taken together, these data extend to male fertility the potential for FAAH to regulate the activity of AEA. Additionally, the finding that Sertoli cells belong to the peripheral endocannabinoid system opens new perspectives to the understanding and treatment of male fertility problems.

4.3 Sex Hormones, Th_1/Th_2 Cytokines, Leukaemia Inhibiting Factor and Endocannabinoids

Human reproductive fluids, such as seminal plasma, mid-cycle oviductal fluid, follicular fluid, amniotic fluid, as well as human amniotic fluid and human milk have been reported to contain AEA, *N*-palmitoylethanolamine (PEA) and *N*-oleoylethanolamine (OEA) in the low nanomolar range, i.e. from 3 nM of AEA in the follicular fluid to 67 nM of OEA in human milk (Schuel et al. 2002b). This suggests that endocannabinoids might regulate multiple physiological and pathological reproductive functions in humans, implying that exogenous cannabinoids delivered by marijuana smoke could impact these processes. Consistent with the hypothesis that endocannabinoids adversely affect human fertility, we have recently found a fall in FAAH activity and expression in the T lymphocytes of women experiencing miscarriage (Maccarrone et al. 2000) and a rise (4-fold) in blood AEA levels of the same subjects, compared to women with normal gestation (Maccarrone et al. 2002). The other components of the endocannabinoid system, like the AEA membrane transporter (AMT) and CB_1 receptors, were not affected (Table 3).

Table 3. FAAH activity, AMT activity and CB_1 receptor binding in women who miscarried and those who did not (data from Maccarrone et al. 2001)

Parameter	Pregnant women	Miscarrying women
FAAH activity[a]	133 ± 9 (100%)	48 ± 5 (36%)*
AMT activity[b]	50 ± 4 (100%)	49 ± 4 (100%)
CB_1 binding[c]	20,380 ± 1,930 (100%)	20,400 ± 1,795 (100%)

AMT, anandamide membrane transporter; FAAH, fatty acid amide hydrolase.
[a]Expressed as $pmol.min^{-1}.mg\ protein^{-1}$.
[b]Expressed as $pmol.min^{-1}.mg\ protein^{-1}$.
[c]Expressed as $cpm.mg\ protein^{-1}$.
*$p<0.0001$ vs pregnant women ($p>0.05$ in all other cases).

Peripheral T lymphocytes regulate fertility at the feto-maternal interface, by producing type 1 T helper (Th_1) and type 2 T helper (Th_2) cytokines (Piccinni et al. 1998). Th_2 cytokines, such as interleukin (IL)-3, IL-4 and IL-10, favour blastocyst implantation and successful pregnancy by promoting trophoblast growth either directly or indirectly through the inhibition of natural killer (NK) cell activity and the stimulation of natural suppressor cells. Conversely, Th_1 cytokines, such as IL-2, IL-12 and interferon-γ (INF-γ), impair gestation by causing direct damage to the trophoblast, by stimulating NK cells and by enhancing tumour necrosis factor-α (TNF-α) secretion by macrophages. Also, trophoblasts

stimulate release of pro-fertility Th_2 cytokines from T lymphocytes (so-called "Th_2 bias"), while suppressing the anti-fertility Th_1 bias. Progesterone (P) favours the Th_2 bias, thus stimulating the release from T lymphocytes of LIF, which in turn favours fetal implantation and survival (Szekenes-Bartho and Wegmann 1996; Stewart and Cullinan 1997; Duval et al. 2000).

P also stimulates FAAH activity and expression in human T lymphocytes (Maccarrone et al. 2001) by enhancing the promoter activity of the *FAAH* gene (Maccarrone et al. 2003c). Regulation of FAAH expression was observed also upon lymphocyte treatment with Th_1/Th_2 cytokines: IL-4 and IL-10 enhanced FAAH, while IL-2 and INF-γ reduced it (Maccarrone et al. 2001). Unlike FAAH, the other proteins of the endocannabinoid system were not affected by P or by any of the cytokines tested (Maccarrone et al. 2001), pointing to FAAH as the "check point" for AEA degradation during pregnancy.

4.4 Perspectives

The reported findings clearly show that in mammals ligand-receptor signalling with endocannabinoids is intimately associated with embryo–uterine interactions during implantation, and that in humans low FAAH in lymphocytes correlates with spontaneous abortion. This calls for attention to AEA-hydrolase as a key point in the control of the endocannabinoid system during pregnancy. Moreover, the results seem to add the endocannabinoids to the hormone-cytokine networks responsible for embryo–uterine interactions, and might represent a useful framework for the interpretation of novel interactions between progesterone, FSH, leptin, Th_1/Th_2 cytokines and (endo)cannabinoids, which appear to regulate both female sexual receptivity and male reproduction.

An interesting possibility raised by the data is that quantitation of FAAH protein in lymphocytes, which is easy to measure in routine analyses, might become an accurate marker of spontaneous abortion in humans. Such markers have long been sought, because of their potential diagnostic value, but they are not yet available or are still restricted to specific clinical situations.

5 General Conclusions

The endocannabinoid system contributes to the control of hypothalamic regulatory mechanisms. We do not know (yet) in which part of the hypothalamus the endocannabinoids are synthesized, but it is possible that cannabinoid receptors, present in the hypothalamus, are activated by AEA or other endocannabinoids released/synthesized quite far away. There is also the possibility that endocannabinoids act on presynaptic membranes to modulate the release of various neurotransmitters. Also of interest is the hypothesis that endocannabinoids may participate in hormone-cytokine networks that regulate reproduction, as this opens new perspectives for the development of novel medicines for human infertility.

Acknowledgements.
M.M. wishes to thank Prof. A. Finazzi-Agrò (Department of Experimental Medicine and Biochemical Sciences, University of Rome "Tor Vergata") for continuing interest and support, and all colleagues who have contributed to the research on endocannabinoids and fertility performed in his laboratory. This investigation was partly supported by Ministero dell'Istruzione, dell'Università e della Ricerca (Cofin 2002 and 2003 to M.M.), Rome. The research grants of Medical Research Council of Hungary (ETT) 003/03 and Hungarian Research Foundation (OTKA) No. T-034365 to T.W. are gratefully acknowledged.

References

Adams IB, Compton DR, Martin BR (1998) Assessment of anandamide interaction with the cannabinoid brain receptor: SR 141716A antagonism studies in mice and autoradiographic analysis of receptor binding in rat brain. J Pharmacol Exp Ther 284:1209–1217

Ambrosini A, Zolese G, Wozniak M, Genga D, Boscaro M, Mantero F, Balercia G (2003) Idiopathic infertility: susceptibility of spermatozoa to in-vitro capacitation, in the presence and the absence of palmitylethanolamide (a homologue of anandamide), is strongly correlated with membrane polarity studied by Laurdan fluorescence. Mol Hum Reprod 9:381–388

Ayalon D, Nir L, Cordova T, Bauminger S, Puder M, Naor Z, Kashi B, Zor U, Harrell A, Lindner HR (1977) Acute effects of Δ9-tetrahydrocannabinol on the hypothalamo-pituitary-ovarian axis in rat. Neuroendocrinology 23:31–42

Bisogno T, Melck D, Bobrov Myu, Grestaya NM, Bezuglov VV, De Petrocellis L, Di Marzo (2000) N-acyl-dopamines: novel synthetic CB1-receptor ligands and inhibitors of anandamide inactivation with cannabimimetic activity in vitro and in vivo. Biochem J 351:817–824

Butrica JL (2002) The medical use of cannabis among the Greek and Romans. J Cannabis Ther 2:51–70

Campfield LA, Smith FJ, Guisez Y, Devos R, Burn P (1995) Recombinant mouse OB protein: evidence for a peripheral signal linking adiposity and central neural networks. Science 269:546–549

Chakravarty I, Sheth AR, Ghosh JJ (1975) Effects of acute Δ9-tetrahydrocannabinol treatment of serum luteinizing hormone and prolactin levels in adult female rats. Fertil Steril 26:947–948

Chakravarty I, Sheth PR, Sheth AR, Ghosh JJ (1982) Delta-9-tetrahydrocannabinol: its effect on hypothalamo-pituitary system in male rats. Arch Androl 8:25–27

Chang MC, Berkery D, Schuel R, Laychock SG, Zimmerman AM, Zimmerman S, Schuel H (1993) Evidence for a cannabinoid receptor in sea urchin sperm and its role in blockade of the acrosome reaction. Mol Reprod Dev 36:507–516

Collu R (1976) Endocrine effects of chronic intraventricular administration of Δ9-tetrahydrocannabinol to prepuberal and adult male rats. Life Sci 18:223–230

Corchero J, Manzanares J, Fuentes JA (2001) Role of gonadal steroids in the corticotropin-releasing hormone and proopiomelanocortin gene expression response to Δ9-tetrahydrocannabinol in the hypothalamus of the rat. Neuroendocrinology 74:185–192

Crawford V (2002) A homelie herbe: medicinal cannabis in early England. J Cannabis Ther 2:71–80

Dalterio SL, Michael SD, Macmillan BT, Bartke A (1981) Differential effects of cannabinoid exposure and stress on plasma prolactin, growth hormone and corticosterone levels in male mice. Life Sci 28:761–766

Das SK, Paria BC, Chakraborty I, Dey SK (1995) Cannabinoid ligand-receptor signaling in the mouse uterus. Proc Natl Acad Sci USA 92:4332–4336

Dewey WI (1986) Cannabinoid pharmacology. Pharmacol Rev 38:151–178

Di Marzo V, Breivogel C, Bisogno T, Melck D, Patrick G, Tao Q, Szallasi A, Razdan RK, Martin BR (2000) Neurobehavioral activity in mice of N-vanillyl-arachidonoyl-amide. Eur J Pharmacol 406:363–374

Di Marzo V, Goparaju SK, Wang L, Liu J, Bátkai S, Járai Z, Fezza F, Miura GI, Palmiter RD, Sugiura T, Kunos G (2001) Leptin-regulated endocannabinoids are involved in maintaining food intake. Nature 410:822–825

Di Marzo V, De Petrocellis L, Fezza F, Ligresti A, Bisogno T (2002) Anandamide receptors. Prostaglandins Leukot Essent Fatty Acids 66:377–391

Doms RW Jr, Nichols SF, Harclerode J (1981) The effects of Δ9-tetrahydrocannabinol and phencyclidine hydrochloride on gonadotrophs in rat anterior pituitary. Fed Proc 40:279

Duval D, Reinhardt B, Kedinger C, Boeuf H (2000) Role of suppressors of cytokine signaling (Socs) in leukemia inhibitory factor (LIF)-dependent embryonic stem cell survival. FASEB J 14:1577–1584

Fernandez-Ruiz JJ, Munoz RM, Romero J, Villanua MA, Makryannis A, Ramos JA (1997) Time course of the effects of different cannabimimetics on prolactin and gonadotrophin secretion: evidence for the presence of CB1 receptors in hypothalamic structures and their involvement in the effects of cannabimimetics. Biochem Pharmacol 53:1919–1927

Friedman JM, Halaas JL (1998) Leptin and regulation of body weight in mammals. Nature 395:763–770

Funahashi H, Yada T, Muroya S, Takigawa M, Ryushi T, Horie S, Nakai Y, Shioda S (1999) The effect of leptin on feeding-regulating neurones in the rat hypothalamus. Neurosci Lett 264:117–120

Geber WF, Schramm LC (1969) Effect of marihuana extract on fetal hamsters and rabbits. Toxicol Appl Pharmacol 14:276

Gonzales S, Bisogno T, Wenger T, Manzanares J, Milone A, Berrendero F, Di Marzo V, Ramos JA, Fernandez-Ruiz JJ (2000) Sex steroid influence on cannabinoid CB1 receptor mRNA and endocannabinoid levels in the anterior pituitary gland. Biochem Biophys Res Commun 270:260–266

Gonzalez S, Manzanares J, Berrendero F, Wenger T, Corchero J, Bisogno T, Romero J, Fuentes JA, Di Marzo V, Ramos JA, Fernandez-Ruiz J (1999) Identification of endocannabinoids and CB1 receptor mRNA in the pituitary gland. Neuroendocrinology 70:137–145

Griswold MD (1995) Interactions between germ cells and Sertoli cells in the testis. Biol Reprod 52:211–216

Hall W, Solowij N (1998) Adverse effects of cannabis. Lancet 352:1611–1616

Hanus L, Abu-Lafi S, Friede E, Breuer A, Vogel Z, Shalev DE, Kustanovich I, Mechoulam R (2001) 2-Arachidonyl glyceryl ether, an endogenous agonist of the CB1 receptor. Proc Natl Acad Sci USA 98:3662–3665

Herkenham M (1995) Localisation of cannabinoid receptors in brain and periphery. In: Pertwee RG (ed) Cannabinoid receptors. Academic Press, London, San Diego, p 145

Kolodney RC, Masters WH, Kolodner RM, Toro G (1974) Depression of plasma testosterone levels after chronic intensive marihuana use. N Engl J Med 290:872–874

Kramer J, Ben David M (1974) Suppression of prolactin secretion by acute administration of Δ9-THC in rats. Proc Soc Exp Biol Med 147:482–484

Kumar MSA, Chen CL (1983) Naloxone blocks the effects of Δ9-tetrahydrocannabinol on serum luteinizing hormone and prolactin in rats. Subst Alcohol Actions Misuse 4:347–353

Levine JE (2000) The hypothalamus as a major integrative center. In: Conn PM, Freeman ME (eds) Neuroendocrinology in physiology and medicine. Humana Press, Totowa, p 75

Lozano I (2001) The therapeutic use of Cannabis sativa (L.) in Arabic medicine. J Cannabis Ther 1:63–70

Maccarrone M, Valensise H, Bari M, Lazzarin N, Romanini C, Finazzi-Agrò A (2000) Relation between decreased anandamide hydrolase concentrations in human lymphocytes and miscarriage. Lancet 355:1326–1329

Maccarrone M, Valensise H, Bari M, Lazzarin N, Romanini C, Finazzi-Agrò A (2001) Progesterone up-regulates anandamide hydrolase in human lymphocytes: role of cytokines and implications for fertility. J Immunol 166:7183–7189

Maccarrone M, Falciglia K, Di Rienzo M, Finazzi-Agrò A (2002) Endocannabinoids, hormone-cytokine networks and human fertility. Prostaglandins Leukot Essent Fatty Acids 66:309–317

Maccarrone M, Di Rienzo M, Finazzi-Agrò A, Rossi A (2003a) Leptin activates the anandamide hydrolase promoter in human T lymphocytes through STAT3. J Biol Chem 278:13318–13324

Maccarrone M, Cecconi S, Rossi G, Battista N, Pauselli R, Finazzi-Agrò A (2003b) Anandamide activity and degradation are regulated by early postnatal ageing and follicle-stimulating hormone in mouse Sertoli cells. Endocrinology 144:20–28

Maccarrone M, Bari M, Di Rienzo M, Finazzi-Agrò A, Rossi A (2003c) Progesterone activates fatty acid amide hydrolase (FAAH) promoter in human T lymphocytes through the transcription factor Ikaros. Evidence for a synergistic effect of leptin. J Biol Chem 278:32726–32732

Mani SK, Mitchell A, O'Malley BW (2001) Progesterone receptor and dopamine receptors are required in Δ9-tetrahydrocannabinol modulation of sexual receptivity in female rats. Proc Natl Acad Sci USA 98:1249–1254

Manzanares J, Corchero J, Fuentes JA (1999) Opioid and cannabinoid receptor-mediated regulation of the increase in adrenocorticotropin hormone and corticosterone plasma concentrations induced by central administration of Δ9-tetrahydrocannabinol in rats. Brain Res 839:173–179

Marks BH (1973) Δ9-TetraHydroCannabinol and luteinizing hormone secretion. Prog Brain Res 39:331–338

Maruta O, Shioda S, Funahashi H, Nakai Y (1999) Immunocytochemical localisation of leptin receptor in rat hypothalamus. Showa Univ J Med Sci 11:75–82

Mechoulam R, Gaoni Y (1965) A total synthesis of dl-delta-tetrahydrocannabinol, the active constituent of hashish. J Am Chem Soc 87:3273–3275

Moldrich G, Wenger T (2000) Localization of the CB1 cannabinoid receptor in the rat brain. An immunohistochemical study. Peptides 21:1735–1742

Moreau JJ (1845) De hachish et de l'alienation mentale: etudes psychologiques. Fortin Masson, Paris

Ness RB, Grisso JA, Hirschinger N, Markovic N, Shaw LM, Day NL, Kline J (1999) Cocaine and tobacco use and the risk of spontaneous abortion. N Engl J Med 340:333–339

Newton SC, Murphy LL, Bartke A (1993) In vitro effects of psychoactive and non-psychoactive cannabinoids on immature rat Sertoli cell function. Life Sci 53:1429–1434

Oomura Y, Ono T, Ooyama H, Wayner MJ (1969) Glucose and osmosensitive neurones of the rat hypothalamus. Nature 222:282–284

Orth JM, McGuinness MP, Qiu J, Jester WF Jr, Li LH (1998) Use of in vitro systems to study male germ cell development in neonatal rats. Theriogenology 49:431–439

Pagotto U, Marsicano G, Fezza F, Theodoropoulou M, Grubler Y Arzberger T, Milone A, Losa M, Di Marzo V, Lutz B, Stalla GK (2001) Normal human pituitary gland and pituitary adenomas express cannabinoid receptor type 1 and synthesise endogenous cannabinoids: first evidence for a direct role of cannabinoids on hormone modulation at the human pituitary level. J Clin Endocrinol Metab 86:2687–2696

Paria BC, Dey SK (2000) Ligand-receptor signaling with endocannabinoids in preimplantation embryo development and implantation. Chem Phys Lipids 108:211–220

Paria BC, Deutsch DD, Dey SK (1996) The uterus is a potential site for anandamide synthesis and hydrolysis: differential profiles of anandamide synthase and hydrolase activities in the mouse uterus during the periimplantation period. Mol Reprod Dev 45:183–192

Paria BC, Song H, Wang X, Schmid PC, Krebsbach RJ, Schmid HH, Bonner TI, Zimmer A, Dey SK (2001) Dysregulated cannabinoid signaling disrupts uterine receptivity for embryo implantation. J Biol Chem 276:20523–20528

Paxinos G, Watson C (1997) The rat brain in steretaxic coordinates. Academic Press, San Diego

Pertwee RG (1985) Effects of cannabinoids on thermoregulation: a brief review. In: Harvey DJ (ed) Marihuana '84. IRL Press, London, p 263

Pertwee RG (1992) In vivo interactions between psychotropic cannabinoids and other drugs involving central and peripheral neurochemical mediators. In: Murphy L, Bartke A (eds) Marijuana/cannabinoids. Neurobiology and neurophysiology. CRC Press, Boca Raton, p 165

Peters H, Nahas G (1999) A brief history of four millennia (B.C. 2000-A.D. 1974). In: Nahas GG, Sutin KM, Harvey DJ, Agurell S (eds) Marihuana and medicine. Humana Press, Totowa, p 3

Piccinni MP, Beloni L, Livi C, Maggi E, Scarselli G, Romagnani S (1998) Defective production of both leukemia inhibitory factor and type 2 T-helper cytokines by decidual T cells in unexplained recurrent abortions. Nat Med 4:1020–1024

Rettori V, Wenger T, Snyder G, Dalterio S, McCann SM (1988) Hypothalamic action of delta9-tetrahydrocannabinol to inhibit the release of prolactin and growth hormone in the rat. Neuroendocrinology 47:498–503

Rodriguez de Fonseca F, Cebeira M, Ramos JA, Martin M, Fernandez-Ruiz JJ (1994) Cannabinoid receptors in the rat brain areas: sexual differences, fluctuations during estrous cycle and changes after gonadectomy and sex steroid replacement. Life Sci 54:159–170

Romero J, Wenger T, deMiguel R, Ramos JA, Fernández-Ruiz JJ (1998) Cannabinoid receptor binding did not vary in several hypothalamic nuclei after hypothalamic deafferentation. Life Sci 63:351–356

Samson WK, Snyder G, Fawcett CP, McCann SM (1980) Chromatographic and biological analysis of ME and OVLT LHRH. Peptides 1:97–102

Schmid PC, Paria BC, Krebsbach RJ, Schmid HHO, Dey SK (1997) Changes in anandamide levels in mouse uterus are associated with uterine receptivity for embryo implantation. Proc Natl Acad Sci USA 94:4188–4192

Schuel H, Goldstein E, Mechoulam R, Zimmerman AM, Zimmerman S (1994) Anandamide (arachidonylethanolamide), a brain cannabinoid receptor agonist, reduces sperm fertilizing capacity in sea urchins by inhibiting the acrosome reaction. Proc Natl Acad Sci USA 91:7678–7682

Schuel H, Burkman LJ, Lippes J, Crickard K, Forester E, Piomelli D, Giuffrida A (2002a) N-acylethanolamines in human reproductive fluids. Chem Phys Lipids 121:211–227

Schuel H, Burkman LJ, Lippes J, Crickard K, Mahony MC, Giuffrida A, Picone RP, Makriyannis A (2002b) Evidence that anandamide-signaling regulates human sperm functions required for fertilization. Mol Reprod Dev 63:376–387

Scorticati C, Mohn C, De Laurentiis A, Vissio P, Fernandez-Solari C, Seilicovich A, McCann SM, Rettori V (2003) The effect of anandamide on prolactin secretion is modulated by estrogen. Proc Natl Acad Sci USA 100:2134–2139

Smith CG, Besch NF, Asch RH (1980) Effects of marihuana on reproductive system. In: Thomas JA (ed) Advances in sex hormone research, vol. 4. Urban and Swarzenberg, Baltimore, p 273

Steger RW, Silverman AY, Siler-Khodr TM, Asch RH (1980) The effects of Δ9-tetrahydrocannabinolon the positive and negative feedback control of luteinizing hormone release. Life Sci 27:1911–1916

Stewart CL, Cullinan EB (1997) Preimplantation development of the mammalian embryo and its regulation by growth factors. Dev Genet 21:91–101

Sugiura T, Kondo S, Sukagawa A, Tonegawa T, Nakane S, Yamashita A, Waku K (1996) Enzymatic synthesis of anandamide an endogenous cannabinoid receptor ligand through N-acylphosphatidylethanolamine pathway in testis: involvement of Ca2+-dependent transacylase and phosphodiesterase activities. Biochem Biophys Res Commun 218:113–117

Sulcova E, Mechoulam R, Friede E (1998) Biphasic effects of anandamide. Pharmacol Biochem Behav 59:347–352

Szekenes-Bartho J, Wegmann TG (1996) A progesterone-dependent immunomodulatory protein alters the Th1/Th2 balance. J Reprod Immunol 31:81–95

Takahashi N, Patel HR, Qi Y, Dushay J, Ahima RS (2002) Divergent effects of leptin in mice susceptible or resistant to obesity. Horm Metab Res 34:691–697

Tartaglia LA, Devos R, Richard GJ, Campfield LA, Clark FT, Deeds J, Muir C, Sanker S, Monarty A, Moore KJ, Smutko JS, Mays GG, Woof EA, Monroe CA, Tepper RI (1995) Identification and expression cloning of a leptin receptor, OB-R. Cell 83:1263–1271

Tyrey C (1980) Δ9-tetrahydrocannabinol: a potent inhibitor of episodic luteinizing hormone secretion. J Pharmacol Exp Ther 213:306–308

Tyrey C (1984) Endocrine aspects of cannabinoid action in female subprimates: search for site of action. In: Braude MC, Ludford JP (eds) Marijuana effects on the endocrine and reproductive systems. NIDA Research Monograph Ser 44, Rockville, p 65

Weidenfeld J, Feldman S, Mechoulam R (1994) Effect of the brain constituent anandamide, a cannabinoid receptor agonist, on the hypothalamo-pituitary-adrenal axis in the rat. Neuroendocrinology 59:110–112

Wenger T, Rettori V, Snyder G, Dalterio S, McCann S (1987) Effects of Δ9-tetrahydrocannabinol (THC) on the hypothalamic-pituitary control of LH and FSH secretion in adult male rats. Neuroendocrinology 46:488–493

Wenger T, Jamali KA, Juaneda C, Leonardelli J, Tramu G (1997) Arachidonoyl ethanolamide (anandamide) activates the parvocellular part of hypothalamic paraventricular nucleus. Biochem Biophys Res Commun 237:724–728

Wenger T, Fernandez-Ruiz J, Ramos J (1999) Immunohistochemical demonstration of CB1 receptors in the anterior lobe of the pituitary gland. J Neuroendocrinol 11:873–878

Wenger T, Ledent C, Csernus V, Gerendai I (2001) The central cannabinoid receptor inactivation suppresses endocrine reproductive functions. Biochem Biophys Res Commun 284:363–368

Williams CM, Kirkham TC (1999) Anandamide induces overeating meditation by central (CB1) receptors. Psychopharmacology (Berl) 143:315–317

Williams CM, Rogers PJ, Kirkham TC (1998) Hyperphagia in prefeed rats following oral Δ9-THC. Physiol Biol Behav 65:343–346

Yang Z-M, Paria BC, Dey SK (1996) Activation of brain-type cannabinoid receptors interferes with preimplantation mouse embryo development. Biol Reprod 55:756–761

HEP (2005) 168:573–598

Cannabinoids and the Digestive Tract

A.A. Izzo[1] · A.A. Coutts[2] (✉)

[1]Department of Experimental Pharmacology, University of Naples Federico II, via D Montesano 49, 80131 Naples, Italy

[2]School of Medical Sciences, College of Life Sciences and Medicine, University of Aberdeen, Institute of Medical Sciences, Foresterhill, Aberdeen AB25 2ZD, UK
a.a.coutts@abdn.ac.uk

Abstract In the digestive tract there is evidence for the presence of high levels of endocannabinoids (anandamide and 2-arachidonoylglycerol) and enzymes involved in the synthesis and metabolism of endocannabinoids. Immunohistochemical studies have shown the presence of CB_1 receptors on myenteric and submucosal nerve plexuses along the alimentary tract. Pharmacological studies have shown that activation of CB_1 receptors produces relaxation of the lower oesophageal sphincter, inhibition of gastric motility and acid secretion, as well as intestinal motility and secretion. In general, CB_1-induced inhibition of intestinal motility and secretion is due to reduced acetylcholine release from enteric nerves. Conversely, endocannabinoids stimulate intestinal primary sensory neurons via the vanilloid VR1 receptor, resulting in enteritis and enhanced motility. The endogenous cannabinoid system has been found to be involved in the physiological control of colonic motility and in some pathophysiological states, including paralytic ileus, intestinal inflammation and cholera toxin-induced diarrhoea. Cannabinoids also possess antiemetic effects mediated by activation of central and peripheral CB_1 receptors. Pharmacological modulation of the endogenous cannabinoid system could provide a new therapeutic target for the treatment of a number of gastrointestinal diseases, including nausea and vomiting, gastric ulcers, secretory diarrhoea, paralytic ileus, inflammatory bowel disease, colon cancer and gastro-oesophageal reflux conditions.

Keywords Cannabinoid receptors · Intestinal motility · Intestinal secretion · Emesis · Intestinal inflammation · Feeding

1
Introduction

Preparations of *Cannabis sativa* (Indian hemp) have been used medicinally for the treatment of a variety of gastrointestinal disorders, including gastrointestinal pain, flatulence, gastroenteritis, Crohn's disease, diarrhoea and diabetic gastroparesis (Di Carlo and Izzo 2003). The main psychotropic constituent of *Cannabis sativa* is Δ^9-tetrahydrocannabinol (Δ^9-THC), which exerts its biological effects mainly by activating two G protein-coupled cannabinoid receptors (Pertwee and Ross 2002). These are CB_1 receptors, present in central and peripheral nerves, including the enteric nervous system, and CB_2 receptors, expressed mainly in immune cells. A general feature of CB_1 activation is the reduction of the release of a variety of neurotransmitters (e.g. acetylcholine from enteric nerves), whereas there is currently no evidence for a role for CB_2 receptors in the gastrointestinal (GI) tract (Di Carlo and Izzo 2003). Endogenous ligands for the cannabinoid receptors have been identified, the best-known being anandamide, 2-arachidonoyl glycerol (2-AG) (non-selective cannabinoid receptor agonists), noladin ether (CB_1 receptor agonist) and virodhamine (CB_1 receptor antagonist/CB_2 receptor agonist) (De Petrocellis et al. 2004). When released, anandamide and 2-AG are removed from extracellular compartments by a carrier-mediated re-uptake process. Once within the cell, endocannabinoids are hydrolysed by the enzyme fatty acid amide hydrolase (FAAH, also named anandamide amidohydrolase) (Sugiura et al. 2002). Also, 2-AG has been shown to be degraded by monoglyceride lipase (monoacyl glycerol lipase). Both FAAH and monoglyceride lipase have been demonstrated in the intestine (Oleinik 1995; Katayama et al. 1997). In addition to the two cannabinoid receptors, anandamide and 2-AG can also activate transient receptor potential vanilloid subtype 1 (VR1, also known as TRPV1) receptors, the molecular target for the pungent plant compound capsaicin (Zygmunt et al. 1999). Cannabinoid receptors, their endogenous ligands (endocannabinoids) and the proteins involved in endocannabinoid inactivation (cellular reuptake and enzymatic degradation) are collectively referred to as the endogenous cannabinoid system (ECS).

Although cannabinoids have a wide variety of biological actions, this article will summarise the main studies dealing with the role of the ECS in the gut, including the effects of cannabinoids on emesis.

2
The Endogenous Cannabinoid System in the Gut

There are several lines of evidence for a functional ECS in the GI tract. The enteric responses to exogenous cannabinoid drugs show all the hallmarks of a receptor-mediated mechanism, namely, high potency, chemical and stereo-selectivity and structure–activity relationships (Coutts et al. 2000; Coutts and Pertwee 1997; Pertwee 2001). This is coupled with the identification of high-affinity specific binding sites that are saturable at low ligand concentrations and whose characteristics resemble those in the brain (Casu et al. 2003; Ross et al. 1998). The presence of CB_1 receptors in rat intestine was demonstrated by radioligand autoradiography with [^{3}H]-CP 55,940 (Lynn and Herkenham 1994) and, more recently, in other species by immunohistochemistry with selective antibodies raised against the N- or C-terminus of the receptor (Casu et al. 2003; Coutts et al. 2002; Kulkarni-Narla and Brown 2000; MacNaughton et al. 2003, 2004; Pinto et al. 2002b; Storr et al. 2004). CB_1 receptor protein was found to be associated with cholinergic neurons in both the submucous and myenteric plexuses in the pig, guinea-pig, rat and mouse (Casu et al. 2003; Pinto et al. 2002b). Cholinergic neurons are identified by the presence of cholinacetyl transferase (ChAT), the enzyme responsible for the synthesis of acetylcholine (ACh). The GI tract of the pig, an omnivorous animal, shares many similarities with that of humans. In cross-sections of the porcine gut, colocalisation experiments indicated that CB_1 receptors were not expressed by nitrergic nor vasoactive intestinal peptide (VIP)-immunoreactive inhibitory neurons (Kulkarni-Narla and Brown 2000). This was also true in guinea-pig tissue, where all CB_1 receptor immunoreactivity was associated with excitatory neurons (Coutts et al. 2002; Kulkarni-Narla and Brown 2000). In primary culture, porcine myenteric CB_1-positive cells also expressed κ- or ∂-opioid receptor-like immunoreactivity, in line with their functional sensitivity to opioid ligands (Poonyachoti et al. 2002). Unlike those from the guinea-pig, pig myenteric neurons do not appear to express μ-opioid receptors (Brown et al. 1998). Analysis of the CB_1 receptor immunoreactivity of myenteric ganglionic neurons in whole mounts of the guinea-pig myenteric plexus-longitudinal muscle preparation (MP-LMP) allowed visualisation of the cellular morphology, unavailable in cross sections. Images showed CB_1 receptor expression in the somata of both Dogiel cell types I and II and punctate expression on neurites of sensory neurons, interneurons and motoneurons, as identified by colocalisation with selective neuronal markers, e.g. calbindin, neurofilament proteins and calretinin (Coutts et al. 2002). There was also a close association with the synaptic protein, synapsin 1, although the limited resolution of the confocal microscope proscribed analysis of the synaptic distribution of these receptors. Similar results were found in guinea-pig colon and rat ileum preparations, though the quantitative distribution of cholinergic subpopulations varied between tissue types (Coutts 2004; Coutts et al. 2002). In mouse intestine, CB_1 receptor labelling was found throughout the GI tract but was most intense in the ileum. In the stomach, the receptors occurred in submucosal ganglia adjacent to the gastric epithelium and also between the smooth muscle layers (Casu et al. 2003; Storr et al. 2004).

CB_1 receptor mRNA was detected in the GI tract of the rat, mouse and guinea-pig (Izzo et al. 2003; Storr et al. 2002). In whole gut homogenates from the guinea-pig, CB_1 receptor and CB_2 receptor-like mRNA transcripts were detected, whereas only CB_1 receptor mRNA was found in the myenteric plexus (Griffin et al. 1997). CB_1 receptor mRNA was also detected in human colon (Shire et al. 1995). Reverse transcription-polymerase chain reaction (RT-PCR) found both CB_1 receptor and CB_2 receptor mRNA in the rat stomach and mouse small intestine (Izzo et al. 2003; Storr et al. 2002). The expression level of CB_1 receptor mRNA in the latter was upregulated after treatment with cholera toxin (Izzo et al. 2003).

Burdyga and colleagues have recently reported that vagal afferent neurons projecting to the rat stomach and duodenum co-express cholecystokinin (CCK)-1 and CB_1 receptors and that the expression of CB_1 receptors was increased by withdrawal of food and decreased after refeeding (Burdyga et al. 2004). Changes in CB_1 expression were blocked by administration of the CCK-1 receptor antagonist lorglumide (i.p.) and mimicked by administration of CCK (a satiety factor). Rat intestinal anandamide levels also increased after food deprivation (with normalisation after refeeding) and peripheral (but not central) administration of the CB_1 antagonist SR141716A-suppressed food intake (Gomez et al. 2002). This is consistent with the observation of an anorexic action of SR141716A in obese humans (Heshmati et al. 2001), suggesting a role for peripheral CB_1 receptors in the regulation of feeding.

Of the endogenous ligands mentioned in the introduction, to date the effects of anandamide and its analogues, 2-AG, which was first isolated from canine ileum, and noladin ether, have been investigated in the GI tract. Noladin ether (i.p.) significantly reduces the defaecation rate in mice (Hanus et al. 2001). Interestingly, intestinal anandamide levels increase after food deprivation (Gomez et al. 2002) or in some pathophysiological states, including experimental ileus (Mascolo et al. 2002), cholera toxin-induced diarrhoea (Izzo et al. 2003) and cancer (patients with adenomatous polyps and carcinomas) (Ligresti et al. 2003). Unlike most hydrophilic neurotransmitters, lipophilic endocannabinoids are not stored in synaptic vesicles, but appear to be synthesised and released on demand. Both anandamide and 2-AG are metabolised by the microsomal enzyme FAAH (Katayama et al. 1997; Ueda and Yamamoto 2000) following uptake by selective membrane uptake processes (Izzo et al. 2001c). This uptake carrier mechanism can be inhibited by AM404 (Pertwee 2001) or VDM11 (Izzo et al. 2003; Mascolo et al. 2002), thus preventing metabolism and potentiating any agonist effect. Although FAAH can catalyse both the synthase and hydrolase reactions, the synthase/hydrolase ratio (5.0) is particularly high in the rat small intestine compared with other rat tissues (Katayama et al. 1997). In the same study, FAAH mRNA was confirmed by Northern blots. This enzyme is thought to exert tonic control of local anandamide levels, and its activity can be reduced by exogenous phenylmethylsulphonyl fluoride (PMSF) (Pertwee et al. 1995) and thus can potentiate the weak agonist activity of anandamide observed in vitro. The presence of specific receptors and endogenous ligands together with their synthetic and catabolic enzymes is strong support for a functional endocannabinoid system in the GI tract.

However, more persuasive evidence for ongoing activity in this system can be derived from the responses to selective CB_1 receptor antagonists, mainly SR141716A, but also AM281 or AM630, in the absence of any exogenous agonist. The direction of these responses is invariably opposite to that which would be expected of a cannabinoid receptor agonist and a useful summary is provided by Pinto et al. (2002a). In mice and rats, SR141716A increased motility, transit, defaecation, fluid accumulation and peristaltic contractions (Casu et al. 2003; Colombo et al. 1998; Izzo et al. 1999b, 2003, 2000a,b; Mancinelli et al. 2001; Pinto et al. 2002b). In the rat stomach, SR141716A increased the occurrence of transient lower oesophageal sphincter relaxations (Lehmann et al. 2002), and AM630 potentiated nonadrenergic–noncholinergic (NANC)-evoked relaxations of the fundus (Storr et al. 2002). SR141716A was first shown to increase neurotransmission and ACh release in the guinea-pig MP-LMP (Coutts et al. 2000; Coutts and Pertwee 1997; Pertwee et al. 1996). SR141716A increased maximal ejection pressure during the emptying phase of peristalsis in the guinea pig ileum (Izzo et al. 2000a) and both tonic and phasic motor activity in the colonic longitudinal smooth muscle in the isolated colon of mouse subjected to electrically evoked peristalsis (Mancinelli et al. 2001). These data suggest that peristaltic activity may be tonically inhibited by the endocannabinoid system. Interestingly, the facilitation of peristalsis in the guinea-pig was not observed by Heinemann (1999), suggesting a possible variability of endocannabinoid tone. Facilitatory effects of SR141716A have also been found on the cholinergic and NANC-mediated contractions of the circular muscle (Izzo et al. 1998). However, in view of the reported inverse agonist properties of SR141716A, it is not possible to determine, conclusively, whether its GI actions are due to antagonism of endocannabinoids or to the presence of CB_1 receptors that are precoupled to their effector mechanisms (inverse agonism). When tested on human innervated longitudinal muscle strips, SR141716A alone appeared to have no discernable effects (Croci et al. 1998; Manara et al. 2002).

3 Gastrointestinal Motility

The predominant action of cannabinoid receptor agonists on the GI tract is an inhibitory effect on gastrointestinal motility, reminiscent of the neuromodulatory response to presynaptic μ-opioid receptor or α_2-adrenoceptor activation of cholinergic, postganglionic parasympathetic neurons. The mechanisms underlying this effect have been studied chiefly in the GI tract of small rodents, but also in man and the pig. Here we shall review the findings of studies carried out in vitro (Sect. 3.1, below) and in vivo (Sect. 3.2).

3.1
In Vitro Studies

3.1.1
Effects on Excitatory Neuronal Pathways

The depressant effects of cannabinoid receptor activation on gastrointestinal motility, as observed in vitro are, principally, the inhibition of evoked cholinergic and NANC contractile responses. Studies have focussed on the inhibition of the peristaltic reflex in segments of whole intestine, on the inhibition of evoked contractions of longitudinal or circular smooth muscle preparations or on the reduction of excitatory neurotransmitter release. Early experiments with Δ^9-THC and some of the more non-polar organic fractions of tincture of *Cannabis* (British Pharmaceutical Codex) indicated the ability of putative cannabinoid receptor agonists to inhibit the contractile responses of the guinea-pig ileum without affecting responses to exogenous ACh (see review by Pertwee 2001). The peristaltic reflex can be reproduced in intestinal segments maintained in vitro. The synthetic cannabinoid receptor agonists WIN 55,212-2 (0.3–300 nM) significantly decreased longitudinal muscle reflex contraction, compliance and maximal ejection pressure, while increasing the threshold pressure and volume required to elicit peristalsis in guinea-pigs (Izzo et al. 2000a). At maximal agonist concentrations, peristalsis was completely prevented. These effects were insensitive to the opioid antagonist naloxone, the α_2-adrenoceptor antagonist, phentolamine or the CB_2 receptor selective antagonist SR144528 (0.1 μM). However, blockade was achieved with the CB_1 receptor-selective antagonist SR141716A (0.1 μM), thus indicating selective activation of cannabinoid CB_1 receptors. Methanandamide, a more stable analogue of anandamide, similarly increased the peristaltic pressure threshold and inhibited the ascending circular muscle contraction (Heinemann et al. 1999). The methanandamide response was antagonised by SR141716A and also by apamin and reduced by the NO synthase inhibitor, *N*-nitro-L-arginine methyl ester (L-NAME) implying a possible involvement of apamin-sensitive Ca^{2+}-activated K^+ channels and nitric oxide (Heinemann et al. 1999). Thus, inhibition by cannabinoids may affect excitatory or inhibitory components of the reflex. These data are consistent with the ability of apamin to reduce cannabinoid CB_1-mediated inhibition of cholinergic transmission in the guinea-pig ileum (Izzo et al. 1998).

Paton and Zar (1968) described the dissection of the MP-LMP of the guinea-pig small intestine. This preparation has been invaluable in the study of neurotransmission from the myenteric plexus to the longitudinal smooth muscle, particularly by opioids and cannabinoids, without the confounding effects of the peristaltic reflex. A similar preparation has been used to study neuromuscular transmission to the circular smooth muscle (Izzo et al. 1998). Contractions of MP-LMP induced by electrical field stimulation (EFS) were potently inhibited in a concentration-dependent fashion by the cannabinoid receptor agonists CP 55,940, CP 50,556, WIN 55,212-2, nabilone, CP 56,667, Δ^9-THC and cannabinol (Coutts and Pertwee 1997; Pertwee 2001). This inhibition was competitively and reversibly antagonised by SR141716A, without any effect on the inhibitory responses to normorphine (μ-opioid receptor agonist) or clonidine (α_2-adrenoceptor agonist) and indicated an involvement of CB_1 receptors. Therefore, electrically stimulated isolated preparations from the guinea-pig ileum have been used to demonstrate the high potency and stereoselectivity of CB_1 receptor agonists (Nye et al. 1985; Pertwee 2001; Pertwee et al. 1992, 1995, 1996). The rank order of potency of agonists correlates well with their affinities for CB_1 receptor binding sites in brain tissue and their known psychotropic effects (Pertwee 1997; Pertwee et al. 1992, 1996). The findings that the cannabinoid-induced inhibition of the guinea-pig MP-LMP was augmented by lowering the extracellular calcium concentration or attenuated by incubating the tissue with forskolin, 8-bromo-cyclic adenosine monophosphate (8-bromo-cAMP) or with the phosphodiesterase inhibitor 3-isobutyl-1-methyl xanthine supports the known signal transduction mechanisms for CB_1 receptors (Coutts and Pertwee 1998). Similar cannabinoid inhibitory effects on evoked responses have been reported for longitudinal strips of human tissue (Croci et al. 1998).

In a single electrophysiological analysis of intracellular recordings from myenteric neurons of the guinea-pig MP-LMP, WIN 55,212-2 or CP 55,940 were found to inhibit fast and slow excitatory synaptic transmission. In a subset of the neurons tested, this effect was reversed by SR141716A (López-Redondo et al. 1997). Both cholinergic and NANC responses of circular smooth muscle due to EFS were presynaptically inhibited by cannabinoids by a mechanism that was sensitive to SR141716A but not L-NAME or naloxone (Izzo et al. 1998). Only the cholinergic component of this response was sensitive to attenuation by apamin, suggesting the involvement of Ca^{2+}-activated K^+ channels. The contractile responses to γ-aminobutyric acid or 5-hydroxytryptamine, agents that release ACh in the intestine, have been shown to be reduced by Δ^9-THC or its analogues (Rosell and Agurell 1975; Rosell et al. 1976). There is some evidence that the release of adenosine, which also inhibits cholinergic neuromuscular transmission in this preparation, is susceptible to modulation via CB_1 receptor activation (Begg et al. 2002a).

3.1.2
Effects on Inhibitory Neurotransmission

There is evidence that cannabinoids affect enteric inhibitory transmission in rodents. Storr and colleagues used standard intracellular recording techniques to study the effect of cannabinoid drugs on enteric transmission (Storr et al. 2004). Focal electrical stimulation of intrinsic neurons of isolated strips of the mouse proximal colon induced a transient excitatory junction potential (EJP, abolished by atropine) followed by a fast (transient) inhibitory junction potential (fIJP, which represents the apamin-sensitive component of inhibitory transmission) and a slow (sustained) inhibitory junction potential (sIJP, which represents the nitric oxide-dependent component of inhibitory transmission). WIN 55,212-2 significantly reduced EJP and the fIJP (an effect sensitive to the CB_1 receptor antagonist SR141716A), but not sIJP; given alone, SR141716A significantly increased EJP, while fIJP and sIJP remained unchanged (Storr et al. 2004). These data suggest that cannabinoids, via CB_1 receptor activation, might reduce the apamin component (which is mediated by ATP or related purines) of the inhibitory transmission in the mouse colon. Other indirect evidence was provided by Heinemann and colleagues, which showed that methanandamide depressed intestinal peristalsis with a mechanism involving, at least in part, facilitation of inhibitory pathways operating via apamin-sensitive K^+ channels and nitric oxide (Heinemann et al. 1999) as mentioned above (Sect. 3.1.1). The effects of cannabinoids on the smooth muscle relaxation of the isolated gastric fundus in response to EFS of NANC innervation are not clear. In rat preparations (Storr et al. 2002), both excitatory cholinergic and NANC transmission were reduced by WIN 55,212-2 and anandamide. Only the anandamide responses were antagonised by the cannabinoid receptor antagonist AM630. By itself, AM630 had no effect on the contractile responses but facilitated the relaxation. This latter effect implied the presence of an ongoing endocannabinoid tone that reduced the NANC neurotransmission. In contrast, Todorov et al. (2003) found no response to anandamide (0.1–10 μM) in the isolated gastric fundus of the guinea-pig. Whether this is due to a species difference or whether the anandamide was metabolised before it could produce a measurable response is unclear. No other, more potent cannabinoid receptor agonist was tested in this study, in which evidence suggested that the NANC response was mediated by nitric oxide and cyclic guanosine monophosphate (cGMP).

3.2 In Vivo Studies

3.2.1 Lower Oesophageal Sphincter

Lower oesophageal sphincter (LOS) relaxation is the chief mechanism for gastro-oesophageal reflux, and thus represents a potential target in the treatment of gastro-oesophageal reflux disease. The principal anatomical components of LOS relaxation are afferent gastric pathways, brainstem integrative centre, and efferent inhibitory pathways to the lower oesophageal sphincter. Functional studies have shown that i.v. administration of the cannabinoid receptor agonists WIN 55,212-2 and Δ^9-THC inhibited (via CB_1 receptor activation) LOS relaxation in dogs (Lehmann et al. 2002) and ferrets (Partosoedarso et al. 2003), the effect being associated, at least in the dog, with the inhibition of gastro-oesophageal reflux (Lehman et al. 2002). The CB_1 receptor antagonist SR141716A, administered alone, stimulated the LOS relaxation incidence and increased the number of reflux episodes and swallowing rate, suggesting an involvement of endocannabinoids in ongoing suppression of LOS relaxation. The most likely site of action is via the CB_1 receptor within the central pattern generator thought to control LOS relaxation. Indeed (1) direct application of Δ^9-THC to the dorsal hindbrain surface attenuated LOS relaxation in ferrets (Partosoedarso et al. 2003) and (2) WIN 55,212-2 reduced the rate of LOS relaxation without altering other characteristics of simultaneous oesophageal contraction in dogs (Lehmann et al. 2002). This is in agreement with the observation that CB_1 receptor staining is present in cell bodies within the area postrema, nucleus tractus solitarius and nodose ganglion (Partosoedarso et al. 2003).

3.2.2
Gastric Motility

Experimental studies performed in the rat have shown that CB_1 receptors modulate gastric motility. A number of cannabinoid receptor agonists, including Δ^9-THC, WIN 55,212-2, CP 55,940 and cannabinol reduced gastric motility, and this effect was antagonised by the CB_1 receptor antagonist SR141716A, but not by the CB_2 receptor antagonist SR144528 (Izzo et al. 1999a; Krowicki et al. 1999; Landi et al. 2002). However, in contrast to the small intestine and the colon, SR141716A, administered alone to the stomach, does not produce any inverse cannabimimetic effects. Most notably, intravenous Δ^9-THC inhibited gastric motility and decreased intragastric pressure in anaesthetised rats. Also, the application of Δ^9-THC directly to the dorsal surface of the medulla evoked very slight changes in gastric motor activity. Both ganglionic blockade and vagotomy, but not spinal cord transection, abolished the gastric motor effects of peripherally administered Δ^9-THC (Krowichi et al. 1999). Taken together, these data indicated that the gastric effects of systemically administered Δ^9-THC depend on intact vagal circuitry.

In agreement with animal data, a double-blind randomised placebo-controlled study performed on 13 healthy volunteers showed that oral Δ^9-THC, at a dose used for preventing chemotherapy-induced nausea and vomiting (10 mg/m^2), significantly delays gastric emptying of solid food in all subjects (McCallum et al. 1999). In contrast, Bateman (1983) found that, in humans, gastric emptying (monitored by a real real-time ultrasound technique) of liquids was unaffected by Δ^9-THC (0.5 and 1 mg/kg i.v., a dose that produced cannabis-like psychomotor and psychological effects). Apart from the different doses and techniques used to measure motility in the two studies, it should be noted that gastric emptying of liquids is mediated by a different mechanism from emptying of solids.

3.2.3
Upper Intestinal Motility

The effect of cannabinoid drugs on upper intestinal motility has been generally studied by evaluating the distance travelled by a non-absorbable marker (e.g. charcoal) from the pylorus to the caecum. Since the marker was given intragastrically, this method does not distinguish between an effect on stomach emptying and transit through the small intestine. Exceptions are the studies by Shook and Burks (1989) and Landi and colleagues (2002) in which the marker was given intraduodenally and motility measured along the small intestine only.

Dewey et al. (1972) first reported that Δ^9-THC delayed gastrointestinal transit in mice. These results were confirmed by Chesher and colleagues (1973) who also showed that Δ^8-THC and three different *Cannabis* extracts dose-dependently reduced the passage of a charcoal meal in mice. Δ^8-THC and Δ^9-THC were shown to be equipotent, while cannabidiol was inactive (Chesher et al. 1973). In a more complete study, Shook and Burks (1989) showed that Δ^9-THC and cannabinol slowed small intestinal transit when injected intravenously in mice and rats, with Δ^9-THC being equipotent to morphine.

More recently, the ability of cannabinoids to reduce intestinal motility has been related to their ability to activate cannabinoid CB_1 receptors. Studies have shown that the endogenous ligand anandamide, the natural agonist cannabinol and the synthetic agonists WIN 55,212-2 and CP 55,940 inhibited gastrointestinal transit motility in mice (Calignano et al. 1997; Colombo et al. 1998b; Izzo et al. 1999b, 2000b, 2001b) an effect counteracted by SR141716A, but not by SR144528. Notably, the inhibitory effect of anandamide was not reduced by the VR1 receptor antagonist capsazepine or by a chronic treatment with capsaicin (a treatment which ablates capsaicin-sensitive afferent neurons) (Izzo et al. 2001a), thus implying that the effect of anandamide on intestinal transit is independent of VR1 receptor activation. SR141716A, but not SR144528, administered alone, increased upper gastrointestinal transit, implying the existence of ongoing background activity of CB_1 receptors due to either tonic release of endocannabinoids or precoupled CB_1 receptors.

WIN 55,212-2 and cannabinol were significantly more effective when administered intracerebroventricularly (i.c.v.) than when administered intraperitoneally (Izzo et al. 2000b), suggesting a central site of action. However, central CB_1 receptors probably contribute little to the effect of peripherally administered cannabinoids, as the effect of i.p.-injected cannabinoid receptor agonists was not modified by the ganglion blocker hexamethonium (Izzo et al. 2000b). The primary role of peripheral CB_1 receptors was emphasised by the observation that i.c.v.-administered SR141716A did not significantly reduce the effect of i.p. WIN 55,212-2 (Landi et al. 2002).

Palmitoylethanolamide (PEA) is an endogenous fatty acid ethanolamide that shares some pharmacological actions with Δ^9-THC and with the endocannabinoids anandamide and 2-AG (Lambert et al. 2002). However PEA does not bind to CB_1 and CB_2 receptors. Capasso and colleagues (2002) reported that i.p.-injected PEA inhibited upper gastrointestinal transit, both in control and in intestine-inflamed mice, and this effect was not antagonised by the cannabinoid receptor antagonists SR141716A or SR144528; moreover, the PEA effect was unaffected by the NO synthase inhibitor L-NAME, the α_2-adrenoceptor antagonist yohimbine, the opioid receptor antagonist naloxone or the nicotinic receptor antagonist hexamethonium.

3.2.4
Motility in the Colon

Pinto and colleagues provided immunohistochemical and pharmacological evidence supporting a role for the endocannabinoids and myenteric CB_1 receptors in regulating colonic motility in vivo in mice (Pinto et al. 2002b). Motility was assessed by measuring the time required for expulsion of a glass bead inserted

2 cm into the distal colon. It was found that the non-selective cannabinoid receptor agonists cannabinol, anandamide and WIN 55,212-2, as well as the selective CB_1 receptor agonist arachidonyl-2-chloroethylamide (ACEA) decreased motility in an SR141716A-sensitive manner. The hypothesis for a local endocannabinoid tone controlling propulsion was strengthened by the following findings: (1) unusually high amounts of endocannabinoids were present in the mouse colon; (2) a stimulatory action on colonic propulsion occurred after selective blockade of the CB_1 receptor with SR141716A; and (3) an inhibitory effect on colonic propulsion occurred after inhibition of endocannabinoid re-uptake with VDM11. Consistent with these in vivo results, CB_1 receptors mediate the antiperistaltic effects of WIN 55,212-2 in the mouse isolated colon (Mancinelli et al. 2001).

4
Intestinal Secretion

Taking short circuit current (Isc) as an indicator of net electrogenic ion transport in Ussing chambers, it was shown that the cannabinoid receptor agonist WIN 55,212-2 reduced (via CB_1 receptor activation) the secretory response to EFS (which is mediated mainly by acetylcholine release from submucosal secretomotor neurons) and capsaicin (which evokes neurotransmitter release such as acetylcholine by activating extrinsic primary afferents) in the rat (Tyler et al. 2000) and guinea-pig ileum (MacNaughton et al. 2004). The inhibitory effect of WIN 55,212-2 was on the enteric nerves, and not on the epithelial cells, since the Isc response to forskolin and carbachol, which act directly on the epithelium to elicit secretion, were unaffected by WIN 55,212-2 pretreatment. Moreover, in extrinsically denervated segments of guinea-pig ileum, the inhibitory effect of WIN 55,212-2 on the response to EFS was completely lost, suggesting a predominant role for capsaicin-sensitive extrinsic primary afferent nerves that innervate submucosal secretomotor neurons (MacNaughton et al. 2004). In agreement, immunohistochemical studies have shown that CB_1 receptors are present on submucosal neurons and extrinsic primary afferent nerves in the submucosa of the small intestine (MacNaughton et al. 2004).

5
Gastrointestinal Signs of Tolerance and Dependence

Chronic treatment with cannabinoids can induce a state of tolerance to their inhibitory effects in the gastrointestinal tract. Studies of this phenomenon have been performed predominantly with pieces of tissue excised from chronically treated animals (ex vivo) or on isolated tissues pretreated in vitro with a cannabinoid receptor agonist. These investigations were comprehensively reviewed by Pertwee (2001) and will be summarised here.

In mice, the inhibition of transit by daily oral Δ^9-THC was reduced for up to 19 days post-treatment (Anderson et al. 1975). Similarly, Δ^9-THC (s.c.) for

3 days reduced the sensitivity of mouse MP-LMP to CP 55,940 compared with vehicle-pretreated littermates, when tested 24–28 h after the final injection (Pertwee et al. 1998). In addition, tolerance to Δ^9-THC and CP 55,940 could be demonstrated in the MP-LMP of guinea-pigs receiving Δ^9-THC (10 mg.kg^{-1}) i.p. daily for 2 days. In tolerant animals, a reduction was observed in the maxima of agonist log concentration–response curves. This was thought to indicate a down-regulation of receptor expression and/or coupling efficiency (Pertwee et al. 1998).

A form of tolerance was induced in guinea-pig ileal segments in vitro by incubation with WIN 55,212-2 (50 nM) for 5 h. At the end of incubation, the size of electrically evoked contractions was not significantly different from untreated preparations (Basilico et al. 1999). MP-LMP from human ileum or distal jejunum, pretreated for 48 h with (+)- or (–)-WIN 55,212 (10 μM), or vehicle alone at 18°C were tested for their sensitivity to subsequent doses of the active isomer, (+)-WIN 55,212 or to SR141716A (Guanini et al. 2000). Those preparations pretreated with (+)-WIN 55,212 but not (–) WIN 55,212 were insensitive to the inhibitory effects of (+)-WIN 55,212 on the evoked contractions. In addition, SR141716A (1 μM) significantly enhanced the contractile responses in (+)-WIN 55,212-pretreated preparations but not those treated with the (–) isomer or the vehicle, dimethylsulfoxide (DMSO). Earlier reports had shown SR141716A not to have inverse agonist effects on human fresh innervated preparations (Croci et al. 1998). This in vitro inverse response to SR141716A supports the "withdrawal" diarrhoea observed on treatment of Δ^9-THC-tolerant dogs with SR141716A. Work in non-GI tissues suggests that selective kinases may be involved in the development of cannabinoid tolerance (Lee et al. 2003).

Opioids and cannabinoids are among the most widely consumed drugs of abuse in humans; therefore, cross-tolerance or interactivity have been investigated with the two drugs in the GI tract. Basilico et al. (1999) found dextral shifts in the log concentration-response curves for the inhibition of electrically evoked contractions for both (+)-WIN 55,212 and morphine in guinea-pig MP-LMP's that had been preincubated for 5 h with either drug. However, in ex vivo preparations from Δ^9-THC-tolerant guinea-pigs (Pertwee et al. 1998), tolerance was not found to the inhibitory responses to normorphine or clonidine (presynaptic α_2-adrenoceptor agonist). Early in vivo studies showed that increases in GI activity (diarrhoea and increased defaecation) and other abstinence signs precipitated by naloxone in morphine-dependent rats could be reduced in a dose-related fashion by Δ^9-THC but not cannabidiol (Hine et al. 1975). Such observations led to hopes for potential treatment of opiate addicts with cannabinoids. An interesting phenomenon observed in the absence of electrical stimulation of morphine-tolerant guinea-pig MP-LMP in vitro is a fast withdrawal contracture in response to naloxone; this is not mimicked by exposure of cannabinoid-tolerant tissues to SR141716A (personal communication). However, the in vitro naloxone "withdrawal" contraction can be significantly reduced by (–)- but not (+)-Δ^9-THC (95 nM) by a presynaptic mechanism (Frederickson et al. 1976). This cross tolerance was confirmed by Morrone et al. (1993) with cannabis extract (equivalent to 5.2 μM Δ^9-THC) in segments of guinea-pig ileum and rabbit jejunum that had been exposed for 5 min to either morphine or the κ-opioid receptor agonist, U-50,488H. The induction of opioid and cannabinoid tolerance by incubation of guinea-pig MP-LMP for 5 h with morphine could be prevented by the addition of (+)-WIN 55,212 (50 nM), as shown by loss of the naloxone-precipitated withdrawal response, which is evident as a slow, sustained contraction. The mechanism responsible for this contraction is thought to be a cannabinoid-sensitive release of endogenous ACh, 5-hydroxytryptamine and/or substance P from myenteric neurons into the neuromuscular space (Basilico et al. 1999; Frederickson et al. 1976).

In the CNS, recent work suggests that the endocannabinoid system is involved in the development of opioid tolerance. In morphine-tolerant rats, autoradiographic binding showed a slight but significant reduction in cannabinoid receptor level in the cerebellum and hippocampus, whereas in the limbic area there was a strong decrease (40%) in receptor/G protein coupling (CP 55,940-stimulated [^{35}S]GTPγS binding). Chronic morphine exposure produced a strong reduction in 2-AG content without changes in anandamide levels in several brain regions (i.e. striatum, cortex, hippocampus, limbic area and hypothalamus) (Vigano et al. 2003).

6 Cannabinoids in Pathological States

6.1 Emesis

Although the antiemetic potential has been recognised for decades, and cannabinoids such as the natural Δ^9-THC or the synthetic cannabinoid nabilone are effectively used in humans (Tramer et al. 2001), the molecular mechanism by which cannabinoids prevent vomiting was only recently ascertained. Immunohistochemistry identified CB_1 receptors and FAAH in areas involved in emesis, including the dorsal vagal complex (DVC) (area postrema and nucleus of the solitary tract, NTS) and the dorsal motor nucleus of the vagus (DMN) (Van Sickle et al. 2001). Functional studies aimed at investigating the role of the endogenous cannabinoid system in nausea and emesis have been performed in both vomiting (i.e. least shrews, ferrets) and non-vomiting (i.e. rats) species. Emesis has been induced mainly by cisplatin or opioids in vomiting species, while conditioned rejection reactions, which may reflect a sensation of nausea, have been elicited in rats mostly by lithium chloride.

A number of cannabinoid receptor agonists (given i.p.), including CP 55,940, Δ^9-THC, WIN 55,212-2 and (–)-11-OH-Δ8-THC dimethylheptyl (HU-210) prevented cisplatin-induced emesis in the least shrew (Darmani 2001a,b; Darmani et al. 2003b), opioid-induced emesis in ferrets (Simoneau et al. 2001; Van Sickle et al. 2001) or lithium-induced conditioned rejection reactions in rats (Parker and Mechoulam 2003; Parker et al. 2003). These effects were mediated by CB_1 receptors, since they were counteracted by selective receptor antagonists such as SR141716A or AM251. Furthermore, the order of potency for reducing both the frequency of emesis and the percentage of shrews vomiting was CP 55,940>WIN 55,212-2>Δ^9-THC, which is consistent with an action on the CB_1 receptor (Darmani 2001a,b).

However, in the least shrew, unlike Δ^9-THC and WIN 55,212-2, the antiemetic activity of CP 55,940 occurs at motor-suppressant doses (Darmani et al. 2003b).

The site of action of cannabinoid receptor agonists has been investigated in ferrets by comparing the effect of Δ^9-THC applied locally to the surface of the brain stem against the emesis induced by intragastric hypertonic saline and, most importantly, by measuring Fos expression induced by cisplatin (Van Sickle et al. 2003). It was found that the anti-emetic effects of cannabinoids are mediated by CB_1 receptors on pathways related to vagal gastric function either centrally, in the area postrema and DVC, or at the peripheral endings of abdominal vagal efferents. Specifically, CB_1 receptors may be involved at three sites: (1) CB_1 receptors on the terminals of primary afferent fibres from the stomach and duodenum could reduce the input indicating intestinal distress and reduce the resulting episodes of emesis, (2) CB_1 receptors on the terminals of interneurons within the NTS could reduce the input to the DMN and therefore reduce emesis, and (3) CB_1 receptors on the terminals of NTS projection neurons could modulate input from the area postrema or directly reduce excitatory transmission to the DMN. Since the chemosensors of the area postrema are located outside the blood–brain barrier, cannabinoids which do not cross this barrier may have antiemetic actions devoid of psychotropic side-effects.

Experimental evidence suggests that an ECS may be present in the brain stem centres that modulate emesis. Indeed, CB_1 receptor antagonists caused emesis when given alone to the least shrews (Darmani 2001a) and also potentiated the emetic response to opioids in the ferret (Van Sickle et al. 2001) as well as lithium-induced nausea in a rat model of nausea (Parker et al. 2003). In the least shrews, the emetic effect of SR141716A was associated with increased forebrain levels of 5-hydroxytryptamine and dopamine (Darmani et al. 2003a). Inconsistent with the putative antiemetic action of the endogenous cannabinoid system is the potent ability of the endocannabinoid 2-AG (but not anandamide) to induce emesis in shrews. This effect is blocked by a non-emetic dose of SR141716A, by the cannabinoid receptor agonist CP 55,940, WIN 55,212-2 or Δ^9-THC and by the cyclo-oxygenase inhibitor indomethacin (Darmani 2002). It has been hypothesised that exogenous 2-AG may elicit its emetic response by acting in brain areas involved in emesis to reduce anti-emetic tone through the displacement from CB_1 receptors of an endogenous CB1 receptor agonist with greater efficacy.

Finally, it should be noted that cannabidiol, a natural cannabinoid that does not activate cannabinoid receptors, suppresses lithium-induced conditioned rejection reactions in a rat model of nausea (Parker et al. 2002) and also potentiates the antiemetic effect of ondansetron and Δ^9-THC in the musk shrew (Kwiatkowska et al. 2004).

6.2 Gastric Ulcer

The gastric antisecretory and antiulcer activity of cannabinoids was first observed in the late 1970s, when it was found that Δ^9-THC reduced gastric juice volume and ulcer formation after ligation of the pylorus (Shay rat test) (Sofia et al. 1978). More recently, it has been shown that the cannabinoid receptor agonist WIN 55,212-2 reduced, in an SR141716A-sensitive manner, stress-induced gastric ulcers in rats (Germanò et al. 2001). The antiulcerative effect of WIN 55,212-2 may well be related to its antisecretory effect (Adami et al. 2002; Coruzzi et al. 1999). Indeed, the non-selective cannabinoid receptor agonists WIN 55,212-2 and HU-210 decreased (via CB_1 activation) the acid secretion induced by indirectly acting secretagogues, such as 2-deoxy-D-glucose (which stimulated acid secretion by increasing the efferent activity of the vagus nerve) and pentagastrin (which acts partly through a cholinergic pathway). These observations were made in anaesthetised rats in which the secretion induced by the activation of parietal cell H_2 receptors by histamine was unaffected, which is consistent with the absence of CB_1 receptors on parietal cells (Adami et al. 2002). Bilateral cervical vagotomy and ganglionic blockade, but not atropine treatment, significantly reduced, but did not abolish, the inhibitory effect of HU-210. These results indicate that gastric antisecretory effects of cannabinoids are mediated by suppression of vagal drive to the stomach through activation of CB_1 receptors, located on pre- and postganglionic cholinergic pathways. In addition, the ineffectiveness of atropine suggests CB_1 receptors may regulate the release of non-cholinergic secretory neurotransmitters.

6.3 Intestinal Inflammation

Many patients with inflammatory bowel disease anecdotally report that they experience relief by smoking marijuana (Di Carlo and Izzo 2003). Furthermore, some cannabinoid-based preparations are already being evaluated in clinical trials for the treatment of inflammatory bowel disease (Di Carlo and Izzo 2003). Experimental evidence indicates that the ECS, via CB_1 activation, mediates protective pathophysiological signals counteracting intestinal inflammatory responses. Enhancement of the cannabinoid signalling, as revealed by the increased expression of enteric CB_1 receptors, has been observed following intestinal inflammation induced by a number of irritants, including intra-colonic dinitrobenzene sulfonic acid (DNBS) (Massa et al. 2004), oral croton oil (Izzo et al. 2001b) and intraperitoneal acetic acid (Mascolo et al. 2002). Massa et al. (2004) showed that colitis induced by intra-colonic DNBS was more severe in CB_1-deficient mice than in wild-types littermates, while FAAH-deficient mice (which are expected to have higher levels of anandamide) showed significant protection against intestinal inflammation. Consistent with experimental results obtained with genetically modified mice, the cannabinoid receptor agonist HU-210 inhibited, while the CB_1 receptor antagonist SR141716A exacerbated, DNBS-induced colonic inflammation (Massa et al. 2004).

The possible involvement of CB_2 receptors in inflammatory bowel disease has been hypothesised on the basis of recent in vitro studies; indeed, cannabinoids exert an inhibitory effect on the expression of tumour necrosis factor (TNF)-α-induced interleukin-release from a human colonic epithelial cell line HT-29, and this effect was reversed by the CB_2 receptor antagonist SR144528 (Ihenetu et al.

2003). Furthermore, Western immunoblotting revealed an immunoreactive protein in this cell line at a region with a size consistent with that of CB_2 receptors (Ihenetu et al. 2003). In contrast with a beneficial role of endocannabinoids, Croci and colleagues (2003) reported that the CB_1 receptor antagonist SR141716A prevented the intestinal ulcers and the rise in TNF-α and myeloperoxidase activity (a marker of inflammation) induced by indomethacin in rats, while the CB_2 receptor antagonist SR144528 reduced the ulcers only (Croci et al. 2003).

Finally, it should be noted that anandamide and 2-AG have been shown to stimulate intestinal primary sensory neurons via the VR1 receptor to release substance P, resulting in ileitis in rats (McVey et al. 2003) and that endocannabinoids may mediate the inflammatory effects of toxin A. Thus, in the intestinal mucosa, endocannabinoids may have both a protective role (via CB_1 receptor activation) and produce deleterious effects (via VR1 receptor activation, presumably at higher concentrations).

6.4 Paralytic Ileus

Paralytic ileus (i.e. a "non-mechanical" bowel obstruction observed in response to nociception initiated at the abdominal level) is a common complication whose pathogenesis is still under debate. Mascolo and colleagues (2002) provided evidence that alterations in the enteric endocannabinoid system contribute to the onset of experimental paralytic ileus induced by peritoneal irritation. Reduced gastrointestinal motility associated with intraperitoneal acetic acid in mice was restored by the CB_1 receptor antagonist SR141716A, while it was worsened by the anandamide cellular re-uptake inhibitor VDM11. Ileus was characterised by increased intestinal levels of anandamide (but not 2-AG) and by an increase in the number and density of CB_1 receptors on acetylcholine- and substance P-containing neurons. Because CB_1 receptor activation reduced excitatory transmission, it was hypothesised that, following peritonitis-induced ileus, overactivity of CB_1 receptors on the enteric cholinergic/substance P neurons lead to a reduced release of both neurotransmitters, with subsequent delayed motility.

6.5 Diarrhoea (Cholera Toxin)

Extracts of *Cannabis* were indicated for the treatment of diarrhoea a century ago in the United States, and there are a number of anecdotal accounts of the effective use of *Cannabis* against dysentery and cholera (Di Carlo and Izzo 2003). Cholera toxin (CT) is the most recognisable enterotoxin causing secretory diarrhoea. The profound dehydrating secretory diarrhoea associated with CT may involve several intestinal secretory mechanisms, including activation of enteric neurons and release and/or synthesis of endogenous secretagogues such as 5-hydroxytryptamine, prostaglandins, tachykinins, vasoactive intestinal peptide, and platelet activating factor (Lundgren 2002). Oral administration of CT to mice increased fluid accumulation in the small intestine, raised anandamide levels and led to overexpression of CB_1 receptor mRNA (Izzo et al. 2003). The non-selective cannabinoid receptor agonist CP 55,940 and the CB_1 selective agonist, ACEA inhibited CT-induced fluid accumulation, and this effect was counteracted by SR141716A (but not by SR144528 or by the vanilloid receptor antagonist capsazepine). The antisecretory effect of cannabinoids may involve peripheral mechanisms, since CP 55,940 still inhibited CT-induced fluid accumulation after ganglionic blockade. Furthermore SR141716A enhanced, while the inhibitor of anandamide uptake VDM11 prevented, CT-induced fluid accumulation. These results indicate that CT, as well as enhancing intestinal secretion, causes overstimulation of endocannabinoid signalling with an antisecretory role in the small intestine.

6.6 Colorectal Cancer

Endocannabinoids are known to inhibit the proliferation of breast cancer cells, prostate cancer cells, and rat thyroid cancer cells (Bifulco and Di Marzo 2002). Ligresti and colleagues (2003) showed that the levels of anandamide and 2-AG were increased relative to controls in adenomatous polyps and carcinomas, but there appeared to be no differences in the expression of CB_1 and CB_2 receptors or FAAH levels among the tissues. To determine if cannabinoids affect colorectal cancer cell growth, the authors used CaCo-2 (which express CB_1 receptor) and DLD-1 cells (which express both CB_1 and CB_2 receptors, with CB_1 receptor less expressed than in CaCo-2 cells). Anandamide, 2-AG and HU-210, as well as an inhibitor of anandamide inactivation, potently inhibited CaCo2 cell proliferation (relative potencies: HU-210>>anandamide≥2-AG), while DLD-1 cells were less responsive to cannabimimetics than CaCo-2 cells (Ligresti et al. 2003). Such data suggest that CB_1 receptors are more important than CB_2 receptors in reducing the proliferation of colorectal carcinoma cells. Consistent with this, in a study performed on SW 480 colon carcinoma cells, Joseph and colleagues (2004) reported that anandamide (via CB_1 activation) inhibited tumour cell migration, which is of paramount importance in metastasis development (Joseph et al. 2004).

7 Anandamide as an Endovanilloid

The unexpected revelation that anandamide is also an agonist at VR1 receptors (Zygmunt et al. 1999) has important implications for the physiological roles of endocannabinoid and VR1 receptor systems. Capsaicin has long been known to affect GI motility (Feher and Vajda 1982; Holzer 2001, 2003). VR1 receptor expression has been associated not only with the oesophagus and GI tract and their related ganglia, but also with areas of the CNS concerned with GI activity. In the rat brain, varicose fibres in the commissural, dorsomedial and gelatinosus subnuclei of the medial solitary tract and lateral area postrema expressed VR1 immunoreactivity that was reduced after vagotomy above the nodose ganglion (Rumessen et al. 2001). A proportion of nodose ganglionic neurons with afferent terminals in the gastric mucosa and vagal afferents from the GI tract overall were found to express VR1 receptors (Rumessen et al. 2001). These fibres were found to traverse both submucous and myenteric plexuses (Akiba et al. 2001) and many individual fibres coexpressed calcitonin gene-related protein (CGRP) (Rumessen et al. 2001). In the pig ileum, some myenteric VR1-positive neurons also expressed δ-opioid and κ-opioid receptors (Poonyachoti et al. 2002); also, in primary cultures of porcine myenteric ileal neurons, some cholinergic cells with δ-opioid-like immunoreactivity were also immunopositive for κ-opioid, cannabinoid or vanilloid receptors (Kulkarni-Narla and Brown 2001).

Anavi-Goffer et al. (2002) identified VR1 immunoreactivity in whole mounts of myenteric plexus preparations from the guinea-pig ileum and colon and rat ileum (Anavi-Goffer and Coutts 2003; Anavi-Goffer et al. 2002). They found VR1 immunoreactivity in a subpopulation (47%) of cholinergic myenteric neurons and fibres in the ganglia, the secondary bundles and tertiary plexus. In guinea-pig myenteric ganglia, intrinsic primary afferent neurons (IPAN's) had the chemical signature ChAT/calbindin/CB_1 receptor/VR1 receptor. In contrast, in rat and human preparations, VR1-immunoreactivity was confined to fibres only, and was increased by inflammation in human tissue (Anavi-Goffer and Coutts 2003; Yiangou et al. 2001).

In a study of hypo- and aganglionic regions of the large bowel in Hirschsprung's disease, hypertrophic extrinsic nerve bundles showed intense VR1 immunoreactivity compared with normoganglionic regions, which were similar to control large intestine (Facer et al. 2001). Aganglionic tissue was also associated with weak purine P2X(3)-receptor immunoreactivity compared with normal specimens. It has been proposed that ATP can lower the threshold for activation of VR1 receptors (Tominaga et al. 2001). It is possible that the relative down-regulation of purinergic receptors in Hirschsprung's disease may be associated with an increased release of ATP and sensitisation of the sensory nerves. Ileitis due to *Clostridium difficile* toxin A could be mimicked by the intraluminal administration of anandamide and 2-AG in rats (McVey et al. 2003): this effect was reduced by pretreatment with the selective VR1 receptor antagonist capsazepine but not the cannabinoid receptor antagonists SR141716A or SR144528. Indeed, toxin A resulted in increased tissue levels of anandamide and 2-AG in the ileum that were further enhanced when their metabolism was reduced by FAAH inhibitors. Responses to both toxin A and anandamide were associated with capsazepine-sensitive substance P release and activation of specific natural killer (NK)-1 receptors and antagonised by the NK-1 antagonist L-733060 (McVey et al. 2003). These results suggest that enteritis due to toxin A involves the release of endocannabinoids that activate VR1 receptors on enteric primary afferent sensory neurons, resulting in the release of inflammatory mediators such as substance P. Clearly, the relevance of vanilloid receptor activation involvement in this field needs further investigation.

It may be of interest that VR1-immunoreactive cells in the rat dorsal root ganglia coexpress CB_1 receptors (Ahluwalia et al. 2000). VR1 mRNA detected by

RT-PCR from rat ileal tissue showed a protein band corresponding to that for VR1 mRNA from rat brain (Anavi-Goffer et al. 2002). Cholinergic VR1 receptor-positive fibres in the tertiary plexus were found to co-express calretinin, substance P and synapsin 1. These findings support results from functional studies indicating that VR1 activation is related to ACh release from motoneurons (Mang et al. 2001). Mang et al. showed that anandamide facilitates spontaneous ACh release from the myenteric plexus by a capsazepine-sensitive mechanism as measured by the release of [^{3}H]-choline. In the same report, Mang et al. demonstrated that SR141716A caused dextral shifts in the log concentration–response curves to CP 55,940 or anandamide for their inhibitory effects on cholinergic transmission. The relative activities of anandamide at CB_1 and VR1 receptors in this tissue are concentration dependent (Begg et al. 2002b). Begg's group found that VR1 receptor activation predominated at higher concentrations, whereas Mang et al. found pEC_{50} values for cannabinoid receptor activation to be less than for vanilloid receptor activation. There is also some controversy as to whether anandamide inhibits ACh release via a CB_1 or a non-CB_1cannabinoid receptor mechanism, since the K_B values differ for the antagonism by SR141716A of CP 55,940 and anandamide (Mang et al. 2001). Whether this difference can be explained by the concomitant effects on ACh release via a VR1-mediated process and/or is due to anandamide metabolism remains to be resolved. There is evidence that VR1 receptor activation by anandamide increases ethylene diamine-induced γ-aminobutyric acid (GABA) release from guinea-pig myenteric plexus by a capsazepine-sensitive mechanism (Begg et al. 2002b). However, it should be noted that no evidence for an activation of capsaicin-sensitive receptors by anandamide has been observed in the human sigmoid colon (Bartho et al. 2002).

Finally, 2-AG has been found to induce contractions in the longitudinal smooth muscle from the guinea-pig distal colon in vitro in a tetrodotoxin-sensitive manner. This response was mimicked by anandamide, but not by the cannabinoid receptor agonist WIN 55,212-2 or the vanilloid receptor agonist AM404 and was not inhibited by antagonists of cannabinoid or vanilloid receptors (Kojima et al. 2002). Since the response to 2-AG was partially reduced by the lipoxygenase inhibitor nordihydroguaiaretic acid, it is possible that leukotrienes may contribute to the neurogenic contractile action of 2-AG.

8 Conclusion

There is now substantial evidence for the presence of endocannabinoid and endovanilloid systems in the GI tract. The anti-inflammatory, anticancer, antiulcerogenic and antiemetic responses to CB_1 receptor activation holds promise for the future management of gastrointestinal diseases. Thus, exploitation of the endocannabinoid system by facilitation at sites of endocannabinoid activity by preventing cellular re-uptake or reducing EC degradation may enhance beneficial endocannabinoid effects without the psychotropic side-effects found with systemic administration of exogenous cannabinoids. Manipulation of the endocannabinoid system, rather than the administration of exogenous cannabinoids, would also lessen the possibility of adverse pharmacokinetic effects or the development of tolerance to or dependence on exogenous cannabinoids. The upregulation of VR1 receptor expression and increased tissue levels of endocannabinoids in inflammatory conditions may have implications for possible therapeutic applications of endovanilloid modulation in a variety of inflammatory gastric (ulceration and oesophageal reflux) and bowel conditions in the future. Clearly, further exploration of the gastrointestinal EC system is likely to produce worthwhile results.

References

Adami M, Frati P, Bertini S, Kulkarni-Narla A, Brown DR, de Caro G, Coruzzi G, Soldani G (2002) Gastric antisecretory role and immunohistochemical localization of cannabinoid receptors in the rat stomach. Br J Pharmacol 135:1598–1606

Ahluwalia J, Urban L, Capogna M, Bevan S, Nagy I (2000) Cannabinoid 1 receptors are expressed in nociceptive primary sensory neurons. Neuroscience 100:685–688

Akiba Y, Nakamura M, Ishii H (2001) Immunolocalization of vanilloid receptor-1 (VR-1) in CGRP-positive neurons and interstitial cells of Cajal in the myenteric plexus of the rat gastrointestinal tract. Gastroenterology 120:1721

Anavi-Goffer S, Coutts AA (2003) Cellular distribution of vanilloid VR1 receptor immunoreactivity in the guinea-pig myenteric plexus. Eur J Pharmacol 458:61–71

Anavi-Goffer S, McKay MG, Ashford MLJ, Coutts AA (2002) Vanilloid receptor type 1-immunoreactivity is expressed by intrinsic afferent neurones in the guinea-pig myenteric plexus. Neurosci Lett 319:53–57

Anderson PF, Jackson DM, Chesher GB, Malor R (1975) Tolerance to the effects of delta-9-tetrahydrocannabinol in mice on intestinal motility, temperature and locomotor activity. Psychopharmacologia 43:31–36

Bartho L, Benko R, Lazar Z, Illenyi L, Horvath OP (2002) Nitric oxide is involved in the relaxant effect of capsaicin in the human sigmoid colon circular muscle. Naunyn Schmiedebergs Arch Pharmacol 366:496–500

Basilico L, Parolaro D, Colleoni M, Costa B, Giagnoni G (1999) Cross-tolerance and convergent dependence between morphine and cannabimimetic agent WIN 55,212-2 in the guinea-pig ileum myenteric plexus. Eur J Pharmacol 376:265–271

Bateman DN (1983) Delta-9-tetrahydrocannabinol and gastric emptying. Br J Clin Pharmacol 15:749–751

Begg M, Dale N, Llaudet E, Molleman A, Parsons ME (2002a) Modulation of the release of endogenous adenosine by cannabinoids in the myenteric preparation of the guinea-pig plexus-longitudinal muscle ileum. Br J Pharmacol 137:1298–1304

Begg M, Molleman A, Parsons M (2002b) Modulation of the release of endogenous gamma-aminobutyric acid by cannabinoids in the guinea pig ileum. Eur J Pharmacol 434:87–94

Bifulco M, Di Marzo V (2002) Targeting the endocannabinoid system in cancer therapy: a call for further research. Nat Med 8:547–550

Brown DR, Poonyachoti S, Osinski MA, Kowalski TR, Pampusch MS, Elde RP, Murtaugh MP (1998) Delta-opioid receptor mRNA expression and immunohistochemical localization in porcine ileum. Dig Dis Sci 43:1402–1410

Burdyga G, Lal S, Varro A, Dimaline R, Thompson DG, Dockray GJ (2004) Expression of cannabinoid CB1 receptors by vagal afferent neurons is inhibited by cholecystokinin. J Neurosci 24:2708–2715

Calignano A, La Rana G, Makriyannis A, Lin SY, Beltramo M, Piomelli D (1997) Inhibition of intestinal motility by anandamide, an endogenous cannabinoid. Eur J Pharmacol 340:R7–R8

Capasso R, Izzo AA, Fezza F, Pinto A, Capasso F, Mascolo N, Di Marzo V (2001) Inhibitory effect of palmitoylethanolamide on gastrointestinal motility in mice. Br J Pharmacol 134:945–950
Casu MA, Porcella A, Ruiu S, Saba P, Marchese G, Carai MA M, Reali R, Gessa GL, Pani L (2003) Differential distribution of functional cannabinoid CB1 receptors in the mouse gastroenteric tract. Eur J Pharmacol 459:97–105
Chesher GB, Dahl CJ, Everingham M, Jackson DM, Marchant Williams H, Starmer GA (1973) The effect of cannabinoids on intestinal motility and their antinociceptive effect in mice. Br J Pharmacol 49:588–594
Colombo G, Agabio R, Lobina C, Reali R, Gessa GL (1998) Cannabinoid modulation of intestinal propulsion in mice. Eur J Pharmacol 344:67–69
Coruzzi G, Adami M, Coppelli G, Frati P, Soldani G (1999) Inhibitory effect of the cannabinoid receptor agonist WIN 55,212–2 on pentagastrin-induced gastric acid secretion in the anaesthetized rat. Naunyn Schmiedebergs Arch Pharmacol 360:715–718
Coutts AA (2004) Cannabinoid receptor activation and the endocannabinoid system in the gastrointestinal tract. Curr Neuropharmacol 2:91–102
Coutts AA, Pertwee RG (1997) Inhibition by cannabinoid receptor agonists of acetylcholine release from the guinea-pig myenteric plexus. Br J Pharmacol 121:1557–1566
Coutts AA, Brewster N, Ingram T, Razdan RK, Pertwee RG (2000) Comparison of novel cannabinoid partial agonists and SR141716A in the guinea-pig small intestine. Br J Pharmacol 129:645–652
Coutts AA, Irving AJ, Mackie K, Pertwee RG, Anavi-Goffer S (2002) Localisation of cannabinoid CB1 receptor immunoreactivity in the guinea pig and rat myenteric plexus. J Comp Neurol 448:410–422
Croci T, Manara L, Aureggi, Guagnini F, Rinaldi-Carmona M, Maffrand J-P, Le Fur G, Mukenge S, Ferla G (1998) In vitro functional evidence of neuronal cannabinoid CB1 receptors in human ileum. Br J Pharmacol 125:1393–1395
Croci T, Landi M, Galzin AM, Marini P (2003) Role of cannabinoid CB1 receptors and tumor necrosis factor-alpha in the gut and systemic anti-inflammatory activity of SR 141716 (Rimonabant) in rodents. Br J Pharmacol 140:115–122
Darmani NA (2001a) Delta(9)-tetrahydrocannabinol and synthetic cannabinoids prevent emesis produced by the cannabinoid CB1 receptor antagonist/inverse agonist SR 141716A. Neuropsychopharmacology 24:198–203
Darmani NA (2001b) The cannabinoid CB1 receptor antagonist SR 141716A reverses the antiemetic and motor depressant actions of WIN 55, 212–2. Eur J Pharmacol 430:49–58
Darmani NA (2002) The potent emetogenic effects of the endocannabinoid, 2-AG (2- arachidonoylglycerol) are blocked by delta(9)- tetrahydrocannabinol and other cannabinoids. J Pharmacol Exp Ther 300:34–42
Darmani NA, Janoyan JJ, Kumar N, Crim JL (2003a) Behaviorally active doses of the CB1 receptor antagonist SR 141716A increase brain serotonin and dopamine levels and turnover. Pharmacol Biochem Behav 75:777–787
Darmani NA, Sim-Selley LJ, Martin BR, Janoyan JJ, Crim JL, Parekh B, Breivogel CS (2003b) Antiemetic and motor-depressive actions of CP55,940: cannabinoid CB1 receptor characterization, distribution, and G- protein activation. Eur J Pharmacol 459:83–95
De Petrocellis L, Cascio MG, Di Marzo V (2004) The endocannabinoid system: a general view and latest additions. Br J Pharmacol 141:765–774
Dewey WL, Harris LS, Kennedy JS (1972) Some pharmacological and toxicological effects of l-trans-delta-8- and l-trans-delta-9-tetrahydrocannabinol in laboratory rodents. Arch Int Pharmacodyn Ther 196:133–145
Di Carlo G, Izzo AA (2003) Cannabinoids for gastrointestinal diseases: potential therapeutic applications. Expert Opin Investig Drugs 12:39–49
Facer P, Knowles CH, Tam PK H, Ford AP, Dyer N, Baecker PA, Anand P (2001) Novel capsaicin (VR1) and purinergic (P2X(3)) receptors in Hirschsprung's intestine. J Pediatr Surg 36:1679–1684

Feher E, Vajda J (1982) Effect of capsaicin on the nerve elements of the small intestine. Acta Morphol Acad Sci Hung 30:57–63

Frederickson RC A, Hewes CR, Aiken JW (1976) Correlation between the in vivo and in vitro expression of opiate withdrawal precipitated by naloxone: their antagonism by l-(–)-delta- 9-tetrahydrocannabinol. J Pharmacol Exp Ther 199:375–384

Germanò MP, D'Angelo V, Mondello R, Pergolizzi S, Capasso F, Capasso R, Izzo AA, Mascolo N, De Pasquale R (2001) Cannabinoid CB1-mediated inhibition of stress-induced gastric ulcers in rats. Naunyn Schmiedebergs Arch Pharmacol 363:241–244

Gomez R, Navarro M, Ferrer B, Trigo JM, Bilbao A, Del Arco I, Cippitelli A, Nava F, Piomelli D, Rodriguez de Fonseca F (2002) A peripheral mechanism for CB1 cannabinoid receptor-dependent modulation of feeding. J Neurosci 22:9612–9617

Griffin G, Fernando SR, Ross RA, McKay NG, Ashford MLJ, Shire D, Huffman JW, Yu S, Lainton JA H, Pertwee RG (1997) Evidence for the presence of CB2-like cannabinoid receptors on peripheral nerve terminals. Eur J Pharmacol 339:53–61

Guanini F, Croci T, Aureggi G, Manara L, Rinaldi-Carmona M, Mukenge S, Aldrighetti L, Ferla G, Maffrand J-P, Le Fur G (2000) Tolerance to (+)WIN55,212–2 inhibitory effect and withdrawal by the cannabinoid CB1 receptor antagonist SR 141716 in isolated strips of small intestine. International Cannabinoid Research Society symposium on the Cannabinoids, Burlington, Vermont

Hanus L, Abu-Lafi S, Fride E, Breuer A, Vogel Z, Shalev DE, Kustanovich I, Mechoulam R (2001) 2-Arachidonyl glyceryl ether, an endogenous agonist of the cannabinoid CB1 receptor. Proc Natl Acad Sci USA 98:3662–3665

Heinemann A, Shahbazian A, Holzer P (1999) Cannabinoid inhibition of guinea-pig intestinal peristalsis via inhibition of excitatory and activation of inhibitory neural pathways. Neuropharmacology 38:1289–1297

Heshmati H, Caplain H, Bellisle F, Mosse M,Fauveau C, Le Fur G (2001) SR141716, a selective CB1 receptor cannabinoid receptor antagonist reduces hunger, caloric intake, and body weight in overweight or obese men. Obes Res 9:70S

Hine B, Friedman E, Torrelio M, Gershon S (1975) Morphine-dependent rats: blockade of precipitated abstinence by tetrahydrocannabinol. Science 187:443–445

Holzer P (2001) Gastrointestinal afferents as targets of novel drugs for the treatment of functional bowel disorders and visceral pain. Eur J Pharmacol 429:177–193

Holzer P (2003) Acid-sensitive ion channels in gastrointestinal function. Curr Opin Pharmacol 3:618–625

Ihenetu K, Molleman A, Parsons ME, Whelan CJ (2003) Inhibition of interleukin-8 release in the human colonic epithelial cell line HT-29 by cannabinoids. Eur J Pharmacol 458:207–215

Izzo AA, Mascolo N, Borrelli F, Capasso F (1998) Excitatory transmission to the circular muscle of the guinea-pig ileum: evidence for the involvement of cannabinoid CB1 receptors. Br J Pharmacol 124:1363–1368

Izzo AA, Mascolo N, Capasso R, Germano MP, DePasquale R, Capasso F (1999a) Inhibitory effect of cannabinoid agonists on gastric emptying in the rat. Naunyn Schmiedebergs Arch Pharmacol 360:221–223

Izzo AA, Mascolo N, Borrelli F, Capasso F (1999b) Defaecation, intestinal fluid accumulation and motility in rodents: implications of cannabinoid CB1 receptors. Naunyn Schmiedebergs Arch Pharmacol 359:65–70

Izzo AA, Mascolo N, Tonini M, Capasso F (2000) Modulation of peristalsis by cannabinoid CB1 ligands in the isolated guinea-pig ileum. Br J Pharmacol 129:984–990

Izzo AA, Pinto L, Borrelli F, Capasso R, Mascolo N, Capasso F (2000b) Central and peripheral cannabinoid modulation of gastrointestinal transit in physiological states or during the diarrhoea induced by croton oil. Br J Pharmacol 129:1627–1632

Izzo AA, Capasso R, Pinto L, Di Carlo G, Mascolo N, Capasso F (2001a) Effect of vanilloid drugs on gastrointestinal transit in mice. Br J Pharmacol 132:1411–1416

Izzo AA, Fezza F, Capasso R, Bisogno T, Pinto L, Iuvone T, Esposito G, Mascolo N, Di Marzo V, Capasso F (2001b) Cannabinoid CB1-receptor mediated regulation of gastrointestinal motility in mice in a model of intestinal inflammation. Br J Pharmacol 134:563–570
Izzo AA, Mascolo N, Capasso F (2001c) The gastrointestinal pharmacology of cannabinoids. Curr Opin Pharmacol 1:597–603
Izzo AA, Capasso F, Costagliola A, Bisogno T, Marsicano G, Ligresti A, Matias I, Capasso R, Pinto L, Borrelli F, Cecio A, Lutz B, Mascolo N, Di Marzo V (2003) An endogenous cannabinoid tone attenuates cholera toxin-induced fluid accumulation in mice. Gastroenterology 125:765–774
Joseph J, Niggemann B, Zaenker KS, Entschladen F (2004) Anandamide is an endogenous inhibitor for the migration of tumor cells and T lymphocytes. Cancer Immunol Immunother 53:723–728
Katayama K, Ueda N, Kurahashi Y, Suzuki H, Yamamoto S, Kato I (1997) Distribution of anandamide amidohydrolase in rat tissues with special reference to small intestine. Biochim Biophys Acta 1347:212–218
Kojima S, Sugiura T, Waku K, Kamikawa Y (2002) Contractile response to a cannabimimetic eicosanoid, 2-arachidonoylglycerol, of longitudinal smooth muscle from the guinea-pig distal colon in vitro. Eur J Pharmacol 444:203–207
Krowicki ZK, Moerschbaecher JM, Winsauer PJ, Digavalli SV, Hornby PJ (1999) Delta9-tetrahydrocannabinol inhibits gastric motility in the rat through cannabinoid CB1 receptors. Eur J Pharmacol 371:187–196
Kulkarni-Narla A, Brown DR (2000) Localization of CB1-cannabinoid receptor immunoreactivity in the porcine enteric nervous system. Cell Tissue Res 302:73–80
Kulkarni-Narla A, Brown DR (2001) Opioid, cannabinoid and vanilloid receptor localization on porcine cultured myenteric neurons. Neurosci Lett 308:153–156
Kwiatkowska M, Parker LA, Burton P, Mechoulam R (2004) A comparative analysis of the potential of cannabinoids and ondansetron to supress cisplatin-induced emesis in the Suncus murinus (house musk shrew). Psychopharmacology (Berl). 174:254–259
Lambert DM, Vandevoorde S, Jonsson KO, Fowler CJ (2002) The palmitoylethanolamide family: A new class of anti-inflammatory agents? Curr Med Chem 9:663–674
Landi M, Croci T, Rinaldi-Carmona M, Maffrand JP, Le Fur G, Manara L (2002) Modulation of gastric emptying and gastrointestinal transit in rats through intestinal cannabinoid CB1 receptors. Eur J Pharmacol 450:77–83
Lee MC, Smith FL, Stevens DL, Welch SP (2003) The role of several kinases in mice tolerant to Delta(9)-tetrahydrocannabinol. J Pharmacol Exp Ther 305:593–599
Lehmann A, Blackshaw LA, Branden L, Carlsson A, Jensen J, Nygren E, Smid SD (2002) Cannabinoid receptor agonism inhibits transient lower esophageal sphincter relaxations and reflux in dogs. Gastroenterology 123:1129–1134
Ligresti A, Bisogno T, Matias I, De Petrocellis L, Cascio MG, Cosenza V, D'Argenio G, Scaglione G, Bifulco M, Sorrentini I, Di Marzo V (2003) Possible endocannabinoid control of colorectal cancer growth. Gastroenterology 125:677–687
López-Redondo F, Lees GM, Pertwee RG (1997) Effects of cannabinoid receptor ligands on electrophysiological properties of myenteric neurones of the guinea-pig ileum. Br J Pharmacol 122:330–334
Lundgren O (2002) Enteric nerves and diarrhoea. Pharmacol Toxicol 90:109–120
Lynn AB, Herkenham M (1994) Localization of cannabinoid receptors and nonsaturable high-density cannabinoid binding sites in peripheral tissues of the rat: implications for receptor-mediated immune modulation by cannabinoids. J Pharmacol Exp Ther 268:1612–1623
MacNaughton WK, Cushing K, Van Sickle MD, Keenan CM, Mackie K, Sharkey KA (2003) Cannabinoid CB1 receptor distribution and function in neurally mediated chloride secretion in the guinea pig ileum. Gastroenterology 124:A342

MacNaughton WK, Van Sickle MD, Keenan CM, Cushing K, Mackie K, Sharkey KA (2004) Distribution and function of the cannabinoid-1 receptor in the modulation of ion transport in the guinea pig ileum: relationship to capsaicin-sensitive nerves. Am J Physiol Gastrointest Liver Physiol (in press)

Manara L, Croci T, Guagnini F, Rinaldi-Carmona M, Maffrand JP, Le Fur G, Mukenge S, Ferla G (2002) Functional assessment of neuronal cannabinoid receptors in the muscular layers of human ileum and colon. Dig Liver Dis 34:262–269

Mancinelli R, Fabrizi A, Del Monaco S, Azzena GB, Vargiu R, Colombo GC, Gessa GL (2001) Inhibition of peristaltic activity by cannabinoids in the isolated distal colon of mouse. Life Sci 69:101–111

Mang CF, Erbelding D, Kilbinger H (2001) Differential effects of anandamide on acetylcholine release in the guinea-pig ileum mediated via vanilloid and non-CB1 cannabinoid receptors. Br J Pharmacol 134:161–167

Mascolo N, Izzo AA, Ligresti A, Costagliola A, Pinto L, Cascio MG, Maffia P, Cecio A, Capasso F, Di Marzo V (2002) The endocannabinoid system and the molecular basis of paralytic ileus in mice. Faseb J 16:1973–1975

Massa F, Marsicano G, Hermann H, Cannich A, Krisztina M, Cravatt BF, Ferri G–L, Sibaev A, Lutz B (2004) The endogenous cannabinoid system protects against colonic inflammation. J Clin Invest 113:1202–1209

McCallum RW, Soykan I, Sridhar KR, Ricci DA, Lange RC, Plankey MW (1999) Delta-9-tetrahydrocannabinol delays the gastric emptying of solid food in humans: a double-blind, randomized study. Aliment Pharmacol Ther 13:77–80

McVey DC, Schmid PC, Schmid HH O, Vigna SR (2003) Endocannabinoids induce ileitis in rats via the capsaicin receptor (VR1). J Pharmacol Exp Ther 304:713–722

Morrone LA, Romanelli L, Mazzanti G, Valeri P, Menichin F (1993) Hashish antagonism on the in vitro development of withdrawal contracture. Pharmacol Res 27 (Suppl 1):63–64

Nye JS, Seltzman HH, Pitt CG, Snyder SH (1985) High affinity cannabinoid binding sites in brain membranes labeled with [H-3]-5′-trimethylammonium delta-8-tetra-hydrocannabinol. J Pharmacol Exp Ther 234:784–791

Oleinik VM (1995) Distribution of digestive enzyme activities along intestine in blue fox, mink, ferret and rat. Comp Biochem Physiol A Physiol 112:55–58

Parker LA, Mechoulam R (2003) Cannabinoid agonists and antagonists modulate lithium-induced conditioned gaping in rats. Integr Physiol Behav Sci 38:134–146

Parker LA, Mechoulam R, Schlievert C (2002) Cannabidiol, a non-psychoactive component of cannabis and its synthetic dimethylheptyl homolog suppress nausea in an experimental model with rats. Neuroreport 13:567–570

Parker LA, Mechoulam R, Schlievert C, Abbott L, Fudge ML, Burton P (2003) Effects of cannabinoids on lithium-induced conditioned rejection reactions in a rat model of nausea. Psychopharmacology (Berl) 166:156–162

Partosoedarso ER, Abrahams TP, Scullion RT, Moerschbaecher JM, Hornby PJ (2003) Cannabinoid1 receptor in the dorsal vagal complex modulates lower oesophageal sphincter relaxation in ferrets. J Physiol 550:149–158

Paton WDM, Zar MA (1968) The origin of acetylcholine released from guinea-pig intestine and longitudinal muscle strips. J Physiol (Lond) 194:13–33

Pertwee RG (1997) Pharmacology of cannabinoid CB1 and CB2 receptors. Pharmacol Ther 74:129–180

Pertwee RG (2001) Cannabinoids and the gastrointestinal tract. Gut 48:859–867

Pertwee RG, Ross RA (2002) Cannabinoid receptors and their ligands. Prostaglandins Leukot Essent Fatty Acids 66:101–121

Pertwee RG, Stevenson LA, Elrick DB, Mechoulam R, Corbett AD (1992) Inhibitory effects of certain enantiomeric cannabinoids in the mouse vas deferens and the myenteric plexus preparation of guinea-pig small intestine. Br J Pharmacol 105:980–984

Pertwee RG, Fernando SR, Griffin G, Abadji V, Makriyannis A (1995) Effect of phenylmethylsulphonyl fluoride on the potency of anandamide as an inhibitor of electrically evoked contractions in two isolated tissue preparations. Eur J Pharmacol 272:73–78

Pertwee RG, Fernando SR, Nash JE, Coutts AA (1996) Further evidence for the presence of cannabinoid CB1 receptors in guinea-pig small intestine. Br J Pharmacol 118:2199–2205

Pertwee RG, Fernando S, Ritchie JEA (1998) Preliminary validation of a novel experimental model for the study of cannabinoid tolerance. International Cannabinoid Research Society symposium on the Cannabinoids, Burlington, Vermont

Pinto L, Capasso R, Di Carlo G, Izzo AA (2002a) Endocannabinoids and the gut. Prostaglandins Leukot Essent Fatty Acids 66:333–341

Pinto L, Izzo AA, Cascio MG, Bisogno T, Hospodar-Scott K, Brown DR, Mascolo N, Di Marzo VCapasso F (2002b) Endocannabinoids as physiological regulators of colonic propulsion in mice. Gastroenterology 123:227–234

Poonyachoti S, Kulkarni-Narla A, Brown DR (2002) Chemical coding of neurons expressing delta- and kappa-opioid receptor and type I vanilloid receptor immunoreactivities in the porcine ileum. Cell Tissue Res 307:23–33

Rosell S, Agurell S (1975) Effects of 7-hydroxy-D6-tetrahydrocannabinol and some related cannabinoids on the guinea-pig isolated ileum. Acta Physiol Scand 94:142–144

Rosell S, Agurell S, Martin BR (1976) Effects of cannabinoids on isolated smooth muscle preparations. New York, Springer-Verlag

Ross RA, Brockie HC, Fernando SR, Saha B, Razdan RK, Pertwee RG (1998) Comparison of cannabinoid binding sites in guinea-pig forebrain and small intestine. Br J Pharmacol 125:1345–1351

Rumessen JJ, d'Exaerde AD, Mignon S, Bernex F, Timmermans JP, Schiffmann SN, Panthier JJ, Vanderwinden JM (2001) Interstitial cells of Cajal in the striated musculature of the mouse esophagus. Cell Tissue Res 306:1–14

Shire D, Carillon C, Kaghad M, Calandra B, Rinaldi-Carmona M, Le Fur G, Caput D, Ferrara P (1995) An amino-terminal variant of the central cannabinoid receptor resulting from alternative splicing. J Biol Chem 270:3726–3731

Shook JE, Burks TF (1989) Psychoactive cannabinoids reduce gastrointestinal propulsion and motility in rodents. J Pharmacol Exp Ther 249:444–449

Simoneau II, Hamza MS, Mata HP, Siegel EM, Vanderah, TW, Porreca F, Makriyannis A, Malan TP (2001) The cannabinoid agonist WIN55,212-2 suppresses opioid-induced emesis in ferrets. Anesthesiology 94:882–887

Sofia RD, Diamantis W, Harrison JE, Melton J (1978) Evaluation of antiulcer activity of delta-9-tetrahydrocannabinol in the Shay rat test. Pharmacology 17:173–177

Storr M, Gaffal E, Saur D, Schusdziarra V, Allescher HD (2002) Effect of cannabinoids on neural transmission in rat gastric fundus. Can J Physiol Pharmacol 80:67–76

Storr M, Sibaev A, Marsicano G, Lutz B, Schusdziarra V, Timmermans JP Allescher HD (2004) Cannabinoid receptor type 1 modulates excitatory and inhibitory neurotransmission in mouse colon. Am J Physiol Gastrointest Liver Physiol 286:G110–G117

Sugiura T, Kobayashi Y, Oka S, Waku K (2002) Biosynthesis and degradation of anandamide and 2-arachidonoylglycerol and their possible physiological significance. Prostaglandins Leukot Essent Fatty Acids 66:173–192

Todorov S, Pozzoli C, Zamfirova R, Poli E (2003) Prejunctional modulation of non-adrenergic non-cholinergic (NANC) inhibitory responses in the isolated guinea-pig gastric fundus. Neurogastroenterol Motil 15:299–306

Tominaga M, Wada M, Masu M (2001) Potentiation of capsaicin receptor activity by metabotropic ATP receptors as a possible mechanism for ATP-evoked pain and hyperalgesia. Proc Natl Acad Sci U S A 98:6951–6956

Tramer MR, Carroll D, Campbell FA, Reynolds DJ M, Moore RA, McQuay HJ (2001) Cannabinoids for control of chemotherapy induced nausea and vomiting: quantitative systematic review. Br Med J 323:16–21

Tyler K, Hillard CJ, Greenwood-Van Meerveld B (2000) Inhibition of small intestinal secretion by cannabinoids is CB1 receptor-mediated in rats. Eur J Pharmacol 409:207–211
Ueda N, Yamamoto S (2000) Anandamide amidohydrolase (fatty acid amide hydrolase). Prostaglandins Other Lipid Mediat 61:19–28
Van Sickle MD, Oland LD, Ho, W, Hillard CJ, Mackie K, Davison JS, Sharkey KA (2001) Cannabinoids inhibit emesis through CB1 receptors in the brainstem of the ferret. Gastroenterology 121:767–774
Van Sickle MD, Oland LD, Mackie K, Davison JS, Sharkey KA (2003) Delta-9-tetrahydrocannabinol selectively acts on cannabinoid 1(CB1) receptors in specific regions of the dorsal vagal complex to inhibit emesis in the ferret. Am J Physiol Gastrointest Liver Physiol 285:G566–G576
Vigano D, Cascio MG, Rubino T, Fezza F, Vaccani A, Di Marzo V, Parolaro D (2003) Chronic morphine modulates the contents of the endocannabinoid, 2-arachidonoyl glycerol, in rat brain. Neuropsychopharmacology 28:1160–1167
Yiangou Y, Facer P, Dyer NH C, Chan CL H, Knowles C, Williams NS, Anand P (2001) Vanilloid receptor 1 immunoreactivity in inflamed human bowel. Lancet 357:1338–1339
Zygmunt PM, Petersson J, Andersson DA, Chuang HH, Sorgard M, Di Marzo V, Julius D, Hogestatt ED (1999) Vanilloid receptors on sensory nerves mediate the vasodilator action of anandamide. Nature 400:452–457

HEP (2005) 168:599–625

Cardiovascular Pharmacology of Cannabinoids

P. Pacher (✉) · S. Bátkai · G. Kunos

Laboratory of Physiologic Studies, National Institute on Alcohol Abuse and Alcoholism, National Institutes of Health, Bethesda MD, 20892-9413, USA
pacher@mail.nih.gov

Abstract Cannabinoids and their synthetic and endogenous analogs affect a broad range of physiological functions, including cardiovascular variables, the most important component of their effect being profound hypotension. The mechanisms of the cardiovascular effects of cannabinoids in vivo are complex and may involve modulation of autonomic outflow in both the central and peripheral nervous systems as well as direct effects on the myocardium and vasculature. Although several lines of evidence indicate that the cardiovascular depressive effects of cannabinoids are mediated by peripherally localized CB_1 receptors, recent studies provide strong support for the existence of as-yet-undefined endothelial and cardiac receptor(s) that mediate certain endocannabinoid-induced cardiovascular effects. The endogenous cannabinoid system has been recently implicated in the mechanism of hypotension associated with hemorrhagic, endotoxic, and cardiogenic shock, and advanced liver cirrhosis. Furthermore, cannabinoids have been considered as novel antihypertensive agents. A protective role of endocannabinoids in myocardial ischemia has also been documented. In this chapter, we summarize current information on the cardiovascular effects of cannabinoids and highlight the importance of these effects in a variety of pathophysiological conditions.

Keywords Cannabinoid · Anandamide · CB_1 receptor · Blood pressure · Cardiac function · Vascular · Ischemia

1 Introduction

The biological effects of marijuana and its main psychoactive ingredient, Δ^9-tetrahydrocannabinol (THC), are mediated by specific, G protein-coupled receptors (GPCRs) (Howlett et al. 1990). To date, two cannabinoid (CB) receptors have been identified by molecular cloning: the CB_1 receptor, which is by far the most abundant of all neurotransmitter receptors in the brain (Matsuda et al. 1990), but is also present in various peripheral tissues including the heart and vasculature (Gebremedhin et al. 1999; Liu et al. 2000; Bonz et al. 2003), and the CB_2 receptor, expressed primarily by immune (Munro et al. 1993) and hematopoietic cells (Valk and Delwel 1998). The natural ligands of these receptors are endogenous, lipid-like substances called endocannabinoids, which include arachidonoyl ethanolamide or anandamide and 2-arachidonoylglycerol (2-AG), as the two most widely studied members of this group (reviewed by Mechoulam et al. 1998).

Cannabinoids and their synthetic and endogenous analogs are best known for their prominent psychoactive properties, but their cardiovascular effects were also recognized as early as the 1960s. The most important component of these effects is a profound decrease in arterial blood pressure, cardiac contractility, and heart rate (Lake et al. 1997a,b; Hillard 2000; Kunos et al. 2000, 2002; Randall et al. 2002; Ralevic et al. 2002; Hiley and Ford 2004). Although several lines of evidence indicate that the cardiovascular depressive effects of cannabinoids are mediated by peripherally localized CB_1 receptors, cannabinoids can also elicit vascular and cardiac effects, which are independent of CB_1 and CB_2 receptors, as discussed in detail later in this chapter.

Recent findings implicate the endogenous cannabinoid system in the pathomechanism of hypotension associated with various forms of shock, including hemorrhagic (Wagner et al. 1997), endotoxic (Varga et al. 1998; Wang et al. 2001; Liu et al. 2003; Bátkai et al. 2004a), and cardiogenic shock (Wagner et al. 2001a, 2003), as well as the hypotension associated with advanced liver cirrhosis (Bátkai et al. 2001; Ros et al. 2002). Furthermore, the possible use of cannabinoids as novel antihypertensive agents has been entertained (Birmingham 1973; Archer 1974; Varma et al. 1975; Crawford and Merritt 1979; Zaugg and Kynel 1983; Lake et al. 1997b; Bátkai et al. 2003 and 2004; Li et al. 2003). In addition, a protective role of endocannabinoids has been described in myocardial ischemia (reviewed in Hiley and Ford 2003, 2004).

The goal of this chapter is to summarize the cardiovascular effects of cannabinoids and to highlight the unique therapeutic potential of the pharmacological manipulation of the endocannabinergic system in a variety of pathological conditions.

2
Cardiovascular Effects of Cannabinoids In Vivo

The in vivo cardiovascular effects of cannabinoids are complex and may involve modulation of the autonomic outflow in both the central and peripheral nervous systems as well as direct effects on the myocardium and vasculature. However, their peripheral actions appear to play the dominant role, at least upon systemic administration at the doses used by most investigators. Moreover, the effects of endocannabinoids are complicated by their rapid metabolism, which may liberate other vasoactive substances and their precursors (reviewed in Mechoulam et al. 1998; Kunos et al. 2002; Randall et al. 2002; Ralevic et al. 2002).

In humans, the acute effect of smoking cannabis usually manifests as an increase in heart rate with no significant change in blood pressure (Kanakis et al. 1976). However, chronic use of cannabis in man, as well as both acute and prolonged administration of THC to experimental animals, elicit a long-lasting decrease in blood pressure and heart rate (Rosenkratz 1974; Benowitz and Jones 1975). Because of the well-known effects of cannabinoids on central nervous system function, early studies of their cardiovascular actions concentrated on the ability of these compounds to inhibit sympathetic tone as the underlying mechanism. Indeed, cross-perfusion experiments in dogs have provided some evidence for a centrally mediated sympatho-inhibitory effect of THC, although additional peripheral sites of action could not be ruled out (Vollmer et al. 1974). Already at this early stage, the potential use of these compounds as antihypertensive agents was considered (Archer 1974), in the hope that their neurobehavioral and cardiovascular effects would turn out to be separable. That this may be possible to achieve was first suggested by a 1977 publication of the biological effects of abnormal cannabidiol, a synthetic analog of the neurobehaviorally inactive, plant-derived cannabinoid, cannabidiol (Adams et al. 1977). However, more than two decades have elapsed before this promising observation was followed up and extended (see below).

2.1
Role of CB_1 Receptors in the Cardiovascular Effects of Cannabinoids

The discovery of anandamide, the first endocannabinoid (Devane et al. 1992), has raised the obvious question whether it possesses cardiovascular activity similar to THC. Upon its intravenous bolus injection into anesthetized rats and mice, anandamide was found to elicit a triphasic blood pressure response and bradycardia (Varga et al. 1995; Pacher et al. 2004; Bátkai et al. 2004b; see Fig. 1) similar to that reported earlier for THC (Siqueira et al. 1979). The first phase of the response

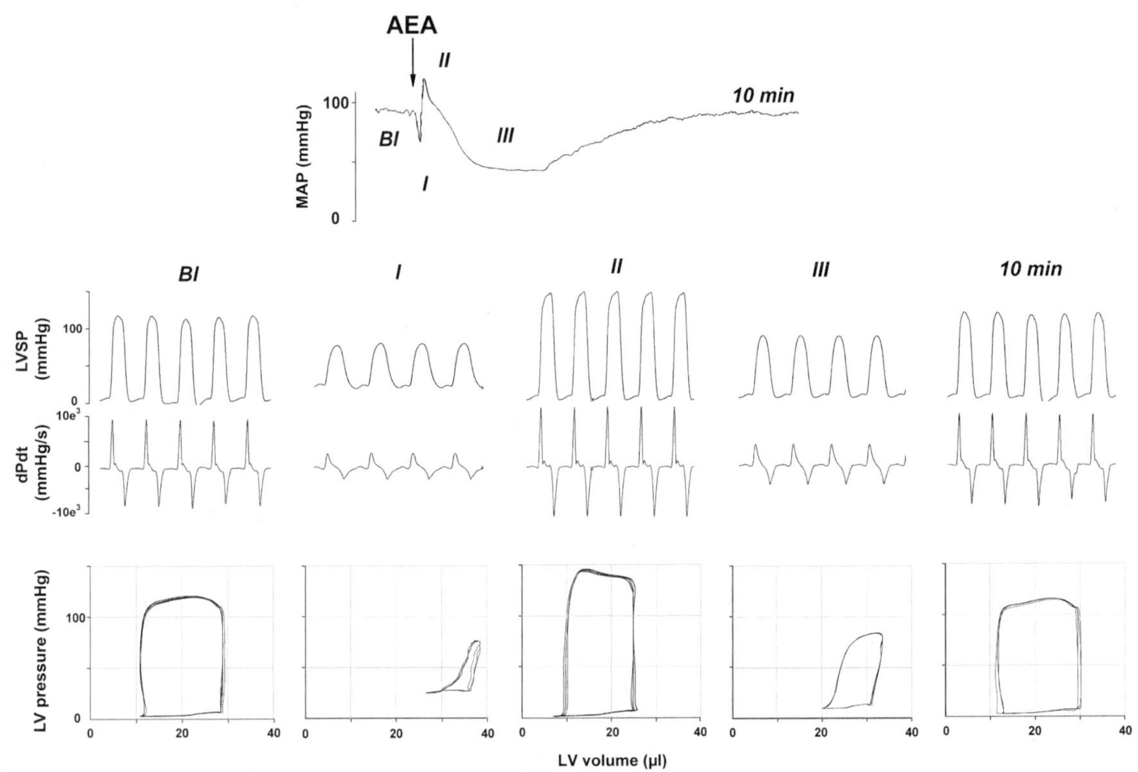

Fig. 1. Hemodynamic effects of anandamide in anesthetized mice. Representative recordings of the effects of anandamide [20 mg/kg i.v., *N*-arachidonoyl-ethanolamine (*AEA*)] on mean arterial pressure (*MAP*, *top panel*), cardiac contractility (left ventricular systolic pressure (*LVSP*) and d*P*/dt (*dPdt*); *middle panel*) and pressure-volume relations (*bottom panel*) in a pentobarbital-anesthetized C57BL6 mouse. The five parts of the *middle* and *bottom panels* represent baseline conditions (*BI*), phase I (*I*), phase II (*II*), and phase III (*III*) of the anandamide response, and recovery *10 min* following the injection. The *arrow* indicates the injection of the drug. The decrease of the amplitude of PV loops and shift to the right indicate decrease of cardiac contractile performance

consists of a precipitous drop in heart rate and blood pressure that lasts for a few seconds only. These effects are vagally mediated, as they are absent in animals after bilateral transection of the cervical vagus nerve, or after pretreatment with methylatropine (Varga et al. 1995). This vagal component is followed by a brief pressor response, which persists in the presence of α-adrenergic blockade and also in rats in which sympathetic tone is abolished by pithing, and is thus not sympathetically mediated (Varga et al. 1995). This pressor component is also unaffected by CB_1 receptor antagonists and it persists in CB_1 knockout mice (Járai et al. 1999; Pacher et al. 2004), indicating the lack of involvement of CB_1 receptors. Recent observations using the radiolabeled microsphere technique in rats suggest that this pressor component may be due to vasoconstriction in certain vascular beds, such as the spleen (Wagner et al. 2001b). The third, and most prominent, phase in the effect of anandamide is hypotension associated with moderate bradycardia that last about 2–10 min. Interestingly, this third phase is absent in conscious normotensive rats (Stein et al. 1996; Lake et al. 1997a), but is present and more prolonged in conscious, spontaneously hypertensive rats (Lake et al. 1997b; Bátkai et al. 2004b). Since sympathetic tone is known to be low in conscious, undisturbed normotensive rats (Carruba et al. 1987), these observations appear to be compatible with a sympatho-inhibitory mechanism underlying anandamide-induced hypotension and bradycardia, as further discussed below. The finding that *R*-methanandamide, a metabolically stable analog of anandamide (Abadji et al. 1994), causes similar but more prolonged hypotension and bradycardia (Kunos et al. 2000) eliminates the possibility that the hypotensive and bradycardic effects of anandamide are mediated indirectly by a metabolite.

Several lines of evidence indicate that cannabinoid-induced hypotension is mediated by CB_1 receptors. First, the hypotension is effectively antagonized by the CB_1-selective antagonist SR141716 (Varga et al. 1995, 1996; Calignano et al. 1997). SR141716 can block hypotension induced by plant-derived and synthetic cannabinoids as well as anandamide (Lake et al. 1997a). However, when tested in anesthetized mice, the hypotensive effect of 2-AG was unexpectedly resistant to inhibition by SR141716, but could be antagonized by indomethacin, suggesting the involvement of a cyclo-oxygenase metabolite (Járai et al. 2000). Indeed, 2-AG was found to be rapidly (<30 s) degraded in mouse blood to generate arachidonic acid. Accordingly, when the metabolically stable analog 2-AG ether was tested, its hypotensive effect was antagonized by SR141716 and it was absent in CB_1-deficient mice indicating that, similar to anandamide, it is an effective agonist of hypotensive CB_1 receptors (Járai et al. 2000). 2-AG ether has been recently identified as an endogenous brain constituent (Hanus et al. 2001), thus it may also be involved in cardiovascular regulation.

The second line of evidence for CB_1 receptor involvement is the strong, positive correlation between the concentrations of various cannabinoid agonists producing half-maximal hypotensive and bradycardic responses (EC_{50}) and their affinity constants for binding to CB_1 receptors in the brain (Lake et al. 1997a). The strongest evidence, however, is the total absence of cannabinoid-induced hypotension and bradycardia in mice lacking the CB_1 receptor (Járai et al. 1999; Ledent et al. 1999). Interestingly, the isolated tachycardia in response to the acute intake of THC by

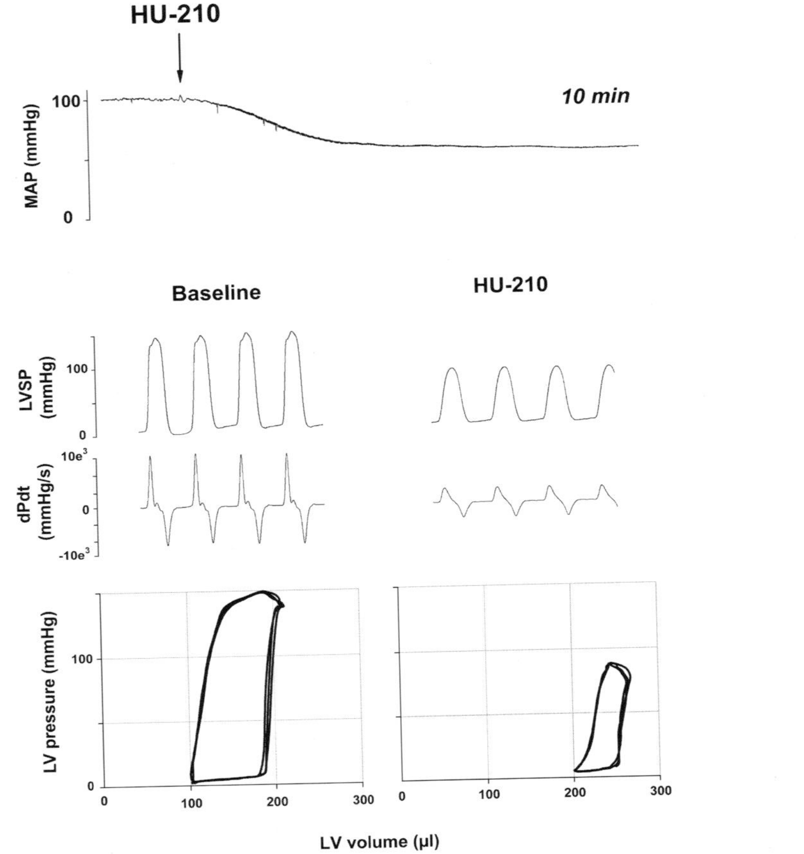

Fig. 2. Hemodynamic effects of HU-210 in anesthetized rat. Representative recordings of the effects of HU-210 (30 µg/kg i.v.) on mean arterial pressure (*MAP*, *top panel*), cardiac contractility (*LVSP* and d*P*/d*t*; *middle panel*), and pressure-volume (*PV*) relations (*bottom panel*) in a pentobarbital-anesthetized rat. The *arrow* indicates the injection of the drug. The decrease of the amplitude of PV loops and their shift to the right are indicative of decreased cardiac contractile performance

human volunteers who are not chronic marijuana users is similarly inhibited by SR141716 (Huestis et al. 2001). In healthy young adults, the heart is under dominant vagal tone, and the tachycardic effect of THC is most likely due to inhibition of acetylcholine release from cardiac vagal efferents via presynaptic CB_1 receptors (Szabo et al. 2001).

Acute drug effects on blood pressure are the result of changes in peripheral vascular resistance, cardiac output, or both. Recent unpublished results in the authors' laboratory, using the pressure-volume conductance system (Pacher et al. 2003 and 2004; Fig. 2) in pentobarbital-anesthetized mice in vivo, indicate that the hypotensive effect of (–)-11-OH-Δ^9-THC dimethylheptyl (HU-210) results primarily from a decrease in cardiac contractility (Fig. 2). In contrast, anandamide reduces both cardiac contractility (Fig. 1) and total peripheral resistance (TPR) (Pacher et al. 2004; Bátkai et al. 2004b), which is in agreement with the cardiodepressant effects of HU-210 and anandamide observed in isolated Langendorff rat hearts and in isolated, electrically stimulated human atrial appendages in vitro (Bonz et al. 2003; Ford et al. 2002; Nahas and Trouve 1985; see below). These cardiodepressor effects may underlie the ability of anandamide and HU-210 to decrease cardiac output observed in studies using the radiolabeled microsphere technique (Wagner et al. 2001b). Wagner et al. also demonstrated that both HU-210 and anandamide produce major vasodilation in the coronary and cerebral circulation, which could be antagonized by SR141716 (Wagner et al. 2001b). Collectively, these findings suggest that cannabinoids cause cardiodepressor effects as well as coronary and cerebral vasodilation via SR141716-sensitive CB_1 receptors.

Despite strong evidence for the exclusive role of CB_1 receptors in the hypotensive effect of cannabinoids, there is growing evidence over the last few years that anandamide-induced vasodilation in the mesenteric, and possibly some other vascular beds, is independent of CB_1 or CB_2 receptors (see below, Sect. 3). This apparent paradox may be resolved by the observation that in anesthetized rats, abnormal cannabidiol (abn-cbd), a non-psychoactive cannabinoid with vasodilator activity (Adams et al. 1977; Járai et al. 1999), can elicit a significant increase in mesenteric blood flow in vivo as measured by Doppler sonography, without having a significant effect on blood pressure (S. Bátkai, P. Pacher, G. Kunos, unpublished observations). It is possible that the vasodilation elicited through local release of endocannabinoids in certain vascular beds is compensated by sympathetic vasoconstriction in others, resulting in no net effect on blood pressure.

2.2 Role of Central Versus Peripheral Mechanisms in the Cardiovascular Effects of Cannabinoids

Early work with THC suggested that cannabinoids lower blood pressure through a centrally mediated sympatho-inhibitory mechanism (Vollmer et al. 1974). However, the hypotension elicited by anandamide in urethane-anesthetized rats is not associated with any change in the activity of sympathetic premotor neurons in the medullary vasomotor center or in the activity of sympathetic postganglionic nerves (Varga et al. 1996), which ruled out centrally mediated sympatho-inhibition or ganglionic blockade as possible underlying mechanisms, at least for anandamide. Intra-cerebroventricular administration in rabbits of the potent synthetic cannabinoid WIN55,212-2 was found to increase rather than decrease sympathetic tone, which also argues against a central mechanism for the hypotensive effect (Niederhoffer and Szabo 2000). Yet, the pressor response triggered by electrical stimulation of the vasomotor center was reversibly inhibited by anandamide, whereas the effect of exogenous phenylephrine was unaffected, suggesting a presynaptic-inhibitory effect of norepinephrine release from peripheral sympathetic nerve terminals (Varga et al. 1996). Indeed, stimulation of presynaptic CB_1 receptors inhibits norepinephrine release both in vitro (Ishac et al. 1996; Deutsch et al. 1997; Schlicker et al. 1997; Christopoulos et al. 2001; Vizi et al. 2001) and in vivo (Malinowska et al. 1997; Niederhoffer and Szabo 2000). However, when sympathetic tone is eliminated by ganglionic blockade and vascular tone is restored by vasopressin infusion, the hypotensive response to the potent synthetic cannabinoid HU-210 remains unchanged, although its bradycardic effect is lost (Wagner et al. 2001b). This suggests that cannabinoid-induced bradycardia may be due to inhibition of sympathetic tone to the heart, but the hypotensive response is due to direct vasodilation, as also indicated by its presence in rats following chemical sympathetic denervation (Vidrio et al. 1996).

3 Cardiovascular Effects of Cannabinoids In Vitro

3.1 Direct Vasorelaxant Effects of Cannabinoids

An early report on the effect of anandamide on cerebral blood flow indicated that the observed vasodilator response could be inhibited by indomethacin (Ellis et al. 1995). The obvious implication of this finding was that anandamide causes vasorelaxation indirectly through the generation of arachidonic acid and its subsequent metabolism by cyclooxygenase. Although THC also produced an indomethacin-sensitive response, subsequent studies could not document an effect of cyclooxygenase inhibition on anandamide-induced vasorelaxation in other blood vessels, including the mesenteric and coronary vasculature (Randall et al. 1996; Randall and Kendall 1997; Plane et al. 1996; White and Hiley 1997), ruling out increased prostanoid formation as a major mechanism for the vasodilatory effect.

The vasorelaxant effect of anandamide displays tissue and interspecies differences. Anandamide has been found to relax rat hepatic and guinea pig basilar arteries (Zygmunt et al. 1999), bovine coronary arteries (Pratt et al. 1998), but not rat carotid arteries (Holland et al. 1999) or the rat aorta (Darker et al. 1998). Anandamide has been reported to mediate vasodilation in kidney afferent arterioles through the endothelial release of nitric oxide (NO) (Deutsch et al. 1997). Anandamide was also found to release NO in a variety of human blood vessels as well as the right atrium (Bilfinger et al. 1998). In contrast, in other studies the anandamide-induced vasorelaxation was insensitive to inhibition of NO synthase (Randall et al. 1996; White and Hiley 1997; Járai et al. 1999). Both anandamide and the CB agonist HU-210 caused up-regulation of the expression and activity of the inducible NO synthase in human umbilical vein endothelial cells (HUVEC), which is unlikely to contribute to the acute vasodilatory effect, but may play an important role in terminating the action of endogenous anandamide by affecting its cellular uptake (Maccarrone et al. 2000).

The interest in the vasodilator action of endocannabinoids was further stimulated by a report in 1996 that the mesenteric vasodilation attributable to an endothelium-derived hyperpolarizing factor (EDHF) is sensitive to inhibition by SR141716 (Randall et al. 1996). The corollary of this finding was that EDHF might be an endocannabinoid released from the vascular endothelium and acting at

SR141716-sensitive CB receptors on vascular smooth muscle cells, which it would hyperpolarize and relax (Randall et al. 1996). Interestingly, carbachol was found to induce 2-AG production by the rat aortic endothelium (Mechoulam et al. 1998b), which is compatible with 2-AG being an EDHF.

Inhibition of EDHF-induced vasorelaxation by SR141716 was confirmed in some (White and Hiley 1997) but not other (Chataigneau et al. 1998; Fulton and Quilley 1998; Niederhoffer and Szabo 1999a; Pratt et al. 1998) studies. A possible source of these discrepancies may be the different species and vascular preparations tested. Additionally, the finding that the vasodilator action of anandamide has both an endothelium-dependent and an endothelium-independent component, and only the former is sensitive to inhibition by SR141716 (Chaytor et al. 1999; Mukhopadhyay et al. 2002; Wagner et al. 1999), also argues against anandamide itself being EDHF, although it leaves open the possibility that anandamide or another endocannabinoid acting at an SR141716-sensitive receptor on vascular endothelial cells may *release* an EDHF (Járai et al. 1999). This latter possibility is compatible with findings that anandamide triggers calcium transients in cultured vascular endothelial cells (Fimiani et al. 1999; Mombouli et al. 1999), and that the anandamide-induced hyperpolarization of the rat hepatic artery is endothelium dependent (Zygmunt et al. 1997). Interestingly, the mesenteric vasodilation caused by the non-psychotropic cannabinoid abn-cbd is inhibited by the same combination of calcium-activated potassium channel toxins (apamin+charybdotoxin; Járai et al. 1999; Ho and Hiley, 2003) that were reported to inhibit EDHF-induced vasodilation (Randall and Kendall 1998), although no such inhibition was observed in rat hepatic arteries (Zygmunt et al. 1997), and in some other preparations the effect of anandamide was inhibited by iberiotoxin, which blocks a different (large conductance) calcium-activated potassium channel (Ishioka et al. 1999; Begg et al. 2001; White et al. 2001). The findings with abn-cbd would suggest that cannabinoids might release EDHF via activation of an endothelial site distinct from CB_1 receptors (which recognizes abn-cbd, see below). Indeed, activation of bona fide CB_1 receptors may have an opposite effect, i.e., inhibition of the release of EDHF, as indicated by the findings of Fleming et al. (1999) in porcine coronary and rabbit carotid and mesenteric arteries.

Another mechanism by which anandamide elicits SR141716-sensitive, endothelium-dependent vasodilation may be through an intracellular site of action at gap junctions. Evidence for this mechanism is the ability of various gap junction inhibitors as well as the anandamide transport inhibitor AM404 to antagonize anandamide-induced mesenteric vasodilation and the ability of SR141716 to inhibit dye transfer through gap junctions (Chaytor et al. 1999). Together, these findings form the basis of the hypothesis that anandamide induces vasodilation at an intracellular site in endothelial cells where it would facilitate the gap junctional transfer of an EDHF to vascular smooth muscle (Chaytor et al. 1999). However, in other studies the vasodilator effect of anandamide was inhibited by some but not other gap junction inhibitors, and the ones that were inhibitory (18α-glycyrrhetinic acid, ouabain) also blocked Na^+, K^+-ATPases at the concentrations used, suggesting a mechanism of action unrelated to gap junction inhibition (Harris et al. 2002). In rabbit aortic rings, which are relaxed by anandamide in a partially endothelium-dependent manner, the anandamide response was unaffected by gap junction inhibitors (Mukhopadhyay et al. 2002), and similar findings were reported for rat isolated coronary arteries (White et al. 2001).

Finally, Zygmunt et al. (1999) described an unusual indirect pathway. They demonstrated that anandamide induces vasorelaxation in rat mesenteric and hepatic arteries, and in guinea pig basilary artery through the activation of TRPV1 receptors on sensory neurons, causing the release of the vasodilatory peptide calcitonin gene-related peptide (CGRP) (see also Sect. 4 below, "Role of Vanilloid TRPV1 Receptors in the Cardiovascular Effects of Cannabinoids").

3.2 Novel Endothelial Endocannabinoid Receptor

The possible existence of cannabinoid receptors distinct from CB_1 or CB_2 was first suggested by findings that potent synthetic cannabinoids as well as THC do not elicit vasodilation in the same rat mesenteric vascular bed preparations in which anandamide and methanandamide have strong vasodilator activity (Wagner et al. 1999). In these experiments, the effects of anandamide and methanandamide could be inhibited by SR141716, but the concentration required was somewhat higher (1–10 μM) than required for inhibition of CB_1 receptors (Járai et al. 1999; Chaytor et al. 1999; Mukhopadhyay et al. 2002; Wagner et al. 2001; White et al. 2001). Also, the ability of SR141716 to inhibit anandamide-induced vasodilation was lost following endothelial denudation (Chaytor et al. 1999; Járai et al. 1999; Mukhopadhyay et al. 2002; Wagner et al. 2001). These findings led to the postulation of an endothelial site, somewhat sensitive to inhibition by SR141716 but distinct from CB_1 or CB_2 receptors, that contributes to anandamide-induced vasodilation in the mesenteric circulation (Járai et al. 1999) and, possibly, in other vascular beds, such as the rat coronary circulation (Ford et al. 2002). In this latter study, a nonCB_1/nonCB_2 mechanism mediating the negative inotropic effect of anandamide has been also identified.

More recently, a non-CB_1/non-CB_2 site was also postulated to exist on glutamatergic terminals in the mouse hippocampus, where its activation by cannabinoids inhibits glutamatergic transmission and excitatory postsynaptic potentials (EPSPs) (Hájos et al. 2001). Similar to the endothelial site, the site in the hippocampus is susceptible to inhibition by SR141716, but it can be activated by the synthetic cannabinoid WIN55,212-2 (Hájos et al. 2001). A similarly WIN55,212-2-sensitive, but SR141716-insensitive, non-CB_1/non-CB_2 site that can activate guanosine triphosphate (GTP)γS labeling in brain membranes has been identified in CB_1 knockout mice (Breivogel et al. 2001) and in astrocytes, where its stimulation inhibits cyclic adenosine monophosphate (cAMP) production (Sagan et al. 1999). Since the endothelial site is insensitive to WIN55,212-2 in the rat mesentery (Wagner et al. 1999) or in the rabbit aorta (Mukhopadhyay et al. 2002), and abn-cbd does not inhibit glutamatergic EPSPs in rat hippocampus (M. Begg, D.M. Lovinger, G. Kunos, unpublished results), it is very likely distinct from the non-CB_1 site described in the rat CNS.

Abn-cbd is a synthetic analog of the behaviorally inactive plant-derived cannabinoid, cannabidiol. Several years ago, abn-cbd was reported to be inactive in two, rather non-specific, behavioral paradigms used to screen cannabinoids in mice, but to cause profound hypotension in dogs (Adams et al. 1977). This prompted us to speculate that abn-cbd may be a selective agonist of the putative vascular endothelial cannabinoid receptor. A detailed study of the pharmacology of abn-cbd supported this possibility (Járai et al. 1999). Abn-cbd does not bind to CB_1 receptors in the rat brain or to the human CB_2 receptor at concentrations up to 100 μM (Offertáler et al. 2003), and is inactive in the Martin behavioral tetrad in mice at doses up to 60 mg/kg (Járai et al. 1999). Yet, abn-cbd (20 mg/kg i.v.) causes SR141716-sensitive hypotension in both wild-type and CB_1 receptor knockout mice. Furthermore, abn-cbd causes endothelium-dependent vasodilation in the buffer-perfused mesenteric vascular bed isolated from rats or from wild-type as well as CB_1 knockout mice, and these effects are also inhibited by 1–5 μM of SR141716 (Járai et al. 1999). The parent compound of abn-cbd, cannabidiol (10 μM), has no vasodilator activity in the rat isolated, perfused mesenteric vascular bed preparation, but is able to inhibit the vasodilation caused by abn-cbd or anandamide (Járai et al. 1999). This suggested that abn-cbd is an agonist and cannabidiol is an antagonist of a novel endothelial cannabinoid receptor mediating vasodilation. Additional experiments indicated that the vasodilator response to abn-cbd is not affected by N^G-nitro-L-arginine methylester (L-NAME)+indomethacin, suggesting that endothelial NO and prostacyclin are not involved. However, a combination of apamin (100 nM) and charybdotoxin (100 nM), inhibitors of calcium-activated potassium channels, significantly attenuated the vasodilation caused by abn-cbd. As the same combination of potassium channel blockers inhibits EDHF-induced mesenteric vasodilation (Randall and Kendall 1998), these findings were compatible with the possible release of EDHF through activation of this novel endothelial site (Járai et al. 1999; also see above).

Capsazepine, an inhibitor of the vanilloid TRPV1 receptor, does not influence the mesenteric vasodilator response to abn-cbd at a concentration that blocks the vasodilator effect of capsaicin or the vasodilator response to anandamide in endothelial-intact preparations (Járai et al. 1999). These observations distinguish the endothelial cannabinoid receptor from TRPV1 receptors, but are compatible with the involvement of the latter in the endothelium-independent, SR141716-insensitive component of the effect of anandamide, as suggested earlier by Zygmunt et al. (1999).

Similar conclusions were reached by Howlett and coworkers, who investigated the vasodilator action of anandamide in isolated aortic rings from rabbits (Mukhopadhyay et al. 2002). In those experiments, the vasorelaxant effect of anandamide had a major (80%) endothelium-dependent and a minor endothelium-independent component, thus making it an attractive model for further exploration of the pharmacological properties of the endothelial site (Mukhopadhyay et al. 2002). The endothelium-dependent component was found to be SR141716-sensitive and also to involve pertussis toxin (PTX)-sensitive G proteins and NO production, whereas the endothelium-independent minor component appeared to be via a PTX-insensitive mechanism involving TRPV1 receptors, CGRP, and NO

(Mukhopadhyay et al. 2002). These findings are in general good agreement with the earlier findings in the rat and suggest, in addition, that the non-CB_1 endothelial receptor is coupled to G_i/G_o.

The possibility that the endothelial cannabinoid receptor is a GPCR is supported by recent findings in rat-isolated mesenteric artery segments (representing small conduit vessels approximately 200 μm in diameter) set up in a wire myograph (Offertáler et al. 2003). The responses of these preparations were similar to those observed using the perfused mesenteric vascular bed, where resistance changes reflect the response of precapillary arterioles (20-30 μm in diameter), in that the vasorelaxant effect of abn-cbd was NO-independent (resistant to inhibition by L-NAME) but sensitive to inhibition by apamin plus charybdotoxin. The observed insensitivity to NO, also observed in the case of anandamide in the same preparation (Harris et al. 2002), is different from the situation in the rabbit aorta (see above) or in the rat renal artery (Deutsch et al. 1997), and may reflect species- and/or vascular region-specific differences. Importantly, the vasorelaxation by abn-cbd was not inhibited by the vanilloid TRPV1 receptor antagonist capsazepine, and the inhibitory effects of the toxin combination implicate calcium-activated potassium channels. However, in agreement with the observations of Mukhopadhyay et al. (2002) with anandamide, mesenteric vasorelaxation by abn-cbd could be inhibited by PTX in endothelium-intact but not in endothelium-denuded preparations, which is compatible with the endothelial cannabinoid receptor being a GPCR coupled to G_i/G_o (Offertáler et al. 2003). Ho and Hiley (2003) reported, however, that the mesenteric vasodilator effect of abn-cbd was unaffected by PTX, even though its other properties, including its endothelium dependence and susceptibility to inhibition by the compound O-1918 (see below) were similar to those reported by Offertáler et al.

An endothelial site of action of abn-cbd is further documented by its ability to activate p42/44 MAP kinase and Akt phosphorylation in cultured HUVEC (Offertáler et al. 2003). As in the earlier studies using the perfused mesenteric vascular bed, in the myograph preparations abn-cbd is a full agonist, i.e., it completely reverses phenylephrine-induced contractions with an EC_{50} of 2–3 μM. Unexpectedly, cannabidiol and SR141716, both of which antagonize abn-cbd-induced vasodilation in the resistance vessels of the mesenteric arterial bed preparation, act as full agonists in the small conduit vessels (EC_{50} of 0.82 μM for cannabidiol and 6.4 μM for SR141716). Thus, these latter compounds are most likely partial agonists rather than pure antagonists at the endothelial cannabinoid receptor. This prompted us to develop structurally modified analogs of cannabidiol to search for a pure antagonist. The compound O-1918 does not relax mesenteric arterial segments at concentrations up to 30 μM, but competitively inhibits the vasodilator response to abn-cbd without affecting vasodilation to carbachol or CGRP (Offertáler et al. 2003). The endothelial site of action of O-1918 is further supported by its ability to antagonize the activation of p42/44 mitogen-activated protein (MAP) kinase and Akt phosphorylation in HUVEC (Offertáler et al. 2003). The finding that O-1918 also inhibits the mesenteric vasorelaxant effect of anandamide strongly suggest that the same endothelial receptor is the site of action of anandamide. Similar to abn-cbd, O-1918 does not bind to CB_1 or CB_2 receptors. Recent studies indicate that O-1918 inhibits the mesenteric vasorelaxant effect of two putative novel endocannabinoids, *N*-arachidonoyl dopamine (O'Sullivan et al. 2004) and virodhamine (C.R. Hiley, personal communication). These findings are important because the relatively low potency of anandamide in eliciting mesenteric vasorelaxation could suggest that the primary endogenous ligand at these receptors is not anandamide.

Inhibition of the vasodilator effect of abn-cbd by charybdotoxin (Ho and Hiley 2003; Offertáler et al. 2003; see also above) has suggested the involvement of a calcium-activated K^+ channel in this effect. In a recent study using the whole cell patch-clamp technique, we have described a voltage-dependent outward current in HUVEC that is carried by K^+ ions and is blocked by charybdotoxin and iberiotoxin, suggesting the involvement of the large conductance calcium-activated potassium (BK_{Ca}) channel (Begg et al. 2003). Although abn-cbd did not elicit a current on its own, it caused a concentration-dependent increase in the voltage-induced K^+ current that was sensitive to PTX, suggesting the involvement of a G_i/G_o-coupled receptor. The increase in K^+ current by abn-cbd was unaffected by relevant concentrations of SR141716 or SR144528, which argues against the involvement of CB_1 and CB_2 receptors (Begg et al. 2003). This was further supported by the lack of effect on K^+ currents of HU-210, a potent CB_1/CB_2 receptor agonist, which is devoid of mesenteric vasodilator activity. On the other hand, the abn-cbd-induced increase in K^+ current was antagonized by O-1918, which produced no effect by itself (Begg et al. 2003). The finding that the iberiotoxin-sensitive current induced by the selective BK_{Ca} opener NS-1619 was unaffected by O-1918 indicates that blockade of the effect of abn-cbd by O-1918 occurs at a site proximal to the channel, most likely at the receptor (Begg et al. 2003). This compound as well as PTX similarly inhibited the endothelium-dependent vasorelaxing effect of abn-cbd in the rat isolated mesenteric artery (Offertáler et al. 2003). This raises the possibility that activation of BK_{Ca} channels is involved in the mesenteric vasodilation mediated by this novel endothelial receptor.

The endogenous cannabinoid anandamide also increased the K^+ current evoked by a single voltage step; however, its effect was only observed at high concentrations. The effect of anandamide was partially inhibited by O-1918 or PTX; thus, part of the effect of anandamide may not be mediated by the same pathway as abn-cbd (Begg et al. 2003). Anandamide acted as a full agonist in the rat isolated mesenteric artery preparation with an EC_{50} comparable to that of abn-cbd (Offertáler et al. 2003), suggesting that there may be subtle differences between the rat and human receptors. The low apparent efficacy of anandamide for the human endothelial receptor could also suggest the existence of an endogenous ligand(s) other than anandamide (see also above).

In HUVEC, intracellular cyclic guanosine monophosphate (cGMP) increased the voltage-induced outward current, comparable with the effect of abn-cbd. A similar increase in outward current was also produced by YC-1, an activator of soluble guanylyl cyclase. The increases evoked by abn-cbd and cGMP were inhibited by a protein kinase G inhibitor KT-5823, and the effect of abn-cbd, but not of cGMP, was also blocked by the guanylyl cyclase inhibitor 1*H*-(1,2,4)oxadiazolo[4,3-*a*]quinozalin-1-one (ODQ). Furthermore, cGMP continued to increase the K^+

current under Ca^{2+}-clamped conditions, indicating that its action is not due to modulation of $[Ca^{2+}]_i$. The effects of abn-cbd, cGMP, and YC-1 on K^+ currents were not additive, suggesting that these compounds utilize a common intracellular pathway (Begg et al. 2003). Finally, abn-cbd was found to increase cellular cGMP levels, and this effect could be inhibited by O-1918 (Begg et al. 2003). Together, these data suggest that the novel, G_i/G_o-coupled receptor activated by abn-cbd is positively coupled to guanylyl cyclase to raise intracellular cGMP, which activates protein kinase G.

Recently, it has been proposed that TRPV4 Ca^{2+} entry channels in vascular endothelial cells contribute to the vasorelaxant effect of anandamide via its enzymatic degradation, yielding arachidonic acid and its subsequent P450 epoxygenase-dependent metabolism (Watanabe et al. 2003). TRPV4 channels are unlikely to be involved in the effects of abn-cbd on the outward current or on vascular tone, because these effects are sensitive to PTX, whereas TRPV4-mediated calcium entry is not. Furthermore, potentiation of the outward current by abn-cbd persisted in the presence of clamped intracellular calcium (Begg et al. 2003). Also, *R*-methanandamide, an anandamide analog resistant to enzymatic degradation and therefore unlikely to give rise to P450 metabolites, has a mesenteric vasodilator effect similar to that of anandamide (Wagner et al. 1999).

Extracellular calcium is known to have a potent vasodilator effect, particularly in the mesenteric circulation, where it is thought to contribute to the postprandial vasodilation associated with the intestinal absorption of nutrients. Bukoski and coworkers found that SR141716 can inhibit Ca^{2+}-induced mesenteric vasodilation through a sensory nerve-dependent mechanism, which led them to suggest that anandamide may be a sensory nerve-derived vasodilator mediator (Ishioka and Bukoski 1999). Interestingly, O-1918 also inhibits calcium-induced mesenteric vasorelaxation, which is similar in wild-type and CB_1 receptor knockout mice (Bukoski et al. 2002), suggesting that the vasodilation by extracellular calcium is most likely mediated by the endothelial abn-cbd-sensitive receptor.

Collectively, the above-mentioned results indicate that the synthetic cannabinoid ligands abn-cbd and O-1918 act as a selective agonist and silent antagonist, respectively, of a novel vascular endothelial receptor distinct from CB_1 and CB_2 that mediates mesenteric vasodilation, and is coupled to a phosphoinositide (PI)3-kinase/Akt-dependent pathway through G_i/G_o.

3.3 Direct Cardiodepressant Effects of Cannabinoids

In contrast to the growing knowledge on the vascular effects of cannabinoids, little is known about cannabinoid-induced direct cardiac effects. The endocannabinoid anandamide (Felder et al. 1996), anandamide amidohydrolase (Bilfinger et al. 1998), and traces of the message for the CB_1 receptor (Galiegue et al. 1995) have all been detected in the human heart. In a more recent study, the existence of CB_1 receptors was confirmed in human atrial myocytes by immunoblotting and immunohistochemistry (Bonz et al. 2003). In the same study, it was demonstrated that anandamide, *R*-methanandamide, and HU-210 dose-dependently decrease contractile performance in isolated, electrically paced human atrial muscle. A selective and potent CB_1 antagonist, AM251 (Gatley et al. 1997), blocked the negative inotropic effect of all three drugs, and the involvement of CB_2 receptor activation, NO, or prostanoid release could all be excluded (Bonz et al. 2003). Consistently with these in vitro results, HU-210 decreases cardiac output in rats in vivo in a CB_1 receptor-dependent manner (Wagner et al. 2001b). Previous studies have also demonstrated that anandamide caused SR141716-sensitive coronary vasorelaxation in isolated perfused rat hearts (Randall and Kendall 1997; Fulton and Quilley 1998), implicating cannabinoid receptors. These effects of anandamide were not mimicked by arachidonic acid, indicating that the vasodilator effect was not mediated by arachidonic acid metabolites.

In isolated, perfused, rat Langendorff heart preparations, anandamide and *R*-methanandamide, but not palmitoylethanolamide or the selective CB_2 receptor agonist JWH015, significantly reduced both left ventricular developed and coronary perfusion pressures, indicating decreased myocardial contractile function and coronary vasodilation (Ford et al. 2002). Interestingly, anandamide-mediated vasodilatation and negative inotropy were both sensitive to inhibition by the CB_1 antagonist SR141716 and the CB_2 antagonist SR144528, but not to the TRPV1 antagonist capsazepine, which led the authors to propose a novel site distinct from classic CB_1 and CB_2 receptors (Ford et al. 2002).

In agreement with the observations on isolated cardiac preparations, our recent unpublished results using the Millar pressure-volume conductance system (see below and also Figs. 1 and 2) to directly measure cardiac performance in vivo strongly suggest the crucial importance of the cardiac component in the hemodynamic effects of cannabinoids. Taken together, the above-mentioned studies suggest that CB_1 receptors are present in cardiomyocytes, and cannabinoids may decrease cardiac contractility through both CB_1-dependent and CB_1-independent mechanisms.

4 Role of Vanilloid TRPV1 Receptors in the Cardiovascular Effects of Cannabinoids

Structural similarities between anandamide and vanilloid compounds such as capsaicin (Di Marzo et al. 1998) raised the possibility of an interplay between these two systems. Indeed, Zygmunt et al. (1999) demonstrated that in rat mesenteric arteries, the endothelium-independent vasodilator effect of anandamide is inhibited by the TRPV1 receptor antagonist capsazepine or by a CGRP receptor antagonist. They further demonstrated that anandamide binds to the cloned TRPV1 receptor with micromolar affinity, and at nanomolar concentrations it releases immunoreactive CGRP from sensory nerve terminals located in the vascular adventitia (Zygmunt et al. 1999). A similar involvement of TRPV1 receptors in the mesenteric vasodilator action of methanandamide has also been suggested (Ralevic et al. 2000). These observations support the hypothesis that anandamide-induced vasodilation involves activation of TRPV1 receptors in sensory nerves and the subsequent release of the potent vasodilator peptide CGRP.

The above observations do not implicate the endothelium in the vasodilator response to anandamide. Other studies, which documented both endothelium-dependent and endothelium-independent components for the vasodilator effect of anandamide, confirmed the role of TRPV1 receptors but only for the endothelium-independent component (Járai et al. 1999; Mukhopadhyay et al. 2002). The endothelium-dependent vasodilator effect of anandamide in the rabbit aorta or the similar effect of abn-cbd in rat mesenteric arteries is unaffected by capsazepine (Mukhopadhyay et al. 2002; Járai et al. 1999; Offertáler et al. 2003; Ho and Hiley 2003). Interestingly, sensory nerve terminals also appear to have CB_1 receptors, stimulation of which by very low doses of anandamide or by the synthetic cannabinoid HU-210, neither of which results in activation of TRPV1 receptors, inhibits sensory neurotransmission (reviewed in Ralevic et al. 2002). Furthermore, a recent study by Zygmunt et al. (2002) indicates that THC and cannabinol, but not other psychotropic cannabinoids, can elicit CGRP release from periarterial sensory nerves by a mechanism that is independent of not only CB_1 and CB_2 receptors, but also of vanilloid TRPV1 receptors. Thus, the sensory nerve-dependent effects of cannabinoids are complex, as interactions with CB_1 and TRPV1 receptors appear to have opposite functional consequences, and there may be additional actions independent of both of these receptors. TRPV1 receptors are not involved in the dilation of isolated coronary arteries by anandamide either in the sheep, where the effect is endothelium dependent (Grainger and Boachie-Ansah 2001), or in the rat, where it is endothelium independent (White et al. 2001). Furthermore, in the rat mesenteric arterial bed, the role of sensory nerves and vanilloid receptors in the dilator effect of anandamide was found to be conditional on the presence of NO (Harris et al. 2002).

TRPV1-containing afferent nerve fibers are present on the epicardial surface of the heart and the activation of these receptors by epicardially injected capsaicin evokes a sympathoexcitatory response with a brief increase in blood pressure (Zahner et al. 2003). Capsaicin infusion also induces a moderate pressor effect in pigs (Kapoor et al. 2003). We have recently found (Pacher et al. 2004) that i.v. injection of 10 μg/kg capsaicin evokes only a brief pressor response in wild-type mice, while at the much higher dose of 100 μg/kg its effect has both a depressor and a pressor component. In contrast, capsaicin elicited no change in blood pressure in TRPV1$^{-/-}$ mice, suggesting that TRPV1 receptors mediate the cardiogenic sympathetic or Bezold-Jarisch reflex in mice, which is in agreement with recent reports in which the cardiovascular effects of capsaicin were inhibited by TRPV1 receptor antagonists (Smith and McQueen 2001; Zahner et al. 2003).

In the absence of anandamide-induced hypotension in CB_1 knockout mice, the physiological relevance of the interaction of anandamide with TRPV1 receptors has been questioned (Szolcsányi 2000). We previously reported that anandamide causes a triphasic blood pressure response where the predominant hypotensive effect is preceded by a transient, vagally mediated drop in heart rate and blood pressure followed by a brief pressor response (Varga et al. 1995, see also Sect. 2 and Fig. 1). This initial component is missing in TRPV1$^{-/-}$ mice (Pacher et al.

2004) and is unaffected by the CB_1 antagonist SR141716 in rats (Varga et al. 1995). Bolus intravenous injections of anandamide may reach high enough plasma concentrations for a few seconds to activate TRPV1 receptors, which may explain the above findings. Indeed, the phase I transient hypotension and bradycardia do not appear when anandamide is injected slowly to limit its peak plasma concentration (Z. Jarai, J.A. Wagner, G. Kunos, unpublished observations). In contrast, the prolonged hypotensive phase of the anandamide response is characterized by decreased cardiac contractility and total peripheral resistance (TPR), which are similar in $TRPV1^{+/+}$ and $TRPV1^{-/-}$ mice and were completely antagonized by SR141716, implicating CB_1 receptors (Pacher et al. 2004). In agreement with this observation, the sustained hypotensive and bradycardic effects of cannabinoids are totally absent in mice lacking the CB_1 receptor (Ledent et al. 1999; Járai et al. 1999). Thus, TRPV1 receptors are not involved in the sustained cardiovascular response to anandamide, but may become transiently activated in response to pharmacological concentrations achieved after bolus i.v. injections. These findings are in agreement with a recent report by Malinowska et al (2001) who found that in rats the transient vagal activation to a bolus injection of anandamide was partially blocked by the TRPV1 antagonists capsazepine or ruthenium red, whereas the CB_1-mediated prolonged hypotension remained unaffected.

Collectively, the above-mentioned studies show that the sustained hypotensive effect of anandamide involves a marked cardiodepressor component in addition to a decrease in TPR, and these effects are mediated by CB_1 but not TRPV1 receptors. The role of TRPV1 receptors is limited to the transient activation of the Bezold-Jarisch reflex by very high plasma concentrations of anandamide.

5 Pathophysiological Role of the Endocannabinergic System in Cardiovascular Disorders

Recent studies indicate that the endogenous cannabinergic system plays an important role in cardiovascular regulation under various pathophysiological conditions, and pharmacological manipulation of this system may offer novel therapeutic approaches in a variety of cardiovascular disorders.

5.1 Role of the Endocannabinergic System in Hemorrhagic, Endotoxic, and Cardiogenic Shock and Liver Cirrhosis

The profound and long-lasting, yet reversible, hypotension elicited by potent synthetic cannabinoids (Lake et al. 1997a) suggested that endocannabinoids may be involved in pathological conditions associated with extreme hypotension such as various forms of shock. Observations over the last decade have provided evidence for a key role of endocannabinoids in the hypotension associated with hemorrhagic (Wagner et al. 1997), endotoxic (Varga et al. 1998; Liu et al. 2003; Bátkai et al. 2004a; Wang et al. 2001; Godlewski et al. 2004), and cardiogenic shock (Wagner et al. 2001a, 2003). The vasodilated state associated with advanced liver cirrhosis appears to be due to a similar mechanism (Bátkai et al. 2001; Ros et al. 2002), which is most likely secondary to the endotoxemia commonly found in patients with late-stage cirrhosis (Lumsden et al. 1988).

In many of these conditions, circulating macrophages and platelets were found to contain elevated levels of endocannabinoids and to elicit SR141716-sensitive hypotension when injected into normal rats, suggesting the involvement of CB_1 receptors. In cirrhosis this was also suggested by the observed increase in CB_1 receptor mRNA and binding sites in vascular endothelial cells from cirrhotic human livers (Bátkai et al. 2001). However, SR141716 can also inhibit a receptor(s) distinct from CB_1 or CB_2 (see above), so the relative role of such receptors versus CB_1 receptors needed to be explored. In a recent study (Bátkai et al. 2004a), we reported that the acute hypotensive response of anesthetized rats to lipopolysaccharide (LPS) is inhibited by SR141716, but not by AM251, an antagonist equipotent with SR141716 at CB_1 receptors (Gatley et al. 1997), but devoid of inhibitory potency at the SR141716-sensitive endothelial and myocardial receptors described above (Ford et al. 2002; Ho and Hiley 2003; O'Sullivan et al. 2004). Furthermore, LPS caused similar, SR141716-sensitive hypotension in wild-type mice and in mice deficient in CB_1 or both CB_1 and CB_2 receptors. Detailed hemodynamic analysis of the effects of LPS also indicated that the hypotension is primarily due to decreased cardiac contractility rather than decreased peripheral vascular resistance. These findings therefore suggest that receptors distinct from CB_1 or CB_2 are primarily responsible for the acute, SR141716-sensitive hypotensive response to LPS (Bátkai et al. 2004a). Anandamide is a ligand for such receptors and LPS stimulates the synthesis of anandamide in macrophages (Liu et al. 2003). Whether LPS may induce the synthesis of additional endogenous ligands for such receptors remains to be determined. Increased target organ sensitivity to anandamide may also play a role in the hemodynamic effects of LPS, as suggested by the potentiation of the vasodilator effect of anandamide in mesenteric beds isolated from endotoxemic rats (Orliac et al. 2003).

Endocannabinoid-mediated cardiovascular effects appear to have survival value, as indicated by the increased mortality, despite the increase in blood pressure, following blockade of CB_1 receptors in hemorrhagic (Wagner et al. 1997) and cardiogenic shock (Wagner et al. 2001a, 2003). In contrast, treatment with the cannabinoid agonists THC or HU-210 improved endothelial function and increased survival in cardiogenic (Wagner et al. 2001a, 2003) and endotoxic shock (Varga et al., 1998). Endocannabinoids may improve tissue oxygenation in these conditions by counteracting the excessive sympathetic vasoconstriction triggered by hemorrhage or myocardial infarction. In addition, endocannabinoids may mediate important protective mechanisms against hypoxic damage in the heart and vasculature and also exert potent anti-inflammatory effects (reviewed in Hiley and Ford 2003, 2004; Walter and Stella 2004, see also below).

5.2
Role of the Endocannabinergic System in Myocardial Reperfusion Damage

Interest in the investigation of potential cardioprotective effects of cannabinoids has recently been rekindled by a study of Lagneux and Lamontagne (2001) in which they compared the effects of a period of 90 min of low-flow ischaemia, followed by 60 min reperfusion at normal flow, in isolated hearts from rats pretreated with bacterial endotoxin or saline. Endotoxin pretreatment reduced infarct size and enhanced functional recovery on reperfusion relative to the saline controls (preconditioning). The beneficial effects of endotoxin-induced preconditioning were blocked by the CB_2 receptor antagonist SR144528 but not by the CB_1 antagonist SR141716 (Lagneux and Lamontagne, 2001). Similarly, Joyeux et al. (2002) found that the infarct-size-reducing effect of heat stress preconditioning was also abolished by SR144528 but not by SR141716. In contrast, in an isolated perfused rat heart model in which preconditioning was induced by a brief period of ischemia (5 min), blockade of either CB_1 receptors with SR141716 or CB_2 receptors with SR1445278 abolished the protective effect of preconditioning. Preconditioning (5 min) also preserved the endothelium-dependent vasodilation evoked by serotonin (5-HT) in another study (Bouchard et al. 2003) in which both CB_1 and CB_2 receptors were implicated. Lepicier et al. (2003) have shown that the CB_2 receptor-mediated cardioprotection by endocannabinoids and synthetic cannabinoid ligands involves p38/ERK1/2 and protein kinase C (PKC) activation.

In an anesthetized rat model of ischemia/reperfusion injury (I/R), induced by coronary occlusion/reocclusion, Krylatov et al. have demonstrated that HU-210 and anandamide reduced the infarct size and the incidence of ventricular arrhythmias through activation of CB_2 but not CB_1 receptors (Krylatov et al. 2001; 2002a,b,c; Ugdyzhenkova et al. 2001; 2002). More recently, in an anesthetized mouse model of myocardial I/R, the mixed CB_1/CB_2 receptor agonist WIN55,212-2 significantly reduced the extent of leukocyte-dependent myocardial damage. The protective effect of WIN55,212-2 was abolished by the selective CB_2 receptor antagonist AM630 and not affected by the selective CB_1 receptor antagonist AM251 (DiFilippo C et al. 2004).

5.3
Role of Endocannabinergic System in Hypertension

As early as in the 1970s, the potential use of cannabinoid ligands as antihypertensive agents had been considered (Archer 1974; Crawford and Merritt 1979; Varma et al. 1975; Zaugg and Kyncl 1983), in the hope that their neurobehavioral and cardiovascular effects could be separated. Although an early study in normotensive rats indicated rapidly developing tolerance to the hypotensive and bradycardic effects of THC (Adams et al. 1976), a subsequent study in spontaneously hypertensive rats (SHR) found no evidence for tolerance for the same effects during a similar, 10-day treatment period (Kosersky 1978).

Following the introduction of selective CB receptor antagonists in the mid 1990s, the finding that treatment of normotensive rats or mice with CB_1 antagonists alone does not affect blood pressure (Lake et al. 1997a; Varga et al. 1995), and that baseline blood pressure is similar in CB_1 knockout mice and their wild-type littermates (Járai et al. 1999; Ledent et al. 1999), indicated the absence of an endocannabinergic "tone" in the maintenance of normal levels of blood pressure. This is also suggested by the lack of significant hypotension following inhibition of anandamide transport (Calignano et al. 1997), in agreement with the relatively modest hypotensive effect of anandamide in normotensive animals (Lake et al. 1997a; Varga et al. 1995). However, SHR respond with greater and longer-lasting hypotension than normotensive rats to both THC (Kosersky 1978) and anandamide (Lake et al. 1997b; Bátkai et al. 2004b). THC inhalation evokes a greater and longer-lasting decrease of arterial blood pressure in hypertensive as compared to normotensive individuals (Crawford and Merritt 1979). Although the mechanism underlying this increased sensitivity has not been explored, it could suggest a role for the endocannabinoid system in regulating cardiovascular functions in hypertension. In a recent study we used three different models of experimental hypertension to explore this possibility (Bátkai et al. 2004b). The results document a significant endocannabinergic tone in hypertension that limits increases in blood pressure and cardiac contractile performance through tonic activation of cardiac and vascular CB_1. They also indicate that upregulation of cardiac and vascular CB_1 contributes to this tone, the potentiation of which, by inhibiting the inactivation of endogenous anandamide, can normalize blood pressure and cardiac contractile performance in hypertension. These findings raise the interesting possibility of the therapeutic use of inhibitors of fatty acid amide hydrolase in the treatment of hypertension.

6 Conclusions

Functional CB_1 receptors are present in vascular tissue as well as the myocardium, and cannabinoid agonists and endocannabinoids exert major hypotensive and cardiodepressor effects in vivo through the stimulation of CB_1 receptors. There is evidence for the existence of an as-yet-undefined endothelial and cardiac receptor or receptors that mediate certain endocannabinoid-induced cardiovascular effects. Vanilloid TRPV1 receptors can be activated by anandamide, but the role of these receptor in in vivo hemodynamic effects appears to be limited to the transient activation of the Bezold-Jarisch reflex by very high initial plasma concentrations of anandamide.

Endocannabinoids play important roles in a variety of pathophysiological conditions including hemorrhagic, endotoxic, and cardiogenic shock, and in the hemodynamic sequelae of advanced liver cirrhosis. Furthermore, pharmacological manipulation of the endocannabinoid system may offer novel therapeutic approaches in hypertension and ischemic heart disease.

References

Abadji V, Lin S, Taha G, Griffin G, Stevenson LA, Pertwee RG, Makriyannis A (1994) (R)-Methanandamide: a chiral novel anandamide possessing higher potency and metabolic stability. J Med Chem 37:1889–1893

Adams MD, Chait LD, Earnhardt JT (1976) Tolerance to the cardiovascular effects of delta9-tetrahydrocannabinol in the rat. Br J Pharmacol 56:43–48

Adams MD, Earnhardt JT, Martin BR, Harris LS, Dewey WL, Razdan RK (1977) A cannabinoid with cardiovascular activity but no overt behavioral effects. Experientia 33:1204–1205

Archer RA (1974) The cannabinoids: therapeutic potentials. Annu Rep Med Chem 9:253–259

Bátkai S, Járai Z, Wagner JA, Goparaju SK, Varga K, Liu J, Wang L, Mirshahi F, Khanolkar AD, Makriyannis A, Urbascheck R, Garcia N Jr, Sanyal AJ Kunos G (2001) Endocannabinoids acting at vascular CB1 receptors mediate the vasodilated state in advanced liver cirrhosis. Nat Med 7:827–832

Bátkai S, Pacher P, Osei-Hyiaman D, Radaeva S, Offertáler L, Bukoski RD, Kunos G (2003) Endocannabinoids are involved in regulating cardiovascular function in spontaneously hypertensive rats. Hypertension 42:A263

Bátkai S, Pacher P, Járai Z, Wagner JA, Kunos G (2004a) Cannabinoid antagonist SR141716 inhibits endotoxic hypotension by a cardiac mechanism not involving CB1 or CB2 receptors. Am J Physiol Heart Circ Physiol 287:H595–H600

Bátkai S, Pacher P, Osei-Hyiaman D, Radaeva S, Liu J, Harvey-White J, Offertaler L, Mackie K, Rudd MA, Bukoski RD, Kunos G (2004b) Endocannabinoids acting at CB1 receptors regulate cardiovascular function in hypertension. Circulation 110(14):1996–2002

Begg M, Baydoun A, Parsons ME, Molleman A (2001) Signal transduction of cannabinoid CB1 receptors in a smooth muscle cell line. J Physiol 531:95–104

Begg M, Mo FM, Offertaler L, Bátkai S, Pacher P, Razdan RK, Lovinger DM, Kunos G (2003) G protein-coupled endothelial receptor for atypical cannabinoid ligands modulates a Ca2+-dependent K+ current. J Biol Chem 278:46188–46194

Benowitz NL, Jones RT (1975) Cardiovascular effects of prolonged delta-9-tetrahydrocannabinol ingestion. Clin Pharmacol Ther 18:287–297

Bilfinger TV, Salzet M, Fimiani C, Deutsch DG, Tramu G, Stefano GB (1998) Pharmacological evidence for anandamide amidase in human cardiac and vascular tissues. Int J Cardiol 64:S15–22

Birmingham MK (1973) Reduction by 9-tetrahydrocannabinol in the blood pressure of hypertensive rats bearing regenerated adrenal glands. Br J Pharmacol 48:169–171

Bisogno T, Hanus L, De Petrocellis L, Tchilibon S, Ponde DE, Brandi I, Moriello AS, Davis JB, Mechoulam R, Di Marzo V (2001) Molecular targets for cannabidiol and its synthetic analogues: effect on vanilloid VR1 receptors and on the cellular uptake and enzymatic hydrolysis of anandamide. Br J Pharmacol 134:845–852

Bonz A, Laser M, Kullmer S, Kniesch S, Babin-Ebell J, Popp V, Ertl G Wagner JA (2003) Cannabinoids acting on CB1 receptors decrease contractile performance in human atrial muscle. J Cardiovasc Pharmacol 41:657–664

Bouchard JF, Lepicier P, Lamontagne D (2003) Contribution of endocannabinoids in the endothelial protection afforded by ischemic preconditioning in the isolated rat heart. Life Sci 72:1859–1870

Breivogel CS, Griffin G, Di Marzo V, Martin BR (2001) Evidence for a new G protein-coupled receptor in mouse brain. Mol Pharmacol 60:155–163

Bukoski RD, Bátkai S, Járai Z, Wang, Y, Offertáler L, Jackson WF, Kunos G (2002) CB(1) receptor antagonist SR141716A inhibits Ca(2+)-induced relaxation in CB(1) receptor-deficient mice. Hypertension 39:251–257

Calignano A, La Rana G, Beltramo M, Makriyannis A, Piomelli D (1997) Potentiation of anandamide hypotension by the transport inhibitor, AM404. Eur J Pharmacol 337:R1–R2

Carruba MO, Bondiolotti G, Picotti GB, Catteruccia N, Da Prada M (1987) Effects of diethyl ether, halothane, ketamine and urethane on sympathetic activity in the rat. Eur J Pharmacol 134:15–24

Chataigneau T, Feletou M, Thollon C, Villeneuve N, Vilaine J-P, Duhault J, Vanhoutte PM (1998) Cannabinoid CB1 receptor and endothelium-dependent hyperpolarization in guinea-pig carotid, rat mesenteric and porcine coronary arteries. Br J Pharmacol 123:968–974

Chaytor AT, Martin PEM, Evans WH, Randall MD, Griffith TM (1999) The endothelial component of cannabinoid-induced relaxation in rabbit mesenteric artery depends on gap junctional communication. J Physiol (Lond) 520:539–550

Christopoulos A, Coles P, Lay L, Lew MJ, Angus JA (2001) Pharmacological analysis of cannabinoid receptor activity in the rat vas deferens. Br J Pharmacol 132:1281–1291

Crawford WJ, Merritt JC (1979) Effects of tetrahydrocannabinol on arterial and intraocular hypertension. Int J Clin Pharmacol Biopharm 17:191–196

Darker IT, Millns PJ, Selbie L, Randall MD, S-Baxter G, Kendall DA (1988) Cannabinoid (CB1) receptor expression is associated with mesenteric resistance vessels but not thoracic aorta in the rat. Br J Pharmacol 125, 95P

Deutsch DG, Goligorsky MS, Schmid PC, Krebsbach RJ, Schmid HH, Das SK, Dey SK, Arreaza G, Thorup C, Stefano G, Moore LC (1997) Production and physiological actions of anandamide in the vasculature of the rat kidney. J Clin Invest 100:1538–1546

Devane WA, Hanus L, Breuer A, Pertwee RG, Stevenson LA, Griffin G, Gibson D, Mandelbaum A, Etinger A, Mechoulam R (1992) Isolation and structure of a brain constituent that binds to the cannabinoid receptor. Science 258:1946–1949

Di Filippo C, Rossi F, Rossi S, DAmico M (2004) Cannabinoid CB2 receptor activation reduces mouse myocardial ischemia-reperfusion injury: involvement of cytokine/chemokines and PMN. J Leukoc Biol 75:453–459

Di Marzo V, Bisogno T, Melck D, Ross R, Brockie H, Stevenson L, Pertwee R, De Petrocellis L (1998) Interactions between synthetic vanilloids and the endogenous cannabinoids system. FEBS Lett 436:449–454

Ellis EF, Moore SF, Willoughby KA (1995) Anandamide and delta 9-THC dilation of cerebral arterioles is blocked by indomethacin. Am J Physiol 269:H1859–H1864

Felder CC, Nielsen A, Briley EM, Palkovits M, Priller J, Axelrod J, Nguyen DN, Richardson JM, Riggin RM, Koppel GA, Paul SM, Becker GW (1996) Isolation and measurement of the endogenous cannabinoid receptor agonist, anandamide, in brain and peripheral tissues of human and rat. FEBS Lett 393:231–235

Fimiani C, Mattocks D, Cavani F, Salzet M, Deutsch DG, Pryor S, Bilfinger TV, Stefano GB (1999) Morphine and anandamide stimulate intracellular calcium transients in human arterial endothelial cells: coupling to nitric oxide release. Cell Signal 11:189–193

Fleming I, Schermer B, Popp R, Busse R (1999) Inhibition of the production of endothelium-derived hyperpolarizing factor by cannabinoid receptor agonists. Br J Pharmacol 126:949–960

Ford WR, Honan SA, White R, Hiley CR (2002) Evidence of a novel site mediating anandamide-induced negative inotropic and coronary vasodilator responses in rat isolated hearts. Br J Pharmacol 135:1191–1198

Fulton DJ, Quilley J (1998) Evidence against anandamide as the hyperpolarizing factor mediating the nitric oxide-independent coronary vasodilator effect of bradykinin in the rat. J Pharmacol Exp Ther 286:1146–1151

Galiegue S, Mary S, Marchand J, Dussossoy D, Carriere D, Carayon P, Bouaboula M, Shire D, Le Fur G, Casellas P (1995) Expression of central and peripheral cannabinoid receptors in human immune tissues and leukocyte subpopulations. Eur J Biochem 232:54–61

Gatley SJ, Lan R, Pyatt B, Gifford AN, Volkow ND, Makriyannis A (1997) Binding of the non-classical cannabinoid CP 55,940, and the diarylpyrazole AM251 to rodent brain cannabinoid receptors. Life Sci 61:PL191–197

Gebremedhin D, Lange AR, Campbell WB, Hillard CJ, Harder DR (1999) Cannabinoid CB1 receptor of cat cerebral arterial muscle functions to inhibit L-type Ca2+ channel current. Am J Physiol 266:H2085–H2093

Godlewski G, Malinowska B, Schlicker E (2004) Presynaptic cannabinoid CB1 receptors are involved in the inhibition of the neurogenic vasopressor response during septic shock in pithed rats. Br J Pharmacol 142:701–708

Grainger J, Boachie-Ansah G (2001) Anandamide-induced relaxation of sheep coronary arteries: the role of the vascular endothelium, arachidonic acid metabolites and potassium channels. Br J Pharmacol 134:1003–1012

Hajos N, Ledent C, Freund TF (2001) Novel cannabinoid-sensitive receptor mediates inhibition of glutamatergic synaptic transmission in the hippocampus. Neuroscience 106:1–4

Hanus L, Abu-Lafi S, Fride E, Breuer A, Vogel Z, Shalev DE, Kustanovich I, Mechoulam R (2001) 2-Arachidonyl glyceryl ether, an endogenous agonist of the cannabinoid CB1 receptor. Proc Natl Acad Sci U S A 98:3662–3665

Harris D, McCulloch AI, Kendall DA, Randall MD (2002) Characterization of vasorelaxant responses to anandamide in the rat mesenteric arterial bed. J Physiol 539:893–902

Hiley CR, Ford WR (2003) Endocannabinoids as mediators in the heart: a potential target for therapy of remodelling after myocardial infarction? Br J Pharmacol 138:1183–1184

Hiley CR, Ford WR (2004) Cannabinoid pharmacology in the cardiovascular system: potential protective mechanisms through lipid signalling. Biol Rev Camb Philos Soc 79:187–205

Hillard CJ (2000) Endocannabinoids and vascular function. J Pharmacol Exp Ther 294:27–32

Ho WS, Hiley CR (2003) Vasodilator actions of abnormal-cannabidiol in rat isolated small mesenteric artery. Br J Pharmacol 138:1320–1332

Holland M, John Challiss RA, Standen NB, Boyle JP (1999) Cannabinoid CB1 receptors fail to cause relaxation, but couple via Gi/Go to the inhibition of adenylyl cyclase in carotid artery smooth muscle. Br J Pharmacol 128:597–604

Howlett AC, Bidaut-Russell M, Devane WA, Melvin LS, Johnson MR, Herkenham M (1990) The cannabinoid receptor: biochemical, anatomical and behavioral characterization. Trends Neurosci 13:420–423

Huestis MA, Gorelick DA, Heishman SJ, Preston KL, Nelson RA, Moolchan ET, Frank RA (2001) Blockade of effects of smoked marijuana by the CB1-selective cannabinoid receptor antagonist SR141716. Arch Gen Psychiatry 58:322–328

Ishac EJ, Jiang L, Lake KD, Varga K, Abood ME, Kunos G (1996) Inhibition of exocytotic noradrenaline release by presynaptic cannabinoid CB1 receptors on peripheral sympathetic nerves. Br J Pharmacol 118:2023–2028

Ishioka N, Bukoski RD (1999) A role for N-arachidonylethanolamine (anandamide) as the mediator of sensory nerve-dependent Ca2+-induced relaxation. J Pharmacol Exp Ther 289:245–250

Járai Z, Wagner JA, Varga K, Lake KD, Compton DR, Martin BR, Zimmer AM, Bonner TI, Buckley NE, Mezey E, Razdan RK, Zimmer A, Kunos G (1999) Cannabinoid-induced mesenteric vasodilation through an endothelial site distinct from CB1 or CB2 receptors. Proc Natl Acad Sci U S A 96:14136–14141

Járai Z, Wagner JA, Goparaju SK, Wang L, Razdan RK, Sugiura T, Zimmer AM, Bonner TI, Zimmer A, Kunos G (2000) Cardiovascular effects of 2-arachidonoyl glycerol in anesthetized mice. Hypertension 35:679–684

Joyeux M, Arnaud C, Godin-Ribuot D, Demenge P, Lamontagne D, Ribuot C (2002) Endocannabinoids are implicated in the infarct size-reducing effect conferred by heat stress preconditioning in isolated rat hearts. Cardiovasc Res 55:619–625

Kanakis C, Pouget JM, Rosen KM (1976) The effects of △9-THC (cannabis) on cardiac performance with or without beta blockade. Circulation 53:703–709

Kapoor K, Arulmani U, Heiligers JP, Garrelds IM, Willems EW, Doods H, Villalon CM, Saxena PR (2003) Effects of the CGRP receptor antagonist BIBN4096BS on capsaicin-induced carotid haemodynamic changes in anaesthetised pigs. Br J Pharmacol 140:329–338

Kosersky DS (1978) Antihypertensive effects of delta9-tetrahydrocannabinol. Arch Int Pharmacodyn Ther 233:76–81
Krylatov AV, Ugdyzhekova DS, Bernatskaya NA, Maslov LN, Mechoulam R, Pertwee RG, Stephano GB (2001) Activation of type II cannabinoid receptors improves myocardial tolerance to arrhythmogenic effects of coronary occlusion and reperfusion. Bull Exp Biol Med 131:523–525
Krylatov AV, Bernatskaia NA, Maslov LN, Pertwee RG, Mechoulam R, Stefano GB, Sharaevskii MA, Sal'nikova OM (2002a) Increase of the heart arrhythmogenic resistance and decrease of the myocardial necrosis zone during activation of cannabinoid receptors. Ross Fiziol Zh Im I M Sechenova 88:560–567
Krylatov AV, Uzhachenko RV, Maslov LN, Bernatskaya NA, Makriyannis A, Mechoulam R, Pertwee RG, Sal'nikova OM, Stefano JB, Lishmanov Y (2002b) Endogenous cannabinoids improve myocardial resistance to arrhythmogenic effects of coronary occlusion and reperfusion: a possible mechanism. Bull Exp Biol Med 133:122–124
Krylatov AV, Uzhachenko RV, Maslov LN, Ugdyzhekova DS, Bernatskaia NA, Pertwee R, Stefano GB, Makriyannis A (2002c) Anandamide and R-(+)-methanandamide prevent development of ischemic and reperfusion arrhythmia in rats by stimulation of CB2-receptors. Eksp Klin Farmakol 65:6–9
Kunos G, Járai Z, Bátkai S, Goparaju SK, Ishac EJ, Liu J, Wang L, Wagner JA (2000) Endocannabinoids as cardiovascular modulators. Chem Phys Lipids 108:159–168
Kunos G, Bátkai S, Offertáler L, Mo F, Liu J, Karcher J, Harvey-White J (2002) The quest for a vascular endothelial cannabinoids receptor. Chem Phys Lipids 121:45–56
Lagneux C, Lamontagne D (2001) Involvement of cannabinoids in the cardioprotection induced by lipopolysaccharide. Br J Pharmacol 132:793–796
Lake KD, Compton DR, Varga K, Martin BR, Kunos G (1997a) Cannabinoid-induced hypotension and bradycardia in rats mediated by CB1-like cannabinoid receptors. J Pharmacol Exp Ther 281:1030–1037
Lake KD, Martin BR, Kunos G, Varga K (1997b) Cardiovascular effects of anandamide in anesthetized and conscious normotensive and hypertensive rats. Hypertension 29:1204–1210
Ledent C, Valverde O, Cossu G, Petitet F, Aubert JF, Beslot F, Bohme GA, Imperato A, Pedrazzini T, Roques BP, Vassart G, Fratta W, Parmentier M (1999) Unresponsiveness to cannabinoids and reduced addictive effects of opiates in CB1 receptor knockout mice. Science 283:401–404
Lepicier P, Bouchard JF, Lagneux C, Lamontagne D (2003) Endocannabinoids protect the rat isolated heart against ischaemia. Br J Pharmacol 139:805–815
Li J, Kaminski NE, Wang DH (2003) Anandamide-induced depressor effect in spontaneously hypertensive rats: role of the vanilloid receptor. Hypertension 41:757–762
Liu J, Gao B, Mirshahi F, Sanyal AJ, Khanolkar AD, Makriyannis A, Kunos G (2000) Functional CB1 cannabinoid receptors in human vascular endothelial cells. Biochem J 346:835–840
Liu J, Bátkai S, Pacher P, Harvey-White J, Wagner JA, Cravatt BF, Gao B, Kunos G (2003) LPS induces anandamide synthesis in macrophages via CD14/MAPK/PI3 K/NF-κB independently of platelet activating factor. J Biol Chem 278:45034–45039
Lumsden AB, Henderson JM, Kutner MH (1988) Endotoxin levels measured by a chromatographic assay in portal, hepatic and peripheral blood in patients with cirrhosis. Hepatology 8:232–236
Maccarrone M, Bari M, Lorenzon T, Bisogno T, Di Marzo V, Finazzi-Agro A (2000) Anandamide uptake by human endothelial cells and its regulation by nitric oxide. Biol Chem 275:13484–13492
Malinowska B, Godlewski G, Bucher B, Schlicker E (1997) Cannabinoid CB1 receptor-mediated inhibition of the neurogenic vasopressor response in the pithed rat. Naunyn Schmiedebergs Arch Pharmacol 356:197–202
Malinowska B, Kwolek G, Gothert M (2001) Anandamide and methanandamide induce both vanilloid VR1- and cannabinoids CB1 receptor-mediated changes in heart rate and blood pressure in anaesthetized rats. Naunyn Schmiedebergs Arch Pharmacol 364:562–569

Matsuda LA, Lolait SJ, Brownstein MJ, Young CA, Bonner TI (1990) Structure of a cannabinoid receptor and functional expression of the cloned cDNA. Nature 346:561–564
Mechoulam R, Fride E, Di Marzo V (1998a) Endocannabinoids. Eur J Pharmacol 359:1–18
Mechoulam R, Fride E, Ben-Shabat S, Meiri U, Horowitz M (1998b) Carbachol, an acetylcholine receptor agonist, enhances production in rat aorta of 2-arachidonoyl glycerol, a hypotensive endocannabinoid. Eur J Pharmacol 362:R1–R3
Mombouli JV, Schaeffer G, Holzmann S, Kostner GM, Graier WF (1999) Anandamide-induced mobilization of cytosolic Ca2+ in endothelial cells. Br J Pharmacol 126:1593–1600
Mukhopadhyay S, Chapnick BM & Howlett AC (2002) Anandamide-induced vasorelaxation in rabbit aortic rings has two components: G protein dependent and independent. Am J Physiol 282:H2046–H2054
Munro S, Thomas KL, Abu-Shaar M (1993) Molecular characterization of a peripheral receptor for cannabinoids. Nature 365:61–65
Nahas G, Trouve R (1985) Effects and interactions of natural cannabinoids on the isolated heart. Proc Soc Exp Biol Med 180:312–316
Niederhoffer N, Szabo B (2000) Cannabinoids cause central sympathoexcitation and bradycardia in rabbits. J Pharmacol Exp Ther 294:707–713
Niederhoffer NB, Szabo B (1999) Involvement of CB1 cannabinoid receptors in the EDHF-dependent vasorelaxation in rabbits. Br J Pharmacol 126:1383–1386
O'Sullivan SE, Kendall DA, Randall MD (2004) Characterisation of the vasorelaxant properties of the novel endocannabinoid N-arachidonoyl-dopamine (NADA). Br J Pharmacol 141:803–812
Offertáler L, Mo FM, Bátkai S, Liu J, Begg M, Razdan RK, Martin BR, Bukoski RD, Kunos G (2003) Selective ligands and cellular effectors of a G protein-coupled endothelial cannabinoid receptor. Mol Pharmacol 63:699–705
Orliac ML, Peroni R, Celuch SM, Adler-Graschinsky E (2003) Potentiation of anandamide effects in mesenteric beds isolated from endotoxemic rats. J Pharmacol Exp Ther 304:179–184
Pacher P, Liaudet L, Bai P, Mabley JG, Kaminski PM, Virág L, Deb A, Szabo E, Ungvári Z, Wolin MS, Groves JT, Szabo C (2003) Potent metalloporphyrin peroxynitrite decomposition catalyst protects against the development of doxorubicin-induced cardiac dysfunction. Circulation 107:896–904
Pacher P, Bátkai S, Kunos G (2004) Haemodynamic profile and responsiveness to anandamide of TRPV1 receptor knock-out mice. J Physiol (Lond) 558:647–657
Plane F, Garland CJ (1996) Influence of contractile agonists on the mechanism of endothelium-dependent relaxation in rat isolated mesenteric artery. Br J Pharmacol 119:191–193
Pratt PF, Hillard CJ, Edgemond WS, Campbell WB (1998) N-Arachidonylethanolamide relaxation of bovine coronary artery is not mediated by CB1 cannabinoid receptor. Am J Physiol 43:H375–H381
Ralevic V, Kendall DA, Randall MD, Zygmunt PM, Movahed P, Hogestatt ED (2000) Vanilloid receptors on capsaicin-sensitive sensory nerves mediate relaxation to methanandamide in the rat isolated mesenteric arterial bed and small mesenteric arteries. Br J Pharmacol 130:1483–1488
Ralevic V, Kendall DA, Randall MD, Smart D (2002) Cannabinoid modulation of sensory neurotransmission via cannabinoid and vanilloid receptors: Roles in regulation of cardiovascular function. Life Sci 71:2577–2594
Randall MD, Kendall DA (1997) Involvement of a cannabinoid in endothelium-derived hyperpolarizing factor-mediated coronary vasorelaxation. Eur J Pharmacol 335:205–209
Randall MD, Kendall DA (1998) Anandamide and endothelium-derived hyperpolarizing factor act via a common vasorelaxant mechanism in rat mesentery. Eur J Pharmacol 346:51–53

Randall MD, Alexander SP, Bennett T, Boyd EA, Fry JR, Gardiner SM, Kemp PA, McCulloch AI, Kendall DA (1996) An endogenous cannabinoid as an endothelium-derived vasorelaxant. Biochem Biophys Res Commun 229:114–120

Randall MD, Harris D, Kendall DA, Ralevic V (2002) Cardiovascular effects of cannabinoids. Pharmacol Ther 95:191–202

Ros J, Claria J, To-Figueras J, Planaguma A, Cejudo-Martin P, Fernandez-Varo G, Martin-Ruiz R, Arroyo V, Rivera F, Rodes J, Jimenez W (2002) Endogenous cannabinoids: a new system involved in the homeostasis of arterial pressure in experimental cirrhosis in the rat. Gastroenterology 122:85–93

Rosenkrantz H, Braude M (1974) Acute, subacute and 23-day chronic marihuana inhalation toxicities in the rat. Toxicol Appl Pharmacol 28:428–441

Sagan S, Venance L, Torrens Y, Cordier J, Glowinski J, Giaume C (1999) Anandamide and WIN55,212-2 inhibit cyclic AMP-formation through G-protein-coupled receptors distinct from CB1 cannabinoid receptors in cultured astrocytes. Eur J Neurosci 11:691–699

Schlicker E, Timm J, Zentner J, Gothert M (1997) Cannabinoid CB1 receptor-mediated inhibition of noradrenaline release in the human and guinea-pig hippocampus. Naunyn Schmiedebergs Arch Pharmacol 356:583–589

Siqueira SW, Lapa AJ, Ribeiro do Valle J (1979) The triple effect induced by delta 9-tetrahydrocannabinol on the rat blood pressure. Eur J Pharmacol 58:351–357

Smith PJ, McQueen DS (2001) Anandamide induces cardiovascular and respiratory reflexes via vasosensory nerves in the anaesthetized rat. Br J Pharmacol 134:655–663

Stein EA, Fuller SA, Edgemond WS, Campbell WB (1996) Physiological and behavioural effects of the endogenous cannabinoid, arachidonylethanolamide (anandamide), in the rat. Br J Pharmacol 119:107–114

Szabo B, Nordheim U, Niederhoffer N (2001) Effects of cannabinoids on sympathetic and parasympathetic neuroeffector transmission in the rabbit heart. J Pharmacol Exp Ther 297:819–826

Szolcsányi J (2000) Are cannabinoids endogenous ligands for the VR1 capsaicin receptor? Trends Pharmacol Sci 21:41–42

Ugdyzhekova DS, Bernatskaya NA, Stefano JB, Graier VF, Tam SW, Mekhoulam R (2001) Endogenous cannabinoid anandamide increases heart resistance to arrhythmogenic effects of epinephrine: role of CB(1) and CB(2) receptors. Bull Exp Biol Med 131:251–253

Ugdyzhekova DS, Krylatov AV, Bernatskaya NA, Maslov LN, Mechoulam R, Pertwee RG (2002) Activation of cannabinoid receptors decreases the area of ischemic myocardial necrosis. Bull Exp Biol Med 133:125–126

Valk PJ, Delwel R (1998) The peripheral cannabinoid receptor, Cb2, in retrovirally-induced leukemic transformation and normal hematopoiesis. Leuk Lymphoma 32:29–43

Varga K, Lake K, Martin BR, Kunos (1995) G Novel antagonist implicates the CB1 cannabinoid receptor in the hypotensive action of anandamide. Eur J Pharmacol 278:279–283

Varga K, Lake KD, Huangfu D, Guyenet PG, Kunos G (1996) Mechanism of the hypotensive action of anandamide in anesthetized rats. Hypertension 28:682–686

Varga K, Wagner JA, Bridgen DT, Kunos G (1998) Platelet- and macrophage-derived endogenous cannabinoids are involved in endotoxin-induced hypotension. FASEB J 12:1035–1044

Varma DR, Goldbaum D (1975) Effect of delta9-tetrahydrocannabinol on experimental hypertension in rats. J Pharm Pharmacol 27:790–791

Vidrio H, Sanchez-Salvatori MA, Medina M (1996) Cardiovascular effects of (−)-11-OH-delta 8-tetrahydrocannabinol-dimethylheptyl in rats. J Cardiovasc Pharmacol 28:332–336

Vizi ES, Katona I, Freund TF (2001) Evidence for presynaptic cannabinoid CB(1) receptor-mediated inhibition of noradrenaline release in the guinea pig lung. Eur J Pharmacol 431:237–244

Vollmer RR, Cavero I, Ertel RJ, Solomon TA, Buckley JP (1974) Role of the central autonomic nervous system in the hypotension and bradycardia induced by (−)-delta 9-trans-tetrahydrocannabinol. J Pharm Pharmacol 26:186–192
Wagner JA, Varga K, Ellis EF, Rzigalinski BA, Martin BR, Kunos G (1997) Activation of peripheral CB1 cannabinoid receptors in haemorrhagic shock. Nature 390:518–521
Wagner JA, Varga K, Járai Z, Kunos G (1999) Mesenteric vasodilation mediated by endothelial anandamide receptors. Hypertension 33:429–434
Wagner JA, Hu K, Bauersachs J, Karcher J, Wiesler M, Goparaju SK, Kunos G, Ertl G (2001a) Endogenous cannabinoids mediate hypotension after experimental myocardial infarction. J Am Coll Cardiol 38:2048–2054
Wagner JA, Járai Z, Bátkai S, Kunos G (2001b) Hemodynamic effects of cannabinoids: coronary and cerebral vasodilation mediated by cannabinoid CB(1) receptors. Eur J Pharmacol 423:203–210
Wagner JA, Hu K, Karcher J, Bauersachs J, Schafer A, Laser M, Han H, Ertl G (2003) CB(1) cannabinoid receptor antagonism promotes remodeling and cannabinoid treatment prevents endothelial dysfunction and hypotension in rats with myocardial infarction. Br J Pharmacol 138:1251–1258
Walter L, Stella N (2004) Cannabinoids and neuroinflammation. Br J Pharmacol 141:775–785
Wang Y, Liu Y, Ito Y, Hashiguchi T, Kitajima I, Yamakuchi M, Shimizu H, Matsuo S, Imaizumi H, Maruyama I (2001) Simultaneous measurement of anandamide and 2-arachidonoylglycerol by polymyxin B-selective adsorption and subsequent high-performance liquid chromatography analysis: increase in endogenous cannabinoids in the sera of patients with endotoxic shock. Anal Biochem 294:73–82
Watanabe H, Vriens J, Prenen J, Droogmans G, Voets T, Nilius B (2003) Anandamide and arachidonic acid use epoxyeicosatrienoic acids to activate TRPV4 channels. Nature 424:434–438
White R, Hiley CR (1997) A comparison of EDHF-mediated and anandamide-induced relaxations in the rat isolated mesenteric artery. Br J Pharmacol 122:1573–1584
White R, Ho WS, Bottrill FE, Ford WR, Hiley CR (2001) Mechanisms of anandamide-induced vasorelaxation in rat isolated coronary arteries. Br J Pharmacol 134:921–929
Zahner MR, Li DP, Chen SR, Pan HL (2003) Cardiac vanilloid receptor 1-expressing afferent nerves and their role in the cardiogenic sympathetic reflex in rats. J Physiol 551:515–523
Zaugg HE, Kyncl J (1983) New antihypertensive cannabinoids. J Med Chem 26:214–217
Zygmunt PM, Högestätt ED, Waldeck K, Edwards G, Kirkup AJ, Weston AH (1997) Studies on the effects of anandamide in rat hepatic artery. Br J Pharmacol 122:1679–1686
Zygmunt PM, Petersson J, Andersson DA, Chuang H-H, Sorgard M, Di Marzo V, Julius D, Högestätt ED (1999) Vanilloid receptors on sensory nerves mediate the vasodilator action of anandamide. Nature 400:452–457
Zygmunt PM, Andersson DA, Högestätt ED (2002) Delta 9-tetrahydrocannabinol and cannabinol activate capsaicin-sensitive sensory nerves via a CB1 and CB2 cannabinoid receptor-independent mechanism. J Neurosci 22:4720–4727

HEP (2005) 168:627–642

Effects on Cell Viability

M. Guzmán

Department of Biochemistry and Molecular Biology I, School of Biology,
Complutense University, 28040 Madrid, Spain
mgp@bbm1.ucm.es

Abstract Cannabinoids are known to control the cell survival/death decision, leading to different outcomes that depend on the nature of the target cell and its proliferative or differentiation status. Cannabinoids induce growth arrest or apoptosis in a number of transformed cells in culture. They do so by modulating key cell signalling pathways involved in the control of tumour cell fate. The best-characterised example is cannabinoid-induced apoptosis of glioma cells, which occurs via sustained ceramide accumulation, extracellular signal-regulated kinase activation and Akt inhibition. In addition, cannabinoid administration inhibits the angiogenesis and slows the growth of different types of tumours in laboratory animals. By contrast, most of the experimental evidence indicates that cannabinoids protect normal neurons and glial cells from apoptosis as induced by toxic insults such as glutamatergic overstimulation, ischaemia and oxidative damage. It is therefore very likely that cannabinoids regulate cell survival and cell death pathways differently in tumour and non-tumour cells. Regarding immune cells, cannabinoids affect proliferation and survival in a complex and still obscure manner that depends on the experimental setting. The findings reviewed here might set the basis for the use of cannabinoids in the treatment of cancer and neurodegenerative diseases.

Keywords Cell death · Apoptosis · Cell proliferation · Cancer · Neuroprotection

1 Cannabinoid Signalling Pathways and Cell Viability

Cannabinoids, the active components of *Cannabis sativa* and their derivatives, act in organisms by mimicking endogenous substances—the endocannabinoids anandamide and 2-arachidonoylglycerol—that bind to and activate specific cannabinoid receptors. So far, two cannabinoid-specific $G_{i/o}$ protein-coupled receptors, CB_1 (Matsuda et al. 1990) and CB_2 (Munro et al. 1993), have been cloned and characterised from mammalian tissues. Most of the effects of cannabinoids rely on CB_1 receptor activation. This receptor is particularly abundant in discrete areas of the brain, but is also expressed in peripheral nerve terminals and various extra-neural sites such as testis, eye, vascular endothelium and spleen. In contrast, the CB_2 receptor is almost exclusively present in the immune system (Howlett et al. 2002).

Before their specific receptors were identified, it was already known that cannabinoids inhibit adenylyl cyclase (AC) with the consequent decrease in intracellular cyclic adenosine monophosphate (cAMP) levels (Howlett 1984). The CB_1 receptor also exerts modulation of ion channels, inducing for example inhibition of N- and P/Q-type voltage-sensitive Ca^{2+} channels (VSCC) and activation of rectifying K^+ channels. These two effects may be responsible for the inhibition of the release of glutamate and other neurotransmitters by blunting membrane depolarisation and exocytosis (Piomelli 2003).

Besides these well-established cannabinoid receptor-coupled signalling events, cannabinoid receptors also modulate several signalling pathways that are more directly involved in the control of cell proliferation and survival (Table 1). Thus, cannabinoid receptors are coupled to activation of extracellular signal-regulated kinase (ERK) (Bouaboula et al. 1995, 1996), c-Jun N-terminal kinase (JNK) and p38 mitogen-activated protein kinase (MAPK) (Liu et al. 2000; Rueda et al. 2000), and phosphatidylinositol 3-kinase (PI-3-K)/Akt (Gómez del Pulgar et al. 2000). Activated Akt can phosphorylate and inhibit nuclear translocation of Forkhead transcription factors, thereby preventing the expression of pro-apoptotic proteins (Samson et al. 2003). The negative coupling of cannabinoid receptors to ERK (Rueda et al. 2002) and Akt (Gómez del Pulgar et al. 2002b) has also been reported.

Cannabinoids modulate sphingolipid-metabolising pathways by inducing sphingomyelin breakdown and increasing acutely the levels of ceramide (Sánchez et al. 1998b; Galve-Roperh et al. 2000), a lipid second messenger that can induce apoptosis and cell cycle arrest. This effect is cannabinoid receptor-dependent but $G_{i/o}$ protein-independent, and seems to involve the participation of the adaptor pro-

Table 1. Control by cannabinoids of intracellular signalling pathways that may affect cell proliferation and survival

Signalling pathway	Cannabinoid effect	Effect on cell number
Ceramide generation	Activation	Decrease
ERK	Activation (sustained)	Decrease
	Activation (acute)	Increase?
PI-3-K/Akt	Activation	Increase
	Inhibition	Decrease
JNK and p38 MAPK	Activation	Decrease
cAMP/PKA	Inhibition	Decrease?

cAMP, cyclic adenosine monophosphate; ERK, extracellular signal-regulated kinase; JNK, c-Jun N-terminal kinase; MAPK, mitogen-activated protein kinase; PI-3-K, phosphoinositide 3-kinase; PKA, protein kinase A.

tein FAN (factor associated with neutral sphingomyelinase activation) (Sánchez et al. 2001b). Cannabinoid receptor activation can also generate a sustained peak of ceramide accumulation via enhanced synthesis de novo that plays an important role in the induction of apoptosis (Galve-Roperh et al. 2000; Gómez del Pulgar et al. 2002b).

2
Tumour Cells

2.1
Antitumoural Effect

A number of plant-derived [for example, Δ^9-tetrahydrocannabinol (THC) and cannabidiol], synthetic (for example, WIN 55,212-2 and HU-210) and endogenous cannabinoids (for example, anandamide and 2-arachidonoylglycerol) have been shown to exert antiproliferative actions on a wide spectrum of tumour cells in culture (Guzmán 2003; Jones and Howl 2003). More importantly, cannabinoid administration to nude mice curbs the growth of various types of tumour xenografts, including lung carcinoma (Munson et al. 1975), glioma (Galve-Roperh et al. 2000), thyroid epithelioma (Bifulco et al. 2001), lymphoma (McKallip et al. 2002a) and skin carcinoma (Casanova et al. 2003).

The antitumoural action of cannabinoids may be exemplified by gliomas. Initial experiments in cultured glioma cells showed that incubation with cannabinoids induces cell death by an apoptotic mechanism (Sánchez et al. 1998a). Further studies with animal models showed that local administration of THC or WIN 55,212-2 reduced the size of tumours generated by intracranial inoculation of C6 glioma cells in rats, leading to complete eradication of gliomas and subsequent survival in one third of the treated rats (Galve-Roperh et al. 2000). Additional studies employed tumour xenografts generated by subcutaneous injection of glioma cells in the flank of immune-deficient mice. Local administration of THC, WIN 55,212-2, the selective CB_2 agonist JWH-133 or cannabidiol decreased the growth of tumours derived from glioma cell lines (Galve-Roperh et al. 2000; Sánchez et al. 2001a; Massi et al. 2004), and in the case of JWH-133 also from cells of tumour biopsies of patients with glioblastoma multiforme (Sánchez et al. 2001a).

2.2 Mechanism of Action

Although the downstream events by which cannabinoids exert their antitumoural action are not completely unravelled, there is certain evidence for the implication of at least three mechanisms: tumour cell apoptosis, tumour cell growth arrest, and inhibition of tumour angiogenesis.

2.2.1 Apoptosis

Glioma cells (Sánchez et al. 1998a; Galve-Roperh et al. 2000) and other cancer cells in culture (Guzmán 2003; Maccarrone and Finazzi-Agro 2003) undergo apoptosis upon long-term cannabinoid challenge. Activation of cannabinoid receptors and accumulation of the proapoptotic sphingolipid ceramide seem necessary for glioma cell apoptosis (Table 1). Of interest, following cannabinoid receptor activation, two peaks of ceramide generation are observed in glioma cells that have different kinetics (minute- versus day-range), magnitude (two- versus fourfold), mechanistic origin (sphingomyelin hydrolysis versus de novo ceramide synthesis) and function (metabolic regulation versus induction of apoptosis) (Guzmán et al. 2001). Studies employing two subclones of C6 glioma cells—one sensitive and the other resistant to THC—indicate that the apoptotic action of cannabinoids relies on the second peak of ceramide generation (Galve-Roperh et al. 2000). It has been suggested that cannabinoids enhance ceramide synthesis de novo via induction of serine palmitoyltransferase, a regulatory enzyme of sphingolipid biosynthesis (Gómez del Pulgar et al. 2002b). It is worth noting that ceramide content has been inversely related with malignant progression and poor prognosis of human astrocytomas. Thus, low-grade astrocytomas have higher ceramide content than high-grade astrocytomas (Riboni et al. 2002). Pharmacological inhibition of de novo ceramide synthesis also prevents cannabinoid-induced death of prostate tumour cells (Mimeault et al. 2003).

The increased ceramide levels observed in glioma cells upon cannabinoid challenge leads to prolonged activation of the Raf-1/MEK/ERK signalling cascade (Galve-Roperh et al. 2000). It is generally accepted that ERK activation leads to cell proliferation. However, the relation between ERK activation and cell fate is complex and depends on many factors, one of which is the duration of the stimulus, as prolonged ERK activation may mediate cell cycle arrest and cell death. Sustained Akt inhibition (Gómez del Pulgar et al. 2002b) and JNK and p38 MAPK induction (Galve-Roperh et al. 2000; Sarker et al. 2003) may also contribute to glioma cell death (Table 1). Nevertheless, further investigation is necessary to clarify the specific downstream targets of these pathways involved in cannabinoid-induced apoptosis of tumour cells and the relative contribution of the apoptotic mechanism in vivo.

2.2.2
Cell Growth Arrest

Cannabinoids have been shown to induce cell cycle arrest in breast carcinoma (De Petrocellis et al. 1998), prostate carcinoma (Melck et al. 2000) and thyroid epithelioma cells (Bifulco et al. 2001). In breast carcinoma cells this has been ascribed to the inhibition of adenylyl cyclase and the cAMP/protein kinase A (PKA) pathway (Table 1). PKA phosphorylates and inhibits Raf-1, so cannabinoids prevent the inhibition of Raf-1 and induce prolonged activation of the Raf-1/MEK/ERK signalling cascade (Melck et al 1999). Cannabinoid-induced inhibition of thyroid epithelioma cell proliferation has been attributed to the induction of the cyclin-dependent kinase inhibitor $p27^{kip1}$ (Portella et al. 2003).

It has been suggested that cannabinoids produce their growth-inhibiting effects on skin and prostate cancer cells at least in part by attenuating epidermal growth factor receptor tyrosine kinase activity (Casanova et al. 2003) and/or by lowering epidermal growth factor receptor expression (Casanova et al. 2003; Mimeault et al. 2003). Furthermore, the antiproliferative action of cannabinoids in breast, prostate and thyroid cancer cells might involve a decrease in the activity and/or expression of prolactin (De Petrocellis et al. 1998), nerve growth factor (Melck et al. 2000) or type 1 vascular endothelial growth factor (VEGF) tyrosine kinase receptors (Portella et al. 2003). In addition, cannabinoids inhibit the type 2 VEGF tyrosine kinase receptor in glioma cells (C. Blázquez and M. Guzmán, unpublished results). Taken together, these observations indicate that attenuation of signalling through tyrosine kinase receptors may constitute a common mechanism of cannabinoid antiproliferative action.

2.2.3
Inhibition of Angiogenesis

To grow beyond minimal size, tumours must generate a new vascular supply (angiogenesis) for purposes of cell nutrition, gas exchange and waste disposal, and therefore blocking the angiogenic process constitutes one of the most promising antitumoural approaches currently available. Immunohistochemical and functional analyses in mouse models of glioma (Blázquez et al. 2003) and skin carcinoma (Casanova et al. 2003) have shown that cannabinoid administration turns the vascular hyperplasia characteristic of actively growing tumours to a pattern of blood vessels characterised by small, differentiated and impermeable capillaries. This is associated with a reduced expression of VEGF and other proangiogenic cytokines (Casanova et al. 2003; Blázquez et al. 2003; Portella et al. 2003), as well as of VEGF receptors (Portella et al. 2003; C. Blázquez and M. Guzmán, unpublished results). Interestingly, pharmacological inhibition of ceramide synthesis de novo abrogates the antitumoural effect of cannabinoids in vivo as well as the cannabinoid-induced inhibition of VEGF production by glioma cells in vitro and by gliomas in vivo (C. Blázquez and M. Guzmán, unpublished results), indicating that ceramide may play a general role in cannabinoid antitumoural action. In addition, activation of cannabinoid receptors in vascular endothelial cells inhibits cell migration and survival, which might contribute as well to the impaired tumour vascularisation observed in cannabinoid-treated gliomas (Blázquez et al. 2003).

Cannabinoids have been recently shown to reduce the number of metastatic nodes produced by paw injection of Lewis lung carcinoma cells in rats (Portella et al. 2003). Moreover, cannabinoid administration to glioma-bearing mice also decreases the activity and expression of matrix metalloproteinase-2 (MMP2), a proteolytic enzyme that allows tissue breakdown and remodelling during angiogenesis and metastasis (Blázquez et al. 2003). Hence it is conceivable that cannabinoids may also control tumour invasiveness.

3 Neural Cells

3.1 Neuroprotective Effect

One of the most exciting aspects of current cannabinoid research is the possibility that cannabinoids play a role as neuroprotective agents, both pharmacologically and physiologically via the endocannabinoid system (Mechoulam et al. 2002; van der Stelt et al. 2002). Thus, most of the experimental evidence indicates that cannabinoids may protect neurons from insults such as ischaemia, glutamatergic excitotoxicity, mechanical trauma and oxidative damage. The neuroprotective action of cannabinoids has been shown in vivo. The neurotoxicity induced by global cerebral ischaemia in vivo following cerebral artery occlusion is reduced in the CA1 region of the hippocampus by WIN 55,212-2 administration via the CB_1 receptor (Nagayama et al. 1999). Cannabinoid treatment also reduces the infarct volume caused by focal cerebral ischaemia, leading to enhanced neuronal survival of penumbral cortical tissue (Nagayama et al. 1999). Likewise, THC exerts a CB_1-mediated reduction of the infarct volume in an in vivo model of ouabain-induced excitotoxicity (van der Stelt et al. 2001). Moreover, the harmful consequences of traumatic brain injury such as edema and neuronal cell death are blunted as a consequence of increased 2-arachidonoylglycerol levels (Panikashvili et al. 2001). The involvement of the CB_1 receptor in endocannabinoid-mediated neuroprotection is further supported by results from experiments with knock-out animals, as CB_1-defective mice are more susceptible to stroke-induced neuronal death (Parmentier-Batteur et al. 2002) and kainate-induced excitotoxicity (Marsicano et al. 2003). The use of CB_1 knock-out mice in which receptor expression is abolished exclusively in principal forebrain neurons, but not in GABAergic interneurons, indicate a direct action of the CB_1 receptor within the same neuronal cells susceptible for primary excitotoxicity damage (Marsicano et al. 2003).

Experiments conducted on cultured neurons have also provided evidence for a cytoprotective action of cannabinoids, although variable results have been obtained regarding the possible involvement of cannabinoid receptors. A neuroprotective action of some cannabinoid agonists in a model of glutamate excitotoxic death was first reported by Skaper et al. (1996) using cerebellar granule neurons. WIN 55,212-2 and CP 55,940 were later shown to protect hippocampal neurons from presynaptically evoked glutamate excitotoxicity by a CB_1-mediated process (Shen and Tayer 1998). Since then, other reports have shown that cannabinoids protect cultured neurons against glutamatergic excitotoxicity via CB_1 receptor activation (e.g. Abood et al. 2001; Hampson and Grimaldi 2001) or independently of the CB_1 receptor (e.g. Nagayama et al. 1999; Sinor et al. 2000). In this respect, various cannabinoids with phenolic structure exert a CB_1-independent neuroprotective effect owing to their intrinsic antioxidant properties (Hampson et al. 1998; Chen and Buck 2000; Marsicano et al. 2002).

3.1.1
Mechanism of Action

Although the precise molecular mechanism for cannabinoid-mediated neuroprotection its not fully understood, probably the best-described mechanism is the cannabinoid inhibitory action on glutamate release, which would protect cells from excitotoxicity (Mechoulam et al. 2002; van der Stelt et al. 2002; Freund et al. 2003). Glutamatergic transmission is regulated by cannabinoids basically at the presynaptic level through inhibition of glutamate release as a consequence of their inhibitory effect on Ca^{2+} influx through different types of voltage-sensitive channels, including N- and P/Q-type channels (Howlett et al. 2002; Freund et al. 2003). Modulation of presynaptic K^+ channels may also contribute to this cannabinoid action. Other cannabinoid effects that may contribute to neuroprotection include, for example, inhibition of nitric oxide production and proinflammatory cytokine release by activated microglial cells (Mechoulam et al. 2002), attenuation of endothelin 1-induced brain vasoconstriction (Mechoulam et al. 2002), and stimulation of brain-derived neurotrophic factor expression (Marsicano et al. 2003).

The observation that anandamide (Di Marzo et al. 1994) and 2-arachidonoylglycerol (Stella et al. 1997) biosyntheses are activated by Ca^{2+} may constitute a mechanism of feedback regulation, with the generated cannabinoids preventing excessive Ca^{2+} influx by their inhibitory action on Ca^{2+} channels. Several brain insults induce a selective increase in certain endocannabinoid species, further suggesting a protective role for these molecules. For example, traumatic brain injury enhances 2-arachidonoylglycerol levels (Panikashvili et al. 2001), glutamatergic excitotoxicity increases anandamide levels (Hansen et al. 2001; Marsicano et al. 2003), and focal cerebral ischaemia raises palmitoylethanolamide levels (Franklin et al. 2003). The biological meaning for this apparent endocannabinoid-species selectivity is, however, unknown.

3.2
Neurotoxic Effect

In contrast to the aforementioned cannabinoid neuroprotection, some studies have reported a neurotoxic effect of THC. Thus, morphological changes in the hippocampus, decreases in the mean volume, synaptic density and dendritic length of CA3 pyramidal neurons, and reduced neuronal density in rat hippocampus associated with chronic THC oral administration have been described (Scallet et al. 1987). However, others could not find any significant histopathological alteration in the brains of rats and mice treated orally with very high doses of THC for 2 years (Chan et al. 1996). Likewise, direct intracranial administration of THC or WIN 55,212-2 to rats for 1 week did not induce neural cell apoptosis (Galve-Roperh et al. 2000).

In vitro studies have also shown that THC induces apoptosis of primary hippocampal (Chan et al. 1998) and cortical (Downer et al. 2003) neurons. In contrast, in another report the viability of cortical neurons in culture did not decrease after prolonged exposure to THC (Sánchez et al. 1998a).

3.2.1
Mechanism of Action

CB_1-dependent apoptosis of primary neurons may involve phospholipase A_2-induced release of arachidonic acid, which promotes the activation of cyclooxygenases and lipoxygenases (Chan et al. 2003). This may generate free radicals and lead to lipid peroxidation and cell death. Activation of the JNK cascade may also contribute to this neurotoxic process (Downer et al. 2003) (Table 1).

3.3
Glioprotective Effect

Cannabinoids protect glial cells from death. Thus, cannabinoids prevent astrocyte apoptosis as induced by ceramide both in vivo and in vitro (Gómez del Pulgar et al. 2002a). Moreover, cannabinoids protect oligodendrocytes from growth factor deprivation-induced cell death (Molina-Holgado et al. 2002), and furthermore they prevent demyelination in vivo in a virus-induced model of multiple sclerosis (Arévalo-Martín et al. 2003). The finding that cannabinoid-induced prevention of neural cell death is not restricted to neurons may be relevant, as protection of more than one type of cell within the brain could result in a synergistic action by increasing the efficiency of the injured-brain response.

3.3.1
Mechanism of Action

—It has been shown that cannabinoids activate two pivotal survival pathways in glial cells: the PI-3-K/Akt pathway (Gómez del Pulgar et al. 2002a; Molina-Holgado et al. 2002) and the ERK pathway (Sánchez et al. 1998b; Gómez del Pulgar 2002a) (Table 1). Both signalling pathways appear to act in concert upon CB_1 receptor activation and G_i protein dissociation, as PI-3-K activation is required for cannabinoid-induced ERK activation, which occurs in a Ras-independent and tyrosine kinase receptor-independent manner (Galve-Roperh et al. 2002). Additionally, adenylyl cyclase inhibition via the CB_1 receptor might supress cAMP-dependent Raf-1 inhibition and cooperate in neural cell ERK activation (Derkinderen et al. 2003). Altogether, these observations show that, most likely, cannabinoids regulate cell survival and cell death pathways differently in transformed and non-transformed neural cells (Fig. 1).

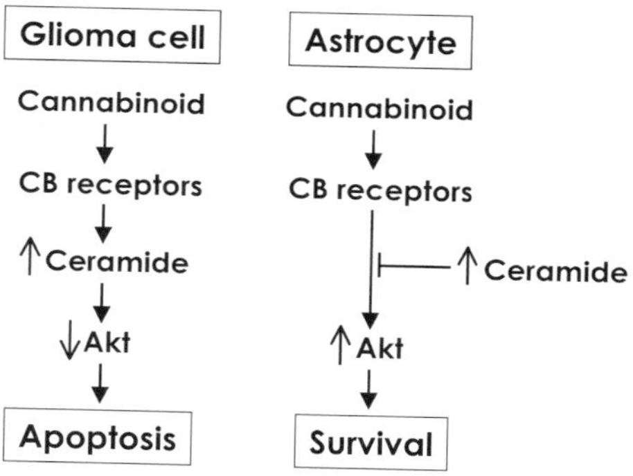

Fig. 1. Dual effect of cannabinoids on the viability of transformed versus non-transformed glial cells. In glioma cells, cannabinoids activate CB_1 and CB_2 receptors, inducing sustained ceramide accumulation and inhibition of the pro-survival kinase Akt, leading in turn to apoptosis. In normal astrocytes, cannabinoids activate CB_1 receptors, preventing ceramide-induced Akt inhibition and leading in turn to cell survival

4
Immune Cells

4.1
Cell Death/Survival Effect

Many in vitro and in vivo studies have shown that cannabinoids are immunosuppressive agents (Roth et al. 2002; Klein et al. 2003). In this context, cannabinoids induce apoptosis of human peripheral blood mononuclear cells (Schwarz et al. 1994) as well as of mouse macrophages and lymphocytes (Zhu et al. 1998). THC also inhibits the proliferation and induces apoptosis of mouse thymocytes and splenocytes in culture and in mice (McKallip et al. 2002b). Moreover, various cannabinoids induce apoptosis of leukaemic and lymphoid cells in vitro and—as mentioned above—in mouse models (McKallip et al. 2002a).

By contrast, some observations indicate that cannabinoids either stimulate or inhibit the function of immune cells. This variation in drug effects may depend on experimental factors such as drug concentration, timing of drug delivery, and type of cell function examined. Thus, Derocq et al. (1995) showed for the first time that human B cell proliferation is stimulated by cannabinoids at nanomolar concentrations. Likewise, Valk et al. (1997b) showed that murine haematopoietic growth factor-dependent cell lines require the presence of anandamide for optimal growth in serum-free medium. The endocannabinoid also enhanced the number and size of interleukin 3-induced myeloid colonies from mouse bone marrow. Moreover, following retroviral insertional mutagenesis these authors have identified a virus integration site, named *Evi11*, within the gene encoding the CB_2 receptor, suggesting that CB_2 is a proto-oncogene involved in leukaemogenesis (Valk et al. 1997a). This is also supported by the observation that, in certain murine leukaemia virus-induced tumours, retroviral integrations occur in the CB_2 receptor gene (Valk et al. 1999).

The possibility that the CB_2 receptor plays a role in the control of immune cell fate is supported by studies on the gene expression profile of human promyelocytic cells, which show that receptor activation up-regulates genes involved in the cell differentiation program (Derocq et al. 2000). These data, together with reports of CB_2-induced blockade of neutrophilic differentiation (Alberich Jordà et al. 2003) and changes in CB_2 receptor expression during the differentiation and activation of human B cells (Carayon et al. 1998) point to an involvement of the CB_2 receptor in the control of immune cell maturation.

4.2 Mechanism of Action

The transduction systems responsible for CB_2 receptor signalling are as yet unclear. It has been traditionally assumed that inhibition of the cAMP pathway may be responsible for the immunosuppressive action of cannabinoids (Kaminski 1998) (Table 1). However, plant-derived and endogenous cannabinoids may behave as partial agonists or even antagonists in the CB_2 receptor-mediated inhibition of adenylyl cyclase (Bayewitch et al. 1996). It is possible therefore that the CB_2 receptor controls immune cell proliferation by coupling positively (Derocq et al. 2000; Alberich Jordà et al. 2003; Samson et al. 2003) or negatively (Faubert Kaplan and Kaminski 2003) to ERK, a process that is dependent on $G_{i/o}$ proteins but independent of cAMP (Bouaboula et al. 1996). Activation of Akt and concomitant phosphorylation of kinase targets have also been shown in mast cells (Samson et al. 2003).

5 Potential Therapeutic Implications

Many studies have dealt with the antiproliferative effect of cannabinoids on different transformed cells. The case of gliomas may be of particular interest because they are one of the most malignant forms of cancer, resulting in the death of affected patients within months after diagnosis. One of the alternative therapeutic approaches for gliomas might be the use of cannabinoid agonists, since these compounds induce apoptosis in vitro and inhibition of tumour angiogenesis and growth without significant collateral effects in vivo (Galve-Roperh et al. 2000; Sánchez et al. 2001a). Based on these observations, a phase I/II clinical trial is currently investigating the effect of local administration of THC on the growth of recurrent glioblastoma multiforme (Guzmán 2003). It would be desirable that other trials on this and other types of tumours are initiated to determine whether cannabinoids can be used as antitumoural agents.

Different neurological and neurodegenerative diseases are accompanied by excitotoxicity, oxidative stress, Ca^{2+} imbalance and/or inflammatory responses, leading to progressive neuronal death. As discussed above, cannabinoids may interfere with these processes, which sets the basis for a potential therapeutic utility of these compounds. Laboratory research has shown that this may be the case with, for example, Huntington's disease (Lastres-Becker et al. 2003), multiple sclerosis (Baker et al. 2000) and brain trauma (Panikashvili et al. 2001). In addition, the existence of an inhibitory loop by which endocannabinoids blunt dopaminergic control of movement (Giuffrida et al. 1999) point to the potential use of CB_1 antagonists for the management of Parkinson's disease (Piomelli 2003) and of CB_1 agonists as agents for preventing symptoms of Tourette's disease (Muller-Vahl 2003). A recent large-scale phase III clinical trial has tested the effect of THC alone or in combination with other cannabinoids for the treatment of spasticity and other muscle-debilitating symptoms of multiple sclerosis, yielding conflicting but encouraging results (Zajicek et al. 2003). Phase III clinical trials are also being conducted with the non-psychoactive cannabinoid HU-211 (dexanabinol), a non-competitive *N*-methyl-D-aspartate receptor antagonist (Feigenbaum et al. 1989), for the treatment of severe head trauma (Knoller et al. 2002). This compound reduces the inflammatory response after closed-head injury or lipopolysaccharide-induced septic shock both in vivo and in vitro, and prevents tremor, seizures and lethality in laboratory animals (Gallily et al. 1997). Ongoing and future clinical trials should determine the usefulness of cannabinoids as therapeutic agents for the management of neurological and neurodegenerative diseases.

As reviewed here, cannabinoids may lead to opposite effects on the cell survival/death decision. For example, in the case of neural cells, cannabinoids may kill tumour cells and protect their non-transformed counterparts from death (Guzmán 2003) (Fig. 1). It is conceivable that different experimental factors may account for this "yin-yang" action, for example : (1) cannabinoid neuroprotection is usually more evident in whole-animal than in cultured-neuron models, which may result from their aforementioned impact on various brain cell types (neurons, astroglia, oligodendroglia, microglia, vascular endothelium); (2) cannabinoids may exert dual effects on neural cell viability depending on signal input (e.g. agonist dose and time of exposure), high inputs usually exerting cell growth inhibition and death; (3) endocannabinoids and exogenous cannabinoids may possess different pharmacological properties (e.g. agonistic potency and stability); and (4) the origin of the neural cell and its stage of differentiation may affect sensitivity to death. Modulation of cell viability by cannabinoids is therefore a complex and still obscure issue that opens exciting biological and clinical issues but requires much further research.

Acknowledgements.
I am indebted to all my laboratory colleagues for their continuous support and for making our research projects possible. Our work is funded by "Fundación Científica de la Asociación Española Contra el Cáncer" and "Ministerio de Ciencia y Tecnología" (SAF2003-00745).

References

Abood ME, Rizvi G, Sallapudi N, McAllister SD (2001) Activation of the CB1 cannabinoid receptor protects cultured spinal neurons against neurotoxicity. Neurosci Lett 309:197–201

Alberich Jordà M, Lowenberg B, Delwel R (2003) The peripheral cannabinoid receptor Cb2, a novel oncoprotein, induces a reversible block in neutrophilic differentiation. Blood 101:1336–1343

Arévalo-Martín A, Vela JM, Molina-Holgado E, Borrell J, Guaza C (2003) Therapeutic action of cannabinoids in a murine model of multiple sclerosis. J Neurosci 23:2511–2516

Baker D, Pryce G, Croxford JL, Brown P, Pertwee RG, Huffman JW, Layward L (2000) Cannabinoids control spasticity and tremor in a multiple sclerosis model. Nature 404:84–87

Bayewitch M, Rhee MH, Avidor-Reiss T, Breuer A, Mechoulam R, Vogel Z (1996) (–)-Delta9-tetrahydrocannabinol antagonizes the peripheral cannabinoid receptor-mediated inhibition of adenylyl cyclase. J Biol Chem 271:9902–9905

Bifulco M, Laezza C, Portella G, Vitale M, Orlando P, De Petrocellis L, Di Marzo V (2001) Control by the endogenous cannabinoid system of ras oncogene-dependent tumor growth. FASEB J 15:2745–2747

Blázquez C, Casanova ML, Planas A, Gómez del Pulgar T, Villanueva C, Fernández-Aceñero MJ, Aragonés J, Huffman JW, Jorcano JL, Guzmán M (2003) Inhibition of tumor angiogenesis by cannabinoids. FASEB J 17:529–531

Bouaboula M, Poinot-Chazel C, Bourrié B, Canat X, Calandra B, Rinaldi-Carmona M, Le Fur G, Casellas P (1995) Activation of mitogen-activated protein kinases by stimulation of the central cannabinoid receptor CB1. Biochem J 312:637–641

Bouaboula M, Poinot-Chazel C, Marchand J, Canat X, Bourrié B, Rinaldi-Carmona M, Calandra B, Le Fur G, Casellas P (1996) Signaling pathway associated with stimulation of CB2 peripheral cannabinoid receptor. Involvement of both mitogen-activated protein kinase and induction of Krox-24 expression. Eur J Biochem 237:704–711

Carayon P, Marchand J, Dussossoy D, Derocq JM, Jbilo O, Bord A, Bouaboula M, Galiègue S, Mondière P, Pénarier G, Le Fur G, Defrance T, Casellas P (1998) Modulation and functional involvement of CB2 peripheral cannabinoid receptors during B-cell differentiation. Blood 92:3605–3615

Casanova ML, Blázquez C, Martínez-Palacio J, Villanueva C, Fernández-Aceñero MJ, Huffman JW, Jorcano JL, Guzmán M (2003) Inhibition of skin tumor growth and angiogenesis in vivo by activation of cannabinoid receptors. J Clin Invest 111:43–50

Chan GC, Hinds TR, Impey S, Storm DR (1998) Hippocampal neurotoxicity of Δ9-tetrahydrocannabinol. J Neurosci 18:5322–5332
Chan PC, Sills RC, Braun AG, Haseman JK, Bucher JR (1996) Toxicity and carcinogenicity of Δ9-tetrahydrocannabinol in Fischer rats and B6C3F1 mice. Fundam Appl Toxicol 30:109–117
Chen Y, Buck J (2000) Cannabinoids protect cells from oxidative cell death: a receptor independent mechanism. J Pharmacol Exp Ther 293:807–812
De Petrocellis L, Melck D, Palmisano A, Bisogno T, Laezza C, Bifulco M, Di Marzo V (1998) The endogenous cannabinoid anandamide inhibits human breast cancer cell proliferation. Proc Natl Acad Sci USA 95:8375–8380
Derkinderen P, Valjent E, Toutant M, Corvol JC, Enslen H, Ledent C, Trzaskos J, Caboche J, Girault JA (2003) Regulation of extracellular signal-regulated kinase by cannabinoids in hippocampus. J Neurosci 23:2371–2382
Derocq JM, Segui M, Marchand J, Le Fur G, Casellas P (1995) Cannabinoids enhance human B-cell growth at low nanomolar concentrations. FEBS Lett 369:177–182
Derocq JM, Jbilo O, Bouaboula M, Ségui M, Clère C, Casellas P (2000) Genomic and functional changes induced by the activation of the peripheral cannabinoid receptor CB2 in the promyelocytic cells HL-60. Possible involvement of the CB2 receptor in cell differentiation. J Biol Chem 275:15621–15628
Di Marzo V, Fontana A, Cadas H, Schinelli S, Cimino G, Schwartz JC, Piomelli D (1994) Formation and inactivation of endogenous cannabinoid anandamide in central neurons. Nature 372:686–691
Downer EJ, Fogarty MP, Campbell VA (2003) TetraHydroCannabinol-induced neurotoxicity depends on CB1 receptor mediated c-Jun N-terminal kinase activation in cultured cortical neurons. Br J Pharmacol 140:547–557
Faubert Kaplan BL, Kaminski NE (2003) Cannabinoids inhibit the activation of ERK MAPK in PMA/Io-stimulated mouse splenocytes. Int Immunopharmacol 3:1503–1510
Feigenbaum JJ, Bergmann F, Richmond SA, Mechoulam R, Nadler V, Kloog Y, Sokolovsky M (1989) Nonpsychotropic cannabinoid acts as a functional N-methyl-D-aspartate receptor blocker. Proc Natl Acad Sci USA 86:9584–9587
Franklin A, Parmentier-Batteur S, Walter L, Greenberg DA, Stella N (2003) Palmitoylethanolamide increases after focal cerebral ischemia and potentiates microglial cell motility. J Neurosci 23:7767–7775
Freund TF, Katona I, Piomelli D (2003) Role of endogenous cannabinoids in synaptic signaling. Physiol Rev 83:1017–1066
Gallily R, Yamin A, Waksmann Y, Ovadia H, Weidenfeld J, Bar-Joseph A, Biegon A, Mechoulam R, Shohami E (1997) Protection against septic shock and suppression of tumor necrosis factor α and nitric oxide production by dexanabinol (HU-211), a nonpsychotropic cannabinoid. J Pharmacol Exp Ther 283:918–924
Galve-Roperh I, Sánchez C, Cortés M, Gómez del Pulgar T, Izquierdo M, Guzmán M (2000) Antitumoral action of cannabinoids: involvement of sustained ceramide accumulation and extracellular signal-regulated kinase activation. Nat Med 6:313–319
Galve-Roperh I, Rueda D, Gómez del Pulgar MT, Velasco G, Guzmán M (2002) Mechanism of extracellular signal-regulated kinase activation by the CB1 cannabinoid receptor. Mol Pharmacol 62:1385–1392
Giuffrida A, Parsons LH, Kerr TM, Rodríguez de Fonseca F, Navarro M, Piomelli D (1999) Dopamine activation of endogenous cannabinoid signaling in dorsal striatum. Nat Neurosci 2:358–363
Gómez del Pulgar T, Velasco G, Guzmán M (2000) The CB1 cannabinoid receptor is coupled to the activation of protein kinase B/Akt. Biochem J 347:369–373
Gómez del Pulgar T, de Ceballos ML, Guzmán M, Velasco G (2002a) Cannabinoids protect astrocytes from ceramide-induced apoptosis through the phosphatidylinositol 3-kinase/protein kinase B pathway. J Biol Chem 277:36527–36533
Gómez del Pulgar T, Velasco G, Sánchez C, Haro A, Guzmán M (2002b) De novo-synthesized ceramide is involved in cannabinoid-induced apoptosis. Biochem J 363:183–188

Guzmán M (2003) Cannabinoids: potential anticancer agents. Nat Rev Cancer 3:745–755
Guzmán M, Galve-Roperh I, Sánchez C (2001) Ceramide: a new second messenger of cannabinoid action. Trends Pharmacol Sci 22:19–22
Hampson AJ, Grimaldi M (2001) Cannabinoid receptor activation and elevated cyclic AMP reduce glutamate neurotoxicity. Eur J Neurosci 13:1529–1536
Hampson AJ, Grimaldi M, Axelrod J, Wink D (1998) Cannabidiol and Δ9-tetrahydrocannabinol are neuroprotective antioxidants. Proc Natl Acad Sci USA 95:8268–8273
Hansen HH, Schmid PC, Bittigau P, Lastres-Becker I, Berrendero F, Manzanares J, Ikonomidou C, Schmid HH, Fernández-Ruiz JJ, Hansen HS (2001) Anandamide, but not 2-arachydonoylglycerol, accumulates during in vivo neurodegeneration. J Neurochem 78:1415–1427
Howlett AC (1984) Inhibition of neuroblastoma adenylate cyclase by cannabinoid and nantradol compounds. Life Sci 35:1803–1810
Howlett AC, Barth F, Bonner TI, Cabral G, Casellas P, Devane WA, Felder CC, Herkenham M, Mackie K, Martin BR, Mechoulam R, Pertwee RG (2002) International Union of Pharmacology. XXVII. Classification of cannabinoid receptors. Pharmacol Rev 54:161–202
Jones S, Howl J (2003) Cannabinoid receptor systems: therapeutic targets for tumour intervention. Expert Opin Ther Targets 7:749–758
Kaminski NE (1998) Regulation of the cAMP cascade, gene expression and immune function by cannabinoid receptors. J Neuroimmunol 83:124–132
Klein TW, Newton C, Larsen K, Lu L, Perkins I, Nong L, Friedman H (2003) The cannabinoid system and immune modulation. J Leukoc Biol 74:486–496
Knoller N, Levi L, Shoshan I, Reichenthal E, Razon N, Rappaport ZH, Biegon A (2002) Dexanabinol (HU-211) in the treatment of severe closed head injury: a randomized, placebo-controlled, phase II clinical trial. Crit Care Med 30:548–554
Lastres-Becker I, De Miguel R, Fernández-Ruiz JJ (2003) The endocannabinoid system and Huntington's disease. Curr Drug Target CNS Neurol Disord 2:335–347
Liu J, Gao B, Mirshahi F, Sanyal AJ, Khanolkar AD, Makriyanis A, Kunos G (2000) Functional CB1 cannabinoid receptors in human vascular endothelial cells. Biochem J 346:835–840
Maccarrone M, Finazzi-Agro A (2003) The endocannabinoid system, anandamide and the regulation of mammalian cell apoptosis. Cell Death Differ 10:946–955
Marsicano G, Moosmann B, Hermann H, Lutz B, Behl C (2002) Neuroprotective properties of cannabinoids against oxidative stress: role of the cannabinoid receptor CB1. J Neurochem 80:448–456
Marsicano G, Goodenough S, Monory K, Hermann H, Eder M, Cannich A, Azad SC, Cascio MG, Gutiérrez SO, van der Stelt M, López-Rodríguez ML, Casanova E, Schutz G, Zieglgansberger W, Di Marzo V, Behl C, Lutz B (2003) CB1 cannabinoid receptors and on-demand defense against excitotoxicity. Science 302:84–88
Massi P, Vaccani A, Ceruti S, Colombo A, Abbracchio MP, Parolaro D (2004) Antitumor effects of cannabidiol, a non-psychotropic cannabinoid, on human glioma cell lines. J Pharmacol Exp Ther 308:838–845
Matsuda LA, Lolait SJ, Brownstein MJ, Young AC, Bonner TI (1990) Structure of a cannabinoid receptor and functional expression of the cloned cDNA. Nature 346:561–564
McKallip RJ, Lombard C, Fisher M, Martin BR, Ryu S, Grant S, Nagarkatti PS, Nagarkatti M (2002a) Targeting CB2 cannabinoid receptors as a novel therapy to treat malignant lymphoblastic disease. Blood 100:627–634
McKallip RJ, Lombard C, Martin BR, Nagarkatti M, Nagarkatti PS (2002b) Δ9-TetraHydro-Cannabinol-induced apoptosis in the thymus and spleen as a mechanism of immunosuppression in vitro and in vivo. J Pharmacol Exp Ther 302:451–465
Mechoulam R, Spatz M, Shohami E (2002) Endocannabinoids and neuroprotection. Sci STKE 129:RE5
Melck D, Rueda D, Galve-Roperh I, De Petrocellis L, Guzmán M, Di Marzo V (1999) Involvement of the cAMP/protein kinase pathway and of mitogen-activated protein kinase in

the anti-proliferative effects of anandamide in human breast cancer cells. FEBS Lett 463:235–240
Melck D, De Petrocellis L, Orlando P, Bisogno T, Laezza C, Bifulco M, Di Marzo V (2000) Suppression of nerve growth factor trk receptors and prolactin receptors by endocannabinoids leads to inhibition of human breast and prostate cancer cell proliferation. Endocrinology 141:118–126
Mimeault M, Pommery N, Wattez N, Bailly C, Henichart JP (2003) Anti-proliferative and apoptotic effects of anandamide in human prostatic cancer cell lines: implication of epidermal growth factor receptor down-regulation and ceramide production. Prostate 56:1–12
Molina-Holgado E, Vela JM, Arévalo-Martín A, Almazán G, Molina-Holgado F, Borrell J, Guaza C (2002) Cannabinoids promote oligodendrocyte progenitor survival: involvement of cannabinoid receptors and phosphatidylinositol 3-kinase/Akt signaling. J Neurosci 22:9742–9753
Muller-Vahl KR (2003) Cannabinoids reduce symptoms of Tourette's syndrome. Expert Opin Pharmacother 4:1717–1725
Munro S, Thomas KL, Abu Shaar M (1993) Molecular characterization of a peripheral receptor for cannabinoids. Nature 365:61–65
Munson AE, Harris LS, Friedman MA, Dewey WL, Carchman RA (1975) Antineoplastic activity of cannabinoids. J Natl Cancer Inst 55:597–602
Nagayama T, Sinor AD, Simon RP, Chen J, Graham SH, Jin K, Greenberg DA (1999) Cannabinoids and neuroprotection in global and focal cerebral ischemia and in neuronal cultures. J Neurosci 19:2987–2995
Panikashvili D, Simeonidou C, Ben-Shabat S, Hanus L, Breuer A, Mechoulam R, Shohami E (2001) An endogenous cannabinoid (2AG) is neuroprotective after brain injury. Nature 413:527–531
Parmentier-Batteur S, Jim K, Mao O, Xie L, Greenberg DA (2002) Increased severity of stroke in CB1 cannabinoid receptor knock-out mice. J Neurosci 22:9771–9775
Piomelli D (2003) The molecular logic of endocannabinoid signalling. Nat Rev Neurosci 4:873–884
Portella G, Laezza C, Laccetti P, De Petrocellis L, Di Marzo V, Bifulco M (2003) Inhibitory effects of cannabinoid CB1 receptor stimulation on tumor growth and metastatic spreading: actions on signals involved in angiogenesis and metastasis. FASEB J 17:1771–1773
Riboni L, Campanella R, Bassi R, Villani R, Gaini SM, Martinelli-Boneschi F, Viani P, Tettamanti G (2002) Ceramide levels are inversely associated with malignant progression of human glial tumors. Glia 39:105–113
Roth MD, Baldwin GC, Tashkin DP (2002) Effects of delta-9-tetrahydrocannabinol on human immune function and host defense. Chem Phys Lipids 31:229–239
Rueda D, Galve-Roperh I, Haro A, Guzmán M (2000) The CB1 cannabinoid receptor is coupled to the activation of c-Jun N-terminal kinase. Mol Pharmacol 58:814–820
Rueda D, Navarro B, Martínez-Serrano A, Guzmán M, Galve-Roperh I (2002) The endocannabinoid anandamide inhibits neuronal progenitor cell differentiation through attenuation of the Rap1/B-Raf/ERK pathway. J Biol Chem 277:46645–46650
Samson MT, Small-Howard A, Shimoda LM, Koblan-Huberson M, Stokes AJ, Turner H (2003) Differential roles of CB1 and CB2 cannabinoid receptors in mast cells. J Immunol 170:4953–4962
Sánchez C, Galve-Roperh I, Canova C, Brachet P, Guzmán M (1998a) Δ9-TetraHydro-Cannabinol induces apoptosis in C6 glioma cells. FEBS Lett 436:6–10
Sánchez C, Galve-Roperh I, Rueda D, Guzmán M (1998b) Involvement of sphingomyelin hydrolysis and the mitogen-activated protein kinase cascade in the Δ9-tetrahydrocannabinol-induced stimulation of glucose metabolism in primary astrocytes. Mol Pharmacol 54:834–843
Sánchez C, de Ceballos ML, Gómez del Pulgar T, Rueda D, Corbacho C, Velasco G, Galve-Roperh I, Huffman JW, Ramón y Cajal S, Guzmán M (2001a) Inhibition of glioma growth in vivo by selective activation of the CB2 cannabinoid receptor. Cancer Res 61:5784–5789

Sánchez C, Rueda D, Ségui B, Galve-Roperh I, Levade T, Guzmán M (2001b) The CB1 cannabinoid receptor of astrocytes is coupled to sphingomyelin hydrolysis through the adaptor protein fan. Mol Pharmacol 59:955–959

Sarker KP, Biswas KK, Yamakuchi M, Lee KY, Hahiguchi T, Kracht M, Kitajima I, Maruyama I (2003) ASK1-p38 MAPK/JNK signaling cascade mediates anandamide-induced PC12 cell death. J Neurochem 85:50–61

Scallet AC, Uemure E, Andrews A, Ali SF, McMillan DE, Paule MC, Brown RM, Slikker W (1987) Morphometric studies of the rat hippocampus following chronic delta-9-tetrahydrocannabinol (THC). Brain Res 436:193–198

Schwarz H, Blanco FJ, Lotz M (1994) Anadamide, an endogenous cannabinoid receptor agonist inhibits lymphocyte proliferation and induces apoptosis. J Neuroimmunol 55:107–115

Shen M, Thayer SA (1998) Cannabinoid receptor agonist protect cultured rat hippocampal neurons from excitotoxicity. Mol Pharmacol 54:459–462

Sinor AD, Irvin SM, Greenberg DA (2000) Endocannabinoids protect cerebral cortical neurons from in vitro ischemia in rats. Neurosci Lett 278:157–160

Skaper SD, Buriani A, Dal Toso R, Petrelli L, Romanello S, Facci L, Leon A (1996) The ALIAmide palmitoylethanolamide and cannabinoids, but not anandamide, are protective in a delayed postglutamate paradigm of excitotoxic death in cerebellar granule neurons. Proc Natl Acad Sci USA 93:3984–3989

Stella N, Schweitzer P, Piomelli D (1997) A second endogenous cannabinoid that modulates long-term potentiation. Nature 388:773–778

Valk PJM, Hol S, Vankan Y, Ihle JN, Askew D, Jenkins NA, Gilbert DJ, Copeland NG, de Both NJ, Löwenberg B, Delwel R (1997a) The genes encoding the peripheral cannabinoid receptor and α-L-fucosidase are located near a newly identified common virus integration site, Evi11. J Virol 71:6796–6804

Valk P, Verbarkel S, Vankan Y, Hol S, Mancham S, Ploemacher R, Mayen A, Löwenberg B, Delwel R (1997b) Anandamide, a natural ligand for the peripheral cannabinoid receptor is a novel synergistic growth factor for hematopoietic cells. Blood 90:1448–1457

Valk PJ, Vankan Y, Joosten M, Jenkins NA, Copeland NG, Löwenberg B, Delwel R (1999) Retroviral insertions in Evi12, a novel common virus integration site upstream of Tra1/Grp94, frequently coincides with insertions in the gene encoding the peripheral cannabinoid receptor Cnr2. J Virol 73:3595–3602

Van der Stelt M, Veldhuis WB, Bär PR, Veldink GA, Vliegenthart JFG, Nikolay K (2001) Neuroprotection by Δ9-tetrahydrocannabinol, the main active compound of marijuana, against ouabain-induced in vivo excitotoxicity. J Neurosci 21:6475–6479

Van der Stelt M, Veldhuis WB, Maccarrone M, Bar PR, Nicolay K, Veldink GA, Di Marzo V, Vliegenthart JF (2002) Acute neuronal injury, excitotoxicity, and the endocannabinoid system. Mol Neurobiol 26:317–346

Zajicek J, Fox P, Sanders H, Wright D, Vickery J, Nunn A, Thompson A, UK MS Research Group (2003) Cannabinoids for treatment of spasticity and other symptoms related to multiple sclerosis (CAMS study): multicentre randomised placebo-controlled trial. Lancet 362:1517–1526

Zhu W, Friedman H, Klein TW (1998) Δ9-TetraHydroCannabinol induces apoptosis in macrophages and lymphocytes: involvement of Bcl-2 and caspase-1. J Pharmacol Exp Ther 286:1103–1109

HEP (2005) 168:643–656

Effects on Development

J.A. Ramos (✉) · M. Gómez · R. de Miguel

Departamento de Bioquímica, Facultad de Medicina, Universidad Complutense, 28040 Madrid, Spain
jara@med.ucm.es

Abstract This chapter will review the effects produced on neural development by maternal consumption of cannabinoids during gestation and lactation, with emphasis in the maturation of several neurotransmitter systems (dopamine, serotonin, opioids, cannabinoids, etc.) and possible modifications in their functional expression at the behavioral or neuroendocrine levels. In addition, we have analyzed the possible existence of a sexual dimorphism in these ontogenic effects of cannabinoids, as well as the possible molecular mechanism underlying such effects. In general, the results discussed support the view that exposure to cannabinoids during critical periods of development produces marked modifications in the functional expression of diverse neuronal systems in adulthood. Furthermore, the functions of endocannabinoids in the brain are large not only in adulthood, but also in the period of prenatal and postnatal development. Thus, endocannabinoids have been reported to be present in early ages and to play a role in the process of brain development: neural proliferation and migration, axonal elongation, synaptogenesis and/or myelogenesis.

Keywords Cannabinoids · Cannabinoid receptor · Endogenous cannabinoids · Perinatal exposure · Brain development

1 Introduction

Marijuana is the most commonly used illicit drug among women of reproductive age. Its use during pregnancy in developed nations is estimated to be approximately 10% (Park et al. 2004). However, reports dealing with the effects of prenatal exposure to this substance of abuse on the length of gestation, fetal growth, and offspring behavior are still controversial (Park et al. 2004).

In animal models, the consumption of marijuana or other *Cannabis sativa* derivatives during pregnancy and/or lactation affects the neurobehavioral development of their litters. The cannabinoids present in these preparations modify the maturation of neurotransmitter systems, including dopamine, serotonin, and opioids and their related-behaviors (Fernández-Ruiz et al. 1999, 2000).

The principal agent responsible for these effects is Δ^9-tetrahydrocannabinol (Δ^9-THC), the major psychoactive component of *Cannabis sativa*. Δ^9-THC can interfere, not only with the activity of classical neurotransmitters but also with the activity of the endogenous cannabinoid system itself. Different studies support a role for this system in brain development and maturation, and several of its components have been characterized (receptors, endogenous ligands, and metabolism pathways). The effects of Δ^9-THC were caused by the activation of cannabinoid receptors, which emerge early in the developing brain (Fernández-Ruiz et al. 1992, 1994, 1996, 1999, 2000).

With regard to the possible influence of cannabis use during human pregnancy, CB_1 receptor immunoreactive labeling has been identified in most major cell types throughout all layers of the human placenta, as well as in the placental villous (Park et al. 2003). Thus, it is not surprising that increased cannabinoid levels may interfere with the materno-fetal process.

Clinical research has been limited to epidemiologic and retrospective studies. In some reports, cannabis use has been correlated with low birth weight, prematurity, intrauterine growth retardation, presence of congenital abnormalities, perinatal death and delayed time for the onset of respiration (for a review see Park et al. 2004).

In some studies, carried out in rats subjected to prenatal exposure to cannabinoids, specific congenital malformations were not produced, even by high doses. However, other studies have reported an increase in embryotoxicity and fetal toxicity at pharmacologically relevant concentrations, resembling the adverse effects of Δ^9-THC on human reproduction. Thus, in rhesus monkeys, Δ^9-THC exposure during early pregnancy produced miscarriage (Asch and Smith 1986). In mice cannabinoids impaired pregnancy and embryo development (Harbison and Mantilla-Plata 1972; Yang et al. 1996).

More complex and less understood are the data concerning the possible long-term consequences of in utero exposure to cannabis derivatives. Some data suggest that prenatal exposure to marijuana could result in a certain impairment of human fetal development (Fried and Smith 2001), albeit restricted to a few executive functions. Thus, a longitudinal cohort study in Ottawa to examine the effects of marijuana consumption during pregnancy upon offspring in the areas of growth, cognitive development, and behavior showed that the consequences of prenatal exposure to marijuana are subtle. Before the age of 3 years, there is little evidence for a prenatal effect either upon growth or behavior. However, beyond this age, there are several findings suggestive of an association between prenatal marijuana exposure and aspects of frontal lobe functioning, including cognitive behavior. Attention/impulsivity and problem-solving situations requiring integration and manipulation of basic visuo-perceptual skills appear to be particularly affected (Fried and Smith 2001).

These results were replicated in a longitudinal cohort study conducted in Pittsburgh to investigate the consequences of prenatal marijuana in offspring beyond early school age. Prenatal marijuana exposure was also found to affect attention and impulsivity, and, thereby, to decrease the ability to plan and execute tasks (Leech et al. 1999). These data are in agreement with the results obtained in animals. In these models, it has been reported that deficits of cognitive functions induced by marijuana use during adulthood could be mainly attributable to the activation of CB_1 receptors located in the hippocampus, a brain region crucial for certain forms of learning and memory. Prenatal exposure to a cannabinoid agonist produced memory deficits linked to dysfunction in hippocampal long-term potentiation and glutamate release (Mereu et al. 2003).

2
Ontogeny of the Endogenous Cannabinoid System

The existence of several components of the endogenous cannabinoid system has been demonstrated in the fetal and neonatal rat brain. The system is active in this period of life and shows significant differences in the expression and/or activity of its components during the consecutive steps of early stages of development.

Most of the studies carried out on the presence and/or the activity of the components of this system during development have focused on the cannabinoid CB_1 receptor (Mailleux and Vanderhaeghen 1992a; Romero et al. 1997; Berrendero et al. 1998; Buckley et al. 1998; Mato et al. 2003). There is also some information about the cannabinoid receptor ligands anandamide (AEA) and 2-arachidonoylglycerol (2-AG) (Berrendero et al. 1999).

2.1
Cannabinoid Receptors

Cannabinoid CB_1 receptor binding and mRNA levels could be detected around gestational day (GD)11–14 in rats, coinciding with the time of phenotypic expression of most of the neurotransmitters (for review see Insel 1995). At these fetal ages, cannabinoid receptors appear to be functional, since they are already coupled to signal transduction mechanisms that involve activation of guanosine triphosphate (GTP)-binding proteins (Berrendero et al. 1998). Pharmacological activation of cannabinoid receptors during the developmental period has been associated with several effects such as induction of key genes, activation of energy metabolism, arachidonic acid mobilization, and other responses potentially related to events of neural development (for review see Ramos et al. 2002).

The levels of these receptors are substantially higher than those seen in the adult rat brain (Berrendero et al. 1999). Moreover, in the fetal and early neonatal brain, there is an atypical distribution of CB_1 receptors compared to the adult brain, particularly with regard to the location of receptor binding in white matter areas (Romero et al. 1997) and mRNA expression in subventricular zones of the forebrain (Berrendero et al. 1998, 1999), areas in which these receptors are scarce or undetectable in the adult brain (Herkenham et al. 1991; Mailleux and Vanderhaeghen 1992b). This atypical location of CB_1 receptors is a transient phenomenon, since during the course of late postnatal development these receptors progressively acquire the classic pattern of distribution observed in the adult brain (Romero et al. 1997; Berrendero et al. 1998).

With regard to humans, data about the appearance and location of CB_1 receptors in the developing brain are still very limited (Mailleux and Vanderhaeghen 1992a; Glass et al. 1997; Biegon and Kerman 2001; Mato et al. 2003). There is a significant population of cannabinoid receptors at week 19 of gestation that is functionally coupled to signal transduction mechanisms. These receptors are present in the same areas as in the adult human brain and seem to increase progressively in number from early prenatal stages to adulthood (Mato et al. 2003).

As in animal models, high densities of cannabinoid receptors have also been detected during human prenatal development in fiber-enriched areas that are practically devoid of these receptors in the adult brain. This atypical distribution of CB_1 receptors, which is similar to that observed in rats, has been interpreted, for both species, as indicating a possible involvement of CB_1 receptors in specific events relating to brain development during fetal and early postnatal periods. As has been suggested for other neurotransmitter receptors (del Olmo and Pazos 2001), the early and transient presence of these receptors could indicate a specific role for the endocannabinoid system in several developmental events, such as metabolic support, cell proliferation and migration, axonal elongation, and later, synaptogenesis and myelogenesis (for review see Fernández-Ruiz et al. 2000).

This hypothesis is supported by the demonstration of a role for CB_1 receptors in neurite remodeling in vitro (Zhou and Song 2001). Support also comes from the findings that endocannabinoids inhibit both cortical neuron differentiation to mature neurons using in vitro cellular models and adult hippocampal neurogenesis in vivo (Rueda et al. 2002). The endocannabinoids interfere with nerve growth factor (NGF) signaling that is responsible for the activation of the differentiation program by acting via CB_1 receptors to inhibit NGF-induced signaling events that ultimately result in inhibition of neural generation (Rueda et al. 2002). Therefore, inhibition of neurogenesis in adult hippocampus, triggered by cannabinoids either during development or in the adult, might help to explain cannabinoid-related disruption of cognitive processes such as learning and short-term memory.

2.2 Cannabinoid Ligands

Both endocannabinoids, AEA and 2-AG, are present in the whole brains of rat fetuses at GD21 (Berrendero et al. 1999). The presence of these cannabinoids in the brain in trace amounts limits how much can be learned about the distribution of these compounds in the different regions of the developing brain and the extent to which this parallels the distribution of the CB_1 receptors that are present at that time. Even so, it has been possible to conclude from the results that have been obtained that the ontogeny of these two endocannabinoids is not the same (Berrendero et al. 1999).

High levels of 2-AG were measured at GD21. These levels peaked at postnatal day (PND)1, with values approximately twofold higher than those found at other ages. In contrast, AEA levels increased during the early postnatal period, reaching their maximum in the adult brain (Berrendero et al. 1999). Cannabinoid receptor binding and mRNA expression in certain brain areas also change with time during development. Interestingly, in the caudate-putamen and cerebellum these changes are similar to the changes that take place in AEA levels, whereas in the hippocampus and cerebral cortex they resemble the changes occurring in 2-AG levels (Berrendero et al. 1999).

The amount of 2-AG in the fetal and early postnatal brain is significantly higher than that of AEA (Berrendero et al. 1999). It is possible that this might indicate a more important role for 2-AG than AEA as an endogenous ligand for the CB_1 receptor in brain development. However, it must be remembered that 2-AG formation in tissues may lead not only to the production of a cannabimimetic signal, but also to the termination of a protein kinase C/diacylglycerol-mediated intracellular signal or to the generation of arachidonic acid. Accordingly, the increase of 2-AG observed at PND1 might be related to an increase in the formation of diacylglycerol, which is an intermediate in the synthesis of 2-AG (Di Marzo 1998). Diacylglycerol has been reported to be significantly involved in the metabolism of phosphoglycerides and sphingolipids during the processes of neurite formation and myelinogenesis, and hence in neural development (Araki and Wurtman 1997; Sillence and Allan 1998).

The physiological significance of these differences between 2-AG and AEA is still unknown, but could be related to regional differences in their distribution and in their access to their receptors, and possibly to their having different roles in brain development. In summary, there is good evidence that the endogenous cannabinoid system plays a functional role in the early stages of brain development. This role changes as the brain develops such that the way in which the endogenous cannabinoid system modulates brain function during its development is not the same as in the adult (i.e., control of movement, nociception, etc.).

3
Effects of Perinatal Exposure to Cannabinoids on Several Neurotransmitters Systems

Endocannabinoids might act as epigenetic factors, through the activation of cannabinoid receptors which emerge early in development. As to plant-derived cannabinoids, by mimicking the effects of natural ligands of cannabinoid receptors these would interfere with the sequence of events that results in the expression of several genes involved in brain development and, in this way, modify the maturation of several neurotransmitter systems (Fernández-Ruiz et al. 1992, 1994, 1996, 1999, 2000).

Also, it is possible that the usual pattern of expression of the endogenous cannabinoid system could be altered by these cannabinoids during brain development (Fernández-Ruiz et al. 1999, 2000). Thus, an acceleration or delay in the expression of the genes implicated in the synthesis of endocannabinoids or their receptors during a particular stage of development could well prevent the endocannabinoid system from functioning normally. Such abnormal functioning might also result from an increase or a decrease in endocannabinoid or cannabinoid receptor concentrations or from a modification in the activity of cannabinoid receptor signaling pathways. Such mechanisms may underlie reported behavioral alterations in adult mice that had been perinatally exposed to anandamide (decreased open-field activity, catalepsy, hypothermia, hypoalgesia, and tolerance to cannabinoid challenges) (Fride and Mechoulam 1996a,b), as well as the interruption of suckling behavior with subsequent inhibition of neonatal growth observed in newborn mice in response to CB_1 receptor blockade (Fride et al. 2001).

To establish the effects of perinatal exposure to Δ^9-THC, studies were conducted in rodents using doses that produce concentrations of this cannabinoid in the body similar to those found in marijuana consumers. The effects observed depended on when the drug treatment was initiated and on the dose used (Dalterio 1986; Fernández-Ruiz et al. 1992, 1994, 1996). These studies demonstrated that cannabinoids may behave as epigenetic factors, modifying normal development of neurotransmission and, most likely, producing neurobehavioral disturbances. Thus, adult animals perinatally exposed to cannabinoids exhibited long-term alterations in male copulatory behavior (Dalterio 1980), open-field activity (Navarro et al. 1994), learning ability (Dalterio 1986), stress response (Mokler et al. 1987), pain sensitivity (Vela et al. 1995), social interaction and sexual motivation (Navarro et al. 1996), drug-seeking behavior (Vela et al. 1998), and neuroendocrine disturbances (Dalterio and Barker 1979; Murphy et al. 1990, 1995), as well as other changes (for review see Dalterio 1986; Fernández-Ruiz et al. 1992, 1994, 1996).

Most of these neurobehavioral effects presumably stemmed from changes in the development of several neurotransmitter systems caused by exposure to cannabinoids during critical prenatal and early postnatal periods of brain development. A large number of studies have demonstrated effects of cannabinoids on the maturation of dopamine (DA) (Fernández-Ruiz et al. 1992, 1994, 1996; Walters and Carr 1988; García-Gil et al. 1998), serotonin (5-HT) (Molina-Holgado et al. 1997),

γ-aminobutyric acid (GABA) (García-Gil et al. 1999b), and opioid peptide systems (Kumar et al. 1990; Corchero et al. 1998; Vela et al. 1998).

3.1 Dopamine

The effects of Δ^9-THC on the development of specific brain dopaminergic pathways appear before the complete differentiation and maturation of dopaminergic projections into their target areas, in particular during the final part of gestation and first week after birth (Bonnin et al. 1996). The activity of tyrosine hydroxylase (TH) represents the rate-limiting step in DA synthesis. This enzyme is present in the growing axons before they make contact with their target neurons and seems to play an important role, together with active receptors located on the target neurons, in the formation of connections between neurons (Insel et al. 1995). In rats, perinatal exposure to Δ^9-THC has been found to cause a marked rise in TH gene expression in the brains of fetuses at GD14, in parallel with a pronounced increase in the levels and activity of this enzyme (Bonnin et al. 1996). These results have been verified in vitro using cultured mesencephalic neurons from fetal brains at GD14 (Hernández et al. 1997). Cultured cells obtained from fetuses that had been exposed to Δ^9-THC daily from day 5 of gestation exhibited higher TH activity compared with cells obtained from vehicle-exposed fetuses (Hernández et al. 2000).

These data suggest that interference of plant-derived cannabinoids with the sequence of events in which the expression of TH gene is involved during brain development might contribute to the abnormal pre- and postnatal development of TH-containing neurons themselves and/or of the different neurons with which they make contact (Walters and Carr 1988; Fernández-Ruiz et al. 1992, 1994, 1996).

Interruption of the exposure of pups to cannabinoids as a consequence of weaning led to an apparent normalization of most of the dopaminergic indices that had altered during development (Bonnin et al. 1996; Rodríguez de Fonseca et al. 1991; Navarro et al. 1996). However, at adulthood, animals that had been perinatally exposed to Δ^9-THC, although mostly having similar basal indices of dopaminergic activity to those of controls (Navarro et al. 1996), exhibited an abnormal ability to respond to a variety of drugs that affect key processes of dopaminergic neurotransmission. Thus, the differences observed in the response of Δ^9-THC-exposed animals to experimental challenges, using D_1 and D_2 antagonists, α-methyl-p-tyrosine combined with reserpine, amphetamine, or NSD1015, support the existence of irreversible, although silent, changes in the adult functionality of dopaminergic neurons due to the early contact with Δ^9-THC (García-Gil et al. 1996, 1999a,b).

A sexual dimorphism is evident in the sensitivity of maturing dopaminergic neurons to Δ^9-THC exposure. This is already apparent at fetal ages (Bonnin et al. 1996), but is particularly evident at early postnatal, immature, and adult ages (Fernández-Ruiz et al. 1992, 1994, 1996; Navarro et al. 1996; García-Gil et al. 1998). Usually, the changes are especially marked and constant in male offspring, whereas in females they are frequently smaller and transient, as has also been reported for other drugs of abuse (Fernández-Ruiz et al. 1992).

3.2
γ-Aminobutyric Acid

Cannabinoid receptors are frequently located on GABA-releasing neurons, for example in the basal ganglia and the hippocampus (Herkenham et al. 1991), or on non-GABAergic neurons that connect with GABA-releasing neurons, for example in the cerebellum, where cannabinoid and GABA receptors colocalize in the same neurons (Pacheco et al. 1993). Perinatal cannabinoid exposure has been found not to produce any measurable change in GABA content or in the activity of glutamic acid decarboxylase (GAD) in motor (caudate-putamen, globus pallidus, and substantia nigra) and limbic (nucleus accumbens and ventral-tegmental area) regions of adult animals. However, both adult male and females that had been perinatally exposed to Δ^9-THC exhibited a higher responsiveness to the $GABA_B$ receptor agonist, baclofen (García-Gil et al. 1999b). This is in concordance with the predominant role proposed for $GABA_B$ receptors in the relationships between GABA and endocannabinoids in the basal ganglia (Romero et al. 1996) and in other brain regions in adult individuals. The changes in the motor responsiveness to baclofen mimicked those observed previously with dopaminergic antagonist, which supports the possibility that the effects on dopaminergic activity are mediated through GABAergic neurons (García-Gil et al. 1996).

3.3
Serotonin

In brain development, 5-HT, like DA, exerts a trophic action, by a reciprocal interaction with other neurotransmitters, in specific development processes (Pares-Herbute et al. 1989). Some studies have reported changes in 5-HT development in rats perinatally exposed to Δ^9-THC. Thus, this cannabinoid produced a decrease in 5-HT content in diencephalic areas but not in other brain areas (Molina-Holgado et al. 1996). When animals perinatally exposed to Δ^9-THC matured, they exhibited an increased 5-HT activity/metabolism in the hypothalamus, neostriatum, hippocampus, septum nuclei, and midbrain raphe nuclei. Some of these effects were only seen in males, indicating a sexually dimorphic response (Molina-Holgado et al. 1997).

3.4
Opioid Peptides

Perinatal exposure to cannabinoids can also alter the development of opioidergic neurons. It has been reported that the administration of cannabinoids increases proenkephalin-mRNA levels in rat fetuses at GD16 and GD18 in motor (caudate-putamen and cerebellum), limbic (septum nuclei), and diencephalic (thalamus and hypothalamus) structures (Pérez-Rosado et al. 2000). In this study, the variations observed after GD21 showed a marked sexual dimorphism with increases in female fetuses and reductions in male fetuses in most of the brain regions analyzed (Pérez-Rosado et al. 2000). Similar results were obtained for two other peptide precursors, proopiomelanocortin (POMC) and prodynorphin, whose levels of mRNA transcripts also increased in female fetuses but were reduced in male fetuses in several brain regions (Pérez-Rosado et al. 2002).

These alterations in the development of opioidergic neurotransmission are likely to produce important long-lasting functional changes in these neurons in the adult brain (Kumar et al. 1990; Vela et al. 1998; Corchero et al. 1998). Indeed, it has been found that adult animals that had been perinatally exposed to cannabinoids exhibit alterations in neuroendocrine control (Kumar et al. 1990), pain sensitivity (Vela et al. 1995), and reward processes (Vela et al. 1998; González et al. 2003). Kumar et al. (1990) reported an increase in both Met-enkephalin and β-endorphin immunoreactivity in the hypothalamus of adult rats perinatally exposed to cannabinoids. This effect could be related to changes in the synthesis and/or release of several anterior pituitary hormones (Dalterio 1986; Murphy et al. 1990, 1995).

Vela et al. (1995) observed that adult animals perinatally exposed to cannabinoids exhibited higher tail-flick latencies at immature ages and were tolerant to the analgesic effect of morphine when they became adults. These effects are sexually dimorphic since both were evident in males but not in females. Animals that had been perinatally exposed to Δ^9-THC also exhibited signs of opiate vulnerability (Vela et al. 1995, 1998; Corchero et al. 1998; Rubio et al. 1998), although the response was again sexually dimorphic. Thus, some somatic signs of withdrawal were evident in rats perinatally exposed to Δ^9-THC after the administration of a single dose of naloxone at day 24 after birth, which coincided with the end of cannabinoid exposure (Vela et al. 1995). These withdrawal signs only appeared in males (Vela et al. 1995). In contrast, adult females, but not males, self-administered morphine more readily if they had been perinatally exposed to Δ^9-THC in a fixed-ratio schedule (Vela et al. 1998). A possible explanation may be that, after treatment, females, but not males exhibited higher μ-opioid receptor binding density in some areas directly or indirectly related to drug reinforcement (Vela et al. 1998), and that in some of these regions the expression of proenkephalin gene decreased in females, but not in males (Corchero et al. 1998). In addition, following perinatal exposure to Δ^9-THC, adult females, but not males, exhibited higher plasma levels of stress hormones, an effect that has also been reported to be an indicator of opiate vulnerability (Rubio et al. 1995). The results obtained using a fixed-ratio schedule (Vela et al. 1998) contrast with those using a progressive ratio schedule, which suggest that perinatal Δ^9-THC exposure does not affect the reinforcing efficacy of morphine and food (González et al. 2003). According to these two schedules we may conclude that when rats are forced to work harder for reward, the response to morphine and food is not influenced by Δ^9-THC exposure.

3.5 Endogenous Cannabinoid System

When adult male, but not female, rats perinatally exposed to Δ^9-THC were subjected to a challenge with this cannabinoid, they showed marked tolerance in the open-field response and in the neurochemical events underlying motor activity (Fernández-Ruiz et al. 1994). However, there were no corresponding changes in CB_1 receptor binding or mRNA expression in the basal ganglia or in other regions that contain large populations of these receptors, such as the hippocampus, anterior limbic structures, and cerebellum (García-Gil et al. 1999). Only in the arcuate nucleus was a change observed in cannabinoid receptor binding levels (an increase of 42.2% in Δ^9-THC-exposed males). This nucleus is only sparsely populated with cannabinoid receptors (Herkenham et al. 1991). Even so, CB_1 receptors seem to play an important role in the modulation of neuroendocrine function. Thus, the administration of cannabinoids has marked effects on this regulation, located in the arcuate nucleus and also in other hypothalamic structures (Fernández-Ruiz et al. 1997).

4 Cannabinoids and Gene Expression of Neural Adhesion Molecules During Brain Development

The ability of plant-derived cannabinoids to affect development does not depend only on their neuromodulatory activity. Recently it has been observed that exposure to Δ^9-THC during critical periods of brain development is associated with an increase in L1 gene expression (Gómez et al. 2003). The protein L1 is a member of the immunoglobulin superfamily and, together with other proteins that can mediate cell–cell and cell–matrix interactions, is involved in various developmental events (Panicker et al. 2003). It plays an important role in various processes that take place during brain development, including cell proliferation and migration, neuritic elongation and guidance, synaptogenesis, and myelogenesis (Burden-Galley et al. 1997).

Exposure to Δ^9-THC increased L1 gene expression in most of the white matter regions analyzed, in particular, transverse commissural tracts, such as the fimbria, the stria terminalis, the stria medullaris, and the corpus callosum (Gómez et al. 2003). It has been reported that CB_1 receptors are present in all these tracts at this fetal age (Berrendero et al. 1998). Δ^9-THC-induced increases in L1-mRNA levels reached statistical significance in males but not in females, yet another example of sexual dimorphism (Gómez et al. 2003). As already discussed, it is likely that this sexual dimorphism stems from male–female differences in the hormonal environment, in particular when the brain is undergoing sexual differentiation, and not from male–female differences in the distribution and/or density of cannabinoid CB_1 receptors, for which there is no evidence at least in the perinatal period (Fernández-Ruiz et al. 2000).

References

Araki W, Wurtman RJ (1997) Control of membrane phosphatidylcholine biosynthesis by diacylglycerol levels in neuronal cells undergoing neurite outgrowth. Proc Natl Acad Sci USA 94:11946–11950

Asch RH, Smith CG (1986) Effects of Δ^9-THC, the principal psychoactive component of marihuana, during pregnancy in the rhesus monkey. J Reprod Med 31:1071–1081

Berrendero F, García-Gil L, Hernández ML, Romero J, Cebeira R, de Miguel R, Ramos JA, Fernández-Ruiz JJ (1998) Localization of mRNA expression and activation of signal transduction mechanisms for cannabinoid receptors in rat brain during fetal development. Development 125:3179–3188

Berrendero F, Sepe N, Ramos JA, Di Marzo V, Fernández-Ruiz JJ (1999) Analysis of cannabinoid receptor binding and mRNA expression and endogenous cannabinoid contents in the developing rat brain during late gestation and early postnatal period. Synapse 33:181–191

Biegon A, Kerman IA (2001) Autoradiographic study of pre- and postnatal distribution of cannabinoids receptors in human brain. Neuroimage 14:1463–1468

Bonnin A, de Miguel R, Castro JG, Ramos JA, Fernández-Ruiz JJ (1996) Effects of perinatal exposure to Δ^9-tetrahydrocannabinol on the fetal and early postnatal development of tyrosine hydroxylase containing neurons in rat brain. J Mol Neurosci 7:291–308

Burden-Galley SM, Pendergast M, Lemmon V (1997) The role of cell adhesion molecule L1 in axonal extension, growth cone motility, and signal transduction. Cell Tissue Res 290:415–422

Corchero J, Garcia-Gil L, Manzanares J, Fernández-Ruiz JJ, Fuentes JA, Ramos JA (1998) Perinatal Δ^9-tetrahydrocannabinol exposure reduces proenkephalin gene expression in the caudate-putamen of adult female rats. Life Sci 63:843–850

Dalterio SL (1980) Perinatal or adult exposure to cannabinoids alter male reproductive functions in mice. Pharmacol Biochem Behav 12:143–153

Dalterio SL (1986) Cannabinoid exposure: effects on development. Neurobehav Toxicol Teratol 8:345–352

Dalterio SL, Barker A (1979) Perinatal exposure to cannabinoids alter male reproductive functions in mice. Science 205:1420–1422

Del Olmo E, Pazos A (2001) Aminergic receptors during the development of the human brain: the contribution of in vitro imaging techniques. J Chem Neuroanat 22:101–114

Di Marzo V (1998) Endocannabinoids and others fatty acid derivatives with cannabimimetics properties: biochemistry and possible physiopathological relevance. Biochim Biophys Acta 1392:153–175

Fernández-Ruiz JJ, Rodriguez de Fonseca F, Navarro M, Ramos JA (1992) Maternal cannabinoid exposure and brain development: changes in the ontogeny of dopaminergic neurons. In: Bartke A, Murphy LL (eds) Neurobiology and neurophysiology of cannabinoids, biochemistry and physiology of substance abuse, vol IV. CRC Press, Boca Raton, pp 119–164

Fernández-Ruiz JJ, Bonnin A, Cebeira M, Ramos JA (1994) Ontogenic and adult changes in the activity of hypothalamic and extrahypothalamic dopaminergic neurons after perinatal cannabinoid exposure. In: Palomo T, Archer T (eds) Strategies for studying brain disorders, vol I. Farrand Press, London, pp 357–390

Fernández-Ruiz JJ, Romero J, García-Gil L, Garcia-Palomero E, Ramos JA (1996) Dopaminergic neurons as neurochemical substrates of neurobehavioral effects of marihuana: developmental and adult studies. In: Beninger RJ, Archer T, Palomo T (eds) Dopamine disease states. CYM Press, Madrid, pp 359–387

Fernández-Ruiz JJ, Muñoz RM, Romero J, Villanúa MA, Makriyannis A, Ramos JA (1997) Time-course of the effects of different cannabimimetics on prolactin and gonadotropin secretion: evidence for the presence of CB1 receptors in hypothalamic structures and their involvement in the effects of cannabimimetics. Biochem Pharmacol 53:1919–1927

Fernández-Ruiz JJ, Bonnin A, de Miguel R, Castro JG, Ramos JA (1998) Perinatal exposure to marihuana or its main psychoactive constituent, Δ^9-tetrahydrocannabinol, affects the development of brain dopaminergic neurons. Arq Med 12:67–77

Fernández-Ruiz JJ, Berrendero F, Hernandez ML, Romero J, Ramos JA (1999) Role of endocannabinoids in brain development. Life Sci 65:725–736

Fernández-Ruiz JJ, Berrendero F, Hernandez ML, Ramos JA (2000) The endogenous cannabinoid system and brain development. Trends Neurosci 23:14–20

Fride E, Mechoulam R (1996a) Developmental aspects of anandamide: ontogeny of response and prenatal exposure. Psychoneuroendocrinology 21:157–172

Fride E, Mechoulam R (1996b) Ontogenic development of the response to anandamide and Δ^9-tetrahydrocannabinol in mice. Brain Res Dev Brain Res 95:131–134

Fride E, Ginzburg Y, Breuer A, Bisogno T, Di Marzo V, Mechoulam R (2001) Critical role of the endogenous cannabinoid system in mouse pup suckling and growth. Eur J Pharmacol 419:207–214

Fried PA, Smith AM (2001) A literature review of the consequences of prenatal marihuana exposure. An emerging theme of a deficiency in aspects of executive function. Neurotoxicol Teratol 23:1–11

García-Gil L, de Miguel R, Ramos JA, Fernández-Ruiz JJ (1996) Perinatal Δ^9-tetrahydrocannabinol exposure in rats modifies the responsiveness of midbrain dopaminergic neurons in adulthood to a variety of challenges with dopaminergic drugs. Drug Alcohol Depend 42:155–166

García-Gil L, Ramos JA, Rubino T, Parolaro D, Fernández-Ruiz JJ (1998) Perinatal Δ^9-tetrahydrocannabinol exposure did not alter dopamine transporter and tyrosine hydroxylase mRNA levels in midbrain dopaminergic neurons of adult male and female rats. Neurotoxicol Teratol 20:549–553

García-Gil L, Romero J, Ramos JA, Fernández-Ruiz JJ (1999a) Cannabinoid receptor binding and mRNA levels in several brain regions of adult male and female rats perinatally exposed to Δ^9-tetrahydrocannabinol. Drug Alcohol Depend 55:127–136

García-Gil L, de Miguel R, Romero J, Perez A, Ramos JA, Fernández-Ruiz JJ (1999b) Perinatal Δ^9-tetrahydrocannabinol exposure augmented the magnitude of motor inhibition caused by GABA-B but not GABA-A, receptor agonists in adult rats. Neurotoxicol Teratol 21:277–283

Glass M, Dragunow M, Faull RLM (1997) Cannabinoid receptors in the human brain: a detailed anatomical and quantitative autoradiographic study in the fetal, neonatal and adult human brain. Neuroscience 77:299–318

Goldschmidt L, Day NL, Richardson GA (2000) Effects of prenatal marihuana exposure on child behavior problems at age 10. Neurobehav Toxicol 22:325–336

Gómez M, Hernández ML, Johansson B, de Miguel R, Ramos JA, Fernández-Ruiz J (2003) Prenatal cannabinoid exposure and gene expression for neural adhesion molecule L1 in the fetal rat brain. Brain Res Dev Brain Res 147:201–207

González B, de Miguel R, Martin S, Perez-Rosado A, Romero J, García-Lecumberri C, Fernández-Ruiz J, Ramos JA, Ambrosio E (2003) Effects of perinatal exposure to Δ^9-tetrahydrocannabinol on operant morphine-reinforced behavior. Pharmacol Biochem Behav 75:577–584

Harbison RD, Mantilla-Plata B (1972) Prenatal toxicity, maternal distribution and placental transfer of tetrahydrocannabinol. J Pharmacol Exp Ther 180:446–453

Herkenham M, Lynn AB, Johnson MR, Melvin LS, de Costa BR, Rice KC (1991) Characterization and localization of cannabinoid receptors in rat brain: a quantitative in vitro autoradiographic study. J Neurosci 11:563–583

Hernández ML, García-Gil L, Berrendero F, Ramos JA, Fernández-Ruiz JJ (1997) Δ^9-tetrahydrocannabinol increases activity of tyrosine hydroxylase in cultured fetal mesencephalic neurons. J Mol Neurosci 8:83–91

Hernández ML, Berrendero F, Suarez I, Garcia-Gil L, Cebeira M, Mackie K, Ramos JA, Fernández-Ruiz JJ (2000) Cannabinoids CB1 receptors colocalize with tyrosine hydroxylase in cultured fetal mesencephalic neurons and their activation increases the levels of this enzyme. Brain Res 857:56–65

Insel TR (1995) The development of brain and behavior. In: Bloom FE, Kupfer DJ (eds) Psychopharmacology: the four generation of progress. Raven Press, New York, pp 683–694

Kumar AM, Haney M, Becker T, Thompson ML, Kream RM, Miczek K (1990) Effect of early exposure to Δ^9-tetrahydrocannabinol on the levels of opioid peptides, gonadotropin-releasing hormone and substance P in the adult male rat brain. Brain Res 525:78–83

Mailleux P, Vanderhaeghen JJ (1992a) Location of cannabinoid receptor in the human developing and adult basal ganglia. Higher levels in the striatonigral neurons. Neurosci Lett 148:173–176

Mailleux P, Vanderhaeghen JJ (1992b) Distribution of neuronal cannabinoid receptor in the adult rat brain: a comparative receptor binding radioautography and in situ hybridization histochemistry. Neuroscience 48:655–668

Mato S, Del Olmo E, Pazos A (2003) Ontogenetic development of cannabinoid receptor expression and signal transduction functionality in the human brain. Eur J Neurosci 17:1747–1754

Mereu G, Fá M, Ferraro L, Cagiano R, Antonelli T, Tattoli M, Ghiglieri V, Tanganelli S, Gessa GL, Cuomo V (2003) Prenatal exposure to a cannabinoid agonist produces memory deficits linked to dysfunction in hippocampal long-term potentiation and glutamate release. Proc Natl Acad Sci USA 100:4915–4920

Mokler DA, Robinson SE, Johnson JH, Hong JS, Rosecrana JA (1987) Neonatal administration of Δ^9-tetrahydrocannabinol alters the neurochemical response to stress in the adult Fischer-344 rat. Neurotoxicol Teratol 9:321–326

Molina-Holgado F, Amaro A, Gonzalez I, Alvarez FJ, Leret ML (1996) Effects of maternal Δ^9-tetrahydrocannabinol on developing serotoninergic neurons. Eur J Pharmacol 316:39–42

Molina-Holgado F, Alvarez FJ, González I, Antonio MT, Leret ML (1997) Maternal exposure to Δ^9-tetrahydrocannabinol (Δ^9-THC) alters indolamine levels and turnover in adult male and female rat brain regions. Brain Res Bull 43:173–178

Murphy LL, Steger RW, Bartke A (1990) Psychoactive and non-psychoactive cannabinoids and their effects on reproductive neuroendocrine parameters. In: Watson RR (ed) Biochemistry and physiology of substance abuse, vol 2. CRC Press, Boca Raton, pp 73–94

Murphy LL, Gher J, Szary A (1995) Effects of prenatal exposure to Δ^9-tetrahydrocannabinol on reproductive, endocrine and immune parameters of the male and the female rat offspring. Endocrine 3:875–881

Navarro M, Rodríguez de Fonseca F, Hernández ML, Ramos JA, Fernández-Ruiz JJ (1994) Motor behavior and nigrostriatal dopaminergic activity in adult rats perinatally exposed to cannabinoids. Pharmacol Biochem Behav 47:47–58

Navarro M, de Miguel R, Rodríguez F, Ramos JA, Fernández JJ (1996) Perinatal cannabinoid exposure modifies the sociosexual approach behavior and the mesolimbic dopaminergic activity of adult male rats. Brain Res Dev Brain Res 75:91–98

Pacheco MA, Ward SJ, Childers SR (1993) Identification of cannabinoids receptors in cultures of rat cerebellar granule cells. Brain Res 603:102–110

Panicker AK, Bulusi M, Theleu K, Maness PF (2003) Cellular signalling mechanisms of neural cell adhesion molecules. Front Biosci 8:900–911

Pares-Herbute N, Tapia-Arancibia L, Artier H (1989) Ontogeny of the metencephalic, mesencephalic and diencephalic content of catecholamines as measured by high performance liquid chromatography with electrochemical detection. Int J Dev Neurosci 7:73–79

Park B, Gibbons HM, Mitchell MD, Glass M (2003) Identification of the CB1 cannabinoid receptor and fatty acid amide hydrolase (FAAH) in the human placenta. Placenta 24:990–995

Park B, McPartland JM, Glass M (2004) Cannabis, cannabinoids and reproduction. Prostaglandins Leukot Essent Fatty Acids 70:189–197

Pérez-Rosado A, Manzanares J, Fernández-Ruiz J, Ramos JA (2000) Prenatal Δ^9-tetrahydrocannabinol exposure modifies proenkephalin gene expression in the fetal rat brain: sex-dependent differences. Brain Res Dev Brain Res 120:77–81

Pérez-Rosado A, Gomez M, Manzanares J, Ramos JA, Fernández-Ruiz J (2002) Changes in prodynorphine and POMC gene expression in several brain regions of rat fetuses prenatally exposed to Δ^9-tetrahydrocannabinol. Neurotox Res 4:211–218

Ramos JA, de Miguel R, Cebeira M, Hernández ML, Fernández-Ruiz J (2002) Exposure to cannabinoids in the development of endogenous cannabinoid system. Neurotox Res 4:363–372

Rodriguez de Fonseca F, Cebeira M, Fernández-Ruiz JJ, Navarro M, Ramos JA (1991) Effects of pre- and perinatal exposure to hashish extracts on the ontogeny of brain dopaminergic neurons. Neuroscience 43:713–723

Romero J, García-Palomero E, Fernández-Ruiz JJ, Ramos JA (1996) Involvement of GABA-B receptors in the motor inhibition produced by agonists of brain cannabinoid receptors. Behav Pharmacol 7:299–302

Romero J, García-Palomero E, Berrendero F, García-Gil L, Hernández ML, Ramos JA, Fernández-Ruiz JJ (1997) Atypical location of cannabinoid receptors in white matter areas during rat brain development. Synapse 26:317–323

Rubio P, Rodríguez de Fonseca F, Muñoz RM, Ariznavarreta C, Martin JL, Navarro M (1995) Long-term behavioral effects of perinatal exposure to Δ^9-tetrahydrocannabinol in rats: possible role of pituitary-adrenal axis. Life Sci 56:2169–2176

Rubio P, Rodríguez de Fonseca F, Martin JL, del Arco I, Bartolomé S, Villanúa MA, Navarro M (1998) Maternal exposure to low doses of Δ^9-tetrahydrocannabinol facilitates morphine-induced place conditioning in adult male offspring. Pharmacol Biochem Behav 61:229–238

Rueda D, Navarro B, Martínez-Serrano A, Guzman M, Galve-Ropert I (2002) The endocannabinoid anandamide inhibits neural progenitor cell differentiation through attenuation of the Rap1/B-Raf/ERK pathway. J Biol Chem 277:46645–46650

Sillence DJ, Allan D (1998) Utilization of phosphatidylcholine and production of diacylglycerol as a consequence of sphingomyelin synthesis. Biochem J 331:251–256

Vela G, Fuentes JA, Bonnin A, Fernández-Ruiz JJ, Ruiz-Gayo M (1995) Perinatal exposure to Δ^9-tetrahydrocannabinol (Δ^9-THC) leads to changes in opioid-related behavioral patterns in rats. Brain Res 680:142–147

Vela G, Martin S, Garcia-Gil L, Crespo JA, Ruiz-Gayo M, Fernández-Ruiz JJ, Garcia Lecumberry C, Pelaprat D, Fuentes JA, Ramos JA, Ambrosio E (1998) Maternal exposure to Δ^9-tetrahydrocannabinol facilitates morphine self-administration behavior and changes regional binding to central μ-opioid receptors in adult offspring female rats. Brain Res 807:101–109

Walters DE, Carr LA (1988) Perinatal exposure to cannabinoids alters neurochemical development in the rat brain. Pharmacol Biochem Behav 29:213–216

Yang ZM, Paria BC, Dey SK (1996) Activation of brain-type cannabinoid receptors interferes with preimplantation mouse embryo development. Biol Reprod 55:756–761

Zhou D, Song ZH (2001) CB1 cannabinoid receptor-mediated neurite remodeling in mouse neuroblastoma N1E-115 cells. J Neurosci Res 65:346–353

HEP (2005) 168:657–690

Pharmacokinetics and Metabolism of the Plant Cannabinoids, Δ^9-Tetrahydrocannabinol, Cannabidiol and Cannabinol

M.A. Huestis

Chemistry and Drug Metabolism, Intramural Research Program,
5500 Nathan Shock Drive, Baltimore MD, 21224, USA
mhuestis@intra.nida.nih.gov

Abstract Increasing interest in the biology, chemistry, pharmacology, and toxicology of cannabinoids and in the development of cannabinoid medications necessitates an understanding of cannabinoid pharmacokinetics and disposition into biological fluids and tissues. A drug's pharmacokinetics determines the onset, magnitude, and duration of its pharmacodynamic effects. This review of cannabinoid pharmacokinetics encompasses absorption following diverse routes of administration and from different drug formulations, distribution of analytes throughout the body, metabolism by different tissues and organs, elimination from the body in the feces, urine, sweat, oral fluid, and hair, and how these processes change over time. Cannabinoid pharmacokinetic research has been especially challenging due to low analyte concentrations, rapid and extensive metabolism, and physicochemical characteristics that hinder the separation of drugs of interest from biological matrices—and from each other—and lower drug recovery due to adsorption of compounds of interest to multiple surfaces. Δ^9-Tetrahydrocannabinol, the primary psychoactive component of *Cannabis sativa*, and its metabolites 11-hydroxy-Δ^9-tetrahydrocannabinol and 11-nor-9-carboxy-tetrahydrocannabinol are the focus of this chapter, although cannabidiol and cannabinol, two other cannabinoids with an interesting array of activities, will also be reviewed. Additional material will be presented on the interpretation of cannabinoid concentrations in human biological tissues and fluids following controlled drug administration.

Keywords Cannabinoids · Pharmacokinetics · Tetrahydrocannabinol · Cannabidiol · Absorption · Distribution · Metabolism · Excretion · Interpretation · Oral fluid · Sweat · Hair · Plasma · Urine · Alternate matrix · Marijuana

1 Introduction

Currently, there is a growing interest in the biology, chemistry, pharmacology, and toxicology of cannabinoids and in the development of potential cannabinoid medications. It is clear that the endogenous cannabinoid system plays a critical role in physiological and behavioral processes. Endogenous cannabinoid neurotransmitters, receptors, and transporters, synthetic cannabinoid agonists and antagonists, and cannabis-based extracts are the subject of extensive research. It is hoped that these agents might provide novel approaches to treat human diseases and disorders. The therapeutic usefulness of oral cannabinoids is being investigated for medicinal applications, including analgesia, treatment of acquired immunodeficiency syndrome (AIDS)-wasting disease, counteracting spasticity of motor diseases, and the prevention of emesis following chemotherapy, among others. Cannabis, also, is one of the oldest and most commonly abused drugs in the world, and its use may have consequences in terms of pathological and behavioral toxicity. For these reasons, it is important to understand cannabinoid pharmacokinetics and the disposition of cannabinoids into biological fluids and tissues. Understanding a drug's pharmacokinetics is essential to understanding the onset, magnitude, and duration of its pharmacodynamic effects.

Pharmacokinetics encompasses the absorption of cannabinoids following diverse routes of administration and from different drug formulations, the distribution of analytes throughout the body, the metabolism of cannabinoids by different tissues and organs, the elimination of cannabinoids from the body in the feces, urine, sweat, oral fluid, and hair, and how these processes change over time. In this chapter, we will review the many contributions to our understanding of cannabinoid pharmacokinetics from the 1970s and 1980s and the more recent research that expands upon this knowledge. Cannabinoid pharmacokinetic research has been especially challenging due to low analyte concentrations, rapid and extensive metabolism, and physicochemical characteristics that (1) hinder the separation of drugs of interest from biological matrices and from each other and (2) lower drug recovery due to adsorption of compounds of interest to multiple surfaces. Much of the earlier data utilized radio-labeled cannabinoids yielding highly sensitive but less specific measurement of individual cannabinoid analytes. Mass spectrometric developments now permit highly sensitive and specific measurement of cannabinoids in a wide variety of biological matrices.

Cannabis sativa contains over 421 different chemical compounds, including over 60 cannabinoids (Claussen and Korte 1968; ElSohly et al. 1984; Turner et al. 1980). Cannabinoid plant chemistry is far more complex than pure Δ^9-tetrahydrocannabinol (THC), and different effects may be expected due to the presence of additional cannabinoids and other chemicals. In all, 18 different classes of chemicals, including nitrogenous compounds, amino acids, hydrocarbons, sugars, terpenes, and simple and fatty acids, contribute to cannabis' known pharmacological and toxicological properties. THC is usually present in cannabis plant material as a mixture of monocarboxylic acids that readily and efficiently decarboxylate upon heating. THC decomposes when exposed to air, heat, or light; exposure to acid can oxidize the compound to cannabinol, a much less potent cannabinoid. In addition, cannabis plants dried in the sun release variable amounts of THC through decarboxylation. During smoking, more than 2,000 compounds may be produced by pyrolysis. The focus of this chapter will be THC, the primary psychoactive component of cannabis, its metabolites, 11-hydroxy-tetrahydrocannabinol (11-OH-THC) and 11-nor-9-carboxy-tetrahydrocannabinol (THCCOOH), and two other cannabinoids present in high concentrations, cannabidiol (CBD), a non-psychoactive agent with an interesting array of other activities, and cannabinol, which is approximately 10% as psychoactive as THC (Perez-Reyes et al. 1982). Mechoulam et al. elucidated the structure of THC after years of effort in 1964, opening the way for studies of the drug's pharmacokinetics (Mechoulam 1970). THC, containing no nitrogen but with two chiral centers in the *trans*-configuration, is described by two different numbering systems, the dibenzopyran or Δ^9, and the monoterpene or Δ^1 system; the dibenzopyran system is used throughout this chapter.

2 Pharmacokinetics of THC

2.1 Absorption

2.1.1 Smoked Administration

Route of drug administration and drug formulation determine the rate of drug absorption. Smoking, the principal route of cannabis administration, provides a rapid and efficient method of drug delivery from the lungs to the brain, contributing to its abuse potential. Intense pleasurable and strongly reinforcing effects may be produced due to almost immediate drug exposure to the central nervous system. Slightly lower peak THC concentrations are achieved after smoking compared to intravenous administration (Ohlsson et al. 1980). Bioavailability following the smoking route was reported as 2% to 56%, due in part to the intra- and inter-subject variability in smoking dynamics that contribute to uncertainty in dose delivery (Agurell et al. 1986; Agurell and Leander 1971; Ohlsson et al. 1982, 1985). The number, duration, and spacing of puffs, hold time and inhalation volume, or smoking topography, greatly influences the degree of drug exposure (Azorlosa et al. 1992; Heishman et al. 1989; Perez-Reyes 1990). Expectation of drug reward also may affect smoking dynamics. Cami et al. noted that subjects were able to change their method of smoking hashish cigarettes to obtain higher plasma concentrations of THC when they expected to receive active drug in comparison to placebo cigarettes (Cami et al. 1991).

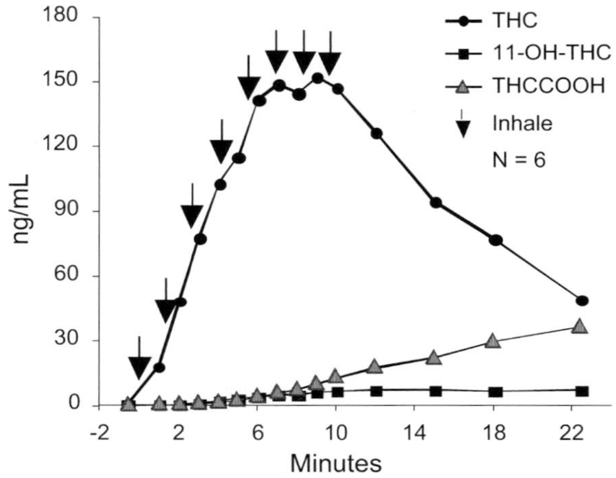

Fig. 1. Mean ($n=6$) plasma concentrations of Δ^9-tetrahydrocannabinol (*THC*), 11-hydroxy-Δ^9-tetrahydrocannabinol (*11-OH-THC*), and 11-nor-9-carboxy-Δ^9-tetrahydrocannabinol (*THCCOOH*) during smoking of a single 3.55% THC cigarette. *Each arrow* represents one inhalation or puff on the cannabis cigarette (M.A. Huestis, unpublished data)

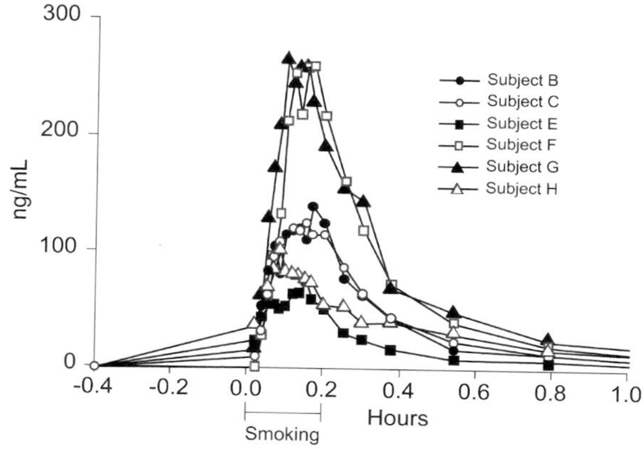

Fig. 2. Individual plasma Δ^9-tetrahydrocannabinol (THC) time course for six subjects following smoking of a single 3.55% THC cigarette. (Reproduced from the *Journal of Analytical Toxicology* by permission of Preston Publications, a division of Preston Industries; Huestis et al. 1992b, Fig. 1d therein)

A continuous blood withdrawal pump was utilized to capture the rapid absorption of THC and formation of 11-OH-THC and THCCOOH during cannabis smoking (Huestis et al. 1992b). The disposition of THC and its metabolites were followed after smoking a single placebo, 1.75%, or 3.55% THC cigarette over 7 days. Plasma concentrations were determined by gas-chromatography mass spectrometry (GC/MS). THC was detected in the plasma immediately after the first cigarette puff (Fig. 1) and was accompanied by the onset of cannabinoid effects (Huestis et al. 1992d). Mean±SD THC concentrations of 7.0±8.1 ng/ml and 18.1±12.0 ng/ml were observed following the first inhalation of a low-dose (1.75% THC, approximately 16 mg) or high-dose (3.55% THC, approximately 30 mg) cigarette, respectively (Huestis et al. 1992b). Concentrations increased rapidly, reaching mean peaks of 84.3 ng/ml (range 50–129) and 162.2 ng/ml (range 76–267) for the low- and high-dose cigarette, respectively. Peak concentrations occurred at 9.0 min, prior to initiation of the last puff sequence at 9.8 min. Despite a computer-paced smoking procedure that controlled the number of puffs, length of inhalation, hold time, and time between puffs, there were large inter-subject differences in plasma THC concentrations due to differences in the depth of inhalation as participants titrated their THC dose (Fig. 2). Mean THC concentrations were approximately 60% and 20% of peak concentrations 15 and 30 min post smoking, respectively. Within 2 h, plasma THC concentrations were at or below 5 ng/ml. The time of detection of THC (GC/MS LOQ = 0.5 ng/ml) varied from 3 to 12 h after the low-dose and from 6 to 27 h after the high-dose cannabis cigarette.

Similar mean THC C_{max} concentrations were reported in specimens collected immediately after cannabis smoking was completed. Mean peak THC concentrations after smoking a single 1.32%, 1.97%, or 2.54% THC cigarette were 94.3, 107.4, and 155.1 ng/ml, respectively (Perez-Reyes et al. 1982). Other reported peak THC concentrations ranged between 45.6 and 187.8 ng/ml following smoking of an approximately 1% THC cigarette (Perez-Reyes et al. 1981) and 33 to 118 ng/ml 3 min after ad lib smoking of an approximate 2% THC cigarette (Ohlsson et al. 1980). Many individuals prefer the smoked route, not only for its rapid drug delivery, but also because it allows them to titrate their dose.

2.1.2 Oral Administration

There are fewer studies on the disposition of THC and metabolites after oral as compared to the smoked route of cannabis administration. THC is readily absorbed due to its high octanol/water coefficient, estimated to be between 6,000 and over 9 million by different technologies (Harder and Rietbrock 1997). The advantages of cannabinoid smoking are offset by the harmful effects of cannabinoid smoke; hence, smoking is generally not recommended for therapeutic applications. Synthetic THC preparations such as dronabinol (Marinol) are usually taken orally but may also be administered rectally. In addition, abuse of cannabis by the oral route also is common. Absorption is slower when cannabinoids are ingested with lower, more delayed peak THC concentrations (Law et al. 1984; Ohlsson et al. 1981). Dose, route of administration, vehicle, and physiological factors such as absorption and rates of metabolism and excretion can influence drug concentrations in the circulation. Perez-Reyes et al. described the efficacy of five different vehicles used in the oral administration of THC in gelatin capsules (Perez-Reyes et al. 1973a). Glycocholate and sesame oil improved the bioavailability of oral THC; however, there was considerable variability in peak concentrations and rates of absorption, even when the drug was administered in the same vehicle. Oral THC bioavailability was reported to be 10% to 20% by Wall et al. (1983). In their study, participants were dosed with either 15 (women) or 20 mg (men) THC dissolved in sesame oil and contained in gelatin capsules. THC plasma concentrations peaked approximately 4 to 6 h after ingestion of 15 to 20 mg of THC in sesame oil. A percentage of the THC was radio-labeled; however, investigators were unable to differentiate labeled THC from its labeled metabolites. Thus, THC concentrations were overestimated.

Possibly a more accurate assessment of oral bioavailability that utilized GC/MS to quantify THC in plasma samples was reported by Ohlsson et al. (1980). Peak THC concentrations ranged from 4.4 to 11 ng/ml and occurred 1 to 5 h following ingestion of 20 mg of THC in a chocolate cookie. Oral bioavailability was estimated to be 6%. Slow rates of absorption and low THC concentrations occur after oral administration of THC or cannabis. Several factors may account for the low oral bioavailability of 4% to 20% (as compared to intravenous drug administration) including variable absorption, degradation of drug in the stomach and significant first-pass metabolism to active 11-OH-THC and inactive metabolites in the liver.

Recently, there has been renewed interest in oral THC pharmacokinetics due to the therapeutic value of orally administered THC. In a study of THC, 11OH-THC, and THCCOOH concentrations in 17 volunteers after a single 10 mg Marinol capsule, mean peak plasma THC concentrations of 3.8 ng/ml (range 1.1–12.7), 11-

OH-THC 3.4 ng/ml (range 1.2–5.6), and THCCOOH 26 ng/ml (range 14–46) were found 1 to 2 h after ingestion (Kim and Yoon 1996). Similar THC and 11-OH-THC concentrations were observed with consistently higher THCCOOH concentrations. Interestingly, two THC peaks were frequently observed due to enterohepatic circulation. The onset, magnitude, and duration of pharmacodynamic effects generally occur later, are lower in magnitude, and have a delayed return to baseline when THC is administered by the oral as compared to the smoked route (Binitie 1975; Meier and Vonesch 1997).

In addition, THC-containing foods, i.e., hemp oil, beer, and other products, are commercially available for oral consumption. Hemp oil is produced from cannabis seed and is an excellent source of essential amino acids and omega-linoleic and linolenic fatty acids. THC content is dependent upon the effectiveness of cannabis seed cleaning and oil filtration processes. Hemp oil of greater than 300 μg THC/g was available in the U.S. and up to 1,500 μg THC/g in Europe. Currently, hemp oil THC concentrations in the U.S. are low, reflecting the efforts of manufacturers to reduce the amount of THC in hemp oil products.

In a recent controlled cannabinoid administration study of THC-containing hemp oils and dronabinol, the pharmacokinetics and pharmacodynamics of oral THC were evaluated. Up to 14.8 mg of THC was ingested by six volunteers each day in three divided doses with meals for five consecutive days (Nebro et al. 2004). There was a 10-day washout phase between each of the five dosing sessions. THC was quantified in plasma by solid-phase extraction followed by positive chemical ionization GC/MS. THC and 11-OH-THC were rarely detected in plasma following the two lowest doses of 0.39 and 0.47 mg/day THC, while peak plasma

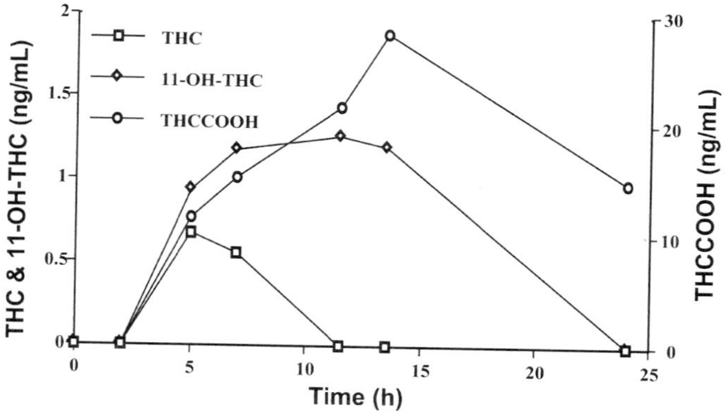

Fig. 3. Plasma Δ^9-tetrahydrocannabinol (*THC*), 11-hydroxy-Δ^9-tetrahydrocannabinol (*11-OH-THC*) and 11-nor-9-carboxy-Δ^9-tetrahydrocannabinol (*THCCOOH*) concentrations in one participant over 24 h following administration of two 2.5-mg dronabinol (synthetic THC) doses. Time zero is the time of the first blood draw at 0730 hours. The 2.5-mg doses were administered with food at 1200 and 1800 (4.5 and 10.5 h after time zero). (Reprinted from Journal of Chromatography B, 789, Gustafson et al., Validated method for the simultaneous determination of delta-9-tetrahydrocannabinol (THC), 11-hydroxy-THC and 11-nor-9-carboxy-THC in human plasma using solid phase extraction and gas chromotography-mass spectrometry with positive chemical ionization, pp. 145, Fig. 2 therein, Copyright (2003) with permission from Elsevier)

concentrations of less than 6.5 ng/ml THC, less than 5.6 ng/ml 11-OH-THC, and less than 43.0 ng/ml THCCOOH were found after the two highest THC doses of 7.5 and 14.8 mg/day (Fig. 3). Interestingly, THCCOOH concentrations after the 7.5 mg/day dronabinol dose were greater than or equal to those of the high potency 14.8 mg/day hemp oil dose. Two possible reasons for the higher bioavailability of THC in dronabinol are greater protection from degradation in the acidic environment of the stomach due to encapsulation and improved absorption of THC from the sesame oil formulation. Plasma THC and 11-OH-THC concentrations fell below the method's limits of quantification of 0.5 ng/ml at 25 h, while THCCOOH was still measurable for more than 50 h after the last dose of the higher concentration hemp oils.

2.1.3
Rectal Administration

Several different suppository formulations were evaluated in monkeys to determine the formulation that maximized bioavailability and reduced first-pass metabolism of THC by the liver (Mattes et al. 1993, 1994); THC-hemisuccinate provided the highest bioavailability of 13.5%. Brenneisen et al. evaluated plasma THC concentrations in two patients who were prescribed THC hemisuccinate suppositories or Marinol for spasticity (Brenneisen et al. 1996). THC did not accumulate in the blood following 10 to 15 mg daily doses. THC concentrations peaked within 1 to 8 h after oral administration and ranged between 2.1 and 16.9 ng/ml. Rectal administration of 2.5 to 5 mg THC produced maximum plasma concentrations of 1.1 to 4.1 ng/ml within 2 to 8 h. The bioavailability of the rectal route was approximately twice that of the oral route due to higher absorption and lower first-pass metabolism.

2.1.4
Sublingual and Dermal Administration

Due to the chemical complexity of cannabis plant material as compared to synthetic THC, cannabis extracts are being explored as therapeutic medications. One reproducible extract of the *Cannabis sativa* plant contains approximately equal amounts of THC and CBD (see Pharmacokinetics of Cannabidiol, Sect. 3). The efficacy of cannabis extracts has been evaluated in clinical trials for analgesia (Holdcroft 1984; Vaughan and Christie 1984), spasticity, and other indications in affected patients (Zajicek et al. 2003). Cannabis extracts can be administered sublingually to avoid first-pass metabolism by the liver.

Another route of drug exposure that avoids first-pass metabolism is topical administration. Although still in the early stages of research, dermal administration of THC also is being explored as a means of improving bioavailability of THC (Stinchcomb et al. 2004).

2.2 Distribution

THC concentrations decrease rapidly after the end of smoking due to its rapid distribution into tissues and metabolism in the liver. THC is highly lipophilic and initially taken up by tissues that are highly perfused, such as the lung, heart, brain, and liver. In animals after i.v. administration of labeled THC, higher levels of radioactivity are present in the lung than in other tissues (Lemberger et al. 1970). Adams and Martin determined that a THC dose of 2 to 22 mg is necessary to produce pharmacological effects in humans (Adams and Martin 1996). Assuming that 10% to 25% of the available THC enters the circulation during smoking, the actual dose required was estimated as 0.2 to 4.4 mg. Furthermore, only about 1% of the dose at peak concentration was found in the brain, indicating that only 2 to 44 μg of THC penetrated the brain. Chiang et al. estimated that equilibration was reached between plasma and tissue THC approximately 6 h after an intravenous THC dose (Chiang and Rapaka 1987).

Metabolism of THC to 11-OH-THC, THCCOOH, and other analytes also contributes to the reduction of THC in the blood. Perez-Reyes et al. compared the pharmacokinetics and pharmacodynamics of tritiated THC and 11-OH-THC in 20 male volunteers (Perez-Reyes et al. 1972). Although equal doses produced equal psychoactive effects, drug effects were perceived more rapidly after 11-OH-THC than after THC. In addition, 11-OH-THC left the intravascular compartment faster than THC. These data suggest that 11-OH-THC diffuses into the brain more readily than THC. Another possible explanation is lower protein binding of 11-OH-THC, as compared to THC, in the blood. Further support for the faster penetration of brain by 11-OH-THC is found in studies documenting a more rapid diffusion of 11-OH-THC than THC into the brains of mice (Perez-Reyes et al. 1972).

THC's volume of distribution (V_d) is large, approximately 10 l/kg, despite the fact that it is 95% to 99% protein bound in plasma, primarily to lipoproteins (Hunt and Jones 1980; Kelly and Jones 1992). More recently, with the benefit of advanced analytical techniques, THC's steady state V_d was found to be 3.4 l/kg (Grotenhermen 2003). Less highly perfused tissues, including fat, accumulate drug more slowly as THC redistributes from the vascular compartment (Harvey 2001). With prolonged drug exposure, THC concentrates in fat and may be retained for extended periods of time (Johansson et al. 1989b; Kreuz and Axelrod 1973). It is suggested that fatty acid conjugates of THC and 11-OH-THC may be formed, increasing the stability of these compounds in fat (Grotenhermen 2003).

Distribution of THC into peripheral organs and brains was found to be similar in THC tolerant and non-tolerant dogs (Dewey et al. 1972). In addition, Dewey et al. found that tolerance to the behavioral effects of THC in pigeons was not due to decreased uptake of cannabinoids into brain (Dewey et al. 1972). Tolerance also was evaluated in humans by Hunt and Jones (1980). Tolerance in humans developed during oral administration of 30 mg of THC every 4 h for 10 to 12 days. Few pharmacokinetic changes were noted during chronic administration, although average total metabolic clearance and initial apparent volume of distribution increased from 605 to 977 ml/min and from 2.6 to 6.4 l/kg, respectively. The pharmacokinetic changes observed after chronic oral THC could not account for the observed behavioral and physiologic tolerance, suggesting rather that tolerance was due to pharmacodynamic adaptation.

THC rapidly crosses the placenta, although concentrations were lower in canine and ovine fetal blood and tissues than in maternal plasma and tissues (Lee and Chiang 1985). THC metabolites 11-OH-THC and THCCOOH crossed the placenta much less efficiently (Bailey et al. 1987; Martin et al. 1977). No THCCOOH was detected in fetal plasma and tissues, indicating a lack of transfer across the placenta and a lack of metabolism of THC in the fetal monkey (Bailey et al. 1987). Blackard and Tennes reported that THC in cord blood was three to six times less than in maternal blood (Blackard and Tennes 1984). Transfer of THC to the fetus was greater in early pregnancy. THC also concentrates into breast milk from maternal plasma due to its high lipophilicity (Atkinson et al. 1988; Perez-Reyes and Wall 1982).

2.3 Metabolism

2.3.1 Hepatic Metabolism

Burstein et al. were the first to show that 11-OH-THC and THCCOOH were primary metabolites of THC in rabbits and rhesus monkeys (Ben-Zvi et al. 1976; Ben-Zvi and Burstein 1974; Burstein et al. 1972). They also documented that THC could be metabolized in the brain. Harvey et al. monitored the metabolism of THC, CBD, and CBN in mice, rats, and guinea pigs and found extensive metabolism, but with inter-species variation (Harvey et al. 1979). Phase I oxidation reactions include allylic and aliphatic hydroxylations, oxidation of alcohols to ketones and acids, beta-oxidation, and degradation of the pentyl side chain. Conjugation with glucuronic acid is a common phase II reaction. 11-OH-THC was the primary metabolite in all three species, followed by 8α-OH-THC concentrations in the mouse and rat, and 8β-OH-THC in guinea pig. Side chain hydroxylation was common in all three species. THCCOOH concentrations were higher in the mouse and rat, while THCCOOH glucuronide concentrations predominated in the guinea pig. THC concentrations accumulated in the liver, lung, heart, and spleen.

The primary metabolic routes and metabolites of THC are depicted in Fig. 4. Hydroxylation of THC at C9 by the hepatic cytochrome P450 enzyme system leads to production of the equipotent metabolite 11-OH-THC (Iribarne et al. 1996; Matsunaga et al. 1995), believed by early investigators to be the true psychoactive analyte (Lemberger et al. 1970). Cytochrome P450 2C9, 2C19, and 3A4 are involved in the oxidation of THC (Matsunaga et al. 1995). More than 100 THC metabolites including di- and tri-hydroxy compounds, ketones, aldehydes, and carboxylic acids have been identified (Grotenhermen 2003; Harvey 2001; Harvey and Paton 1986). Although 11-OH-THC predominates as the first oxidation product, significant amounts of 8β-OH-THC and lower amounts of the 8α-OH-THC are formed. Much

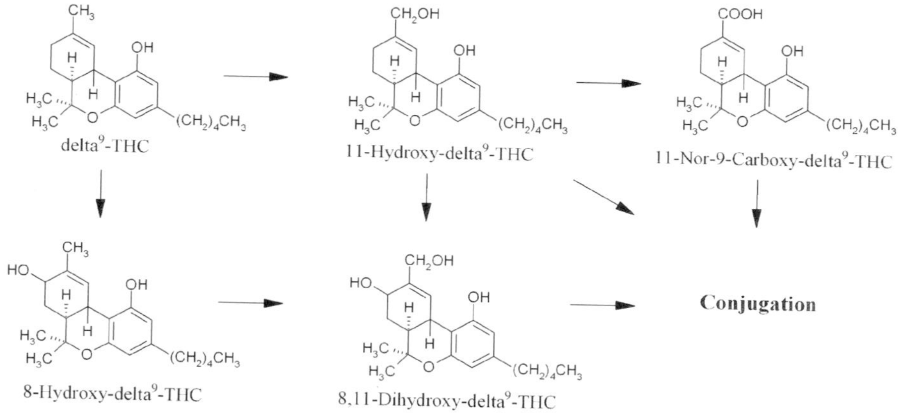

Fig. 4. Major metabolic routes for Δ^9-tetrahydrocannabinol (*THC*) in humans (M.A. Huestis, unpublished data)

lower plasma 11-OH-THC concentrations (approximately 10% of THC concentrations) are found after cannabis smoking than after oral administration (Wall et al. 1983). Peak 11-OH-THC concentrations occurred approximately 13 min after the start of smoking (Huestis et al. 1992b). Bornheim et al. reported that 11-OH-THC and 8-β-OH-THC were formed at the same rate in human liver microsomes, with smaller amounts of epoxy-hexahydrocannabinol, 8α-OH-THC and 8-keto-THC (Bornheim et al. 1992). Cytochrome P450 2C9 is believed to be primarily responsible for the formation of 11-OH-THC, whereas P450 3A catalyzes the formation of 8-β-OH-THC, epoxy hexahydrocannabinol, and other minor metabolites. Less than a fivefold variability in 2C9 rates of activity were observed, while much higher variability was noted for 3A. Dihydroxylation of THC yields 8β-11-di-OH-THC. Excretion of 8β-11-di-OH-THC in urine was reported to be a good biomarker for recent cannabis use (McBurney et al. 1986).

Oxidation of the active 11-OH-THC produces the inactive metabolite 11-nor-9-carboxy-Δ^9-tetrahydrocannabinol (THCCOOH) (Lemberger et al. 1970; Mechoulam et al. 1973). THCCOOH and its glucuronide conjugate are the major end products of biotransformation in most species, including man (Halldin et al. 1982; Harvey and Paton 1986). THCCOOH concentrations gradually increase and are greater than THC concentrations 30 to 45 min after the end of smoking (Mason and McBay 1985). After ingestion of a single 10 mg oral dose of Marinol, plasma THCCOOH concentrations were higher than THC and 11-OH-THC concentrations as early as 1 h after dosing (Sporkert et al. 2001). Unlike after smoking, THC and 11-OH-THC concentrations are similar after oral THC administration. Phase II metabolism of THCCOOH involves addition of glucuronic acid, and less commonly, sulfate, glutathione, amino acids, and fatty acids via the C11 carboxyl group. The phenolic hydroxyl group may be a target as well. It is also possible to have two glucuronic acid moieties attached to THCCOOH, although steric hindrance at the phenolic hydroxyl group could be a factor. Addition of the glucuronide group improves water solubility facilitating excretion, but renal clearance of these polar metabolites is low due to extensive protein binding (Hunt and Jones 1980). No significant differences in metabolism between men and women have been reported (Wall et al. 1983).

After the initial distribution phase, the rate-limiting step in the metabolism of THC is its redistribution from lipid depots into blood (Garrett and Hunt 1977). Lemberger et al. suggested that frequent cannabis smoking could induce THC metabolism (Lemberger et al. 1971). However, later studies did not replicate this finding (Agurell et al. 1986; Harvey and Paton 1986).

2.3.2 Extrahepatic Metabolism

Other tissues, including brain, intestine, and lung, may contribute to the metabolism of THC, although alternate hydroxylation pathways may be more prominent (Ben-Zvi et al. 1976; Greene and Saunders 1974; Krishna and Klotz 1994; Watanabe et al. 1988; Widman et al. 1975). An extrahepatic metabolic site should be suspected whenever total body clearance exceeds blood flow to the liver, or if severe liver dysfunction does not affect metabolic clearance (Krishna and Klotz 1994). Of the ten mammalian classes of cytochrome P450 systems, the cytochrome 1, 2, 3, and 4 families primarily metabolize xenobiotics and are found in the liver, small intestine, peripheral blood, bone marrow, and mast cells in decreasing concentrations, with the lowest concentrations in the brain, pancreas, gall bladder, kidney, skin, salivary glands, and testes. Within the brain, higher concentrations of cytochrome P450 enzymes are found in the brain stem and cerebellum (Krishna and Klotz 1994). The hydrolyzing enzymes, non-specific esterases, β-glucuronidases, and sulfatases, are primarily found in the gastrointestinal tract. Side chain hydroxylation of THC is prominent in THC metabolism by the lung. Metabolism of THC by fresh biopsies of human intestinal mucosa yielded polar hydroxylated metabolites that directly correlated with time and the amount of intestinal tissue (Greene and Saunders 1974).

In a study of the metabolism of THC in the brains of mice, rats, guinea pigs, and rabbits, Watanabe et al. found that brain microsomes oxidized THC to monohydroxylated metabolites (Watanabe et al. 1988). Hydroxylation of C4 of the pentyl side chain produced the most common THC metabolite in the brains of these animals, similar to THC metabolites produced in the lung. These metabolites are pharmacologically active, but their relative activity is unknown.

2.4 Elimination

Within 5 days, a total of 80% to 90% of a THC dose is excreted, mostly as hydroxylated and carboxylated metabolites (Halldin et al. 1982; Harvey 2001). More than 65% is excreted in the feces, with approximately 20% eliminated in the urine (Wall et al. 1983). Numerous acidic metabolites are found in the urine, many of which are conjugated with glucuronic acid to increase their water solubility. The primary urinary metabolite is the acid-linked THCCOOH glucuronide conjugate (Williams and Moffat 1980), while 11-OH-THC predominates in the feces (Harvey 2001). The concentration of free THCCOOH and the cross-reactivity of glucuronide-bound THCCOOH enable cannabinoid immunoassays to be performed directly on non-hydrolyzed urine, but confirmation and quantification of THCCOOH is usually performed after alkaline hydrolysis or β-glucuronidase hydrolysis to free THC-COOH for measurement by GC/MS. It is generally thought that little to no THC or 11-OH-THC is excreted in the urine.

2.4.1 Terminal Elimination Half-Lives of THCCOOH

Another common problem with studying the pharmacokinetics of cannabinoids in humans is the need for highly sensitive procedures to measure low cannabinoid concentrations in the terminal phase of excretion, and the requirement for monitoring plasma concentrations over an extended period to adequately determine cannabinoid half-lives. Many studies utilized short sampling intervals of 24 to 72 h that underestimate terminal THC and THCCOOH half-lives. The slow release of THC from lipid storage compartments and significant enterohepatic circulation contribute to THC's long terminal half-life in plasma, reported as greater than 4.1 days in chronic cannabis users (Johansson et al. 1988). Isotopically labeled THC and sensitive analytical procedures were used to obtain this drug half-life. Garrett and Hunt reported that 10% to 15% of the THC dose is enterohepatically circulated in dogs (Garrett and Hunt 1977). Johansson et al. reported a THC-COOH plasma elimination half-life up to 12.6 days in a chronic cannabis user when monitoring THCCOOH concentrations for 4 weeks (Johansson et al. 1989a). Mean plasma THCCOOH elimination half-lives were 5.2±0.8 and 6.2±6.7 days for frequent and infrequent cannabis users, respectively. Similarly, when sensitive analytical procedures and sufficient sampling periods were employed for determining the terminal urinary excretion half-life of THCCOOH, it was estimated to be 3 to 4 days (Johansson and Halldin 1989). Urinary THCCOOH concentrations drop rapidly until approximately 20 to 50 ng/ml, and then decrease at a much slower rate. No significant pharmacokinetic differences between chronic and occasional users have been substantiated (Chiang and Rapaka 1987).

2.4.2 Percentage THC Dose Excreted as Urinary THCCOOH

An average of 93.9±24.5 µg THCCOOH (range 34.6–171.6) was measured in urine over a 7-day period following smoking of a single 1.75% THC cigarette containing approximately 18 mg THC (Huestis et al. 1996). The average amount of THCCOOH excreted in the same time period following the high dose (3.55% THC containing approximately 34 mg THC) was 197.4±33.6 µg (range 107.5–305.0). This represented an average of only 0.54±0.14% and 0.53±0.09% of the original amount of

THC in the low- and high-dose cigarettes, respectively. The small percentage of the total dose found in the urine as THCCOOH is not surprising considering the many factors that influence THCCOOH excretion after smoking. Prior to harvesting, cannabis plant material contains little active THC. When smoked, THC carboxylic acids spontaneously decarboxylate to produce THC with nearly complete conversion upon heating. Pyrolysis of THC during smoking destroys additional drug. Drug availability is further reduced by loss of drug in the side-stream smoke and drug remaining in the unsmoked cigarette butt. These factors contribute to high variability in drug delivery by the smoked route. It is estimated that the systemic availability of smoked THC is approximately 8% to 24% and that bioavailability depends strongly upon the experience of the cannabis user (Lindgren et al. 1981; Ohlsson et al. 1980; Perez-Reyes et al. 1981). THC bioavailability is reduced due to the combined effect of these factors; the actual available dose is much lower than the amount of THC and THC precursor present in the cigarette. Most of the THC dose is excreted in the feces (30%–65%), rather than in the urine (20%) (Wall and Perez-Reyes 1981; Wall et al. 1983). Another factor affecting the low amount of recovered dose is measurement of a single metabolite. Numerous cannabinoid metabolites are produced in humans as a result of THC metabolism, most of which are not measured or included in the percentage-of-dose-excreted calculations when utilizing GC/MS.

2.4.3
Cannabinoid Glucuronide Conjugates

Specimen preparation for cannabinoid testing frequently includes a hydrolysis step to free cannabinoids from their glucuronide conjugates. Most GC/MS confirmation procedures in urine measure total THCCOOH following either an enzymatic hydrolysis with β-glucuronidase, or more commonly, an alkaline hydrolysis with sodium hydroxide. Alkaline hydrolysis appears to efficiently hydrolyze the ester THCCOOH glucuronide linkage.

2.4.4
Urinary Biomarkers of Recent Cannabis Use

Significantly higher concentrations of THC and 11-OH-THC in urine are observed when *Escherichia coli* β-glucuronidase is employed in the hydrolysis method compared to either *Helix pomatia* β-glucuronidase or base (Kemp et al. 1995a,b). THC and 11-OH-THC are primarily excreted in urine as glucuronide conjugates that are resistant to cleavage by alkaline hydrolysis and by enzymatic hydrolysis procedures employing some types of β-glucuronidase. Kemp et al. demonstrated that β-glucuronidase from *E. coli* was needed to hydrolyze the ether glucuronide linkages of the active cannabinoid analytes. Mean THC concentration in urine specimens from seven subjects collected after each had smoked a single 3.58% marijuana cigarette was 22 ng/ml using the *E. coli* β-glucuronidase hydrolysis method, while THC concentrations using either *H. pomatia* β-glucuronidase or base hydrolysis methods were near zero (Kemp et al. 1995a,b). Similar differences were found for 11-OH-THC with a mean concentration of 72 ng/ml from the *E. coli* method and concentrations less than 10 ng/ml from the other methods. The authors suggested that finding THC and/or 11-OH-THC in the urine might provide a reliable marker of recent cannabis use, but adequate data from controlled drug administration studies were not yet available to support or refute this observation. Using a modified analytical method with *E. coli* β-glucuronidase, we have analyzed hundreds of urine specimens collected following controlled THC administration. We found that 11-OH-THC may be excreted in the urine of chronic cannabis users for a much longer period of time, beyond the period of pharmacodynamic effects and performance impairment. However, it does appear that THC is only present in urine for a short period after use. Additional research is necessary to determine the validity of estimating time of cannabis use from THC and 11-OH-THC concentrations in urine.

3 Pharmacokinetics of Cannabidiol

Cannabidiol (CBD) is a natural constituent of *Cannabis sativa* that is not psychoactive (Benowitz et al. 1980; Perez-Reyes et al. 1973b; Pertwee 2004), but possesses pharmacological activity that is being explored for therapeutic applications (Pertwee 2004). CBD has been reported to be neuroprotective (Hampson et al. 1998), analgesic (Holdcroft 1984; Karst et al. 2003; Vaughan and Christie 1984), sedating (Holdcroft 1984; Melamede 1984; Plasse 1984; Vaughan and Christie 1984), anti-emetic (Plasse 1984), anti-spasmodic (Baker et al. 2000), and anti-inflammatory (Malfait et al. 2000). In addition, it has been reported that CBD blocks the anxiety produced by THC (Zuardi et al. 1982) and is useful in the treatment of autoimmune diseases (Melamede 1984). These potential therapeutic applications alone warrant investigation of CBD pharmacokinetics, but also, the controversy over whether CBD alters the pharmacokinetics of THC in a clinically significant manner needs to be resolved (Agurell et al. 1984; McArdle et al. 2001).

Cannabidiol metabolism is similar to that of THC, with primary oxidation of C9 to the hydroxy and carboxylic acid moieties (Agurell et al. 1986; Harvey and Mechoulam 1990) and side chain oxidation (Harvey et al. 1979; Harvey and Mechoulam 1990). Like THC, CBD is subjected to a significant first-pass effect; however, unlike THC, a large proportion of the dose is excreted unchanged in the feces (Wall et al. 1976). Benowitz et al. reported that CBD was an in vitro inhibitor of liver microsomal drug-metabolizing enzymes and inhibited hexobarbital metabolism in humans (Benowitz et al. 1980). Others have reported that CBD selectively inhibits THC metabolite formation in vitro (McArdle et al. 2001). Hunt et al. reported that THC's pharmacokinetic properties were not affected by CBD, except for a slight slowing of the metabolism of 11-OH-THC to THCCOOH (Hunt et al. 1981). Co-administration of CBD did not significantly affect the total clearance, volume of distribution, and terminal elimination half-lives of THC metabolites.

The bioavailability of CBD following the smoked route averaged 31% (range 11%–45%) as compared to intravenously administered drug (Ohlsson et al. 1986).

Similar results were obtained when comparing the sublingual administration of 25 mg THC to 25 mg THC and 25 mg CBD in cannabis-based medicinal extracts (Guy and Robson 2004a). There were no statistically significant differences in mean THC C_{max}, half-life, or AUC for THC and 11-OH-THC following administration of these two compounds. The only statistically significant difference was in the time of maximum THC concentration. Despite administration of equivalent amounts of THC and CBD, lower plasma concentrations of CBD were always observed. In a separate evaluation of 10 mg THC and 10 mg CBD from a cannabis-based medicine extract, the pharmacokinetics of THC, 11-OH-THC, and CBD were determined after sublingual, buccal, oro-pharyngeal, and oral administration (Guy and Robson 2004b). All three analytes were measurable approximately 30 min after dosing with higher THC than CBD concentrations. 11-OH-THC generally exceeded THC concentrations within 45 min of dosing. Mean C_{max} concentrations for THC, CBD, and 11-OH-THC were less than 5, less than 2, and less than 7 ng/ml across all administration routes. High intra- and inter-subject variability was noted.

4 Pharmacokinetics of Cannabinol

Cannabinol (CBN) is a natural constituent of *Cannabis sativa* with approximately 10% of the activity of THC (Perez-Reyes 1985; Perez-Reyes et al. 1973b). CBN metabolism is also similar to that of THC with the hydroxylation of C9 yielding the primary metabolite (Wall et al. 1976). Due to the fact that one additional ring is aromatic, CBN is metabolized less extensively and more slowly than THC (Harvey et al. 1979). The average bioavailability of a smoked CBN dose, as compared to intravenous CBN, was 41% with a range of 8% to 77% (Ohlsson et al. 1985).

5 Interpretation of Cannabinoid Concentrations in Biological Fluids

5.1 Plasma Concentrations of THC, 11-OH-THC, and THCCOOH

Compared to other drugs of abuse, analysis of cannabinoids presents some difficult challenges. THC and 11-OH-THC are highly lipophilic and present in low concentrations in body fluids. Complex specimen matrices, i.e., blood, sweat, and hair, may require multi-step extractions to separate cannabinoids from endogenous lipids and proteins. Care must be taken to avoid low recoveries of cannabinoids due to their high affinity to glass and plastic containers, and to collection devices for alternate matrices (Blanc et al. 1993; Bloom 1982; Christophersen 1986; Joern 1992). THC and THCCOOH are predominantly found in the plasma fraction of blood, where 95% to 99% are bound to lipoproteins. Only about 10% of either compound is found in the erythrocytes (Garrett and Hunt 1974; Widman et al. 1974). Whole blood cannabinoid concentrations are approximately one-half the concentrations found in plasma specimens, due to the low partition coefficient of drug into erythrocytes (Huang et al. 2001; Mason and McBay 1985; Owens et al. 1981; Widman et al. 1974).

5.1.1 Following Intravenous THC Administration

Kelly et al. intravenously administered 5 mg of THC to eight males and periodically monitored THC, THCCOOH, and THCCOOH-glucuronide conjugates by GC/MS [limit of detection (LOD) 1 ng/ml for THC and THCCOOH] in plasma with and without alkaline hydrolysis for up to 10 h, and then once daily for up to 12 days (Kelly and Jones 1992). The elimination half-lives of THC, THCCOOH, and THCCOOH-glucuronide in the plasma of frequent cannabis users were 116.8 min, 5.2 days, and 6.8 days, respectively, and 93.3 min, 6.2 days and 3.7 days in infrequent users. Conjugated THCCOOH was detected in the plasma of 75% of the frequent and 25% of the infrequent users at day 12.

5.1.2 Following Smoked Cannabis Administration

THC detection times in plasma of 3.5 to 5.5 h were reported in individuals who smoked two cannabis cigarettes containing a total of approximately 10 mg of THC (GC/MS LOD 0.8 ng/ml) (McBurney et al. 1986) and up to 13 days for deuterated THC in the blood of chronic cannabis users who smoked four deuterium-labeled THC cigarettes (GC/MS LOD = 0.02 ng/ml) (Johansson et al. 1988). In the latter study, the terminal half-life of THC in plasma was determined to be approximately 4.1 days, as compared to frequent estimates of 24 to 36 h in several other studies (Agurell et al. 1984; Lemberger et al. 1972; Wall et al. 1983) that lacked the sensitivity and the lengthy monitoring window of the radio-labeled protocol.

Few controlled drug administration studies have monitored active 11-OH-THC plasma concentrations. Huestis et al. found plasma 11-OH-THC concentrations to be approximately 6% to 10% of the concurrent THC concentrations for up to 45 min after the start of smoking (Huestis et al. 1992b). Mean peak 11-OH-THC concentrations occurred 13.5 min (range 9.0–22.8) after the start of smoking and were 6.7 ng/ml (range 3.3–10.4) and 7.5 ng/ml (range 3.8–16) after one 1.75% or 3.55% THC cigarette, respectively. 11-OH-THC concentrations decreased gradually with mean detection times of 4.5 h and 11.2 h after the two doses.

THCCOOH concentrations were monitored in human plasma for 7 days after controlled cannabis smoking (Huestis et al. 1992b). This inactive metabolite was detected in all subjects' plasma by 8 min after the start of smoking. THC-COOH concentrations in plasma increased slowly and plateaued for up to 4 h. Peak concentrations were consistently lower than peak THC concentrations, but were higher than peak 11-OH-THC concentrations. Mean peak THCCOOH concentrations were 24.5 ng/ml (range 15–54) and 54.0 ng/ml (range 22–101) after the 1.75% and 3.55% THC cigarettes, respectively. Following smoking of the lower dose, THCCOOH was detected from 48 to 168 h, with a mean of 84 h. Detection times ranged from 72 to 168 h with a mean of 152 h following smoking of the higher dose. The time course of detection of THCCOOH is much longer than either that of THC or 11-OH-THC. The area under the curve for the mean data from 0 to 168 h was 36.5 and 72.2 ng-h/ml, respectively, for the low- and high-dose conditions, demonstrating a dose–response relationship for the mean data (Huestis et al. 1992b). Figure 2 shows individual THC concentration time profiles for six subjects and demonstrates the large inter-subject variability of the smoked route of drug administration. Moeller et al. measured serum THC and THCCOOH concentrations in 24 experienced users from 40 to 220 min after smoking 300-μg/kg cannabis cigarettes (Moeller et al. 1992). Mean serum THC and THCCOOH concentrations were approximately 13 and 22 ng/ml at 40 min and 1 and 13 ng/ml at 220 min after smoking. The half-life of the rapid distribution phase of THC was estimated to be 55 min over this short sampling interval.

Most plasma or whole blood cannabinoid analytical methods have not included measurement of the glucuronide conjugates of THC, 11-OH-THC, or THCCOOH. The relative percentages of free and conjugated cannabinoids in plasma after different routes of drug administration are unclear. Even the efficacy of alkaline and enzymatic hydrolysis procedures to release analytes from their conjugates is not fully understood (Feng et al. 2000; Foltz 1984; Green et al. 1997; Kelly and Jones 1992; Kemp et al. 1995a,b; Law et al. 1984; Manno et al. 2001; McBurney et al. 1986; Wall and Perez-Reyes 1981; Wall and Taylor 1984; Widman et al. 1974). In general, conjugate concentrations are believed to be lower in plasma following intravenous or smoked cannabis, but may be of much greater magnitude after oral drug administration. There is no indication that the glucuronide conjugates are active, although supporting data are lacking.

5.1.3
Oral THC

After oral and sublingual administration of THC, THC-containing food products, or cannabis-based extracts, concentrations of THC and 11-OH-THC are much lower than after smoked administration. Plasma concentrations of THC in patients receiving 10 to 15 mg of Marinol as an anti-emetic were low to non-measurable in 57 patients (Shaw et al. 1991). Brenneisen et al. found peak plasma concentrations of THC and THCCOOH after daily oral 10 to 15 mg Marinol doses of 2.1 to 16.9 ng/ml within 1 to 8 h and 74.5 to 244 ng/ml within 2 to 8 h, respectively (Brenneisen et al. 1996). In our oral THC controlled administration studies, peak plasma THC, 11-OH-THC, and THCCOOH concentrations were less than 6.5, 5.6, and 24.4 ng/ml, respectively, following up to 14.8 mg/day of THC in the form of THC-containing food products or Marinol (Nebro et al. 2004). Peak concentrations and time to peak concentrations varied, sometimes considerably, between subjects. Plasma

THC and 11-OH-THC were negative for all participants and for all doses by 16 h after the last THC dose. Plasma THCCOOH persisted for a longer period of time following the two highest doses of 7.5 mg/day dronabinol and 14.8 mg/day THC in hemp oil. Ohlsson et al. reported that orally administered THC (20 mg in a cookie) yielded low and irregular plasma concentrations compared to intravenous and inhaled THC (Ohlsson et al. 1980).

5.1.4
Cannabinoid Concentrations After Frequent Use

Most THC plasma data have been collected following acute exposure; less is known of plasma THC concentrations in frequent users. Peat reported THC, 11-OH-THC, and THCCOOH plasma concentrations in frequent cannabis users of 0.86±0.22, 0.46±0.17, and 45.8±3.1 ng/ml, respectively, a minimum of 12 h after the last smoked dose (Peat 1989). No difference in terminal half-life in frequent or infrequent users was observed. Johansson et al. administered radiolabeled THC to frequent cannabis users and found a terminal elimination half-life of 4.1 days for THC in plasma due to extensive storage and release from body fat (Johansson et al. 1988).

5.1.5
Prediction Models for Estimation of Cannabis Exposure

Although there continues to be controversy in the interpretation of blood cannabinoid results, some general concepts have wide support. A dose–response relationship has been demonstrated for smoked THC and THC plasma concentrations (Perez-Reyes et al. 1981, 1982). It is well established that plasma THC concentrations begin to decline prior to the time of peak effects, although it has been shown that THC effects appear rapidly after initiation of smoking (Huestis et al. 1992d). Individual drug concentrations and ratios of cannabinoid metabolite to parent drug concentration have been suggested as potentially useful indicators of recent drug use (Hanson et al. 1983; Law et al. 1984). The ratio of plasma THCCOOH to THC was found to exceed 1 at 45 min after cannabis smoking (Kelly and Jones 1992). This is in agreement with results reported by Mason and McBay (1985) and Huestis et al. (1992d) who found that peak effects occurred when THC and THC-COOH concentrations reached equivalency, within 30 to 45 min after initiation of smoking. Measurement of cannabinoid analytes with short time courses of detection (e.g., 8β, 11-dihydroxy-tetrahydrocannabinol) as a marker of recent exposure has not found widespread use (Mason and McBay 1985). Recent exposure (6 to 8 h) and possible impairment have been linked to plasma THC concentrations in excess of 2 to 3 ng/ml (Huestis et al. 1992b; Mason and McBay 1985). Gjerde et al. (1993) suggested that 1.6 ng/ml THC in whole blood might indicate possible impairment. This correlates well with the suggested concentration of plasma THC, due to the fact that THC in hemolyzed blood is approximately one-half the concentration of

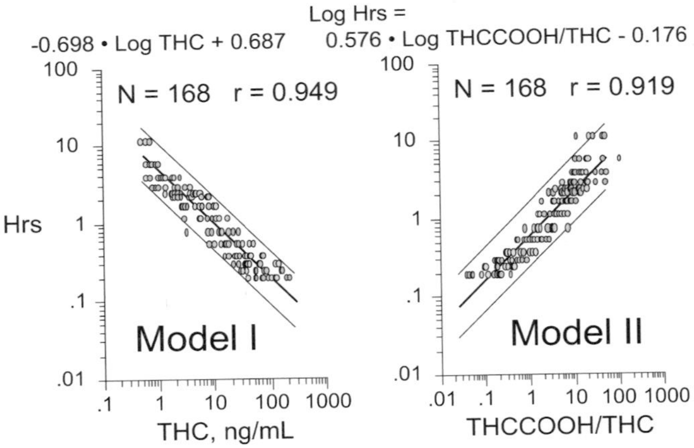

Fig. 5. Predictive mathematical models for estimating the elapsed time in hours (*Hrs*) of last cannabis use based on plasma Δ^9-tetrahydrocannabinol (*THC*) and 11-nor-9-carboxy-Δ^9-tetrahydrocannabinol (*THCCOOH*) concentrations. (Reproduced from the *Journal of Analytic Toxicology*, by permission of Preston Publications, a division of Preston Industries; Huestis et al. 1992c, Fig. 1 therein)

plasma THC (Mason and McBay 1984). Interpretation is further complicated by residual THC and THCCOOH concentrations found in blood of frequent cannabis users. In general, it is suggested that chronic cannabis smokers may have residual plasma THC concentrations of less than 2 ng/ml 12 h after smoking cannabis (Peat 1989). Significantly higher residual concentrations of THCCOOH may be found.

Having an accurate prediction of the time of cannabis exposure would provide valuable information in establishing the role of cannabis as a contributing factor to events under investigation. Two mathematical models for the prediction of time of cannabis use from the analysis of a single plasma specimen for cannabinoids were developed (Huestis et al. 1992c). Model I was based on THC concentrations and model II was based on the ratio of THCCOOH to THC in plasma (Fig. 5). Both correctly predicted the times of exposure within the 95% confidence interval for more than 90% of the specimens evaluated. Furthermore, plasma THC and THCCOOH concentrations reported in the literature following oral and smoked cannabis exposure, in frequent and infrequent cannabis smokers, and with measurements obtained by a wide variety of methods, including radioimmunoassay and GC/MS, were evaluated with the models. Plasma THC concentrations less than 2.0 ng/ml were excluded from use in both models due to the possibility of residual THC concentrations in frequent smokers. Manno et al. evaluated the models' usefulness in predicting the time of cannabis use in a controlled cannabis smoking study (Manno et al. 2001). The models were found to accurately predict the time of last use within the 95% confidence intervals. Due to the limited distribution of THC and THCCOOH into red blood cells, it is important to remember that when comparing whole blood THC and/or THCCOOH concentrations to plasma concentrations, it is necessary to double the whole blood concentration prior to comparison.

5.2 Urinary THCCOOH Concentrations

Detection of cannabinoids in urine is indicative of prior cannabis exposure, but the long excretion half-life of THCCOOH in the body, especially in chronic cannabis users, makes it difficult to predict the timing of past drug use. In a single extreme case, one individual's urine was positive at a concentration greater than 20 ng/ml by immunoassay up to 67 days after last drug exposure (Ellis et al. 1985). This individual had used cannabis heavily for more than 10 years. However, a naïve user's urine may be found negative by immunoassay after only a few hours following the smoking of a single cannabis cigarette (Huestis et al. 1995). Assay cutoff concentrations and the sensitivity and specificity of the immunoassay affect drug detection times. A positive urine test for cannabinoids indicates only that drug exposure has occurred. The result does not provide information on the route of administration, the amount of drug exposure, when drug exposure occurred, or the degree of impairment.

To date, there are too few urinary THC and 11-OH-THC data to guide interpretation of positive urine cannabinoid tests; however, data are available for guiding interpretation of total urinary THCCOOH concentrations. Total THCCOOH concentrations include both the free THCCOOH and THCCOOH-glucuronide concentrations that are obtained after alkaline or enzymatic hydrolysis. Substantial intra- and inter-subject variability occurs in patterns of THCCOOH excretion. THCCOOH concentration in the first specimen after smoking is indicative of how rapidly the metabolite can appear in urine. Mean first urine THCCOOH concentrations were 47±22.3 ng/ml and 75.3±48.9 ng/ml after smoking one 1.75% or 3.55% THC cigarette, respectively (Huestis et al. 1996). Of the subjects' first urine specimens, 50% after the low dose and 83% after the high dose were positive by GC/MS at a 15 ng/ml THCCOOH cutoff concentration. Thus, THCCOOH concentrations in the first urine specimen are dependent upon the relative potency of the cigarette, the elapsed time following drug administration, smoking efficiency, and individual differences in drug metabolism and excretion. Mean peak urine THCCOOH concentrations averaged 89.8±31.9 ng/ml (range 20.6–234.2) and 153.4±49.2 ng/ml (range 29.9–355.2) following smoking of approximately 15.8 and 33.8 mg THC, respectively. The mean times of peak urine concentration were 7.7±0.8 h after the 1.75% THC and 13.9±3.5 h after the 3.55% THC dose. Although peak concentrations appeared to be dose related, there was a 12-fold variation between individuals.

5.2.1 THCCOOH Detection Windows in Urine

Drug detection time, or the duration of time after drug administration that an individual's urine tests positive for cannabinoids, is an important factor in the interpretation of urine drug results. Detection time is dependent on pharmacological factors (e.g., drug dose, route of administration, rates of metabolism and excretion) and analytical factors (e.g., assay sensitivity, specificity, accuracy). Mean detection times in urine following smoking vary considerably between subjects, even in controlled smoking studies where cannabis dosing is standardized and smoking is computer-paced. During the terminal elimination phase, consecutive urine specimens may fluctuate between positive and negative as THCCOOH concentrations approach the cutoff concentration. It may be important in drug treatment settings or in clinical trials to differentiate between new drug use and residual excretion of previously used cannabinoids. After smoking a 1.75% THC cigarette, three of six subjects had additional positive urine samples interspersed between negative urine samples (Huestis et al. 1995). This had the effect of producing much longer detection times for the last positive specimen. Using the 15 ng/ml confirmation cutoff for THCCOOH currently used for most urine drug testing, the mean GC/MS THCCOOH detection times for the last positive urine sample following the smoking of a single 1.75% or 3.55% THC cigarette were 33.7±9.2 h (range 8–68.5) and 88.6±23.2 h (range 57–122.3), respectively.

5.2.2 Normalization of Cannabinoid Urine Concentrations to Urine Creatinine Concentrations

Normalization of the cannabinoid drug concentration to the urine creatinine concentration aids in the differentiation of new vs prior cannabis use and reduces the variability of drug measurement due to urine dilution. Due to the long half-life of drug in the body, especially in chronic cannabis users, toxicologists and practitioners are frequently asked to determine if a positive urine test represents a new episode of drug use or represents continued excretion of residual drug. Random urine specimens contain varying amounts of creatinine depending on the degree of concentration of the urine. Hawks first suggested creatinine normalization of urine test results to account for variations in urine volume in the bladder (Hawks 1983). Whereas urine volume is highly variable due to changes in liquid, salt, and protein intake, exercise, and age, creatinine excretion is much more stable. Manno et al. recommended that an increase of 150% in the creatinine normalized cannabinoid concentration above the previous specimen be considered indicative of a new episode of drug exposure (Manno et al. 1984). If the increase is greater than or equal to the threshold selected, then new use is predicted. This approach has received wide attention for potential use in treatment and employee assistance programs, but there has been limited evaluation of the usefulness of this ratio under controlled dosing conditions. Huestis et al. conducted a controlled clinical study of the excretion profile of creatinine and cannabinoid metabolites in a group of six cannabis users who smoked two different doses of cannabis separated by weekly intervals (Huestis and Cone 1998b). As seen in Fig. 6, normalization of urinary THCCOOH concentration to the urinary creatinine concentration produces a smoother excretion pattern and facilitates interpretation of consecutive urine drug test results. A relative operating characteristic (ROC) curve was constructed from sensitivity and specificity data for 26 different cutoffs ranging from 10% to

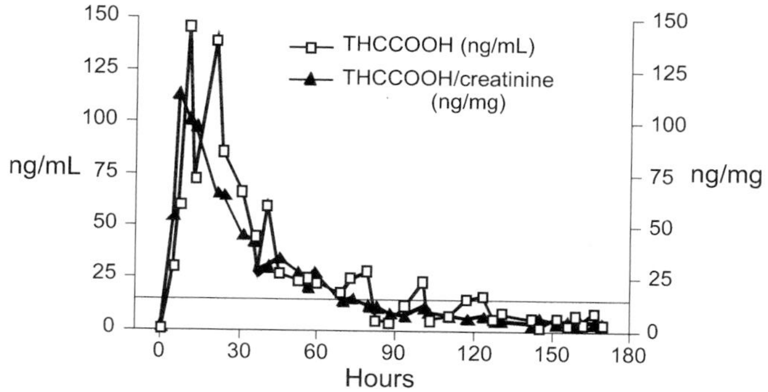

Fig. 6. Urine concentrations of 11-nor-9-carboxy-Δ^9-tetrahydrocannabinol (*THCCOOH*; ng/ml, and ng/mg creatinine) for one subject after smoking a single 3.55% THC cigarette. (Reproduced from the *Journal of Analytic Toxicology* by permission of Preston Publications, a division of Preston Industries; Huestis and Cone 1998b, Fig. 3 therein)

200%. The most accurate ratio (85.4%) was 50%, with a sensitivity of 80.1% and a specificity of 90.2%, with 5.6% false-positive and 7.4% false-negative predictions. If the previously recommended increase of 150% was used as the threshold for new use, sensitivity of detecting new use was only 33.4%, specificity was high at 99.8%, and there was an overall accuracy prediction of 74.2%. To further substantiate the validity of the derived ROC curve, urine cannabinoid metabolite and creatinine data from another controlled clinical trial that specifically addressed water dilution as a means of specimen adulteration were evaluated (Cone et al. 1998). Sensitivity, specificity, accuracy, percentage false positives, and percentage false negatives were 71.9%, 91.6%, 83.9%, 5.4%, and 10.7%, respectively, when the 50% criterion was applied. These data indicate selection of a threshold to evaluate sequential creatinine-normalized urine drug concentrations can improve the ability to distinguish residual excretion from new drug usage.

5.3 Oral Fluid Testing

Oral fluid is also a suitable specimen for monitoring cannabinoid exposure and is being evaluated for driving under the influence of drugs, drug treatment, workplace drug testing, and for clinical trials (Cairns et al. 1990; Gross and Soares 1978; Gross et al. 1985; Mura et al. 1999; Soares et al. 1976, 1982). Adequate sensitivity is best achieved with an assay directed toward detection of THC, rather than 11-OH-THC or THCCOOH. The oral mucosa is exposed to high concentrations of THC during smoking and serves as the source of THC found in oral fluid. Only minor amounts of drug and metabolites diffuse from the plasma into oral fluid (Hawks 1983). Following intravenous administration of radiolabeled THC, no radioactivity could be demonstrated in oral fluid (Hawks 1982). No measurable 11-OH-THC or THCCOOH was found in oral fluid collected immediately following and up to

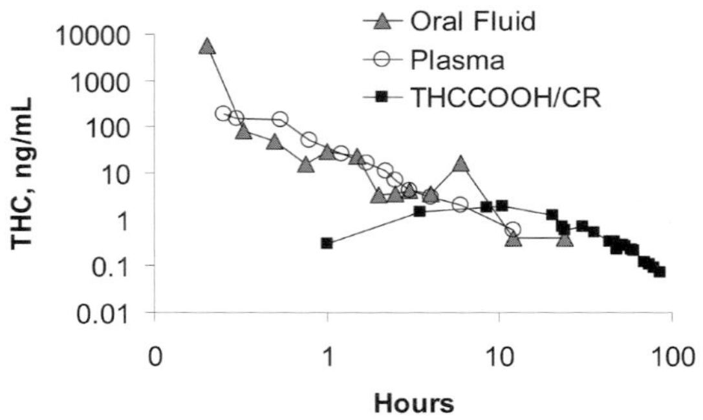

Fig. 7. Excretion patterns of Δ^9-tetrahydrocannabinol (THC) concentrations (ng/ml) in oral fluid and plasma, and urinary 11-nor-9-carboxy-Δ^9-tetrahydrocannabinol (ng THCCOOH/mg creatinine) in one human subject following smoking of a single cannabis cigarette (3.55%). The ng THCCOOH/mg creatinine ratio is illustrated for all urine specimens collected through the last positive specimen. Analyses were performed by GC-MS at cutoff concentrations of 0.5 ng/ml for oral fluid and plasma and 15 ng/ml for urine. (Reproduced from the *Journal of Analytic Toxicology* by permission of Preston Publications, a division of Preston Industries; Huestis and Cone 2004, Fig. 2 therein)

7 days after cannabis smoking with a GC/MS LOQ of 0.5 ng/ml (Huestis and Cone 1998a). Similarly, 11-OH-THC and THCCOOH were not detected in the oral fluid of 22 subjects who were documented cannabis users (Kintz et al. 2000). Oral fluid collected with the Salivette collection device was positive for THC in 14 of these 22 participants. Although no 11-OH-THC or THCCOOH was identified by GC/MS, cannabinol and cannabidiol were found in addition to THC. Hours after smoking, the oral mucosa serves as a depot for release of THC into the oral fluid. In addition, as detection limits continue to decrease with the development of new analytical instrumentation, it may be possible to measure low concentrations of THCCOOH in oral fluid.

Detection times of cannabinoids in oral fluid are shorter than in urine, and more indicative of recent cannabis use (Cairns et al. 1990; Gross et al. 1985). Oral fluid THC concentrations temporally correlate with plasma cannabinoid concentrations and behavioral and physiological effects, but wide intra- and inter-individual variation precludes the use of oral fluid concentrations as indicators of drug impairment (Huestis and Cone 1998a; Huestis et al. 1992a). THC may be detected at low concentrations by radioimmunoassay for up to 24 h after use. Figure 7 depicts excretion of THC in oral fluid and plasma and creatinine-normalized THCCOOH excretion in urine in one subject after smoking a single 3.55% cannabis cigarette (Huestis and Cone 2004). After smoking cannabis, oral fluid cannabinoid tests were positive for THC by GC/MS/MS with a cutoff of 0.5 ng/ml for 13±3 h (range 1–24) (Niedbala et al. 2001). After these times, occasional positive oral fluid results were interspersed with negative tests for up to 34 h. Peel et al. tested oral fluid samples from 56 drivers suspected of being under the influence of cannabis with the enzyme-multiplied immunoassay test (EMIT) screening test and GC/MS confirmation (Peel et al. 1984). They suggested that the ease and non-invasiveness of sample collection made oral fluid a useful alternative matrix for detection of recent cannabis use. Oral fluid samples also are being evaluated in the European Union's Roadside Testing Assessment (ROSITA) Project to reduce the number of individuals driving under the influence of drugs and to improve road safety. The ease and non-invasiveness of oral fluid collection, reduced hazards in specimen handling and testing, and shorter detection window are attractive attributes of this method for identifying the presence of potential performance-impairing drugs.

In a recent study of smoked and oral cannabis use, the Intercept DOA Oral Specimen Collection Device and GC/MS/MS (cutoff 0.5 ng/ml) were paired to monitor oral fluid cannabinoids in ten participants (Niedbala et al. 2001). Oral fluid specimens tested positive following smoked cannabis for an average of 13±3 h (range 1–24). After these times, occasional positive oral fluid results were interspersed with negative tests for up to 34 h. A different oral fluid collection device, the Cozart RapiScan device, utilizes a 10 ng/ml cannabinoid cutoff to screen for cannabis use (Jehanli et al. 2001). Positive oral fluid cannabinoid tests were not obtained more than 2 h after last use, suggesting that much lower cutoff concentrations were needed to improve sensitivity. A procedure for direct analysis of cannabinoids in oral fluid with solid-phase microextraction and ion trap GC/MS has been developed with a limit of detection of 1.0 ng/ml (Hall et al. 1998). Detection of cannabinoids in oral fluid is a rapidly developing field; however, there are many scientific issues to resolve. One of the most important is the degree of absorption of the drug to oral fluid collection devices.

5.4 Cannabinoids in Sweat

To date, there are no published data on the excretion of cannabinoids in sweat following controlled THC administration. Sweat testing is being applied to monitor cannabis use in drug treatment, criminal justice, workplace drug testing, and clinical studies (Huestis and Cone 1998a; Kidwell et al. 1998). In 1989, Balabanova and Schneider utilized radioimmunoassay to detect cannabinoids in apocrine sweat (Balabanova and Schneider 1989). Currently, there is a single commercially available sweat collection device, the PharmCheck patch, offered by PharmChem Laboratories in Texas, USA. Generally, the patch is worn for 7 days and exchanged for a new patch once each week during visits to the treatment clinic or parole officer. Theoretically, this permits constant monitoring of drug use throughout the week, extending the window of drug detection and improving test sensitivity. As with oral fluid testing, this is a developing analytical technique with much to be learned about the pharmacokinetics of cannabinoid excretion in sweat, potential reabsorption of THC by the skin, possible degradation of THC on the patch, and adsorption of THC onto the patch collection device. It is known that THC is the primary analyte detected in sweat, with little 11-OH-THC and THCCOOH. Several investigators have evaluated the sensitivity and specificity of different screening assays for detecting cannabinoids in sweat (Mura et al. 1999; Samyn and van Haeren 2000). Kintz et al. identified THC (4–38 ng/patch) in 20 known heroin abusers who wore the PharmChek patch for 5 days while attending a detoxification center (Kintz et al. 1997). Sweat was extracted with methanol and analyzed by GC/MS. The same investigators also evaluated forehead swipes with cosmetic pads for monitoring cannabinoids in sweat from individuals suspected of driving under the influence of drugs (Kintz et al. 2000). THC, but not 11-OH-THC or THCCOOH, was detected (4 to 152 ng/pad) by electron impact GC/MS in the sweat of 16 of 22 individuals who tested positive for cannabinoids in urine. Ion trap tandem mass spectrometry has also been used to measure cannabinoids in sweat collected with the PharmChek sweat patch with a limit of detection of 1 ng/patch (Ehorn et al. 1994).

5.5 Cannabinoids in Hair

There are multiple mechanisms for the incorporation of cannabinoids in hair. THC and metabolites may be incorporated into the hair bulb that is surrounded by capillaries. Drug may also diffuse into hair from sebum that is secreted onto the hair shaft and from sweat that is excreted onto the skin surface. Drug may also be incorporated into hair from the environment. Cannabis is primarily smoked, providing an opportunity for environmental contamination of hair with THC in cannabis smoke. Basic drugs such as cocaine and methamphetamine concentrate in hair due to ionic bonding to melanin, the pigment in hair that determines hair color. The more neutral and lipophilic THC is not highly bound to melanin, resulting in much lower concentrations of THC in hair as compared to other drugs of abuse. Usually THC is present in hair at a higher concentration than its THCCOOH metabolite (Cairns et al. 1995; Cirimele 1996; Kintz et al. 1995; Moore et al. 2001). An advantage of measuring THCCOOH in hair is that THCCOOH is not present in cannabis smoke, avoiding the issue of passive exposure from the environment. Analysis of cannabinoids in hair is challenging due to the high analytical sensitivity required. THCCOOH is present in the femtogram to picogram per milligram of hair range. GC/MS/MS is required in most analytical techniques. A novel approach to the screening of hair specimens for the presence of cannabinoids in hair was proposed by Cirimele et al. (1996). They developed a rapid, simple GC/MS screening method for THC, cannabinol, and cannabidiol in hair that did not require derivatization prior to analysis. The method was found to be a sensitive screen for cannabis detection with GC/MS identification of THCCOOH recommended as a confirmatory procedure.

It is difficult to conduct controlled cannabinoid administration studies on the disposition of cannabinoids in hair because of the inability to differentiate administered drug from previously self-administered cannabis. If isotopically labeled drug were administered, it would be possible to identify newly administered drug in hair. There are advantages to monitoring drug use with hair testing including a wide window of drug detection, a less invasive specimen collection procedure, and the ability to collect a second specimen at a later time. However, one of the weakest aspects of testing for cannabinoids in hair is the low sensitivity of drug detection in this alternate matrix. In controlled cannabinoid administration studies conducted by Huestis et al., only about one-third of non-daily users and two-thirds of daily cannabis users had positive cannabinoid hair tests by GC/MS/MS with detection limits of 1 ng/mg for THC and 0.1 ng/mg for THCCOOH. All participants had positive urine cannabinoid tests at the time of hair collection (unpublished data).

References

Adams IB, Martin BR (1996) Cannabis: pharmacology and toxicology in animals and humans. Addiction 91:1585–1614

Agurell S, Leander K (1971) Stability, transfer and absorption of cannabinoid constituents of cannabis (hashish) during smoking. Acta Pharm Suec 8:391–402

Agurell S, Gillespie H, Halldin M, Hollister LE, Johansson E, Lindgren JE, Ohlsson A, Szirmai M, Widman M (1984) A review of recent studies on the pharmacokinetics and metabolism of delta-1-tetrahydrocannabinol, cannabidiol and cannabinol in man. In: Paton SW, Nahas GG (eds) Marijuana '84. IRL Press, Oxford, pp 49–62

Agurell S, Halldin M, Lindgren JE, Ohlsson A, Widman M, Gillespie H, Hollister L (1986) Pharmacokinetics and metabolism of delta-tetrahydrocannabinol and other cannabinoids with emphasis on man. Pharmacol Rev 38:21–43

Atkinson HC, Begg EJ, Darrlow BA (1988) Drugs in human milk clinical pharmacokinetic considerations. Clin Pharmacokinet 14:217–240

Azorlosa JL, Heishman SJ, Stitzer ML, Mahaffey JM (1992) Marijuana smoking: effect of varying delta 9-tetrahydrocannabinol content and number of puffs. J Pharmacol Exp Ther 261:114–122

Bailey JR, Cunny HC, Paule M, Slikker W Jr (1987) Fetal disposition of delta-9-tetrahydrocannbinol (THC) during late pregnancy in the rhesus monkey. Toxicol Appl Pharmacol 90:315–321

Baker D, Pryce G, Croxford JL, Brown P, Pertwee RG, Huffman JW, Layward L (2000) Cannabinoids control spasticity and tremor in a multiple sclerosis model. Nature 404:84–87

Balabanova S, Schneider E (1989) Nachweis von drogen im schweis. Beitr Gerichtl Med 48:45–49

Ben-Zvi Z, Burstein S (1974) 7-Oxo-delta-1-tetrahydrocannabinol: a novel metabolite of delta 1 tetrahydrocannabinol. Int J Clin Pharmacol Ther 8:223–229

Ben-Zvi Z, Bergen JR, Burstein S, Sehgal PK, Varanelli C (1976) The metabolism of delta-tetrahydrocannabinol in the rhesus monkey. In: Braude MC, Szara S (eds) The pharmacology of marijuana. Raven Press, New York, pp 63–75

Benowitz NL, Nguyen TL, Jones RT, Herning RI, Bachman J (1980) Metabolic and psychophysiologic studies of cannabidiol-hexobarbital interaction. Clin Pharmacol Ther 28:115–117

Binitie A (1975) Psychosis following ingestion of hemp in children. Psychopharmacologia 44:301–302

Blackard C, Tennes K (1984) Human placental transfer of cannabinoids. N Engl J Med 311:797

Blanc JA, Manneh VA, Ernst R, Berger DE, de Keczer SA, Chase C, Centofanti JM, DeLizza AJ (1993) Adsorption losses from urine-bases cannabinoid calibrators during routine use. Clin Chem 39:1705–1717

Bloom AS (1982) Effect of delta-9-tetrahydrocannabinol on the synthesis of dopamine and norepinephrine in mouse brain synaptosomes. J Pharmacol Exp Ther 221:97–103

Bornheim LM, Lasker JM, Raucy JL (1992) Human hepatic microsomal metabolism of delta-1-tetrahydrocannabinol. Drug Metab Dispos 20:241–246

Brenneisen R, Egli A, ElSohly MA, Henn V, Spiess Y (1996) The effect of orally and rectally administered delta-9-tetrahydrocannabinol on spasticity: a pilot study with 2 patients. Int J Clin Pharmacol Ther 34:446–452

Burstein S, Rosenfeld J, Wittstruck T (1972) Isolation and characterization of two major urinary metabolites of delta-1-tetrahydrocannabinol. Science 176:422–424

Cairns ER, Howard RC, Hung CT, Menkes DB (1990) Saliva THC levels and cannabis intoxication. Report No. 2411, Chemistry Division, Division of Scientific and Industrial Research, Lower Hutt, New Zealand, pp 1–71

Cairns T, Kippenberger DJ, Scholtz H, Baumgartner WA (1995) determination of carboxy-THC in hair by mass spectrometry/mass spectrometry. In: de Zeeuw RA, Al Hosani I, Al Munthiri S, Maqbool A (eds) Hair Analysis in Forensic Toxicology: proceedings of the 1995 international conference and workshop. The Organizing Committee of the conference, Abu Dhabi, pp 185–193

Cami J, Guerra D, Ugena B, Segura J, de La Torre R (1991) Effect of subject expectancy on the THC intoxication and disposition from smoked hashish cigarettes. Pharmacol Biochem Behav 40:115–119

Chiang CN, Rapaka RS (1987) Pharmacokinetics and disposition of cannabinoids. In: Rapaka RS, Makriyannis A (eds) NIDA research monograph—structure-activity relationships of the cannabinoids. National Institute on Drug Abuse, Rockville, pp 173–188

Christophersen AS (1986) TetraHydroCannabinol stability in whole blood: plastic versus glass containers. J Anal Toxicol 10:129–131

Cirimele V (1996) Cannabis and amphetamine determination in human hair. In: Kintz P (ed) Drug testing in hair. CRC Press, Boca Raton, pp 181–189

Cirimele V, Sachs H, Kintz P, Mangin P (1996) Testing human hair for cannabis. III. Rapid screening procedure for the simultaneous identification of delta9-tetrahydrocannabinol, cannabinol, and cannabidiol. J Anal Toxicol 20:13–16

Claussen U, Korte F (1968) Uber das verhalten von hanf und von delta-9-6a, 10a-trans-tetrahydrocannabinol biem rauchen. Justus Liebigs Ann Chem 713:162–165

Cone EJ, Lange R, Darwin WD (1998) In vivo adulteration: excess fluid ingestion causes false-negative marijuana and cocaine urine test results. J Anal Toxicol 22:460–473

Dewey WL, McMillan DE, Harris LS, Turk RF (1972) Distribution of radioactivity in brain of tolerant and nontolerant pigeons treated with 3-H-delta-9-tetrahydrocannabinol. Biochem Pharmacol 22:399–405

Ehorn C, Fretthold D, Maharaj M (1994) Ion trap GC/MS/MS for analysis of THC from sweat. Abstract presented at the Society of Forensic Toxicologists annual meeting, Tampa, 31 October–4 November, p 11

Ellis GM, Mann MA, Judson BA, Schramm NT, Tashchian A (1985) Excretion patterns of cannabinoid metabolites after last use in a group of chronic users. Clin Pharmacol Ther 38:572–578

ElSohly HN, Boeren EG, Turner CE, ElSohly MA (1984) Constituents of cannabis sativa L. XXIV: Cannabitetrol, a new polyhydroxylated cannabinoid. In: Agurell S, Dewey WL, Willette RE (eds) The cannabinoids: chemical, pharmacologic and therapeutic aspects. Academic Press, Orlando, pp 89–96

Feng S, ElSohly MA, Salamone S, Salem MY (2000) Simultaneous analysis of delta-9-THC and its major metabolites in urine, plasma, and meconium by GC-MS using an immunoaffinity extraction procedure. J Anal Toxicol 24:395–402

Foltz RL (1984) Analysis of cannabinoids in physiological specimens by gas chromatography/mass spectrometry. In: Baselt RC (ed) Advances in analytical toxicology. Biomedical Publications, Foster City, pp 125–157

Garrett ER, Hunt CA (1974) Physiochemical properties, solubility, and protein binding of delta-9-tetrahydrocannabinol. J Pharm Pharm Sci 63:1056–1064

Garrett ER, Hunt CA (1977) Pharmacokinetics of delta-9-tetrahydrocannabinol in dogs. J Pharm Pharm Sci 66:395–406

Gjerde H, Beylich KM, Morland J (1993) Incidence of alcohol and drugs in fatally injured car drivers in Norway. Accid Anal Prev 25:479–483

Green MD, Belanger G, Hum DW, Belanger A, Tephly TR (1997) Glucuronidation of opioids, carboxylic acid-containing drugs, and hydroxylated xenobiotics catalyzed by expressed monkey UDP-glucuronosyltransferase 2B9 protein. Drug Metab Dispos 25:1389–1394

Greene ML, Saunders DR (1974) Metabolism of tetrahydrocannabinol by the small intestine. Gastroenterology 66:365–372

Gross SJ, Soares JR (1978) Validated direct blood delta-9-THC radioimmune quantitation. J Anal Toxicol 2:98–100

Gross SJ, Worthy TE, Nerder L, Zimmermann EG, Soares JR, Lomax P (1985) Detection of recent cannabis use by saliva delta-9-THC radioimmunoassay. J Anal Toxicol 9:1–5

Grotenhermen F (2003) Pharmacokinetics and pharmacodynamics of cannabinoids. Clin Pharmacokinet 42:327–360

Gustafson RA, Moolchan ET, Barnes A, Levine B, Huestis MA (2003) Validated method for the simultaneous determination of delta-9-tetrahydrocannabinol (THC), 11-hydroxy-THC and 11-nor-9-carboxy-THC in human plasma using solid phase extraction and gas chromatography-mass spectrometry with positive chemical ionization. J Chromatogr B Analyt Technol Biomed Life Sci 798:145–154

Guy GW, Robson PJ (2004a) A phase I, double blind, three-way crossover study to assess the pharmacokinetic profile of cannabis based medicine extract (CBME) administered sublingually in variant cannabinoid ratios in normal healthy male volunteers (GWPK0215). J Cannabis Ther 3:121–152

Guy GW, Robson PJ (2004b) A phase I, open label, four-way crossover study to compare the pharmacokinetic profiles of a single dose of 20 mg of a cannabis based medicine extract (CBME) administered on 3 different areas of the buccal mucosa and to investigate the pharmacokinetics of CBME per oral in healthy male and female volunteers (GWPK0112). J Cannabis Ther 3:79–120

Hall BJ, Satterfield-Doerr M, Parikh AR, Brodbelt JS (1998) Determination of cannabinoids in water and human saliva by solid-phase microextraction and quadrupole ion trap gas chromatography-mass spectrometry. Anal Chem 70:1788–1796

Halldin MM, Widman M, Bahr CV, Lindgren JE, Martin BR (1982) Identification of in vitro metabolites of delta-tetrahydrocannabinol formed by human livers. Drug Metab Dispos 10:297–301

Hampson AJ, Grimaldi M, Axelrod J, Wink D (1998) Cannabidiol and (–) delta-9-tetrahydrocannabinol are neuroprotective antioxidants. Proc Natl Acad Sci U S A 95:8268–8273

Hanson VW, Buonarati MH, Baselt RC, Wade NA, Yep C, Biasotti AA, Reeve VC, Wong AS, Orbanowsky MW (1983) Comparison of 3H-and 125I-radioimmunoassay and gas chromatography/mass spectrometry for the determination of delta-9-tetrahydrocannabinol and cannabinoids in blood and serum. J Anal Toxicol 7:96–102

Harder S, Rietbrock S (1997) Concentration-effect relationship of delta-9-tetrahydrocannabiol and prediction of psychotropic effects after smoking marijuana. Int J Clin Pharmacol Ther 35:155–159

Harvey DJ (2001) Absorption, distribution, and biotransformation of the cannabinoids. In: Nahas GG, Sutin KM, Harvey DJ, Agurell S (eds) Marijuana and medicine. Humana Press, Totowa, pp 91–103

Harvey DJ, Mechoulam R (1990) Metabolites of cannabidiol identified in human urine. Xenobiotica 20:303–320

Harvey DJ, Paton WDM (1986) Metabolism of the cannabinoids. Rev Biochem Toxicol 6:221–264

Harvey DJ, Martin BR, Paton WDM (1979) Identification and measurement of cannabinoids and their in vivo metabolites in liver by gas chromatography-mass spectrometry. In: Nahas GG, Paton WDM (eds) Marijuana: biological effects. Pergamon Press, Oxford, pp 45–62

Hawks RL (1982) The constituents of cannabis and the disposition and metabolism of cannabinoids. In: Hawks R (ed) The analysis of cannabinoids in biological Fluids. Research Monograph 42. National Institute on Drug Abuse, Rockville, pp 125–137

Hawks RL (1983) Developments in cannabinoid analyses of body fluids: implications for forensic applications. In: Agurell S, Dewey W, Willette R (eds) The cannabinoids: chemical, pharmacologic, and therapeutic aspects. Academic Press, Rockville, pp 1–12

Heishman SJ, Stitzer ML, Yingling JE (1989) Effects of tetrahydrocannabinol content on marijuana smoking behavior, subjective reports, and performance. Pharmacol Biochem Behav 34:173–179

Holdcroft A (1984) Pain therapy. In: Grotenhermen F, Russo E (eds) Cannabis and cannabinoids. Pharmacology, toxicology, and therapeutic potential. The Haworth Integrative Healing Press, New York, pp 181–186

Huang W, Moody DE, Andrenyak DM, Smith EK, Foltz RL, Huestis MA, Newton JF (2001) Simultaneous determination of delta-9-tetrahydrocannabinol and 11-nor-9-carboxy-delta-9-tetrahydrocannabinol in human plasma by solid phase extraction and gas chromatography—negative ion chemical ionization—mass spectrometry. J Anal Toxicol 25:531–537

Huestis MA, Cone EJ (1998a) Alternative testing matrices. In: Karch SB (ed) Drug abuse handbook. CRC Press, Boca Raton, pp 799–857

Huestis MA, Cone EJ (1998b) Differentiating new marijuana use from residual drug excretion in occasional marijuana users. J Anal Toxicol 22:445–454

Huestis MA, Cone EJ (2004) Relationship of Delta-9-tetrahydrocannabinol concentrations in oral fluid and plasma after controlled administration of smoked cannabis. J Anal Toxicol 28:394–399

Huestis MA, Dickerson S, Cone EJ (1992a) Can saliva THC levels be correlated to behavior? American Academy of Forensic Sciences Annual Meeting, New Orleans, p 190

Huestis MA, Henningfield JE, Cone EJ (1992b) Blood cannabinoids. I. Absorption of THC and formation of 11-OH-THC and THCCOOH during and after smoking marijuana. J Anal Toxicol 16:276–282

Huestis MA, Henningfield JE, Cone EJ (1992c) Blood cannabinoids. II. Models for the prediction of time of marijuana exposure from plasma concentrations of delta-9-tetrahydrocannabinol (THC) and 11-nor-9-carboxy-delta-9-tetrahydrocannabinol (THCCOOH). J Anal Toxicol 16:283–290

Huestis MA, Sampson AH, Holicky BJ, Henningfield JE, Cone EJ (1992d) Characterization of the absorption phase of marijuana smoking. Clin Pharmacol Ther 52:31–41

Huestis MA, Mitchell JM, Cone EJ (1995) Detection times of marijuana metabolites in urine by immunoassay and GC-MS. J Anal Toxicol 19:443–449

Huestis MA, Mitchell JM, Cone EJ (1996) Urinary excretion profiles of 11-nor-9-carboxy-Δ9-tetrahydrocannabinol in humans after single smoked doses of marijuana. J Anal Toxicol 20:441–452

Hunt CA, Jones RT (1980) Tolerance and disposition of tetrahydrocannabinol in man. J Pharmacol Exp Ther 215:35–44

Hunt CA, Jones RT, Herning RI, Bachman J (1981) Evidence that cannabidiol does not significantly alter the pharmacokinetics of tetrahydrocannabinol in man. J Pharmacokinet Biopharm 9:245–260

Iribarne C, Berthou F, Baird S, Dreano Y, Picart D, Bail JP, Beaune P, Menez JF (1996) Involvement of cytochrome P450 3A4 enzyme in the N-demethylation of methadone in human liver microsomes. Chem Res Toxicol 9:365–373

Jehanli A, Brannan S, Moore L, Spiehler VR (2001) Blind trials of an onsite saliva drug test for marijuana and opiates. J Forensic Sci 46:1214–1220

Joern WA (1992) Surface adsorption of the urinary marijuana carboxy metabolite: the problem and a partial solution. J Anal Toxicol 16:401

Johansson E, Halldin MM (1989) Urinary excretion half-life of delta1-tetrahydrocannabinol-7-oic acid in heavy marijuana users after smoking. J Anal Toxicol 13:218–223

Johansson E, Agurell S, Hollister LE, Halldin MM (1988) Prolonged apparent half-life of delta-1-tetrahydrocannabinol in plasma of chronic marijuana users. J Pharm Pharmacol 40:374–375

Johansson E, Halldin MM, Agurell S, Hollister LE, Gillespie HK (1989a) Terminal elimination plasma half-life of delta-1-tetrahydrocannabinol (selta-1-THC) in heavy users of marijuana. Eur J Clin Pharmacol 37:273–277

Johansson E, Noren K, Sjovall J, Halldin MM (1989b) Determination of delta-1-tetrahydrocannabinol in human fat biopsies from marihuana users by gas chromatography-mass spectrometry. Biomed Chromatogr 3:35–38

Karst M, Salim K, Burstein S, Conrad I, Hoy L, Schneider U (2003) Analgesic effect of the synthetic cannabinoid CT-3 on chronic neuropathic pain. J Am Med Assoc 290:1757–1762

Kelly P, Jones RT (1992) Metabolism of tetrahydrocannabinol in frequent and infrequent marijuana users. J Anal Toxicol 16:228–235

Kemp PM, Abukhalaf IK, Manno JE, Manno BR, Alford DD, Abusada GA (1995a) Cannabinoids in humans. I. Analysis of delta-9-tetrahydrocannabinol and six metabolites in plasma and urine using GC-MS. J Anal Toxicol 19:285–291

Kemp PM, Abukhalaf IK, Manno JE, Manno BR, Alford DD, McWilliams ME, Nixon FE, Fitzgerald MJ, Reeves RR, Wood MJ (1995b) Cannabinoids in humans. II. The influence of three methods of hydrolysis on the concentration of THC and two metabolites in urine. J Anal Toxicol 19:292–298

Kidwell DA, Holland JC, Athanaselis S (1998) Testing for drugs of abuse in saliva and sweat. J Chromatogr 713:111–135

Kim KR, Yoon HR (1996) Rapid screening for acidic non-steroidal anti-inflammatory drugs in urine by gas chromatography-mass spectrometry in the selected-ion monitoring mode. J Chromatogr B Biomed Appl 682:55–66

Kintz P, Cirimele V, Mangin P (1995) Testing human hair for cannabis II. Identification of THC-COOH by GC-MS-NCI as a unique proof. J Forensic Sci 40:619–622

Kintz P, Brenniesen R, Bundeli P, Mangin P (1997) Sweat testing for heroin and metabolites in a heroin maintenance program. Clin Chem 43:736–739

Kintz P, Cirimele V, Ludes B (2000) Detection of cannabis in oral fluid (saliva) and forehead wipes (sweat) from impaired drivers. J Anal Toxicol 24:557–561

Kreuz DS, Axelrod J (1973) Delta-9-tetrahydrocannabinol: localization in body fat. Science 179:391–393

Krishna DR, Klotz U (1994) Extrahepatic metabolism of drugs in humans. Clin Pharmacokinet 26:144–160

Law B, Mason PA, Moffat AC, Gleadle RI, King LJ (1984) Forensic aspects of the metabolism and excretion of cannabinoids following oral ingestion of cannabis resin. J Pharm Pharmacol 36:289–294

Lee CC, Chiang CN (1985) Appendix. Maternal-fetal transfer of abused substances: pharmacokinetic and pharmacodynamic data. In: Chiang CN, Lee CC (eds) NIDA Research monograph 60. U.S. Department of Health and Human Services, Rockville, pp 110–147

Lemberger L, Silberstein SD, Axelrod J, Kopin IJ (1970) Marihuana: studies on the disposition and metabolism of delta-9-tetrahydrocannabinol in man. Science 170:1320–1322

Lemberger L, Tamarkin NR, Axelrod J (1971) Delta-9-tetrahydrocannabinol: metabolism and disposition in long-term marihuana smokers. Science 173:72–74

Lemberger L, Crabtree RE, Rowe HM (1972) 11-hydroxy-delta-9-tetrahydrocannabinol: pharmacology, disposition, and metabolism of a major metabolite of marijuana in man. Science 177:62–64

Lindgren JE, Ohlsson A, Agurell S, Hollister L, Gillespie H (1981) Clinical effects and plasma levels of delta-9-tetrahydrocannabinol (delta-9-THC) in heavy and light users of cannabis. Psychopharmacology (Berl) 74:208–212

Malfait AM, Galllily R, Sumariwalla PF, Malik AS, Andreakos E, Mechoulam R, Feldmann M (2000) The nonpsychoactive cannabis constituent cannabidiol is an oral anti-arthritic therapeutic in murine collagen-induced arthritis. Proc Natl Acad Sci U S A 97:9561–9566

Manno JE, Ferslew KE, Manno BR (1984) Urine excretion patterns of cannabinoids and the clinical application of the EMIT-d.a.u. cannabinoid urine assay for substance abuse treatment. In: Agurell S, Dewey WL, Willette RE (eds) The cannabinoids: chemical, pharmacologic, and therapeutic aspects. Harcourt Brace Jonanovich, Orlando, pp 281–290

Manno JE, Manno BR, Kemp PM, Alford DD, Abukhalaf IK, McWilliams ME, Hagaman FN, Fitzgerald MJ (2001) Temporal indication of marijuana use can be estimated from plasma and urine concentrations of delta-9-tetrahydrocannabinol, 11-hydroxy-delta-9-tetrahydrocannabinol, and 11-nor-delta-9-tetrahydrocannabinol-9-carboxylic acid. J Anal Toxicol 25:538–549

Martin BR, Dewey WL, Harris LS, Beckner JS (1977) 3H-Delta-9-tetrahydrocannbinol distribution in pregnant dogs and their fetuses. Res Commun Chem Pathol Pharmacol 17:457–470

Mason AP, McBay AJ (1984) Ethanol, marijuana, and other drug use in 600 drivers killed in single-vehicle crashers in North Carolina, 1978–1981. J Forensic Sci 29:987–1026

Mason AP, McBay AJ (1985) Cannabis: pharmacology and interpretation of effects. J Forensic Sci 30:615–631

Matsunaga T, Iwawaki Y, Watanabe K, Yamamoto I, Kageyama T, Yoshimura H (1995) Metabolism of delta-9-tetrahydrocannabinol by cytochrome P450 isozymes purified from hepatic microsomes of monkeys. Life Sci 56:2089–2095

Mattes RD, Shaw LM, Edling-Owens J, Engelman K, ElSohly MA (1993) Bypassing the first-pass effect for the therapeutic use of cannabinoids. Pharmacol Biochem Behav 44:745–747

Mattes RD, Engelman K, Shaw LM, ElSohly MA (1994) Cannabinoids and appetite stimulation. Pharmacol Biochem Behav 49:187–195

McArdle K, Mackie P, Pertwee R, Guy G, Whittle B, Hawksworth G (2001) Selective inhibition of delta-9-tetrahydrocannabinol metabolite formation by cannabidiol in vitro (abstract) in Proceedings of the BTS Annual Congress. Toxicology 168:133–134

McBurney LJ, Bobbie BA, Sepp LA (1986) GC/MS and Emit analyses for delta-9-tetrahydrocannabinol metabolites in plasma and urine of human subjects. J Anal Toxicol 10:56–64

Mechoulam R (1970) Marihuana chemistry. Science 168:1159–1166

Mechoulam R, Zvi ZB, Agurell S, Nilsson IM, Nilsson JLG, Edery H, Grunfeld Y (1973) Delta-6 tetrahydrocannabinol-7-oic acid, a urinary delta-6-THC metabolite: isolation and synthesis. Experientia 29:1193–1195

Meier HJ, Vonesch HJ (1997) Cannabis-intoxikation nach salatgenuss. Schweiz Med Wochenschr 127:214–218

Melamede R (1984) Possible mechanisms in autoimmune diseases. In: Grotenhermen F, Russo E (eds) Cannabis and cannabinoids. Pharmacology, toxicology, and therapeutic potential. The Haworth Integrative Healing Press, New York, pp 111–122

Moeller MR, Doerr G, Warth S (1992) Simultaneous quantitation of delta-9-tetrahydrocannabinol (THC) and 11-nor-9-carboxy-delta-9-tetrahydrocannabinol (THC-COOH) in serum by GC/MS using deuterated internal standards and its application to a smoking study and forensic cases. J Forensic Sci 37:969–983

Moore C, Guzaldo F, Donahue T (2001) The determination of 11-nor-delta-9-tetrahydrocannabinol-9-carboxylic acid (THC-COOH) in hair using negative ion gas chromatography-mass spectrometry and high-volume injection. J Anal Toxicol 25:555–558

Mura P, Kintz P, Papet Y, Ruesch G, Piriou A (1999) Evaluation of six rapid tests for screening of cannabinoids in sweat, saliva and urine. Acta Clin Belg 1:35–38

Nebro W, Barnes A, Gustafson R, Moolchan E, Huestis M (2004) Delta-9-tetrahydrocannabinol (THC), 11-hydroxy-delta-9-tetrahydrocannabinol (11-OH-THC) and 11-nor-9-carboxy-delta-9-tetrahydrocannabinol (THCCOOH) in human plasma following oral administration of hemp oil. 2004 FBI/SOFT/TIAFT Meeting

Niedbala RS, Kardos KW, Fritch DF, Kardos S, Fries T, Waga J (2001) Detection of marijuana use by oral fluid and urine analysis following single-dose administration of smoked and oral marijuana. J Anal Toxicol 25:289–303

Ohlsson A, Lindgren JE, Wahlen A, Agurell S, Hollister LE, Gillespie HK (1980) Plasma delta-9-tetrahydrocannabinol concentrations and clinical effects after oral and intravenous administration and smoking. Clin Pharmacol Ther 28:409–416

Ohlsson A, Lindgren JE, Wahlen A, Agurell S, Hollister LE, Gillespie HK (1981) Plasma levels of delta-9-tetrahydrocannabinol after intravenous, oral and smoke administration. NIDA Res Monogr 34:250–256

Ohlsson A, Lindgren JE, Wahlen A, Agurell S, Hollister LE, Gillespie HK (1982) Single dose kinetics of deuterium labelled delta-1-tetrahydrocannabinol in heavy and light cannabis users. Biomed Environ Mass Spectrom 9:6–10

Ohlsson A, Agurell S, Londgren JE, Gillespie HK, Hollister LE (1985) Pharmacokinetic studies of delta-1-tetrahydrocannabinol in man. In: Barnett G, Chiang CN (eds) Pharmacokinetics and pharmacodynamics of psychoactive drugs. Mosby Yearbook, St Louis, pp 75–92

Ohlsson A, Lindgren JE, Andersson S, Agurell S, Gillespie H, Hollister LE (1986) Single-dose kinetics of deuterium-labelled cannabidiol in man after smoking and intravenous administration. Biomed Environ Mass Spectrom 13:77–83

Owens SM, McBay AJ, Reisner HM, Perez-Reyes M (1981) 125 l radioimmunoassay of delta-9-tetrahydrocannabinol in blood and plasma with a solid-phase second-antibody separation method. Clin Chem 27:619–624

Peat MA (1989) Distribution of delta-9-tetrahydrocannabinol and its metabolites. In: Baselt RC (ed) Advances in analytical toxicology II. Year Book Medical Publishers, Chicago, pp 186–217

Peel HW, Perrigo BJ, Mikhael NZ (1984) Detection of drugs in saliva of impaired drivers. J Forensic Sci 29:185–189

Perez-Reyes M (1985) Pharmacodynamics of certain drugs of abuse. In: Barnett G, Chiang CN (eds) Pharmacokinetics and pharmacodynamics of psychoactive drugs. Biomedical Publications, Foster City, pp 287–310

Perez-Reyes M (1990) Marijuana smoking: factors that influence the bioavailability of tetrahydrocannabinol. In: Chiang CN, Hawks RL (eds) Research findings on smoking of abused substances. NIDA Research Monograph 99, Rockville, pp 42–62

Perez-Reyes M, Wall ME (1982) Presence of delta-9-tetrahydrocannabinol in human milk. N Engl J Med 307:819–820

Perez-Reyes M, Timmons MC, Lipton MA, Davis KH, Wall ME (1972) Intravenous injection in man of delta-9-tetrahydrocannabinol and 11-OH-delta-9-tetrahydrocannabinol. Science 177:633–635

Perez-Reyes M, Lipton MA, Timmons MC, Wall ME, Brine DR, Davis KH (1973a) Pharmacology of orally administered delta-9-tetrahydrocannabinol. Clin Pharmacol Ther 14:48–55

Perez-Reyes M, Timmons MC, Davis KH, Wall EM (1973b) A comparison of the pharmacological activity in man of intravenously administered delta-9-tetrahydrocannabinol, cannabinol and cannabidiol. Experientia 29:1368–1369

Perez-Reyes M, Owens SM, Di Guiseppi S (1981) The clinical pharmacology and dynamics of marijuana cigarette smoking. J Clin Pharmacol 21:201S–207S

Perez-Reyes M, Di Guiseppi S, Davis KH, Schindler VH, Cook CE (1982) Comparison of effects of marijuana cigarettes of three different potencies. Clin Pharmacol Ther 31:617–624

Pertwee RG (2004) The pharmacology and therapeutic potential of cannabidiol. In: Di Marzo V (ed) Cannabinoids. Kluwer Academic/Plenum Publishers, New York, pp 32–83

Plasse T (1984) Antiemetic effects of cannabinoids. In: Grotenhermen F, Russo E (eds) Cannabis and cannabinoids. Pharmacology, toxicology, and therapeutic potential. The Haworth Integrative Healing Press, New York, pp 165–180

Samyn N, van Haeren C (2000) On-site testing of saliva and sweat with Drugwipe and determination of concentrations of drugs of abuse in saliva, plasma and urine of suspected users. Int J Legal Med 113:150–154

Shaw LM, Edling-Owens J, Mattes R (1991) Ultrasensitive measurement of delta-9-tetrahydrocannabinol with a high energy dynode detector and electron-capture negative chemical-ionization mass spectrometry. Clin Chem 37:2062–2068

Soares JR, Gross SJ (1976) Separate radioimmune measurements of body fluid delta-9THC and 11-nor-9-carboxy-delta-THC. Life Sci 19:1711–1717

Soares JR, Grant JD, Gross SJ (1982) Significant developments in radioimmune methods applied to delta-9-THC and its 9-substituted metabolites. In: Hawks R (ed) NIDA Research Monograph 42: National Institute on Drug Abuse, Bethesda, pp 44–55

Sporkert F, Pragst F, Ploner CJ, Tschirch A, Stadelmann AM (2001) Pharmacokinetic investigations and delta-9-tetrahydrocannabinol and its metabolites after single administration of 10 mg Marinol in attendance of a psychiatric study. The Annual Meeting of The International Association of Forensic Toxicologists, Prague, pp 1–62

Stinchcomb AL, Valiveti S, Hammell DC, Ramsey DR (2004) Human skin permeation of delta-8-tetrahydrocannabinol, cannabidiol and cannabinol. J Pharm Pharmacol 56:291–297

Turner CE, ElSohly MA, Boeren EG (1980) Constituents of cannabis sativa L. XVII. A review of the natural constituents. J Nat Prod 43:169–234

Vaughan CW, Christie MJ (1984) Mechanisms of cannabinoid analgesia. In: Grotenhermen F, Russo E (eds) Cannabis and cannabinoids. Pharmacology, toxicology, and therapeutic potential. The Haworth Integrative Healing Press, New York, pp 89–100

Wall ME, Perez-Reyes M (1981) The metabolism of delta-9-tetrahydrocannabinol and related cannabinoids in man. J Clin Pharmacol 21:178S–189S

Wall ME, Taylor HL (1984) Conjugation of acidic metabolites of delta-8 and delta-9-THC in man. In: Harvey DJ, Paton SW, Nahas GG (eds) Marihuana 84 Proceedings of the Oxford Symposium on Cannabis. IRL Press Limited, Oxford, pp 69–76

Wall ME, Brine DR, Perez-Reyes M (1976) Metabolism of cannabinoids in man. In: Braude MC, Szara S (eds) The Pharmacology of marihuana. Raven Press, New York, pp 93–113

Wall ME, Sadler BM, Brine D, Taylor H, Perez-Reyes M (1983) Metabolism, disposition, and kinetics of delta-9-tetrahydrocannabinol in men and women. Clin Pharmacol Ther 34:352–363

Watanabe K, Tanaka T, Yamamoto I, Yoshimura H (1988) Brain microsomal oxidation of delta-8- and delta-9-tetrahydrocannabinol. Biochem Biophys Res Commun 157:75–80

Widman M, Agurell S, Ehrnebo M, Jones G (1974) Binding of (+) and (−)-delta1-tetrahydrocannabinols and (−)-7-hydroxy-delta1-tetrahydrocannabinol to blood cells and plasma proteins in man. J Pharm Pharmacol 26:914–916

Widman M, Nordqvist M, Dollery CT, Briant RH (1975) Metabolism of delta-1-tetrahydrocannabinol by the isolated perfused dog lung. Comparison with in vitro liver metabolism. J Pharm Pharmacol 27:842–848

Williams PL, Moffat AC (1980) Identification in human urine of delta-9-tetrahydrocannabinol-11-oic glucuronide: a tetrahydrocannabinol metabolite. J Pharm Pharmacol 32:445–448

Zajicek J, Fox P, Sanders H, Wright D, Vickery J, Nunn A, Thompson A (2003) Cannabinoids for treatment of spasticity and other symptoms related to multiple sclerosis (CAMS study): multicentre randomised placebo-controlled trial. Lancet 362:1517–1526

Zuardi AW, Shirakawa I, Finkelfarb E, Karniol IG (1982) Action of cannabidiol on the anxiety and other effects produced by delta-9-THC in normal subjects. Psychopharmacology (Berl) 76:245–250

HEP (2005) 168:691–717

Cannabinoid Tolerance and Dependence

A.H. Lichtman (✉) · B.R. Martin

Department of Pharmacology and Toxicology, Virginia Commonwealth University, 410 North 12th Street, P.O. Box 980613, Richmond VA, 23298-0613, USA
alichtma@hsc.vcu.edu

Abstract The use of marijuana for recreational and medicinal purposes has resulted in a large prevalence of chronic marijuana users. Consequences of chronic cannabinoid administration include profound behavioral tolerance and withdrawal symptoms upon drug cessation. A marijuana withdrawal syndrome is only recently gaining acceptance as being clinically significant. Similarly, laboratory animals exhibit both tolerance and dependence following chronic administration of cannabinoids. These animal models are being used to evaluate the high degree of plasticity that occurs at the molecular level in various brain regions following chronic cannabinoid exposure. In this review, we describe recent advances that have increased our understanding of the impact of chronic cannabinoid administration on cannabinoid receptors and their signal transduction pathways. Additionally, we discuss several potential pharmacotherapies that have been examined to treat marijuana dependence.

Keywords Cannabinoid · Dependence · Marijuana · THC · Withdrawal · Tolerance · Rimonabant · Cannabis

1 Introduction

The worldwide use of cannabis for recreational as well as for medicinal purposes has resulted in a large population of individuals chronically using this drug. Consequently, many recent reviews have focused on the long-term consequences of chronic marijuana use, particularly as it relates to tolerance and dependence (Lichtman et al. 2002; Maldonado 2002; Maldonado and Rodriguez De Fonseca 2002; Tanda and Goldberg 2003; Martin et al. 2004). Although similar antecedents (i.e., prolonged exposure to a particular drug) lead to tolerance and withdrawal, both processes are mediated by different neurochemical and neuroanatomical substrates. The profound behavioral tolerance that occurs during repeated administration of Δ^9-tetrahydrocannabinol (THC), the primary active constituent of marijuana, or other cannabinoid agonists illustrates the high degree of plasticity that can occur within the endocannabinoid system. This plasticity can be attributed, in part, to changes that occur to the CB_1 receptor, cell signaling processes, and possibly changes in levels of endogenous cannabinoids in the CNS. In contrast, termination of chronic cannabinoids results in a variety of withdrawal symptoms in humans as well as laboratory animals. In humans, abstinence from continual marijuana use leads to delayed withdrawal symptoms manifested as physiological symptoms of decreased appetite and weight loss, as well as emotional changes, which include irritability, anxiety, restlessness, and strange dreams. Acceptance that these symptoms reflect a clinically significant withdrawal syndrome is gaining. In addition, the availability of several laboratory animal models of cannabinoid withdrawal is playing an important role in understanding cannabinoid dependence and may contribute to the development of pharmacotherapies. In this review, we will discuss published research that has been conducted on (1) characterizing cannabinoid tolerance and dependence, (2) providing insight into the mechanisms of action underlying cannabinoid tolerance and dependence, and (3) evaluating potential pharmacotherapies to treat cannabinoid withdrawal.

2 Reinforcing Effects of Cannabinoids in Animals

Despite the fact that marijuana has consistently been the most commonly used illicit drug for more than 25 years (Johnston et al. 2004), it is only relatively recently that cannabinoids have been shown to elicit rewarding effects in animal models of addiction, including drug self-administration, conditioned place-preference, and intracranial self-stimulation paradigms. As this issue is the topic of several recent reviews (Gardner 2002; Maldonado 2002; Maldonado and Rodriguez de Fonseca 2002; Tanda and Goldberg 2003; Varvel et al. 2004), it will only be briefly discussed here.

The most compelling preclinical evidence suggesting that a drug has reinforcing properties comes from drug-self administration studies. Most early studies attempting to determine whether THC is self-administered in laboratory animals met with failure, except for studies by Takahashi and Singer showing that THC self-administration occurs in food-deprived rats when a fixed-time 1-min non-contingent food delivery schedule is operating (Takahashi and Singer 1979, 1980). However, the refinement of procedural issues, particularly the use of sufficiently low doses that lacked both motor and aversive effects, has led to several recent papers demonstrating that mice (Martellotta et al. 1998; Ledent et al. 1999), rats (Braida et al. 2001b; Fattore et al. 2001), and squirrel monkeys (Tanda et al. 2000; Justinova et al. 2003) will self-administer Δ^9-THC as well as exogenous cannabinoids (e.g., WIN 55,212-2 and CP 55,940).

In the conditioned place preference (CPP) paradigm, subjective effects of a given drug are repeatedly paired with stimuli associated with one of two experimental chambers, while the other chamber is repeatedly paired with vehicle injections. On test days, the subjects are given free access to both chambers, and the relative amount of time spent in the drug-paired chamber is taken as an indicator of the rewarding/aversive properties of that drug. As in the case of self-administration, the occurrence of preference for the chamber paired with drug is generally inversely related to drug dose: low doses lead to a place preference (Lepore et al. 1995; Valjent and Maldonado 2000; Braida et al. 2001a,b; Ghozland et al. 2002) while high doses elicit conditioned place aversions (Sanudo-Pena et al. 1997; Hutcheson et al. 1998; Zimmer et al. 2001).

The third general technique that has been used to infer the rewarding properties of drugs is the propensity of a drug to decrease the threshold for electrical self-stimulation of neural reward circuits (see Wise 1996). Substantial work demonstrating that Δ^9-THC decreases brain-stimulation reward thresholds comes from the laboratory of Gardner and colleagues (Gardner et al. 1988, 1989; Gardner and Lowinson 1991). Interestingly, the positive effects of Δ^9-THC in this paradigm are strain dependent, with the greatest efficacy in Lewis rats, followed by Sprague-Dawley rats; there is a complete lack of efficacy in Fischer rats (Lepore et al. 1996).

3
Overview of Cannabinoid Tolerance in Whole Animals

THC and a variety of cannabinoid agonists have been reliably shown to produce a constellation of pharmacological effects as evaluated in a variety of animal behavioral models, including the dog-static ataxia test, rat and monkey drug discrimination, and the tetrad test in mice (i.e., depression of spontaneous activity, antinociception, hypothermia, and catalepsy) (Dewey et al. 1972; Chaperon and Thiebot 1999; Martin 2002). These behavioral effects are well known to undergo tolerance upon subchronic dosing of THC and potent synthetic cannabinoid agonists, such as WIN 55,212-2, CP 55,940, and HU-210. Not only do these behaviors undergo tolerance at different rates (Fan et al. 1994; De Vry et al. 2004), but also they recover at different rates following cessation of cannabinoid administration (Bass and Martin 2000). These findings suggest that the various behaviors may be subserved by distinct mechanisms for the production and maintenance of tolerance.

THC has also been shown to produce dose-dependent decreases in cerebral glucose utilization, especially in structures subserving limbic and sensory functions (Freedland et al. 2002). After subchronic dosing of THC, the rates of glucose utilization in the majority of brain structures were similar to control levels, indicating that tolerance had taken place (Whitlow et al. 2003). Remarkably, THC continued to produce acute alterations in functional activity in mesolimbic and amygdalar regions, despite repeated THC administration, suggesting the provocative possibility that behaviors subserved by these structures (e.g. anxiety, stress, reward, and memory) may continue to be affected by THC, even after chronic dosing (Whitlow et al. 2003).

Below, we will discuss the cellular changes that occur following repeated cannabinoid administration, and which may underlie behavioral tolerance. However, it is important to note that associative learning is also known to contribute to drug tolerance and dependence (Siegel and Ramos 2002). Using a Pavlovian conditioning paradigm, it was shown that repeated administration of HU-210 resulted in a more rapid development of tolerance to the decreased ambulatory behavior when the drug was administered in the testing environment than when it was given in a separate context (Hill et al. 2004).

3.1 Cellular and Molecular Changes Associated with Cannabinoid Tolerance

The profound tolerance that occurs following repeated administration of cannabinoids illustrates the high degree of plasticity that can occur in the endocannabinoid system. This plasticity can be attributed, in part, to changes that occur to the CB_1 receptor, which include sequestration into an intracellular vesicle (internalization) and either receptor degradation (downregulation) or recycling to the cell membrane (Ferguson and Caron 1998; Krupnick and Benovic 1998). Several significant changes have been demonstrated to occur to CB_1 signaling pathways. Acute stimulation of CB_1 receptors has been demonstrated to activate the pertussis toxin-sensitive G proteins (Howlett et al. 2002), the consequence of which includes inhibition of adenylyl cyclase, decreased Ca^{2+} conductance, and increased K^+ conductance. Further downstream, many intracellular kinases are activated including the mitogen-activated protein kinases (MAPK), extracellular signal-regulated kinases type 1 and 2 (ERK1/2) (Bouaboula et al. 1995; Rueda et al. 2000), protein kinase B (PKB, also, known as Akt) (Gomez Del Pulgar et al. 2000), and focal adhesion kinase (FAK) (Derkinderen et al. 1996). Repeated administration of cannabinoids leads to receptor/G protein uncoupling and desensitization. Interestingly, these changes appear to vary between brain regions, which is likely to account for differences in tolerance to various cannabinoid-mediated behaviors. These processes are discussed in the following sections.

3.2
CB_1 Receptor Downregulation

Several groups have demonstrated that repeated cannabinoid administration consistently leads to a reduction in CB_1 receptor levels in brain. Specifically, repetitive treatment with THC, CP 55,940, or WIN 55,212-2 results in a decrease in CB_1 receptor binding in brain sections or membrane homogenates from whole brain (Oviedo et al. 1993; Romero et al. 1997; Breivogel et al. 1999; Rubino et al. 2000b; Sim-Selley and Martin 2002). Autoradiographic studies have allowed analysis of a large number of brain regions and have revealed that downregulation occurs in all CB_1 receptor-containing brain regions following subchronic administration of WIN 55,212-2 or THC, including cerebellum, hippocampus, caudate-putamen, globus pallidus, substantia nigra, prefrontal, cingulate and entorhinal cortices, nucleus accumbens, amygdala, hypothalamus, thalamus, and periaqueductal gray (Sim-Selley and Martin 2002). However, the magnitude of downregulation varies in different brain regions. For example, comparatively small changes are found in the basal ganglia output nuclei (globus pallidus, entopeduncular nucleus, and substantia nigra). In addition, regional differences in CB_1 receptor downregulation are not constant throughout the period of tolerance development (Romero et al. 1998b; Breivogel et al. 1999). Downregulation occurs more rapidly and with greater magnitude in the hippocampus and cerebellum compared with the basal ganglia.

Although the mechanisms that regulate CB_1 receptor synthesis, posttranslational modification, degradation, and internalization remain largely unknown, studies are beginning to address this area. In hippocampal neuronal cultures and in a Xenopus oocyte expression system, CB_1 receptor desensitization has been shown to require G protein-coupled receptor kinase (GRK) and β-arrestin (Jin et al. 1999; Kouznetsova et al. 2002). Internalization of CB_1 receptors following agonist treatment in CB_1 receptor-transfected cells (Hsieh et al. 1999) and hippocampal cultures (Coutts et al. 2001) has been shown to lead to internalization of CB_1 receptors.

Another possible explanation for decreased CB_1 receptor binding after chronic cannabinoid treatment is that receptor synthesis has been attenuated. However, studies investigating CB_1 receptor mRNA following repeated cannabinoid administration have resulted in mixed results. In one study, CB_1 receptor mRNA was decreased only in the caudate-putamen of animals treated with THC or CP 55,940 (Rubino et al. 1994). The duration of THC treatment as well as the interval between drug injection and sacrifice appear to contribute to the direction and magnitude of change in CB_1 receptor mRNA (Zhuang et al. 1998). However, the results of this study also suggested that alterations in CB_1 receptor synthesis underlie adaptations in the caudate-putamen, but not in the hippocampus and cerebellum.

3.3 CB_1 Receptor-Activated G Proteins and Effectors

Agonist-stimulated [^{35}S]GTPγS binding has been a very useful assay to investigate the consequences of receptor downregulation and desensitization. Following THC administration for 3 weeks, a high level of CB_1 receptor desensitization occurs in many brain areas of the rat, including the hippocampus, cerebellum, caudate-putamen, globus pallidus, substantia nigra, septum, and cortex (Sim et al. 1996a). The pattern of desensitization was similar to that of downregulation, with the globus pallidus and substantia nigra exhibiting the smallest magnitude of change. Repeated administration of THC, WIN 55212-2, CP 55,940 or anandamide results in CB_1 receptor desensitization (Breivogel et al. 1999; Rubino et al. 2000a,b; Sim-Selley and Martin 2002). As in the case of downregulation, desensitization develops at different rates and magnitudes in different brain regions (Breivogel et al. 1999). Desensitization in the hippocampus occurs rapidly and with a large magnitude, while desensitization in the cerebellum develops at a slightly slower rate, but to the same magnitude, as compared to the hippocampus. However, the caudate putamen and globus pallidus exhibit slower rates of development and smaller magnitudes of desensitization than found in hippocampus and cerebellum. Whereas changes in G protein mRNA have been found and could contribute to loss of CB_1 receptor-mediated G protein activity, these alterations were not accompanied by concomitant changes in protein expression (Rubino et al. 1997). Thus, it is unclear whether the loss of CB_1 receptor-mediated G protein activity is the result of CB_1 receptor downregulation, receptor–G protein uncoupling, altered G protein expression, or some combination of these adaptations.

The results of cell culture models have yielded convincing evidence demonstrating that subchronic cannabinoid administration leads to desensitization of adenylyl cyclase and ERK1/2 (Rinaldi-Carmona et al. 1998). In contrast, the results have been somewhat mixed when attempts have been made to examine this signaling pathway in whole animals. Mice treated repeatedly with CP 55,940 did not exhibit altered CB_1-mediated inhibition of adenylyl cyclase in the cerebellum (Fan et al. 1996). In contrast, several other studies have reported increases in basal, forskolin or Ca^{2+}-stimulated adenylyl cyclase activity in the cortex, striatum, and cerebellum of mice treated subchronically with THC (Hutcheson et al. 1998; Rubino et al. 2000b; Tzavara et al. 2000).

Tolerance procedures have great utility for identifying signaling events that are uniquely specific to the pharmacological effects of cannabinoids. In particular, regional brain differences in the magnitude and time-course of downregulation and desensitization appear to occur concomitantly (Sim-Selley 2003). For example, the time-courses for development and recovery of tolerance that develops to cannabinoid hypothermia, hypomotility, antinociception, and memory impairment appear to be associated with CB_1 receptor adaptation in hypothalamus, striatum, cerebellum, periaqueductal gray, spinal cord, and hippocampus.

CB_1 receptor-mediated G protein activation leads to activation of $G\alpha_i$, attenuation of adenylyl cyclase activity, decreased cyclic AMP (cAMP) synthesis, and ultimately decreased protein kinase A (PKA) activity (Howlett et al. 2002). Chronic cannabinoid administration elevates cAMP that consequently increases PKA activity (Rubino et al. 2000b), an effect that is augmented by administration of a CB_1 receptor antagonist (Hutcheson et al. 1998; Tzavara et al. 2000). Free $\beta\gamma$-dimers also regulate other cellular signaling events, such as ion channel conductance, phospholipid metabolism, and the activity of several intracellular kinases. Cannabinoids stimulate ERK1/2 activation in cultured cell lines (Bouaboula et al. 1995; Sanchez et al. 1998) and brain (Derkinderen et al. 2003). Interestingly, in hippocampus and neuroblastoma cells, ERK1/2 activation by cannabinoids was downstream of inhibition of the cAMP/PKA pathway (Davis et al. 2003; Derkinderen et al. 2003). In astrocytoma cells, however, this activation required $G\beta\gamma$-sensitive phosphoinositide 3-kinase (PI3K), similarly to cannabinoid activation of PKB/Akt (Galve-Roperh et al. 2002). Cannabinoids also activate p38 MAPK and c-Jun N-terminal kinase (JNK) in several cell types (Liu et al. 2000; Rueda et al. 2000), though only p38 was activated in hippocampus (Derkinderen et al. 2001a). There is also evidence for cannabinoid activation of non-receptor tyrosine kinases in brain, including a neuronal FAK isoform that was activated via recruitment of the Src family kinase Fyn (Derkinderen et al. 2001b). Similarly to ERK1/2 activation, evidence suggested that these signaling events were a result of inhibition of cAMP/PKA signaling.

Plasticity of an endogenous system often occurs via phosphorylation events. It appears that the endocannabinoid system is no exception. Src tyrosine kinase and PKA are involved in tolerance to spinally mediated cannabinoid analgesia (Lee et al. 2003). Tolerance to THC was reversed with either an Src family tyrosine kinase inhibitor or a PKA inhibitor. Inhibitors of protein kinases C and G (PKC and PKG) and of PI3K were ineffective in altering tolerance to this effect, despite an earlier report that PKC activation disrupted CB_1 receptor signaling through direct phosphorylation of the receptor (Garcia et al. 1998). Underscoring the importance of PKA in the maintenance of THC tolerance, another study also reported that intracerebroventricular administration of a PKA inhibitor reversed tolerance to the antinociceptive, cataleptic, and hypomotility, but not hypothermic, effects of THC (Bass et al. 2004). However, Src tyrosine kinase may also play a role in the THC tolerance, albeit a less prominent one than PKA; its inhibition reversed tolerance to decreased locomotor activity, but failed to affect other in vivo actions of THC. In contrast, PKG and PKC inhibitors failed to affect any measures of tolerance. Collectively, these data suggest that PKA activity plays a major role in THC-induced tolerance, and that THC produces its multiple effects through different signaling pathways. If direct involvement of specific protein kinases in cannabinoid tolerance is confirmed, future studies will need to determine the targets of these protein kinases.

3.4
Changes in Endogenous Cannabinoid Levels

In addition to altering CB_1 receptors and the signal transduction pathways, repeated cannabinoid administration has been associated with differential changes of endogenous cannabinoids in a regionally dependent manner. Reliable decreases of both anandamide and 2-arachidonoyl glycerol (2-AG) were found in striatum, while increases of only anandamide were found in limbic forebrain of THC-tolerant rats (Di Marzo et al. 2000; Gonzalez et al. 2004). Additionally, significant increases of anandamide were found in limbic forebrain following repeated THC dosing (Gonzalez et al. 2004). No significant differences of either endogenous cannabinoid were found in cerebral cortex of THC-tolerant rats (Di Marzo et al. 2000; Gonzalez et al. 2004). However, the effects of repeated THC administration on 2-AG in cerebellum, brain stem, and hippocampus are less certain. Whereas a recent study reported increased levels of 2-AG in these brain areas (Gonzalez et al. 2004), an earlier study by the same group found no differences (Di Marzo et al. 2000). Nonetheless, intriguing possibilities concerning the consequences of altered levels of endogenous cannabinoid content in specific brain regions of THC-dependent animals have been proposed. Specifically, the altered levels of endogenous cannabinoids may be part of a homeostatic mechanism and have important ramifications for motor behavior and emotional states, as well as for cannabinoid dependence (Gonzalez et al. 2004).

4 Characterization of Cannabinoid Dependence

Although chronic cannabis users have long been known to undergo withdrawal upon abrupt discontinuation of the drug (Fraser 1949; Wikler 1976), a marijuana withdrawal syndrome is not yet included in the Diagnostic and Statistical Manuel of Mental Disorders (DSM-IV 1994). Two factors may contribute to the reluctance of accepting the notion that marijuana dependence is clinically relevant. First, the likelihood of progressing from occasional drug use to daily use is considerably lower for marijuana than for other drugs of abuse such as nicotine, cocaine, or heroin (Anthony et al. 1994). However, given the fact that marijuana has consistently been the most frequently used illicit drug in the United States (Johnston et al. 2004), it should not be surprising that the estimated proportion of Americans dependent on marijuana was 4.2%, which was higher than all other illicit drugs, including those that have a greater abuse potential such as cocaine (2.7%) and heroin (0.4%) (Anthony et al. 1994). Second, the delayed onset of cannabis withdrawal symptoms due to the long half-life of THC is also likely to contribute to the lingering doubts concerning the clinical relevance of a cannabinoid withdrawal symptom. However, a growing body of research indicates that abrupt discontinuation following prolonged cannabinoid administration can lead to physical withdrawal symptoms in humans as well as in laboratory animals.

4.1 Clinical Significance of Cannabis Withdrawal

Although it has been contended that further controlled research is needed to diagnose a withdrawal syndrome in human marijuana users (Smith 2002), criteria for cannabinoid withdrawal have been proposed (Budney et al. 2003). Moreover, converging lines of evidence from retrospective, outpatient, and inpatient studies indicate a pattern of cannabis withdrawal signs. In an early inpatient study, Jones and his colleagues reported that subjects reported a variety of subjective effects upon abrupt discontinuation from chronic oral THC, which included strange dreams, decreased appetite, restlessness, irritability, sweating, chills, and nausea (Jones and Benowitz 1976; Jones et al. 1976). More recently, similar abstinence symptoms were reported by subjects following abrupt withdrawal from continued administration of either oral THC (Haney et al. 1999a) or inhalation of smoked marijuana (Haney et al. 1999b). In these studies, subjects were found to experience increased anxiety, irritability, and stomach pain, as well as exhibit decreases in food intake. The authors of these studies suggested that daily marijuana use might be sustained, in part, to alleviate or avoid withdrawal effects.

Although a cannabis withdrawal syndrome can be obtained in a controlled laboratory setting, these findings do not address whether a cannabis withdrawal syndrome represents a clinically significant malady. The results of both retrospective and outpatient studies addressing this issue support the hypothesis that a cannabis withdrawal syndrome is indeed clinically relevant. In one study, marijuana users who were identified from a population of alcohol-dependent subjects recalled a variety of symptoms from when they had previously abstained from marijuana smoking (Wiesbeck et al. 1996). These symptoms included nervousness, sleep disturbances, and changes in appetite. Despite the inherent limitations associated with retrospective self-report data in polysubstance abuse subjects, this pattern of withdrawal symptoms was similar to those described in the laboratory studies investigating cannabis and THC and distinct from those associated with other drugs. In another retrospective study, adults seeking treatment for marijuana dependence recalled similar symptoms upon their most recent period of abstinence that included craving for marijuana, irritability, nervousness, restlessness, depressed mood, increased anger, sleep difficulties, strange dreams, decreased appetite, and headaches (Budney et al. 1999). In addition, the amount of marijuana smoked per day yielded a positive correlation with withdrawal severity.

These findings reported in retrospective and laboratory studies have been corroborated in an outpatient study of regular marijuana users (Budney et al. 2001). In this experiment, subjects were instructed to smoke marijuana for 5 days, followed by a 3-day abstinence period, and then this cycle was repeated. Statistically significant withdrawal symptoms reported during each abstinence period included marijuana craving, decreased appetite, sleep difficulty, and a global withdrawal discomfort score that consisted of the other three measures as well as self-reported data of anger, depressed mood, headaches, irritability, nervousness, restlessness, and strange dreams. Additionally, the subjects lost a significant amount of weight during each abstinence phase. The fact that the withdrawal symptoms increased during abstinence from marijuana smoking, returned to baseline when smoking was reinitiated, and increased again during the second abstinence period suggests that the effects were caused by cessation of marijuana use.

In an outpatient study of marijuana users, Budney and his colleagues attempted to characterize the time course and recovery of marijuana withdrawal symptoms (Budney et al. 2003). In addition to the observations that the onset (i.e., 1–3 days) and peak effects (i.e., 2–6 days) were quite gradual, the duration of these effects was quite prolonged (i.e., 4–14 days). The protracted nature of marijuana withdrawal may contribute to the difficulties in maintaining abstinence. Collectively, this pattern of results is consistent with the hypothesis that cannabis withdrawal symptoms contribute to continued marijuana use.

4.2 Investigation of Cannabinoid Withdrawal in Laboratory Animal Models

Two general categories of dependent measures are used to assess withdrawal in laboratory animals: (1) recording the occurrence of behavioral and physiological changes and (2) the use of operant procedures. In the former approach, subjects are observed for alterations in behavior, and these behaviors are generally either scored as quantal responses or quantified. In the latter approach, animals that had been previously trained to emit an operant response (e.g., lever pressing) for food reinforcement will exhibit decreases in response rates during withdrawal. In both cases, readministration of the drug results in response rates returning to normal. Below, we will first discuss the results of abstinence withdrawal studies conducted in laboratory animals followed by precipitated withdrawal studies.

There are two types of procedures that are used to induce withdrawal in drug-dependent organisms, abstinence withdrawal and precipitated withdrawal. Abstinence withdrawal occurs when drug administration is abruptly discontinued or reduced, following prolonged exposure to the drug. As the body metabolizes and/or excretes the agent, physiological symptoms ranging from mild rebound to severe life-threatening effects can emerge. The pharmacokinetic and pharmacodynamic characteristics of the drug, as well as the degree of dependence, influence the specific withdrawal syndrome, its intensity, and the onset of withdrawal responses. In contrast, a second procedure used to induce withdrawal is the use of a receptor antagonist that precipitates withdrawal in a drug-dependent organism. The antagonist displaces the agonist from the receptor, immediately eliciting withdrawal effects. A typical clinical example of precipitated withdrawal is the treatment of an opioid overdose with naloxone or other opioid receptor antagonist. Upon near instantaneous reversal of respiratory depression and other overdose symptoms, an opioid-dependent individual will present with opioid withdrawal effects. The precipitated withdrawal procedure has been particularly useful in investigating cannabinoid withdrawal symptoms in laboratory animals; however, there are currently no published reports in which this procedure has been used in humans. Below, we will review studies examining cannabinoid dependence in humans and laboratory animals, as well discuss potential pharmacotherapies to alleviate cannabinoid withdrawal symptoms.

4.2.1 Abstinence Withdrawal Versus Precipitated Withdrawal in Laboratory Animals

Research investigating abstinence withdrawal following repeated cannabinoid administration in laboratory animals has led to mixed results. A variety of unconditional behavioral effects including hyperirritability, tremors, and anorexia have been reported (Kaymakcalan and Deneau 1972), though other studies failed to observe abrupt withdrawal effects following subchronic THC administration to dogs (Mcmillan et al. 1971) or rats (Leite and Carlini 1974; Aceto et al. 1996). Indeed, rhesus monkeys that received chronic THC and trained to press a lever for food exhibit a marked suppression in response rates; an effect that was reversed upon readministration of drug (Beardsley et al. 1986). In addition, rats have been observed to exhibit small, but significant increases in wet-dog shakes and facial rubbing 24 h following discontinuation of continuous infusion of a medium ramping dose, but not a low or high ramping dose, of WIN 55,212-2 (Aceto et al. 2001). Similarly, cessation following repeated CP 55,940 treatment in mice led to subtle behavioral changes that included increases in motor activity and rearing, but decreases in grooming, wet-dog shakes, and rubbing behavior that were interpreted as a withdrawal syndrome (Oliva et al. 2004). The long half-life of these drugs further contributes to the difficulty in studying cannabinoid withdrawal in non-human animals. Other related factors that complicate the investigation of abrupt cannabinoid withdrawal in laboratory animals include species differences, strain differences, the time at which withdrawal is assessed, the particular withdrawal measures that are scored, and other methodological issues.

The development of SR 141716 and other selective CB_1 receptor antagonists resulted in powerful pharmacological tools to investigate cannabinoid pharmacology (Rinaldi-Carmona et al. 1994). In addition to its usefulness in determining whether the acute pharmacological actions of an agent are mediated through a CB_1 receptor mechanism and to infer endogenous cannabinoid tone, SR 141716 has been demonstrated to precipitate withdrawal reactions following subchronic administration of cannabinoid agonists in mice (Cook et al. 1998; Rubino et al. 1998), rats (Aceto et al. 1995; Tsou et al. 1995), and dogs (Lichtman et al. 1998). The specific withdrawal effects are species specific, and even within a species a variety of factors affects the withdrawal behavior, including strain, dosing regimen, and test conditions.

Rats exhibit a variety of somatic withdrawal signs that include wet-dog shakes, facial rubs, horizontal and vertical activity, forepaw fluttering, chewing, tongue rolling, head shakes, retropulsion, myoclonic spasms, front paw treading, and eyelid ptosis (Aceto et al. 1995; Tsou et al. 1995). Additionally, SR 141716 dose-dependently reduced food-maintained response rates of THC-tolerant rats, but had no effect in non-tolerant rats (Beardsley and Martin 2000). This disruption of response rates occurred in the absence of somatic withdrawal signs, suggesting that antagonist-precipitated disruptions of operant behavior may be a more sensitive measure of THC withdrawal than the occurrence of unconditional withdrawal behaviors. Others have reported increases in the rate-suppressing effects of SR 141716 in both naïve and THC-tolerant rats, though significantly greater suppression occurred in the tolerant rats compared to naïve animals (Freedland et al. 2003). Again, this effect occurred in the absence of profound somatic signs of withdrawal, further indicating that food-maintained responding is a sensitive measure of the effects of SR 141716-precipitated cannabinoid withdrawal.

SR 141716 was found to precipitate a constellation of withdrawal signs in THC-tolerant dogs that included excessive salivation, vomiting, diarrhea, restless behavior (e.g., circling), trembling, and decreases in social behavior (Lichtman et al. 1998). Several of these signs bore some resemblance to those reported in humans undergoing abrupt withdrawal from THC, including restlessness, nausea, and loose stools (Jones et al. 1981).

Notwithstanding the influence that species and strain differences contribute to the specific withdrawal effects that are observed, the utility of the animal models is verified by the fact that SR 141716 reliably precipitates withdrawal in animals treated repetitively with cannabinoids.

4.2.2 SR 141716 Inverse Activity

One complication in interpreting the results of experiments using SR 141716 to precipitate withdrawal in cannabinoid-dependent animals is that it has inverse agonist properties, in addition to its antagonist actions. Specifically, SR 141716 was found to decrease [^{35}S]GTPγS binding in membranes isolated from human cannabinoid CB_1 receptor-transfected CHO cells (Landsman et al. 1997; Pan et al. 1998), an effect opposite that of cannabinoid agonists (Sim et al. 1996b; Burkey et al. 1997). The observation that SR 141716 is over 7,000-fold more potent as a CB_1 receptor antagonist than as an inverse agonist indicates a high degree of a selectivity as antagonist (Sim-Selley et al. 2001). However, it is difficult to determine whether SR 141716's inverse agonist properties play a role in precipitating cannabinoid withdrawal.

Another important consideration in the design and interpretation of SR 141716-precipitated withdrawal studies is that SR 141716 given alone has been shown to produce mild withdrawal-like effects in naïve (Rodriguez De Fonseca et al. 1997) or vehicle-treated animals (Aceto et al. 1995), such as scratching of the face and body (Aceto et al. 1996; Rubino et al. 1998), head shakes (Cook et al. 1998; Lichtman et al. 2001a), and forepaw fluttering (Rubino et al. 1998). Some of these intrinsic effects of SR 141716 have been linked to other receptor systems. For example SR 141716-induced head twitches were completely blocked by a serotonergic 5-HT_{2A}/5-HT_{2C} receptor antagonist and were partially blocked by an α-amino-3-hydroxy-5-methyl-4-isoxazolepropionic acid (AMPA)/kainate receptor antagonist as well as by a tachykinin natural killer (NK)1 antagonist (Darmani and Pandya 2000). On the other hand, because SR 141716 lacks affinity for these receptors (Rinaldi-Carmona et al. 1994), it is likely that the serotonergic 5-HT_{2A}/5-HT_{2C}, AMPA/kainate,, and tachykinin NK1 receptor antagonists acted "downstream" from the CB_1 receptor.

It remains to be determined whether SR 141716-induced behaviors are due to inverse agonist activity, blockade of endogenous cannabinoid tone, or noncannabinoid sites of action. However, these intrinsic behavioral effects of SR 141716 do not occur universally. Moreover, the magnitude of SR 141716-induced head shakes and paw tremors in naïve animals is generally significantly less than that found in cannabinoid-dependent animals (Aceto et al. 1995, 1996; Aceto et al. 1998; Cook et al. 1998). Nonetheless, the fact that SR 141716 can produce "withdrawal-like" behavior in naïve animals underscores the importance of including an appropriate control group comprising animals that receive repeated doses of vehicle, in order to control for the intrinsic effects of SR 141716 at testing.

4.2.3
SR 141716 Precipitated Anandamide Withdrawal?

SR 141716 reliably precipitates mild to moderate withdrawal responses following subchronic dosing of a variety of cannabinoid agonists including THC, WIN 55,212-2, CP 55,940, and HU-210. However, studies evaluating the effect of this antagonist in rats treated repeatedly with anandamide have led to mixed results. SR 141716 failed to precipitate withdrawal in rats that were infused continuously with anandamide over a 4-day period (Aceto et al. 1998). On the other hand, both abstinence withdrawal and SR 141716-precipitated withdrawal were reported to occur in rats treated repeatedly with daily i.p. injections of anandamide (20 mg/kg) (Costa et al. 2000). In addition to the procedural differences between the two studies, the short half-life of anandamide (Willoughby et al. 1997) contributes to the difficulty of evaluating this endocannabinoid in the whole animal. However, the use of mice deficient in fatty acid amide hydrolase (FAAH), the primary enzyme responsible for anandamide catabolism (Cravatt et al. 2001), and selective inhibitors of this enzyme (Kathuria et al. 2003; Lichtman et al. 2004) will be of value for investigating both the dependence liability and other behavioral effects of anandamide. Interestingly, a single nucleotide polymorphism found in the human gene that encodes FAAH, which produces a variant that displays an enhanced sensitivity to proteolytic degradation, was found to be associated with both street drug use and problem drug/alcohol use (Sipe et al. 2002). This finding suggests the intriguing possibility that the FAAH-endocannabinoid system may play a regulatory role in addictive behavior.

4.3
Neuroadaptive Changes Underlying Cannabinoid Dependence

While it is clear that repeated stimulation of CB_1 receptors by cannabinoid agonists is necessary for the development of cannabinoid dependence, recent research is just beginning to shed light on the underlying cellular mechanisms of action as well as brain regions that mediate these effects. Most of the research directed toward these issues has employed an approach in which cannabinoid-dependent animals are subjected to either a precipitated withdrawal or abstinence withdrawal. The subjects are then euthanized and the biochemical and molecular indices are examined. Several measures of interest include assessing changes of the CB_1 receptor, the signal transduction pathway of this receptor, and other neurochemical systems that affect or are affected by this process. Many studies use a strategy in which behavioral withdrawal signs are also assessed in an attempt to elucidate the relationship between cannabinoid dependence and the underlying neuroadaptations.

4.3.1 Role of the CB_1 Receptor in Cannabinoid Dependence

Repeated administration of a cannabinoid agonist generally results in decreases in CB_1 receptor density in a variety of brain regions as measured by radioligand binding (Romero et al. 1998a; Breivogel et al. 1999). At the level of the G protein, daily injections of THC for 21 days produced significant decreases of CB_1 receptor-stimulated G protein activity in various brain regions, including hippocampus, cerebellum, caudate-putamen, globus pallidus, substantia nigra, septum, and various regions of cortex. In addition to being region-dependent, this desensitization was time-dependent and appeared to be specific for CB_1 receptors and not other G protein-coupled receptors (Breivogel et al. 1999). Although these biochemical correlates are believed to play an important role in the development of tolerance, their role in dependence is less clear. CP 55,940-dependent mice during abstinence were found to undergo subtle behavioral changes that included increases in motor activity and rearing, but decreases in grooming, wet-dog shakes, and rubbing behavior, which were associated with upregulation of CB_1 gene expression in caudate-putamen, ventromedial hypothalamic nucleus, central amygdaloid nucleus, and in the CA1 field of hippocampus, but a decrease in the CA3 field of hippocampus (Oliva et al. 2004).

4.3.2 Cellular Mechanisms Underlying Cannabinoid Dependence

Recent studies have linked alterations in the cAMP second messenger cascade with cannabinoid withdrawal. SR 141716 administered to THC-dependent mice resulted in significant increases of both basal and forskolin-stimulated adenylyl cyclase activity in the cerebellum, but not in other brain regions including the cortex, hippocampus, striatum, and periaqueductal gray (Hutcheson et al. 1998). Similarly, significantly higher levels of calcium–calmodulin-stimulated adenylyl cyclase were found in the cerebellum of THC-dependent rats undergoing withdrawal than in non-dependent rats treated with SR 141716. In another well-designed study (Rubino et al. 2000c), G protein, adenylyl cyclase, and PKA activity were assessed in cerebral cortex, striatum, hippocampus, and cerebellum of rats undergoing precipitated withdrawal. Significant increases of adenylyl cyclase and PKA activity were found in the cerebellum of these animals. However, no effects were found on either receptor density or G protein activity. These findings further implicate the involvement of the cerebellum in cannabinoid dependence and suggest that changes are occurring downstream from the CB_1 receptor.

Functional evidence also suggests that the adenylyl cyclase second messenger cascade in the cerebellum may be involved in cannabinoid withdrawal. An intracerebellar infusion of the cAMP blocker Rp-8Br-cAMPs reduced several behavioral signs of withdrawal including tremors, ataxia, mastication, front paw tremors, ptosis, piloerection, and wet-dog shakes in THC-dependent mice following SR 141716 challenge (Tzavara et al. 2000). Interestingly, Sp-8Br-cAMPs, a cAMP analog, actually induced each of these behavioral effects in vehicle-treated mice. Taken together with the biochemical data, these intriguing findings suggest that upregulation of cAMP signal transduction in the cerebellum may represent a critical biochemical event underlying precipitated withdrawal.

In addition, there is evidence accumulating that implicates the involvement of corticotropin-releasing factor (CRF) and other hormones associated with stress in cannabinoid dependence. SR 141716 significantly elevated plasma corticosterone levels in rats treated subchronically with HU-210 compared with non-dependent rats that were administered SR 141716 or dependent rats not going through precipitated withdrawal (Rodriguez De Fonseca et al. 1997). In another study, however, SR 141716 failed to significantly increase plasma levels of corticosterone in THC-dependent rats (Gonzalez et al. 2004). However, it is unclear whether this apparent failure to replicate was due to a great deal of variability associated with measuring endocrine levels or other methodological differences, such as the use of different cannabinoid agonists to induce dependence (i.e., HU-210 versus THC).

4.3.3
Brain Areas Implicated in Cannabinoid Dependence

As described above (see Sect. 4.3.2) inhibition of adenylyl cyclase in the cerebellum was found to significantly decrease the expression of cannabinoid withdrawal behaviors (Tzavara et al. 2000). Other compelling evidence supporting the notion that CB_1 receptors in the cerebellum play a predominant role in cannabinoid dependence is recent work from Valverde and colleagues. They found that intracerebral injections of SR 141716 into the cerebellum of mice treated repeatedly with WIN 55,212-2 precipitated robust withdrawal responses, which included significant increases in wet-dog shakes, body tremor, paw tremor, piloerection, mastication, genital licks, and sniffing (Castane et al. 2004). Microinjection of SR 141716 into the hippocampus and the amygdala precipitated a moderate but still significant withdrawal syndrome, suggesting the involvement of these brain regions as well. However, SR 141716 infusion into the striatum failed to elicit any significant increases in withdrawal signs. Collectively, these exciting findings strongly implicate the involvement of the cerebellum and possibly the hippocampus and amygdala in cannabinoid withdrawal (Castane et al. 2004).

The role of CRF during cannabinoid withdrawal appears to involve the amygdala. Specifically, SR 141716 challenge to cannabinoid-dependent rats led to significant concomitant increases in CRF and Fos-immunopositive cell activity in the central nucleus of the amygdala (Rodriguez De Fonseca et al. 1997). Similar alterations in amygdaloid CRF function have also been found following ethanol, cocaine, and opioid withdrawal (for review see Weiss et al. 2001). Moreover, SR 141716 challenge to HU-210-tolerant rats led to increases in mRNA CRF in the central amygdala compared to that of HU-210-tolerant subjects not administered SR 141716 (Caberlotto et al. 2004). An interpretation of this finding is that increased gene expression in the amygdala contributes to increased CRF release and behavioral withdrawal signs during precipitated withdrawal. It should be noted that the increase in mRNA CRF levels was only a relative increase, as levels did not significantly differ from those of untreated control subjects. It was proposed that repeated HU-210 administration activated a hitherto unknown counter-regulatory process, thereby returning the system to equilibrium in the presence of the drug (Caberlotto et al. 2004). In this provocative model, withdrawal would occur when SR 141716 rapidly displaces the agonist from the receptor so as to cause the purported counter-regulatory process to go unchecked, despite the apparently normal CRF mRNA levels.

Many other brain regions are also affected by SR 141716-precipitated withdrawal. SR 141716 significantly increased CRF-mRNA levels in the paraventricular hypothalamic nucleus of THC-tolerant rats compared to a non-tolerant group (Gonzalez et al. 2004). Additionally, Fos-immunopositive activity has been reported to occur in the accumbens shell, piriform cortex, hippocampus, caudate putamen, ventral pallidum, ventral tegmental area, locus coeruleus solitary tract, and area postrema (Rodriguez De Fonseca et al. 1997). It remains to be established whether these biochemical changes represent an underlying mechanism of action for cannabinoid dependence or are merely correlated with this phenomenon.

4.4
Influences of Other Neurochemical Systems on Cannabinoid Dependence

Considerable evidence is emerging that supports an interaction between endocannabinoid and opioid systems on many physiological processes (for a review see Varvel et al. 2004). For example, an acute injection of morphine significantly reduced the magnitude of SR 141716-precipitated THC withdrawal (Lichtman et al. 2001b). Genetic alteration of the opioid system has also been found to ameliorate SR 141716-precipitated cannabinoid withdrawal. Cannabinoid withdrawal symptoms, as well as tolerance, were significantly diminished in pre-proenkephalin-deficient mice compared to the wild-type mice (Valverde et al. 2000). In contrast, dynorphin knockout mice failed to exhibit statistically significant changes in either THC tolerance or THC withdrawal (Zimmer et al. 2001). On the other hand, assessing cannabinoid withdrawal in μ-opioid receptor knockout mice has led to contradictory results. While these mice exhibited a significant attenuation of SR 141716-precipitated THC paw tremors and head shakes compared to the wild-type controls in one study (Lichtman et al. 2001b), another study found that THC withdrawal was unaffected by deletion of μ, δ, or κ receptors (Ghozland et al. 2002). Still, another study found that THC withdrawal effects were significantly decreased in mutant mice deficient of both μ- and δ-opioid receptors, suggesting the involvement of multiple opioid receptors in the expression of cannabinoid withdrawal (Castane et al. 2003). A variety of methodological factors could contribute to the apparent discrepant results among these studies, including the background strain and dosing regimen.

Studies examining acute blockade of opioid receptors in cannabinoid-tolerant animals have also reported inconsistent findings, with some finding that naloxone precipitated withdrawal (Hirschhorn and Rosecrans 1974; Kaymakcalan et al. 1977). Similarly, naloxone precipitated withdrawal in rats following repeated injections of the potent cannabinoid analog HU-210 for 15 days (Navarro et al. 1998). It should be noted that considerable toxicity occurred following chronic high doses of THC (Kaymakcalan et al. 1977), though no such toxicity was reported in the other studies. Conversely, naloxone was ineffective in precipitating withdrawal in THC-dependent monkeys (Beardsley et al. 1986), pigeons (Mcmillan et al. 1971), or mice (Lichtman et al. 2001b). Considerable methodological differences used among the studies, including the selection of agonist, species, dosing regimen, and the dependence measures, make it difficult to account for the differential effectiveness of the antagonist in precipitating withdrawal. In general, however, the cannabinoid dosing regimen employed in studies that find naloxone precipitated withdrawal tend to be greater than those in which naloxone failed to elicit any withdrawal effects. Nonetheless, naltrexone failed to precipitate any apparent withdrawal effects in marijuana smokers (Haney et al. 2003a).

It has been well established that clonidine, as well as other α_2-agonists, abrogates many of the withdrawal effects in morphine-dependent animals (Fielding et al. 1978) as well as in human opioid addicts (Gold et al. 1978). Clonidine also ameliorated SR 141716-precipitated paw tremors in THC-dependent mice independently of motor depressive or motor impairment effects (Lichtman et al. 2001a). Although clonidine may hold some promise for treating withdrawal, its hypotensive side effects (Gossop 1988) must be considered before any potential development for its use in alleviating drug withdrawal.

4.5 Pharmacotherapies for Cannabis Dependence

Despite a great need, there are currently no efficacious pharmacotherapies for the treatment of cannabinoid dependence (for a review see McRae et al. 2003). Haney and her colleagues have made important inroads into this area by evaluating several potential treatments, including two antidepressants, bupropion (Zyban) (Haney et al. 2001) and nefazodone (Serzone) (Haney et al. 2003b), the mood stabilizer divalproex, and oral THC (Haney et al. 2004) in double-blind studies (see Table 1). During abstinence from marijuana, the placebo treatment group reliably reported significant increased ratings of marijuana craving, misery, anxiety, trouble sleeping, "strange dreams," chills, irritability, and muscle pain, but decreased ratings of feeling high, content, friendly, social, mellow, and self-confident. Oral THC decreased many of these withdrawal measures such as marijuana craving, feelings of anxiety, misery, trouble sleeping, and chills, and increased the ratings of self-confidence (Haney et al. 2004). In addition, total daily caloric intake was significantly decreased during the withdrawal period in the placebo group and increased in the oral THC group.

Table 1. Potential pharmacotherapies to treat marijuana dependence

Drug/class	Positive effects for marijuana withdrawal	Potential problems	Reference
Oral THC/ cannabinoid agonist	Decreased marijuana craving and withdrawal symptoms	Potential dependence to THC	Haney et al. 2004
Divalproex/ mood stabilizer	Decreased marijuana craving	Worsened anxiety and other withdrawal indices	Haney et al. 2004
Nefazodone (Serzone)/ antidepressant	Decreased anxiety	Ineffective on most withdrawal measures; may cause liver failure	Haney et al. 2003b
Naltrexone/opioid antagonist	No apparent effect	Increased "positive" subjective effects of oral THC	Haney et al. 2003a
Bupropion (Zyban)/ antidepressant	No apparent benefit	Worsened anxiety and other withdrawal indices	Haney et al. 2001
Fluoxetine (Prozac)/ antidepressant	Reduced marijuana use in a subgroup of depressed alcoholics	Controlled studies needed to evaluate efficacy during marijuana withdrawal	Cornelius et al. 1999

The results of other potential pharmacotherapies were less encouraging than those found for THC. Divalproex decreased marijuana craving during marijuana abstinence; however, it increased ratings of anxiety, irritability, bad effect, and sleepiness during marijuana abstinence (Haney et al. 2004), thus limiting its utility. Although bupropion has been well established to be an effective treatment for nicotine withdrawal (Ferry and Johnston 2003; Richmond and Zwar 2003), it not only failed to alleviate symptoms associated with marijuana withdrawal, but also exacerbated them (Haney et al. 2001). Specifically, bupropion worsened mood during marijuana abstinence as reflected by increased self-reports of irritability, misery, restlessness, depression, and lack of motivation compared to placebo. It also worsened subjective reports of sleep quality during marijuana withdrawal compared to placebo. One concern with bupropion is that its stimulatory effects were the cause of the increased severity of withdrawal effects. Therefore, a subsequent study examined whether nefazodone, an antidepressant with sedative properties would attenuate symptoms of marijuana withdrawal. Indeed, the effects of nefazodone maintenance on marijuana were more promising than those of bupropion. Nefazodone decreased subjective ratings of anxiety and muscle pain, but failed to ameliorate the increased ratings of irritability and misery, and decreased rating of sleep quality during marijuana withdrawal (Haney et al. 2003b). In any event, the recent finding that nefazodone is associated with the risk of hepatic failure (FDA 2004) would certainly diminish enthusiasm to develop this drug to treat cannabinoid dependence.

The results of a study conducted in a subgroup of depressed alcoholic marijuana users suggest that fluoxetine (Prozac) may also have some promise for treating marijuana dependence (Cornelius et al. 1999). Specifically, this serotonin re-uptake inhibitor significantly reduced marijuana use as well as depression and alcohol consumption in this subgroup, suggesting that daily marijuana use may result from an attempt to self-medicate for depression (see Table 1). Further controlled studies will be needed to assess the efficacy of fluoxetine in treating marijuana withdrawal symptoms.

Another study by Haney and her colleagues was designed to examine the effects of naltrexone in heavy marijuana users. The goals of this experiment were to examine whether naltrexone would (1) block marijuana's pharmacological effects, and (2) precipitate withdrawal in heavy marijuana users (Haney et al. 2003a). Naltrexone failed to elicit these actions; however, it did increase many of the "positive" subjective effects of oral THC in heavy marijuana smokers. An implication of these findings is that naltrexone may be expected to increase marijuana use.

Currently, oral THC appears to be the best candidate agent to treat marijuana dependence. Of significance, THC was also shown to ameliorate withdrawal symptoms in both monkeys and mice (Beardsley et al. 1986; Lichtman et al. 2001a). However, none of the other agents listed in Table 1 has been evaluated in preclinical models of cannabinoid withdrawal. Thus, it will be important to establish whether the various animal models of cannabinoid dependence are relevant to marijuana-dependent humans. Clearly, the availability of a viable animal model could facilitate the development of effective pharmacotherapies to treat cannabis dependence.

5
Conclusions

The availability of laboratory animal models of cannabinoid tolerance and dependence has greatly increased the ability to investigate the mechanisms underlying these processes. A substantial effort has been focused on characterizing the adaptive changes that occur to the CB_1 receptor and its signal transduction pathways in response to repeated stimulation by cannabinoid agonists. A compelling body of research has led to the wide acceptance that tolerance to the pharmacological effects of cannabinoids is strongly associated with downregulation and desensitization of this receptor. Moreover, a growing body of in vitro and in vivo evidence is beginning to establish that PKA may play an integral role in the maintenance of cannabinoid tolerance.

Cannabinoid-tolerant mice have been demonstrated to exhibit a constellation of somatic withdrawal signs as well as decreases in food-reinforced behavior upon abrupt discontinuation of drug or challenge with a CB_1 receptor antagonist. Additionally, cannabinoid withdrawal has been shown to elicit a variety of physiological responses associated with stress, such as elevations in corticotropin releasing factor and Fos-immunopositive cell activity. The preponderance of evidence from human research supports the notion that cannabinoid dependence is clinically significant and that a need for treatment is warranted. It will be important to establish whether the animal models of dependence will be of value in developing pharmacotherapies for the treatment of cannabis dependence. A multidisciplinary approach examining molecular and cellular changes in conjunction with animal models of dependence, and clinical trials will undoubtedly further our basic understanding of cannabinoid dependence as well as develop pharmacotherapies for cannabinoid dependence disorders.

Acknowledgements.
This work was supported by NIH grants DA03672, DA005274, DA14277, and DA15197.

References

Aceto M, Scates S, Lowe J, Martin B (1995) Cannabinoid precipitated withdrawal by the selective cannabinoid receptor antagonist, SR 141716A. Eur J Pharmacol 282:R1–R2

Aceto M, Scates S, Lowe J, Martin B (1996) Dependence on Δ9-tetrahydrocannabinol: studies on precipitated and abrupt withdrawal. J Pharmacol Exp Ther 278:1290–1295

Aceto MD, Scates SM, Razdan RK, Martin BR (1998) Anandamide, an endogenous cannabinoid, has a very low physical dependence potential. J Pharmacol Exp Ther 287:598–605

Aceto MD, Scates SM, Martin BB (2001) Spontaneous and precipitated withdrawal with a synthetic cannabinoid, WIN 55212-2. Eur J Pharmacol 416:75–81

Anthony J, Warner L, Kessler R (1994) Comparative epidemiology of dependence on tobacco, alcohol, controlled substances and inhalants: basic findings from the National Comorbidity Survey. Exp Clin Psychopharmacol 2:244–268

Bass C, Welch S, Martin B (2004) Reversal of delta(9)-tetrahydrocannabinol-induced tolerance by specific kinase inhibitors. Eur J Pharmacol 496:99–108

Bass CE, Martin BR (2000) Time course for the induction and maintenance of tolerance to delta(9)-tetrahydrocannabinol in mice. Drug Alcohol Depend 60:113–119

Beardsley PM, Martin BR (2000) Effects of the cannabinoid CB(1) receptor antagonist, SR141716A, after delta(9)-tetrahydrocannabinol withdrawal. Eur J Pharmacol 387:47–53

Beardsley PM, Balster RL, Harris LS (1986) Dependence on tetrahydrocannabinol in rhesus monkeys. J Pharmacol Exp Ther 239:311–319

Bouaboula M, Poinot-Chazel C, Bourrie B, Canat X, Calandra B, Rinaldi-Carmona M, Le Fur G, Casellas P (1995) Activation of mitogen-activated protein kinases by stimulation of the central cannabinoid receptor CB1. Biochem J 312:637–641

Braida D, Pozzi M, Cavallini R, Sala M (2001a) Conditioned place preference induced by the cannabinoid agonist CP 55,940: interaction with the opioid system. Neuroscience 104:923–926

Braida D, Pozzi M, Parolaro D, Sala M (2001b) Intracerebral self-administration of the cannabinoid receptor agonist CP 55,940 in the rat: interaction with the opioid system. Eur J Pharmacol 413:227–234

Breivogel CS, Childers SR, Deadwyler SA, Hampson RE, Vogt LJ, Sim-Selley LJ (1999) Chronic delta9-tetrahydrocannabinol treatment produces a time-dependent loss of cannabinoid receptors and cannabinoid receptor-activated G proteins in rat brain. J Neurochem 73:2447–2459

Budney AJ, Novy PL, Hughes JR (1999) Marijuana withdrawal among adults seeking treatment for marijuana dependence. Addiction 94:1311–1322

Budney AJ, Hughes JR, Moore BA, Novy PL (2001) Marijuana abstinence effects in marijuana smokers maintained in their home environment. Arch Gen Psychiatry 58:917–924

Budney AJ, Moore BA, Vandrey RG, Hughes JR (2003) The time course and significance of cannabis withdrawal. J Abnorm Psychol 112:393–402

Burkey TH, Quock RM, Consroe P, Roeske WR, Yamamura HI (1997) Δ9-TetraHydro-Cannabinol is a partial agonist of cannabinoid receptors in mouse brain. Eur J Pharmacol 323:R3–R4
Caberlotto L, Rimondini R, Hansson A, Eriksson S, Heilig M (2004) Corticotropin-releasing hormone (CRH) mRNA expression in rat central amygdala in cannabinoid tolerance and withdrawal: evidence for an allostatic shift? Neuropsychopharmacology 29:15–22
Castane A, Robledo P, Matifas A, Kieffer BL, Maldonado R (2003) Cannabinoid withdrawal syndrome is reduced in double mu and delta opioid receptor knockout mice. Eur J Neurosci 17:155–159
Castane A, Maldonado R, Valverde O (2004) Role of different brain structures in the behavioural expression of WIN 55,212-2 withdrawal in mice. Br J Pharmacol
Chaperon F, Thiebot MH (1999) Behavioral effects of cannabinoid agents in animals. Crit Rev Neurobiol 13:243–281
Cook SA, Lowe JA, Martin BR (1998) CB1 receptor antagonist precipitates withdrawal in mice exposed to Delta9-tetrahydrocannabinol. J Pharmacol Exp Ther 285:1150–1156
Cornelius JR, Salloum IM, Haskett RF, Ehler JG, Jarrett PJ, Thase ME, Perel JM (1999) Fluoxetine versus placebo for the marijuana use of depressed alcoholics. Addict Behav 24:111–114
Costa B, Giagnoni G, Colleoni M (2000) Precipitated and spontaneous withdrawal in rats tolerant to anandamide. Psychopharmacology (Berl) 149:121–128
Coutts AA, Anavi-Goffer S, Ross RA, MacEwan DJ, Mackie K, Pertwee RG, Irving AJ (2001) Agonist-induced internalization and trafficking of cannabinoid CB1 receptors in hippocampal neurons. J Neurosci 21:2425–2433
Cravatt BF, Demarest K, Patricelli MP, Bracey MH, Giang DK, Martin BR, Lichtman AH (2001) Supersensitivity to anandamide and enhanced endogenous cannabinoid signaling in mice lacking fatty acid amide hydrolase. Proc Natl Acad Sci U S A 98:9371–9376
Darmani NA, Pandya DK (2000) Involvement of other neurotransmitters in behaviors induced by the cannabinoid CB1 receptor antagonist SR 141716A in naïve mice. J Neural Transm 107:931–945
Davis MI, Ronesi J, Lovinger DM (2003) A predominant role for inhibition of the adenylate cyclase/protein kinase A pathway in ERK activation by cannabinoid receptor 1 in N1E-115 neuroblastoma cells. J Biol Chem 278:48973–48980
De Vry J, Jentzsch KR, Kuhl E, Eckel G (2004) Behavioral effects of cannabinoids show differential sensitivity to cannabinoid receptor blockade and tolerance development. Behav Pharmacol 15:1–12
Derkinderen P, Toutant M, Burgaya F, Le Bert M, Siciliano JC, de Franciscis V, Gelman M, Girault JA (1996) Regulation of a neuronal form of focal adhesion kinase by anandamide. Science 273:1719–1722
Derkinderen P, Ledent C, Parmentier M, Girault JA (2001a) Cannabinoids activate p38 mitogen-activated protein kinases through CB1 receptors in hippocampus. J Neurochem 77:957–960
Derkinderen P, Toutant M, Kadare G, Ledent C, Parmentier M, Girault JA (2001b) Dual role of Fyn in the regulation of FAK+6,7 by cannabinoids in hippocampus. J Biol Chem 276:38289–38296
Derkinderen P, Valjent E, Toutant M, Corvol JC, Enslen H, Ledent C, Trzaskos J, Caboche J, Girault JA (2003) Regulation of extracellular signal-regulated kinase by cannabinoids in hippocampus. J Neurosci 23:2371–2382
Dewey WL, Jenkins J, Rourke T, Harris LS (1972) The effects of chronic administration of trans-Δ9-tetrahydrocannabinol on behavior and the cardiovascular system. Arch Int Pharmacodyn Ther 198:118–131
Di Marzo V, Berrendero F, Bisogno T, Gonzalez S, Cavaliere P, Romero J, Cebeira M, Ramos JA, Fernandez-Ruiz JJ (2000) Enhancement of anandamide formation in the limbic forebrain and reduction of endocannabinoid contents in the striatum of delta-9- tetrahydrocannabinol-tolerant rats. J Neurochem 74:1627–1635

DSM-IV (1994) Diagnostic and statistical manual of mental disorders: DSM-IV. American Psychiatric Association, Washington
Fan F, Compton DR, Ward S, Melvin L, Martin BR (1994) Development of cross-tolerance between delta 9-tetrahydrocannabinol, CP 55,940 and WIN 55,212. J Pharmacol Exp Ther 271:1383–1390
Fan F, Tao Q, Abood M, Martin BR (1996) Cannabinoid receptor down-regulation without alteration of the inhibitory effect of CP 55,940 on adenylyl cyclase in the cerebellum of CP 55,940-tolerant mice. Brain Res 706:13–20
Fattore L, Cossu G, Martellotta CM, Fratta W (2001) Intravenous self-administration of the cannabinoid CB1 receptor agonist WIN 55,212-2 in rats. Psychopharmacology (Berl) 156:410–416
FDA (2004) http://www.fda.gov/medwatch/SAFETY/2004/safety04.htm serzone. Cited 12 September 2004
Ferguson SS, Caron MG (1998) G protein-coupled receptor adaptation mechanisms. Semin Cell Dev Biol 9:119–127
Ferry L, Johnston JA (2003) Efficacy and safety of bupropion SR for smoking cessation: data from clinical trials and five years of postmarketing experience. Int J Clin Pract 57:224–230
Fielding S, Wilker J, Hynes M, Szewczak M, Novick WJ Jr, Lal H (1978) A comparison of clonidine with morphine for antinociceptive and antiwithdrawal actions. J Pharmacol Exp Ther 207:899–905
Fraser JD (1949) Withdrawal symptoms in cannabis-indica addicts. Lancet 257:747–748
Freedland CS, Whitlow CT, Miller MD, Porrino LJ (2002) Dose-dependent effects of delta9-tetrahydrocannabinol on rates of local cerebral glucose utilization in rat. Synapse 45:134–142
Freedland CS, Whitlow CT, Smith HR, Porrino LJ (2003) Functional consequences of the acute administration of the cannabinoid receptor antagonist, SR141716A, in cannabinoid-naïve and -tolerant animals: a quantitative 2-[14C]deoxyglucose study. Brain Res 962:169–179
Galve-Roperh I, Rueda D, Gomez del Pulgar T, Velasco G, Guzman M (2002) Mechanism of extracellular signal-regulated kinase activation by the CB(1) cannabinoid receptor. Mol Pharmacol 62:1385–1392
Garcia DE, Brown S, Hille B, Mackie K (1998) Protein kinase C disrupts cannabinoid actions by phosphorylation of the CB1 cannabinoid receptor. J Neurosci 18:2834–2841
Gardner EL (2002) Addictive potential of cannabinoids: the underlying neurobiology. Chem Phys Lipids 121:267–290
Gardner EL, Lowinson JH (1991) Marijuana's interaction with brain reward systems—update 1991. Pharmacol Biochem Behav 40:571–580
Gardner EL, Paredes W, Smith D, Donner A, Milling C, Cohen A, Morrison D (1988) Facilitation of brain stimulation reward by D9-tetrahydrocannabinol. Psychopharmacology (Berl) 96:142–144
Gardner EL, Paredes W, Smith D, Zukin RS (1989) Facilitation of brain stimulation reward by D9-tetrahydrocannabinol is mediated by an endogenous opioid mechanism. Adv Biosci 75:671–674
Ghozland S, Matthes HW, Simonin F, Filliol D, Kieffer BL, Maldonado R (2002) Motivational effects of cannabinoids are mediated by mu-opioid and kappa-opioid receptors. J Neurosci 22:1146–1154
Gold MS, Redmond DE Jr, Kleber HD (1978) Clonidine blocks acute opiate-withdrawal symptoms. Lancet 2:599–602
Gomez del Pulgar T, Velasco G, Guzman M (2000) The CB1 cannabinoid receptor is coupled to the activation of protein kinase B/Akt. Biochem J 347:369–373
Gonzalez S, Fernandez-Ruiz J, Marzo VD, Hernandez M, Arevalo C, Nicanor C, Cascio MG, Ambrosio E, Ramos JA (2004) Behavioral and molecular changes elicited by acute administration of SR141716 to Delta9-tetrahydrocannabinol-tolerant rats: an experimental model of cannabinoid abstinence. Drug Alcohol Depend 74:159–170

Gossop M (1988) Clonidine and the treatment of the opiate withdrawal syndrome. Drug Alcohol Depend 21:253–259

Haney M, Ward AS, Comer SD, Foltin RW, Fischman MW (1999a) Abstinence symptoms following oral THC administration to humans. Psychopharmacology (Berl) 141:385–394

Haney M, Ward AS, Comer SD, Foltin RW, Fischman MW (1999b) Abstinence symptoms following smoked marijuana in humans. Psychopharmacology (Berl) 141:395–404

Haney M, Ward AS, Comer SD, Hart CL, Foltin RW, Fischman MW (2001) Bupropion SR worsens mood during marijuana withdrawal in humans. Psychopharmacology (Berl) 155:171–179

Haney M, Bisaga A, Foltin RW (2003a) Interaction between naltrexone and oral THC in heavy marijuana smokers. Psychopharmacology (Berl) 166:77–85

Haney M, Hart CL, Ward AS, Foltin RW (2003b) Nefazodone decreases anxiety during marijuana withdrawal in humans. Psychopharmacology (Berl) 165:157–165

Haney M, Hart CL, Vosburg SK, Nasser J, Bennett A, Zubaran C, Foltin RW (2004) Marijuana withdrawal in humans: effects of oral THC or divalproex. Neuropsychopharmacology 29:158–170

Hill MN, Gorzalka BB, Choi JW (2004) Augmentation of the development of behavioral tolerance to cannabinoid administration through pavlovian conditioning. Neuropsychobiology 49:94–100

Hirschhorn ID, Rosecrans JA (1974) Morphine and Δ9-tetrahydrocannabinol: tolerance to the stimulus effects. Psychopharmacologia 36:243–253

Howlett AC, Barth F, Bonner TI, Cabral G, Casellas P, Devane WA, Felder CC, Herkenham M, Mackie K, Martin BR, Mechoulam R, Pertwee RG (2002) International Union of Pharmacology. XXVII. Classification of cannabinoid receptors. Pharmacol Rev 54:161–202

Hsieh C, Brown S, Derleth C, Mackie K (1999) Internalization and recycling of the CB1 cannabinoid receptor. J Neurochem 73:493–501

Hutcheson DM, Tzavara ET, Smadja C, Valjent E, Roques BP, Hanoune J, Maldonado R (1998) Behavioural and biochemical evidence for signs of abstinence in mice chronically treated with delta-9-tetrahydrocannabinol. Br J Pharmacol 125:1567–1577

Jin W, Brown S, Roche JP, Hsieh C, Celver JP, Kovoor A, Chavkin C, Mackie K (1999) Distinct domains of the CB1 cannabinoid receptor mediate desensitization and internalization. J Neurosci 19:3773–3780

Johnston LD, O'Malley PM, Bachman JG, Schulenberg JE (2004) Monitoring the Future national results on adolescent drug use: overview of key findings, 2003. National Institute on Drug Abuse, Bethdesda

Jones RT, Benowitz N (1976) The 30-day trip—clinical studies of cannabis tolerance and dependence. In: Braude MC, Szara S (eds) Pharmacology of Marihuana. Raven Press, New York, pp 627–642

Jones RT, Benowitz N, Bachman J (1976) Clinical studies of cannabis tolerance and dependence. Ann NY Acad Sci 282:221–239

Jones RT, Benowitz NL, Herning RI (1981) Clinical relevance of cannabis tolerance and dependence. J Clin Pharmacol 21:143S–152S

Justinova Z, Tanda G, Redhi GH, Goldberg SR (2003) Self-administration of delta9-tetrahydrocannabinol (THC) by drug naive squirrel monkeys. Psychopharmacology (Berl) 169:135–140

Kathuria S, Gaetani S, Fegley D, Valino F, Duranti A, Tontini A, Mor M, Tarzia G, Rana GL, Calignano A, Giustino A, Tattoli M, Palmery M, Cuomo V, Piomelli D (2003) Modulation of anxiety through blockade of anandamide hydrolysis. Nat Med 9:76–81

Kaymakcalan K, Ayhan IH, Tulunay FC (1977) Naloxone-induced or postwithdrawal abstinence signs in Δ9-tetrahydrocannabinol-tolerant rats. Psychopharmacology (Berl) 55:243–249

Kaymakcalan S, Deneau GA (1972) Some pharmacologic properties of synthetic Δ9-tetrahydrocannabinol. Acta Medica Turcica Suppl 1:5

Kouznetsova M, Kelley B, Shen M, Thayer SA (2002) Desensitization of cannabinoid-mediated presynaptic inhibition of neurotransmission between rat hippocampal neurons in culture. Mol Pharmacol 61:477–485

Krupnick JG, Benovic JL (1998) The role of receptor kinases and arrestins in G protein-coupled receptor regulation. Annu Rev Pharmacol Toxicol 38:289–319

Landsman RS, Burkey TH, Consroe P, Roeske WR, Yamamura HI (1997) SR141716A is an inverse agonist at the human cannabinoid CB1 receptor. Eur J Pharmacol 334:R1–2

Ledent C, Valverde O, Cossu G, Petitet F, Aubert JF, Beslot F, Bohme GA, Imperato A, Pedrazzini T, Roques BP, Vassart G, Fratta W, Parmentier M (1999) Unresponsiveness to cannabinoids and reduced addictive effects of opiates in CB1 receptor knockout mice. Science 283:401–404

Lee MC, Smith FL, Stevens DL, Welch SP (2003) The role of several kinases in mice tolerant to delta 9-tetrahydrocannabinol. J Pharmacol Exp Ther 305:593–599

Leite JR, Carlini EA (1974) Failure to obtain "cannabis-directed behavior" and abstinence syndrome in rats chronically treated with cannabis sativa extracts. Psychopharmacologia 36:133–145

Lepore M, Vorel SR, Lowinson J, Gardner EL (1995) Conditioned place preference induced by delta 9-tetrahydrocannabinol: comparison with cocaine, morphine, and food reward. Life Sci 56:2073–2080

Lepore M, Liu X, Savage V, Matalon D, Gardner EL (1996) Genetic differences in D9-tetrahydrocannabinol-induced facilitation of brain stimulation reward as measured by a rate-frequency curve-shift electrical brain stimulation paradigm in three different rat strains. Life Sci 58:365–0372

Lichtman AH, Wiley JL, LaVecchia KL, Neviaser ST, Arthrur DB, Wilson DM, Martin BR (1998) Acute and chronic cannabinoid effects: characterization of precipitated withdrawal in dogs. Eur J Pharmacol 357:139–148

Lichtman AH, Fisher J, Martin BR (2001a) Precipitated cannabinoid withdrawal is reversed by Delta(9)- tetrahydrocannabinol or clonidine. Pharmacol Biochem Behav 69:181–188

Lichtman AH, Sheikh SM, Loh HH, Martin BR (2001b) Opioid and cannabinoid modulation of precipitated withdrawal in delta(9)-tetrahydrocannabinol and morphine-dependent mice. J Pharmacol Exp Ther 298:1007–1014

Lichtman AH, Varvel SA, Martin BR (2002) Endocannabinoids in cognition and dependence. Prostaglandins Leukot Essent Fatty Acids 66:269–285

Lichtman AH, Leung D, Shelton C, Saghatelian A, Hardouin C, Boger D, Cravatt BF (2004) Reversible inhibitors of fatty acid amide hydrolase that promote analgesia: evidence for an unprecedented combination of potency and selectivity. J Pharmacol Exp Ther

Liu J, Gao B, Mirshahi F, Sanyal AJ, Khanolkar AD, Makriyannis A, Kunos G (2000) Functional CB1 cannabinoid receptors in human vascular endothelial cells. Biochem J 346:835–840

Maldonado R (2002) Study of cannabinoid dependence in animals. Pharmacol Ther 95:153–164

Maldonado R, Rodriguez de Fonseca F (2002) Cannabinoid addiction: behavioral models and neural correlates. J Neurosci 22:3326–3331

Martellotta MC, Cossu G, Fattore L, Gessa GL, Fratta W (1998) Self-administration of the cannabinoid receptor agonist WIN 55,212-2 in drug-naive mice. Neuroscience 85:327–330

Martin BR (2002) Identification of the endogenous cannabinoid system through integrative pharmacological approaches. J Pharmacol Exp Ther 301:790–796

Martin BR, Sim-Selley LJ, Selley DE (2004) Signaling pathways involved in the development of cannabinoid tolerance. Trends Pharmacol Sci 25:325–330

McMillan DE, Dewey WL, Harris LS (1971) Characteristics of tetrahydrocannabinol tolerance. Ann NY Acad Sci 191:83–99

McRae AL, Budney AJ, Brady KT (2003) Treatment of marijuana dependence: a review of the literature. J Subst Abuse Treat 24:369–376

Navarro M, Chowen J, Carrera MRA, del Arco I, Vallanua MA, Martin Y, Roberts AJ, Koob GF, Rodriguez de Fonseca F (1998) CB1 cannabinoid receptor antagonist-induced opiate withdrawal in morphine-dependent rats. Neuroreport 9:3397–3402

Oliva JM, Ortiz S, Palomo T, Manzanares J (2004) Spontaneous cannabinoid withdrawal produces a differential time-related responsiveness in cannabinoid CB1 receptor gene expression in the mouse brain. J Psychopharmacol 18:59–65

Oviedo A, Glowa J, Herkenham M (1993) Chronic cannabinoid administration alters cannabinoid receptor binding in rat brain: a quantitative autoradiographic study. Brain Res 616:293–302

Pan X, Ikeda SR, Lewis DL (1998) SR 141716A acts as an inverse agonist to increase neuronal voltage-dependent Ca2+ currents by reversal of tonic CB1 cannabinoid receptor activity. Mol Pharmacol 54:1064–1072

Richmond R, Zwar N (2003) Review of bupropion for smoking cessation. Drug Alcohol Rev 22:203–220

Rinaldi-Carmona M, Barth F, Héaulme M, Shire D, Calandra B, Congy C, Martinez S, Maruani J, Néliat G, Caput D, Ferrara P, Soubrié P, Brelière JC, Le Fur G (1994) SR141716A, a potent and selective antagonist of the brain cannabinoid receptor. FEBS Lett 350:240–244

Rinaldi-Carmona M, Le Duigou A, Oustric D, Barth F, Bouaboula M, Carayon P, Casellas P, Le Fur G (1998) Modulation of CB1 cannabinoid receptor functions after a long-term exposure to agonist or inverse agonist in the Chinese hamster ovary cell expression system. J Pharmacol Exp Ther 287:1038–1047

Rodriguez de Fonseca F, Carrera M, Navarro M, Koob K, Weiss F (1997) Activation of corticotropin-releasing factor in the limbic system during cannabinoid withdrawal. Science 276:2050–2054

Romero J, Garcia-Palomero E, Castro JG, Garcia-Gil L, Ramos JA, Fernandez-Ruiz JJ (1997) Effects of chronic exposure to delta9-tetrahydrocannabinol on cannabinoid receptor binding and mRNA levels in several rat brain regions. Brain Res Mol Brain Res 46:100–108

Romero J, Berrendero F, Garcia-Gil L, De La Cruz P, Ramos A, Fernandez-Ruiz JJ (1998a) Loss of cannabinoid receptor binding and messenger RNA levels and cannabinoid agonist-stimulated [35 s]guanylyl-5'-O-(thio)-triphosphate binding in the basal ganglia of rats. Neuroscience 84:1075–1083

Romero J, Berrendero F, Manzanares J, Perez A, Corchero J, Fuentes JA, Fernandez-Ruiz JJ, Ramos JA (1998b) Time-course of the cannabinoid receptor down-regulation in the adult rat brain caused by repeated exposure to delta9-tetrahydrocannabinol. Synapse 30:298–308

Rubino T, Massi P, Patrini G, Venier I, Giagnoni G, Parolaro D (1994) Chronic CP-55,940 alters cannabinoid receptor mRNA in the rat brain: An in situ hybridization study. NeuroReport 5:2493–2496

Rubino T, Patrini G, Parenti M, Massi P, Parolaro D (1997) Chronic treatment with a synthetic cannabinoid CP-55,940 alters G-protein expression in the rat central nervous system. Brain Res Mol Brain Res 44:191–197

Rubino T, Patrini G, Massi P, Fuzio D, Vigano D, Giagnoni G, Parolaro D (1998) Cannabinoid-precipitated withdrawal: a time-course study of the behavioral aspect and its correlation with cannabinoid receptors and G protein expression. J Pharmacol Exp Ther 285:813–819

Rubino T, Vigano D, Costa B, Colleoni M, Parolaro D (2000a) Loss of cannabinoid-stimulated guanosine 5'-O-(3-[(35)S]thiotriphosphate) binding without receptor down-regulation in brain regions of anandamide-tolerant rats. J Neurochem 75:2478–2484

Rubino T, Vigano D, Massi P, Spinello M, Zagato E, Giagnoni G, Parolaro D (2000b) Chronic delta-9-tetrahydrocannabinol treatment increases cAMP levels and cAMP-dependent protein kinase activity in some rat brain regions. Neuropharmacology 39:1331–1336

Rubino T, Vigano D, Zagato E, Sala M, Parolaro D (2000c) In vivo characterization of the specific cannabinoid receptor antagonist, SR141716A: behavioral and cellular responses after acute and chronic treatments. Synapse 35:8–14

Rueda D, Galve-Roperh I, Haro A, Guzman M (2000) The CB(1) cannabinoid receptor is coupled to the activation of c-Jun N-terminal kinase. Mol Pharmacol 58:814–820
Sanchez C, Galve-Roperh I, Rueda D, Guzman M (1998) Involvement of sphingomyelin hydrolysis and the mitogen-activated protein kinase cascade in the delta9-tetrahydrocannabinol-induced stimulation of glucose metabolism in primary astrocytes. Mol Pharmacol 54:834–843
Sanudo-Pena MC, Tsou K, Delay ER, Hohman AG, Force M, Walker JM (1997) Endogenous cannabinoids as an aversive or counter-rewarding system in the rat. Neurosci Lett 223:125–128
Siegel S, Ramos BM (2002) Applying laboratory research: drug anticipation and the treatment of drug addiction. Exp Clin Psychopharmacol 10:162–183
Sim LJ, Hampson RE, Deadwyler SA, Childers SR (1996a) Effects of chronic treatment with Δ9-tetrahydrocannabinol on cannabinoid-stimulated [35]GTPγS autoradiography in rat brain. J Neurosci 16:8057–8066
Sim LJ, Selley DE, Xiao R, Childers SR (1996b) Differences in G-protein activation by μ- and δ-opioid, and cannabinoid, receptors in rat striatum. Eur J Pharmacol 307:97–105
Sim-Selley LJ (2003) Regulation of cannabinoid CB1 receptors in the central nervous system by chronic cannabinoids. Crit Rev Neurobiol 15:91–119
Sim-Selley LJ, Martin BR (2002) Effect of chronic administration of R-(+)-[2,3-dihydro-5-methyl-3-[(morpholinyl)methyl]pyrrolo[1,2,3-de]-1,4-b enzoxazinyl]-(1-naphthalenyl)methanone mesylate (WIN55,212-2) or delta(9)-tetrahydrocannabinol on cannabinoid receptor adaptation in mice. J Pharmacol Exp Ther 303:36–44
Sim-Selley LJ, Brunk LK, Selley DE (2001) Inhibitory effects of SR141716A on G-protein activation in rat brain. Eur J Pharmacol 414:135–143
Sipe JC, Chiang K, Gerber AL, Beutler E, Cravatt BF (2002) A missense mutation in human fatty acid amide hydrolase associated with problem drug use. Proc Natl Acad Sci U S A 99:8394–8399
Smith NT (2002) A review of the published literature into cannabis withdrawal symptoms in human users. Addiction 97:621–632
Takahashi RN, Singer G (1979) Self-administration of D9-tetrahydrocannabinol by rats. Pharmacol Biochem Behav 11:737–740
Takahashi RN, Singer G (1980) Effects of body weight levels on cannabis self-injection. Pharmacol Biochem Behav 13:877–881
Tanda G, Goldberg SR (2003) Cannabinoids: reward, dependence, and underlying neurochemical mechanisms—a review of recent preclinical data. Psychopharmacology (Berl) 169:115–134
Tanda G, Munzar P, Goldberg SR (2000) Self-administration behavior is maintained by the psychoactive ingredient of marijuana in squirrel monkeys. Nat Neurosci 3:1073–1074
Tsou K, Patrick S, Walker JM (1995) Physical withdrawal in rats tolerant to Δ9-tetrahydrocannabinol precipitated by a cannabinoid receptor antagonist. Eur J Pharmacol 280:R13–R15
Tzavara ET, Valjent E, Firmo C, Mas M, Beslot F, Defer N, Roques BP, Hanoune J, Maldonado R (2000) Cannabinoid withdrawal is dependent upon PKA activation in the cerebellum. Eur J Neurosci 12:1038–1046
Valjent E, Maldonado R (2000) A behavioural model to reveal place preference to delta 9-tetrahydrocannabinol in mice. Psychopharmacology (Berl) 147:436–438
Valverde O, Maldonado R, Valjent E, Zimmer AM, Zimmer A (2000) Cannabinoid withdrawal syndrome is reduced in pre-proenkephalin knock-out mice. J Neurosci 20:9284–9289
Varvel SA, Cichewicz DL, Lichtman AH (2004) Interactions between cannabinoids and opioids. In: Wenger T (ed) Recent advances on pharmacology and physiology of cannabinoids research. Signpost, Keraka, p 157–182
Weiss F, Ciccocioppo R, Parsons LH, Katner S, Liu X, Zorrilla EP, Valdez GR, Ben-Shahar O, Angeletti S, Richter RR (2001) Compulsive drug-seeking behavior and relapse. Neuroadaptation, stress, and conditioning factors. Ann N Y Acad Sci 937:1–26

Whitlow CT, Freedland CS, Porrino LJ (2003) Functional consequences of the repeated administration of delta9-tetrahydrocannabinol in the rat. Drug Alcohol Depend 71:169–177

Wiesbeck GA, Schuckit MA, Kalmijn JA, Tipp JE, Bucholz KK, Smith TL (1996) An evaluation of the history of a marijuana withdrawal syndrome in a large population. Addiction 91:1469–1478

Wikler A (1976) Aspects of tolerance to and dependence on cannabis. Ann N Y Acad Sci 282:126–147

Willoughby KA, Moore SF, Martin BR, Ellis EF (1997) The biodisposition and metabolism of anandamide in mice. J Pharmacol Exp Ther 282:243–247

Wise RA (1996) Addictive drugs and brain stimulation reward. Annu Rev Neurosci 19:319–340

Zhuang S, Kittler J, Grigorenko EV, Kirby MT, Sim LJ, Hampson RE, Childers SR, Deadwyler SA (1998) Effects of long-term exposure to delta9-THC on expression of cannabinoid receptor (CB1) mRNA in different rat brain regions. Brain Res Mol Brain Res 62:141–149

Zimmer A, Valjent E, Konig M, Zimmer AM, Robledo P, Hahn H, Valverde O, Maldonado R (2001) Absence of delta-9-tetrahydrocannabinol dysphoric effects in dynorphin-deficient mice. J Neurosci 21:9499–9505

HEP (2005) 168:719–756

Human Studies of Cannabinoids and Medicinal Cannabis

P. Robson

Department of Psychiatry, Oxford University, Warneford Hospital, Oxford OX3 7JX, UK
pjr@gwpharm.com

Abstract Cannabis has been known as a medicine for several thousand years across many cultures. It reached a position of prominence within Western medicine in the nineteenth century but became mired in disrepute and legal controls early in the twentieth century. Despite unremitting world-wide suppression, recreational cannabis exploded into popular culture in the 1960s and has remained easily obtainable on the black market in most countries ever since. This ready availability has allowed many thousands of patients to rediscover the apparent power of the drug to alleviate symptoms of some of the most cruel and refractory diseases known to humankind. Pioneering clinical research in the last quarter of the twentieth century has given some support to these anecdotal reports, but the methodological challenges to human research involving a pariah drug are formidable. Studies have tended to be small, imperfectly controlled, and have often incorporated unsatisfactory synthetic cannabinoid analogues or smoked herbal material of uncertain composition and irregular bioavailability. As a result, the scientific evaluation of medicinal cannabis in humans is still in its infancy. New possibilities in human research have been opened up by the discovery of the endocannabinoid system, a rapidly expanding knowledge of cannabinoid pharmacology, and a more sympathetic political environment in several countries. More and more scientists and clinicians are becoming interested in exploring the potential of cannabis-based medicines. Future targets will extend beyond symptom relief into disease modification, and already cannabinoids seem to offer particular promise in the treatment of certain inflammatory and neurodegenerative conditions. This chapter will begin with an outline of the development and current status of legal controls pertaining to cannabis, following which the existing human research will be reviewed. Some key safety issues will then be considered, and the chapter will conclude with some suggestions as to future directions for human research.

Keywords Cannabinoids · Medicinal cannabis · Human research · Therapeutic potential

1 Introduction

The pariah status of cannabis is a relatively modern phenomenon. Cultivation of the plant for hemp extends back to the Stone Age, and medicinal use dates back at least 4,000 years (reviewed by Mechoulam 1986). In China a medical treatise dating from around 2600 B.C.E. recommends its use for relieving the symptoms of malaria, constipation, rheumatic pains and dysmenorrhoea (Grinspoon and Bakalar 1993). There are subsequent records of medicinal use throughout Asia, the Middle East, Southern Africa and South America. Known to European physicians as Indian hemp until christened *Cannabis sativa* by Linnaeus in 1753, it was not until the mid-nineteenth century that it emerged as a mainstream medicine in Britain. The Irish scientist and physician W.B. O'Shaughnessy had observed its use in India as an analgesic, anti-spasmodic, anti-emetic and hypnotic. After testing its safety on dogs, goats and himself he went on to administer cannabis resin in an ethanolic solution to patients with a range of maladies. His report (O'Shaughnessy 1843) of these experiments generated considerable interest, and medicinal use expanded rapidly. By 1854 it had found its way into the U.S. Dispensatory, and "over-the-counter" preparations were soon available in pharmacies throughout England and Scotland. Establishment status was fully achieved through the enthusiastic endorsement of one of Queen Victoria's physicians (Reynolds 1890), but by the end of the century cannabis had passed its zenith as a prescribed medicine and home remedy. Although Sir William Osler was still recommending it for migraine sufferers in 1913, its popularity was in steep decline for a number of reasons: variable potency of herbal preparations, unreliable sources of supply, poor storage stability, unpredictable response to oral administration, the growing availability of potent synthetic medicines, and commercial pressures. An increasingly influential factor was increasing concern in some countries about recreational use, notably South Africa, Egypt and the U.S.

These concerns were brought to the 1923 meeting of the League of Nations, and thence referred for consideration at the 1925 Geneva Convention on the manufacture, sale and movement of dangerous drugs. Signatory nations agreed to enforce a limitation of the use of cannabis solely for medical or scientific purposes. In 1928 the UK government ratified this convention, but prescription of cannabis remained possible until the Misuse of Drugs Act (1971) brought down the final curtain. This Act provides rules for the manufacture, supply and possession of a long list of controlled drugs. For the purposes of determining penalties for malefactors it places them in three classes according to the "harmfulness attributable to a drug when it is misused". On this basis, cannabis and cannabis resin were assigned to Class B along with amphetamines, barbiturates, codeine and dihydrocodeine. In 2001, the British Home Secretary asked a leading committee of experts [Advisory Council on the Misuse of Drugs (ACMD)] to review the classification of cannabis in the light of current scientific evidence. The ACMD carried out a detailed scrutiny of all the relevant literature and in 2002 concluded that, though certainly not innocuous, cannabis

> ... is less harmful than other substances (amphetamines, barbiturates, codeine-like compounds) within Class B of Schedule 2 to the Misuse of Drugs Act 1971. The continuing juxtaposition of cannabis with these more harmful Class B drugs erroneously (and dangerously) suggests that their harmful effects are equivalent. This may lead to the belief, among cannabis users, that if they have had no harmful effects from cannabis then other Class B substances will be equally safe.

ACMD recommended reclassification of all cannabis preparations to Class C, and in February 2004, despite hostile media comment, the Home Secretary implemented this advice.

An important issue for medicinal cannabis in Britain is its inclusion in schedule 1 of the Misuse of Drugs Regulations (1985). This means that it belongs to a group of controlled drugs [alongside lysergic acid diethylamide (LSD), raw opium and coca leaf] that have no recognised medicinal use, and which are totally prohibited for possession or supply unless authorised by a special licence from the Home Office. However, the Home Secretary is on record as saying in 2001: "Should, as I believe it will, this programme (of trials) be proved to be successful, I will recommend to the Medicines Control Agency that they should go ahead with authorising the medical use" (UK Parliament 2002).

In the U.S., concern about the recreational use of cannabis had reached fever pitch by the 1930s (for a full review, see Mead 2004). This was fuelled by some lurid propaganda largely instigated by the chief of the Federal Bureau of Narcotics, Harry J. Anslinger (Abel 1980). This highly effective campaign, which generated some baseless myths that survive to the present day, culminated in the Marihuana Tax Act (1937) that effectively ruled out both recreational and medicinal use. In 1941 cannabis was removed from the U.S. Pharmacopoeia. Scientific reports that challenged claims that cannabis use was closely associated with insanity, addiction, violence and crime were ignored by politicians, regulators and the American Medical Association. Cannabis continued to be portrayed as a dangerous, addictive drug that also acted as a "gateway" into opiate or cocaine addiction. In the late 1940s the confused international situation regarding drug control led the United Nations Commission on Narcotic Drugs (CND) to seek an international agreement. In the resulting 1961 Single Convention on Narcotic Drugs, cannabis and cannabis resin were placed in one of the most restricted categories (along with heroin). Signatory nations were obliged to impose complete prohibition and "adequate punishment" for transgressors. The 1971 Convention on Psychotropic Substances and the 1988 Convention against Illicit Traffic in Narcotic Drugs and Psychotropic Substances were subsequent developments. The 1971 convention placed dronabinol (Marinol), a synthetic formulation of Δ^9-tetrahydrocannabinol (THC) for oral use, in a less restrictive category. Following research funded by the U.S. National Cancer Institute, dronabinol was approved by the U.S. regulatory authority for the treatment of nausea and vomiting associated with cancer chemotherapy.

U.S. Advocacy groups such as the National Organisation for the Reform of Marijuana Laws (NORML) and Alliance for Cannabis Therapeutics (ACT) have vigorously opposed the suppression of medicinal cannabis (Mead 2004). Rescheduling litigation was not, in the end, successful at a national level, but many individual states enacted legislation to make cannabis available to specific patients. Numerous cannabis buyers' clubs sprang up to provide supplies, but these are certainly not immune from prosecution by the federal authorities. California has been a particular focus for activity, and a Center for Medicinal Cannabis Research has been established within the University of California at San Diego.

Nations have some flexibility in implementing the 1961 and 1971 conventions (Mead 2004). For example, if a national court ruled that an individual had a constitutional right to use medicinal cannabis, that nation would be relieved of any obligation to punish such activity. This elasticity has resulted in a marked disparity in approach between countries (for a full review, see Mead 2004).

Unfortunately, the blossoming of recreational cannabis during a period of social turmoil in the 1960s has hardened its image as an agent of alienation and subversion in the eyes of many politicians and regulators. Rigorous prohibition has remained the central policy, despite inescapable evidence that the "War on Drugs" is a futile approach that wastes billions of dollars every year (Robson 1999). The price of black market cannabis continues to fall in real terms, and it remains easily accessible in virtually every country in the world to anyone who wishes to consume it. However, medicinal research involving such a pariah drug presents profound methodological challenges, and this is reflected in the scientific limitations inherent in many of the clinical trials conducted during the last quarter of the twentieth century.

Partly as a result of the discovery of the endocannabinoid system and a growing realisation of its importance in both normal and pathological function, the final years of the twentieth century have seen renewed interest in exploring the potential of cannabis-based medicines among scientists and politicians in a number of countries. In the UK this has led to pioneering work in developing whole plant medicinal cannabis extracts containing different ratios of active ingredients targeted at different medical conditions (Whittle et al. 2001; Robson and Guy 2004). Whole plant extracts may have advantages over single chemical entities (such as synthetic THC) for several reasons (McPartland and Russo 2001). The non-psychoactive cannabinoid, cannabidiol (CBD), shows therapeutic promise in its own right (Pertwee 2004), and may modulate some of the less desirable actions of THC by both pharmacodynamic and pharmacokinetic mechanisms (Karniol 1973; McPartland and Russo 2001). Other cannabinoids and plant components such as terpenes, flavonoids and phenols may also have medicinal potential (McPartland and Russo 2001). Oromucosal sprays and vapourisers are promising delivery systems which provide greater flexibility for self-titration than the oral route (Whittle et al. 2001).

Conditions have never been more propitious for the rigorous scientific evaluation in humans of many of the hitherto anecdotal accounts summarised below.

2 Review of Clinical Research

2.1 Symptomatic Relief in Multiple Sclerosis and Spinal Cord Injury

Spasticity is a central feature of multiple sclerosis (MS) and spinal cord injury (SCI). It consists of a velocity-dependent increase in tonic stretch reflexes with exaggerated tendon jerks, resulting from hyperexcitability of the stretch reflex as one component of the upper motor syndrome (Young 1994). Existing drug therapy is far from satisfactory in terms of efficacy and unwanted effects (Panegyres 1992). Tremor, ataxia and lower urinary tract symptoms are frequently troublesome in MS. Both neuropathic and nociceptive pain (dealt with in Sect. 2.3) are also common in MS and SCI, and dozens of very painful muscle spasms can occur each day. Small wonder that there is also a high incidence of anxiety and depression in these conditions.

THC and other cannabinoids have been shown (Baker et al. 2000) to improve both tremor and spasticity in a well-validated animal model of MS (experimental allergic encephalomyelitis). Antagonism of the CB_1 receptor aggravated these signs, indicating a role for endogenous cannabinoids in the control of tremor and spasticity.

Many patients have reported anecdotally that cannabis can relieve some of the most distressing symptoms of MS and SCI, including spasticity, muscle pain, tremor, spasms on walking, paraesthesiae, leg weakness, trunk numbness, facial pain, impaired balance, nystagmus, anxiety and depression (Grinspoon and Bakalar 1993; Consroe et al. 1997). Hodges (1992) described the severe progression of her MS from its onset in 1983. Prescribed medicine was only moderately effective and produced unpleasant side-effects. Having with reluctance and no small difficulty established an illicit supply of cannabis, she wrote:

> When I smoke it, my body completely relaxes, which relieves the tension and spasms I have. It has had other beneficial effects. I am now more efficient at controlling my bladder, so I don't get the recurrent urinary infections that I was having before. It relieves my nausea and I can now sleep much better, so that I am not tired all the time.

Malec (1982) reported that 21 out of 24 SCI patients with spasticity who had tried cannabis found it had alleviated their symptoms. A recent survey of MS patients in the UK and USA found that between 30% and 97% experienced relief in symptoms with cannabis, depending on the particular symptoms (Consroe et al. 1997). In descending order of improvement, these were: spasticity, chronic pain, acute paroxysmal phenomena, tremor, emotional problems, anorexia/weight loss, fatigue states, double vision, sexual dysfunction, bowel and bladder symptoms, vision dimness, difficulty with walking and balance, and memory loss.

Open or single-blind observations of small numbers of patients on the effects of synthetic THC given orally have provided some support to these reports (Dunn and Davis 1974; Petro 1980; Clifford 1983; Meinck et al. 1989; Brenneissen et al. 1996). Subjective improvements in spasticity are a consistent finding, with some studies also indicating benefits for tremor, bladder control, mobility and mood. Unwanted effects do not seem to have been prominent. Schon et al. (1999) reported amplitude reduction of pendular nystagmus and improved visual acuity in an MS patient following smoked cannabis, but no effect following cannabis capsules or nabilone (a synthetic THC analogue). Of related interest is a report from Russo et al. (2003) describing improved night vision following both THC and cannabis in a single subject.

Brady et al. (2003) carried out an open pilot study in 15 MS patients with refractory lower urinary tract symptoms. They each received whole plant cannabis medicinal extracts (CBME) containing either predominantly THC or an equal proportion of THC and CBD for consecutive 8-week periods. Incontinence episodes, nocturia episodes, incidence of urinary urgency and frequency all decreased significantly, whilst the number of planned or normal voids significantly increased. Most patients experienced mild intoxication during the initial titration phases and two had short-lived hallucinations that disappeared on dose reduction. The authors concluded that CBME may prove to be a safe and effective additional treatment for this harrowing condition. A pilot open label study in 15 patients with overactive bladders as a result of SCI also showed symptomatic improvement following 10 mg THC by either oral or rectal routes (Hagenbach et al. 2001).

The first double-blind placebo-controlled study in MS patients was reported by Petro and Ellenberger (1981). Oral THC in a single dose of 5 or 10 mg was compared with placebo in a crossover design in 9 subjects. Both doses of THC were significantly superior to placebo in relieving spasticity measured by clinical examination or, where feasible, electromyography during quadriceps stretching. One patient receiving THC 10 mg and one receiving placebo reported feeling "high". Ungerleider and colleagues (1987) found in a randomised double-blind crossover study with 5-day treatment periods that THC 7.5 mg produced significantly improved patient ratings of spasticity in comparison with placebo. In a double-blind, placebo-controlled crossover trial Hanigan et al. (1985) reported that THC 30 mg/day for 20 days significantly improved objective measures of spasticity in 2 out of 5 patients with traumatic paraplegia. Martyn (1995) reported that nabilone 1 mg on alternate days for 1 month was better than placebo in a double-blind crossover study in a single MS patient. Improvement in nocturia, muscle spasm and general well-being were also noted in this patient, with mild sedation the only unwanted effect. On the negative side, a single dose of smoked cannabis (THC content 1.54%) impaired both posture and balance in comparison with placebo in 10 MS patients and 10 normal subjects (Greenberg et al. 1994), a not-unexpected occurrence with any skeletal muscle relaxant.

More recent trials of cannabis-based medicines in MS have given mixed results. Vaney and colleagues (2002) enrolled 57 MS patients in a randomised, crossover comparison of 15 mg THC daily in divided doses for 15 days with placebo. A significant improvement in a subjective rating of spasm frequency was not accompanied by objective improvement as represented by the Ashworth Score (Ashworth 1964). This is a measure of biological impairment, as opposed to disability or handicap, and relies upon an estimation by a clinician. A trend towards improvement in mobility was noted, but no effect on tremor, sleep quality, or lower urinary tract symptoms. Adverse events occurred with similar frequency in the active and control groups, but were more severe in the former. Killestein et al. (2002) reported an unambiguously negative study in 16 MS patients. In a randomised, double-blind crossover design, they compared synthetic THC with a cannabis plant extract containing the same amount of THC and placebo over 4 weeks of treatment. Starting dose was 2.5 mg orally twice daily, with the option to increase this to 5 mg twice daily after 2 weeks if the first dose was well tolerated. There was no improvement in spasticity as represented by the Ashworth Score. Both active medicines were well tolerated, but were inferior to placebo in terms of the patients' subjective global impression of change. An accompanying editorial (Thompson and Baker 2002) pointed out that the study was not powered to detect efficacy, and the writers drew attention to the difficulty in achieving the most appropriate individual dose by the oral route.

The very low water solubility of key cannabis constituents aggravates still further the well-known variability of absorption from the gastro-intestinal tract, resulting in poor predictability of both the timing and intensity of peak effects by the oral route. Titration of dose against symptom relief, as is the norm for most individuals who smoke cannabis medicinally, is very difficult in these circumstances. An additional drawback is the production of larger quantities of the reputedly psychoactive metabolite 11-OH-THC as a result of the hepatic first-pass phenomenon. The use of whole plant cannabis-based medicinal extracts in liquid form delivered by a pump action oromucosal spray (Whittle et al. 2001) represents an attempt to overcome these problems and permit the patient to self-titrate to an optimal individualised daily dose.

This mode of delivery was utilised in a consecutive series of double-blind, randomised, placebo-controlled single patient crossover trials with 2-week treatment periods (Wade et al. 2003). Twenty-four patients received whole plant extracts by oromucosal spray containing primarily THC, primarily CBD, an equal proportion of THC and CBD, or matched placebo at doses determined by titration against symptom relief or unwanted effects within the range 2.5–120 mg/24 h (1–48 sprays). Eligible patients had neurogenic symptoms which had responded poorly to standard treatments, and the majority had MS or SCI. Patients recorded symptom, well-being and intoxication scores on a daily basis using visual analogue scales (VAS), completed standard measures of disability, mood and cognition on regular clinic visits, and recorded adverse events. Average dose following self-titration in the active treatment groups was around 9 sprays/24 h. At the nursing assessments, all three CBMEs significantly improved the subjective measure of spasticity in comparison with placebo, and both THC CBME and THC: CBD CBME improved muscle spasm. Patients' daily diaries showed that THC CBME significantly improved VAS scores of pain, muscle spasm and spasticity, THC: CBD CBME significantly improved spasm and sleep, and CBD CBME significantly improved pain. Four patients withdrew due to unwanted effects, and the percentage of patients with at least one adverse event was considerably higher when THC was not accompanied by an equal proportion of CBD (55% vs 30%). The authors concluded that CBME can improve neurogenic symptoms unresponsive to standard treatments, and that unwanted effects were predictable and generally well tolerated.

An important trial funded by the UK Medical Research Council ("CAMS" study) has explored the effects of synthetic THC (Marinol) and a cannabis extract ("Cannador") given orally on spasticity and other symptoms related to multiple sclerosis (Zajicek et al. 2003). This was a randomised, placebo-controlled trial involving 33 centres and 630 patients, and the primary outcome measure was change in overall spasticity score as represented by the Ashworth scale.

The results of the study were mixed, and a large placebo effect was noted. There was no change in Ashworth score following 15 weeks of treatment with either THC or Cannador, but both active treatments demonstrated significant improvements in subjective measures of spasticity, muscle spasms, pain and sleep, and also in an objective measure of mobility. No effect was apparent on irritability, depression, tiredness, tremor or loss of energy. The authors noted an unexpected reduction in hospital admissions for relapse in the two active treatment groups. The known interaction of cannabinoids with the immune system, and the fact that MS is still regarded as an auto-immune condition led them to comment that this finding was worthy of further investigation. Minor unwanted effects were frequently reported in all three treatment groups, with a higher prevalence for the active treatments. The small number of serious adverse events were evenly spread across the three groups.

The limitations of the Ashworth scale in measuring such a complex phenomenon as spasticity is well known (Hinderer and Gupta 1996) and is acknowledged by the authors. They also noted that the evidence in support of currently available standard drug treatments for spasticity (and many other MS-related symptoms) is weak. Although the study incorporated a titration phase, the fixed twice daily dosing routine was not ideal in seeking to allow patients to optimise the balance between positive and negative effects given the known variations in individual response. An accompanying Lancet editorial (Metz and Page 2003) drew attention to the high variability in degree of spasticity among the trial patients and commented that the primary outcome measure does not correlate with function or other measures of spasticity. It recommended that "future studies should consider the potential confounding effect of including ... patients with severe spinal cord disease and should not rely totally on the Ashworth scale". It was also noted that poor bioavailability of oral cannabinoids may have influenced the outcome.

A significant effect upon a subjective measure of spasticity was the principal finding in another large study of cannabis-based medicine in MS (Wade et al. 2004). The effects of a whole plant extract containing an equal proportion of THC and CBD (Sativex) was compared with placebo in a parallel-group, double-blind, randomised study in 160 MS patients. Eligible patients were experiencing one of the following symptoms which had proved refractory to standard treatment: spasticity, muscle spasms, lower urinary tract symptoms, neuropathic pain or tremor. An oromucosal spray delivered 2.5 mg of each cannabinoid or matched placebo on each activation. After initial standardised dosing in an outpatient clinic, patients gradually titrated the dose upwards at home to a maximum of 48 sprays/24 h, aiming for an optimal balance between symptom relief and unwanted effects. Treatment period was 6 weeks, and the primary outcome measure was a composite derived from the VAS score of each patient's most troublesome symptom. Secondary measures were individual symptom VAS scores, and standardised measures of disability, cognition, mood, sleep and fatigue.

Once again, there was a strikingly large placebo effect. The composite score (max 100) following Sativex fell from a mean (SE) of 74.4 (11.1) to 48.9 (22.0) and from 74.3 (12.5) to 54.8 (26.3) following placebo (ns). Spasticity VAS scores fell by 31.2 following Sativex and by 8.4 following placebo [difference = −22.8; 95% confidence interval (CI): −35.52 to −10.07, $p = 0.001$]. Statistically non-significant improvements were also seen for spasms, bladder control and tremor. A similar pattern of responses was also noted from diary symptom VAS scores recorded by patients on a daily basis. Patients using Sativex assessed the quality of their sleep as significantly improved ($p = 0.047$). No significant adverse effects on cognition or mood were noted. Sativex was generally well tolerated. In particular, intoxication was usually mild, and largely avoidable with careful dose titration.

Clearly, further work is required to clarify the exact role of cannabis-based medicine in the symptomatic treatment of MS and SCI. Perhaps the position at the time of writing is best summarised by the comments of the Chief Executive of the Multiple Sclerosis Trust on the results of the CAMS study. In a press release on 7 November 2003, he stated:

> It is frustrating that the results of the study are somewhat equivocal. We are pleased that the CAMS study confirms the strong anecdotal evidence of the benefit of cannabis for some people with MS. It is particularly encouraging that patients receiving cannabis perceived an improvement in both spasticity and pain, when compared with those on placebo, and that no significant side-effects were reported. However, it is clear that the primary assessment tool used to measure spasticity, the Ashworth Scale, has failed to capture the full impact of this aspect of MS. Spasticity is a complex collection of symptoms encompassing pain and stiffness, some of which can only accurately be as-

sessed using subjective measures. However, overall, we believe that this study, combined with others which demonstrate symptomatic improvement, provides convincing evidence that cannabis may be clinically useful in treating some of the symptoms of MS.

2.2
Symptomatic Relief in Other Neurological Conditions

Stimulated by anecdotal reports that smoked cannabis improved a variety of movement disorders, Consroe and colleagues (1986) gave CBD 100–600 mg daily for 6 weeks to five patients with a variety of dystonic movement disorders. Dose-related improvements in dystonia were noted in all the patients, with maximal improvements ranging from 20% to 50%. Side-effects, described as mild, consisted of hypotension, dry mouth, sedation and light-headedness. However, CBD was "neither symptomatically beneficial nor toxic" in 10 patients with Huntington's disease at a dosage of 10 mg/kg/day for 6-week treatment periods (Consroe et al. 1991).

L-Dopa-induced dyskinesia (LDD) in Parkinson's disease (PD) presents a formidable therapeutic challenge. Overactivity in the lateral globus pallidus has been identified as a possible mechanism (Brotchie 2000) and, noting that this structure is rich in CB_1 receptors, Sieradzan et al. (2001) compared the synthetic THC analogue nabilone (0.03 mg/kg) with placebo in a double-blind, crossover trial in 7 patients. Mean total LDD score was significantly reduced following nabilone in comparison with placebo (17 vs 22, $p < 0.05$). Two patients were withdrawn following nabilone, one complaining of vertigo and the other because of postural hypotension. However, a further placebo-controlled study of 13 patients with primary dystonia (Fox et al. 2001) revealed no beneficial effect of nabilone. A recent survey (Venderova et al. 2003) identified 85 PD patients who had tried illicit cannabis for symptom relief, of whom 39 (45.9%) reported some improvement in rest tremor, bradykinesia, muscle rigidity or LDD. Interestingly, it took an average of 1.7 months for the benefit to appear, and improvement was recorded significantly more frequently by patients using cannabis for 3 months or more, and on a regular basis—at least once daily.

In a study primarily investigating possible appetite-stimulating effects of oral THC (dronabinol) in 12 patients with Alzheimer's disease (Volicer et al. 1997), the prevalence of disturbed behaviour measured by the Cohen-Mansfield Agitation Inventory (CMAI) was also assessed. Patients received THC 2.5 mg twice daily and placebo in a randomised, crossover design with 6-week treatment periods. THC significantly improved CMAI scores in comparison with placebo ($p = 0.05$). Unwanted effects included tiredness, somnolence and euphoria, and one patient experienced an epileptic convulsion (type not specified) soon after receiving the first dose of THC.

A few case studies have suggested that cannabis may produce beneficial effects in Tourette's syndrome (Sandyk et al. 1988; Hemming et al. 1993), although no clear rationale for a mechanism of action has been established. Muller-Vahl et al.

(1999) reported marked amelioration of both vocal and motor tics in an open trial of THC 10 mg in a 25-year-old patient. Improvement began 30 min after dosing, total tic severity was down from 41 at baseline to 7 at 2 h post dose, and benefit lasted for about 7 h. No adverse effects were reported. In a preliminary randomised, double-blind, placebo-controlled study (Muller-Vahl et al. 2003) THC in dosages up to 10 mg/day over a 6-week treatment period were compared with placebo in 24 patients with Tourette's syndrome. Seven patients dropped out, but even so there were some significant benefits for the active treatment using standardised outcome measures (e.g. Tourette Syndrome Symptom List). No serious adverse events were reported. On the basis of these findings the authors hypothesised that central cannabinoid receptors may play a part in the pathology of the syndrome.

2.3 Chronic Pain

Relief of intractable pain is one of the core historical applications of cannabis. There are many modern anecdotes as to its utility in cancer pain, bone and joint pain, migraine, menstrual cramps and labour pain (Grinspoon and Bakalar 1993). Cannabis has been shown to have a dose-dependent antinociceptive effect on experimental pain in healthy subjects (Greenwald and Stitzer 2000).

Unfortunately, scientific evidence for analgesic utility in humans remains scanty. Early studies evaluated oral THC or other synthetic cannabinoids in severe cancer-related, postoperative, or neurogenic pain. Noyes et al. (1975a) compared oral THC in single doses of 5, 10, 15 and 20 mg with placebo in a randomised, crossover design in 10 patients with cancer pain whose regular medication was withheld. A dose-related effect was observed, and the two higher doses gave significantly better pain relief than placebo, but these doses were associated with marked sedation. Other unwanted effects included slurred speech, blurred vision, mental clouding, dizziness and ataxia. Noyes' group went on to compare the efficacy of oral THC 10 and 20 mg with codeine 60 and 120 mg and placebo in a randomised, double-blind trial in 36 patients with cancer pain (Noyes et al. 1975b). A dose-related and equivalent analgesic effect was noted for both drugs, with the higher doses of both significantly superior to placebo. The effect of THC was maximal at 5 h (compared with 3 h for codeine) but 20 mg caused sedation and mental clouding in most patients. THC 10 mg was well tolerated but suitable only for mild pain.

Jain and colleagues (1981) compared intramuscular levonantradol (a synthetic cannabinoid) at several doses with placebo in a randomised, double-blind trial in 56 patients with severe postoperative or trauma pain. There was no apparent dose–effect relationship, but all doses of levonantradol were significantly superior to placebo. Unwanted effects were common but generally mild, with drowsiness occurring in almost half the subjects receiving active drug. Levonantradol subsequently disappeared without trace.

Two detailed single case studies were published in the 1990s. Maurer et al. (1990) compared the effects of oral THC (5 mg), codeine (50 mg) and placebo in a randomised, double-blind crossover study in a patient suffering severe pain related to muscle spasticity. Analgesic effects of both active drugs were similar and both superior to placebo. It was noted that THC also significantly improved the spasticity. No adverse effects were reported. Holdcroft et al. (1997) compared oral THC (50 mg daily in divided doses) with placebo in a 6-week, double-blind, crossover trial in a patient who required daily morphine to control chronic pain associated with familial Mediterranean fever. The patient was allowed to take morphine tablets as required, and although VAS pain scores remained similar in the THC and placebo conditions, the morphine consumption was significantly reduced ($p < 0.001$) in the THC period.

This limited body of work was subjected to meta-analysis (Campbell et al. 2001), and the authors reached the following conclusion:

> Cannabinoids are no more effective than codeine in controlling pain and have depressant effects on the central nervous system that limit their use. Their widespread introduction into clinical practice for pain management is therefore undesirable. In acute postoperative pain they should not be used. Before cannabinoids can be considered for treating spasticity and neuropathic pain, further valid randomised controlled studies are needed.

The validity of this conclusion was challenged by several correspondents to the editor of the journal. For example, Iversen (2001), noting that "a wealth of animal data support a role for cannabinoids in pain modulation" in contrast to the paucity of controlled human studies available for review, criticised the authors for "coming to a series of emphatic but ill-founded conclusions".

A further study of oral THC in postoperative pain has also given negative results (Buggy et al. 2003). THC 5 mg was compared with placebo in a randomised, double-blind, single-dose study in 40 women who had undergone abdominal hysterectomy. Measurement of summed pain-intensity difference at 6 h post dose revealed no difference between THC and placebo. However, there was also no difference between the groups in the incidence of adverse events, so the negative findings may be the result of a sub-therapeutic dose of THC.

Emerging evidence from basic science (e.g. Bridges et al. 2001; Fox et al. 2001; and Walker and Hohmann, this volume) implies that cannabis may benefit neuropathic pain. The 1997 National Institute of Health workshop on medical cannabis concluded: "Neuropathic pain represents a treatment problem for which currently available analgesics are, at best, marginally effective. Since Δ^9-THC is not acting by the same mechanism as either opioids or NSAIDs [nonsteroidal anti-inflammatory drugs], it may be useful in this inadequately treated type of pain." The UK House of Lords Science and Technology Committee (1998) came to a similar conclusion: "... pain which originates from damaged nerves might respond to cannabinoids.... An example of such pain is phantom limb pain following amputation.... [There is] anecdotal evidence that cannabis can relieve this pain [and] ... trials of cannabis should be undertaken in such patients."

Notcutt and colleagues (1997) reported their qualitative experience of the use of nabilone (synthetic THC analogue) in the treatment of 43 patients with severe pain resulting from MS, SCI and other sources of peripheral or central nerve damage, or malignancy. Of 43 patients, 25 were deemed to have benefited, and the main unwanted effects of nabilone were drowsiness and dysphoria.

More recent human studies focusing primarily on neuropathic pain have generally provided positive results. Wade et al. (2003) investigated the effects of three whole plant cannabis extracts (CBME) in a series of 24 single-case, double-blind, placebo-controlled crossover studies (see MS section above, Sect. 2.1, for details of the design) in patients with intractable neurogenic symptoms including pain. Significant analgesic effects in comparison to placebo were seen with both THC CBME and CBD CBME. The latter finding was considered particularly notable since CBD is a non-psychoactive cannabinoid. Using a similar design and the same extracts, Notcutt et al. (2002) reported the results of a series of trials involving 29 patients who were experiencing refractory pain as a result of MS or nerve damage following surgery or trauma. Significant improvements were seen in pain, sleep, depression, activity and general health. Three patients experienced postural hypotension during the initial self-titration, and some degree of intoxication was reported by several patients. One of these extracts (Sativex), containing equal proportions of THC and CBD, was compared with placebo in a double-blind, randomised trial over 3 weeks of treatment in 70 patients with chronic refractory neuropathic pain due to MS or other defects of neurological function (Sharief et al. 2004). Treatment difference in pain scores (BS-11) was 0.39 boxes in favour of Sativex ($p = 0.332$; 95% CI: −1.18, 0.4). Median percentage of days on which escape medication was used was 5% for Sativex and 45% for placebo ($p = 0.006$; 95% CI: −47.62, 0.00). Treatment was generally well tolerated, withdrawals were similar in both groups. Sleep disturbance was improved following Sativex ($p = 0.052$). The authors concluded that, on the basis of a reduced need for rescue medication, Sativex was efficacious in the treatment of chronic neuropathic pain.

Sativex has been the focus of two further controlled trials. Young and Rog (2003) compared it to placebo in a randomised, double-blind, parallel-group trial over 4 weeks of treatment in 64 patients with intractable central neuropathic pain due to MS. Patients were allowed to self-titrate their dose over a period of 1 week to a maximum of 48 sprays daily. At the end of the 4-week treatment period, pain relief following Sativex was significantly superior to placebo on both a BS-11 scale ($p = 0.005$) and the Neuropathic Pain Scale ($p = 0.039$). A subjective measure of sleep disturbance was also improved by Sativex ($p = 0.003$), and patients reported a greater overall impression of benefit following the CBME ($p = 0.005$). Most patients (88%) experienced at least one adverse effect on CBME (placebo = 69%) and one patient in the Sativex group withdrew from the study. Cognitive function was tested using the Brief Repeatable Battery of Neurological Tests. CBME showed a small but statistically significant difference ($p = 0.009$) in favour of placebo in one of the five components of the battery (the long-term storage component of the Selective Reminding Test). The authors concluded that Sativex was effective in reducing pain and sleep disturbance in MS-related central neuropathic pain, and is mostly well tolerated.

The effect of Sativex and THC CBME in treating refractory pain due to traction injuries of the brachial plexus has been studied in a randomised, placebo-controlled, crossover trial in 45 patients (McKerral et al. 2003). This injury produces a highly characteristic pain syndrome that is particularly difficult to treat. The authors note that opioids, anticonvulsants and tricyclic antidepressants are routinely used in the treatment of this pain, but are partially effective at best. Eligible patients continued on previously stabilised medicines, and received each test medicine for 2 weeks. During the first week of each treatment period, they were instructed cautiously to self-titrate to an optimal individualised dose within a daily maximum of 48 sprays. Both CBMEs produced moderate but highly statistically significant improvements (Sativex: $p < 0.002$; THC CBME: $p < 0.005$) in BS-11 pain scores in comparison with placebo. Sleep quality was also significantly improved by both CBMEs. Average number of sprays/24 h was 9.2 (placebo), 7.3 (THC CBME) and 6.9 (Sativex). The authors speculated that these relatively low doses might have been a result of the relatively short treatment periods limiting scope for self-titration, the fact that patients remained on their pre-existing analgesics, and patients' need to avoid dosing if they intended to drive. Taking into account the low doses achieved and the refractory nature of this type of neuropathic pain, the authors concluded that CBME "may represent a significant advance in treatment."

A small controlled study (Svendsen et al. 2003) suggests that dronabinol (synthetic THC) may also be useful in MS-related pain. THC (maximum dose of 10 mg/day) was compared with placebo in a randomised, double-blind, crossover trial with 3-week treatment periods in 24 patients with central neuropathic pain. Spontaneous pain intensity and pain relief were both significantly improved by THC. There was no comment on unwanted effects in this conference abstract.

Abrams et al. (2003) reported the effects of smoked cannabis in painful peripheral neuropathy secondary to human immunodeficiency virus (HIV) and/or antiretroviral treatment. In a preliminary uncontrolled pilot study (in preparation for a planned placebo-controlled trial) "excellent" correlation was reported between cannabis dosing and pain improvement, with 10 of 16 participants experiencing a greater than 30% reduction in pain. These results provide the ethical justification to proceed with the controlled trial.

Finally, the synthetic cannabinoid CT-3 was compared (Karst et al. 2003) with placebo in a randomised, double-blind, crossover trial in 21 patients with chronic neuropathic pain (cause unspecified). In 1-week treatment periods, patients received 4 capsules (10 mg CT-3 or placebo) daily in divided doses for the first 4 days and 8 daily for the following 3 days. Pain VAS scores were significantly improved by CT-3 in comparison with placebo ($p = 0.02$), although there was no dose–response relationship. Unwanted effects (most commonly dry mouth and tiredness) occurred more frequently following CT-3. The authors concluded that this preliminary evaluation suggested that CT-3 was effective in reducing chronic neuropathic pain.

2.4 Effects on Nausea and Vomiting

Many cytotoxic drugs used in the treatment of malignant disease are powerful emetics, and the distress caused by drug-induced nausea and vomiting is the major limiting factor in determining patients' acceptance of cancer chemotherapy (Carmichael 1992). Premedication with anti-emetics is routine, but severe vomiting induced by such drugs as cisplatin, dacarbazine or cyclophosphamide can be very difficult to control.

The anti-emetic properties of cannabis were rediscovered in the 1960s, when recreational users receiving cancer chemotherapy told their doctors it relieved their nausea. Anecdotal reports (e.g. Grinspoon and Bakalar 1993) preceded a range of controlled clinical trials in the 1970s and 1980s. These established that natural and synthetic forms of THC were invariably superior to placebo (Chang et al. 1979; Orr and Mckernan 1981; Jones et al. 1982). Controlled comparisons of THC with the anti-emetics available at the time suggested that it is either equivalent (Ungerleider et al. 1982) or superior (Formukong et al. 1989; Plasse et al. 1991; Orr and Mckernan 1981; Einhorn et al. 1981; Niiranen and Mattson 1985; Dalzell et al. 1986; Niederle et al. 1986; Pomery et al. 1986; Chan et al. 1987; Penta et al. 1981; Levitt 1986) to such drugs as prochlorperazine, domperidone, alizapride, dexamethasone and metoclopramide. Commonest unwanted effects included somnolence, dry mouth, ataxia, dizziness, dysphoria, and postural hypotension. Oral THC and nabilone often produced more unwanted effects than comparison drugs, yet THC was usually preferred by patients (Ungerleider et al. 1982; Einhorn et al. 1981; Niiranen and Mattson 1985; Dalzell et al. 1986).

Penta and colleagues (1981) reviewed 12 studies that examined the anti-emetic effects of THC (9) or nabilone (3) involving 600 patients. They reported that THC was "effective" in 8/9 and nabilone in 3/3. Levitt (1986) reviewed 55 studies, of which 32 were of randomised, double-blind design. Low-dose preventative treatment gave better results than targeting established vomiting. Levonantradol produced a higher frequency of dysphoric effects than nabilone or THC. A review by Formukong et al. (1989) suggested that the emesis produced by certain drugs (e.g. methotrexate, doxorubicin, cyclophosphamide, fluorouracil) responded better to THC than others (e.g. nitrosoureas, mustine, cisplatin). Younger patients responded better than older. Plasse and colleagues (1991) reviewed clinical experience with dronabinol (capsules of THC in sesame oil), which was first marketed in the U.S. in 1987. Meta-analysis suggested that an optimal balance of efficacy and unwanted effects is achieved with relatively modest doses of THC (i.e. 7 mg/m^2 or less). Sedation and psychotropic effects were commonly reported but were usually only of mild to moderate intensity and resolved rapidly on discontinuation.

Children seemed to respond well to nabilone (Dalzell et al. 1986; Chan et al. 1987) and to be tolerant of adverse effects, but confirmation is required. A small pilot study (Abrahamov et al. 1995) indicated a positive response to Δ^8-THC in 8 children receiving highly emetic antineoplastic therapy for various blood cancers. Vomiting was reported in 60% children receiving metoclopramide, but when Δ^8-THC was given orally 2 h before chemotherapy and repeated every 6 h for 24 h, no vomiting occurred on any of the 480 occasions this strategy was applied. Two children reported unwanted effects: both were "slightly irritable" and one (age 4) showed "slight euphoria". Surprisingly, this very promising result has not been followed up with a more definitive study.

The introduction of the highly effective (though expensive) 5-HT_3 antagonists including granisetron, ondansetron and tropisetron seems to have undermined interest in cannabis-based medicines for this indication. There have been no recent trials, so no information is available as to how they may compare with these newer and highly effective treatments. However, the combination of an anti-emetic effect alongside other attributes (e.g. analgesia, muscle relaxation, sedation) still provides a compelling case for exploration of a potential role for cannabinoids in conditions such as acquired immunodeficiency syndrome (AIDS), cancer, or perioperative pain. Of additional interest is the emerging evidence that non-psychoactive cannabinoids such as CBD may have anti-emetic properties (Parker et al. 2002; Javid et al. 2002).

2.5 Appetite Stimulation

Recreational users are familiar with the appetite-stimulating effect of cannabis ("the munchies"), and controlled studies in healthy subjects have confirmed this (Hollister 1971; Foltin et al. 1986). Kirkham and Williams (2001) have provided a comprehensive review of the effects of exogenous and endogenous cannabinoids on appetite and weight in animals and humans. There appears to be a link to the reward mechanisms that mediate the incentive value of food.

Open studies in cancer patients (Plasse et al. 1991; Nelson et al. 1994) suggested that THC has a positive effect on appetite and weight. In a double-blind study in 54 patients with various cancers, Regelson et al. (1976) found that oral THC (0.1–0.4 mg/kg four times daily) produced a significant ($p < 0.05$) gain or preservation of weight in comparison with placebo. THC also improved depression and "tranquillity" scores, but somnolence, dizziness and disassociation were troublesome in a quarter of the patients and led to 9 dropouts. A more recent study (Jatoi et al. 2002) compared dronabinol alone (2.5 mg BD) or in combination with megestrol acetate (MA: 800 mg/day) with MA alone in 469 patients with advanced cancer who were troubled with recent poor appetite or weight loss of at least 2.268 kg (5 lb). MA alone was significantly superior to dronabinol alone ($p = 0.0001$ for appetite; $p = 0.02$ for weight gain), and the addition of dronabinol to MA resulted in no significant improvements in appetite or weight over those that occurred with MA alone. Impotence was a significant problem for MA-treated men. The relative absence of typical THC-related unwanted effects suggests a sub-optimal dose.

Progressive weight loss is a major problem in AIDS. Beal and colleagues (1995) carried out a randomised, controlled trial of dronabinol in 139 late-stage AIDS patients (of whom 88 were "evaluable") who had experienced at least 2.5 kg reduction from their normal weight. Oral THC 5 mg daily significantly improved appetite in comparison with placebo ($p < 0.015$) and also reduced nausea ($p = 0.05$). There was a trend towards mood improvement in the dronabinol group ($p = 0.06$) and there was a tendency toward weight gain. THC produced significantly more adverse effects than placebo ($p < 0.001$), the most frequent being euphoria, dizziness, "thinking abnormalities", and sedation, but three quarters of these fell into the mild or moderate categories. Drop out rates between active and placebo groups were similar. Beal et al. (1997) followed up 94 patients from this study for a further 12 months. These subjects continued to receive dronabinol 2.5 mg once or twice daily, and consistent improvement in appetite was noted, typically at least twice baseline levels. Unwanted effects were as expected from a THC-containing medicine but were generally well tolerated.

Apart from appetite improvement, AIDS patients have reported a number of other benefits from cannabis including reduction in nausea, reduced anxiety, relief of aches and pains, improved sleep, and inhibition of oral candidiasis (Grinspoon and Bakalar 1993; Plasse et al. 1991). Commonest reasons for smoking cannabis given in a recently published survey (Sidney 2001) of HIV-positive subjects were to feel better mentally or reduce stress (79%), improve appetite or gain weight (67%) and decrease nausea (66%).

The study team who conducted the U.S. Institute of Medicine Review (1999) concluded (page 177), "For patients such as those with AIDS or who are undergoing chemotherapy, and who suffer simultaneously from severe pain, nausea, and appetite loss, cannabinoid drugs might offer broad-spectrum relief not found in any other single medication."

Concern has been expressed that HIV-infected individuals may be more vulnerable to the immunosuppressive effects of cannabis or THC. Kaslow and colleagues (1989) monitored the progress of nearly 5,000 HIV-positive men for 18 months and found no evidence that use of psychoactive substances (including cannabis) had any discernable effect upon T helper lymphocyte counts or progression to AIDS. A randomised controlled trial (Bredt et al. 2002) compared the effects of marijuana cigarettes (0.9 g, 3.95% THC, up to 3 daily), dronabinol (2.5 mg up to 3 times daily) and placebo over a 3-week treatment period in 62 HIV-positive subjects being treated with protease inhibitor anti-retroviral drugs. Neither active treatment produced any significant effects on the percentage of $CD4^+$ and $CD8^+$ T cells, T cell activation, changes in cytokine flow cytometry, natural killer cell number and function, or in a lymphoproliferation assay. Within the limitations of a short-term study, the authors concluded that there were no detrimental effects of cannabinoids on any of the immune parameters measured. A separate analysis of the same patient group (Abrams et al. 2003) revealed no significant effects on viral load as represented by HIV RNA levels.

Another condition frequently associated with decreased appetite and malnutrition is senile dementia of Alzheimer type. Eleven patients with Alzheimer's disease were treated for 12 weeks on an alternating schedule of dronabinol (THC: 2.5 mg twice daily) and placebo (6 weeks of each treatment). The dronabinol treatment resulted in substantial weight gains and a decline in disturbed behaviour (Volicer et al. 1997). No serious side-effects were observed. One patient had a seizure and was removed from the study, but the investigators were unsure whether this was attributable to dronabinol. Patel and colleagues (2003) recently reported an open study in this population. Forty-eight patients with Alzheimer's disease with uncontrolled agitation and anorexia were given dronabinol 5–10 mg daily for a month. The authors reported weight gain in all patients.

2.6 Appetite Suppression in Obesity

A growing understanding of the role of central cannabinoid systems in the regulation of appetite (Williams and Kirkham 1999) has raised the possibility that blocking CB_1 receptors might inhibit appetite (Kirkham 2003). Testing this hypothesis has become a possibility with the development of the selective CB_1 receptor antagonist SR141716A (rimonabant). Studies in various animal models have demonstrated that this produces marked reduction of food intake, body weight and adiposity (e.g. Ravinet et al. 2002).

At the time of writing, seven phase III clinical trials are in progress focusing on rimonabant's effect on weight loss and smoking cessation. None of these has yet been published in peer-reviewed journals, but two have been completed and the results presented at a U.S. cardiology conference in 2004. According to information supplied by the manufacturer, overweight patients treated with rimonabant 20 mg daily for 1 year lost significantly more weight than placebo patients ($p < 0.001$). Improvement in some associated cardiovascular risk factors (e.g. waist circumference, HDL cholesterol and triglyceride plasma levels, C-reactive protein levels) were also reported. Unwanted effects were described as consisting mainly of mild and transient nausea and dizziness, though twice as many patients dropped out on rimonabant 20 mg than placebo. A second study suggested that smokers seeking abstinence were twice as likely to be successful when treated with rimonabant 20 mg for 10 weeks in comparison with placebo ($p = 0.002$). Rimonabant also appeared to protect against the weight gain commonly seen following smoking cessation. Once again, however, there were twice as many dropouts on active treatment. It must be noted that these results await peer review.

2.7 Glaucoma

Glaucoma is the commonest cause of blindness in the Western World. Raised intraocular pressure (IOP) is usually due to an obstruction to the outflow of aqueous humour at the front of the eye, and by far the commonest deficit is primary open-angle (chronic simple) glaucoma. A range of topical and systemic drugs are used to treat this, but efficacy is variable and there are many possible unwanted effects.

The discovery that cannabis lowers IOP was first reported by Hepler and Frank (1971), and the mechanism by which this is achieved still remains to be clarified. Controlled studies in healthy subjects (Hepler et al. 1976; Perez-Reyes et al. 1976; Jones et al. 1981) have shown that oral, injected or smoked THC produces dose-related reductions of IOP as much as 30% below baseline, though tolerance may occur on chronic dosing.

In the 1970s, anecdotal reports of symptom relief by smoked marijuana appeared and a small number of glaucoma patients successfully argued in the U.S. for legal access to the drug (Grinspoon and Bakalar 1993). Hepler and colleagues (1976) carried out a pilot study of smoked marijuana and oral THC (15 mg) in 11 patients. IOP reductions averaging 30% were seen in 7, whilst 4 had no response.

Two small placebo-controlled studies of smoked and topical THC confirmed a significant IOP reduction in glaucoma patients. Merrit and colleagues (1980) compared smoked THC (2%) with placebo in a double-blind parallel-group study in 18 patients. IOP was significantly reduced in comparison with placebo between 1.5 and 2.5 h after dosing. Unfortunately, these effects were accompanied by reductions in blood pressure, increases in heart rate, and "alterations in mental status" which were not propitious for clinical utility. Merritt (1981) went on to investigate THC eye-drops in a double-blind, placebo-controlled study in 8 patients. Dose-related reductions in IOP were recorded using 0.05% and 1% drops with minimal unwanted effects. Parallel reductions were noted in the untreated eye, suggesting a systemic rather than a local mode of action.

It is now apparent that raised IOP is not the only pathological mechanism in glaucoma. Impaired auto-regulation in arteries supplying the optic nerve head may interfere with perfusion and cause neural damage (Prunte et al. 1998). The discovery that CB_1 receptors are present in micro-vasculature (Sugiura et al. 1998) and the ability of endogenous cannabinoids to produce vasodilation (Sugiura et al. 1998) suggests the possibility that exogenous cannabinoids may alleviate this deficit. Antioxidant and *N*-methyl-D-aspartate (NMDA) receptor neuroprotective properties of cannabinoids (Hampson et al. 1998) raise the hope that they might improve survival of ischaemic retinal ganglion cells. Future prospects have been reviewed by Jarvinen et al. (2002). Non-irritant local delivery using cyclodextrins and non-psychoactive cannabinoids offers considerable promise.

2.8 Epilepsy

Epilepsy afflicts around 1% of the world's population, and historically was an important target for medicinal cannabis (O'Shaugnessy 1843; Reynolds 1890). Modern anti-epileptic drugs fail to provide satisfactory control in up to 30% of patients, and all can produce disabling or even life-threatening unwanted effects.

A confusing picture emerges when cannabinoids are evaluated in animal models of epilepsy (Karler and Turkanis 1981; Consroe and Snider 1986). CBD has anticonvulsant properties with a spectrum distinct from standard anticonvulsants, apparently not hampered by the development of tolerance but with a varying profile according to the species tested. THC can produce seizures in some circumstances but is anticonvulsant in others. In a recent study, THC (10 mg/kg) completely abolished spontaneous seizures in the rat pilocarpine model of epilepsy (Wallace et al. 2003). The results also indicated that endogenous cannabinoid tone may modulate seizure termination and duration via the CB_1 receptor.

Human research data are almost non-existent. There are anecdotal reports of beneficial effects of cannabis in human epileptics (Grinspoon and Bakalar 1993) and a couple of published single case reports. A man with grand mal epilepsy stopped taking his anticonvulsants and suffered no fits for 6 months. He then smoked cannabis on seven occasions over a 3-week period and suffered three fits during this time, though these were unrelated to periods of intoxication (Keeler and Reifler 1967). In contrast, a young man whose seizure control was poor began smoking 2–5 cannabis cigarettes nightly in addition to his conventional medication and found this terminated his seizures (Consroe et al. 1975).

One solitary controlled trial is on record, comparing CBD (200–300 mg daily for up to 4.5 months added to standard therapy) with placebo in a double-blind, parallel-group design in 15 poorly controlled patients with "secondary generalised epilepsy" (Cunha et al. 1980). Half the CBD patients remained "almost free" of fits throughout the experiment and all but one of the others showed "partial improvement". With a single exception, the placebo patients remained unchanged. Drowsiness in a quarter of the patients was the only unwanted effect associated with CBD.

In view of the continuing uncertainty as to whether cannabis and its constituents pose a risk to individuals with past or present epilepsy or on the contrary offer a novel mode of treatment, a properly powered controlled trial is urgently required.

2.9 Psychiatric Disorders

There is some evidence that nabilone may have an anxiolytic effect. Fabre and McLendon (1981) compared nabilone 3 mg daily with placebo in a randomised, double-blind, parallel-group study in 20 anxious patients. "Dramatic improvements" in anxiety scores were reported for nabilone relative to placebo ($p < 0.001$). Commonest unwanted effects were dry mouth, dry eyes and drowsiness. Ilaria et al. (1981) compared nabilone 2–5 mg daily with placebo in a double-blind crossover study over a 2-week period in 11 anxious patients. Significant improvements in outcome scores were accompanied by postural hypotension in most patients, though this tended to tolerate out over time.

Cannabis and THC are known in certain circumstances to induce anxiety or panic, and Zuardi and colleagues (1982) reported that CBD antagonises anxiogenic effects of THC along with some other marijuana-like effects in healthy volunteers. At a dose of 300 mg orally it reduced anxiety in comparison with placebo in a simulated public speaking task (Zuardi et al. 1993). CBD was also found to behave like an atypical antipsychotic in the apomorphine-induced stereotypy model in rodents (Zuardi et al. 1991). In a report of a single case, CBD (in doses up to 1500 mg/day) was found to improve psychotic symptoms without toxic effects in a psychotic patient who had experienced intolerable unwanted effects with haloperidol (Zuardi et al. 1995).

A controlled trial in 15 insomniac volunteers suggested that CBD (160 mg) may be an effective hypnotic (Carlini and Cunha 1981), but in a more recent sleep laboratory study in healthy subjects (Nicholson et al. 2004) much smaller doses of CBD (5 and 15 mg) appeared to have alerting properties. When CBD (15 mg) was given in combination with THC (15 mg) at 10 pm, it counteracted the morning-after sedative effects seen when THC was given alone and increased wakeful activity during sleep. Effects on sleep architecture were modest, but some effects of both cannabinoids on slow wave sleep were reported. Overall, these results suggest that the improvement in sleep quality frequently reported in clinical trials is mainly due to nocturnal symptom relief rather than a primary hypnotic effect.

THC (0.1 mg/kg) was reported to have anti-depressant properties in cancer patients (Regelson et al. 1976). There are anecdotal reports that cannabis may act as a mood stabiliser in bipolar affective disorder (Grinspoon and Bakalar 1998).

The discovery that the endogenous cannabinoid system has a central function in extinction of aversive memories (Marsicano et al. 2002) raises the fascinating possibility that CB_1 agonists may prove therapeutic in phobias or post-traumatic stress disorder.

2.10 Asthma

Although cannabis was used as a bronchodilator in the nineteenth century, modern human research seems to have been limited to a brief period in the 1970s. Small controlled studies in asthmatic volunteers (Tashkin et al. 1974; Williams et al. 1976; Tashkin et al. 1977) showed that oral, smoked and aerosolised THC had significant bronchodilator activity comparable to that of salbutamol, though slower in onset. Dose-related tachycardia and intoxication occurred at higher doses. An inhaled aerosol avoided systemic absorption of THC but induced cough and chest discomfort, which limited its usefulness.

3 Safety Issues with Cannabis-Based Medicines

Cannabis is known to demonstrate very low acute toxicity. To the best of this author's knowledge, it remains the case that no human death has been reliably ascribed to cannabis toxicity alone. It has been estimated, based on extrapolation from mouse to man, that the lethal dose to effective dose ratio is around 40,000:1 (Grinspoon and Bakalar 1993, p 138). Minor adverse events (AEs) including intoxication, dizziness and dry mouth occur frequently with THC-containing medicines, but are generally mild or moderate in intensity and well tolerated by patients. In a recent large study (Zajicek et al. 2003), out of 417 patients allocated to THC or cannabis extract, only 9 patients discontinued treatment because of intolerable AEs, and serious or life-threatening AEs were no more frequent following active treatments than placebo.

Cannabis and THC are known to increase heart rate, cardiac output and supine blood pressure, and can cause orthostatic hypotension (Jones 2002). Because of the resulting increase in cardiac work, cannabis and THC are probably best avoided by patients with clinically significant cardiovascular disorders. Cardiovascular effects tend to tolerate out over chronic dosing (Benowitz and Jones 1981). A survey of myocardial infarction survivors set out to investigate whether smoking marijuana might have been a trigger for this event (Mittleman et al. 2001). Unfortunately, only 124 of the 3,882 patients surveyed admitted to smoking marijuana. The risk of infarction appeared to be elevated 4.8 times over baseline during the 60 min following marijuana use but decreased rapidly thereafter. However, this conclusion has been much criticised, not least because the sample of subjects upon which it is based (those who had smoked cannabis within 1 h of infarction) amounted to only 9 individuals, of whom 3 admitted at least one other "triggering activity" (e.g. cocaine use or sexual intercourse). Epidemiological data on 65,000 patients in the San Francisco Bay Area do not support an increased risk of cardiac events in cannabis smokers (Sidney et al. 1997).

Animal and human data regarding effects of recreational cannabis on fertility, pregnancy and birth outcomes, teratogenicity, and possible neurodevelopment effects on the infant are conflicting and no clear conclusions are possible. In these circumstances, it would be prudent for couples seeking to conceive and pregnant women to avoid cannabis-based medicines. THC is transferred into breast milk and may reach concentrations eight times higher than those in maternal plasma (Astley and Little 1990).

3.1 Cognitive/Motor Effects

Any medicine containing THC may produce similar acute cognitive effects to recreational cannabis if taken in sufficient dosage. These effects include: euphoria, sensory enhancement, increased social conviviality, and a sense of relaxation and contentment; perceptual effects including distorted time and space estimation and alteration in sensory modalities; impairment in both sustained and divided attention; impairment in reaction time, motor control and dexterity; impairment in various aspects of memory and higher cognitive function including associative and abstractive processes, planning and organisational strategies (reviewed by Solowij 1998: pp 29–40). The possible implications for those receiving cannabis-based medicines who wish to continue driving have been reviewed by Hadorn (2004). Interestingly, analysis of responsibility for traffic collisions has repeatedly indicated that drivers with only cannabis in their systems (and especially no alcohol) were, if anything, less culpable than drug-free drivers. In prospective studies using driving simulators or road tests, cannabis does impair subjects' ability to maintain road position and constant following distances. However, cannabis users generally seem aware of being impaired and compensate by driving more cautiously. Alcohol consistently produced greater impairment than cannabis in comparable social doses, tended to induce more aggressive driving and, in contrast to cannabis smokers, alcohol subjects lacked insight into their impairment and thus made no attempt to compensate. These studies suggest that, as should be the case with many other prescribed drugs, patients receiving cannabis-based medicines should simply be warned to avoid driving and other potentially hazardous tasks at any time they feel impaired.

Do any of these acute deficits persist after cannabis has been discontinued and fully metabolised? A large and expanding scientific literature has still not fully resolved this question. Recognising the methodological shortcomings that have dogged much of this research, Gonzalez et al. (2002) proposed seven "minimal criteria" which should be applied to any study purporting to explore non-acute cognitive effects of cannabis: only 13 out of 40 eligible studies met these basic criteria. The authors point out that negative results have been disseminated in the media without any acknowledgement of these serious shortcomings.

A major problem lies in distinguishing long-lasting but reversible residual effects (due to slow metabolism of cannabis components or withdrawal phenomena) from irreversible effects. Pope et al. (2002) tested 77 current heavy users and 87 controls. The former showed significant memory deficits at 0, 1 and 7 days of abstinence, but by day 28 were virtually indistinguishable from control subjects. There was no association between duration of cannabis use and cognitive performance after 28 days of abstinence. This conflicts with the finding of Solowij et al. (2002) that deficits on several neuropsychological measures were correlated with lifetime duration of cannabis exposure. In seeking to explain this, Pope et al. (2002) point out that even well-controlled studies depend on the assumption that, after adjustment for more obvious confounding factors, cannabis users and non-users are comparable on all factors other than exposure to cannabis. Additionally, heavy use of an illegal drug may produce non-pharmacological deficits such as family alienation or school drop-out that impact outcome measures. Grant et al. (2003) carried out a meta-analysis of studies examining non-acute cognitive effects, and found no substantial, systematic or detrimental effect of recreational cannabis on neuropsychological performance. They concluded:

> The small magnitude of effect sizes from observations of chronic users of cannabis suggests that cannabis compounds, if found to have therapeutic value, should have a good margin of safety from a neurocognitive standpoint under the more limited conditions of exposure that would likely obtain in a medical setting.

3.2 Dependency/Abuse

Properties of THC that may have a bearing on its dependency and abuse potential have been investigated in numerous animal models, but how reliable these may be in predicting human behaviour is open to question. Despite the cripplingly expensive War on Drugs, recreational cannabis is still easily available, cheaper in real terms and used extensively throughout the world, so it seems sensible to examine what actually happens outside the laboratory.

The evidence for cannabis dependence in humans has been reviewed by Johns (2001). Characteristic components of a dependence syndrome are the need over time to take more of the drug to maintain the desired effect (tolerance), a predictable group of symptoms and signs over a consistent time course when the drug is withdrawn, difficulty in keeping consumption under control, and a preoccupation with the drug that interferes with the normal activities of living.

Tolerance to the subjective effects of marijuana has been reported (Georgotas and Zeidenberg 1979), and a minority (16%) of regular smokers experienced at least one of the following symptoms following abrupt withdrawal of cannabis: irritability, insomnia, tremor, sweating, gastro-intestinal disturbance or appetite change (Wisbeck et al. 1996). These effects peak between 2 and 6 days after abrupt withdrawal (Budney et al. 2003). It has been reported that a third of regular users experienced some difficulty in controlling their use of the drug (Thomas 1996). All research in this area is dogged by serious methodological problems, including highly selected samples, non-validated measures, poor response rates in community surveys, and the existence of many confounding variables. However, it seems reasonable to accept that psychological dependence will occur in a small minority of cannabis smokers. The existence of a clear-cut physical dependence syndrome is much less convincing on the basis of the published literature. If it exists at all, it is probably mild and transient, and is likely to consist of a few days of sleep disturbance and somatic symptoms of anxiety in heavy daily users who abstain abruptly.

In an interview study (Robson and Bruce 1997), the dependence potential of various street drugs was assessed in 201 problem and 380 "social" users of heroin, cocaine or amphetamine using the well-validated Severity of Dependence Scale (SDS). Scores (maximum = 15) in the problem group were 12.9 for heroin, 9.6 for other opioids, 6.1 for amphetamine and 5.5 for crack cocaine. All of these scores were consistent with findings in other studies. Cannabis SDS score was 2.6 and comparable with those of LSD (3.1) and ecstasy (1.3), two drugs that are generally not associated with physical or psychological dependence. In the parallel sample of social users, the cannabis SDS was similar at 3.4.

Attempting to define and investigate cannabis dependence in patients is still more challenging, especially if the individual is experiencing a beneficial therapeutic effect. Developing an emotional attachment or preoccupation with a drug that has helped with previously intractable, life-impairing symptoms is a very different matter from becoming over-preoccupied with a recreational drug. It would hardly be surprising for a patient abruptly denied such a medicine to yearn for it and become preoccupied with re-establishing a supply. Is the diabetic addicted to insulin? Experience in the therapeutic setting with much more powerfully addictive drugs than cannabis is encouraging. For example, the abuse of opiates is extremely unusual among patients treated appropriately for pain and other symptoms (Porter and Jick 1980; Portenoy 1990), and this is very likely to be the case with cannabis-based medicines. Support for this is provided by the intoxication data from a recent study (Wade et al. 2004) comparing a THC-containing cannabis extract (Sativex) with placebo for a 6-week treatment period in patients with MS. At the end of the trial, all patients re-titrated on to the active medicine for a further 4 weeks. Intoxication scores were recorded in a daily diary on a 100 mm VAS scale shown in Fig. 1. Average peak scores reached only around 20/100, and levels appeared to diminish over time. There was no evidence that Sativex was abused by any of these patients.

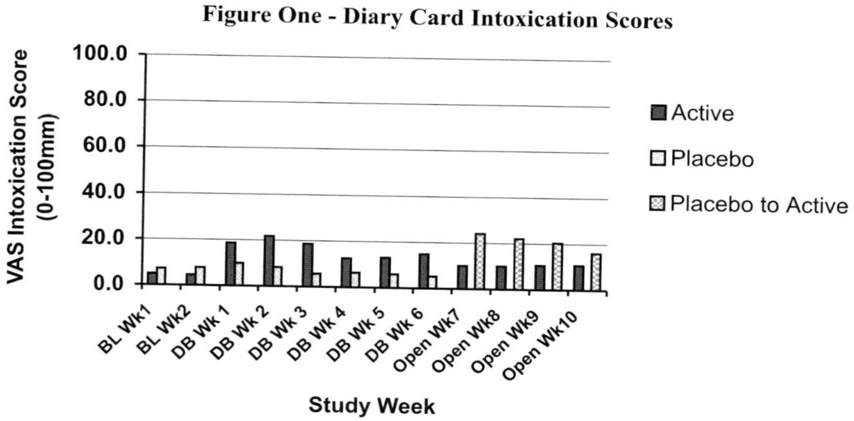

Fig. 1. Diary Card intoxication scores. *BL*, baseline. *DB*, double-blind. Patients self-titrated active medication (THC:CBD) or placebo against symptom relief or intolerable unwanted effects. Doses reached a plateau after 4 weeks. Open patients from both arms re-titrated onto active medication. (Reproduced with kind permission from Arnold Publishers)

3.3 Effects on Mental Health

All the following considerations refer to information derived from recreational cannabis smokers, and the implications for medicinal users are unknown. However, an obvious difference exists between the two groups: the primary intention of the former is to experience intoxication, whilst the vast majority of the latter seek to avoid it.

Cannabis smoking is known to produce anxiety, dysphoria, panic, paranoia, tiredness and low motivation in a proportion of users, particularly younger people and those with unusual personalities or social disadvantage (Hall et al. 1994). Large doses can produce a transient "toxic psychosis" with hallucinations and delusions that generally resolves within a week or so of abstinence (Johns 2001). Although there are exceptions to this, a consensus view among psychiatrists would be that recreational cannabis use is likely to aggravate the symptoms and behavioural consequences of pre-existing psychiatric illness (Johns 2001). This would suggest that patients with existing psychiatric illness or a strong family history should avoid cannabis-based medicines. Intriguingly, raised concentrations of endocannabinoids were discovered in the cerebrospinal fluid of schizophrenia patients in comparison with normal controls (Leweke et al. 1999), leading the authors to speculate that an imbalance in endocannabinoid signalling may contribute to the pathogenesis of schizophrenia.

A much more controversial question is whether cannabis might actually be an independent risk factor for schizophrenia in previously healthy subjects. Undoubtedly, cannabis smoking is more prevalent in psychiatric populations (Regier et al. 1990), but retrospective or cross-sectional studies are of no help in evaluating the presence or direction of causality. Five prospective studies have been subjected to critical review (Arsenault et al. 2004). The authors' conclusion was that cannabis smoking by young adolescents confers an overall twofold increase in the risk of developing schizophrenia. However, they state that "cannabis use appears to be neither a sufficient nor a necessary cause for psychosis. It is a component cause, part of a complex constellation of factors leading to psychosis". They further conclude:

> Although the majority of young people are able to use cannabis in adolescence without harm, a vulnerable minority experiences harmful outcomes. The epidemiological evidence suggests that cannabis use among psychologically vulnerable adolescents should be strongly discouraged by parents, teachers and health practitioners alike.

However, the five studies reviewed in this paper have been criticised elsewhere for methodological shortcomings including: presence of clinical or sub-clinical psychiatric illness prior to cannabis consumption; lack of a clear temporal link between cannabis use and subsequent psychiatric illness; poor reliability of the diagnosis of schizophrenia; confusion between acute toxic states and functional mental illness; confusion of association with causation; confounding effects of other recreational drugs and environmental risk factors for mental illness; unreliability of self-report of an illegal activity; and a lack of a correlation in epidemiological studies between prevalence of cannabis consumption and schizophrenia. The UK Advisory Council on the Misuse of Drugs (ACMD) reviewed the evidence in depth and concluded (2002, p. 8) "... no clear causal link has been demonstrated." Degenhardt and Hall reached a similar conclusion (2002): "Time trends in schizophrenia and cannabis use are not consistent with the hypothesis that cannabis use causes schizophrenia de novo."

In conclusion, the link between functional mental illness and recreational cannabis use in previously healthy subjects with no psychiatric history remains controversial, and a causative link has not yet been established. However, it would seem advisable for individuals with existing psychiatric illness or a strong family history to avoid THC-containing medicines.

4 Future Directions

Notwithstanding all the hard work summarised above, the scientific evaluation of medicinal cannabis in humans is in its infancy. The role of cannabis-based medicines in all the clinical indications so far discussed requires clarification through further well-controlled, adequately powered randomised trials. The rapidly expanding knowledge of the structure and function of the endocannabinoid system raises the hope of exciting new pharmacological entities. To give a few examples: It may be possible to enhance the activity of endocannabinoids by inhibiting degradation mechanisms such as fatty acid amide hydrolase, and since there appears to be local up-regulation of endocannabinoids in certain pathological conditions, this gives the added possibility of site selectivity (Baker et al. 2001); the discovery that the CB_2-selective cannabinoid agonist AM1241 suppresses capsaicin-evoked thermal and mechanical hyperalgesia and allodynia (Hohmann et al. 2004) along with associated pain behaviour in rats raises the possibility of novel treatments for pain, free from unwanted psychoactive effects; it may be possible to develop CB_1 agonists that do not cross the blood–brain barrier (Chaperon and Thiebot 1999). Other possibilities are discussed elsewhere in this book, but these developments are all for the future. Of more immediate concern is the question as to which new directions are worthy of clinical pursuit with the synthetic and plant-derived materials available right now?

The answer to that question will reflect to some extent the personal interests of the respondent, but it seems logical that target conditions should satisfy at least one of the following two requirements: historical or anecdotal evidence which suggests that cannabis may be helpful, and currently available treatment is unsatisfactory either because of limited efficacy or unacceptable toxicity; the activity profile of cannabis or its components in some relevant in vitro or in vivo models indicates a potentially beneficial effect on symptoms/signs or disease progression. Given the rapid expansion in basic research involving both exogenous and endogenous cannabinoids over recent years, there are many conditions that satisfy both requirements. The following is by no means an exhaustive list.

4.1 Inflammatory Conditions

These disorders certainly satisfy both the above categories. Musculoskeletal pain features prominently in historical accounts. In a recent survey (Ware et al. 2003) of 2,969 people who agreed to fill in a questionnaire about medicinal cannabis, nearly a quarter gave symptom relief for arthritis as the reason for smoking cannabis. This was the fourth-commonest indication after chronic pain, MS and depression. Elucidation of the anti-inflammatory and immunomodulatory effects of several cannabis constituents (see chapters by Cabral and Staab, this volume, and Pertwee, also in this volume) has provided a strong scientific rationale for clinical evaluation. Of particular relevance was the discovery (Malfait et al. 2000) that CBD given either intraperitoneally or orally inhibited disease progression in a murine model of rheumatoid arthritis (RA). Clinical improvement and joint protection were related to a combination of lymphocyte and granulocyte suppression and inhibition of the inflammatory cytokine tumour necrosis factor (TNF). RA is the commonest form of inflammatory arthritis and afflicts up to 3% of the population of Western countries. Non-steroidal anti-inflammatory drugs and corticosteroids form the backbone of treatment, but are often seriously toxic. TNF antagonism looks a promising approach (Taylor 2001) but available agents (e.g. etanercept, infliximab) are expensive and have to be given by injection.

The combination of analgesic and anti-inflammatory effects is also highly relevant for inflammatory bowel conditions such as Crohn's disease. Dysregulation of immune mechanisms are strongly implicated in the disease process with excess production of inflammatory cytokines, particularly TNF, by lymphocytes and macrophages in the gut wall. Disruption of mucosal function leads to chronic diarrhoea and weight loss. In these circumstances certain cannabinoids may produce beneficial symptomatic effects by depressing gastrointestinal motility, delaying gastric emptying, and inhibiting peristalsis by both central and peripheral mechanisms (Pertwee 2001). Examination of human biopsy specimens has demonstrated the presence of CB_1 receptors in the epithelium and smooth muscle of both normal and diseased colon, implying a role for the endocannabinoid system in gastrointestinal physiology (Wright et al. 2003).

4.2 Chronic Nociceptive Pain

Existing (albeit flawed) research reviewed above suggests that cannabis and THC offer few advantages over standard treatments for nociceptive pain, but recent research has indicated that a combination of THC with opioids may provide benefits greater than the sum of the two parts. This synergy was certainly recognised by nineteenth century physicians.

The combination of analgesic agents with different modes of action is a well-accepted principle (Dahl and Raeder 2000), and the anti-emetic activity of THC is important since nausea and vomiting are the most troublesome and dose-limiting unwanted effects of opioids. However, the important work of Welch, Cichewicz and colleagues shows that the advantages go well beyond this. Small doses of THC not only enhance the analgesic effects of opioids (Cichewicz and McCarthy 2003) but also prevent the development of tolerance and physical dependence (Cichewicz and Welch 2003) and extend the duration of action of both morphine and codeine (Cichewicz et al. 2003). Clinical research to explore the exciting potential of this combination in humans is urgently required, and at the time of writing a large multi-centre study of THC in combination with patient-controlled morphine analgesia in postoperative patients is getting underway in the UK.

4.3 Neuroprotection

Brain trauma or ischaemia and a range of neurodegenerative disorders including MS, Parkinson's disease, Huntington's disease, Alzheimer's disease and motor neuron disease share common mechanisms of neuron damage. These include excitotoxic effects resulting from excessive release of glutamate, which massively increases intracellular calcium concentration through overstimulation of NMDA, *S*-α-amino-3-hydroxy-5-methyl-4-isoxazolepropionic acid (AMPA) and kainate receptors, and damage from reactive oxygen species. Following the demonstration by Hampson and colleagues (1998) that both THC and CBD could protect against these effects in vitro, there is now a considerable literature in this area (see chapters by Pertwee and Guzmán, this volume). Encouraging results have been found in animal models of cerebral ischaemia, closed head injury, Huntington's disease, Parkinson's disease, amyotrophic lateral sclerosis (SOD_1 model), and soman-induced seizures. Vulnerability to excitotoxicity is probably a major factor in the progression of MS, so the discovery that CB_1 agonists limit neurodegeneration in an animal model of MS (Pryce et al. 2003) is of considerable interest. This gives potential significance to the observation by Zajicek et al. (2003) that MS patients receiving THC or a cannabis extract experienced fewer hospital admissions for relapse than placebo patients.

The investigation of neuroprotective activity in humans poses daunting ethical, financial and methodological challenges. Timely enrollment of stroke and trauma patients is difficult, and the inherent variability in progression of neurodegenerative conditions means large numbers of subjects are needed. Outcome measures are often unreliable or expensive. Brain imaging techniques are likely to be central. These include structural and function magnetic resonance imaging, magnetic resonance spectroscopy, positron emission tomography, and single-photon emission computerised tomography. Unfortunately, in many conditions lesions revealed by these techniques show little relation to clinical disease progression, and still more focused measures may be required such as imaging the MRS neuronal marker *N*-acetylaspartate in MS (Mathews et al. 1998).

Dexanabinol (HU-211), a non-psychoactive synthetic cannabinoid, has been the subject of the only controlled study yet to be reported in humans (Knoller et al. 2002). In a randomised, double-blind comparison with placebo, single doses of either 48 or 150 mg dexanabinol were given intravenously to neurosurgical inpatients within 6 h of severe closed head injury. Since outcome measures did not indicate a dose-related response, comparisons were made between combined active dose groups and placebo. Significant beneficial effects on intracranial pressure and cerebral perfusion pressure independent of systemic blood pressure were seen in the active treatment groups. Neurological outcome as assessed by the Glasgow scale was better ($p = 0.04$) in the combined active groups at 3 months, but this was no longer significant ($p = 0.14$) at 6 months. Dexanabinol appeared well tolerated and there was no significant difference between placebo and active groups in the incidence of unwanted effects.

4.4
Anti-cancer Effects

The symptomatic benefits of cannabis and its derivatives in patients with cancer has been discussed above, but considerable evidence has accumulated from in vitro and in vivo animal studies that cannabinoids may inhibit the growth of various types of tumour cell (For a review see Guzmán's contribution in this volume and Guzmán 2003). Possible mechanisms include the selective promotion of cancer cell apoptosis and inhibition of tumour vascularisation. Preliminary clinical studies have been initiated but no results reported at the time of writing. An issue to be determined is whether effects will be apparent at the tissue levels achievable in humans by systemic dosing—in some circumstances it may be preferable to seek ways to deliver the cannabinoid direct to the target site (Guzmán 2003).

4.5 Drug Withdrawal Treatments

In contrast to contemporary concerns about the addictive potential of cannabis, the drug was used in the nineteenth century in the treatment of dependencies on various other substances including alcohol, cocaine, chloral hydrate and morphine. Anyone who discusses the problems of opiate withdrawal with a modern heroin addict is likely to be told of the beneficial effects of marijuana in allaying withdrawal symptoms, and this anecdotal evidence is given some scientific credibility by a number of studies in animals (Hine et al. 1975; Bhargava 1976; Chesher and Jackson 1985). In animal pain models THC inhibits the development of opioid tolerance and physical dependence (Chichewicz and Welch 2003). At the time of writing, the efficacy of a combination of THC and CBD (Sativex) in alleviating the opioid withdrawal syndrome is being explored in a double-blind, placebo-controlled study.

There are anecdotal reports that cannabis is useful in countering both the withdrawal symptoms (Labigalini et al. 1999) and paranoia and weight loss (Dreher 2002) associated with smoking crack cocaine.

See above (Sect. 2.6) for the promising preliminary outcome of a trial evaluating the CB_1 receptor antagonist rimonabant as an aid to abstaining from tobacco smoking.

4.6 Migraine

This is a common disorder in which attacks, sometimes preceded by an aura, consist of intense headache along with nausea and sensitivity to light and sound lasting anywhere from a few hours to several days. In historical times, cannabis was widely used in the treatment of headache, and there are numerous modern anecdotes (Grinspoon and Bakalar 1993). The pathology underlying the disorder remains controversial, but serotonergic, dopaminergic, inflammatory and brain stem mechanisms have been implicated.

In a detailed review, Russo (2001) considers how cannabinoids may impact on these systems and makes a compelling case for initiating controlled clinical trials.

4.7 Intractable Breathlessness

A number of lung diseases (e.g. chronic bronchitis and emphysema) are capable of producing shortness of breath that is often extremely distressing to the patient. Many of these conditions are irreversible, so it becomes necessary to target the symptom itself. The sensation of breathlessness is a complicated phenomenon that seems to depend upon central processing through respiratory and non-respiratory mechanisms (Guz 1996). Ideally, a treatment would relieve the unpleasant sensation without further compromising respiratory function. Opioids and benzodiazepines produce some relief but may have the dangerous side-effect of depressing respiration.

Patients have reported anecdotally that cannabis can relieve breathlessness by relieving anxiety and promoting relaxation. CB_1 receptors are virtually absent from the part of the brain-stem which drives respiration (Herkenham et al. 1990), so it seems possible that symptom relief may be achieved without negative effects upon breathing. THC has been shown to have anxiety-reducing and sedating effects (Fabre and McLendon 1981; Nicholson et al. 2004), as has CBD in larger doses (Zuardi et al. 1997). CBD is also thought to have useful modulating effects on some of the undesirable effects of THC (McPartland and Russo 2001).

At the time of writing, exploratory research of THC/CBD combinations in refractory breathlessness is getting underway, incorporating careful monitoring of respiratory function.

References

Abel EL (1980) Outlawing marihuana. In: Marihuana: the first twelve thousand years. Plenum Press, New York, pp 237–247

Abrahamov A, Abrahamov A, Mechoulam R (1995) An efficient new cannabinoids antiemetic in pediatric oncology. Life Sci 56:2097–2102

Abrams DI, Hilton JF, Leiser RJ, et al (2003) Short-term effects of cannabinoids in patients with HIV-1 infection. A randomized, placebo-controlled clinical trial. Ann Intern Med 139:258–266

Advisory Council on the Misuse of Drugs (2002) The classification of cannabis under the Misuse of Drugs Act 1971. Home Office, London

Arseneault L, Cannon M, Witton J, et al (2004) Causal association between cannabis and psychosis: examination of the evidence. Br J Psychiatry 184:110–117

Ashworth B (1964) Preliminary trial of carisoprodol in multiple sclerosis. Practitioner 192:540–542

Astley SJ, Little RE (1990) Maternal marijuana use during lactation and infant development at one year. Neurotoxicol Teratol 12:161–168

Baker D, Pryce G, Croxford JL, et al (2000) Cannabinoids control spasticity and tremor in a multiple sclerosis model. Nature 404:84–87

Baker D, Pryce G, Croxford JL, et al (2001) Endocannabinoids control spasticity in a multiple sclerosis model. FASEB J 15:300–302

Beal JE, Olson R, Laubenstein L, et al (1995) Dronabinol as a treatment for anorexia associated with weight loss in patients with AIDS. J Pain Symptom Manage 10:89–97

Beal JE, Olson R, Lefkowitz L, et al (1997) Long-term efficacy and safety of dronabinol for acquired immunodeficiency syndrome-associated nausea. J Pain Symptom Manage 14:7–14

Benowitz NL, Jones RT (1981) Cardiovascular and metabolic considerations in prolonged cannabinoid administration in man. J Clin Pharmacol 21:214–223S

Brady C, DasGupta R, Wiseman O, et al (2003) Sublingual cannabis based medicinal extracts for bladder dysfunction in advanced multiple sclerosis. Int Cannabinoid Res Soc 13th Annu Symp Cannabinoids, Abstr 56

Bredt BM, Higuera-Alhino D, Shade SB, et al (2002) Short-term effects of cannabinoids on immune phenotype and function in HIV-1-infected patients. J Clin Pharmacol 42:82S–89S

Brenneissen R, Egli A, Elsohly MA, Henn V, Spiess Y (1996) The effect of orally and rectally administered delta-9-THC on spasticity: a pilot study with two patients. Int J Clin Pharmacol Ther 34:446–452

Bridges D, Ahmad K, Rice, AS (2001) The synthetic cannabinoid WIN55,212–2 attenuates hyperalgesia and allodynia in a rat model of neuropathic pain. Br J Pharmacol 133:586–594

Brotchie JM (2000) The neural mechanisms underlying levodopa-induced dyskinesia. Ann Neurol 47:105–114S

Budney AJ, Moore BA, Vandrey RG, et al (2003) The time course and significance of cannabis withdrawal. J Abnorm Psychol 112:393–402

Buggy DJ, Toogood L, Maric S, et al (2003) Lack of analgesic efficacy of oral delta-9-tetrahydrocannabinol in postoperative pain. Pain 106:169–172

Campbell FA, Tramer MR, Carroll D, et al (2001) Are cannabinoids an effective and safe option in the management of pain? A qualitative systematic review. Br Med J 323:13–16

Carmichael J (1992) The principles of cancer chemotherapy. In: Grahame-Smith DG, Aronson JK (eds) Oxford textbook of clinical pharmacology, 2nd edn. Oxford University Press, Oxford, 505–515

Chan HSL, Correia JA, MacLeod SM (1987) Nabilone versus prochlorperazine for control of cancer chemotherapy-induced emesis in children: a double-blind crossover trial. Pediatrics 79:946–952

Chang AE, Shiling DJ, Stillman RC, et al (1979) Delta-9-THC as an antiemetic in cancer patients receiving high-dose methotrexate. Ann Intern Med 91:819–830

Chaperon F, Thiebot MH (1999) Behavioural effects of cannabinoid agents in animals. Crit Rev Neurobiol 13:243–281

Cichewicz DL, McCarthy EA (2003) Antinociceptive synergy between delta-9-tetrahydrocannabinol and opioids after oral administration. J Pharmacol Exp Ther 304:1010–1015

Cichewicz DL, Welch SP (2003) Modulation of oral morphine antinociceptive tolerance and naloxone-precipitated withdrawal signs by oral delta 9-tetrahydrocannabinol. J Pharmacol Exp Ther 305:812–817

Cichewicz DL, Rubo A, Welch SP (2003) Recovery of morphine- and codeine-induced antinociception by delta-9-tetrahydrocannabinol. Int Cannabinoid Res Soc 13th Annu Symp Cannabinoids, Abstr 40

Clifford DB (1983) THC for tremor in multiple sclerosis. Ann Neurol 13:669–671

Consroe P, Snider SR (1986) Therapeutic potential of cannabinoids in neurological disorders. In: Mechoulam E (ed) Cannabinoids as therapeutic agents. CRC Press, Boca Raton, 21–49

Consroe P, Sandyk R, Snider SR (1986) Open label evaluation of cannabidiol in dystonic movement disorders. Int J NeuroSci 30:277–282C

Consroe P, Laguna J, Allender J, et al (1991) Controlled clinical trial of cannabidiol in Huntington's disease. Pharmacol Biochem Behav 40:701–708C

Consroe P, Musty R, Rein J, Tillery W, Pertwee RG (1997) The Perceived effects of smoked cannabis on patients with multiple sclerosis. Eur Neurol 38:44–48

Consroe PF, Wood GC, Buchsbaum H (1975) Anticonvulsant nature of marihuana smoking. J Am Med Assoc 234:306–307

Cunha JM, Carlini EA, Pereira AE, et al (1980) Chronic administration of cannabidiol to healthy volunteers and epileptic patients. Pharmacology 21:175–185

Dahl V, Raeder JC (2000) Non-opioid postoperative analgesia. Acta Anaesthesiol Scand 44:1191–1203

Dalzell AM, Bartlett H, Lilleyman JS (1986) Nabilone: an alternative antiemetic for cancer chemotherapy. Arch Dis Child 61:502–505

Degenhardt L, Hall W (2002) Cannabis and psychosis. Curr Psychiatry Rep 4:191–196

Dunn M, Davis R (1974) The perceived effects of marijuana on spinal cord injured males. Paraplegia 12:175

Einhorn LH, Nagy C, Furnas B, et al (1981) Nabilone: an effective antiemetic in patients receiving cancer chemotherapy. J Clin Pharmacol 21:64S–69S

Fabre LF, McLendon D (1981) The efficacy and safety of nabilone (a synthetic cannabinoid) in the treatment of anxiety. J Clin Pharmacol 21:377S–382S
Foltin RW, Fischman MW, Byrne MF (1986) Behavioural analysis of marijuana effects on food intake in humans. Pharmacol Biochem Behav 25:577–582
Formukong EA, Evans AT, Evans FJ (1989) The medicinal use of cannabis and its constituents. PhytoTher Res 3:219–231
Fox A, Kesingland A, Gentry, et al (2001) The role of central and peripheral Cannabinoid1 receptors in the antihyperalgesic activity of cannabinoids in a model of neuropathic pain. Pain 92:91–100
Fox SH, Kellett M, Moore AP, et al (2002) Randomised, double-blind, placebo-controlled trial to assess the potential of cannabinoid receptor stimulation in the treatment of dystonia. Mov Disord 17:145–149
Georgotas A, Zeidenberg P (1979) Observations on the effects of heavy marijuana smoking on group interaction and individual behaviour. Compr Psychiatry 20:427–432
Gonzalez R, Carey C, Grant I (2002) Nonacute (residual) effects of cannabis use: a qualitative analysis and systematic review. J Clin Pharmacol 42:48–57S
Grant I, Gonzalez R, Carey CL, et al (2003) Non-acute (residual) neurocognitive effects of cannabis use: a meta-analytic study. J Int Neuropsychol Soc 9:679–689
Greenberg HS, Werness SA, Pugh JE, et al (1994) Short-term effects of smoking marijuana on balance in patients with multiple sclerosis and normal volunteers. Clin Pharmacol Ther 55:324–328
Greenwald MK, Stitzer ML (2000) Antinociceptive, subjective and behavioural effects of smoked marijuana in humans. Drug Alcohol Depend 59:261–275
Grinspoon L, Bakalar JB (1993) The history of cannabis. In: Marihuana: the forbidden medicine. Yale University Press, New Haven, pp 1–23
Grinspoon L, Bakalar JB (1998) The use of cannabis as a mood stabilizer in bipolar disorder: anecdotal evidence and the need for clinical research. J Psychoactive Drugs 30:171–177
Guz A (1996) Respiratory sensations: some clinical perspectives. In: Eds Abrams L, Guz A (eds) Lung biology in health and disease, vol 90. Marcel Dekker, New York, 389–395
Guzmán M (2003) Cannabinoids: potential anticancer agents. Nat Rev Cancer 3:745–755
Hadorn D (2004) A review of cannabis and driving skills. In: Guy G, Whittle B, Robson P (eds) The medicinal use of cannabis and cannabinoids. Pharmaceutical Press, London, 329–365
Hagenbach U, Gafoor N, Brenneisen R, et al (2001) Clinical investigation of delta-9-tetrahydrocannabinol (THC) as an alternative therapy for overactive bladders in spinal cord injury patients. Congress on Cannabis and Cannabinoids, Cologne, Germany. International Association for Cannabis as Medicine, p 10
Hall W, Solowij N, Lemon J (1994) The health and social consequences of cannabis use. Monograph series No. 25. Australian Government Publishing Service, Canberra
Hampson A, Grimaldi M, Axelrod J, et al (1998) Cannabidiol and delta-9-tetrahydrocannabinol are neuroprotective antioxidants. Proc Natl Acad Sci USA 95:8268–8273
Hanigan WC, Destree R, Truong XT (1985) The effect of delta-9-THC on human spasticity. Clin Pharmacol Ther 35:198
Hemming M, Yellowlees PM (1993) Effective treatment of Tourette's syndrome with marijuana. J Clin Pharmacol 7:389–391
Hepler RS, Frank IM (1971) Marihuana smoking and intraocular pressure. JAMA 217:1392
Hepler RS, Frank IM, Petrus R (1976) Ocular effects of marihuana smoking. In: Braude MC, Szara S (eds) The pharmacology of marihuana. Raven Press, New York, pp 815–824
Herkenham M, Lynn AB, Little MD, et al (1990) Cannabinoid receptor localization in brain. Proc Natl Acad Sci USA 87:1932–1936
Hinderer SR, Gupta S (1996) Functional outcome measures to assess interventions for spasticity. Arch Phys Med Rehabil 77:1083–1089
Hodges C (1992) Very alternative medicine. Spectator, 1 August 1992, p 18

Hohmann AG, Farthing JN, Zvonok AM, Makriyannis A (2004) Selective activation of cannabinoid CB2 receptors suppresses hyperalgesia evoked by intradermal capsaicin. J Pharmacol Exp Ther 308:446–453
Holdcroft A, Smith M, Jacklin A, et al (1997) Pain relief with oral cannabinoids in familial Mediterranean fever. Anaesthesia 5:483–486
Hollister LE (1971) Hunger and appetite after single doses of marihuana, alcohol, and dextroamphetamine. Clin Pharmacol Ther 12:44–49
House of Lords Select Committee on Science and Technology (1998) Ninth Report. London
Ilaria RL, Thornby JI, Fann WE (1981) Nabilone, a cannabinoid derivative, in the treatment of anxiety neurosis. Curr Ther Res Clin Exp 29:943–949
Institute of Medicine (1999) The medical value of marijuana and related substances. In: Joy JE, Watson SJ, Benson JA (eds) Marijuana and medicine: assessing the science base. National Academic Press, Washington, 137–191
Iversen L (2001) Few well controlled trials of cannabis exist for systematic review. Br Med J 323:1250
Jain AK, Ryan JR, McMahon FG, Smith G (1981) Evaluation of intramuscular levonantradol and placebo in acute postoperative pain. J Clin Pharmacol 21:320S–326S
Jarvinen T, Pate D, Laine K (2002) Cannabinoids in the treatment of glaucoma. Pharmacol Ther 95:203–220
Jatoi A, Windshitl HE, Loprinzi CL, et al (2002) Dronabinol versus megestrol acetate versus combination therapy for cancer-associated anorexia. J Clin Oncol 20:567–573
Javid FA, Wright C, Naylor RJ, et al (2002) An inhibitory role for cannabinoids in the control of motion sickness in Suncus murinus. In: Symposium on the Cannabinoids; Asilomar Conference Center, Pacific Grove, CA. International Cannabinoid Research Society, 13 July 2002: abstr 141
Johns A (2001) Psychiatric effects of cannabis. Br J Psychiatry 178:116–122
Jones RT (2002) Cardiovascular system effects of marijuana. J Clin Pharmacol 42:58–63S
Jones RT, Benowitz NL, Herning RL (1981) Clinical relevance of cannabis tolerance and dependence. J Clin Pharmacol 21:143S–152S
Jones SE, Durant JR, Greco FA, et al (1982) A multi-institutional phase III study of nabilone vs placebo in chemotherapy-induced nausea and vomiting. Cancer Treat Rev 9:45–48S
Karler R, Turkanis SA (1981) The cannabinoids as potential antiepileptics. J Clin Pharmacol 21:437S–448S
Karniol IG, Carlini EA (1973) Pharmacological interaction between cannabidiol and delta-9-THC. Psychopharmacologia 33:53–70
Karst M, Salim K, Burstein S, et al (2003) Analgesic effect of the synthetic cannabinoid CT-3 on chronic neuropathic pain: a randomised controlled trial. JAMA 290:1757–1762
Kaslow RA, Blackwelder WC, Ostrow DG, et al (1989) No evidence for a role of alcohol or other psychoactive drugs in accelerating immunodeficiency in HIV-1-positive individuals. JAMA 261:3424–3429
Keeler MH, Reifler CB (1967) Grand mal convulsions subsequent to marihuana smoking. Dis Nerv Syst 18:474–475
Killestein J, Hoogervorst EL, Reif M, et al (2002) Safety, tolerability, and efficacy of orally administered cannabinoids in MS. Neurology 58:1404–1407
Kirkham TC (2003) Endogenous cannabinoids: a new target in the treatment of obesity. Am J Physiol Regul Integr Comp Physiol 284:R343–R344
Kirkham TC, Williams CM (2001) Endogenous cannabinoids and appetite. Nutr Res Rev 14:65–86
Knoller N, Levi L, Shoshan I, et al (2002) Dexanabinol (HU-211) in the treatment of severe closed head injury: a randomised, placebo-controlled, phase II clinical trial. Crit Care Med 30:548–554
Levitt M (1986) Cannabinoids as antiemetics in cancer chemotherapy. In: Mechoulam R (ed) Cannabinoids as therapeutic agents. CRC Press, Boca Raton, 71–83

Leweke FM, Giuffrida A, Wurster U, et al (1999) Elevated endogenous cannabinoids in schizophrenia. Neuroreport 10:1665–1669
Malec J, Harvey RF, Cayner JJ (1982) Cannabis effect on spasticity in spinal cord injury. Arch Phys Med Rehabil 63:116–118
Malfait AM, Gallily R, Sumariwalla PF, et al (2000) The non-psychoactive cannabis-constituent cannabidiol is an oral anti-arthritic therapeutic in murine collagen-induced arthritis. Proc Natl Acad Sci USA 97:9561–9566
Marsicano G, Wotjak CT, Azad SC, et al (2002) Endogenous cannabinoid system controls extinction aversive memories. Nature 418:530–534
Martyn CN, Illis LS, Thom J (1995) Nabilone in the treatment of multiple sclerosis. Lancet 345:579
Mathews PM, De Stefano N, Narayanan S, et al (1998) Putting magnetic resonance spectroscopy studies in context: axonal damage and disability in multiple sclerosis. Semin Neurol 18:327–336
Maurer M, Henn V, Dittrich A, Hoffman A (1990) Delta-9-THC shows antispastic and analgesic effects in a single case double blind trial. Eur Arch Psychiatry Clin Neurosci 240:1–4
McKerral S, Berman J, Lee J, et al (2003) Efficacy of two cannabis based medicinal extracts for relief of central neuropathic pain from brachial plexus avulsion: results of a randomised controlled trial. Int Cannabinoid Res Soc 13th Annu Symp Cannabinoids, Abstr 45
McPartland J, Russo E (2001) Cannabis and cannabis extracts: greater than the sum of their parts? J Cannabis Ther 1:103–132
Mead A (2004) International control of cannabis: changing attitudes. In: Guy G, Whittle B, Robson P (eds) The medicinal use of cannabis and cannabinoids. Pharmaceutical Press, London
Mechoulam R (1986) The pharmacohistory of cannabis sativa. In: Mechoulam R (ed) Cannabinoids as therapeutic agents. CRC Press, Boca Raton, pp 1–19
Meinck HM, Schonle PW, Conrad B (1989) Effect of cannabinoids on spasticity and ataxia in multiple scleroris. J Neurol 236:120–122
Merritt JC, Crawford WJ, Alexander PC, et al (1980) Effect of marihuana on intraocular and blood pressure in glaucoma. Ophthalmology 87:222–228
Merritt JC, Olsen JL, Armstrong PC, et al (1981) Topical delta-9-THC in hypertensive glaucomas. J Pharm Pharmacol 33:40–41
Metz L, Page S (2003) Oral cannabinoids for spasticity in multiple sclerosis: will attitude continue to limit use? (editorial). Lancet 362:1513
Mittleman MA, Lewis RA, Maclure M, et al (2001) Triggering myocardial infarction by marijuana. Circulation 103:2805–2809
Muller-Vahl K, Schneider U, Kolbe H, et al (1999) Treatment of Tourette's syndrome with delta-9-tetrahydrocannabinol. Am J Psychiatry 156:495
Muller-Vahl KR, Schneider U, Prevedel H, et al (2003) Delta-9-tetrahydrocannabinol (THC) is effective in the treatment of Tourette syndrome: a 6-week randomised trial. J Clin Psychiatry 64:459–465
National Institute of Health (1997) Report from a workshop on the medical utility of marijuana, 19–20 February 1997: p 6. http://www.parklandtrading.com/users/thc4ms/pdf/0211.pdf
Nelson K, Walsh D, Deeter P, et al (1994) A phase-II study of delta-9-THC for appetite-stimulation in cancer-associated anorexia. J Palliat Care 10:14–18
Nicholson AN, Turner C, Stone BM, Robson P (2004) Effect of delta-9-THC and cannabidiol on nocturnal sleep and early morning behaviour in young adults. J Clin PsychoPharmacol (in press)
Niederle N, Schutte J, Schmidt CG (1986) Crossover comparison of the antiemetic efficacy of nabilone and alizapride in patients with nonseminomatous testicular cancer receiving cisplatin therapy. Klin Wochenschr 64:362–365
Niiranen A, Mattson K (1985) A cross-over comparison of nabilone and prochlorperazine for emesis induced by cancer chemotherapy. Am J Clin Oncol 8:336–340

Notcutt W, Price M, Chapman G (1997) Clinical experience with nabilone for chronic pain. Pharm Sci 3:551–555
Notcutt W, Price M, Sansom C, et al (2002) Medicinal cannabis extract in chronic pain: overall results of 29 "n of 1" studies (CBME-1) In: Symposium on the Cannabinoids; Asilomar Conference Center, Pacific Grove, CA: International Cannabinoid Research Society; 13 July 2002: abstr 55
Noyes R, Brunk SF, Baram DA, Canter A (1975a) Analgesic effects of delta-9-THC. J Clin Pharmacol 15:139–143
Noyes R, Brunk SF, Avery DH, Canter A (1975b) The analgesic properties of delta-9-THC and codeine. Clin Pharmacol Ther 18:84–89
O'Shaughnessy WB (1843) On the cannabis indica or Indian hemp. Pharmacol J 2:594
Orr LE, McKernan JF (1981) Antiemetic effect of delta-9-THC in chemotherapy-associated nausea and emesis as compared to placebo and compazine. J Clin Pharmacol 21:76S–80S
Panegyres PK (1992) The drug therapy of neurological disorders. In: Grahame-Smith DG, Aronson JK (eds) Oxford textbook of clinical pharmacology, 2nd edn. Oxford University Press, 441–442
Parker LA, Mechoulam R, Schlievert C (2002) Cannabidiol, a non-psychoactive component of cannabis and its synthetic dimethylheptyl homolog suppress nausea in an experimental model with rats. Neuroreport 13:567–570
Patel S, Shua-Haim JR, Pass M (2003) Safety and efficacy of dronabinol in the treatment of agitation in patients with Alzheimer's disease with anorexia: a retrospective chart review. International Psychogeriatric Association, Eleventh International Congress
Penta JS, Poster DS, Bruno S, et al (1981) Clinical trials with anti-emetic agents in cancer patients receiving chemotherapy. J Clin Pharmacol 21:11S–22S
Perez-Reyes M, Wagner D, Wall ME, et al (1976) Intravenous administration of cannabinoids and intraocular pressure. In: Braude MC, Szara S (eds) The pharmacology of marihuana. Raven Press, New York, pp 829–832
Pertwee RG (2001) Cannabinoids and the gastrointestinal tract. Gut 48:859–867
Pertwee RG (2004) The pharmacology and therapeutic potential of cannabidiol. In: Di Marzo V (ed) Cannabinoids. Kluwer Academic/Plenum Publishers (http://www.eurekah.com/)
Petro DJ (1980) Marihuana as a therapeutic agent for muscle spasm or spasticity. Psychosomatics 21:81–85C
Petro DJ, Ellenberger C (1981) Treatment of human spasticity with delta-9-THC. J Clin Pharmacol 21:413S–416S
Plasse TF, Gorter RW, Krasnow SH, et al (1991) Recent clinical experience with dronabinol. Pharmacol Biochem Behav 40:695–700
Pomeroy M, Fennelly JJ, Towers M (1986) Prospective randomised double-blind trial of nabilone versus domperidone in the treatment of cytotoxic-induced emesis. Cancer ChemoTher Pharmacol 17:285–288
Pope HG, Gruber AJ, Hudson JI, et al (2002) Cognitive measures in long-term cannabis users. J Clin Pharmacol 42:41–47S
Portenoy RK (1990) Chronic opioid therapy in non-malignant pain. J Pain Symptom Manage 5 (Suppl 1):46–62
Porter J, Jick H (1980) Addiction rare in patients treated with narcotics (letter). N Engl J Med 302:123
Prunte C, Orgul S, Flammer J (1998) Abnormalities of microcirculation in glaucoma: facts and hints. Curr Opin Ophthalmol 9:50–55
Pryce G, Ahmed Z, Hankey DJR, et al (2003) Cannabinoids inhibit neurodegeneration in models of multiple sclerosis. Brain 126:2191–2202
Ravinet Trillou C, Arnone M, Delgorge C, et al (2002) Anti-obesity effect of SR141716, a CB1 receptor antagonist, in diet-induced obese mice. Am J Physiol Regul Integr Comp Physiol 284:R345–R353
Regelson W, Butler JR, Schulz J, et al (1976) Delta-9-THC as an effective antidepressant and appetite-stimulating agent in advanced cancer patients. In: Braude MC, Szara S (eds) The pharmacology of marijuana. Raven Press, New York, pp 763–776

Regier DA, Farmer ME, Swift W, et al (1990) Comorbidity of mental disorders with alcohol and other drug abuse. Results from the Epidemiologic Catchment Area (ECA) study. J Am Med Assoc 264:2511–2518

Reynolds JR (1890) Therapeutic uses and toxic effects of cannabis indica. Lancet 1:637–638

Robson P, Bruce M (1997) A comparison of 'visible' and 'invisible' users of amphetamine, cocaine and heroin: two distinct populations? Addiction 92:1729–1736

Robson P, Guy GW (2004) Clinical studies of cannabis based medicines. In: Guy G, Whittle B, Robson P (eds) The medicinal use of cannabis and cannabinoids. Pharmaceutical Press, London

Robson PJ (1999) Drug policy—a time for change? In: Forbidden drugs. Oxford University Press, Oxford, pp 239–259

Russo E (2001) Hemp for headache: an in-depth historical and scientific review of cannabis in migraine treatment. J Cannabis Ther 1:21–92

Russo E, Merzouki A, Mesa JM, Frey K (2003) Cannabis improves night vision: a pilot study of visual threshold and dark adaptometry in kif smokers in the Rif region of Northern Morocco. International Cannabinoid Research Society 13th Annual Symposium on the Cannabinoids, abstr 61

Sandyk R, Awerbuch G (1988) Marijuana and Tourette's syndrome. J Clin PsychoPharmacol 8:444–445

Sharief MK, Notcutt WG, Mutiboko I, et al (2004) Sativex in the treatment of patients with chronic refractory pain due to MS or other defects of neurological function. Association of British Neurologists Spring Scientific Meeting, 14–16 April 2004, London

Sidney S (2001) Marijuana use in HIV-positive and AIDS patients: results of an anonymous mail survey. J Cannabis Ther 1:35–43

Sidney S, Beck JE, Tekawa IS, et al (1997) Marijuana use and mortality. Am J Public Health 87:585–590

Sieradzan KA, Fox SH, Hill M, et al (2001) Cannabinoids reduce levodopa-induced dyskinesia in Parkinson's disease: a pilot study. Neurology 57:2108–2111

Solowij N (1998) Acute effects of cannabis on cognitive functioning In: Cannabis and cognitive functioning. Cambridge University Press, Cambridge, pp 29–38

Solowij N, Stephens RS, Roffman RA, et al (2002) Cognitive functioning of long-term heavy cannabis users seeking treatment. JAMA 287:1123–1131

Sugiura T, Kodaka T, Nakane S (1998) Detection of an endogenous cannabimimetic molecule, 2-arachidonoylglycerol, and cannabinoid CB1 receptor mRNA in human vascular cells: Is 2-arachidonoylglycerol a possible vasomodulator? Biochem Biophys Res Commun 243:838–843

Svendsen KB, Jensen TS, Bach FW (2003) Dronabinol (delta-9-THC) alleviates pain in multiple sclerosis. Congress of the European Federation of IASP Chapters (EFIC), 2–6 September 2003, Prague

Tashkin DP, Shapiro BJ, Frank IM (1974) Acute effects of smoked marihuana and oral delta-9-THC on specific airway conductance in asthmatic subjects. Am Rev Respir Dis 109:420–428

Tashkin DP, Reiss S, Shapiro BJ, et al (1977) Bronchial effects of aerosolised delta-9-THC in healthy and asthmatic subjects. Am Rev Respir Dis 115:57–65

Taylor PC (2001) Anti-TNF therapy for rheumatoid arthritis and other inflammatory diseases. Mol Biotechnol 19:153–168

Thomas H (1993) Psychiatric symptoms in cannabis users. Br J Psychiatry 163:141–149

Thompson AJ, Baker D (2002) Cannabinoids in MS: potentially useful but not just yet? Editorial. Neurology 58:1323–1324

UK Parliament Select Committee on Home Affairs (Third Report) (2002) The government's drug policy: is it working: different controls for different drugs; cannabis paragraph 108. HC318:1; ISBN 0-10-500334-9 http://www.publications.parliament.uk/pa/cm200102/cmselect/cmhaff/318/31807.htm a25. Cited 12 Oct 2004

Ungerleider T, Andrysiak T, Fairbanks L, et al (1982) Cannabis and cancer chemotherapy: a comparison of oral delta-9-THC and prochlorperazine. Cancer 50:636–645
Vaney C, Jobin P, Tschopp F, Heinzel M, Schnelle M (2002) Efficacy, safety and tolerability of an orally administered cannabis extract in the treatment of spasticity in patients with multiple sclerosis. In: Symposium on the Cannabinoids; Asilomar Conference Center, Pacific Grove, CA: International Cannabinoid Research Society; 13 July 2002, p 57
Venderova K, Ruzicka E, Vorisek V, Visnovsky P (2003) Cannabis and Parkinson's disease: subjective improvement of symptoms and levodopa-induced dyskinesias. International Cannabinoid Research Society 13th Annual Symposium on the Cannabinoids, abstr 145
Volicer L, Stelly M, Morris J, McLaughlin J, Volicer BJ (1997) Effects of dronabinol on anorexia and disturbed behavior in patients with Alzheimer's disease. Int J Geriatr Psychiatry 12:913–919
Wade DT, Robson PJ, House H, et al (2003) A preliminary controlled study to determine whether whole-plant cannabis extracts can improve intractable neurogenic symptoms. Clin Rehabil 17:18–26
Wade DT, Makela P, Robson P, et al (2004) Do cannabis-based medicinal extracts have general or specific effects on symptoms in multiple sclerosis? A double-blind, randomised, placebo-controlled study in 160 patients. Mult Scler 10:434–442
Wallace MJ, Blair RE, Falenski KW, et al (2003) The endogenous cannabinoid system regulates seizure frequency and duration in a model of temporal lobe epilepsy. J Pharmacol Exp Ther 307:129–137
Ware MA, Adams H, Guy G (2003) The medicinal use of cannabis in the United Kingdom. International Cannabinoid Research Society 13th Annual Symposium on the Cannabinoids, abstr 139
Whittle BA, Guy GW, Robson P (2001) Prospects for new cannabis-based prescription medicines. J Cannabis Ther 1:183–205
Williams CM, Kirkham TC (1999) Anandamide induces overeating: mediation by central cannabinoid (CB1) receptors. Psychopharmacology (Berl) 143:315–317
Williams SJ, Hartley JPR, Graham JDP (1976) Bronchodilator effect of delta-9-THC administered by aerosol to asthmatic patients. Thorax 31:720–723
Wisbeck GA, Schuckit MA, Kalmijn JA (1996) An evaluation of the history of a marijuana withdrawal syndrome in a large population. Addiction 91:1469–1478
Wright K, Rooney N, Tate J, et al (2003) Functional cannabinoid receptor expression in human colonic epithelium. International Cannabinoid Research Society 13th Annual Symposium on the Cannabinoids, abstr 25
Young CA, Rog DJ (2003) Randomised controlled trial of cannabis based medicinal extracts (CBME) in central neuropathic pain due to multiple sclerosis. Congress of the European Federation of IASP Chapters (EFIC), 2–6 September 2003, Prague
Young RR (1994) Spasticity: a review. Neurology 44:S12–S20
Zajicek J, Fox P, Sanders H, et al (2003) Cannabinoids for treatment of spasticity and other symptoms related to multiple sclerosis (CAMS study): multicentre randomised placebo-controlled trial. Lancet 362:1517–1526
Zuardi AW, Guimaraes FS (1997) Cannabidiol as an anxiolytic and antipsychotic. In: Mathre ML (ed) Cannabis in medical practice: a legal, historical and pharmacological overview of the therapeutic use of marijuana. McFarland, Jefferson, pp 133–141
Zuardi AW, Shirakawa I, Finkelfarb E, Karniol IG (1982) Action of cannabidiol on the anxiety and other effects produced by delta-9-THC in normal subjects. Psychopharmacology (Berl) 76:245–250
Zuardi AW, Rogdrigues JA, Cunha JM (1991) Effects of cannabidiol in animal models predictive of antipsychotic activity. Psychopharmacology (Berl) 104:260–264
Zuardi AW, Cosme RA, Graeff FG, Guimaraes FS (1993) Effects of ipsapirone and cannabidiol on human experimental anxiety. J PsychoPharmacol 7:82–88
Zuardi AW, Morais SL, Guimaraes FS, Mechoulam R (1995) Antipsychotic effect of cannabidiol. J Clin Psychiatry 56:485–486

Subject Index